Prepare for Your Tests With

The Chapter Test Prep Video included in this text will help you prepare for tests and use your study time more efficiently. Step-by-step solutions presented by Elayn Martin-Gay are included for every problem in the Chapter Tests in *Beginning Algebra, Fifth Edition*. Now captioned in English and Spanish.

Good study skills are key to success in mathematics. Use the resources available to help you make the most of your study time and get the practice you need to understand the concepts.

Three Steps to Success

To make the most of this resource when studying for a test, follow these three steps:

1. Take the Chapter Test found at the end of the text chapter.
2. Check your answers in the back of the text.
3. Use this video to review every step of the worked-out solution to those specific questions you answered incorrectly or need to review.

The easy-to-use navigation gives you direct access to the specific questions you need to review.

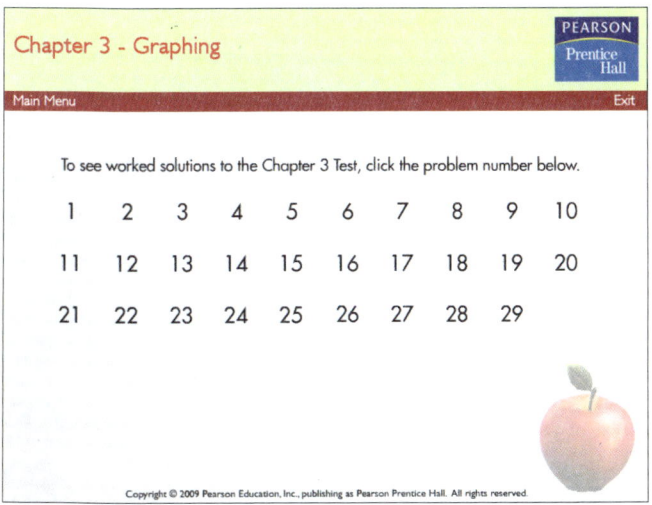

Chapter Test Menu: Lists all questions contained in the Chapter Test. Students select the test questions they wish to review.

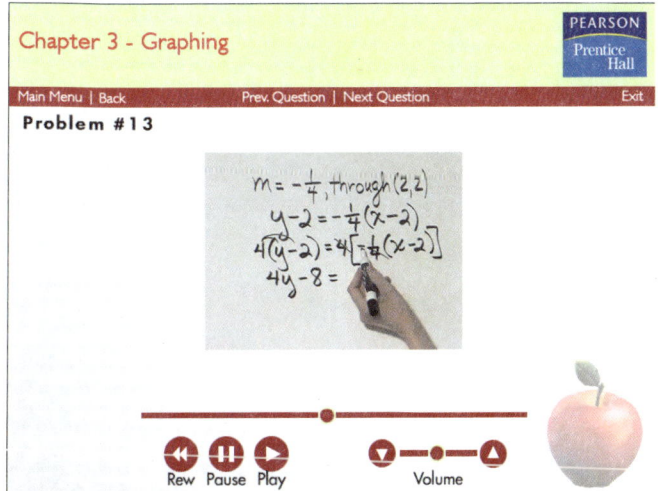

Previous and Next buttons allow easy navigation within tests. Rewind, Pause, and Play buttons let you view the solutions at your own pace.

How to Get Started

1. Insert the Video CD into your CD-ROM drive. The video should start automatically. If it does not, or if you have auto-launch turned off, you will need to open the program manually.

   To open the program manually:

   ### Windows users

   - Double-click on "My Computer," and locate your CD-ROM drive. It is usually the D: Drive.
   - Double-click on the CD-ROM drive to open the CD Video.
   - Find the file named "Start."
   - Double-click on "Start" to begin the video.

   ### Macintosh users

   - Locate the CD-ROM icon on your desktop and double-click it to reveal the contents of the CD.
   - Locate "Start," and double-click on it to start the program.

2. The opening screen lists your text chapter titles.
3. Click on the chapter title for the Chapter Test you are working on. You will be taken to the Chapter Test for that chapter.
4. The Chapter Test menu lists all the problems contained in your Chapter Test in the text.
5. Double-click on the problem number(s) to view the solution(s) you need to review. You can also navigate to other problems on the test by using the Previous and Next buttons.
6. If you do not have QuickTime, refer to the ReadMe file included on this Video CD. The ReadMe includes instructions on how to download and install the latest version of QuickTime.
7. **English and Spanish Captioning.** To activate the captioning feature, select the appropriate logo on the main menu, and click the "off" button to "on." Choosing the CC logo will activate English captions, while selecting the ESP logo will activate Spanish captions. Captioning will then be enabled for all video segments on this CD. To disable captioning, click the selected button on the main menu to "off."

Technical Support: To obtain support, please visit us online at http://247pearsoned.custhelp.com where you can search our knowledge base for common solutions, view product alerts, and review all options for additional assistance.

SINGLE PC LICENSE AGREEMENT AND LIMITED WARRANTY

READ THIS LICENSE CAREFULLY BEFORE OPENING THIS PACKAGE. BY OPENING THIS PACKAGE, YOU ARE AGREEING TO THE TERMS AND CONDITIONS OF THIS LICENSE. IF YOU DO NOT AGREE, DO NOT OPEN THE PACKAGE. PROMPTLY RETURN THE UNOPENED PACKAGE AND ALL ACCOMPANYING ITEMS TO THE PLACE YOU OBTAINED THEM [[FOR A FULL REFUND OF ANY SUMS YOU HAVE PAID FOR THE SOFTWARE]]. *THESE TERMS APPLY TO ALL LICENSED SOFTWARE ON THE DISK EXCEPT THAT THE TERMS FOR USE OF ANY SHAREWARE OR FREEWARE ON THE DISKETTES ARE AS SET FORTH IN THE ELECTRONIC LICENSE LOCATED ON THE DISK:*

1. **GRANT OF LICENSE AND OWNERSHIP:** The enclosed computer programs and data ("Software") are licensed, not sold, to you by Pearson Education, Inc. publishing as Prentice-Hall, Inc. ("We" or the "Company") in consideration of your purchase or adoption of the accompanying Company textbooks and/or other materials, and your agreement to these terms. We reserve any rights not granted to you. You own only the disk(s) but we and/or our licensors own the Software itself. This license allows individuals who have purchased the accompanying Company textbook to use and display their copy of the Software on a single computer (i.e., with a single CPU) at a single location for *academic* use only, so long as you comply with the terms of this Agreement. You may make one copy for back up, or transfer your copy to another CPU, provided that the Software is usable on only one computer.

2. **RESTRICTIONS:** You may *not* transfer or distribute the Software or documentation to anyone else. Except for backup, you may *not* copy the documentation or the Software. You may *not* network the Software or otherwise use it on more than one computer or computer terminal at the same time. You may *not* reverse engineer, disassemble, decompile, modify, adapt, translate, or create derivative works based on the Software or the Documentation. You may be held legally responsible for any copying or copyright infringement that is caused by your failure to abide by the terms of these restrictions.

3. **TERMINATION:** This license is effective until terminated. This license will terminate automatically without notice from the Company if you fail to comply with any provisions or limitations of this license. Upon termination, you shall destroy the Documentation and all copies of the Software. All provisions of this Agreement as to limitation and disclaimer of warranties, limitation of liability, remedies or damages, and our ownership rights shall survive termination.

4. **LIMITED WARRANTY AND DISCLAIMER OF WARRANTY:** Company warrants that for a period of 60 days from the date you purchase this SOFTWARE (or purchase or adopt the accompanying textbook), the Software, when properly installed and used in accordance with the Documentation, will operate in substantial conformity with the description of the Software set forth in the Documentation, and that for a period of 30 days the disk(s) on which the Software is delivered shall be free from defects in materials and workmanship under normal use. The Company does *not* warrant that the Software will meet your requirements or that the operation of the Software will be uninterrupted or error-free. Your only remedy and the Company's only obligation under these limited warranties is, at the Company's option, return of the disk for a refund of any amounts paid for it by you or replacement of the disk. THIS LIMITED WARRANTY IS THE ONLY WARRANTY PROVIDED BY THE COMPANY AND ITS LICENSORS, AND THE COMPANY AND ITS LICENSORS DISCLAIM ALL OTHER WARRANTIES, EXPRESS OR IMPLIED, INCLUDING WITHOUT LIMITATION, THE IMPLIED WARRANTIES OF MERCHANTABILITY AND FITNESS FOR A PARTICULAR PURPOSE. THE COMPANY DOES NOT WARRANT, GUARANTEE OR MAKE ANY REPRESENTATION REGARDING THE ACCURACY, RELIABILITY, CURRENTNESS, USE, OR RESULTS OF USE, OF THE SOFTWARE.

5. **LIMITATION OF REMEDIES AND DAMAGES:** IN NO EVENT, SHALL THE COMPANY OR ITS EMPLOYEES, AGENTS, LICENSORS, OR CONTRACTORS BE LIABLE FOR ANY INCIDENTAL, INDIRECT, SPECIAL, OR CONSEQUENTIAL DAMAGES ARISING OUT OF OR IN CONNECTION WITH THIS LICENSE OR THE SOFTWARE, INCLUDING FOR LOSS OF USE, LOSS OF DATA, LOSS OF INCOME OR PROFIT, OR OTHER LOSSES, SUSTAINED AS A RESULT OF INJURY TO ANY PERSON, OR LOSS OF OR DAMAGE TO PROPERTY, OR CLAIMS OF THIRD PARTIES, EVEN IF THE COMPANY OR AN AUTHORIZED REPRESENTATIVE OF THE COMPANY HAS BEEN ADVISED OF THE POSSIBILITY OF SUCH DAMAGES. IN NO EVENT SHALL THE LIABILITY OF THE COMPANY FOR DAMAGES WITH RESPECT TO THE SOFTWARE EXCEED THE AMOUNTS ACTUALLY PAID BY YOU, IF ANY, FOR THE SOFTWARE OR THE ACCOMPANYING TEXTBOOK. BECAUSE SOME JURISDICTIONS DO NOT ALLOW THE LIMITATION OF LIABILITY IN CERTAIN CIRCUMSTANCES, THE ABOVE LIMITATIONS MAY NOT ALWAYS APPLY TO YOU.

6. **GENERAL:** THIS AGREEMENT SHALL BE CONSTRUED IN ACCORDANCE WITH THE LAWS OF THE UNITED STATES OF AMERICA AND THE STATE OF NEW YORK, APPLICABLE TO CONTRACTS MADE IN NEW YORK, AND SHALL BENEFIT THE COMPANY, ITS AFFILIATES AND ASSIGNEES. THIS AGREEMENT IS THE COMPLETE AND EXCLUSIVE STATEMENT OF THE AGREEMENT BETWEEN YOU AND THE COMPANY AND SUPERSEDES ALL PROPOSALS OR PRIOR AGREEMENTS, ORAL, OR WRITTEN, AND ANY OTHER COMMUNICATIONS BETWEEN YOU AND THE COMPANY OR ANY REPRESENTATIVE OF THE COMPANY RELATING TO THE SUBJECT MATTER OF THIS AGREEMENT. If you are a U.S. Government user, this Software is licensed with "restricted rights" as set forth in subparagraphs (a)–(d) of the Commercial Computer-Restricted Rights clause at FAR 52.227-19 or in subparagraphs (c)(1)(ii) of the Rights in Technical Data and Computer Software clause at DFARS 252.227-7013, and similar clauses, as applicable.

Windows System Requirements:
Intel® Pentium® 700-MHz processor
Windows 2000 (Service Pack 4), XP, or Vista
800 x 600 resolution
8x CD drive
QuickTime 7.x
Sound Card
Internet browser

Macintosh System Requirements:
Power PC® or Intel 300 MHz processor
Mac 10.x
800 x 600 resolution monitor
8x CD drive
QuickTime 7
Internet browser

Should you have any questions concerning this agreement or if you wish to contact the Company for any reason, please contact in writing:
Director, Media Production
Pearson Education
1 Lake Street
Upper Saddle River, NJ 07458

Beginning Algebra

Beginning Algebra

Fifth Edition

Elayn Martin-Gay
University of New Orleans

Upper Saddle River, New Jersey 07458

Library of Congress Cataloging-in-Publication Data

Martin-Gay, K. Elayn
  Beginning algebra/K. Elayn Martin-Gay—5th ed.
    p. cm.
  Includes index.
  ISBN 0-13-600702-3
  1. Algebra I. Title

President: *Greg Tobin*
Editor in Chief: *Paul Murphy*
Vice President and Editorial Director: *Christine Hoag*
Sponsoring Editor: *Mary Beckwith*
Assistant Editor: *Christine Whitlock*
Editorial Assistant: *Georgina Brown*
Production Management: *Elm Street Publishing Services*
Senior Managing Editor: *Linda Mihatov Behrens*
Operations Specialist: *Ilene Kahn*
Senior Operations Supervisor: *Diane Peirano*
Vice President and Executive Director of Development: *Carol Trueheart*
Development Editor: *Lisa Collette*
Media Producer: *Audra J. Walsh*
Lead Media Project Manager: *Richard Bretan*
Software Development: *MyMathLab: Jennifer Sparkes, Media Producer*
MathXL: Janet Szykolony, Software Editor
TestGen: Ted Hartman, Software Editor
Vice President and Director of Marketing: *Amy Cronin*
Executive Marketing Manager: *Kate Valentine*
Senior Marketing Manager: *Michelle Renda*
Marketing Manager: *Marlana Voerster*
Marketing Assistants: *Jill Kapinus and Nathaniel Koven*
Senior Art Director: *Juan R. López*
Interior Designer: *Mike Fruhbeis*
Cover Art Direction, Design, and Illustration: *Kenny Beck*
AV Project Manager: *Thomas Benfatti*
Director, Image Resource Center: *Melinda Patelli*
Manager, Rights and Permissions: *Zina Arabia*
Manager, Visual Research: *Beth Brenzel*
Image Permission Coordinator: *Craig Jones*
Photo Researcher: *David Tietz, Editorial Image, LLC*
Art Studios: *Scientific Illustrators/Laserwords*
Compositor: *ICC Macmillan Inc.*

© 2009, 2005, 2001, 1997, 1993 Pearson Education, Inc.
Pearson Prentice Hall
Pearson Education, Inc.
Upper Saddle River, New Jersey 07458

All rights reserved. No part of this book may be reproduced, in any form or by any means, without permission in writing from the publisher.

Pearson Prentice Hall™ is a trademark of Pearson Education, Inc.
Printed in the United States of America

10 9 8 7 6 5 4 3

ISBN-10 0-13-600702-3
ISBN-13 978-0-13-600702-9

Pearson Education LTD., *London*
Pearson Education Australia PTY. Limited, *Sydney*
Pearson Education Singapore PTE. LTD.
Pearson Education North Asia, LTD., *Hong Kong*
Pearson Education Canada, LTD., *Toronto*
Pearson Educacíon de Mexico, S.A., de C.V.
Pearson Education, Japan, *Tokyo*
Pearson Education Malaysia, PTE. LTD.

This book is dedicated in memory of
Dennis Lee Wood

Some of his many wise sayings were . . .
"Sometimes the 'Hokey Pokey' really
is what it's all about."
"No matter what it is, somebody will find
a way to take it way too seriously."
"When I start to believe that all my problems
are behind me, that means I have
overlooked something."

Contents

Preface xv
Application Index xxv

CHAPTER 1 — REVIEW OF REAL NUMBERS 1

1.1 Tips for Success in Mathematics 2
1.2 Symbols and Sets of Numbers 7
1.3 Fractions 16
1.4 Introduction to Variable Expressions and Equations 24
1.5 Adding Real Numbers 34
1.6 Subtracting Real Numbers 41
Integrated Review—Operations on Real Numbers 48
1.7 Multiplying and Dividing Real Numbers 49
1.8 Properties of Real Numbers 58
Chapter 1 Group Activity 64
Chapter 1 Vocabulary Check 65
Chapter 1 Highlights 65
Chapter 1 Review 69
Chapter 1 Test 72

CHAPTER 2 — EQUATIONS, INEQUALITIES, AND PROBLEM SOLVING 73

2.1 Simplifying Algebraic Expressions 74
2.2 The Addition Property of Equality 82
2.3 The Multiplication Property of Equality 90
2.4 Solving Linear Equations 97
Integrated Review—Solving Linear Equations 105
2.5 An Introduction to Problem Solving 105
2.6 Formulas and Problem Solving 115
2.7 Percent and Mixture Problem Solving 127
2.8 Further Problem Solving 138
2.9 Solving Linear Inequalities 145
Chapter 2 Group Activity 157
Chapter 2 Vocabulary Check 158
Chapter 2 Highlights 158
Chapter 2 Review 163
Chapter 2 Test 166
Chapter 2 Cumulative Review 167

CHAPTER 3 — GRAPHING 169

3.1 Reading Graphs and the Rectangular Coordinate System 170
3.2 Graphing Linear Equations 185
3.3 Intercepts 194
3.4 Slope and Rate of Change 203
Integrated Review—Summary on Slope and Graphing Linear Equations 217
3.5 Equations of Lines 218
3.6 Functions 227
Chapter 3 Group Activity 237

vii

Chapter 3 Vocabulary Check 238
Chapter 3 Highlights 238
Chapter 3 Review 242
Chapter 3 Test 246
Chapter 3 Cumulative Review 247

CHAPTER 4 SOLVING SYSTEMS OF LINEAR EQUATIONS AND INEQUALITIES 249

4.1 Solving Systems of Linear Equations by Graphing 250
4.2 Solving Systems of Linear Equations by Substitution 258
4.3 Solving Systems of Linear Equations by Addition 265
Integrated Review—Solving Systems of Equations 271
4.4 Systems of Linear Equations and Problem Solving 272
4.5 Graphing Linear Inequalities 282
4.6 Systems of Linear Inequalities 288
Chapter 4 Group Activity 291
Chapter 4 Vocabulary Check 292
Chapter 4 Highlights 293
Chapter 4 Review 296
Chapter 4 Test 298
Chapter 4 Cumulative Review 299

CHAPTER 5 EXPONENTS AND POLYNOMIALS 301

5.1 Exponents 302
5.2 Adding and Subtracting Polynomials 313
5.3 Multiplying Polynomials 323
5.4 Special Products 330
Integrated Review—Exponents and Operations on Polynomials 336
5.5 Negative Exponents and Scientific Notation 337
5.6 Dividing Polynomials 345
Chapter 5 Group Activity 352
Chapter 5 Vocabulary Check 353
Chapter 5 Highlights 354
Chapter 5 Review 356
Chapter 5 Test 359
Chapter 5 Cumulative Review 360

CHAPTER 6 FACTORING POLYNOMIALS 362

6.1 The Greatest Common Factor and Factoring by Grouping 363
6.2 Factoring Trinomials of the Form $x^2 + bx + c$ 371
6.3 Factoring Trinomials of the Form $ax^2 + bx + c$ and Perfect Square Trinomials 377
6.4 Factoring Trinomials of the Form $ax^2 + bx + c$ by Grouping 386
6.5 Factoring Binomials 391
Integrated Review—Choosing a Factoring Strategy 398
6.6 Solving Quadratic Equations by Factoring 402
6.7 Quadratic Equations and Problem Solving 411
Chapter 6 Group Activity 419
Chapter 6 Vocabulary Check 420
Chapter 6 Highlights 420
Chapter 6 Review 424
Chapter 6 Test 426
Chapter 6 Cumulative Review 427

Contents ix

CHAPTER 7

RATIONAL EXPRESSIONS 428

7.1 Simplifying Rational Expressions 429
7.2 Multiplying and Dividing Rational Expressions 437
7.3 Adding and Subtracting Rational Expressions with Common Denominators and Least Common Denominator 444
7.4 Adding and Subtracting Rational Expressions with Unlike Denominators 452
7.5 Solving Equations Containing Rational Expressions 459
Integrated Review—Summary on Rational Expressions 465
7.6 Proportion and Problem Solving with Rational Equations 466
7.7 Variation and Problem Solving 478
7.8 Simplifying Complex Fractions 488
Chapter 7 Group Activity 493
Chapter 7 Vocabulary Check 494
Chapter 7 Highlights 494
Chapter 7 Review 499
Chapter 7 Test 501
Chapter 7 Cumulative Review 501

CHAPTER 8

ROOTS AND RADICALS 503

8.1 Introduction to Radicals 504
8.2 Simplifying Radicals 511
8.3 Adding and Subtracting Radicals 518
8.4 Multiplying and Dividing Radicals 522
Integrated Review—Simplifying Radicals 530
8.5 Solving Equations Containing Radicals 531
8.6 Radical Equations and Problem Solving 537
8.7 Rational Exponents 543
Chapter 8 Group Activity 547
Chapter 8 Vocabulary Check 547
Chapter 8 Highlights 547
Chapter 8 Review 551
Chapter 8 Test 553
Chapter 8 Cumulative Review 554

CHAPTER 9

QUADRATIC EQUATIONS 556

9.1 Solving Quadratic Equations by the Square Root Property 557
9.2 Solving Quadratic Equations by Completing the Square 562
9.3 Solving Quadratic Equations by the Quadratic Formula 568
Integrated Review—Summary on Solving Quadratic Equations 577
9.4 Complex Solutions of Quadratic Equations 579
9.5 Graphing Quadratic Equations 584
Chapter 9 Group Activity 591
Chapter 9 Vocabulary Check 592
Chapter 9 Highlights 592
Chapter 9 Review 595
Chapter 9 Test 598
Chapter 9 Cumulative Review 598

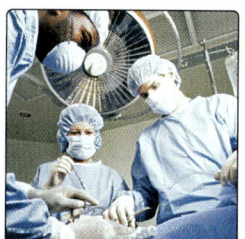

APPENDICES 601

A THE BIGGER PICTURE/PRACTICE FINAL EXAM 601
B GEOMETRY 606
C ADDITIONAL EXERCISES ON PROPORTION AND PROPORTION APPLICATIONS 620
D OPERATIONS ON DECIMALS 622
E MEAN, MEDIAN, AND MODE 624
F TABLES 626

Answers to Selected Exercises A1
Index I1
Photo Credits P1

Tools to Help Students Succeed

Your textbook includes a number of features designed to help you succeed in this math course—as well as the next math course you take. These features include:

Feature	Benefit	Page
Well-crafted Exercise Sets: We learn math by doing math	The exercise sets in your text offer an ample number of exercises carefully ordered so you can master basic mathematical skills and concepts while developing all-important problem solving skills. Exercise sets include Mixed Practice exercises to help you master multiple key concepts, as well as Vocabulary and Readiness Check, Writing, Applications, Concept Check, Concept Extension, and Review and Preview Exercises.	102–104
Study Skills Builders: Maximize your chances for success	Study Skills Builders reinforce the material in *Section 1.1—Tips for Success in Mathematics.* Study Skills Builders are a great resource for study ideas and self-assessment to maximize your opportunity for success in this course. Take your new study skills with you to help you succeed in your next math course.	97
The Bigger Picture: Succeed in this math course and the next one you take	The Bigger Picture focuses on the key concepts of this course—simplifying expressions and solving equations and inequalities—and asks you to keep an ongoing study guide so you can simplify expressions and solve equations and inequalities, and recognize the difference between them. A strong foundation in simplifying expressions and solving equations and inequalities will help you succeed in this algebra course, as well as the next math course you take.	157
Examples: Step-by-step instruction for you	Examples in the text provide you with clear, concise step-by-step instructions to help you learn. Annotations in the examples provide additional instruction.	77
Helpful Hints: Help where you'll need it most	Helpful Hints provide tips and advice at exact locations where students need it most. Strategically placed where you might have the most difficulty, Helpful Hints will help you work through common trouble spots.	77
Practice Exercises: Immediate reinforcement	New Practice exercises offer immediate reinforcement after every example. Try each Practice exercise after studying the corresponding example to make sure you have a good working knowledge of the concept.	77
Integrated Review: Mid-chapter progress check	To ensure you understand the key concepts covered in the first sections of the chapter, work the exercises in the Integrated Review before you continue with the rest of the chapter.	105
Vocabulary and Readiness Check, Vocabulary Check: Key terms and vocabulary	Use the Vocabulary and Readiness Checks to build your vocabulary and warm-up on concepts in the section. Make sure you understand key terms and vocabulary in each chapter with the end-of-chapter Vocabulary Check.	102, 158
Chapter Highlights: Study smart	Chapter Highlights outline the key concepts of the chapter along with examples to help you focus your studying efforts as you prepare for your test.	158–163
Chapter Test: Take a practice test	In preparation for your classroom test, take this practice test to make sure you understand the key topics in the chapter. Be sure to use the **Chapter Test Prep Video** included with this text to see the author present a fully worked-out solution to each exercise in the Chapter Test.	166
Practice Final Exam: Take a practice final	In preparation for your final, take the practice final exam found in Appendix A.2. **Martin-Gay's Interactive DVD/CD Lecture Series** includes the Practice Final Exam and provides you with full video solutions to each exercise. Overview clips provide a brief overview on how to approach different problem types.	604

Martin-Gay's VIDEO RESOURCES
Help Students Succeed

MARTIN-GAY'S CHAPTER TEST PREP VIDEO (AVAILABLE WITH THIS TEXT) TEST PREP VIDEO

- Provides students with help during their most "teachable moment"—while they are studying for a test.
- Text author Elayn Martin-Gay presents step-by-step solutions to the exact exercises found in each Chapter Test in the book.
- Easy video navigation allows students to instantly access the worked-out solutions to the exercises they want to review.
- Close captioned in English and Spanish.

NEW MARTIN-GAY'S INTERACTIVE DVD/CD LECTURE SERIES

Martin-Gay's video series has been comprehensively updated to address the way today's students study and learn. The new videos offer students active learning at their pace, with the following resources and more:

- **A complete lecture** for each section of the text, presented by Elayn Martin-Gay. Students can easily review a section or a specific topic before a homework assignment, quiz, or test. Exercises in the text marked with the are worked on the video.
- A **new interface** with menu and navigation features helps students quickly find and focus on the examples and exercises they need to review.
- Martin-Gay's "pop-ups" reinforce key terms and definitions and are a great support for multiple learning styles.
- A new **Practice Final Exam Video** helps students prepare for the final exam. This Practice Final Exam is included in the text in Appendix A.2. At the click of a button, students can watch the full solutions to each exercise on the exam when they need help. Overview clips provide a brief overview on how to approach different problem types—just as they will need to do on a Final Exam.
- **Interactive Concept Checks** allow students to check their understanding of essential concepts. Like the concept checks in the text, these multiple choice exercises focus on common misunderstandings. After making their answer selection, students are told whether they're correct or not, and why! Elayn also presents the full solution.
- **Study Skills Builders** help students develop effective study habits and reinforce the advice provided in Section 1.1, Tips for Success in Mathematics, found in the text and video.
- **Close-captioned in Spanish and English**
- Ask your bookstore for information about Martin-Gay's *Beginning Algebra,* Fifth Edition Interactive DVD/CD Lecture Series or visit www.mypearsonstore.com.

You will find Interactive Concept Checks and Study Skills Builders in the following sections on the Interactive DVD/CD Lecture Series:

Interactive Concept Checks Section		Study Skills Builders Section
1.3	5.6	1.2 Time Management
1.4	6.1	2.5 Are You Familiar with the Resources Available with Your Textbook?
1.5	6.2	
1.6	6.3	2.6 Have You Decided to Complete This Course Successfully?
1.7	6.5	
1.8	6.6	2.7 How Well Do You Know Your Textbook?
2.1	Ch 6 Integrated Review	
2.2	7.1	2.8 Tips for Studying for an Exam
2.3	7.2	3.1 How Are Your Homework Assignments Going?
2.4	7.3	3.2 Doing Your Homework Online
2.8	7.5	
3.3	7.6	4.6 What to Do the Day of an Exam?
3.4	7.7	5.1 Are You Satisfied with Your Performance on a Particular Quiz?
3.5	7.8	
3.6	8.1	5.2 Are You Organized?
4.1	8.3	5.5 Are You Familiar with the Resources Available with Your Textbook?
4.2	8.4	
4.3	8.7	6.4 Have You Decided to Complete This Course Successfully?
4.4	9.2	
4.5	9.3	6.7 How Well Do You Know Your Textbook?
5.3	9.4	7.4 Tips for Studying for an Exam
5.4	9.5	8.2 Are You Satisfied with Your Performance on a Particular Quiz or Exam?
		8.5 What to Do the Day of an Exam?
		8.6 How Are Your Homework Assignment Going?
		9.1 Preparing for Your Final Exam

Additional Resources to Help You Succeed

Student Study Pack

A single, easy-to-use package—available bundled with your textbook or by itself—for purchase through your bookstore. This package contains the following resources to help you succeed:

Student Solutions Manual
- Contains worked-out solutions to odd-numbered exercises from each section exercise set, all Practice exercises, Vocabulary and Readiness Check Exercises, and all exercises found in the Chapter Review Chapter Tests, and Integrated Reviews.

Martin-Gay's Interactive Video Lectures
- Text author Elayn Martin-Gay presents the key concepts from every section of the text with 15–20 minute mini-lectures. Students can easily review a section or a specific topic before a homework assignment, quiz, or test.
- A new interface allows easy navigation through the lesson.
- Includes fully worked-out solutions to exercises marked with a icon in each section. Also includes *Section 1.1, Tips for Success in Mathematics.*
- Close-captioned in English and Spanish.

Pearson Tutor Center

Online Homework and Tutorial Resources

MyMathLab® MyMathLab

MyMathLab is a series of text-specific, easily customizable online courses for Pearson Education's textbooks in mathematics and statistics. Powered by CourseCompass™ (our online teaching and learning environment) and MathXL® (our online homework, tutorial, and assessment system), MyMathLab gives you the tools you need to deliver all or a portion of your course online, whether your students are in a lab setting or working from home. MyMathLab provides a rich and flexible set of course materials, featuring free-response exercises that are algorithmically generated for unlimited practice and mastery. Students can also use online tools, such as video lectures, animations, and a multimedia textbook, to independently improve their understanding and performance. Instructors can use MyMathLab's homework and test managers to select and assign online exercises correlated directly to the textbook, and they can also create and assign their own online exercises and import TestGen tests for added flexibility. MyMathLab's online gradebook—designed specifically for mathematics and statistics—automatically tracks students' homework and test results and gives the instructor control over how to calculate final grades. Instructors can also add offline (paper-and-pencil) grades to the gradebook. Includes access to the Pearson Tutor Center. MyMathLab is available to qualified adopters. For more information, visit our website at www.mythlab.com or contact your sales representative.

MathXL® PRACTICE

MathXL® is a powerful online homework, tutorial, and assessment system that accompanies Pearson Education's textbooks in mathematics or statistics. With MathXL, instructors can create, edit, and assign online homework and tests using algorithmically generated exercises correlated at the objective level to the textbook. They can also create and assign their own online exercises and import TestGen tests for added flexibility. All student work is tracked in MathXL's online gradebook. Students can take chapter tests in MathXL and receive personalized study plans based on their test results. The study plan diagnoses weaknesses and links students directly to tutorial exercises for the objectives they need to study and retest. Students can also access supplemental animations and video clips directly from selected exercises. MathXL is available to qualified adopters. For more information, visit our website at www.mathxl.com, or contact your sales representative.

Preface

ABOUT THE BOOK

Beginning Algebra, Fifth Edition was written to provide a **solid foundation in algebra** for students who might not have had previous experience in algebra. Specific care has been taken to ensure that students have the most **up-to-date and relevant** text preparation for their next mathematics course, as well as to help students succeed in nonmathematical courses that require a grasp of algebraic fundamentals. I have tried to achieve this by writing a user-friendly text that is keyed to objectives and contains many worked-out examples. The basic concepts of graphing are introduced early, and problem solving techniques, real-life and real-data applications, data interpretation, appropriate use of technology, mental mathematics, number sense, critical thinking, decision-making, and geometric concepts are emphasized and integrated throughout the book.

The many factors that contributed to the success of the previous editions have been retained. In preparing this edition, I considered the comments and suggestions of colleagues throughout the country, students, and many users of the prior editions. The AMATYC Crossroads in Mathematics: Standards for Introductory College Mathematics before Calculus and the MAA and NCTM standards (plus Addenda), together with advances in technology, also influenced the writing of this text.

Throughout the series, pedagogical features are designed to develop student proficiency in algebra and problem solving, and to prepare students for future courses.

WHAT'S NEW IN THE FIFTH EDITION

New Martin-Gay's Interactive DVD/CD Lecture Series, featuring Elayn Martin-Gay, provides students with active learning at their pace. The new videos offer the following resources and more:

- A complete lecture for each section of the text. A new interface with menu and navigation features allows students to quickly find and focus on the examples and exercises they need to review.
- The new **Practice Final Exam** helps students prepare for an end of course final. Students can watch full video solutions to each exercise. Overview clips give students a brief overview on how to approach different problem types.
- **Interactive Concept Check exercises** that allow students to check their understanding of essential concepts

A New Three-Point System to support students and instructors:

- **Examples** in the text prepare students for class and homework.
- New **Practice Exercises** are paired with each text example and are ready-made for use in the classroom and can be assigned as homework.
- **Classroom Examples,** found only in the Annotated Instructor's Edition, are also paired with each example to give instructors a convenient way to further illustrate skills and concepts.

ENHANCED EXERCISE SETS

- **New Vocabulary and Readiness exercises** appear at the beginning of exercise sets. The **Vocabulary** exercises reinforce a student's understanding of new terms so that forthcoming instructions in the exercise set are clear. The **Readiness** exercises serve as a warm-up on the skills and concepts necessary to complete the exercise set.

- **NEW! Concept Check exercises** have been added to the section exercise sets. These exercises are related to the Concept Check(s) found within the section. They help students measure their understanding of key concepts by focusing on common trouble areas. These exercises may ask students to identify a common error, and/or provide an explanation.
- **New Mixed Review** exercises are included at the end of the Chapter Review. These exercises require students to determine the problem type and strategy needed in order to solve it.
- **New Practice Final Exam**—included in Appendix A.2 helps students prepare for an end of course final. Martin-Gay's Interactive DVD/CD Lecture Series provides students with solutions to all exercises on the final. Overview clips give a brief introduction on how to approach different types of problems.

INCREASED EMPHASIS ON STUDY SKILLS AND STUDENT SUCCESS

- **NEW! Study Skills Builders** (formerly Study Skill Reminders) Found at the end of many exercise sets, Study Skills Builders allow instructors to assign exercises that will help students improve their study skills and take responsibility for their part of the learning process. Study Skills Builders reinforce the material found in Section 1.1, "Tips for Success in Mathematics" and serve as an excellent tool for self-assessment.
- **NEW! The Bigger Picture** is a recurring feature, starting in Section 1.7, that focuses on the key concepts of the course—simplifying expressions, and solving equations and inequalities. Students develop an on going study guide to help them be able to simplify expressions and solve equations and inequalities, and to know the difference between them. By working the exercises and developing this study guide throughout the text, students can begin to transition from thinking "section by section" to thinking about how the mathematics in this course is part of the "bigger picture" of mathematics in general. A completed outline is provided in Appendix A so students have a model for their work. It's great preparation for the Practice Final in A.2.

CONTINUING SUPPORT FOR TEST PREPARATION

- **TEST PREP VIDEO** **Chapter Test Prep Video** provides students with help during their most "teachable moment"—while they are studying for a test. Included with every copy of the student edition of the text, this video provides fully worked-out solutions by the author to every exercise from each Chapter Test in the text. The easy video navigation allows students to instantly access the solutions to the exercises they want to review. The problems are solved by the author in the same manner as in the text.
- **Chapter Test files in TestGen®** provide algorithms specific to each exercise from each Chapter Test in the text. Allows for easy replication of Chapter Tests with consistent, algorithmically generated problem types for additional assignments or assessments purposes.

CONTENT CHANGES IN THE FIFTH EDITION

The sections on problem solving in Chapter 2 have been revised to make them more accessible to students. The changes include:

- Section 2.5 has been reorganized into three objectives: solving problems involving direct translation; solving problems involving relationships among unknown quantities; and finding consecutive integers.
- A new Section 2.7, Percent and Mixture Problem Solving, covers percent applications, discount and mark up, percent increase and decrease, and mixture applications.

- Section 2.8 now covers distance, money, and interest applications.
- Appendix C contains additional exercises on solving proportion and proportion applications. This appendix may be used to augment Section 7.6 on this material or used to place proportions at any appropriate place during the course.
- There is now one section on Equations of Lines, Section 3.5, that reviews the slope-intercept form and introduces the point-slope form. This section now contains mixed exercises that help students decide what equation form to use.
- Section 6.4, Factoring Trinomials of the Form $ax^2 + bx + c$ by Grouping, is new for those instructors who prefer this method of factoring trinomials.
- Appendix A.2 is a Practice Final Exam that is an excellent review for students. The entire exam is worked by the author on the Interactive DVD/CD Lecture Series.
- Reading graphs has been moved from Chapter 1 to Chapter 3. Although simple bar and circle graphs with data labels remain in Chapters 1 and 2, the formal introduction of reading graphs is now in Section 3.1.

KEY PEDAGOGICAL FEATURES

The following key features have been retained and/or updated for the Fifth Edition of the text:

Problem Solving Process This is formally introduced in Chapter 2 with a four-step process that is integrated throughout the text. The four steps are **Understand, Translate, Solve,** and **Interpret.** The repeated use of these steps in a variety of examples shows their wide applicability. Reinforcing the steps can increase students' comfort level and confidence in tackling problems.

Exercise Sets Revised and Updated The exercise sets have been carefully examined and extensively revised. Special focus was placed on making sure that even- and odd-numbered exercises are paired.

Each text section ends with an exercise set, usually divided into two parts. Both parts contain graded exercises. The **first part is carefully keyed** to at least one worked example in the text. Once a student has gained confidence in a skill, **the second part contains exercises not keyed to examples.** Exercises and examples marked with a (🎬) have been worked out step-by-step by the author in the interactive videos that accompany this text.

Throughout the text exercises there is an emphasis on data and graphical interpretation via tables, charts, and graphs. The ability to interpret data and read and create a variety of types of graphs is developed gradually so students become comfortable with it. Similarly, throughout the text there is integration of geometric concepts, such as perimeter and area. Exercises and examples marked with a geometry icon (△) have been identified for convenience.

Examples Detailed step-by-step examples were added, deleted, replaced, or updated as needed. Many of these reflect real life. Additional instructional support is provided in the annotated examples.

Helpful Hints Helpful Hints contain practical advice on applying mathematical concepts. Strategically placed where students are most likely to need immediate reinforcement, Helpful Hints help students avoid common trouble areas and mistakes.

Concept Checks This feature allows students to gauge their grasp of an idea as it is being presented in the text. Concept Checks stress conceptual understanding at the point-of-use and help suppress misconceived notions before they start. Answers appear at the bottom of the page. Exercises related to Concept Checks are now included in the exercise sets.

Mixed Practice Exercises These exercises combine objectives within a section. They require students to determine the problem type and strategy needed in order to solve it. In doing so, students need to think about key concepts to proceed with a correct method of solving—just as they would need to do on a test.

Concept Extensions These exercises require students to combine several skills of concepts to solve exercises in the section.

Integrated Reviews A unique, mid-chapter exercise set that helps students assimilate new skills and concepts that they have learned separately over several sections. These reviews provide yet another opportunity for students to work with "mixed" exercises as they master the topics.

Vocabulary Check Provides an opportunity for students to become more familiar with the use of mathematical terms as they strengthen their verbal skills. These appear at the end of each chapter before the Chapter Highlights.

Chapter Highlights Found at the end of every chapter, these contain key definitions and concepts with examples to help students understand and retain what they have learned and help them organize their notes and study for tests.

Chapter Review The end of every chapter contains a comprehensive review of topics introduced in the chapter. The Chapter Review offers exercises keyed to every section in the chapter, as well as **Mixed Review (NEW!)** exercises that are not keyed to sections.

Cumulative Review Follows every chapter in the text except Chapter 1. Each odd-numbered exercise contained in the Cumulative Review is an earlier worked example in the text that is referenced in the back of the book along with the answer.

Writing Exercises ＼ These exercises occur in almost every exercise set and require students to provide a written response to explain concepts or justify their thinking.

Applications Real-world and real-data applications have been thoroughly updated and many new applications are included. These exercises occur in almost every exercise set and show the relevance of mathematics and help students gradually, and continuously develop their problem solving skills.

Review and Preview Exercises These exercises occur in each exercise set (except Chapter 1) and are keyed to earlier sections. They review concepts learned earlier in the text that will be needed in the next section or chapter.

Exercise Set Resource Icons at the opening of each exercise set remind students of the resources available for extra practice and support:

See Student Resource descriptions pages xxi–xxii for details on the individual resources available.

Exercise Icons These icons facilitate the assignment of specialized exercises and let students know what resources can support them.

 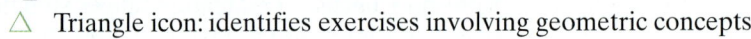

 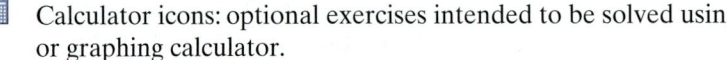

Group Activities Found at the end of each chapter, these activities are for individual or group completion, and are usually hands-on or data-based activities that extend the concepts found in the chapter allowing students to make decisions and interpretations and to think and write about algebra.

Optional: Calculator Exploration Boxes and Calculator Exercises The optional Calculator Explorations provide key strokes and exercises at appropriate points to provide an opportunity for students to become familiar with these tools. Section exercises that are best completed by using a calculator are identified by 🖩 or 🖩 for ease of assignment.

A Word about Textbook Design and Student Success

The design of developmental mathematics textbooks has become increasingly important. As students and instructors have told Pearson in focus groups and market research surveys, these textbooks cannot look "cluttered" or "busy." A "busy" design can distract a student from what is most important in the text. It can also heighten math anxiety.

As a result of the conversations and meetings we have had with students and instructors, we concluded the design of this text should be understated and focused on the most important pedagogical elements. Students and instructors helped us to identify the primary elements that are central to student success. These primary elements include:

- Exercise Sets
- Examples and Practice Problems
- Helpful Hints
- Rules, Property, and Definition boxes

As you will notice in this text, these primary features are the most prominent elements in the design. We have made every attempt to make sure these elements are the features the eye is drawn to. The remaining features, the secondary elements in the design, blend into the "fabric" or "grain" of the overall design. These secondary elements complement the primary elements without becoming distractions.

Pearson's thanks goes to all of the students and instructors (as noted by the author in Acknowledgments) who helped us develop the design of this text. At every step in the design process, their feedback proved valuable in helping us to make the right decisions. Thanks to your input, we're confident the design of this text will be both practical and engaging as it serves its educational and learning purposes.

Sincerely,

Paul Murphy
Editor-in-Chief
Developmental Mathematics
Pearson Arts & Sciences

INSTRUCTOR AND STUDENT RESOURCES

The following resources are available to help instructors and students use this text more effectively.

INSTRUCTOR RESOURCES

Annotated Instructor's Edition

- Answers to all exercises printed on the same text page or in the Graphing Answer Section.
- Teaching Tips throughout the text placed at key points.

- Classroom Examples paired with each text example for use as an additional resource during class.
- General tips and suggestions for classroom or group activities.
- Graphing Answer Section includes graphical answers and answers to the Group Activities.

Instructor's Solutions Manual

- Solutions to the even numbered exercises
- Solutions to every Vocabulary Check exercise
- Solutions to every Practice exercise
- Solutions to every exercise in the Integrated Reviews, Chapter Reviews, Chapter Tests, and Cumulative Reviews

Instructor's Resource Manual with Tests and Mini-Lectures

- **NEW!** Includes Mini-Lectures for every section from the text
- Additional Exercises now 3 forms per section, to help instructors support students at different skill and ability levels.
- Free Response Test Forms, Multiple Choice Test Forms, Cumulative Tests, Group Activities
- Answers to all items

Martin-Gay's Instructor to Instructor Videos

- Text author Elayn Martin-Gay presents tips, hints, and suggestions for engaging students and presenting key topics.
- Available as part of the Instructor Resource Kit.
- Contact your sales representative for more information.

Instructor Resource Kit

The Martin-Gay Instructor Resource Kit contains tools and resources to help instructors succeed in the classroom. The kit includes:

- Instructor-to-Instructor CD Videos that offer tips, suggestions, and strategies for engaging students and presenting key topics
- PDF files of the Instructor Solutions Manual and the Instructor's Resource Manual
- Powerpoint Lecture Slides and Active Learning Questions

MYMATHLAB®

(Instructor Version 0-13-147898-2)

MyMathLab is a series of text-specific, easily customizable, online courses for Pearson Education's textbooks in mathematics and statistics. Powered by CourseCompass™ (our online teaching and learning environment) and MathXL® (our online homework, tutorial, and assessment system), MyMathLab gives you the tools you need to deliver all or a portion of your course online, whether your students are in a lab setting or working from home. MyMathLab provides a rich and flexible set of course materials, featuring free-response exercises that are algorithmically generated for unlimited practice and mastery. Students can also use online tools, such as video lectures, animations, and a multimedia textbook, to independently improve their understanding and performance. Instructors can use MyMathLab's homework and test managers to select and assign online exercises correlated directly to the textbook, and they can also create and assign their own online exercises and import TestGen tests for added flexibility. MyMathLab's online gradebook—designed specifically for mathematics and statistics—automatically tracks students' homework and test results and gives the instructor control over how to calculate final grades. Instructors can also add offline (paper-and-pencil) grades to the gradebook. **Includes access to the Pearson Tutor Center. Students**

can receive tutoring via toll free phone, fax, email, and Internet. MyMathLab is available to qualified adopters. For more information, visit our website at www.mymathlab.com or contact your sales representative.

MATHXL®

(Instructor version 0-13-147895-8)

MathXL® is a powerful online homework, tutorial, and assessment system that accompanies Pearson Education's textbooks in mathematics or statistics. With MathXL, instructors can create, edit, and assign online homework and tests using algorithmically generated exercises correlated at the objective level to the textbook. They can also create and assign their own online exercises and import TestGen tests for added flexibility. All student work is tracked in MathXL's online gradebook. Students can take chapter tests in MathXL and receive personalized study plans based on their test results. The study plan diagnoses weaknesses and links students directly to tutorial exercises for the objectives they need to study and retest. Students can also access supplemental animations and video clips directly from selected exercises MathXL is available to qualified adopters. For more information, visit our website at www.mathxl.com, or contact your sales representative.

INTERACT MATH TUTORIAL WEBSITE: WWW.INTERACTMATH.COM

Get practice and tutorial help online! This interactive tutorial website provides algorithmically generated practice exercises that correlate directly to the exercises in the textbook. Students can retry an exercise as many times as they like with new values each time for unlimited practice and mastery. Every exercise is accompanied by an interactive guided solution that provides helpful feedback for incorrect answers, and students can also view a worked-out sample problem that steps them through an exercise similar to the one they're working on.

TESTGEN®

TestGen enables instructors to build, edit, print, and administer tests using a computerized bank of questions developed to cover all the objectives of the text. TestGen is algorithmically based, allowing instructors to create multiple but equivalent versions of the same question or test with the click of a button. Insturctors can also modify test bank questions or add new questions. Tests can be printed or administered online. The software and testbank are available for download from Pearson Education's online catalog.

STUDENT RESOURCES

Student Solutions Manual

- Solutions to the odd-numbered section exercises
- Solutions to the Practice exercises
- Solutions to the Vocabulary and Readiness Check
- Solutions to every exercise found in the Chapter Reviews and Chapter Tests, cumulative Reviews and Integrated Reviews

MARTIN-GAY'S INTERACTIVE DVD/CD LECTURE SERIES

Martin-Gay's video series has been comprehensively updated to address the way today's students study and learn. The new videos offer students active learning at their pace, with the following resources and more:

- **A complete lecture** for each section of the text, presented by Elayn Martin-Gay. Students can easily review a section or a specific topic before a homework assignment, quiz, or test. Exercises in the text marked with the 📽 are worked on the video.

- A **new interface** with menu and navigation features allows students to quickly find and focus on examples and exercises they need to review.
- Pop-ups reinforce key terms and definitions.
- A new **Practice Final Exam Video** helps students prepare for the final exam. This Practice Final Exam is included in the text in Appendix A.2. At the click of a button, students can watch the full solutions to each exercise on the exam that they need help with. Overview clips provide a brief overview on how to approach different problem types.
- **Interactive Concept Checks** allow students to check their understanding of essential concepts. These multiple choice exercises focus on common misunderstandings. After making their answer selection, students are told whether they're correct or not, and why! Elayn also presents the full solution.
- **Study Skills Builders** help students develop effective study habits and reinforce the advice provided in Section 1.1, Tips for Success in Mathematics, found in the text and video.
- **Close-captioned in Spanish and English**

Beginning Algebra Fifth Edition *Student Study Pack*

The Student Study Pack includes:

- Martin-Gay's Interactive DVD/CD Lecture Series
- Student Solutions Manual
- Pearson Tutor Center

Chapter Test Prep Video CD — Standalone TEST PREP VIDEO

- Includes fully worked-out solutions to every problem from each Chapter Test in the text.

MATHXL® TUTORIALS ON CD

This interactive tutorial CD-ROM provides algorithmically generated practice exercises that are correlated at the objective level to the exercises in the textbook. Every practice exercise is accompanied by an example and a guided solution designed to involve students in the solution process. Selected exercises may also include a video clip to help students visualize concepts. The software provides helpful feedback for incorrect answers and can generate printed summaries of students' progress.

INTERACT MATH TUTORIAL WEBSITE: WWW.INTERACTMATH.COM

Get practice and tutorial help online! This interactive tutorial website provides algorithmically generated practice exercises that correlate directly to the exercises in the textbook. Students can retry an exercise as many times as they like with new values each time for unlimited practice and mastery. Every exercise is accompanied by an interactive guided solution that provides helpful feedback for incorrect answers, and student can also view a worked-out sample problem that steps them through an exercise similar to the one they're working on.

ACKNOWLEDGMENTS

There are many people who helped me develop this text, and I will attempt to thank some of them here. Carrie Green and Edutorial Services were *invaluable* for contributing to the overall accuracy of the text. Suellen Robinson, Lisa Collette, and Kim Lane were *invaluable* for their many suggestions and contributions during the development and writing of this Fifth Edition. Ingrid Benson provided guidance throughout the production process.

A special thanks to my Editor-in-Chief, Paul Murphy, for all of his assistance, support, and contributions to this project. A very special thank you goes to my Sponsoring Editor, Mary Beckwith, for being there 24/7/365, as my students say. Last, my thanks to the staff at Pearson Education for all their support: Linda Behrens, Ilene Kahn, Juan López, Mike Fruhbeis, Kenny Beck, Richard Bretan, Tom Benfatti, Kate Valentine, Michelle Renda, Chris Hoag, and Greg Tobin.

I would like to thank the following reviewers for their input and suggestions:

Barbara Little, *Central Texas College*
Ellen Falkenstein, *Central Texas College*
Teresa Hasenauer, *Indian River Community College*
Maria Witheral, *Pasco Hernandez Community College*
Marie Caruso and students, *Middlesex Community College*

I would also like to thank the following dedicated group of instructors who participated in our focus groups, Martin-Gay Summits, and our design review. Their feedback and insights have helped to strengthen this text. These instructors include:

Cedric Atkins, *Mott Community College*
Michelle Beermann, *Pasco Hernandez Community College*
Laurel Berry, *Bryant & Stratton*
John Beyers, *University of Maryland University College*
Lisa and Bob Brown, *Community College Baltimore County–Essex*
Gail Burkett, *Palm Beach Community College*
Cheryl Cantwell, *Seminole Community College*
Jackie Cohen, *Augusta State*
Julie Dewan, *Mohawk Community College*
Janice Ervin, *Central Piedmont Community College*
Karen Estes, *St. Petersburg College*
Cindy Gaddis, *Tyler Junior College*
Pauline Hall, *Iowa State*
Sonya Johnson, *Central Piedmont Community College*
Irene Jones, *Fullerton College*
Paul Jones, *University of Cincinnati*
Nancy Lange, *Inver Hills Community College*
Sandy Lofstock, *St. Petersburg College*
Jean McArthur, *Joliet Junior College*
Marcia Molle, *Metropolitan Community College*
Greg Nguyen, *Fullerton College*
Linda Padilla, *Joliet Junior College*
Rena Petrello, *Moorpark College*
Ena Salter, *Manatee Community College*
Carole Shapero, *Oakton Community College*
Ann Smallen, *Mohawk Community College*
Jennifer Strehler, *Oakton Community College*
Tanomo Taguchi, *Fullerton College*
Sam Tinsley, *Richland College*
Linda Tucker, *Rose State College*
Leigh Ann Wheeler, *Greenville Technical Community College*
Jenny Wilson, *Tyler Junior College*
Valerie Wright, *Central Piedmont Community College*

A special thank you to those students who participated in our design review: Katherine Browne, Mike Bulfin, Nancy Canipe, Ashley Carpenter, Jeff Chojnachi, Roxanne Davis, Mike Dieter, Amy Dombrowski, Kay Herring, Todd Jaycox, Kaleena Levan, Matt Montgomery, Tony Plese, Abigail Polkinghorn, Harley Price, Eli Robinson, Avery Rosen, Robyn Schott, Cynthia Thomas, and Sherry Ward.

ADDITIONAL ACKNOWLEDGMENTS

As usual, I would like to thank my husband, Clayton, for his constant encouragement. I would also like to thank my children, Eric and Bryan. They are now both attending college and I miss them dearly. I would also like to thank my extended family for their help and wonderful sense of humor. Their contributions are too numerous to list. They are Rod and Karen Pasch; Peter, Michael, Christopher, and Matthew Callac; Jessica and Matt Chavez; Perry and Melissa Landrum; Stuart and Earline Martin; Josh, Mandy, Bailey, Ethan, Avery, and Mia Barnes; Mark, Sabrina, and Madison Martin; Leo and Barbara Miller; and Jewett Gay.

Elayn Martin-Gay

ABOUT THE AUTHOR

Elayn Martin-Gay has taught mathematics at the University of New Orleans for more than 25 years. Her numerous teaching awards include the local University Alumni Association's Award for Excellence in Teaching, and Outstanding Developmental Educator at University of New Orleans, presented by the Louisiana Association of Developmental Educators.

Prior to writing textbooks, Elayn Martin-Gay developed an acclaimed series of lecture videos to support developmental mathematics students in their quest for success. These highly successful videos originally served as the foundation material for her texts. Today, the videos are specific to each book in the Martin-Gay series. The author has also created Chapter Test Prep Videos to help students during their most "teachable moment"—as they prepare for a test, along with Instructor-to-Instructor videos that provide teaching tips, hints, and suggestions for each developmental mathematics course, including basic mathematics, prealgebra, beginning algebra, and intermediate algebra.

Elayn is the author of 12 published textbooks as well as multimedia interactive mathematics, all specializing in developmental mathematics courses. She has participated as an author across the broadest range of educational materials: textbooks, videos, tutorial software, and courseware. This offers an opportunity of various combinations for an integrated teaching and learning package offering great consistency for the student.

Application Index

A

Agriculture
- area of wood needed for water trough, 521
- cattle on Kansas farms, 562
- corn production, 134
- farm size averages, 182
- farms in the U.S., 136–137, 298
- farm land operations, 183
- horse pasture fencing, 281
- price of farmland, 216
- soybean production rate, 562
- U.S. beef production, 246

Animals & Insects
- animal pen dimensions, 124
- animal running speeds, 477
- beetles species, 113
- dog fences, 118
- dog medicine dosages, 237
- flying fish travel distance, 125
- goldfish tank sizes, 123
- piranha fish tank dimensions, 123

Astronomy & Space
- light travel, 125
- magnitude of stars, 1, 16
- meteorite size and weight, 89, 111
- revolutions of planets around the sun, 451
- space plane travel, 123
- temperatures on other planets, 125

Automobiles
- auto manufacturing quality control tours, 471–472
- average fuel economy for autos, 216
- best-seller sports car speeds, 475
- car rental expenses, 287
- car travel, 500
- cost of owning/operating a compact car, 215
- cost of owning/operating a standard truck, 215
- gas prices, 135
- gasoline-electric hybrid cars registered, 225
- number of cars sold in U.S., 134
- unleaded gasoline price per gallon, 181
- used car dealership sales, 134
- used car values, 179
- vehicle sales worldwide, 135

Aviation
- airport elevations in California cities, 47
- corporate jet travel, 475
- jet airplane travel, 138
- jet vs. propeller plane speeds, 143
- plane speed with/against headwinds/tailwinds, 279, 475
- runway lengths, 124

B

Business
- advertising costs, 291–292
- airlines net income/losses, 40
- apparel and accessory stores, 225
- balancing company books, 476
- bookstore sales, 567
- car manufacturers net incomes/losses, 40, 57
- catalog shoppers, 135
- coat sales, 135
- company downsizing, layoffs, 136
- computer disc and notebook sales, 278
- cost of revenue, 292
- costs of manufacturing, 487, 500
- defective light bulbs, 501
- delivery service cost of daily operation, 517
- digital camera average prices, 201
- earnings calculations, 479, 487
- efficiency of experienced vs. apprentice surveyors, 474
- employee hourly wages and production numbers, 183
- fax machine manufacturing costs, 436
- Gap-owned retail stores, 23
- hourly minimum wage in U.S., 235–236
- job completion rates, 473, 477
- large companies, net incomes, losses, 40, 56, 57
- machine processing rates, 500
- music store advertising, 134
- online advertising revenue, 169, 217
- percentages of citizens who shop, 135
- proofreading rates, 476
- retail chains net sales, 174
- retail home improvement stores, 193
- revenues and profit margins, 436
- sales revenues from music CDs, 436
- tourism budgets, 114–115
- T-shirt sale predictions, 221–222
- word processing charges, 287
- World Wide Web use, 301

C

Chemistry
- acid solution mixtures, 134, 137, 161, 165, 248, 281, 297, 298
- antibiotic solutions mixtures, 134
- antifreeze solution, 136
- boiling and freezing points of water, 14
- bug spray concentrates, 475
- copper alloy mixtures, 136
- dosage formula comparisons, 493
- ethanol fuel production, 136
- eye wash solutions and chemicals, 133
- hydrochloric acid solutions, 279
- pharmacy alcohols solutions, 276–277
- saline solution mixtures, 276–277, 278–280, 554
- tanning lotion mixtures, 136
- weed killer mixtures, 475

Communications & Technology
- cell phone prices, 129
- cell phone ranges and distances, 144
- computer desk production costs, 183
- computer software application engineers, 190
- cost of long distance phone call, 211
- effects of cell phone use on driving, 165
- international phone calls, 112
- long distance calling revenue, 243
- radio stations in U.S., 264
- Smart Cards issued, 136
- software packages for computers, 191
- telephone rates, 34
- U.S. households with computers, 584
- value of computers, 178–179
- value of fax machines, 178
- walkie-talkie ranges and distances, 144

Construction & Home Improvement
- architectural plans for a deck, 476
- baseboard and carpeting measurements, 122
- board lengths, 86, 88, 104, 114, 164, 457
- bookcase materials needed, 114
- decking materials needed, 117
- efficiency of experienced vs. apprentice bricklayers, 474
- fertilizer solutions, 280
- flower bed dimensions, 123
- framing dimensions, 122
- garden fencing needed, 117, 248, 281
- lawn manicuring, 500
- lawn shapes, 123
- length of connecting pipe, 540
- length of diagonal wall brace, 540
- lumber needed for fencing, 122
- lumber price, 242
- molding cuts, 70
- painting times, 475, 500
- pipe and concrete estimates, 475
- pitch of a roof, 210, 214, 216
- rope lengths, 113, 124
- sewer pipe slopes, 215
- siding pieces, 114
- steel section lengths, 112
- washer circumferences, 156
- water sealant paints, 166
- wood stick lengths, 107–108

D

Demographics
- birth rate in U.S., 136, 225
- population and health insurance estimates, 475
- population per square mile, 225

E

Economics & Finance
- crude oil production, 226
- ethanol fuel production, 181
- gold prices, 567, 596
- health insurance providers net income, 72
- interest earned on investments, 142–144, 155
- international stocks gains/losses, 37
- investment accounts, 278
- investments of inherited money, 142, 502
- job growth, 249
- large retail chain growth, 183
- McDonald's stock values, 37
- occupation increase predictions, 271
- pharmaceutical industry net income, 237
- retirement pensions, 165
- salary increases, 134
- sales commissions, 155
- silver production trends, 193
- stock gains and losses, 37, 72
- stock investments, 166
- stocks, 279
- Tyson Industries stocks gains and losses, 47

Education
- algebra student assistance, 450
- algebra test scores, 155
- bachelor's degrees earned, 264
- biology books purchased, 136
- college education costs, 14, 130, 134
- college enrollment, 181
- college entrance exam scores, 165
- community college average costs and fees, 153
- doctorates awarded in U.S., 23
- Internet access in college classrooms, 137
- student survey of desired employment benefits, 137
- student: teacher ratio in public schools, 180

Entertainment & Recreation
- amusement park admission prices, 136
- birthday celebration meals, 134
- cable TV news viewers, 281
- card game scores, 46
- CD and cassette tape purchases, 278
- CD holder purchases, 243
- cinema screens in U.S., 111, 137
- cost of recordable CDs, 468
- digital movie screens worldwide, 131–132
- DVD and CD purchases, 279
- eating establishments (number of), 225
- feature films made, 132
- Internet use, 165
- movie theatre admissions, 201–202, 217
- newspaper annual expenditure per person, 156
- overnight stays in national parks, 242
- purpose of Americans traveling, 128–129
- recorded music expenditures, 183
- retirement dinner for two, 134
- retirement party catering fees, 152–153, 155
- Rubik's cube dimensions, 517
- specialty entertainment productions admission prices, 273–274
- television stations available, 111

xxv

Application Index

Entertainment & Recreation (*continued*)
- theater ticket sales, 144, 297
- theme park and resort revenues, 201
- top five Internet users, 170
- top tourist destinations, 180
- vacation trip lengths, 123
- visitors to U.S. National Parks, 15
- wedding anniversary reception costs, 155
- wedding costs, 152, 287
- zoo ticket prices, 144

F

Food & Nutrition
- average daily water use per person, 247
- bottles of Pepsi consumed, 468
- breakfast item charges in delicatessens, 297
- calories in Eagle Brand Milk, 476
- calories per grams, 474
- candy mixtures, 281
- chocolate shop products, 136
- coffee blends, 135, 277, 280
- fish and shellfish consumption, 257
- food manufacturing plants, 165
- food purchases for a barbecue, 451
- lettuce sales and consumption, 135–136
- nut mixtures, 135, 280
- nutrition label elements, 137
- pepper hotness (Scoville units), 137
- pie preparation time, 476
- pineapple sauce sales, 225
- spaghetti supper attendance, 280
- trail mix mixture ingredients, 137

G

Geography
- currents in the bayou, 500
- desert areas, 89, 111
- elevation differences of localities, 44, 45
- elevations of Africa and Tanzania, 47
- Grand Canyon depth, 40
- New Orleans, Louisiana elevation, 46
- volcano heights, 160
- wildfires in U.S., 175

Geology
- glacier flow rates, 116, 125
- production of diamonds in carats, 113
- stalactites and stalagmites, 126
- volcanic lava flow rates, 117, 124

Geometry
- angle measures in geodesic dome, 114
- base of square-based pyramid, 535
- billboard perimeters, 164
- conference table dimensions, 575
- cube surface area, 484
- cylinder dimensions, 547
- cylinder radius, 535
- distance across a lake, 600
- flag dimensions, 111, 112, 113
- parallelogram heights, 502
- Pentagon dimensions, 104
- sphere radius and surface area, 552
- triangle dimensions, 136
- triangle heights, 502
- Vietnam Veterans Memorial angles, 109

H

Health & Medicine
- administering medicine, 89, 96
- attention deficit hyperactivity disorder (ADHD) drugs prescribed, 215
- body mass index, 436
- body surface area formula, 516
- cephalic index for classification of skulls, 436
- cholesterol levels, 155
- forensics, 237
- kidney transplant numbers per year, 575, 596
- kidney transplants performed in U.S., 216
- medical assistants in U.S., 190
- medicine dosages, 436, 457
- smoking and pulse rate, 171–172

M

Miscellaneous
- architects' blueprints, 474
- beach cleaning rate, 464
- consecutive number problems, 96–97, 110, 112, 113, 166
- dry ice temperatures, 125
- electricity needs of cities/towns, 477
- electronic billing statements, 226
- filling problems, 464, 473, 476, 500, 501
- floor space in office buildings, 114
- length of flagpole guywire, 540
- magic squares, 64, 65
- money problems, 144, 165, 599
- sight distance from a height, 543
- superlatives
  - fastest trains, 114
  - largest chocolate bar, 575
  - longest river, 87
  - most cars produced in the world, 475
  - most visited Web sites, 301
  - smallest diamond ring, 165
  - world's first surviving octuplets, 70
  - world's largest pink ribbon, 123
  - world's largest sign, 123
- swimming pool volumes, 164
- weight on balance scales, 80, 81, 82, 90
- wheelchair ramp grades, 215

P

Personal Finances
- bank account balances, 45
- bank statements deposits and withdrawals, 37
- commission sales, 165
- information technologist's salary, 575
- interest on loan, 34
- monthly day care costs, 244
- new mothers returning to work, percentage of, 246
- nonbusiness bankruptcies, 502

Physics
- gas volume, 485
- height of fireball from Roman candle, 590
- Hoberman Sphere expansion/contraction, 125
- metal expansion in heat, 543
- modeling parabolic path of water from fountain, 591
- objects falling/thrown from a height, 225, 487, 541, 561
- parachutist time in free fall, 560, 561
- period of a pendulum, 488, 510
- rocket trajectory, 575
- spring stretch distances with weights attached, 487

Politics & Government
- Democratic and Republican Representatives, 108–109, 248
- electoral votes, 108
- mayoral elections, 88
- Supreme Court decisions, 137
- U.S. government's projected spending, 244
- votes cast, 113

R

Real Estate
- condominium sales and price relationships, 222
- prices of homes, 131

S

Sports
- ballgame admissions, 275
- baseball hits, 436
- baseball home runs, 278
- basketball player heights, 155
- bicycle workouts, 474
- bowling scores, 155
- charity 10K races, 165
- disc throwing women's world records, 137
- football
  - NFL players' salaries, 193
  - ratings, 436
  - yards gained/lost, 46, 72
- golf scores, 41, 71, 165
- hang glider flights, 123
- high school sports records, 475
- ice hockey statistics, 457
- jogging workouts, 474
- pole vault record, 23
- race car competitions, 124–125
- race track distance, 23
- sail heights, 124
- skateboarding vs. inline roller skating, 271
- skateboards manufactured, 144
- snowboarding as a growing sport, 192–193
- sports played in college, 157
- Superbowl
  - attendance, 180
  - regular-season games scores of Superbowl winners, 181
  - scores, 114
- surfboard prices, 129
- Tour de France Championships distances traveled, 165
- treadmill prices, 130
- Water Cube swimming center dimensions, 561
- Winter Olympics medals won, 113
- Women's National Basketball Association scores, 278

T

Temperature & Weather
- Alaska sunset times, 235
- average daily/monthly temperatures, 41, 46, 47, 48, 118, 126, 231, 247
- highest temperatures, 165
- lightning bolt temperatures, 125
- rainfall, 436
- snowfall at distances from the equator, 181–182
- sunrise times, 230–231
- temperature conversions, Fahrenheit to Celsius, 123, 124
- temperatures in cold climates, 37
- thermometer patterns, 35
- tornado wind speed scales, 155
- wind speed, 475

Time & Distance
- bicycling and walking times, 279
- bicycling average speeds, 138–139
- bus vs. car speeds, 143
- car speeds, 275
- distance traveled before and after speeding ticket, 144
- driving distance and time, 116, 274, 481
- driving distance between cities, 34
- hiking and walking speeds, 139, 144, 276, 280, 298
- jogging and walking, 297
- rate of travel, 487
- rowing/paddling with/against currents, 144, 279
- speed of vehicles/persons walking traveling toward each other, 554, 599
- swimming distance against current, 516–517
- train speed records, 73, 87
- train travel speeds, 139–140, 165, 166
- vehicle speeds, 472–473, 476

Transportation & Safety
- bullet train world records, 124
- car stopping distance, 487
- car travel on highways, 280
- catamaran auto ferry operations, 123
- railroad fares, 278
- railroad track grades, 211, 214
- riverboat speeds, 297
- road grades, 210–211, 214, 215
- safe speed on cloverleaf exit, 541
- skid distance on roadway, 541
- suspension bridge length, 87
- taxi cab rides, 287
- traffic signs, 281
- traffic/road sign dimensions, 119, 123, 281
- yield sign shapes, 123

CHAPTER 1

Review of Real Numbers

1.1 Tips for Success in Mathematics
1.2 Symbols and Sets of Numbers
1.3 Fractions
1.4 Introduction to Variable Expressions and Equations
1.5 Adding Real Numbers
1.6 Subtracting Real Numbers
Integrated Review—Operations on Real Numbers
1.7 Multiplying and Dividing Real Numbers
1.8 Properties of Real Numbers

The apparent magnitude of a star is the measure of its brightness as seen by someone on Earth. The smaller the apparent magnitude, the brighter the star. Below, the apparent magnitudes of some stars are listed.

Around 150 B.C., a Greek astronomer, Hipparchus, devised a system of classifying the brightness of stars. Hipparchus's system is the basis of the apparent magnitude scale used by modern astronomers. In Exercises 77 through 82, Section 1.2, we shall see how this scale is used to describe the brightness of objects such as the sun, the moon, and some planets.

Star	Apparent Magnitude	Star	Apparent Magnitude
Arcturus	−0.04	Spica	0.98
Sirius	−1.46	Rigel	0.12
Vega	0.03	Regulus	1.35
Antares	0.96	Canopus	−0.72
Sun	−26.7	Hadar	0.61

(*Source: Norton's 2000.0: Star Atlas and Reference Handbook,* 18th ed., Longman Group, UK, 1989)

The power of mathematics is its flexibility. We apply numbers to almost every aspect of our lives, from an ordinary trip to the grocery store to a rocket launched into space. The power of algebra is its generality. Using letters to represent numbers, we tie together the trip to the grocery store and the launched rocket.

In this chapter we review the basic symbols and words—the language—of arithmetic and introduce using variables in place of numbers. This is our starting place in the study of algebra.

1.1 TIPS FOR SUCCESS IN MATHEMATICS

OBJECTIVES

1. Get ready for this course.
2. Understand some general tips for success.
3. Understand how to use this text.
4. Get help as soon as you need it.
5. Learn how to prepare for and take an exam.
6. Develop good time management.

Before reading this section, remember that your instructor is your best source for information. Please see your instructor for any additional help or information.

OBJECTIVE 1 ▶ Getting ready for this course. Now that you have decided to take this course, remember that a *positive attitude* will make all the difference in the world. Your belief that you can succeed is just as important as your commitment to this course. Make sure you are ready for this course by having the time and positive attitude that it takes to succeed.

Next, make sure you have scheduled your math course at a time that will give you the best chance for success. For example, if you are also working, you may want to check with your employer to make sure that your work hours will not conflict with your course schedule. Also, schedule your class during a time of day when you are more attentive and do your best work.

On the day of your first class period, double-check your schedule and allow yourself extra time to arrive in case of traffic problems or difficulty locating your classroom. Make sure that you bring at least your textbook, paper, and a writing instrument. Are you required to have a lab manual, graph paper, calculator, or some other supply besides this text? If so, also bring this material with you.

OBJECTIVE 2 ▶ General tips for success. Below are some general tips that will increase your chance for success in a mathematics class. Many of these tips will also help you in other courses you may be taking.

Exchange names and phone numbers with at least one other person in class. This contact person can be a great help if you miss an assignment or want to discuss math concepts or exercises that you find difficult.

Choose to attend all class periods and be on time. If possible, sit near the front of the classroom. This way, you will see and hear the presentation better. It may also be easier for you to participate in classroom activities.

Do your homework. You've probably heard the phrase "practice makes perfect" in relation to music and sports. It also applies to mathematics. You will find that the more time you spend solving mathematics problems, the easier the process becomes. Be sure to schedule enough time to complete your assignments before the next class period.

Check your work. Review the steps you made while working a problem. Learn to check your answers in the original problems. You may also compare your answers with the answers to selected exercises section in the back of the book. If you have made a mistake, try to figure out what went wrong. Then correct your mistake. If you can't find what went wrong, don't erase your work or throw it away. Bring your work to your instructor, a tutor in a math lab, or a classmate. It is easier for someone to find where you had trouble if they look at your original work.

Learn from your mistakes and be patient with yourself. Everyone, even your instructor, makes mistakes. (That definitely includes me—Elayn Martin-Gay.) Use your errors to learn and to become a better math student. The key is finding and understanding your errors.

Was your mistake a careless one, or did you make it because you can't read your own math writing? If so, try to work more slowly or write more neatly and make a conscious effort to carefully check your work.

Did you make a mistake because you don't understand a concept? Take the time to review the concept or ask questions to better understand it.

Did you skip too many steps? Skipping steps or trying to do too many steps mentally may lead to preventable mistakes.

Know how to get help if you need it. It's all right to ask for help. In fact, it's a good idea to ask for help whenever there is something that you don't understand. Make sure you know when your instructor has office hours and how to find his or her office. Find out whether math tutoring services are available on your campus. Check out the hours, location, and requirements of the tutoring service. Videotapes and software are available with this text. Learn how to access these resources.

Organize your class materials, including homework assignments, graded quizzes and tests, and notes from your class or lab. All of these items will be valuable references throughout your course especially when studying for upcoming tests and the final exam. Make sure that you can locate these materials when you need them.

Read your textbook before class. Reading a mathematics textbook is unlike leisure reading such as reading a novel or newspaper. Your pace will be much slower. It is helpful to have paper and a pencil with you when you read. Try to work out examples on your own as you encounter them in your text. You should also write down any questions that you want to ask in class. When you read a mathematics textbook, some of the information in a section may be unclear. But when you hear a lecture or watch a videotape on that section, you will understand it much more easily than if you had not read your text beforehand.

Don't be afraid to ask questions. Instructors are not mind readers. Many times we do not know a concept is unclear until a student asks a question. You are not the only person in class with questions. Other students are normally grateful that someone has spoken up.

Hand in assignments on time. This way you can be sure that you will not lose points for being late. Show every step of a problem and be neat and organized. Also be sure that you understand which problems are assigned for homework. You can always double-check this assignment with another student in your class.

OBJECTIVE 3 ▶ Using this text. There are many helpful resources that are available to you in this text. It is important that you become familiar with and use these resources. They should increase your chances for success in this course.

- The main section of exercises in each exercise set is referenced by an example(s). Use this referencing if you have trouble completing an assignment from the exercise set.
- If you need extra help in a particular section, look at the beginning of the section to see what videotapes and software are available.
- Make sure that you understand the meaning of the icons that are beside many exercises. The video icon 📹 tells you that the corresponding exercise may be viewed on the videotape that corresponds to that section. The pencil icon ✎ tells you that this exercise is a writing exercise in which you should answer in complete sentences. The △ icon tells you that the exercise involves geometry.
- Practice exercises are located immediately after a worked-out Example in the text. This exercise is similar to the previous Example and is a great way for you to immediately reinforce a learned concept. All answers to the practice exercises are found in the back of the text.
- Integrated Reviews in each chapter offer you a chance to practice—in one place—the many concepts that you have learned separately over several sections.
- There are many opportunities at the end of each chapter to help you understand the concepts of the chapter.

    Chapter Highlights contain chapter summaries and examples.

    Chapter Reviews contain review problems organized by section.

    Chapter Tests are sample tests to help you prepare for an exam.

    Cumulative Reviews are reviews consisting of material from the beginning of the book to the end of that particular chapter.

- *The Bigger Picture.* This feature contains the directions for building an outline to be used throughout the course. The purpose of this outline is to help you make the transition from thinking "section by section" to thinking about how the mathematics in this course is part of a bigger picture.

- *Study Skills Builder*. This feature is found at the end of many exercise sets. In order to increase your chance of success in this course, please read and answer the questions in the Study Skills Builder. For your convenience, the table below contains selected Study Skills Builder titles and their location.

Study Skills Builder Title	Page of First Occurrence
Learning New Terms	Page 57
What to Do the Day of an Exam?	Page 145
Have You Decided to Complete This Course Successfully?	Page 81
Organizing a Notebook	Page 126
How Are Your Homework Assignments Going?	Page 97
Are You Familiar with Your Textbook Supplements?	Page 202
How Well Do You Know Your Textbook?	Page 312
Are You Organized?	Page 329
Are You Satisfied with Your Performance on a Particular Quiz or Exam?	Page 386
Tips for Studying for an Exam	Page 288
Are You Getting All the Mathematics Help That You Need?	Page 398
How Are You Doing?	Page 451
Are You Preparing for Your Final Exam?	Page 562

See the Preface at the beginning of this text for a more thorough explanation of the features of this text.

OBJECTIVE 4 ▶ Getting help. If you have trouble completing assignments or understanding the mathematics, get help as soon as you need it! This tip is presented as an objective on its own because it is so important. In mathematics, usually the material presented in one section builds on your understanding of the previous section. This means that if you don't understand the concepts covered during a class period, there is a good chance that you will not understand the concepts covered during the next class period. If this happens to you, get help as soon as you can.

Where can you get help? Many suggestions have been made in the section on where to get help, and now it is up to you to do it. Try your instructor, a tutoring center, or a math lab, or you may want to form a study group with fellow classmates. If you do decide to see your instructor or go to a tutoring center, make sure that you have a neat notebook and are ready with your questions.

OBJECTIVE 5 ▶ Preparing for and taking an exam. Make sure that you allow yourself plenty of time to prepare for a test. If you think that you are a little "math anxious," it may be that you are not preparing for a test in a way that will ensure success. The way that you prepare for a test in mathematics is important. To prepare for a test,

1. Review your previous homework assignments. You may also want to rework some of them.
2. Review any notes from class and section-level quizzes you have taken. (If this is a final exam, also review chapter tests you have taken.)
3. Review concepts and definitions by reading the Highlights at the end of each chapter.
4. Practice working out exercises by completing the Chapter Review found at the end of each chapter. (If this is a final exam, go through a Cumulative Review. There is

one found at the end of each chapter except Chapter 1. Choose the review found at the end of the latest chapter that you have covered in your course.) *Don't stop here!*

5. It is important that you place yourself in conditions similar to test conditions to find out how you will perform. In other words, as soon as you feel that you know the material, get a few blank sheets of paper and take a sample test. There is a Chapter Test available at the end of each chapter. During this sample test, do not use your notes or your textbook. Once you complete the Chapter Test, check your answers in the back of the book. If any answer is incorrect, there is a CD available with each exercise of each chapter test worked. Use this CD or your instructor to correct your sample test. Your instructor may also provide you with a review sheet. If you are not satisfied with the results, study the areas that you are weak in and try again.

6. Get a good night's sleep before the exam.

7. On the day of the actual test, allow yourself plenty of time to arrive at where you will be taking your exam.

When taking your test,

1. Read the directions on the test carefully.

2. Read each problem carefully as you take the test. Make sure that you answer the question asked.

3. Pace yourself by first completing the problems you are most confident with. Then work toward the problems you are least confident with. Watch your time so you do not spend too much time on one particular problem.

4. If you have time, check your work and answers.

5. Do not turn your test in early. If you have extra time, spend it double-checking your work.

OBJECTIVE 6 ▶ Managing your time. As a college student, you know the demands that classes, homework, work, and family place on your time. Some days you probably wonder how you'll ever get everything done. One key to managing your time is developing a schedule. Here are some hints for making a schedule:

1. Make a list of all of your weekly commitments for the term. Include classes, work, regular meetings, extracurricular activities, etc. You may also find it helpful to list such things as laundry, regular workouts, grocery shopping, etc.

2. Next, estimate the time needed for each item on the list. Also make a note of how often you will need to do each item. Don't forget to include time estimates for reading, studying, and homework you do outside of your classes. You may want to ask your instructor for help estimating the time needed.

3. In the following exercise set, you are asked to block out a typical week on the schedule grid given. Start with items with fixed time slots like classes and work.

4. Next, include the items on your list with flexible time slots. Think carefully about how best to schedule some items such as study time.

5. Don't fill up every time slot on the schedule. Remember that you need to allow time for eating, sleeping, and relaxing! You should also allow a little extra time in case some items take longer than planned.

6. If you find that your weekly schedule is too full for you to handle, you may need to make some changes in your workload, classload, or in other areas of your life. You may want to talk to your advisor, manager or supervisor at work, or someone in your college's academic counseling center for help with such decisions.

Note: Don't forget in this chapter we begin a feature called Study Skills Builder. The purpose of this feature is to remind you of some of the information given in this section and to further expand on some topics in this section.

1.1 EXERCISE SET

1. What is your instructor's name?
2. What are your instructor's office location and office hours?
3. What is the best way to contact your instructor?
4. What does the ⬊ icon mean?
5. What does the 🎧 icon mean?
6. What does the △ icon mean?
7. Where are answers located in this text?
8. What Exercise Set answers are available to you in the answers section?
9. What Chapter Review, Chapter Test, and Cumulative Review answers are available to you in the answer section?
10. Search after the solution of any worked example in this text. What are Practice exercises?
11. When might be the best time to work a Practice exercise?
12. Where are the answers to Practice exercises?
13. List any similarities between a Practice exercise and the worked example before it.
14. Go to the Highlights section at the end of this chapter. Describe how this section may be helpful to you when preparing for a test.
15. Do you have the name and contact information of at least one other student in class?
16. Will your instructor allow you to use a calculator in this class?
17. Are videotapes, CDs, and/or tutorial software available to you? If so, where?
18. Is there a tutoring service available? If so, what are its hours?
19. Have you attempted this course before? If so, write down ways that you might improve your chances of success during this second attempt.
20. List some steps that you can take if you begin having trouble understanding the material or completing an assignment.
21. Read or reread objective 6 and fill out the schedule grid below.
22. Study your filled-out grid from Exercise 21. Decide whether you have the time necessary to successfully complete this course and any other courses you may be registered for.

	Monday	Tuesday	Wednesday	Thursday	Friday	Saturday	Sunday
4:00 A.M.							
5:00 A.M.							
6:00 A.M.							
7:00 A.M.							
8:00 A.M.							
9:00 A.M.							
10:00 A.M.							
11:00 A.M.							
12:00 P.M.							
1:00 P.M.							
2:00 P.M.							
3:00 P.M.							
4:00 P.M.							
5:00 P.M.							
6:00 P.M.							
7:00 P.M.							
8:00 P.M.							
9:00 P.M.							
10:00 P.M.							
11:00 P.M.							
Midnight							
1:00 A.M.							
2:00 A.M.							
3:00 A.M.							

1.2 SYMBOLS AND SETS OF NUMBERS

OBJECTIVES

1. Use a number line to order numbers.
2. Translate sentences into mathematical statements.
3. Identify natural numbers, whole numbers, integers, rational numbers, irrational numbers, and real numbers.
4. Find the absolute value of a real number.

OBJECTIVE 1 ▶ Using a number line to order numbers. We begin with a review of the set of natural numbers and the set of whole numbers and how we use symbols to compare these numbers. A **set** is a collection of objects, each of which is called a **member** or **element** of the set. A pair of brace symbols { } encloses the list of elements and is translated as "the set of" or "the set containing."

> **Natural Numbers**
> The set of **natural numbers** is $\{1, 2, 3, 4, 5, 6, \dots\}$.

> **Whole Numbers**
> The set of **whole numbers** is $\{0, 1, 2, 3, 4, \dots\}$.

The three dots (an ellipsis) at the end of the list of elements of a set means that the list continues in the same manner indefinitely.

These numbers can be pictured on a **number line.** We will use number lines often to help us visualize distance and relationships between numbers. Visualizing mathematical concepts is an important skill and tool, and later we will develop and explore other visualizing tools.

To draw a number line, first draw a line. Choose a point on the line and label it 0. To the right of 0, label any other point 1. Being careful to use the same distance as from 0 to 1, mark off equally spaced distances. Label these points 2, 3, 4, 5, and so on. Since the whole numbers continue indefinitely, it is not possible to show every whole number on this number line. The arrow at the right end of the line indicates that the pattern continues indefinitely.

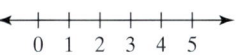

Picturing whole numbers on a number line helps us to see the order of the numbers. Symbols can be used to describe concisely in writing the order that we see.

The **equal symbol** = means "is equal to."

The symbol ≠ means "is not equal to."

These symbols may be used to form a **mathematical statement.** The statement might be true or it might be false. The two statements below are both true.

$2 = 2$ states that "two is equal to two"

$2 \neq 6$ states that "two is not equal to six"

If two numbers are not equal, then one number is larger than the other. The symbol > means "is greater than." The symbol < means "is less than." For example,

$2 > 0$ states that "two is greater than zero"

$3 < 5$ states that "three is less than five"

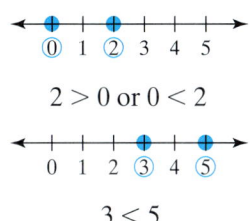

$2 > 0$ or $0 < 2$

$3 < 5$

On a number line, we see that a number **to the right of** another number is **larger.** Similarly, a number **to the left of** another number is smaller. For example, 3 is to the left of 5 on a number line, which means that 3 is less than 5, or $3 < 5$. Similarly, 2 is to the right of 0 on a number line, which means that 2 is greater than 0, or $2 > 0$. Since 0 is to the left of 2, we can also say that 0 is less than 2, or $0 < 2$.

The symbols ≠, <, and > are called **inequality symbols.**

> ▶ **Helpful Hint**
>
> Notice that $2 > 0$ has exactly the same meaning as $0 < 2$. Switching the order of the numbers and reversing the "direction of the inequality symbol" does not change the meaning of the statement.
>
> $5 > 3$ has the same meaning as $3 < 5$.
>
> Also notice that, when the statement is true, the inequality arrow points to the smaller number.

EXAMPLE 1 Insert $<$, $>$, or $=$ in the space between each pair of numbers to make each statement true.

a. 2   3 **b.** 7   4 **c.** 72   27

Solution

a. $2 < 3$ since 2 is to the left of 3 on the number line.
b. $7 > 4$ since 7 is to the right of 4 on the number line.
c. $72 > 27$ since 72 is to the right of 27 on the number line.

PRACTICE 1 Insert $<$, $>$, or $=$ in the space between each pair of numbers to make each statement true.

a. 5   8 **b.** 6   4 **c.** 16   82

Two other symbols are used to compare numbers. The symbol $\leq$ means "is less than or equal to." The symbol $\geq$ means "is greater than or equal to." For example,

$$7 \leq 10 \text{ states that "seven is less than or equal to ten"}$$

This statement is true since $7 < 10$ is true. If either $7 < 10$ or $7 = 10$ is true, then $7 \leq 10$ is true.

$$3 \geq 3 \text{ states that "three is greater than or equal to three"}$$

This statement is true since $3 = 3$ is true. If either $3 > 3$ or $3 = 3$ is true, then $3 \geq 3$ is true.

The statement $6 \geq 10$ is false since neither $6 > 10$ nor $6 = 10$ is true. The symbols $\leq$ and $\geq$ are also called **inequality symbols.**

EXAMPLE 2 Tell whether each statement is true or false.

a. $8 \geq 8$ **b.** $8 \leq 8$ **c.** $23 \leq 0$ **d.** $23 \geq 0$

Solution

a. True, since $8 = 8$ is true.
b. True, since $8 = 8$ is true.
c. False, since neither $23 < 0$ nor $23 = 0$ is true.
d. True, since $23 > 0$ is true.

PRACTICE 2 Tell whether each statement is true or false.

a. $9 \geq 3$ **b.** $3 \geq 8$ **c.** $25 \leq 25$ **d.** $4 \leq 14$

OBJECTIVE 2 ▶ Translating sentences. Now, let's use the symbols discussed above to translate sentences into mathematical statements.

EXAMPLE 3 Translate each sentence into a mathematical statement.

a. Nine is less than or equal to eleven.
b. Eight is greater than one.
c. Three is not equal to four.

Solution

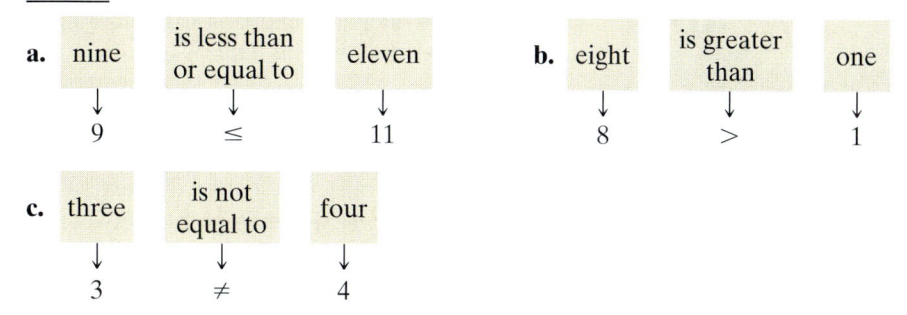

a. nine is less than or equal to eleven
   ↓   ↓   ↓
   9   ≤   11

b. eight is greater than one
   ↓   ↓   ↓
   8   >   1

c. three is not equal to four
   ↓   ↓   ↓
   3   ≠   4

PRACTICE 3 Translate each sentence into a mathematical statement.

a. Three is less than eight.
b. Fifteen is greater than or equal to nine.
c. Six is not equal to seven.

OBJECTIVE 3 ▶ Identifying common sets of numbers. Whole numbers are not sufficient to describe many situations in the real world. For example, quantities smaller than zero must sometimes be represented, such as temperatures less than 0 degrees.

We can picture numbers less than zero on a number line as follows:

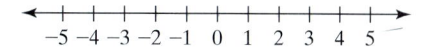

$$\longleftarrow | \ | \ | \ | \ | \ | \ | \ | \ | \ | \ | \longrightarrow$$
$$-5 \ -4 \ -3 \ -2 \ -1 \ 0 \ 1 \ 2 \ 3 \ 4 \ 5$$

Numbers less than 0 are to the left of 0 and are labeled $-1, -2, -3,$ and so on. A $-$ sign, such as the one in -1, tells us that the number is to the left of 0 on a number line. In words, -1 is read "negative one." A $+$ sign or no sign tells us that a number lies to the right of 0 on the number line. For example, 3 and $+3$ both mean positive three.

The numbers we have pictured are called the set of **integers**. Integers to the left of 0 are called **negative integers**; integers to the right of 0 are called **positive integers**. The integer **0 is neither positive nor negative.**

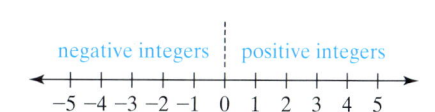

negative integers | positive integers
$$-5 \ -4 \ -3 \ -2 \ -1 \ 0 \ 1 \ 2 \ 3 \ 4 \ 5$$

Integers

The set of **integers** is $\{\ldots, -3, -2, -1, 0, 1, 2, 3, \ldots\}$.

Notice the ellipses (three dots) to the left and to the right of the list for the integers. This indicates that the positive integers and the negative integers continue indefinitely.

10 CHAPTER 1 Review of Real Numbers

EXAMPLE 4 Use an integer to express the number in the following. "Pole of Inaccessibility, Antarctica, is the coldest location in the world, with an average annual temperature of 72 degrees below zero." (*Source: The Guinness Book of Records*)

Solution The integer -72 represents 72 degrees below zero.

PRACTICE 4 Use an integer to express the number in the following: Fred overdrew his checking account and now owes his bank 52 dollars.

A problem with integers in real-life settings arises when quantities are smaller than some integer but greater than the next smallest integer. On a number line, these quantities may be visualized by points between integers. Some of these quantities between integers can be represented as a quotient of integers. For example,

The point on a number line halfway between 0 and 1 can be represented by $\frac{1}{2}$, a quotient of integers.

The point on a number line halfway between 0 and -1 can be represented by $-\frac{1}{2}$. Other quotients of integers and their graphs are shown.

These numbers, each of which can be represented as a quotient of integers, are examples of **rational numbers.** It's not possible to list the set of rational numbers using the notation that we have been using. For this reason, we will use a different notation.

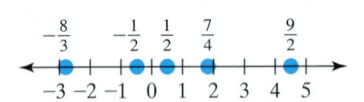

> **Rational Numbers**
> $$\left\{\frac{a}{b}\,\middle|\, a \text{ and } b \text{ are integers and } b \neq 0\right\}$$

We read this set as "the set of all numbers $\frac{a}{b}$ such that a and b are integers and ***b* is not equal to 0.**" Notice that every integer is also a rational number since each integer can be expressed as a quotient of integers. For example, the integer 5 is also a rational number since $5 = \frac{5}{1}$.

The number line also contains points that cannot be expressed as quotients of integers. These numbers are called **irrational numbers** because they cannot be represented by rational numbers. For example, $\sqrt{2}$ and π are irrational numbers.

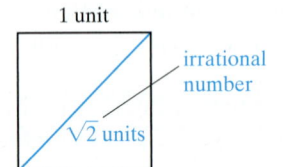

> **Irrational Numbers**
> The set of **irrational numbers** is
>
> {Nonrational numbers that correspond to points on the number line}.
>
> That is, an irrational number is a number that cannot be expressed as a quotient of integers.

Both rational numbers and irrational numbers can be written as decimal numbers. The decimal equivalent of a rational number will either terminate or repeat in a pattern. For example, upon dividing we find that

$$\frac{3}{4} = 0.75 \text{ (decimal number terminates or ends) and}$$

$$\frac{2}{3} = 0.66666\ldots \text{ (decimal number repeats in a pattern)}$$

The decimal representation of an irrational number will neither terminate nor repeat. For example, the decimal representations of irrational numbers $\sqrt{2}$ and π are

$\sqrt{2} = 1.414213562\ldots$ (decimal number does not terminate or repeat in a pattern)

$\pi = 3.141592653\ldots$ (decimal number does not terminate or repeat in a pattern)

(For further review of decimals, see the Appendix.)

Combining the natural numbers with the irrational numbers gives the set of **real numbers**. One and only one point on a number line corresponds to each real number.

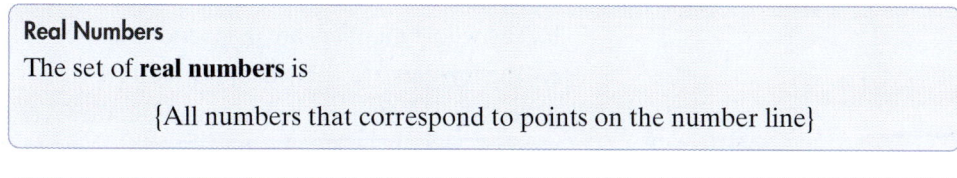

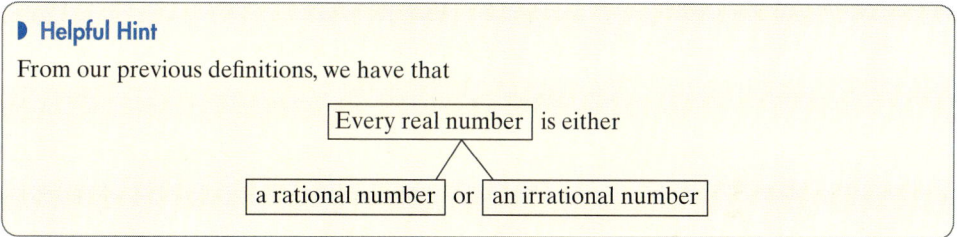

> **Real Numbers**
> The set of **real numbers** is
>
> {All numbers that correspond to points on the number line}

> **▶ Helpful Hint**
> From our previous definitions, we have that
>
> Every real number is either a rational number or an irrational number

On the following number line, we see that real numbers can be positive, negative, or 0. Numbers to the left of 0 are called **negative numbers**; numbers to the right of 0 are called **positive numbers**. Positive and negative numbers are also called **signed numbers**.

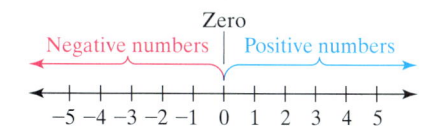

Several different sets of numbers have been discussed in this section. The following diagram shows the relationships among these sets of real numbers.

Common Sets of Numbers

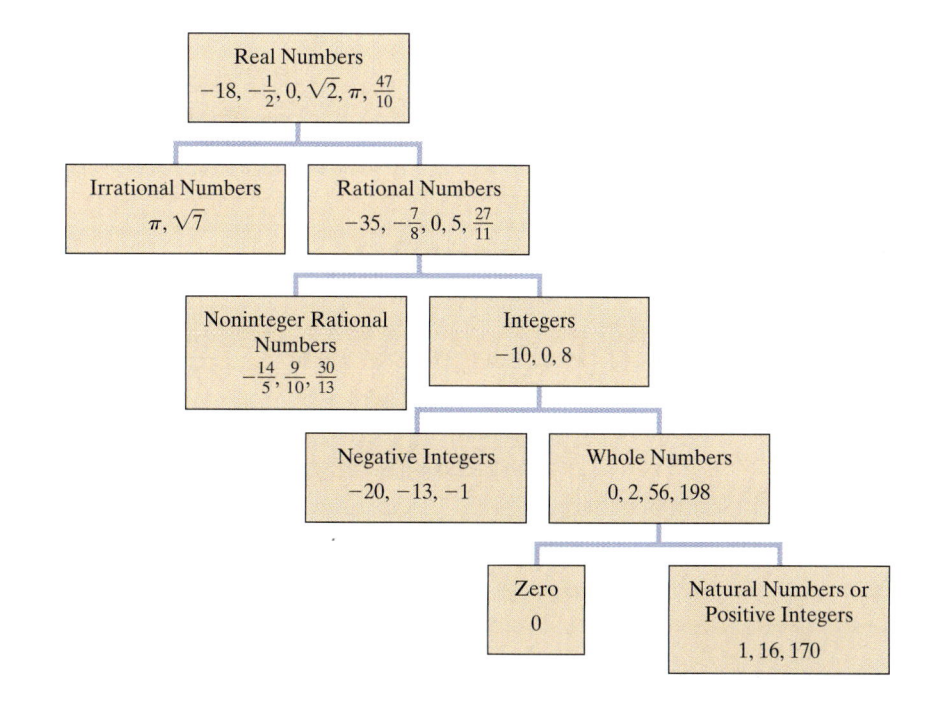

Real Numbers: $-18, -\frac{1}{2}, 0, \sqrt{2}, \pi, \frac{47}{10}$

Irrational Numbers: $\pi, \sqrt{7}$

Rational Numbers: $-35, -\frac{7}{8}, 0, 5, \frac{27}{11}$

Noninteger Rational Numbers: $-\frac{14}{5}, \frac{9}{10}, \frac{30}{13}$

Integers: $-10, 0, 8$

Negative Integers: $-20, -13, -1$

Whole Numbers: $0, 2, 56, 198$

Zero: 0

Natural Numbers or Positive Integers: $1, 16, 170$

EXAMPLE 5 Given the set $\left\{-2, 0, \frac{1}{4}, -1.5, 112, -3, 11, \sqrt{2}\right\}$, list the numbers in this set that belong to the set of:

a. Natural numbers
b. Whole numbers
c. Integers
d. Rational numbers
e. Irrational numbers
f. Real numbers

Solution

a. The natural numbers are 11 and 112.
b. The whole numbers are 0, 11, and 112.
c. The integers are $-3, -2, 0, 11$, and 112.
d. Recall that integers are rational numbers also. The rational numbers are $-3, -2, -1.5, 0, \frac{1}{4}, 11$, and 112.
e. The irrational number is $\sqrt{2}$.
f. The real numbers are all numbers in the given set.

PRACTICE 5 Given the set $\left\{25, \frac{7}{3}, -15, \frac{-3}{4}, \sqrt{5}, -3.7, 8.8, -99\right\}$, list the numbers in this set that belong to the set of:

a. Natural numbers
b. Whole numbers
c. Integers
d. Rational numbers
e. Irrational numbers
f. Real numbers

We can now extend the meaning and use of inequality symbols such as $<$ and $>$ to apply to all real numbers.

Order Property for Real Numbers

Given any two real numbers a and b, $a < b$ if a is to the left of b on a number line. Similarly, $a > b$ if a is to the right of b on a number line.

$a < b$

$a > b$

EXAMPLE 6 Insert $<, >,$ or $=$ in the appropriate space to make each statement true.

a. $-1 \quad 0$
b. $7 \quad \frac{14}{2}$
c. $-5 \quad -6$

Solution

a. $-1 < 0$ since -1 is to the left of 0 on a number line.

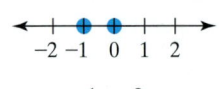

$-1 < 0$

Section 1.2   Symbols and Sets of Numbers   13

b. $7 = \dfrac{14}{2}$ since $\dfrac{14}{2}$ simplifies to 7.

c. $-5 > -6$ since -5 is to the right of -6 on the number line.

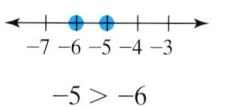

$$-5 > -6$$

PRACTICE 6   Insert $<$, $>$, or $=$ in the appropriate space to make each statement true.

a. 0   3   **b.** 15 -5 **c.** 3 $\dfrac{12}{4}$

OBJECTIVE 4 ▶ Finding the absolute value of a real number. A number line not only gives us a picture of the real numbers, it also helps us visualize the distance between numbers. The distance between a real number a and 0 is given a special name called the **absolute value** of a. "The absolute value of a" is written in symbols as $|a|$.

Absolute Value

The absolute value of a real number a, denoted by $|a|$, is the distance between a and 0 on a number line.

For example, $|3| = 3$ and $|-3| = 3$ since both 3 and -3 are a distance of 3 units from 0 on a number line.

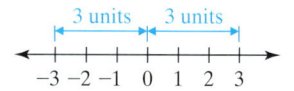

▶ **Helpful Hint**

Since $|a|$ is a distance, $|a|$ is always either positive or 0, never negative. That is, **for any real number a, $|a| \geq 0$.**

EXAMPLE 7   Find the absolute value of each number.

a. $|4|$ **b.** $|-5|$ **c.** $|0|$ **d.** $\left|-\dfrac{1}{2}\right|$ **e.** $|5.6|$

Solution

a. $|4| = 4$ since 4 is 4 units from 0 on a number line.
b. $|-5| = 5$ since -5 is 5 units from 0 on a number line.
c. $|0| = 0$ since 0 is 0 units from 0 on a number line.
d. $\left|-\dfrac{1}{2}\right| = \dfrac{1}{2}$ since $-\dfrac{1}{2}$ is $\dfrac{1}{2}$ unit from 0 on a number line.
e. $|5.6| = 5.6$ since 5.6 is 5.6 units from 0 on a number line.

PRACTICE 7   Find the absolute value of each number.

a. $|-8|$ **b.** $|9|$ **c.** $|-2.5|$ **d.** $\left|\dfrac{5}{11}\right|$ **e.** $|\sqrt{3}|$

EXAMPLE 8 Insert <, >, or = in the appropriate space to make each statement true.

a. $|0|$ 2   b. $|-5|$ 5   c. $|-3|$ $|-2|$ d. $|5|$ $|6|$ e. $|-7|$ $|6|$

Solution

a. $|0| < 2$ since $|0| = 0$ and $0 < 2$.   b. $|-5| = 5$ since $5 = 5$.
c. $|-3| > |-2|$ since $3 > 2$.   d. $|5| < |6|$ since $5 < 6$.
e. $|-7| > |6|$ since $7 > 6$.

PRACTICE 8 Insert <, >, or = in the appropriate space to make each statement true.

a. $|8|$ $|-8|$ b. $|-3|$ 0   c. $|-7|$ $|-11|$ d. $|3|$ $|2|$ e. $|0|$ $|-4|$

VOCABULARY & READINESS CHECK

Use the choices below to fill in each blank.

real   natural   whole   irrational
$|b|$ inequality   integers   rational

1. The _____ numbers are {0, 1, 2, 3, 4, …}.
2. The _____ numbers are {1, 2, 3, 4, 5, …}.
3. The symbols ≠, ≤, and > are called _____ symbols.
4. The _____ are {…, −3, −2, −1, 0, 1, 2, 3, …}.
5. The _____ numbers are {all numbers that correspond to points on the number line}.
6. The _____ numbers are $\left\{ \dfrac{a}{b} \,\middle|\, a \text{ and } b \text{ are integers}, b \neq 0 \right\}$.
7. The _____ numbers are {nonrational numbers that correspond to points on the number line}.
8. The distance between a number b and 0 on a number line is _____.

1.2 EXERCISE SET

Insert <, >, or = in the appropriate space to make the statement true. See Example 1.

1. 7   3
2. 9   15
3. 6.26   6.26
4. 2.13   1.13
5. 0   7
6. 20   0
7. −2   2
8. −4   −6

9. The freezing point of water is 32° Fahrenheit. The boiling point of water is 212° Fahrenheit. Write an inequality statement using < or > comparing the numbers 32 and 212.

10. The freezing point of water is 0° Celsius. The boiling point of water is 100° Celsius. Write an inequality statement using < or > comparing the numbers 0 and 100.

11. The spring 2007 tuition and fees for a Texas resident undergraduate student at University of Texas at El Paso were approximately $2631 for a 15-credit load. At the same time the tuition and fees for a Florida resident attending University of Florida were approximately $2456. Write an inequality statement using < or > comparing the numbers 2631 and 2456. (*Source:* UTEP and UF)

12. The average salary in the San Jose, California, area for a chemical engineer is $67,841. The average salary for a database administrator in the same area is $75,657. Write an inequality statement using < or > comparing the numbers 67,841 and 75,657. (*Source: The Wall Street Journal*)

Are the following statements true or false? See Example 2.

13. 11 ≤ 11
14. 4 ≥ 7
15. 10 > 11
16. 17 > 16
17. 3 + 8 ≥ 3(8)
18. 8 · 8 ≤ 8 · 7
19. 7 > 0
20. 4 < 7

△ 21. An angle measuring 30° is shown and an angle measuring 45° is shown. Use the inequality symbol ≤ or ≥ to write a statement comparing the numbers 30 and 45.

△ 22. The sum of the measures of the angles of a triangle is 180°. The sum of the measures of the angles of a parallelogram is 360°. Use the inequality symbol ≤ or ≥ to write a statement comparing the numbers 360 and 180.

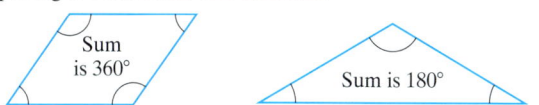

Write each sentence as a mathematical statement. See Example 3.

23. Eight is less than twelve.
24. Fifteen is greater than five.
25. Five is greater than or equal to four.
26. Negative ten is less than or equal to thirty-seven.
27. Fifteen is not equal to negative two.
28. Negative seven is not equal to seven.

Use integers to represent the values in each statement. See Example 4.

29. Driskill Mountain, in Louisiana, has an altitude of 535 feet. New Orleans, Louisiana, lies 8 feet below sea level. (*Source:* U.S. Geological Survey)
30. During a Green Bay Packers football game, the team gained 23 yards and then lost 12 yards on consecutive plays.
31. From 2005 to 2010, the population of Washington, D.C., is expected to decrease by approximately 21,350. (*Source:* U.S. Census Bureau)
32. From 2005 to 2010, the population of Alaska is expected to grow by about 33,000 people. (*Source:* U.S. Census Bureau)
33. Aaron Miller deposited $350 in his savings account. He later withdrew $126.
34. Aris Peña was deep-sea diving. During her dive, she ascended 30 feet and later descended 50 feet.

The graph below is called a bar graph. This particular graph shows the annual numbers of recreational visitors to U.S. National Parks. Each bar represents a different year, and the height of the bar represents the number of visitors (in millions) in that year.

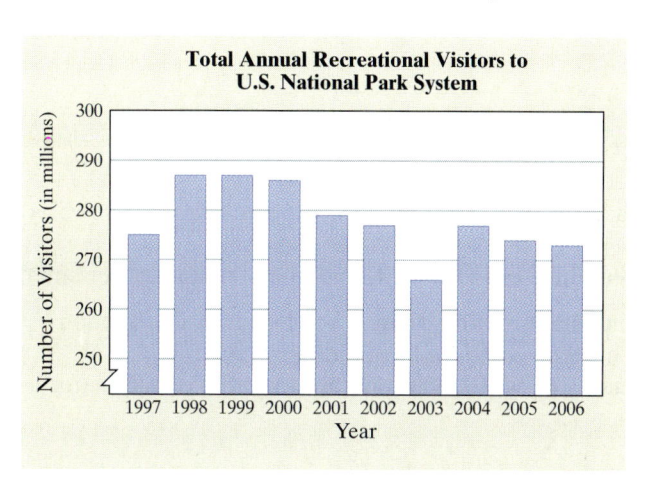

35. In which year(s) was the number of visitors the greatest?
36. What was the greatest number of visitors shown?
37. In what year(s) was the number of visitors greater than 280 million?
38. In what year(s) was the number of visitors less than 270 million?
39. Write an inequality statement comparing the number of annual visitors in 2001 and 2006.
40. Do you notice any trends shown by this bar graph?

Tell which set or sets each number belongs to: natural numbers, whole numbers, integers, rational numbers, irrational numbers, and real numbers. See Example 5.

41. 0
42. $\frac{1}{4}$
43. -2
44. $-\frac{1}{2}$
45. 6
46. 5
47. $\frac{2}{3}$
48. $\sqrt{3}$
49. $-\sqrt{5}$
50. $-1\frac{5}{9}$

Tell whether each statement is true or false.

51. Every rational number is also an integer.
52. Every negative number is also a rational number.
53. Every natural number is positive.
54. Every rational number is also a real number.
55. 0 is a real number.
56. Every real number is also a rational number.
57. Every whole number is an integer.
58. $\frac{1}{2}$ is an integer.
59. A number can be both rational and irrational.
60. Every whole number is positive.

Insert <, >, or = in the appropriate space to make a true statement. See Examples 6 through 8.

61. -10 -100
62. -200 -20
63. 32 5.2
64. 7.1 -7
65. $\frac{18}{3}$ $\frac{24}{3}$
66. $\frac{8}{2}$ $\frac{12}{3}$
67. -51 -50
68. $|-20|$ -200
69. $|-5|$ -4
70. 0 $|0|$
71. $|-1|$ $|1|$
72. $\left|\frac{2}{5}\right|$ $\left|-\frac{2}{5}\right|$
73. $|-2|$ $|-3|$
74. -500 $|-50|$
75. $|0|$ $|-8|$
76. $|-12|$ $\frac{24}{2}$

16　CHAPTER 1　Review of Real Numbers

CONCEPT EXTENSIONS

The apparent magnitude of a star is the measure of its brightness as seen by someone on Earth. The smaller the apparent magnitude, the brighter the star. Use the apparent magnitudes in the table to answer Exercises 77 through 82.

Star	Apparent Magnitude	Star	Apparent Magnitude
Arcturus	−0.04	Spica	0.98
Sirius	−1.46	Rigel	0.12
Vega	0.03	Regulus	1.35
Antares	0.96	Canopus	−0.72
Sun	−26.7	Hadar	0.61

(*Source: Norton's 2000: Star Atlas and Reference Handbook,* 18th ed., Longman Group, UK, 1989)

77. The apparent magnitude of the sun is −26.7. The apparent magnitude of the star Arcturus is −0.04. Write an inequality statement comparing the numbers −0.04 and −26.7.

78. The apparent magnitude of Antares is 0.96. The apparent magnitude of Spica is 0.98. Write an inequality statement comparing the numbers 0.96 and 0.98.

79. Which is brighter, the sun or Arcturus?

80. Which is dimmer, Antares or Spica?

81. Which star listed is the brightest?

82. Which star listed is the dimmest?

Rewrite the following inequalities so that the inequality symbol points in the opposite direction and the resulting statement has the same meaning as the given one.

83. $25 \geq 20$　　　　**84.** $-13 \leq 13$

85. $0 < 6$　　　　**86.** $5 > 3$

87. $-10 > -12$　　　　**88.** $-4 < -2$

89. In your own words, explain how to find the absolute value of a number.

90. Give an example of a real-life situation that can be described with integers but not with whole numbers.

1.3　FRACTIONS

OBJECTIVES

1. Write fractions in simplest form.
2. Multiply and divide fractions.
3. Add and subtract fractions.

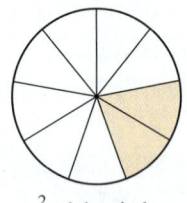

$\frac{2}{9}$ of the circle is shaded.

OBJECTIVE 1 ▶ Writing fractions in simplest form. A quotient of two numbers such as $\frac{2}{9}$ is called a **fraction**. In the fraction $\frac{2}{9}$, the top number, 2, is called the **numerator** and the bottom number, 9, is called the **denominator**.

A fraction may be used to refer to part of a whole. For example, $\frac{2}{9}$ of the circle to the left is shaded. The denominator 9 tells us how many equal parts the whole circle is divided into and the numerator 2 tells us how many equal parts are shaded.

To simplify fractions, we can factor the numerator and the denominator. In the statement $3 \cdot 5 = 15$, 3 and 5 are called **factors** and 15 is the **product**. (The raised dot symbol indicates multiplication.)

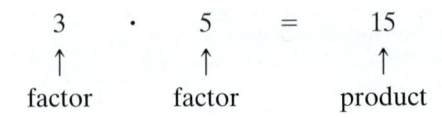

$$\underset{\text{factor}}{3} \cdot \underset{\text{factor}}{5} = \underset{\text{product}}{15}$$

To **factor** 15 means to write it as a product. The number 15 can be factored as $3 \cdot 5$ or as $1 \cdot 15$.

A fraction is said to be **simplified** or in **lowest terms** when the numerator and the denominator have no factors in common other than 1. For example, the fraction $\frac{5}{11}$ is in lowest terms since 5 and 11 have no common factors other than 1.

To help us simplify fractions, we write the numerator and the denominator as a product of **prime numbers.**

> **Prime Number**
> A prime number is a natural number, other than 1, whose only factors are 1 and itself. The first few prime numbers are
>
> $$2, 3, 5, 7, 11, 13, 17, 19, 23, 29, \text{ and so on.}$$

A natural number, other than 1, that is not a prime number is called a **composite number.** Every composite number can be written as a product of prime numbers. We call this product of prime numbers the prime factorization of the composite number.

EXAMPLE 1 Write each of the following numbers as a product of primes.

a. 40   **b.** 63

Solution

a. First, write 40 as the product of any two whole numbers, other than 1.

$$40 = 4 \cdot 10$$

Next, factor each of these numbers. Continue this process until all of the factors are prime numbers.

$$40 = 4 \cdot 10 = 2 \cdot 2 \cdot 2 \cdot 5$$

All the factors are now prime numbers. Then 40 written as a product of primes is

$$40 = 2 \cdot 2 \cdot 2 \cdot 5$$

b. $63 = 9 \cdot 7 = 3 \cdot 3 \cdot 7$

PRACTICE 1 Write each of the following numbers as a product of primes.

a. 36   **b.** 75

To use prime factors to write a fraction in lowest terms, apply the fundamental principle of fractions.

> **Fundamental Principle of Fractions**
> If $\dfrac{a}{b}$ is a fraction and c is a nonzero real number, then
>
> $$\frac{a \cdot c}{b \cdot c} = \frac{a}{b}$$

To understand why this is true, we use the fact that since c is not zero, then $\dfrac{c}{c} = 1$.

$$\frac{a \cdot c}{b \cdot c} = \frac{a}{b} \cdot \frac{c}{c} = \frac{a}{b} \cdot 1 = \frac{a}{b}$$

We will call this process dividing out the common factor of c.

EXAMPLE 2 Write each fraction in lowest terms.

a. $\dfrac{42}{49}$ b. $\dfrac{11}{27}$ c. $\dfrac{88}{20}$

Solution

a. Write the numerator and the denominator as products of primes; then apply the fundamental principle to the common factor 7.

$$\dfrac{42}{49} = \dfrac{2 \cdot 3 \cdot 7}{7 \cdot 7} = \dfrac{2 \cdot 3}{7} = \dfrac{6}{7}$$

b. $\dfrac{11}{27} = \dfrac{11}{3 \cdot 3 \cdot 3}$

There are no common factors other than 1, so $\dfrac{11}{27}$ is already in lowest terms.

c. $\dfrac{88}{20} = \dfrac{2 \cdot 2 \cdot 2 \cdot 11}{2 \cdot 2 \cdot 5} = \dfrac{22}{5}$

PRACTICE 2 Write each fraction in lowest terms.

a. $\dfrac{63}{72}$ b. $\dfrac{64}{12}$ c. $\dfrac{7}{25}$

Concept Check ✓

Explain the error in the following steps.

a. $\dfrac{15}{55} = \dfrac{1\cancel{5}}{5\cancel{5}} = \dfrac{1}{5}$ b. $\dfrac{6}{7} = \dfrac{5+1}{5+2} = \dfrac{1}{2}$

OBJECTIVE 2 ▶ Multiplying and dividing fractions. To multiply two fractions, multiply numerator times numerator to obtain the numerator of the product; multiply denominator times denominator to obtain the denominator of the product.

Multiplying Fractions

$$\dfrac{a}{b} \cdot \dfrac{c}{d} = \dfrac{a \cdot c}{b \cdot d}, \quad \text{if } b \neq 0 \text{ and } d \neq 0$$

EXAMPLE 3 Multiply $\dfrac{2}{15}$ and $\dfrac{5}{13}$. Write the product in lowest terms.

Solution

$$\dfrac{2}{15} \cdot \dfrac{5}{13} = \dfrac{2 \cdot 5}{15 \cdot 13} \quad \text{Multiply numerators.} \\ \text{Multiply denominators.}$$

Next, simplify the product by dividing the numerator and the denominator by any common factors.

$$= \dfrac{2 \cdot 5}{3 \cdot 5 \cdot 13}$$

$$= \dfrac{2}{39}$$

PRACTICE 3 Multiply $\dfrac{3}{8}$ and $\dfrac{7}{9}$. Write the product in lowest terms.

Before dividing fractions, we first define **reciprocals.** Two fractions are reciprocals of each other if their product is 1. For example $\dfrac{2}{3}$ and $\dfrac{3}{2}$ are reciprocals since $\dfrac{2}{3} \cdot \dfrac{3}{2} = 1$. Also, the reciprocal of 5 is $\dfrac{1}{5}$ since $5 \cdot \dfrac{1}{5} = \dfrac{5}{1} \cdot \dfrac{1}{5} = 1$.

Answers to Concept Check:
answers may vary

Section 1.3 Fractions 19

To divide fractions, multiply the first fraction by the reciprocal of the second fraction.

> **Dividing Fractions**
>
> $$\frac{a}{b} \div \frac{c}{d} = \frac{a}{b} \cdot \frac{d}{c}, \quad \text{if } b \neq 0, d \neq 0, \text{ and } c \neq 0$$

EXAMPLE 4 Divide. Write all quotients in lowest terms.

a. $\dfrac{4}{5} \div \dfrac{5}{16}$ b. $\dfrac{7}{10} \div 14$ c. $\dfrac{3}{8} \div \dfrac{3}{10}$

Solution

a. $\dfrac{4}{5} \div \dfrac{5}{16} = \dfrac{4}{5} \cdot \dfrac{16}{5} = \dfrac{4 \cdot 16}{5 \cdot 5} = \dfrac{64}{25}$

b. $\dfrac{7}{10} \div 14 = \dfrac{7}{10} \div \dfrac{14}{1} = \dfrac{7}{10} \cdot \dfrac{1}{14} = \dfrac{7 \cdot 1}{2 \cdot 5 \cdot 2 \cdot 7} = \dfrac{1}{20}.$

c. $\dfrac{3}{8} \div \dfrac{3}{10} = \dfrac{3}{8} \cdot \dfrac{10}{3} = \dfrac{3 \cdot 2 \cdot 5}{2 \cdot 2 \cdot 2 \cdot 3} = \dfrac{5}{4}$

PRACTICE 4 Divide. Write all quotients in lowest terms.

a. $\dfrac{3}{4} \div \dfrac{4}{9}$ b. $\dfrac{5}{12} \div 15$ c. $\dfrac{7}{6} \div \dfrac{7}{15}$

OBJECTIVE 3 ▶ Adding and subtracting fractions. To add or subtract fractions with the same denominator, combine numerators and place the sum or difference over the common denominator.

> **Adding and Subtracting Fractions with the Same Denominator**
>
> $$\frac{a}{b} + \frac{c}{b} = \frac{a + c}{b}, \quad \text{if } b \neq 0$$
>
> $$\frac{a}{b} - \frac{c}{b} = \frac{a - c}{b}, \quad \text{if } b \neq 0$$

EXAMPLE 5 Add or subtract as indicated. Write each result in lowest terms.

a. $\dfrac{2}{7} + \dfrac{4}{7}$ b. $\dfrac{3}{10} + \dfrac{2}{10}$ c. $\dfrac{9}{7} - \dfrac{2}{7}$ d. $\dfrac{5}{3} - \dfrac{1}{3}$

Solution

a. $\dfrac{2}{7} + \dfrac{4}{7} = \dfrac{2 + 4}{7} = \dfrac{6}{7}$ b. $\dfrac{3}{10} + \dfrac{2}{10} = \dfrac{3 + 2}{10} = \dfrac{5}{10} = \dfrac{5}{2 \cdot 5} = \dfrac{1}{2}$

c. $\dfrac{9}{7} - \dfrac{2}{7} = \dfrac{9 - 2}{7} = \dfrac{7}{7} = 1$ d. $\dfrac{5}{3} - \dfrac{1}{3} = \dfrac{5 - 1}{3} = \dfrac{4}{3}$

PRACTICE 5 Add or subtract as indicated. Write each result in lowest terms.

a. $\dfrac{8}{5} - \dfrac{3}{5}$ b. $\dfrac{8}{5} - \dfrac{2}{5}$ c. $\dfrac{3}{5} + \dfrac{1}{5}$ d. $\dfrac{5}{12} + \dfrac{1}{12}$

To add or subtract fractions without the same denominator, first write the fractions as **equivalent fractions** with a common denominator. Equivalent fractions are fractions that represent the same quantity. For example, $\frac{3}{4}$ and $\frac{12}{16}$ are equivalent fractions since they represent the same portion of a whole, as the diagram shows. Count the larger squares and the shaded portion is $\frac{3}{4}$. Count the smaller squares and the shaded portion is $\frac{12}{16}$. Thus, $\frac{3}{4} = \frac{12}{16}$.

We can write equivalent fractions by multiplying a given fraction by 1, as shown in the next example. Multiplying a fraction by 1 does not change the value of the fraction.

EXAMPLE 6 Write $\frac{2}{5}$ as an equivalent fraction with a denominator of 20.

Solution Since $5 \cdot 4 = 20$, multiply the fraction by $\frac{4}{4}$. Multiplying by $\frac{4}{4} = 1$ does not change the value of the fraction.

$$\frac{2}{5} = \frac{2}{5} \cdot \frac{4}{4} = \frac{2 \cdot 4}{5 \cdot 4} = \frac{8}{20}$$

Multiply by $\frac{4}{4}$ or 1.

PRACTICE 6 Write $\frac{2}{3}$ as an equivalent fraction with a denominator of 21.

EXAMPLE 7 Add or subtract as indicated. Write each answer in lowest terms.

a. $\frac{2}{5} + \frac{1}{4}$ **b.** $\frac{1}{2} + \frac{17}{22} - \frac{2}{11}$ **c.** $3\frac{1}{6} - 1\frac{11}{12}$

Solution

a. Fractions must have a common denominator before they can be added or subtracted. Since 20 is the smallest number that both 5 and 4 divide into evenly, 20 is the **least common denominator.** Write both fractions as equivalent fractions with denominators of 20. Since

$$\frac{2}{5} \cdot \frac{4}{4} = \frac{2 \cdot 4}{5 \cdot 4} = \frac{8}{20} \quad \text{and} \quad \frac{1}{4} \cdot \frac{5}{5} = \frac{1 \cdot 5}{4 \cdot 5} = \frac{5}{20}$$

then

$$\frac{2}{5} + \frac{1}{4} = \frac{8}{20} + \frac{5}{20} = \frac{13}{20}$$

b. The least common denominator for denominators 2, 22, and 11 is 22. First, write each fraction as an equivalent fraction with a denominator of 22. Then add or subtract from left to right.

$$\frac{1}{2} = \frac{1}{2} \cdot \frac{11}{11} = \frac{11}{22}, \quad \frac{17}{22} = \frac{17}{22}, \quad \text{and} \quad \frac{2}{11} = \frac{2}{11} \cdot \frac{2}{2} = \frac{4}{22}$$

Then

$$\frac{1}{2} + \frac{17}{22} - \frac{2}{11} = \frac{11}{22} + \frac{17}{22} - \frac{4}{22} = \frac{24}{22} = \frac{12}{11}$$

c. To find $3\frac{1}{6} - 1\frac{11}{12}$, let's use a vertical format.

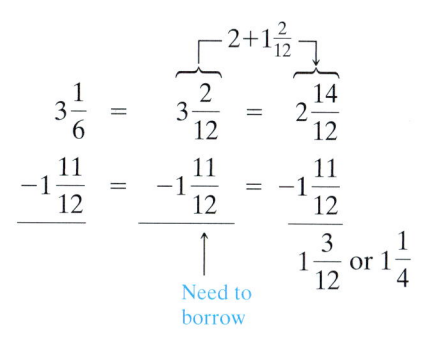

$$3\frac{1}{6} = 3\frac{2}{12} = 2\frac{14}{12}$$
$$-1\frac{11}{12} = -1\frac{11}{12} = -1\frac{11}{12}$$
$$\phantom{-1\frac{11}{12} = -1\frac{11}{12} = } 1\frac{3}{12} \text{ or } 1\frac{1}{4}$$

(Need to borrow; $2 + 1\frac{2}{12}$)

PRACTICE 7 Add or subtract as indicated. Write answers in lowest terms.

a. $\dfrac{5}{11} + \dfrac{1}{7}$ **b.** $9\dfrac{1}{13} - 5\dfrac{1}{2}$ **c.** $\dfrac{1}{3} + \dfrac{29}{30} - \dfrac{4}{5}$

VOCABULARY & READINESS CHECK

Use the choices below to fill in each blank. Some choices may be used more than once.

| simplified | reciprocals | equivalent | denominator |
| product | factors | fraction | numerator |

1. A quotient of two numbers, such as $\dfrac{5}{8}$, is called a _____ .
2. In the fraction $\dfrac{3}{11}$, the number 3 is called the _____ and the number 11 is called the _____ .
3. To factor a number means to write it as a _____ .
4. A fraction is said to be _____ when the numerator and the denominator have no common factors other than 1.
5. In $7 \cdot 3 = 21$, the numbers 7 and 3 are called _____ and the number 21 is called the _____ .
6. The fractions $\dfrac{2}{9}$ and $\dfrac{9}{2}$ are called _____ .
7. Fractions that represent the same quantity are called _____ fractions.

Represent the shaded part of each geometric figure by a fraction

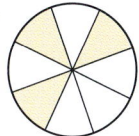

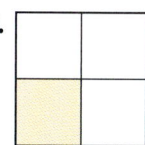

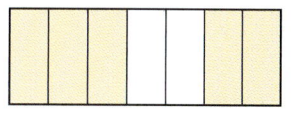

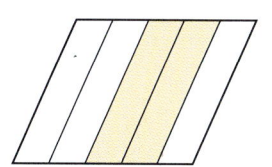

8.   9.   10.   11.

1.3 EXERCISE SET

Write each number as a product of primes. See Example 1.

1. 33
2. 60
3. 98
4. 27
5. 20
6. 56
7. 75
8. 32
9. 45
10. 24

Write the fraction in lowest terms. See Example 2.

11. $\dfrac{2}{4}$
12. $\dfrac{3}{6}$
13. $\dfrac{10}{15}$
14. $\dfrac{15}{20}$
15. $\dfrac{3}{7}$
16. $\dfrac{5}{9}$
17. $\dfrac{18}{30}$
18. $\dfrac{42}{45}$

22 CHAPTER 1 Review of Real Numbers

Multiply or divide as indicated. Write the answer in lowest terms. See Examples 3 and 4.

19. $\dfrac{1}{2} \cdot \dfrac{3}{4}$ 20. $\dfrac{1}{8} \cdot \dfrac{3}{5}$ 21. $\dfrac{2}{3} \cdot \dfrac{3}{4}$

22. $\dfrac{7}{8} \cdot \dfrac{3}{21}$ 23. $\dfrac{1}{2} \div \dfrac{7}{12}$ 24. $\dfrac{7}{12} \div \dfrac{1}{2}$

25. $\dfrac{3}{4} \div \dfrac{1}{20}$ 26. $\dfrac{3}{5} \div \dfrac{9}{10}$ 27. $\dfrac{7}{10} \cdot \dfrac{5}{21}$

28. $\dfrac{3}{35} \cdot \dfrac{10}{63}$ 29. $2\dfrac{7}{9} \cdot \dfrac{1}{3}$ 30. $\dfrac{1}{4} \cdot 5\dfrac{5}{6}$

The area of a plane figure is a measure of the amount of surface of the figure. Find the area of each figure below. (The area of a rectangle is the product of its length and width. The area of a triangle is $\dfrac{1}{2}$ the product of its base and height.)

△ 31. Rectangle: $\dfrac{3}{5}$ mile by $\dfrac{11}{12}$ mile

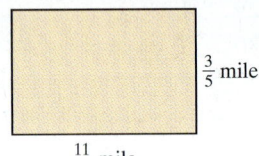

△ 32. Triangle: height $\dfrac{1}{2}$ meter, base $1\dfrac{1}{4}$ meters

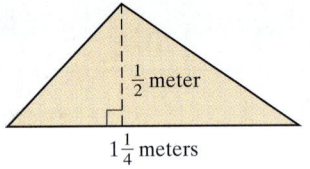

Add or subtract as indicated. Write the answer in lowest terms. See Example 5.

33. $\dfrac{4}{5} - \dfrac{1}{5}$ 34. $\dfrac{6}{7} - \dfrac{1}{7}$

35. $\dfrac{4}{5} + \dfrac{1}{5}$ 36. $\dfrac{6}{7} + \dfrac{1}{7}$

37. $\dfrac{17}{21} - \dfrac{10}{21}$ 38. $\dfrac{18}{35} - \dfrac{11}{35}$

39. $\dfrac{23}{105} + \dfrac{4}{105}$ 40. $\dfrac{13}{132} + \dfrac{35}{132}$

Write each fraction as an equivalent fraction with the given denominator. See Example 6.

41. $\dfrac{7}{10}$ with a denominator of 30

42. $\dfrac{2}{3}$ with a denominator of 9

43. $\dfrac{2}{9}$ with a denominator of 18

44. $\dfrac{8}{7}$ with a denominator of 56

45. $\dfrac{4}{5}$ with a denominator of 20

46. $\dfrac{4}{5}$ with a denominator of 25

Add or subtract as indicated. Write the answer in lowest terms. See Example 7.

47. $\dfrac{2}{3} + \dfrac{3}{7}$ 48. $\dfrac{3}{4} + \dfrac{1}{6}$

49. $2\dfrac{13}{15} - 1\dfrac{1}{5}$ 50. $5\dfrac{2}{9} - 3\dfrac{1}{6}$

51. $\dfrac{5}{22} - \dfrac{5}{33}$ 52. $\dfrac{7}{10} - \dfrac{8}{15}$

53. $\dfrac{12}{5} - 1$ 54. $2 - \dfrac{3}{8}$

Each circle below represents a whole, or 1. Use subtraction to determine the unknown part of the circle.

55. Circle with parts $\dfrac{3}{10}$, $\dfrac{5}{10}$, and ?

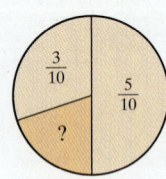

56. Circle with parts $\dfrac{3}{11}$, $\dfrac{2}{11}$, and ?

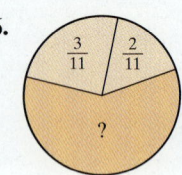

57. Circle with parts ?, $\dfrac{1}{4}$, and $\dfrac{3}{8}$

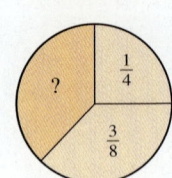

58. Circle with parts ?, $\dfrac{3}{5}$, and $\dfrac{1}{10}$

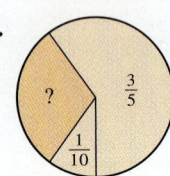

59. Circle with parts $\dfrac{1}{2}$, ?, $\dfrac{2}{9}$, and $\dfrac{1}{6}$

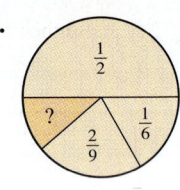

60. Circle with parts $\dfrac{5}{12}$, $\dfrac{1}{3}$, $\dfrac{1}{6}$, and ?

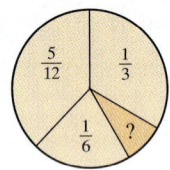

MIXED PRACTICE

Perform the following operations. Write answers in lowest terms.

61. $\dfrac{10}{21} + \dfrac{5}{21}$ 62. $\dfrac{11}{35} + \dfrac{3}{35}$ 63. $\dfrac{10}{3} - \dfrac{5}{21}$

64. $\dfrac{11}{7} - \dfrac{3}{35}$ 65. $\dfrac{2}{3} \cdot \dfrac{3}{5}$ 66. $\dfrac{2}{3} \div \dfrac{3}{5}$

67. $\dfrac{3}{4} \div \dfrac{7}{12}$ 68. $\dfrac{3}{4} \cdot \dfrac{7}{12}$ 69. $\dfrac{5}{12} + \dfrac{4}{12}$

70. $\dfrac{2}{7} + \dfrac{4}{7}$ 71. $5 + \dfrac{2}{3}$ 72. $7 + \dfrac{1}{10}$

73. $\dfrac{7}{8} \div 3\dfrac{1}{4}$ 74. $3 \div \dfrac{3}{4}$ 75. $\dfrac{7}{18} \div \dfrac{14}{36}$

76. $4\dfrac{3}{7} \div \dfrac{31}{7}$ 77. $\dfrac{23}{105} - \dfrac{2}{105}$ 78. $\dfrac{57}{132} - \dfrac{13}{132}$

79. $1\dfrac{1}{2} + 3\dfrac{2}{3}$ 80. $2\dfrac{3}{5} + 4\dfrac{7}{10}$

81. $\dfrac{2}{3} - \dfrac{5}{9} + \dfrac{5}{6}$ 82. $\dfrac{8}{11} - \dfrac{1}{4} + \dfrac{1}{2}$

The perimeter of a plane figure is the total distance around the figure. Find the perimeter of each figure in Exercises 83 and 84.

△ 83. Floor plan with Breakfast Area and Kitchen: top sides $4\dfrac{1}{8}$ feet, 5 feet, $4\dfrac{1}{8}$ feet; left and right sides $15\dfrac{3}{4}$ feet; bottom $10\dfrac{1}{2}$ feet

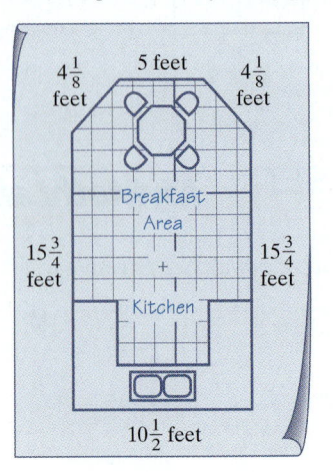

84. [Figure: Bedroom with dimensions $12\frac{3}{8}$ feet, $16\frac{1}{2}$ feet, $2\frac{3}{8}$ feet, $9\frac{1}{4}$ feet]

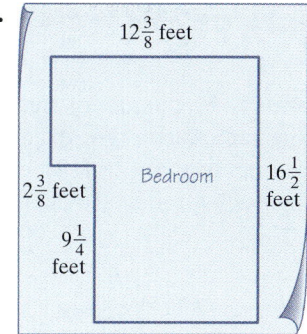

85. Yelena Isinbaeva currently holds the women's pole vault world record at $5\frac{1}{50}$ meters. The men's pole vault world record is currently held by Sergei Bubka, at $1\frac{3}{25}$ meters higher than the Women's record. What is the current men's pole vault record? (*Source: International Association of Athletics Federations*)

86. The Preakness, one of the horse races in the Triple Crown, is a $1\frac{3}{16}$-mile race. The Belmont, another of the three races, is $\frac{5}{16}$ of a mile longer than the Preakness. How long is the Belmont? (*Source: Sports Illustrated*)

87. In your own words, explain how to add two fractions with different denominators.

88. In your own words, explain how to multiply two fractions.

The following trail chart is given to visitors at the Lakeview Forest Preserve.

Trail Name	Distance (miles)
Robin Path	$3\frac{1}{2}$
Red Falls	$5\frac{1}{2}$
Green Way	$2\frac{1}{8}$
Autumn Walk	$1\frac{3}{4}$

89. How much longer is Red Falls Trail than Green Way Trail?

90. Find the total distance traveled by someone who hiked along all four trails.

CONCEPT EXTENSIONS

The breakdown of science and engineering doctorate degrees awarded in the United States is summarized in the graph on the next column, called a circle graph or a pie chart. Use the graph to answer the questions. (Source: National Science Foundation)

91. What fraction of science and engineering doctorates are awarded in the physical sciences?

92. Engineering doctorates make up what fraction of all science and engineering doctorates awarded in the United States?

Science and Engineering Doctorates Awarded, by Field of Study

- Social sciences, 4/25
- Biological/agricultural sciences
- Psychology, 7/50
- Earth, atmospheric, ocean sciences, 3/100
- Physical sciences, 7/50
- Engineering, 21/100
- Mathematical/computer sciences, 7/100

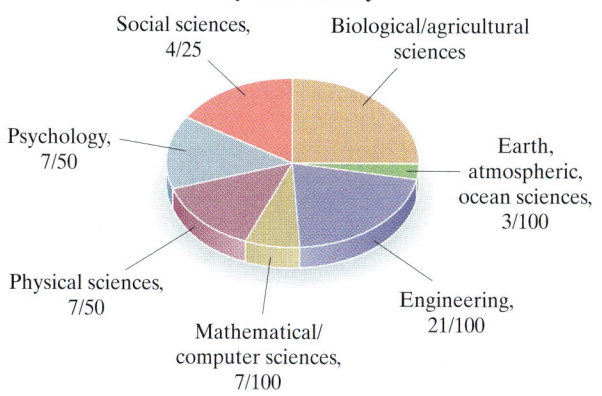

93. What fraction of all science and engineering doctorates are awarded in the biological and agricultural sciences?

94. Social sciences and psychology doctorates together make up what fraction of all science and engineering doctorates awarded in the United States?

In 2006, Gap Inc. operated a total of 3054 stores worldwide. The following chart shows the store breakdown by brand. (Source: Gap Inc.)

Brand	Number of Stores
Gap (Domestic)	1335
Gap (International)	256
Banana Republic	498
Old Navy	960
Forth & Towne	5
Total	3054

95. What fraction of Gap-brand stores were Old Navy stores? Simplify this fraction.

96. What fraction of Gap-brand stores were either domestic or international Gap stores or Forth & Towne stores? Simplify this fraction.

The area of a plane figure is a measure of the amount of surface of the figure. Find the area of each figure. (The area of a triangle is $\frac{1}{2}$ the product of its base and height. The area of a rectangle is the product of its length and width. Recall that area is measured in square units.)

97. [Triangle with height $\frac{4}{9}$ foot and base $\frac{7}{8}$ foot]

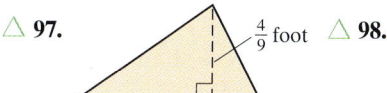

98. [Rectangle with sides $\frac{3}{11}$ meter and $\frac{2}{5}$ meter]

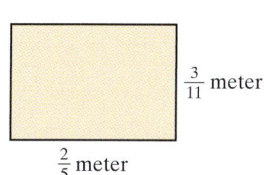

1.4 INTRODUCTION TO VARIABLE EXPRESSIONS AND EQUATIONS

OBJECTIVES

1. Define and use exponents and the order of operations.
2. Evaluate algebraic expressions, given replacement values for variables.
3. Determine whether a number is a solution of a given equation.
4. Translate phrases into expressions and sentences into equations.

OBJECTIVE 1 ▶ Using exponents and the order of operations. Frequently in algebra, products occur that contain repeated multiplication of the same factor. For example, the volume of a cube whose sides each measure 2 centimeters is $(2 \cdot 2 \cdot 2)$ cubic centimeters. We may use **exponential notation** to write such products in a more compact form. For example,

$$2 \cdot 2 \cdot 2 \quad \text{may be written as} \quad 2^3.$$

The 2 in 2^3 is called the **base**; it is the repeated factor. The 3 in 2^3 is called the **exponent** and is the number of times the base is used as a factor. The expression 2^3 is called an **exponential expression.**

$$\text{base} \searrow 2^{\overset{\text{exponent}}{3}} = 2 \cdot 2 \cdot 2 = 8$$

2 is a factor 3 times

Volume is $(2 \cdot 2 \cdot 2)$ cubic centimeters.

EXAMPLE 1 Evaluate the following:

a. 3^2 [read as "3 squared" or as "3 to the second power"]
b. 5^3 [read as "5 cubed" or as "5 to the third power"]
c. 2^4 [read as "2 to the fourth power"]
d. 7^1 e. $\left(\dfrac{3}{7}\right)^2$

Solution

a. $3^2 = 3 \cdot 3 = 9$ b. $5^3 = 5 \cdot 5 \cdot 5 = 125$
c. $2^4 = 2 \cdot 2 \cdot 2 \cdot 2 = 16$ d. $7^1 = 7$
e. $\left(\dfrac{3}{7}\right)^2 = \left(\dfrac{3}{7}\right)\left(\dfrac{3}{7}\right) = \dfrac{9}{49}$

PRACTICE 1 Evaluate:

a. 1^3 b. 5^2 c. $\left(\dfrac{1}{10}\right)^2$ d. 9^1 e. $\left(\dfrac{2}{5}\right)^3$

▶ **Helpful Hint**
$2^3 \neq 2 \cdot 3$ since 2^3 indicates repeated **multiplication** of the same factor.
$$2^3 = 2 \cdot 2 \cdot 2 = 8, \text{ whereas } 2 \cdot 3 = 6.$$

Using symbols for mathematical operations is a great convenience. However, the more operation symbols presented in an expression, the more careful we must be when performing the indicated operation. For example, in the expression $2 + 3 \cdot 7$, do we add first or multiply first? To eliminate confusion, **grouping symbols** are used. Examples of grouping symbols are parentheses ( ), brackets [ ], braces { }, and the fraction bar. If we wish $2 + 3 \cdot 7$ to be simplified by adding first, we enclose $2 + 3$ in parentheses.

$$(2 + 3) \cdot 7 = 5 \cdot 7 = 35$$

If we wish to multiply first, $3 \cdot 7$ may be enclosed in parentheses.

$$2 + (3 \cdot 7) = 2 + 21 = 23$$

Section 1.4 Introduction to Variable Expressions and Equations **25**

To eliminate confusion when no grouping symbols are present, use the following agreed upon order of operations.

> **Order of Operations**
> Simplify expressions using the order below. If grouping symbols such as parentheses are present, simplify expressions within those first, starting with the innermost set. If fraction bars are present, simplify the numerator and the denominator separately.
> 1. Evaluate exponential expressions.
> 2. Perform multiplications or divisions in order from left to right.
> 3. Perform additions or subtractions in order from left to right.

Now simplify $2 + 3 \cdot 7$. There are no grouping symbols and no exponents, so we multiply and then add.

$$2 + 3 \cdot 7 = 2 + 21 \quad \text{Multiply.}$$
$$= 23 \quad \text{Add.}$$

EXAMPLE 2 Simplify each expression.

a. $6 \div 3 + 5^2$ **b.** $\dfrac{2(12 + 3)}{|-15|}$ **c.** $3 \cdot 10 - 7 \div 7$ **d.** $3 \cdot 4^2$ **e.** $\dfrac{3}{2} \cdot \dfrac{1}{2} - \dfrac{1}{2}$

Solution

a. Evaluate 5^2 first.

$$6 \div 3 + 5^2 = 6 \div 3 + 25$$

Next divide, then add.

$$= 2 + 25 \quad \text{Divide.}$$
$$= 27 \quad \text{Add.}$$

b. First, simplify the numerator and the denominator separately.

$$\dfrac{2(12 + 3)}{|-15|} = \dfrac{2(15)}{15} \quad \text{Simplify numerator and denominator separately.}$$
$$= \dfrac{30}{15}$$
$$= 2 \quad \text{Simplify.}$$

c. Multiply and divide from left to right. Then subtract.

$$3 \cdot 10 - 7 \div 7 = 30 - 1$$
$$= 29 \quad \text{Subtract.}$$

d. In this example, only the 4 is squared. The factor of 3 is not part of the base because no grouping symbol includes it as part of the base.

$$3 \cdot 4^2 = 3 \cdot 16 \quad \text{Evaluate the exponential expression.}$$
$$= 48 \quad \text{Multiply.}$$

e. The order of operations applies to operations with fractions in exactly the same way as it applies to operations with whole numbers.

$$\frac{3}{2} \cdot \frac{1}{2} - \frac{1}{2} = \frac{3}{4} - \frac{1}{2} \quad \text{Multiply.}$$

$$= \frac{3}{4} - \frac{2}{4} \quad \text{The least common denominator is 4.}$$

$$= \frac{1}{4} \quad \text{Subtract.}$$

PRACTICE 2 Simplify each expression

a. $6 + 3 \cdot 9$ **b.** $4^3 \div 8 + 3$ **c.** $\left(\frac{2}{3}\right)^2 \cdot |-8|$

d. $\dfrac{9(14 - 6)}{|-2|}$ **e.** $\dfrac{7}{4} \cdot \dfrac{1}{4} - \dfrac{1}{4}$

> **Helpful Hint**
> Be careful when evaluating an exponential expression. In $3 \cdot 4^2$, the exponent 2 applies only to the base 4. In $(3 \cdot 4)^2$, we multiply first because of parentheses, so the exponent 2 applies to the product $3 \cdot 4$.
>
> $$3 \cdot 4^2 = 3 \cdot 16 = 48 \qquad (3 \cdot 4)^2 = (12)^2 = 144$$

Expressions that include many grouping symbols can be confusing. When simplifying these expressions, keep in mind that grouping symbols separate the expression into distinct parts. Each is then simplified separately.

EXAMPLE 3 Simplify $\dfrac{3 + |4 - 3| + 2^2}{6 - 3}$.

Solution The fraction bar serves as a grouping symbol and separates the numerator and denominator. Simplify each separately. Also, the absolute value bars here serve as a grouping symbol. We begin in the numerator by simplifying within the absolute value bars.

$$\frac{3 + |4 - 3| + 2^2}{6 - 3} = \frac{3 + |1| + 2^2}{6 - 3} \quad \text{Simplify the expression inside the absolute value bars.}$$

$$= \frac{3 + 1 + 2^2}{3} \quad \text{Find the absolute value and simplify the denominator.}$$

$$= \frac{3 + 1 + 4}{3} \quad \text{Evaluate the exponential expression.}$$

$$= \frac{8}{3} \quad \text{Simplify the numerator.}$$

PRACTICE 3 Simplify $\dfrac{6^2 - 5}{3 + |6 - 5| \cdot 8}$

EXAMPLE 4 Simplify $3[4 + 2(10 - 1)]$.

Solution Notice that both parentheses and brackets are used as grouping symbols. Start with the innermost set of grouping symbols.

> **Helpful Hint**
> Be sure to follow order of operations and resist the temptation to incorrectly add 4 and 2 first.

$$3[4 + 2(10 - 1)] = 3[4 + 2(9)] \quad \text{Simplify the expression in parentheses.}$$
$$= 3[4 + 18] \quad \text{Multiply.}$$
$$= 3[22] \quad \text{Add.}$$
$$= 66 \quad \text{Multiply.}$$

PRACTICE 4 Simplify $4[25 - 3(5 + 3)]$.

EXAMPLE 5 Simplify $\dfrac{8 + 2 \cdot 3}{2^2 - 1}$.

Solution

$$\frac{8 + 2 \cdot 3}{2^2 - 1} = \frac{8 + 6}{4 - 1} = \frac{14}{3}$$

PRACTICE 5 Simplify $\dfrac{36 \div 9 + 5}{5^2 - 3}$.

OBJECTIVE 2 **Evaluating algebraic expressions.** In algebra, we use symbols, usually letters such as x, y, or z, to represent unknown numbers. A symbol that is used to represent a number is called a **variable.** An **algebraic expression** is a collection of numbers, variables, operation symbols, and grouping symbols. For example,

$$2x, \quad -3, \quad 2x + 10, \quad 5(p^2 + 1), \quad \text{and} \quad \frac{3y^2 - 6y + 1}{5}$$

are algebraic expressions. The expression $2x$ means $2 \cdot x$. Also, $5(p^2 + 1)$ means $5 \cdot (p^2 + 1)$ and $3y^2$ means $3 \cdot y^2$. If we give a specific value to a variable, we can **evaluate an algebraic expression.** To evaluate an algebraic expression means to find its numerical value once we know the values of the variables.

Algebraic expressions often occur during problem solving. For example, the expression

$$16t^2$$

gives the distance in feet (neglecting air resistance) that an object will fall in t seconds. (See Exercise 63 in this section.)

EXAMPLE 6 Evaluate each expression if $x = 3$ and $y = 2$.

a. $2x - y$ **b.** $\dfrac{3x}{2y}$ **c.** $\dfrac{x}{y} + \dfrac{y}{2}$ **d.** $x^2 - y^2$

Solution

a. Replace x with 3 and y with 2.

$$\begin{aligned} 2x - y &= 2(3) - 2 &&\text{Let } x = 3 \text{ and } y = 2. \\ &= 6 - 2 &&\text{Multiply.} \\ &= 4 &&\text{Subtract.} \end{aligned}$$

b. $\dfrac{3x}{2y} = \dfrac{3 \cdot 3}{2 \cdot 2} = \dfrac{9}{4}$ Let $x = 3$ and $y = 2$.

c. Replace x with 3 and y with 2. Then simplify.

$$\dfrac{x}{y} + \dfrac{y}{2} = \dfrac{3}{2} + \dfrac{2}{2} = \dfrac{5}{2}$$

d. Replace x with 3 and y with 2.

$$x^2 - y^2 = 3^2 - 2^2 = 9 - 4 = 5$$

PRACTICE 6 Evaluate each expression if $x = 2$ and $y = 5$.

a. $2x + y$ **b.** $\dfrac{4x}{3y}$ **c.** $\dfrac{3}{x} + \dfrac{x}{y}$ **d.** $x^3 + y^2$

OBJECTIVE 3 ▶ Determining whether a number is a solution of an equation. Many times a problem-solving situation is modeled by an equation. An **equation** is a mathematical statement that two expressions have equal value. The equal symbol "=" is used to equate the two expressions. For example, $3 + 2 = 5$, $7x = 35$, $\dfrac{2(x - 1)}{3} = 0$, and $I = PRT$ are all equations.

> **▶ Helpful Hint**
> An equation contains the equal symbol "=". An algebraic expression does not.

Concept Check ✓

Which of the following are equations? Which are expressions?

a. $5x = 8$ **b.** $5x - 8$ **c.** $12y + 3x$ **d.** $12y = 3x$

When an equation contains a variable, deciding which values of the variable make an equation a true statement is called **solving** an equation for the variable. A **solution** of an equation is a value for the variable that makes the equation true. For example, 3 is a solution of the equation $x + 4 = 7$, because if x is replaced with 3 the statement is true.

$$\begin{aligned} x + 4 &= 7 \\ \downarrow& \\ 3 + 4 &= 7 &&\text{Replace } x \text{ with 3.} \\ 7 &= 7 &&\text{True} \end{aligned}$$

Similarly, 1 is not a solution of the equation $x + 4 = 7$, because $1 + 4 = 7$ is **not** a true statement.

Answers to Concept Check:
equations: a, d; expressions: b, c.

EXAMPLE 7 Decide whether 2 is a solution of $3x + 10 = 8x$.

Solution Replace x with 2 and see if a true statement results.

$$3x + 10 = 8x \quad \text{Original equation}$$
$$3(2) + 10 \stackrel{?}{=} 8(2) \quad \text{Replace } x \text{ with 2.}$$
$$6 + 10 \stackrel{?}{=} 16 \quad \text{Simplify each side.}$$
$$16 = 16 \quad \text{True}$$

Since we arrived at a true statement after replacing x with 2 and simplifying both sides of the equation, 2 is a solution of the equation.

PRACTICE 7 Decide whether 4 is a solution of $9x - 6 = 7x$.

OBJECTIVE 4 ▶ Translating phrases to expressions and sentences to equations. Now that we know how to represent an unknown number by a variable, let's practice translating phrases into algebraic expressions and sentences into equations. Oftentimes solving problems requires the ability to translate word phrases and sentences into symbols. Below is a list of some key words and phrases to help us translate.

> **▶ Helpful Hint**
> Order matters when subtracting and also dividing, so be especially careful with these translations.

Addition (+)	*Subtraction* (−)	*Multiplication* (·)	*Division* (÷)	*Equality* (=)
Sum	Difference of	Product	Quotient	Equals
Plus	Minus	Times	Divide	Gives
Added to	Subtracted from	Multiply	Into	Is/was/should be
More than	Less than	Twice	Ratio	Yields
Increased by	Decreased by	Of	Divided by	Amounts to
Total	Less			Represents/ Is the same as

EXAMPLE 8 Write an algebraic expression that represents each phrase. Let the variable x represent the unknown number.

a. The sum of a number and 3
b. The product of 3 and a number
c. Twice a number
d. 10 decreased by a number
e. 5 times a number, increased by 7

Solution

a. $x + 3$ since "sum" means to add
b. $3 \cdot x$ and $3x$ are both ways to denote the product of 3 and x
c. $2 \cdot x$ or $2x$
d. $10 - x$ because "decreased by" means to subtract
e. $\underbrace{5x}_{\text{5 times a number}} + 7$

PRACTICE 8 Write an algebraic expression that represents each phase. Let the variable x represent the unknown number.

a. Six times a number
b. A number decreased by 8
c. The product of a number and 9
d. Two times a number, plus 3
e. The sum of 7 and a number

> **Helpful Hint**
> Make sure you understand the difference when translating phrases containing "decreased by," "subtracted from," and "less than."
>
> | Phrase | Translation |
> |---|---|
> | A number decreased by 10 | $x - 10$ |
> | A number subtracted from 10 | $10 - x$ |
> | 10 less than a number | $x - 10$ |
> | A number less 10 | $x - 10$ |
>
> Notice the order.

Now let's practice translating sentences into equations.

EXAMPLE 9 Write each sentence as an equation or inequality. Let x represent the unknown number.

a. The quotient of 15 and a number is 4.
b. Three subtracted from 12 is a number.
c. Four times a number, added to 17, is not equal to 21.
d. Triple a number is less than 48.

Solution

a. In words: the quotient of 15 and a number | is | 4

Translate: $\dfrac{15}{x} = 4$

b. In words: three subtracted **from** 12 | is | a number

Translate: $12 - 3 = x$

Care must be taken when the operation is subtraction. The expression $3 - 12$ would be incorrect. Notice that $3 - 12 \neq 12 - 3$.

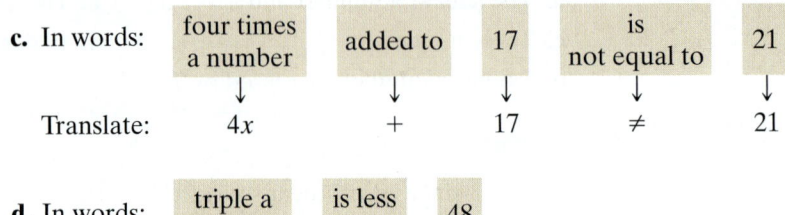

c. In words: four times a number | added to | 17 | is not equal to | 21

Translate: $4x + 17 \neq 21$

d. In words: triple a number | is less than | 48

Translate: $3x < 48$

PRACTICE 9 Write each sentence as an equation or inequality. Let x represent the unknown number.

a. A number increased by 7 is equal to 13.

b. Two less than a number is 11.

c. Double a number, added to 9, is not equal to 25.

d. Five times 11 is greater than or equal to an unknown number.

Calculator Explorations

Exponents

To evaluate exponential expressions on a scientific calculator, find the key marked y^x or $\wedge$. To evaluate, for example, 3^5, press the following keys: 3 y^x 5 = or 3 $\wedge$ 5 = .

↕ or ENTER

The display should read 243 or $3\wedge 5$ / 243

Order of Operations

Some calculators follow the order of operations, and others do not. To see whether or not your calculator has the order of operations built in, use your calculator to find $2 + 3 \cdot 4$. To do this, press the following sequence of keys:

2 + 3 × 4 = .

↕ or ENTER

The correct answer is 14 because the order of operations is to multiply before we add. If the calculator displays 14, then it has the order of operations built in.

Even if the order of operations is built in, parentheses must sometimes be inserted. For example, to simplify $\dfrac{5}{12-7}$, press the keys

5 ÷ (1 2 − 7) = .

↕ or ENTER

The display should read 1 or $5/(12-7)$ / 1

Use a calculator to evaluate each expression.

1. 5^4
2. 7^4
3. 9^5
4. 8^6
5. $2(20 - 5)$
6. $3(14 - 7) + 21$
7. $24(862 - 455) + 89$
8. $99 + (401 + 962)$
9. $\dfrac{4623 + 129}{36 - 34}$
10. $\dfrac{956 - 452}{89 - 86}$

VOCABULARY & READINESS CHECK

Use the choices below to fill in each blank.

add	multiply	equation	variable	base	grouping
subtract	divide	expression	solution	solving	exponent

1. In the expression 5^2, the 5 is called the _____ and the 2 is called the _____.
2. The symbols ( ), [ ], and { } are examples of _____ symbols.
3. A symbol that is used to represent a number is called a(n) _____.
4. A collection of numbers, variables, operation symbols, and grouping symbols is called a(n) _____.
5. A mathematical statement that two expressions are equal is called a(n) _____.
6. A value for the variable that makes an equation a true statement is called a(n) _____.
7. Deciding what values of a variable make an equation a true statement is called _____ the equation.
8. To simplify the expression $1 + 3 \cdot 6$, first _____.
9. To simplify the expression $(1 + 3) \cdot 6$, first _____.
10. To simplify the expression $(20 - 4) \cdot 2$, first _____.
11. To simplify the expression $20 - 4 \div 2$, first _____.

1.4 EXERCISE SET

Evaluate. See Example 1.

1. 3^5
2. 2^5
3. 3^3
4. 4^4
5. 1^5
6. 1^8
7. 5^1
8. 8^1
9. $\left(\dfrac{1}{5}\right)^3$
10. $\left(\dfrac{6}{11}\right)^2$
11. $\left(\dfrac{2}{3}\right)^4$
12. $\left(\dfrac{1}{2}\right)^5$
13. 7^2
14. 9^2
15. 4^2
16. 4^3
17. $(1.2)^2$
18. $(0.07)^2$

MIXED PRACTICE

Simplify each expression. See Examples 2 through 5.

19. $5 + 6 \cdot 2$
20. $8 + 5 \cdot 3$
21. $4 \cdot 8 - 6 \cdot 2$
22. $12 \cdot 5 - 3 \cdot 6$
23. $2(8 - 3)$
24. $5(6 - 2)$
25. $2 + (5 - 2) + 4^2$
26. $6 - 2 \cdot 2 + 2^5$
27. $5 \cdot 3^2$
28. $2 \cdot 5^2$
29. $\dfrac{1}{4} \cdot \dfrac{2}{3} - \dfrac{1}{6}$
30. $\dfrac{3}{4} \cdot \dfrac{1}{2} + \dfrac{2}{3}$
31. $\dfrac{6-4}{9-2}$
32. $\dfrac{8-5}{24-20}$
33. $2[5 + 2(8 - 3)]$
34. $3[4 + 3(6 - 4)]$
35. $\dfrac{19 - 3 \cdot 5}{6 - 4}$
36. $\dfrac{4 \cdot 3 + 2}{4 + 3 \cdot 2}$
37. $\dfrac{|6 - 2| + 3}{8 + 2 \cdot 5}$
38. $\dfrac{15 - |3 - 1|}{12 - 3 \cdot 2}$
39. $\dfrac{3 + 3(5 + 3)}{3^2 + 1}$
40. $\dfrac{3 + 6(8 - 5)}{4^2 + 2}$
41. $\dfrac{6 + |8 - 2| + 3^2}{18 - 3}$
42. $\dfrac{16 + |13 - 5| + 4^2}{17 - 5}$

43. Are parentheses necessary in the expression $2 + (3 \cdot 5)$? Explain your answer.
44. Are parentheses necessary in the expression $(2 + 3) \cdot 5$? Explain your answer.

For Exercises 45 and 46, match each expression in the first column with its value in the second column.

45.
   a. $(6 + 2) \cdot (5 + 3)$ 19
   b. $(6 + 2) \cdot 5 + 3$ 22
   c. $6 + 2 \cdot 5 + 3$ 64
   d. $6 + 2 \cdot (5 + 3)$ 43

46.
   a. $(1 + 4) \cdot 6 - 3$ 15
   b. $1 + 4 \cdot (6 - 3)$ 13
   c. $1 + 4 \cdot 6 - 3$ 27
   d. $(1 + 4) \cdot (6 - 3)$ 22

Evaluate each expression when $x = 1$, $y = 3$, and $z = 5$. See Example 6.

47. $3y$
48. $4x$
49. $\dfrac{z}{5x}$
50. $\dfrac{y}{2z}$

51. $3x - 2$
52. $6y - 8$
53. $|2x + 3y|$
54. $|5z - 2y|$
55. $5y^2$
56. $2z^2$

Evaluate each expression if $x = 12$, $y = 8$, and $z = 4$. See Example 6.

57. $\dfrac{x}{z} + 3y$
58. $\dfrac{y}{z} + 8x$
59. $x^2 - 3y + x$
60. $y^2 - 3x + y$
61. $\dfrac{x^2 + z}{y^2 + 2z}$
62. $\dfrac{y^2 + x}{x^2 + 3y}$

Neglecting air resistance, the expression $16t^2$ gives the distance in feet an object will fall in t seconds.

63. Complete the chart below. To evaluate $16t^2$, remember to first find t^2, then multiply by 16.

Time t (in seconds)	Distance $16t^2$ (in feet)
1	
2	
3	
4	

64. Does an object fall the same distance *during* each second? Why or why not? (See Exercise 63.)

Decide whether the given number is a solution of the given equation. See Example 7.

65. Is 5 a solution of $3x + 30 = 9x$?
66. Is 6 a solution of $2x + 7 = 3x$?
67. Is 0 a solution of $2x + 6 = 5x - 1$?
68. Is 2 a solution of $4x + 2 = x + 8$?
69. Is 8 a solution of $2x - 5 = 5$?
70. Is 6 a solution of $3x - 10 = 8$?
71. Is 2 a solution of $x + 6 = x + 6$?
72. Is 10 a solution of $x + 6 = x + 6$?
73. Is 0 a solution of $x = 5x + 15$?
74. Is 1 a solution of $4 = 1 - x$?

Write each phrase as an algebraic expression. Let x represent the unknown number. See Example 8.

75. Fifteen more than a number
76. One-half times a number
77. Five subtracted from a number
78. The quotient of a number and 9
79. Three times a number, increased by 22
80. The product of 8 and a number, decreased by 10

Write each sentence as an equation or inequality. Use x to represent any unknown number. See Example 9.

81. One increased by two equals the quotient of nine and three.
82. Four subtracted from eight is equal to two squared.
83. Three is not equal to four divided by two.
84. The difference of sixteen and four is greater than ten.
85. The sum of 5 and a number is 20.
86. Twice a number is 17.
87. Thirteen minus three times a number is 13.
88. Seven subtracted from a number is 0.
89. The quotient of 12 and a number is $\dfrac{1}{2}$.
90. The sum of 8 and twice a number is 42.
91. In your own words, explain the difference between an expression and an equation.
92. Determine whether each is an expression or an equation.
    a. $3x^2 - 26$
    b. $3x^2 - 26 = 1$
    c. $2x - 5 = 7x - 5$
    d. $9y + x - 8$

CONCEPT EXTENSIONS

93. Insert parentheses so that the following expression simplifies to 32.
$$20 - 4 \cdot 4 \div 2$$

94. Insert parentheses so that the following expression simplifies to 28.
$$2 \cdot 5 + 3^2$$

Solve the following.

95. The perimeter of a figure is the distance around the figure. The expression $2l + 2w$ represents the perimeter of a rectangle when l is its length and w is its width. Find the perimeter of the following rectangle by substituting 8 for l and 6 for w.

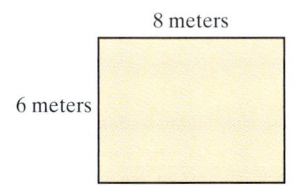

8 meters
6 meters

96. The expression $a + b + c$ represents the perimeter of a triangle when a, b, and c are the lengths of its sides. Find the perimeter of the following triangle.

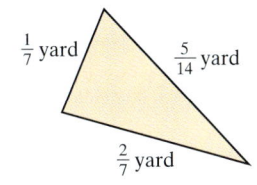

$\frac{1}{7}$ yard
$\frac{5}{14}$ yard
$\frac{2}{7}$ yard

▲ **97.** The area of a figure is the total enclosed surface of the figure. Area is measured in square units. The expression lw represents the area of a rectangle when l is its length and w is its width. Find the area of the following rectangular-shaped lot.

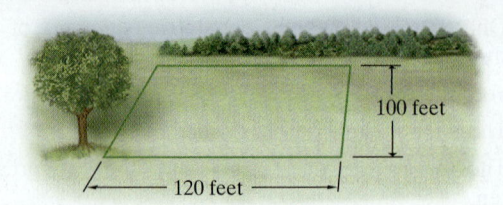

100 feet

120 feet

▲ **98.** A trapezoid is a four-sided figure with exactly one pair of parallel sides. The expression $\frac{1}{2}h(B + b)$ represents its area, when B and b are the lengths of the two parallel sides and h is the height between these sides. Find the area if $B = 15$ inches, $b = 7$ inches, and $h = 5$ inches.

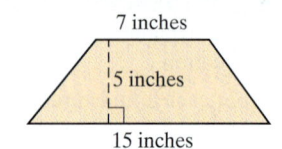

7 inches

5 inches

15 inches

99. The expression $\frac{I}{PT}$ represents the rate of interest being charged if a loan of P dollars for T years required I dollars in interest to be paid. Find the interest rate if a $650 loan for 3 years to buy a used IBM personal computer requires $126.75 in interest to be paid.

100. The expression $\frac{d}{t}$ represents the average speed r in miles per hour if a distance of d miles is traveled in t hours. Find the rate to the nearest whole number if the distance between Dallas, Texas, and Kaw City, Oklahoma, is 432 miles, and it takes Peter Callac 8.5 hours to drive the distance.

101. Verizon long-distance service offers a "Talk to the World Plan," which offers lower rates on international phone calls to Verizon subscribers. This plan charges $3.00 per month, plus $0.12 per minute for calls to Japan. The expression $3.00 + 0.12m$ represents the long-distance charge for a call to Japan last month by a Verizon customer enrolled in this plan. Find the monthly bill for the customer whose only call to Japan lasted 84 minutes.

102. In forensics, the density of a substance is used to help identify it. The expression $\frac{M}{V}$ represents the density of an object with a mass of M grams and a volume of V milliliters. Find the density of an object having a mass of 29.76 grams and a volume of 12 milliliters.

1.5 ADDING REAL NUMBERS

OBJECTIVES

1. Add real numbers with the same sign.
2. Add real numbers with unlike signs.
3. Solve problems that involve addition of real numbers.
4. Find the opposite of a number.

OBJECTIVE 1 ▶ Adding real numbers with the same sign. Real numbers can be added, subtracted, multiplied, divided, and raised to powers, just as whole numbers can. We use a number line to help picture the addition of real numbers.

EXAMPLE 1   Add: $3 + 2$

Solution   Recall that 3 and 2 are called addends. We start at 0 on a number line, and draw an arrow representing the addend 3. This arrow is three units long and points to the right since 3 is positive. From the tip of this arrow, we draw another arrow representing the addend 2. The number below the tip of this arrow is the sum, 5.

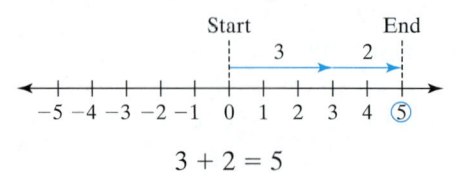

$3 + 2 = 5$

PRACTICE 1   Add using a number line: $2 + 4$.

EXAMPLE 2   Add: $-1 + (-2)$

Solution   Here, -1 and -2 are addends. We start at 0 on a number line, and draw an arrow representing -1. This arrow is one unit long and points to the left since -1 is negative. From the tip of this arrow, we draw another arrow representing -2. The number below the tip of this arrow is the sum, -3.

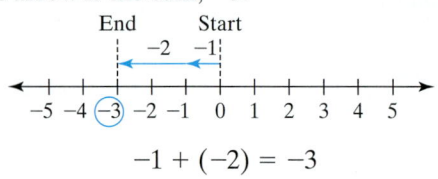

$-1 + (-2) = -3$

PRACTICE
2   Add using a number line: $-2 + (-3)$.

Thinking of signed numbers as money earned or lost might help make addition more meaningful. Earnings can be thought of as positive numbers. If $1 is earned and later another $3 is earned, the total amount earned is $4. In other words, $1 + 3 = 4$.

On the other hand, losses can be thought of as negative numbers. If $1 is lost and later another $3 is lost, a total of $4 is lost. In other words, $(-1) + (-3) = -4$.

Using a number line each time we add two numbers can be time consuming. Instead, we can notice patterns in the previous examples and write rules for adding signed numbers. When adding two numbers with the same sign, notice that the sign of the sum is the same as the sign of the addends.

> **Adding Two Numbers with the Same Sign**
> Add their absolute values. Use their common sign as the sign of the sum.

EXAMPLE 3   Add.

a. $-3 + (-7)$ **b.** $-1 + (-20)$ **c.** $-2 + (-10)$

Solution   Notice that each time, we are adding numbers with the same sign.

a. $-3 + (-7) = -10$ ⟵ Add their absolute values: $3 + 7 = 10$.
      Use their common sign.

b. $-1 + (-20) = -21$ ⟵ Add their absolute values: $1 + 20 = 21$.
      Common sign.

c. $-2 + (-10) = -12$ ⟵ Add their absolute values.
      Common sign.

PRACTICE
3   Add.   **a.** $-5 + (-8)$ **b.** $-31 + (-1)$

OBJECTIVE 2 ▶ Adding real numbers with unlike signs. Adding numbers whose signs are not the same can also be pictured on a number line.

EXAMPLE 4   Add: $-4 + 6$

Solution

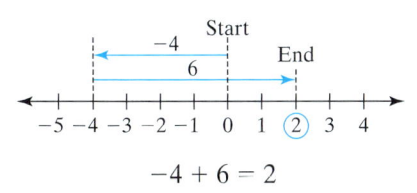

$-4 + 6 = 2$

PRACTICE
4   Add using a number line: $-3 + 8$.

Using temperature as an example, if the thermometer registers 4 degrees below 0 degrees and then rises 6 degrees, the new temperature is 2 degrees above 0 degrees. Thus, it is reasonable that $-4 + 6 = 2$.

Once again, we can observe a pattern: when adding two numbers with different signs, the sign of the sum is the same as the sign of the addend whose absolute value is larger.

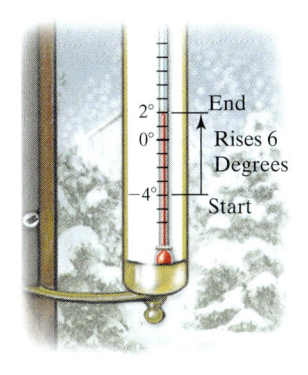

Adding Two Numbers with Different Signs
Subtract the smaller absolute value from the larger absolute value. Use the sign of the number whose absolute value is larger as the sign of the sum.

EXAMPLE 5 Add.

a. $3 + (-7)$ **b.** $-2 + 10$ **c.** $0.2 + (-0.5)$

Solution Notice that each time, we are adding numbers with **different signs**.

a. $3 + (-7) = -4$ ← Subtract their absolute values: $7 - 3 = 4$.
    ↑ The negative number, -7, has the larger absolute value so the sum is negative.

b. $-2 + 10 = 8$ ← Subtract their absolute values: $10 - 2 = 8$.
    ↑ The positive number, 10, has the larger absolute value so the sum is positive.

c. $0.2 + (-0.5) = -0.3$ ← Subtract their absolute values: $0.5 - 0.2 = 0.3$.
    ↑ The negative number, -0.5, has the larger absolute value so the sum is negative.

PRACTICE 5 Add.

a. $15 + (-18)$ **b.** $-19 + 20$ **c.** $-0.6 + 0.4$

EXAMPLE 6 Add.

a. $-8 + (-11)$ **b.** $-5 + 35$ **c.** $0.6 + (-1.1)$

d. $-\dfrac{7}{10} + \left(-\dfrac{1}{10}\right)$ **e.** $11.4 + (-4.7)$ **f.** $-\dfrac{3}{8} + \dfrac{2}{5}$

Solution

a. $-8 + (-11) = -19$ Same sign. Add absolute values and use the common sign.

b. $-5 + 35 = 30$ Different signs. Subtract absolute values and use the sign of the number with the larger absolute value.

c. $0.6 + (-1.1) = -0.5$ Different signs.

d. $-\dfrac{7}{10} + \left(-\dfrac{1}{10}\right) = -\dfrac{8}{10} = -\dfrac{4}{5}$ Same sign.

e. $11.4 + (-4.7) = 6.7$

f. $-\dfrac{3}{8} + \dfrac{2}{5} = -\dfrac{15}{40} + \dfrac{16}{40} = \dfrac{1}{40}$

> **Helpful Hint**
> Don't forget that a common denominator is needed when adding or subtracting fractions. The common denominator here is 40.

PRACTICE 6 Add.

a. $-\dfrac{3}{5} + \left(-\dfrac{2}{5}\right)$ **b.** $3 + (-9)$

c. $2.2 + (-1.7)$ **d.** $-\dfrac{2}{7} + \dfrac{3}{10}$

EXAMPLE 7 Add.

a. $3 + (-7) + (-8)$ **b.** $[7 + (-10)] + [-2 + |-4|]$

Solution

a. Perform the additions from left to right.

$3 + (-7) + (-8) = -4 + (-8)$ Adding numbers with different signs.
$ = -12$ Adding numbers with like signs.

b. Simplify inside brackets first.

$[7 + (-10)] + [-2 + |-4|] = [-3] + [-2 + 4]$
$ = [-3] + [2]$
$ = -1$ Add.

> **Helpful Hint**
> Don't forget that brackets are grouping symbols. We simplify within them first.

PRACTICE 7 Add.

a. $8 + (-5) + (-9)$ **b.** $[-8 + 5] + [-5 + |-2|]$

OBJECTIVE 3 ▶ Solving problems by adding real numbers. Positive and negative numbers are often used in everyday life. Stock market returns show gains and losses as positive and negative numbers. Temperatures in cold climates often dip into the negative range, commonly referred to as "below zero" temperatures. Bank statements report deposits and withdrawals as positive and negative numbers.

EXAMPLE 8 Finding the Gain or Loss of a Stock

During a three-day period, a share of Fremont General Corporation stock recorded the following gains and losses:

Monday	Tuesday	Wednesday
a gain of $2	a loss of $1	a loss of $3

Find the overall gain or loss for the stock for the three days.

Solution Gains can be represented by positive numbers. Losses can be represented by negative numbers. The overall gain or loss is the sum of the gains and losses.

In words: gain plus loss plus loss
Translate: $2 \; + \; (-1) \; + \; (-3) = -2$

The overall loss is $2.

PRACTICE 8 During a three-day period, a share of McDonald's stock recorded the following gain and losses:

Monday	Tuesday	Wednesday
a loss of $5	a gain of $8	a loss of $2

Find the overall gain or loss for the stock for the three days.

OBJECTIVE 4 ▶ Finding the opposite of a number. To help us subtract real numbers in the next section, we first review the concept of opposites. The graphs of 4 and −4 are shown on a number line below.

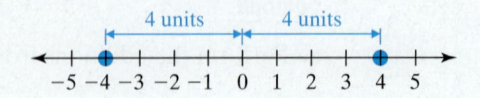

[Number line showing points at −4 and 4, each 4 units from 0]

Notice that 4 and −4 lie on opposite sides of 0, and each is 4 units away from 0.

This relationship between −4 and +4 is an important one. Such numbers are known as **opposites** or **additive inverses** of each other.

> **Opposites or Additive Inverses**
> Two numbers that are the same distance from 0 but lie on opposite sides of 0 are called opposites or additive inverses of each other.

Let's discover another characteristic about opposites. Notice that the sum of a number and its opposite is 0.

$$10 + (-10) = 0$$
$$-3 + 3 = 0$$
$$\frac{1}{2} + \left(-\frac{1}{2}\right) = 0$$

In general, we can write the following:

> The sum of a number a and its opposite $-a$ is 0.
> $$a + (-a) = 0$$

This is why opposites are also called additive inverses. Notice that this also means that the opposite of 0 is then 0 since $0 + 0 = 0$.

EXAMPLE 9 Find the opposite or additive inverse of each number.

a. 5 **b.** −6 **c.** $\frac{1}{2}$ **d.** −4.5

Solution

a. The opposite of 5 is −5. Notice that 5 and −5 are on opposite sides of 0 when plotted on a number line and are equal distances away.
b. The opposite of −6 is 6.
c. The opposite of $\frac{1}{2}$ is $-\frac{1}{2}$.
d. The opposite of −4.5 is 4.5.

PRACTICE 9 Find the opposite or additive inverse of each number.

a. $-\frac{5}{9}$ **b.** 8 **c.** 6.2 **d.** −3

We use the symbol "−" to represent the phrase "the opposite of" or "the additive inverse of." In general, if a is a number, we write the opposite or additive inverse of a as $-a$. We know that the opposite of -3 is 3. Notice that this translates as

the opposite of -3 is 3
$$-(-3) = 3$$

This is true in general.

If a is a number, then $-(-a) = a$.

EXAMPLE 10   Simplify each expression.

a. $-(-10)$ b. $-\left(-\dfrac{1}{2}\right)$ c. $-(-2x)$ d. $-|-6|$

Solution

a. $-(-10) = 10$ b. $-\left(-\dfrac{1}{2}\right) = \dfrac{1}{2}$ c. $-(-2x) = 2x$

d. Since $|-6| = 6$, then $-|-6| = -6$.

PRACTICE 10   Simplify each expression.

a. $-|-15|$ b. $-\left(-\dfrac{3}{5}\right)$

c. $-(-5y)$ d. $-(-8)$

VOCABULARY & READINESS CHECK

Use the choices below to fill in each blank.

     positive number n opposites
     negative number 0 $-n$

1. Two numbers that are the same distance from 0 but lie on opposite sides of 0 are called _____.
2. The sum of a number and its opposite is always _____.
3. If n is a number, then $-(-n) = $ _____.

Tell whether the sum is a positive number, a negative number, or 0. Do not actually find the sum.

4. $-80 + (-127) = $ _____.
5. $-162 + 164 = $ _____.
6. $-162 + 162 = $ _____.
7. $-1.26 + (-8.3) = $ _____.
8. $-3.68 + 0.27 = $ _____.
9. $-\dfrac{2}{3} + \dfrac{2}{3} = $ _____.

1.5 EXERCISE SET

MIXED PRACTICE

Add. See Examples 1 through 7.

1. $6 + 3$
2. $9 + (-12)$
3. $-6 + (-8)$
4. $-6 + (-14)$
5. $8 + (-7)$
6. $6 + (-4)$
7. $-14 + 2$
8. $-10 + 5$
9. $-2 + (-3)$
10. $-7 + (-4)$
11. $-9 + (-3)$
12. $7 + (-5)$
13. $-7 + 3$
14. $-5 + 9$
15. $10 + (-3)$
16. $8 + (-6)$
17. $5 + (-7)$
18. $3 + (-6)$
19. $-16 + 16$
20. $23 + (-23)$
21. $27 + (-46)$
22. $53 + (-37)$
23. $-18 + 49$
24. $-26 + 14$
25. $-33 + (-14)$
26. $-18 + (-26)$
27. $6.3 + (-8.4)$
28. $9.2 + (-11.4)$
29. $|-8| + (-16)$
30. $|-6| + (-61)$
31. $117 + (-79)$
32. $144 + (-88)$
33. $-9.6 + (-3.5)$
34. $-6.7 + (-7.6)$
35. $-\frac{3}{8} + \frac{5}{8}$
36. $-\frac{5}{12} + \frac{7}{12}$
37. $-\frac{7}{16} + \frac{1}{4}$
38. $-\frac{5}{9} + \frac{1}{3}$
39. $-\frac{7}{10} + \left(-\frac{3}{5}\right)$
40. $-\frac{5}{6} + \left(-\frac{2}{3}\right)$
41. $-15 + 9 + (-2)$
42. $-9 + 15 + (-5)$
43. $-21 + (-16) + (-22)$
44. $-18 + (-6) + (-40)$
45. $-23 + 16 + (-2)$
46. $-14 + (-3) + 11$
47. $|5 + (-10)|$
48. $|7 + (-17)|$
49. $6 + (-4) + 9$
50. $8 + (-2) + 7$
51. $[-17 + (-4)] + [-12 + 15]$
52. $[-2 + (-7)] + [-11 + 22]$
53. $|9 + (-12)| + |-16|$
54. $|43 + (-73)| + |-20|$
55. $-1.3 + [0.5 + (-0.3) + 0.4]$
56. $-3.7 + [0.1 + (-0.6) + 8.1]$

Solve. See Example 8.

57. The low temperature in Anoka, Minnesota, was $-15°$ last night. During the day it rose only $9°$. Find the high temperature for the day.

58. On January 2, 1943, the temperature was $-4°$ at 7:30 A.M. in Spearfish, South Dakota. Incredibly, it got $49°$ warmer in the next 2 minutes. To what temperature did it rise by 7:32?

59. The deepest canyon in the world is the Great Canyon of the Yarlung Tsangpo in Tibet. The bottom of the canyon is 17,657 feet below the surrounding terrain, called the rim. If you are standing 1230 feet above the bottom of the canyon, how far from the rim are you?

60. The lowest point in Africa is -512 feet at Lake Assal in Djibouti. If you are standing at a point 658 feet above Lake Assal, what is your elevation? (*Source:* Microsoft Encarta)

A negative net income results when a company's expenses are more than the money brought in.

61. The table below shows net incomes for Ford Motor Company's Automotive sector for the years 2004, 2005, and 2006. Find the total net income for three years.

Year	Net Income (in millions)
2004	$-\$155$
2005	$-\$3895$
2006	$-\$5200$

(*Source:* Ford Motor Company)

62. The table below shows net incomes for Continental Airlines for the years 2004, 2005, and 2006. Find the total net income for these years.

Year	Net Income (in millions)
2004	$-\$409$
2005	$-\$68$
2006	$\$343$

(*Source:* Continental Airlines)

In golf, scores that are under par for the entire round are shown as negative scores; positive scores are shown for scores that are over par, and 0 is par.

63. Paula Creamer was the winner of the 2007 LPGA SBS Open at Turtle Bay. Her scores were $-5, -2,$ and -2. What was her overall score? (*Source:* Ladies Professional Golf Association)
64. During the 2007 PGA Buick Invitational Golf Tournament, Tiger Woods won with scores of $-6, 0, -3,$ and -6. What was his overall score? (*Source:* Professional Golf Association)

Find each additive inverse or opposite. See Example 9.

65. 6
66. 4
67. -2
68. -8
69. 0
70. $-\dfrac{1}{4}$
71. $|-6|$
72. $|-11|$

73. In your own words, explain how to find the opposite of a number.
74. In your own words, explain why 0 is the only number that is its own opposite.

Simplify each of the following. See Example 10.

75. $-|-2|$
76. $-(-3)$
77. $-|0|$
78. $\left|-\dfrac{2}{3}\right|$
79. $-\left|-\dfrac{2}{3}\right|$
80. $-(-7)$

81. Explain why adding a negative number to another negative number always gives a negative sum.
82. When a positive and a negative number are added, sometimes the sum is positive, sometimes it is zero, and sometimes it is negative. Explain why and when this happens.

Decide whether the given number is a solution of the given equation.

83. Is -4 a solution of $x + 9 = 5$?
84. Is 10 a solution of $7 = -x + 3$?
85. Is -1 a solution of $y + (-3) = -7$?
86. Is -6 a solution of $1 = y + 7$?

CONCEPT EXTENSIONS

The following bar graph shows each month's average daily low temperature in degrees Fahrenheit for Barrow, Alaska. Use this graph to answer Exercises 87 through 92.

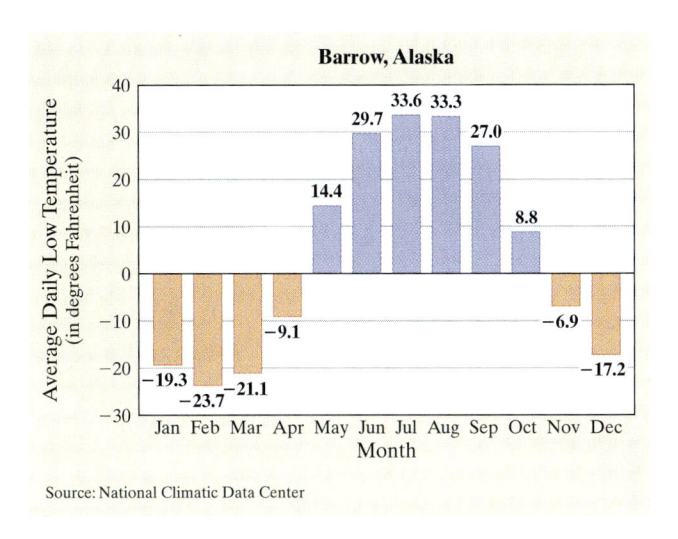

Barrow, Alaska

Jan -19.3, Feb -23.7, Mar -21.1, Apr -9.1, May 14.4, Jun 29.7, Jul 33.6, Aug 33.3, Sep 27.0, Oct 8.8, Nov -6.9, Dec -17.2

Source: National Climatic Data Center

87. For what month is the graphed temperature the highest?
88. For what month is the graphed temperature the lowest?
89. For what month is the graphed temperature positive *and* closest to 0°?
90. For what month is the graphed temperature negative *and* closest to 0°?
91. Find the average of the temperatures shown for the months of April, May, and October. (To find the average of three temperatures, find their sum and divide by 3.)
92. Find the average of the temperatures shown for the months of January, September, and October.

If a is a positive number and b is a negative number, fill in the blanks with the words positive or negative.

93. $-a$ is _____.
94. $-b$ is _____.
95. $a + a$ is _____.
96. $b + b$ is _____.

1.6 SUBTRACTING REAL NUMBERS

OBJECTIVES

1. Subtract real numbers.
2. Add and subtract real numbers.
3. Evaluate algebraic expressions using real numbers.
4. Solve problems that involve subtraction of real numbers.

OBJECTIVE 1 ▶ Subtracting real numbers. Now that addition of signed numbers has been discussed, we can explore subtraction. We know that $9 - 7 = 2$. Notice that $9 + (-7) = 2$, also. This means that

$$9 - 7 = 9 + (-7)$$

Notice that the difference of 9 and 7 is the same as the sum of 9 and the opposite of 7. In general, we have the following.

Subtracting Two Real Numbers

If a and b are real numbers, then $a - b = a + (-b)$.

In other words, to find the difference of two numbers, add the first number to the opposite of the second number.

EXAMPLE 1 Subtract.

a. $-13 - 4$ **b.** $5 - (-6)$ **c.** $3 - 6$ **d.** $-1 - (-7)$

Solution

a. $-13 - 4 = -13 + (-4)$ Add -13 to the opposite of $+4$, which is -4.

$= -17$

b. $5 - (-6) = 5 + (6)$ Add 5 to the opposite of -6, which is 6.

$= 11$

c. $3 - 6 = 3 + (-6)$ Add 3 to the opposite of 6, which is -6.

$= -3$

d. $-1 - (-7) = -1 + (7) = 6$

PRACTICE 1 Subtract.

a. $-7 - 6$ **b.** $-8 - (-1)$ **c.** $9 - (-3)$ **d.** $5 - 7$

▶ **Helpful Hint**

Study the patterns indicated.

No change ──────── Change to addition.
              Change to opposite.

$5 - 11 = 5 + (-11) = -6$
$-3 - 4 = -3 + (-4) = -7$
$7 - (-1) = 7 + (1) = 8$

EXAMPLE 2 Subtract.

a. $5.3 - (-4.6)$ **b.** $-\dfrac{3}{10} - \dfrac{5}{10}$ **c.** $-\dfrac{2}{3} - \left(-\dfrac{4}{5}\right)$

Solution

a. $5.3 - (-4.6) = 5.3 + (4.6) = 9.9$

b. $-\dfrac{3}{10} - \dfrac{5}{10} = -\dfrac{3}{10} + \left(-\dfrac{5}{10}\right) = -\dfrac{8}{10} = -\dfrac{4}{5}$

c. $-\dfrac{2}{3} - \left(-\dfrac{4}{5}\right) = -\dfrac{2}{3} + \left(\dfrac{4}{5}\right) = -\dfrac{10}{15} + \dfrac{12}{15} = \dfrac{2}{15}$ The common denominator is 15.

PRACTICE 2 Subtract.

a. $8.4 - (-2.5)$ **b.** $-\dfrac{5}{8} - \left(-\dfrac{1}{8}\right)$ **c.** $-\dfrac{3}{4} - \dfrac{1}{5}$

EXAMPLE 3 Subtract 8 from -4.

Solution Be careful when interpreting this: The order of numbers in subtraction is important. 8 is to be subtracted **from** -4.
$$-4 - 8 = -4 + (-8) = -12$$

PRACTICE 3 Subtract 5 from -2.

OBJECTIVE 2 ▶ **Adding and subtracting real numbers.** If an expression contains additions and subtractions, just write the subtractions as equivalent additions. Then simplify from left to right.

EXAMPLE 4 Simplify each expression.

a. $-14 - 8 + 10 - (-6)$ b. $1.6 - (-10.3) + (-5.6)$

Solution

a. $-14 - 8 + 10 - (-6) = -14 + (-8) + 10 + 6$
$\qquad = -6$

b. $1.6 - (-10.3) + (-5.6) = 1.6 + 10.3 + (-5.6)$
$\qquad = 6.3$

PRACTICE 4 Simplify each expression.

a. $-15 - 2 - (-4) + 7$ b. $3.5 + (-4.1) - (-6.7)$

When an expression contains parentheses and brackets, remember the order of operations. Start with the innermost set of parentheses or brackets and work your way outward.

EXAMPLE 5 Simplify each expression.

a. $-3 + [(-2 - 5) - 2]$ b. $2^3 - |10| + [-6 - (-5)]$

Solution

a. Start with the innermost sets of parentheses. Rewrite $-2 - 5$ as a sum.

$-3 + [(-2 - 5) - 2] = -3 + [(-2 + (-5)) - 2]$
$\qquad = -3 + [(-7) - 2]$ Add: $-2 + (-5)$.
$\qquad = -3 + [-7 + (-2)]$ Write $-7 - 2$ as a sum.
$\qquad = -3 + [-9]$ Add.
$\qquad = -12$ Add.

b. Start simplifying the expression inside the brackets by writing $-6 - (-5)$ as a sum.

$2^3 - |10| + [-6 - (-5)] = 2^3 - |10| + [-6 + 5]$
$\qquad = 2^3 - |10| + [-1]$ Add.
$\qquad = 8 - 10 + (-1)$ Evaluate 2^3 and $|10|$.
$\qquad = 8 + (-10) + (-1)$ Write $8 - 10$ as a sum.
$\qquad = -2 + (-1)$ Add.
$\qquad = -3$ Add.

PRACTICE 5 Simplify each expression.

a. $-4 + [(-8 - 3) - 5]$ b. $|-13| - 3^2 + [2 - (-7)]$

OBJECTIVE 3 ▶ **Evaluating algebraic expressions.** Knowing how to evaluate expressions for given replacement values is helpful when checking solutions of equations and when solving problems whose unknowns satisfy given expressions. The next example illustrates this.

EXAMPLE 6 Find the value of each expression when $x = 2$ and $y = -5$.

a. $\dfrac{x - y}{12 + x}$ **b.** $x^2 - 3y$

Solution

a. Replace x with 2 and y with -5. Be sure to put parentheses around -5 to separate signs. Then simplify the resulting expression.

$$\frac{x - y}{12 + x} = \frac{2 - (-5)}{12 + 2}$$
$$= \frac{2 + 5}{14}$$
$$= \frac{7}{14} = \frac{1}{2}$$

b. Replace the x with 2 and y with -5 and simplify.

$$x^2 - 3y = 2^2 - 3(-5)$$
$$= 4 - 3(-5)$$
$$= 4 - (-15)$$
$$= 4 + 15$$
$$= 19$$

PRACTICE 6 Find the value of each expression when $x = -3$ and $y = 4$.

a. $\dfrac{7 - x}{2y + x}$ **b.** $y^2 + x$

OBJECTIVE 4 ▶ **Solving problems by subtracting real numbers.** One use of positive and negative numbers is in recording altitudes above and below sea level, as shown in the next example.

EXAMPLE 7 **Finding the Difference in Elevations**

The lowest point on the surface of the Earth is the Dead Sea, at an elevation of 1349 feet below sea level. The highest point is Mt. Everest, at an elevation of 29,035 feet. How much of a variation in elevation is there between these two world extremes? (*Source:* National Geographic Society)

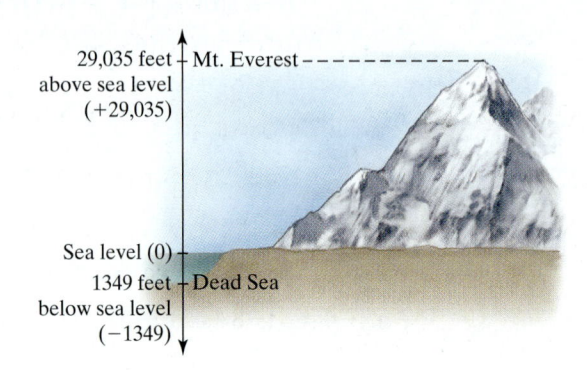

Solution To find the variation in elevation between the two heights, find the difference of the high point and the low point.

In words: high point minus low point

Translate: $29{,}035 \quad - \quad (-1349) = 29{,}035 + 1349$
$$= 30{,}384 \text{ feet}$$

Thus, the variation in elevation is 30,384 feet.

PRACTICE 7 On Tuesday morning, a bank account balance was $282. On Thursday the account balance had dropped to $-\$75$. Find the overall change in this account balance.

A knowledge of geometric concepts is needed by many professionals, such as doctors, carpenters, electronic technicians, gardeners, machinists, and pilots, just to name a few. With this in mind, we review the geometric concepts of **complementary** and **supplementary angles**.

Complementary and Supplementary Angles
Two angles are **complementary** if their sum is 90°.

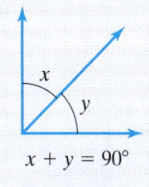

$x + y = 90°$

Two angles are **supplementary** if their sum is 180°.

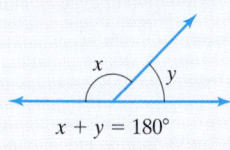

$x + y = 180°$

EXAMPLE 8 Find each unknown complementary or supplementary angle.

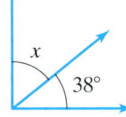

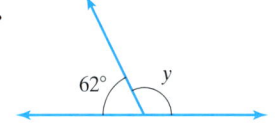

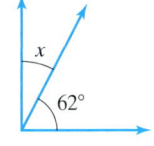

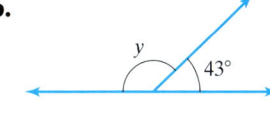

a. (x, 38°)   b. (62°, y)

Solution

a. These angles are complementary, so their sum is $90°$. This means that x is $90° - 38°$.
$$x = 90° - 38° = 52°$$

b. These angles are supplementary, so their sum is $180°$. This means that y is $180° - 62°$.
$$y = 180° - 62° = 118°$$

PRACTICE 8 Find each unknown complementary or supplementary angle.

a. (x, 62°)   b. (y, 43°)

VOCABULARY & READINESS CHECK

Translate each phrase. Let x represent "a number." Use the choices below to fill in each blank.

$7 - x$ $x - 7$

1. 7 minus a number _____
2. 7 subtracted from a number _____
3. A number decreased by 7 _____
4. 7 less a number _____
5. A number less than 7 _____
6. A number subtracted from 7 _____

1.6 EXERCISE SET

MIXED PRACTICE

Subtract. See Examples 1 through 5.

1. $-6 - 4$
2. $-12 - 8$
3. $4 - 9$
4. $8 - 11$
5. $16 - (-3)$
6. $12 - (-5)$
7. $\dfrac{1}{2} - \dfrac{1}{3}$
8. $\dfrac{3}{4} - \dfrac{7}{8}$
9. $-16 - (-18)$
10. $-20 - (-48)$
11. $-6 - 5$
12. $-8 - 4$
13. $7 - (-4)$
14. $3 - (-6)$
15. $-6 - (-11)$
16. $-4 - (-16)$
17. $16 - (-21)$
18. $15 - (-33)$
19. $9.7 - 16.1$
20. $8.3 - 11.2$
21. $-44 - 27$
22. $-36 - 51$
23. $-21 - (-21)$
24. $-17 - (-17)$
25. $-2.6 - (-6.7)$
26. $-6.1 - (-5.3)$
27. $-\dfrac{3}{11} - \left(-\dfrac{5}{11}\right)$
28. $-\dfrac{4}{7} - \left(-\dfrac{1}{7}\right)$
29. $-\dfrac{1}{6} - \dfrac{3}{4}$
30. $-\dfrac{1}{10} - \dfrac{7}{8}$
31. $8.3 - (-0.62)$
32. $4.3 - (-0.87)$

Perform the operation. See Example 3.

33. Subtract -5 from 8.
34. Subtract 3 from -2.
35. Subtract -1 from -6.
36. Subtract 17 from 1.
37. Subtract 8 from 7.
38. Subtract 9 from -4.
39. Decrease -8 by 15.
40. Decrease 11 by -14.

41. In your own words, explain why $5 - 8$ simplifies to a negative number.
42. Explain why $6 - 11$ is the same as $6 + (-11)$.

Simplify each expression. (Remember the order of operations.) See Examples 4 and 5.

43. $-10 - (-8) + (-4) - 20$
44. $-16 - (-3) + (-11) - 14$
45. $5 - 9 + (-4) - 8 - 8$
46. $7 - 12 + (-5) - 2 + (-2)$
47. $-6 - (2 - 11)$
48. $-9 - (3 - 8)$
49. $3^3 - 8 \cdot 9$
50. $2^3 - 6 \cdot 3$
51. $2 - 3(8 - 6)$
52. $4 - 6(7 - 3)$
53. $(3 - 6) + 4^2$
54. $(2 - 3) + 5^2$
55. $-2 + [(8 - 11) - (-2 - 9)]$
56. $-5 + [(4 - 15) - (-6) - 8]$
57. $|-3| + 2^2 + [-4 - (-6)]$
58. $|-2| + 6^2 + (-3 - 8)$

Evaluate each expression when $x = -5$, $y = 4$, and $t = 10$. See Example 6.

59. $x - y$
60. $y - x$
61. $|x| + 2t - 8y$
62. $|x + t - 7y|$
63. $\dfrac{9 - x}{y + 6}$
64. $\dfrac{15 - x}{y + 2}$
65. $y^2 - x$
66. $t^2 - x$
67. $\dfrac{|x - (-10)|}{2t}$
68. $\dfrac{|5y - x|}{6t}$

Solve. See Example 7.

69. Within 24 hours in 1916, the temperature in Browning, Montana, fell from 44 degrees to -56 degrees. How large a drop in temperature was this?

70. Much of New Orleans is below sea level. If George descends 12 feet from an elevation of 5 feet above sea level, what is his new elevation?

71. In a series of plays, the San Francisco 49ers gain 2 yards, lose 5 yards, and then lose another 20 yards. What is their total gain or loss of yardage?

72. In some card games, it is possible to have a negative score. Lavonne Schultz currently has a score of 15 points. She then loses 24 points. What is her new score?

73. Pythagoras died in the year −475 (or 475 B.C.). When was he born, if he was 94 years old when he died?
74. The Greek astronomer and mathematician Geminus died in 60 A.D. at the age of 70. When was he born?
75. A commercial jet liner hits an air pocket and drops 250 feet. After climbing 120 feet, it drops another 178 feet. What is its overall vertical change?
76. Tyson Industries stock posted a loss of 1.625 points yesterday. If it drops another 0.75 point today, find its overall change for the two days.
77. The highest point in Africa is Mt. Kilimanjaro, Tanzania, at an elevation of 19,340 feet. The lowest point is Lake Assal, Djibouti, at 512 feet below sea level. How much higher is Mt. Kilimanjaro than Lake Assal? (*Source:* National Geographic Society)

78. The airport in Bishop, California, is at an elevation of 4101 feet above sea level. The nearby Furnace Creek Airport in Death Valley, California, is at an elevation of 226 feet below sea level. How much higher in elevation is the Bishop Airport than the Furnace Creek Airport? (*Source:* National Climatic Data Center)

Find each unknown complementary or supplementary angle. See Example 8.

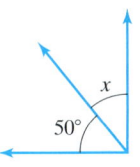

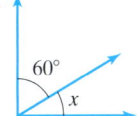

79. [angle diagram: y, 50°]
80. [angle diagram: x, 50°]
81. [angle diagram: 60°, x]
82. [angle diagram: y, 105°]

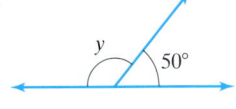

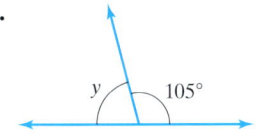

Decide whether the given number is a solution of the given equation.

83. Is −4 a solution of $x - 9 = 5$?
84. Is 3 a solution of $x - 10 = -7$?
85. Is −2 a solution of $-x + 6 = -x - 1$?
86. Is −10 a solution of $-x - 6 = -x - 1$?
87. Is 2 a solution of $-x - 13 = -15$?
88. Is 5 a solution of $4 = 1 - x$?

CONCEPT EXTENSIONS

Recall from the last section the bar graph below that shows each month's average daily low temperature in degrees Fahrenheit for Barrow, Alaska. Use this graph to answer Exercises 89 through 91.

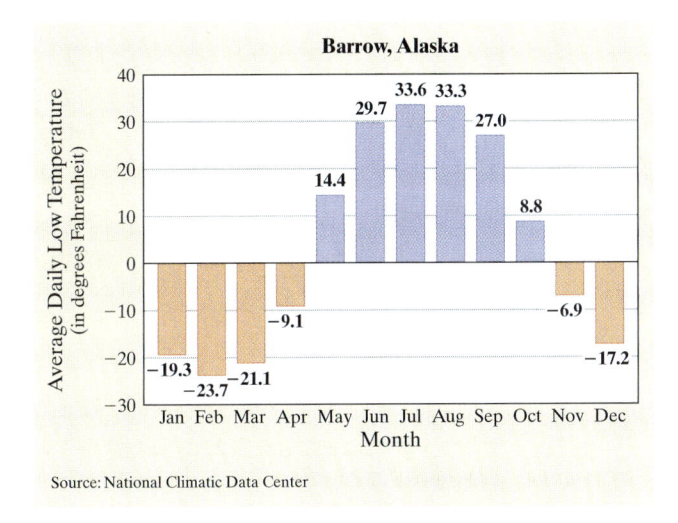

Barrow, Alaska

Average Daily Low Temperature (in degrees Fahrenheit)

Month	Temperature
Jan	−19.3
Feb	−23.7
Mar	−21.1
Apr	−9.1
May	14.4
Jun	29.7
Jul	33.6
Aug	33.3
Sep	27.0
Oct	8.8
Nov	−6.9
Dec	−17.2

Source: National Climatic Data Center

89. Record the monthly increases and decreases in the low temperature from the previous month.

Month	Monthly Increase or Decrease
February	
March	
April	
May	
June	
July	
August	
September	
October	
November	
December	

48 CHAPTER 1 Review of Real Numbers

90. Which month had the greatest increase in temperature?
91. Which month had the greatest decrease in temperature?

If a is a positive number and b is a negative number, determine whether each statement is true or false.

92. $a - b$ is always a positive number.
93. $b - a$ is always a negative number.
94. $|b| - |a|$ is always a positive number.
95. $|b - a|$ is always a positive number.

Without calculating, determine whether each answer is positive or negative. Then use a calculator to find the exact difference.

96. $56{,}875 - 87{,}262$
97. $4.362 - 7.0086$

INTEGRATED REVIEW   OPERATIONS ON REAL NUMBERS
Sections 1.1–1.6

Answer the following with positive, negative, or 0.

1. The opposite of a positive number is a _____ number.
2. The sum of two negative numbers is a _____ number.
3. The absolute value of a negative number is a _____ number.
4. The absolute value of zero is _____.
5. The reciprocal of a positive number is a _____ number.
6. The sum of a number and its opposite is _____.
7. The absolute value of a positive number is a _____ number.
8. The opposite of a negative number is a _____ number.

Fill in the chart:

Number	Opposite	Absolute Value
9. $\frac{1}{7}$		
10. $-\frac{12}{5}$		
11.	-3	
12.		$\frac{9}{11}$

Perform each indicated operation and simplify.

13. $-19 + (-23)$
14. $7 - (-3)$
15. $-15 + 17$
16. $-8 - 10$
17. $18 + (-25)$
18. $-2 + (-37)$
19. $-14 - (-12)$
20. $5 - 14$
21. $4.5 - 7.9$
22. $-8.6 - 1.2$
23. $-\frac{3}{4} - \frac{1}{7}$
24. $\frac{2}{3} - \frac{7}{8}$
25. $-9 - (-7) + 4 - 6$
26. $11 - 20 + (-3) - 12$
27. $24 - 6(14 - 11)$
28. $30 - 5(10 - 8)$
29. $(7 - 17) + 4^2$
30. $9^2 + (10 - 30)$
31. $|-9| + 3^2 + (-4 - 20)$
32. $|-4 - 5| + 5^2 + (-50)$
33. $-7 + [(1 - 2) + (-2 - 9)]$
34. $-6 + [(-3 + 7) + (4 - 15)]$
35. Subtract 5 from 1.
36. Subtract -2 from -3.
37. Subtract $-\frac{2}{5}$ from $\frac{1}{4}$.
38. Subtract $\frac{1}{10}$ from $-\frac{5}{8}$.
39. $2(19 - 17)^3 - 3(-7 + 9)^2$
40. $3(10 - 9)^2 + 6(20 - 19)^3$

Evaluate each expression when $x = -2$, $y = -1$, and $z = 9$.

41. $x - y$
42. $x + y$
43. $y + z$
44. $z - y$
45. $\dfrac{|5z - x|}{y - x}$
46. $\dfrac{|-x - y + z|}{2z}$

1.7 MULTIPLYING AND DIVIDING REAL NUMBERS

OBJECTIVES

1. Multiply and divide real numbers.
2. Evaluate algebraic expressions using real numbers.

OBJECTIVE 1 ▶ Multiplying and dividing real numbers. In this section, we discover patterns for multiplying and dividing real numbers. To discover sign rules for multiplication, recall that multiplication is repeated addition. Thus $3 \cdot 2$ means that 2 is an addend 3 times. That is,

$$2 + 2 + 2 = 3 \cdot 2$$

which equals 6. Similarly, $3 \cdot (-2)$ means -2 is an addend 3 times. That is,

$$(-2) + (-2) + (-2) = 3 \cdot (-2)$$

Since $(-2) + (-2) + (-2) = -6$, then $3 \cdot (-2) = -6$. This suggests that the product of a positive number and a negative number is a negative number.

What about the product of two negative numbers? To find out, consider the following pattern.

Factor decreases by 1 each time

$$\left.\begin{array}{l} -3 \cdot 2 = -6 \\ -3 \cdot 1 = -3 \\ -3 \cdot 0 = 0 \end{array}\right\} \text{Product increases by 3 each time.}$$

This pattern continues as

Factor decreases by 1 each time

$$\left.\begin{array}{l} -3 \cdot -1 = 3 \\ -3 \cdot -2 = 6 \end{array}\right\} \text{Product increases by 3 each time.}$$

This suggests that the product of two negative numbers is a positive number.

Multiplying Real Numbers
1. The product of two numbers with the *same* sign is a positive number.
2. The product of two numbers with *different* signs is a negative number.

EXAMPLE 1 Multiply.

a. $(-8)(4)$ b. $14(-1)$ c. $-9(-10)$

Solution

a. $-8(4) = -32$ b. $14(-1) = -14$ c. $-9(-10) = 90$

PRACTICE 1 Multiply.

a. $8(-5)$ b. $(-3)(-4)$ c. $(-6)(9)$

We know that every whole number multiplied by zero equals zero. This remains true for real numbers.

Zero as a Factor
If b is a real number, then $b \cdot 0 = 0$. Also, $0 \cdot b = 0$.

EXAMPLE 2 Perform the indicated operations.

a. $(7)(0)(-6)$ **b.** $(-2)(-3)(-4)$ **c.** $(-1)(5)(-9)$ **d.** $(-4)(-11) - (5)(-2)$

Solution

a. By the order of operations, we multiply from left to right. Notice that, because one of the factors is 0, the product is 0.

$$(7)(0)(-6) = 0(-6) = 0$$

b. Multiply two factors at a time, from left to right.

$$(-2)(-3)(-4) = (6)(-4) \quad \text{Multiply } (-2)(-3).$$
$$= -24$$

c. Multiply from left to right.

$$(-1)(5)(-9) = (-5)(-9) \quad \text{Multiply } (-1)(5).$$
$$= 45$$

d. Follow the rules for order of operation.

$$(-4)(-11) - (5)(-2) = 44 - (-10) \quad \text{Find each product.}$$
$$= 44 + 10 \quad \text{Add 44 to the opposite of } -10.$$
$$= 54 \quad \text{Add.}$$

PRACTICE 2 Perform the indicated operations.

a. $(-1)(-5)(-6)$ **b.** $(-3)(-2)(4)$
c. $(-4)(0)(5)$ **d.** $(-2)(-3) - (-4)(5)$

> **Helpful Hint**
> You may have noticed from the example that if we multiply:
>
> - an *even* number of negative numbers, the product is *positive*.
> - an *odd* number of negative numbers, the product is *negative*.

Multiplying signed decimals or fractions is carried out exactly the same way as multiplying by integers.

EXAMPLE 3 Multiply.

a. $(-1.2)(0.05)$ **b.** $\dfrac{2}{3} \cdot \left(-\dfrac{7}{10}\right)$ **c.** $\left(-\dfrac{4}{5}\right)(-20)$

Solution

a. The product of two numbers with different signs is negative.

$$(-1.2)(0.05) = -[(1.2)(0.05)]$$
$$= -0.06$$

b. $\dfrac{2}{3} \cdot \left(-\dfrac{7}{10}\right) = -\dfrac{2 \cdot 7}{3 \cdot 10} = -\dfrac{2 \cdot 7}{3 \cdot 2 \cdot 5} = -\dfrac{7}{15}$

c. $\left(-\dfrac{4}{5}\right)(-20) = \dfrac{4 \cdot 20}{5 \cdot 1} = \dfrac{4 \cdot 4 \cdot 5}{5 \cdot 1} = \dfrac{16}{1} \text{ or } 16$

PRACTICE 3 Multiply.

a. $(0.23)(-0.2)$ **b.** $\left(-\dfrac{3}{5}\right) \cdot \left(\dfrac{4}{9}\right)$ **c.** $\left(-\dfrac{7}{12}\right)(-24)$

Now that we know how to multiply positive and negative numbers, let's see how we find the values of $(-4)^2$ and -4^2, for example. Although these two expressions look similar, the difference between the two is the parentheses. In $(-4)^2$, the parentheses tell us that the base, or repeated factor, is -4. In -4^2, only 4 is the base. Thus,

$$(-4)^2 = (-4)(-4) = 16 \quad \text{The base is } -4.$$

$$-4^2 = -(4 \cdot 4) = -16 \quad \text{The base is } 4.$$

EXAMPLE 4 Evaluate.

a. $(-2)^3$ **b.** -2^3 **c.** $(-3)^2$ **d.** -3^2

Solution

a. $(-2)^3 = (-2)(-2)(-2) = -8$ The base is -2.
b. $-2^3 = -(2 \cdot 2 \cdot 2) = -8$ The base is 2.
c. $(-3)^2 = (-3)(-3) = 9$ The base is -3.
d. $-3^2 = -(3 \cdot 3) = -9$ The base is 3.

PRACTICE 4 Evaluate.

a. $(-6)^2$ **b.** -6^2 **c.** $(-4)^3$ **d.** -4^3

> **Helpful Hint**
> Be careful when identifying the base of an exponential expression.
>
> $(-3)^2$ $\qquad\qquad$ -3^2
> Base is -3 $\qquad$ Base is 3
> $(-3)^2 = (-3)(-3) = 9$ $\quad$ $-3^2 = -(3 \cdot 3) = -9$

Just as every difference of two numbers $a - b$ can be written as the sum $a + (-b)$, so too every quotient of two numbers can be written as a product. For example, the quotient $6 \div 3$ can be written as $6 \cdot \dfrac{1}{3}$. Recall that the pair of numbers 3 and $\dfrac{1}{3}$ has a special relationship. Their product is 1 and they are called reciprocals or **multiplicative inverses** of each other.

> **Reciprocals or Multiplicative Inverses**
> Two numbers whose product is 1 are called reciprocals or multiplicative inverses of each other.

Notice that **0 has no multiplicative inverse** since 0 multiplied by any number is never 1 but always 0.

EXAMPLE 5 Find the reciprocal of each number.

a. 22   b. $\dfrac{3}{16}$ c. -10 d. $-\dfrac{9}{13}$

Solution

a. The reciprocal of 22 is $\dfrac{1}{22}$ since $22 \cdot \dfrac{1}{22} = 1$.

b. The reciprocal of $\dfrac{3}{16}$ is $\dfrac{16}{3}$ since $\dfrac{3}{16} \cdot \dfrac{16}{3} = 1$.

c. The reciprocal of -10 is $-\dfrac{1}{10}$.

d. The reciprocal of $-\dfrac{9}{13}$ is $-\dfrac{13}{9}$.

PRACTICE 5 Find the reciprocal of each number.

a. $\dfrac{8}{3}$ b. 15   c. $-\dfrac{2}{7}$ d. -5

We may now write a quotient as an equivalent product.

Quotient of Two Real Numbers

If a and b are real numbers and b is not 0, then

$$a \div b = \dfrac{a}{b} = a \cdot \dfrac{1}{b}$$

In other words, the quotient of two real numbers is the product of the first number and the multiplicative inverse or reciprocal of the second number.

EXAMPLE 6 Use the definition of the quotient of two numbers to divide.

a. $-18 \div 3$ b. $\dfrac{-14}{-2}$ c. $\dfrac{20}{-4}$

Solution

a. $-18 \div 3 = -18 \cdot \dfrac{1}{3} = -6$ b. $\dfrac{-14}{-2} = -14 \cdot -\dfrac{1}{2} = 7$

c. $\dfrac{20}{-4} = 20 \cdot -\dfrac{1}{4} = -5$

PRACTICE 6 Use the definition of the quotient of two numbers to divide.

a. $\dfrac{16}{-2}$ b. $24 \div (-6)$ c. $\dfrac{-35}{-7}$

Since the quotient $a \div b$ can be written as the product $a \cdot \dfrac{1}{b}$, it follows that sign patterns for dividing two real numbers are the same as sign patterns for multiplying two real numbers.

Multiplying and Dividing Real Numbers

1. The product or quotient of two numbers with the *same* sign is a positive number.
2. The product or quotient of two numbers with *different* signs is a negative number.

Section 1.7 Multiplying and Dividing Real Numbers   53

EXAMPLE 7 Divide.

a. $\dfrac{-24}{-4}$ b. $\dfrac{-36}{3}$ c. $\dfrac{2}{3} \div \left(-\dfrac{5}{4}\right)$ d. $-\dfrac{3}{2} \div 9$

Solution

a. $\dfrac{-24}{-4} = 6$ b. $\dfrac{-36}{3} = -12$ c. $\dfrac{2}{3} \div \left(-\dfrac{5}{4}\right) = \dfrac{2}{3} \cdot \left(-\dfrac{4}{5}\right) = -\dfrac{8}{15}$

d. $-\dfrac{3}{2} \div 9 = -\dfrac{3}{2} \cdot \dfrac{1}{9} = -\dfrac{3 \cdot 1}{2 \cdot 9} = -\dfrac{3 \cdot 1}{2 \cdot 3 \cdot 3} = -\dfrac{1}{6}$

PRACTICE 7   Divide.

a. $\dfrac{-18}{-6}$ b. $\dfrac{-48}{3}$ c. $\dfrac{3}{5} \div \left(-\dfrac{1}{2}\right)$ d. $-\dfrac{4}{9} \div 8$

The definition of the quotient of two real numbers does not allow for division by 0 because 0 does not have a multiplicative inverse. There is no number we can multiply 0 by to get 1. How then do we interpret $\dfrac{3}{0}$? We say that division by 0 is not allowed or not defined and that $\dfrac{3}{0}$ does not represent a real number. The denominator of a fraction can never be 0.

Can the numerator of a fraction be 0? Can we divide 0 by a number? Yes. For example,

$$\dfrac{0}{3} = 0 \cdot \dfrac{1}{3} = 0$$

In general, the quotient of 0 and any nonzero number is 0.

Zero as a Divisor or Dividend
1. The quotient of any nonzero real number and 0 is undefined. In symbols, if $a \neq 0$, $\dfrac{a}{0}$ is **undefined.**
2. The quotient of 0 and any real number except 0 is 0. In symbols, if $a \neq 0$, $\dfrac{0}{a} = 0$.

EXAMPLE 8   Perform the indicated operations.

a. $\dfrac{1}{0}$ b. $\dfrac{0}{-3}$ c. $\dfrac{0(-8)}{2}$

Solution

a. $\dfrac{1}{0}$ is undefined   b. $\dfrac{0}{-3} = 0$ c. $\dfrac{0(-8)}{2} = \dfrac{0}{2} = 0$

PRACTICE 8   Perform the indicated operations.

a. $\dfrac{0}{-2}$ b. $\dfrac{-4}{0}$ c. $\dfrac{-5}{6(0)}$

Notice that $\dfrac{12}{-2} = -6$, $-\dfrac{12}{2} = -6$, and $\dfrac{-12}{2} = -6$. This means that

$$\dfrac{12}{-2} = -\dfrac{12}{2} = \dfrac{-12}{2}$$

In words, a single negative sign in a fraction can be written in the denominator, in the numerator, or in front of the fraction without changing the value of the fraction. Thus,

$$\frac{1}{-7} = \frac{-1}{7} = -\frac{1}{7}$$

> In general, if a and b are real numbers, $b \neq 0$, $\frac{a}{-b} = \frac{-a}{b} = -\frac{a}{b}$.

Examples combining basic arithmetic operations along with the principles of order of operations help us to review these concepts.

EXAMPLE 9 Simplify each expression.

a. $\dfrac{(-12)(-3) + 3}{-7 - (-2)}$ b. $\dfrac{2(-3)^2 - 20}{-5 + 4}$

Solution

a. First, simplify the numerator and denominator separately, then divide.

$$\frac{(-12)(-3) + 3}{-7 - (-2)} = \frac{36 + 3}{-7 + 2}$$

$$= \frac{39}{-5} \text{ or } -\frac{39}{5}$$

b. Simplify the numerator and denominator separately, then divide.

$$\frac{2(-3)^2 - 20}{-5 + 4} = \frac{2 \cdot 9 - 20}{-5 + 4} = \frac{18 - 20}{-5 + 4} = \frac{-2}{-1} = 2$$

PRACTICE 9 Simplify each expression.

a. $\dfrac{(-8)(-11) - 4}{-9 - (-4)}$ b. $\dfrac{3(-2)^3 - 9}{-6 + 3}$

OBJECTIVE 2 ▶ Evaluating algebraic expressions using real numbers. Using what we have learned about multiplying and dividing real numbers, we continue to practice evaluating algebraic expressions.

EXAMPLE 10 If $x = -2$ and $y = -4$, evaluate each expression.

a. $5x - y$ b. $x^4 - y^2$ c. $\dfrac{3x}{2y}$

Solution

a. Replace x with -2 and y with -4 and simplify.

$$5x - y = 5(-2) - (-4) = -10 - (-4) = -10 + 4 = -6$$

b. Replace x with -2 and y with -4.

$$x^4 - y^2 = (-2)^4 - (-4)^2 \quad \text{Substitute the given values for the variables.}$$
$$= 16 - (16) \quad \text{Evaluate exponential expressions.}$$
$$= 0 \quad \text{Subtract.}$$

c. Replace x with -2 and y with -4 and simplify.

$$\frac{3x}{2y} = \frac{3(-2)}{2(-4)} = \frac{-6}{-8} = \frac{3}{4}$$

PRACTICE 10
If $x = -5$ and $y = -2$, evaluate each expression.

a. $7y - x$ b. $x^2 - y^3$ c. $\dfrac{2x}{3y}$

Calculator Explorations

Entering Negative Numbers on a Scientific Calculator

To enter a negative number on a scientific calculator, find a key marked $\boxed{+/-}$. (On some calculators, this key is marked $\boxed{\text{CHS}}$ for "change sign.") To enter -8, for example, press the keys $\boxed{8}$ $\boxed{+/-}$. The display will read $\boxed{-8}$.

Entering Negative Numbers on a Graphing Calculator

To enter a negative number on a graphing calculator, find a key marked $\boxed{(-)}$. Do not confuse this key with the key $\boxed{-}$, which is used for subtraction. To enter -8, for example, press the keys $\boxed{(-)}$ $\boxed{8}$. The display will read $\boxed{-8}$.

Operations with Real Numbers

To evaluate $-2(7 - 9) - 20$ on a calculator, press the keys
$\boxed{2}$ $\boxed{+/-}$ $\boxed{\times}$ $\boxed{(}$ $\boxed{7}$ $\boxed{-}$ $\boxed{9}$ $\boxed{)}$ $\boxed{-}$ $\boxed{2}$ $\boxed{0}$ $\boxed{=}$, or
$\boxed{(-)}$ $\boxed{2}$ $\boxed{(}$ $\boxed{7}$ $\boxed{-}$ $\boxed{9}$ $\boxed{)}$ $\boxed{-}$ $\boxed{2}$ $\boxed{0}$ $\boxed{\text{ENTER}}$.

The display will read $\boxed{-16}$ or $\boxed{\begin{array}{r}-2(7-9)-20\\-16\end{array}}$.

Use a calculator to simplify each expression.

1. $-38(26 - 27)$
2. $-59(-8) + 1726$
3. $134 + 25(68 - 91)$
4. $45(32) - 8(218)$
5. $\dfrac{-50(294)}{175 - 265}$
6. $\dfrac{-444 - 444.8}{-181 - 324}$
7. $9^5 - 4550$
8. $5^8 - 6259$
9. $(-125)^2$ (Be careful.)
10. -125^2 (Be careful.)

VOCABULARY & READINESS CHECK

Use the choices below to fill in each blank.

positive   0   negative   undefined

1. If n is a real number, then $n \cdot 0 =$ ___ and $0 \cdot n =$ ___.
2. If n is a real number, but not 0, then $\dfrac{0}{n} =$ ___ and we say $\dfrac{n}{0}$ is _____.
3. The product of two negative numbers is a _____ number.
4. The quotient of two negative numbers is a _____ number.
5. The quotient of a positive number and a negative number is a _____ number.
6. The product of a positive number and a negative number is a _____ number.
7. The reciprocal of a positive number is a _____ number.
8. The opposite of a positive number is a _____ number.

1.7 EXERCISE SET

Multiply. See Examples 1 through 3.

1. $-6(4)$
2. $-8(5)$
3. $2(-1)$
4. $7(-4)$
5. $-5(-10)$
6. $-6(-11)$
7. $-3 \cdot 4$
8. $-2 \cdot 8$
9. $-7 \cdot 0$
10. $-6 \cdot 0$
11. $2(-9)$
12. $3(-5)$
13. $-\dfrac{1}{2}\left(-\dfrac{3}{5}\right)$
14. $-\dfrac{1}{8}\left(-\dfrac{1}{3}\right)$
15. $-\dfrac{3}{4}\left(-\dfrac{8}{9}\right)$
16. $-\dfrac{5}{6}\left(-\dfrac{3}{10}\right)$
17. $5(-1.4)$
18. $6(-2.5)$
19. $-0.2(-0.7)$
20. $-0.5(-0.3)$
21. $-10(80)$
22. $-20(60)$
23. $4(-7)$
24. $5(-9)$
25. $(-5)(-5)$
26. $(-7)(-7)$
27. $\dfrac{2}{3}\left(-\dfrac{4}{9}\right)$
28. $\dfrac{2}{7}\left(-\dfrac{2}{11}\right)$
29. $-11(11)$
30. $-12(12)$
31. $-\dfrac{20}{25}\left(\dfrac{5}{16}\right)$
32. $-\dfrac{25}{36}\left(\dfrac{6}{15}\right)$
33. $(-1)(2)(-3)(-5)$
34. $(-2)(-3)(-4)(-2)$

Perform the indicated operations. See Example 2.

35. $(-2)(5) - (-11)(3)$
36. $8(-3) - 4(-5)$
37. $(-6)(-1)(-2) - (-5)$
38. $20 - (-4)(3)(-2)$

Decide whether each statement is true or false.

39. The product of three negative integers is negative.
40. The product of three positive integers is positive.
41. The product of four negative integers is negative.
42. The product of four positive integers is positive.

Evaluate. See Example 4.

43. $(-2)^4$
44. -2^4
45. -1^5
46. $(-1)^5$
47. $(-5)^2$
48. -5^2
49. -7^2
50. $(-7)^2$

Find each reciprocal or multiplicative inverse. See Example 5.

51. 9
52. 100
53. $\dfrac{2}{3}$
54. $\dfrac{1}{7}$
55. -14
56. -8
57. $-\dfrac{3}{11}$
58. $-\dfrac{6}{13}$
59. 0.2
60. 1.5
61. $\dfrac{1}{-6.3}$
62. $\dfrac{1}{-8.9}$

Divide. See Examples 6 through 8.

63. $\dfrac{18}{-2}$
64. $\dfrac{20}{-10}$
65. $\dfrac{-16}{-4}$
66. $\dfrac{-18}{-6}$
67. $\dfrac{-48}{12}$
68. $\dfrac{-60}{5}$
69. $\dfrac{0}{-4}$
70. $\dfrac{0}{-9}$
71. $-\dfrac{15}{3}$
72. $-\dfrac{24}{8}$
73. $\dfrac{5}{0}$
74. $\dfrac{3}{0}$
75. $\dfrac{-12}{-4}$
76. $\dfrac{-45}{-9}$
77. $\dfrac{30}{-2}$
78. $\dfrac{14}{-2}$
79. $\dfrac{6}{7} \div \left(-\dfrac{1}{3}\right)$
80. $\dfrac{4}{5} \div \left(-\dfrac{1}{2}\right)$
81. $-\dfrac{5}{9} \div \left(-\dfrac{3}{4}\right)$
82. $-\dfrac{1}{10} \div \left(-\dfrac{8}{11}\right)$
83. $-\dfrac{4}{9} \div \dfrac{4}{9}$
84. $-\dfrac{5}{12} \div \dfrac{5}{12}$

MIXED PRACTICE

Simplify. See Example 9.

85. $\dfrac{-9(-3)}{-6}$
86. $\dfrac{-6(-3)}{-4}$
87. $\dfrac{12}{9 - 12}$
88. $\dfrac{-15}{1 - 4}$
89. $\dfrac{-6^2 + 4}{-2}$
90. $\dfrac{3^2 + 4}{5}$
91. $\dfrac{8 + (-4)^2}{4 - 12}$
92. $\dfrac{6 + (-2)^2}{4 - 9}$
93. $\dfrac{22 + (3)(-2)}{-5 - 2}$
94. $\dfrac{-20 + (-4)(3)}{1 - 5}$
95. $\dfrac{-3 - 5^2}{2(-7)}$
96. $\dfrac{-2 - 4^2}{3(-6)}$
97. $\dfrac{6 - 2(-3)}{4 - 3(-2)}$
98. $\dfrac{8 - 3(-2)}{2 - 5(-4)}$
99. $\dfrac{-3 - 2(-9)}{-15 - 3(-4)}$
100. $\dfrac{-4 - 8(-2)}{-9 - 2(-3)}$
101. $\dfrac{|5 - 9| + |10 - 15|}{|2(-3)|}$
102. $\dfrac{|-3 + 6| + |-2 + 7|}{|-2 \cdot 2|}$

If $x = -5$ and $y = -3$, evaluate each expression. See Example 10.

103. $3x + 2y$
104. $4x + 5y$
105. $2x^2 - y^2$
106. $x^2 - 2y^2$
107. $x^3 + 3y$
108. $y^3 + 3x$
109. $\dfrac{2x - 5}{y - 2}$
110. $\dfrac{2y - 12}{x - 4}$
111. $\dfrac{-3 - y}{x - 4}$
112. $\dfrac{4 - 2x}{y + 3}$

113. At the end of 2006, Delta Airlines posted a net loss of $203 million, which we will write as $-$6203 million. If this continues, what will Delta's income be after four years? (*Source:* Delta Airlines)

114. At the end of the third quarter of 2006, General Motors reported a net loss of $115 million. If this continued, what would General Motor's income be after four more quarters? (*Source:* General Motors)

Decide whether the given number is a solution of the given equation.

115. Is 7 a solution of $-5x = -35$?
116. Is -4 a solution of $2x = x - 1$?
117. Is -20 a solution of $\dfrac{x}{10} = 2$?
118. Is -3 a solution of $\dfrac{45}{x} = -15$?
119. Is 5 a solution of $-3x - 5 = -20$?
120. Is -4 a solution of $2x + 4 = x + 8$?

CONCEPT EXTENSIONS

121. Explain why the product of an even number of negative numbers is a positive number.
122. If a and b are any real numbers, is the statement $a \cdot b = b \cdot a$ always true? Why or why not?
123. Find any real numbers that are their own reciprocal.
124. Explain why 0 has no reciprocal.

If q is a negative number, r is a negative number, and t is a positive number, determine whether each expression simplifies to a positive or negative number. If it is not possible to determine, state so.

125. $\dfrac{q}{r \cdot t}$
126. $q^2 \cdot r \cdot t$
127. $q + t$
128. $t + r$
129. $t(q + r)$
130. $r(q - t)$

Write each of the following as an expression and evaluate.

131. The sum of -2 and the quotient of -15 and 3
132. The sum of 1 and the product of -8 and -5
133. Twice the sum of -5 and -3
134. 7 subtracted from the quotient of 0 and 5

STUDY SKILLS BUILDER

Learning New Terms

Many of the terms used in this text may be new to you. It will be helpful to make a list of new mathematical terms and symbols as you encounter them and to review them frequently. Placing these new terms (including page references) on 3 × 5 index cards might help you later when you're preparing for a quiz.

Answer the following.

1. Name one way you might place a word and its definition on a 3 × 5 card.
2. How do new terms stand out in this text so that they can be found?

THE BIGGER PICTURE   SIMPLIFYING EXPRESSIONS

This is a special feature that we introduce in this section. Among other concepts introduced later in this text, it is very important for you to be able to simplify expressions and solve equations and to know the difference between the two. To help with this, we began an outline below and expand this outline throughout the text. Although suggestions are given, this outline should be in your own words. Once you complete the new portion of your outline, try the exercises to the right. Remember: Study your outline often as you proceed through this text.

I. **Simplifying Expressions**
   A. **Real Numbers**
      1. **Add:**
         $-1.7 + (-0.21) = -1.91$ Adding like signs. Add absolute values. Attach the common sign.
         $-7 + 3 = -4$ Adding unlike signs. Subtract absolute values. Attach the sign of the number with the larger absolute value.
      2. **Subtract:** Add the first number to the opposite of the second number.
         $\dfrac{1}{7} - \dfrac{1}{3} = \dfrac{3}{21} + \left(-\dfrac{7}{21}\right) = -\dfrac{4}{21}$

   3. **Multiply or Divide:** Multiply or divide as usual. If the signs of the two numbers are the same, the answer is positive. If the signs of the two numbers are different, the answer is negative.
      $-\dfrac{3}{8} \cdot \dfrac{7}{11} = -\dfrac{21}{88}, \quad -42 \div (-10) = 4.2$

Perform the indicated operations.

1. $-0.2(25)$
2. $86 - 100$
3. $-\dfrac{1}{7} + \left(-\dfrac{3}{5}\right)$
4. $\dfrac{-40}{-5}$
5. $(-7)^2$
6. -7^2
7. $\dfrac{|-42|}{-|-2|}$
8. $\dfrac{8.6}{0}$
9. $\dfrac{0}{8.6}$
10. $-25 - (-13)$
11. $-8.3 - 8.3$
12. $-\dfrac{8}{9}\left(-\dfrac{3}{16}\right)$
13. $2 + 3(8 - 11)^3$
14. $-2\dfrac{1}{2} \div \left(-3\dfrac{1}{4}\right)$
15. $20 \div 2 \cdot 5$
16. $-2[(1 - 5) - (7 - 17)]$

1.8 PROPERTIES OF REAL NUMBERS

OBJECTIVES

1. Use the commutative and associative properties.
2. Use the distributive property.
3. Use the identity and inverse properties.

OBJECTIVE 1 ▶ Using the commutative and associative properties. In this section we give names to properties of real numbers with which we are already familiar. Throughout this section, the variables a, b, and c represent real numbers.

We know that order does not matter when adding numbers. For example, we know that $7 + 5$ is the same as $5 + 7$. This property is given a special name—the **commutative property of addition.** We also know that order does not matter when multiplying numbers. For example, we know that $-5(6) = 6(-5)$. This property means that multiplication is commutative also and is called the **commutative property of multiplication.**

Commutative Properties

Addition: $\qquad a + b = b + a$

Multiplication: $\qquad a \cdot b = b \cdot a$

These properties state that the *order* in which any two real numbers are added or multiplied does not change their sum or product. For example, if we let $a = 3$ and $b = 5$, then the commutative properties guarantee that

$$3 + 5 = 5 + 3 \quad \text{and} \quad 3 \cdot 5 = 5 \cdot 3$$

> **▶ Helpful Hint**
>
> Is subtraction also commutative? Try an example. Does $3 - 2 = 2 - 3$? **No!** The left side of this statement equals 1; the right side equals -1. There is no commutative property of subtraction. Similarly, there is no commutative property for division. For example, $10 \div 2$ does not equal $2 \div 10$.

EXAMPLE 1 Use a commutative property to complete each statement.

a. $x + 5 = $ _____
b. $3 \cdot x = $ _____

Solution

a. $x + 5 = 5 + x \qquad$ By the commutative property of addition
b. $3 \cdot x = x \cdot 3 \qquad$ By the commutative property of multiplication

PRACTICE 1 Use a commutative property to complete each statement.

a. $x \cdot 8 = $ ____
b. $x + 17 = $ ____

Concept Check ✓

Which of the following pairs of actions are commutative?

a. "raking the leaves" and "bagging the leaves"
b. "putting on your left glove" and "putting on your right glove"
c. "putting on your coat" and "putting on your shirt"
d. "reading a novel" and "reading a newspaper"

Answers to Concept Check:
b, d

Let's now discuss grouping numbers. We know that when we add three numbers, the way in which they are grouped or associated does not change their sum. For example, we know that $2 + (3 + 4) = 2 + 7 = 9$. This result is the same if we group the numbers differently. In other words, $(2 + 3) + 4 = 5 + 4 = 9$, also. Thus, $2 + (3 + 4) = (2 + 3) + 4$. This property is called the **associative property of addition.**

We also know that changing the grouping of numbers when multiplying does not change their product. For example, $2 \cdot (3 \cdot 4) = (2 \cdot 3) \cdot 4$ (check it). This is the **associative property of multiplication.**

Associative Properties

Addition: $\qquad (a + b) + c = a + (b + c)$

Multiplication: $\qquad (a \cdot b) \cdot c = a \cdot (b \cdot c)$

These properties state that the way in which three numbers are *grouped* does not change their sum or their product.

EXAMPLE 2 Use an associative property to complete each statement.

a. $5 + (4 + 6) =$ _____   **b.** $(-1 \cdot 2) \cdot 5 =$ _____

Solution

a. $5 + (4 + 6) = (5 + 4) + 6$ By the associative property of addition
b. $(-1 \cdot 2) \cdot 5 = -1 \cdot (2 \cdot 5)$ By the associative property of multiplication

PRACTICE 2 Use an associative property to complete each statement.

a. $(2 + 9) + 7 =$ _____   **b.** $-4 \cdot (2 \cdot 7) =$ _____

▶ **Helpful Hint**

Remember the difference between the commutative properties and the associative properties. The commutative properties have to do with the *order* of numbers, and the associative properties have to do with the *grouping* of numbers.

Let's now illustrate how these properties can help us simplify expressions.

EXAMPLE 3 Simplify each expression.

a. $10 + (x + 12)$ **b.** $-3(7x)$

Solution

a. $10 + (x + 12) = 10 + (12 + x)$ By the commutative property of addition
$\qquad \qquad \qquad = (10 + 12) + x$ By the associative property of addition
$\qquad \qquad \qquad = 22 + x$ Add.
b. $-3(7x) = (-3 \cdot 7)x$ By the associative property of multiplication
$\qquad \qquad = -21x$ Multiply.

PRACTICE 3 Simplify each expression.

a. $(5 + x) + 9$ **b.** $5(-6x)$

OBJECTIVE 2 ▶ Using the distributive property. The **distributive property of multiplication over addition** is used repeatedly throughout algebra. It is useful because it allows us to write a product as a sum or a sum as a product.

We know that $7(2 + 4) = 7(6) = 42$. Compare that with $7(2) + 7(4) = 14 + 28 = 42$. Since both original expressions equal 42, they must equal each other, or

$$7(2 + 4) = 7(2) + 7(4)$$

This is an example of the distributive property. The product on the left side of the equal sign is equal to the sum on the right side. We can think of the 7 as being distributed to each number inside the parentheses.

Distributive Property of Multiplication Over Addition

$$a(b + c) = ab + ac$$

Since multiplication is commutative, this property can also be written as

$$(b + c)a = ba + ca$$

The distributive property can also be extended to more than two numbers inside the parentheses. For example,

$$3(x + y + z) = 3(x) + 3(y) + 3(z)$$
$$= 3x + 3y + 3z$$

Since we define subtraction in terms of addition, the distributive property is also true for subtraction. For example

$$2(x - y) = 2(x) - 2(y)$$
$$= 2x - 2y$$

EXAMPLE 4 Use the distributive property to write each expression without parentheses. Then simplify if possible.

a. $2(x + y)$
b. $-5(-3 + 2z)$
c. $5(x + 3y - z)$
d. $-1(2 - y)$
e. $-(3 + x - w)$
f. $4(3x + 7) + 10$

Solution

a. $2(x + y) = 2 \cdot x + 2 \cdot y$
$$= 2x + 2y$$

b. $-5(-3 + 2z) = -5(-3) + (-5)(2z)$
$$= 15 - 10z$$

c. $5(x + 3y - z) = 5(x) + 5(3y) - 5(z)$
$$= 5x + 15y - 5y$$

d. $-1(2 - y) = (-1)(2) - (-1)(y)$
$= -2 + y$

e. $-(3 + x - w) = -1(3 + x - w)$
$= (-1)(3) + (-1)(x) - (-1)(w)$
$= -3 - x + w$

> **Helpful Hint**
> Notice in part (**e**) that $-(3 + x - w)$ is first rewritten as $-1(3 + x - w)$.

f. $4(3x + 7) + 10 = 4(3x) + 4(7) + 10$ Apply the distributive property.
$= 12x + 28 + 10$ Multiply.
$= 12x + 38$ Add.

PRACTICE 4 Use the distributive property to write each expression without parentheses. Then simplify, if possible.

a. $5(x - y)$
b. $-6(4 + 2t)$
c. $2(3x - 4y - z)$
d. $(3 - y) \cdot (-1)$
e. $-(x - 7 + 2s)$
f. $2(7x + 4) + 6$

We can use the distributive property in reverse to write a sum as a product.

EXAMPLE 5 Use the distributive property to write each sum as a product.

a. $8 \cdot 2 + 8 \cdot x$
b. $7s + 7t$

Solution

a. $8 \cdot 2 + 8 \cdot x = 8(2 + x)$
b. $7s + 7t = 7(s + t)$

PRACTICE 5 Use the distributive property to write each sum as a product.

a. $5 \cdot w + 5 \cdot 3$
b. $9w + 9z$

OBJECTIVE 3 ▶ Using the identity and inverse properties. Next, we look at the **identity properties**.

The number 0 is called the identity for addition because when 0 is added to any real number, the result is the same real number. In other words, the *identity* of the real number is not changed.

The number 1 is called the identity for multiplication because when a real number is multiplied by 1, the result is the same real number. In other words, the *identity* of the real number is not changed.

Identities for Addition and Multiplication

0 is the identity element for addition.

$$a + 0 = a \quad \text{and} \quad 0 + a = a$$

1 is the identity element for multiplication.

$$a \cdot 1 = a \quad \text{and} \quad 1 \cdot a = a$$

Notice that 0 is the *only* number that can be added to any real number with the result that the sum is the same real number. Also, 1 is the *only* number that can be multiplied by any real number with the result that the product is the same real number.

Additive inverses or **opposites** were introduced in Section 1.5. Two numbers are called additive inverses or opposites if their sum is 0. The additive inverse or opposite of 6 is -6 because $6 + (-6) = 0$. The additive inverse or opposite of -5 is 5 because $-5 + 5 = 0$.

Reciprocals or **multiplicative inverses** were introduced in Section 1.3. Two nonzero numbers are called reciprocals or multiplicative inverses if their product is 1. The reciprocal or multiplicative inverse of $\frac{2}{3}$ is $\frac{3}{2}$ because $\frac{2}{3} \cdot \frac{3}{2} = 1$. Likewise, the reciprocal of -5 is $-\frac{1}{5}$ because $-5\left(-\frac{1}{5}\right) = 1$.

Concept Check ✓

Which of the following, $1, -\frac{10}{3}, \frac{3}{10}, 0, \frac{10}{3}, -\frac{3}{10}$, is the

a. opposite of $-\frac{3}{10}$?

b. reciprocal of $-\frac{3}{10}$?

Additive or Multiplicative Inverses

The numbers a and $-a$ are additive inverses or opposites of each other because their sum is 0; that is,

$$a + (-a) = 0$$

The numbers b and $\frac{1}{b}$ (for $b \neq 0$) are reciprocals or multiplicative inverses of each other because their product is 1; that is,

$$b \cdot \frac{1}{b} = 1$$

EXAMPLE 6 Name the property or properties illustrated by each true statement.

Solution

a. $3 \cdot y = y \cdot 3$ Commutative property of multiplication (order changed)
b. $(x + 7) + 9 = x + (7 + 9)$ Associative property of addition (grouping changed)
c. $(b + 0) + 3 = b + 3$ Identity element for addition
d. $0.2 \cdot (z \cdot 5) = 0.2 \cdot (5 \cdot z)$ Commutative property of multiplication (order changed)
e. $-2 \cdot \left(-\frac{1}{2}\right) = 1$ Multiplicative inverse property
f. $-2 + 2 = 0$ Additive inverse property
g. $-6 \cdot (y \cdot 2) = (-6 \cdot 2) \cdot y$ Commutative and associative properties of multiplication (order and grouping changed)

PRACTICE 6 Name the property or properties illustrated by each true statement.

a. $(7 \cdot 3x) \cdot 4 = (3x \cdot 7) \cdot 4$
b. $6 + (3 + y) = (6 + 3) + y$
c. $8 + (t + 0) = 8 + t$
d. $-\frac{3}{4} \cdot \left(-\frac{4}{3}\right) = 1$
e. $(2 + x) + 5 = 5 + (2 + x)$
f. $3 + (-3) = 0$
g. $(-3b) \cdot 7 = (-3 \cdot 7) \cdot b$

Answers to Concept Check:
a. $\frac{3}{10}$ **b.** $-\frac{10}{3}$

VOCABULARY & READINESS CHECK

Use the choices below to fill in each blank.

distributive property   associative property of multiplication   commutative property of addition
opposites or additive inverses   associative property of addition
reciprocals or multiplicative inverses   commutative property of multiplication

1. $x + 5 = 5 + x$ is a true statement by the _____.
2. $x \cdot 5 = 5 \cdot x$ is a true statement by the _____.
3. $3(y + 6) = 3 \cdot y + 3 \cdot 6$ is true by the _____.
4. $2 \cdot (x \cdot y) = (2 \cdot x) \cdot y$ is a true statement by the _____.
5. $x + (7 + y) = (x + 7) + y$ is a true statement by the _____.
6. The numbers $-\frac{2}{3}$ and $-\frac{3}{2}$ are called _____.
7. The numbers $-\frac{2}{3}$ and $\frac{2}{3}$ are called _____.

1.8 EXERCISE SET

Use a commutative property to complete each statement. See Example 1.

1. $x + 16 =$ _____
2. $4 + y =$ _____
3. $-4 \cdot y =$ _____
4. $-2 \cdot x =$ _____
5. $xy =$ _____
6. $ab =$ _____
7. $2x + 13 =$ _____
8. $19 + 3y =$ _____

Use an associative property to complete each statement. See Example 2.

9. $(xy) \cdot z =$ _____
10. $3 \cdot (xy) =$ _____
11. $2 + (a + b) =$ _____
12. $(y + 4) + z =$ _____
13. $4 \cdot (ab) =$ _____
14. $(-3y) \cdot z =$ _____
15. $(a + b) + c =$ _____
16. $6 + (r + s) =$ _____

Use the commutative and associative properties to simplify each expression. See Example 3.

17. $8 + (9 + b)$
18. $(r + 3) + 11$
19. $4(6y)$
20. $2(42x)$
21. $\frac{1}{5}(5y)$
22. $\frac{1}{8}(8z)$
23. $(13 + a) + 13$
24. $7 + (x + 4)$
25. $-9(8x)$
26. $-3(12y)$
27. $\frac{3}{4}\left(\frac{4}{3}s\right)$
28. $\frac{2}{7}\left(\frac{7}{2}r\right)$

29. Write an example that shows that division is not commutative.
30. Write an example that shows that subtraction is not commutative.

Use the distributive property to write each expression without parentheses. Then simplify the result. See Example 4.

31. $4(x + y)$
32. $7(a + b)$
33. $9(x - 6)$
34. $11(y - 4)$
35. $2(3x + 5)$
36. $5(7 + 8y)$
37. $7(4x - 3)$
38. $3(8x - 1)$
39. $3(6 + x)$
40. $2(x + 5)$
41. $-2(y - z)$
42. $-3(z - y)$
43. $-7(3y + 5)$
44. $-5(2r + 11)$
45. $5(x + 4m + 2)$
46. $8(3y + z - 6)$
47. $-4(1 - 2m + n)$
48. $-4(4 + 2p + 5)$
49. $-(5x + 2)$
50. $-(9r + 5)$
51. $-(r - 3 - 7p)$
52. $-(q - 2 + 6r)$
53. $\frac{1}{2}(6x + 8)$
54. $\frac{1}{4}(4x - 2)$
55. $-\frac{1}{3}(3x - 9y)$
56. $-\frac{1}{5}(10a - 25b)$
57. $3(2r + 5) - 7$
58. $10(4s + 6) - 40$
59. $-9(4x + 8) + 2$
60. $-11(5x + 3) + 10$
61. $-4(4x + 5) - 5$
62. $-6(2x + 1) - 1$

Use the distributive property to write each sum as a product. See Example 5.

63. $4 \cdot 1 + 4 \cdot y$
64. $14 \cdot z + 14 \cdot 5$
65. $11x + 11y$
66. $9a + 9b$
67. $(-1) \cdot 5 + (-1) \cdot x$
68. $(-3)a + (-3)b$
69. $30a + 30b$
70. $25x + 25y$

Name the properties illustrated by each true statement. See Example 6.

71. $3 \cdot 5 = 5 \cdot 3$
72. $4(3 + 8) = 4 \cdot 3 + 4 \cdot 8$
73. $2 + (x + 5) = (2 + x) + 5$
74. $(x + 9) + 3 = (9 + x) + 3$
75. $9(3 + 7) = 9 \cdot 3 + 9 \cdot 7$
76. $1 \cdot 9 = 9$
77. $(4 \cdot y) \cdot 9 = 4 \cdot (y \cdot 9)$
78. $6 \cdot \dfrac{1}{6} = 1$
79. $0 + 6 = 6$
80. $(a + 9) + 6 = a + (9 + 6)$
81. $-4(y + 7) = -4 \cdot y + (-4) \cdot 7$
82. $(11 + r) + 8 = (r + 11) + 8$
83. $-4 \cdot (8 \cdot 3) = (8 \cdot -4) \cdot 3$
84. $r + 0 = r$

CONCEPT EXTENSIONS

Fill in the table with the opposite (additive inverse), and the reciprocal (multiplicative inverse). Assume that the value of each expression is not 0.

	Expression	Opposite	Reciprocal
85.	8		
86.	$-\dfrac{2}{3}$		
87.	x		
88.	$4y$		
89.			$\dfrac{1}{2x}$
90.		$7x$	

Determine which pairs of actions are commutative.

91. "taking a test" and "studying for the test"
92. "putting on your shoes" and "putting on your socks"
93. "putting on your left shoe" and "putting on your right shoe"
94. "reading the sports section" and "reading the comics section"
95. Explain why 0 is called the identity element for addition.
96. Explain why 1 is called the identity element for multiplication.

CHAPTER 1 GROUP ACTIVITY

Sections 1.3, 1.4, 1.5

Magic Squares

A magic square is a set of numbers arranged in a square table so that the sum of the numbers in each column, row, and diagonal is the same. For instance, in the magic square below, the sum of each column, row, and diagonal is 15. Notice that no number is used more than once in the magic square.

2	9	4
7	5	3
6	1	8

The properties of magic squares have been known for a very long time and once were thought to be good luck charms. The ancient Egyptians and Greeks understood their patterns. A magic square even made it into a famous work of art. The engraving titled *Melencolia I*, created by German artist Albrecht Dürer in 1514, features the following four-by-four magic square on the building behind the central figure.

16	3	2	13
5	10	11	8
9	6	7	12
4	15	14	1

Group Exercises

1. Verify that what is shown in the Dürer engraving is, in fact, a magic square. What is the common sum of the columns, rows, and diagonals?

2. Negative numbers can also be used in magic squares. Complete the following magic square:

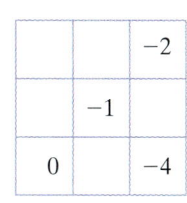

		−2
	−1	
0	−4	

3. Use the numbers −12, −9, −6, −3, 0, 3, 6, 9, and 12 to form a magic square.

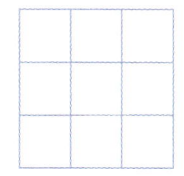

CHAPTER 1 VOCABULARY CHECK

Fill in each blank with one of the words or phrases listed below.

set　　　　　　inequality symbols　　opposites　　　absolute value　　numerator
denominator　　grouping symbols　　exponent　　　base　　　　　　　reciprocals
variable　　　　equation　　　　　　solution

1. The symbols ≠, <, and > are called _____.
2. A mathematical statement that two expressions are equal is called an _____.
3. The _____ of a number is the distance between that number and 0 on the number line.
4. A symbol used to represent a number is called a _____.
5. Two numbers that are the same distance from 0 but lie on opposite sides of 0 are called _____.
6. The number in a fraction above the fraction bar is called the _____.
7. A _____ of an equation is a value for the variable that makes the equation a true statement.
8. Two numbers whose product is 1 are called _____.
9. In 2^3, the 2 is called the _____ and the 3 is called the _____.
10. The number in a fraction below the fraction bar is called the _____.
11. Parentheses and brackets are examples of _____.
12. A ___ is a collection of objects.

> **Helpful Hint**
> Are you preparing for your test? Don't forget to take the Chapter 1 Test on page 72. Then check your answers at the back of the text and use the Chapter Test Prep Video CD to see the fully worked-out solutions to any of the exercises you want to review.

CHAPTER 1 HIGHLIGHTS

DEFINITIONS AND CONCEPTS	EXAMPLES
SECTION 1.2　SYMBOLS AND SETS OF NUMBERS	
A **set** is a collection of objects, called **elements**, enclosed in braces.	$\{a, c, e\}$
Natural Numbers: $\{1, 2, 3, 4, \ldots\}$ **Whole Numbers:** $\{0, 1, 2, 3, 4, \ldots\}$	Given the set $\left\{-3.4, \sqrt{3}, 0, \frac{2}{3}, 5, -4\right\}$, list the numbers that belong to the set of
Integers: $\{\ldots, -3, -2, -1, 0, 1, 2, 3, \ldots\}$	Natural numbers: 5
Rational Numbers: {real numbers that can be expressed as a quotient of integers}	Whole numbers: 0, 5 Integers: −4, 0, 5
Irrational Numbers: {real numbers that cannot be expressed as a quotient of integers}	Rational numbers: $-4, -3.4, 0, \frac{2}{3}, 5$ Irrational Numbers: $\sqrt{3}$

(continued)

66 CHAPTER 1 Review of Real Numbers

DEFINITIONS AND CONCEPTS	EXAMPLES
SECTION 1.2 SYMBOLS AND SETS OF NUMBERS (continued)	
Real Numbers: {all numbers that correspond to a point on the number line}	Real numbers: $-4, -3.4, 0, \frac{2}{3}, \sqrt{3}, 5$
A line used to picture numbers is called a **number line**.	
The **absolute value** of a real number a, denoted by $\|a\|$, is the distance between a and 0 on the number line.	$\|5\| = 5 \quad \|0\| = 0 \quad \|-2\| = 2$
Symbols: $=$ is equal to	$-7 = -7$
$\neq$ is not equal to	$3 \neq -3$
$>$ is greater than	$4 > 1$
$<$ is less than	$1 < 4$
$\leq$ is less than or equal to	$6 \leq 6$
$\geq$ is greater than or equal to	$18 \geq -\frac{1}{3}$
Order Property for Real Numbers	
For any two real numbers a and b, a is less than b if a is to the left of b on a number line.	$-3 < 0 \quad 0 > -3 \quad 0 < 2.5 \quad 2.5 > 0$
SECTION 1.3 FRACTIONS	
A quotient of two integers is called a **fraction**. The **numerator** of a fraction is the top number. The **denominator** of a fraction is the bottom number.	$\frac{13}{17}$ ← numerator ← denominator
If $a \cdot b = c$, then a and b are **factors** and c is the **product**.	$\underset{\text{factor}}{7} \cdot \underset{\text{factor}}{9} = \underset{\text{product}}{63}$
A fraction is in **lowest terms** when the numerator and the denominator have no factors in common other than 1.	$\frac{13}{17}$ is in lowest terms.
To write a fraction in lowest terms, factor the numerator and the denominator; then apply the fundamental principle.	Write in lowest terms. $\frac{6}{14} = \frac{2 \cdot 3}{2 \cdot 7} = \frac{3}{7}$
Two fractions are **reciprocals** if their product is 1. The reciprocal of $\frac{a}{b}$ is $\frac{b}{a}$.	The reciprocal of $\frac{6}{25}$ is $\frac{25}{6}$.
To multiply fractions, numerator times numerator is the numerator of the product and denominator times denominator is the denominator of the product.	Perform the indicated operations. $\frac{2}{5} \cdot \frac{3}{7} = \frac{6}{35}$
To divide fractions, multiply the first fraction by the reciprocal of the second fraction.	$\frac{5}{9} \div \frac{2}{7} = \frac{5}{9} \cdot \frac{7}{2} = \frac{35}{18}$
To add fractions with the same denominator, add the numerators and place the sum over the common denominator.	$\frac{5}{11} + \frac{3}{11} = \frac{8}{11}$
To subtract fractions with the same denominator, subtract the numerators and place the difference over the common denominator.	$\frac{13}{15} - \frac{3}{15} = \frac{10}{15} = \frac{2}{3}$
Fractions that represent the same quantity are called **equivalent fractions**.	$\frac{1}{5} = \frac{1 \cdot 4}{5 \cdot 4} = \frac{4}{20}$ $\frac{1}{5}$ and $\frac{4}{20}$ are equivalent fractions.

DEFINITIONS AND CONCEPTS	EXAMPLES
SECTION 1.4 INTRODUCTION TO VARIABLE EXPRESSIONS AND EQUATIONS	

The expression a^n is an **exponential expression.** The number a is called the **base;** it is the repeated factor. The number n is called the **exponent;** it is the number of times that the base is a factor.

$$4^3 = 4 \cdot 4 \cdot 4 = 64$$
$$7^2 = 7 \cdot 7 = 49$$

Order of Operations

Simplify expressions in the following order. If grouping symbols are present, simplify expressions within those first, starting with the innermost set. Also, simplify the numerator and the denominator of a fraction separately.

1. Simplify exponential expressions.
2. Multiply or divide in order from left to right.
3. Add or subtract in order from left to right.

$$\frac{8^2 + 5(7-3)}{3 \cdot 7} = \frac{8^2 + 5(4)}{21}$$
$$= \frac{64 + 5(4)}{21}$$
$$= \frac{64 + 20}{21}$$
$$= \frac{84}{21}$$
$$= 4$$

A symbol used to represent a number is called a **variable.**

Examples of variables are:
$$q, x, z$$

An **algebraic expression** is a collection of numbers, variables, operation symbols, and grouping symbols.

Examples of algebraic expressions are:
$$5x, 2(y-6), \frac{q^2 - 3q + 1}{6}$$

To evaluate an algebraic expression containing a variable, substitute a given number for the variable and simplify.

Evaluate $x^2 - y^2$ if $x = 5$ and $y = 3$.
$$x^2 - y^2 = (5)^2 - 3^2$$
$$= 25 - 9$$
$$= 16$$

A mathematical statement that two expressions are equal is called an **equation.**

Equations:
$$3x - 9 = 20$$
$$A = \pi r^2$$

A **solution** of an equation is a value for the variable that makes the equation a true statement.

Determine whether 4 is a solution of $5x + 7 = 27$.
$$5x + 7 = 27$$
$$5(4) + 7 \stackrel{?}{=} 27$$
$$20 + 7 \stackrel{?}{=} 27$$
$$27 = 27 \quad \text{True}$$
4 is a solution.

SECTION 1.5 ADDING REAL NUMBERS	

To Add Two Numbers with the Same Sign

1. Add their absolute values.
2. Use their common sign as the sign of the sum.

Add.
$$10 + 7 = 17$$
$$-3 + (-8) = -11$$

To Add Two Numbers with Different Signs

1. Subtract their absolute values.
2. Use the sign of the number whose absolute value is larger as the sign of the sum.

$$-25 + 5 = -20$$
$$14 + (-9) = 5$$

(continued)

68 CHAPTER 1 Review of Real Numbers

DEFINITIONS AND CONCEPTS	EXAMPLES
SECTION 1.5 ADDING REAL NUMBERS (continued)	
Two numbers that are the same distance from 0 but lie on opposite sides of 0 are called **opposites** or **additive inverses**. The opposite of a number a is denoted by $-a$.	The opposite of -7 is 7. The opposite of 123 is -123.
The sum of a number a and its opposite, $-a$, is 0. $$a + (-a) = 0$$ If a is a number, then $-(-a) = a$.	$-4 + 4 = 0$ $12 + (-12) = 0$ $-(-8) = 8$ $-(-14) = 14$
SECTION 1.6 SUBTRACTING REAL NUMBERS	
To subtract two numbers a and b, add the first number a to the opposite of the second number b. $$a - b = a + (-b)$$	Subtract. $3 - (-44) = 3 + 44 = 47$ $-5 - 22 = -5 + (-22) = -27$ $-30 - (-30) = -30 + 30 = 0$
SECTION 1.7 MULTIPLYING AND DIVIDING REAL NUMBERS	
Quotient of two real numbers $$\frac{a}{b} = a \cdot \frac{1}{b}$$	Multiply or divide. $\dfrac{42}{2} = 42 \cdot \dfrac{1}{2} = 21$
Multiplying and Dividing Real Numbers The product or quotient of two numbers with the same sign is a positive number. The product or quotient of two numbers with different signs is a negative number.	$7 \cdot 8 = 56 \qquad -7 \cdot (-8) = 56$ $-2 \cdot 4 = -8 \qquad 2 \cdot (-4) = -8$ $\dfrac{90}{10} = 9 \qquad \dfrac{-90}{-10} = 9$ $\dfrac{42}{-6} = -7 \qquad \dfrac{-42}{6} = -7$
Products and Quotients Involving Zero The product of 0 and any number is 0. $$b \cdot 0 = 0 \quad \text{and} \quad 0 \cdot b = 0$$	$-4 \cdot 0 = 0 \qquad 0 \cdot \left(-\dfrac{3}{4}\right) = 0$
The quotient of a nonzero number and 0 is undefined. $$\dfrac{b}{0} \text{ is undefined.}$$	$\dfrac{-85}{0}$ is undefined.
The quotient of 0 and any nonzero number is 0. $$\dfrac{0}{b} = 0$$	$\dfrac{0}{18} = 0 \qquad \dfrac{0}{-47} = 0$
SECTION 1.8 PROPERTIES OF REAL NUMBERS	
Commutative Properties Addition: $a + b = b + a$ Multiplication: $a \cdot b = b \cdot a$	$3 + (-7) = -7 + 3$ $-8 \cdot 5 = 5 \cdot (-8)$
Associative Properties Addition: $(a + b) + c = a + (b + c)$ Multiplication: $(a \cdot b) \cdot c = a \cdot (b \cdot c)$	$(5 + 10) + 20 = 5 + (10 + 20)$ $(-3 \cdot 2) \cdot 11 = -3 \cdot (2 \cdot 11)$

DEFINITIONS AND CONCEPTS	EXAMPLES
SECTION 1.8 PROPERTIES OF REAL NUMBERS (continued)	
Two numbers whose product is 1 are called **multiplicative inverses** or **reciprocals.** The reciprocal of a nonzero number a is $\frac{1}{a}$ because $a \cdot \frac{1}{a} = 1$.	The reciprocal of 3 is $\frac{1}{3}$. The reciprocal of $-\frac{2}{5}$ is $-\frac{5}{2}$.
Distributive Property $a(b + c) = a \cdot b + a \cdot c$	$5(6 + 10) = 5 \cdot 6 + 5 \cdot 10$ $-2(3 + x) = -2 \cdot 3 + (-2)(x)$
Identities $\quad a + 0 = a \quad 0 + a = a$ $\qquad\qquad a \cdot 1 = a \quad 1 \cdot a = a$	$5 + 0 = 5 \quad 0 + (-2) = -2$ $-14 \cdot 1 = -14 \quad 1 \cdot 27 = 27$
Inverses Addition or opposite: $\quad a + (-a) = 0$ Multiplication or reciprocal: $\quad b \cdot \frac{1}{b} = 1$	$7 + (-7) = 0$ $3 \cdot \frac{1}{3} = 1$

STUDY SKILLS BUILDER

Are You Preparing for a Test on Chapter 1?

Below I have listed some *common trouble areas* for topics covered in Chapter 1. After studying for your test—but before taking your test—read these.

- Do you know the difference between $|-3|$, $-|-3|$, and $-(-3)$?

  $|-3| = 3; \quad -|-3| = -3; \quad \text{and} \quad -(-3) = 3$ (Section 1.2)

- Evaluate $x - y$ if $x = 7$ and $y = -3$.

  $x - y = 7 - (-3) = 10$ (Sections 1.4 and 1.6)

- Make sure you are familiar with order of operations. Sometimes the simplest-looking expressions can give you the most trouble.

  $1 + 2(3 + 6) = 1 + 2(9) = 1 + 18 = 19$ (Section 1.4)

- Do you know the difference between $(-3)^2$ and -3^2?

  $(-3)^2 = 9 \quad \text{and} \quad -3^2 = -9$ (Section 1.7)

- Do you know that these fractions are equivalent?

  $-\frac{1}{3} = \frac{-1}{3} = \frac{1}{-3}$ (Section 1.7)

- Do you know the difference between opposite and reciprocal? If not, study the table below.

Number	Opposite	Reciprocal
5	-5	$\frac{1}{5}$
$-\frac{4}{7}$	$\frac{4}{7}$	$-\frac{7}{4}$
$-\frac{1}{3}$	$\frac{1}{3}$	-3

  (Sections 1.5 and 1.7)

Remember: This is simply a checklist of selected topics given to check your understanding. For a review of Chapter 1 in the text, see the material at the end of Chapter 1.

CHAPTER 1 REVIEW

(1.2) Insert $<, >,$ or $=$ in the appropriate space to make the following statements true.

1. 8   10
2. 7   2
3. -4 -5
4. $\frac{12}{2}$ -8
5. $|-7|$ $|-8|$
6. $|-9|$ -9
7. $-|-1|$ -1
8. $|-14|$ $-(-14)$
9. 1.2   1.02
10. $-\frac{3}{2}$ $-\frac{3}{4}$

Translate each statement into symbols.

11. Four is greater than or equal to negative three.
12. Six is not equal to five.
13. 0.03 is less than 0.3.
14. New York City has 155 museums and 400 art galleries. Write an inequality comparing the numbers 155 and 400. (*Source: Absolute Trivia.com*)

70 CHAPTER 1 Review of Real Numbers

Given the following sets of numbers, list the numbers in each set that also belong to the set of:

a. Natural numbers b. Whole numbers
c. Integers d. Rational numbers
e. Irrational numbers f. Real numbers

15. $\left\{-6, 0, 1, 1\frac{1}{2}, 3, \pi, 9.62\right\}$

16. $\left\{-3, -1.6, 2, 5, \frac{11}{2}, 15.1, \sqrt{5}, 2\pi\right\}$

The following chart shows the gains and losses in dollars of Density Oil and Gas stock for a particular week.

Day	Gain or Loss in Dollars
Monday	+1
Tuesday	−2
Wednesday	+5
Thursday	+1
Friday	−4

17. Which day showed the greatest loss?
18. Which day showed the greatest gain?

(1.3) *Write the number as a product of prime factors.*

19. 36
20. 120

Perform the indicated operations. Write results in lowest terms.

21. $\dfrac{8}{15} \cdot \dfrac{27}{30}$
22. $\dfrac{7}{8} \div \dfrac{21}{32}$
23. $\dfrac{7}{15} + \dfrac{5}{6}$
24. $\dfrac{3}{4} - \dfrac{3}{20}$
25. $2\dfrac{3}{4} + 6\dfrac{5}{8}$
26. $7\dfrac{1}{6} - 2\dfrac{2}{3}$
27. $5 \div \dfrac{1}{3}$
28. $2 \cdot 8\dfrac{3}{4}$

29. Determine the unknown part of the given circle.

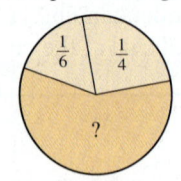

Find the area and the perimeter of each figure.

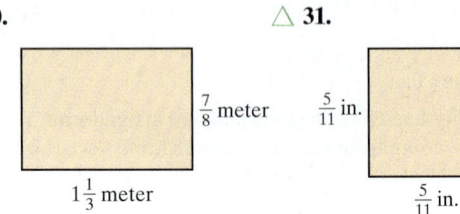

△ 30. (rectangle: $\frac{7}{8}$ meter by $1\frac{1}{3}$ meter)

△ 31. (figure with sides $\frac{5}{11}$ in., $\frac{3}{11}$ in., $\frac{3}{11}$ in., $\frac{5}{11}$ in.)

△ 32. A trim carpenter needs a piece of quarter round molding $6\frac{1}{8}$ feet long for a bathroom. She finds a piece $7\frac{1}{2}$ feet long. How long a piece does she need to cut from the $7\frac{1}{2}$-foot-long molding in order to use it in the bathroom?

In December 1998, Nkem Chukwu gave birth to the world's first surviving octuplets in Houston, Texas. The following chart gives the octuplets' birthweights. The babies are listed in order of birth.

Baby's Name	Gender	Birthweight (pounds)
Ebuka	girl	$1\frac{1}{2}$
Chidi	girl	$1\frac{11}{16}$
Echerem	girl	$1\frac{3}{4}$
Chima	girl	$1\frac{5}{8}$
Odera	girl	$1\frac{11}{16}$
Ikem	boy	$1\frac{1}{8}$
Jioke	boy	$1\frac{13}{16}$
Gorom	girl	$1\frac{1}{8}$

(*Source:* Texas Children's Hospital, Houston, Texas)

33. What was the total weight of the boy octuplets?
34. What was the total weight of the girl octuplets?
35. Find the combined weight of all eight octuplets.
36. Which baby weighed the most?
37. Which baby weighed the least?
38. How much more did the heaviest baby weigh than the lightest baby?
39. By March 1999, Chima weighed $5\frac{1}{2}$ pounds. How much weight had she gained since birth?
40. By March 1999, Ikem weighed $4\frac{5}{32}$ pounds. How much weight had he gained since birth?

(1.4) *Simplify each expression.*

41. 2^4
42. 5^2
43. $\left(\dfrac{2}{7}\right)^2$
44. $\left(\dfrac{3}{4}\right)^3$
45. $6 \cdot 3^2 + 2 \cdot 8$
46. $68 - 5 \cdot 2^3$
47. $3(1 + 2 \cdot 5) + 4$
48. $8 + 3(2 \cdot 6 - 1)$
49. $\dfrac{4 + |6 - 2| + 8^2}{4 + 6 \cdot 4}$
50. $5[3(2 + 5) - 5]$

Translate each word statement to symbols.

51. The difference of twenty and twelve is equal to the product of two and four.
52. The quotient of nine and two is greater than negative five.

Evaluate each expression if $x = 6$, $y = 2$, and $z = 8$.

53. $2x + 3y$

54. $x(y + 2z)$

55. $\dfrac{x}{y} + \dfrac{z}{2y}$

56. $x^2 - 3y^2$

△ **57.** The expression $180 - a - b$ represents the measure of the unknown angle of the given triangle. Replace a with 37 and b with 80 to find the measure of the unknown angle.

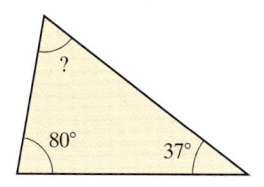

Decide whether the given number is a solution to the given equation.

58. Is $x = 3$ a solution of $7x - 3 = 18$?

59. Is $x = 1$ a solution of $3x^2 + 4 = x - 1$?

(1.5) *Find the additive inverse or the opposite.*

60. -9

61. $\dfrac{2}{3}$

62. $|-2|$

63. $-|-7|$

Find the following sums.

64. $-15 + 4$

65. $-6 + (-11)$

66. $\dfrac{1}{16} + \left(-\dfrac{1}{4}\right)$

67. $-8 + |-3|$

68. $-4.6 + (-9.3)$

69. $-2.8 + 6.7$

70. The lowest elevation in North America is -282 feet at Death Valley in California. If you are standing at a point 728 feet above Death Valley, what is your elevation? (*Source:* National Geographic Society)

(1.6) *Perform the indicated operations.*

71. $6 - 20$

72. $-3.1 - 8.4$

73. $-6 - (-11)$

74. $4 - 15$

75. $-21 - 16 + 3(8 - 2)$

76. $\dfrac{11 - (-9) + 6(8 - 2)}{2 + 3 \cdot 4}$

If $x = 3$, $y = -6$, and $z = -9$, evaluate each expression.

77. $2x^2 - y + z$

78. $\dfrac{y - x + 5x}{2x}$

(1.7) *Find the multiplicative inverse or reciprocal.*

79. -6

80. $\dfrac{3}{5}$

Simplify each expression.

81. $6(-8)$

82. $(-2)(-14)$

83. $\dfrac{-18}{-6}$

84. $\dfrac{42}{-3}$

85. $\dfrac{4(-3) + (-8)}{2 + (-2)}$

86. $\dfrac{3(-2)^2 - 5}{-14}$

87. $\dfrac{-6}{0}$

88. $\dfrac{0}{-2}$

89. $-4^2 - (-3 + 5) \div (-1) \cdot 2$

90. $-5^2 - (2 - 20) \div (-3) \cdot 3$

If $x = -5$ and $y = -2$, evaluate each expression.

91. $x^2 - y^4$

92. $x^2 - y^3$

93. During the 1999 LPGA Sara Lee Classic, Michelle McGann had scores of -9, -7, and $+1$ in three rounds of golf. Find her average score per round. (*Source:* Ladies Professional Golf Association)

94. During the 1999 PGA Masters Tournament, Bob Estes had scores of -1, 0, -3, and 0 in four rounds of golf. Find his average score per round. (*Source:* Professional Golf Association)

(1.8) *Name the property illustrated.*

95. $-6 + 5 = 5 + (-6)$

96. $6 \cdot 1 = 6$

97. $3(8 - 5) = 3 \cdot 8 + 3 \cdot (-5)$

98. $4 + (-4) = 0$

99. $2 + (3 + 9) = (2 + 3) + 9$

100. $2 \cdot 8 = 8 \cdot 2$

101. $6(8 + 5) = 6 \cdot 8 + 6 \cdot 5$

102. $(3 \cdot 8) \cdot 4 = 3 \cdot (8 \cdot 4)$

103. $4 \cdot \dfrac{1}{4} = 1$

104. $8 + 0 = 8$

Use the distributive property to write each expression without parentheses.

105. $5(y - 2)$

106. $-3(z + y)$

107. $-(7 - x + 4z)$

108. $\dfrac{1}{2}(6z - 10)$

109. $-4(3x + 5) - 7$

110. $-8(2y + 9) - 1$

MIXED REVIEW

Insert $<$, $>$, or $=$ in the space between each pair of numbers.

111. $-|-11|$ $\quad$ $|11.4|$

112. $-1\dfrac{1}{2}$ $\quad$ $-2\dfrac{1}{2}$

Perform the indicated operations.

113. $-7.2 + (-8.1)$

114. $14 - 20$

115. $4(-20)$

116. $\dfrac{-20}{4}$

117. $-\dfrac{4}{5}\left(\dfrac{5}{16}\right)$

118. $-0.5(-0.3)$

119. $8 \div 2 \cdot 4$

120. $(-2)^4$

121. $\dfrac{-3 - 2(-9)}{-15 - 3(-4)}$

122. $5 + 2[(7 - 5)^2 + (1 - 3)]$

123. $-\dfrac{5}{8} \div \dfrac{3}{4}$

124. $\dfrac{-15 + (-4)^2 + |-9|}{10 - 2 \cdot 5}$

CHAPTER 1 TEST

Translate the statement into symbols.

1. The absolute value of negative seven is greater than five.

2. The sum of nine and five is greater than or equal to four.

Simplify the expression.

3. $-13 + 8$
4. $-13 - (-2)$
5. $12 \div 4 \cdot 3 - 6 \cdot 2$
6. $(13)(-3)$
7. $(-6)(-2)$
8. $\dfrac{|-16|}{-8}$
9. $\dfrac{-8}{0}$
10. $\dfrac{|-6| + 2}{5 - 6}$
11. $\dfrac{1}{2} - \dfrac{5}{6}$
12. $-1\dfrac{1}{8} + 5\dfrac{3}{4}$
13. $(2 - 6) \div \dfrac{-2 - 6}{-3 - 1} - \dfrac{1}{2}$
14. $3(-4)^2 - 80$
15. $6[5 + 2(3 - 8) - 3]$
16. $\dfrac{-12 + 3 \cdot 8}{4}$
17. $\dfrac{(-2)(0)(-3)}{-6}$

Insert $<$, $>$, or $=$ in the appropriate space to make each of the following statements true.

18. -3 $\quad$ -7
19. 4 $\quad$ -8
20. 2 $\quad$ $|-3|$
21. $|-2|$ $\quad$ $-1 - (-3)$

22. In the state of Massachusetts, there are 2221 licensed child care centers and 10,993 licensed home-based child care providers. Write an inequality statement comparing the numbers 2221 and 10,993. (*Source:* Children's Foundation)

23. Given $\left\{-5, -1, 0, \dfrac{1}{4}, 1, 7, 11.6, \sqrt{7}, 3\pi\right\}$, list the numbers in this set that also belong to the set of:
    a. Natural numbers
    b. Whole numbers
    c. Integers
    d. Rational numbers
    e. Irrational numbers
    f. Real numbers

If $x = 6$, $y = -2$, and $z = -3$, evaluate each expression.

24. $x^2 + y^2$
25. $x + yz$
26. $2 + 3x - y$
27. $\dfrac{y + z - 1}{x}$

Identify the property illustrated by each expression.

28. $8 + (9 + 3) = (8 + 9) + 3$
29. $6 \cdot 8 = 8 \cdot 6$
30. $-6(2 + 4) = -6 \cdot 2 + (-6) \cdot 4$
31. $\dfrac{1}{6}(6) = 1$
32. Find the opposite of -9.
33. Find the reciprocal of $-\dfrac{1}{3}$.

The New Orleans Saints were 22 yards from the goal when the following series of gains and losses occurred.

Gains and Losses in Yards	
First Down	5
Second Down	-10
Third Down	-2
Fourth Down	29

34. During which down did the greatest loss of yardage occur?

35. Was a touchdown scored?

36. The temperature at the Winter Olympics was a frigid 14 degrees below zero in the morning, but by noon it had risen 31 degrees. What was the temperature at noon?

37. United HealthCare is a health insurance provider. In 3 consecutive recent years, it had net incomes of $356 million, $460 million, and $-$166 million. What was United HealthCare's total net income for these three years? (*Source:* United HealthCare Corp.)

38. Jean Avarez decided to sell 280 shares of stock, which decreased in value by $1.50 per share yesterday. How much money did she lose?

CHAPTER 2

Equations, Inequalities, and Problem Solving

2.1 Simplifying Algebraic Expressions
2.2 The Addition Property of Equality
2.3 The Multiplication Property of Equality
2.4 Solving Linear Equations
    Integrated Review—Solving Linear Equations
2.5 An Introduction to Problem Solving
2.6 Formulas and Problem Solving
2.7 Percent and Mixture Problem Solving
2.8 Further Problem Solving
2.9 Solving Linear Inequalities

There is an expanding market for high-speed trains as more countries, such as China, turn to bullet trains. In April 2007, a French TGV, short for "train à grande vitesse" or "very fast train," broke the previous world speed record on rails. (It should be noted that the current record for overall train speed is held by a magnetically levitated train called the Maglev, from Japan. This train hovers above the rails.)

The bar graph below shows a history of train speed records. In Section 2.5, Exercise 49, you will have the opportunity to calculate the speeds of the Maglev and the TGV.

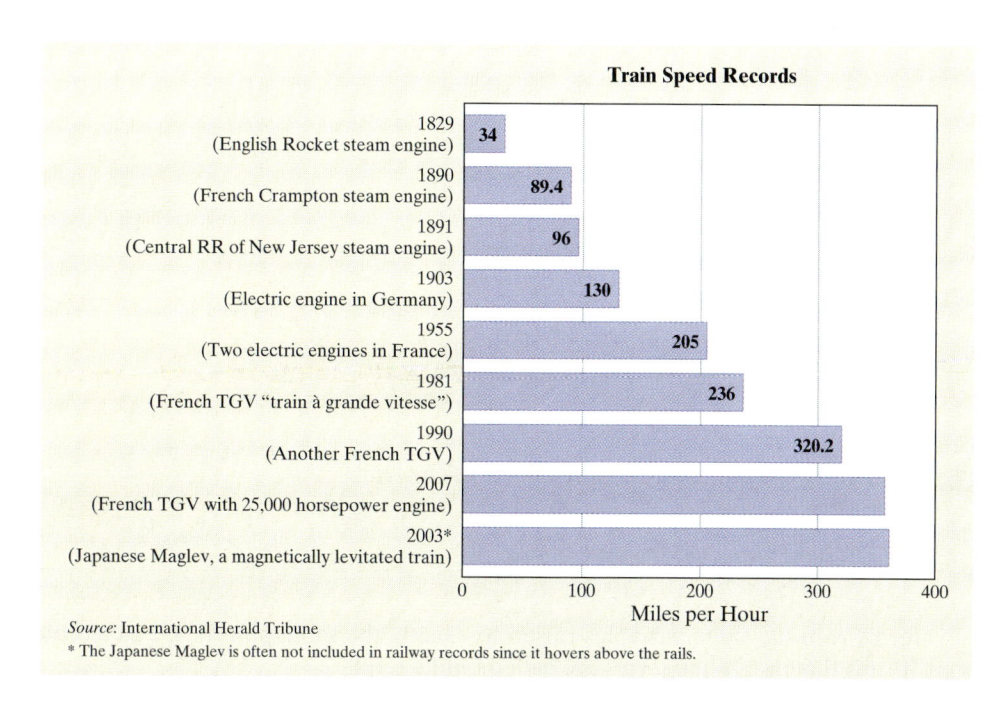

Much of mathematics relates to deciding which statements are true and which are false. For example, the statement $x + 7 = 15$ is an equation stating that the sum $x + 7$ has the same value as 15. Is this statement true or false? It is false for some values of x and true for just one value of x, namely 8. Our purpose in this chapter is to learn ways of deciding which values make an equation or an inequality true.

Train Speed Records

Year (Train)	Miles per Hour
1829 (English Rocket steam engine)	34
1890 (French Crampton steam engine)	89.4
1891 (Central RR of New Jersey steam engine)	96
1903 (Electric engine in Germany)	130
1955 (Two electric engines in France)	205
1981 (French TGV "train à grande vitesse")	236
1990 (Another French TGV)	320.2
2007 (French TGV with 25,000 horsepower engine)	
2003* (Japanese Maglev, a magnetically levitated train)	

Source: International Herald Tribune
* The Japanese Maglev is often not included in railway records since it hovers above the rails.

73

2.1 SIMPLIFYING ALGEBRAIC EXPRESSIONS

OBJECTIVES

1. Identify terms, like terms, and unlike terms.
2. Combine like terms.
3. Use the distributive property to remove parentheses.
4. Write word phrases as algebraic expressions.

As we explore in this section, an expression such as $3x + 2x$ is not as simple as possible, because — even without replacing x by a value — we can perform the indicated addition.

OBJECTIVE 1 ▶ Identifying terms, like terms, and unlike terms. Before we practice simplifying expressions, some new language of algebra is presented. A **term** is a number or the product of a number and variables raised to powers.

Terms

$$-y, \quad 2x^3, \quad -5, \quad 3xz^2, \quad \frac{2}{y}, \quad 0.8z$$

The **numerical coefficient** (sometimes also simply called the **coefficient**) of a term is the numerical factor. The numerical coefficient of $3x$ is 3. Recall that $3x$ means $3 \cdot x$.

Term	Numerical Coefficient
$3x$	3
$\dfrac{y^3}{5}$	$\dfrac{1}{5}$ since $\dfrac{y^3}{5}$ means $\dfrac{1}{5} \cdot y^3$
$0.7ab^3c^5$	0.7
z	1
$-y$	-1
-5	-5

▶ **Helpful Hint**
The term $-y$ means $-1y$ and thus has a numerical coefficient of -1. The term z means $1z$ and thus has a numerical coefficient of 1.

EXAMPLE 1 Identify the numerical coefficient in each term.

a. $-3y$ b. $22z^4$ c. y d. $-x$ e. $\dfrac{x}{7}$

Solution

a. The numerical coefficient of $-3y$ is -3.
b. The numerical coefficient of $22z^4$ is 22.
c. The numerical coefficient of y is 1, since y is $1y$.
d. The numerical coefficient of $-x$ is -1, since $-x$ is $-1x$.
e. The numerical coefficient of $\dfrac{x}{7}$ is $\dfrac{1}{7}$, since $\dfrac{x}{7}$ means $\dfrac{1}{7} \cdot x$.

PRACTICE 1 Identify the numerical coefficients in each term.

a. t b. $-7x$ c. $-\dfrac{w}{5}$ d. $43x^4$ e. $-b$

Terms with the same variables raised to exactly the same powers are called **like terms**. Terms that aren't like terms are called **unlike terms**.

Like Terms	Unlike Terms	
$3x, 2x$	$5x, 5x^2$	Why? Same variable x, but different powers x and x^2
$-6x^2y, 2x^2y, 4x^2y$	$7y, 3z, 8x^2$	Why? Different variables
$2ab^2c^3, ac^3b^2$	$6abc^3, 6ab^2$	Why? Different variables and different powers

> **Helpful Hint**
> In like terms, each variable and its exponent must match exactly, but these factors don't need to be in the same order.
>
> $2x^2y$ and $3yx^2$ are like terms.

EXAMPLE 2 Determine whether the terms are like or unlike.

a. $2x, 3x^2$ b. $4x^2y, x^2y, -2x^2y$ c. $-2yz, -3zy$ d. $-x^4, x^4$

Solution

a. Unlike terms, since the exponents on x are not the same.
b. Like terms, since each variable and its exponent match.
c. Like terms, since $zy = yz$ by the commutative property.
d. Like terms.

PRACTICE 2 Determine whether the terms are like or unlike.

a. $-4xy, 5yx$ b. $5q, -3q^2$

c. $3ab^2, -2ab^2, 43ab^2$ d. $y^5, \dfrac{y^5}{2}$

OBJECTIVE 2 ▶ Combining like terms. An algebraic expression containing the sum or difference of like terms can be simplified by applying the distributive property. For example, by the distributive property, we rewrite the sum of the like terms $3x + 2x$ as

$$3x + 2x = (3 + 2)x = 5x$$

Also,

$$-y^2 + 5y^2 = (-1 + 5)y^2 = 4y^2$$

Simplifying the sum or difference of like terms is called **combining like terms.**

EXAMPLE 3 Simplify each expression by combining like terms.

a. $7x - 3x$ b. $10y^2 + y^2$ c. $8x^2 + 2x - 3x$

Solution

a. $7x - 3x = (7 - 3)x = 4x$
b. $10y^2 + y^2 = 10y^2 + 1y^2 = (10 + 1)y^2 = 11y^2$
c. $8x^2 + 2x - 3x = 8x^2 + (2 - 3)x = 8x^2 - x$

PRACTICE 3 Simplify each expression by combining like terms.

a. $4x^2 + 3x^2$ b. $-3y + y$ c. $5x - 3x^2 + 8x^2$

EXAMPLE 4 Simplify each expression by combining like terms.

a. $2x + 3x + 5 + 2$
b. $-5a - 3 + a + 2$
c. $4y - 3y^2$
d. $2.3x + 5x - 6$
e. $-\frac{1}{2}b + b$

Solution Use the distributive property to combine like terms.

a. $2x + 3x + 5 + 2 = (2 + 3)x + (5 + 2)$
$= 5x + 7$

b. $-5a - 3 + a + 2 = -5a + 1a + (-3 + 2)$
$= (-5 + 1)a + (-3 + 2)$
$= -4a - 1$

c. $4y - 3y^2$ These two terms cannot be combined because they are unlike terms.

d. $2.3x + 5x - 6 = (2.3 + 5)x - 6$
$= 7.3x - 6$

e. $-\frac{1}{2}b + b = -\frac{1}{2}b + 1b = \left(-\frac{1}{2} + 1\right)b = \frac{1}{2}b$

PRACTICE 4 Use the distributive property to combine like terms.

a. $3y + 8y - 7 + 2$
b. $6x - 3 - x - 3$
c. $\frac{3}{4}t - t$
d. $9y + 3.2y + 10 + 3$
e. $5z - 3z^4$

The examples above suggest the following:

> **Combining Like Terms**
> To **combine like terms,** add the numerical coefficients and multiply the result by the common variable factors.

OBJECTIVE 3 ▶ Using the distributive property. Simplifying expressions makes frequent use of the distributive property to also remove parentheses.

EXAMPLE 5 Find each product by using the distributive property to remove parentheses.

a. $5(x + 2)$
b. $-2(y + 0.3z - 1)$
c. $-(x + y - 2z + 6)$

Solution

a. $5(x + 2) = 5 \cdot x + 5 \cdot 2$ Apply the distributive property.
$= 5x + 10$ Multiply.

b. $-2(y + 0.3z - 1) = -2(y) + (-2)(0.3z) + (-2)(-1)$ Apply the distributive property.
$= -2y - 0.6z + 2$ Multiply.

c. $-(x + y - 2z + 6) = -1(x + y - 2z + 6)$ Distribute -1 over each term.
$= -1(x) - 1(y) - 1(-2z) - 1(6)$
$= -x - y + 2z - 6$

PRACTICE 5 Find each product by using the distributive property to remove parentheses.

a. $3(2x - 7)$
b. $-5(3x - 4z - 5)$
c. $-(2x - y + z - 2)$

> **Helpful Hint**
> If a "−" sign precedes parentheses, the sign of each term inside the parentheses is changed when the distributive property is applied to remove parentheses.
>
> **Examples:**
>
> $-(2x + 1) = -2x - 1 \qquad -(-5x + y - z) = 5x - y + z$
> $-(x - 2y) = -x + 2y \qquad -(-3x - 4y - 1) = 3x + 4y + 1$

When simplifying an expression containing parentheses, we often use the distributive property in both directions—first to remove parentheses and then again to combine any like terms.

EXAMPLE 6 Simplify the following expressions.

a. $3(2x - 5) + 1$ **b.** $-2(4x + 7) - (3x - 1)$ **c.** $9 - 3(4x + 10)$

Solution

a. $3(2x - 5) + 1 = 6x - 15 + 1$ Apply the distributive property.
$\qquad\qquad\qquad = 6x - 14$ Combine like terms.

b. $-2(4x + 7) - (3x - 1) = -8x - 14 - 3x + 1$ Apply the distributive property.
$\qquad\qquad\qquad\qquad\qquad = -11x - 13$ Combine like terms.

c. $9 - 3(4x + 10) = 9 - 12x - 30$ Apply the distributive property.
$\qquad\qquad\qquad = -21 - 12x$ Combine like terms.

> **Helpful Hint**
> Don't forget to use the distributive property to multiply before adding or subtracting like terms.

PRACTICE 6 Simplify the following expressions.

a. $4(9x + 1) + 6$ **b.** $-7(2x - 1) - (6 - 3x)$ **c.** $8 - 5(6x + 5)$

EXAMPLE 7 Write the phrase below as an algebraic expression. Then simplify if possible.

"Subtract $4x - 2$ from $2x - 3$."

Solution "Subtract $4x - 2$ **from** $2x - 3$" translates to $(2x - 3) - (4x - 2)$. Next, simplify the algebraic expression.

$(2x - 3) - (4x - 2) = 2x - 3 - 4x + 2$ Apply the distributive property.
$\qquad\qquad\qquad\qquad = -2x - 1$ Combine like terms.

PRACTICE 7 Write the phrase below as an algebraic expression. Then simplify if possible.

"Subtract $7x - 1$ from $2x + 3$."

OBJECTIVE 4 ▶ Writing word phrases as algebraic expressions. Next, we practice writing word phrases as algebraic expressions.

EXAMPLE 8 Write the following phrases as algebraic expressions and simplify if possible. Let x represent the unknown number.

a. Twice a number, added to 6
b. The difference of a number and 4, divided by 7
c. Five added to 3 times the sum of a number and 1
d. The sum of twice a number, 3 times the number, and 5 times the number

Solution

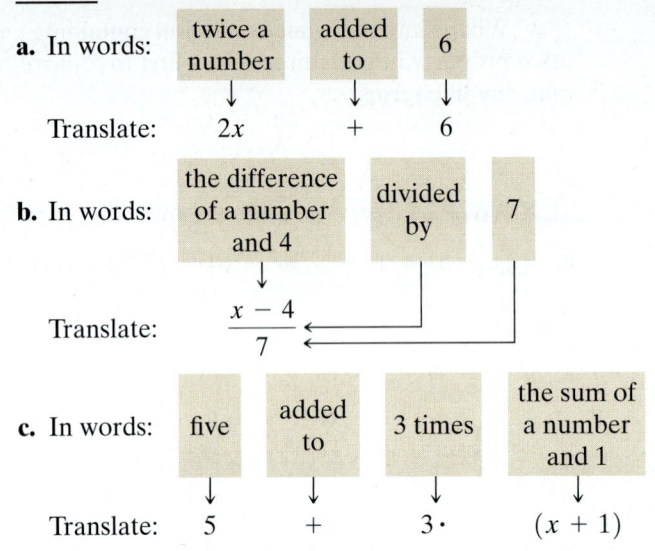

a. In words: twice a number | added to | 6
Translate: $2x \quad + \quad 6$

b. In words: the difference of a number and 4 | divided by | 7
Translate: $\dfrac{x-4}{7}$

c. In words: five | added to | 3 times | the sum of a number and 1
Translate: $5 \quad + \quad 3 \cdot \quad (x+1)$

Next, we simplify this expression.

$5 + 3(x+1) = 5 + 3x + 3$ Use the distributive property.
$ = 8 + 3x$ Combine like terms.

d. The phrase "the sum of" means that we add.

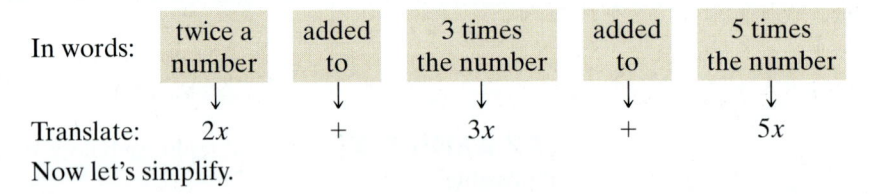

In words: twice a number | added to | 3 times the number | added to | 5 times the number
Translate: $2x \quad + \quad 3x \quad + \quad 5x$

Now let's simplify.

$2x + 3x + 5x = 10x$ Combine like terms.

PRACTICE 8 Write the following phrases as algebraic expressions and simplify if possible. Let x represent the unknown number.

a. Three added to double a number
b. Six subtracted from the sum of 5 and a number
c. Two times the sum of 3 and a number, increased by 4
d. The sum of a number, half the number, and 5 times the number

Section 2.1 Simplifying Algebraic Expressions 79

VOCABULARY & READINESS CHECK

Use the choices below to fill in each blank. Some choices may be used more than once.

| like | numerical coefficient | term | distributive |
| unlike | combine like terms | expression | |

1. $23y^2 + 10y - 6$ is called a(n) _____ while $23y^2$, $10y$, and -6 are each called a(n) _____.
2. To simplify $x + 4x$, we _____.
3. The term y has an understood _____ of 1.
4. The terms $7z$ and $7y$ are _____ terms and the terms $7z$ and $-z$ are _____ terms.
5. For the term $-\frac{1}{2}xy^2$, the number $-\frac{1}{2}$ is the _____.
6. $5(3x - y)$ equals $15x - 5y$ by the _____ property.

Fill in the blank with the numerical coefficient of each term. See Example 1.

7. $-7y$ _____   8. $3x$ _____   9. x _____   10. $-y$ _____   11. $-\frac{5y}{3}$ _____   12. $-\frac{2}{3}z$ _____

Indicate whether the following lists of terms are like or unlike. See Example 2.

13. $5y, -y$ _____   14. $-2x^2y, 6xy$ _____   15. $2z, 3z^2$ _____   16. $b^2a, -\frac{7}{8}ab^2$ _____

2.1 EXERCISE SET

Simplify each expression by combining any like terms. See Examples 3 and 4.

1. $7y + 8y$
2. $3x + 2x$
3. $8w - w + 6w$
4. $c - 7c + 2c$
5. $3b - 5 - 10b - 4$
6. $6g + 5 - 3g - 7$
7. $m - 4m + 2m - 6$
8. $a + 3a - 2 - 7a$
9. $5g - 3 - 5 - 5g$
10. $8p + 4 - 8p - 15$
11. $6.2x - 4 + x - 1.2$
12. $7.9y - 0.7 - y + 0.2$
13. $6x - 5x + x - 3 + 2x$
14. $8h + 13h - 6 + 7h - h$
15. $7x^2 + 8x^2 - 10x^2$
16. $8x^3 + x^3 - 11x^3$
17. $6x + 0.5 - 4.3x - 0.4x + 3$
18. $0.4y - 6.7 + y - 0.3 - 2.6y$
19. In your own words, explain how to combine like terms.
20. Do like terms contain the same numerical coefficients? Explain your answer.

Simplify each expression. First use the distributive property to remove any parentheses. See Examples 5 and 6.

21. $5(y - 4)$
22. $7(r - 3)$
23. $-2(x + 2)$
24. $-4(y + 6)$
25. $7(d - 3) + 10$
26. $9(z + 7) - 15$
27. $-5(2x - 3y + 6)$
28. $-2(4x - 3z - 1)$
29. $-(3x - 2y + 1)$
30. $-(y + 5z - 7)$
31. $5(x + 2) - (3x - 4)$
32. $4(2x - 3) - 2(x + 1)$

Write each of the following as an algebraic expression. Simplify if possible. See Example 7.

33. Add $6x + 7$ to $4x - 10$.
34. Add $3y - 5$ to $y + 16$.
35. Subtract $7x + 1$ from $3x - 8$.
36. Subtract $4x - 7$ from $12 + x$.
37. Subtract $5m - 6$ from $m - 9$.
38. Subtract $m - 3$ from $2m - 6$.

MIXED PRACTICE

Simplify each expression. See Examples 3 through 7.

39. $2k - k - 6$
40. $7c - 8 - c$
41. $-9x + 4x + 18 - 10x$
42. $5y - 14 + 7y - 20y$
43. $-4(3y - 4) + 12y$
44. $-3(2x + 5) - 6x$
45. $3(2x - 5) - 5(x - 4)$
46. $2(6x - 1) - (x - 7)$
47. $-2(3x - 4) + 7x - 6$
48. $8y - 2 - 3(y + 4)$
49. $5k - (3k - 10)$
50. $-11c - (4 - 2c)$
51. Subtract $6x - 1$ from $3x + 4$
52. Subtract $4 + 3y$ from $8 - 5y$
53. $3.4m - 4 - 3.4m - 7$
54. $2.8w - 0.9 - 0.5 - 2.8w$
55. $\frac{1}{3}(7y - 1) + \frac{1}{6}(4y + 7)$
56. $\frac{1}{5}(9y + 2) + \frac{1}{10}(2y - 1)$
57. $2 + 4(6x - 6)$
58. $8 + 4(3x - 4)$
59. $0.5(m + 2) + 0.4m$
60. $0.2(k + 8) - 0.1k$
61. $10 - 3(2x + 3y)$
62. $14 - 11(5m + 3n)$
63. $6(3x - 6) - 2(x + 1) - 17x$
64. $7(2x + 5) - 4(x + 2) - 20x$
65. $\frac{1}{2}(12x - 4) - (x + 5)$
66. $\frac{1}{3}(9x - 6) - (x - 2)$

Write each phrase as an algebraic expression and simplify if possible. Let x represent the unknown number. See Examples 7 and 8.

67. Twice a number, decreased by four
68. The difference of a number and two, divided by five
69. Seven added to double a number
70. Eight more than triple a number
71. Three-fourths of a number, increased by twelve
72. Eleven, increased by two-thirds of a number
73. The sum of 5 times a number and -2, added to 7 times a number
74. The sum of 3 times a number and 10, **subtracted from** 9 times a number
75. Eight times the sum of a number and six
76. Six times the difference of a number and five
77. Double a number, minus the sum of the number and ten
78. Half a number, minus the product of the number and eight
79. Seven, multiplied by the quotient of a number and six
80. The product of a number and ten, less twenty
81. The sum of 2, three times a number, -9, and four times a number
82. The sum of twice a number, -1, five times a number, and -12

REVIEW AND PREVIEW

Evaluate the following expressions for the given values. See Section 1.7.

83. If $x = -1$ and $y = 3$, find $y - x^2$.
84. If $g = 0$ and $h = -4$, find $gh - h^2$.
85. If $a = 2$ and $b = -5$, find $a - b^2$.
86. If $x = -3$, find $x^3 - x^2 + 4$.
87. If $y = -5$ and $z = 0$, find $yz - y^2$.
88. If $x = -2$, find $x^3 - x^2 - x$.

CONCEPT EXTENSIONS

△ 89. Recall that the perimeter of a figure is the total distance around the figure. Given the following rectangle, express the perimeter as an algebraic expression containing the variable x.

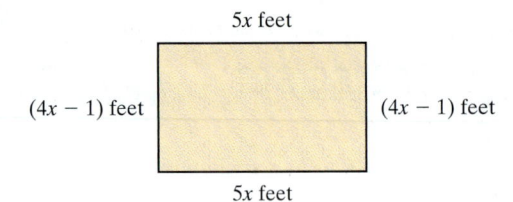

5x feet

$(4x - 1)$ feet $(4x - 1)$ feet

5x feet

△ 90. Given the following triangle, express its perimeter as an algebraic expression containing the variable x.

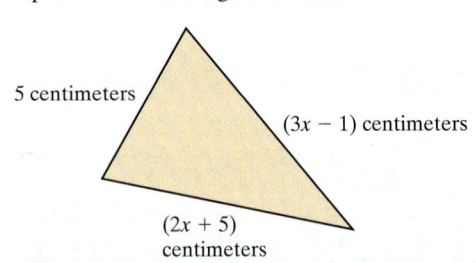

5 centimeters

$(3x - 1)$ centimeters

$(2x + 5)$ centimeters

Given the following two rules, determine whether each scale in Exercises 91 through 94 is balanced or not.

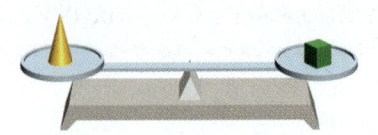

1 cone balances 1 cube

1 cylinder balances 2 cubes

91.

92.

93.

94.

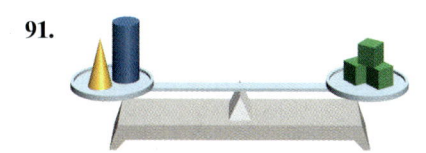

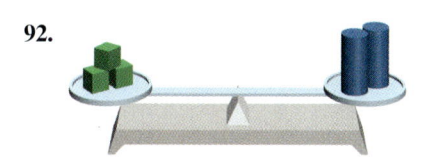

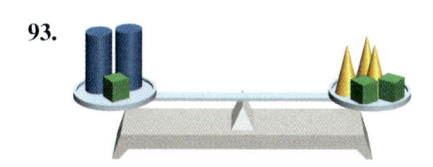

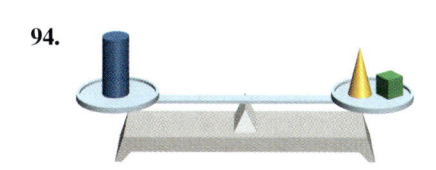

Write each algebraic expression described.

95. Write an expression with 4 terms that simplifies to $3x - 4$.
96. Write an expression of the form ___ (___ + ___) whose product is $6x + 24$.

97. To convert from feet to inches, we multiply by 12. For example, the number of inches in 2 feet is $12 \cdot 2$ inches. If one board has a length of $(x + 2)$ *feet* and a second board has a length of $(3x - 1)$ *inches*, express their total length in inches as an algebraic expression.

98. The value of 7 nickels is $5 \cdot 7$ cents. Likewise, the value of x nickels is $5x$ cents. If the money box in a drink machine contains x *nickels*, $3x$ *dimes*, and $(30x - 1)$ *quarters*, express their total value in cents as an algebraic expression.

For Exercises 99 through 104, see the example below.

Example

Simplify $-3xy + 2x^2y - (2xy - 1)$.

Solution

$$-3xy + 2x^2y - (2xy - 1)$$
$$= -3xy + 2x^2y - 2xy + 1 = -5xy + 2x^2y + 1$$

Simplify each expression.

99. $5b^2c^3 + 8b^3c^2 - 7b^3c^2$
100. $4m^4p^2 + m^4p^2 - 5m^2p^4$
101. $3x - (2x^2 - 6x) + 7x^2$
102. $9y^2 - (6xy^2 - 5y^2) - 8xy^2$
103. $-(2x^2y + 3z) + 3z - 5x^2y$
104. $-(7c^3d - 8c) - 5c - 4c^3d$

STUDY SKILLS BUILDER

Have You Decided to Complete This Course Successfully?

Ask yourself if one of your current goals is to complete this course successfully.

If it is not a goal of yours, ask yourself why? One common reason is fear of failure. Amazingly enough, fear of failure alone can be strong enough to keep many of us from doing our best in any endeavor.

Another common reason is that you simply haven't taken the time to think about or write down your goals for this course. To help accomplish this, answer the questions below.

Self-Check

1. Write down your goal(s) for this course.
2. Now list steps you will take to make sure your goal(s) in Question 1 are accomplished.

3. Rate your commitment to this course with a number between 1 and 5. Use the diagram below to help.

High Commitment		Average Commitment		Not committed at all
5	4	3	2	1

4. If you have rated your personal commitment level (from the exercise above) as a 1, 2, or 3, list the reasons why this is so. Then determine whether it is possible to increase your commitment level to a 4 or 5.

Good luck, and don't forget that a positive attitude will make a big difference.

82 CHAPTER 2 Equations, Inequalities, and Problem Solving

2.2 THE ADDITION PROPERTY OF EQUALITY

OBJECTIVES

1. Define linear equations and use the addition property of equality to solve linear equations.
2. Write word phrases as algebraic expressions.

OBJECTIVE 1 ▶ Defining linear equations and using the addition property. Recall from Section 1.4 that an equation is a statement that two expressions have the same value. Also, a value of the variable that makes an equation a true statement is called a solution or root of the equation. The process of finding the solution of an equation is called **solving** the equation for the variable. In this section we concentrate on solving **linear equations** in one variable.

> **Linear Equation in One Variable**
> **A linear equation in one variable** can be written in the form
> $$ax + b = c$$
> where a, b, and c are real numbers and $a \neq 0$.

Evaluating a linear equation for a given value of the variable, as we did in Section 1.4, can tell us whether that value is a solution, but we can't rely on evaluating an equation as our method of solving it.

Instead, to solve a linear equation in x, we write a series of simpler equations, all *equivalent* to the original equation, so that the final equation has the form

$$x = \text{number} \quad \text{or} \quad \text{number} = x$$

Equivalent equations are equations that have the same solution. This means that the "number" above is the solution to the original equation.

The first property of equality that helps us write simpler equivalent equations is the **addition property of equality.**

> **Addition Property of Equality**
> If a, b, and c are real numbers, then
> $$a = b \quad \text{and} \quad a + c = b + c$$
> are equivalent equations.

This property guarantees that adding the same number to both sides of an equation does not change the solution of the equation. Since subtraction is defined in terms of addition, we may also **subtract the same number from both sides** without changing the solution.

A good way to picture a true equation is as a balanced scale. Since it is balanced, each side of the scale weighs the same amount.

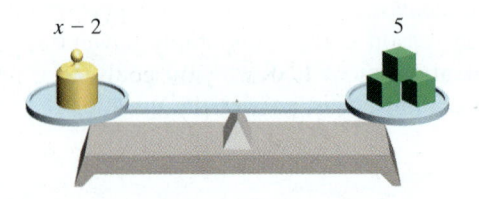

If the same weight is added to or subtracted from each side, the scale remains balanced.

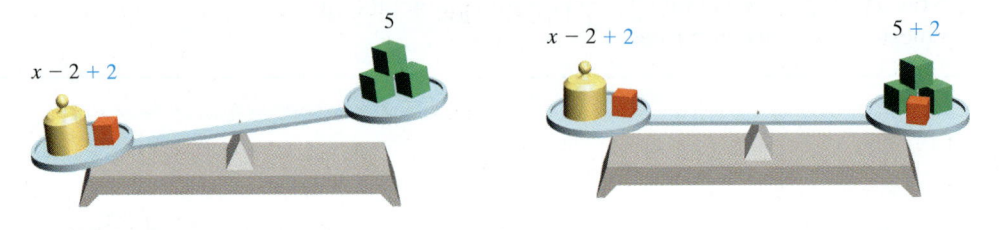

We use the addition property of equality to write equivalent equations until the variable is by itself on one side of the equation, and the equation looks like "$x =$ number" or "number $= x$."

EXAMPLE 1 Solve $x - 7 = 10$ for x.

Solution To solve for x, we want x alone on one side of the equation. To do this, we add 7 to both sides of the equation.

$$x - 7 = 10$$
$$x - 7 + 7 = 10 + 7 \quad \text{Add 7 to both sides.}$$
$$x = 17 \quad \text{Simplify.}$$

The solution of the equation $x = 17$ is obviously 17. Since we are writing equivalent equations, the solution of the equation $x - 7 = 10$ is also 17.

Check: To check, replace x with 17 in the original equation.

$$x - 7 = 10$$
$$17 - 7 \stackrel{?}{=} 10 \quad \text{Replace } x \text{ with 17 in the original equation.}$$
$$10 = 10 \quad \text{True}$$

Since the statement is true, 17 is the solution or we can say that the solution set is $\{17\}$.

PRACTICE 1 Solve: $x + 3 = -5$ for x.

Concept Check ✓

Use the addition property to fill in the blank so that the middle equation simplifies to the last equation.

$$x - 5 = 3$$
$$x - 5 + \underline{} = 3 + \underline{}$$
$$x = 8$$

EXAMPLE 2 Solve $y + 0.6 = -1.0$ for y.

Solution To get y alone on one side of the equation, subtract 0.6 from both sides of the equation.

$$y + 0.6 = -1.0$$
$$y + 0.6 - 0.6 = -1.0 - 0.6 \quad \text{Subtract 0.6 from both sides.}$$
$$y = -1.6 \quad \text{Combine like terms.}$$

Check: To check the proposed solution, -1.6, replace y with -1.6 in the original equation.

$$y + 0.6 = -1.0$$
$$-1.6 + 0.6 \stackrel{?}{=} -1.0 \quad \text{Replace } y \text{ with } -1.6 \text{ in the original equation.}$$
$$-1.0 = -1.0 \quad \text{True}$$

The solution is -1.6 or we can say that the solution set is $\{-1.6\}$.

PRACTICE 2 Solve: $y - 0.3 = -2.1$ for y.

Answer to Concept Check: 5

84 CHAPTER 2 Equations, Inequalities, and Problem Solving

EXAMPLE 3 Solve: $\dfrac{1}{2} = x - \dfrac{3}{4}$

Solution To get x alone, we add $\dfrac{3}{4}$ to both sides.

$$\dfrac{1}{2} = x - \dfrac{3}{4}$$

$$\dfrac{1}{2} + \dfrac{3}{4} = x - \dfrac{3}{4} + \dfrac{3}{4} \quad \text{Add } \dfrac{3}{4} \text{ to both sides.}$$

$$\dfrac{1}{2} \cdot \dfrac{2}{2} + \dfrac{3}{4} = x \quad \text{The LCD is 4.}$$

$$\dfrac{2}{4} + \dfrac{3}{4} = x \quad \text{Add the fractions.}$$

$$\dfrac{5}{4} = x$$

Check:

$$\dfrac{1}{2} = x - \dfrac{3}{4} \quad \text{Original equation}$$

$$\dfrac{1}{2} \stackrel{?}{=} \dfrac{5}{4} - \dfrac{3}{4} \quad \text{Replace } x \text{ with } \dfrac{5}{4}.$$

$$\dfrac{1}{2} \stackrel{?}{=} \dfrac{2}{4} \quad \text{Subtract.}$$

$$\dfrac{1}{2} = \dfrac{1}{2} \quad \text{True}$$

The solution is $\dfrac{5}{4}$.

PRACTICE 3 Solve: $\dfrac{2}{5} = x + \dfrac{3}{10}$

> **Helpful Hint**
> We may solve an equation so that the variable is alone on *either* side of the equation. For example, $\dfrac{5}{4} = x$ is equivalent to $x = \dfrac{5}{4}$.

EXAMPLE 4 Solve $5t - 5 = 6t + 2$ for t.

Solution To solve for t, we first want all terms containing t on one side of the equation and all other terms on the other side of the equation. To do this, first subtract $5t$ from both sides of the equation.

$$5t - 5 = 6t + 2$$

$$5t - 5 - 5t = 6t + 2 - 5t \quad \text{Subtract } 5t \text{ from both sides.}$$

$$-5 = t + 2 \quad \text{Combine like terms.}$$

Next, subtract 2 from both sides and the variable t will be isolated.

$$-5 = t + 2$$

$$-5 - 2 = t + 2 - 2 \quad \text{Subtract 2 from both sides.}$$

$$-7 = t$$

Check: Check the solution, -7, in the original equation. The solution is -7.

PRACTICE 4 Solve: $4t + 7 = 5t - 3$

Many times, it is best to simplify one or both sides of an equation before applying the addition property of equality.

EXAMPLE 5 Solve: $2x + 3x - 5 + 7 = 10x + 3 - 6x - 4$

Solution First we simplify both sides of the equation.

$2x + 3x - 5 + 7 = 10x + 3 - 6x - 4$
$5x + 2 = 4x - 1$ Combine like terms on each side of the equation.

Next, we want all terms with a variable on one side of the equation and all numbers on the other side.

$5x + 2 - 4x = 4x - 1 - 4x$ Subtract $4x$ from both sides.
$x + 2 = -1$ Combine like terms.
$x + 2 - 2 = -1 - 2$ Subtract 2 from both sides to get x alone.
$x = -3$ Combine like terms.

Check:

$2x + 3x - 5 + 7 = 10x + 3 - 6x - 4$ Original equation
$2(-3) + 3(-3) - 5 + 7 \stackrel{?}{=} 10(-3) + 3 - 6(-3) - 4$ Replace x with -3.
$-6 - 9 - 5 + 7 \stackrel{?}{=} -30 + 3 + 18 - 4$ Multiply.
$-13 = -13$ True

The solution is -3.

PRACTICE 5 Solve: $8x - 5x - 3 + 9 = x + x + 3 - 7$

If an equation contains parentheses, we use the distributive property to remove them, as before. Then we combine any like terms.

EXAMPLE 6 Solve: $6(2a - 1) - (11a + 6) = 7$

Solution $6(2a - 1) - 1(11a + 6) = 7$
$6(2a) + 6(-1) - 1(11a) - 1(6) = 7$ Apply the distributive property.
$12a - 6 - 11a - 6 = 7$ Multiply.
$a - 12 = 7$ Combine like terms.
$a - 12 + 12 = 7 + 12$ Add 12 to both sides.
$a = 19$ Simplify.

Check: Check by replacing a with 19 in the original equation.

PRACTICE 6 Solve: $4(2a - 3) - (7a + 4) = 2$

EXAMPLE 7 Solve: $3 - x = 7$

Solution First we subtract 3 from both sides.

$3 - x = 7$
$3 - x - 3 = 7 - 3$ Subtract 3 from both sides.
$-x = 4$ Simplify.

We have not yet solved for x since x is not alone. However, this equation does say that the opposite of x is 4. If the opposite of x is 4, then x is the opposite of 4, or $x = -4$. If $-x = 4$, then $x = -4$.

Check:
$$3 - x = 7 \quad \text{Original equation}$$
$$3 - (-4) \stackrel{?}{=} 7 \quad \text{Replace } x \text{ with } -4.$$
$$3 + 4 \stackrel{?}{=} 7 \quad \text{Add.}$$
$$7 = 7 \quad \text{True}$$

The solution is -4.

PRACTICE 7 Solve: $12 - x = 20$

OBJECTIVE 2 ▶ Writing word phrases as algebraic expressions. Next, we practice writing word phrases as algebraic expressions.

EXAMPLE 8

a. The sum of two numbers is 8. If one number is 3, find the other number.

b. The sum of two numbers is 8. If one number is x, write an expression representing the other number.

c. An 8-foot board is cut into two pieces. If one piece is x feet, express the length of the other piece in terms of x.

Solution

a. If the sum of two numbers is 8 and one number is 3, we find the other number by subtracting 3 from 8. The other number is $8 - 3$ or 5.

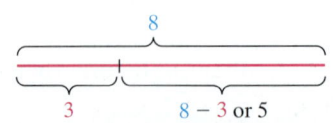

b. If the sum of two numbers is 8 and one number is x, we find the other number by subtracting x from 8. The other number is represented by $8 - x$.

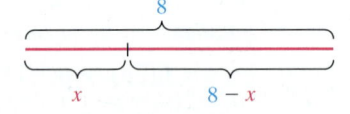

c. If an 8-foot board is cut into two pieces and one piece is x feet, we find the other length by subtracting x from 8. The other piece is $(8 - x)$ feet.

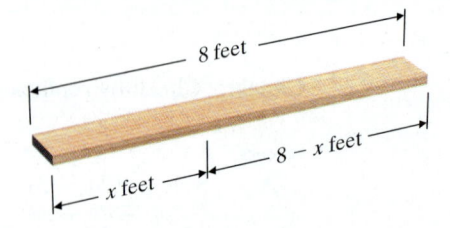

PRACTICE 8

a. The sum of two numbers is 9. If one number is 2, find the other number.

b. The sum of two numbers is 9. If one number is x, write an expression representing the other number.

c. A 9-foot rope is cut into two pieces. If one piece is x feet, express the length of the other piece in terms of x.

EXAMPLE 9

The Verrazano-Narrows Bridge in New York City is the longest suspension bridge in North America. The Golden Gate Bridge in San Francisco is 60 feet shorter than the Verrazano-Narrows Bridge. If the length of the Verrazano-Narrows Bridge is m feet, express the length of the Golden Gate Bridge as an algebraic expression in m. (*Source:* Survey of State Highway Engineers)

Solution Since the Golden Gate is 60 feet shorter than the Verrazano-Narrows Bridge, we have that its length is

	length of Verrazano-Narrows Bridge	minus	60
In words:			
Translate:	m	$-$	60

The Golden Gate Bridge is $(m - 60)$ feet long.

PRACTICE 9 Currently, the fastest train is the Japanese Maglev. The French TGV is 3.8 mph slower than the Maglev (see the Chapter 2 opener). If the speed of the Maglev is s miles per hour, express the speed of the French TGV as an algebraic expression in s.

VOCABULARY & READINESS CHECK

Use the choices below to fill in each blank. Not all choices will be used.

| addition | solving | expression | true |
| equivalent | equation | solution | false |

1. The difference between an equation and an expression is that a(n) _____ contains an equal sign, whereas an _____ does not.
2. _____ equations are equations that have the same solution.
3. A value of the variable that makes the equation a true statement is called a(n) _____ of the equation.
4. The process of finding the solution of an equation is called _____ the equation for the variable.
5. By the _____ property of equality, $x = -2$ and $x + 10 = -2 + 10$ are equivalent equations.
6. True or false: The equations $x = \frac{1}{2}$ and $\frac{1}{2} = x$ are equivalent equations. _____

2.2 EXERCISE SET

Solve each equation. Check each solution. See Examples 1 through 3.

1. $x + 7 = 10$
2. $x + 14 = 25$
3. $x - 2 = -4$
4. $y - 9 = 1$
5. $-2 = t - 5$
6. $-17 = x + 3$
7. $r - 8.6 = -8.1$
8. $t - 9.2 = -6.8$
9. $\frac{3}{4} = \frac{1}{3} + f$
10. $\frac{3}{8} = c + \frac{1}{6}$
11. $5b - 0.7 = 6b$
12. $9x + 5.5 = 10x$
13. $7x - 3 = 6x$
14. $18x - 9 = 19x$
15. In your own words, explain what is meant by the solution of an equation.

16. In your own words, explain how to check a solution of an equation.

Solve each equation. Don't forget to first simplify each side of the equation, if possible. Check each solution. See Examples 4 through 7.

17. $7x + 2x = 8x - 3$
18. $3n + 2n = 7 + 4n$
19. $\frac{5}{6}x + \frac{1}{6}x = -9$
20. $\frac{13}{11}y - \frac{2}{11}y = -3$
21. $2y + 10 = 5y - 4y$
22. $4x - 4 = 10x - 7x$
23. $-5(n - 2) = 8 - 4n$
24. $-4(z - 3) = 2 - 3z$
25. $\frac{3}{7}x + 2 = -\frac{4}{7}x - 5$
26. $\frac{1}{5}x - 1 = -\frac{4}{5}x - 13$
27. $5x - 6 = 6x - 5$
28. $2x + 7 = x - 10$
29. $8y + 2 - 6y = 3 + y - 10$
30. $4p - 11 - p = 2 + 2p - 20$
31. $-3(x - 4) = -4x$
32. $-2(x - 1) = -3x$
33. $\frac{3}{8}x - \frac{1}{6} = -\frac{5}{8}x - \frac{2}{3}$
34. $\frac{2}{5}x - \frac{1}{12} = \frac{3}{5}x - \frac{3}{4}$
35. $2(x - 4) = x + 3$
36. $3(y + 7) = 2y - 5$
37. $3(n - 5) - (6 - 2n) = 4n$
38. $5(3 + z) - (8z + 9) = -4z$
39. $-2(x + 6) + 3(2x - 5) = 3(x - 4) + 10$
40. $-5(x + 1) + 4(2x - 3) = 2(x + 2) - 8$

MIXED PRACTICE

Solve. See Examples 1 through 7.

41. $-11 = 3 + x$
42. $-8 = 8 + z$
43. $x - \frac{2}{5} = -\frac{3}{20}$
44. $y - \frac{4}{7} = -\frac{3}{14}$
45. $3x - 6 = 2x + 5$
46. $7y + 2 = 6y + 2$
47. $13x - 9 + 2x - 5 = 12x - 1 + 2x$
48. $15x + 20 - 10x - 9 = 25x + 8 - 21x - 7$
49. $7(6 + w) = 6(2 + w)$
50. $6(5 + c) = 5(c - 4)$
51. $n + 4 = 3.6$
52. $m + 2 = 7.1$
53. $10 - (2x - 4) = 7 - 3x$
54. $15 - (6 - 7k) = 2 + 6k$
55. $\frac{1}{3} = x + \frac{2}{3}$
56. $\frac{1}{11} = y + \frac{10}{11}$
57. $-6.5 - 4x - 1.6 - 3x = -6x + 9.8$
58. $-1.4 - 7x - 3.6 - 2x = -8x + 4.4$
59. $-3\left(x - \frac{1}{4}\right) = -4x$
60. $-2\left(x - \frac{1}{7}\right) = -3x$
61. $7(m - 2) - 6(m + 1) = -20$
62. $-4(x - 1) - 5(2 - x) = -6$
63. $0.8t + 0.2(t - 0.4) = 1.75$
64. $0.6v + 0.4(0.3 + v) = 2.34$

See Examples 8 and 9.

65. Two numbers have a sum of 20. If one number is p, express the other number in terms of p.

66. Two numbers have a sum of 13. If one number is y, express the other number in terms of y.

67. A 10-foot board is cut into two pieces. If one piece is x feet long, express the other length in terms of x.

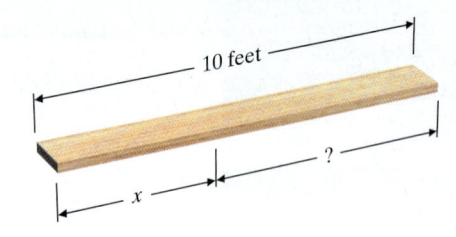

68. A 5-foot piece of string is cut into two pieces. If one piece is x feet long, express the other length in terms of x.

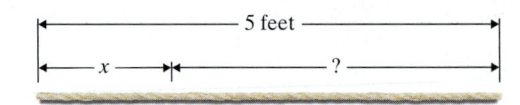

69. Two angles are *supplementary* if their sum is 180°. If one angle measures $x°$, express the measure of its supplement in terms of x.

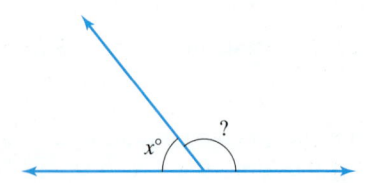

70. Two angles are *complementary* if their sum is 90°. If one angle measures $x°$, express the measure of its complement in terms of x.

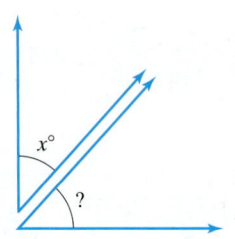

71. The length of the top of a computer desk is $1\frac{1}{2}$ feet longer than its width. If its width measures m feet, express its length as an algebraic expression in m.

72. In a mayoral election, April Catarella received 284 more votes than Charles Pecot. If Charles received n votes, how many votes did April receive?

73. The area of the Sahara Desert in Africa is 7 times the area of the Gobi Desert in Asia. If the area of the Gobi Desert is x square miles, express the area of the Sahara Desert as an algebraic expression in x.

74. The largest meteorite in the world is the Hoba West located in Namibia. Its weight is 3 times the weight of the Armanty meteorite located in Outer Mongolia. If the weight of the Armanty meteorite is y kilograms, express the weight of the Hoba West meteorite as an algebraic expression in y.

REVIEW AND PREVIEW

Find the reciprocal or multiplicative inverse of each. See Section 1.7.

75. $\dfrac{5}{8}$ **76.** $\dfrac{7}{6}$ **77.** 2

78. 5   **79.** $-\dfrac{1}{9}$ **80.** $-\dfrac{3}{5}$

Perform each indicated operation and simplify. See Section 1.7.

81. $\dfrac{3x}{3}$ **82.** $\dfrac{-2y}{-2}$ **83.** $-5\left(-\dfrac{1}{5}y\right)$

84. $7\left(\dfrac{1}{7}r\right)$ **85.** $\dfrac{3}{5}\left(\dfrac{5}{3}x\right)$ **86.** $\dfrac{9}{2}\left(\dfrac{2}{9}x\right)$

CONCEPT EXTENSIONS

87. The sum of the angles of a triangle is 180°. If one angle of a triangle measures $x°$ and a second angle measures $(2x + 7)°$, express the measure of the third angle in terms of x. Simplify the expression.

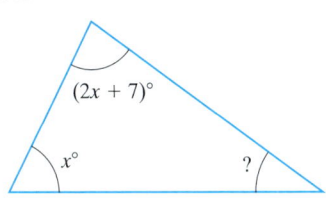

88. A quadrilateral is a four-sided figure like the one shown below whose angle sum is 360°. If one angle measures $x°$, a second angle measures $3x°$, and a third angle measures $5x°$, express the measure of the fourth angle in terms of x. Simplify the expression.

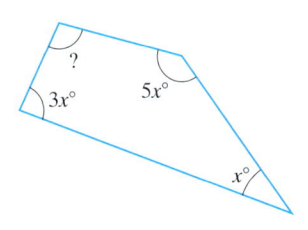

89. Write two terms whose sum is $-3x$.

90. Write four terms whose sum is $2y - 6$.

Use the addition property to fill in the blank so that the middle equation simplifies to the last equation. See the Concept Check in this section.

91. $x - 4 = -9$
$x - 4 + (\ \ ) = -9 + (\ \ )$
$x = -5$

92. $a + 9 = 15$
$a + 9 + (\ \ ) = 15 + (\ \ )$
$a = 6$

Fill in the blanks with numbers of your choice so that each equation has the given solution. Note: Each blank may be replaced with a different number.

93. ____ $+ x =$ ____ ; Solution: -3

94. $x -$ ____ $=$ ____ ; Solution: -10

95. A nurse's aide recorded the following fluid intakes for a patient on her night shift: 200 ml, 150 ml, 400 ml. If the patient's doctor requested that a total of 1000 ml of fluid be taken by the patient overnight, how much more fluid must the nurse give the patient? To solve this problem, solve the equation $200 + 150 + 400 + x = 1000$.

96. Let $x = 1$ and then $x = 2$ in the equation $x + 5 = x + 6$. Is either number a solution? How many solutions do you think this equation has? Explain your answer.

97. Let $x = 1$ and then $x = 2$ in the equation $x + 3 = x + 3$. Is either number a solution? How many solutions do you think this equation has? Explain your answer.

Use a calculator to determine whether the given value is a solution of the given equation.

98. $1.23x - 0.06 = 2.6x - 0.1285$; $x = 0.05$

99. $8.13 + 5.85y = 20.05y - 8.91$; $y = 1.2$

100. $3(a + 4.6) = 5a + 2.5$; $a = 6.3$

101. $7(z - 1.7) + 9.5 = 5(z + 3.2) - 9.2$; $z = 4.8$

2.3 THE MULTIPLICATION PROPERTY OF EQUALITY

OBJECTIVES

1. Use the multiplication property of equality to solve linear equations.
2. Use both the addition and multiplication properties of equality to solve linear equations.
3. Write word phrases as algebraic expressions.

OBJECTIVE 1 ▶ Using the multiplication property. As useful as the addition property of equality is, it cannot help us solve every type of linear equation in one variable. For example, adding or subtracting a value on both sides of the equation does not help solve

$$\frac{5}{2}x = 15.$$

Instead, we apply another important property of equality, the **multiplication property of equality.**

Multiplication Property of Equality

If a, b, and c are real numbers and $c \neq 0$, then

$$a = b \quad \text{and} \quad ac = bc$$

are equivalent equations.

This property guarantees that multiplying both sides of an equation by the same nonzero number does not change the solution of the equation. Since division is defined in terms of multiplication, we may also **divide both sides of the equation by the same nonzero number** without changing the solution.

EXAMPLE 1 Solve: $\frac{5}{2}x = 15$.

Solution To get x alone, multiply both sides of the equation by the reciprocal of $\frac{5}{2}$, which is $\frac{2}{5}$.

$$\frac{5}{2}x = 15$$

$$\frac{2}{5} \cdot \frac{5}{2}x = \frac{2}{5} \cdot 15 \quad \text{Multiply both sides by } \frac{2}{5}.$$

$$\left(\frac{2}{5} \cdot \frac{5}{2}\right)x = \frac{2}{5} \cdot 15 \quad \text{Apply the associative property.}$$

$$1x = 6 \quad \text{Simplify.}$$

or

$$x = 6$$

Check: Replace x with 6 in the original equation.

$$\frac{5}{2}x = 15 \quad \text{Original equation}$$

$$\frac{5}{2}(6) \stackrel{?}{=} 15 \quad \text{Replace } x \text{ with 6.}$$

$$15 = 15 \quad \text{True}$$

The solution is 6 or we say that the solution set is $\{6\}$.

PRACTICE 1 Solve: $\frac{4}{5}x = 16$

In the equation $\frac{5}{2}x = 15$, $\frac{5}{2}$ is the coefficient of x. When the coefficient of x is a *fraction*, we will get x alone by multiplying by the reciprocal. When the coefficient of x is an integer or a decimal, it is usually more convenient to divide both sides by the coefficient. (Dividing by a number is, of course, the same as multiplying by the reciprocal of the number.)

EXAMPLE 2 Solve: $-3x = 33$

Solution Recall that $-3x$ means $-3 \cdot x$. To get x alone, we divide both sides by the coefficient of x, that is, -3.

$$-3x = 33$$
$$\frac{-3x}{-3} = \frac{33}{-3} \quad \text{Divide both sides by } -3.$$
$$1x = -11 \quad \text{Simplify.}$$
$$x = -11$$

Check:
$$-3x = 33 \quad \text{Original equation}$$
$$-3(-11) \stackrel{?}{=} 33 \quad \text{Replace } x \text{ with } -11.$$
$$33 = 33 \quad \text{True}$$

The solution is -11, or the solution set is $\{-11\}$.

PRACTICE 2 Solve: $8x = -96$

EXAMPLE 3 Solve: $\frac{y}{7} = 20$

Solution Recall that $\frac{y}{7} = \frac{1}{7}y$. To get y alone, we multiply both sides of the equation by 7, the reciprocal of $\frac{1}{7}$.

$$\frac{y}{7} = 20$$
$$\frac{1}{7}y = 20$$
$$7 \cdot \frac{1}{7}y = 7 \cdot 20 \quad \text{Multiply both sides by 7.}$$
$$1y = 140 \quad \text{Simplify.}$$
$$y = 140$$

Check:
$$\frac{y}{7} = 20 \quad \text{Original equation}$$
$$\frac{140}{7} \stackrel{?}{=} 20 \quad \text{Replace } y \text{ with 140.}$$
$$20 = 20 \quad \text{True}$$

The solution is 140.

PRACTICE 3 Solve: $\frac{x}{5} = 13$

CHAPTER 2 Equations, Inequalities, and Problem Solving

EXAMPLE 4 Solve: $3.1x = 4.96$

Solution
$$3.1x = 4.96$$
$$\frac{3.1x}{3.1} = \frac{4.96}{3.1} \quad \text{Divide both sides by 3.1.}$$
$$1x = 1.6 \quad \text{Simplify.}$$
$$x = 1.6$$

Check: Check by replacing x with 1.6 in the original equation. The solution is 1.6.

PRACTICE 4 Solve: $2.7x = 4.05$

EXAMPLE 5 Solve: $-\frac{2}{3}x = -\frac{5}{2}$

Solution To get x alone, we multiply both sides of the equation by $-\frac{3}{2}$, the reciprocal of the coefficient of x.

$$-\frac{2}{3}x = -\frac{5}{2}$$
$$-\frac{3}{2} \cdot -\frac{2}{3}x = -\frac{3}{2} \cdot -\frac{5}{2} \quad \text{Multiply both sides by } -\frac{3}{2}, \text{ the reciprocal of } -\frac{2}{3}.$$
$$x = \frac{15}{4} \quad \text{Simplify.}$$

▶ **Helpful Hint**
Don't forget to multiply *both* sides by $-\frac{3}{2}$.

Check: Check by replacing x with $\frac{15}{4}$ in the original equation. The solution is $\frac{15}{4}$.

PRACTICE 5 Solve: $-\frac{5}{3}x = \frac{4}{7}$

OBJECTIVE 2 ▶ **Using both the addition and multiplication properties.** We are now ready to combine the skills learned in the last section with the skills learned from this section to solve equations by applying more than one property.

EXAMPLE 6 Solve: $-z - 4 = 6$

Solution First, get $-z$, the term containing the variable alone on one side. To do so, add 4 to both sides of the equation.

$$-z - 4 + 4 = 6 + 4 \quad \text{Add 4 to both sides.}$$
$$-z = 10 \quad \text{Simplify.}$$

Next, recall that $-z$ means $-1 \cdot z$. To get z alone, either multiply or divide both sides of the equation by -1. In this example, we divide.

$$-z = 10$$
$$\frac{-z}{-1} = \frac{10}{-1} \quad \text{Divide both sides by the coefficient } -1.$$
$$z = -10 \quad \text{Simplify.}$$

Check: To check, replace z with -10 in the original equation. The solution is -10.

PRACTICE 6 Solve: $-y + 3 = -8$

Don't forget to simplify one or both sides of an equation, if possible.

EXAMPLE 7 Solve: $12a - 8a = 10 + 2a - 13 - 7$

Solution First, simplify both sides of the equation by combining like terms.

$$12a - 8a = 10 + 2a - 13 - 7$$
$$4a = 2a - 10 \quad \text{Combine like terms.}$$

To get all terms containing a variable on one side, subtract $2a$ from both sides.

$$4a - 2a = 2a - 10 - 2a \quad \text{Subtract } 2a \text{ from both sides.}$$
$$2a = -10 \quad \text{Simplify.}$$
$$\frac{2a}{2} = \frac{-10}{2} \quad \text{Divide both sides by 2.}$$
$$a = -5 \quad \text{Simplify.}$$

Check: Check by replacing a with -5 in the original equation. The solution is -5.

PRACTICE 7 Solve: $6b - 11b = 18 + 2b - 6 + 9$

EXAMPLE 8 Solve: $7x - 3 = 5x + 9$

Solution To get x alone, let's first use the addition property to get variable terms on one side of the equation and numbers on the other side. One way to get variable terms on one side is to subtract $5x$ from both sides.

$$7x - 3 = 5x + 9$$
$$7x - 3 - 5x = 5x + 9 - 5x \quad \text{Subtract } 5x \text{ from both sides.}$$
$$2x - 3 = 9 \quad \text{Simplify.}$$

Now, to get numbers on the other side, let's add 3 to both sides.

$$2x - 3 + 3 = 9 + 3 \quad \text{Add 3 to both sides.}$$
$$2x = 12 \quad \text{Simplify.}$$

Use the multiplication property to get x alone.

$$\frac{2x}{2} = \frac{12}{2} \quad \text{Divide both sides by 2.}$$
$$x = 6 \quad \text{Simplify.}$$

Check: To check, replace x with 6 in the original equation to see that a true statement results. The solution is 6.

PRACTICE 8 Solve: $10x - 4 = 7x + 14$

If an equation has parentheses, don't forget to use the distributive property to remove them. Then combine any like terms.

EXAMPLE 9 Solve: $5(2x + 3) = -1 + 7$

Solution

$5(2x + 3) = -1 + 7$

$5(2x) + 5(3) = -1 + 7$ Apply the distributive property.

$10x + 15 = 6$ Multiply and write $-1 + 7$ as 6.

$10x + 15 - 15 = 6 - 15$ Subtract 15 from both sides.

$10x = -9$ Simplify.

$\dfrac{10x}{10} = -\dfrac{9}{10}$ Divide both sides by 10.

$x = -\dfrac{9}{10}$ Simplify.

Check: To check, replace x with $-\dfrac{9}{10}$ in the original equation to see that a true statement results. The solution is $-\dfrac{9}{10}$.

PRACTICE 9 Solve: $4(3x - 2) = -1 + 4$

OBJECTIVE 3 ▶ Writing word phrases as algebraic expressions. Next, we continue to sharpen our problem-solving skills by writing word phrases as algebraic expressions.

EXAMPLE 10 If x is the first of three consecutive integers, express the sum of the three integers in terms of x. Simplify if possible.

Solution An example of three consecutive integers is

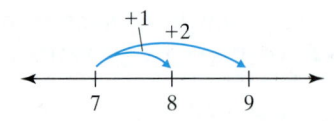

```
       +1
        \  +2
         \  \
    ←--+--+--+--→
       7  8  9
```

The second consecutive integer is always 1 more than the first, and the third consecutive integer is 2 more than the first. If x is the first of three consecutive integers, the three consecutive integers are

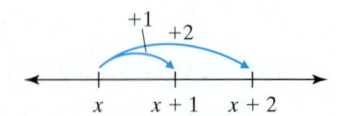

```
       +1
        \  +2
         \  \
    ←--+--+--+--→
       x  x+1 x+2
```

Their sum is

In words: first integer + second integer + third integer

Translate: $x + (x + 1) + (x + 2)$

which simplifies to $3x + 3$.

PRACTICE 10 If x is the first of three consecutive *even* integers, express their sum in terms of x.

Below are examples of consecutive even and odd integers.

Consecutive Even integers:

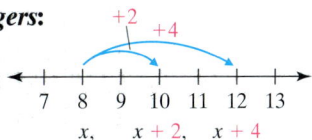

7 8 9 10 11 12 13
x,   x + 2,   x + 4

Consecutive Odd integers:

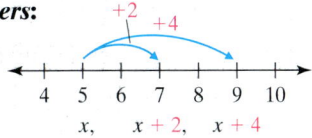

4 5 6 7 8 9 10
x,   x + 2,   x + 4

> **Helpful Hint**
> If x is an odd integer, then $x + 2$ is the next odd integer. This 2 simply means that odd integers are always 2 units from each other. (The same is true for even integers. They are always 2 units from each other.)

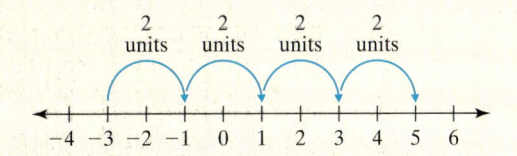

VOCABULARY & READINESS CHECK

Use the choices below to fill in each blank. Not all choices will be used.

true   addition   false   multiplication

1. By the _____ property of equality, $y = \frac{1}{2}$ and $5 \cdot y = 5 \cdot \frac{1}{2}$ are equivalent equations.

2. True or false: The equations $\frac{z}{4} = 10$ and $4 \cdot \frac{z}{4} = 10$ are equivalent equations. _____

3. True or false: The equations $-7x = 30$ and $\frac{-7x}{-7} = \frac{30}{7}$ are equivalent equations. _____

4. By the _____ property of equality, $9x = -63$ and $\frac{9x}{9} = \frac{-63}{9}$ are equivalent equations.

Solve each equation mentally.

5. $3a = 27$
6. $9c = 54$
7. $5b = 10$
8. $7t = 14$

2.3 EXERCISE SET

Solve each equation. Check each solution. See Examples 1 through 5.

1. $-5x = -20$
2. $-7x = -49$
3. $3x = 0$
4. $2x = 0$
5. $-x = -12$
6. $-y = 8$
7. $\frac{2}{3}x = -8$
8. $\frac{3}{4}n = -15$
9. $\frac{1}{6}d = \frac{1}{2}$
10. $\frac{1}{8}v = \frac{1}{4}$
11. $\frac{a}{2} = 1$
12. $\frac{d}{15} = 2$
13. $\frac{k}{-7} = 0$
14. $\frac{f}{-5} = 0$
15. $1.7x = 10.71$
16. $8.5y = 19.55$

Solve each equation. Check each solution. See Examples 6 and 7.

17. $2x - 4 = 16$
18. $3x - 1 = 26$
19. $-x + 2 = 22$
20. $-x + 4 = -24$

21. $6a + 3 = 3$
22. $8t + 5 = 5$
23. $\frac{x}{3} - 2 = -5$
24. $\frac{b}{4} - 1 = -7$
25. $6z - z = -2 + 2z - 1 - 6$
26. $4a + a = -1 + 3a - 1 - 2$
27. $1 = 0.4x - 0.6x - 5$
28. $19 = 0.4x - 0.9x - 6$
29. $\frac{2}{3}y - 11 = -9$
30. $\frac{3}{5}x - 14 = -8$
31. $\frac{3}{4}t - \frac{1}{2} = \frac{1}{3}$
32. $\frac{2}{7}z - \frac{1}{5} = \frac{1}{2}$

Solve each equation. See Examples 8 and 9.

33. $8x + 20 = 6x + 18$
34. $11x + 13 = 9x + 9$
35. $3(2x + 5) = -18 + 9$
36. $2(4x + 1) = -12 + 6$
37. $2x - 5 = 20x + 4$
38. $6x - 4 = -2x - 10$
39. $2 + 14 = -4(3x - 4)$
40. $8 + 4 = -6(5x - 2)$
41. $-6y - 3 = -5y - 7$
42. $-17z - 4 = -16z - 20$
43. $\frac{1}{2}(2x - 1) = -\frac{1}{7} - \frac{3}{7}$
44. $\frac{1}{3}(3x - 1) = -\frac{1}{10} - \frac{2}{10}$
45. $-10z - 0.5 = -20z + 1.6$
46. $-14y - 1.8 = -24y + 3.9$
47. $-4x + 20 = 4x - 20$
48. $-3x + 15 = 3x - 15$

MIXED PRACTICE

See Examples 1 through 9.

49. $42 = 7x$
50. $81 = 3x$
51. $4.4 = -0.8x$
52. $6.3 = -0.6x$
53. $6x + 10 = -20$
54. $10y + 15 = -5$
55. $5 - 0.3k = 5$
56. $2 - 0.4p = 2$
57. $13x - 5 = 11x - 11$
58. $20x - 20 = 16x - 40$
59. $9(3x + 1) = 4x - 5x$
60. $7(2x + 1) = 18x - 19x$
61. $-\frac{3}{7}p = -2$
62. $-\frac{4}{5}r = -5$
63. $-\frac{4}{3}x = 12$
64. $-\frac{10}{3}x = 30$
65. $-2x - \frac{1}{2} = \frac{7}{2}$
66. $-3n - \frac{1}{3} = \frac{8}{3}$
67. $10 = 2x - 1$
68. $12 = 3j - 4$
69. $10 - 3x - 6 - 9x = 7$
70. $12x + 30 + 8x - 6 = 10$
71. $z - 5z = 7z - 9 - z$
72. $t - 6t = -13 + t - 3t$
73. $-x - \frac{4}{5} = x + \frac{1}{2} + \frac{2}{5}$
74. $x + \frac{3}{7} = -x + \frac{1}{3} + \frac{4}{7}$
75. $-15 + 37 = -2(x + 5)$
76. $-19 + 74 = -5(x + 3)$

Write each algebraic expression described. Simplify if possible. See Example 10.

77. If x represents the first of two consecutive odd integers, express the sum of the two integers in terms of x.
78. If x is the first of four consecutive even integers, write their sum as an algebraic expression in x.
79. If x is the first of four consecutive integers, express the sum of the first integer and the third integer as an algebraic expression containing the variable x.
80. If x is the first of two consecutive integers, express the sum of 20 and the second consecutive integer as an algebraic expression containing the variable x.
81. Classrooms on one side of the science building are all numbered with consecutive even integers. If the first room on this side of the building is numbered x, write an expression in x for the sum of five classroom numbers in a row. Then simplify this expression.

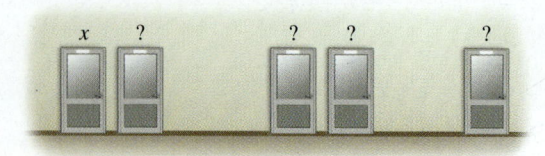

82. Two sides of a quadrilateral have the same length, x, while the other two sides have the same length, both being the next consecutive odd integer. Write the sum of these lengths. Then simplify this expression.

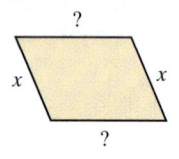

REVIEW AND PREVIEW

Simplify each expression. See Section 2.1.

83. $5x + 2(x - 6)$
84. $-7y + 2y - 3(y + 1)$
85. $-(x - 1) + x$
86. $-(3a - 3) + 2a - 6$

Insert $<$, $>$, or $=$ in the appropriate space to make each statement true. See Sections 1.2 and 1.7.

87. $(-3)^2$ ___ -3^2
88. $(-2)^4$ ___ -2^4
89. $(-2)^3$ ___ -2^3
90. $(-4)^3$ ___ -4^3
91. $-|-6|$ ___ 6
92. $-|-0.7|$ ___ -0.7

CONCEPT EXTENSIONS

Fill in the blank with a number so that each equation has the given solution.

93. $6x = $ ___ ; solution: -8
94. ___ $x = 10$; solution: $\frac{1}{2}$
95. The equation $3x + 6 = 2x + 10 + x - 4$ is true for all real numbers. Substitute a few real numbers for x to see that this is so and then try solving the equation. Describe what happens.
96. The equation $6x + 2 - 2x = 4x + 1$ has no solution. Try solving this equation for x and describe what happens.
97. From the results of Exercises 95 and 96, when do you think an equation has all real numbers as its solutions?
98. From the results of Exercises 95 and 96, when do you think an equation has no solution?
99. A licensed nurse practitioner is instructed to give a patient 2100 milligrams of an antibiotic over a period of 36 hours. If the antibiotic is to be given every 4 hours starting immediately,

how much antibiotic should be given in each dose? To answer this question, solve the equation $9x = 2100$.

100. Suppose you are a pharmacist and a customer asks you the following question. His child is to receive 13.5 milliliters of a nausea medicine over a period of 54 hours. If the nausea medicine is to be administered every 6 hours starting immediately, how much medicine should be given in each dose?

Solve each equation.

101. $-3.6x = 10.62$

102. $4.95y = -31.185$

103. $7x - 5.06 = -4.92$

104. $0.06y + 2.63 = 2.5562$

STUDY SKILLS BUILDER

How Are Your Homework Assignments Going?

It is very important in mathematics to keep up with homework. Why? Many concepts build on each other. Often your understanding of a day's concepts depends on an understanding of the previous day's material.

Remember that completing your homework assignment involves a lot more than attempting a few of the problems assigned.

To complete a homework assignment, remember these four things:

- Attempt all of it.
- Check it.
- Correct it.
- If needed, ask questions about it.

Self-Check

Take a moment and review your completed homework assignments. Answer the questions below based on this review.

1. Approximate the fraction of your homework you have attempted.
2. Approximate the fraction of your homework you have checked (if possible).
3. If you are able to check your homework, have you corrected it when errors have been found?
4. When working homework, if you do not understand a concept, what do you do?

2.4 SOLVING LINEAR EQUATIONS

OBJECTIVES

1. Apply a general strategy for solving a linear equation.
2. Solve equations containing fractions.
3. Solve equations containing decimals.
4. Recognize identities and equations with no solution.

OBJECTIVE 1 ▶ Apply a general strategy for solving a linear equation. We now present a general strategy for solving linear equations. One new piece of strategy is a suggestion to "clear an equation of fractions" as a first step. Doing so makes the equation more manageable, since operating on integers is more convenient than operating on fractions.

Solving Linear Equations in One Variable

STEP 1. Multiply on both sides by the LCD to clear the equation of fractions if they occur.

STEP 2. Use the distributive property to remove parentheses if they occur.

STEP 3. Simplify each side of the equation by combining like terms.

STEP 4. Get all variable terms on one side and all numbers on the other side by using the addition property of equality.

STEP 5. Get the variable alone by using the multiplication property of equality.

STEP 6. Check the solution by substituting it into the original equation.

EXAMPLE 1 Solve: $4(2x - 3) + 7 = 3x + 5$

Solution There are no fractions, so we begin with Step 2.

$$4(2x - 3) + 7 = 3x + 5$$

STEP 2. $\qquad 8x - 12 + 7 = 3x + 5 \quad$ Apply the distributive property.

STEP 3. $\qquad 8x - 5 = 3x + 5 \quad$ Combine like terms.

STEP 4. Get all variable terms on the same side of the equation by subtracting $3x$ from both sides, then adding 5 to both sides.

$$8x - 5 - 3x = 3x + 5 - 3x \quad \text{Subtract } 3x \text{ from both sides.}$$
$$5x - 5 = 5 \quad \text{Simplify.}$$
$$5x - 5 + 5 = 5 + 5 \quad \text{Add 5 to both sides.}$$
$$5x = 10 \quad \text{Simplify.}$$

STEP 5. Use the multiplication property of equality to get x alone.

$$\frac{5x}{5} = \frac{10}{5} \quad \text{Divide both sides by 5.}$$
$$x = 2 \quad \text{Simplify.}$$

STEP 6. Check.

> **Helpful Hint**
> When checking solutions, remember to use the original written equation.

$$4(2x - 3) + 7 = 3x + 5 \quad \text{Original equation}$$
$$4[2(2) - 3] + 7 \stackrel{?}{=} 3(2) + 5 \quad \text{Replace } x \text{ with 2.}$$
$$4(4 - 3) + 7 \stackrel{?}{=} 6 + 5$$
$$4(1) + 7 \stackrel{?}{=} 11$$
$$4 + 7 \stackrel{?}{=} 11$$
$$11 = 11 \quad \text{True}$$

The solution is 2 or the solution set is $\{2\}$.

PRACTICE 1 Solve: $2(4a - 9) + 3 = 5a - 6$

EXAMPLE 2 Solve: $8(2 - t) = -5t$

Solution First, we apply the distributive property.

$$8(2 - t) = -5t$$

STEP 2. $\qquad 16 - 8t = -5t \quad$ Use the distributive property.

STEP 4. $\quad 16 - 8t + 8t = -5t + 8t \quad$ To get variable terms on one side, add $8t$ to both sides.
$$16 = 3t \quad \text{Combine like terms.}$$

STEP 5. $\qquad \dfrac{16}{3} = \dfrac{3t}{3} \quad$ Divide both sides by 3.
$$\dfrac{16}{3} = t \quad \text{Simplify.}$$

STEP 6. Check.

$$8(2 - t) = -5t \quad \text{Original equation}$$
$$8\left(2 - \dfrac{16}{3}\right) \stackrel{?}{=} -5\left(\dfrac{16}{3}\right) \quad \text{Replace } t \text{ with } \dfrac{16}{3}.$$
$$8\left(\dfrac{6}{3} - \dfrac{16}{3}\right) \stackrel{?}{=} -\dfrac{80}{3} \quad \text{The LCD is 3.}$$

Section 2.4 Solving Linear Equations 99

$$8\left(-\frac{10}{3}\right) \stackrel{?}{=} -\frac{80}{3} \quad \text{Subtract fractions.}$$

$$-\frac{80}{3} = -\frac{80}{3} \quad \text{True}$$

The solution is $\frac{16}{3}$.

PRACTICE 2 Solve: $7(x - 3) = -6x$

OBJECTIVE 2 ▶ Solving equations containing fractions. If an equation contains fractions, we can clear the equation of fractions by multiplying both sides by the LCD of all denominators. By doing this, we avoid working with time-consuming fractions.

EXAMPLE 3 Solve: $\frac{x}{2} - 1 = \frac{2}{3}x - 3$

Solution We begin by clearing fractions. To do this, we multiply both sides of the equation by the LCD of 2 and 3, which is 6.

$$\frac{x}{2} - 1 = \frac{2}{3}x - 3$$

STEP 1. $6\left(\frac{x}{2} - 1\right) = 6\left(\frac{2}{3}x - 3\right)$ Multiply both sides by the LCD, 6.

▶ **Helpful Hint**
Don't forget to multiply *each* term by the LCD.

STEP 2. $6\left(\frac{x}{2}\right) - 6(1) = 6\left(\frac{2}{3}x\right) - 6(3)$ Apply the distributive property.

$$3x - 6 = 4x - 18 \quad \text{Simplify.}$$

There are no longer grouping symbols and no like terms on either side of the equation, so we continue with Step 4.

$$3x - 6 = 4x - 18$$

STEP 4. $3x - 6 - 3x = 4x - 18 - 3x$ To get variable terms on one side, subtract $3x$ from both sides.
$-6 = x - 18$ Simplify.
$-6 + 18 = x - 18 + 18$ Add 18 to both sides.
$12 = x$ Simplify.

STEP 5. The variable is now alone, so there is no need to apply the multiplication property of equality.

STEP 6. Check.

$$\frac{x}{2} - 1 = \frac{2}{3}x - 3 \quad \text{Original equation}$$

$$\frac{12}{2} - 1 \stackrel{?}{=} \frac{2}{3} \cdot 12 - 3 \quad \text{Replace } x \text{ with 12.}$$

$$6 - 1 \stackrel{?}{=} 8 - 3 \quad \text{Simplify.}$$

$$5 = 5 \quad \text{True}$$

The solution is 12.

PRACTICE 3 Solve: $\frac{3}{5}x - 2 = \frac{2}{3}x - 1$

EXAMPLE 4 Solve: $\dfrac{2(a+3)}{3} = 6a + 2$

Solution We clear the equation of fractions first.

$$\dfrac{2(a+3)}{3} = 6a + 2$$

STEP 1. $\quad 3 \cdot \dfrac{2(a+3)}{3} = 3(6a+2) \quad$ Clear the fraction by multiplying both sides by the LCD, 3.

$$2(a+3) = 3(6a+2)$$

STEP 2. Next, we use the distributive property and remove parentheses.

$\qquad\qquad\qquad 2a + 6 = 18a + 6 \qquad$ Apply the distributive property.

STEP 4. $\qquad 2a + 6 - 6 = 18a + 6 - 6 \qquad$ Subtract 6 from both sides.

$\qquad\qquad\qquad 2a = 18a$

$\qquad\qquad 2a - 18a = 18a - 18a \qquad$ Subtract 18a from both sides.

$\qquad\qquad\qquad -16a = 0$

STEP 5. $\qquad \dfrac{-16a}{-16} = \dfrac{0}{-16} \qquad$ Divide both sides by -16.

$\qquad\qquad\qquad a = 0 \qquad$ Write the fraction in simplest form.

STEP 6. To check, replace a with 0 in the original equation. The solution is 0.

PRACTICE 4 Solve: $\dfrac{4(y+3)}{3} = 5y - 7$

OBJECTIVE 3 ▶ Solving equations containing decimals. When solving a problem about money, you may need to solve an equation containing decimals. If you choose, you may multiply to clear the equation of decimals.

EXAMPLE 5 Solve: $0.25x + 0.10(x - 3) = 0.05(22)$

Solution First we clear this equation of decimals by multiplying both sides of the equation by 100. Recall that multiplying a decimal number by 100 has the effect of moving the decimal point 2 places to the right.

$$0.25x + 0.10(x - 3) = 0.05(22)$$

▶ **Helpful Hint**
By the distributive property, 0.10 is multiplied by x and -3. Thus to multiply each term here by 100, we only need to multiply 0.10 by 100.

STEP 1. $\qquad 0.25x + 0.10(x - 3) = 0.05(22) \qquad$ Multiply both sides by 100.

$\qquad\qquad 25x + 10(x - 3) = 5(22)$

STEP 2. $\qquad 25x + 10x - 30 = 110 \qquad$ Apply the distributive property.

STEP 3. $\qquad\qquad 35x - 30 = 110 \qquad$ Combine like terms.

STEP 4. $\qquad 35x - 30 + 30 = 110 + 30 \qquad$ Add 30 to both sides.

$\qquad\qquad\qquad 35x = 140 \qquad$ Combine like terms.

STEP 5. $\qquad \dfrac{35x}{35} = \dfrac{140}{35} \qquad$ Divide both sides by 35.

$\qquad\qquad\qquad x = 4$

STEP 6. To check, replace x with 4 in the original equation. The solution is 4.

PRACTICE 5 Solve: $0.35x + 0.09(x + 4) = 0.03(12)$

OBJECTIVE 4 ▶ Recognizing identities and equations with no solution. So far, each equation that we have solved has had a single solution. However, not every equation in one variable has a single solution. Some equations have no solution, while others have an infinite number of solutions. For example,

$$x + 5 = x + 7$$

has no solution since no matter which **real number** we replace x with, the equation is false.

real number $+ 5 =$ same **real number** $+ 7$ **FALSE**

On the other hand,

$$x + 6 = x + 6$$

has infinitely many solutions since x can be replaced by any real number and the equation is always true.

real number $+ 6 =$ same **real number** $+ 6$ **TRUE**

The equation $x + 6 = x + 6$ is called an **identity**. The next few examples illustrate special equations like these.

EXAMPLE 6   Solve: $-2(x - 5) + 10 = -3(x + 2) + x$

Solution

$$-2(x - 5) + 10 = -3(x + 2) + x$$
$$-2x + 10 + 10 = -3x - 6 + x \quad \text{Apply the distributive property on both sides.}$$
$$-2x + 20 = -2x - 6 \quad \text{Combine like terms.}$$
$$-2x + 20 + 2x = -2x - 6 + 2x \quad \text{Add } 2x \text{ to both sides.}$$
$$20 = -6 \quad \text{Combine like terms.}$$

The final equation contains no variable terms, and there is no value for x that makes $20 = -6$ a true equation. We conclude that there is **no solution** to this equation. In set notation, we can indicate that there is no solution with the empty set, $\{\ \}$, or use the empty set or null set symbol, $\varnothing$. In this chapter, we will simply write *no solution*.

PRACTICE 6   Solve: $4(x + 4) - x = 2(x + 11) + x$

EXAMPLE 7   Solve: $3(x - 4) = 3x - 12$

Solution
$$3(x - 4) = 3x - 12$$
$$3x - 12 = 3x - 12 \quad \text{Apply the distributive property.}$$

The left side of the equation is now identical to the right side. Every real number may be substituted for x and a true statement will result. We arrive at the same conclusion if we continue.

$$3x - 12 = 3x - 12$$
$$3x - 12 + 12 = 3x - 12 + 12 \quad \text{Add 12 to both sides.}$$
$$3x = 3x \quad \text{Combine like terms.}$$
$$3x - 3x = 3x - 3x \quad \text{Subtract } 3x \text{ from both sides.}$$
$$0 = 0$$

Again, one side of the equation is identical to the other side. Thus, $3(x - 4) = 3x - 12$ is an **identity** and **all real numbers** are solutions. In set notation, this is $\{$all real numbers$\}$.

PRACTICE 7   Solve: $12x - 18 = 9(x - 2) + 3x$

Answers to Concept Check:
a. Every real number is a solution.
b. The solution is 0.
c. There is no solution.

Concept Check ✓

Suppose you have simplified several equations and obtain the following results. What can you conclude about the solutions to the original equation?

a. $7 = 7$ **b.** $x = 0$ **c.** $7 = -4$

Calculator Explorations

Checking Equations

We can use a calculator to check possible solutions of equations. To do this, replace the variable by the possible solution and evaluate both sides of the equation separately.

Equation: $\quad 3x - 4 = 2(x + 6) \quad$ *Solution:* $x = 16$

$3x - 4 = 2(x + 6) \quad$ Original equation

$3(16) - 4 \stackrel{?}{=} 2(16 + 6) \quad$ Replace x with 16.

Now evaluate each side with your calculator.

Evaluate left side:

$\boxed{3} \boxed{\times} \boxed{16} \boxed{-} \boxed{4} \boxed{=}$ or $\boxed{\text{ENTER}}$ Display: $\boxed{44}$ or $\begin{array}{c} 3*16 - 4 \\ \hfill 44 \end{array}$

Evaluate right side:

$\boxed{2} \boxed{(} \boxed{16} \boxed{+} \boxed{6} \boxed{)} \boxed{=}$ or $\boxed{\text{ENTER}}$ Display: $\boxed{44}$ or $\begin{array}{c} 2(16 + 6) \\ \hfill 44 \end{array}$

Since the left side equals the right side, the equation checks.

Use a calculator to check the possible solutions to each equation.

1. $2x = 48 + 6x; \quad x = -12$ **2.** $-3x - 7 = 3x - 1; \quad x = -1$

3. $5x - 2.6 = 2(x + 0.8); \quad x = 4.4$ **4.** $-1.6x - 3.9 = -6.9x - 25.6; \quad x = 5$

5. $\dfrac{564x}{4} = 200x - 11(649); \quad x = 121$ **6.** $20(x - 39) = 5x - 432; \quad x = 23.2$

VOCABULARY & READINESS CHECK

Throughout algebra, it is important to be able to identify equations and expressions.

Remember,
- an equation contains an equals sign and
- an expression does not.

Among other things,
- we solve equations and
- we simplify or perform operations on expressions.

Identify each as an equation or an expression.

1. $x = -7$ _____   **2.** $x - 7$ _____   **3.** $4y - 6 + 9y + 1$ _____

4. $4y - 6 = 9y + 1$ _____   **5.** $\dfrac{1}{x} - \dfrac{x-1}{8}$ _____   **6.** $\dfrac{1}{x} - \dfrac{x-1}{8} = 6$ _____

7. $0.1x + 9 = 0.2x$ _____   **8.** $0.1x^2 + 9y - 0.2x^2$ _____

2.4 EXERCISE SET

Solve each equation. See Examples 1 and 2.

1. $-4y + 10 = -2(3y + 1)$
2. $-3x + 1 = -2(4x + 2)$
3. $15x - 8 = 10 + 9x$
4. $15x - 5 = 7 + 12x$
5. $-2(3x - 4) = 2x$
6. $-(5x - 10) = 5x$
7. $5(2x - 1) - 2(3x) = 1$
8. $3(2 - 5x) + 4(6x) = 12$
9. $-6(x - 3) - 26 = -8$
10. $-4(n - 4) - 23 = -7$
11. $8 - 2(a + 1) = 9 + a$
12. $5 - 6(2 + b) = b - 14$
13. $4x + 3 = -3 + 2x + 14$
14. $6y - 8 = -6 + 3y + 13$
15. $-2y - 10 = 5y + 18$
16. $-7n + 5 = 8n - 10$

Solve each equation. See Examples 3 through 5.

17. $\frac{2}{3}x + \frac{4}{3} = -\frac{2}{3}$
18. $\frac{4}{5}x - \frac{8}{5} = -\frac{16}{5}$
19. $\frac{3}{4}x - \frac{1}{2} = 1$
20. $\frac{2}{9}x - \frac{1}{3} = 1$
21. $0.50x + 0.15(70) = 35.5$
22. $0.40x + 0.06(30) = 9.8$
23. $\frac{2(x + 1)}{4} = 3x - 2$
24. $\frac{3(y + 3)}{5} = 2y + 6$
25. $x + \frac{7}{6} = 2x - \frac{7}{6}$
26. $\frac{5}{2}x - 1 = x + \frac{1}{4}$
27. $0.12(y - 6) + 0.06y = 0.08y - 0.7$
28. $0.60(z - 300) + 0.05z = 0.70z - 205$

Solve each equation. See Examples 6 and 7.

29. $4(3x + 2) = 12x + 8$
30. $14x + 7 = 7(2x + 1)$
31. $\frac{x}{4} + 1 = \frac{x}{4}$
32. $\frac{x}{3} - 2 = \frac{x}{3}$
33. $3x - 7 = 3(x + 1)$
34. $2(x - 5) = 2x + 10$
35. $-2(6x - 5) + 4 = -12x + 14$
36. $-5(4y - 3) + 2 = -20y + 17$

MIXED PRACTICE

Solve. See Examples 1 through 7.

37. $\frac{6(3 - z)}{5} = -z$
38. $\frac{4(5 - w)}{3} = -w$
39. $-3(2t - 5) + 2t = 5t - 4$
40. $-(4a - 7) - 5a = 10 + a$
41. $5y + 2(y - 6) = 4(y + 1) - 2$
42. $9x + 3(x - 4) = 10(x - 5) + 7$
43. $\frac{3(x - 5)}{2} = \frac{2(x + 5)}{3}$
44. $\frac{5(x - 1)}{4} = \frac{3(x + 1)}{2}$
45. $0.7x - 2.3 = 0.5$
46. $0.9x - 4.1 = 0.4$
47. $5x - 5 = 2(x + 1) + 3x - 7$
48. $3(2x - 1) + 5 = 6x + 2$
49. $4(2n + 1) = 3(6n + 3) + 1$
50. $4(4y + 2) = 2(1 + 6y) + 8$
51. $x + \frac{5}{4} = \frac{3}{4}x$
52. $\frac{7}{8}x + \frac{1}{4} = \frac{3}{4}x$
53. $\frac{x}{2} - 1 = \frac{x}{5} + 2$
54. $\frac{x}{5} - 7 = \frac{x}{3} - 5$
55. $2(x + 3) - 5 = 5x - 3(1 + x)$
56. $4(2 + x) + 1 = 7x - 3(x - 2)$
57. $0.06 - 0.01(x + 1) = -0.02(2 - x)$
58. $-0.01(5x + 4) = 0.04 - 0.01(x + 4)$
59. $\frac{9}{2} + \frac{5}{2}y = 2y - 4$
60. $3 - \frac{1}{2}x = 5x - 8$
61. $-2y - 10 = 5y + 18$
62. $7n + 5 = 10n - 10$
63. $0.6x - 0.1 = 0.5x + 0.2$
64. $0.2x - 0.1 = 0.6x - 2.1$
65. $0.02(6t - 3) = 0.12(t - 2) + 0.18$
66. $0.03(2m + 7) = 0.06(5 + m) - 0.09$

REVIEW AND PREVIEW

Write each phrase as an algebraic expression. Use x for the unknown number. See Section 2.1.

67. A number subtracted from -8
68. Three times a number
69. The sum of -3 and twice a number
70. The difference of 8 and twice a number

71. The product of 9 and the sum of a number and 20
72. The quotient of -12 and the difference of a number and 3

See Section 2.1.

73. A plot of land is in the shape of a triangle. If one side is x meters, a second side is $(2x - 3)$ meters and a third side is $(3x - 5)$ meters, express the perimeter of the lot as a simplified expression in x.

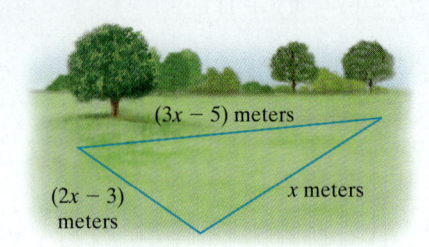

74. A portion of a board has length x feet. The other part has length $(7x - 9)$ feet. Express the total length of the board as a simplified expression in x.

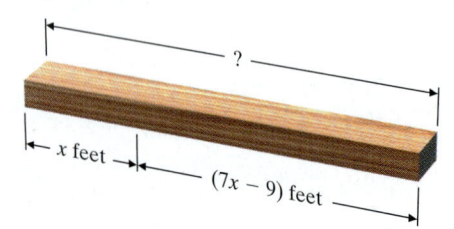

CONCEPT EXTENSIONS

See the Concept Check in this section.

75. **a.** Solve: $x + 3 = x + 3$
    b. If you simplify an equation and get $0 = 0$, what can you conclude about the solution(s) of the original equation?
    c. On your own, construct an equation for which every real number is a solution.

76. **a.** Solve: $x + 3 = x + 5$
    b. If you simplify an equation and get $3 = 5$, what can you conclude about the solution(s) of the original equation?
    c. On your own, construct an equation that has no solution.

Match each equation in the first column with its solution in the second column. Items in the second column may be used more than once.

77. $5x + 1 = 5x + 1$
78. $3x + 1 = 3x + 2$
79. $2x - 6x - 10 = -4x + 3 - 10$
80. $x - 11x - 3 = -10x - 1 - 2$
81. $9x - 20 = 8x - 20$
82. $-x + 15 = x + 15$

a. all real numbers
b. no solution
c. 0

83. Explain the difference between simplifying an expression and solving an equation.

84. On your own, write an expression and then an equation. Label each.

For Exercises 85 and 86, **a.** *Write an equation for perimeter.* **b.** *Solve the equation in part (a).* **c.** *Find the length of each side.*

85. The perimeter of a geometric figure is the sum of the lengths of its sides. The perimeter of the following pentagon (five-sided figure) is 28 centimeters.

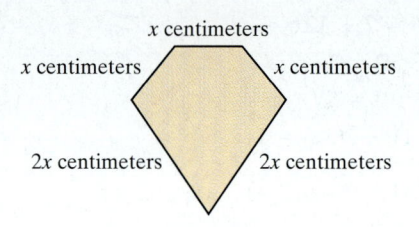

86. The perimeter of the following triangle is 35 meters.

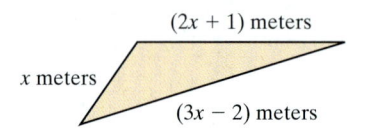

Fill in the blanks with numbers of your choice so that each equation has the given solution. Note: Each blank may be replaced by a different number.

87. $x + \underline{\quad} = 2x - \underline{\quad}$; solution: 9
88. $-5x - \underline{\quad} = \underline{\quad}$; solution: 2

Solve.

89. $1000(7x - 10) = 50(412 + 100x)$
90. $1000(x + 40) = 100(16 + 7x)$
91. $0.035x + 5.112 = 0.010x + 5.107$
92. $0.127x - 2.685 = 0.027x - 2.38$

For Exercises 93 through 96, see the example below.

Example

Solve: $t(t + 4) = t^2 - 2t + 6$.

Solution
$$t(t + 4) = t^2 - 2t + 6$$
$$t^2 + 4t = t^2 - 2t + 6$$
$$t^2 + 4t - t^2 = t^2 - 2t + 6 - t^2$$
$$4t = -2t + 6$$
$$4t + 2t = -2t + 6 + 2t$$
$$6t = 6$$
$$t = 1$$

Solve each equation.

93. $x(x - 3) = x^2 + 5x + 7$
94. $t^2 - 6t = t(8 + t)$
95. $2z(z + 6) = 2z^2 + 12z - 8$
96. $y^2 - 4y + 10 = y(y - 5)$

INTEGRATED REVIEW SOLVING LINEAR EQUATIONS

Sections 2.1–2.4

Solve. Feel free to use the steps given in Section 2.4.

1. $x - 10 = -4$
2. $y + 14 = -3$
3. $9y = 108$
4. $-3x = 78$
5. $-6x + 7 = 25$
6. $5y - 42 = -47$
7. $\frac{2}{3}x = 9$
8. $\frac{4}{5}z = 10$
9. $\frac{r}{-4} = -2$
10. $\frac{y}{-8} = 8$
11. $6 - 2x + 8 = 10$
12. $-5 - 6y + 6 = 19$
13. $2x - 7 = 2x - 27$
14. $3 + 8y = 8y - 2$
15. $-3a + 6 + 5a = 7a - 8a$
16. $4b - 8 - b = 10b - 3b$
17. $-\frac{2}{3}x = \frac{5}{9}$
18. $-\frac{3}{8}y = -\frac{1}{16}$
19. $10 = -6n + 16$
20. $-5 = -2m + 7$
21. $3(5c - 1) - 2 = 13c + 3$
22. $4(3t + 4) - 20 = 3 + 5t$
23. $\frac{2(z + 3)}{3} = 5 - z$
24. $\frac{3(w + 2)}{4} = 2w + 3$
25. $-2(2x - 5) = -3x + 7 - x + 3$
26. $-4(5x - 2) = -12x + 4 - 8x + 4$
27. $0.02(6t - 3) = 0.04(t - 2) + 0.02$
28. $0.03(m + 7) = 0.02(5 - m) + 0.03$
29. $-3y = \frac{4(y - 1)}{5}$
30. $-4x = \frac{5(1 - x)}{6}$
31. $\frac{5}{3}x - \frac{7}{3} = x$
32. $\frac{7}{5}n + \frac{3}{5} = -n$
33. $\frac{1}{10}(3x - 7) = \frac{3}{10}x + 5$
34. $\frac{1}{7}(2x - 5) = \frac{2}{7}x + 1$
35. $5 + 2(3x - 6) = -4(6x - 7)$
36. $3 + 5(2x - 4) = -7(5x + 2)$

2.5 AN INTRODUCTION TO PROBLEM SOLVING

OBJECTIVES

Apply the steps for problem solving as we

1. Solve problems involving direct translations.
2. Solve problems involving relationships among unknown quantities.
3. Solve problems involving consecutive integers.

OBJECTIVE 1 ▶ Solving direct translation problems. In previous sections, you practiced writing word phrases and sentences as algebraic expressions and equations to help prepare for problem solving. We now use these translations to help write equations that model a problem. The problem-solving steps given next may be helpful.

General Strategy for Problem Solving

1. UNDERSTAND the problem. During this step, become comfortable with the problem. Some ways of doing this are:

    Read and reread the problem.

    Choose a variable to represent the unknown.

    Construct a drawing, whenever possible.

    Propose a solution and check. Pay careful attention to how you check your proposed solution. This will help when writing an equation to model the problem.

2. TRANSLATE the problem into an equation.
3. SOLVE the equation.
4. INTERPRET the results: *Check* the proposed solution in the stated problem and state your conclusion.

Much of problem solving involves a direct translation from a sentence to an equation.

106 CHAPTER 2 Equations, Inequalities, and Problem Solving

> **EXAMPLE 1** Finding an Unknown Number
>
> Twice a number, added to seven, is the same as three subtracted from the number. Find the number.

Solution Translate the sentence into an equation and solve.

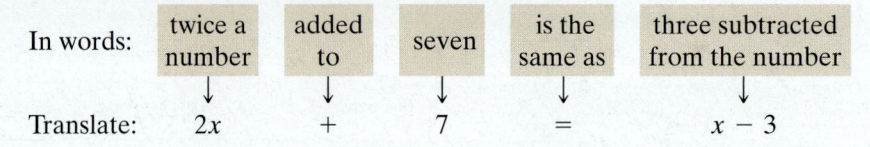

In words:	twice a number	added to	seven	is the same as	three subtracted from the number
Translate:	$2x$	$+$	7	$=$	$x - 3$

To solve, begin by subtracting x from both sides to isolate the variable term.

$$2x + 7 = x - 3$$
$$2x + 7 - x = x - 3 - x \quad \text{Subtract } x \text{ from both sides.}$$
$$x + 7 = -3 \quad \text{Combine like terms.}$$
$$x + 7 - 7 = -3 - 7 \quad \text{Subtract 7 from both sides.}$$
$$x = -10 \quad \text{Combine like terms.}$$

Check the solution in the problem as it was originally stated. To do so, replace "number" in the sentence with -10. Twice "-10" added to 7 is the same as 3 subtracted from "-10."

$$2(-10) + 7 = -10 - 3$$
$$-13 = -13$$

The unknown number is -10.

PRACTICE 1 Three times a number, minus 6, is the same as two times a number, plus 3. Find the number.

> ▶ **Helpful Hint**
> When checking solutions, go back to the original stated problem, rather than to your equation in case errors have been made in translating to an equation.

> **EXAMPLE 2** Finding an Unknown Number
>
> Twice the sum of a number and 4 is the same as four times the number, decreased by 12. Find the number.

Solution

1. UNDERSTAND. Read and reread the problem. If we let

$$x = \text{the unknown number, then}$$

"the sum of a number and 4" translates to "$x + 4$" and "four times the number" translates to "$4x$."

2. TRANSLATE.

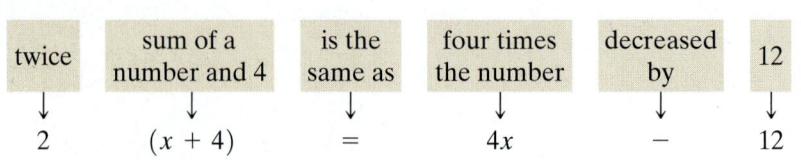

twice	sum of a number and 4	is the same as	four times the number	decreased by	12
2	$(x + 4)$	$=$	$4x$	$-$	12

3. SOLVE.

$$2(x + 4) = 4x - 12$$
$$2x + 8 = 4x - 12 \quad \text{Apply the distributive property.}$$
$$2x + 8 - 4x = 4x - 12 - 4x \quad \text{Subtract } 4x \text{ from both sides.}$$
$$-2x + 8 = -12$$
$$-2x + 8 - 8 = -12 - 8 \quad \text{Subtract 8 from both sides.}$$
$$-2x = -20$$
$$\frac{-2x}{-2} = \frac{-20}{-2} \quad \text{Divide both sides by } -2.$$
$$x = 10$$

4. INTERPRET.

Check: Check this solution in the problem as it was originally stated. To do so, replace "number" with 10. Twice the sum of "10" and 4 is 28, which is the same as 4 times "10" decreased by 12.

State: The number is 10.

PRACTICE
2 Three times a number, decreased by 4, is the same as double the difference of the number and 1.

OBJECTIVE 2 ▶ Solving problems involving relationships among unknown quantities.
The next three examples have to do with relationships among unknown quantities.

EXAMPLE 3 Finding the Length of a Board

Balsa wood sticks are commonly used for building models (for example, bridge models). A 48-inch Balsa wood stick is to be cut into two pieces so that the longer piece is 3 times the shorter. Find the length of each piece.

Solution

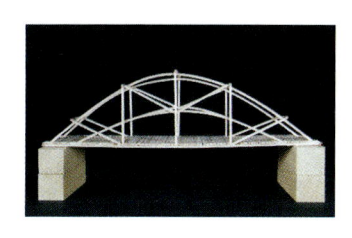

1. UNDERSTAND the problem. To do so, read and reread the problem. You may also want to propose a solution. For example, if 10 inches represents the length of the shorter piece, then $3(10) = 30$ inches is the length of the longer piece, since it is 3 times the length of the shorter piece. This guess gives a total board length of 10 inches + 30 inches = 40 inches, too short. However, the purpose of proposing a solution is not to guess correctly, but to help better understand the problem and how to model it.

   Since the length of the longer piece is given in terms of the length of the shorter piece, let's let

   $$x = \text{length of shorter piece, then}$$
   $$3x = \text{length of longer piece}$$

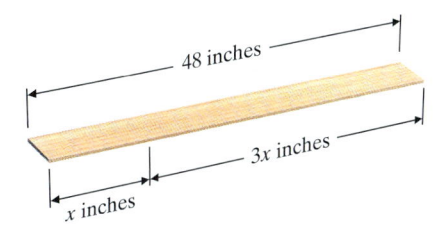

2. TRANSLATE the problem. First, we write the equation in words.

length of shorter piece	added to	length of longer piece	equals	total length of board
↓	↓	↓	↓	↓
x	$+$	$3x$	$=$	48

3. SOLVE.

$$x + 3x = 48$$
$$4x = 48 \quad \text{Combine like terms.}$$
$$\frac{4x}{4} = \frac{48}{4} \quad \text{Divide both sides by 4.}$$
$$x = 12$$

4. INTERPRET.

Check: Check the solution in the stated problem. If the shorter piece of board is 12 inches, the longer piece is $3 \cdot (12 \text{ inches}) = 36$ inches and the sum of the two pieces is 12 inches + 36 inches = 48 inches.

State: The shorter piece of Balsa wood is 12 inches and the longer piece of Balsa wood is 36 inches.

> **Helpful Hint**
> Make sure that units are included in your answer, if appropriate.

PRACTICE 3 A 45-inch board is to be cut into two pieces so that the longer piece is 4 times the shorter. Find the length of each piece.

EXAMPLE 4 Finding the Number of Democratic and Republican Representatives

In a recent year, the U.S. House of Representatives had a total of 435 Democrats and Republicans. There were 31 more Democratic representatives than Republican representatives. Find the number of representatives from each party. (*Source:* Office of the Clerk of the U.S. House of Representatives)

Solution

1. UNDERSTAND. Read and reread the problem. Let's suppose that there were 200 Republican representatives. Since there were 31 more Democrats than Republicans, there must have been 200 + 31 = 231 Democrats. The total number of Democrats and Republicans was then 200 + 231 = 431. This is incorrect since the total should be 435, but now we have a better understanding of the problem.

In general, if we let

$$x = \text{number of Republicans, then}$$
$$x + 31 = \text{number of Democrats}$$

2. TRANSLATE. First we write the equation in words.

Number of Republicans	added to	number of Democrats	equals	435
↓	↓	↓	↓	↓
x	$+$	$(x + 31)$	$=$	435

3. SOLVE.

$$x + (x + 31) = 435$$
$$2x + 31 = 435$$
$$2x + 31 - 31 = 435 - 31$$
$$2x = 404$$
$$\frac{2x}{2} = \frac{404}{2}$$
$$x = 202$$

Section 2.5 An Introduction to Problem Solving **109**

4. INTERPRET.

Check: If there were 202 Republican representatives, then there were 202 + 31 = 233 Democratic representatives. The total number of Democratic or Republican representatives is 202 + 233 = 435. The results check.

State: There were 202 Republican and 233 Democratic representatives in Congress.

PRACTICE

4 In a recent year, there were 6 more Democratic State Governors than Republican State Governors. Find the number of State Governors from each party. (We are only counting the 50 states.) (*Source:* National Conference of State Legislatures).

EXAMPLE 5 Finding Angle Measures

If the two walls of the Vietnam Veterans Memorial in Washington, D.C., were connected, an isosceles triangle would be formed. The measure of the third angle is 97.5° more than the measure of either of the other two equal angles. Find the measure of the third angle. (*Source:* National Park Service)

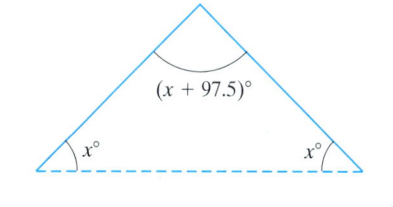

Solution

1. UNDERSTAND. Read and reread the problem. We then draw a diagram (recall that an isosceles triangle has two angles with the same measure) and let

   x = degree measure of one angle

   x = degree measure of the second equal angle

   $x + 97.5$ = degree measure of the third angle

2. TRANSLATE. Recall that the sum of the measures of the angles of a triangle equals 180.

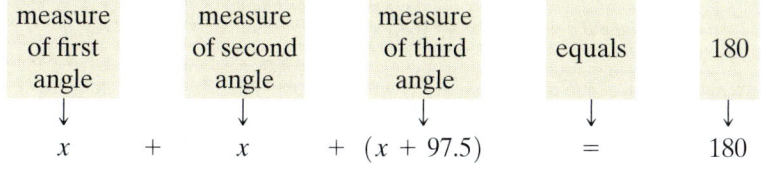

measure of first angle	measure of second angle	measure of third angle	equals	180
↓	↓	↓	↓	↓
x +	x +	$(x + 97.5)$	=	180

3. SOLVE.

$$x + x + (x + 97.5) = 180$$
$$3x + 97.5 = 180 \quad \text{Combine like terms.}$$
$$3x + 97.5 - 97.5 = 180 - 97.5 \quad \text{Subtract 97.5 from both sides.}$$
$$3x = 82.5$$
$$\frac{3x}{3} = \frac{82.5}{3} \quad \text{Divide both sides by 3.}$$
$$x = 27.5$$

4. **INTERPRET.**

Check: If $x = 27.5$, then the measure of the third angle is $x + 97.5 = 125$. The sum of the angles is then $27.5 + 27.5 + 125 = 180$, the correct sum.

State: The third angle measures 125°.*

PRACTICE

5 The second angle of a triangle measures three times as large as the first. If the third angle measures 55° more than the first, find the measures of all three angles.

---

*The two walls actually meet at an angle of 125 degrees 12 minutes. The measurement of 97.5° given in the problem is an approximation.

OBJECTIVE 3 ▶ **Solving consecutive integer problems.** The next example has to do with consecutive integers. Recall what we have learned thus far about these integers.

	Example		General Representation
Consecutive Integers	11, 12, 13 (+1, +1)	Let x be an integer.	x, $x + 1$, $x + 2$ (+1, +1)
Consecutive Even Integers	38, 40, 42 (+2, +2)	Let x be an even integer.	x, $x + 2$, $x + 4$ (+2, +2)
Consecutive Odd Integers	57, 59, 61 (+2, +2)	Let x be an odd integer.	x, $x + 2$, $x + 4$ (+2, +2)

EXAMPLE 6 Some states have a single area code for the entire state. Two such states have area codes that are consecutive odd integers. If the sum of these integers is 1208, find the two area codes. (*Source:* North American Numbering Plan Administration)

Solution:

1. UNDERSTAND. Read and reread the problem. If we let

$$x = \text{the first odd integer, then}$$
$$x + 2 = \text{the next odd integer}$$

▶ **Helpful Hint**
Remember, the 2 here means that odd integers are 2 units apart, for example, the odd integers 13 and 13 + 2 = 15.

2. TRANSLATE.

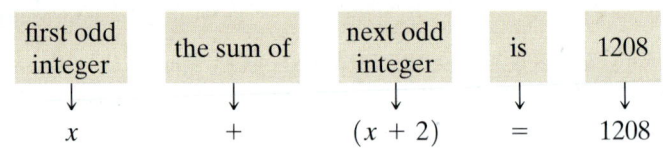

first odd integer	the sum of	next odd integer	is	1208
↓	↓	↓	↓	↓
x	$+$	$(x + 2)$	$=$	1208

3. SOLVE.

$$x + x + 2 = 1208$$
$$2x + 2 = 1208$$
$$2x + 2 - 2 = 1208 - 2$$
$$2x = 1206$$
$$\frac{2x}{2} = \frac{1206}{2}$$
$$x = 603$$

4. INTERPRET.

Check: If $x = 603$, then the next odd integer $x + 2 = 603 + 2 = 605$. Notice their sum, $603 + 605 = 1208$, as needed.

State: The area codes are 603 and 605.

Note: New Hampshire's area code is 603 and South Dakota's area code is 605.

PRACTICE 6 The sum of three consecutive even integers is 144. Find the integers.

VOCABULARY & READINESS CHECK

Fill in the table.

	A number:	→	Operation:	→	Further operation:
1.	A number: x	→	Double the number:	→	Double the number, decreased by 31:
2.	A number: x	→	Three times the number:	→	Three times the number, increased by 17:
3.	A number: x	→	The sum of the number and 5:	→	Twice the sum of the number and 5:
4.	A number: x	→	The difference of the number and 11:	→	Seven times the difference of a number and 11:
5.	A number: y	→	The difference of 20 and the number:	→	The difference of 20 and the number, divided by 3:
6.	A number: y	→	The sum of -10 and the number:	→	The sum of -10 and the number, divided by 9:

2.5 EXERCISE SET

Write each of the following as equations. Then solve. See Examples 1 and 2.

1. The sum of twice a number, and 7, is equal to the sum of a number and 6. Find the number.

2. The difference of three times a number, and 1, is the same as twice a number. Find the number.

3. Three times a number, minus 6, is equal to two times a number, plus 8. Find the number.

4. The sum of 4 times a number, and -2, is equal to the sum of 5 times a number, and -2. Find the number.

5. Twice the difference of a number and 8 is equal to three times the sum of the number and 3. Find the number.

6. Five times the sum of a number and -1 is the same as 6 times the number. Find the number.

7. Four times the sum of -2 and a number is the same as five times the number increased by $\frac{1}{2}$. Find the number.

8. If the difference of a number and four is doubled, the result is $\frac{1}{4}$ less than the number. Find the number.

Solve. See Examples 3 through 5.

9. A 17-foot piece of string is cut into two pieces so that the longer piece is 2 feet longer than twice the shorter piece. Find the lengths of both pieces.

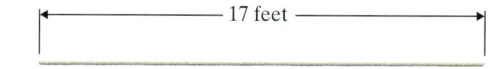

     |←——— 17 feet ———→|

10. A 25-foot wire is to be cut so that the longer piece is one foot longer than 5 times the shorter piece. Find the length of each piece.

11. The largest meteorite in the world is the Hoba West located in Namibia. Its weight is 3 times the weight of the Armanty meteorite located in Outer Mongolia. If the sum of their weights is 88 tons, find the weight of each.

12. The area of the Sahara Desert is 7 times the area of the Gobi Desert. If the sum of their areas is 4,000,000 square miles, find the area of each desert.

13. The countries with the most cinema screens in the world are China and the United States. China has 5806 more cinema screens than the United States whereas the total screens for both countries is 78,994. Find the number of cinema screens for both countries. (*Source:* Film Distributor's Association)

14. The countries with the most television stations in the world are Russia and China. Russia has 4066 more television stations than China whereas the total stations for both countries is 10,546. Find the number of television stations for both countries. (*Source:* Central Intelligence Agency, *The World Factbook 2006*)

15. The flag of Equatorial Guinea contains an isosceles triangle. (Recall that an isosceles triangle contains two angles with the

112 CHAPTER 2 Equations, Inequalities, and Problem Solving

same measure.) If the measure of the third angle of the triangle is 30° more than twice the measure of either of the other two angles, find the measure of each angle of the triangle. (*Hint:* Recall that the sum of the measures of the angles of a triangle is 180°.)

16. Recall that the sum of the measures of the angles of a triangle is 180°. In the triangle below, angle *C* has the same measure as angle *B*, and angle *A* measures 42° less than angle *B*. Find the measure of each angle.

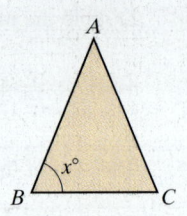

Solve. See Example 6. Fill in the table. Most of the first row has been completed for you.

	First Integer →	Next Integers →	Indicated Sum
17. Three consecutive integers:	Integer: x	$x+1$, $x+2$	Sum of the three consecutive integers, simplified:
18. Three consecutive integers:	Integer: x		Sum of the second and third consecutive integers, simplified:
19. Three consecutive even integers:	Even integer: x		Sum of the first and third even consecutive integers, simplified:
20. Three consecutive odd integers:	Odd integer: x		Sum of the three consecutive odd integers, simplified:
21. Four consecutive integers:	Integer: x		Sum of the four consecutive integers, simplified:
22. Four consecutive integers:	Integer: x		Sum of the first and fourth consecutive integers, simplified:
23. Three consecutive odd integers:	Odd integer: x		Sum of the second and third consecutive odd integers, simplified:
24. Three consecutive even integers:	Even integer: x		Sum of the three consecutive even integers, simplified:

25. The left and right page numbers of an open book are two consecutive integers whose sum is 469. Find these page numbers.

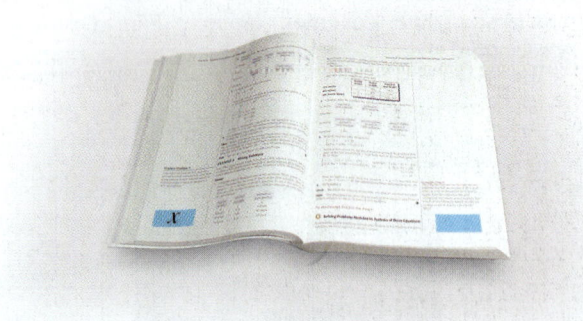

26. The room numbers of two adjacent classrooms are two consecutive even numbers. If their sum is 654, find the classroom numbers.

27. To make an international telephone call, you need the code for the country you are calling. The codes for Belgium, France, and Spain are three consecutive integers whose sum is 99. Find the code for each country. (*Source: The World Almanac and Book of Facts, 2007*)

28. To make an international telephone call, you need the code for the country you are calling. The codes for Mali Republic, Côte d'Ivoire, and Niger are three consecutive odd integers whose sum is 675. Find the code for each country.

MIXED PRACTICE

Solve. See Examples 1 through 6.

29. A 25-inch piece of steel is cut into three pieces so that the second piece is twice as long as the first piece, and the third piece is one inch more than five times the length of the first piece. Find the lengths of the pieces.

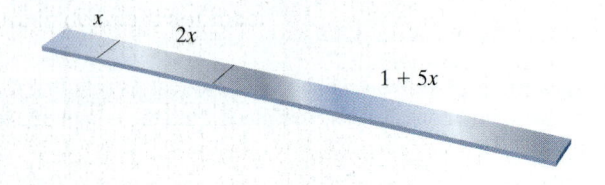

Section 2.6 Formulas and Problem Solving 121

Next, solve for *x* by dividing both sides by *m*.

$$\frac{y-b}{m} = \frac{mx}{m}$$

$$\frac{y-b}{m} = x \quad \text{Simplify.}$$

PRACTICE 6 Solve $H = 5as + 10a$ for *s*.

Concept Check ✓

Solve:

a. 🟡 = 🟥 − 🟩 for 🟥

b. 🟡 = 🟥 · 🔺 − 🟩 for 🟥

▲ EXAMPLE 7 Solve $P = 2l + 2w$ for *w*.

Solution This formula relates the perimeter of a rectangle to its length and width. Find the term containing the variable *w*. To get this term, 2*w*, alone subtract 2*l* from both sides.

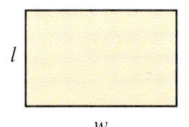

> ▶ **Helpful Hint**
> The 2's may *not* be divided out here. Although 2 is a factor of the denominator, 2 is *not* a factor of the numerator since it is not a factor of both terms in the numerator.

$$P = 2l + 2w$$
$$P - 2l = 2l + 2w - 2l \quad \text{Subtract } 2l \text{ from both sides.}$$
$$P - 2l = 2w \quad \text{Combine like terms.}$$
$$\frac{P - 2l}{2} = \frac{2w}{2} \quad \text{Divide both sides by 2.}$$
$$\frac{P - 2l}{2} = w \quad \text{Simplify.}$$

PRACTICE 7 Solve $N = F + d(n - 1)$ for *d*.

The next example has an equation containing a fraction. We will first clear the equation of fractions and then solve for the specified variable.

EXAMPLE 8 Solve $F = \frac{9}{5}C + 32$ for *C*.

Solution

$$F = \frac{9}{5}C + 32$$

$$5(F) = 5\left(\frac{9}{5}C + 32\right) \quad \text{Clear the fraction by multiplying both sides by the LCD.}$$

$$5F = 9C + 160 \quad \text{Distribute the 5.}$$

$$5F - 160 = 9C + 160 - 160 \quad \text{To get the term containing the variable } C \text{ alone, subtract 160 from both sides.}$$

$$5F - 160 = 9C \quad \text{Combine like terms.}$$

$$\frac{5F - 160}{9} = \frac{9C}{9} \quad \text{Divide both sides by 9.}$$

$$\frac{5F - 160}{9} = C \quad \text{Simplify.}$$

Note: Another equivalent way to write this formula is $C = \frac{5}{9}(F - 32)$.

Answers to Concept Check:

a. 🟡 + 🟩

b. $\frac{🟡 + 🟩}{🔺}$

PRACTICE 8 Solve $A = \frac{1}{2}a(b + B)$ for *B*.

2.6 EXERCISE SET

Substitute the given values into each given formula and solve for the unknown variable. If necessary, round to one decimal place. See Examples 1 through 3.

1. $A = bh$; $A = 45, b = 15$ (Area of a parallelogram)
2. $d = rt$; $d = 195, t = 3$ (Distance formula)
3. $S = 4lw + 2wh$; $S = 102, l = 7, w = 3$ (Surface area of a special rectangular box)
4. $V = lwh$; $l = 14, w = 8, h = 3$ (Volume of a rectangular box)
5. $A = \frac{1}{2}h(B + b)$; $A = 180, B = 11, b = 7$ (Area of a trapezoid)
6. $A = \frac{1}{2}h(B + b)$; $A = 60, B = 7, b = 3$ (Area of a trapezoid)
7. $P = a + b + c$; $P = 30, a = 8, b = 10$ (Perimeter of a triangle)
8. $V = \frac{1}{3}Ah$; $V = 45, h = 5$ (Volume of a pyramid)
9. $C = 2\pi r$; $C = 15.7$ (use the approximation 3.14 or a calculator approximation for π) (Circumference of a circle)
10. $A = \pi r^2$; $r = 4.5$ (use the approximation 3.14 or a calculator approximation for π) (Area of a circle)
11. $I = PRT$; $I = 3750, P = 25,000, R = 0.05$ (Simple interest formula)
12. $I = PRT$; $I = 1,056,000, R = 0.055, T = 6$ (Simple interest formula)
13. $V = \frac{1}{3}\pi r^2 h$; $V = 565.2, r = 6$ (use a calculator approximation for π) (Volume of a cone)
14. $V = \frac{4}{3}\pi r^3$; $r = 3$ (use a calculator approximation for π) (Volume of a sphere)

Solve each formula for the specified variable. See Examples 5 through 8.

15. $f = 5gh$ for h
16. $A = \pi ab$ for b
17. $V = lwh$ for w
18. $T = mnr$ for n
19. $3x + y = 7$ for y
20. $-x + y = 13$ for y
21. $A = P + PRT$ for R
22. $A = P + PRT$ for T
23. $V = \frac{1}{3}Ah$ for A
24. $D = \frac{1}{4}fk$ for k
25. $P = a + b + c$ for a
26. $PR = x + y + z + w$ for z
27. $S = 2\pi rh + 2\pi r^2$ for h
28. $S = 4lw + 2wh$ for h

Solve. See Examples 1 through 4.

29. For the purpose of purchasing new baseboard and carpet,
    a. Find the area and perimeter of the room below (neglecting doors).
    b. Identify whether baseboard has to do with area or perimeter and the same with carpet.

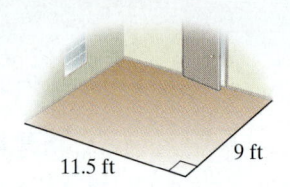

    11.5 ft, 9 ft

30. For the purpose of purchasing lumber for a new fence and seed to plant grass,
    a. Find the area and perimeter of the yard below.
    b. Identify whether a fence has to do with area or perimeter and the same with grass seed.

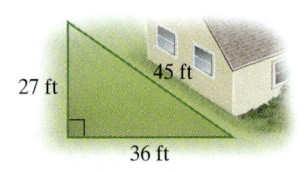

    27 ft, 45 ft, 36 ft

31. A frame shop charges according to both the amount of framing needed to surround the picture and the amount of glass needed to cover the picture.
    a. Find the area and perimeter of the trapezoid-shaped framed picture below.
    b. Identify whether the amount of framing has to do with perimeter or area and the same with the amount of glass.

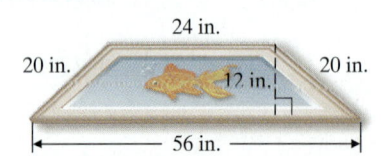

    24 in., 20 in., 12 in., 20 in., 56 in.

32. A decorator is painting and placing a border completely around the parallelogram-shaped wall.
    a. Find the area and perimeter of the wall below.
    b. Identify whether the border has to do with perimeter or area and the same with paint.

    11.7 ft, 7 ft, 9.3 ft

33. The world's largest pink ribbon, the sign of the fight against breast cancer, was erected out of pink post-it notes on a billboard in New York City in October, 2004. If the area of the rectangular billboard covered by the ribbon is approximately 3990 square feet, and the width of the billboard was approximately 57 feet, what was the height of this billboard?

34. The world's largest sign for Coca-Cola is located in Arica, Chile. The rectangular sign has a length of 400 feet and has an area of 52,400 square feet. Find the width of the sign. (*Source:* Fabulous Facts about Coca-Cola, Atlanta, GA)

35. Convert Nome, Alaska's 14°F high temperature to Celsius.

36. Convert Paris, France's low temperature of −5°C to Fahrenheit.

37. The X-30 is a "space plane" that skims the edge of space at 4000 miles per hour. Neglecting altitude, if the circumference of the Earth is approximately 25,000 miles, how long will it take for the X-30 to travel around the Earth?

38. In the United States, a notable hang glider flight was a 303-mile, $8\frac{1}{2}$ hour flight from New Mexico to Kansas. What was the average rate during this flight?

39. An architect designs a rectangular flower garden such that the width is exactly two-thirds of the length. If 260 feet of antique picket fencing are to be used to enclose the garden, find the dimensions of the garden.

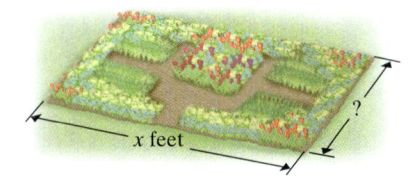

40. If the length of a rectangular parking lot is 10 meters less than twice its width, and the perimeter is 400 meters, find the length of the parking lot.

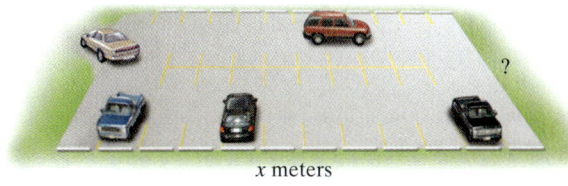

41. A flower bed is in the shape of a triangle with one side twice the length of the shortest side, and the third side is 30 feet more than the length of the shortest side. Find the dimensions if the perimeter is 102 feet.

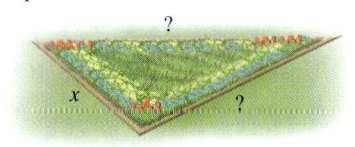

42. The perimeter of a yield sign in the shape of an isosceles triangle is 22 feet. If the shortest side is 2 feet less than the other two sides, find the length of the shortest side. (*Hint:* An isosceles triangle has two sides the same length.)

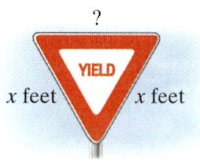

43. The Cat is a high-speed catamaran auto ferry that operates between Bar Harbor, Maine, and Yarmouth, Nova Scotia. The Cat can make the 138-mile trip in about $2\frac{1}{2}$ hours. Find the catamaran speed for this trip. (*Source:* Bay Ferries)

44. A family is planning their vacation to Disney World. They will drive from a small town outside New Orleans, Louisiana, to Orlando, Florida, a distance of 700 miles. They plan to average a rate of 55 mph. How long will this trip take?

45. Piranha fish require 1.5 cubic feet of water per fish to maintain a healthy environment. Find the maximum number of piranhas you could put in a tank measuring 8 feet by 3 feet by 6 feet.

46. Find the maximum number of goldfish you can put in a cylindrical tank whose diameter is 8 meters and whose height is 3 meters if each goldfish needs 2 cubic meters of water.

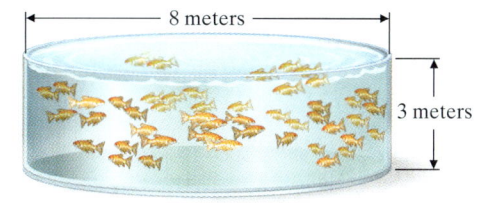

47. A lawn is in the shape of a trapezoid with a height of 60 feet and bases of 70 feet and 130 feet. How many whole bags of fertilizer must be purchased to cover the lawn if each bag covers 4000 square feet?

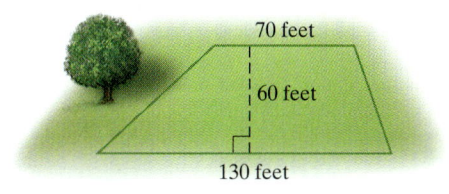

48. If the area of a right-triangularly shaped sail is 20 square feet and its base is 5 feet, find the height of the sail.

49. Maria's Pizza sells one 16-inch cheese pizza or two 10-inch cheese pizzas for $9.99. Determine which size gives more pizza.

50. Find how much rope is needed to wrap around the Earth at the equator, if the radius of the Earth is 4000 miles. (*Hint:* Use 3.14 for π and the formula for circumference.)

51. The perimeter of a geometric figure is the sum of the lengths of its sides. If the perimeter of the following pentagon (five-sided figure) is 48 meters, find the length of each side.

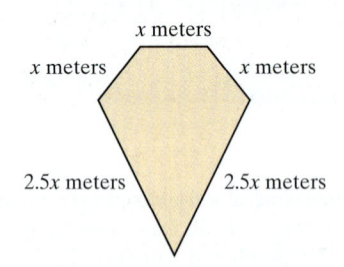

x meters

x meters ... *x* meters

2.5*x* meters ... 2.5*x* meters

52. The perimeter of the following triangle is 82 feet. Find the length of each side.

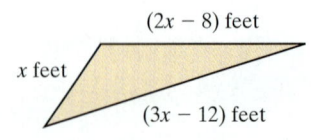

$(2x - 8)$ feet

x feet

$(3x - 12)$ feet

53. A Japanese "bullet" train set a new world record for train speed at 361 miles per hour during a manned test run on the Yamanashi Maglev Test Line in 2003. How long does it take this train to travel 72.2 miles at this speed? Give the result in hours; then convert to minutes.

54. In 1983, the Hawaiian volcano Kilauea began erupting in a series of episodes still occurring at the time of this writing. At times, the lava flows advanced at speeds of up to 0.5 kilometer per hour. In 1983 and 1984 lava flows destroyed 16 homes in the Royal Gardens subdivision, about 6 km away from the eruption site. Roughly how long did it take the lava to reach Royal Gardens? Assume that the lava traveled at its fastest rate, 0.5 kph. (*Source:* U.S. Geological Survey Hawaiian Volcano Observatory)

55. The perimeter of an equilateral triangle is 7 inches more than the perimeter of a square, and the side of the triangle is 5 inches longer than the side of the square. Find the side of the triangle. (*Hint:* An equilateral triangle has three sides the same length.)

56. A square animal pen and a pen shaped like an equilateral triangle have equal perimeters. Find the length of the sides of each pen if the sides of the triangular pen are fifteen less than twice a side of the square pen.

57. Find how long it takes a person to drive 135 miles on I-10 if she merges onto I-10 at 10 a.m. and drives nonstop with her cruise control set on 60 mph.

58. Beaumont, Texas, is about 150 miles from Toledo Bend. If Leo Miller leaves Beaumont at 4 a.m. and averages 45 mph, when should he arrive at Toledo Bend?

59. The longest runway at Los Angeles International Airport has the shape of a rectangle and an area of 1,813,500 square feet. This runway is 150 feet wide. How long is the runway? (*Source:* Los Angeles World Airports)

60. Normal room temperature is about 78°F. Convert this temperature to Celsius.

61. The highest temperature ever recorded in Europe was 122°F in Seville, Spain, in August of 1881. Convert this record high temperature to Celsius. (*Source:* National Climatic Data Center)

62. The lowest temperature ever recorded in Oceania was −10°C at the Haleakala Summit in Maui, Hawaii, in January 1961. Convert this record low temperature to Fahrenheit. (*Source:* National Climatic Data Center)

63. The CART FedEx Championship Series is an open-wheeled race car competition based in the United States. A CART car has a maximum length of 199 inches, a maximum width of 78.5 inches, and a maximum height of 33 inches. When the CART series travels to another country for a grand prix, teams must ship their cars. Find the volume of the smallest shipping crate needed to ship a CART car of maximum dimensions. (*Source:* Championship Auto Racing Teams, Inc.)

CART Racing Car

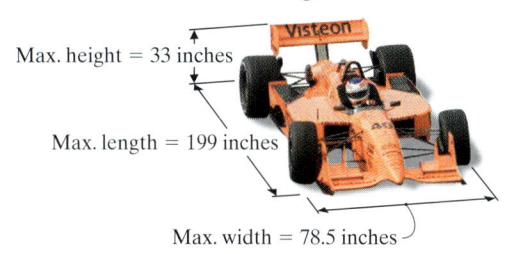

Max. height = 33 inches
Max. length = 199 inches
Max. width = 78.5 inches

64. On a road course, a CART car's speed can average up to around 105 mph. Based on this speed, how long would it take a CART driver to travel from Los Angeles to New York City, a distance of about 2810 miles by road, without stopping? Round to the nearest tenth of an hour.

65. The Hoberman Sphere is a toy ball that expands and contracts. When it is completely closed, it has a diameter of 9.5 inches. Find the volume of the Hoberman Sphere when it is completely closed. Use 3.14 for π. Round to the nearest whole cubic inch. (*Source:* Hoberman Designs, Inc.)

66. When the Hoberman Sphere (see Exercise 65) is completely expanded, its diameter is 30 inches. Find the volume of the Hoberman Sphere when it is completely expanded. Use 3.14 for π. Round to the nearest whole cubic inch. (*Source:* Hoberman Designs, Inc.)

67. The average temperature on the planet Mercury is 167°C. Convert this temperature to degrees Fahrenheit. (*Source:* National Space Science Data Center)

68. The average temperature on the planet Jupiter is −227°F. Convert this temperature to degrees Celsius. Round to the nearest degree. (*Source:* National Space Science Data Center)

REVIEW AND PREVIEW

Write the following phrases as algebraic expressions. See Section 2.1.

69. Nine divided by the sum of a number and 5
70. Half the product of a number and five
71. Three times the sum of a number and four
72. Double the sum of ten and four times a number
73. Triple the difference of a number and twelve
74. A number minus the sum of the number and six

CONCEPT EXTENSIONS

Solve. See the Concept Check in this section.

75. ▢ − ● · ▮ = ▲ for ●

76. ⬠ · ▮ + ▲ = ● for ▮

77. Dry ice is a name given to solidified carbon dioxide. At −78.5° Celsius it changes directly from a solid to a gas. Convert this temperature to Fahrenheit.

78. Lightning bolts can reach a temperature of 50,000° Fahrenheit. Convert this temperature to Celsius.

79. The distance from the sun to the Earth is approximately 93,000,000 miles. If light travels at a rate of 186,000 miles per second, how long does it take light from the sun to reach us?

80. Light travels at a rate of 186,000 miles per second. If our moon is 238,860 miles from the Earth, how long does it take light from the moon to reach us? (Round to the nearest tenth of a second.)

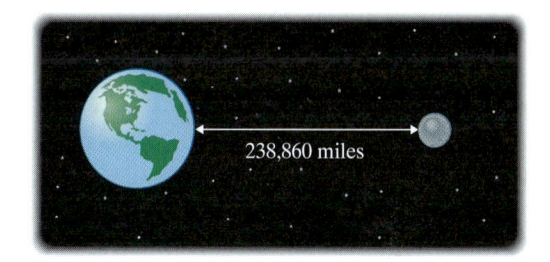

238,860 miles

81. A glacier is a giant mass of rocks and ice that flows downhill like a river. Exit Glacier, near Seward, Alaska, moves at a rate of 20 inches a day. Find the distance in feet the glacier moves in a year. (Assume 365 days a year. Round to 2 decimal places.)

82. Flying fish do not *actually* fly, but glide. They have been known to travel a distance of 1300 feet at a rate of 20 miles per hour. How many seconds did it take to travel this distance? (*Hint:* First convert miles per hour to feet per second. Recall that 1 mile = 5280 feet. Round to the nearest tenth of a second.)

126 CHAPTER 2 Equations, Inequalities, and Problem Solving

83. Stalactites join stalagmites to form columns. A column found at Natural Bridge Caverns near San Antonio, Texas, rises 15 feet and has a *diameter* of only 2 inches. Find the volume of this column in cubic inches. (*Hint:* Use the formula for volume of a cylinder and use a calculator approximation for π. Round to the nearest tenth of an inch.)

84. Find the temperature at which the Celsius measurement and Fahrenheit measurement are the same number.

85. The formula $A = bh$ is used to find the area of a parallelogram. If the base of a parallelogram is doubled and its height is doubled, how does this affect the area?

86. The formula $V = LWH$ is used to find the volume of a box. If the length of a box is doubled, the width is doubled, and the height is doubled, how does this affect the volume?

STUDY SKILLS BUILDER

Organizing a Notebook

It's never too late to get organized. If you need ideas about organizing a notebook for your mathematics course, try some of these:

- Use a spiral or ring binder notebook with pockets and use it for mathematics only.
- Start each page by writing the book's section number you are working on at the top.
- When your instructor is lecturing, take notes. *Always* include any examples your instructor works for you.
- Place your worked-out homework exercises in your notebook immediately after the lecture notes from that section. This way, a section's worth of material is together.
- Homework exercises: Attempt and check all assigned homework.
- Place graded quizzes in the pockets of your notebook or a special section of your binder.

Self-Check

Check your notebook organization by answering the following questions.

1. Do you have a spiral or ring binder notebook for your mathematics course only?
2. Have you ever had to flip through several sheets of notes and work in your mathematics notebook to determine what section's work you are in?
3. Are you now writing the textbook's section number at the top of each notebook page?
4. Have you ever lost or had trouble finding a graded quiz or test?
5. Are you now placing all your graded work in a dedicated place in your notebook?
6. Are you attempting all of your homework and placing all of your work in your notebook?
7. Are you checking and correcting your homework in your notebook? If not, why not?
8. Are you writing in your notebook the examples your instructor works for you in class?

2.7 PERCENT AND MIXTURE PROBLEM SOLVING

OBJECTIVES

1. Solve percent equations.
2. Solve discount and mark-up problems.
3. Solve percent increase and percent decrease problems.
4. Solve mixture problems.

This section is devoted to solving problems in the categories listed. The same problem-solving steps used in previous sections are also followed in this section. They are listed below for review.

> **General Strategy for Problem Solving**
>
> 1. UNDERSTAND the problem. During this step, become comfortable with the problem. Some ways of doing this are as follows:
>   Read and reread the problem.
>   Choose a variable to represent the unknown.
>   Construct a drawing, whenever possible.
>   Propose a solution and check. Pay careful attention to how you check your proposed solution. This will help writing an equation to model the problem.
> 2. TRANSLATE the problem into an equation.
> 3. SOLVE the equation.
> 4. INTERPRET the results: *Check* the proposed solution in the stated problem and *state* your conclusion.

OBJECTIVE 1 ▶ Solving percent equations. Many of today's statistics are given in terms of percent: a basketball player's free throw percent, current interest rates, stock market trends, and nutrition labeling, just to name a few. In this section, we first explore percent, percent equations, and applications involving percents. See Appendix F.2 if a further review of percents is needed.

EXAMPLE 1 The number 63 is what percent of 72?

Solution

1. UNDERSTAND. Read and reread the problem. Next, let's suppose that the percent is 80%. To check, we find 80% of 72.

$$80\% \text{ of } 72 = 0.80(72) = 57.6$$

This is close, but not 63. At this point, though, we have a better understanding of the problem, we know the correct answer is close to and greater than 80%, and we know how to check our proposed solution later.

Let x = the unknown percent.

2. TRANSLATE. Recall that "is" means "equals" and "of" signifies multiplying. Let's translate the sentence directly.

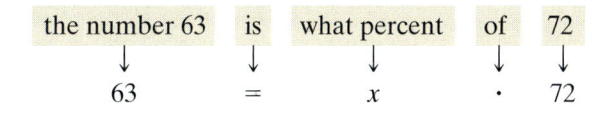

the number 63	is	what percent	of	72
↓	↓	↓	↓	↓
63	=	x	·	72

3. SOLVE.

$$63 = 72x$$
$$0.875 = x \quad \text{Divide both sides by 72.}$$
$$87.5\% = x \quad \text{Write as a percent.}$$

4. INTERPRET.

Check: Verify that 87.5% of 72 is 63.

State: The number 63 is 87.5% of 72.

PRACTICE 1 The number 35 is what percent of 56?

EXAMPLE 2 The number 120 is 15% of what number?

Solution

1. UNDERSTAND. Read and reread the problem.

   Let x = the unknown number.

2. TRANSLATE.

the number 120	is	15%	of	what number
120	=	15%	·	x

3. SOLVE.

   $120 = 0.15x$ Write 15% as 0.15.

   $800 = x$ Divide both sides by 0.15.

4. INTERPRET.

Check: Check the proposed solution by finding 15% of 800 and verifying that the result is 120.

State: Thus, 120 is 15% of 800.

PRACTICE 2 The number 198 is 55% of what number?

The next example contains a circle graph. This particular circle graph shows percents of American travelers in certain categories. Since the circle graph represents all American travelers, the percents should add to 100%.

> **Helpful Hint**
> The percents in a circle graph should have a sum of 100%.

EXAMPLE 3 The circle graph below shows the purpose of trips made by American travelers. Use this graph to answer the questions below.

Purpose of American Travelers

- Personal/Other, 13%
- Combined Business/Pleasure, 4%
- Business, 17%
- Pleasure, 66%

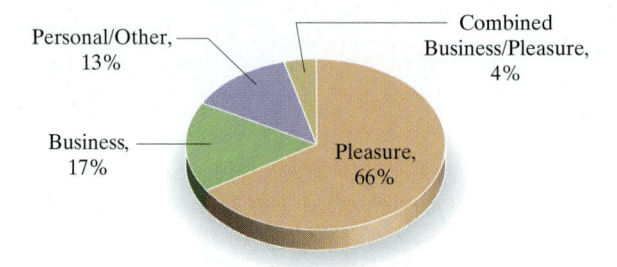

Source: Travel Industry Association of America

a. What percent of trips made by American travelers are solely for the purpose of business?
b. What percent of trips made by American travelers are for the purpose of business or combined business/pleasure?
c. On an airplane flight of 253 Americans, how many of these people might we expect to be traveling solely for business?

Solution

a. From the circle graph, we see that 17% of trips made by American travelers are solely for the purpose of business.
b. From the circle graph, we know that 17% of trips are solely for business and 4% of trips are for combined business/pleasure. The sum 17% + 4% or 21% of trips made by American travelers are for the purpose of business or combined business/pleasure.
c. Since 17% of trips made by American travelers are for business, we find 17% of 253. Remember that "of" translates to "multiplication."

$$17\% \text{ of } 253 = 0.17(253) \quad \text{Replace "of" with the operation of multiplication.}$$
$$= 43.01$$

We might then expect that about 43 American travelers on the flight are traveling solely for business.

PRACTICE 3 Use the Example 3 circle graph to answer each question.

a. What percent of trips made by American travelers are for combined business/pleasure?
b. What percent of trips made by American travelers are for the purpose of business, pleasure, or combined business/pleasure?
c. On a flight of 325 Americans, how many of these people might we expect to be traveling for business/pleasure?

OBJECTIVE 2 ▶ Solving discount and mark-up problems. The next example has to do with discounting the price of a cell phone.

EXAMPLE 4 Cell Phones Unlimited recently reduced the price of a $140 phone by 20%. What is the discount and the new price?

Solution

1. UNDERSTAND. Read and reread the problem. Make sure you understand the meaning of the word "discount." Discount is the amount of money by which the cost of an item has been decreased. To find the discount, we simply find 20% of $140. In other words, we have the formulas,

$$\text{discount} = \text{percent} \cdot \text{original price} \quad \text{Then}$$

$$\text{new price} = \text{original price} - \text{discount}$$

2, 3. TRANSLATE and SOLVE.

$$\text{discount} = \text{percent} \cdot \text{original price}$$
$$= 20\% \cdot \$140$$
$$= 0.20 \cdot \$140$$
$$= \$28$$

Thus, the discount in price is $28.

$$\text{new price} = \text{original price} - \text{discount}$$
$$= \$140 - \$28$$
$$= \$112$$

4. INTERPRET.

Check: Check your calculations in the formulas, and also see if our results are reasonable. They are.

State: The discount in price is $28 and the new price is $112.

PRACTICE 4 A used treadmill, originally purchased for $480, was sold at a garage sale at a discount of 85% of the original price. What was the discount and the new price?

A concept similar to discount is mark-up. What is the difference between the two? A discount is subtracted from the original price while a mark-up is added to the original price. For mark-ups,

$$\text{mark-up} = \text{percent} \cdot \text{original price}$$

$$\text{new price} = \text{original price} + \text{mark-up}$$

▶ **Helpful Hint**
Discounts are subtracted from the original price while mark-ups are added.

Mark-up exercises can be found in Exercise Set 2.7.

OBJECTIVE 3 ▶ Solving percent increase and percent decrease problems. Percent increase or percent decrease is a common way to describe how some measurement has increased or decreased. For example, crime increased by 8%, teachers received a 5.5% increase in salary, or a company decreased its employees by 10%. The next example is a review of percent increase.

EXAMPLE 5 The cost of attending a public college rose from $9258 in 1996 to $12,796 in 2006. Find the percent increase, rounded to the nearest tenth of a percent. (*Source:* The College Board)

Solution

1. UNDERSTAND. Read and reread the problem. Let's guess that the percent increase is 20%. To see if this is the case, we find 20% of $9258 to find the *increase* in cost. Then we add this increase to $9258 to find the *new cost*. In other words, 20% ($9258) = 0.20($9258) = $1851.60, the *increase* in cost. The new cost then would be $9258 + $1851.60 = $11,109.60, less than the actual new cost of $12,976. We now know that the increase is greater than 20% and we know how to check our proposed solution.

Let x = the percent increase.

2. TRANSLATE. First, find the **increase,** and then the **percent increase.** The increase in cost is found by

In words: increase = new cost − old cost   or

Translate: increase = $12,796 − $9258
                      = $3538

Next, find the percent increase. The percent increase or percent decrease is always a percent of the original number or, in this case, the old cost.

In words: increase   is   what percent increase   of   old cost

Translate: $3538   =   x   · $9258

3. SOLVE.

$$3538 = x \cdot 9258 \quad \text{Divide both sides by 9258.}$$
$$0.382 \approx x \quad \text{Round to 3 decimal places.}$$
$$38.2\% \approx x \quad \text{Write as a percent.}$$

4. INTERPRET.

Check: Check the proposed solution, as shown in Step 1.

State: The percent increase in cost is approximately 38.2%.

PRACTICE 5   The average price of a single family home in the United States rose from $198,900 in 2000 to $299,800 in 2005. Find the percent increase. Round to the nearest tenth of a percent. (*Source:* Federal Housing Finance Board)

Percent decrease is found using a similar method. First find the decrease, then determine what percent of the original or first amount is that decrease.

Read the next example carefully. For Example 5, we were asked to find percent increase. In Example 6, we are given the percent increase and asked to find the number before the increase.

EXAMPLE 6   Most of the movie screens globally project analog film, but the number of cinemas using digital is increasing. Find the number of digital screens worldwide last year if, after a 153% increase the number this year is 849. Round to the nearest whole number. (*Source:* Motion Picture Association of America)

Solution

1. UNDERSTAND. Read and reread the problem. Let's guess a solution and see how we would check our guess. If the number of digital screens worldwide last year was 400, we would see if 400 plus the increase is 849; that is,

$$400 + 153\%(400) = 400 + 1.53(400) = 2.53(400) = 1012$$

Since 1012 is too large, we know that our guess of 400 is too large. We also have a better understanding of the problem. Let

$$x = \text{number of digital screens last year}$$

2. TRANSLATE. To translate an equation, we remember that

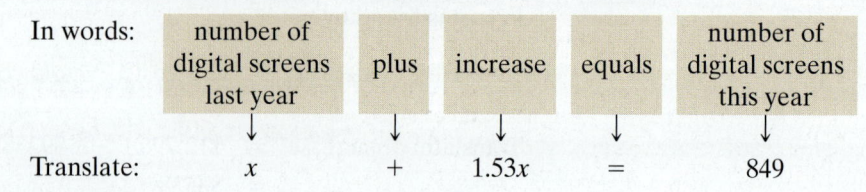

In words:	number of digital screens last year	plus	increase	equals	number of digital screens this year
	↓	↓	↓	↓	↓
Translate:	x	$+$	$1.53x$	$=$	849

3. SOLVE.

$$2.53x = 849 \quad \text{Add like terms.}$$
$$x = \frac{849}{2.53}$$
$$x \approx 336$$

4. INTERPRET.

Check: Recall that x represents the number of digital screens worldwide last year. If this number is approximately 336, let's see if 336 plus the increase is close to 849. (We use the word "close" since 336 is rounded.)

$$336 + 153\%(336) = 336 + 1.53(336) = 2.53(336) = 850.08$$

which is close to 849.

State: There were approximately 336 digital screens worldwide last year. □

PRACTICE
6 In 2005, 535 new feature films were released in the United States. This was an increase of 2.8% over the number of new feature films released in 2004. Find the number of new feature films released in 2004. (*Source:* Motion Picture Association of America)

OBJECTIVE 4 ▶ Solving mixture problems. Mixture problems involve two or more different quantities being combined to form a new mixture. These applications range from Dow Chemical's need to form a chemical mixture of a required strength to Planter's Peanut Company's need to find the correct mixture of peanuts and cashews, given taste and price constraints.

EXAMPLE 7 Calculating Percent for a Lab Experiment

A chemist working on his doctoral degree at Massachusetts Institute of Technology needs 12 liters of a 50% acid solution for a lab experiment. The stockroom has only 40% and 70% solutions. How much of each solution should be mixed together to form 12 liters of a 50% solution?

Solution:

1. **UNDERSTAND.** First, read and reread the problem a few times. Next, guess a solution. Suppose that we need 7 liters of the 40% solution. Then we need $12 - 7 = 5$ liters of the 70% solution. To see if this is indeed the solution, find the amount of pure acid in 7 liters of the 40% solution, in 5 liters of the 70% solution, and in 12 liters of a 50% solution, the required amount and strength.

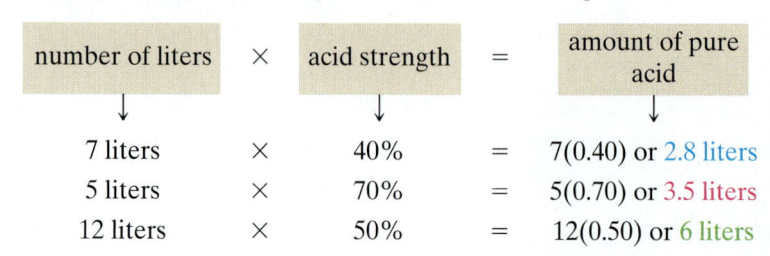

number of liters	×	acid strength	=	amount of pure acid
↓		↓		↓
7 liters	×	40%	=	7(0.40) or 2.8 liters
5 liters	×	70%	=	5(0.70) or 3.5 liters
12 liters	×	50%	=	12(0.50) or 6 liters

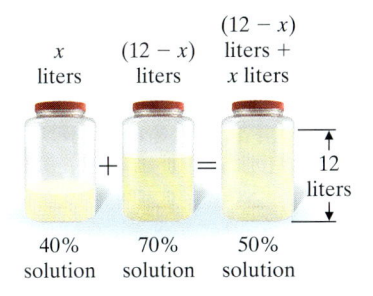

Since 2.8 liters + 3.5 liters = 6.3 liters and not 6, our guess is incorrect, but we have gained some valuable insight into how to model and check this problem.

Let

x = number of liters of 40% solution; then
$12 - x$ = number of liters of 70% solution.

2. **TRANSLATE.** To help us translate to an equation, the following table summarizes the information given. Recall that the amount of acid in each solution is found by multiplying the acid strength of each solution by the number of liters.

	No. of Liters ·	Acid Strength =	Amount of Acid
40% Solution	x	40%	$0.40x$
70% Solution	$12 - x$	70%	$0.70(12 - x)$
50% Solution Needed	12	50%	$0.50(12)$

The amount of acid in the final solution is the sum of the amounts of acid in the two beginning solutions.

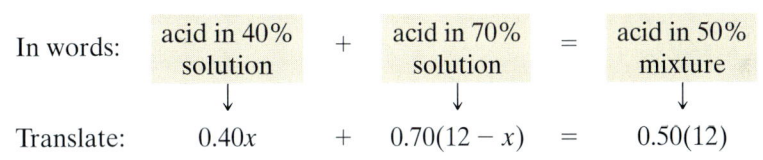

In words: acid in 40% solution + acid in 70% solution = acid in 50% mixture

Translate: $0.40x + 0.70(12 - x) = 0.50(12)$

3. **SOLVE.**

$$0.40x + 0.70(12 - x) = 0.50(12)$$
$$0.4x + 8.4 - 0.7x = 6 \quad \text{Apply the distributive property.}$$
$$-0.3x + 8.4 = 6 \quad \text{Combine like terms.}$$
$$-0.3x = -2.4 \quad \text{Subtract 8.4 from both sides.}$$
$$x = 8 \quad \text{Divide both sides by } -0.3.$$

4. **INTERPRET.**

Check: To check, recall how we checked our guess.

State: If 8 liters of the 40% solution are mixed with $12 - 8$ or 4 liters of the 70% solution, the result is 12 liters of a 50% solution.

PRACTICE 7 Hamida Barash was responsible for refilling the eye wash stations in the chemistry lab. She needed 6 liters of 3% strength eyewash to refill the dispensers. The supply room only had 2% and 5% eyewash in stock. How much of each solution should she mix to produce the needed 3% strength eyewash?

VOCABULARY & READINESS CHECK

Tell whether the percent labels in the circle graphs are correct.

1. (circle graph: 25%, 40%, 25%)
2. (circle graph: 30%, 30%, 30%)
3. (circle graph: 25%, 25%, 25%, 25%)
4. (circle graph: 40%, 50%, 10%)

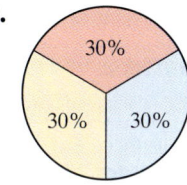

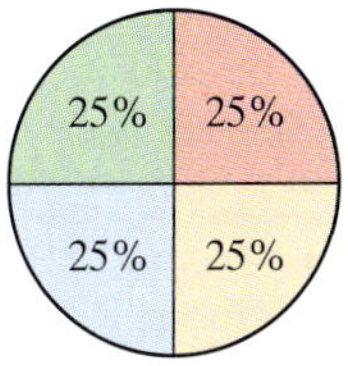

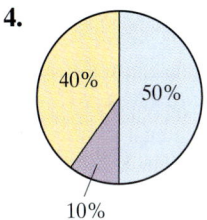

2.7 EXERCISE SET

Find each number described. See Examples 1 and 2.

1. What number is 16% of 70?
2. What number is 88% of 1000?
3. The number 28.6 is what percent of 52?
4. The number 87.2 is what percent of 436?
5. The number 45 is 25% of what number?
6. The number 126 is 35% of what number?

The circle graph below shows the uses of U.S. corn production. Use this graph for Exercises 7 through 10. See Example 3.

U.S. Corn Production Use

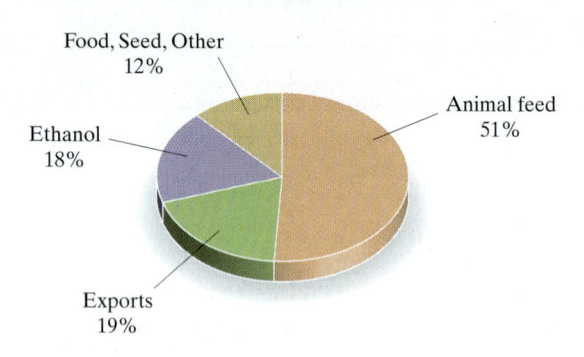

- Food, Seed, Other 12%
- Ethanol 18%
- Exports 19%
- Animal feed 51%

Source: USDA, American Farm Bureau Federation

7. What percent of corn production is used for animal feed or ethanol?
8. What percent of corn production is *not* used for exports?
9. The U.S. corn production in 2006–2007 was 10,535 million bushels. How many bushels were used to make ethanol?
10. How many bushels of the 2006–2007 corn production was used for food, seed, or other? (See Exercise 9.)

Solve. If needed, round answers to the nearest cent. See Example 4.

11. A used automobile dealership recently reduced the price of a used compact car by 8%. If the price of the car before discount was $18,500, find the discount and the new price.
12. A music store is advertising a 25%-off sale on all new releases. Find the discount and the sale price of a newly released CD that regularly sells for $12.50.
13. A birthday celebration meal is $40.50 including tax. Find the total cost if a 15% tip is added to the cost.
14. A retirement dinner for two is $65.40 including tax. Find the total cost if a 20% tip is added to the cost.

Solve. See Example 5.

15. The number of different cars sold in the United States rose from 208 in 1997 to 280 in 2007. Find the percent increase. Round to the nearest whole percent. (*Source: New York Times*, May 2007)
16. The cost of attending a private college rose from $19,000 in 2000 to $22,200 in 2006. Find the percent increase. Round to the nearest whole percent.

17. By decreasing each dimension by 1 unit, the area of a rectangle decreased from 40 square feet (on the left) to 28 square feet (on the right). Find the percent decrease in area.

8 ft — Area: 40 sq ft — 5 ft

7 ft — Area: 28 sq ft — 4 ft

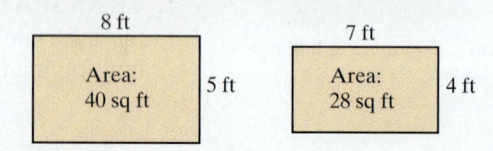

18. By decreasing the length of the side by one unit, the area of a square decreased from 100 square meters to 81 square meters. Find the percent decrease in area.

10 m — Area: 100 sq m

9 m — Area: 81 sq m

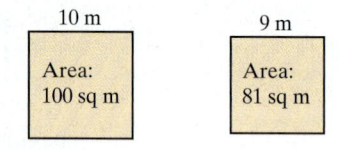

Solve. See Example 6.

19. Find the original price of a pair of shoes if the sale price is $78 after a 25% discount.
20. Find the original price of a popular pair of shoes if the increased price is $80 after a 25% increase.
21. Find last year's salary if after a 4% pay raise, this year's salary is $44,200.
22. Find last year's salary if after a 3% pay raise, this year's salary is $55,620.

Solve. For each exercise, a table is given for you to complete and use to write an equation that models the situation. See Example 7.

23. How much pure acid should be mixed with 2 gallons of a 40% acid solution in order to get a 70% acid solution?

	Number of Gallons	·	Acid Strength	=	Amount of Acid
Pure Acid			100%		
40% Acid Solution					
70% Acid Solution Needed					

24. How many cubic centimeters (cc) of a 25% antibiotic solution should be added to 10 cubic centimeters of a 60% antibiotic solution in order to get a 30% antibiotic solution?

	Number of Cubic cm	·	Antibiotic Strength	=	Amount of Antibiotic
25% Antibiotic Solution					
60% Antibiotic Solution					
30% Antibiotic Solution Needed					

25. Community Coffee Company wants a new flavor of Cajun coffee. How many pounds of coffee worth $7 a pound should be added to 14 pounds of coffee worth $4 a pound to get a mixture worth $5 a pound?

	Number of Pounds	·	Cost per Pound	=	Value
$7 per lb Coffee					
$4 per lb Coffee					
$5 per lb Coffee Wanted					

26. Planter's Peanut Company wants to mix 20 pounds of peanuts worth $3 a pound with cashews worth $5 a pound in order to make an experimental mix worth $3.50 a pound. How many pounds of cashews should be added to the peanuts?

	Number of Pounds	·	Cost per Pound	=	Value
$3 per lb Peanuts					
$5 per lb Cashews					
$3.50 per lb Mixture Wanted					

MIXED PRACTICE

Solve. If needed, round money amounts to two decimal places and all other amounts to one decimal place. See Examples 1 through 7.

27. Find 23% of 20.
28. Find 140% of 86.
29. The number 40 is 80% of what number?
30. The number 56.25 is 45% of what number?
31. The number 144 is what percent of 480?
32. The number 42 is what percent of 35?

The graph shows the communities in the United States that have the highest percents of citizens that shop by catalog. Use the graph to answer Exercises 33 through 36.

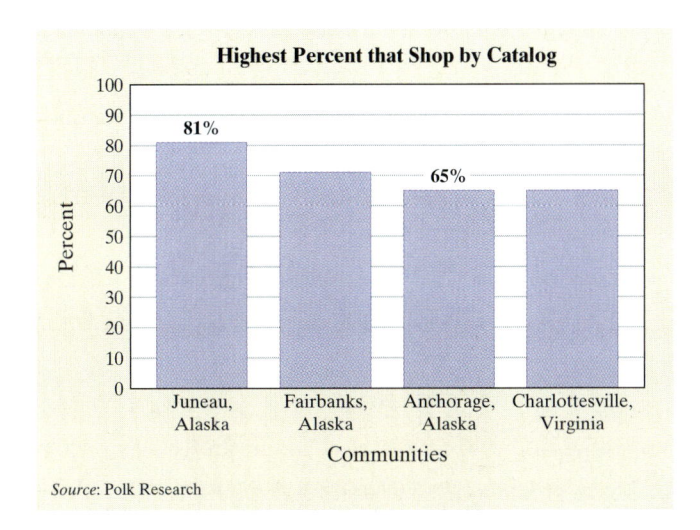

Highest Percent that Shop by Catalog

- Juneau, Alaska: 81%
- Fairbanks, Alaska: 71%
- Anchorage, Alaska: 65%
- Charlottesville, Virginia: 65%

Source: Polk Research

33. Estimate the percent of the population in Fairbanks, Alaska, who shops by catalog.
34. Estimate the percent of the population in Charlottesville, Virginia, who shops by catalog.
35. According to the *World Almanac*, Anchorage has a population of 275,043. How many catalog shoppers might we predict live in Anchorage? Round to the nearest whole number.
36. According to the *World Almanac*, Juneau has a population of 30,987. How many catalog shoppers might we predict live in Juneau? Round to the nearest whole number.

For Exercises 37 and 38, fill in the percent column in each table. Each table contains a worked-out example.

37.

Ford Motor Company
Model Year 2006 Vehicle Sales Worldwide

	Thousands of Vehicles	Percent of Total (Rounded to Nearest Percent)
North America	3051	
Europe	1846	
Asia-Pacific-Africa	589	
South America	381	
Rest of the World	730	Example: $\frac{730}{6597} \approx 11\%$
Total	6597	

Source: Ford Motor Company

38.

Kraft Foods North America
Volume Food Produced in a year

Food Group	Volume (in pounds)	Percent (Round to Nearest Percent)
Cheese, Meals, and Enhancers	6183	
Biscuits, Snacks, and Confectionaries	2083	Example: $\frac{2083}{13,741} \approx 15\%$
Beverages, Desserts, and Cereals	3905	
Oscar Mayer and Pizza	1570	
Total	13,741	

Source: Kraft Foods, North America

39. Nordstrom advertised a 25%-off sale. If a London Fog coat originally sold for $256, find the decrease in price and the sale price.
40. A gasoline station decreased the price of a $0.95 cola by 15%. Find the decrease in price and the new price.
41. Iceberg lettuce is grown and shipped to stores for about 40 cents a head, and consumers purchase it for about 86 cents a head. Find the percent increase. (*Source: Statistical Abstract of the United States*)

42. The lettuce consumption per capita in 1980 was about 25.6 pounds, and in 2005 the consumption dropped to about 22.4 pounds. Find the percent decrease. (*Source: Statistical Abstract of the United States.*)

43. Smart Cards (cards with an embedded computer chip) have been growing in popularity in recent years. In 2006, about 1900 million Smart Cards were expected to be issued. This represents a 726% increase from the number of cards that were issued in 2001. How many Smart Cards were issued in 2001? Round to the nearest million. (*Source:* The Freedonia Group)

44. Fuel ethanol production is projected to be 10,800 million gallons in 2009. This represents a 44% increase from the number of gallons produced in 2007. How many millions of gallons were produced in 2007? (*Source:* Renewable Fuels Association)

45. How much of an alloy that is 20% copper should be mixed with 200 ounces of an alloy that is 50% copper in order to get an alloy that is 30% copper?

46. How much water should be added to 30 gallons of a solution that is 70% antifreeze in order to get a mixture that is 60% antifreeze?

47. A junior one-day admission to Hershey Park amusement park in Hershey, Pennsylvania, is $27. This price is increased by 70% for (nonsenior) adults. Find the mark-up and the adult price. (*Note:* Prices given are approximations.)

48. The price of a biology book recently increased by 10%. If this book originally cost $99.90, find the mark-up and the new price.

49. By doubling each dimension, the area of a parallelogram increased from 36 square centimeters to 144 square centimeters. Find the percent increase in area.

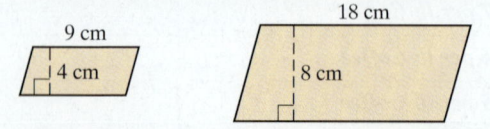

50. By doubling each dimension, the area of a triangle increased from 6 square miles to 24 square miles. Find the percent increase in area.

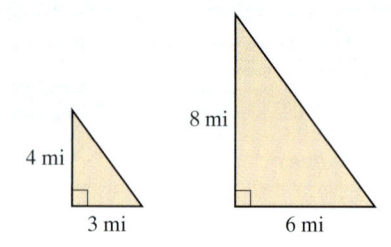

51. A company recently downsized its number of employees by 35%. If there are still 78 employees, how many employees were there prior to the layoffs?

52. The average number of children born to each U.S. woman has decreased by 44% since 1920. If this average is now 1.9, find the average in 1920. Round to the nearest tenth.

53. The owner of a local chocolate shop wants to develop a new trail mix. How many pounds of chocolate-covered peanuts worth $5 a pound should be mixed with 10 pounds of granola bites worth $2 a pound to get a mixture worth $3 per pound?

54. A new self-tanning lotion for everyday use is to be sold. First, an experimental lotion mixture is made by mixing 800 ounces of everyday moisturizing lotion worth $0.30 an ounce with self-tanning lotion worth $3 per ounce. If the experimental lotion is to cost $1.20 per ounce, how many ounces of the self-tanning lotion should be in the mixture?

55. The number of farms in the United States was 2.19 million in 2000. By 2006, the number had dropped to 2.09 million. What was the percent of decrease? Round to the nearest tenth of a percent. (*Source:* USDA: National Agricultural Statistical Service)

56. The average size of farms in the United States was 436 acres in 2000. By 2005, the average size had increased to 444 acres. What was the percent increase? Round to the nearest tenth of a percent. (*Source:* USDA: National Agricultural Statistical Service)

57. The number of Supreme Court decisions has been decreasing in recent years. During the 2005–2006 term, 182 decisions were announced. This is a 45.7% decrease from the number of decisions announced during the 1982–1983 term. How many decisions were announced during 1982–1983? Round to the nearest whole. (*Source: World Almanac*)

58. The total number of movie screens in the United States has been increasing in recent years. In 2005, there were 37,092 indoor movie screens. This is a 4.3% increase from the number of indoor movie screens in 2000. How many movie screens were operating in 2000? Round to the nearest whole. (*Source:* National Association of Theater Owners)

59. Scoville units are used to measure the hotness of a pepper. Measuring 577 thousand Scoville units, the "Red Savina" habañero pepper was known as the hottest chili pepper. That has recently changed with the discovery of Naga Jolokia pepper from India. It measures 48% hotter than the habañero. Find the measure of the Naga Jolokia pepper. Round to the nearest thousand units.

60. At this writing, the women's world record for throwing a disc (like a heavy Frisbee) was set by Jennifer Griffin of the United States in 2000. Her throw was 138.56 meters. The men's world record was set by Christian Sandstrom of Sweden in 2002. His throw was 80.4% farther than Jennifer's. Find the distance of his throw. Round to the nearest meter. (*Source:* World Flying Disc Federation)

61. A recent survey showed that 42% of recent college graduates named flexible hours as their most desired employment benefit. In a graduating class of 860 college students, how many would you expect to rank flexible hours as their top priority in job benefits? (Round to the nearest whole.) (*Source:* JobTrak.com)

62. A recent survey showed that 64% of U.S. colleges have Internet access in their classrooms. There are approximately 9800 post-secondary institutions in the United States. How many of these would you expect to have Internet access in their classrooms? (*Source:* Market Data Retrieval, National Center for Education Statistics)

REVIEW AND PREVIEW

Place $<, >,$ *or* $=$ *in the appropriate space to make each a true statement. See Sections 1.2, 1.3, and 1.6.*

63. -5 $\quad$ -7

64. $\dfrac{12}{3}$ $\quad$ 2^2

65. $|-5|$ $\quad$ $-(-5)$

66. -3^3 $\quad$ $(-3)^3$

67. $(-3)^2$ $\quad$ -3^2

68. $|-2|$ $\quad$ $-|-2|$

CONCEPT EXTENSIONS

69. Is it possible to mix a 10% acid solution and a 40% acid solution to obtain a 60% acid solution? Why or why not?

70. Must the percents in a circle graph have a sum of 100%? Why or why not?

71. A trail mix is made by combining peanuts worth $3 a pound, raisins worth $2 a pound, and M & M's worth $4 a pound. Would it make good business sense to sell the trail mix for $1.98 a pound? Why or why not?

72. a. Can an item be marked-up by more than 100%? Why or why not?

  b. Can an item be discounted by more than 100%? Why or why not?

Standardized nutrition labels like the one below have been displayed on food items since 1994. The percent column on the right shows the percent of daily values (based on a 2000-calorie diet) shown at the bottom of the label. For example, a serving of this food contains 4 grams of total fat, where the recommended daily fat based on a 2000-calorie diet is less than 65 grams of fat. This means that $\dfrac{4}{65}$ *or approximately 6% (as shown) of your daily recommended fat is taken in by eating a serving of this food. Use this nutrition label to answer Exercises 73 through 75.*

Nutrition Facts

Serving Size 18 Crackers (31g)
Servings Per Container About 9

Amount Per Serving

Calories 130   Calories from Fat 35

% Daily Value*

Total Fat 4g	6%
Saturated Fat 0.5g	3%
Polyunsaturated Fat 0g	
Monounsaturated Fat 1.5g	
Cholesterol 0mg	0%
Sodium 230mg	x
Total Carbohydrate 23g	y
Dietary Fiber 2g	8%
Sugars 3g	
Protein 2g	

Vitamin A 0% • Vitamin C 0%
Calcium 2% • Iron 6%

* Percent Daily Values are based on a 2,000 calorie diet. Your daily values may be higher or lower depending on your calorie needs.

		Calories	2,000	2,500
Total Fat	Less than		65g	80g
Sat. Fat	Less than		20g	25g
Cholesterol	Less than		300mg	300mg
Sodium	Less than		2400mg	2400mg
Total Carbohydrate			300g	375g
Dietary Fiber			25g	30g

73. Based on a 2000-calorie diet, what percent of daily value of sodium is contained in a serving of this food? In other words, find x in the label. (Round to the nearest tenth of a percent.)

74. Based on a 2000-calorie diet, what percent of daily value of total carbohydrate is contained in a serving of this food? In other words, find y in the label. (Round to the nearest tenth of a percent.)

75. Notice on the nutrition label that one serving of this food contains 130 calories and 35 of these calories are from fat. Find the percent of calories from fat. (Round to the nearest tenth of a percent.) It is recommended that no more than 30% of calorie intake come from fat. Does this food satisfy this recommendation?

Use the nutrition label below to answer Exercises 76 through 78.

NUTRITIONAL INFORMATION PER SERVING	
Serving Size: 9.8 oz.	Servings Per Container: 1
Calories 280	Polyunsaturated Fat 1g
Protein 12g	Saturated Fat 3g
Carbohydrate 45g	Cholesterol 20mg
Fat 6g	Sodium 520mg
Percent of Calories from Fat....?	Potassium 220mg

76. If fat contains approximately 9 calories per gram, find the percent of calories from fat in one serving of this food. (Round to the nearest tenth of a percent.)

77. If protein contains approximately 4 calories per gram, find the percent of calories from protein from one serving of this food. (Round to the nearest tenth of a percent.)

78. Find a food that contains more than 30% of its calories per serving from fat. Analyze the nutrition label and verify that the percents shown are correct.

2.8 FURTHER PROBLEM SOLVING

OBJECTIVES

1. Solve problems involving distance.
2. Solve problems involving money.
3. Solve problems involving interest.

This section is devoted to solving problems in the categories listed. The same problem-solving steps used in previous sections are also followed in this section. They are listed below for review.

> **General Strategy for Problem Solving**
>
> 1. UNDERSTAND the problem. During this step, become comfortable with the problem. Some ways of doing this are:
>
>   Read and reread the problem.
>
>   Choose a variable to represent the unknown.
>
>   Construct a drawing, whenever possible.
>
>   Propose a solution and check. Pay careful attention to how you check your proposed solution. This will help writing an equation to model the problem.
> 2. TRANSLATE the problem into an equation.
> 3. SOLVE the equation.
> 4. INTERPRET the results: *Check* the proposed solution in the stated problem and *state* your conclusion.

OBJECTIVE 1 ▶ Solving distance problems. Our first example involves distance. For a review of the distance formula, $d = r \cdot t$, see Section 2.6, Example 1 and the table before the example.

EXAMPLE 1   Finding Time Given Rate and Distance

Marie Antonio, a bicycling enthusiast, rode her 21-speed at an average speed of 18 miles per hour on level roads and then slowed down to an average of 10 mph on the hilly roads of the trip. If she covered a distance of 98 miles, how long did the entire trip take if traveling the level roads took the same time as traveling the hilly roads?

Solution

1. UNDERSTAND the problem. To do so, read and reread the problem. The formula $d = r \cdot t$ is needed. At this time, let's guess a solution. Suppose that she spent 2 hours traveling on the level roads. This means that she also spent 2 hours traveling on the hilly roads, since the times spent were the same. What is her total distance? Her distance on the level road is rate · time = 18(2) = 36 miles. Her distance on the hilly roads is rate · time = 10(2) = 20 miles. This gives a total distance of 36 miles + 20 miles = 56 miles, not the correct distance of 98 miles. Remember that the purpose of guessing a solution is not to guess correctly (although this may

happen) but to help better understand the problem and how to model it with an equation. We are looking for the length of the entire trip, so we begin by letting

x = the time spent on level roads.

Because the same amount of time is spent on hilly roads, then also

x = the time spent on hilly roads.

2. **TRANSLATE.** To help us translate to an equation, we now summarize the information from the problem on the following chart. Fill in the rates given, the variables used to represent the times, and use the formula $d = r \cdot t$ to fill in the distance column.

	Rate ·	Time =	Distance
Level	18	x	$18x$
Hilly	10	x	$10x$

Since the entire trip covered 98 miles, we have that

In words: total distance = level distance + hilly distance

Translate: $98 = 18x + 10x$

3. **SOLVE.**

$$98 = 28x \quad \text{Add like terms.}$$
$$\frac{98}{28} = \frac{28x}{28} \quad \text{Divide both sides by 28.}$$
$$3.5 = x$$

4. **INTERPRET** the results.

Check: Recall that x represents the time spent on the level portion of the trip and also the time spent on the hilly portion. If Marie rides for 3.5 hours at 18 mph, her distance is $18(3.5) = 63$ miles. If Marie rides for 3.5 hours at 10 mph, her distance is $10(3.5) = 35$ miles. The total distance is 63 miles + 35 miles = 98 miles, the required distance.

State: The time of the entire trip is then 3.5 hours + 3.5 hours or 7 hours.

PRACTICE

1 Sat Tranh took a short hike with his friends up Mt. Wachusett. They hiked uphill at a steady pace of 1.5 miles per hour, and downhill at a rate of 4 miles per hour. If the time to climb the mountain took an hour more than the time to hike down, how long was the entire hike?

EXAMPLE 2 Finding Train Speeds

The Kansas City Southern Railway operates in 10 states and Mexico. Suppose two trains leave Neosho, Missouri, at the same time. One travels north and the other travels south at a speed that is 15 miles per hour faster. In 2 hours, the trains are 230 miles apart. Find the speed of each train.

Kansas City Southern Railway

140 CHAPTER 2 Equations, Inequalities, and Problem Solving

Solution

1. UNDERSTAND the problem. Read and reread the problem. Guess a solution and check. Let's let

$$x = \text{speed of train traveling north}$$

Because the train traveling south is 15 mph faster, we have

$$x + 15 = \text{speed of train traveling south}$$

2. TRANSLATE. Just as for Example 1, let's summarize our information on a chart. Use the formula $d = r \cdot t$ to fill in the distance column.

	r	$\cdot$	t	$=$	d
North Train	x		2		$2x$
South Train	$x + 15$		2		$2(x + 15)$

Since the total distance between the trains is 230 miles, we have

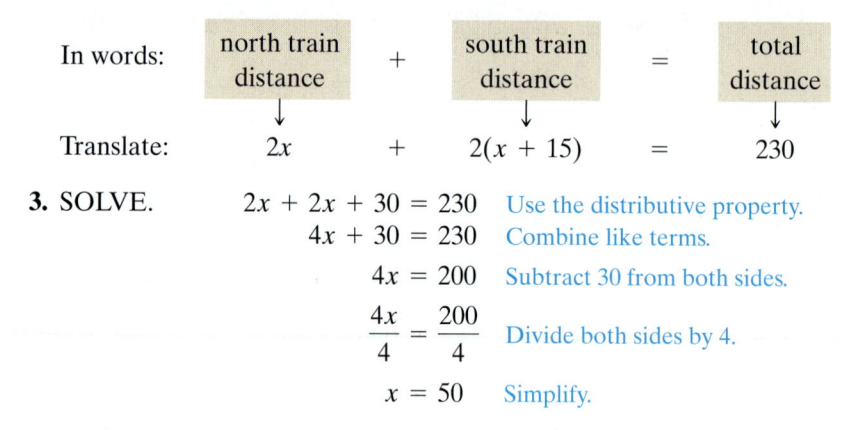

In words: north train distance + south train distance = total distance

Translate: $2x$ + $2(x + 15)$ = 230

3. SOLVE.

$$2x + 2x + 30 = 230 \quad \text{Use the distributive property.}$$
$$4x + 30 = 230 \quad \text{Combine like terms.}$$
$$4x = 200 \quad \text{Subtract 30 from both sides.}$$
$$\frac{4x}{4} = \frac{200}{4} \quad \text{Divide both sides by 4.}$$
$$x = 50 \quad \text{Simplify.}$$

4. INTERPRET the results.

Check: Recall that x is the speed of the train traveling north, or 50 mph. In 2 hours, this train travels a distance of $2(50) = 100$ miles. The speed of the train traveling south is $x + 15$ or $50 + 15 = 65$ mph. In 2 hours, this train travels $2(65) = 130$ miles. The total distance of the trains is 100 miles + 130 miles = 230 miles, the required distance.

State: The northbound train's speed is 50 mph and the southbound train's speed is 65 mph. □

PRACTICE

2 The Kansas City Southern Railway has a station in Mexico City, Mexico. Suppose two trains leave Mexico City at the same time. One travels east and the other west at a speed that is 10 mph slower. In 1.5 hours, the trains are 171 miles apart. Find the speed of each train.

OBJECTIVE 2 ▶ Solving money problems. The next example has to do with finding an unknown number of a certain denomination of coin or bill. These problems are extremely useful in that they help you understand the difference between the number of coins or bills and the total value of the money.

For example, suppose there are seven $5-bills. The *number* of $5-bills is 7 and the *total value* of the money is $5(7) = $35.

Study the table below for more examples.

Denomination or Coin or Bill	Number of Coins or Bills	Value of Coins or Bills
20-dollar bills	17	$20(17) = $340
nickels	31	$0.05(31) = $1.55
quarters	x	$0.25(x) = $0.25x

EXAMPLE 3 Finding Numbers of Denominations

Part of the proceeds from a local talent show was $2420 worth of $10 and $20 bills. If there were 37 more $20 bills than $10 bills, find the number of each denomination.

Solution

1. UNDERSTAND the problem. To do so, read and reread the problem. If you'd like, let's guess a solution. Suppose that there are 25 $10 bills. Since there are 37 more $20 bills, we have $25 + 37 = 62$ $20 bills. The total amount of money is $10(25) + $20(62) = $1490, below the given amount of $2420. Remember that our purpose for guessing is to help us better understand the problem.

   We are looking for the number of each denomination, so we let

   $$x = \text{number of } \$10 \text{ bills}$$

   There are 37 more $20 bills, so

   $$x + 37 = \text{number of } \$20 \text{ bills}$$

2. TRANSLATE. To help us translate to an equation, study the table below

Denomination	Number of Bills	Value of Bills (in dollars)
$10 bills	x	$10x$
$20 bills	$x + 37$	$20(x + 37)$

Since the total value of these bills is $2420, we have

In words: value of $10 bills plus value of $20 bills is 2420

Translate: $10x \quad + \quad 20(x + 37) \quad = \quad 2420$

3. SOLVE:

$$10x + 20x + 740 = 2420 \quad \text{Use the distributive property.}$$
$$30x + 740 = 2420 \quad \text{Add like terms.}$$
$$30x = 1680 \quad \text{Subtract 740 from both sides.}$$
$$\frac{30x}{30} = \frac{1680}{30} \quad \text{Divide both sides by 30.}$$
$$x = 56$$

4. INTERPRET the results.

Check: Since x represents the number of $10 bills, we have 56 $10 bills and $56 + 37$, or 93 $20 bills. The total amount of these bills is $10(56) + $20(93) = $2420, the correct total.

State: There are 56 $10 bills and 93 $20 bills.

PRACTICE 3 A stack of $5 and $20 bills was counted by the treasurer of an organization. The total value of the money was $1710 and there were 47 more $5 bills than $20 bills. Find the number of each type of bill.

OBJECTIVE 3 ▶ Solving interest problems. The next example is an investment problem. For a review of the simple interest formula, $I = PRT$, see the table at the beginning of Section 2.6 and also Exercises 11 and 12 in that exercise set.

EXAMPLE 4 Finding the Investment Amount

Rajiv Puri invested part of his $20,000 inheritance in a mutual funds account that pays 7% simple interest yearly and the rest in a certificate of deposit that pays 9% simple interest yearly. At the end of one year, Rajiv's investments earned $1550. Find the amount he invested at each rate.

Solution

1. UNDERSTAND. Read and reread the problem. Next, guess a solution. Suppose that Rajiv invested $8000 in the 7% fund and the rest, $12,000, in the fund paying 9%. To check, find his interest after one year. Recall the formula, $I = PRT$, so the interest from the 7% fund = $8000(0.07)(1) = $560. The interest from the 9% fund = $12,000(0.09)(1) = $1080. The sum of the interests is $560 + $1080 = $1640. Our guess is incorrect, since the sum of the interests is not $1550, but we now have a better understanding of the problem.

   Let

   x = amount of money in the account paying 7%.

   The rest of the money is $20,000 less x or

   $20{,}000 - x$ = amount of money in the account paying 9%.

2. TRANSLATE. We apply the simple interest formula $I = PRT$ and organize our information in the following chart. Since there are two different rates of interest and two different amounts invested, we apply the formula twice.

	Principal ·	Rate ·	Time =	Interest
7% Fund	x	0.07	1	$x(0.07)(1)$ or $0.07x$
9% Fund	$20{,}000 - x$	0.09	1	$(20{,}000 - x)(0.09)(1)$ or $0.09(20{,}000 - x)$
Total	20,000			1550

The total interest earned, $1550, is the sum of the interest earned at 7% and the interest earned at 9%.

In words: [interest at 7%] + [interest at 9%] = [total interest]

Translate: $0.07x$ + $0.09(20{,}000 - x)$ = 1550

3. SOLVE.

$$0.07x + 0.09(20{,}000 - x) = 1550$$
$$0.07x + 1800 - 0.09x = 1550 \quad \text{Apply the distributive property.}$$
$$1800 - 0.02x = 1550 \quad \text{Combine like terms.}$$
$$-0.02x = -250 \quad \text{Subtract 1800 from both sides.}$$
$$x = 12{,}500 \quad \text{Divide both sides by } -0.02.$$

4. INTERPRET.

Check: If $x = 12,500$, then $20,000 - x = 20,000 - 12,500$ or 7500. These solutions are reasonable, since their sum is $20,000 as required. The annual interest on $12,500 at 7% is $875; the annual interest on $7500 at 9% is $675, and $875 + $675 = $1550.

State: The amount invested at 7% is $12,500. The amount invested at 9% is $7500.

PRACTICE 4 Suzanne Scarpulla invested $30,000, part of it in a high-risk venture that yielded 11.5% per year, and the rest in a secure mutual fund paying interest of 6% per year. At the end of one year, Suzanne's investments earned $2790. Find the amount she invested at each rate.

2.8 EXERCISE SET

Solve. See Examples 1 and 2.

1. A jet plane traveling at 500 mph overtakes a propeller plane traveling at 200 mph that had a 2-hour head start. How far from the starting point are the planes?

2. How long will it take a bus traveling at 60 miles per hour to overtake a car traveling at 40 mph if the car had a 1.5-hour head start?

3. A bus traveled on a level road for 3 hours at an average speed 20 miles per hour faster than it traveled on a winding road. The time spent on the winding road was 4 hours. Find the average speed on the level road if the entire trip was 305 miles.

4. The Jones family drove to Disneyland at 50 miles per hour and returned on the same route at 40 mph. Find the distance to Disneyland if the total driving time was 7.2 hours.

Complete the table. The first and sixth rows have been completed for you. See Example 3.

		Number of Coins or Bills	Value of Coins or Bills (in dollars)
	pennies	x	$0.01x$
5.	dimes	y	
6.	quarters	z	
7.	nickels	$(x + 7)$	
8.	half-dollars	$(20 - z)$	
	$5 bills	$9x$	$5(9x)$
9.	$20 bills	$4y$	
10.	$100 bills	$97z$	
11.	$50 bills	$(35 - x)$	
12.	$10 bills	$(15 - y)$	

13. Part of the proceeds from a garage sale was $280 worth of $5 and $10 bills. If there were 20 more $5 bills than $10 bills, find the number of each denomination.

	Number of Bills	Value of Bills
$5 bills		
$10 bills		
Total		

14. A bank teller is counting $20 and $50-dollar bills. If there are six times as many $20 bills as $50 bills and the total amount of money is $3910, find the number of each denomination.

	Number of Bills	Value of Bills
$20 bills		
$50 bills		
Total		

Solve. See Example 4.

15. Zoya Lon invested part of her $25,000 advance at 8% annual simple interest and the rest at 9% annual simple interest. If her total yearly interest from both accounts was $2135, find the amount invested at each rate.

16. Karen Waugtal invested some money at 9% annual simple interest and $250 more than that amount at 10% annual simple interest. If her total yearly interest was $101, how much was invested at each rate?

17. Sam Mathius invested part of his $10,000 bonus in a fund that paid an 11% profit and invested the rest in stock that suffered a 4% loss. Find the amount of each investment if his overall net profit was $650.

18. Bruce Blossum invested a sum of money at 10% annual simple interest and invested twice that amount at 12% annual

simple interest. If his total yearly income from both investments was $2890, how much was invested at each rate?

19. The Concordia Theatre contains 500 seats and the ticket prices for a recent play were $43 for adults and $28 for children. For one matinee, if the total proceeds were $16,805, how many of each type of ticket were sold?

20. A zoo in Oklahoma charged $22 for adults and $15 for children. During a summer day, 732 zoo tickets were sold and the total receipts were $12,912. How many children and how many adult tickets were sold?

MIXED PRACTICE

21. How can $54,000 be invested, part at 8% annual simple interest and the remainder at 10% annual simple interest, so that the interest earned by the two accounts will be equal?

22. Ms. Mills invested her $20,000 bonus in two accounts. She took a 4% loss on one investment and made a 12% profit on another investment, but ended up breaking even. How much was invested in each account?

23. Alan and Dave Schaferkötter leave from the same point driving in opposite directions, Alan driving at 55 miles per hour and Dave at 65 mph. Alan has a one-hour head start. How long will they be able to talk on their car phones if the phones have a 250-mile range?

24. Kathleen and Cade Williams leave simultaneously from the same point hiking in opposite directions, Kathleen walking at 4 miles per hour and Cade at 5 mph. How long can they talk on their walkie-talkies if the walkie-talkies have a 20-mile radius?

25. A youth organization collected nickels and dimes for a charity drive. By the end of the 1-day drive, the youth had collected $56.35. If there were three times as many dimes as nickels, how many of each type of coin was collected?

26. A collection of dimes and quarters are retrieved from a soft drink machine. There are five times as many dimes as quarters and the total value of the coins is $27.75. Find the number of dimes and the number of quarters.

27. If $3000 is invested at 6% annual simple interest, how much should be invested at 9% annual simple interest so that the total yearly income from both investments is $585?

28. Trudy Waterbury, a financial planner, invested a certain amount of money at 9% annual simple interest, twice that amount at 10% annual simple interest, and three times that amount at 11% annual simple interest. Find the amount invested at each rate if her total yearly income from the investments was $2790.

29. Two hikers are 11 miles apart and walking toward each other. They meet in 2 hours. Find the rate of each hiker if one hiker walks 1.1 mph faster than the other.

30. Nedra and Latonya Dominguez are 12 miles apart hiking toward each other. How long will it take them to meet if Nedra walks at 3 mph and Latonya walks 1 mph faster?

31. Mark Martin can row upstream at 5 mph and downstream at 11 mph. If Mark starts rowing upstream until he gets tired and then rows downstream to his starting point, how far did Mark row if the entire trip took 4 hours?

32. On a 255-mile trip, Gary Alessandrini traveled at an average speed of 70 mph, got a speeding ticket, and then traveled at 60 mph for the remainder of the trip. If the entire trip took 4.5 hours and the speeding ticket stop took 30 minutes, how long did Gary speed before getting stopped?

REVIEW AND PREVIEW

Perform the indicated operations. See Sections 1.5 and 1.6.

33. $3 + (-7)$

34. $(-2) + (-8)$

35. $\dfrac{3}{4} - \dfrac{3}{16}$

36. $-11 + 2.9$

37. $-5 - (-1)$

38. $-12 - 3$

CONCEPT EXTENSIONS

39. A stack of $20, $50, and $100 bills was retrieved as part of an FBI investigation. There were 46 more $50 bills than $100 bills. Also, the number of $20 bills was 7 times the number of $100 bills. If the total value of the money was $9550, find the number of each type of bill.

40. A man places his pocket change in a jar every day. The jar is full and his children have counted the change. The total value is $44.86. Let x represent the number of quarters and use the information below to find the number of each type of coin.

    There are: 136 more dimes than quarters
    8 times as many nickels as quarters
    32 more than 16 times as many pennies as quarters

To "break even" in a manufacturing business, revenue R (income) must equal the cost C of production, or R = C.

41. The cost C to produce x number of skateboards is given by $C = 100 + 20x$. The skateboards are sold wholesale for $24 each, so revenue R is given by $R = 24x$. Find how many skateboards the manufacturer needs to produce and sell to break even. (*Hint:* Set the expression for R equal to the expression for C, then solve for x.)

42. The revenue R from selling x number of computer boards is given by $R = 60x$, and the cost C of producing them is given by $C = 50x + 5000$. Find how many boards must be sold to break even. Find how much money is needed to produce the break-even number of boards.

43. The cost C of producing x number of paperback books is given by $C = 4.50x + 2400$. Income R from these books is given by $R = 7.50x$. Find how many books should be produced and sold to break even.

44. Find the break-even quantity for a company that makes x number of computer monitors at a cost C given by $C = 870 + 70x$ and receives revenue R given by $R = 105x$.

45. Exercises 41 through 44 involve finding the break-even point for manufacturing. Discuss what happens if a company makes and sells fewer products than the break-even point. Discuss what happens if more products than the break-even point are made and sold.

STUDY SKILLS BUILDER

What to Do the Day of an Exam?

Your first exam may be soon. On the day of an exam, don't forget to try the following:

- Allow yourself plenty of time to arrive.
- Read the directions on the test carefully.
- Read each problem carefully as you take your test. Make sure that you answer the question asked.
- Watch your time and pace yourself so that you may attempt each problem on your test.
- Check your work and answers.
- **Do not turn your test in early.** If you have extra time, spend it double-checking your work.

Good luck!

Answer the following questions based on your most recent mathematics exam, whenever that was.

1. How soon before class did you arrive?
2. Did you read the directions on the test carefully?
3. Did you make sure you answered the question asked for each problem on the exam?
4. Were you able to attempt each problem on your exam?
5. If your answer to question 4 is no, list reasons why.
6. Did you have extra time on your exam?
7. If your answer to question 6 is yes, describe how you spent that extra time.

2.9 SOLVING LINEAR INEQUALITIES

OBJECTIVES

1. Define linear inequality in one variable, graph solution sets on a number line, and use interval notation.
2. Solve linear inequalities.
3. Solve compound inequalities.
4. Solve inequality applications.

OBJECTIVE 1 ▶ Graphing solution sets to linear inequalities and using interval notation.

In Chapter 1, we reviewed these inequality symbols and their meanings:

$<$ means "is less than" $\qquad$ $\leq$ means "is less than or equal to"

$>$ means "is greater than" $\qquad$ $\geq$ means "is greater than or equal to"

Equations	Inequalities
$x = 3$	$x \leq 3$
$5n - 6 = 14$	$5n - 6 > 14$
$12 = 7 - 3y$	$12 \leq 7 - 3y$
$\frac{x}{4} - 6 = 1$	$\frac{x}{4} - 6 > 1$

A linear inequality is similar to a linear equation except that the equality symbol is replaced with an inequality symbol.

Linear Inequality in One Variable

A **linear inequality in one variable** is an inequality that can be written in the form

$$ax + b < c$$

where a, b, and c are real numbers and a is not 0.

This definition and all other definitions, properties, and steps in this section also hold true for the inequality symbols, $>$, $\geq$, and $\leq$.

A **solution of an inequality** is a value of the variable that makes the inequality a true statement. The solution set is the set of all solutions. For the inequality $x < 3$, replacing x with any number less than 3, that is, to the left of 3 on a number line, makes the resulting inequality true. This means that any number less than 3 is a solution of the inequality $x < 3$.

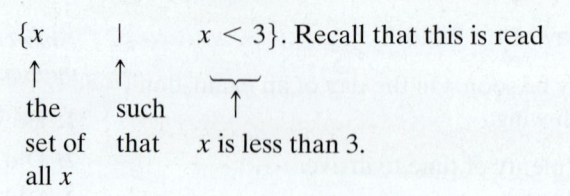

Since there are infinitely many such numbers, we cannot list all the solutions of the inequality. We *can* use set notation and write

$\{x \mid x < 3\}$. Recall that this is read

↑   ↑
the   such
set of   that   ↑
all x x is less than 3.

We can also picture the solutions on a number line. If we use open/closed-circle notation, the graph of $\{x \mid x < 3\}$ looks like the following.

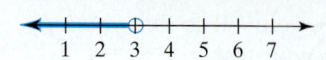

In this text, a convenient notation, called **interval notation,** will be used to write solution sets of inequalities. To help us understand this notation, a different graphing notation will be used. Instead of an open circle, we use a parenthesis; instead of a closed circle, we use a bracket. With this new notation, the graph of $\{x \mid x < 3\}$ now looks like

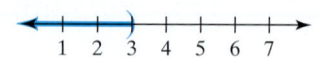

and can be represented in interval notation as $(-\infty, 3)$. The symbol $-\infty$, read as "negative infinity," does not indicate a number, but does indicate that the shaded arrow to the left never ends. In other words, the interval $(-\infty, 3)$ includes *all* numbers less than 3.

Picturing the solutions of an inequality on a number line is called **graphing** the solutions or graphing the inequality, and the picture is called the **graph** of the inequality.

To graph $\{x \mid x \leq 3\}$ or simply $x \leq 3$, shade the numbers to the left of 3 and place a bracket at 3 on the number line. The bracket indicates that 3 **is** a solution: 3 **is** less than or equal to 3. In interval notation, we write $(-\infty, 3]$.

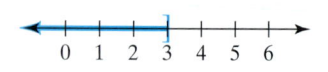

> ▶ **Helpful Hint**
>
> When writing an inequality in interval notation, it may be easier to first graph the inequality, then write it in interval notation. To help, think of the number line as approaching $-\infty$ to the left and $+\infty$ or ∞ to the right. Then simply write the interval notation by following your shading from left to right.
>
> $x > 5$ $x \leq -7$
>
> $(5, \infty)$ $(-\infty, -7]$

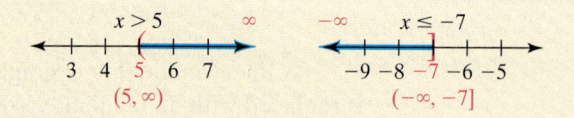

EXAMPLE 1   Graph $x \geq -1$. Then write the solutions in interval notation.

Solution   We place a bracket at -1 since the inequality symbol is $\geq$ and -1 is greater than or equal to -1. Then we shade to the right of -1.

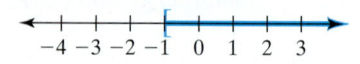

In interval notation, this is $[-1, \infty)$.

PRACTICE 1   Graph $x < 5$. Then write the solutions in interval notation.

OBJECTIVE 2 ▶ Solving linear inequalities.

When solutions of a linear inequality are not immediately obvious, they are found through a process similar to the one used to solve a linear equation. Our goal is to get the variable alone, and we use properties of inequality similar to properties of equality.

> **Addition Property of Inequality**
> If a, b, and c are real numbers, then
> $$a < b \quad \text{and} \quad a + c < b + c$$
> are equivalent inequalities.

This property also holds true for subtracting values, since subtraction is defined in terms of addition. In other words, adding or subtracting the same quantity from both sides of an inequality does not change the solution of the inequality.

EXAMPLE 2 Solve $x + 4 \leq -6$ for x. Graph the solution set and write it in interval notation.

Solution To solve for x, subtract 4 from both sides of the inequality.

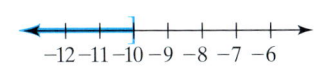

$$x + 4 \leq -6 \quad \text{Original inequality}$$
$$x + 4 - 4 \leq -6 - 4 \quad \text{Subtract 4 from both sides.}$$
$$x \leq -10 \quad \text{Simplify.}$$

The solution set is $(-\infty, -10]$.

PRACTICE 2 Solve $x + 11 \geq 6$. Graph the solution set and write it in interval notation.

> ▶ **Helpful Hint**
> Notice that any number less than or equal to -10 is a solution to $x \leq -10$. For example, solutions include
> $$-10, \; -200, \; -11\frac{1}{2}, \; -7\pi, \; -\sqrt{130}, \; -50.3$$

An important difference between linear equations and linear inequalities is shown when we multiply or divide both sides of an inequality by a nonzero real number. For example, start with the true statement $6 < 8$ and multiply both sides by 2. As we see below, the resulting inequality is also true.

$$6 < 8 \quad \text{True}$$
$$2(6) < 2(8) \quad \text{Multiply both sides by 2.}$$
$$12 < 16 \quad \text{True}$$

But if we start with the same true statement $6 < 8$ and multiply both sides by -2, the resulting inequality is not a true statement.

$$6 < 8 \quad \text{True}$$
$$-2(6) < -2(8) \quad \text{Multiply both sides by } -2.$$
$$-12 < -16 \quad \text{False}$$

Notice, however, that if we reverse the direction of the inequality symbol, the resulting inequality is true.

$$-12 < -16 \quad \text{False}$$
$$-12 > -16 \quad \text{True}$$

This demonstrates the multiplication property of inequality.

> **Multiplication Property of Inequality**
> 1. If a, b, and c are real numbers, and c is **positive,** then
> $$a < b \quad \text{and} \quad ac < bc$$
> are equivalent inequalities.
> 2. If a, b, and c are real numbers, and c is **negative,** then
> $$a < b \quad \text{and} \quad ac > bc$$
> are equivalent inequalities.

Because division is defined in terms of multiplication, this property also holds true when dividing both sides of an inequality by a nonzero number. If we multiply or divide both sides of an inequality by a negative number, **the direction of the inequality sign must be reversed for the inequalities to remain equivalent.**

> ▶ **Helpful Hint**
> Whenever both sides of an inequality are multiplied or divided by a negative number, the direction of the inequality symbol **must be** reversed to form an equivalent inequality.

EXAMPLE 3 Solve $-2x \leq -4$. Graph the solution set and write it in interval notation.

Solution Remember to reverse the direction of the inequality symbol when dividing by a negative number.

▶ **Helpful Hint**
Don't forget to reverse the direction of the inequality sign.

$$-2x \leq -4$$
$$\frac{-2x}{-2} \geq \frac{-4}{-2} \quad \text{Divide both sides by } -2 \text{ and reverse the direction of the inequality sign.}$$
$$x \geq 2 \quad \text{Simplify.}$$

The solution set $[2, \infty)$ is graphed as shown.

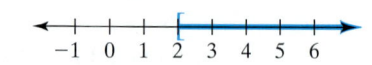

$$\longleftarrow \mid \; \mid \; \mid \; [\; \mid \; \mid \; \mid \; \mid \longrightarrow$$
$$ -1 \; 0 \; 1 \; 2 \; 3 \; 4 \; 5 \; 6$$

PRACTICE 3 Solve $-5x \geq -15$. Graph the solution set and write it in interval notation.

EXAMPLE 4 Solve $2x < -4$. Graph the solution set and write it in interval notation.

Solution

▶ **Helpful Hint**
Do not reverse the inequality sign.

$$2x < -4$$
$$\frac{2x}{2} < \frac{-4}{2} \quad \text{Divide both sides by 2.}$$
$$ \text{Do not reverse the direction of the inequality sign.}$$
$$x < -2 \quad \text{Simplify.}$$

The solution set $(-\infty, -2)$ is graphed as shown.

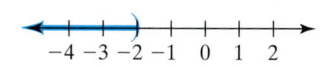

$$\longleftarrow \mid \; \mid \;) \; \mid \; \mid \; \mid \; \mid \longrightarrow$$
$$ -4 \; -3 \; -2 \; -1 \; 0 \; 1 \; 2$$

Section 2.9 Solving Linear Inequalities 149

PRACTICE 4 Solve $3x > -9$. Graph the solution set and write it in interval notation.

Concept Check ✓

Fill in the blank with $<$, $>$, $\leq$, or $\geq$.

a. Since $-8 < -4$, then $3(-8)$ _____ $3(-4)$.

b. Since $5 \geq -2$, then $\dfrac{5}{-7}$ _____ $\dfrac{-2}{-7}$.

c. If $a < b$, then $2a$ _____ $2b$.

d. If $a \geq b$, then $\dfrac{a}{-3}$ _____ $\dfrac{b}{-3}$.

The following steps may be helpful when solving inequalities. Notice that these steps are similar to the ones given in Section 2.4 for solving equations.

> **Solving Linear Inequalities in One Variable**
>
> **STEP 1.** Clear the inequality of fractions by multiplying both sides of the inequality by the lowest common denominator (LCD) of all fractions in the inequality.
>
> **STEP 2.** Remove grouping symbols such as parentheses by using the distributive property.
>
> **STEP 3.** Simplify each side of the inequality by combining like terms.
>
> **STEP 4.** Write the inequality with variable terms on one side and numbers on the other side by using the addition property of inequality.
>
> **STEP 5.** Get the variable alone by using the multiplication property of inequality.

▶ **Helpful Hint**

Don't forget that if both sides of an inequality are multiplied or divided by a negative number, the direction of the inequality sign must be reversed.

EXAMPLE 5 Solve $-4x + 7 \geq -9$. Graph the solution set and write it in interval notation.

Solution

$-4x + 7 \geq -9$

$-4x + 7 - 7 \geq -9 - 7$ Subtract 7 from both sides.

$-4x \geq -16$ Simplify.

$\dfrac{-4x}{-4} \leq \dfrac{-16}{-4}$ Divide both sides by -4 and reverse the direction of the inequality sign.

$x \leq 4$ Simplify.

The solution set $(-\infty, 4]$ is graphed as shown.

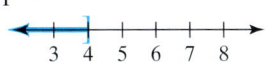

PRACTICE 5 Solve $45 - 7x \leq -4$. Graph the solution set and write it in interval notation.

Answers to Concept Check:
a. $<$ **b.** $\leq$ **c.** $<$ **d.** $\leq$

150 CHAPTER 2 Equations, Inequalities, and Problem Solving

EXAMPLE 6 Solve $2x + 7 \leq x - 11$. Graph the solution set and write it in interval notation.

Solution

$$2x + 7 \leq x - 11$$
$$2x + 7 - x \leq x - 11 - x \quad \text{Subtract } x \text{ from both sides.}$$
$$x + 7 \leq -11 \quad \text{Combine like terms.}$$
$$x + 7 - 7 \leq -11 - 7 \quad \text{Subtract 7 from both sides.}$$
$$x \leq -18 \quad \text{Combine like terms.}$$

The graph of the solution set $(-\infty, -18]$ is shown.

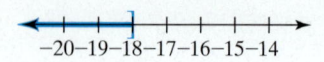

$$\overset{\longleftarrow}{\underset{-20\ -19\ -18\ -17\ -16\ -15\ -14}{\vphantom{|}}}$$

PRACTICE 6 Solve $3x + 20 \leq 2x + 13$. Graph the solution set and write it in interval notation.

EXAMPLE 7 Solve $-5x + 7 < 2(x - 3)$. Graph the solution set and write it in interval notation.

Solution

$$-5x + 7 < 2(x - 3)$$
$$-5x + 7 < 2x - 6 \quad \text{Apply the distributive property.}$$
$$-5x + 7 - 2x < 2x - 6 - 2x \quad \text{Subtract } 2x \text{ from both sides.}$$
$$-7x + 7 < -6 \quad \text{Combine like terms.}$$
$$-7x + 7 - 7 < -6 - 7 \quad \text{Subtract 7 from both sides.}$$
$$-7x < -13 \quad \text{Combine like terms.}$$
$$\frac{-7x}{-7} > \frac{-13}{-7} \quad \text{Divide both sides by } -7 \text{ and reverse the direction of the inequality sign.}$$
$$x > \frac{13}{7} \quad \text{Simplify.}$$

The graph of the solution set $\left(\frac{13}{7}, \infty\right)$ is shown.

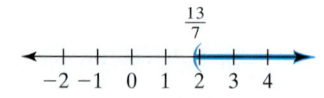

$$\overset{\longleftarrow}{\underset{-2\ -1\ \ 0\ \ 1\ \ 2\ \ 3\ \ 4}{\vphantom{|}}} \overset{\frac{13}{7}}{\ }$$

PRACTICE 7 Solve $6 - 5x > 3(x - 4)$. Graph the solution set and write it in interval notation.

EXAMPLE 8 Solve $2(x - 3) - 5 \leq 3(x + 2) - 18$. Graph the solution set and write it in interval notation.

Solution

$$2(x - 3) - 5 \leq 3(x + 2) - 18$$
$$2x - 6 - 5 \leq 3x + 6 - 18 \quad \text{Apply the distributive property.}$$
$$2x - 11 \leq 3x - 12 \quad \text{Combine like terms.}$$
$$-x - 11 \leq -12 \quad \text{Subtract } 3x \text{ from both sides.}$$
$$-x \leq -1 \quad \text{Add 11 to both sides.}$$

$$\frac{-x}{-1} \geq \frac{-1}{-1} \qquad \text{Divide both sides by } -1 \text{ and reverse the direction of the inequality sign.}$$

$$x \geq 1 \qquad \text{Simplify.}$$

The graph of the solution set $[1, \infty)$ is shown.

PRACTICE 8 Solve $3(x - 4) - 5 \leq 5(x - 1) - 12$. Graph the solution set and write it in interval notation.

OBJECTIVE 3 ▶ Solving compound inequalities. Inequalities containing one inequality symbol are called **simple inequalities,** while inequalities containing two inequality symbols are called **compound inequalities.** A compound inequality is really two simple inequalities in one. The compound inequality

$$3 < x < 5 \quad \text{means} \quad 3 < x \text{ and } x < 5$$

This can be read "x is greater than 3 and less than 5."

A solution of a compound inequality is a value that is a solution of both of the simple inequalities that make up the compound inequality. For example,

$$4\frac{1}{2} \text{ is a solution of } 3 < x < 5 \text{ since } 3 < 4\frac{1}{2} \text{ and } 4\frac{1}{2} < 5.$$

To graph $3 < x < 5$, place parentheses at both 3 and 5 and shade between.

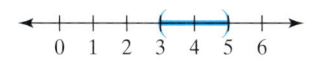

EXAMPLE 9 Graph $2 < x \leq 4$. Write the solutions in interval notation.

Solution Graph all numbers greater than 2 and less than or equal to 4. Place a parenthesis at 2, a bracket at 4, and shade between.

In interval notation, this is $(2, 4]$.

PRACTICE 9 Graph $-3 \leq x < 1$. Write the solutions in interval notation.

When we solve a simple inequality, we isolate the variable on one side of the inequality. When we solve a compound inequality, we isolate the variable in the middle part of the inequality. Also, when solving a compound inequality, we must perform the same operation to all **three** parts of the inequality: left, middle, and right.

EXAMPLE 10 Solve $-1 \leq 2x - 3 < 5$. Graph the solution set and write it in interval notation.

Solution

$$-1 \leq 2x - 3 < 5$$
$$-1 + 3 \leq 2x - 3 + 3 < 5 + 3 \qquad \text{Add 3 to all three parts.}$$
$$2 \leq 2x < 8 \qquad \text{Combine like terms.}$$
$$\frac{2}{2} \leq \frac{2x}{2} < \frac{8}{2} \qquad \text{Divide all three parts by 2.}$$
$$1 \leq x < 4 \qquad \text{Simplify.}$$

The graph of the solution set [1, 4) is shown.

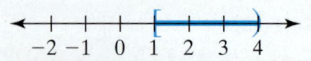

PRACTICE 10 Solve $-4 < 3x + 2 \leq 8$. Graph the solution set and write it in interval notation.

EXAMPLE 11 Solve $3 \leq \dfrac{3x}{2} + 4 \leq 5$. Graph the solution set and write it in interval notation.

Solution

$$3 \leq \dfrac{3x}{2} + 4 \leq 5$$

$$2(3) \leq 2\left(\dfrac{3x}{2} + 4\right) \leq 2(5) \quad \text{Multiply all three parts by 2 to clear the fraction.}$$

$$6 \leq 3x + 8 \leq 10 \quad \text{Distribute.}$$

$$-2 \leq 3x \leq 2 \quad \text{Subtract 8 from all three parts.}$$

$$\dfrac{-2}{3} \leq \dfrac{3x}{3} \leq \dfrac{2}{3} \quad \text{Divide all three parts by 3.}$$

$$-\dfrac{2}{3} \leq x \leq \dfrac{2}{3} \quad \text{Simplify.}$$

The graph of the solution set $\left[-\dfrac{2}{3}, \dfrac{2}{3}\right]$ is shown.

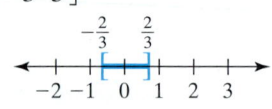

PRACTICE 11 Solve $1 < \dfrac{3}{4}x + 5 < 6$. Graph the solution set and write it in interval notation.

OBJECTIVE 4 ▶ Solving inequality applications. Problems containing words such as "at least," "at most," "between," "no more than," and "no less than" usually indicate that an inequality should be solved instead of an equation. In solving applications involving linear inequalities, use the same procedure you use to solve applications involving linear equations.

EXAMPLE 12 Staying within Budget

Marie Chase and Jonathan Edwards are having their wedding reception at the Gallery Reception Hall. They may spend at most $2000 for the reception. If the reception hall charges a $100 cleanup fee plus $36 per person, find the greatest number of people that they can invite and still stay within their budget.

Solution

1. UNDERSTAND. Read and reread the problem. Next, guess a solution. If 40 people attend the reception, the cost is $100 + $36(40) = $100 + $1440 = $1540. Let x = the number of people who attend the reception.

2. TRANSLATE.

In words:	cleanup fee	+	cost per person		must be less than or equal to		$2000
	↓		↓		↓		↓
Translate:	100	+	36x		≤		2000

3. SOLVE.

$$100 + 36x \leq 2000$$
$$36x \leq 1900 \quad \text{Subtract 100 from both sides.}$$
$$x \leq 52\frac{7}{9} \quad \text{Divide both sides by 36.}$$

4. INTERPRET.

Check: Since x represents the number of people, we round down to the nearest whole, or 52. Notice that if 52 people attend, the cost is

$$\$100 + \$36(52) = \$1972. \text{ If 53 people attend, the cost is}$$
$$\$100 + \$36(53) = \$2008, \text{ which is more than the given \$2000.}$$

State: Marie Chase and Jonathan Edwards can invite at most 52 people to the reception.

PRACTICE 12 Kasonga is eager to begin his education at his local community college. He has budgeted $1500 for college this semester. His local college charges a $300 matriculation fee and costs an average of $375 for tuition, fees, and books for each three-credit course. Find the greatest number of classes Kasonga can afford to take this semester.

VOCABULARY & READINESS CHECK

Use the choices below to fill in each blank.

    expression   inequality   equation

1. $6x - 7(x + 9)$ _____
2. $6x = 7(x + 9)$ _____
3. $6x < 7(x + 9)$ _____
4. $5y - 2 \geq -38$ _____
5. $\frac{9}{7} = \frac{x + 2}{14}$ _____
6. $\frac{9}{7} - \frac{x + 2}{14}$ _____

Decide which number listed is not a solution to each given inequality.

7. $x \geq -3; \quad -3, 0, -5, \pi$ _____
8. $x < 6; \quad -6, |-6|, 0, -3.2$ _____
9. $x < 4.01; \quad 4, -4.01, 4.1, -4.1$ _____
10. $x \geq -3; \quad -4, -3, -2, -(-2)$ _____

2.9 EXERCISE SET

Graph each set of numbers given in interval notation. Then write an inequality statement in x describing the numbers graphed.

1. $[2, \infty)$
2. $(-3, \infty)$
3. $(-\infty, -5)$
4. $(-\infty, 4]$

Graph each inequality on a number line. Then write the solutions in interval notation. See Example 1.

5. $x \leq -1$
6. $y < 0$
7. $x < \frac{1}{2}$
8. $z < -\frac{2}{3}$

9. $y \geq 5$

10. $x > 3$

Solve each inequality. Graph the solution set and write it in interval notation. See Examples 2 through 4.

11. $2x < -6$
12. $3x > -9$
13. $x - 2 \geq -7$
14. $x + 4 \leq 1$
15. $-8x \leq 16$
16. $-5x < 20$

Solve each inequality. Graph the solution set and write it in interval notation. See Examples 5 and 6.

17. $3x - 5 > 2x - 8$
18. $3 - 7x \geq 10 - 8x$
19. $4x - 1 \leq 5x - 2x$
20. $7x + 3 < 9x - 3x$

Solve each inequality. Graph the solution set and write it in interval notation. See Examples 7 and 8.

21. $x - 7 < 3(x + 1)$
22. $3x + 9 \leq 5(x - 1)$
23. $-6x + 2 \geq 2(5 - x)$
24. $-7x + 4 > 3(4 - x)$
25. $4(3x - 1) \leq 5(2x - 4)$
26. $3(5x - 4) \leq 4(3x - 2)$
27. $3(x + 2) - 6 > -2(x - 3) + 14$
28. $7(x - 2) + x \leq -4(5 - x) - 12$

MIXED PRACTICE

Solve the following inequalities. Graph each solution set and write it in interval notation.

29. $-2x \leq -40$
30. $-7x > 21$
31. $-9 + x > 7$
32. $y - 4 \leq 1$
33. $3x - 7 < 6x + 2$
34. $2x - 1 \geq 4x - 5$
35. $5x - 7x \geq x + 2$
36. $4 - x < 8x + 2x$

37. $\dfrac{3}{4}x > 2$
38. $\dfrac{5}{6}x \geq -8$
39. $3(x - 5) < 2(2x - 1)$
40. $5(x + 4) < 4(2x + 3)$
41. $4(2x + 1) < 4$
42. $6(2 - x) \geq 12$
43. $-5x + 4 \geq -4(x - 1)$
44. $-6x + 2 < -3(x + 4)$
45. $-2(x - 4) - 3x < -(4x + 1) + 2x$
46. $-5(1 - x) + x \leq -(6 - 2x) + 6$
47. $-3x + 6 \geq 2x + 6$
48. $-(x - 4) < 4$

49. Explain how solving a linear inequality is similar to solving a linear equation.

50. Explain how solving a linear inequality is different from solving a linear equation.

Graph each inequality. Then write the solutions in interval notation. See Example 9.

51. $-1 < x < 3$
52. $2 \leq y \leq 3$
53. $0 \leq y < 2$
54. $-1 \leq x \leq 4$

Solve each inequality. Graph the solution set and write it in interval notation. See Examples 10 and 11.

55. $-3 < 3x < 6$
56. $-5 < 2x < -2$
57. $2 \leq 3x - 10 \leq 5$
58. $4 \leq 5x - 6 \leq 19$
59. $-4 < 2(x - 3) \leq 4$
60. $0 < 4(x + 5) \leq 8$
61. $-2 < 3x - 5 < 7$
62. $1 < 4 + 2x \leq 7$
63. $-6 < 3(x - 2) \leq 8$
64. $-5 \leq 2(x + 4) < 8$

65. Explain how solving a linear inequality is different from solving a compound inequality.

66. Explain how solving a linear inequality is similar to solving a compound inequality.

Solve. See Example 12.

67. Six more than twice a number is greater than negative fourteen. Find all numbers that make this statement true.

68. Five times a number, increased by one, is less than or equal to ten. Find all such numbers.

69. Dennis and Nancy Wood are celebrating their 30th wedding anniversary by having a reception at Tiffany Oaks reception hall. They have budgeted $3000 for their reception. If the reception hall charges a $50.00 cleanup fee plus $34 per person, find the greatest number of people that they may invite and still stay within their budget.

70. A surprise retirement party is being planned for Pratep Puri. A total of $860 has been collected for the event, which is to be held at a local reception hall. This reception hall charges a cleanup fee of $40 and $15 per person for drinks and light snacks. Find the greatest number of people that may be invited and still stay within $860.

71. Find the values for x so that the perimeter of this rectangle is no greater than 100 centimeters.

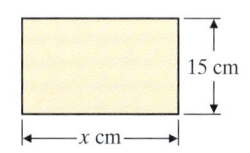

72. Find the values for x so that the perimeter of this triangle is no longer than 87 inches.

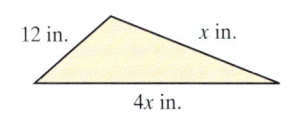

73. A financial planner has a client with $15,000 to invest. If he invests $10,000 in a certificate of deposit paying 11% annual simple interest, at what rate does the remainder of the money need to be invested so that the two investments together yield at least $1600 in yearly interest?

74. Alex earns $600 per month plus 4% of all his sales over $1000. Find the minimum sales that will allow Alex to earn at least $3000 per month.

75. Ben Holladay bowled 146 and 201 in his first two games. What must he bowl in his third game to have an average of at least 180?

76. On an NBA team the two forwards measure 6'8" and 6'6" and the two guards measure 6'0" and 5'9" tall. How tall a center should they hire if they wish to have a starting team average height of at least 6'5"?

77. High blood cholesterol levels increase the risk of heart disease in adults. Doctors recommend that total blood cholesterol be less than 200 milligrams per deciliter. Total cholesterol levels from 200 up to 240 milligrams per deciliter are considered borderline. Any total cholesterol reading above 240 milligrams per deciliter is considered high.

Letting x represent a patient's total blood cholesterol level, write a series of three inequalities that describe the ranges corresponding to recommended, borderline, and high levels of total blood cholesterol.

78. In 1971, T. Theodore Fujita created the Fujita Scale (or F Scale), which uses ratings from 0 to 5 to classify tornadoes, based on the damage that wind intensity causes to structures as the tornado passes through. This scale was updated by meteorologists and wind engineers working for the National Oceanic and Atmospheric Administration (NOAA) and implemented in February 2007. This new scale is called the Enhanced F Scale. An EF-0 tornado has wind speeds between 65 and 85 mph, inclusive. The winds in an EF-1 tornado range from 86 to 110 mph. In an EF-2 tornado, winds are from 111 to 135 mph. An EF-3 tornado has wind speeds ranging from 136 to 165 mph. Wind speeds in an EF-4 tornado are clocked at 166 to 200 mph. The most violent tornadoes are ranked at EF-5, with wind speeds of at least 201 mph. (*Source:* Storm Prediction Center, NOAA)

Letting y represent a tornado's wind speed, write a series of six inequalities that describe the wind speed ranges corresponding to each Enhanced Fujita Scale rank.

79. Twice a number, increased by one, is between negative five and seven. Find all such numbers.

80. Half a number, decreased by four, is between two and three. Find all such numbers.

81. The temperatures in Ohio range from $-39°C$ to $45°C$. Use a compound inequality to convert these temperatures to Fahrenheit temperatures. (*Hint:* Use $C = \frac{5}{9}(F - 32)$.)

82. Mario Lipco has scores of 85, 95, and 92 on his algebra tests. Use a compound inequality to find the range of scores he can make on his final exam in order to receive an A in the course. The final exam counts as three tests, and an A is received if the final course average is from 90 to 100. (*Hint:* The average of a list of numbers is their sum divided by the number of numbers in the list.)

REVIEW AND PREVIEW

Evaluate the following. See Section 1.4.

83. $(2)^3$

84. $(3)^3$

85. $(1)^{12}$

86. 0^5

87. $\left(\frac{4}{7}\right)^2$

88. $\left(\frac{2}{3}\right)^3$

156 CHAPTER 2 Equations, Inequalities, and Problem Solving

This broken line graph shows the average annual per person expenditure on newspapers for the given years. Use this graph for Exercises 89 through 92. See Section 3.1. (Source: Veronis Suhler Stevenson)

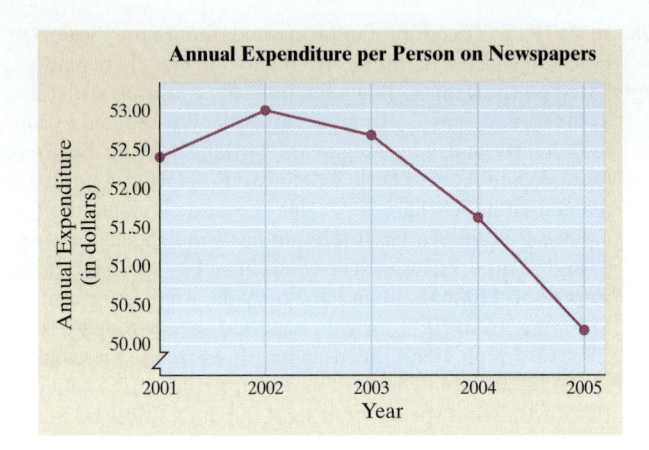

Annual Expenditure per Person on Newspapers

89. What was the average per person expenditure on newspapers in 2003?

90. What was the average per person expenditure on newspapers in 2005?

91. What year had the greatest drop in newspaper expenditures?

92. What years had per person newspaper expenditures over $52?

CONCEPT EXTENSIONS

93. The formula $C = 3.14d$ can be used to approximate the circumference of a circle given its diameter. Waldo Manufacturing manufactures and sells a certain washer with an outside circumference of 3 centimeters. The company has decided that a washer whose actual circumference is in the interval $2.9 \leq C \leq 3.1$ centimeters is acceptable. Use a compound inequality and find the corresponding interval for diameters of these washers. (Round to 3 decimal places.)

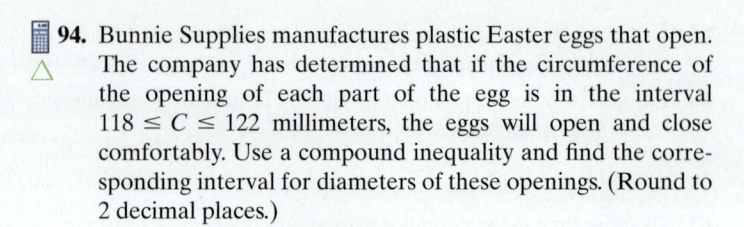

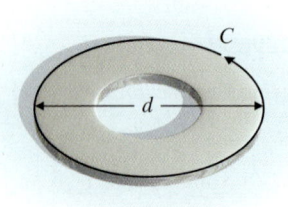

94. Bunnie Supplies manufactures plastic Easter eggs that open. The company has determined that if the circumference of the opening of each part of the egg is in the interval $118 \leq C \leq 122$ millimeters, the eggs will open and close comfortably. Use a compound inequality and find the corresponding interval for diameters of these openings. (Round to 2 decimal places.)

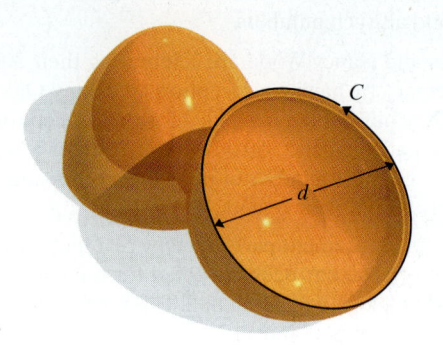

For Exercises 95 through 98, see the example below.

Solve $x(x - 6) > x^2 - 5x + 6$. Graph the solution set and write it in interval notation.

Solution

$$x(x - 6) > x^2 - 5x + 6$$
$$x^2 - 6x > x^2 - 5x + 6$$
$$x^2 - 6x - x^2 > x^2 - 5x + 6 - x^2$$
$$-6x > -5x + 6$$
$$-x > 6$$
$$\frac{-x}{-1} < \frac{6}{-1}$$
$$x < -6$$

The solution set $(-\infty, -6)$ is graphed as shown.

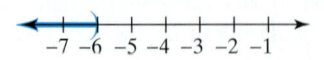

Solve each inequality. Graph the solution set and write it in interval notation.

95. $x(x + 4) > x^2 - 2x + 6$

96. $x(x - 3) \geq x^2 - 5x - 8$

97. $x^2 + 6x - 10 < x(x - 10)$

98. $x^2 - 4x + 8 < x(x + 8)$

THE BIGGER PICTURE   SIMPLIFYING EXPRESSIONS AND SOLVING EQUATIONS

Now we continue our outline started in Section 1.7. Although suggestions are given, this outline should be in your own words. Once you complete this new portion, try the exercises below.

I. Simplifying Expressions
   A. Real Numbers
      1. Add (Section 1.5)
      2. Subtract (Section 1.6)
      3. Multiply or Divide (Section 1.7)

II. Solving Equations and Inequalities
   A. **Linear Equations:** power on variable is 1 and there are no variables in the denominator

$7(x - 3) = 4x + 6$	Linear equation. Simplify both sides, then get variable terms on one side, numbers on the other side.
$7x - 21 = 4x + 6$	Use the distributive property.
$7x = 4x + 27$	Add 21 to both sides.
$3x = 27$	Subtract $4x$ from both sides.
$x = 9$	Divide both sides by 3.

   B. **Linear Inequalities:** same as linear equation, except there are inequality symbols, $\leq, <, \geq, >$ Remember, if you multiply or divide by a negative number, then reverse the direction of the inequality symbol.

$-4x - 11 \leq 1$	Linear inequality.
$-4x \leq 12$	Add 11 to both sides.
$\dfrac{-4x}{-4} \geq \dfrac{12}{-4}$	Divide both sides by -4 and reverse the direction of the inequality symbol.
$x \geq -3$	Simplify.

Solve each equation or inequality.

1. $-5x = 15$
2. $-5x > 15$
3. $9y - 14 = -12$
4. $9x - 3 = 5x - 4$
5. $4(x - 2) \leq 5x + 7$
6. $5(4x - 1) = 2(10x - 1)$
7. $-5.4 = 0.6x - 9.6$
8. $\dfrac{1}{3}(x - 4) < \dfrac{1}{4}(x + 7)$
9. $3y - 5(y - 4) = -2(y - 10)$
10. $\dfrac{7(x - 1)}{3} = \dfrac{2(x + 1)}{5}$

CHAPTER 2 GROUP ACTIVITY

Investigating Averages

Sections 2.1–2.9
Materials:
- small rubber ball or crumpled paper ball
- bucket or waste can

This activity may be completed by working in groups or individually.

1. Try shooting the ball into the bucket or waste can 5 times. Record your results below.

   Shots Made   Shots Missed

2. Find your shooting percent for the 5 shots (that is, the percent of the shots you actually made out of the number you tried).

3. Suppose you are going to try an additional 5 shots. How many of the next 5 shots will you have to make to have a 50% shooting percent for all 10 shots? An 80% shooting percent?

4. Did you solve an equation in Question 3? If so, explain what you did. If not, explain how you could use an equation to find the answers.

5. Now suppose you are going to try an additional 22 shots. How many of the next 22 shots will you have to make to have at least a 50% shooting percent for all 27 shots? At least a 70% shooting percent?

6. Choose one of the sports played at your college that is currently in season. How many regular-season games are scheduled? What is the team's current percent of games won?

7. Suppose the team has a goal of finishing the season with a winning percent better than 110% of their current wins. At least how many of the remaining games must they win to achieve their goal?

CHAPTER 2 VOCABULARY CHECK

Fill in each blank with one of the words or phrases listed below.

like terms
equivalent equations
linear equation in one variable
numerical coefficient
formula
linear inequality in one variable
compound inequalities

1. Terms with the same variables raised to exactly the same powers are called _____.
2. A _____ can be written in the form $ax + b = c$.
3. Equations that have the same solution are called _____.
4. Inequalities containing two inequality symbols are called _____.
5. An equation that describes a known relationship among quantities is called a _____.
6. A _____ can be written in the form $ax + b < c$, (or $>$, $\leq$, $\geq$).
7. The _____ of a term is its numerical factor.

> **Helpful Hint**
> Are you preparing for your test? Don't forget to take the Chapter 2 Test on page 166. Then check your answers at the back of the text and use the Chapter Test Prep Video CD to see the fully worked-out solutions to any of the exercises you want to review.

CHAPTER 2 HIGHLIGHTS

DEFINITIONS AND CONCEPTS	EXAMPLES

SECTION 2.1 SIMPLIFYING ALGEBRAIC EXPRESSIONS

DEFINITIONS AND CONCEPTS	EXAMPLES
The **numerical coefficient** of a **term** is its numerical factor.	**Term** **Numerical Coefficient** $-7y$ -7 x 1 $\frac{1}{5}a^2b$ $\frac{1}{5}$
Terms with the same variables raised to exactly the same powers are **like terms**.	**Like Terms** **Unlike Terms** $12x, -x$ $3y, 3y^2$ $-2xy, 5yx$ $7a^2b, -2ab^2$
To combine like terms, add the numerical coefficients and multiply the result by the common variable factor.	$9y + 3y = 12y$ $-4z^2 + 5z^2 - 6z^2 = -5z^2$
To remove parentheses, apply the distributive property.	$-4(x + 7) + 10(3x - 1)$ $= -4x - 28 + 30x - 10$ $= 26x - 38$

DEFINITIONS AND CONCEPTS	EXAMPLES
SECTION 2.2 THE ADDITION PROPERTY OF EQUALITY	
A **linear equation in one variable** can be written in the form $ax + b = c$ where $a, b,$ and c are real numbers and $a \neq 0$.	**Linear Equations** $$-3x + 7 = 2$$ $$3(x - 1) = -8(x + 5) + 4$$
Equivalent equations are equations that have the same solution.	$x - 7 = 10$ and $x = 17$ are equivalent equations.
Addition Property of Equality Adding the same number to or subtracting the same number from both sides of an equation does not change its solution.	$$y + 9 = 3$$ $$y + 9 - 9 = 3 - 9$$ $$y = -6$$
SECTION 2.3 THE MULTIPLICATION PROPERTY OF EQUALITY	
Multiplication Property of Equality Multiplying both sides or dividing both sides of an equation by the same nonzero number does not change its solution.	$$\frac{2}{3}a = 18$$ $$\frac{3}{2}\left(\frac{2}{3}a\right) = \frac{3}{2}(18)$$ $$a = 27$$
SECTION 2.4 SOLVING LINEAR EQUATIONS	
To Solve Linear Equations	Solve: $\dfrac{5(-2x + 9)}{6} + 3 = \dfrac{1}{2}$
1. Clear the equation of fractions.	1. $6 \cdot \dfrac{5(-2x + 9)}{6} + 6 \cdot 3 = 6 \cdot \dfrac{1}{2}$ $$5(-2x + 9) + 18 = 3$$
2. Remove any grouping symbols such as parentheses.	2. $-10x + 45 + 18 = 3$ Distributive property
3. Simplify each side by combining like terms.	3. $-10x + 63 = 3$ Combine like terms.
4. Write variable terms on one side and numbers on the other side using the addition property of equality.	4. $-10x + 63 - 63 = 3 - 63$ Subtract 63. $$-10x = -60$$
5. Get the variable alone using the multiplication property of equality.	5. $\dfrac{-10x}{-10} = \dfrac{-60}{-10}$ Divide by -10. $$x = 6$$
6. Check by substituting in the original equation.	6. $\dfrac{5(-2x + 9)}{6} + 3 = \dfrac{1}{2}$ $$\dfrac{5(-2 \cdot 6 + 9)}{6} + 3 \stackrel{?}{=} \dfrac{1}{2}$$ $$\dfrac{5(-3)}{6} + 3 \stackrel{?}{=} \dfrac{1}{2}$$ $$-\dfrac{5}{2} + \dfrac{6}{2} \stackrel{?}{=} \dfrac{1}{2}$$ $$\dfrac{1}{2} = \dfrac{1}{2} \quad \text{True}$$

DEFINITIONS AND CONCEPTS

EXAMPLES

SECTION 2.5 AN INTRODUCTION TO PROBLEM SOLVING

Problem-Solving Steps

The height of the Hudson volcano in Chili is twice the height of the Kiska volcano in the Aleutian Islands. If the sum of their heights is 12,870 feet, find the height of each.

1. UNDERSTAND the problem.

1. Read and reread the problem. Guess a solution and check your guess.
    Let x be the height of the Kiska volcano. Then $2x$ is the height of the Hudson volcano.

    x — Kiska $2x$ — Hudson

2. TRANSLATE the problem.

2. In words: height of Kiska + height of Hudson is 12,870

    Translate: $x + 2x = 12{,}870$

3. SOLVE.

3.
$$x + 2x = 12{,}870$$
$$3x = 12{,}870$$
$$x = 4290$$

4. INTERPRET the results.

4. *Check:* If x is 4290 then $2x$ is $2(4290)$ or 8580. Their sum is $4290 + 8580$ or 12,870, the required amount.

    State: Kiska volcano is 4290 feet high and Hudson volcano is 8580 feet high.

SECTION 2.6 FORMULAS AND PROBLEM SOLVING

An equation that describes a known relationship among quantities is called a **formula**.

To solve a formula for a specified variable, use the same steps as for solving a linear equation. Treat the specified variable as the only variable of the equation.

Formulas

$A = lw$ (area of a rectangle)
$I = PRT$ (simple interest)

Solve: $P = 2l + 2w$ for l.
$$P = 2l + 2w$$
$$P - 2w = 2l + 2w - 2w \quad \text{Subtract } 2w.$$
$$P - 2w = 2l$$
$$\frac{P - 2w}{2} = \frac{2l}{2} \quad \text{Divide by 2.}$$
$$\frac{P - 2w}{2} = l \quad \text{Simplify.}$$

If all values for the variables in a formula are known except for one, this unknown value may be found by substituting in the known values and solving.

If $d = 182$ miles and $r = 52$ miles per hour in the formula $d = r \cdot t$, find t.

$$d = r \cdot t$$
$$182 = 52 \cdot t \quad \text{Let } d = 182 \text{ and } r = 52.$$
$$3.5 = t$$

The time is 3.5 hours.

DEFINITIONS AND CONCEPTS	EXAMPLES
SECTION 2.7 PERCENT AND MIXTURE PROBLEM SOLVING	

Use the same problem-solving steps to solve a problem containing percents.

32% of what number is 36.8?

1. UNDERSTAND.

1. Read and reread. Propose a solution and check. Let x = the unknown number.

2. TRANSLATE.

2. 32%   of   what number   is   36.8
   ↓   ↓   ↓   ↓   ↓
   32%   · x =   36.8

3. SOLVE.

3. Solve: $32\% \cdot x = 36.8$
$$0.32x = 36.8$$
$$\frac{0.32x}{0.32} = \frac{36.8}{0.32} \quad \text{Divide by 0.32.}$$
$$x = 115 \quad \text{Simplify.}$$

4. INTERPRET.

4. Check, then state: 32% of 115 is 36.8.

How many liters of a 20% acid solution must be mixed with a 50% acid solution in order to obtain 12 liters of a 30% solution?

1. UNDERSTAND.

1. Read and reread. Guess a solution and check.
Let x = number of liters of 20% solution.
Then $12 - x$ = number of liters of 50% solution.

2. TRANSLATE.

2.

	No. of Liters	·	Acid Strength	=	Amount of Acid
20% Solution	x		20%		$0.20x$
50% Solution	$12 - x$		50%		$0.50(12 - x)$
30% Solution Needed	12		30%		$0.30(12)$

In words:

acid in 20% solution + acid in 50% solution = acid in 30% solution

Translate: $0.20x + 0.50(12 - x) = 0.30(12)$

3. SOLVE.

3. Solve: $0.20x + 0.50(12 - x) = 0.30(12)$
$$0.20x + 6 - 0.50x = 3.6 \quad \text{Apply the distributive property.}$$
$$-0.30x + 6 = 3.6$$
$$-0.30x = -2.4 \quad \text{Subtract 6.}$$
$$x = 8 \quad \text{Divide by } -0.30.$$

4. INTERPRET.

4. Check, then state:
If 8 liters of a 20% acid solution are mixed with $12 - 8$ or 4 liters of a 50% acid solution, the result is 12 liters of a 30% solution.

162 CHAPTER 2 Equations, Inequalities, and Problem Solving

DEFINITIONS AND CONCEPTS	EXAMPLES
SECTION 2.8 FURTHER PROBLEM SOLVING	
Problem-Solving Steps	A collection of dimes and quarters has a total value of $19.55. If there are three times as many quarters as dimes, find the number of quarters.
1. UNDERSTAND.	1. Read and reread. Propose a solution and check. Let x = number of dimes and $3x$ = number of quarters.
2. TRANSLATE.	2. In words: value of dimes + value of quarters = 19.55 Translate: $0.10x + 0.25(3x) = 19.55$
3. SOLVE.	3. Solve: $0.10x + 0.75x = 19.55$ Multiply. $0.85x = 19.55$ Add like terms. $x = 23$ Divide by 0.85.
4. INTERPRET.	4. Check, then state. The number of dimes is 23 and the number of quarters is 3(23) or 69. The total value of this money is $0.10(23) + 0.25(69) = 19.55$, so our result checks. The number of quarters is 69.
SECTION 2.9 SOLVING LINEAR INEQUALITIES	
A **linear inequality in one variable** is an inequality that can be written in one of the forms: $ax + b < c$ $ax + b \le c$ $ax + b > c$ $ax + b \ge c$ where $a, b,$ and c are real numbers and a is not 0.	**Linear Inequalities** $2x + 3 < 6$ $5(x - 6) \ge 10$ $\dfrac{x - 2}{5} > \dfrac{5x + 7}{2}$ $\dfrac{-(x + 8)}{9} \le \dfrac{-2x}{11}$
Addition Property of Inequality Adding the same number to or subtracting the same number from both sides of an inequality does not change the solutions.	$y + 4 \le -1$ $y + 4 - 4 \le -1 - 4$ Subtract 4. $y \le -5$
Multiplication Property of Inequality Multiplying or dividing both sides of an inequality by the same positive number does not change its solutions.	$\dfrac{1}{3}x > -2$ $3\left(\dfrac{1}{3}x\right) > 3 \cdot -2$ Multiply by 3. $x > -6$
Multiplying or dividing both sides of an inequality by the same **negative number and reversing the direction of the inequality sign** does not change its solutions.	$-2x \le 4$ $\dfrac{-2x}{-2} \ge \dfrac{4}{-2}$ Divide by -2, reverse inequality sign. $x \ge -2$

DEFINITIONS AND CONCEPTS	EXAMPLES

SECTION 2.9 SOLVING LINEAR INEQUALITIES (continued)

To Solve Linear Inequalities
1. Clear the equation of fractions.
2. Remove grouping symbols.
3. Simplify each side by combining like terms.
4. Write variable terms on one side and numbers on the other side using the addition property of inequality.
5. Get the variable alone using the multiplication property of inequality.
	Solve: $3(x + 2) \leq -2 + 8$
	1. No fractions to clear. $3(x + 2) \leq -2 + 8$
	2. $3x + 6 \leq -2 + 8$ Distributive property
	3. $3x + 6 \leq 6$ Combine like terms.
	4. $3x + 6 - 6 \leq 6 - 6$ Subtract 6.
	    $3x \leq 0$
	5. $\dfrac{3x}{3} \leq \dfrac{0}{3}$ Divide by 3.
	    $x \leq 0$

Inequalities containing two inequality symbols are called **compound inequalities.**	**Compound Inequalities** $-2 < x < 6$ $5 \leq 3(x - 6) < \dfrac{20}{3}$
To solve a compound inequality, isolate the variable in the middle part of the inequality. Perform the same operation to all three parts of the inequality: left, middle, right.	Solve: $-2 < 3x + 1 < 7$ $-2 - 1 < 3x + 1 - 1 < 7 - 1$ Subtract 1. $-3 < 3x < 6$ $\dfrac{-3}{3} < \dfrac{3x}{3} < \dfrac{6}{3}$ Divide by 3. $-1 < x < 2$

CHAPTER 2 REVIEW

(2.1) *Simplify the following expressions.*

1. $5x - x + 2x$
2. $0.2z - 4.6x - 7.4z$
3. $\dfrac{1}{2}x + 3 + \dfrac{7}{2}x - 5$
4. $\dfrac{4}{5}y + 1 + \dfrac{6}{5}y + 2$
5. $2(n - 4) + n - 10$
6. $3(w + 2) - (12 - w)$
7. Subtract $7x - 2$ from $x + 5$.
8. Subtract $1.4y - 3$ from $y - 0.7$.

Write each of the following as algebraic expressions.

9. Three times a number decreased by 7
10. Twice the sum of a number and 2.8 added to 3 times a number

(2.2) *Solve each equation.*

11. $8x + 4 = 9x$
12. $5y - 3 = 6y$
13. $\dfrac{2}{7}x + \dfrac{5}{7}x = 6$
14. $3x - 5 = 4x + 1$
15. $2x - 6 = x - 6$
16. $4(x + 3) = 3(1 + x)$
17. $6(3 + n) = 5(n - 1)$
18. $5(2 + x) - 3(3x + 2) = -5(x - 6) + 2$

Use the addition property to fill in the blank so that the middle equation simplifies to the last equation.

19. $\quad x - 5 = 3$
$x - 5 + \underline{} = 3 + \underline{}$
$x = 8$

20. $\quad x + 9 = -2$
$x + 9 - \underline{} = -2 - \underline{}$
$x = -11$

Choose the correct algebraic expression.

21. The sum of two numbers is 10. If one number is x, express the other number in terms of x.
    a. $x - 10$ b. $10 - x$
    c. $10 + x$ d. $10x$

22. Mandy is 5 inches taller than Melissa. If x inches represents the height of Mandy, express Melissa's height in terms of x.
    a. $x - 5$ b. $5 - x$
    c. $5 + x$ d. $5x$

△ 23. If one angle measures $x°$, express the measure of its complement in terms of x.
   a. $(180 - x)°$
   b. $(90 - x)°$
   c. $(x - 180)°$
   d. $(x - 90)°$

△ 24. If one angle measures $(x + 5)°$, express the measure of its supplement in terms of x.
   a. $(185 + x)°$
   b. $(95 + x)°$
   c. $(175 - x)°$
   d. $(x - 170)°$

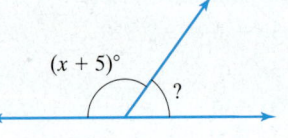

(2.3) Solve each equation.

25. $\frac{3}{4}x = -9$
26. $\frac{x}{6} = \frac{2}{3}$
27. $-5x = 0$
28. $-y = 7$
29. $0.2x = 0.15$
30. $\frac{-x}{3} = 1$
31. $-3x + 1 = 19$
32. $5x + 25 = 20$
33. $7(x - 1) + 9 = 5x$
34. $7x - 6 = 5x - 3$
35. $-5x + \frac{3}{7} = \frac{10}{7}$
36. $5x + x = 9 + 4x - 1 + 6$

37. Write the sum of three consecutive integers as an expression in x. Let x be the first integer.

38. Write the sum of the first and fourth of four consecutive even integers. Let x be the first even integer.

(2.4) Solve each equation.

39. $\frac{5}{3}x + 4 = \frac{2}{3}x$
40. $\frac{7}{8}x + 1 = \frac{5}{8}x$
41. $-(5x + 1) = -7x + 3$
42. $-4(2x + 1) = -5x + 5$
43. $-6(2x - 5) = -3(9 + 4x)$
44. $3(8y - 1) = 6(5 + 4y)$
45. $\frac{3(2 - z)}{5} = z$
46. $\frac{4(n + 2)}{5} = -n$
47. $0.5(2n - 3) - 0.1 = 0.4(6 + 2n)$
48. $-9 - 5a = 3(6a - 1)$
49. $\frac{5(c + 1)}{6} = 2c - 3$
50. $\frac{2(8 - a)}{3} = 4 - 4a$
51. $200(70x - 3560) = -179(150x - 19,300)$
52. $1.72y - 0.04y = 0.42$

(2.5) Solve each of the following.

53. The height of the Washington Monument is 50.5 inches more than 10 times the length of a side of its square base. If the sum of these two dimensions is 7327 inches, find the height of the Washington Monument. (*Source:* National Park Service)

54. A 12-foot board is to be divided into two pieces so that one piece is twice as long as the other. If x represents the length of the shorter piece, find the length of each piece.

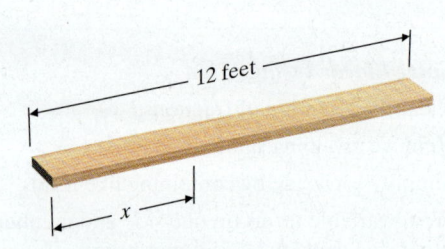

55. In a recent year, Kellogg Company acquired Keebler Foods Company. After the merger, the total number of Kellogg and Keebler manufacturing plants was 53. The number of Kellogg plants was one less than twice the number of Keebler plants. How many of each type of plant were there? (*Source: Kellogg Company 2000 Annual Report*)

56. Find three consecutive integers whose sum is -114.

57. The quotient of a number and 3 is the same as the difference of the number and two. Find the number.

58. Double the sum of a number and 6 is the opposite of the number. Find the number.

(2.6) Substitute the given values into the given formulas and solve for the unknown variable.

59. $P = 2l + 2w$; $P = 46, l = 14$
60. $V = lwh$; $V = 192, l = 8, w = 6$

Solve each equation for the indicated variable.

61. $y = mx + b$ for m
62. $r = vst - 5$ for s
63. $2y - 5x = 7$ for x
64. $3x - 6y = -2$ for y
△ 65. $C = \pi D$ for π
△ 66. $C = 2\pi r$ for π
△ 67. A swimming pool holds 900 cubic meters of water. If its length is 20 meters and its height is 3 meters, find its width.

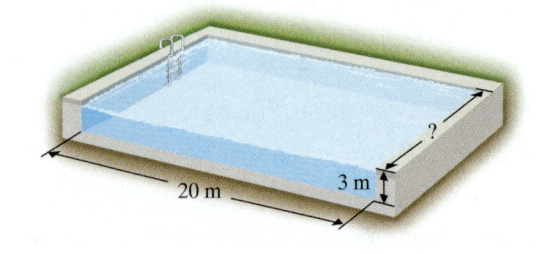

68. The perimeter of a rectangular billboard is 60 feet and has a length 6 feet longer than its width. Find the dimensions of the billboard.

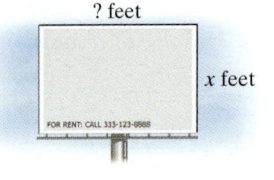

69. A charity 10K race is given annually to benefit a local hospice organization. How long will it take to run/walk a 10K race (10 kilometers or 10,000 meters) if your average pace is 125 **meters** per minute? Give your time in hours and minutes.

70. On April 28, 2001, the highest temperature recorded in the United States was 104°F, which occurred in Death Valley, California. Convert this temperature to degrees Celsius. (*Source:* National Weather Service)

(2.7) Find each of the following.

71. The number 9 is what percent of 45?
72. The number 59.5 is what percent of 85?
73. The number 137.5 is 125% of what number?
74. The number 768 is 60% of what number?
75. The price of a small diamond ring was recently increased by 11%. If the ring originally cost $1900, find the mark-up and the new price of the ring.
76. A recent survey found that 66.9% of Americans use the Internet. If a city has a population of 76,000 how many people in that city would you expect to use the Internet? (*Source:* UCLA Center for Communication Policy)
77. Thirty gallons of a 20% acid solution is needed for an experiment. Only 40% and 10% acid solutions are available. How much of each should be mixed to form the needed solution?
78. The ACT Assessment is a college entrance exam taken by about 60% of college-bound students. The national average score was 20.7 in 1993 and rose to 21.0 in 2001. Find the percent increase. (Round to the nearest hundredth of a percent.)

The graph below shows the percent(s) of cell phone users who have engaged in various behaviors while driving and talking on their cell phones. Use this graph to answer Exercises 79 through 82.

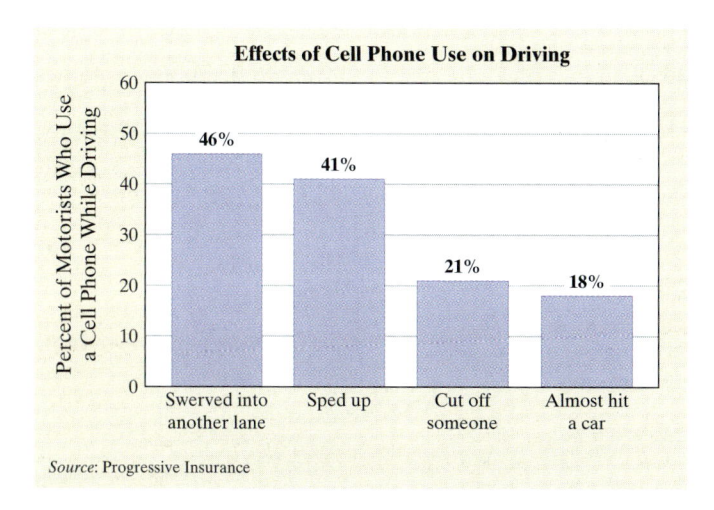

Effects of Cell Phone Use on Driving

- Swerved into another lane: 46%
- Sped up: 41%
- Cut off someone: 21%
- Almost hit a car: 18%

Source: Progressive Insurance

79. What percent of motorists who use a cell phone while driving have almost hit another car?
80. What is the most common effect of cell phone use on driving?
81. If a cell-phone service has an estimated 4600 customers who use their cell phones while driving, how many of these customers would you expect to have cut someone off while driving and talking on their cell phones?
82. Do the percents in the graph have a sum of 100%? Why or why not?
83. In 2005, Lance Armstrong incredibly won his seventh Tour de France, the first man in history to win more than five Tour de France Championships. Suppose he rides a bicycle up a category 2 climb at 10 km/hr and rides down the same distance at a speed of 50 km/hr. Find the distance traveled if the total time on the mountain was 3 hours.

(2.8) Solve.

84. A $50,000 retirement pension is to be invested into two accounts: a money market fund that pays 8.5% and a certificate of deposit that pays 10.5%. How much should be invested at each rate in order to provide a yearly interest income of $4550?
85. A pay phone is holding its maximum number of 500 coins consisting of nickels, dimes, and quarters. The number of quarters is twice the number of dimes. If the value of all the coins is $88.00, how many nickels were in the pay phone?
86. How long will it take an Amtrak passenger train to catch up to a freight train if their speeds are 60 and 45 mph and the freight train had an hour and a half head start?

(2.9) Solve and graph the solution of each of the following inequalities.

87. $x > 0$
88. $x \leq -2$
89. $0.5 \leq y < 1.5$
90. $-1 < x < 1$
91. $-3x > 12$
92. $-2x \geq -20$
93. $x + 4 \geq 6x - 16$
94. $5x - 7 > 8x + 5$
95. $-3 < 4x - 1 < 2$
96. $2 \leq 3x - 4 < 6$
97. $4(2x - 5) \leq 5x - 1$
98. $-2(x - 5) > 2(3x - 2)$
99. Tina earns $175 per week plus a 5% commission on all her sales. Find the minimum amount of sales to ensure that she earns at least $300 per week.
100. Ellen Catarella shot rounds of 76, 82, and 79 golfing. What must she shoot on her next round so that her average will be below 80?

MIXED REVIEW

Solve each equation.

101. $6x + 2x - 1 = 5x + 11$
102. $2(3y - 4) = 6 + 7y$
103. $4(3 - a) - (6a + 9) = -12a$
104. $\dfrac{x}{3} - 2 = 5$
105. $2(y + 5) = 2y + 10$
106. $7x - 3x + 2 = 2(2x - 1)$

Solve.

107. The sum of six and twice a number is equal to seven less than the number. Find the number.
108. A 23-inch piece of string is to be cut into two pieces so that the length of the longer piece is three more than four times the shorter piece. If x represents the length of the shorter piece, find the lengths of both pieces.

Solve for the specified variable.

109. $V = \dfrac{1}{3} Ah$ for h
110. What number is 26% of 85?
111. The number 72 is 45% of what number?
112. A company recently increased their number of employees from 235 to 282. Find the percent increase.

Solve each inequality. Graph the solution set.

113. $4x - 7 > 3x + 2$
114. $-5x < 20$
115. $-3(1 + 2x) + x \geq -(3 - x)$

CHAPTER 2 TEST

Simplify each of the following expressions.

1. $2y - 6 - y - 4$
2. $2.7x + 6.1 + 3.2x - 4.9$
3. $4(x - 2) - 3(2x - 6)$
4. $7 + 2(5y - 3)$

Solve each of the following equations.

5. $-\dfrac{4}{5}x = 4$
6. $4(n - 5) = -(4 - 2n)$
7. $5y - 7 + y = -(y + 3y)$
8. $4z + 1 - z = 1 + z$
9. $\dfrac{2(x + 6)}{3} = x - 5$
10. $\dfrac{1}{2} - x + \dfrac{3}{2} = x - 4$
11. $-0.3(x - 4) + x = 0.5(3 - x)$
12. $-4(a + 1) - 3a = -7(2a - 3)$
13. $-2(x - 3) = x + 5 - 3x$

Solve each of the following applications.

14. A number increased by two-thirds of the number is 35. Find the number.
15. A gallon of water seal covers 200 square feet. How many gallons are needed to paint two coats of water seal on a deck that measures 20 feet by 35 feet?
16. Some states have a single area code for the entire state. Two such states have area codes where one is double the other. If the sum of these integers is 1203, find the two area codes. (*Source:* North American Numbering Plan Administration)
17. Sedric Angell invested an amount of money in Amoxil stock that earned an annual 10% return, and then he invested twice the original amount in IBM stock that earned an annual 12% return. If his total return from both investments was $2890, find how much he invested in each stock.
18. Two trains leave Los Angeles simultaneously traveling on the same track in opposite directions at speeds of 50 and 64 mph. How long will it take before they are 285 miles apart?
19. Find the value of x if $y = -14$, $m = -2$, and $b = -2$ in the formula $y = mx + b$.

Solve each of the following equations for the indicated variable.

20. $V = \pi r^2 h$ for h
21. $3x - 4y = 10$ for y

Solve and graph each of the following inequalities.

22. $3x - 5 \geq 7x + 3$
23. $x + 6 > 4x - 6$
24. $-2 < 3x + 1 < 8$
25. $\dfrac{2(5x + 1)}{3} > 2$

CHAPTER 2 CUMULATIVE REVIEW

1. Given the set $\left\{-2, 0, \frac{1}{4}, -1.5, 112, -3, 11, \sqrt{\sqrt{2}}\right\}$, list the numbers in this set that belong to the set of:
   a. Natural numbers
   b. Whole numbers
   c. Integers
   d. Rational numbers
   e. Irrational numbers
   f. Real numbers

2. Given the set $\left\{7, 2, -\frac{1}{5}, 0, \sqrt{3}, -185, 8\right\}$, list the numbers in this set that belong to the set of:
   a. Natural numbers
   b. Whole Numbers
   c. Integers
   d. Rational Numbers
   e. Irrational numbers
   f. Real numbers

3. Find the absolute value of each number.
   a. $|4|$
   b. $|-5|$
   c. $|0|$
   d. $\left|-\frac{1}{2}\right|$
   e. $|5.6|$

4. Find the absolute value of each number.
   a. $|5|$
   b. $|-8|$
   c. $\left|-\frac{2}{3}\right|$

5. Write each of the following numbers as a product of primes.
   a. 40
   b. 63

6. Write each number as a product of primes.
   a. 44
   b. 90

7. Write $\frac{2}{5}$ as an equivalent fraction with a denominator of 20.

8. Write $\frac{2}{3}$ as an equivalent fraction with a denominator of 24.

9. Simplify $3[4 + 2(10 - 1)]$.

10. Simplify $5[16 - 4(2 + 1)]$.

11. Decide whether 2 is a solution of $3x + 10 = 8x$.

12. Decide whether 3 is a solution of $5x - 2 = 4x$.

Add.

13. $-1 + (-2)$
14. $(-2) + (-8)$
15. $-4 + 6$
16. $-3 + 10$

17. Simplify each expression.
    a. $-(-10)$
    b. $-\left(-\frac{1}{2}\right)$
    c. $-(-2x)$
    d. $-|-6|$

18. Simplify each expression.
    a. $-(-5)$
    b. $-\left(-\frac{2}{3}\right)$
    c. $-(-a)$
    d. $-|-3|$

19. Subtract.
    a. $5.3 - (-4.6)$
    b. $-\frac{3}{10} - \frac{5}{10}$
    c. $-\frac{2}{3} - \left(-\frac{4}{5}\right)$

20. Subtract
    a. $-2.7 - 8.4$
    b. $-\frac{4}{5} - \left(-\frac{3}{5}\right)$
    c. $\frac{1}{4} - \left(-\frac{1}{2}\right)$

21. Find each unknown complementary or supplementary angle.
    a. (38°, x)
    b. (62°, y)

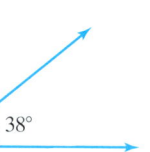

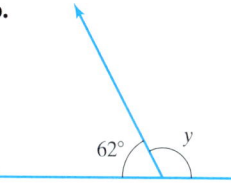

22. Find each unknown complementary or supplementary angle.
    a. (72°, x)
    b. (47°, y)

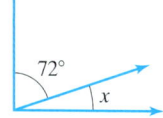

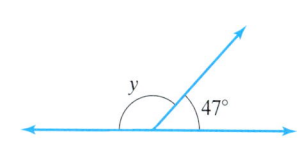

23. Find each product.
    a. $(-1.2)(0.05)$
    b. $\frac{2}{3} \cdot \left(-\frac{7}{10}\right)$
    c. $\left(-\frac{4}{5}\right)(-20)$

24. Find each product.
    a. $(4.5)(-0.08)$
    b. $-\frac{3}{4} \cdot -\frac{8}{17}$

25. Find each quotient.
    a. $\frac{-24}{-4}$
    b. $\frac{-36}{3}$
    c. $\frac{2}{3} \div \left(-\frac{5}{4}\right)$
    d. $-\frac{3}{2} \div 9$

26. Find each quotient.
    a. $\frac{-32}{8}$
    b. $\frac{-108}{-12}$
    c. $\frac{-5}{7} \div \left(\frac{-9}{2}\right)$

27. Use a commutative property to complete each statement.
    a. $x + 5 = $ _____
    b. $3 \cdot x = $ _____

28. Use a commutative property to complete each statement.
    a. $y + 1 = $ _____
    b. $y \cdot 4 = $ _____

29. Use the distributive property to write each sum as a product.
   a. $8 \cdot 2 + 8 \cdot x$
   b. $7s + 7t$

30. Use the distributive property to write each sum as a product.
   a. $4 \cdot y + 4 \cdot \frac{1}{3}$
   b. $0.10x + 0.10y$

31. Subtract $4x - 2$ from $2x - 3$.

32. Subtract $10x + 3$ from $-5x + 1$.

Solve.

33. $\frac{1}{2} = x - \frac{3}{4}$

34. $\frac{5}{6} + x = \frac{2}{3}$

35. $6(2a - 1) - (11a + 6) = 7$

36. $-3x + 1 - (-4x - 6) = 10$

37. $\frac{y}{7} = 20$

38. $\frac{x}{4} = 18$

39. $4(2x - 3) + 7 = 3x + 5$

40. $6x + 5 = 4(x + 4) - 1$

41. Twice the sum of a number and 4 is the same as four times the number, decreased by 12. Find the number.

42. A number increased by 4 is the same as 3 times the number decreased by 8. Find the number.

43. Solve $V = lwh$ for l.

44. Solve $C = 2\pi r$ for r.

45. Solve $x + 4 \leq -6$ for x. Graph the solution set and write it in interval notation.

46. Solve $x - 3 > 2$ for x. Graph the solution set and write it in interval notation.

CHAPTER

3 Graphing

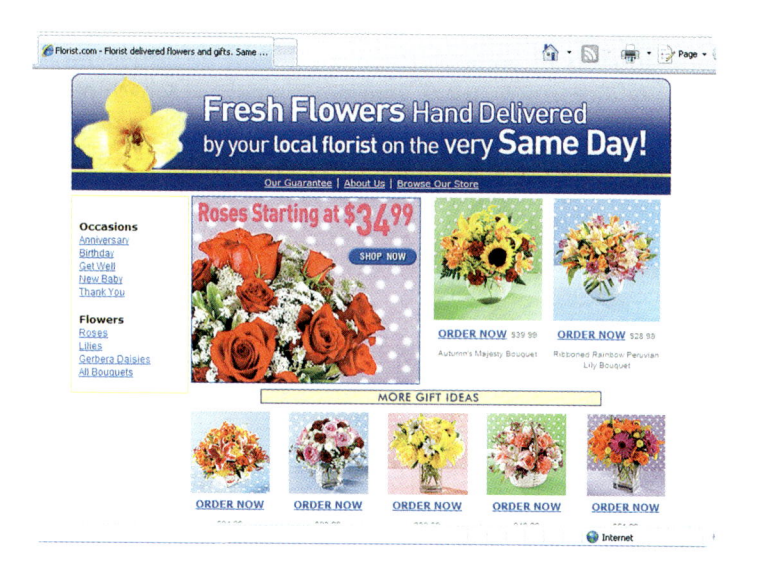

3.1	Reading Graphs and the Rectangular Coordinate System
3.2	Graphing Linear Equations
3.3	Intercepts
3.4	Slope and Rate of Change
	Integrated Review— Summary on Slope and Graphing Linear Equations
3.5	Equations of Lines
3.6	Functions

Online advertising is a way to promote services and products via the Internet. This fairly new way of advertising is quickly growing in popularity as more of the world is looking to the Internet for news and information. The broken-line graph below shows the yearly revenue generated by online advertising.

In Chapter 3's Integrated Review, Exercise 16, you will have the opportunity to use a linear equation, generated by the years 2003–2010, to predict online advertising revenue.

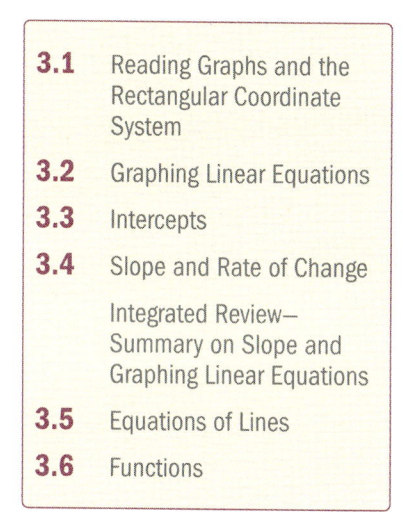

In the previous chapter we learned to solve and graph the solutions of linear equations and inequalities in one variable. Now we define and present techniques for solving and graphing linear equations and inequalities in two variables.

Online Advertising Revenue

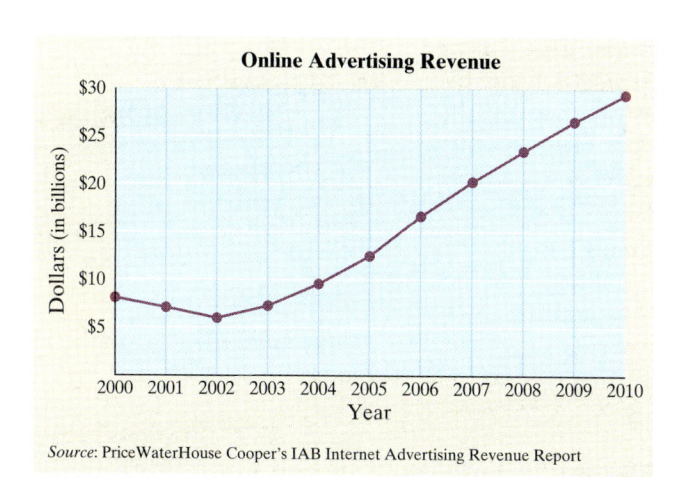

Source: PriceWaterHouse Cooper's IAB Internet Advertising Revenue Report

3.1 READING GRAPHS AND THE RECTANGULAR COORDINATE SYSTEM

OBJECTIVES

1. Read bar and line graphs.
2. Define the rectangular coordinate system and plot ordered pairs of numbers.
3. Graph paired data to create a scatter diagram.
4. Determine whether an ordered pair is a solution of an equation in two variables.
5. Find the missing coordinate of an ordered pair solution, given one coordinate of the pair.

In today's world, where the exchange of information must be fast and entertaining, graphs are becoming increasingly popular. They provide a quick way of making comparisons, drawing conclusions, and approximating quantities.

OBJECTIVE 1 ▶ Reading bar and line graphs. A **bar graph** consists of a series of bars arranged vertically or horizontally. The bar graph in Example 1 shows a comparison of worldwide Internet users by country. The names of the countries are listed vertically and a bar is shown for each country. Corresponding to the length of the bar for each country is a number along a horizontal axis. These horizontal numbers are number of Internet users in millions.

EXAMPLE 1

The following bar graph shows the estimated number of Internet users worldwide by country, as of a recent year.

a. Find the country that has the most Internet users and approximate the number of users.

b. How many more users are in the United States than in China?

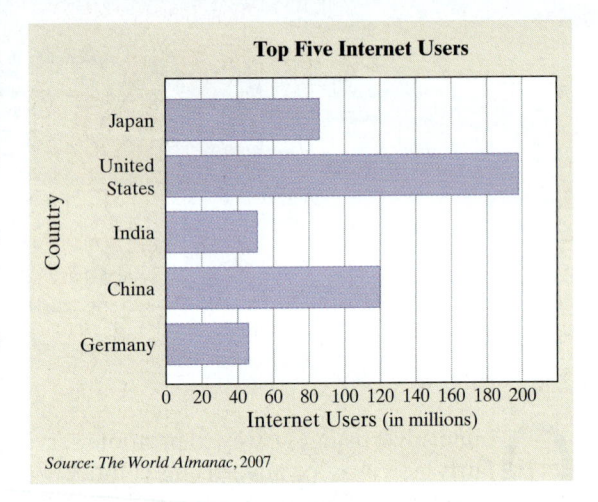

Solution

a. Since these bars are arranged horizontally, we look for the longest bar, which is the bar representing the United States. To approximate the number associated with this country, we move from the right edge of this bar vertically downward to the Internet user axis. This country has approximately 198 million Internet users.

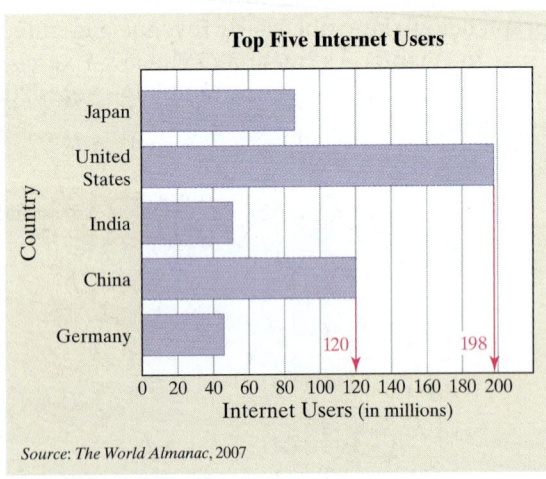

b. The United States has approximately 198 million Internet users. China has approximately 120 million Internet users. To find how many more users are in the United States, we subtract 198 − 120 = 78 or 78 million more Internet users. □

PRACTICE 1 Use the graph from Example 1 to answer the following.

a. Find the country shown with the fewest Internet users and approximate the number of users.

b. How many more users are in India than in Germany?

A **line graph** consists of a series of points connected by a line. The next graph is an example of a line graph. It is also sometimes called a **broken line graph.**

EXAMPLE 2 The line graph shows the relationship between time spent smoking a cigarette and pulse rate. Time is recorded along the horizontal axis in minutes, with 0 minutes being the moment a smoker lights a cigarette. Pulse is recorded along the vertical axis in heartbeats per minute.

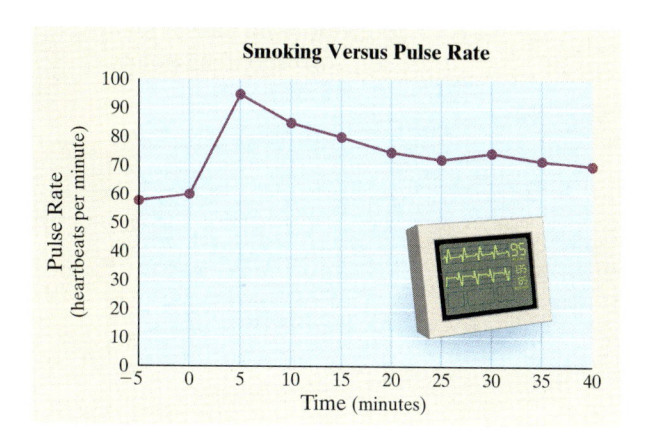

a. What is the pulse rate 15 minutes after a cigarette is lit?
b. When is the pulse rate the lowest?
c. When does the pulse rate show the greatest change?

Solution

a. We locate the number 15 along the time axis and move vertically upward until the line is reached. From this point on the line, we move horizontally to the left until the pulse rate axis is reached. Reading the number of beats per minute, we find that the pulse rate is 80 beats per minute 15 minutes after a cigarette is lit.

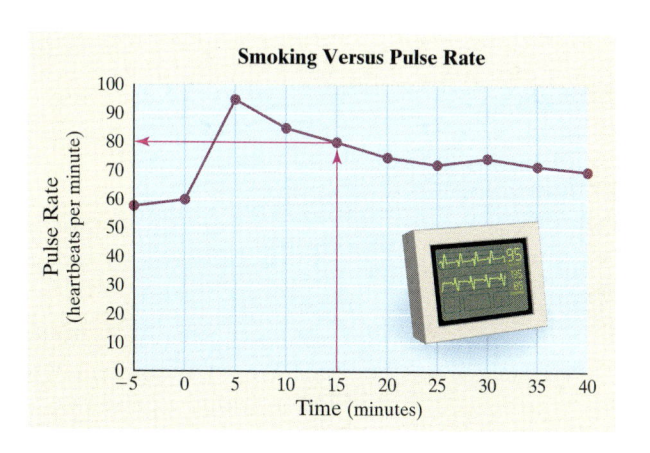

b. We find the lowest point of the line graph, which represents the lowest pulse rate. From this point, we move vertically downward to the time axis. We find that the pulse rate is the lowest at −5 minutes, which means 5 minutes *before* lighting a cigarette.

c. The pulse rate shows the greatest change during the 5 minutes between 0 and 5. Notice that the line graph is *steepest* between 0 and 5 minutes.

PRACTICE 2 Use the graph from Example 2 to answer the following.

a. What is the pulse rate 40 minutes after lighting a cigarette?
b. What is the pulse rate when the cigarette is being lit?
c. When is the pulse rate the highest?

OBJECTIVE 2 ▶ Defining the rectangular coordinate system and plotting ordered pairs of numbers. Notice in the previous graph that there are two numbers associated with each point of the graph. For example, we discussed earlier that 15 minutes after lighting a cigarette, the pulse rate is 80 beats per minute. If we agree to write the time first and the pulse rate second, we can say there is a point on the graph corresponding to the **ordered pair** of numbers (15, 80). A few more ordered pairs are listed alongside their corresponding points.

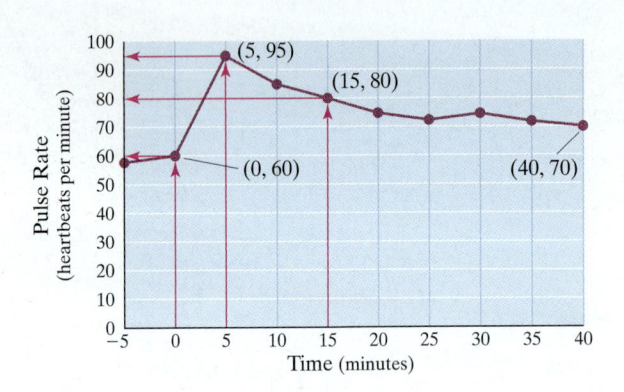

In general, we use this same ordered pair idea to describe the location of a point in a plane (such as a piece of paper). We start with a horizontal and a vertical axis. Each axis is a number line, and for the sake of consistency we construct our axes to intersect at the 0 coordinate of both. This point of intersection is called the **origin.** Notice that these two number lines or axes divide the plane into four regions called **quadrants.** The quadrants are usually numbered with Roman numerals as shown. The axes are not considered to be in any quadrant.

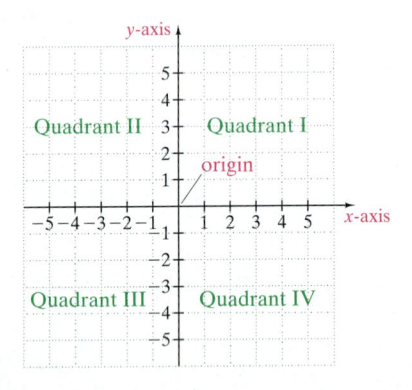

It is helpful to label axes, so we label the horizontal axis the ***x*-axis** and the vertical axis the ***y*-axis.** We call the system described above the **rectangular coordinate system.**

Just as with the pulse rate graph, we can then describe the locations of points by ordered pairs of numbers. We list the horizontal ***x*-axis** measurement first and the vertical ***y*-axis** measurement second.

To plot or graph the point corresponding to the ordered pair

$$(a, b)$$

we start at the origin. We then move a units left or right (right if a is positive, left if a is negative). From there, we move b units up or down (up if b is positive, down if b is negative). For example, to plot the point corresponding to the ordered pair (3, 2), we start at the origin, move 3 units right, and from there move 2 units up. (See the figure to the left.) The x-value, 3, is called the ***x*-coordinate** and the y-value, 2, is called the ***y*-coordinate.** From now on, we will call the point with coordinates (3, 2) simply the point (3, 2). The point (−2, 5) is graphed to the left also.

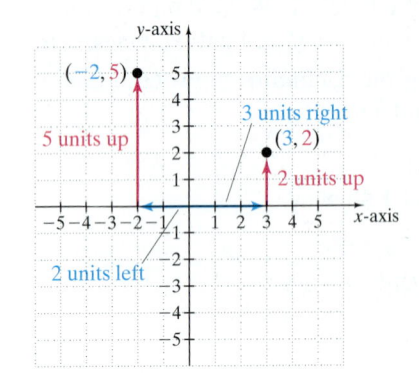

Does the order in which the coordinates are listed matter? Yes! Notice that the point corresponding to the ordered pair (2, 3) is in a different location than the point corresponding to (3, 2). These two ordered pairs of numbers describe two different points of the plane.

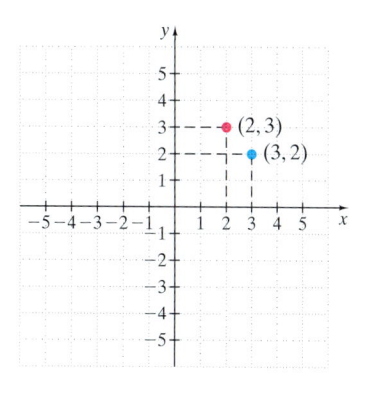

Concept Check ✓

Is the graph of the point (−5, 1) in the same location as the graph of the point (1, −5)? Explain.

> **Helpful Hint**
> Don't forget that **each ordered pair corresponds to exactly one point in the plane and that each point in the plane corresponds to exactly one ordered pair.**

EXAMPLE 3 On a single coordinate system, plot each ordered pair. State in which quadrant, if any, each point lies.

a. (5, 3)   **b.** (−5, 3)   **c.** (−2, −4)   **d.** (1, −2)

e. (0, 0)   **f.** (0, 2)   **g.** (−5, 0)   **h.** $\left(0, -5\frac{1}{2}\right)$

Solution

Point (5, 3) lies in quadrant I.
Point (−5, 3) lies in quadrant II.
Point (−2, −4) lies in quadrant III.
Point (1, −2) lies in quadrant IV.

Points (0, 0), (0, 2), (−5, 0), and $\left(0, -5\frac{1}{2}\right)$ lie on axes, so they are not in any quadrant.

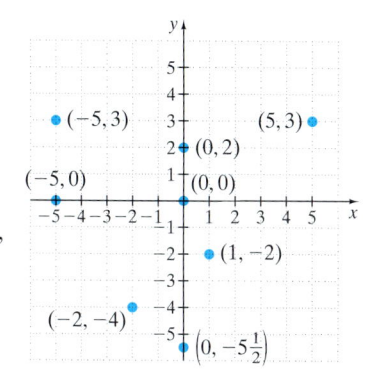

From Example 1, notice that the y-coordinate of any point on the x-axis is 0. For example, the point (−5, 0) lies on the x-axis. Also, the x-coordinate of any point on the y-axis is 0. For example, the point (0, 2) lies on the y-axis.

PRACTICE 3 On a single coordinate system, plot each ordered pair. State in which quadrant, if any, each point lies.

a. (4, −3)   **b.** (−3, 5)   **c.** (0, 4)   **d.** (−6, 1)

e. (−2, 0)   **f.** (5, 5)   **g.** $\left(3\frac{1}{2}, 1\frac{1}{2}\right)$ **h.** (−4, −5)

Answer to Concept Check:
The graph of point (−5, 1) lies in quadrant II and the graph of point (1, −5) lies in quadrant IV. They are *not* in the same location.

174 CHAPTER 3 Graphing

Concept Check ✓

For each description of a point in the rectangular coordinate system, write an ordered pair that represents it.

a. Point A is located three units to the left of the y-axis and five units above the x-axis.

b. Point B is located six units below the origin.

OBJECTIVE 3 ▶ Graphing paired data. Data that can be represented as an ordered pair is called **paired data.** Many types of data collected from the real world are paired data. For instance, the annual measurement of a child's height can be written as an ordered pair of the form (year, height in inches) and is paired data. The graph of paired data as points in the rectangular coordinate system is called a **scatter diagram.** Scatter diagrams can be used to look for patterns and trends in paired data.

EXAMPLE 4 The table gives the annual net sales for Wal-Mart Stores for the years shown. (*Source:* Wal-Mart Stores, Inc.)

Year	Wal-Mart Net Sales (in billions of dollars)
2000	181
2001	204
2002	230
2003	256
2004	285
2005	312
2006	345

a. Write this paired data as a set of ordered pairs of the form (year, sales in billions of dollars).

b. Create a scatter diagram of the paired data.

c. What trend in the paired data does the scatter diagram show?

Solution

a. The ordered pairs are (2000, 181), (2001, 204), (2002, 230), (2003, 256), (2004, 285), (2005, 312), and (2006, 345).

b. We begin by plotting the ordered pairs. Because the x-coordinate in each ordered pair is a year, we label the x-axis "Year" and mark the horizontal axis with the years given. Then we label the y-axis or vertical axis "Net Sales (in billions of dollars)." In this case it is convenient to mark the vertical axis in multiples of 20. Since no net sale is less than 180, we use the notation ⌇ to skip to 180, then proceed by multiples of 20.

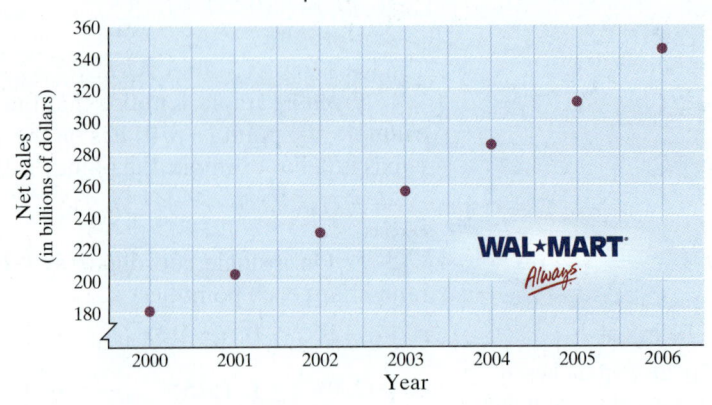

c. The scatter diagram shows that Wal-Mart net sales steadily increased over the years 2000–2006.

Answers to Concept Check:
a. $(-3, 5)$ **b.** $(0, -6)$

PRACTICE 4 The table gives the approximate annual number of wildfires (in the thousands) that have occurred in the United States for the years shown. (*Source: National Interagency Fire Center*)

Year	Wildfires (in thousands)
2000	92
2001	84
2002	73
2003	64
2004	65
2005	67
2006	96

a. Write this paired data as a set of ordered pairs of the form (year, number of wildfires in thousands).

b. Create a scatter diagram of the paired data.

OBJECTIVE 4 ▶ Determining whether an ordered pair is a solution. Let's see how we can use ordered pairs to record solutions of equations containing two variables. An equation in one variable such as $x + 1 = 5$ has one solution, which is 4: the number 4 is the value of the variable x that makes the equation true.

An equation in two variables, such as $2x + y = 8$, has solutions consisting of two values, one for x and one for y. For example, $x = 3$ and $y = 2$ is a solution of $2x + y = 8$ because, if x is replaced with 3 and y with 2, we get a true statement.

$$2x + y = 8$$
$$2(3) + 2 = 8$$
$$8 = 8 \quad \text{True}$$

The solution $x = 3$ and $y = 2$ can be written as $(3, 2)$, an **ordered pair** of numbers. The first number, 3, is the x-value and the second number, 2, is the y-value.

In general, an ordered pair is a **solution** of an equation in two variables if replacing the variables by the values of the ordered pair results in a true statement.

EXAMPLE 5 Determine whether each ordered pair is a solution of the equation $x - 2y = 6$.

a. $(6, 0)$ b. $(0, 3)$ c. $\left(1, -\frac{5}{2}\right)$

Solution

a. Let $x = 6$ and $y = 0$ in the equation $x - 2y = 6$.

$$x - 2y = 6$$
$$6 - 2(0) = 6 \quad \text{Replace } x \text{ with 6 and } y \text{ with 0.}$$
$$6 - 0 = 6 \quad \text{Simplify.}$$
$$6 = 6 \quad \text{True}$$

$(6, 0)$ is a solution, since $6 = 6$ is a true statement.

b. Let $x = 0$ and $y = 3$.

$$x - 2y = 6$$
$$0 - 2(3) = 6 \quad \text{Replace } x \text{ with 0 and } y \text{ with 3.}$$
$$0 - 6 = 6$$
$$-6 = 6 \quad \text{False}$$

$(0, 3)$ is *not* a solution, since $-6 = 6$ is a false statement.

c. Let $x = 1$ and $y = -\dfrac{5}{2}$ in the equation.

$$x - 2y = 6$$
$$1 - 2\left(-\dfrac{5}{2}\right) = 6 \quad \text{Replace } x \text{ with 1 and } y \text{ with } -\dfrac{5}{2}.$$
$$1 + 5 = 6$$
$$6 = 6 \quad \text{True}$$

$\left(1, -\dfrac{5}{2}\right)$ is a solution, since $6 = 6$ is a true statement. □

PRACTICE 5 Determine whether each ordered pair is a solution of the equation $x + 3y = 6$.

a. $(3, 1)$ **b.** $(6, 0)$ **c.** $\left(-2, \dfrac{2}{3}\right)$

OBJECTIVE 5 ▶ Completing ordered pair solutions. If one value of an ordered pair solution of an equation is known, the other value can be determined. To find the unknown value, replace one variable in the equation by its known value. Doing so results in an equation with just one variable that can be solved for the variable using the methods of Chapter 2.

EXAMPLE 6 Complete the following ordered pair solutions for the equation $3x + y = 12$.

a. $(0,)$ **b.** $(, 6)$ **c.** $(-1,)$

Solution

a. In the ordered pair $(0,)$, the x-value is 0. Let $x = 0$ in the equation and solve for y.

$$3x + y = 12$$
$$3(0) + y = 12 \quad \text{Replace } x \text{ with 0.}$$
$$0 + y = 12$$
$$y = 12$$

The completed ordered pair is $(0, 12)$.

b. In the ordered pair $(, 6)$, the y-value is 6. Let $y = 6$ in the equation and solve for x.

$$3x + y = 12$$
$$3x + 6 = 12 \quad \text{Replace } y \text{ with 6.}$$
$$3x = 6 \quad \text{Subtract 6 from both sides.}$$
$$x = 2 \quad \text{Divide both sides by 3.}$$

The ordered pair is $(2, 6)$.

Section 3.1 Reading Graphs and the Rectangular Coordinate System **177**

c. In the ordered pair $(-1,\ )$, the x-value is -1. Let $x = -1$ in the equation and solve for y.

$$3x + y = 12$$
$$3(-1) + y = 12 \quad \text{Replace } x \text{ with } -1.$$
$$-3 + y = 12$$
$$y = 15 \quad \text{Add 3 to both sides.}$$

The ordered pair is $(-1, 15)$.

PRACTICE 6 Complete the following ordered pair solutions for the equation $2x - y = 8$.

a. $(0,\ )$ **b.** $(\ , 4)$ **c.** $(-3,\ )$

Solutions of equations in two variables can also be recorded in a **table of values**, as shown in the next example.

EXAMPLE 7 Complete the table for the equation $y = 3x$.

	x	y
a.	-1	
b.		0
c.		-9

Solution

a. Replace x with -1 in the equation and solve for y.

$$y = 3x$$
$$y = 3(-1) \quad \text{Let } x = -1.$$
$$y = -3$$

The ordered pair is $(-1, -3)$.

b. Replace y with 0 in the equation and solve for x.

$$y = 3x$$
$$0 = 3x \quad \text{Let } y = 0.$$
$$0 = x \quad \text{Divide both sides by 3.}$$

The ordered pair is $(0, 0)$.

c. Replace y with -9 in the equation and solve for x.

$$y = 3x$$
$$-9 = 3x \quad \text{Let } y = -9.$$
$$-3 = x \quad \text{Divide both sides by 3.}$$

The ordered pair is $(-3, -9)$. The completed table is shown to the left.

x	y
-1	-3
0	0
-3	-9

PRACTICE 7 Complete the table for the equation $y = -4x$.

	x	y
a.	-2	
b.		-12
c.		0

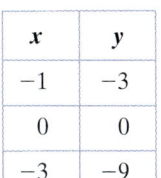

EXAMPLE 8 Complete the table for the equation

$$y = \frac{1}{2}x - 5.$$

	x	y
a.	-2	
b.	0	
c.		0

Solution

a. Let $x = -2$.

$y = \frac{1}{2}x - 5$

$y = \frac{1}{2}(-2) - 5$

$y = -1 - 5$

$y = -6$

b. Let $x = 0$.

$y = \frac{1}{2}x - 5$

$y = \frac{1}{2}(0) - 5$

$y = 0 - 5$

$y = -5$

c. Let $y = 0$.

$y = \frac{1}{2}x - 5$

$0 = \frac{1}{2}x - 5$ Now, solve for x.

$5 = \frac{1}{2}x$ Add 5.

$10 = x$ Multiply by 2.

Ordered Pairs: $(-2, -6)$ $(0, -5)$ $(10, 0)$

The completed table is

x	y
-2	-6
0	-5
10	0

PRACTICE 8 Compute the table for the equation $y = \frac{1}{5}x - 2$.

	x	y
a.	-10	
b.	0	
c.		0

EXAMPLE 9 Finding the Value of a Computer

A computer was recently purchased for a small business for $2000. The business manager predicts that the computer will be used for 5 years and the value in dollars y of the computer in x years is $y = -300x + 2000$. Complete the table.

x	0	1	2	3	4	5
y						

Solution To find the value of y when x is 0, replace x with 0 in the equation. We use this same procedure to find y when x is 1 and when x is 2.

When $x = 0$,

$y = -300x + 2000$

$y = -300 \cdot 0 + 2000$

$y = 0 + 2000$

$y = 2000$

When $x = 1$,

$y = -300x + 2000$

$y = -300 \cdot 1 + 2000$

$y = -300 + 2000$

$y = 1700$

When $x = 2$,

$y = -300x + 2000$

$y = -300 \cdot 2 + 2000$

$y = -600 + 2000$

$y = 1400$

We have the ordered pairs (0, 2000), (1, 1700), and (2, 1400). This means that in 0 years the value of the computer is $2000, in 1 year the value of the computer is $1700, and in

Section 3.1 Reading Graphs and the Rectangular Coordinate System 179

2 years the value is $1400. To complete the table of values, we continue the procedure for $x = 3$, $x = 4$, and $x = 5$.

When $x = 3$,
$y = -300x + 2000$
$y = -300 \cdot 3 + 2000$
$y = -900 + 2000$
$y = 1100$

When $x = 4$,
$y = -300x + 2000$
$y = -300 \cdot 4 + 2000$
$y = -1200 + 2000$
$y = 800$

When $x = 5$,
$y = -300x + 2000$
$y = -300 \cdot 5 + 2000$
$y = -1500 + 2000$
$y = 500$

The completed table is

x	0	1	2	3	4	5
y	2000	1700	1400	1100	800	500

PRACTICE

9 A college student purchased a used car for $12,000. The student predicted that she would need to use the car for four years and the value in dollars y of the car in x years is $y = -1800x + 12{,}000$. Complete this table.

x	0	1	2	3	4
y					

The ordered pair solutions recorded in the completed table for the example above are graphed below. Notice that the graph gives a visual picture of the decrease in value of the computer.

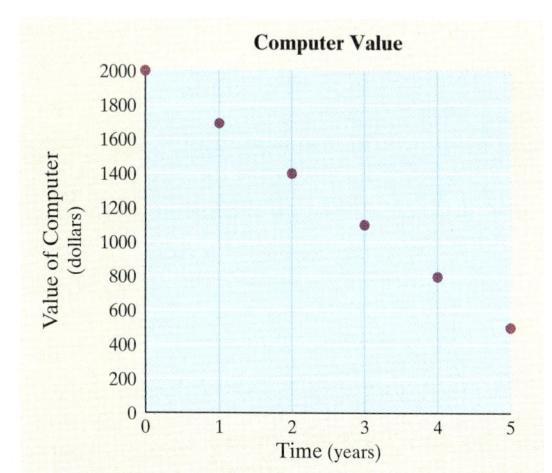

Computer Value

x	y
0	2000
1	1700
2	1400
3	1100
4	800
5	500

VOCABULARY & READINESS CHECK

Use the choices below to fill in each blank. The exercises below all have to do with the rectangular coordinate system.

origin   x-coordinate   x-axis   one   four
quadrants   y-coordinate   y-axis   solution

1. The horizontal axis is called the _____.
2. The vertical axis is called the _____.
3. The intersection of the horizontal axis and the vertical axis is a point called the _____.
4. The axes divide the plane into regions, called _____. There are _____ of these regions.
5. In the ordered pair of numbers $(-2, 5)$, the number -2 is called the _____ and the number 5 is called the _____.
6. Each ordered pair of numbers corresponds to _____ point in the plane.
7. An ordered pair is a _____ of an equation in two variables if replacing the variables by the coordinates of the ordered pair results in a true statement.

3.1 EXERCISE SET

The following bar graph shows the top 10 tourist destinations and the number of tourists that visit each country per year. Use this graph to answer Exercises 1 through 6. See Example 1.

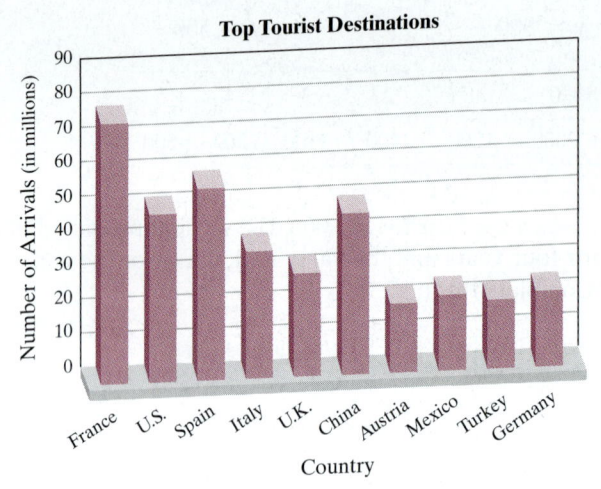

Top Tourist Destinations

Source: World Tourism Organization

1. Which country shown is the most popular tourist destination?
2. Which country shown is the least popular tourist destination?
3. Which countries shown have more than 40 million tourists per year?
4. Which countries shown have between 40 and 50 million tourists per year?
5. Estimate the number of tourists per year whose destination is the United Kingdom.
6. Estimate the number of tourists per year whose destination is Turkey.

The following line graph shows the attendance at each Super Bowl game from 2000 through 2007. Use this graph to answer Exercises 7 through 10. See Example 2.

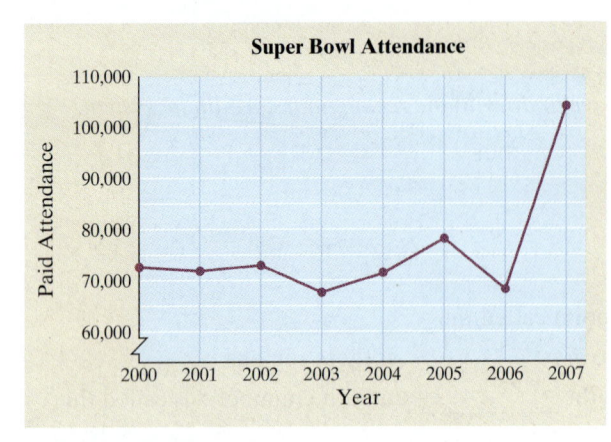

Super Bowl Attendance

7. Estimate the Super Bowl attendance in 2000.
8. Estimate the Super Bowl attendance in 2004.
9. Find the year on the graph with the greatest Super Bowl attendance and approximate that attendance.
10. Find the year on the graph with the least Super Bowl attendance and approximate that attendance.

The line graph below shows the number of students per teacher in U.S. public elementary and secondary schools. Use this graph for Exercises 11 through 16. See Example 2.

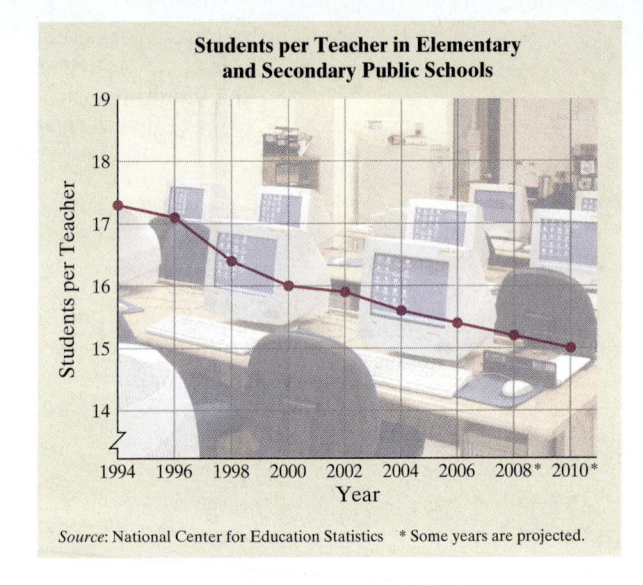

Students per Teacher in Elementary and Secondary Public Schools

*Source: National Center for Education Statistics *Some years are projected.*

11. Approximate the number of students per teacher in 2002.
12. Approximate the number of students per teacher in 2010.
13. Between what years shown did the greatest decrease in number of students per teacher occur?
14. What was the first year shown that the number of students per teacher fell below 17?
15. What was the first year shown that the number of students per teacher fell below 16?
16. Discuss any trends shown by this line graph.

Plot each ordered pair. State in which quadrant or on which axis each point lies. See Example 3.

17. a. $(1, 5)$ b. $(-5, -2)$
    c. $(-3, 0)$ d. $(0, -1)$
    e. $(2, -4)$ f. $\left(-1, 4\frac{1}{2}\right)$
    g. $(3.7, 2.2)$ h. $\left(\frac{1}{2}, -3\right)$

18. a. $(2, 4)$ b. $(0, 2)$
    c. $(-2, 1)$ d. $(-3, -3)$
    e. $\left(3\frac{3}{4}, 0\right)$ f. $(5, -4)$
    g. $(-3.4, 4.8)$ h. $\left(\frac{1}{3}, -5\right)$

Find the x- and y-coordinates of each labeled point. See Example 3.

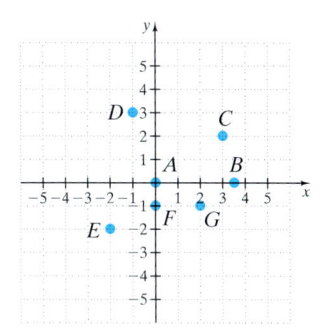

19. A
20. B
21. C
22. D
23. E
24. F
25. G

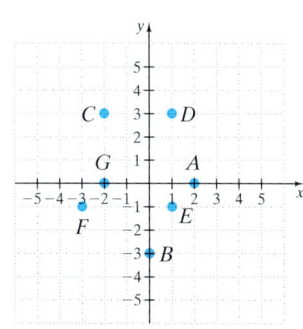

26. A
27. B
28. C
29. D
30. E
31. F
32. G

Solve. See Example 4.

33. The table shows the number of regular-season NFL football games won by the winner of the Super Bowl for the years shown. (*Source:* National Football League)

Year	Regular-Season Games Won by Super Bowl Winner
2002	12
2003	14
2004	14
2005	11
2006	12

   a. Write each paired data as an ordered pair of the form (year, games won).

   b. Draw a grid such as the one in Example 4 and create a scatter diagram of the paired data.

34. The table shows the average price of a gallon of regular unleaded gasoline (in dollars) for the years shown. (*Source:* Energy Information Administration)

Year	Price per Gallon of Unleaded Gasoline (in dollars)
2001	1.38
2002	1.31
2003	1.52
2004	1.81
2005	2.24
2006	2.53

   a. Write each paired data as an ordered pair of the form (year, gasoline price).

   b. Draw a grid such as the one in Example 4 and create a scatter diagram of the paired data.

35. The table shows the ethanol fuel production in the United States. (*Source:* Renewable Fuels Association; *some years projected)

Year	Ethanol Fuel Production (in millions of gallons)
2001	1770
2003	2800
2005	3904
2007*	7500
2009*	10,800

   a. Write each paired data as an ordered pair of the form (year, millions of gallons produced)

   b. Draw a grid such as the one in Example 4 and create a scatter diagram of the paired data.

   c. What trend in the paired data does the scatter diagram show?

36. The table shows the enrollment in college in the United States for the years shown. (*Source:* U.S. Department of Education)

Year	Enrollment in College (in millions)
1970	8.6
1980	12.1
1990	13.8
2000	15.3
2010*	18.7

*projected

   a. Write each paired data as an ordered pair of the form (year, number of institutions).

   b. Draw a grid such as the one in Example 4 and create a scatter diagram of the paired data.

   c. What trend in the paired data does the scatter diagram show?

37. The table shows the distance from the equator (in miles) and the average annual snowfall (in inches) for each of eight

selected U.S. cities. (*Sources:* National Climatic Data Center, Wake Forest University Albatross Project)

City	Distance from Equator (in miles)	Average Annual Snowfall (in inches)
1. Atlanta, GA	2313	2
2. Austin, TX	2085	1
3. Baltimore, MD	2711	21
4. Chicago, IL	2869	39
5. Detroit, MI	2920	42
6. Juneau, AK	4038	99
7. Miami, FL	1783	0
8. Winston-Salem, NC	2493	9

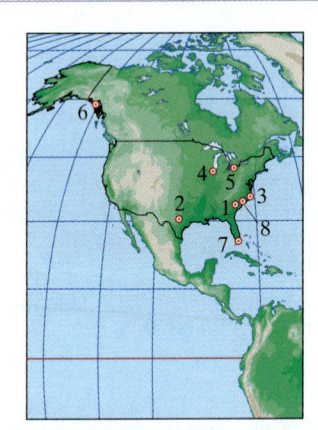

a. Write this paired data as a set of ordered pairs of the form (distance from equator, average annual snowfall).

b. Create a scatter diagram of the paired data. Be sure to label the axes appropriately.

c. What trend in the paired data does the scatter diagram show?

38. The table shows the average farm size (in acres) in the United States during the years shown. (*Source:* National Agricultural Statistics Service)

Year	Average Farm Size (in acres)
2001	438
2002	440
2003	441
2004	443
2005	445
2006	446

a. Write this paired data as a set of ordered pairs of the form (year, average farm size).

b. Create a scatter diagram of the paired data. Be sure to label the axes appropriately.

Determine whether each ordered pair is a solution of the given linear equation. See Example 5.

39. $2x + y = 7$; $(3, 1), (7, 0), (0, 7)$
40. $3x + y = 8$; $(2, 3), (0, 8), (8, 0)$
41. $x = -\frac{1}{3}y$; $(0, 0), (3, -9)$
42. $y = -\frac{1}{2}x$; $(0, 0), (4, 2)$
43. $x = 5$; $(4, 5), (5, 4), (5, 0)$
44. $y = -2$; $(-2, 2), (2, -2), (0, -2)$

Complete each ordered pair so that it is a solution of the given linear equation. See Examples 6 through 8.

45. $x - 4y = 4$; $(\ \ , -2), (4, \ \ )$
46. $x - 5y = -1$; $(\ \ , -2), (4, \ \ )$
47. $y = \frac{1}{4}x - 3$; $(-8, \ \ ), (\ \ , 1)$
48. $y = \frac{1}{5}x - 2$; $(-10, \ \ ), (\ \ , 1)$

Complete the table of ordered pairs for each linear equation. See Examples 6 through 8.

49. $y = -7x$

x	y
0	
-1	
	2

50. $y = -9x$

x	y
0	0
-3	
	2

51. $y = -x + 2$

x	y
0	
	0
-3	

52. $x = -y + 4$

x	y
	0
0	
	-3

53. $y = \frac{1}{2}x$

x	y
0	
-6	
	1

54. $y = \frac{1}{3}x$

x	y
0	
-6	
	1

55. $x + 3y = 6$

x	y
0	
	0
	1

56. $2x + y = 4$

x	y
	4
2	
	2

57. $y = 2x - 12$

x	y
0	
	-2
3	

58. $y = 5x + 10$

x	y
0	
	5
	0

59. $2x + 7y = 5$

x	y
0	
	0
	1

60. $x - 6y = 3$

x	y
0	
	1
	-1

MIXED PRACTICE

Complete the table of ordered pairs for each equation. Then plot the ordered pair solutions. See Examples 1 through 7.

61. $x = -5y$

x	y
	0
	1
10	

62. $y = -3x$

x	y
0	
-2	
	9

63. $y = \frac{1}{3}x + 2$

x	y
0	
-3	
	0

64. $y = \frac{1}{2}x + 3$

x	y
0	
-4	
	0

Solve. See Example 9.

65. The cost in dollars y of producing x computer desks is given by $y = 80x + 5000$.

a. Complete the table.

x	100	200	300
y			

b. Find the number of computer desks that can be produced for $8600. (*Hint:* Find x when $y = 8600$.)

66. The hourly wage y of an employee at a certain production company is given by $y = 0.25x + 9$ where x is the number of units produced by the employee in an hour.

a. Complete the table.

x	0	1	5	10
y				

b. Find the number of units that an employee must produce each hour to earn an hourly wage of $12.25. (*Hint:* Find x when $y = 12.25$.)

67. The average amount of money y spent per person on recorded music from 2001 to 2005 is given by $y = -2.35x + 55.92$. In this equation, x represents the number of years after 2001. (*Source:* Veronis Suhler Stevenson)

a. Complete the table.

x	1	3	5
y			

b. Find the year in which the yearly average amount of money per person spent on recorded music was approximately $46. (*Hint:* Find x when $y = 46$ and round to the nearest whole number.)

68. The amount y of land operated by farms in the United States (in million acres) from 2000 through 2006 is given by $y = -2.18x + 944.68$. In the equation, x represents the number of years after 2000. (*Source:* National Agricultural Statistics Service)

a. Complete the table.

x	2	4	6
y			

b. Find the year in which there were approximately 933 million acres of land operated by farms. (*Hint:* Find x when $y = 933$ and round to the nearest whole number.)

The graph below shows the number of Target stores for each year. Use this graph to answer Exercises 69 through 72.

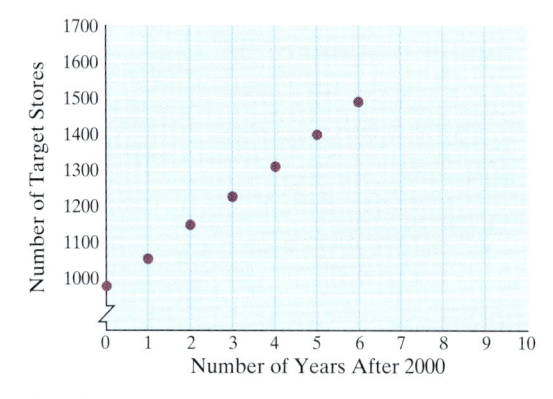

Source: Target

69. The ordered pair (4, 1308) is a point of the graph. Write a sentence describing the meaning of this ordered pair.

70. The ordered pair (6, 1488) is a point of the graph. Write a sentence describing the meaning of this ordered pair.

71. Estimate the increase in Target stores for years 1, 2, and 3.

72. Use a straightedge or ruler and this graph to predict the number of Target stores in the year 2009.

73. When is the graph of the ordered pair (a, b) the same as the graph of the ordered pair (b, a)?

74. In your own words, describe how to plot an ordered pair.

REVIEW AND PREVIEW

Solve each equation for y. See Section 2.6.

75. $x + y = 5$
76. $x - y = 3$
77. $2x + 4y = 5$
78. $5x + 2y = 7$
79. $10x = -5y$
80. $4y = -8x$
81. $x - 3y = 6$
82. $2x - 9y = -20$

CONCEPT EXTENSIONS

Answer each exercise with true or false.

83. Point $(-1, 5)$ lies in quadrant IV.

84. Point $(3, 0)$ lies on the y-axis.

85. For the point $\left(-\frac{1}{2}, 1.5\right)$, the first value, $-\frac{1}{2}$, is the x-coordinate and the second value, 1.5, is the y-coordinate.

86. The ordered pair $\left(2, \frac{2}{3}\right)$ is a solution of $2x - 3y = 6$.

For Exercises 87 through 91, fill in each blank with "0," "positive," or "negative." For Exercises 92 and 93, fill in each blank with "x" or "y."

	Point	Location
87.	(_____, _____)	quadrant III
88.	(_____, _____)	quadrant I
89.	(_____, _____)	quadrant IV
90.	(_____, _____)	quadrant II
91.	(_____, _____)	origin
92.	(number, 0)	___-axis
93.	(0, number)	___-axis

94. Give an example of an ordered pair whose location is in (or on)

a. quadrant I
b. quadrant II
c. quadrant III
d. quadrant IV
e. x-axis
f. y-axis

Solve. See the Concept Check in this section.

95. Is the graph of $(3, 0)$ in the same location as the graph of $(0, 3)$? Explain why or why not.

96. Give the coordinates of a point such that if the coordinates are reversed, their location is the same.

97. In general, what points can have coordinates reversed and still have the same location?

98. In your own words, describe how to plot or graph an ordered pair of numbers.

Write an ordered pair for each point described.

99. Point C is four units to the right of the y-axis and seven units below the x-axis.

100. Point D is three units to the left of the origin.

101. Three vertices of a rectangle are $(-2, -3), (-7, -3)$, and $(-7, 6)$.

a. Find the coordinates of the fourth vertex of a rectangle.
b. Find the perimeter of the rectangle.
c. Find the area of the rectangle.

102. Three vertices of a square are $(-4, -1), (-4, 8)$, and $(5, 8)$.

a. Find the coordinates of the fourth vertex of the square.
b. Find the perimeter of the square.
c. Find the area of the square.

STUDY SKILLS BUILDER

Are You Satisfied with Your Performance in This Course Thus Far?

To see if there is room for improvement, answer these questions:

1. Am I attending all classes and arriving on time?
2. Am I working and checking my homework assignments on time?
3. Am I getting help (from my instructor or a campus learning resource lab) when I need it?
4. In addition to my instructor, am I using the text supplements that might help me?
5. Am I satisfied with my performance on quizzes and exams?

If you answered no to any of these questions, read or reread Section 1.1 for suggestions in these areas. Also, you might want to contact your instructor for additional feedback.

3.2 GRAPHING LINEAR EQUATIONS

OBJECTIVES

1. Identify linear equations.
2. Graph a linear equation by finding and plotting ordered pair solutions.

OBJECTIVE 1 ▶ Identifying linear equations. In the previous section, we found that equations in two variables may have more than one solution. For example, both $(6, 0)$ and $(2, -2)$ are solutions of the equation $x - 2y = 6$. In fact, this equation has an infinite number of solutions. Other solutions include $(0, -3)$, $(4, -1)$, and $(-2, -4)$. If we graph these solutions, notice that a pattern appears.

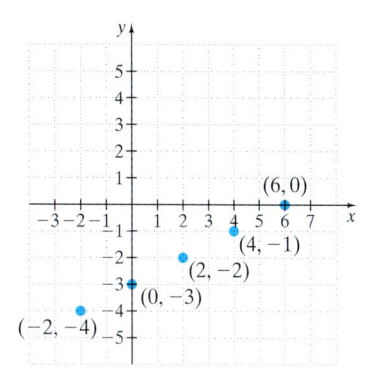

These solutions all appear to lie on the same line, which has been filled in below. It can be shown that every ordered pair solution of the equation corresponds to a point on this line, and every point on this line corresponds to an ordered pair solution. Thus, we say that this line is the **graph of the equation** $x - 2y = 6$.

▶ **Helpful Hint**

Notice that we can only show a part of a line on a graph. The arrowheads on each end of the line remind us that the line actually extends indefinitely in both directions.

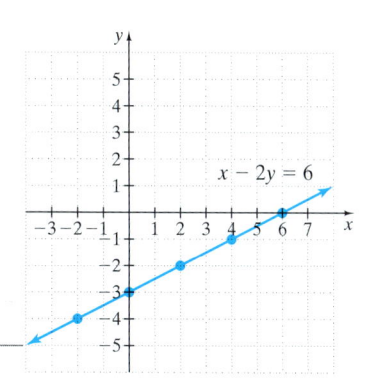

The equation $x - 2y = 6$ is called a **linear equation in two variables** and **the graph of every linear equation in two variables is a line.**

Linear Equation in Two Variables
A linear equation in two variables is an equation that can be written in the form

$$Ax + By = C$$

where A, B, and C are real numbers and A and B are not both 0. **The graph of a linear equation in two variables is a straight line.**

The form $Ax + By = C$ is called **standard form.**

▶ **Helpful Hint**
Notice in the form $Ax + By = C$, the understood exponent on both x and y is 1.

Examples of Linear Equations in Two Variables

$$2x + y = 8 \qquad -2x = 7y \qquad y = \frac{1}{3}x + 2 \qquad y = 7$$
(Standard Form)

Before we graph linear equations in two variables, let's practice identifying these equations.

EXAMPLE 1 Determine whether each equation is a linear equation in two variables.

a. $x - 1.5y = -1.6$ **b.** $y = -2x$ **c.** $x + y^2 = 9$ **d.** $x = 5$

Solution

a. This is a linear equation in two variables because it is written in the form $Ax + By = C$ with $A = 1$, $B = -1.5$, and $C = -1.6$.

b. This is a linear equation in two variables because it can be written in the form $Ax + By = C$.

$$y = -2x$$
$$2x + y = 0 \qquad \text{Add } 2x \text{ to both sides.}$$

c. This is *not* a linear equation in two variables because y is squared.

d. This is a linear equation in two variables because it can be written in the form $Ax + By = C$.

$$x = 5$$
$$x + 0y = 5 \qquad \text{Add } 0 \cdot y.$$

PRACTICE 1 Determine whether each equation is a linear equation in two variables.

a. $3x + 2.7y = -5.3$ **b.** $x^2 + y = 8$ **c.** $y = 12$ **d.** $5x = -3y$

OBJECTIVE 2 ▶ Graphing linear equations by plotting ordered pair solutions. From geometry, we know that a straight line is determined by just two points. Graphing a linear equation in two variables, then, requires that we find just two of its infinitely many solutions. Once we do so, we plot the solution points and draw the line connecting the points. Usually, we find a third solution as well, as a check.

EXAMPLE 2 Graph the linear equation $2x + y = 5$.

Solution Find three ordered pair solutions of $2x + y = 5$. To do this, choose a value for one variable, x or y, and solve for the other variable. For example, let $x = 1$. Then $2x + y = 5$ becomes

$$2x + y = 5$$
$$2(1) + y = 5 \qquad \text{Replace } x \text{ with 1.}$$
$$2 + y = 5 \qquad \text{Multiply.}$$
$$y = 3 \qquad \text{Subtract 2 from both sides.}$$

Since $y = 3$ when $x = 1$, the ordered pair $(1, 3)$ is a solution of $2x + y = 5$. Next, let $x = 0$.

$$2x + y = 5$$
$$2(0) + y = 5 \qquad \text{Replace } x \text{ with 0.}$$
$$0 + y = 5$$
$$y = 5$$

The ordered pair $(0, 5)$ is a second solution.

Section 3.2 Graphing Linear Equations   187

The two solutions found so far allow us to draw the straight line that is the graph of all solutions of $2x + y = 5$. However, we find a third ordered pair as a check. Let $y = -1$.

$$2x + y = 5$$
$$2x + (-1) = 5 \quad \text{Replace } y \text{ with } -1.$$
$$2x - 1 = 5$$
$$2x = 6 \quad \text{Add 1 to both sides.}$$
$$x = 3 \quad \text{Divide both sides by 2.}$$

The third solution is $(3, -1)$. These three ordered pair solutions are listed in table form as shown. The graph of $2x + y = 5$ is the line through the three points.

x	y
1	3
0	5
3	-1

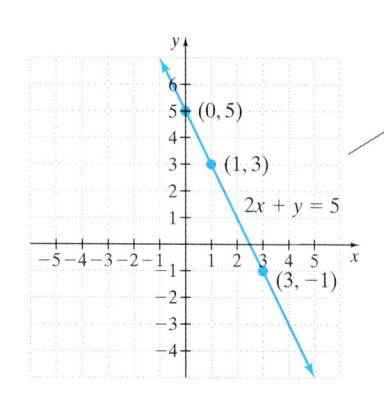

> **Helpful Hint**
> All three points should fall on the same straight line. If not, check your ordered pair solutions for a mistake.

PRACTICE 2   Graph the linear equation $x + 3y = 9$.

EXAMPLE 3   Graph the linear equation $-5x + 3y = 15$.

Solution   Find three ordered pair solutions of $-5x + 3y = 15$.

Let $x = 0$.	Let $y = 0$.	Let $x = -2$.
$-5x + 3y = 15$	$-5x + 3y = 15$	$-5x + 3y = 15$
$-5 \cdot 0 + 3y = 15$	$-5x + 3 \cdot 0 = 15$	$-5(-2) + 3y = 15$
$0 + 3y = 15$	$-5x + 0 = 15$	$10 + 3y = 15$
$3y = 15$	$-5x = 15$	$3y = 5$
$y = 5$	$x = -3$	$y = \dfrac{5}{3}$

The ordered pairs are $(0, 5)$, $(-3, 0)$, and $\left(-2, \dfrac{5}{3}\right)$. The graph of $-5x + 3y = 15$ is the line through the three points.

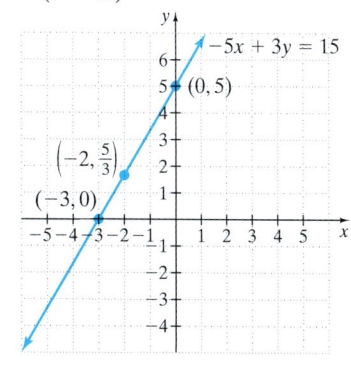

x	y
0	5
-3	0
-2	$\dfrac{5}{3} = 1\dfrac{2}{3}$

PRACTICE 3   Graph the linear equation $3x - 4y = 12$.

EXAMPLE 4 Graph the linear equation $y = 3x$.

Solution To graph this linear equation, we find three ordered pair solutions. Since this equation is solved for y, choose three x values.

If $x = 2$, $y = 3 \cdot 2 = 6$.
If $x = 0$, $y = 3 \cdot 0 = 0$.
If $x = -1$, $y = 3 \cdot -1 = -3$.

x	y
2	6
0	0
-1	-3

Next, graph the ordered pair solutions listed in the table above and draw a line through the plotted points as shown on the next page. The line is the graph of $y = 3x$. Every point on the graph represents an ordered pair solution of the equation and every ordered pair solution is a point on this line.

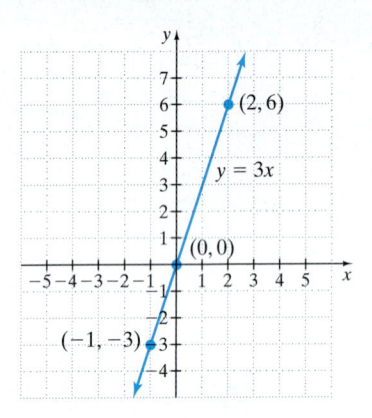

PRACTICE 4 Graph the linear equation $y = -2x$.

EXAMPLE 5 Graph the linear equation $y = -\frac{1}{3}x + 2$.

Solution Find three ordered pair solutions, graph the solutions, and draw a line through the plotted solutions. To avoid fractions, choose x values that are multiples of 3 to substitute in the equation. When a multiple of 3 is multiplied by $-\frac{1}{3}$, the result is an integer. See the calculations shown above the table.

If $x = 6$, then $y = -\frac{1}{3} \cdot 6 + 2 = -2 + 2 = 0$

If $x = 0$, then $y = -\frac{1}{3} \cdot 0 + 2 = 0 + 2 = 2$

If $x = -3$, then $y = -\frac{1}{3} \cdot -3 + 2 = 1 + 2 = 3$

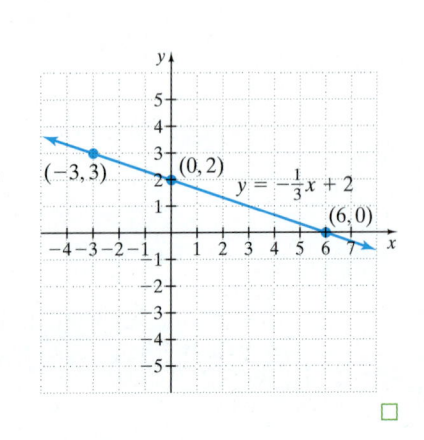

x	y
6	0
0	2
-3	3

PRACTICE 5 Graph the linear equation $y = \frac{1}{2}x + 3$.

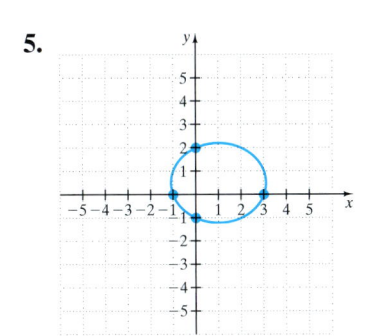

5.

Solution
x-intercepts: $(-1, 0), (3, 0)$

y-intercepts: $(0, 2), (0, -1)$

PRACTICES

1–5 Identify the *x*- and *y*-intercepts.

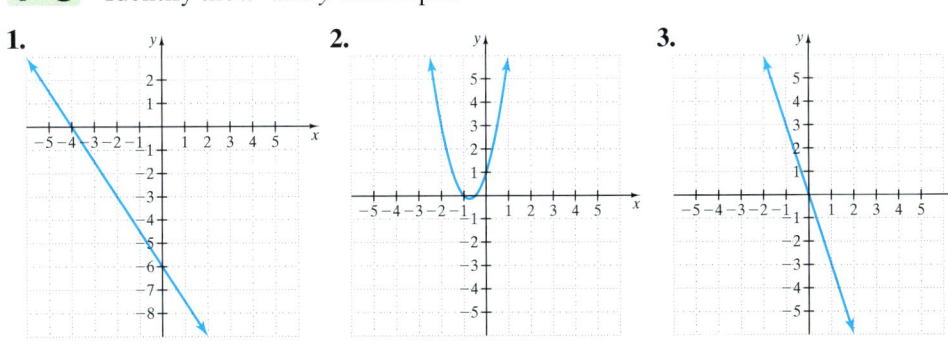

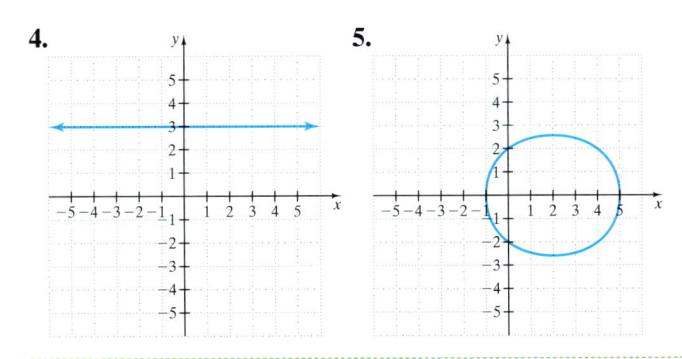

1.

2.

3.

4.

5.

OBJECTIVE 2 ▶ Using intercepts to graph a linear equation. Given the equation of a line, intercepts are usually easy to find since one coordinate is 0.

One way to find the *y*-intercept of a line, given its equation, is to let $x = 0$, since a point on the *y*-axis has an *x*-coordinate of 0. To find the *x*-intercept of a line, let $y = 0$, since a point on the *x*-axis has a *y*-coordinate of 0.

> **Finding x- and y-intercepts**
> To find the *x*-intercept, let $y = 0$ and solve for *x*.
> To find the *y*-intercept, let $x = 0$ and solve for *y*.

EXAMPLE 6 Graph $x - 3y = 6$ by finding and plotting intercepts.

Solution Let $y = 0$ to find the x-intercept and let $x = 0$ to find the y-intercept.

$$
\begin{array}{ll}
\text{Let } y = 0 & \text{Let } x = 0 \\
x - 3y = 6 & x - 3y = 6 \\
x - 3(0) = 6 & 0 - 3y = 6 \\
x - 0 = 6 & -3y = 6 \\
x = 6 & y = -2
\end{array}
$$

The x-intercept is $(6, 0)$ and the y-intercept is $(0, -2)$. We find a third ordered pair solution to check our work. If we let $y = -1$, then $x = 3$. Plot the points $(6, 0)$, $(0, -2)$, and $(3, -1)$. The graph of $x - 3y = 6$ is the line drawn through these points, as shown.

x	y
6	0
0	-2
3	-1

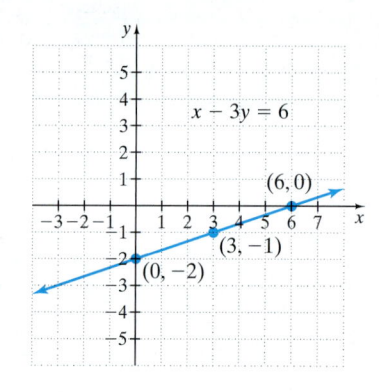

PRACTICE 6 Graph $x + 2y = -4$ by finding and plotting intercepts.

EXAMPLE 7 Graph $x = -2y$ by plotting intercepts.

Solution Let $y = 0$ to find the x-intercept and $x = 0$ to find the y-intercept.

$$
\begin{array}{ll}
\text{Let } y = 0 & \text{Let } x = 0 \\
x = -2y & x = -2y \\
x = -2(0) & 0 = -2y \\
x = 0 & 0 = y
\end{array}
$$

Both the x-intercept and y-intercept are $(0, 0)$. In other words, when $x = 0$, then $y = 0$, which gives the ordered pair $(0, 0)$. Also, when $y = 0$, then $x = 0$, which gives the same ordered pair $(0, 0)$. This happens when the graph passes through the origin. Since two points are needed to determine a line, we must find at least one more ordered pair that satisfies $x = -2y$. Let $y = -1$ to find a second ordered pair solution and let $y = 1$ as a checkpoint.

$$
\begin{array}{ll}
\text{Let } y = -1 & \text{Let } y = 1 \\
x = -2(-1) & x = -2(1) \\
x = 2 & x = -2
\end{array}
$$

The ordered pairs are $(0, 0)$, $(2, -1)$, and $(-2, 1)$. Plot these points to graph $x = -2y$.

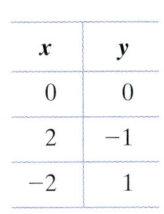

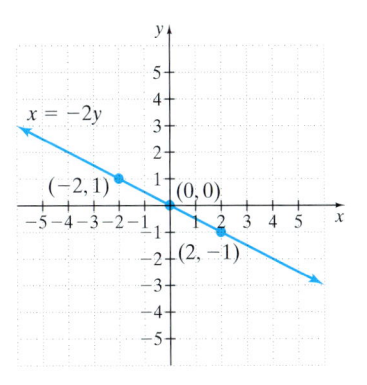

x	y
0	0
2	-1
-2	1

PRACTICE 7 Graph $x = 3y$ by plotting intercepts.

EXAMPLE 8 Graph $4x = 3y - 9$.

Solution Find the x- and y-intercepts, and then choose $x = 2$ to find a third checkpoint.

Let $y = 0$
$4x = 3(0) - 9$
$4x = -9$
Solve for x.
$x = -\dfrac{9}{4}$ or $-2\dfrac{1}{4}$

Let $x = 0$
$4 \cdot 0 = 3y - 9$
$9 = 3y$
Solve for y.
$3 = y$

Let $x = 2$
$4(2) = 3y - 9$
$8 = 3y - 9$
Solve for y.
$17 = 3y$
$\dfrac{17}{3} = y$ or $y = 5\dfrac{2}{3}$

The ordered pairs are $\left(-2\dfrac{1}{4}, 0\right)$, $(0, 3)$, and $\left(2, 5\dfrac{2}{3}\right)$. The equation $4x = 3y - 9$ is graphed as follows.

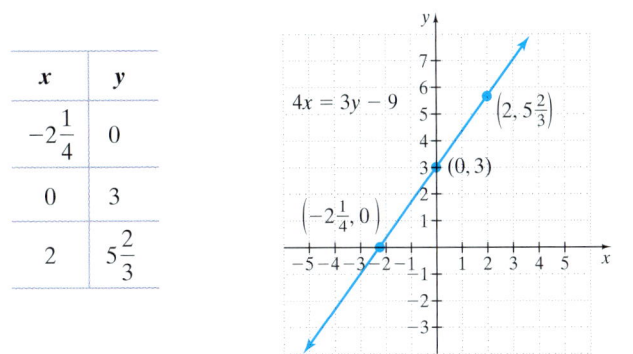

x	y
$-2\dfrac{1}{4}$	0
0	3
2	$5\dfrac{2}{3}$

PRACTICE 8 Graph $3x = 2y + 4$.

OBJECTIVE 3 ▶ Graphing vertical and horizontal lines. The equation $x = c$, where c is a real number constant, is a linear equation in two variables because it can be written in the form $x + 0y = c$. The graph of this equation is a vertical line as shown in the next example.

EXAMPLE 9 Graph $x = 2$.

Solution The equation $x = 2$ can be written as $x + 0y = 2$. For any y-value chosen, notice that x is 2. No other value for x satisfies $x + 0y = 2$. Any ordered pair whose x-coordinate is 2 is a solution of $x + 0y = 2$. We will use the ordered pair solutions $(2, 3)$, $(2, 0)$, and $(2, -3)$ to graph $x = 2$.

x	y
2	3
2	0
2	-3

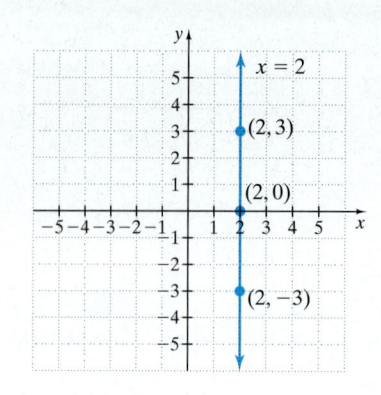

The graph is a vertical line with x-intercept $(2, 0)$. Note that this graph has no y-intercept because x is never 0.

PRACTICE 9 Graph $y = 2$.

Vertical Lines

The graph of $x = c$, where c is a real number, is a vertical line with x-intercept $(c, 0)$.

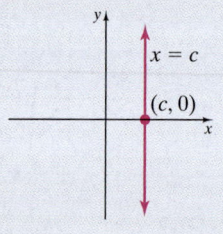

EXAMPLE 10 Graph $y = -3$.

Solution The equation $y = -3$ can be written as $0x + y = -3$. For any x-value chosen, y is -3. If we choose 4, 1, and -2 as x-values, the ordered pair solutions are $(4, -3)$, $(1, -3)$, and $(-2, -3)$. Use these ordered pairs to graph $y = -3$. The graph is a horizontal line with y-intercept $(0, -3)$ and no x-intercept.

x	y
4	-3
1	-3
-2	-3

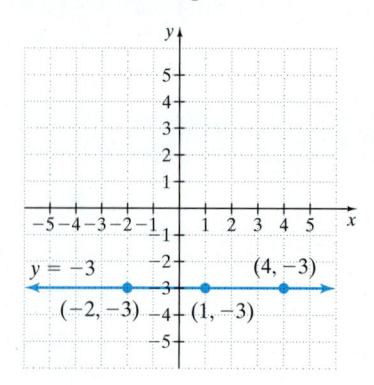

PRACTICE 10 Graph $x = -2$.

Horizontal Lines

The graph of $y = c$, where c is a real number, is a horizontal line with y-intercept $(0, c)$.

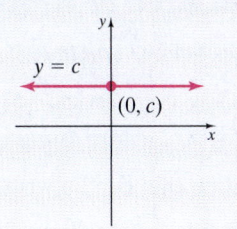

Graphing Calculator Explorations

You may have noticed that to use the $\boxed{Y=}$ key on a grapher to graph an equation, the equation must be solved for y. For example, to graph $2x + 3y = 7$, we solve this equation for y.

$$2x + 3y = 7$$
$$3y = -2x + 7 \quad \text{Subtract } 2x \text{ from both sides.}$$
$$\frac{3y}{3} = -\frac{2x}{3} + \frac{7}{3} \quad \text{Divide both sides by 3.}$$
$$y = -\frac{2}{3}x + \frac{7}{3} \quad \text{Simplify.}$$

To graph $2x + 3y = 7$ or $y = -\frac{2}{3}x + \frac{7}{3}$, press the $\boxed{Y=}$ key and enter

$$Y_1 = -\frac{2}{3}x + \frac{7}{3}$$

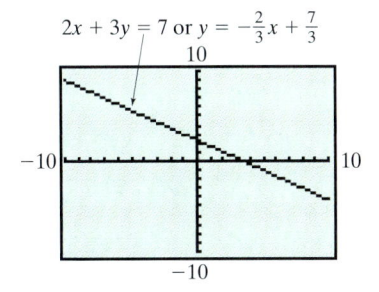

Graph each linear equation.

1. $x = 3.78y$ **2.** $-2.61y = x$ **3.** $3x + 7y = 21$

4. $-4x + 6y = 12$ **5.** $-2.2x + 6.8y = 15.5$ **6.** $5.9x - 0.8y = -10.4$

VOCABULARY & READINESS CHECK

Use the choices below to fill in each blank. Some choices may be used more than once. Exercises 1 and 2 come from Section 3.2.

x	vertical	x-intercept	linear
y	horizontal	y-intercept	standard

1. An equation that can be written in the form $Ax + By = C$ is called a _____ equation in two variables.
2. The form $Ax + By = C$ is called _____ form.
3. The graph of the equation $y = -1$ is a _____ line.
4. The graph of the equation $x = 5$ is a _____ line.
5. A point where a graph crosses the y-axis is called a(n) _____.
6. A point where a graph crosses the x-axis is called a(n) _____.
7. Given an equation of a line, to find the x-intercept (if there is one), let _____ = 0 and solve for _____.
8. Given an equation of a line, to find the y-intercept (if there is one), let _____ = 0 and solve for _____.

Answer the following true or false.

9. All lines have an x-intercept *and* a y-intercept.
10. The graph of $y = 4x$ contains the point $(0, 0)$.
11. The graph of $x + y = 5$ has an x-intercept of $(5, 0)$ and a y-intercept of $(0, 5)$.
12. The graph of $y = 5x$ contains the point $(5, 1)$.

3.3 EXERCISE SET

Identify the intercepts. See Examples 1 through 5.

1. 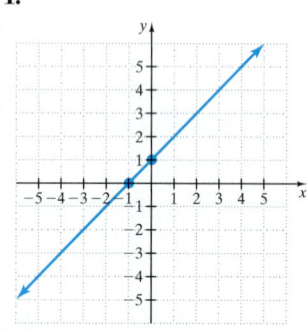
2. 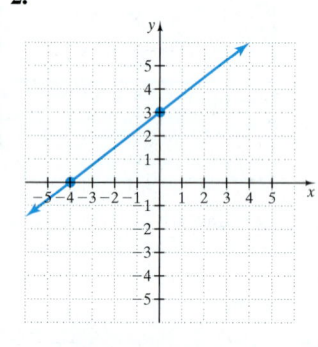
3. 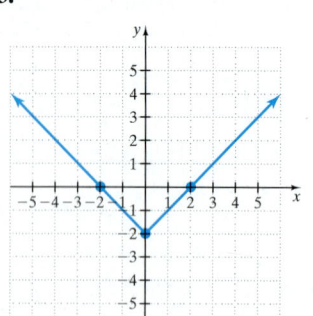
4. 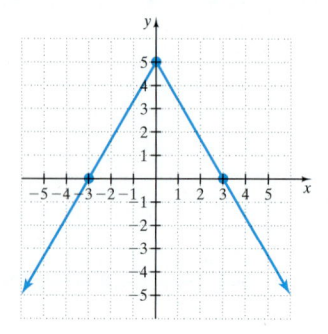
5. 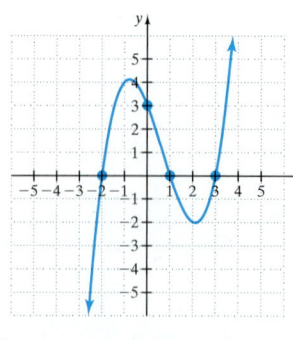
6. 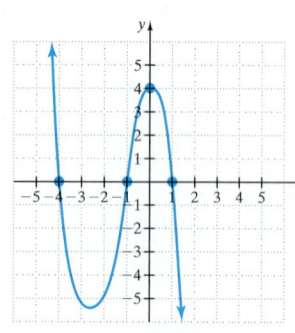
7. 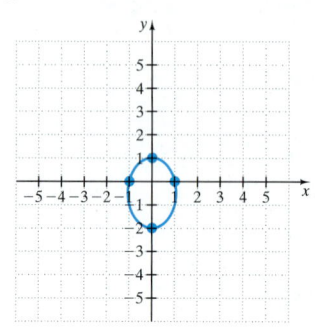
8.

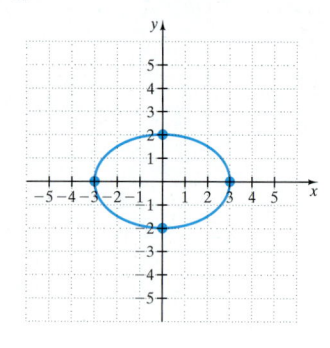

Solve. See Example 1.

9. What is the greatest number of intercepts for a line?
10. What is the least number of intercepts for a line?
11. What is the least number of intercepts for a circle?
12. What is the greatest number of intercepts for a circle?

Graph each linear equation by finding and plotting its intercepts. See Examples 6 through 8.

13. $x - y = 3$
14. $x - y = -4$
15. $x = 5y$
16. $x = 2y$
17. $-x + 2y = 6$
18. $x - 2y = -8$
19. $2x - 4y = 8$
20. $2x + 3y = 6$
21. $y = 2x$
22. $y = -2x$
23. $y = 3x + 6$
24. $y = 2x + 10$

Graph each linear equation. See Examples 9 and 10.

25. $x = -1$
26. $y = 5$
27. $y = 0$
28. $x = 0$
29. $y + 7 = 0$
30. $x - 2 = 0$
31. $x + 3 = 0$
32. $y - 6 = 0$

MIXED PRACTICE

Graph each linear equation. See Examples 6 through 10.

33. $x = y$
34. $x = -y$
35. $x + 8y = 8$
36. $x + 3y = 9$
37. $5 = 6x - y$
38. $4 = x - 3y$
39. $-x + 10y = 11$
40. $-x + 9y = 10$
41. $x = -4\frac{1}{2}$
42. $x = -1\frac{3}{4}$
43. $y = 3\frac{1}{4}$
44. $y = 2\frac{1}{2}$
45. $y = -\frac{2}{3}x + 1$
46. $y = -\frac{3}{5}x + 3$
47. $4x - 6y + 2 = 0$
48. $9x - 6y + 3 = 0$

For Exercises 49 through 54, match each equation with its graph.

A. 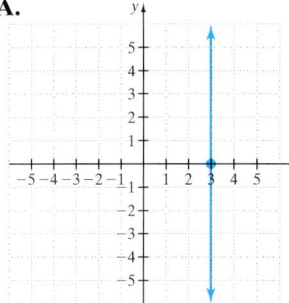
B. 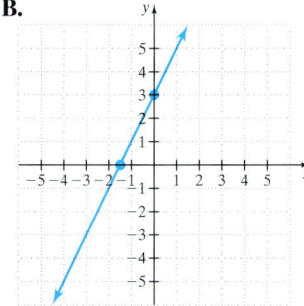
C. 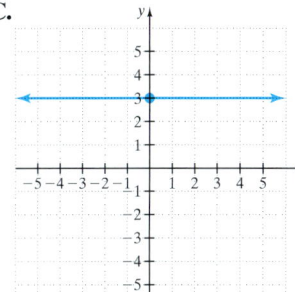
D. 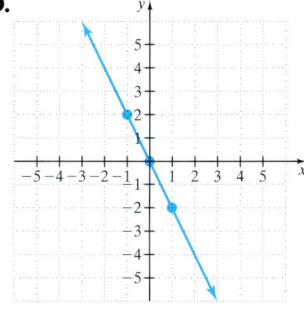
E. 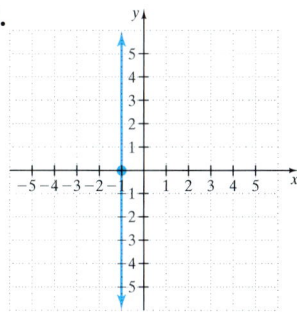
F.

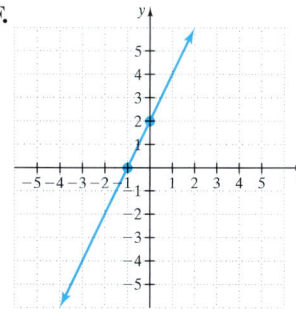

49. $y = 3$
50. $y = 2x + 2$
51. $x = -1$
52. $x = 3$
53. $y = 2x + 3$
54. $y = -2x$

REVIEW AND PREVIEW

Simplify.

55. $\dfrac{-6 - 3}{2 - 8}$
56. $\dfrac{4 - 5}{-1 - 0}$
57. $\dfrac{-8 - (-2)}{-3 - (-2)}$
58. $\dfrac{12 - 3}{10 - 9}$
59. $\dfrac{0 - 6}{5 - 0}$
60. $\dfrac{2 - 2}{3 - 5}$

CONCEPT EXTENSIONS

61. The revenue for the Disney Parks and Resorts y (in millions) for the years 2003–2006 can be approximated by the equation $y = 1181x + 6505$, where x represents the number of years after 2003. (*Source:* Based on data from The Walt Disney Company)
    a. Find the y-intercept of this equation.
    b. What does the y-intercept mean?

62. The average price of a digital camera y (in dollars) can be modeled by the linear equation $y = -78.1x + 491.8$ where x represents the number of years after 2000. (*Source*: NPDTechworld)
    a. Find the y-intercept of this equation.
    b. What does this y-intercept mean?

63. Since 2002, admissions at movie theaters have been in a decline. The number of people y (in billions) who go to movie theaters each year can be estimated by the equation $y = -0.075x + 1.65$, where x represents the number of years

since 2002. (*Source:* Based on data from Motion Picture Association of America)

a. Find the *x*-intercept of this equation.
b. What does this *x*-intercept mean?
c. Use part (b) to comment on the limitations of using equations to model real data.

64. The price of admission to a movie theater has been steadily increasing. The price of regular admission *y* (in dollars) to a movie theater may be represented by the equation $y = 0.2x + 5.42$, where *x* is the number of years after 2000. (*Source:* Based on data from Motion Picture Association of America)

a. Find the *x*-intercept of this equation.
b. What does this *x*-intercept mean?
c. Use part (b) to comment on the limitation of using equations to model real data.

65. The production supervisor at Alexandra's Office Products finds that it takes 3 hours to manufacture a particular office chair and 6 hours to manufacture a computer desk. A total of 1200 hours is available to produce office chairs and desks of this style. The linear equation that models this situation is $3x + 6y = 1200$, where *x* represents the number of chairs produced and *y* the number of desks manufactured.

a. Complete the ordered pair solution (0,) of this equation. Describe the manufacturing situation that corresponds to this solution.
b. Complete the ordered pair solution (,0) of this equation. Describe the manufacturing situation that corresponds to this solution.
c. Use the ordered pairs found above and graph the equation $3x + 6y = 1200$.
d. If 50 computer desks are manufactured, find the greatest number of chairs that they can make.

Two lines in the same plane that do not intersect are called **parallel lines.**

66. Draw a line parallel to the line $x = 5$ that intersects the *x*-axis at $(1, 0)$. What is the equation of this line?
67. Draw a line parallel to the line $y = -1$ that intersects the *y*-axis at $(0, -4)$. What is the equation of this line?
68. Discuss whether a vertical line ever has a *y*-intercept.
69. Explain why it is a good idea to use three points to graph a linear equation.
70. Discuss whether a horizontal line ever has an *x*-intercept.
71. Explain how to find intercepts.

STUDY SKILLS BUILDER

Are You Familiar with Your Textbook Supplements?

Below is a review of some of the student supplements available for additional study. Check to see if you are using the ones most helpful to you.

- Chapter Test Prep Videos on CD. This CD is found with your textbook and contains video clip solutions to the Chapter Test exercises in this text. You will find this extremely useful when studying for chapter tests.
- Lecture Videos on CD-ROM. These are keyed to each section of the text. The material is presented by me, Elayn Martin-Gay, and I have placed a 💿 by the exercises in the text that I have worked on the video.
- The *Student Solutions Manual.* This contains worked out solutions to odd-numbered exercises as well as every exercise in the Integrated Reviews, Chapter Reviews, Chapter Tests, and Cumulative Reviews.
- Pearson Tutor Center. Mathematics questions may be phoned, faxed, or emailed to this center.
- MyMathLab is a text-specific online course. MathXL is an online homework, tutorial, and assessment system. Take a moment and determine whether these are available to you.

As usual, your instructor is your best source of information.

Let's see how you are doing with textbook supplements.

1. Name one way the Lecture Videos can be helpful to you.
2. Name one way the Chapter Test Prep Video can help you prepare for a chapter test.
3. List any textbook supplements that you have found useful.
4. Have you located and visited a learning resource lab located on your campus?
5. List the textbook supplements that are currently housed in your campus' learning resource lab.

3.4 SLOPE AND RATE OF CHANGE

OBJECTIVES

1. Find the slope of a line given two points of the line.
2. Find the slope of a line given its equation.
3. Find the slopes of horizontal and vertical lines.
4. Compare the slopes of parallel and perpendicular lines.
5. Slope as a rate of change.

OBJECTIVE 1 ▶ Finding the slope of a line given two points of the line. Thus far, much of this chapter has been devoted to graphing lines. You have probably noticed by now that a key feature of a line is its slant or steepness. In mathematics, the slant or steepness of a line is formally known as its **slope**. We measure the slope of a line by the ratio of vertical change to the corresponding horizontal change as we move along the line.

On the line below, for example, suppose that we begin at the point (1, 2) and move to the point (4, 6). The vertical change is the change in y-coordinates: $6 - 2$ or 4 units. The corresponding horizontal change is the change in x-coordinates: $4 - 1 = 3$ units. The ratio of these changes is

$$\text{slope} = \frac{\text{change in } y \text{ (vertical change)}}{\text{change in } x \text{ (horizontal change)}} = \frac{4}{3}$$

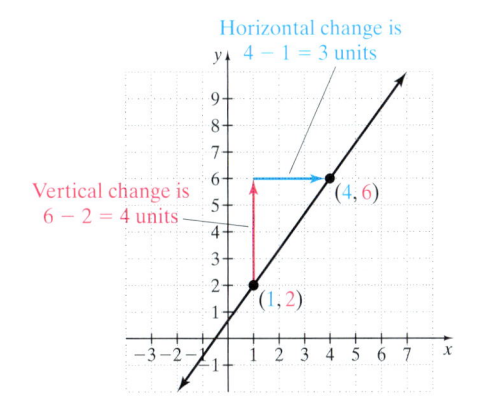

The slope of this line, then, is $\frac{4}{3}$. This means that for every 4 units of change in y-coordinates, there is a corresponding change of 3 units in x-coordinates.

▶ **Helpful Hint**
It makes no difference what two points of a line are chosen to find its slope. The slope of a line is the same everywhere on the line.

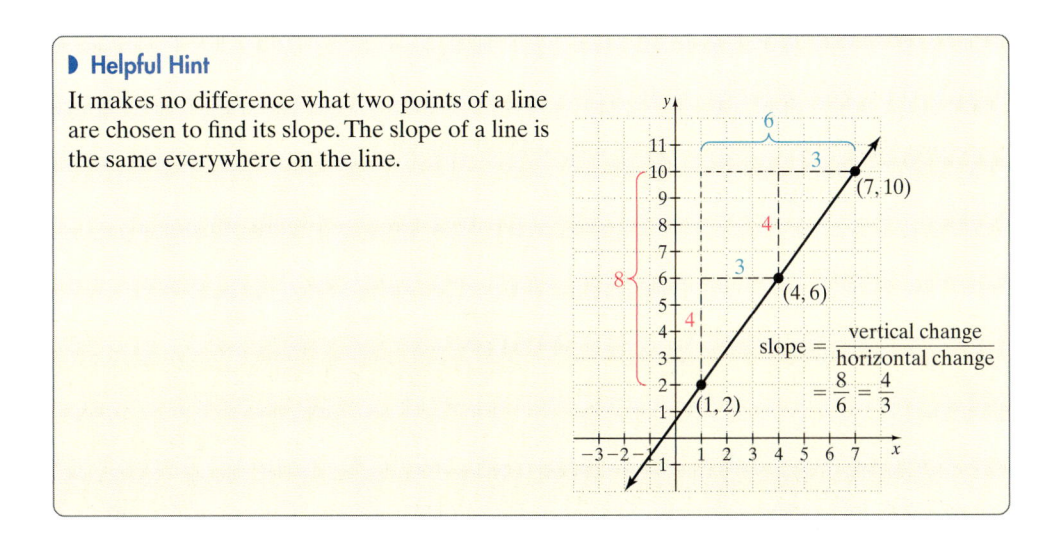

$$\text{slope} = \frac{\text{vertical change}}{\text{horizontal change}} = \frac{8}{6} = \frac{4}{3}$$

To find the slope of a line, then, choose two points of the line. Label the two x-coordinates of two points, x_1 and x_2 (read "x sub one" and "x sub two"), and label the corresponding y-coordinates y_1 and y_2.

The vertical change or **rise** between these points is the difference in the y-coordinates: $y_2 - y_1$. The horizontal change or **run** between the points is the difference of the x-coordinates: $x_2 - x_1$. The slope of the line is the ratio of $y_2 - y_1$ to $x_2 - x_1$, and we traditionally use the letter m to denote slope $m = \dfrac{y_2 - y_1}{x_2 - x_1}$.

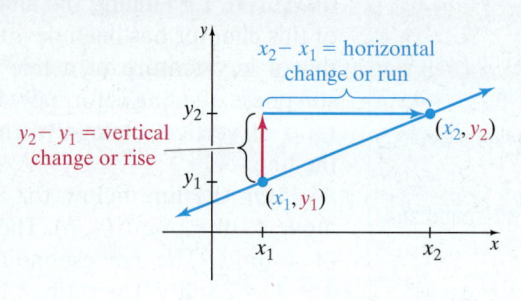

> **Slope of a Line**
>
> The slope m of the line containing the points (x_1, y_1) and (x_2, y_2) is given by
>
> $$m = \dfrac{\text{rise}}{\text{run}} = \dfrac{\text{change in } y}{\text{change in } x} = \dfrac{y_2 - y_1}{x_2 - x_1}, \quad \text{as long as } x_2 \neq x_1$$

EXAMPLE 1 Find the slope of the line through $(-1, 5)$ and $(2, -3)$. Graph the line.

Solution If we let (x_1, y_1) be $(-1, 5)$, then $x_1 = -1$ and $y_1 = 5$. Also, let (x_2, y_2) be $(2, -3)$ so that $x_2 = 2$ and $y_2 = -3$. Then, by the definition of slope,

$$m = \dfrac{y_2 - y_1}{x_2 - x_1}$$

$$= \dfrac{-3 - 5}{2 - (-1)}$$

$$= \dfrac{-8}{3} = -\dfrac{8}{3}$$

The slope of the line is $-\dfrac{8}{3}$.

PRACTICE 1 Find the slope of the line through $(-4, 11)$ and $(2, 5)$.

> **Helpful Hint**
>
> When finding slope, it makes no difference which point is identified as (x_1, y_1) and which is identified as (x_2, y_2). Just remember that whatever y-value is first in the numerator, its corresponding x-value is first in the denominator. Another way to calculate the slope in Example 1 is:
>
> $$m = \dfrac{y_2 - y_1}{x_2 - x_1} = \dfrac{5 - (-3)}{-1 - 2} = \dfrac{8}{-3} \quad \text{or} \quad -\dfrac{8}{3} \quad \leftarrow \text{Same slope as found in Example 1.}$$

Answer to Concept Check:

$$m = \dfrac{3}{2}$$

Concept Check ✓

The points $(-2, -5)$, $(0, -2)$, $(4, 4)$, and $(10, 13)$ all lie on the same line. Work with a partner and verify that the slope is the same no matter which points are used to find slope.

Section 3.4 Slope and Rate of Change 205

EXAMPLE 2 Find the slope of the line through $(-1, -2)$ and $(2, 4)$. Graph the line.

Solution Let (x_1, y_1) be $(2, 4)$ and (x_2, y_2) be $(-1, -2)$.

$$m = \frac{y_2 - y_1}{x_2 - x_1}$$

$$= \frac{-2 - 4}{-1 - 2} \quad \begin{array}{l} y\text{-value} \\ \text{corresponding } x\text{-value} \end{array}$$

$$= \frac{-6}{-3} = 2$$

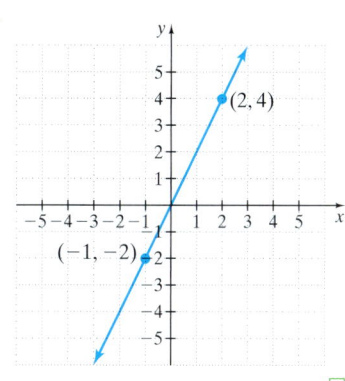

> **Helpful Hint**
> The slope for Example 2 is the same if we let (x_1, y_1) be $(-1, -2)$ and (x_2, y_2) be $(2, 4)$.
>
> $$m = \frac{\overset{y\text{-value}}{4 - (-2)}}{\underset{\text{corresponding } x\text{-value}}{2 - (-1)}} = \frac{6}{3} = 2$$

PRACTICE 2 Find the slope of the line through $(-3, -1)$ and $(3, 1)$.

Concept Check ✓

What is wrong with the following slope calculation for the points $(3, 5)$ and $(-2, 6)$?

$$m = \frac{5 - 6}{-2 - 3} = \frac{-1}{-5} = \frac{1}{5}$$

Notice that the slope of the line in Example 1 is negative, whereas the slope of the line in Example 2 is positive. Let your eye follow the line with negative slope from left to right and notice that the line "goes down." Following the line with positive slope from left to right, notice that the line "goes up." This is true in general.

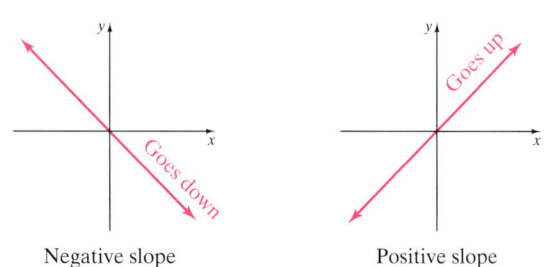

Negative slope   Positive slope

> **Helpful Hint**
> To decide whether a line "goes up" or "goes down," always follow the line from left to right.

OBJECTIVE 2 ▶ Finding the slope of a line given its equation. As we have seen, the slope of a line is defined by two points on the line. Thus, if we know the equation of a line, we can find its slope by finding two of its points. For example, let's find the slope of the line

$$y = 3x + 2$$

To find two points, we can choose two values for x and substitute to find corresponding y-values. If $x = 0$, for example, $y = 3 \cdot 0 + 2$ or $y = 2$. If $x = 1$, $y = 3 \cdot 1 + 2$ or $y = 5$. This gives the ordered pairs $(0, 2)$ and $(1, 5)$. Using the definition for slope, we have

$$m = \frac{5 - 2}{1 - 0} = \frac{3}{1} = 3 \quad \text{The slope is 3.}$$

Notice that the slope, 3, is the same as the coefficient of x in the equation $y = 3x + 2$.

Also, recall from Section 3.2 that the graph of an equation of the form $y = mx + b$ has y-intercept $(0, b)$.

This means that the y-intercept of the graph of $y = 3x + 2$ is $(0, 2)$. This is true in general.

Answer to Concept Check:
The order in which the x- and y-values are used must be the same.

$$m = \frac{5 - 6}{3 - (-2)} = \frac{-1}{5} = -\frac{1}{5}$$

When a linear equation is written in the form $y = mx + b$, not only is $(0, b)$ the y-intercept of the line, but m is its slope. The form $y = mx + b$ is appropriately called the **slope-intercept form**.

$$y = \underset{\uparrow}{m}x + \underset{\uparrow}{b}$$
$$\text{slope} \quad \text{y-intercept}$$
$$(0, b)$$

> **Slope-Intercept Form**
>
> When a linear equation in two variables is written in slope-intercept form,
>
> $$y = mx + b$$
>
> m is the slope of the line and $(0, b)$ is the y-intercept of the line.

EXAMPLE 3 Find the slope and y-intercept of the line whose equation is $y = \frac{3}{4}x + 6$.

Solution The equation is in slope-intercept form, $y = mx + b$.

$$y = \frac{3}{4}x + 6$$

The coefficient of x, $\frac{3}{4}$, is the slope and the constant term, 6 is the y-value of the y-intercept, $(0, 6)$.

PRACTICE 3 Find the slope and y-intercept of the line whose equation is $y = \frac{2}{3}x - 2$.

EXAMPLE 4 Find the slope and the y-intercept of the line whose equation is $5x + y = 2$.

Solution Write the equation in slope-intercept form by solving the equation for y.

$$5x + y = 2$$
$$y = -5x + 2 \quad \text{Subtract } 5x \text{ from both sides.}$$

The coefficient of x, -5, is the slope and the constant term, 2, is the y-value of the y-intercept, $(0, 2)$.

PRACTICE 4 Find the slope and y-intercept of the line whose equation is $6x - y = 5$.

EXAMPLE 5 Find the slope and the y-intercept of the line whose equation is $3x - 4y = 4$.

Solution Write the equation in slope-intercept form by solving for y.

$$3x - 4y = 4$$
$$-4y = -3x + 4 \quad \text{Subtract } 3x \text{ from both sides.}$$
$$\frac{-4y}{-4} = \frac{-3x}{-4} + \frac{4}{-4} \quad \text{Divide both sides by } -4.$$
$$y = \frac{3}{4}x - 1 \quad \text{Simplify.}$$

The coefficient of x, $\frac{3}{4}$, is the slope, and the y-intercept is $(0, -1)$.

PRACTICE 5 Find the slope and the y-intercept of the line whose equation is $5x + 2y = 8$.

OBJECTIVE 3 ▶ Finding slopes of horizontal and vertical lines. Recall that if a line tilts upward from left to right, its slope is positive. If a line tilts downward from left to right, its slope is negative. Let's now find the slopes of two special lines, horizontal and vertical lines.

EXAMPLE 6 Find the slope of the line $y = -1$.

Solution Recall that $y = -1$ is a horizontal line with y-intercept $(0, -1)$. To find the slope, find two ordered pair solutions of $y = -1$. Solutions of $y = -1$ must have a y-value of -1. Let's use points $(2, -1)$ and $(-3, -1)$, which are on the line.

$$m = \frac{y_2 - y_1}{x_2 - x_1} = \frac{-1 - (-1)}{-3 - 2} = \frac{0}{-5} = 0$$

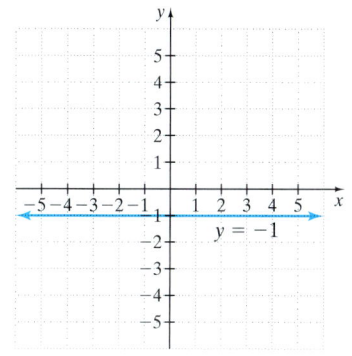

The slope of the line $y = -1$ is 0 and its graph is shown.

PRACTICE 6 Find the slope of the line $y = 3$.

Any two points of a horizontal line will have the same y-values. This means that the y-values will always have a difference of 0 for all horizontal lines. Thus, **all horizontal lines have a slope 0.**

EXAMPLE 7 Find the slope of the line $x = 5$.

Solution Recall that the graph of $x = 5$ is a vertical line with x-intercept $(5, 0)$.

To find the slope, find two ordered pair solutions of $x = 5$. Solutions of $x = 5$ must have an x-value of 5. Let's use points $(5, 0)$ and $(5, 4)$, which are on the line.

$$m = \frac{y_2 - y_1}{x_2 - x_1} = \frac{4 - 0}{5 - 5} = \frac{4}{0}$$

Since $\frac{4}{0}$ is undefined, we say the slope of the vertical line $x = 5$ is undefined, and its graph is shown.

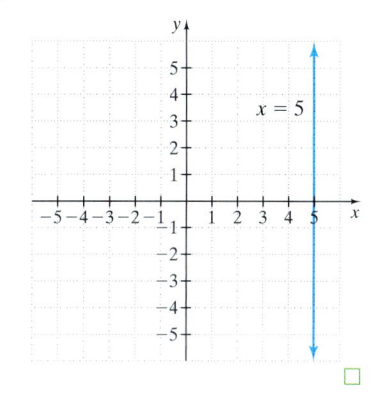

PRACTICE 7 Find the slope of the line $x = -4$.

Any two points of a vertical line will have the same x-values. This means that the x-values will always have a difference of 0 for all vertical lines. Thus **all vertical lines have undefined slope.**

▶ **Helpful Hint**

Slope of 0 and undefined slope are not the same. Vertical lines have undefined slope or no slope, while horizontal lines have a slope of 0.

Here is a general review of slope.

Summary of Slope

Slope m of the line through (x_1, y_1) and (x_2, y_2) is given by the equation $m = \dfrac{y_2 - y_1}{x_2 - x_1}$.

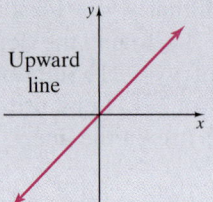

Upward line
Positive slope: $m > 0$

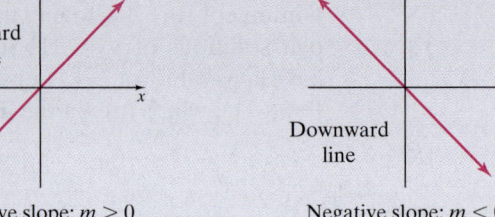

Downward line
Negative slope: $m < 0$

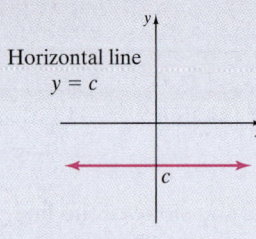

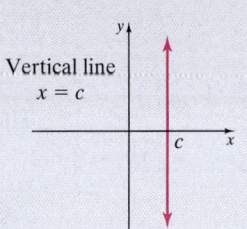

Horizontal line
$y = c$
Zero slope: $m = 0$

Vertical line
$x = c$
Undefined slope or no slope

OBJECTIVE 4 ▶ Slopes of parallel and perpendicular lines. Two lines in the same plane are **parallel** if they do not intersect. Slopes of lines can help us determine whether lines are parallel. Parallel lines have the same steepness, so it follows that they have the same slope.

For example, the graphs of

$$y = -2x + 4$$

and

$$y = -2x - 3$$

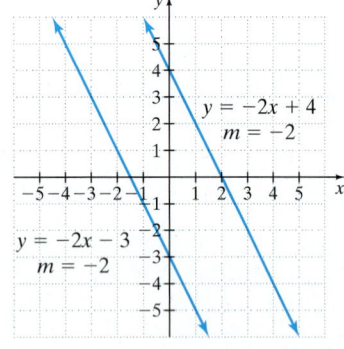

$y = -2x + 4$
$m = -2$

$y = -2x - 3$
$m = -2$

are shown. These lines have the same slope, -2. They also have different y-intercepts, so the lines are parallel. (If the y-intercepts were the same also, the lines would be the same.)

Parallel Lines

Nonvertical parallel lines have the same slope and different y-intercepts.

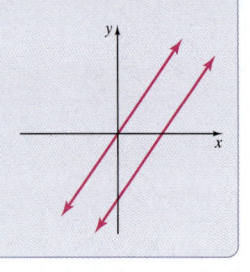

Two lines are **perpendicular** if they lie in the same plane and meet at a 90° (right) angle. How do the slopes of perpendicular lines compare? The product of the slopes of two perpendicular lines is -1.

For example, the graphs of
$$y = 4x + 1$$
and
$$y = -\frac{1}{4}x - 3$$
are shown. The slopes of the lines are 4 and $-\frac{1}{4}$. Their product is $4\left(-\frac{1}{4}\right) = -1$, so the lines are perpendicular.

> **Perpendicular Lines**
>
> If the product of the slopes of two lines is -1, then the lines are perpendicular.
>
> (Two nonvertical lines are perpendicular if the slopes of one is the negative reciprocal of the slope of the other.)

> ▶ **Helpful Hint**
>
> Here are examples of numbers that are negative (opposite) reciprocals.
>
> | Number | Negative Reciprocal | Their Product Is -1. |
> |---|---|---|
> | $\frac{2}{3}$ | $-\frac{3}{2}$ | $\frac{2}{3} \cdot -\frac{3}{2} = -\frac{6}{6} = -1$ |
> | -5 or $-\frac{5}{1}$ | $\frac{1}{5}$ | $-5 \cdot \frac{1}{5} = -\frac{5}{5} = -1$ |

> ▶ **Helpful Hint**
>
> Here are a few important facts about vertical and horizontal lines.
>
> - Two distinct vertical lines are parallel.
> - Two distinct horizontal lines are parallel.
> - A horizontal line and a vertical line are always perpendicular.

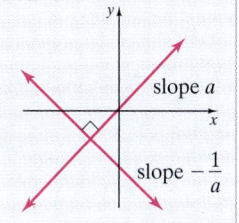

△ **EXAMPLE 8** Determine whether each pair of lines is parallel, perpendicular, or neither.

a. $y = -\frac{1}{5}x + 1$
$2x + 10y = 3$

b. $x + y = 3$
$-x + y = 4$

c. $3x + y = 5$
$2x + 3y = 6$

Solution

a. The slope of the line $y = -\frac{1}{5}x + 1$ is $-\frac{1}{5}$. We find the slope of the second line by solving its equation for y.

$$2x + 10y = 3$$
$$10y = -2x + 3 \quad \text{Subtract } 2x \text{ from both sides.}$$
$$y = \frac{-2}{10}x + \frac{3}{10} \quad \text{Divide both sides by 10.}$$
$$y = -\frac{1}{5}x + \frac{3}{10} \quad \text{Simplify.}$$

The slope of this line is $-\frac{1}{5}$ also. Since the lines have the same slope and different y-intercepts, they are parallel, as shown in the figure on the next page.

b. To find each slope, we solve each equation for y.

$$x + y = 3 \qquad\qquad -x + y = 4$$
$$y = -x + 3 \qquad\qquad y = x + 4$$

The slope is −1.   The slope is 1.

The slopes are not the same, so the lines are not parallel. Next we check the product of the slopes: $(-1)(1) = -1$. Since the product is −1, the lines are perpendicular, as shown in the figure.

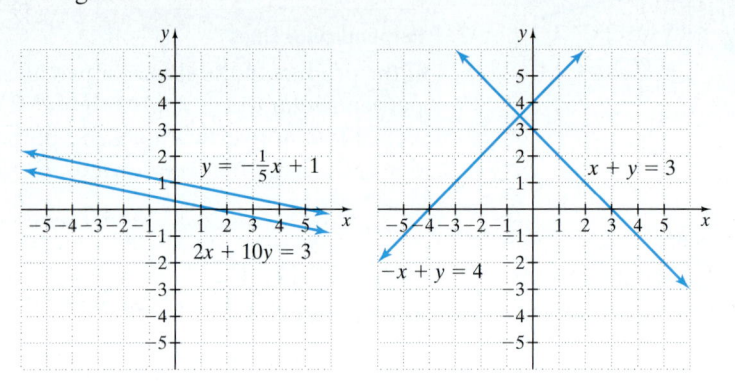

c. We solve each equation for y to find each slope. The slopes are -3 and $-\frac{2}{3}$. The slopes are not the same and their product is not −1. Thus, the lines are neither parallel nor perpendicular.

PRACTICE 8 Determine whether each pair of lines is parallel, perpendicular, or neither.

a. $y = -5x + 1$
   $x - 5y = 10$

b. $x + y = 11$
   $2x + y = 11$

c. $2x + 3y = 21$
   $6y = -4x - 2$

Concept Check ✓

Consider the line $-6x + 2y = 1$.

a. Write the equations of two lines parallel to this line.
b. Write the equations of two lines perpendicular to this line.

OBJECTIVE 5 ▶ Slope as a rate of change. Slope can also be interpreted as a rate of change. In other words, slope tells us how fast y is changing with respect to x. To see this, let's look at a few of the many real-world applications of slope. For example, the pitch of a roof, used by builders and architects, is its slope. The pitch of the roof on the left is $\frac{7}{10}\left(\frac{\text{rise}}{\text{run}}\right)$. This means that the roof rises vertically 7 feet for every horizontal 10 feet. The rate of change for the roof is 7 vertical feet (y) per 10 horizontal feet (x).

The grade of a road is its slope written as a percent. A 7% grade, as shown below, means that the road rises (or falls) 7 feet for every horizontal 100 feet. $\left(\text{Recall that } 7\% = \frac{7}{100}.\right)$ Here, the slope of $\frac{7}{100}$ gives us the rate of change. The road rises (in our diagram) 7 vertical feet (y) for every 100 horizontal feet (x).

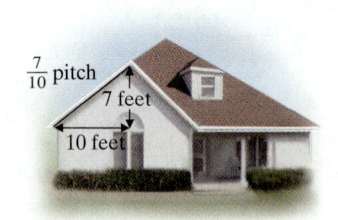

Answers to Concept Check:

a. any two lines with $m = 3$ and y-intercept not $\left(0, \frac{1}{2}\right)$

b. any two lines with $m = -\frac{1}{3}$

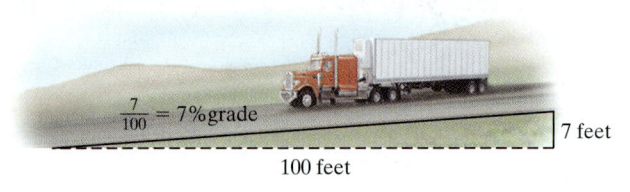

Section 3.4 Slope and Rate of Change **211**

EXAMPLE 9   Finding the Grade of a Road

At one part of the road to the summit of Pikes Peak, the road rises at a rate of 15 vertical feet for a horizontal distance of 250 feet. Find the grade of the road.

Solution   Recall that the grade of a road is its slope written as a percent.

$$\text{grade} = \frac{\text{rise}}{\text{run}} = \frac{15}{250} = 0.06 = 6\%$$

15 feet

250 feet

The grade is 6%.

PRACTICE
9   One part of the Mt. Washington (New Hampshire) cog railway rises about 1794 feet over a horizontal distance of 7176 feet. Find the grade of this part of the railway.

EXAMPLE 10   Finding the Slope of a Line

The following graph shows the cost y (in cents) of a nationwide long-distance telephone call from Texas with a certain telephone-calling plan, where x is the length of the call in minutes. Find the slope of the line and attach the proper units for the rate of change. Then write a sentence explaining the meaning of slope in this application.

Solution   Use (2, 34) and (6, 62) to calculate slope.

$$m = \frac{62 - 34}{6 - 2} = \frac{28}{4} = \frac{7 \text{ cents}}{1 \text{ minute}}$$

This means that the rate of change of a phone call is 7 cents per 1 minute or the cost of the phone call is 7 cents per minute.

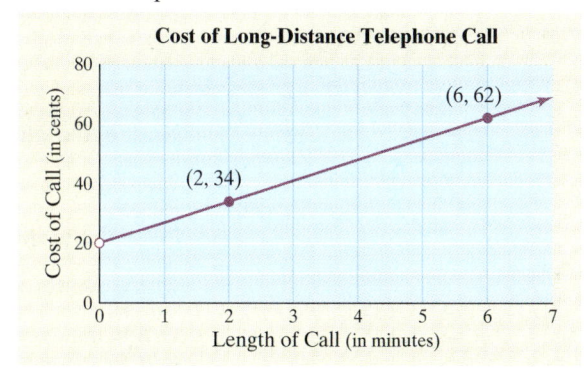

Cost of Long-Distance Telephone Call

PRACTICE
10   The following graph shows the cost y (in dollars) of having laundry done at the Wash-n-Fold, where x is the number of pounds of laundry. Find the slope of the line, and attach the proper units for the rate of change.

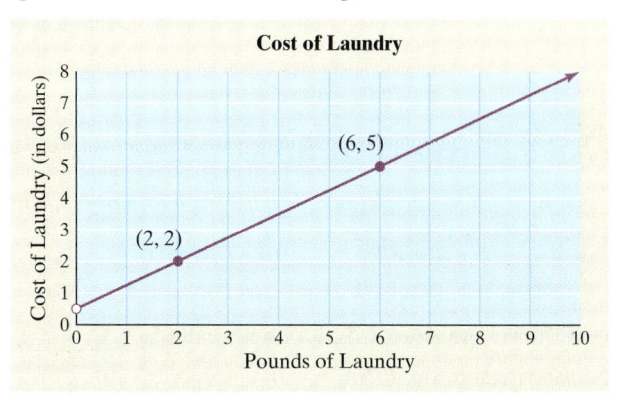

Cost of Laundry

Graphing Calculator Explorations

It is possible to use a grapher to sketch the graph of more than one equation on the same set of axes. This feature can be used to confirm our findings from Section 3.2 when we learned that the graph of an equation written in the form $y = mx + b$ has a y-intercept of b. For example, graph the equations $y = \frac{2}{5}x$, $y = \frac{2}{5}x + 7$, and $y = \frac{2}{5}x - 4$ on the same set of axes. To do so, press the $\boxed{Y=}$ key and enter the equations on the first three lines.

$$Y_1 = \left(\frac{2}{5}\right)x$$

$$Y_2 = \left(\frac{2}{5}\right)x + 7$$

$$Y_3 = \left(\frac{2}{5}\right)x - 4$$

The screen should look like:

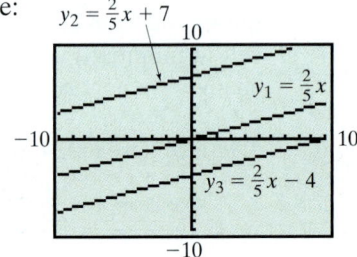

Notice that all three graphs appear to have the same positive slope. The graph of $y = \frac{2}{5}x + 7$ is the graph of $y = \frac{2}{5}x$ moved 7 units upward with a y-intercept of 7. Also, the graph of $y = \frac{2}{5}x - 4$ is the graph of $y = \frac{2}{5}x$ moved 4 units downward with a y-intercept of -4.

Graph the equations on the same set of axes. Describe the similarities and differences in their graphs.

1. $y = 3.8x$, $y = 3.8x - 3$, $y = 3.8x + 9$
2. $y = -4.9x$, $y = -4.9x + 1$, $y = -4.9x + 8$
3. $y = \frac{1}{4}x$; $y = \frac{1}{4}x + 5$, $y = \frac{1}{4}x - 8$
4. $y = -\frac{3}{4}x$, $y = -\frac{3}{4}x - 5$, $y = -\frac{3}{4}x + 6$

VOCABULARY & READINESS CHECK

Use the choices below to fill in each blank. Not all choices will be used.

    m   x   0   positive   undefined
    b   y   slope   negative

1. The measure of the steepness or tilt of a line is called _____.
2. If an equation is written in the form $y = mx + b$, the value of the letter _____ is the value of the slope of the graph.
3. The slope of a horizontal line is _____.
4. The slope of a vertical line is _____.
5. If the graph of a line moves upward from left to right, the line has _____ slope.

6. If the graph of a line moves downward from left to right, the line has _____ slope.

7. Given two points of a line, slope = $\dfrac{\text{change in } \_\_}{\text{change in } \_\_}$.

State whether the slope of the line is positive, negative, 0, or is undefined.

8.   9. 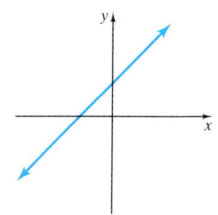   10. 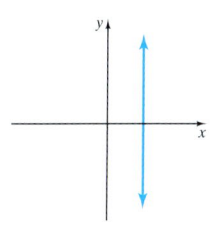   11.

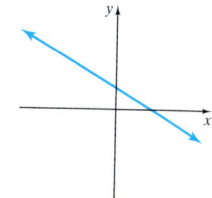

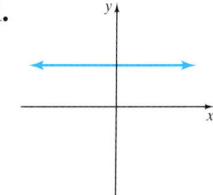

Decide whether a line with the given slope is upward, downward, horizontal, or vertical.

12. $m = \dfrac{7}{6}$ _____   13. $m = -3$ _____   14. $m = 0$ _____   15. m is undefined. _____

3.4 EXERCISE SET

Find the slope of the line that passes through the given points. See Examples 1 and 2.

1. $(-1, 5)$ and $(6, -2)$
2. $(3, 1)$ and $(2, 6)$
3. $(-4, 3)$ and $(-4, 5)$
4. $(6, -6)$ and $(6, 2)$
5. $(-2, 8)$ and $(1, 6)$
6. $(4, -3)$ and $(2, 2)$
7. $(5, 1)$ and $(-2, 1)$
8. $(0, 13)$ and $(-4, 13)$

Find the slope of each line if it exists. See Examples 1 and 2.

9.   10.

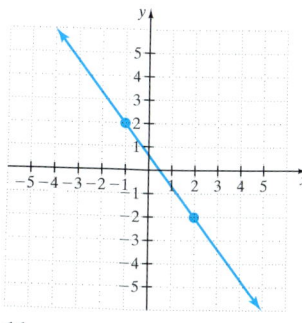

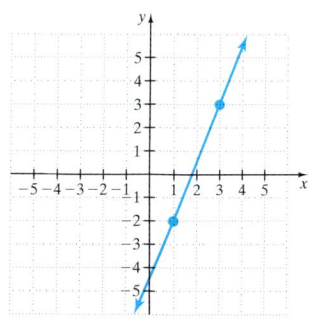

11. 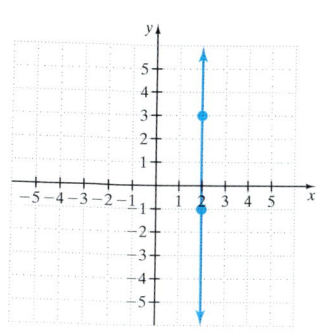   12.

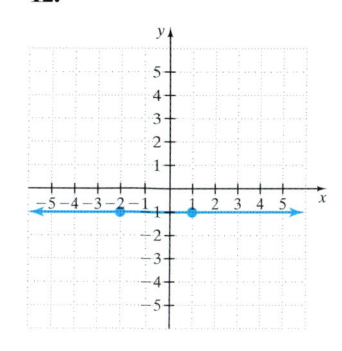

13. 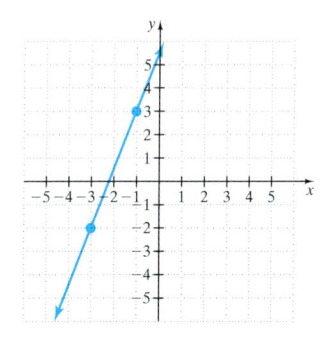   14.

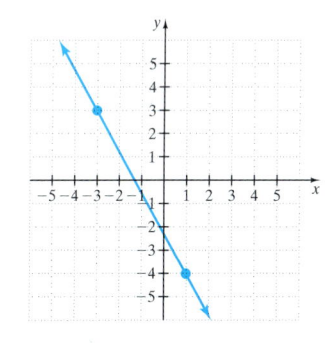

For each graph, determine which line has the greater slope.

15.   16.

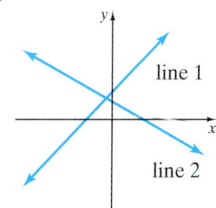

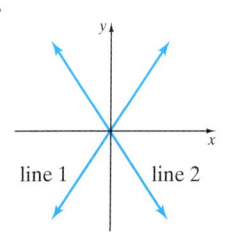

17. 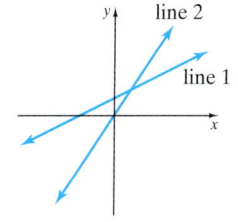   18.

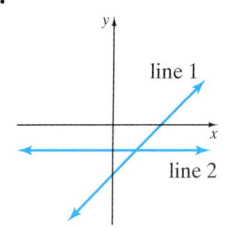

214   CHAPTER 3   Graphing

In Exercises 25 through 30, match each line with its slope.

A. $m = 0$ **B.** undefined slope   **C.** $m = 3$

D. $m = 1$ **E.** $m = -\dfrac{1}{2}$ **F.** $m = -\dfrac{3}{4}$

19.

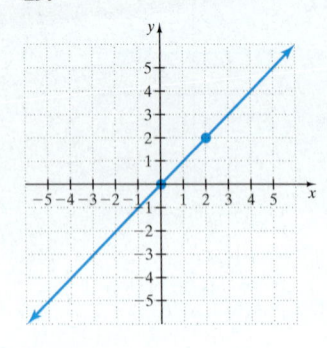

20.

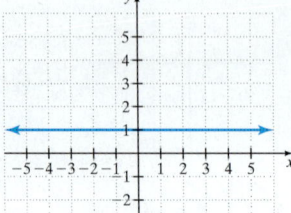

21.

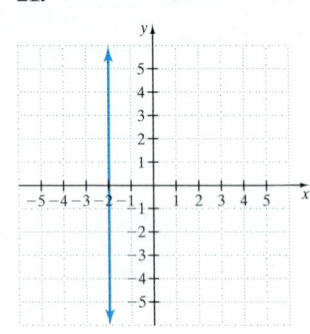

22.

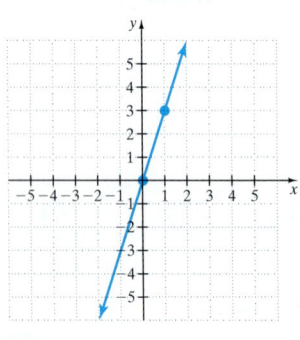

23.

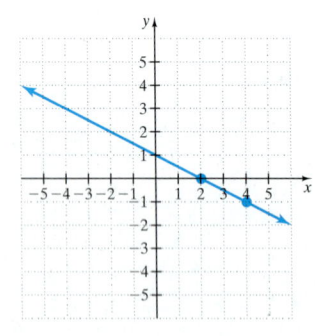

24.

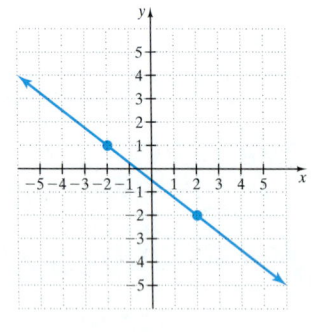

Find the slope of each line. See Examples 6 and 7.

25. $x = 6$
26. $y = 4$
27. $y = -4$
28. $x = 2$
29. $x = -3$
30. $y = -11$
31. $y = 0$
32. $x = 0$

MIXED PRACTICE

Find the slope of each line. See Examples 3 through 7.

33. $y = 5x - 2$
34. $y = -2x + 6$
35. $y = -0.3x + 2.5$
36. $y = -7.6x - 0.1$
37. $2x + y = 7$
38. $-5x + y = 10$
39. $2x - 3y = 10$
40. $3x - 5y = 1$
41. $x = 1$
42. $y = -2$
43. $x = 2y$
44. $x = -4y$
45. $y = -3$
46. $x = 5$

47. $-3x - 4y = 6$
48. $-4x - 7y = 9$
49. $20x - 5y = 1.2$
50. $24x - 3y = 5.7$

△ *Determine whether each pair of lines is parallel, perpendicular, or neither. See Example 8.*

51. $y = \dfrac{2}{9}x + 3$

    $y = -\dfrac{2}{9}x$

52. $y = \dfrac{1}{5}x + 20$

    $y = -\dfrac{1}{5}x$

53. $x - 3y = -6$

    $y = 3x - 9$

54. $y = 4x - 2$

    $4x + y = 5$

55. $6x = 5y + 1$

    $-12x + 10y = 1$

56. $-x + 2y = -2$

    $2x = 4y + 3$

57. $6 + 4x = 3y$

    $3x + 4y = 8$

58. $10 + 3x = 5y$

    $5x + 3y = 1$

The pitch of a roof is its slope. Find the pitch of each roof shown. See Example 9.

59.

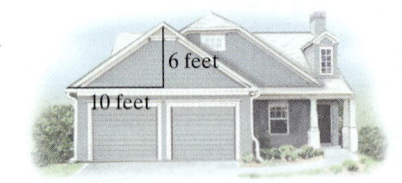

6 feet
10 feet

60.

5
10

The grade of a road is its slope written as a percent. Find the grade of each road shown. See Example 9.

61.

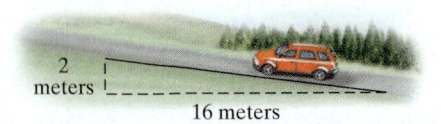

2 meters
16 meters

62.

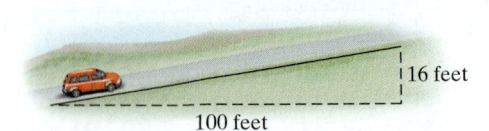

16 feet
100 feet

63. One of Japan's superconducting "bullet" trains is researched and tested at the Yamanashi Maglev Test Line near Otsuki City. The steepest section of the track has a rise of 2580 meters for a horizontal distance of 6450 meters. What is the grade of this section of track? (*Source:* Japan Railways Central Co.)

2580 meters
6450 meters

64. Professional plumbers suggest that a sewer pipe should rise 0.25 inch for every horizontal foot. Find the recommended slope for a sewer pipe. Round to the nearest hundredth.

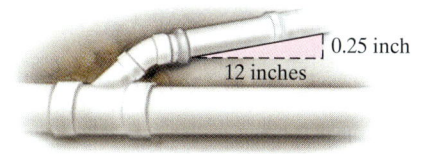

65. The steepest street is Baldwin Street in Dunedin, New Zealand. It has a maximum rise of 10 meters for a horizontal distance of 12.66 meters. Find the grade of this section of road. Round to the nearest whole percent. (*Source: The Guinness Book of Records*)

66. According to federal regulations, a wheelchair ramp should rise no more than 1 foot for a horizontal distance of 12 feet. Write the slope as a grade. Round to the nearest tenth of a percent.

Find the slope of each line and write the slope as a rate of change. Don't forget to attach the proper units. See Example 10.

67. This graph approximates the number of U.S. households that have personal computers y (in millions) for year x.

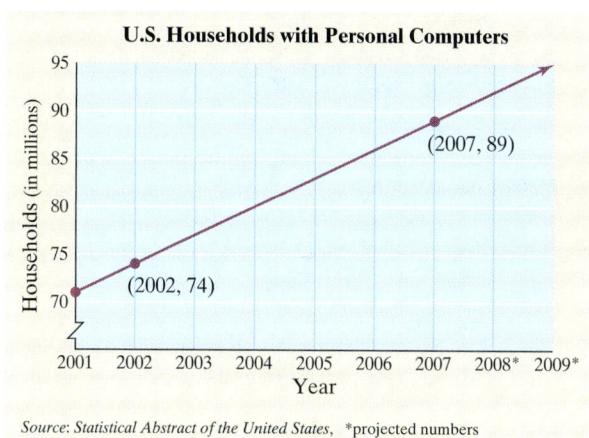

U.S. Households with Personal Computers
(2002, 74), (2007, 89)
Source: Statistical Abstract of the United States, *projected numbers

68. This graph approximates the number y (per hundred population) of Attention Deficit Hyperactivity Disorder (ADHD) prescriptions for children under 18 for the year x. (*Source: Centers for Disease Control and Prevention (CDC) National Center for Health Statistics*)

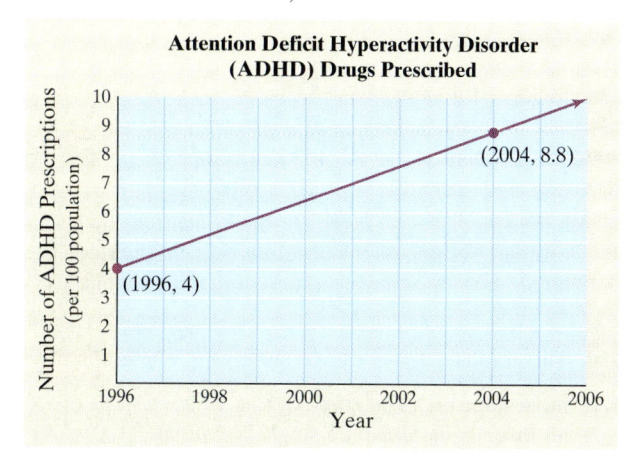

Attention Deficit Hyperactivity Disorder (ADHD) Drugs Prescribed
(1996, 4), (2004, 8.8)

69. The graph below shows the total cost y (in dollars) of owning and operating a compact car where x is the number of miles driven.

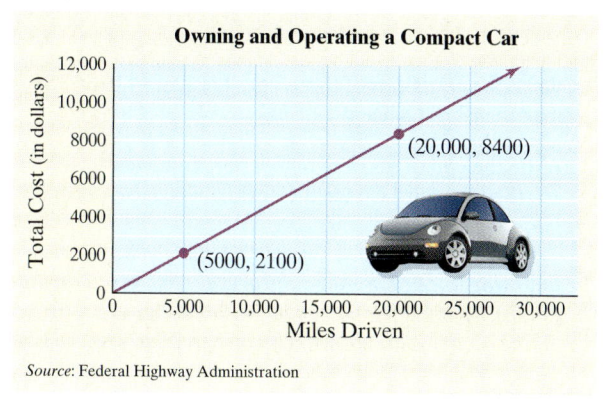

Owning and Operating a Compact Car
(5000, 2100), (20,000, 8400)
Source: Federal Highway Administration

70. The graph below shows the total cost y (in dollars) of owning and operating a standard pickup truck, where x is the number of miles driven.

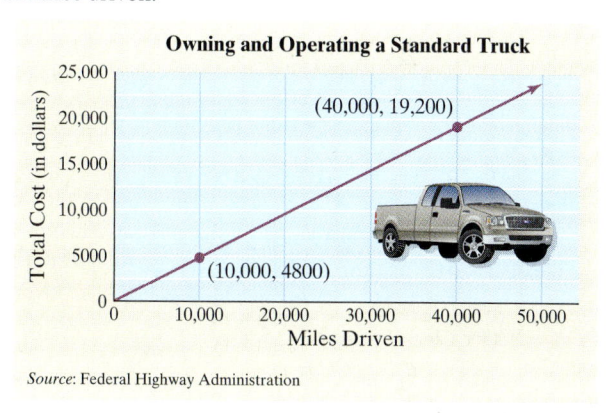

Owning and Operating a Standard Truck
(10,000, 4800), (40,000, 19,200)
Source: Federal Highway Administration

REVIEW AND PREVIEW

Solve each equation for y. See Section 2.5.

71. $y - (-6) = 2(x - 4)$
72. $y - 7 = -9(x - 6)$
73. $y - 1 = -6(x - (-2))$
74. $y - (-3) = 4(x - (-5))$

CONCEPT EXTENSIONS

△ *Find the slope of the line that is (a) parallel and (b) perpendicular to the line through each pair of points.*

75. $(-3, -3)$ and $(0, 0)$

76. $(6, -2)$ and $(1, 4)$

77. $(-8, -4)$ and $(3, 5)$

78. $(6, -1)$ and $(-4, -10)$

Solve. See a Concept Check in this section.

79. Verify that the points $(2, 1), (0, 0), (-2, -1)$ and $(-4, -2)$ are all on the same line by computing the slope between each pair of points. (See the first Concept Check.)

80. Given the points $(2, 3)$ and $(-5, 1)$, can the slope of the line through these points be calculated by $\dfrac{1-3}{2-(-5)}$? Why or why not? (See the second Concept Check.)

81. Write the equations of three lines parallel to $10x - 5y = -7$. (See the third Concept Check.)

82. Write the equations of two lines perpendicular to $10x - 5y = -7$. (See the third Concept Check.)

The following line graph shows the average fuel economy (in miles per gallon) by mid-size passenger automobiles produced during each of the model years shown. Use this graph to answer Exercises 83 through 88.

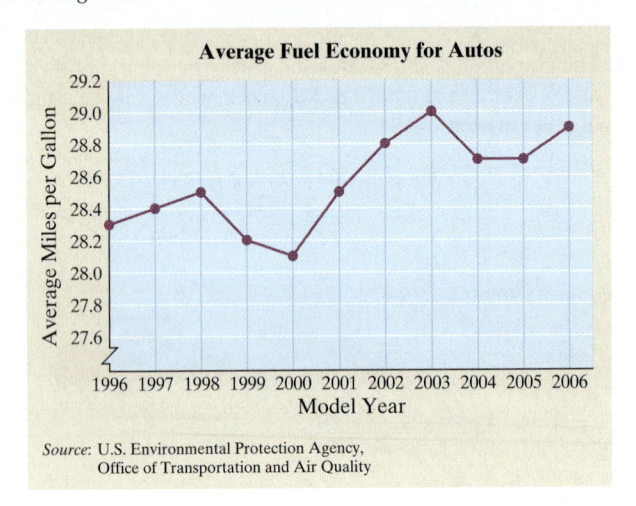

83. What was the average fuel economy (in miles per gallon) for automobiles produced during 2001?

84. Find the decrease in average fuel economy for automobiles between the years 2003 to 2004.

85. During which of the model years shown was average fuel economy the lowest?
What was the average fuel economy for that year?

86. During which of the model years shown was average fuel economy the highest?
What was the average fuel economy for that year?

87. What line segment has the greatest slope?

88. What line segment has the least positive slope?

Solve.

89. Find x so that the pitch of the roof is $\dfrac{1}{3}$.

90. Find x so that the pitch of the roof is $\dfrac{2}{5}$.

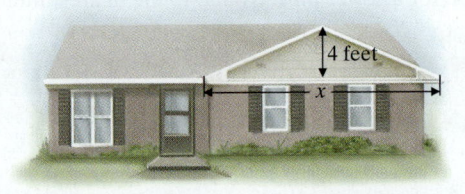

91. The average price of an acre of U.S. farmland was $1132 in 2001. In 2006, the price of an acre rose to approximately $1657. (*Source:* National Agricultural Statistics Service)
    a. Write two ordered pairs of the form (year, price of acre)
    b. Find the slope of the line through the two points.
    c. Write a sentence explaining the meaning of the slope as a rate of change.

92. There were approximately 14,774 kidney transplants performed in the United States in 2002. In 2006, the number of kidney transplants performed in the United States rose to 15,722. (*Source:* Organ Procurement and Transplantation Network)
    a. Write two ordered pairs of the form (year, number of kidney transplants).
    b. Find the slope of the line between the two points.
    c. Write a sentence explaining the meaning of the slope as a rate of change.

93. Show that a triangle with vertices at the points $(1, 1)$, $(-4, 4)$, and $(-3, 0)$ is a right triangle.

94. Show that the quadrilateral with vertices $(1, 3), (2, 1), (-4, 0)$, and $(-3, -2)$ is a parallelogram.

Find the slope of the line through the given points.

95. $(2.1, 6.7)$ and $(-8.3, 9.3)$

96. $(-3.8, 1.2)$ and $(-2.2, 4.5)$

97. $(2.3, 0.2)$ and $(7.9, 5.1)$

98. $(14.3, -10.1)$ and $(9.8, -2.9)$

99. The graph of $y = -\dfrac{1}{3}x + 2$ has a slope of $-\dfrac{1}{3}$. The graph of $y = -2x + 2$ has a slope of -2. The graph of $y = -4x + 2$ has a slope of -4. Graph all three equations on a single coordinate system. As the absolute value of the slope becomes larger, how does the steepness of the line change?

100. The graph of $y = \dfrac{1}{2}x$ has a slope of $\dfrac{1}{2}$. The graph of $y = 3x$ has a slope of 3. The graph of $y = 5x$ has a slope of 5. Graph all three equations on a single coordinate system. As slope becomes larger, how does the steepness of the line change?

INTEGRATED REVIEW — SUMMARY ON SLOPE & GRAPHING LINEAR EQUATIONS

Sections 3.1–3.4

Find the slope of each line.

1. 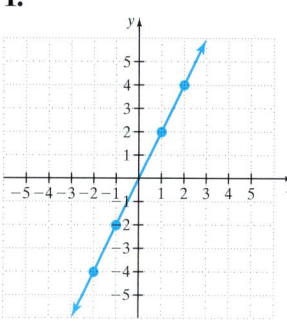
2. 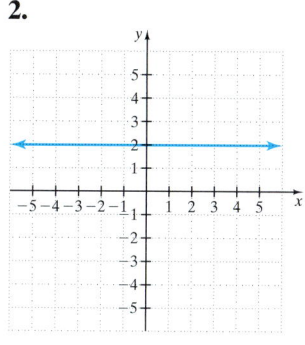
3. 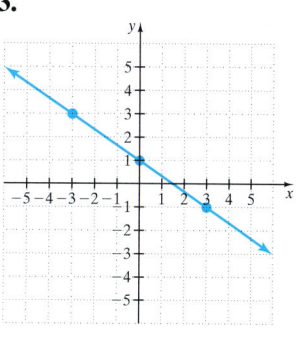
4.

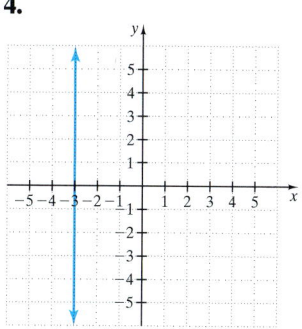

Graph each linear equation.

5. $y = -2x$

6. $x + y = 3$

7. $x = -1$

8. $y = 4$

9. $x - 2y = 6$

10. $y = 3x + 2$

11. $5x + 3y = 15$

12. $2x - 4y = 8$

Determine whether the lines through the points are parallel, perpendicular, or neither.

13. $y = -\dfrac{1}{5}x + \dfrac{1}{3}$
    $3x = -15y$

14. $x - y = \dfrac{1}{2}$
    $3x - y = \dfrac{1}{2}$

15. In the years 2002 through 2005 the number of admissions to movie theaters in the United States can be modeled by the linear equation $y = -75x + 1650$ where x is years after 2002 and y is admissions in millions. (*Source:* Motion Picture Assn. of America)
    a. Find the y-intercept of this line.
    b. Write a sentence explaining the meaning of this intercept.
    c. Find the slope of this line.
    d. Write a sentence explaining the meaning of the slope as a rate of change.

16. Online advertising is a means of promoting products and services using the Internet. The revenue (in billions of dollars) for online advertising for the years 2003 through a projected 2010 is given by $y = 3.3x - 3.1$, where x is the number of years after 2000.
    a. Use this equation to complete the ordered pair (9,   ).
    b. Write a sentence explaining the meaning of the answer to part (a).

3.5 EQUATIONS OF LINES

OBJECTIVES

1. Use the slope-intercept form to write an equation of a line.
2. Use the slope-intercept form to graph a linear equation.
3. Use the point-slope form to find an equation of a line given its slope and a point of the line.
4. Use the point-slope form to find an equation of a line given two points of the line.
5. Find equations of vertical and horizontal lines.
6. Use the point-slope form to solve problems.

Recall that the form $y = mx + b$ is appropriately called the *slope-intercept form* of a linear equation.

$y = \underset{\text{slope}}{m}x + \underset{\text{y-intercept is }(0,b)}{b}$

> **Slope-Intercept Form**
> When a linear equation in two variables is written in **slope-intercept form**,
> $$y = mx + b$$
> where m is slope and $(0, b)$ is the y-intercept,
> then m is the slope of the line and $(0, b)$ is the y-intercept of the line.

OBJECTIVE 1 ▶ **Using the slope-intercept form to write an equation.** As we know from the previous section, writing an equation in slope-intercept form is a way to find the slope and y-intercept of its graph. The slope-intercept form can be used to write the equation of a line when we know its slope and y-intercept.

EXAMPLE 1 Find an equation of the line with y-intercept $(0, -3)$ and slope of $\frac{1}{4}$.

Solution We are given the slope and the y-intercept. We let $m = \frac{1}{4}$ and $b = -3$ and write the equation in slope-intercept form, $y = mx + b$.

$y = mx + b$

$y = \frac{1}{4}x + (-3)$ Let $m = \frac{1}{4}$ and $b = -3$.

$y = \frac{1}{4}x - 3$ Simplify.

PRACTICE 1 Find an equation of the line with y-intercept $(0, 7)$ and slope of $\frac{1}{2}$.

OBJECTIVE 2 ▶ **Using the slope-intercept form to graph an equation.** We also can use the slope-intercept form of the equation of a line to graph a linear equation.

EXAMPLE 2 Use the slope-intercept form to graph the equation

$$y = \frac{3}{5}x - 2$$

Solution Since the equation $y = \frac{3}{5}x - 2$ is written in slope-intercept form $y = mx + b$, the slope of its graph is $\frac{3}{5}$ and the y-intercept is $(0, -2)$. To graph this equation, we begin by plotting the point $(0, -2)$. From this point, we can find another point of the graph by using the slope $\frac{3}{5}$ and recalling that slope is $\frac{\text{rise}}{\text{run}}$. We start at the y-intercept and move 3 units up since the numerator of the slope is 3; then we move 5 units to the right since the denominator of the slope is 5. We stop at the point $(5, 1)$. The line through $(0, -2)$ and $(5, 1)$ is the graph of $y = \frac{3}{5}x - 2$.

PRACTICE 2 Graph $y = \frac{2}{3}x - 5$.

Section 3.5 Equations of Lines **219**

EXAMPLE 3 Use the slope-intercept form to graph the equation $4x + y = 1$.

Solution First we write the given equation in slope-intercept form.

$$4x + y = 1$$
$$y = -4x + 1$$

The graph of this equation will have slope -4 and y-intercept $(0, 1)$. To graph this line, we first plot the point $(0, 1)$. To find another point of the graph, we use the slope -4, which can be written as $\frac{-4}{1}$ ($\frac{4}{-1}$ could also be used). We start at the point $(0, 1)$ and move 4 units down (since the numerator of the slope is -4), and then 1 unit to the right (since the denominator of the slope is 1).

We arrive at the point $(1, -3)$. The line through $(0, 1)$ and $(1, -3)$ is the graph of $4x + y = 1$.

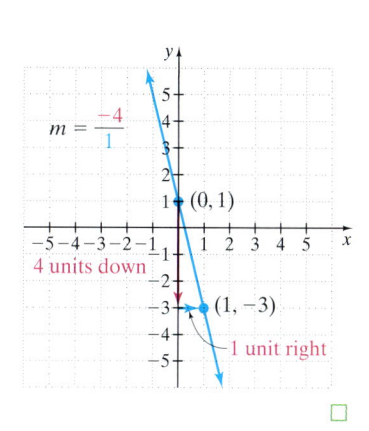

> **Helpful Hint**
> In Example 3, if we interpret the slope of -4 as $\frac{4}{-1}$, we arrive at $(-1, 5)$ for a second point. Notice that this point is also on the line.

PRACTICE 3 Use the slope-intercept form to graph the equation $3x - y = 2$.

OBJECTIVE 3 ▶ Writing an equation given slope and a point. Thus far, we have seen that we can write an equation of a line if we know its slope and y-intercept. We can also write an equation of a line if we know its slope and any point on the line. To see how we do this, let m represent slope and (x_1, y_1) represent the point on the line. Then if (x, y) is any other point of the line, we have that

$$\frac{y - y_1}{x - x_1} = m$$

$$y - y_1 = m(x - x_1) \quad \text{Multiply both sides by } (x - x_1).$$
$$\uparrow$$
$$\text{slope}$$

This is the *point-slope form* of the equation of a line.

Point-Slope Form of the Equation of a Line

The **point-slope form** of the equation of a line is

$$y - y_1 = m(x - x_1)$$
$$\text{slope} \quad (x_1, y_1) \text{ point of the line}$$

where m is the slope of the line and (x_1, y_1) is a point on the line.

EXAMPLE 4 Find an equation of the line with slope -2 that passes through $(-1, 5)$. Write the equation in slope-intercept form, $y = mx + b$, and in standard form, $Ax + By = C$.

Solution Since the slope and a point on the line are given, we use point-slope form $y - y_1 = m(x - x_1)$ to write the equation. Let $m = -2$ and $(-1, 5) = (x_1, y_1)$.

$$y - y_1 = m(x - x_1)$$
$$y - 5 = -2[x - (-1)] \quad \text{Let } m = -2 \text{ and } (x_1, y_1) = (-1, 5).$$
$$y - 5 = -2(x + 1) \quad \text{Simplify.}$$
$$y - 5 = -2x - 2 \quad \text{Use the distributive property.}$$

To write the equation in slope-intercept form, $y = mx + b$, we simply solve the equation for y. To do this, we add 5 to both sides.

$$y - 5 = -2x - 2$$
$$y = -2x + 3 \quad \text{Slope-intercept form.}$$
$$2x + y = 3 \quad \text{Add } 2x \text{ to both sides and we have standard form.} \qquad \square$$

PRACTICE

4 Find an equation of the line passing through $(2, 3)$ with slope 4. Write the equation in standard form: $Ax + By = C$.

OBJECTIVE 4 ▶ Writing an equation given two points. We can also find the equation of a line when we are given any two points of the line.

EXAMPLE 5 Find an equation of the line through $(2, 5)$ and $(-3, 4)$. Write the equation in standard form.

Solution First, use the two given points to find the slope of the line.

$$m = \frac{4 - 5}{-3 - 2} = \frac{-1}{-5} = \frac{1}{5}$$

Next we use the slope $\frac{1}{5}$ and either one of the given points to write the equation in point-slope form. We use $(2, 5)$. Let $x_1 = 2$, $y_1 = 5$, and $m = \frac{1}{5}$.

$$y - y_1 = m(x - x_1) \quad \text{Use point-slope form.}$$
$$y - 5 = \frac{1}{5}(x - 2) \quad \text{Let } x_1 = 2, y_1 = 5, \text{ and } m = \frac{1}{5}.$$
$$5(y - 5) = 5 \cdot \frac{1}{5}(x - 2) \quad \text{Multiply both sides by 5 to clear fractions.}$$
$$5y - 25 = x - 2 \quad \text{Use the distributive property and simplify.}$$
$$-x + 5y - 25 = -2 \quad \text{Subtract } x \text{ from both sides.}$$
$$-x + 5y = 23 \quad \text{Add 25 to both sides.} \qquad \square$$

PRACTICE

5 Find an equation of the line through $(-1, 6)$ and $(3, 1)$. Write the equation in standard form.

> ▶ **Helpful Hint**
>
> Multiply both sides of the equation $-x + 5y = 23$ by -1, and it becomes $x - 5y = -23$. Both $-x + 5y = 23$ and $x - 5y = -23$ are in standard form, and they are equations of the same line.

OBJECTIVE 5 ▶ Finding equations of vertical and horizontal lines. Recall from Section 3.3 that:

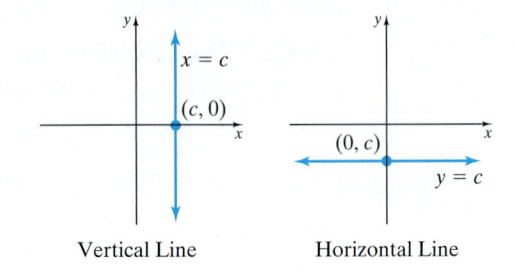

Vertical Line: $x = c$, point $(c, 0)$

Horizontal Line: $y = c$, point $(0, c)$

EXAMPLE 6 Find an equation of the vertical line through $(-1, 5)$.

Solution The equation of a vertical line can be written in the form $x = c$, so an equation for a vertical line passing through $(-1, 5)$ is $x = -1$.

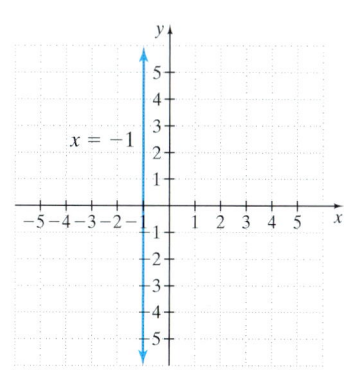

PRACTICE 6 Find an equation of the vertical line through $(3, -2)$.

EXAMPLE 7 Find an equation of the line parallel to the line $y = 5$ and passing through $(-2, -3)$.

Solution Since the graph of $y = 5$ is a horizontal line, any line parallel to it is also horizontal. The equation of a horizontal line can be written in the form $y = c$. An equation for the horizontal line passing through

$(-2, -3)$ is $y = -3$.

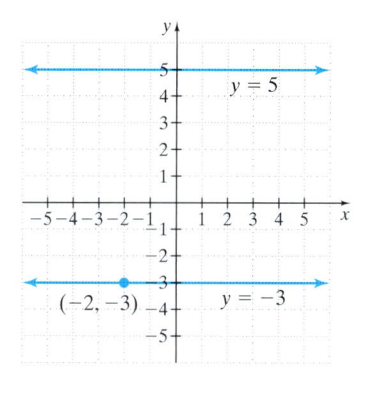

PRACTICE 7 Find an equation of the line parallel to the line $y = -2$ and passing through $(4, 3)$.

OBJECTIVE 6 ▶ Using the point-slope form to solve problems. Problems occurring in many fields can be modeled by linear equations in two variables. The next example is from the field of marketing and shows how consumer demand of a product depends on the price of the product.

EXAMPLE 8 Predicting the Sales of T-Shirts

A web-based T-shirt company has learned that by pricing a clearance-sale T-shirt at $6, sales will reach 2000 T-shirts per day. Raising the price to $8 will cause the sales to fall to 1500 T-shirts per day.

a. Assume that the relationship between sales price and number of T-shirts sold is linear and write an equation describing this relationship. Write the equation in slope-intercept form.

b. Predict the daily sales of T-shirts if the price is $7.50.

Solution

a. First, use the given information and write two ordered pairs. Ordered pairs will be in the form (sales price, number sold) so that our ordered pairs are (6, 2000) and

(8, 1500). Use the point-slope form to write an equation. To do so, we find the slope of the line that contains these points.

$$m = \frac{2000 - 1500}{6 - 8} = \frac{500}{-2} = -250$$

Next, use the slope and either one of the points to write the equation in point-slope form. We use (6, 2000).

$y - y_1 = m(x - x_1)$	Use point-slope form.
$y - 2000 = -250(x - 6)$	Let $x_1 = 6$, $y_1 = 2000$, and $m = -250$.
$y - 2000 = -250x + 1500$	Use the distributive property.
$y = -250x + 3500$	Write in slope-intercept form.

b. To predict the sales if the price is $7.50, we find y when $x = 7.50$.

$y = -250x + 3500$	
$y = -250(7.50) + 3500$	Let $x = 7.50$.
$y = -1875 + 3500$	
$y = 1625$	

If the price is $7.50, sales will reach 1625 T-shirts per day.

PRACTICE

8 The new *Camelot* condos were selling at a rate of 30 per month when they were priced at $150,000 each. Lowering the price to $120,000 caused the sales to rise to 50 condos per month.

a. Assume that the relationship between number of condos sold and price is linear, and write an equation describing this relationship. Write the equation in slope-intercept form.

b. What should the condos be priced at if the developer wishes to sell 60 condos per month?

The preceding example may also be solved by using ordered pairs of the form (number sold, sales price).

Forms of Linear Equations

$Ax + By = C$	**Standard form** of a linear equation. A and B are not both 0.
$y = mx + b$	**Slope-intercept form** of a linear equation. The slope is m and the y-intercept is $(0, b)$.
$y - y_1 = m(x - x_1)$	**Point-slope form** of a linear equation. The slope is m and (x_1, y_1) is a point on the line.
$y = c$	**Horizontal line** The slope is 0 and the y-intercept is $(0, c)$.
$x = c$	**Vertical line** The slope is undefined and the x-intercept is $(c, 0)$.

Parallel and Perpendicular Lines

Nonvertical parallel lines have the same slope.

The product of the slopes of two nonvertical perpendicular lines is -1.

Graphing Calculator Explorations

A grapher is a very useful tool for discovering patterns. To discover the change in the graph of a linear equation caused by a change in slope, try the following. Use a standard window and graph a linear equation in the form $y = mx + b$. Recall that the graph of such an equation will have slope m and y-intercept b.

First graph $y = x + 3$. To do so, press the $\boxed{Y=}$ key and enter $Y_1 = x + 3$. Notice that this graph has slope 1 and that the y-intercept is 3. Next, on the same set of axes, graph $y = 2x + 3$ and $y = 3x + 3$ by pressing $\boxed{Y=}$ and entering $Y_2 = 2x + 3$ and $Y_3 = 3x + 3$.

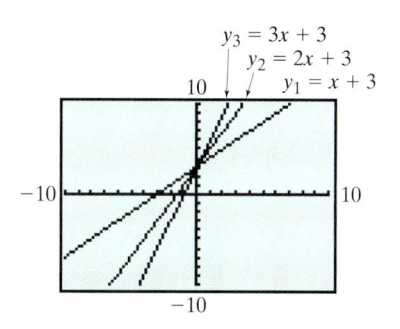

Notice the difference in the graph of each equation as the slope changes from 1 to 2 to 3. How would the graph of $y = 5x + 3$ appear? To see the change in the graph caused by a change in negative slope, try graphing $y = -x + 3$, $y = -2x + 3$, and $y = -3x + 3$ on the same set of axes.

Use a grapher to graph the following equations. For each exercise, graph the first equation and use its graph to predict the appearance of the other equations. Then graph the other equations on the same set of axes and check your prediction.

1. $y = x;\ y = 6x,\ y = -6x$

2. $y = -x;\ y = -5x,\ y = -10x$

3. $y = \dfrac{1}{2}x + 2;\ y = \dfrac{3}{4}x + 2,\ y = x + 2$

4. $y = x + 1;\ y = \dfrac{5}{4}x + 1,\ y = \dfrac{5}{2}x + 1$

5. $y = -7x + 5;\ y = 7x + 5$

6. $y = 3x - 1;\ y = -3x - 1$

VOCABULARY & READINESS CHECK

Use the choices below to fill in each blank. Some choices may be used more than once and some not at all.

b	(y_1, x_1)	point-slope	vertical	standard
m	(x_1, y_1)	slope-intercept	horizontal	

1. The form $y = mx + b$ is called _____ form. When a linear equation in two variables is written in this form, _____ is the slope of its graph and (0, _____) is its y-intercept.

2. The form $y - y_1 = m(x - x_1)$ is called _____ form. When a linear equation in two variables is written in this form, _____ is the slope of its graph and _____ is a point on the graph.

For Exercises 3, 4, and 7, identify the form that the linear equation in two variables is written in. For Exercises 5 and 6, identify the appearance of the graph of the equation.

3. $y - 7 = 4(x + 3)$; _____ form

4. $5x - 9y = 11$; _____ form

5. $y = \dfrac{1}{2}$; _____ line

6. $x = -17$; _____ line

7. $y = \dfrac{3}{4}x - \dfrac{1}{3}$; _____ form

3.5 EXERCISE SET

Write an equation of the line with each given slope, m, and y-intercept, (0, b). See Example 1.

1. $m = 5, b = 3$
2. $m = -3, b = -3$
3. $m = -4, b = -\frac{1}{6}$
4. $m = 2, b = \frac{3}{4}$
5. $m = \frac{2}{3}, b = 0$
6. $m = -\frac{4}{5}, b = 0$
7. $m = 0, b = -8$
8. $m = 0, b = -2$
9. $m = -\frac{1}{5}, b = \frac{1}{9}$
10. $m = \frac{1}{2}, b = -\frac{1}{3}$

Use the slope-intercept form to graph each equation. See Examples 2 and 3.

11. $y = 2x + 1$
12. $y = -4x - 1$
13. $y = \frac{2}{3}x + 5$
14. $y = \frac{1}{4}x - 3$
15. $y = -5x$
16. $y = -6x$
17. $4x + y = 6$
18. $-3x + y = 2$
19. $4x - 7y = -14$
20. $3x - 4y = 4$
21. $x = \frac{5}{4}y$
22. $x = \frac{3}{2}y$

Find an equation of each line with the given slope that passes through the given point. Write the equation in the form $Ax + By = C$. See Example 4.

23. $m = 6$; $(2, 2)$
24. $m = 4$; $(1, 3)$
25. $m = -8$; $(-1, -5)$
26. $m = -2$; $(-11, -12)$
27. $m = \frac{3}{2}$; $(5, -6)$
28. $m = \frac{2}{3}$; $(-8, 9)$
29. $m = -\frac{1}{2}$; $(-3, 0)$
30. $m = -\frac{1}{5}$; $(4, 0)$

Find an equation of the line passing through each pair of points. Write the equation in the form $Ax + By = C$. See Example 5.

31. $(3, 2)$ and $(5, 6)$
32. $(6, 2)$ and $(8, 8)$
33. $(-1, 3)$ and $(-2, -5)$
34. $(-4, 0)$ and $(6, -1)$
35. $(2, 3)$ and $(-1, -1)$
36. $(7, 10)$ and $(-1, -1)$
37. $(0, 0)$ and $\left(-\frac{1}{8}, \frac{1}{13}\right)$
38. $(0, 0)$ and $\left(-\frac{1}{2}, \frac{1}{3}\right)$

Find an equation of each line. See Example 6.

39. Vertical line through $(0, 2)$
40. Horizontal line through $(1, 4)$
41. Horizontal line through $(-1, 3)$
42. Vertical line through $(-1, 3)$
43. Vertical line through $\left(-\frac{7}{3}, -\frac{2}{5}\right)$
44. Horizontal line through $\left(\frac{2}{7}, 0\right)$

Find an equation of each line. See Example 7.

45. Parallel to $y = 5$, through $(1, 2)$
46. Perpendicular to $y = 5$, through $(1, 2)$
47. Perpendicular to $x = -3$, through $(-2, 5)$
48. Parallel to $y = -4$, through $(0, -3)$
49. Parallel to $x = 0$, through $(6, -8)$
50. Perpendicular to $x = 7$, through $(-5, 0)$

MIXED PRACTICE

See Examples 1 through 7. Find an equation of each line described. Write each equation in slope-intercept form (solved for y), when possible.

51. With slope $-\frac{1}{2}$, through $\left(0, \frac{5}{3}\right)$
52. With slope $\frac{5}{7}$, through $(0, -3)$
53. Through $(10, 7)$ and $(7, 10)$
54. Through $(5, -6)$ and $(-6, 5)$
55. With undefined slope, through $\left(-\frac{3}{4}, 1\right)$
56. With slope 0, through $(6.7, 12.1)$
57. Slope 1, through $(-7, 9)$
58. Slope 5, through $(6, -8)$
59. Slope -5, y-intercept $(0, 7)$
60. Slope -2; y-intercept $(0, -4)$
61. Through $(6, 7)$, parallel to the x-axis
62. Through $(1, -5)$, parallel to the y-axis
63. Through $(2, 3)$ and $(0, 0)$
64. Through $(4, 7)$ and $(0, 0)$

65. Through $(-2, -3)$, perpendicular to the y-axis

66. Through $(0, 12)$, perpendicular to the x-axis

67. Slope $-\dfrac{4}{7}$, through $(-1, -2)$

68. Slope $-\dfrac{3}{5}$, through $(4, 4)$

Solve. Assume each exercise describes a linear relationship. Write the equations in slope-intercept form. See Example 8.

69. A rock is dropped from the top of a 400-foot cliff. After 1 second, the rock is traveling 32 feet per second. After 3 seconds, the rock is traveling 96 feet per second.

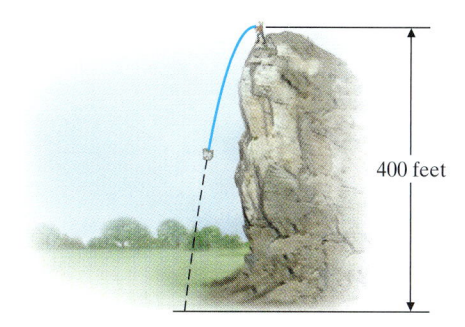

  a. Assume that the relationship between time and speed is linear and write an equation describing this relationship. Use ordered pairs of the form (time, speed).

  b. Use this equation to determine the speed of the rock 4 seconds after it was dropped.

70. A Hawaiian fruit company is studying the sales of a pineapple sauce to see if this product is to be continued. At the end of its first year, profits on this product amounted to $30,000. At the end of the fourth year, profits were $66,000.

  a. Assume that the relationship between years on the market and profit is linear and write an equation describing this relationship. Use ordered pairs of the form (years on the market, profit).

  b. Use this equation to predict the profit at the end of 7 years.

71. In January 2007, there were 71,000 registered gasoline-electric hybrid cars in the United States. In 2004, there were only 29,000 registered gasoline-electric hybrids. (*Source:* U.S. Energy Information Administration)

  a. Write an equation describing the relationship between time and number of registered gasoline-hybrid cars. Use ordered pairs of the form (years past 2004, number of cars).

  b. Use this equation to predict the number of gasoline-electric hybrids in the year 2010.

72. In 2006, there were 935 thousand eating establishments in the United States. In 1996, there were 457 thousand eating establishments. (*Source:* National Restaurant Association)

  a. Write an equation describing the relationship between time and number of eating establishments. Use ordered pairs of the form (years past 1996, number of eating establishments in thousands).

  b. Use this equation to predict the number of eating establishments in 2010.

73. In 2006, the U.S. population per square mile of land area was 85. In 2000, the person per square mile population was 79.6.

  a. Write an equation describing the relationship between year and persons per square mile. Use ordered pairs of the form (years past 2000, persons per square mile).

  b. Use this equation to predict the person per square mile population in 2010.

74. In 2001, there were a total of 152 thousand apparel and accessory stores. In 2005, there were a total of 150 thousand apparel and accessory stores. (*Source:* U.S. Bureau of the Census. *County Business Patterns, annual*)

  a. Write an equation describing this relationship. Use ordered pairs of the form (years past 2001, numbers of stores in thousand).

  b. Use this equation to predict the number of apparel and accessory stores in 2011.

75. The birth rate in the United States in 1996 was 14.7 births per thousand population. In 2006, the birth rate was 14.14 births per thousand. (*Source:* Department of Health and Human Services, National Center for Health Statistics)

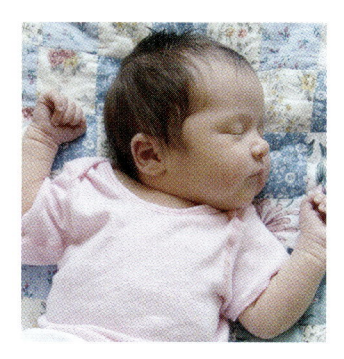

  a. Write two ordered pairs of the form (years after 1996, birth rate per thousand population).

  b. Assume that the relationship between years after 1996 and birth rate per thousand is linear over this period. Use the ordered pairs from part (a) to write an equation of the line relating years to birth rate.

  c. Use the linear equation from part (b) to estimate the birth rate in the United States in the year 2016.

76. In 2002, crude oil production by OPEC countries was about 28.7 million barrels per day. In 2007, crude oil production had risen to about 34.5 million barrels per day. (*Source:* OPEC)
   a. Write two ordered pairs of the form (years after 2002, crude oil production) for this situation.
   b. Assume that crude oil production is linear between the years 2002 and 2007. Use the ordered pairs from part (a) to write an equation of the line relating year and crude oil production.
   c. Use the linear equation from part (b) to estimate the crude oil production by OPEC countries in 2004.

77. Better World Club is a relatively new automobile association which prides itself on its "green" philosophy. In 2003, the membership totaled 5 thousand. By 2006, there were 20 thousand members of this ecologically minded club. (*Source:* Better World Club)
   a. Write two ordered pairs of the form (years after 2003, membership in thousands)
   b. Assume that the membership is linear between the years 2003 and 2006. Use the ordered pairs from part (a) to write an equation of the line relating year and Better World membership.
   c. Use the linear equation from part (b) to predict the Better World Club membership in 2012.

78. In 2002, 9.9 million electronic bill statements were delivered and payment occurred. In 2005, that number rose to 26.9 million. (*Source:* Forrester Research)

   a. Write two ordered pairs of the form (years after 2002, millions of electronic bills).
   b. Assume that this method of delivery and payment between the years 2002 and 2005 is linear. Use the ordered pairs from part (a) to write an equation of the line relating year and number of electronic bills.
   c. Use the linear equation from part (b) to predict the number of electronic bills to be delivered and paid in 2011.

REVIEW AND PREVIEW

Find the value of $x^2 - 3x + 1$ for each given value of x. See Section 1.7.

79. 2   80. 5   81. −1   82. −3

For each graph, determine whether any x-values correspond to two or more y-values. See Section 3.1.

83.

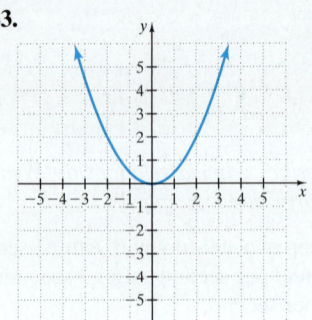

84.

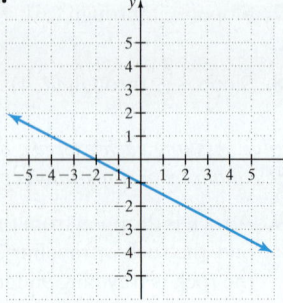

85.

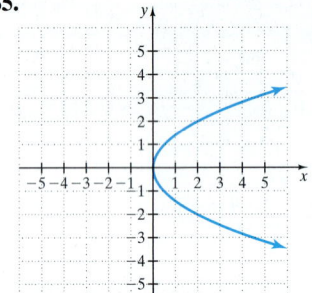

86.

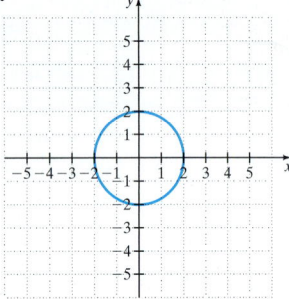

CONCEPT EXTENSIONS

87. Given the equation of a nonvertical line, explain how to find the slope without finding two points on the line.

88. Given two points on a nonvertical line, explain how to use the point-slope form to find the equation of the line.

89. Write an equation in standard form of the line that contains the point $(-1, 2)$ and is
   a. parallel to the line $y = 3x - 1$.
   b. perpendicular to the line $y = 3x - 1$.

90. Write an equation in standard form of the line that contains the point $(4, 0)$ and is
   a. parallel to the line $y = -2x + 3$.
   b. perpendicular to the line $y = -2x + 3$.

91. Write an equation in standard form of the line that contains the point $(3, -5)$ and is
   a. parallel to the line $3x + 2y = 7$.
   b. perpendicular to the line $3x + 2y = 7$.

92. Write an equation in standard form of the line that contains the point $(-2, 4)$ and is
   a. parallel to the line $x + 3y = 6$.
   b. perpendicular to the line $x + 3y = 6$.

3.6 FUNCTIONS

OBJECTIVES

1. Identify relations, domains, and ranges.
2. Identify functions.
3. Use the vertical line test.
4. Use function notation.

OBJECTIVE 1 ▶ Identifying relations, domains, and ranges. In previous sections, we have discussed the relationships between two quantities. For example, the relationship between the length of the side of a square x and its area y is described by the equation $y = x^2$. Ordered pairs can be used to write down solutions of this equation. For example, $(2, 4)$ is a solution of $y = x^2$, and this notation tells us that the x-value 2 is related to the y-value 4 for this equation. In other words, when the length of the side of a square is 2 units, its area is 4 square units.

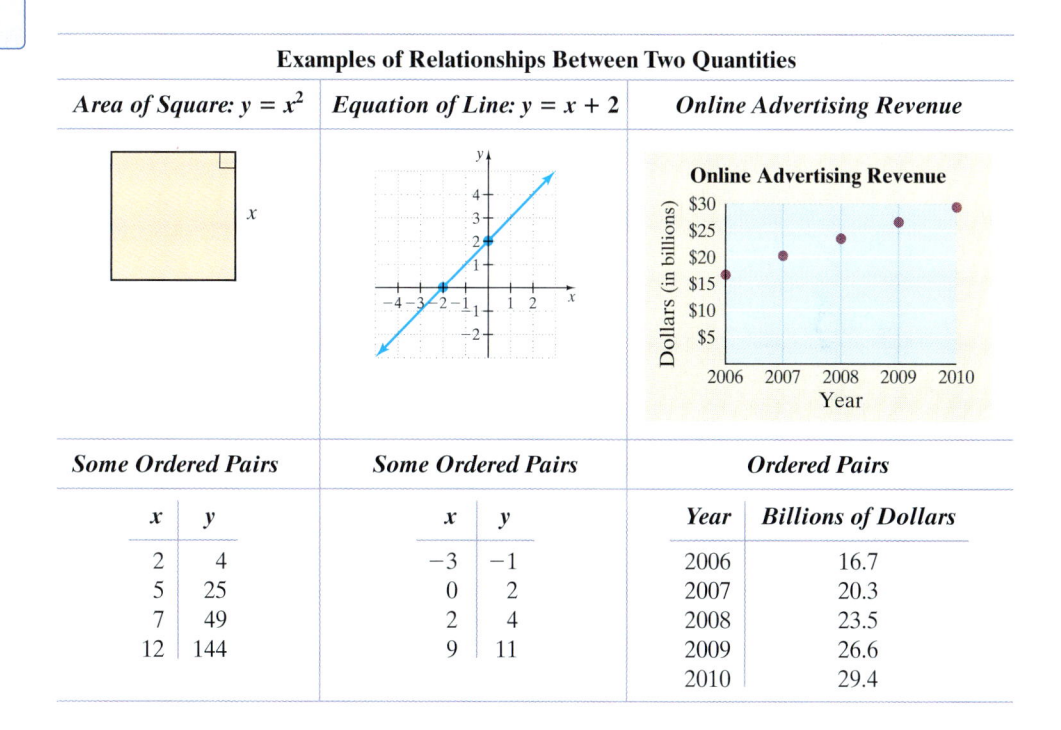

Examples of Relationships Between Two Quantities

Area of Square: $y = x^2$	Equation of Line: $y = x + 2$	Online Advertising Revenue

Some Ordered Pairs

x	y
2	4
5	25
7	49
12	144

Some Ordered Pairs

x	y
-3	-1
0	2
2	4
9	11

Ordered Pairs

Year	Billions of Dollars
2006	16.7
2007	20.3
2008	23.5
2009	26.6
2010	29.4

A set of ordered pairs is called a **relation**. The set of all x-coordinates is called the **domain** of a relation, and the set of all y-coordinates is called the **range** of a relation. Equations such as $y = x^2$ are also called relations since equations in two variables define a set of ordered pair solutions.

EXAMPLE 1 Find the domain and the range of the relation $\{(0, 2), (3, 3), (-1, 0), (3, -2)\}$.

Solution The domain is the set of all x-values or $\{-1, 0, 3\}$, and the range is the set of all y-values, or $\{-2, 0, 2, 3\}$.

PRACTICE 1 Find the domain and the range of the relation $\{(1, 3)(5, 0)(0, -2)(5, 4)\}$.

OBJECTIVE 2 ▶ Identifying functions. Some relations are also functions.

> **Function**
> A function is a set of ordered pairs that assigns to each x-value exactly one y-value.

228 CHAPTER 3 Graphing

EXAMPLE 2 Which of the following relations are also functions?

a. $\{(-1, 1), (2, 3), (7, 3), (8, 6)\}$ **b.** $\{(0, -2), (1, 5), (0, 3), (7, 7)\}$

Solution

a. Although the ordered pairs (2, 3) and (7, 3) have the same *y*-value, each *x*-value is assigned to only one *y*-value so this set of ordered pairs is a function.

b. The *x*-value 0 is assigned to two *y*-values, −2 and 3, so this set of ordered pairs is not a function.

PRACTICE 2 Which of the following relations are also functions?

a. $\{(4, 1)(3, -2)(8, 5)(-5, 3)\}$ **b.** $\{(1, 2)(-4, 3)(0, 8)(1, 4)\}$

Relations and functions can be described by a graph of their ordered pairs.

EXAMPLE 3 Which graph is the graph of a function?

a.

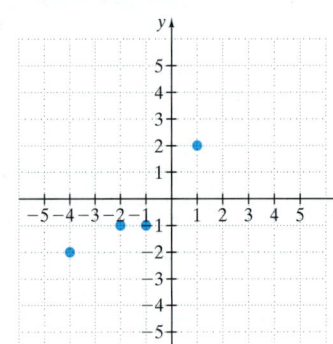

b.

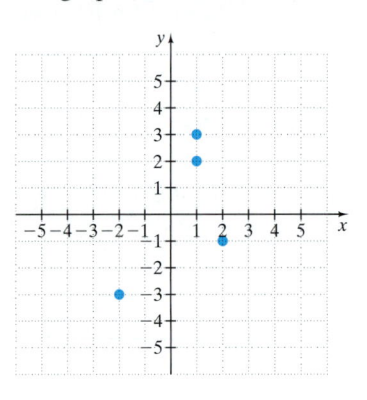

Solution

a. This is the graph of the relation $\{(-4, -2), (-2, -1)(-1, -1), (1, 2)\}$. Each *x*-coordinate has exactly one *y*-coordinate, so this is the graph of a function.

b. This is the graph of the relation $\{(-2, -3), (1, 2), (1, 3), (2, -1)\}$. The *x*-coordinate 1 is paired with two *y*-coordinates, 2 and 3, so this is not the graph of a function.

PRACTICE 3 Which graph is the graph of a function?

a.

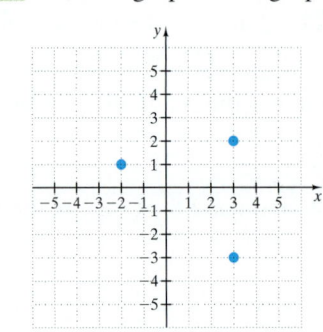

b.

x-coordinate 1 paired with two *y*-coordinates, 2 and 3.

(1, 3)
(1, 2)

Not the graph of a function.

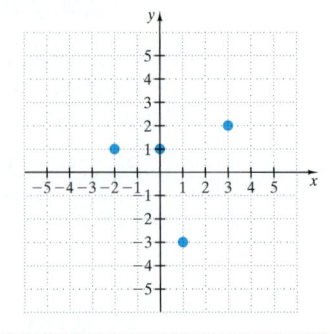

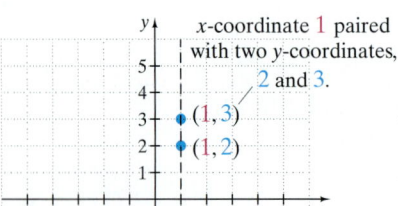

OBJECTIVE 3 ▶ Using the vertical line test. The graph in Example 3(b) was not the graph of a function because the *x*-coordinate 1 was paired with two *y*-coordinates, 2 and 3. Notice that when an *x*-coordinate is paired with more than one *y*-coordinate, a vertical line can be drawn that will intersect the graph at more than one point. We can use this fact to determine whether a relation is also a function. We call this the **vertical line test.**

> **Vertical Line Test**
> If a vertical line can be drawn so that it intersects a graph more than once, the graph is not the graph of a function.

This vertical line test works for all types of graphs on the rectangular coordinate system.

EXAMPLE 4 Use the vertical line test to determine whether each graph is the graph of a function.

a.   b.   c.   d.

Solution

a. This graph is the graph of a function since no vertical line will intersect this graph more than once.

b. This graph is also the graph of a function; no vertical line will intersect it more than once.

c. This graph is not the graph of a function. Vertical lines can be drawn that intersect the graph in two points. An example of one is shown.

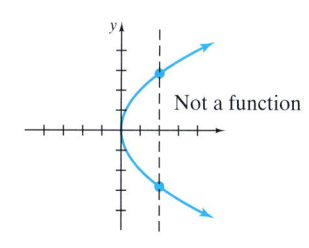

Not a function

d. This graph is not the graph of a function. A vertical line can be drawn that intersects this line at every point.

PRACTICE 4 Use the vertical line test to determine whether each graph is the graph of a function.

a.   b.   c.   d.

Recall that the graph of a linear equation is a line, and a line that is not vertical will pass the vertical line test. **Thus, all linear equations are functions except those of the form $x = c$, which are vertical lines.**

EXAMPLE 5 Which of the following linear equations are functions?

a. $y = x$ **b.** $y = 2x + 1$ **c.** $y = 5$ **d.** $x = -1$

Solution a, b, and c are functions because their graphs are nonvertical lines. d is not a function because its graph is a vertical line. □

PRACTICE 5 Which of the following linear equations are functions?

a. $y = 2x$ **b.** $y = -3x - 1$ **c.** $y = 8$ **d.** $x = 2$

Examples of functions can often be found in magazines, newspapers, books, and other printed material in the form of tables or graphs such as that in Example 6.

EXAMPLE 6 The graph shows the sunrise time for Indianapolis, Indiana, for the year. Use this graph to answer the questions.

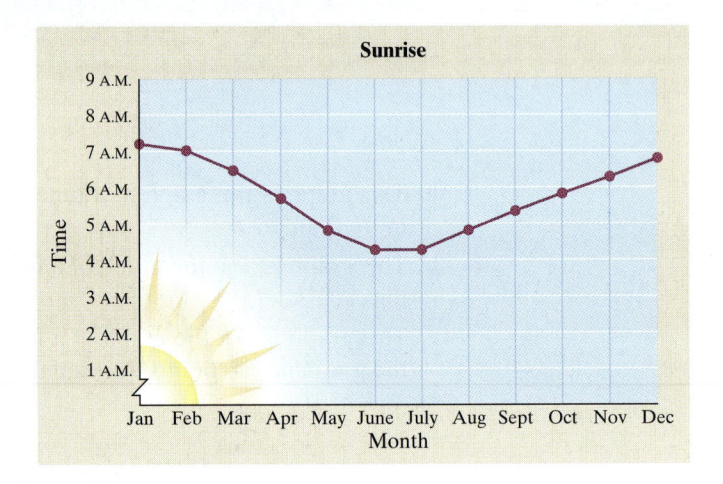

a. Approximate the time of sunrise on February 1.
b. Approximately when does the sun rise at 5 A.M.?
c. Is this the graph of a function?

Solution

a. To approximate the time of sunrise on February 1, we find the mark on the horizontal axis that corresponds to February 1. From this mark, we move vertically upward until the graph is reached. From that point on the graph, we move horizontally to the left until the vertical axis is reached. The vertical axis there reads 7 A.M.

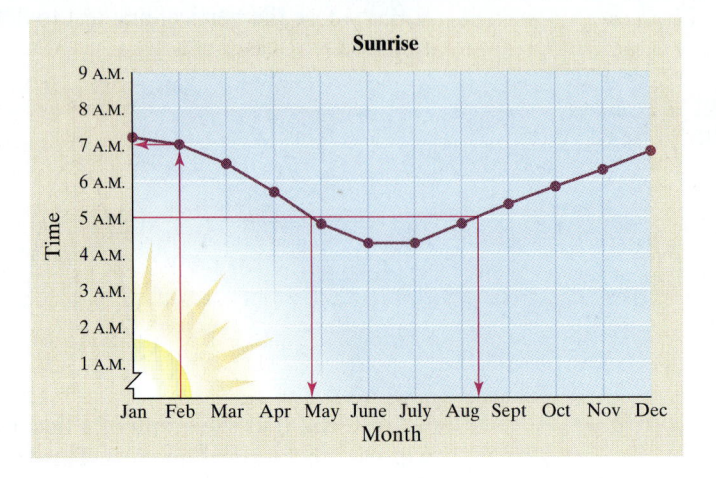

b. To approximate when the sun rises at 5 A.M., we find 5 A.M. on the time axis and move horizontally to the right. Notice that we will reach the graph twice, corresponding to two dates for which the sun rises at 5 A.M. We follow both points on the graph vertically downward until the horizontal axis is reached. The sun rises at 5 A.M. at approximately the end of the month of April and the middle of the month of August.

c. The graph is the graph of a function since it passes the vertical line test. In other words, for every day of the year in Indianapolis, there is exactly one sunrise time. □

PRACTICE
6   The graph shows the average monthly temperature for Chicago, Illinois, for the year. Use this graph to answer the questions.

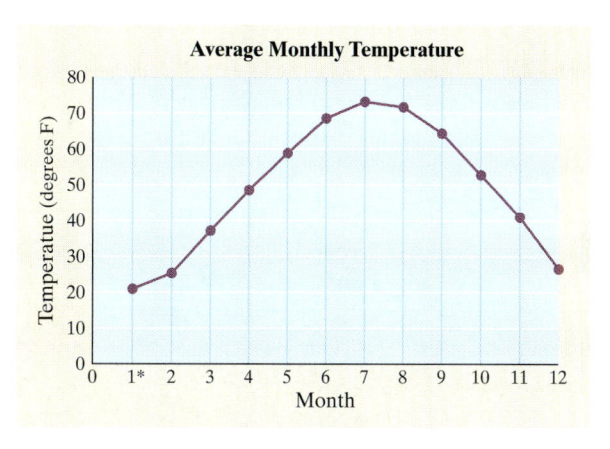

Average Monthly Temperature

*(1 is Jan., 12 is Dec.)

a. Approximate the average monthly temperature for June.
b. Approximately when is the average monthly temperature 40°?
c. Is this the graph of a function?

OBJECTIVE 4 ▶ Using function notation. The graph of the linear equation $y = 2x + 1$ passes the vertical line test, so we say that $y = 2x + 1$ is a function. In other words, $y = 2x + 1$ gives us a rule for writing ordered pairs where every x-coordinate is paired with one y-coordinate.

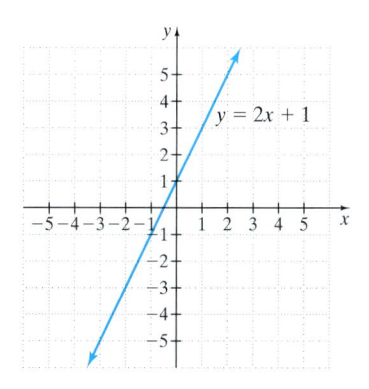

We often use letters such as f, g, and h to name functions. For example, the symbol $f(x)$ means *function of x* and is read "f of x." This notation is called **function notation.** The equation $y = 2x + 1$ can be written as $f(x) = 2x + 1$ using function notation, and these equations mean the same thing. In other words, $y = f(x)$.

The notation $f(1)$ means to replace x with 1 and find the resulting y or function value. Since

$$f(x) = 2x + 1$$

then

$$f(1) = 2(1) + 1 = 3$$

This means that, when $x = 1$, y or $f(x) = 3$, and we have the ordered pair $(1, 3)$. Now let's find $f(2), f(0)$, and $f(-1)$.

$f(x) = 2x + 1$	$f(x) = 2x + 1$	$f(x) = 2x + 1$
$f(2) = 2(2) + 1$	$f(0) = 2(0) + 1$	$f(-1) = 2(-1) + 1$
$= 4 + 1$	$= 0 + 1$	$= -2 + 1$
$= 5$	$= 1$	$= -1$

Ordered
Pair: $(2, 5)$ $(0, 1)$ $(-1, -1)$

> **Helpful Hint**
> Note that, for example, if $f(2) = 5$, the corresponding ordered pair is $(2, 5)$.

> **Helpful Hint**
> Note that $f(x)$ is a special symbol in mathematics used to denote a function. The symbol $f(x)$ is read "f of x." It does **not** mean $f \cdot x$ (f times x).

EXAMPLE 7 Given $g(x) = x^2 - 3$, find the following. Then write down the corresponding ordered pairs generated.

a. $g(2)$ **b.** $g(-2)$ **c.** $g(0)$

Solution

a. $g(x) = x^2 - 3$	**b.** $g(x) = x^2 - 3$	**c.** $g(x) = x^2 - 3$
$g(2) = 2^2 - 3$	$g(-2) = (-2)^2 - 3$	$g(0) = 0^2 - 3$
$= 4 - 3$	$= 4 - 3$	$= 0 - 3$
$= 1$	$= 1$	$= -3$

Ordered Pairs:	$g(2) = 1$ gives $(2, 1)$	$g(-2) = 1$ gives $(-2, 1)$	$g(0) = -3$ gives $(0, -3)$

PRACTICE 7 Given $h(x) = x^2 + 5$, find the following. Then write the corresponding ordered pairs generated.

a. $h(2)$ **b.** $h(-5)$ **c.** $h(0)$

We now practice finding the domain and the range of a function. The domain of our functions will be the set of all possible real numbers that x can be replaced by. The range is the set of corresponding y-values.

EXAMPLE 8 Find the domain of each function.

a. $g(x) = \dfrac{1}{x}$ **b.** $f(x) = 2x + 1$

Solution

a. Recall that we cannot divide by 0 so that the domain of $g(x)$ is the set of all real numbers except 0. In interval notation, we can write $(-\infty, 0) \cup (0, \infty)$.

b. In this function, x can be any real number. The domain of $f(x)$ is the set of all real numbers, or $(-\infty, \infty)$ in interval notation.

PRACTICE 8 Find the domain of each function.

a. $h(x) = 6x + 3$ **b.** $f(x) = \dfrac{1}{x^2}$

Concept Check ✓

Suppose that the value of f is -7 when the function is evaluated at 2. Write this situation in function notation.

EXAMPLE 9 Find the domain and the range of each function graphed. Use interval notation.

a.

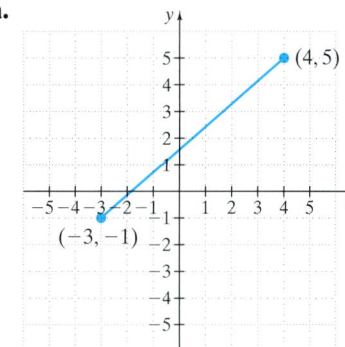

b.

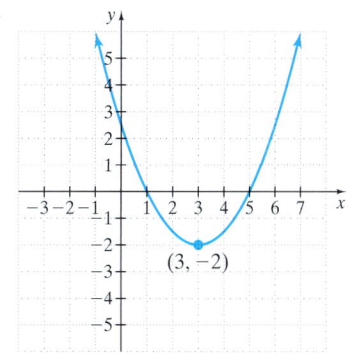

Solution

a. Range: $[-1, 5]$ Domain $[-3, 4]$

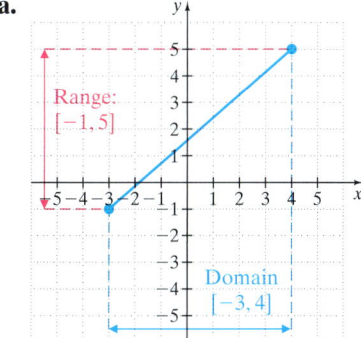

b. Range: $[-2, \infty)$ Domain: All real numbers or $(-\infty, \infty)$

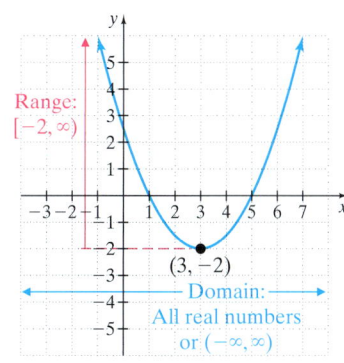

Answer to Concept Check:
$f(2) = -7$

PRACTICE 9 Find the domain and the range of each function graphed. Use interval notation.

a.

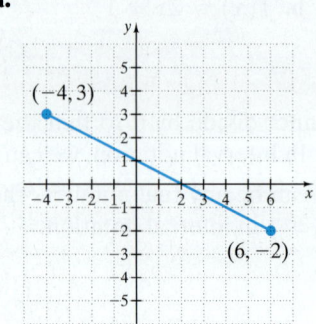

b.

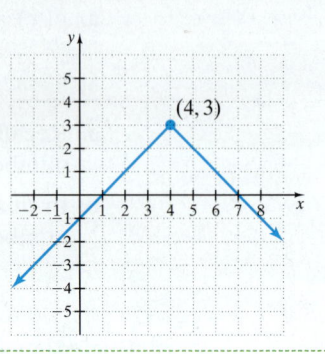

VOCABULARY & READINESS CHECK

Use the choices below to fill in each blank. Some choices may not be used.

| $x = c$ | horizontal | domain | relation | (7, 3) | $(-\infty, 5]$ |
| $y = c$ | vertical | range | function | (3, 7) | $(-\infty, \infty)$ |

1. A set of ordered pairs is called a _____ .
2. A set of ordered pairs that assigns to each x-value exactly one y-value is called a _____ .
3. The set of all y-coordinates of a relation is called the _____ .
4. The set of all x-coordinates of a relation is called the _____ .
5. All linear equations are functions except those whose graphs are _____ lines.
6. All linear equations are functions except those whose equations are of the form _____ .
7. If $f(3) = 7$, the corresponding ordered pair is _____ .
8. The domain of $f(x) = x + 5$ is _____ .

3.6 EXERCISE SET

Find the domain and the range of each relation. See Example 1.

1. $\{(2, 4), (0, 0), (-7, 10), (10, -7)\}$
2. $\{(3, -6), (1, 4), (-2, -2)\}$
3. $\{(0, -2), (1, -2), (5, -2)\}$
4. $\{(5, 0), (5, -3), (5, 4), (5, 3)\}$

Determine whether each relation is also a function. See Example 2.

5. $\{(1, 1), (2, 2), (-3, -3), (0, 0)\}$
6. $\{(11, 6), (-1, -2), (0, 0), (3, -2)\}$
7. $\{(-1, 0), (-1, 6), (-1, 8)\}$
8. $\{(1, 2), (3, 2), (1, 4)\}$

MIXED PRACTICE

Determine whether each graph is the graph of a function. See Examples 3 and 4.

9.

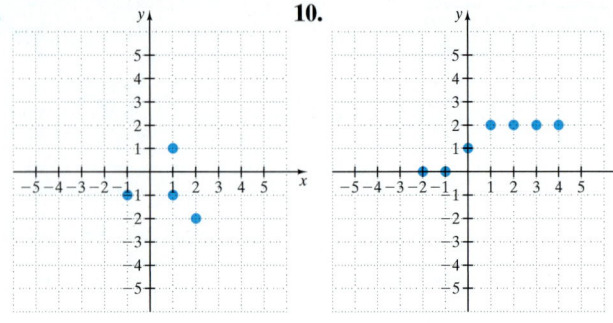

10.

11. [graph]   12. [graph]

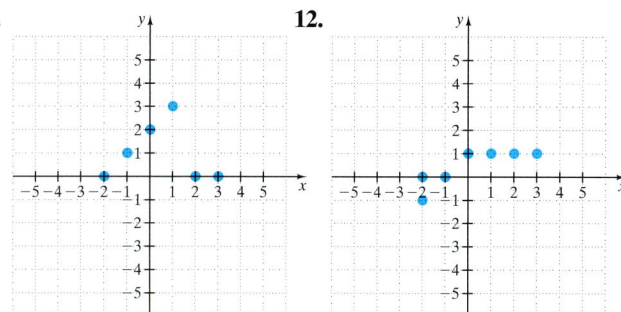

Decide whether the equation describes a function. See Example 5.

17. $y = x + 1$ 18. $y = x - 1$
19. $y - x = 7$ 20. $2x - 3y = 9$
21. $y = 6$ 22. $x = 3$
23. $x = -2$ 24. $y = -9$
25. $x = y^2$ 26. $y = x^2 - 3$

The graph shows the sunset times for Seward, Alaska. Use this graph to answer Exercises 27 through 32. See Example 6.

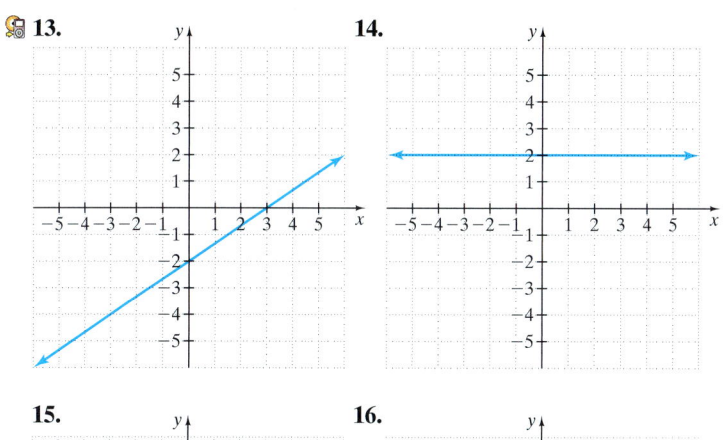

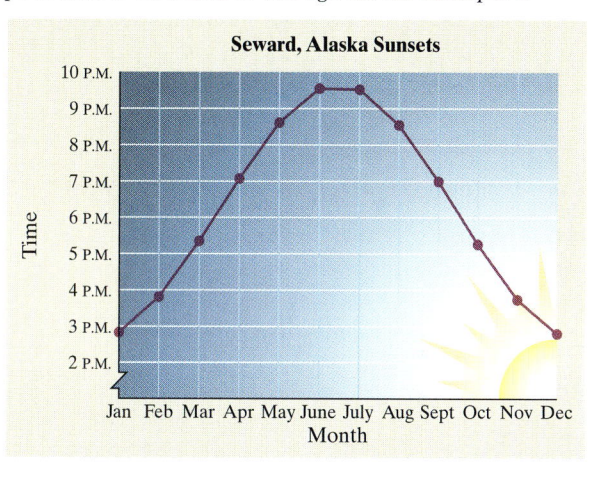

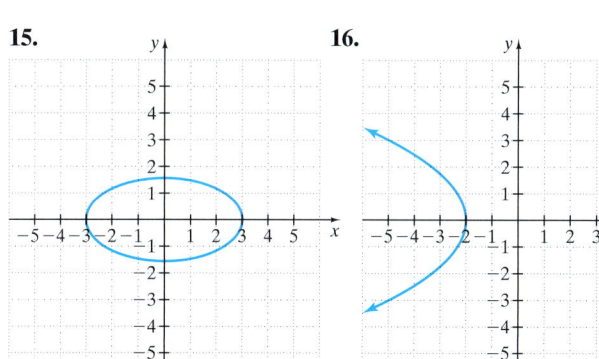

13. [graph]   14. [graph]

[Seward, Alaska Sunsets graph]

15. [graph]   16. [graph]

27. Approximate the time of sunset on June 1.
28. Approximate the time of sunset on November 1.
29. Approximate the date(s) when the sunset is at 3 P.M.
30. Approximate the date(s) when the sunset is at 9 P.M.
31. Is this graph the graph of a function? Why or why not?
32. Do you think a graph of sunset times for any location will always be a function? Why or why not?

This graph shows the U.S. hourly minimum wage for each year shown. Use this graph to answer Exercises 33 through 38. See Example 6.

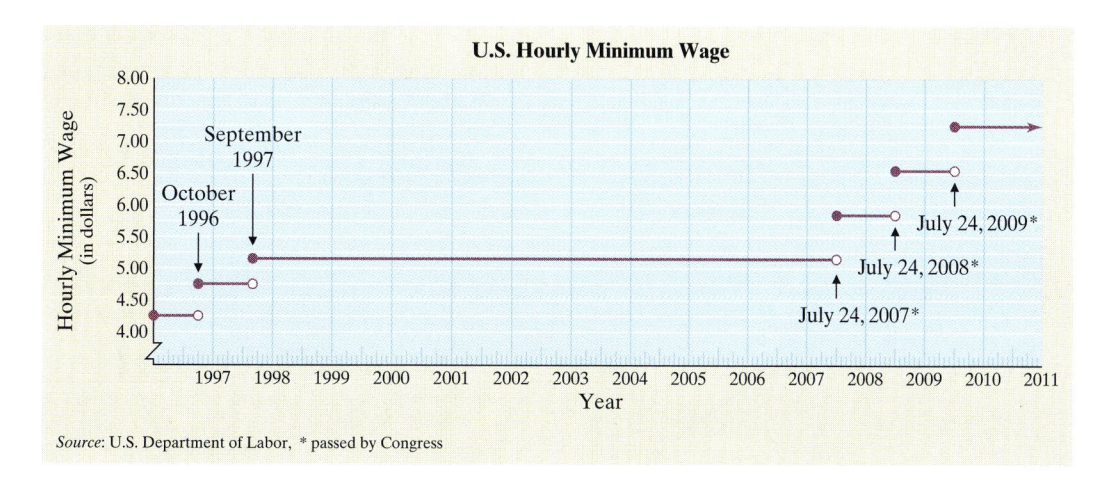

[U.S. Hourly Minimum Wage graph]

Source: U.S. Department of Labor, * passed by Congress

33. Approximate the minimum wage before October, 1996.
34. Approximate the minimum wage in 2006.
35. Approximate the year when the minimum wage will increase to over $7.00 per hour.

236 CHAPTER 3 Graphing

36. According to the graph, what hourly wage was in effect for the greatest number of years?
37. Is this graph the graph of a function? Why or why not?
38. Do you think that a similar graph of your hourly wage on January 1 of every year (whether you are working or not) will be the graph of a function? Why or why not?

Find $f(-2)$, $f(0)$, and $f(3)$ for each function. See Example 7.

39. $f(x) = 2x - 5$
40. $f(x) = 3 - 7x$
41. $f(x) = x^2 + 2$
42. $f(x) = x^2 - 4$
43. $f(x) = 3x$
44. $f(x) = -3x$
45. $f(x) = |x|$
46. $f(x) = |2 - x|$

Find $h(-1)$, $h(0)$, and $h(4)$ for each function. See Example 7.

47. $h(x) = -5x$
48. $h(x) = -3x$
49. $h(x) = 2x^2 + 3$
50. $h(x) = 3x^2$

For each given function value, write a corresponding ordered pair.

51. $f(3) = 6$,
52. $f(7) = -2$,
53. $g(0) = -\dfrac{1}{2}$
54. $g(0) = -\dfrac{7}{8}$
55. $h(-2) = 9$
56. $h(-10) = 1$

Find the domain of each function. See Example 8.

57. $f(x) = 3x - 7$
58. $g(x) = 5 - 2x$
59. $h(x) = \dfrac{1}{x + 5}$
60. $f(x) = \dfrac{1}{x - 6}$
61. $g(x) = |x + 1|$
62. $h(x) = |2x|$

Find the domain and the range of each relation graphed. See Example 9.

63. 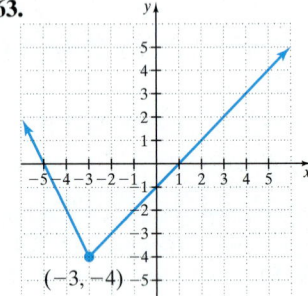
64. 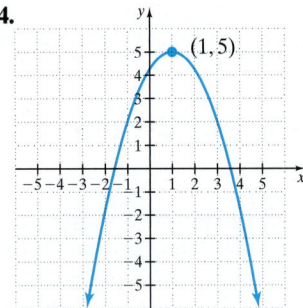
65. 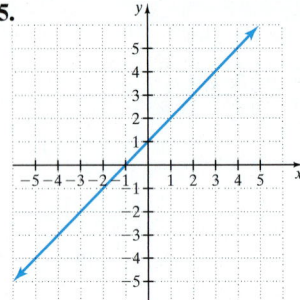
66. 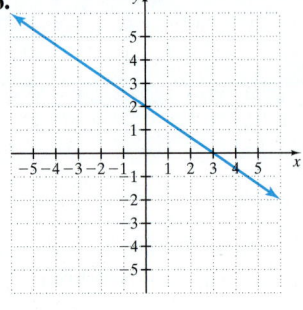
67.
68.

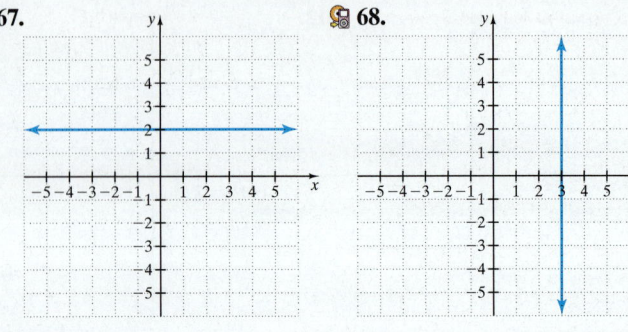

REVIEW AND PREVIEW

Find the coordinates of the point of intersection. See Section 3.1.

69. 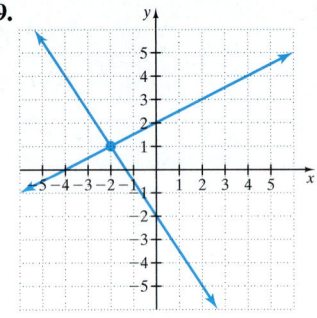
70. 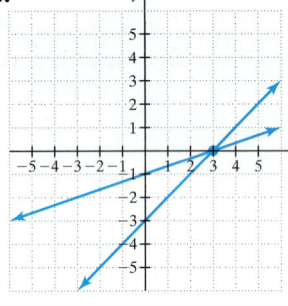
71. 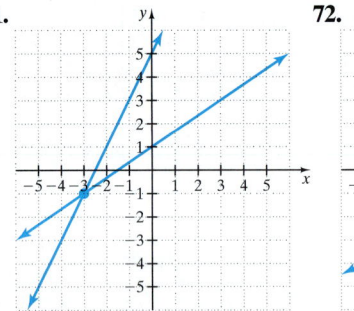
72.

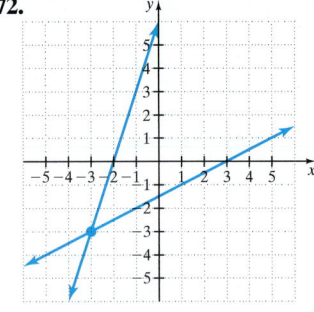

CONCEPT EXTENSIONS

Solve. See the Concept Check in this section.

73. If a function f is evaluated at -5, the value of the function is 12. Write this situation using function notation.
74. Suppose $(9, 20)$ is an ordered-pair solution for the function g. Write this situation using function notation.

The graph of the function, f, is below. Use this graph to answer Exercises 75 through 78.

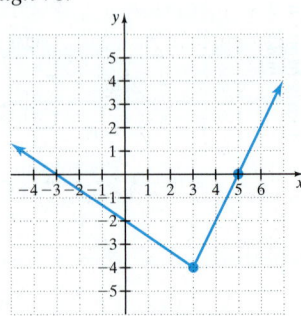

75. Write the coordinates of the lowest point of the graph.
76. Write the answer to Exercise 73 in function notation.

77. An x-intercept of this graph is $(5,0)$. Write this using function notation.

78. Write the other x-intercept of this graph (see Exercise 77) using function notation.

79. Forensic scientists use the function
$$H(x) = 2.59x + 47.24$$
to estimate the height of a woman in centimeters given the length x of her femur bone.

  a. Estimate the height of a woman whose femur measures 46 centimeters.

  b. Estimate the height of a woman whose femur measures 39 centimeters.

80. The dosage in milligrams D of Ivermectin, a heartworm preventive for a dog who weighs x pounds, is given by the function
$$D(x) = \frac{136}{25}x$$

  a. Find the proper dosage for a dog that weighs 35 pounds.

  b. Find the proper dosage for a dog that weighs 70 pounds.

81. In your own words define **(a)** function; **(b)** domain; **(c)** range.

82. Explain the vertical line test and how it is used.

83. Since $y = x + 7$ is a function, rewrite the equation using function notation.

See the example below for Exercises 84 through 87.

Example

If $f(x) = x^2 + 2x + 1$, find $f(\pi)$.

Solution:
$$f(x) = x^2 + 2x + 1$$
$$f(\pi) = \pi^2 + 2\pi + 1$$

Given the following functions, find the indicated values.

84. $f(x) = 2x + 7$
  a. $f(2)$ **b.** $f(a)$

85. $g(x) = -3x + 12$
  a. $g(s)$ **b.** $g(r)$

86. $h(x) = x^2 + 7$
  a. $h(3)$ **b.** $h(a)$

87. $f(x) = x^2 - 12$
  a. $f(12)$ **b.** $f(a)$

CHAPTER 3 GROUP ACTIVITY

Financial Analysis

Investment analysts investigate a company's sales, net profit, debt, and assets to decide whether investing in it is a wise choice. One way to analyze this data is to graph it and look for trends over time. Another way is to find algebraically the rate at which the data changes over time.

The following table gives the net incomes in millions of dollars for some of the leading U.S. businesses in the pharmaceutical industry for the years 2004 and 2005. In this project, you will analyze the performances of these companies and, based on this information alone, make an investment recommendation. This project may be completed by working in groups or individually.

Pharmaceutical Industry Net Income (In Millions of Dollars)

Company	2004	2005
Merck	$5813.4	$4631.3
Pfizer	$11,361	$8085
Johnson & Johnson	$8509	$10411
Bristol-Myers Squibb	$2378	$2992
Abbot Laboratories	$3175.8	$3372.1
Eli Lilly	$1819.1	$1979.6
Schering-Plough	$269	-$947
Wyeth	$1234	$3656.3

Source: The 2006 annual report for each of the companies listed.

1. Scan the table. Did any of the companies have a loss during the years shown? If so, which company and when? What does this mean?

2. Write the data for each company as two ordered pairs of the form (year, net income). Assuming that the trends in net income are linear, use graph paper to graph the line represented by the ordered pairs for each company. Describe the trends shown by each graph.

3. Find the slope of the line for each company.

4. Which of the lines, if any, have positive slopes? What does that mean in this context? Which of the lines have negative slopes? What does that mean in this context?

5. Of these pharmaceutical companies, which one(s) would you recommend as an investment choice? Why?

6. Do you think it is wise to make a decision after looking at only two years of net profits? What other factors do you think should be taken into consideration when making an investment choice?

(Optional) Use financial magazines, company annual reports, or online investing information to find net income information for two different years for two to four companies in the same industry. Analyze the net income and make an investment recommendation.

CHAPTER 3 VOCABULARY CHECK

Fill in each blank with one of the words listed below.

relation	function	domain	range	standard	slope-intercept
y-axis	x-axis	solution	linear	slope	point-slope
x-intercept	y-intercept	y	x		

1. An ordered pair is a _____ of an equation in two variables if replacing the variables by the coordinates of the ordered pair results in a true statement.
2. The vertical number line in the rectangular coordinate system is called the _____.
3. A _____ equation can be written in the form $Ax + By = C$.
4. A(n) _____ is a point of the graph where the graph crosses the x-axis.
5. The form $Ax + By = C$ is called _____ form.
6. A(n) _____ is a point of the graph where the graph crosses the y-axis.
7. The equation $y = 7x - 5$ is written in _____ form.
8. The equation $y + 1 = 7(x - 2)$ is written in _____ form.
9. To find an x-intercept of a graph, let ____ = 0.
10. The horizontal number line in the rectangular coordinate system is called the _____.
11. To find a y-intercept of a graph, let ____ = 0.
12. The _____ of a line measures the steepness or tilt of a line.
13. A set of ordered pairs that assigns to each x-value exactly one y-value is called a _____.
14. The set of all x-coordinates of a relation is called the _____ of the relation.
15. The set of all y-coordinates of a relation is called the _____ of the relation.
16. A set of ordered pairs is called a _____.

> **Helpful Hint**
> Are you preparing for your test? Don't forget to take the Chapter 3 Test on page 246. Then check your answers at the back of the text and use the Chapter Test Prep Video CD to see the fully worked-out solutions to any of the exercises you want to review.

CHAPTER 3 HIGHLIGHTS

DEFINITIONS AND CONCEPTS	EXAMPLES
SECTION 3.1 READING GRAPHS AND THE RECTANGULAR COORDINATE SYSTEM	
The **rectangular coordinate system** consists of a plane and a vertical and a horizontal number line intersecting at their 0 coordinates. The vertical number line is called the **y-axis** and the horizontal number line is called the **x-axis**. The point of intersection of the axes is called the **origin**.	

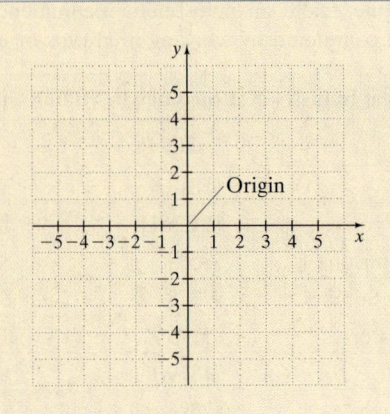

DEFINITIONS AND CONCEPTS	EXAMPLES

SECTION 3.1 READING GRAPHS AND THE RECTANGULAR COORDINATE SYSTEM (continued)

To **plot** or **graph** an ordered pair means to find its corresponding point on a rectangular coordinate system.

To plot or graph an ordered pair such as $(3, -2)$, start at the origin. Move 3 units to the right and from there, 2 units down.

To plot or graph $(-3, 4)$ start at the origin. Move 3 units to the left and from there, 4 units up.

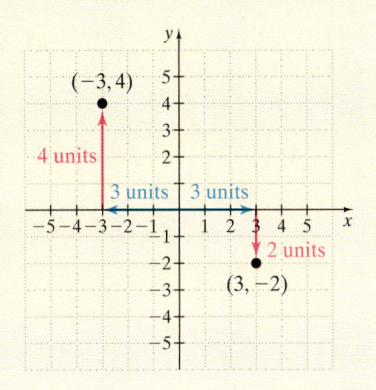

An ordered pair is a **solution** of an equation in two variables if replacing the variables by the coordinates of the ordered pair results in a true statement.

Determine whether $(-1, 5)$ is a solution of $2x + 3y = 13$.

$$2x + 3y = 13$$
$$2(-1) + 3 \cdot 5 = 13 \quad \text{Let } x = -1, y = 5$$
$$-2 + 15 = 13$$
$$13 = 13 \quad \text{True}$$

If one coordinate of an ordered pair solution is known, the other value can be determined by substitution.

Complete the ordered pair solution $(0,)$ for the equation $x - 6y = 12$.

$$x - 6y = 12$$
$$0 - 6y = 12 \quad \text{Let } x = 0.$$
$$\frac{-6y}{-6} = \frac{12}{-6} \quad \text{Divide by } -6.$$
$$y = -2$$

The ordered pair solution is $(0, -2)$.

SECTION 3.2 GRAPHING LINEAR EQUATIONS

A **linear equation in two variables** is an equation that can be written in the form $Ax + By = C$ where A and B are not both 0. The form $Ax + By = C$ is called **standard form**.

Linear Equations

$$3x + 2y = -6 \qquad x = -5$$
$$y = 3 \qquad y = -x + 10$$

$x + y = 10$ is in standard form.

To graph a linear equation in two variables, find three ordered pair solutions. Plot the solution points and draw the line connecting the points.

Graph $x - 2y = 5$.

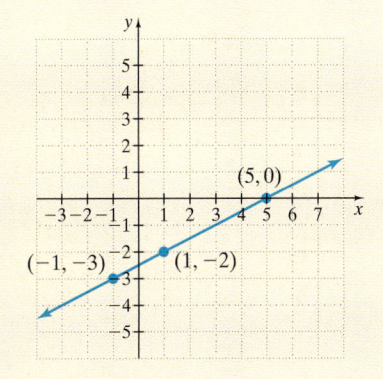

x	y
5	0
1	-2
-1	-3

DEFINITIONS AND CONCEPTS

EXAMPLES

SECTION 3.3 INTERCEPTS

An **intercept** of a graph is a point where the graph intersects an axis. If a graph intersects the *x*-axis at *a*, then $(a, 0)$ is the **x-intercept.** If a graph intersects the *y*-axis at *b*, then $(0, b)$ is the **y-intercept.**

The *y*-intercept is $(0, 3)$
The *x*-intercept is $(5, 0)$

To find the *x*-intercept, let $y = 0$ and solve for *x*.

To find the *y*-intercept, let $x = 0$ and solve for *y*.

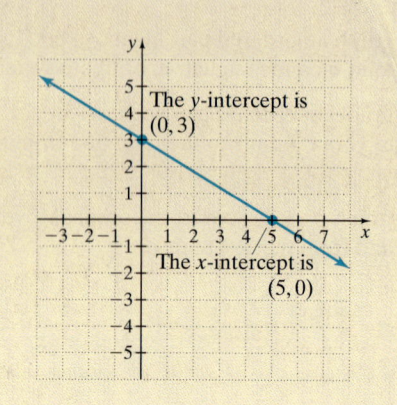

Graph $2x - 5y = -10$ by finding intercepts.

If $y = 0$, then
$2x - 5 \cdot 0 = -10$
$2x = -10$
$\dfrac{2x}{2} = \dfrac{-10}{2}$
$x = -5$

If $x = 0$, then
$2 \cdot 0 - 5y = -10$
$-5y = -10$
$\dfrac{-5y}{-5} = \dfrac{-10}{-5}$
$y = 2$

The *x*-intercept is $(-5, 0)$. The *y*-intercept is $(0, 2)$.

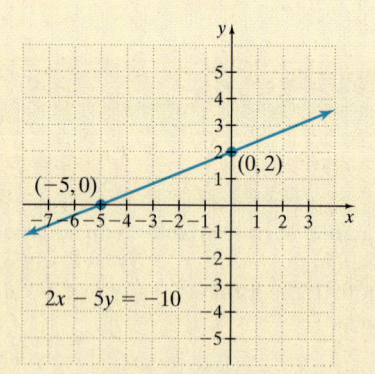

$2x - 5y = -10$

The graph of $x = c$ is a vertical line with *x*-intercept $(c, 0)$.

The graph of $y = c$ is a horizontal line with *y*-intercept $(0, c)$.

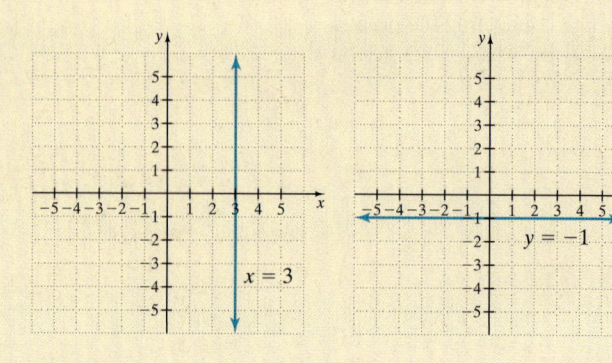

$x = 3$

$y = -1$

DEFINITIONS AND CONCEPTS	EXAMPLES

SECTION 3.4 SLOPE AND RATE OF CHANGE

The **slope** m of the line through points (x_1, y_1) and (x_2, y_2) is given by

$$m = \frac{y_2 - y_1}{x_2 - x_1} \quad \text{as long as } x_2 \neq x_1$$

The slope of the line through points $(-1, 6)$ and $(-5, 8)$ is

$$m = \frac{y_2 - y_1}{x_2 - x_1} = \frac{8 - 6}{-5 - (-1)} = \frac{2}{-4} = -\frac{1}{2}$$

The slope of the line $y = -5$ is 0.

The line $x = 3$ has undefined slope.

A horizontal line has slope 0.

The slope of a vertical line is undefined.

Nonvertical parallel lines have the same slope.

Two nonvertical lines are perpendicular if the slope of one is the negative reciprocal of the slope of the other.

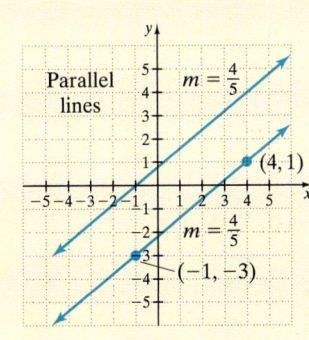

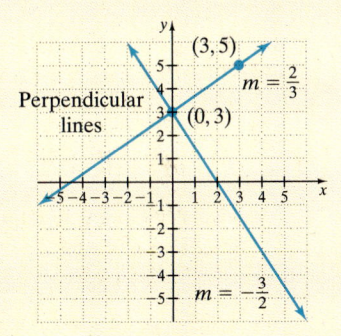

SECTION 3.5 EQUATIONS OF LINES

Slope-Intercept Form

$$y = mx + b$$

m is the slope of the line.
$(0, b)$ is the y-intercept.

Find the slope and the y-intercept of the line whose equation is $2x + 3y = 6$.

Solve for y:

$$2x + 3y = 6$$
$$3y = -2x + 6 \quad \text{Subtract } 2x.$$
$$y = -\frac{2}{3}x + 2 \quad \text{Divide by 3.}$$

The slope of the line is $-\frac{2}{3}$ and the y-intercept is $(0, 2)$.

Find an equation of the line with slope 3 and y-intercept $(0, -1)$.

The equation is $y = 3x - 1$.

Point-Slope Form

$$y - y_1 = m(x - x_1)$$

m is the slope.
(x_1, y_1) is a point on the line.

Find an equation of the line with slope $\frac{3}{4}$ that contains the point $(-1, 5)$.

$$y - 5 = \frac{3}{4}[x - (-1)]$$

$$4(y - 5) = 3(x + 1) \quad \text{Multiply by 4.}$$
$$4y - 20 = 3x + 3 \quad \text{Distribute.}$$
$$-3x + 4y = 23 \quad \text{Subtract } 3x \text{ and add 20.}$$

DEFINITIONS AND CONCEPTS	EXAMPLES
SECTION 3.6 FUNCTIONS	
A set of ordered pairs is a **relation**. The set of all *x*-coordinates is called the **domain** of the relation and the set of all *y*-coordinates is called the **range** of the relation.	The domain of the relation $\{(0,5),(2,5),(4,5),(5,-2)\}$ is $\{0,2,4,5\}$. The range is $\{-2,5\}$.
A **function** is a set of ordered pairs that assigns to each *x*-value exactly one *y*-value.	Which are graphs of functions?
Vertical Line Test If a vertical line can be drawn so that it intersects a graph more than once, the graph is not the graph of a function.	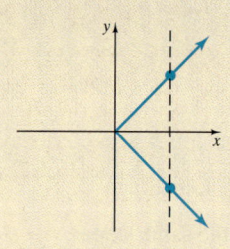 This graph is not the graph of a function. 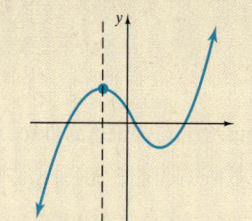 This graph is the graph of a function.
The symbol $f(x)$ means **function of *x*.** This notation is called **function** notation.	If $f(x) = 2x^2 + 6x - 1$, find $f(3)$. $$f(3) = 2(3)^2 + 6 \cdot 3 - 1$$ $$= 2 \cdot 9 + 18 - 1$$ $$= 18 + 18 - 1$$ $$= 35$$

CHAPTER 3 REVIEW

(3.1) Plot the following ordered pairs on a Cartesian coordinate system.

1. $(-7, 0)$
2. $\left(0, 4\frac{4}{5}\right)$
3. $(-2, -5)$
4. $(1, -3)$
5. $(0.7, 0.7)$
6. $(-6, 4)$

7. A local lumberyard uses quantity pricing. The table shows the price per board for different amounts of lumber purchased.

Price per Board (in dollars)	Number of Boards Purchased
8.00	1
7.50	10
6.50	25
5.00	50
2.00	100

a. Write each paired data as an ordered pair of the form (price per board, number of boards purchased).

b. Create a scatter diagram of the paired data. Be sure to label the axes appropriately.

8. The table shows the annual overnight stays in national parks. (*Source:* National Park Service)

Year	Overnight Stays in National Parks (in millions)
2001	9.8
2002	15.1
2003	14.6
2004	14.0
2005	13.8
2006	13.6

a. Write each paired data as an ordered pair of the form (year, number of overnight stays).

b. Create a scatter diagram of the paired data. Be sure to label the axes properly.

Determine whether each ordered pair is a solution of the given equation.

9. $7x - 8y = 56; (0, 56), (8, 0)$
10. $-2x + 5y = 10; (-5, 0), (1, 1)$
11. $x = 13; (13, 5), (13, 13)$
12. $y = 2; (7, 2), (2, 7)$

Complete the ordered pairs so that each is a solution of the given equation.

13. $-2 + y = 6x$; $(7, \ )$ **14.** $y = 3x + 5$; $(\ , -8)$

Complete the table of values for each given equation; then plot the ordered pairs. Use a single coordinate system for each exercise.

15. $9 = -3x + 4y$

x	y
0	
3	
9	

16. $y = 5$

x	y
	7
	-7
	0

17. $x = 2y$

x	y
	0
	5
	-5

18. The cost in dollars of producing x compact disk holders is given by $y = 5x + 2000$.

a. Complete the following table.

x	y
1	
100	
1000	

b. Find the number of compact disk holders that can be produced for $6430.

(3.2) Graph each linear equation.

19. $x - y = 1$ **20.** $x + y = 6$
21. $x - 3y = 12$ **22.** $5x - y = -8$
23. $x = 3y$ **24.** $y = -2x$
25. $2x - 3y = 6$ **26.** $4x - 3y = 12$

27. The projected U.S. long-distance revenue (in billions of dollars) from 1999 to 2004 is given by the equation, $y = 3x + 111$ where x is the number of years after 1999. Graph this equation and use it to estimate the amount of long-distance revenue in 2007. (*Source*: Giga Information Group)

(3.3) Identify the intercepts.

28.

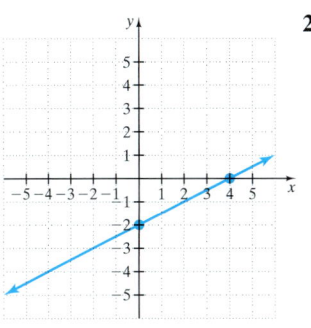

29.

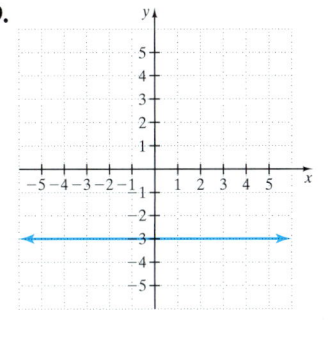

30.

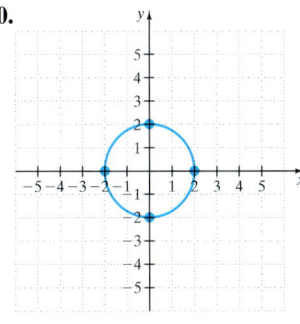

31.

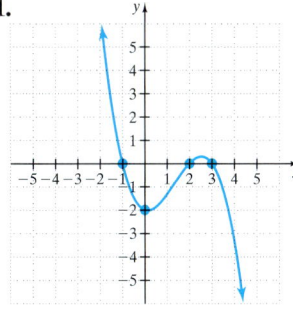

Graph each linear equation by finding its intercepts.

32. $x - 3y = 12$ **33.** $-4x + y = 8$
34. $y = -3$ **35.** $x = 5$
36. $y = -3x$ **37.** $x = 5y$
38. $x - 2 = 0$ **39.** $y + 6 = 0$

(3.4) Find the slope of each line.

40.

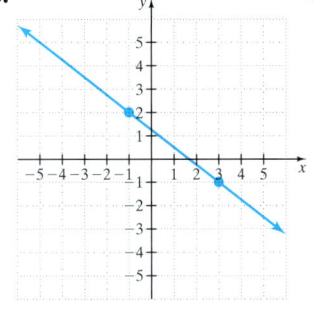

41.

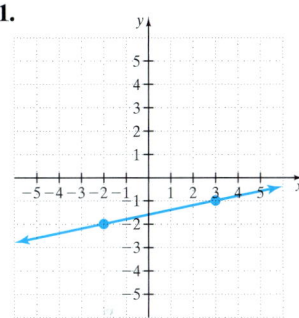

In Exercises 42 through 45, match each line with its slope.

a.

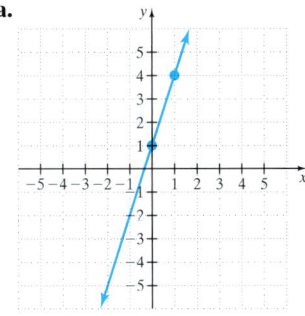

b.

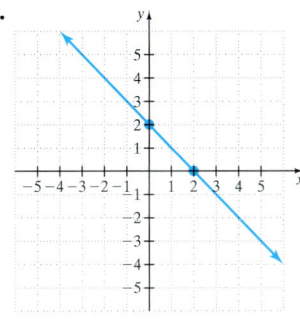

c.

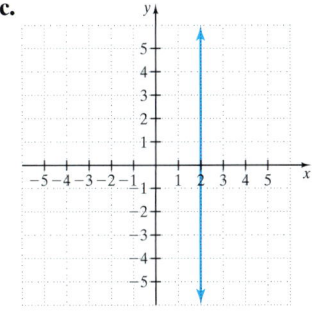

d.

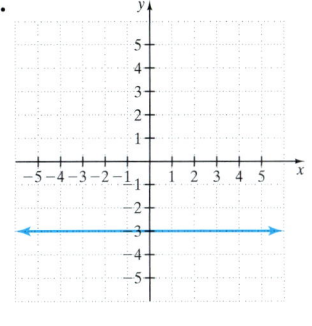

244 CHAPTER 3 Graphing

e.

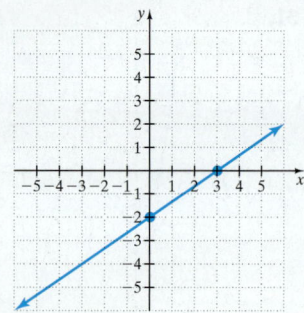

42. $m = 0$
43. $m = -1$
44. undefined slope
45. $m = 3$
46. $m = \dfrac{2}{3}$

Find the slope of the line that goes through the given points.

47. $(2, 5)$ and $(6, 8)$
48. $(4, 7)$ and $(1, 2)$
49. $(1, 3)$ and $(-2, -9)$
50. $(-4, 1)$ and $(3, -6)$

Find the slope of each line.

51. $y = 3x + 7$
52. $x - 2y = 4$
53. $y = -2$
54. $x = 0$

△ *Determine whether each pair of lines is parallel, perpendicular, or neither.*

55. $x - y = -6$
   $x + y = 3$

56. $3x + y = 7$
   $-3x - y = 10$

57. $y = 4x + \dfrac{1}{2}$
   $4x + 2y = 1$

58. $x = 4$
   $y = -2$

Find the slope of each line and write the slope as a rate of change. Don't forget to attach the proper units.

59. The graph below shows the average monthly day care cost for a 3-year-old attending 8 hours a day, 5 days a week.

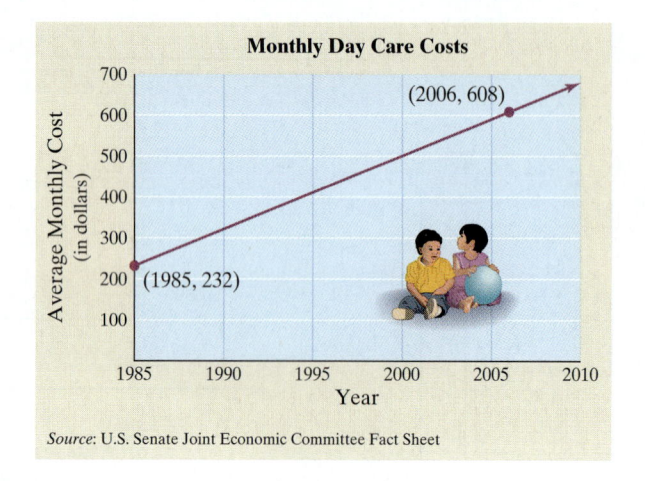

Monthly Day Care Costs
(1985, 232), (2006, 608)

Source: U.S. Senate Joint Economic Committee Fact Sheet

60. The graph below shows the U.S. government's projected spending (in billions of dollars) on technology. (Some years projected.)

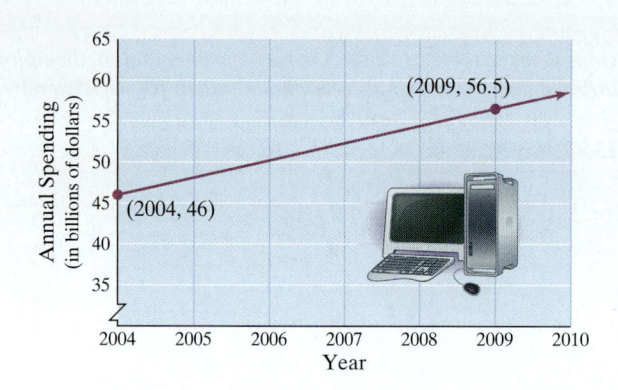

(2004, 46), (2009, 56.5)

(3.5) *Determine the slope and the y-intercept of the graph of each equation.*

61. $3x + y = 7$
62. $x - 6y = -1$
63. $y = 2$
64. $x = -5$

Write an equation of each line in slope-intercept form.

65. slope -5; y-intercept $\dfrac{1}{2}$
66. slope $\dfrac{2}{3}$; y-intercept 6

Use the slope-intercept form to graph each equation.

67. $y = 3x - 1$
68. $y = -3x$
69. $5x - 3y = 15$
70. $-x + 2y = 8$

Match each equation with its graph.

71. $y = -4x$
72. $y = -2x + 1$
73. $y = 2x - 1$
74. $y = 2x$

a.

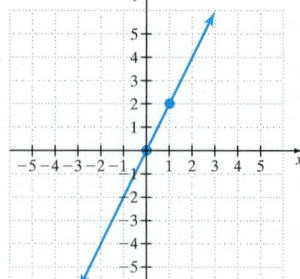

b.

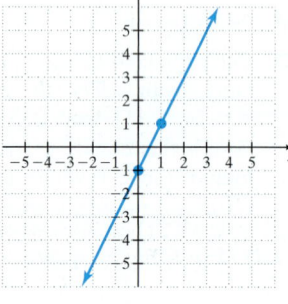

c.

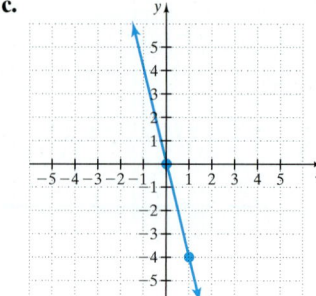

d.

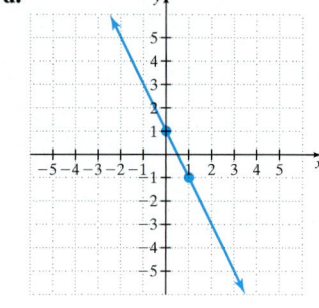

Write an equation of each line in standard form.

75. With slope -3, through $(0, -5)$
76. With slope $\frac{1}{2}$, through $\left(0, -\frac{7}{2}\right)$
77. With slope 0, through $(-2, -3)$
78. With 0 slope, through the origin
79. With slope -6, through $(2, -1)$
80. With slope 12, through $\left(\frac{1}{2}, 5\right)$
81. Through $(0, 6)$ and $(6, 0)$
82. Through $(0, -4)$ and $(-8, 0)$
83. Vertical line, through $(5, 7)$
84. Horizontal line, through $(-6, 8)$
85. Through $(6, 0)$, perpendicular to $y = 8$
86. Through $(10, 12)$, perpendicular to $x = -2$

(3.6) Determine which of the following are functions

87. $\{(7, 1), (7, 5), (2, 6)\}$
88. $\{(0, -1), (5, -1), (2, 2)\}$
89. $7x - 6y = 1$
90. $y = 7$
91. $x = 2$
92. $y = x^3$
93.
94.

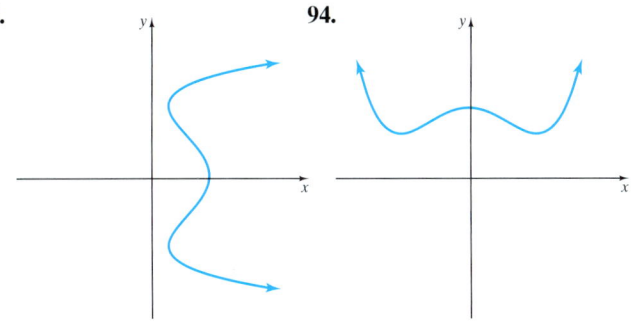

Given the following functions, find the indicated function values.

95. Given $f(x) = -2x + 6$, find
    a. $f(0)$
    b. $f(-2)$
    c. $f\left(\frac{1}{2}\right)$

96. Given $h(x) = -5 - 3x$, find
    a. $h(2)$
    b. $h(-3)$
    c. $h(0)$

97. Given $g(x) = x^2 + 12x$, find
    a. $g(3)$
    b. $g(-5)$
    c. $g(0)$

98. Given $h(x) = 6 - |x|$, find
    a. $h(-1)$
    b. $h(1)$
    c. $h(-4)$

Find the domain of each function.

99. $f(x) = 2x + 7$
100. $g(x) = \dfrac{7}{x - 2}$

Find the domain and the range of each function graphed.

101. 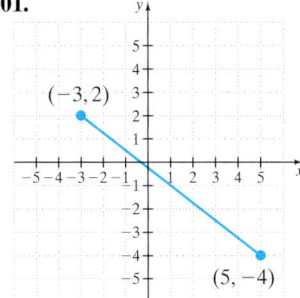
102. 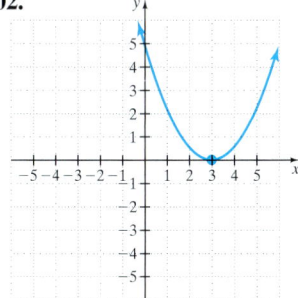
103. 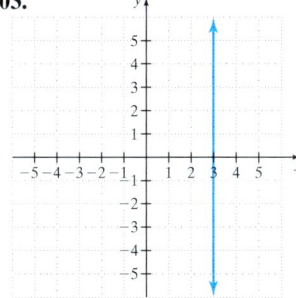
104.

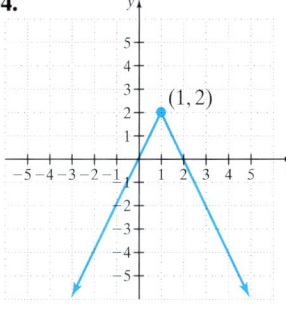

MIXED REVIEW

Complete the table of values for each given equation.

105. $2x - 5y = 9$

x	y
	1
2	
	-3

106. $x = -3y$

x	y
	0
	1
6	

Find the intercepts for each equation.

107. $2x - 3y = 6$
108. $-5x + y = 10$

Graph each linear equation.

109. $x - 5y = 10$
110. $x + y = 4$
111. $y = -4x$
112. $2x + 3y = -6$
113. $x = 3$
114. $y = -2$

Find the slope of the line that passes through each pair of points.

115. $(3, -5)$ and $(-4, 2)$
116. $(1, 3)$ and $(-6, -8)$

Find the slope of each line.

117.

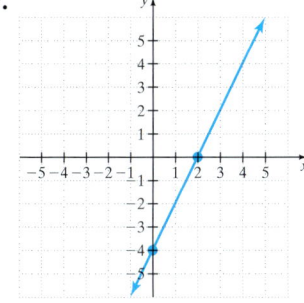

118.

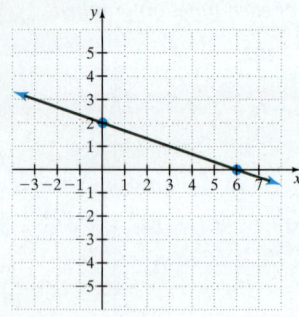

Determine the slope and y-intercept of the graph of each equation.

119. $-2x + 3y = -15$

120. $6x + y - 2 = 0$

Write an equation of the line with the given slope that passes through the given point. Write the equation in the form $Ax + By = C$.

121. $m = -5; (3, -7)$

122. $m = 3; (0, 6)$

Write an equation of the line passing through each pair of points. Write the equation in the form $Ax + By = C$.

123. $(-3, 9)$ and $(-2, 5)$

124. $(3, 1)$ and $(5, -9)$

Use the line graph to answer Exercises 125 through 128.

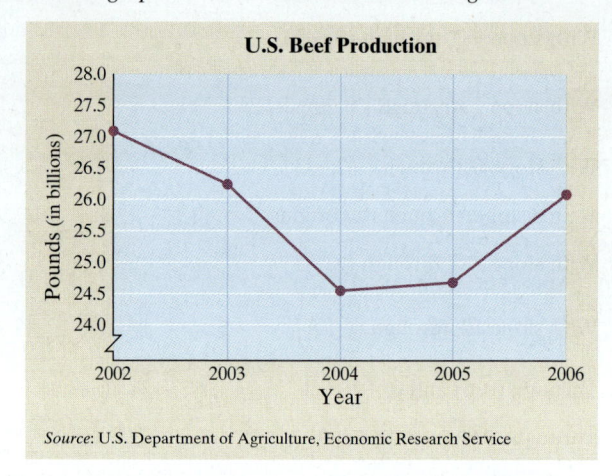

U.S. Beef Production

Source: U.S. Department of Agriculture, Economic Research Service

125. Which year shows the greatest production of beef? Estimate production for that year.

126. Which year shows the least production of beef? Estimate production for that year.

127. Which years had beef production greater then 25 billion pounds?

128. Which year shows the greatest increase in beef production?

CHAPTER 3 TEST

Graph the following.

1. $y = \dfrac{1}{2}x$

2. $2x + y = 8$

3. $5x - 7y = 10$

4. $y = -1$

5. $x - 3 = 0$

Find the slopes of the following lines.

6.

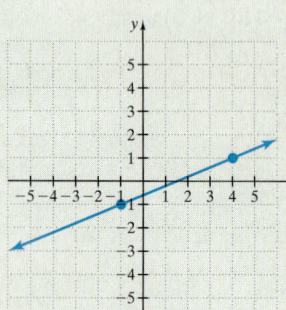

7.

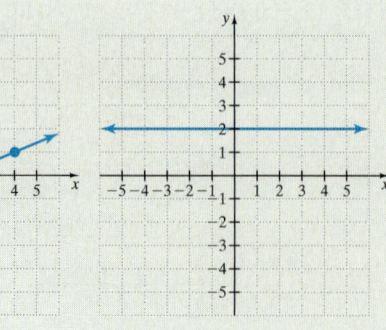

8. Through $(6, -5)$ and $(-1, 2)$

9. $-3x + y = 5$ **10.** $x = 6$

11. Determine the slope and the y-intercept of the graph of $7x - 3y = 2$.

12. Determine whether the graphs of $y = 2x - 6$ and $-4x = 2y$ are parallel lines, perpendicular lines, or neither.

Find equations of the following lines. Write the equation in standard form.

13. With slope of $-\dfrac{1}{4}$, through $(2, 2)$

14. Through the origin and $(6, -7)$

15. Through $(2, -5)$ and $(1, 3)$

16. Through $(-5, -1)$ and parallel to $x = 7$

17. With slope $\dfrac{1}{8}$ and y-intercept $(0, 12)$

Which of the following are functions?

18.

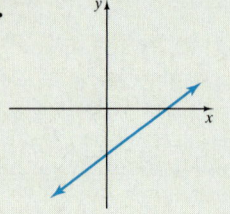

19.

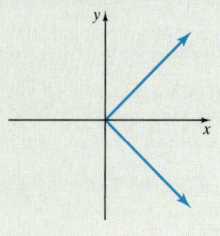

Given the following functions, find the indicated function values.

20. $h(x) = x^3 - x$

    a. $h(-1)$ **b.** $h(0)$ **c.** $h(4)$

21. Find the domain of $y = \dfrac{1}{x+1}$.

Find the domain and the range of each function graphed.

22.

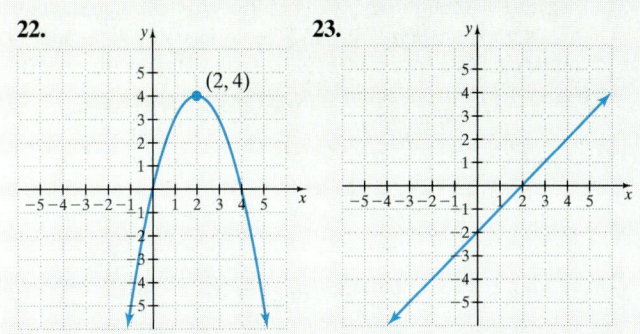

23.

24. If $f(7) = 20$, write the corresponding ordered pair.

Use the bar graph below to answer Exercises 25 and 26.

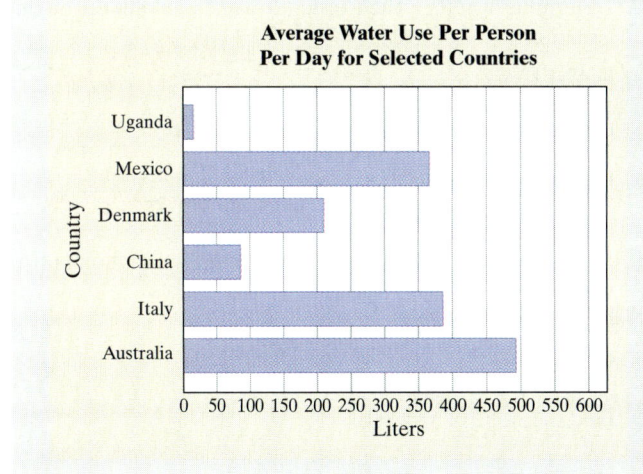

Average Water Use Per Person Per Day for Selected Countries

25. Estimate the average water use per person per day in Denmark.

26. Estimate the average water use per person per day in Australia.

Use this graph to answer Exercises 27 through 29.

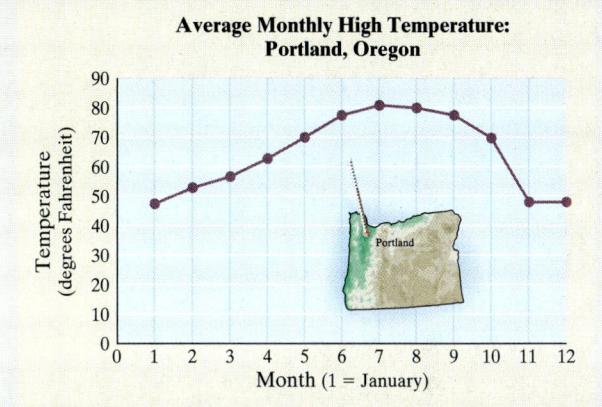

Average Monthly High Temperature: Portland, Oregon

Source: The Weather Channel Enterprises, Inc.

27. During what month is the average high temperature the greatest?

28. Approximate the average high temperature for the month of April.

29. During what month(s) is the average high temperature below 60°F?

CHAPTER 3 CUMULATIVE REVIEW

1. Insert $<$, $>$, or $=$ in the space between each pair of numbers to make each statement true.

    a. 2   3   **b.** 7   4   **c.** 72   27

2. Write the fraction $\dfrac{56}{64}$ in lowest terms.

3. Multiply $\dfrac{2}{15}$ and $\dfrac{5}{13}$. Write the product in lowest terms

4. Add: $\dfrac{10}{3} + \dfrac{5}{21}$

5. Simplify: $\dfrac{3 + |4 - 3| + 2^2}{6 - 3}$

6. Simplify: $16 - 3 \cdot 3 + 2^4$

7. Add.

    a. $-8 + (-11)$ **b.** $-5 + 35$

    c. $0.6 + (-1.1)$ **d.** $-\dfrac{7}{10} + \left(-\dfrac{1}{10}\right)$

    e. $11.4 + (-4.7)$ **f.** $-\dfrac{3}{8} + \dfrac{2}{5}$

8. Simplify: $|9 + (-20)| + |-10|$

9. Simplify each expression.

    a. $-14 - 8 + 10 - (-6)$

    b. $1.6 - (-10.3) + (-5.6)$

10. Simplify: $-9 - (3 - 8)$

11. If $x = -2$ and $y = -4$, evaluate each expression.
    a. $5x - y$
    b. $x^4 - y^2$
    c. $\dfrac{3x}{2y}$

12. Is -20 a solution of $\dfrac{x}{-10} = 2$?

13. Simplify each expression.
    a. $10 + (x + 12)$
    b. $-3(7x)$

14. Simplify: $(12 + x) - (4x - 7)$

15. Identify the numerical coefficient in each term.
    a. $-3y$
    b. $22z^4$
    c. y
    d. $-x$
    e. $\dfrac{x}{7}$

16. Multiply: $-5(x - 7)$

17. Solve $y + 0.6 = -1.0$ for y.

18. Solve: $5(3 + z) - (8z + 9) = -4$

19. Solve: $-\dfrac{2}{3}x = -\dfrac{5}{2}$

20. Solve: $\dfrac{x}{4} - 1 = -7$

21. If x is the first of three consecutive integers, express the sum of the three integers in terms of x. Simplify if possible.

22. Solve: $\dfrac{x}{3} - 2 = \dfrac{x}{3}$

23. Solve: $\dfrac{2(a + 3)}{3} = 6a + 2$

24. Solve: $x + 2y = 6$ for y.

25. In a recent year, the U.S. House of Representatives had a total of 435 Democrats and Republicans. There were 31 more Democratic representatives than Republican representatives. Find the number of representatives from each party. (*Source:* Office of the Clerk of the U.S. House of Representatives)

26. Solve $5(x + 4) \geq 4(2x + 3)$. Write the solution set in interval notation.

27. Charles Pecot can afford enough fencing to enclose a rectangular garden with a perimeter of 140 feet. If the width of his garden is to be 30 feet, find the length.

28. Solve $-3 < 4x - 1 \leq 2$. Write the solution set in interval notation.

29. Solve $y = mx + b$ for x.

30. Complete the table for $y = -5x$.

x	y
0	
-1	
	-10

31. A chemist working on his doctoral degree at Massachusetts Institute of Technology needs 12 liters of a 50% acid solution for a lab experiment. The stockroom has only 40% and 70% solutions. How much of each solution should be mixed together to form 12 liters of a 50% solution?

32. Graph: $y = -3x + 5$

33. Graph $x \geq -1$.

34. Find the x- and y-intercepts of $2x + 4y = -8$.

35. Solve $-1 \leq 2x - 3 < 5$. Graph the solution set and write it in interval notation.

36. Graph $x = 2$ on a rectangular coordinate system.

37. Determine whether each ordered pair is a solution of the equation $x - 2y = 6$.
    a. $(6, 0)$
    b. $(0, 3)$
    c. $\left(1, -\dfrac{5}{2}\right)$

38. Find the slope of the line through $(0, 5)$ and $(-5, 4)$.

39. Determine whether each equation is a linear equation in two variables.
    a. $x - 1.5y = -1.6$
    b. $y = -2x$
    c. $x + y^2 = 9$
    d. $x = 5$

40. Find the slope of $x = -10$.

41. Find the slope of the line $y = -1$.

42. Find the slope and y-intercept of the line whose equation is $2x - 5y = 10$.

43. Find an equation of the line with y-intercept $(0, -3)$ and slope of $\dfrac{1}{4}$.

44. Write an equation of the line through $(2, 3)$ and $(0, 0)$. Write the equation in standard form.

CHAPTER

4 Solving Systems of Linear Equations and Inequalities

4.1 Solving Systems of Linear Equations by Graphing

4.2 Solving Systems of Linear Equations by Substitution

4.3 Solving Systems of Linear Equations by Addition

Integrated Review—Solving Systems of Equations

4.4 Systems of Linear Equations and Problem Solving

4.5 Graphing Linear Inequalities

4.6 Systems of Linear Inequalities

Many of the occupations predicted to have the largest percent increase in number of jobs are in the fields of medicine and computer science. For example, from 2004 to 2014, the job growth predicted for pharmacy technicians is 28.6%, and for network systems and data communication analysts it is 54.5%. Although the demand for both jobs is growing, these jobs are growing at different rates. In Section 4.3, Exercise 73, we will predict when these occupations might have the same number of jobs.

In Chapter 3, we graphed equations containing two variables. Equations like these are often needed to represent relationships between two different values. There are also many real-life opportunities to compare and contrast two such equations, called a system of equations. This chapter presents linear systems and ways we solve these systems and apply them to real-life situations.

Job Title	Job Description	Employment (in thousands) 2004	2014
Pharmacy technician	Prepare medications under the direction of a pharmacist	258	332
Network systems and data communications analyst	Analyze, design, test, and evaluate network systems, Internet, intranet, and other data communication systems	231	357

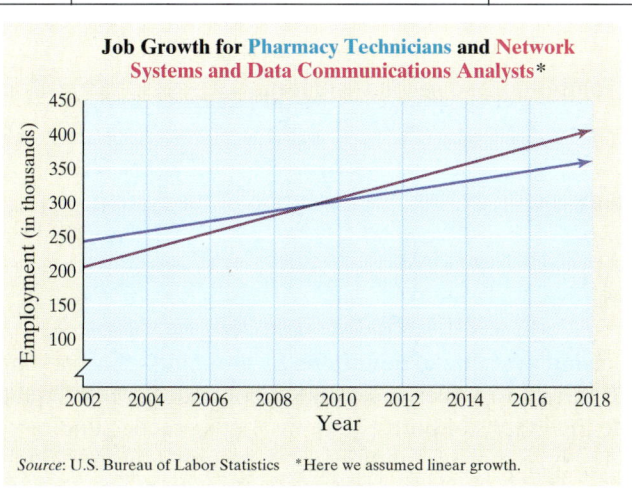

Job Growth for Pharmacy Technicians and Network Systems and Data Communications Analysts*

Source: U.S. Bureau of Labor Statistics *Here we assumed linear growth.

4.1 SOLVING SYSTEMS OF LINEAR EQUATIONS BY GRAPHING

OBJECTIVES

1. Determine if an ordered pair is a solution of a system of equations in two variables.
2. Solve a system of linear equations by graphing.
3. Without graphing, determine the number of solutions of a system.

OBJECTIVE 1 ▶ **Deciding whether an ordered pair is a solution.** A **system of linear equations** consists of two or more linear equations. In this section, we focus on solving systems of linear equations containing two equations in two variables. Examples of such linear systems are

$$\begin{cases} 3x - 3y = 0 \\ x = 2y \end{cases} \quad \begin{cases} x - y = 0 \\ 2x + y = 10 \end{cases} \quad \begin{cases} y = 7x - 1 \\ y = 4 \end{cases}$$

A **solution** of a system of two equations in two variables is an ordered pair of numbers that is a solution of both equations in the system.

EXAMPLE 1 Determine whether $(12, 6)$ is a solution of the system

$$\begin{cases} 2x - 3y = 6 \\ x = 2y \end{cases}$$

Solution To determine whether $(12, 6)$ is a solution of the system, we replace x with 12 and y with 6 in both equations.

$$\begin{array}{llll}
2x - 3y = 6 & \text{First equation} & x = 2y & \text{Second equation} \\
2(12) - 3(6) \stackrel{?}{=} 6 & \text{Let } x = 12 \text{ and } y = 6. & 12 \stackrel{?}{=} 2(6) & \text{Let } x = 12 \text{ and } y = 6. \\
24 - 18 \stackrel{?}{=} 6 & \text{Simplify.} & 12 = 12 & \text{True} \\
6 = 6 & \text{True} & &
\end{array}$$

Since $(12, 6)$ is a solution of both equations, it is a solution of the system. □

PRACTICE 1 Determine whether $(4, 12)$ is a solution of the system.

$$\begin{cases} 4x - y = 2 \\ y = 3x \end{cases}$$

EXAMPLE 2 Determine whether $(-1, 2)$ is a solution of the system

$$\begin{cases} x + 2y = 3 \\ 4x - y = 6 \end{cases}$$

Solution We replace x with -1 and y with 2 in both equations.

$$\begin{array}{llll}
x + 2y = 3 & \text{First equation} & 4x - y = 6 & \text{Second equation} \\
-1 + 2(2) \stackrel{?}{=} 3 & \text{Let } x = -1 \text{ and } y = 2. & 4(-1) - 2 \stackrel{?}{=} 6 & \text{Let } x = -1 \text{ and } y = 2. \\
-1 + 4 \stackrel{?}{=} 3 & \text{Simplify.} & -4 - 2 \stackrel{?}{=} 6 & \text{Simplify.} \\
3 = 3 & \text{True} & -6 = 6 & \text{False}
\end{array}$$

$(-1, 2)$ is not a solution of the second equation, $4x - y = 6$, so it is not a solution of the system. □

PRACTICE 2 Determine whether $(-4, 1)$ is a solution of the system.

$$\begin{cases} x - 3y = -7 \\ 2x + 9y = 1 \end{cases}$$

OBJECTIVE 2 ▶ **Solving systems of equations by graphing.** Since a solution of a system of two equations in two variables is a solution common to both equations, it is also a point common to the graphs of both equations. Let's practice finding solutions of both equations in a system—that is, solutions of a system—by graphing and identifying points of intersection.

EXAMPLE 3 Solve the system of equations by graphing.

$$\begin{cases} -x + 3y = 10 \\ x + y = 2 \end{cases}$$

Solution On a single set of axes, graph each linear equation.

$-x + 3y = 10$

x	y
0	$\frac{10}{3}$
-4	2
2	4

$x + y = 2$

x	y
0	2
2	0
1	1

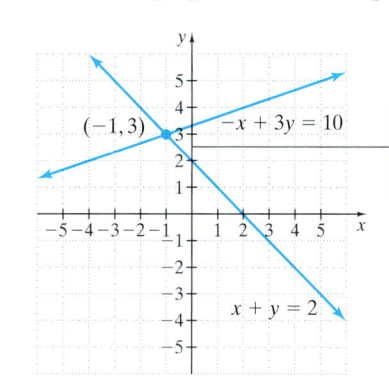

> **Helpful Hint**
> The point of intersection gives the solution of the system.

The two lines appear to intersect at the point $(-1, 3)$. To check, we replace x with -1 and y with 3 in both equations.

$-x + 3y = 10$ First equation $\qquad$ $x + y = 2$ Second equation
$-(-1) + 3(3) \stackrel{?}{=} 10$ Let $x = -1$ and $y = 3$. $\qquad$ $-1 + 3 \stackrel{?}{=} 2$ Let $x = -1$ and $y = 3$.
$1 + 9 \stackrel{?}{=} 10$ Simplify. $\qquad\qquad\qquad\qquad$ $2 = 2$ True
$10 = 10$ True

$(-1, 3)$ checks, so it is the solution of the system.

PRACTICE 3 Solve the system of equations by graphing:

$$\begin{cases} x - y = 3 \\ x + 2y = 18 \end{cases}$$

> **Helpful Hint**
> Neatly drawn graphs can help when you are estimating the solution of a system of linear equations by graphing.
>   In the example above, notice that the two lines intersected in a point. This means that the system has 1 solution.

EXAMPLE 4 Solve the system of equations by graphing.

$$\begin{cases} 2x + 3y = -2 \\ x = 2 \end{cases}$$

Solution We graph each linear equation on a single set of axes.

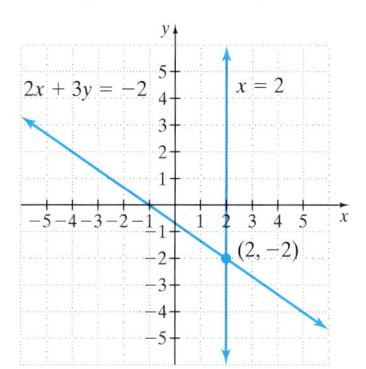

The two lines appear to intersect at the point $(2, -2)$. To determine whether $(2, -2)$ is the solution, we replace x with 2 and y with -2 in both equations.

$2x + 3y = -2$	First equation	$x = 2$	Second equation
$2(2) + 3(-2) \stackrel{?}{=} -2$	Let $x = 2$ and $y = -2$.	$2 \stackrel{?}{=} 2$	Let $x = 2$.
$4 + (-6) \stackrel{?}{=} -2$	Simplify.	$2 = 2$	True
$-2 = -2$	True		

Since a true statement results in both equations, $(2, -2)$ is the solution of the system.

PRACTICE 4 Solve the system of equations by graphing.

$$\begin{cases} -4x + 3y = -3 \\ y = -5 \end{cases}$$

A system of equations that has at least one solution as in Examples 3 and 4 is said to be a **consistent system**. A system that has no solution is said to be an **inconsistent system**.

EXAMPLE 5 Solve the following system of equations by graphing.

$$\begin{cases} 2x + y = 7 \\ 2y = -4x \end{cases}$$

Solution Graph the two lines in the system.

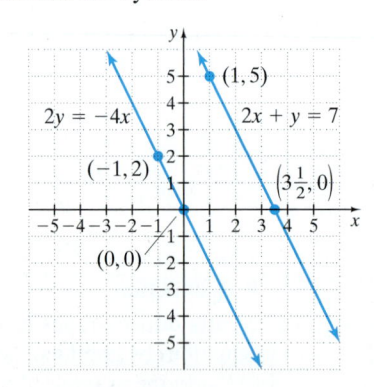

The lines **appear** to be parallel. To confirm this, write both equations in slope-intercept form by solving each equation for y.

$2x + y = 7$	First equation	$2y = -4x$	Second equation
$y = -2x + 7$	Subtract $2x$ from both sides.	$\dfrac{2y}{2} = \dfrac{-4x}{2}$	Divide both sides by 2.
		$y = -2x$	

Recall that when an equation is written in slope-intercept form, the coefficient of x is the slope. Since both equations have the same slope, -2, but different y-intercepts, the lines are parallel and have no points in common. Thus, there is no solution of the system and the system is inconsistent.

PRACTICE 5 Solve the system of equations by graphing.

$$\begin{cases} 3y = 9x \\ 6x - 2y = 12 \end{cases}$$

In Examples 3, 4, and 5, the graphs of the two linear equations of each system are different. When this happens, we call these equations **independent equations.** If the graphs of the two equations in a system are identical, we call the equations **dependent equations.**

EXAMPLE 6 Solve the system of equations by graphing.

$$\begin{cases} x - y = 3 \\ -x + y = -3 \end{cases}$$

Solution Graph each line.

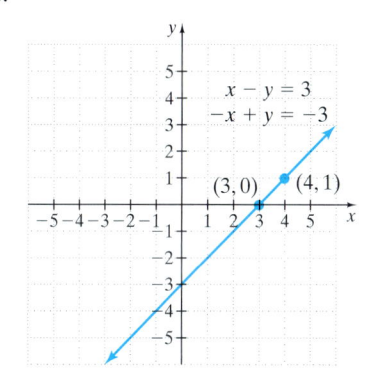

These graphs **appear** to be identical. To confirm this, write each equation in slope-intercept form.

$x - y = 3$ First equation $-x + y = -3$ Second equation

$-y = -x + 3$ Subtract x from both sides. $y = x - 3$ Add x to both sides.

$\dfrac{-y}{-1} = \dfrac{-x}{-1} + \dfrac{3}{-1}$ Divide both sides by -1.

$y = x - 3$

The equations are identical and so must be their graphs. The lines have an infinite number of points in common. Thus, there is an infinite number of solutions of the system and this is a consistent system. The equations are dependent equations.

PRACTICE 6 Solve the system of equations by graphing.

$$\begin{cases} x - y = 4 \\ -2x + 2y = -8 \end{cases}$$

As we have seen, three different situations can occur when graphing the two lines associated with the equations in a linear system:

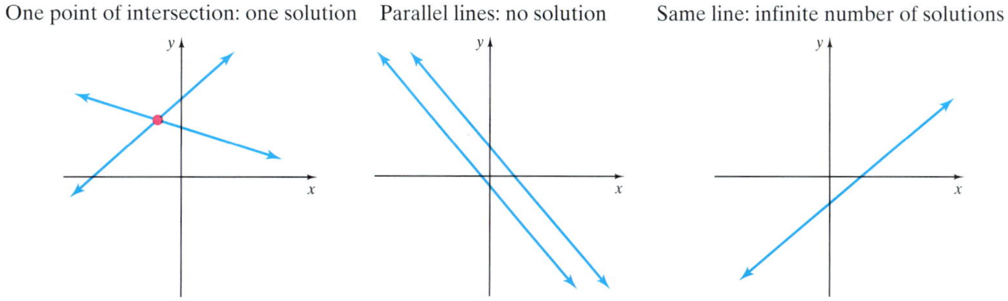

One point of intersection: one solution   Parallel lines: no solution   Same line: infinite number of solutions

Consistent system   **Inconsistent system**   **Consistent system**
(at least one solution)   (no solution)   (at least one solution)
Independent equations   **Independent equations**   **Dependent equations**
(graphs of equations differ)   (graphs of equations differ)   (graphs of equations identical)

OBJECTIVE 3 ▶ Finding the number of solutions of a system without graphing. You may have suspected by now that graphing alone is not an accurate way to solve a system of linear equations. For example, a solution of $\left(\dfrac{1}{2}, \dfrac{2}{9}\right)$ is unlikely to be read correctly from a graph. The next two sections present two accurate methods of solving these systems. In the meantime, we can decide how many solutions a system has by writing each equation in the slope-intercept form.

EXAMPLE 7 Without graphing, determine the number of solutions of the system.

$$\begin{cases} \dfrac{1}{2}x - y = 2 \\ x = 2y + 5 \end{cases}$$

Solution First write each equation in slope-intercept form.

$\dfrac{1}{2}x - y = 2$	First equation	$x = 2y + 5$	Second equation
$\dfrac{1}{2}x = y + 2$	Add y to both sides.	$x - 5 = 2y$	Subtract 5 from both sides.
$\dfrac{1}{2}x - 2 = y$	Subtract 2 from both sides.	$\dfrac{x}{2} - \dfrac{5}{2} = \dfrac{2y}{2}$	Divide both sides by 2.
		$\dfrac{1}{2}x - \dfrac{5}{2} = y$	Simplify.

The slope of each line is $\dfrac{1}{2}$, but they have different y-intercepts. This tells us that the lines representing these equations are parallel. Since the lines are parallel, the system has no solution and is inconsistent. □

PRACTICE 7 Without graphing, determine the number of solutions of the system.

$$\begin{cases} 5x + 4y = 6 \\ x - y = 3 \end{cases}$$

EXAMPLE 8 Without graphing, determine the number of solutions of the system.

$$\begin{cases} 3x - y = 4 \\ x + 2y = 8 \end{cases}$$

Solution Once again, the slope-intercept form helps determine how many solutions this system has.

$3x - y = 4$	First equation	$x + 2y = 8$	Second equation
$3x = y + 4$	Add y to both sides.	$x = -2y + 8$	Subtract $2y$ from both sides.
$3x - 4 = y$	Subtract 4 from both sides.	$x - 8 = -2y$	Subtract 8 from both sides.
		$\dfrac{x}{-2} - \dfrac{8}{-2} = \dfrac{-2y}{-2}$	Divide both sides by -2.
		$-\dfrac{1}{2}x + 4 = y$	Simplify.

The slope of the second line is $-\dfrac{1}{2}$, whereas the slope of the first line is 3. Since the slopes are not equal, the two lines are neither parallel nor identical and must intersect. Therefore, this system has one solution and is consistent. □

PRACTICE 8 Without graphing, determine the number of solutions of the system.

$$\begin{cases} -\dfrac{2}{3}x + y = 6 \\ 3y = 2x + 5 \end{cases}$$

Graphing Calculator Explorations

A graphing calculator may be used to approximate solutions of systems of equations. For example, to approximate the solution of the system

$$\begin{cases} y = -3.14x - 1.35 \\ y = 4.88x + 5.25, \end{cases}$$

first graph each equation on the same set of axes. Then use the intersect feature of your calculator to approximate the point of intersection.

The approximate point of intersection is $(-0.82, 1.23)$.

Solve each system of equations. Approximate the solutions to two decimal places.

1. $\begin{cases} y = -2.68x + 1.21 \\ y = 5.22x - 1.68 \end{cases}$
2. $\begin{cases} y = 4.25x + 3.89 \\ y = -1.88x + 3.21 \end{cases}$
3. $\begin{cases} 4.3x - 2.9y = 5.6 \\ 8.1x + 7.6y = -14.1 \end{cases}$
4. $\begin{cases} -3.6x - 8.6y = 10 \\ -4.5x + 9.6y = -7.7 \end{cases}$

VOCABULARY & READINESS CHECK

Fill in each blank with one of the words or phrases listed below.

system of linear equations   solution   consistent

dependent   inconsistent   independent

1. In a system of linear equations in two variables, if the graphs of the equations are the same, the equations are _____ equations.
2. Two or more linear equations are called a _____.
3. A system of equations that has at least one solution is called a(n) _____ system.
4. A _____ of a system of two equations in two variables is an ordered pair of numbers that is a solution of both equations in the system.
5. A system of equations that has no solution is called a(n) _____ system.
6. In a system of linear equations in two variables, if the graphs of the equations are different, the equations are _____ equations.

Each rectangular coordinate system shows the graph of the equations in a system of equations. Use each graph to determine the number of solutions for each associated system. If the system has only one solution, give its coordinates.

7.

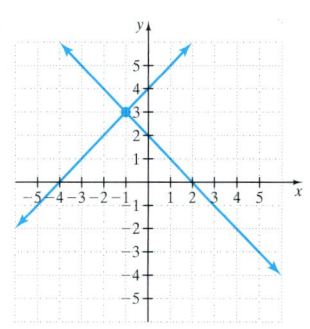

8.

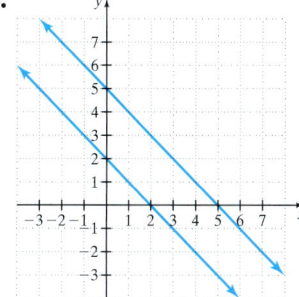

9.

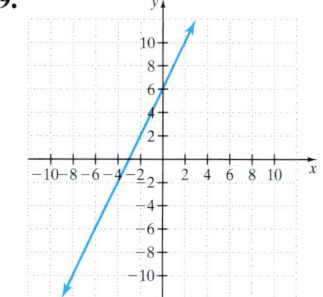

10.

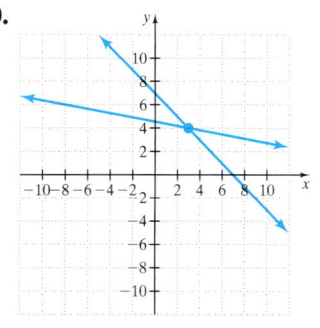

4.1 EXERCISE SET

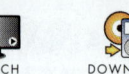

Determine whether each ordered pair is a solution of the system of linear equations. See Examples 1 and 2.

1. $\begin{cases} x + y = 8 \\ 3x + 2y = 21 \end{cases}$
   a. $(2, 4)$
   b. $(5, 3)$

2. $\begin{cases} 2x + y = 5 \\ x + 3y = 5 \end{cases}$
   a. $(5, 0)$
   b. $(2, 1)$

3. $\begin{cases} 3x - y = 5 \\ x + 2y = 11 \end{cases}$
   a. $(3, 4)$
   b. $(0, -5)$

4. $\begin{cases} 2x - 3y = 8 \\ x - 2y = 6 \end{cases}$
   a. $(-2, -4)$
   b. $(7, 2)$

5. $\begin{cases} 2y = 4x + 6 \\ 2x - y = -3 \end{cases}$
   a. $(-3, -3)$
   b. $(0, 3)$

6. $\begin{cases} x + 5y = -4 \\ -2x = 10y + 8 \end{cases}$
   a. $(-4, 0)$
   b. $(6, -2)$

7. $\begin{cases} -2 = x - 7y \\ 6x - y = 13 \end{cases}$
   a. $(-2, 0)$
   b. $\left(\dfrac{1}{2}, \dfrac{5}{14}\right)$

8. $\begin{cases} 4x = 1 - y \\ x - 3y = -8 \end{cases}$
   a. $(0, 1)$
   b. $\left(\dfrac{1}{6}, \dfrac{1}{3}\right)$

MIXED PRACTICE

Solve each system of linear equations by graphing. See Examples 3 through 6.

9. $\begin{cases} x + y = 4 \\ x - y = 2 \end{cases}$

10. $\begin{cases} x + y = 3 \\ x - y = 5 \end{cases}$

11. $\begin{cases} x + y = 6 \\ -x + y = -6 \end{cases}$

12. $\begin{cases} x + y = 1 \\ -x + y = -3 \end{cases}$

13. $\begin{cases} y = 2x \\ 3x - y = -2 \end{cases}$

14. $\begin{cases} y = -3x \\ 2x - y = -5 \end{cases}$

15. $\begin{cases} y = x + 1 \\ y = 2x - 1 \end{cases}$

16. $\begin{cases} y = 3x - 4 \\ y = x + 2 \end{cases}$

17. $\begin{cases} 2x + y = 0 \\ 3x + y = 1 \end{cases}$

18. $\begin{cases} 2x + y = 1 \\ 3x + y = 0 \end{cases}$

19. $\begin{cases} y = -x - 1 \\ y = 2x + 5 \end{cases}$

20. $\begin{cases} y = x - 1 \\ y = -3x - 5 \end{cases}$

21. $\begin{cases} x + y = 5 \\ x + y = 6 \end{cases}$

22. $\begin{cases} x - y = 4 \\ x - y = 1 \end{cases}$

23. $\begin{cases} 2x - y = 6 \\ y = 2 \end{cases}$

24. $\begin{cases} x + y = 5 \\ x = 4 \end{cases}$

25. $\begin{cases} x - 2y = 2 \\ 3x + 2y = -2 \end{cases}$

26. $\begin{cases} x + 3y = 7 \\ 2x - 3y = -4 \end{cases}$

27. $\begin{cases} 2x + y = 4 \\ 6x = -3y + 6 \end{cases}$

28. $\begin{cases} y + 2x = 3 \\ 4x = 2 - 2y \end{cases}$

29. $\begin{cases} y - 3x = -2 \\ 6x - 2y = 4 \end{cases}$

30. $\begin{cases} x - 2y = -6 \\ -2x + 4y = 12 \end{cases}$

31. $\begin{cases} x = 3 \\ y = -1 \end{cases}$

32. $\begin{cases} x = -5 \\ y = 3 \end{cases}$

33. $\begin{cases} y = x - 2 \\ y = 2x + 3 \end{cases}$

34. $\begin{cases} y = x + 5 \\ y = -2x - 4 \end{cases}$

35. $\begin{cases} 2x - 3y = -2 \\ -3x + 5y = 5 \end{cases}$

36. $\begin{cases} 4x - y = 7 \\ 2x - 3y = -9 \end{cases}$

37. $\begin{cases} 6x - y = 4 \\ \dfrac{1}{2}y = -2 + 3x \end{cases}$

38. $\begin{cases} 3x - y = 6 \\ \dfrac{1}{3}y = -2 + x \end{cases}$

Without graphing, decide.

a. Are the graphs of the equations identical lines, parallel lines, or lines intersecting at a single point?

b. How many solutions does the system have? See Examples 7 and 8.

39. $\begin{cases} 4x + y = 24 \\ x + 2y = 2 \end{cases}$

40. $\begin{cases} 3x + y = 1 \\ 3x + 2y = 6 \end{cases}$

41. $\begin{cases} 2x + y = 0 \\ 2y = 6 - 4x \end{cases}$

42. $\begin{cases} 3x + y = 0 \\ 2y = -6x \end{cases}$

43. $\begin{cases} 6x - y = 4 \\ \dfrac{1}{2}y = -2 + 3x \end{cases}$

44. $\begin{cases} 3x - y = 2 \\ \dfrac{1}{3}y = -2 + 3x \end{cases}$

45. $\begin{cases} x = 5 \\ y = -2 \end{cases}$

46. $\begin{cases} y = 3 \\ x = -4 \end{cases}$

47. $\begin{cases} 3y - 2x = 3 \\ x + 2y = 9 \end{cases}$

48. $\begin{cases} 2y = x + 2 \\ y + 2x = 3 \end{cases}$

49. $\begin{cases} 6y + 4x = 6 \\ 3y - 3 = -2x \end{cases}$

50. $\begin{cases} 8y + 6x = 4 \\ 4y - 2 = 3x \end{cases}$

51. $\begin{cases} x + y = 4 \\ x + y = 3 \end{cases}$

52. $\begin{cases} 2x + y = 0 \\ y = -2x + 1 \end{cases}$

REVIEW AND PREVIEW

Solve each equation. See Section 2.4.

53. $5(x - 3) + 3x = 1$

54. $-2x + 3(x + 6) = 17$

55. $4\left(\dfrac{y+1}{2}\right) + 3y = 0$

56. $-y + 12\left(\dfrac{y-1}{4}\right) = 3$

57. $8a - 2(3a - 1) = 6$

58. $3z - (4z - 2) = 9$

CONCEPT EXTENSIONS

59. Draw a graph of two linear equations whose associated system has the solution $(-1, 4)$.

60. Draw a graph of two linear equations whose associated system has the solution $(3, -2)$.

61. Draw a graph of two linear equations whose associated system has no solution.

62. Draw a graph of two linear equations whose associated system has an infinite number of solutions.

63. Explain how to use a graph to determine the number of solutions of a system.

64. The ordered pair $(-2, 3)$ is a solution of all three independent equations:

$$x + y = 1$$
$$2x - y = -7$$
$$x + 3y = 7$$

Describe the graph of all three equations on the same axes.

The double line graph below shows the number of pounds of fish and shellfish consumed per person in the United States for the years shown. Use the graph for Exercises 65 and 66. (Source: Economic Research Service, U.S. Department of Agriculture)

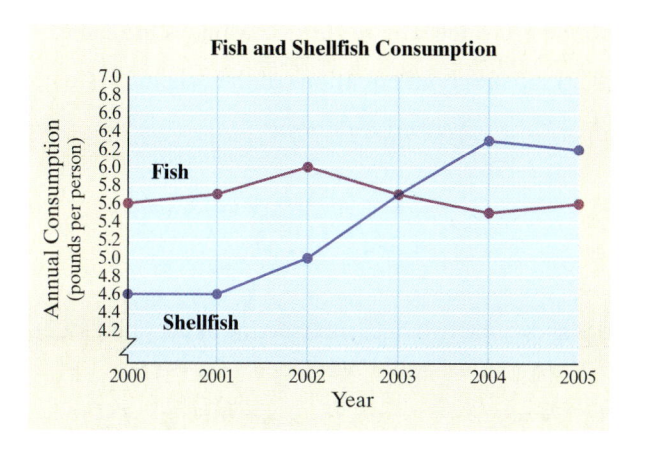

65. In what year(s) was the pounds per person of fish greater than the pounds per person of shellfish?

66. In what year(s) was the pounds per person of shellfish greater than or equal to the pounds per person of fish?

The double line graph below shows the annual number of Toyota cars and General Motors cars sold in the United States for the years shown. Use this graph to answer Exercises 67–70. (Sources: Toyota Corporation, General Motors Corporation)

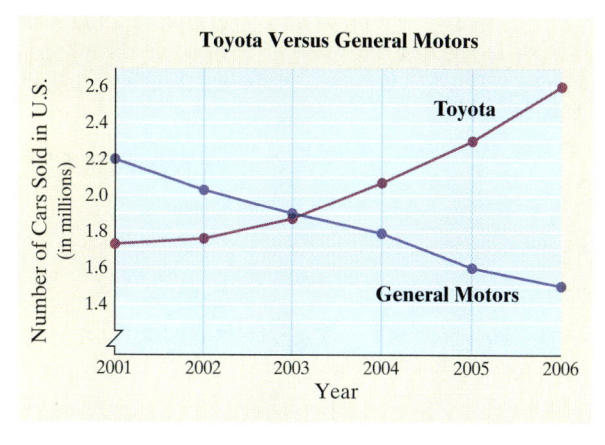

67. In what year(s) was the number of Toyota cars sold in the United States less than the number of GM cars sold in the United States?

68. In what year(s) was the number of GM cars sold in the United States less than the number of Toyota cars sold in the United States?

69. Describe any trends you see in this graph.

70. Approximate how many more cars Toyota sold in the United States than GM in 2005.

71. Construct a system of two linear equations that has $(1, 3)$ as a solution.

72. Construct a system of two linear equations that has $(0, 7)$ as a solution.

73. Below are two tables of values for two linear equations. Using the tables,
    a. find a solution of the corresponding system.
    b. graph several ordered pairs from each table and sketch the two lines.
    c. Does your graph confirm the solution from part a?

x	y
1	3
2	5
3	7
4	9
5	11

x	y
1	6
2	7
3	8
4	9
5	10

74. Explain how writing each equation in a linear system in slope-intercept form helps determine the number of solutions of a system.

75. Is it possible for a system of two linear equations in two variables to be inconsistent, but with dependent equations? Why or why not?

4.2 SOLVING SYSTEMS OF LINEAR EQUATIONS BY SUBSTITUTION

OBJECTIVE

1. Use the substitution method to solve a system of linear equations.

OBJECTIVE 1 ▶ Using the substitution method. As we stated in the preceding section, graphing alone is not an accurate way to solve a system of linear equations. In this section, we discuss a second, more accurate method for solving systems of equations. This method is called the **substitution method** and is introduced in the next example.

EXAMPLE 1   Solve the system:

$$\begin{cases} 2x + y = 10 & \text{First equation} \\ x = y + 2 & \text{Second equation} \end{cases}$$

Solution   The second equation in this system is $x = y + 2$. This tells us that x and $y + 2$ have the same value. This means that we may substitute $y + 2$ for x in the first equation.

$2x + y = 10$ First equation

$2(y + 2) + y = 10$ Substitute $y + 2$ for x since $x = y + 2$.

Notice that this equation now has one variable, y. Let's now solve this equation for y.

$2(y + 2) + y = 10$

$2y + 4 + y = 10$ Use the distributive property.

$3y + 4 = 10$ Combine like terms.

$3y = 6$ Subtract 4 from both sides.

$y = 2$ Divide both sides by 3.

▶ **Helpful Hint**

Don't forget the distributive property.

Now we know that the y-value of the ordered pair solution of the system is 2. To find the corresponding x-value, we replace y with 2 in the equation $x = y + 2$ and solve for x.

$x = y + 2$

$x = 2 + 2$ Let $y = 2$.

$x = 4$

The solution of the system is the ordered pair $(4, 2)$. Since an ordered pair solution must satisfy both linear equations in the system, we could have chosen the equation $2x + y = 10$ to find the corresponding x-value. The resulting x-value is the same.

Check: We check to see that $(4, 2)$ satisfies both equations of the original system.

First Equation	*Second Equation*
$2x + y = 10$	$x = y + 2$
$2(4) + 2 \stackrel{?}{=} 10$	$4 \stackrel{?}{=} 2 + 2$ Let $x = 4$ and $y = 2$.
$10 = 10$ True	$4 = 4$ True

The solution of the system is $(4, 2)$.

A graph of the two equations shows the two lines intersecting at the point $(4, 2)$.

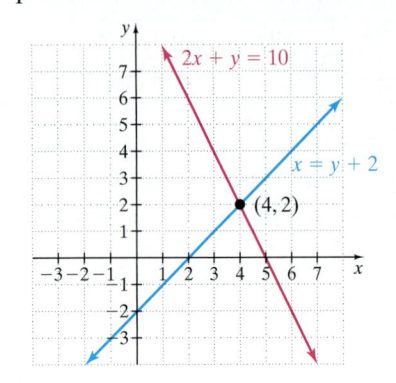

PRACTICE

1 Solve the system:
$$\begin{cases} 2x - y = 9 \\ x = y + 1 \end{cases}$$

EXAMPLE 2 Solve the system:
$$\begin{cases} 5x - y = -2 \\ y = 3x \end{cases}$$

Solution The second equation is solved for y in terms of x. We substitute $3x$ for y in the first equation.

$5x - y = -2$ First equation

$5x - (3x) = -2$ Substitute $3x$ for y.

Now we solve for x.

$5x - 3x = -2$

$2x = -2$ Combine like terms.

$x = -1$ Divide both sides by 2.

The x-value of the ordered pair solution is -1. To find the corresponding y-value, we replace x with -1 in the second equation $y = 3x$.

$y = 3x$ Second equation

$y = 3(-1)$ Let $x = -1$.

$y = -3$

Check to see that the solution of the system is $(-1, -3)$.

PRACTICE

2 Solve the system:
$$\begin{cases} 7x - y = -15 \\ y = 2x \end{cases}$$

To solve a system of equations by substitution, we first need an equation solved for one of its variables, as in Examples 1 and 2. If neither equation in a system is solved for x or y, this will be our first step.

EXAMPLE 3 Solve the system:
$$\begin{cases} x + 2y = 7 \\ 2x + 2y = 13 \end{cases}$$

Solution We choose one of the equations and solve for x or y. We will solve the first equation for x by subtracting $2y$ from both sides.

$x + 2y = 7$ First equation

$x = 7 - 2y$ Subtract $2y$ from both sides.

Since $x = 7 - 2y$, we now substitute $7 - 2y$ for x in the second equation and solve for y.

$2x + 2y = 13$ Second equation

$2(7 - 2y) + 2y = 13$ Let $x = 7 - 2y$.

$14 - 4y + 2y = 13$ Use the distributive property.

$14 - 2y = 13$ Simplify.

$-2y = -1$ Subtract 14 from both sides.

$y = \dfrac{1}{2}$ Divide both sides by -2.

▶ **Helpful Hint**

Don't forget to insert parentheses when substituting $7 - 2y$ for x.

260 CHAPTER 4 Solving Systems of Linear Equations and Inequalities

To find x, we let $y = \dfrac{1}{2}$ in the equation $x = 7 - 2y$.

$$x = 7 - 2y$$
$$x = 7 - 2\left(\dfrac{1}{2}\right) \quad \text{Let } y = \dfrac{1}{2}.$$
$$x = 7 - 1$$
$$x = 6$$

> **Helpful Hint**
> To find x, any equation in two variables equivalent to the original equations of the system may be used. We used this equation since it is solved for x.

The solution is $\left(6, \dfrac{1}{2}\right)$. Check the solution in both equations of the original system. □

PRACTICE 3 Solve the system:
$$\begin{cases} x + 3y = 6 \\ 2x + 3y = 10 \end{cases}$$

The following steps may be used to solve a system of equations by the substitution method.

Solving a System of Two Linear Equations by the Substitution Method
STEP 1. Solve one of the equations for one of its variables.
STEP 2. Substitute the expression for the variable found in Step 1 into the other equation.
STEP 3. Solve the equation from Step 2 to find the value of one variable.
STEP 4. Substitute the value found in Step 3 in any equation containing both variables to find the value of the other variable.
STEP 5. Check the proposed solution in the original system.

Concept Check ✓

As you solve the system $\begin{cases} 2x + y = -5 \\ x - y = 5 \end{cases}$ you find that $y = -5$. Is this the solution of the system?

EXAMPLE 4 Solve the system:
$$\begin{cases} 7x - 3y = -14 \\ -3x + y = 6 \end{cases}$$

Solution To avoid introducing fractions, we will solve the second equation for y.

$$-3x + y = 6 \qquad \text{Second equation}$$
$$y = 3x + 6$$

Next, substitute $3x + 6$ for y in the first equation.

$$7x - 3y = -14 \qquad \text{First equation}$$
$$7x - 3(3x + 6) = -14 \qquad \text{Let } y = 3x + 6.$$
$$7x - 9x - 18 = -14 \qquad \text{Use the distributive property.}$$
$$-2x - 18 = -14 \qquad \text{Simplify.}$$
$$-2x = 4 \qquad \text{Add 18 to both sides.}$$
$$x = -2 \qquad \text{Divide both sides by } -2.$$

Answer to Concept Check:
No, the solution will be an ordered pair.

To find the corresponding y-value, substitute -2 for x in the equation $y = 3x + 6$. Then $y = 3(-2) + 6$ or $y = 0$. The solution of the system is $(-2, 0)$. Check this solution in both equations of the system.

PRACTICE 4 Solve the system:
$$\begin{cases} 5x + 3y = -9 \\ -2x + y = 8 \end{cases}$$

> **Helpful Hint**
> When solving a system of equations by the substitution method, begin by solving an equation for one of its variables. If possible, solve for a variable that has a coefficient of 1 or -1. This way, we avoid working with time-consuming fractions.

EXAMPLE 5 Solve the system:
$$\begin{cases} \dfrac{1}{2}x - y = 3 \\ x = 6 + 2y \end{cases}$$

Solution The second equation is already solved for x in terms of y. Thus we substitute $6 + 2y$ for x in the first equation and solve for y.

$\dfrac{1}{2}x - y = 3$ First equation

$\dfrac{1}{2}(6 + 2y) - y = 3$ Let $x = 6 + 2y$.

$3 + y - y = 3$ Use the distributive property.

$3 = 3$ Simplify.

Arriving at a true statement such as $3 = 3$ indicates that the two linear equations in the original system are equivalent. This means that their graphs are identical and there are an infinite number of solutions of the system. Any solution of one equation is also a solution of the other.

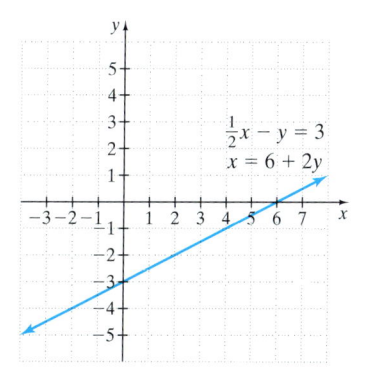

PRACTICE 5 Solve the system:
$$\begin{cases} \dfrac{1}{4}x - y = 2 \\ x = 4y + 8 \end{cases}$$

EXAMPLE 6 Use substitution to solve the system.

$$\begin{cases} 6x + 12y = 5 \\ -4x - 8y = 0 \end{cases}$$

Solution Choose the second equation and solve for y.

$$-4x - 8y = 0 \quad \text{Second equation}$$
$$-8y = 4x \quad \text{Add } 4x \text{ to both sides.}$$
$$\frac{-8y}{-8} = \frac{4x}{-8} \quad \text{Divide both sides by } -8.$$
$$y = -\frac{1}{2}x \quad \text{Simplify.}$$

Now replace y with $-\frac{1}{2}x$ in the first equation.

$$6x + 12y = 5 \quad \text{First equation}$$
$$6x + 12\left(-\frac{1}{2}x\right) = 5 \quad \text{Let } y = -\frac{1}{2}x.$$
$$6x + (-6x) = 5 \quad \text{Simplify.}$$
$$0 = 5 \quad \text{Combine like terms.}$$

The false statement $0 = 5$ indicates that this system has no solution and is inconsistent. The graph of the linear equations in the system is a pair of parallel lines.

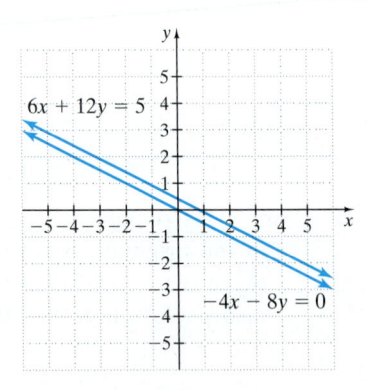

PRACTICE 6 Use substitution to solve the system.

$$\begin{cases} 4x - 3y = 12 \\ -8x + 6y = -30 \end{cases}$$

Concept Check ✓

Describe how the graphs of the equations in a system appear if the system has

a. no solution
b. one solution
c. an infinite number of solutions

Answers to Concept Check:
a. parallel lines
b. intersect at one point
c. identical graphs

VOCABULARY & READINESS CHECK

Give the solution of each system. If the system has no solution or an infinite number of solutions, say so. If the system has one solution, find it.

1. $\begin{cases} y = 4x \\ -3x + y = 1 \end{cases}$
   When solving, you obtain $x = 1$

2. $\begin{cases} 4x - y = 17 \\ -8x + 2y = 0 \end{cases}$
   When solving, you obtain $0 = 34$

3. $\begin{cases} 4x - y = 17 \\ -8x + 2y = -34 \end{cases}$
   When solving, you obtain $0 = 0$

4. $\begin{cases} 5x + 2y = 25 \\ x = y + 5 \end{cases}$
   When solving, you obtain $y = 0$

5. $\begin{cases} x + y = 0 \\ 7x - 7y = 0 \end{cases}$
   When solving, you obtain $x = 0$

6. $\begin{cases} y = -2x + 5 \\ 4x + 2y = 10 \end{cases}$
   When solving, you obtain $0 = 0$

4.2 EXERCISE SET

Solve each system of equations by the substitution method. See Examples 1 and 2.

1. $\begin{cases} x + y = 3 \\ x = 2y \end{cases}$

2. $\begin{cases} x + y = 20 \\ x = 3y \end{cases}$

3. $\begin{cases} x + y = 6 \\ y = -3x \end{cases}$

4. $\begin{cases} x + y = 6 \\ y = -4x \end{cases}$

5. $\begin{cases} y = 3x + 1 \\ 4y - 8x = 12 \end{cases}$

6. $\begin{cases} y = 2x + 3 \\ 5y - 7x = 18 \end{cases}$

7. $\begin{cases} y = 2x + 9 \\ y = 7x + 10 \end{cases}$

8. $\begin{cases} y = 5x - 3 \\ y = 8x + 4 \end{cases}$

MIXED PRACTICE

Solve each system of equations by the substitution method. See Examples 1 through 6.

9. $\begin{cases} 3x - 4y = 10 \\ y = x - 3 \end{cases}$

10. $\begin{cases} 4x - 3y = 10 \\ y = x - 5 \end{cases}$

11. $\begin{cases} x + 2y = 6 \\ 2x + 3y = 8 \end{cases}$

12. $\begin{cases} x + 3y = -5 \\ 2x + 2y = 6 \end{cases}$

13. $\begin{cases} 3x + 2y = 16 \\ x = 3y - 2 \end{cases}$

14. $\begin{cases} 2x + 3y = 18 \\ x = 2y - 5 \end{cases}$

15. $\begin{cases} 2x - 5y = 1 \\ 3x + y = -7 \end{cases}$

16. $\begin{cases} 3y - x = 6 \\ 4x + 12y = 0 \end{cases}$

17. $\begin{cases} 4x + 2y = 5 \\ -2x = y + 4 \end{cases}$

18. $\begin{cases} 2y = x + 2 \\ 6x - 12y = 0 \end{cases}$

19. $\begin{cases} 4x + y = 11 \\ 2x + 5y = 1 \end{cases}$

20. $\begin{cases} 3x + y = -14 \\ 4x + 3y = -22 \end{cases}$

21. $\begin{cases} x + 2y + 5 = -4 + 5y - x \\ 2x + x = y + 4 \end{cases}$
    (Hint: First simplify each equation.)

22. $\begin{cases} 5x + 4y - 2 = -6 + 7y - 3x \\ 3x + 4x = y + 3 \end{cases}$
    (Hint: See Exercise 21.)

23. $\begin{cases} 6x - 3y = 5 \\ x + 2y = 0 \end{cases}$

24. $\begin{cases} 10x - 5y = -21 \\ x + 3y = 0 \end{cases}$

25. $\begin{cases} 3x - y = 1 \\ 2x - 3y = 10 \end{cases}$

26. $\begin{cases} 2x - y = -7 \\ 4x - 3y = -11 \end{cases}$

27. $\begin{cases} -x + 2y = 10 \\ -2x + 3y = 18 \end{cases}$

28. $\begin{cases} -x + 3y = 18 \\ -3x + 2y = 19 \end{cases}$

29. $\begin{cases} 5x + 10y = 20 \\ 2x + 6y = 10 \end{cases}$

30. $\begin{cases} 6x + 3y = 12 \\ 9x + 6y = 15 \end{cases}$

31. $\begin{cases} 3x + 6y = 9 \\ 4x + 8y = 16 \end{cases}$

32. $\begin{cases} 2x + 4y = 6 \\ 5x + 10y = 16 \end{cases}$

33. $\begin{cases} \dfrac{1}{3}x - y = 2 \\ x - 3y = 6 \end{cases}$

34. $\begin{cases} \dfrac{1}{4}x - 2y = 1 \\ x - 8y = 4 \end{cases}$

35. $\begin{cases} x = \dfrac{3}{4}y - 1 \\ 8x - 5y = -6 \end{cases}$

36. $\begin{cases} x = \dfrac{5}{6}y - 2 \\ 12x - 5y = -9 \end{cases}$

Solve each system by the substitution method. First simplify each equation by combining like terms.

37. $\begin{cases} -5y + 6y = 3x + 2(x - 5) - 3x + 5 \\ 4(x + y) - x + y = -12 \end{cases}$

38. $\begin{cases} 5x + 2y - 4x - 2y = 2(2y + 6) - 7 \\ 3(2x - y) - 4x = 1 + 9 \end{cases}$

REVIEW AND PREVIEW

Write equivalent equations by multiplying both sides of the given equation by the given nonzero number. See Section 2.3.

39. $3x + 2y = 6$ by -2

40. $-x + y = 10$ by 5

41. $-4x + y = 3$ by 3

42. $5a - 7b = -4$ by -4

Add the binomials. See Section 2.1.

43. $\begin{array}{r}3n + 6m\\2n - 6m\end{array}$ 44. $\begin{array}{r}-2x + 5y\\2x + 11y\end{array}$ 45. $\begin{array}{r}-5a - 7b\\5a - 8b\end{array}$ 46. $\begin{array}{r}9q + p\\-9q - p\end{array}$

CONCEPT EXTENSIONS

47. Explain how to identify a system with no solution when using the substitution method.

48. Occasionally, when using the substitution method, we obtain the equation $0 = 0$. Explain how this result indicates that the graphs of the equations in the system are identical.

Solve. See a Concept Check in this section.

49. As you solve the system $\begin{cases}3x - y = -6\\-3x + 2y = 7\end{cases}$, you find that $y = 1$. Is this the solution to the system?

50. As you solve the system $\begin{cases}x = 5y\\y = 2x\end{cases}$, you find that $x = 0$ and $y = 0$. What is the solution to this system?

51. To avoid fractions, which of the equations below would you use if solving for y? Explain why.
    a. $\frac{1}{2}x - 4y = \frac{3}{4}$ b. $8x - 5y = 13$
    c. $7x - y = 19$

52. Give the number of solutions for a system if the graphs of the equations in the system are
    a. lines intersecting in one point
    b. parallel lines
    c. same line

53. The number of men and women receiving bachelor's degrees each year has been steadily increasing. For the years 1970 through the projection of 2014, the number of men receiving degrees (in thousands) is given by the equation $y = 3.9x + 443$, and for women, the equation is $y = 14.2x + 314$ where x is the number of years after 1970. (*Source:* National Center for Education Statistics)
    a. Use the substitution method to solve this system of equations. (Round your final results to the nearest whole numbers.)
    b. Explain the meaning of your answer to part (a).
    c. Sketch a graph of the system of equations. Write a sentence describing the trends for men and women receiving bachelor degrees.

54. The number of Adult Contemporary Music radio stations in the United States from 2000 to 2006 is given by the equation $y = -6.17x + 719$, where x is the number of years after 2000. The number of Spanish radio stations is given by $y = 33.9x + 534$ for the same time period. (*Source:* M Street Corporation)
    a. Use the substitution method to solve this system of equations. (Round your numbers to the nearest tenth.)
    b. Explain the meaning of your answer to part (a).
    c. Sketch a graph of the system of equations. Write a sentence describing the trends in the popularity of these two types of music format.

Solve each system by substitution. When necessary, round answers to the nearest hundredth.

55. $\begin{cases}y = 5.1x + 14.56\\y = -2x - 3.9\end{cases}$

56. $\begin{cases}y = 3.1x - 16.35\\y = -9.7x + 28.45\end{cases}$

57. $\begin{cases}3x + 2y = 14.05\\5x + y = 18.5\end{cases}$

58. $\begin{cases}x + y = -15.2\\-2x + 5y = -19.3\end{cases}$

STUDY SKILLS BUILDER

How Are Your Homework Assignments Going?

Remember that it is important to keep up with homework. Why? Many concepts in mathematics build on each other. Often, your understanding of a day's lecture depends on an understanding of the previous day's material.

To complete a homework assignment, remember these 4 things:

- Attempt all of it.
- Check it.
- Correct it.
- If needed, ask questions about it.

Take a moment and review your completed homework assignments. Answer the exercises below based on this review.

1. Approximate the fraction of your homework you have attempted.
2. Approximate the fraction of your homework you have checked (if possible).
3. If you are able to check your homework, have you corrected it when errors have been found?
4. When working homework, if you do not understand a concept, what do you personally do?

4.3 SOLVING SYSTEMS OF LINEAR EQUATIONS BY ADDITION

OBJECTIVE

1. Use the addition method to solve a system of linear equations.

OBJECTIVE 1 ▶ Using the addition method. We have seen that substitution is an accurate way to solve a linear system. Another method for solving a system of equations accurately is the **addition** or **elimination method.** The addition method is based on the addition property of equality: adding equal quantities to both sides of an equation does not change the solution of the equation. In symbols,

$$\text{if } A = B \text{ and } C = D, \text{ then } A + C = B + D.$$

EXAMPLE 1 Solve the system:
$$\begin{cases} x + y = 7 \\ x - y = 5 \end{cases}$$

Solution Since the left side of each equation is equal to the right side, we add equal quantities by adding the left sides of the equations together and the right sides of the equations together. This adding eliminates the variable y and gives us an equation in one variable, x. We can then solve for x.

▶ **Helpful Hint**

Our goal when solving a system of equations by the addition method is to eliminate a variable when adding the equations.

$$\begin{array}{ll} x + y = 7 & \text{First equation} \\ \underline{x - y = 5} & \text{Second equation} \\ 2x = 12 & \text{Add the equations.} \\ x = 6 & \text{Divide both sides by 2.} \end{array}$$

The x-value of the solution is 6. To find the corresponding y-value, let $x = 6$ in either equation of the system. We will use the first equation.

$$\begin{array}{ll} x + y = 7 & \text{First equation} \\ 6 + y = 7 & \text{Let } x = 6. \\ y = 7 - 6 & \text{Solve for } y. \\ y = 1 & \text{Simplify.} \end{array}$$

Check: The solution is (6, 1). Check this in both equations.

First Equation	Second Equation
$x + y = 7$	$x - y = 5$
$6 + 1 \stackrel{?}{=} 7$	$6 - 1 \stackrel{?}{=} 5$ Let $x = 6$ and $y = 1$.
$7 = 7$ True	$5 = 5$ True

Thus, the solution of the system is (6, 1) and the graphs of the two equations intersect at the point (6, 1) as shown next.

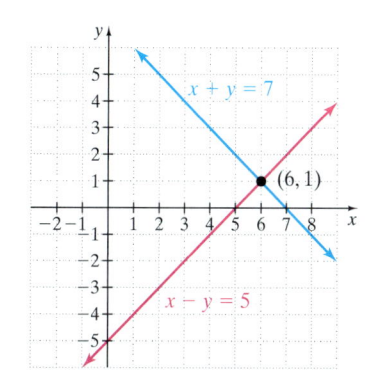

PRACTICE 1 Solve the system: $\begin{cases} x - y = 2 \\ x + y = 8 \end{cases}$

EXAMPLE 2 Solve the system: $\begin{cases} -2x + y = 2 \\ -x + 3y = -4 \end{cases}$

Solution If we simply add the two equations, the result is still an equation in two variables. However, our goal is to eliminate one of the variables. Notice what happens if we multiply *both sides* of the first equation by -3, which we are allowed to do by the multiplication property of equality. The system

$\begin{cases} -3(-2x + y) = -3(2) \\ -x + 3y = -4 \end{cases}$ simplifies to $\begin{cases} 6x - 3y = -6 \\ -x + 3y = -4 \end{cases}$

Now add the resulting equations and the y-variable is eliminated.

$$\begin{aligned} 6x - 3y &= -6 \\ -x + 3y &= -4 \\ \hline 5x &= -10 \quad \text{Add.} \\ x &= -2 \quad \text{Divide both sides by 5.} \end{aligned}$$

To find the corresponding y-value, let $x = -2$ in any of the preceding equations containing both variables. We use the first equation of the original system.

$$\begin{aligned} -2x + y &= 2 &&\text{First equation} \\ -2(-2) + y &= 2 &&\text{Let } x = -2. \\ 4 + y &= 2 \\ y &= -2 &&\text{Subtract 4 from both sides.} \end{aligned}$$

The solution is $(-2, -2)$. Check this ordered pair in both equations of the original system. □

PRACTICE 2 Solve the system: $\begin{cases} x - 2y = 11 \\ 3x - y = 13 \end{cases}$

In Example 2, the decision to multiply the first equation by -3 was no accident. **To eliminate a variable** when adding two equations, **the coefficient of the variable in one equation must be the opposite of its coefficient in the other equation.**

> **Helpful Hint**
> Be sure to multiply *both sides* of an equation by a chosen number when solving by the addition method. A common mistake is to multiply only the side containing the variables.

EXAMPLE 3 Solve the system: $\begin{cases} 2x - y = 7 \\ 8x - 4y = 1 \end{cases}$

Solution Multiply both sides of the first equation by -4 and the resulting coefficient of x is -8, the opposite of 8, the coefficient of x in the second equation. The system becomes

> **Helpful Hint**
> Don't forget to multiply both sides by -4.

$\begin{cases} -4(2x - y) = -4(7) \\ 8x - 4y = 1 \end{cases}$ simplifies to $\begin{cases} -8x + 4y = -28 \\ 8x - 4y = 1 \end{cases}$

Now add the resulting equations.

$$\begin{aligned} -8x + 4y &= -28 \\ 8x - 4y &= 1 \\ \hline 0 &= -27 \quad \text{Add the equations.} \\ & \text{False} \end{aligned}$$

When we add the equations, both variables are eliminated and we have $0 = -27$, a false statement. This means that the system has no solution. The graphs of these equations are parallel lines. □

PRACTICE 3 Solve the system: $\begin{cases} x - 3y = 5 \\ 2x - 6y = -3 \end{cases}$

Section 4.3 Solving Systems of Linear Equations by Addition 267

EXAMPLE 4 Solve the system: $\begin{cases} 3x - 2y = 2 \\ -9x + 6y = -6 \end{cases}$

Solution First we multiply both sides of the first equation by 3, then we add the resulting equations.

$\begin{cases} 3(3x - 2y) = 3(2) \\ -9x + 6y = -6 \end{cases}$ simplifies to $\begin{cases} 9x - 6y = 6 \\ -9x + 6y = -6 \end{cases}$

$ 0 = 0$ Add the equations. True

Both variables are eliminated and we have $0 = 0$, a true statement. Whenever you eliminate a variable and get the equation $0 = 0$, the system has an infinite number of solutions. The graphs of these equations are identical.

PRACTICE 4 Solve the system: $\begin{cases} 4x - 3y = 5 \\ -8x + 6y = -10 \end{cases}$

Concept Check ✓

Suppose you are solving the system

$$\begin{cases} 3x + 8y = -5 \\ 2x - 4y = 3 \end{cases}$$

You decide to use the addition method and begin by multiplying both sides of the first equation by -2. In which of the following was the multiplication performed correctly? Explain.

a. $-6x - 16y = -5$ **b.** $-6x - 16y = 10$

EXAMPLE 5 Solve the system: $\begin{cases} 3x + 4y = 13 \\ 5x - 9y = 6 \end{cases}$

Solution We can eliminate the variable y by multiplying the first equation by 9 and the second equation by 4.

$\begin{cases} 9(3x + 4y) = 9(13) \\ 4(5x - 9y) = 4(6) \end{cases}$ simplifies to $\begin{cases} 27x + 36y = 117 \\ 20x - 36y = 24 \end{cases}$

$ 47x = 141$ Add the equations.
$ x = 3$ Divide both sides by 47.

To find the corresponding y-value, we let $x = 3$ in any equation in this example containing two variables. Doing so in any of these equations will give $y = 1$. The solution to this system is $(3, 1)$. Check to see that $(3, 1)$ satisfies each equation in the original system.

PRACTICE 5 Solve the system: $\begin{cases} 4x + 3y = 14 \\ 3x - 2y = 2 \end{cases}$

If we had decided to eliminate x instead of y in Example 5, the first equation could have been multiplied by 5 and the second by -3. Try solving the original system this way to check that the solution is $(3, 1)$.

The following steps summarize how to solve a system of linear equations by the addition method.

Answer to Concept Check: b

Solving a System of Two Linear Equations by the Addition Method

STEP 1. Rewrite each equation in standard form $Ax + By = C$.

STEP 2. If necessary, multiply one or both equations by a nonzero number so that the coefficients of a chosen variable in the system are opposites.

STEP 3. Add the equations.

STEP 4. Find the value of one variable by solving the resulting equation from Step 3.

STEP 5. Find the value of the second variable by substituting the value found in Step 4 into either of the original equations.

STEP 6. Check the proposed solution in the original system.

Concept Check ✓

Suppose you are solving the system

$$\begin{cases} -4x + 7y = 6 \\ x + 2y = 5 \end{cases}$$

by the addition method.

a. What step(s) should you take if you wish to eliminate x when adding the equations?
b. What step(s) should you take if you wish to eliminate y when adding the equations?

EXAMPLE 6 Solve the system:

$$\begin{cases} -x - \dfrac{y}{2} = \dfrac{5}{2} \\ -\dfrac{x}{2} + \dfrac{y}{4} = 0 \end{cases}$$

Solution We begin by clearing each equation of fractions. To do so, we multiply both sides of the first equation by the LCD 2 and both sides of the second equation by the LCD 4. Then the system

$$\begin{cases} 2\left(-x - \dfrac{y}{2}\right) = 2\left(\dfrac{5}{2}\right) \\ 4\left(-\dfrac{x}{2} + \dfrac{y}{4}\right) = 4(0) \end{cases} \quad \text{simplifies to} \quad \begin{cases} -2x - y = 5 \\ -2x + y = 0 \end{cases}$$

Now we add the resulting equations in the simplified system.

$$\begin{array}{r} -2x - y = 5 \\ -2x + y = 0 \\ \hline -4x = 5 \end{array}$$ Add the equations.

$$x = -\dfrac{5}{4}$$

To find y, we could replace x with $-\dfrac{5}{4}$ in one of the equations with two variables.

Answers to Concept Check:
a. multiply the second equation by 4
b. possible answer: multiply the first equation by −2 and the second equation by 7

Instead, let's go back to the simplified system and multiply by appropriate factors to eliminate the variable x and solve for y. To do this, we multiply the first equation in the simplified system by -1. Then the system

$$\begin{cases} -1(-2x - y) = -1(5) \\ -2x + y = 0 \end{cases} \text{ simplifies to } \begin{cases} 2x + y = -5 \\ \underline{-2x + y = 0} \\ 2y = -5 \quad \text{Add.} \\ y = -\dfrac{5}{2} \quad \text{Solve for } y. \end{cases}$$

Check the ordered pair $\left(-\dfrac{5}{4}, -\dfrac{5}{2}\right)$ in both equations of the original system. The solution is $\left(-\dfrac{5}{4}, -\dfrac{5}{2}\right)$.

PRACTICE 6 Solve the system: $\begin{cases} -2x + \dfrac{3y}{2} = 5 \\ -\dfrac{x}{2} - \dfrac{y}{4} = \dfrac{1}{2} \end{cases}$

4.3 EXERCISE SET

Solve each system of equations by the addition method. See Example 1.

1. $\begin{cases} 3x + y = 5 \\ 6x - y = 4 \end{cases}$

2. $\begin{cases} 4x + y = 13 \\ 2x - y = 5 \end{cases}$

3. $\begin{cases} x - 2y = 8 \\ -x + 5y = -17 \end{cases}$

4. $\begin{cases} x - 2y = -11 \\ -x + 5y = 23 \end{cases}$

Solve each system of equations by the addition method. If a system contains fractions or decimals, you may want to first clear each equation of fractions or decimals. See Examples 2 through 6.

5. $\begin{cases} 3x + y = -11 \\ 6x - 2y = -2 \end{cases}$

6. $\begin{cases} 4x + y = -13 \\ 6x - 3y = -15 \end{cases}$

7. $\begin{cases} 3x + 2y = 11 \\ 5x - 2y = 29 \end{cases}$

8. $\begin{cases} 4x + 2y = 2 \\ 3x - 2y = 12 \end{cases}$

9. $\begin{cases} x + 5y = 18 \\ 3x + 2y = -11 \end{cases}$

10. $\begin{cases} x + 4y = 14 \\ 5x + 3y = 2 \end{cases}$

11. $\begin{cases} x + y = 6 \\ x - y = 6 \end{cases}$

12. $\begin{cases} x - y = 1 \\ -x + 2y = 0 \end{cases}$

13. $\begin{cases} 2x + 3y = 0 \\ 4x + 6y = 3 \end{cases}$

14. $\begin{cases} 3x + y = 4 \\ 9x + 3y = 6 \end{cases}$

15. $\begin{cases} -x + 5y = -1 \\ 3x - 15y = 3 \end{cases}$

16. $\begin{cases} 2x + y = 6 \\ 4x + 2y = 12 \end{cases}$

17. $\begin{cases} 3x - 2y = 7 \\ 5x + 4y = 8 \end{cases}$

18. $\begin{cases} 6x - 5y = 25 \\ 4x + 15y = 13 \end{cases}$

19. $\begin{cases} 8x = -11y - 16 \\ 2x + 3y = -4 \end{cases}$

20. $\begin{cases} 10x + 3y = -12 \\ 5x = -4y - 16 \end{cases}$

21. $\begin{cases} 4x - 3y = 7 \\ 7x + 5y = 2 \end{cases}$

22. $\begin{cases} -2x + 3y = 10 \\ 3x + 4y = 2 \end{cases}$

23. $\begin{cases} 4x - 6y = 8 \\ 6x - 9y = 12 \end{cases}$

24. $\begin{cases} 9x - 3y = 12 \\ 12x - 4y = 18 \end{cases}$

25. $\begin{cases} 2x - 5y = 4 \\ 3x - 2y = 4 \end{cases}$

26. $\begin{cases} 6x - 5y = 7 \\ 4x - 6y = 7 \end{cases}$

27. $\begin{cases} \dfrac{x}{3} + \dfrac{y}{6} = 1 \\ \dfrac{x}{2} - \dfrac{y}{4} = 0 \end{cases}$

28. $\begin{cases} \dfrac{x}{2} + \dfrac{y}{8} = 3 \\ x - \dfrac{y}{4} = 0 \end{cases}$

29. $\begin{cases} \dfrac{10}{3}x + 4y = -4 \\ 5x + 6y = -6 \end{cases}$

30. $\begin{cases} \dfrac{3}{2}x + 4y = 1 \\ 9x + 24y = 5 \end{cases}$

31. $\begin{cases} x - \dfrac{y}{3} = -1 \\ -\dfrac{x}{2} + \dfrac{y}{8} = \dfrac{1}{4} \end{cases}$

32. $\begin{cases} 2x - \dfrac{3y}{4} = -3 \\ x + \dfrac{y}{9} = \dfrac{13}{3} \end{cases}$

33. $-4(x+2) = 3y$
    $2x - 2y = 3$

34. $-9(x+3) = 8y$
    $3x - 3y = 8$

35. $\begin{cases} \dfrac{x}{3} - y = 2 \\ -\dfrac{x}{2} + \dfrac{3y}{2} = -3 \end{cases}$

36. $\begin{cases} \dfrac{x}{2} + \dfrac{y}{4} = 1 \\ -\dfrac{x}{4} - \dfrac{y}{8} = 1 \end{cases}$

37. $\begin{cases} \dfrac{3}{5}x - y = -\dfrac{4}{5} \\ 3x + \dfrac{y}{2} = -\dfrac{9}{5} \end{cases}$

38. $\begin{cases} 3x + \dfrac{7}{2}y = \dfrac{3}{4} \\ -\dfrac{x}{2} + \dfrac{5}{3}y = -\dfrac{5}{4} \end{cases}$

39. $\begin{cases} 3.5x + 2.5y = 17 \\ -1.5x - 7.5y = -33 \end{cases}$

40. $\begin{cases} -2.5x - 6.5y = 47 \\ 0.5x - 4.5y = 37 \end{cases}$

41. $\begin{cases} 0.02x + 0.04y = 0.09 \\ -0.1x + 0.3y = 0.8 \end{cases}$

42. $\begin{cases} 0.04x - 0.05y = 0.105 \\ 0.2x - 0.6y = 1.05 \end{cases}$

MIXED PRACTICE

Solve each system by either the addition method or the substitution method.

43. $\begin{cases} 2x - 3y = -11 \\ y = 4x - 3 \end{cases}$

44. $\begin{cases} 4x - 5y = 6 \\ y = 3x - 10 \end{cases}$

45. $\begin{cases} x + 2y = 1 \\ 3x + 4y = -1 \end{cases}$

46. $\begin{cases} x + 3y = 5 \\ 5x + 6y = -2 \end{cases}$

47. $\begin{cases} 2y = x + 6 \\ 3x - 2y = -6 \end{cases}$

48. $\begin{cases} 3y = x + 14 \\ 2x - 3y = -16 \end{cases}$

49. $\begin{cases} y = 2x - 3 \\ y = 5x - 18 \end{cases}$

50. $\begin{cases} y = 6x - 5 \\ y = 4x - 11 \end{cases}$

51. $\begin{cases} x + \dfrac{1}{6}y = \dfrac{1}{2} \\ 3x + 2y = 3 \end{cases}$

52. $\begin{cases} x + \dfrac{1}{3}y = \dfrac{5}{12} \\ 8x + 3y = 4 \end{cases}$

53. $\begin{cases} \dfrac{x+2}{2} = \dfrac{y+11}{3} \\ \dfrac{x}{2} = \dfrac{2y+16}{6} \end{cases}$

54. $\begin{cases} \dfrac{x+5}{2} = \dfrac{y+14}{4} \\ \dfrac{x}{3} = \dfrac{2y+2}{6} \end{cases}$

55. $\begin{cases} 2x + 3y = 14 \\ 3x - 4y = -69.1 \end{cases}$

56. $\begin{cases} 5x - 2y = -19.8 \\ -3x + 5y = -3.7 \end{cases}$

REVIEW AND PREVIEW

Rewrite the following sentences using mathematical symbols. Do not solve the equations. See Sections 2.4 and 2.5.

57. Twice a number, added to 6, is 3 less than the number.
58. The sum of three consecutive integers is 66.
59. Three times a number, subtracted from 20, is 2.
60. Twice the sum of 8 and a number is the difference of the number and 20.
61. The product of 4 and the sum of a number and 6 is twice the number.
62. The quotient of twice a number and 7 is subtracted from the reciprocal of the number.

CONCEPT EXTENSIONS

Solve. See a Concept Check in this section.

63. To solve this system by the addition method and eliminate the variable y,
    $$\begin{cases} 4x + 2y = -7 \\ 3x - y = -12 \end{cases}$$
    by what value would you multiply the second equation? What do you get when you complete the multiplication?

Given the system of linear equations $\begin{cases} 3x - y = -8 \\ 5x + 3y = 2 \end{cases}$

64. Use the addition method and
    a. Solve the system by eliminating x.
    b. Solve the system by eliminating y.

65. Suppose you are solving the system
    $$\begin{cases} 3x + 8y = -5 \\ 2x - 4y = 3. \end{cases}$$
    You decide to use the addition method by multiplying both sides of the second equation by 2. In which of the following was the multiplication performed correctly? Explain.
    a. $4x - 8y = 3$
    b. $4x - 8y = 6$

66. Suppose you are solving the system
    $$\begin{cases} -2x - y = 0 \\ -2x + 3y = 6. \end{cases}$$
    You decide to use the addition method by multiplying both sides of the first equation by 3, then adding the resulting equation to the second equation. Which of the following is the correct sum? Explain.
    a. $-8x = 6$
    b. $-8x = 9$

67. When solving a system of equations by the addition method, how do we know when the system has no solution?

68. Explain why the addition method might be preferred over the substitution method for solving the system $\begin{cases} 2x - 3y = 5 \\ 5x + 2y = 6. \end{cases}$

69. Use the system of linear equations below to answer the questions.
    $$\begin{cases} x + y = 5 \\ 3x + 3y = b \end{cases}$$
    a. Find the value of b so that the system has an infinite number of solutions.
    b. Find a value of b so that there are no solutions to the system.

70. Use the system of linear equations below to answer the questions.

$$\begin{cases} x + y = 4 \\ 2x + by = 8 \end{cases}$$

  a. Find the value of b so that the system has an infinite number of solutions.

  b. Find a value of b so that the system has a single solution.

Solve each system by the addition method.

71. $\begin{cases} 1.2x + 3.4y = 27.6 \\ 7.2x - 1.7y = -46.56 \end{cases}$ **72.** $\begin{cases} 5.1x - 2.4y = 3.15 \\ -15.3x + 1.2y = 27.75 \end{cases}$

73. Two occupations predicted to greatly increase in number of jobs are pharmacy technicians and network systems and data communication analysts. The number of pharmacy technician jobs predicted for 2004 through 2014 can be approximated by $7.4x - y = -258$. The number of network and data analyst jobs for the same years can be approximated by $12.6x - y = -231$. For both equations, x is the number of years since 2004 and y is the number of jobs in thousands.

  a. Use the addition method to solve this system of equations:

  $$\begin{cases} 7.4x - y = -258 \\ 12.6x - y = -231 \end{cases}$$

  (Round answer to the nearest whole number.)

  b. Use your result from part (a) and estimate the year in which the number of both jobs is equal.

  c. Use your result from part (a) and estimate the number of pharmacy technician jobs (or the number of analyst jobs since they should be equal).

74. In recent years, the number of Americans (in millions) who have skateboarded at least once in a year has been increasing, while the number of Americans who have in-line roller skated has been decreasing. The number of skateboarders (people who have skateboarded at least once a year) from 1996 to 2006 is given by $-0.7x + y = 4.6$ and the number of in-line roller skaters can be given by the equation $1.3x + y = 26.9$. For both equations, x is the number of years after 1996. (*Source:* National Sporting Goods Association)

  a. Use the addition method to solve this system of equations:

  $$\begin{cases} -0.7x + y = 4.6 \\ 1.3x + y = 26.9 \end{cases}$$

  (Round to the nearest whole number. Because of rounding, the y-value of your ordered pair solution may vary.)

  b. Use your result from part (a) and estimate the year in which the number of skate boarders equals the number of in-line skaters.

  c. Use your result from part (a) and estimate the number of skateboarders (or the number of in-line skaters since they should be equal).

INTEGRATED REVIEW   SOLVING SYSTEMS OF EQUATIONS

Sections 4.1–4.3

Solve each system by either the addition method or the substitution method.

1. $\begin{cases} 2x - 3y = -11 \\ y = 4x - 3 \end{cases}$ **2.** $\begin{cases} 4x - 5y = 6 \\ y = 3x - 10 \end{cases}$ **3.** $\begin{cases} x + y = 3 \\ x - y = 7 \end{cases}$ **4.** $\begin{cases} x - y = 20 \\ x + y = -8 \end{cases}$

5. $\begin{cases} x + 2y = 1 \\ 3x + 4y = -1 \end{cases}$ **6.** $\begin{cases} x + 3y = 5 \\ 5x + 6y = -2 \end{cases}$ **7.** $\begin{cases} y = x + 3 \\ 3x - 2y = -6 \end{cases}$ **8.** $\begin{cases} y = -2x \\ 2x - 3y = -16 \end{cases}$

9. $\begin{cases} y = 2x - 3 \\ y = 5x - 18 \end{cases}$ **10.** $\begin{cases} y = 6x - 5 \\ y = 4x - 11 \end{cases}$ **11.** $\begin{cases} x + \dfrac{1}{6}y = \dfrac{1}{2} \\ 3x + 2y = 3 \end{cases}$ **12.** $\begin{cases} x + \dfrac{1}{3}y = \dfrac{5}{12} \\ 8x + 3y = 4 \end{cases}$

13. $\begin{cases} x - 5y = 1 \\ -2x + 10y = 3 \end{cases}$
14. $\begin{cases} -x + 2y = 3 \\ 3x - 6y = -9 \end{cases}$
15. $\begin{cases} 0.2x - 0.3y = -0.95 \\ 0.4x + 0.1y = 0.55 \end{cases}$
16. $\begin{cases} 0.08x - 0.04y = -0.11 \\ 0.02x - 0.06y = -0.09 \end{cases}$

17. $\begin{cases} x = 3y - 7 \\ 2x - 6y = -14 \end{cases}$
18. $\begin{cases} y = \dfrac{x}{2} - 3 \\ 2x - 4y = 0 \end{cases}$
19. $\begin{cases} 2x + 5y = -1 \\ 3x - 4y = 33 \end{cases}$
20. $\begin{cases} 7x - 3y = 2 \\ 6x + 5y = -21 \end{cases}$

21. Which method, substitution or addition, would you prefer to use to solve the system below? Explain your reasoning.

$$\begin{cases} 3x + 2y = -2 \\ y = -2x \end{cases}$$

22. Which method, substitution or addition, would you prefer to use to solve the system below? Explain your reasoning.

$$\begin{cases} 3x - 2y = -3 \\ 6x + 2y = 12 \end{cases}$$

4.4 SYSTEMS OF LINEAR EQUATIONS AND PROBLEM SOLVING

OBJECTIVE

1 Use a system of equations to solve problems.

OBJECTIVE 1 ▶ Using a system of equations for problem solving. Many of the word problems solved earlier using one-variable equations can also be solved using two equations in **two** variables. We use the same problem-solving steps that have been used throughout this text. The only difference is that two variables are assigned to represent the two unknown quantities and that the stated problem is translated into **two** equations.

Problem-Solving Steps

1. UNDERSTAND the problem. During this step, become comfortable with the problem. Some ways of doing this are to

   Read and reread the problem.

   Choose two variables to represent the two unknowns.

   Construct a drawing, if possible.

   Propose a solution and check. Pay careful attention to how you check your proposed solution. This will help when writing equations to model the problem.

2. TRANSLATE the problem into two equations.
3. SOLVE the system of equations.
4. INTERPRET the results: **Check** the proposed solution in the stated problem and **state** your conclusion.

EXAMPLE 1 Finding Unknown Numbers

Find two numbers whose sum is 37 and whose difference is 21.

Solution

1. UNDERSTAND. Read and reread the problem. Suppose that one number is 20. If their sum is 37, the other number is 17 because 20 + 17 = 37. Is their difference 21? No; 20 − 17 = 3. Our proposed solution is incorrect, but we now have a better understanding of the problem.

   Since we are looking for two numbers, we let

   x = first number

   y = second number

Section 4.4 Systems of Linear Equations and Problem Solving   273

2. TRANSLATE. Since we have assigned two variables to this problem, we translate our problem into two equations.

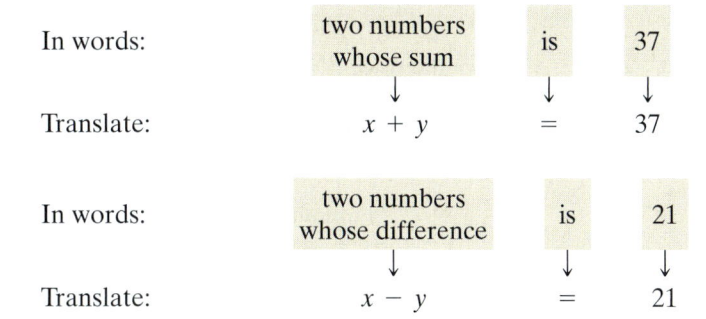

In words: two numbers whose sum is 37
Translate: $x + y$ = 37

In words: two numbers whose difference is 21
Translate: $x - y$ = 21

3. SOLVE. Now we solve the system

$$\begin{cases} x + y = 37 \\ x - y = 21 \end{cases}$$

Notice that the coefficients of the variable y are opposites. Let's then solve by the addition method and begin by adding the equations.

$$\begin{aligned} x + y &= 37 \\ \underline{x - y} &= \underline{21} \\ 2x &= 58 \end{aligned}$$ Add the equations.

$$x = \frac{58}{2} = 29$$ Divide both sides by 2.

Now we let $x = 29$ in the first equation to find y.

$x + y = 37$ First equation
$29 + y = 37$
$y = 8$ Subtract 29 from both sides.

4. INTERPRET. The solution of the system is (29, 8).

Check: Notice that the sum of 29 and 8 is $29 + 8 = 37$, the required sum. Their difference is $29 - 8 = 21$, the required difference.

State: The numbers are 29 and 8.

PRACTICE
1   Find two numbers whose sum is 30 and whose difference is 6.

EXAMPLE 2   Solving a Problem about Prices

The Cirque du Soleil show Corteo is performing locally. Matinee admission for 4 adults and 2 children is $374, while admission for 2 adults and 3 children is $285.

a. What is the price of an adult's ticket?
b. What is the price of a child's ticket?
c. Suppose that a special rate of $1000 is offered for groups of 20 persons. Should a group of 4 adults and 16 children use the group rate? Why or why not?

Solution

1. UNDERSTAND. Read and reread the problem and guess a solution. Let's suppose that the price of an adult's ticket is $50 and the price of a child's ticket is $40. To check our proposed solution, let's see if admission for 4 adults and 2 children is $374.

Admission for 4 adults is 4($50) or $200 and admission for 2 children is 2($40) or $80. This gives a total admission of $200 + $80 = $280, not the required $374. Again though, we have accomplished the purpose of this process. We have a better understanding of the problem. To continue, we let

A = the price of an adult's ticket and

C = the price of a child's ticket

2. TRANSLATE. We translate the problem into two equations using both variables.

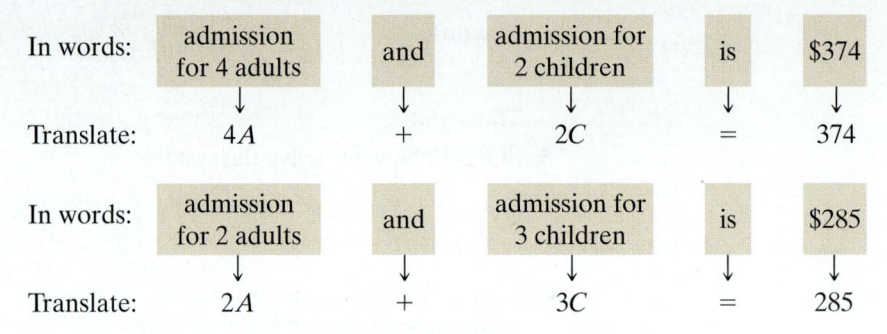

In words:	admission for 4 adults	and	admission for 2 children	is	$374
Translate:	$4A$	$+$	$2C$	$=$	374

In words:	admission for 2 adults	and	admission for 3 children	is	$285
Translate:	$2A$	$+$	$3C$	$=$	285

3. SOLVE. We solve the system.

$$\begin{cases} 4A + 2C = 374 \\ 2A + 3C = 285 \end{cases}$$

Since both equations are written in standard form, we solve by the addition method. First we multiply the second equation by -2 so that when we add the equations we eliminate the variable A. Then the system

$$\begin{cases} 4A + 2C = 374 \\ -2(2A + 3C) = -2(285) \end{cases} \text{ simplifies to } \begin{cases} 4A + 2C = 374 \\ -4A - 6C = -570 \end{cases}$$

$$-4C = -196 \quad \text{Add the equations.}$$
$$C = 49 \quad \text{Divide by } -4.$$

or $49, the children's ticket price.

To find A, we replace C with 49 in the first equation.

$$4A + 2C = 374 \quad \text{First equation}$$
$$4A + 2(49) = 374 \quad \text{Let } C = 49.$$
$$4A + 98 = 374$$
$$4A = 276$$
$$A = 69$$

or $69, the adult's ticket price.

4. INTERPRET.

Check: Notice that 4 adults and 2 children will pay $4(\$69) + 2(\$49) = \$276 + \$98 = \$374$, the required amount. Also, the price for 2 adults and 3 children is $2(\$69) + 3(\$49) = \$138 + \$147 = \$285$, the required amount.

State: Answer the three original questions.

a. Since $A = 69$, the price of an adult's ticket is $69.

b. Since $C = 49$, the price of a child's ticket is $49.

c. The regular admission price for 4 adults and 16 children is

$$4(\$69) + 16(\$49) = \$276 + \$784$$
$$= \$1060$$

This is $60 more than the special group rate of $1000, so they should request the group rate. □

Section 4.4 Systems of Linear Equations and Problem Solving   275

PRACTICE 2 It is considered a premium game when the Red Sox or the Yankees come to Texas to play the Rangers. Admission for one of these games for three adults and three children under 14 is $75, while admission for two adults and 4 children is $62. (*Source: MLB.com, Texas Rangers*)

a. What is the price of an adult admission at Ameriquest Park?

b. What is the price of a child's admission?

c. Suppose that a special rate of $200 is offered for groups of 20 persons. Should a group of 5 adults and 15 children use the group rate? Why or why not?

EXAMPLE 3   Finding Rates

As part of an exercise program, Albert and Louis started walking each morning. They live 15 miles away from each other and decided to meet one day by walking toward one another. After 2 hours they meet. If Louis walks one mile per hour faster than Albert, find both walking speeds.

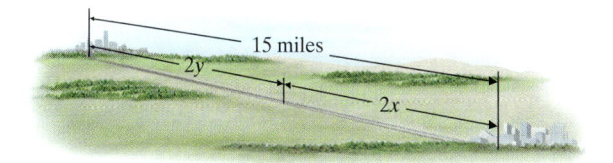

Solution

1. UNDERSTAND. Read and reread the problem. Let's propose a solution and use the formula $d = r \cdot t$ to check. Suppose that Louis's rate is 4 miles per hour. Since Louis's rate is 1 mile per hour faster, Albert's rate is 3 miles per hour. To check, see if they can walk a total of 15 miles in 2 hours. Louis's distance is rate · time = 4(2) = 8 miles and Albert's distance is rate time = 3(2) = 6 miles. Their total distance is 8 miles + 6 miles = 14 miles, not the required 15 miles. Now that we have a better understanding of the problem, let's model it with a system of equations.

   First, we let

   x = Albert's rate in miles per hour

   y = Louis's rate in miles per hour

   Now we use the facts stated in the problem and the formula $d = rt$ to fill in the following chart.

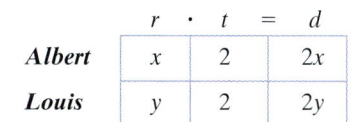

	r	·	t	=	d
Albert	x		2		$2x$
Louis	y		2		$2y$

2. TRANSLATE. We translate the problem into two equations using both variables.

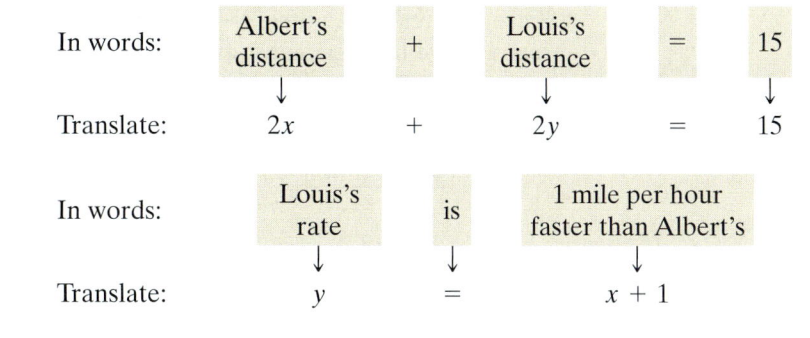

   In words: Albert's distance + Louis's distance = 15

   Translate: $2x + 2y = 15$

   In words: Louis's rate is 1 mile per hour faster than Albert's

   Translate: $y = x + 1$

3. SOLVE. The system of equations we are solving is

$$\begin{cases} 2x + 2y = 15 \\ y = x + 1 \end{cases}$$

Let's use substitution to solve the system since the second equation is solved for y.

$$2x + 2y = 15 \quad \text{First equation}$$

$$2x + 2(x + 1) = 15 \quad \text{Replace } y \text{ with } x + 1.$$
$$2x + 2x + 2 = 15$$
$$4x = 13$$
$$x = 3.25 \quad \text{Divide both sides by 4 and write the result as a decimal.}$$
$$y = x + 1 = 3.25 + 1 = 4.25$$

4. INTERPRET. Albert's proposed rate is 3.25 miles per hour and Louis's proposed rate is 4.25 miles per hour.

Check: Use the formula $d = rt$ and find that in 2 hours, Albert's distance is $(3.25)(2)$ miles or 6.5 miles. In 2 hours, Louis's distance is $(4.25)(2)$ miles or 8.5 miles. The total distance walked is 6.5 miles + 8.5 miles or 15 miles, the given distance.

State: Albert walks at a rate of 3.25 miles per hour and Louis walks at a rate of 4.25 miles per hour.

PRACTICE

3 Two hikers on a straight trail are 22 miles apart and walking toward each other. After 4 hours they meet. If one hiker is a nature lover and walks 2 miles per hour slower than the other hiker, find both walking speeds.

EXAMPLE 4 Finding Amounts of Solutions

Eric Daly, a chemistry teaching assistant, needs 10 liters of a 20% saline solution (salt water) for his 2 p.m. laboratory class. Unfortunately, the only mixtures on hand are a 5% saline solution and a 25% saline solution. How much of each solution should he mix to produce the 20% solution?

Solution

1. UNDERSTAND. Read and reread the problem. Suppose that we need 4 liters of the 5% solution. Then we need $10 - 4 = 6$ liters of the 25% solution. To see if this gives us 10 liters of a 20% saline solution, let's find the amount of pure salt in each solution.

	concentration rate	×	amount of solution	=	amount of pure salt
5% solution:	0.05	×	4 liters	=	0.2 liters
25% solution:	0.25	×	6 liters	=	1.5 liters
20% solution:	0.20	×	10 liters	=	2 liters

Since 0.2 liters + 1.5 liters = 1.7 liters, not 2 liters, our proposed solution is incorrect. But we have gained some insight into how to model and check this problem.

We let

$x = $ number of liters of 5% solution

$y = $ number of liters of 25% solution

Now we use a table to organize the given data.

	Concentration Rate	Liters of Solution	Liters of Pure Salt
First Solution	5%	x	$0.05x$
Second Solution	25%	y	$0.25y$
Mixture Needed	20%	10	$(0.20)(10)$

2. **TRANSLATE.** We translate into two equations using both variables.

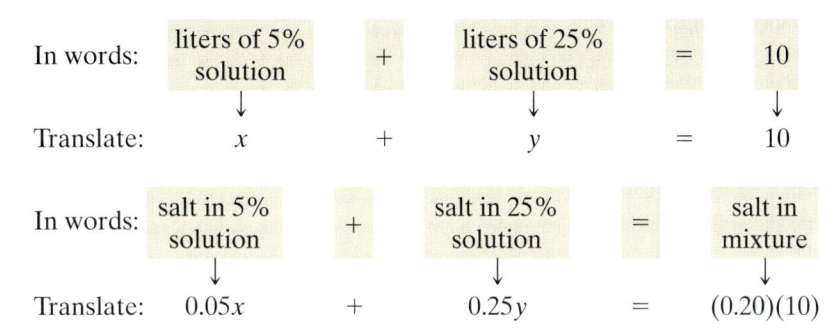

In words: liters of 5% solution + liters of 25% solution = 10

Translate: $\quad x \quad + \quad y \quad = \quad 10$

In words: salt in 5% solution + salt in 25% solution = salt in mixture

Translate: $\quad 0.05x \quad + \quad 0.25y \quad = \quad (0.20)(10)$

3. **SOLVE.** Here we solve the system

$$\begin{cases} x + y = 10 \\ 0.05x + 0.25y = 2 \end{cases}$$

To solve by the addition method, we first multiply the first equation by -25 and the second equation by 100. Then the system

$$\begin{cases} -25(x + y) = -25(10) \\ 100(0.05x + 0.25y) = 100(2) \end{cases} \text{ simplifies to } \begin{cases} -25x - 25y = -250 \\ 5x + 25y = 200 \end{cases}$$

$$-20x = -50 \quad \text{Add.}$$
$$x = 2.5 \quad \text{Divide by } -20.$$

To find y, we let $x = 2.5$ in the first equation of the original system.

$$x + y = 10$$
$$2.5 + y = 10 \quad \text{Let } x = 2.5.$$
$$y = 7.5$$

4. **INTERPRET.** Thus, we propose that Eric needs to mix 2.5 liters of 5% saline solution with 7.5 liters of 25% saline solution.

Check: Notice that $2.5 + 7.5 = 10$, the required number of liters. Also, the sum of the liters of salt in the two solutions equals the liters of salt in the required mixture:

$$0.05(2.5) + 0.25(7.5) = 0.20(10)$$
$$0.125 + 1.875 = 2$$

State: Eric needs 2.5 liters of the 5% saline solution and 7.5 liters of the 25% solution.

PRACTICE 4 Jemima Juarez owns the Sola Café in southern California. She is known for her interesting coffee blends. To create a new blend, she has decided to use Hawaiian Kona and Jamaica Blue Mountain coffee. To test her new blend, she intends to create 20 pounds of the mix. If the Hawaiian Kona costs $20 per pound and the Jamaica Blue Mountain costs $28 per pound, how much of each coffee type should she use to create a new blend that costs $22 per pound?

Concept Check ✓

Suppose you mix an amount of a 30% acid solution with an amount of a 50% acid solution. Which of the following acid strengths would be possible for the resulting acid mixture?

a. 22%   **b.** 44%   **c.** 63%

Answer to Concept Check: b

4.4 EXERCISE SET

Without actually solving each problem, choose each correct solution by deciding which choice satisfies the given conditions.

1. The length of a rectangle is 3 feet longer than the width. The perimeter is 30 feet. Find the dimensions of the rectangle.
   a. length = 8 feet; width = 5 feet
   b. length = 8 feet; width = 7 feet
   c. length = 9 feet; width = 6 feet

2. An isosceles triangle, a triangle with two sides of equal length, has a perimeter of 20 inches. Each of the equal sides is one inch longer than the third side. Find the lengths of the three sides.
   a. 6 inches, 6 inches, and 7 inches
   b. 7 inches, 7 inches, and 6 inches
   c. 6 inches, 7 inches, and 8 inches

3. Two computer disks and three notebooks cost $17. However, five computer disks and four notebooks cost $32. Find the price of each.
   a. notebook = $4; computer disk = $3
   b. notebook = $3; computer disk = $4
   c. notebook = $5; computer disk = $2

4. Two music CDs and four music cassette tapes cost a total of $40. However, three music CDs and five cassette tapes cost $55. Find the price of each.
   a. CD = $12; cassette = $4   b. CD = $15; cassette = $2
   c. CD = $10; cassette = $5

5. Kesha has a total of 100 coins, all of which are either dimes or quarters. The total value of the coins is $13.00. Find the number of each type of coin.
   a. 80 dimes; 20 quarters   b. 20 dimes; 44 quarters
   c. 60 dimes; 40 quarters

6. Samuel has 28 gallons of saline solution available in two large containers at his pharmacy. One container holds three times as much as the other container. Find the capacity of each container.
   a. 15 gallons; 5 gallons   b. 20 gallons; 8 gallons
   c. 21 gallons; 7 gallons

Write a system of equations in x and y describing each situation. Do not solve the system. See Example 1.

7. A smaller number and a larger number add up to 15 and have a difference of 7. (Let x be the larger number.)

8. The total of two numbers is 16. The first number plus 2 more than 3 times the second equals 18. (Let x be the first number.)

9. Keiko has a total of $6500, which she has invested in two accounts. The larger account is $800 greater than the smaller account. (Let x be the amount of money in the larger account.)

10. Dominique has four times as much money in his savings account as in his checking account. The total amount is $2300. (Let x be the amount of money in his checking account.)

MIXED PRACTICE

Solve. See Examples 1 through 4.

11. Two numbers total 83 and have a difference of 17. Find the two numbers.

12. The sum of two numbers is 76 and their difference is 52. Find the two numbers.

13. A first number plus twice a second number is 8. Twice the first number, plus the second totals 25. Find the numbers.

14. One number is 4 more than twice the second number. Their total is 25. Find the numbers.

15. The highest scorer during the WNBA 2006 regular season was Diana Taurasi of the Phoenix Mercury. Over the season, Taurasi scored 116 more points than Seimone Augustus of the Minnesota Lynx. Together, Taurasi and Augustus scored 1604 points during the 2006 regular season. How many points did each player score over the course of the season? (*Source:* Women's National Basketball Association)

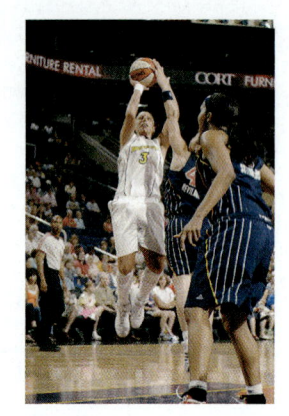

16. During the 2006 regular MLB season, Ryan Howard of the Philadelphia Phillies hit the most home runs of any player in the major leagues. Over the course of the season, he hit 4 more home runs than David Ortiz of the Boston Red Sox. Together, these batting giants hit 112 home runs. How many home runs did each player hit? (*Source:* Major League Baseball)

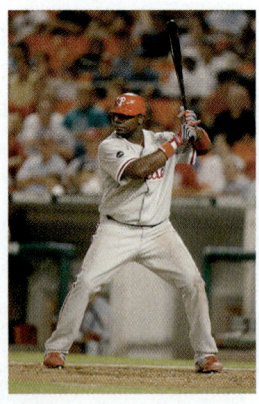

17. Ann Marie Jones has been pricing Amtrak train fares for a group trip to New York. Three adults and four children must pay $159. Two adults and three children must pay $112. Find the price of an adult's ticket, and find the price of a child's ticket.

18. Last month, Jerry Papa purchased five DVDs and two CDs at Wall-to-Wall Sound for $65. This month he bought three DVDs and four CDs for $81. Find the price of each DVD, and find the price of each CD.

19. Johnston and Betsy Waring have a jar containing 80 coins, all of which are either quarters or nickels. The total value of the coins is $14.60. How many of each type of coin do they have?

20. Sarah and Keith Robinson purchased 40 stamps, a mixture of 39¢ and 24¢ stamps. Find the number of each type of stamp if they spent $14.85.

21. Davie and Judi Mihaly own 50 shares of Apple stock and 60 shares of Microsoft stock. At the close of the markets on March 9, 2007, their stock portfolio was worth $6035.90. The closing price of the Microsoft stock was $60.68 less than the closing price of Apple stock on that day. What was the price of each stock on March 9, 2007? (*Source:* New York Stock Exchange)

22. Pho Lin has investments in EBay and Amazon stock. On March 9, 2007, EBay stock closed at $30.82 per share and Amazon stock closed at $38.84 per share. Pho's stock portfolio was worth $2866.60 at the end of the day. If Pho owns 20 more shares of Amazon stock than EBay stock, how many of each type of stock does she own?

23. Twice last month, Judy Carter rented a car from Enterprise in Fresno, California, and traveled around the Southwest on business. Enterprise rents its cars for a daily fee, plus an additional charge per mile driven. Judy recalls that her first trip lasted 4 days, she drove 450 miles, and the rental cost her $240.50. On her second business trip she drove 200 miles in 3 days, and paid $146.00 for the rental. Find the daily fee and the mileage charge.

24. Joan Gundersen rented a car from Hertz, which rents its cars for a daily fee plus an additional charge per mile driven. Joan recalls that a car rented for 5 days and driven for 300 miles cost her $178, while a car rented for 4 days and driven for 500 miles cost $197. Find the daily fee, and find the mileage charge.

25. Pratap Puri rowed 18 miles down the Delaware River in 2 hours, but the return trip took him $4\frac{1}{2}$ hours. Find the rate Pratap can row in still water, and find the rate of the current.
Let x = rate Pratap can row in still water and
    y = rate of the current

	d =	r ·	t
Downstream		$x + y$	
Upstream		$x - y$	

26. The Jonathan Schultz family took a canoe 10 miles down the Allegheny River in $1\frac{1}{4}$ hours. After lunch it took them 4 hours to return. Find the rate of the current.
Let x = rate the family can row in still water and
    y = rate of the current

	d =	r ·	t
Downstream		$x + y$	
Upstream		$x - y$	

27. Dave and Sandy Hartranft are frequent flyers with Delta Airlines. They often fly from Philadelphia to Chicago, a distance of 780 miles. On one particular trip they fly into the wind, and the flight takes 2 hours. The return trip, with the wind behind them, only takes $1\frac{1}{2}$ hours. If the wind speed is the same on each trip, find the speed of the wind and find the speed of the plane in still air.

28. With a strong wind behind it, a United Airlines jet flies 2400 miles from Los Angeles to Orlando in $4\frac{3}{4}$ hours. The return trip takes 6 hours, as the plane flies into the wind. If the wind speed is the same on each trip, find the speed of the plane in still air, and find the wind speed to the nearest tenth of a mile per hour.

29. Jim Williamson began a 96-mile bicycle trip to build up stamina for a triathlete competition. Unfortunately, his bicycle chain broke, so he finished the trip walking. The whole trip took 6 hours. If Jim walks at a rate of 4 miles per hour and rides at 20 miles per hour, find the amount of time he spent on the bicycle.

30. In Canada, eastbound and westbound trains travel along the same track, with sidings to pull onto to avoid accidents. Two trains are now 150 miles apart, with the westbound train traveling twice as fast as the eastbound train. A warning must be issued to pull one train onto a siding or else the trains will crash in $1\frac{1}{4}$ hours. Find the speed of the eastbound train and the speed of the westbound train.

31. Doreen Schmidt is a chemist with Gemco Pharmaceutical. She needs to prepare 12 ounces of a 9% hydrochloric acid solution. Find the amount of a 4% solution and the amount of a 12% solution she should mix to get this solution.

Concentration Rate	Ounces of Solution	Ounces of Pure Acid
0.04	x	0.04x
0.12	y	?
0.09	12	?

32. Elise Everly is preparing 15 liters of a 25% saline solution. Elise has two other saline solutions with strengths of 40% and

10%. Find the amount of 40% solution and the amount of 10% solution she should mix to get 15 liters of a 25% solution.

Concentration Rate	Liters of Solution	Liters of Pure Salt
0.40	x	$0.40x$
0.10	y	?
0.25	15	?

33. Wayne Osby blends coffee for a local coffee café. He needs to prepare 200 pounds of blended coffee beans selling for $3.95 per pound. He intends to do this by blending together a high-quality bean costing $4.95 per pound and a cheaper bean costing $2.65 per pound. To the nearest pound, find how much high-quality coffee bean and how much cheaper coffee bean he should blend.

34. Macadamia nuts cost an astounding $16.50 per pound, but research by an independent firm says that mixed nuts sell better if macadamias are included. The standard mix costs $9.25 per pound. Find how many pounds of macadamias and how many pounds of the standard mix should be combined to produce 40 pounds that will cost $10 per pound. Find the amounts to the nearest tenth of a pound.

35. Recall that two angles are complementary if the sum of their measures is 90°. Find the measures of two complementary angles if one angle is twice the other.

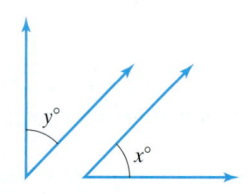

36. Recall that two angles are supplementary if the sum of their measures is 180°. Find the measures of two supplementary angles if one angle is 20° more than four times the other.

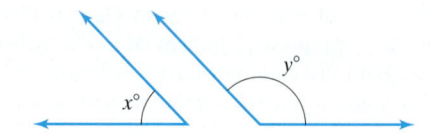

37. Find the measures of two complementary angles if one angle is 10° more than three times the other.

38. Find the measures of two supplementary angles if one angle is 18° more than twice the other.

39. Kathi and Robert Hawn had a pottery stand at the annual Skippack Craft Fair. They sold some of their pottery at the original price of $9.50 each, but later decreased the price of each by $2. If they sold all 90 pieces and took in $721, find how many they sold at the original price and how many they sold at the reduced price.

40. A charity fund-raiser consisted of a spaghetti supper where a total of 387 people were fed. They charged $6.80 for adults and half-price for children. If they took in $2444.60, find how many adults and how many children attended the supper.

41. The Santa Fe National Historic Trail is approximately 1200 miles between Old Franklin, Missouri, and Santa Fe, New Mexico. Suppose that a group of hikers start from each town and walk the trail toward each other. They meet after a total hiking time of 240 hours. If one group travels $\frac{1}{2}$ mile per hour slower than the other group, find the rate of each group. (*Source:* National Park Service)

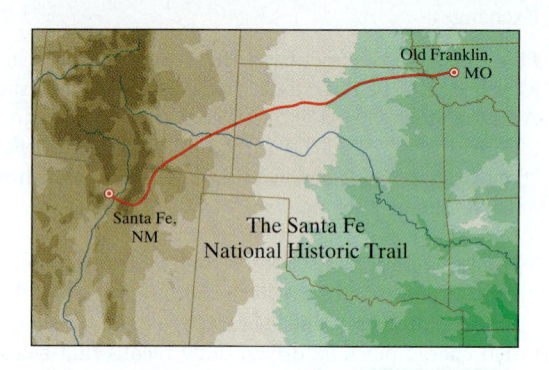

42. California 1 South is a historic highway that stretches 123 miles along the coast from Monterey to Morro Bay. Suppose that two antique cars start driving this highway, one from each town. They meet after 3 hours. Find the rate of each car if one car travels 1 mile per hour faster than the other car. (*Source:* National Geographic)

43. A 30% solution of fertilizer is to be mixed with a 60% solution of fertilizer in order to get 150 gallons of a 50% solution. How many gallons of the 30% solution and 60% solution should be mixed?

44. A 10% acid solution is to be mixed with a 50% acid solution in order to get 120 ounces of a 20% acid solution. How many ounces of the 10% solution and 50% solution should be mixed?

45. Traffic signs are regulated by the *Manual on Uniform Traffic Control Devices* (MUTCD). According to this manual, if the sign below is placed on a freeway, its perimeter must be 144 inches. Also, its length is 12 inches longer than its width. Find the dimensions of this sign.

46. According to the MUTCD (see Exercise 45), this sign must have a perimeter of 60 inches. Also, its length must be 6 inches longer than its width. Find the dimensions of this sign.

REVIEW AND PREVIEW

Solve each linear inequality. See Section 2.8.

47. $-3x < -9$
48. $2x - 7 \leq 5x + 11$
49. $4(2x - 1) \geq 0$
50. $\frac{2}{3}x < \frac{1}{3}$

CONCEPT EXTENSIONS

Solve. See the Concept Check in the section.

51. Suppose you mix an amount of candy costing $0.49 a pound with candy costing $0.65 a pound. Which of the following costs per pound could result?
    a. $0.58   b. $0.72   c. $0.29

52. Suppose you mix a 50% acid solution with pure acid (100%). Which of the following acid strengths are possible for the resulting acid mixture?
    a. 25%   b. 150%   c. 62%   d. 90%

△ 53. Dale and Sharon Mahnke have decided to fence off a garden plot behind their house, using their house as the "fence" along one side of the garden. The length (which runs parallel to the house) is 3 feet less than twice the width. Find the dimensions if 33 feet of fencing is used along the three sides requiring it.

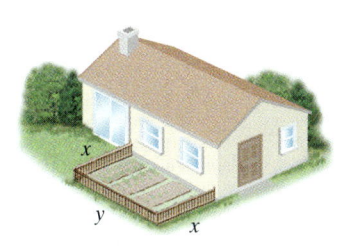

△ 54. Judy McElroy plans to erect 152 feet of fencing around her rectangular horse pasture. A river bank serves as one side length of the rectangle. If each width is 4 feet longer than half the length, find the dimensions.

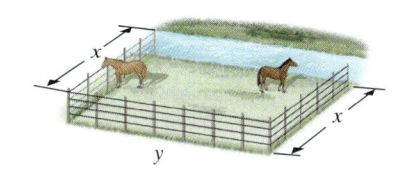

55. The percent of viewers who watch nightly network news can be approximated by the equation $y = 0.82x + 17.2$, where x is the years of age over 18 of the viewer. The percent of viewers who watch cable TV news is approximated by the equation $y = 0.33x + 30.5$ where x is also the years of age over 18 of the viewer. (*Source:* The Pew Research Center for The People & The Press)

   a. Solve the system of equations: $\begin{cases} y = 0.82x + 17.2 \\ y = 0.33x + 30.5 \end{cases}$

   Round x and y to the nearest tenth.

   b. Explain what the point of intersection means in terms of the context of the exercise.

   c. Look at the slopes of both equations of the system. What type of news attracts older viewers more? What type of news attracts younger viewers more?

56. In the triangle below, the measure of angle x is 6 times the measure of angle y. Find the measure of x and y by writing a system of two equations in two unknowns and solving the system.

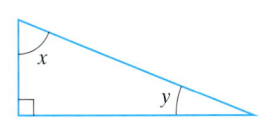

4.5 GRAPHING LINEAR INEQUALITIES

OBJECTIVE

1 Graph a linear inequality in two variables.

In the next section, we continue our work with systems by solving systems of linear inequalities. Before that section, we first need to learn to graph a single linear inequality in two variables.

Recall that a linear equation in two variables is an equation that can be written in the form $Ax + By = C$ where A, B, and C are real numbers and A and B are not both 0. The definition of a linear inequality is the same except that the equal sign is replaced with an inequality sign.

A **linear inequality in two variables** is an inequality that can be written in one of the forms:

$$Ax + By < C \qquad Ax + By \leq C$$
$$Ax + By > C \qquad Ax + By \geq C$$

where A, B, and C are real numbers and A and B are not both 0. Just as for linear equations in x and y, an ordered pair is a **solution** of an inequality in x and y if replacing the variables by coordinates of the ordered pair results in a true statement.

OBJECTIVE 1 ▶ Graphing linear inequalities in two variables. The linear equation $x - y = 1$ is graphed next. Recall that all points on the line correspond to ordered pairs that satisfy the equation $x - y = 1$.

Notice the line defined by $x - y = 1$ divides the rectangular coordinate system plane into 2 sides. All points on one side of the line satisfy the inequality $x - y < 1$ and all points on the other side satisfy the inequality $x - y > 1$. The graph below shows a few examples of this.

Check	$x - y < 1$
$(1, 3)$	$1 - 3 < 1$ True
$(-2, 1)$	$-2 - 1 < 1$ True
$(-4, -4)$	$-4 - (-4) < 1$ True

Check	$x - y > 1$
$(4, 1)$	$4 - 1 > 1$ True
$(2, -2)$	$2 - (-2) > 1$ True
$(0, -4)$	$0 - (-4) > 1$ True

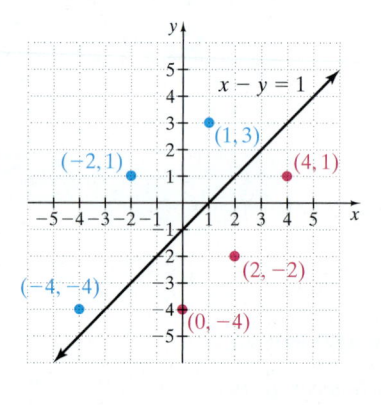

The graph of $x - y < 1$ is the region shaded blue and the graph of $x - y > 1$ is the region shaded red below.

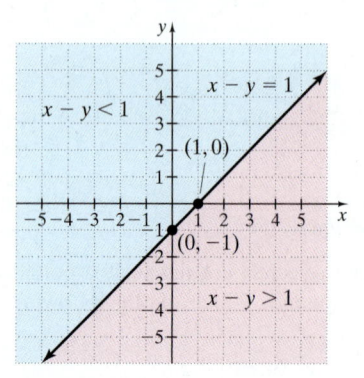

The region to the left of the line and the region to the right of the line are called **half-planes.** Every line divides the plane (similar to a sheet of paper extending indefinitely in all directions) into two half-planes; the line is called the **boundary.**

Recall that the inequality $x - y \leq 1$ means

$$x - y = 1 \quad \text{or} \quad x - y < 1$$

Thus, the graph of $x - y \leq 1$ is the half-plane $x - y < 1$ along with the boundary line $x - y = 1$.

> **Graphing a Linear Inequality in Two Variables**
>
> **STEP 1.** Graph the boundary line found by replacing the inequality sign with an equal sign. If the inequality sign is $>$ or $<$, graph a dashed boundary line (indicating that the points on the line are not solutions of the inequality). If the inequality sign is $\geq$ or $\leq$, graph a solid boundary line (indicating that the points on the line are solutions of the inequality).
>
> **STEP 2.** Choose a point, *not* on the boundary line, as a test point. Substitute the coordinates of this test point into the *original* inequality.
>
> **STEP 3.** If a true statement is obtained in Step 2, shade the half-plane that contains the test point. If a false statement is obtained, shade the half-plane that does not contain the test point.

EXAMPLE 1 Graph: $x + y < 7$

Solution

STEP 1. First we graph the boundary line by graphing the equation $x + y = 7$. We graph this boundary as a dashed line because the inequality sign is $<$, and thus the points on the line are not solutions of the inequality $x + y < 7$.

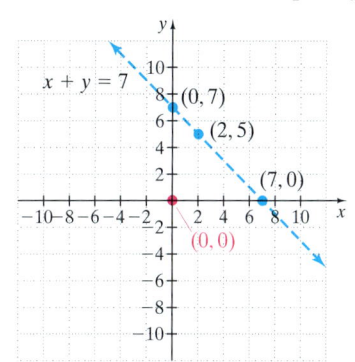

STEP 2. Next, choose a test point, being careful not to choose a point on the boundary line. We choose $(0, 0)$. Substitute the coordinates of $(0, 0)$ into $x + y < 7$.

$$x + y < 7 \quad \text{Original inequality}$$
$$0 + 0 \stackrel{?}{<} 7 \quad \text{Replace } x \text{ with 0 and } y \text{ with 0.}$$
$$0 < 7 \quad \text{True}$$

STEP 3. Since the result is a true statement, $(0, 0)$ is a solution of $x + y < 7$, and every point in the same half-plane as $(0, 0)$ is also a solution. To indicate this, shade the entire half-plane containing $(0, 0)$, as shown.

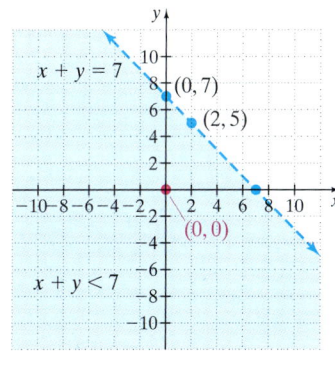

PRACTICE

1 Graph: $x + y > 5$

Concept Check ✓

Determine whether $(0, 0)$ is included in the graph of

a. $y \geq 2x + 3$ b. $x < 7$ c. $2x - 3y < 6$

EXAMPLE 2 Graph: $2x - y \geq 3$

Solution

STEP 1. We graph the boundary line by graphing $2x - y = 3$. We draw this line as a solid line because the inequality sign is $\geq$, and thus the points on the line are solutions of $2x - y \geq 3$.

STEP 2. Once again, $(0, 0)$ is a convenient test point since it is not on the boundary line. We substitute 0 for x and 0 for y into the original inequality.

$$2x - y \geq 3$$
$$2(0) - 0 \geq 3 \quad \text{Let } x = 0 \text{ and } y = 0.$$
$$0 \geq 3 \quad \text{False}$$

STEP 3. Since the statement is false, no point in the half-plane containing $(0, 0)$ is a solution. Therefore, we shade the half-plane that does not contain $(0, 0)$. Every point in the shaded half-plane and every point on the boundary line is a solution of $2x - y \geq 3$.

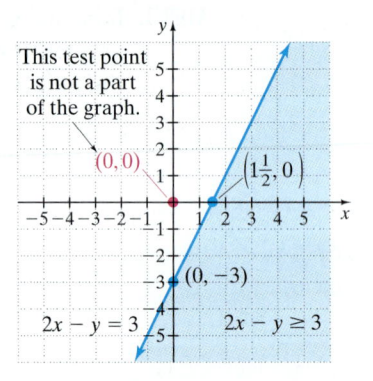

PRACTICE 2 Graph: $3x - y \geq 4$

> **Helpful Hint**
>
> When graphing an inequality, make sure the test point is substituted into the **original inequality**. For Example 2, we substituted the test point $(0, 0)$ into the **original inequality** $2x - y \geq 3$, not $2x - y = 3$.

EXAMPLE 3 Graph: $x > 2y$

Solution

STEP 1. We find the boundary line by graphing $x = 2y$. The boundary line is a dashed line since the inequality symbol is $>$.

STEP 2. We cannot use $(0, 0)$ as a test point because it is a point on the boundary line. We choose instead $(0, 2)$.

$$x > 2y$$
$$0 > 2(2) \quad \text{Let } x = 0 \text{ and } y = 2.$$
$$0 > 4 \quad \text{False}$$

Answers to Concept Check:
a. no b. yes c. yes

Section 4.5 Graphing Linear Inequalities 285

STEP 3. Since the statement is false, we shade the half-plane that does not contain the test point $(0, 2)$, as shown.

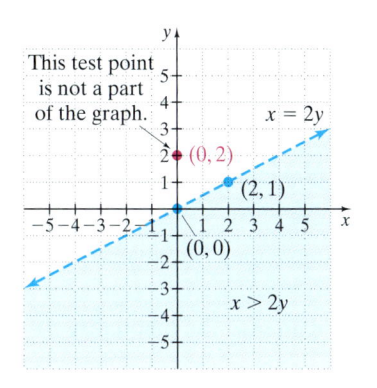

PRACTICE 3 Graph: $x > 3y$

EXAMPLE 4 Graph: $5x + 4y \leq 20$

Solution We graph the solid boundary line $5x + 4y = 20$ and choose $(0, 0)$ as the test point.

$$5x + 4y \leq 20$$
$$5(0) + 4(0) \stackrel{?}{\leq} 20 \quad \text{Let } x = 0 \text{ and } y = 0.$$
$$0 \leq 20 \quad \text{True}$$

We shade the half-plane that contains $(0, 0)$, as shown.

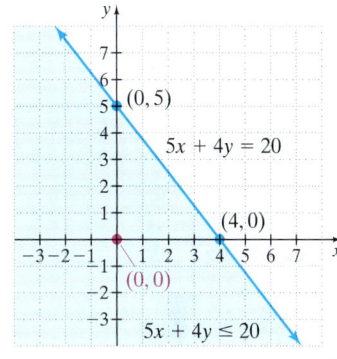

PRACTICE 4 Graph: $3x + 4y \geq 12$

EXAMPLE 5 Graph: $y > 3$

Solution We graph the dashed boundary line $y = 3$ and choose $(0, 0)$ as the test point. (Recall that the graph of $y = 3$ is a horizontal line with y-intercept 3.)

$$y > 3$$
$$0 \stackrel{?}{>} 3 \quad \text{Let } y = 0.$$
$$0 > 3 \quad \text{False}$$

We shade the half-plane that does not contain $(0, 0)$, as shown.

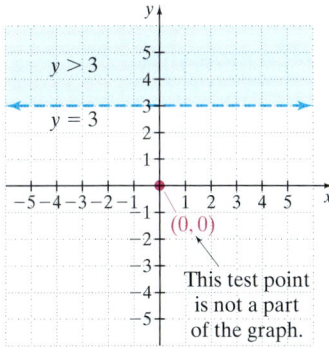

PRACTICE 5 Graph: $x > 3$

VOCABULARY & READINESS CHECK

Use the choices below to fill in each blank. Some choices may be used more than once, and some not at all.

| true | $x < 3$ | $y < 3$ | half-planes | yes |
| false | $x \leq 3$ | $y \leq 3$ | linear inequality in two variables | no |

1. The statement $5x - 6y < 7$ is an example of a(n) _____.
2. A boundary line divides a plane into two regions called _____.
3. True or false: The graph of $5x - 6y < 7$ includes its corresponding boundary line. _____
4. True or false: When graphing a linear inequality, to determine which side of the boundary line to shade, choose a point *not* on the boundary line. _____
5. True or false: The boundary line for the inequality $5x - 6y < 7$ is the graph of $5x - 6y = 7$. _____
6. The graph of _____ is

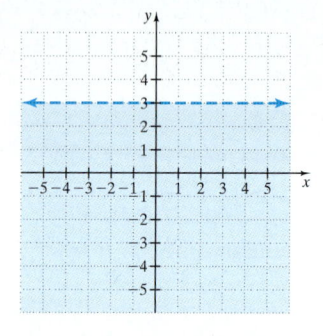

State whether the graph of each inequality includes its corresponding boundary line.

7. $y \geq x + 4$ 8. $x - y > -7$ 9. $y \geq x$ 10. $x > 0$

Decide whether $(0, 0)$ is a solution of each given inequality.

11. $x + y > -5$ 12. $2x + 3y < 10$ 13. $x - y \leq -1$ 14. $\frac{2}{3}x + \frac{5}{6}y > 4$

4.5 EXERCISE SET

Determine which ordered pairs given are solutions of the linear inequality in two variables. See Example 1.

1. $x - y > 3; (2, -1), (5, 1)$
2. $y - x < -2; (2, 1), (5, -1)$
3. $3x - 5y \leq -4; (-1, -1), (4, 0)$
4. $2x + y \geq 10; (-1, -4), (5, 0)$
5. $x < -y; (0, 2), (-5, 1)$
6. $y > 3x; (0, 0), (-1, -4)$

MIXED PRACTICE

Graph each inequality. See Examples 2 through 5.

7. $x + y \leq 1$
8. $x + y \geq -2$
9. $2x + y > -4$
10. $x + 3y \leq 3$
11. $x + 6y \leq -6$
12. $7x + y > -14$
13. $2x + 5y > -10$
14. $5x + 2y \leq 10$
15. $x + 2y \leq 3$
16. $2x + 3y > -5$
17. $2x + 7y > 5$
18. $3x + 5y \leq -2$
19. $x - 2y \geq 3$
20. $4x + y \leq 2$
21. $5x + y < 3$
22. $x + 2y > -7$
23. $4x + y < 8$
24. $9x + 2y \geq -9$
25. $y \geq 2x$
26. $x < 5y$
27. $x \geq 0$
28. $y \leq 0$
29. $y \leq -3$
30. $x > -\frac{2}{3}$
31. $2x - 7y > 0$
32. $5x + 2y \leq 0$
33. $3x - 7y \geq 0$
34. $-2x - 9y > 0$
35. $x > y$
36. $x \leq -y$
37. $x - y \leq 6$
38. $x - y > 10$
39. $-\frac{1}{4}y + \frac{1}{3}x > 1$
40. $\frac{1}{2}x - \frac{1}{3}y \leq -1$
41. $-x < 0.4y$
42. $0.3x \geq 0.1y$

In Exercises 43 through 48, match each inequality with its graph.

a. $x > 2$ b. $y < 2$ c. $y < 2x$
d. $y \leq -3x$ e. $2x + 3y < 6$ f. $3x + 2y > 6$

43.

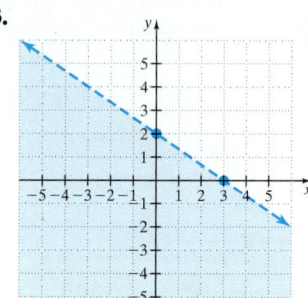

44.

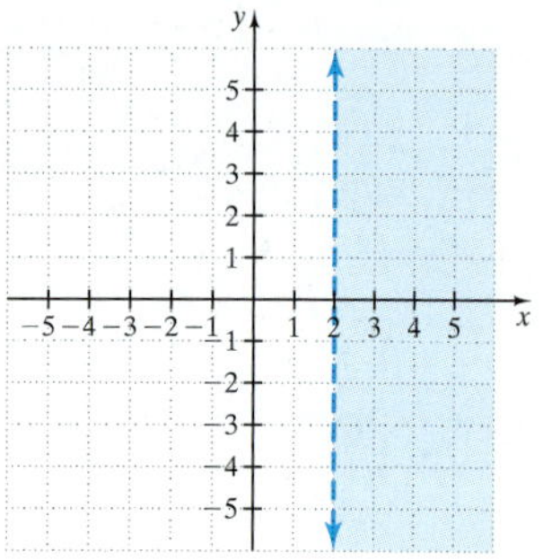

45.

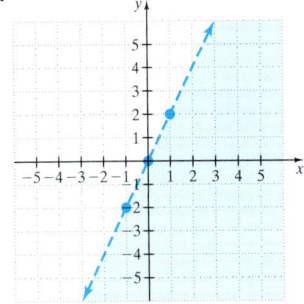

46.

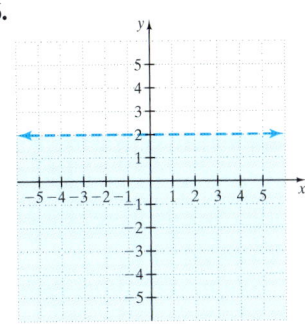

47.

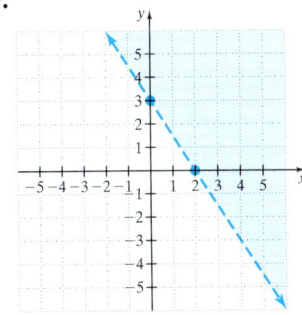

48.

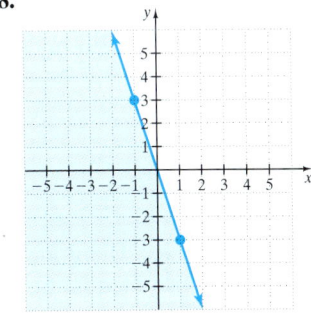

REVIEW AND PREVIEW

Evaluate. See Section 1.4.

49. 2^3
50. 3^4
51. $(-2)^5$
52. -2^5
53. $3 \cdot 4^2$
54. $4 \cdot 3^3$

Evaluate each expression for the given replacement value. See Section 1.4.

55. x^2 if x is -5
56. x^3 if x is -5
57. $2x^3$ if x is -1
58. $3x^2$ if x is -1

CONCEPT EXTENSIONS

Determine whether (1, 1) is included in each graph. See the Concept Check in this section.

59. $3x + 4y < 8$
60. $y > 5x$
61. $y \geq -\dfrac{1}{2}x$
62. $x > 3$

63. Write an inequality whose solutions are all pairs of numbers x and y whose sum is at least 13. Graph the inequality.

64. Write an inequality whose solutions are all the pairs of numbers x and y whose sum is at most -4. Graph the inequality.

65. Explain why a point on the boundary line should not be chosen as the test point.

66. Describe the graph of a linear inequality.

67. The price for a taxi cab in a small city is $2.50 per mile, x, while traveling, and $.25 every minute, y, while waiting. If you have $20 to spend on a cab ride, the inequality

$$2.5x + 0.25y \leq 20$$

represents your situation. Graph this inequality in the first quadrant only.

68. A word processor charges $22 per hour, x, for typing a first draft, and $15 per hour, y, for making changes and typing a second draft. If you need a document typed and have $100, the inequality

$$22x + 15y \leq 100$$

represents your situation. Graph the inequality in the first quadrant only.

69. In Exercises 67 and 68, why were you instructed to graph each inequality in the first quadrant only?

70. Scott Sambracci and Sara Thygeson are planning their wedding. They have calculated that they want the cost of their wedding ceremony x plus the cost of their reception y to be no more than $5000.
   a. Write an inequality describing this relationship.
   b. Graph this inequality.
   c. Why should we be interested in only quadrant I of this graph?

71. It's the end of the budgeting period for Dennis Fernandes and he has $500 left in his budget for car rental expenses. He plans to spend this budget on a sales trip throughout southern Texas. He will rent a car that costs $30 per day and $0.15 per mile and he can spend no more than $500.
   a. Write an inequality describing this situation. Let x = number of days and let y = number of miles.
   b. Graph this inequality.
   c. Why should we be interested in only quadrant I of this graph?

STUDY SKILLS BUILDER

Tips for Studying for an Exam

To prepare for an exam, try the following study techniques:

- Start the study process days before your exam.
- Make sure that you are up-to-date on your assignments.
- If there is a topic that you are insure of, use one of the many resources that are available to you. For example,
  See your instructor.
  View a lecture video on the topic.
  Visit a learning resource center on campus.
  Read the textbook material and examples on the topic.
- Reread your notes and carefully review the Chapter Highlights at the end of any chapter.
- Work the review exercises at the end of the chapter.
- Find a quiet place to take the Chapter Test found at the end of the chapter. Do not use any resources when taking this sample test. This way, you will have a clear indication of how prepared you are for your exam. Check your answers and use the Test Prep CD to make sure that you correct any missed exercises.

Good luck, and keep a positive attitude.

Let's see how you did on your last exam.

1. How many days before your last exam did you start studying for that exam?
2. Were you up-to-date on your assignments at that time or did you need to catch up on assignments?
3. List the most helpful text supplement (if you used one).
4. List the most helpful campus supplement (if you used one).
5. List your process for preparing for a mathematics test.
6. Was this process helpful? In other words, were you satisfied with your performance on your exam?
7. If not, what changes can you make in your process that will make it more helpful to you?

4.6 SYSTEMS OF LINEAR INEQUALITIES

OBJECTIVE

1 Solve a system of linear inequalities.

OBJECTIVE 1 ▶ Solving systems of linear inequalities. In Section 4.5, we graphed linear inequalities in two variables. Just as two linear equations make a system of linear equations, two linear inequalities make a **system of linear inequalities.** Systems of inequalities are very important in a process called linear programming. Many businesses use linear programming to find the most profitable way to use limited resources such as employees, machines, or buildings.

A **solution of a system of linear inequalities** is an ordered pair that satisfies each inequality in the system. The set of all such ordered pairs is the solution set of the system. Graphing this set gives us a picture of the solution set. We can graph a system of inequalities by graphing each inequality in the system and identifying the region of overlap.

EXAMPLE 1 Graph the solution of the system: $\begin{cases} 3x \geq y \\ x + 2y \leq 8 \end{cases}$

Solution We begin by graphing each inequality on the same set of axes. The graph of the solution of the system is the region contained in the graphs of both inequalities. It is their intersection.

First, graph $3x \geq y$. The boundary line is the graph of $3x = y$. Sketch a solid boundary line since the inequality $3x \geq y$ means $3x > y$ or $3x = y$. The test point $(1, 0)$ satisfies the inequality, so shade the half-plane that includes $(1, 0)$.

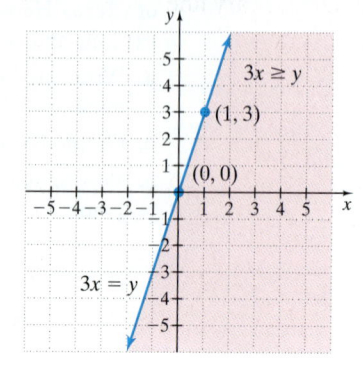

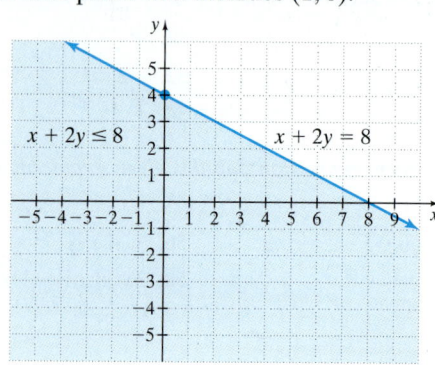

Next, sketch a solid boundary line $x + 2y = 8$ on the same set of axes. The test point $(0, 0)$ satisfies the inequality $x + 2y \leq 8$, so shade the half-plane that includes $(0, 0)$. (For clarity, the graph of $x + 2y \leq 8$ is shown on a separate set of axes.)

An ordered pair solution of the system must satisfy both inequalities. These solutions are points that lie in both shaded regions. The solution of the system is the purple shaded region as seen below. This solution includes parts of both boundary lines.

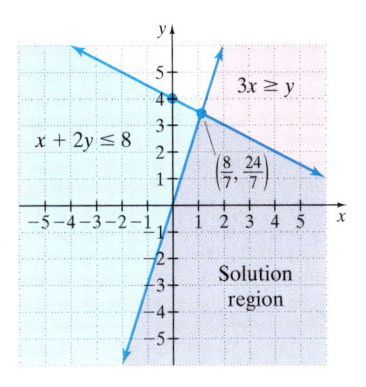

PRACTICE 1 Graph the solution of the system: $\begin{cases} 4x \leq y \\ x + 3y \geq 9 \end{cases}$

In linear programming, it is sometimes necessary to find the coordinates of the **corner point:** the point at which the two boundary lines intersect. To find the point of intersection, solve the related linear system

$$\begin{cases} 3x = y \\ x + 2y = 8 \end{cases}$$

by the substitution method or the addition method. The lines intersect at $\left(\dfrac{8}{7}, \dfrac{24}{7}\right)$, the corner point of the graph.

> **Graphing the Solution of a System of Linear Inequalities**
>
> **STEP 1.** Graph each inequality in the system on the same set of axes.
>
> **STEP 2.** The solutions of the system are the points common to the graphs of all the inequalities in the system.

EXAMPLE 2 Graph the solution of the system: $\begin{cases} x - y < 2 \\ x + 2y > -1 \end{cases}$

Solution Graph both inequalities on the same set of axes. Both boundary lines are dashed lines since the inequality symbols are $<$ and $>$. The solution of the system is the region shown by the purple shading. In this example, the boundary lines are not a part of the solution.

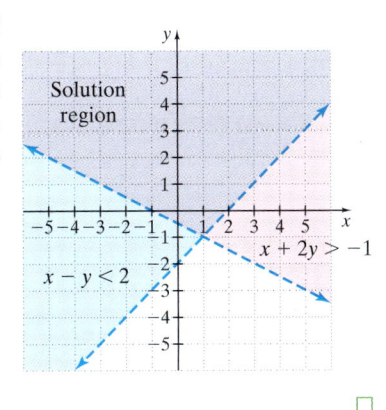

PRACTICE 2 Graph the solution of the system: $\begin{cases} x - y > 4 \\ x + 3y < -4 \end{cases}$

EXAMPLE 3 Graph the solution of the system: $\begin{cases} -3x + 4y < 12 \\ x \geq 2 \end{cases}$

Solution Graph both inequalities on the same set of axes.

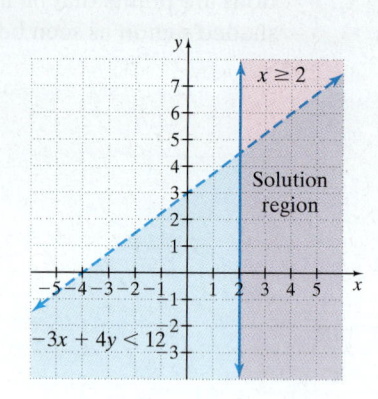

The solution of the system is the purple shaded region, including a portion of the line $x = 2$.

PRACTICE 3 Graph the solution of the system: $\begin{cases} y \leq 6 \\ -2x + 5y > 10 \end{cases}$

4.6 EXERCISE SET

MIXED PRACTICE

Graph the solution of each system of linear inequalities. See Examples 1 through 3.

1. $\begin{cases} y \geq x + 1 \\ y \geq 3 - x \end{cases}$

2. $\begin{cases} y \geq x - 3 \\ y \geq -1 - x \end{cases}$

3. $\begin{cases} y < 3x - 4 \\ y \leq x + 2 \end{cases}$

4. $\begin{cases} y \leq 2x + 1 \\ y > x + 2 \end{cases}$

5. $\begin{cases} y \leq -2x - 2 \\ y \geq x + 4 \end{cases}$

6. $\begin{cases} y \leq 2x + 4 \\ y \geq -x - 5 \end{cases}$

7. $\begin{cases} y \geq -x + 2 \\ y \leq 2x + 5 \end{cases}$

8. $\begin{cases} y \geq x - 5 \\ y \leq -3x + 3 \end{cases}$

9. $\begin{cases} x \geq 3y \\ x + 3y \leq 6 \end{cases}$

10. $\begin{cases} -2x < y \\ x + 2y < 3 \end{cases}$

11. $\begin{cases} y + 2x \geq 0 \\ 5x - 3y \leq 12 \end{cases}$

12. $\begin{cases} y + 2x \leq 0 \\ 5x + 3y \geq -2 \end{cases}$

13. $\begin{cases} 3x - 4y \geq -6 \\ 2x + y \leq 7 \end{cases}$

14. $\begin{cases} 4x - y \geq -2 \\ 2x + 3y \leq -8 \end{cases}$

15. $\begin{cases} x \leq 2 \\ y \geq -3 \end{cases}$

16. $\begin{cases} x \geq -3 \\ y \geq -2 \end{cases}$

17. $\begin{cases} y \geq 1 \\ x < -3 \end{cases}$

18. $\begin{cases} y > 2 \\ x \geq -1 \end{cases}$

19. $\begin{cases} 2x + 3y < -8 \\ x \geq -4 \end{cases}$

20. $\begin{cases} 3x + 2y \leq 6 \\ x < 2 \end{cases}$

21. $\begin{cases} 2x - 5y \leq 9 \\ y \leq -3 \end{cases}$

22. $\begin{cases} 2x + 5y \leq -10 \\ y \geq 1 \end{cases}$

23. $\begin{cases} y \geq \frac{1}{2}x + 2 \\ y \leq \frac{1}{2}x - 3 \end{cases}$

24. $\begin{cases} y \geq \frac{-3}{2}x + 3 \\ y < \frac{-3}{2}x + 6 \end{cases}$

REVIEW AND PREVIEW

Find the square of each expression. For example, the square of 7 is 7^2 or 49. The square of 5x is $(5x)^2$ or $25x^2$. See Section 1.4.

25. 4

26. 3

27. 6x

28. 11y

29. $10y^3$

30. $8x^5$

CONCEPT EXTENSIONS

For each system of inequalities, choose the corresponding graph.

31. $\begin{cases} y < 5 \\ x > 3 \end{cases}$ 32. $\begin{cases} y > 5 \\ x < 3 \end{cases}$

33. $\begin{cases} y \leq 5 \\ x < 3 \end{cases}$ 34. $\begin{cases} y > 5 \\ x \geq 3 \end{cases}$

A.   B.

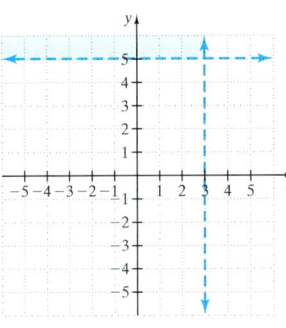

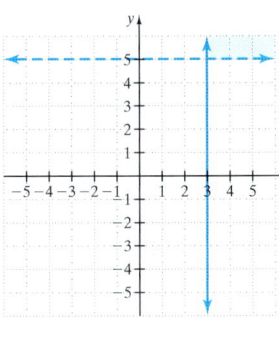

C. 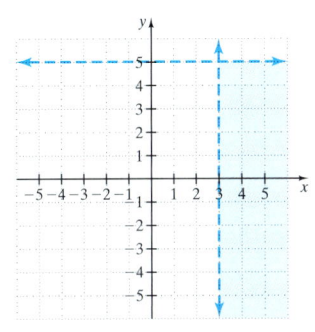   D.

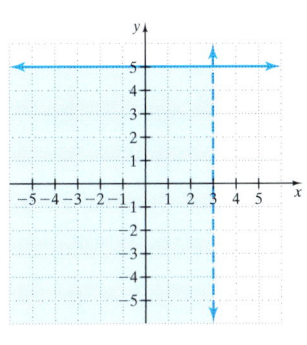

35. Explain how to decide which region to shade to show the solution region of the following system.
$$\begin{cases} x \geq 3 \\ y \geq -2 \end{cases}$$

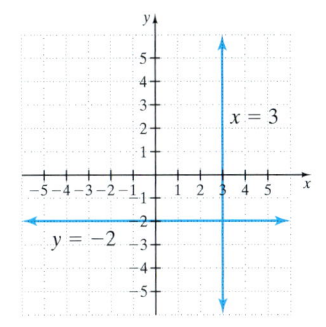

36. Describe the location of the solution region of the system
$$\begin{cases} x > 0 \\ y > 0. \end{cases}$$

37. Graph the solution of $\begin{cases} 2x - y \leq 6 \\ x \geq 3 \\ y > 2 \end{cases}$

38. Graph the solution of $\begin{cases} x + y < 5 \\ y < 2x \\ x \geq 0 \\ y \geq 0 \end{cases}$

CHAPTER 4 GROUP ACTIVITY

Break-Even Point
Sections 4.1, 4.2, 4.3, 4.4

When a business sells a new product, it generally does not start making a profit right away. There are usually many expenses associated with creating a new product. These expenses might include an advertising blitz to introduce the product to the public. These start-up expenses might also include the cost of market research and product development or any brand-new equipment needed to manufacture the product. Start-up costs like these are generally called *fixed costs* because they don't depend on the number of items manufactured. Expenses that depend on the number of items manufactured, such as the cost of materials and shipping, are called *variable costs*. The total cost of manufacturing the new product is given by the cost equation: Total cost = Fixed costs + Variable costs.

For instance, suppose a greeting card company is launching a new line of greeting cards. The company spent $7000 doing product research and development for the new line and spent $15,000 on advertising the new line. The company does not need to buy any new equipment to manufacture the cards, but the paper and ink needed to make each card will cost $0.20 per card. The total cost y in dollars for manufacturing x cards is $y = 22,000 + 0.20x$.

Once a business sets a price for the new product, the company can find the product's expected *revenue*. Revenue is the amount of money the company takes in from the sales of its product. The revenue from selling a product is given by the revenue equation: Revenue = Price per item × Number of items sold.

For instance, suppose that the card company plans to sell its new cards for $1.50 each. The revenue y, in dollars, that the company can expect to receive from the sales of x cards is $y = 1.50x$.

If the total cost and revenue equations are graphed on the same coordinate system, the graphs should intersect. The point of intersection is where total cost equals revenue and is called the *break-even point*. The break-even point gives the number of items x that must be manufactured and sold for the company to recover its expenses. If fewer than this number of items are produced and sold, the company loses money. If more than this number of items are produced and sold, the company makes a profit. In the case of the greeting card company, approximately 16,923 cards must be manufactured and sold for the company to break even on this new card line. The total cost and revenue of producing and selling 16,923 cards is the same. It is approximately $25,385.

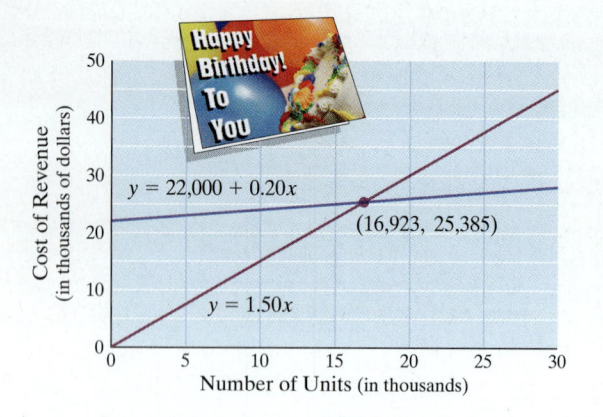

Group Activity

Suppose your group is starting a small business near your campus.

a. Choose a business and decide what campus-related product or service you will provide.
b. Research the fixed costs of starting up such a business.
c. Research the variable costs of producing such a product or providing such a service.
d. Decide how much you would charge per unit of your product or service.
e. Find a system of equations for the total cost and revenue of your product or service.
f. How many units of your product or service must be sold before your business will break even?

CHAPTER 4 VOCABULARY CHECK

Fill in each blank with one of the words or phrases listed below.

system of linear equations   solution   consistent   independent
dependent   inconsistent   substitution   addition
system of linear inequalities

1. In a system of linear equations in two variables, if the graphs of the equations are the same, the equations are _____ equations.
2. Two or more linear equations are called a _____.
3. A system of equations that has at least one solution is called a(n) _____ system.
4. A _____ of a system of two equations in two variables is an ordered pair of numbers that is a solution of both equations in the system.
5. Two algebraic methods for solving systems of equations are _____ and _____.
6. A system of equations that has no solution is called a(n) _____ system.
7. In a system of linear equations in two variables, if the graphs of the equations are different, the equations are _____ equations.
8. Two or more linear inequalities are called a _____.

STUDY SKILLS BUILDER

Are You Preparing for a Test on Chapter 4?

Below I have listed some common trouble areas for topics covered in Chapter 4. After studying for your test—but before taking your test—read these.

- If you are having trouble drawing a neat graph, remember to ask your instructor if you can use graph paper on your test. This will save your time and keep your graphs neat.
- Do you remember how to check solutions of systems of equations? If $(-1, 5)$ is a solution of the system
$$\begin{cases} 3x - y = -8 \\ -x + y = 6, \end{cases}$$

then the ordered pair will make *both* equations a true statement.

$3x - y = -8$ $-x + y = 6$
$3(-1) - 5 = -8$ $-(-1) + 5 = 6$ Let $x = -1$ and $y = 5$.
$-8 = -8$ True $6 = 6$ True

Remember: This is simply a list of a few common trouble areas. For a review of Chapter 4, see the Highlights and Chapter Review at the end of this chapter.

CHAPTER 4 HIGHLIGHTS

> **Helpful Hint**
>
> Are you preparing for your test? Don't forget to take the Chapter 4 Test on page 298. Then check your answers at the back of the text and use the Chapter Test Prep Video CD to see the fully worked-out solutions to any of the exercises you want to review.

DEFINITIONS AND CONCEPTS	EXAMPLES

SECTION 4.1 SOLVING SYSTEMS OF LINEAR EQUATIONS BY GRAPHING

A **solution** of a system of two equations in two variables is an ordered pair of numbers that is a solution of both equations in the system.

Determine whether $(-1, 3)$ is a solution of the system:
$$\begin{cases} 2x - y = -5 \\ x = 3y - 10 \end{cases}$$

Replace x with -1 and y with 3 in both equations.

$$2x - y = -5 \qquad x = 3y - 10$$
$$2(-1) - 3 \stackrel{?}{=} -5 \qquad -1 \stackrel{?}{=} 3 \cdot 3 - 10$$
$$-5 = -5 \text{ True} \qquad -1 = -1 \text{ True}$$

$(-1, 3)$ is a solution of the system.

Graphically, a solution of a system is a point common to the graphs of both equations.

Solve by graphing. $\begin{cases} 3x - 2y = -3 \\ x + y = 4 \end{cases}$

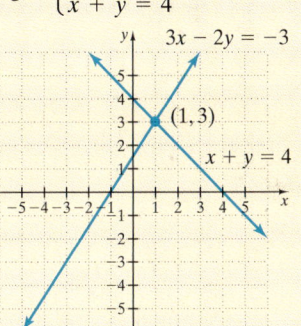

A system of equations with at least one solution is a **consistent system**. A system that has no solution is an **inconsistent system**.

If the graphs of two linear equations are identical, the equations are **dependent**. If their graphs are different, the equations are **independent**.

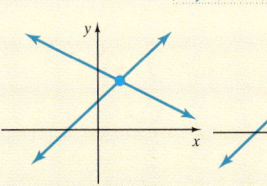

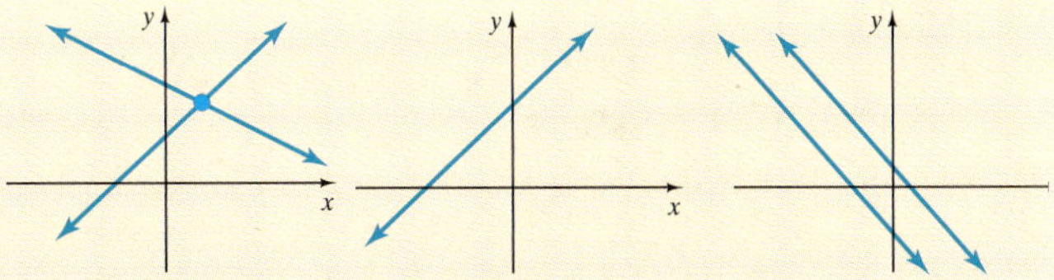

Consistent and independent | Consistent and dependent | Inconsistent and independent

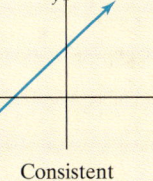

SECTION 4.2 SOLVING SYSTEMS OF LINEAR EQUATIONS BY SUBSTITUTION

To solve a system of linear equations by the substitution method.

Step 1. Solve one equation for a variable.

Step 2. Substitute the expression for the variable into the other equation.

Step 3. Solve the equation from Step 2 to find the value of one variable.

Step 4. Substitute the value from Step 3 in either original equation to find the value of the other variable.

Solve by substitution.
$$\begin{cases} 3x + 2y = 1 \\ x = y - 3 \end{cases}$$

Substitute $y - 3$ for x in the first equation.

$$3x + 2y = 1$$
$$3(y - 3) + 2y = 1$$
$$3y - 9 + 2y = 1$$
$$5y = 10$$
$$y = 2 \qquad \text{Divide by 5.}$$

(continued)

DEFINITIONS AND CONCEPTS	EXAMPLES

SECTION 4.2 SOLVING SYSTEMS OF LINEAR EQUATIONS BY SUBSTITUTION (continued)

Step 5. Check the solution in both equations.	To find x, substitute 2 for y in $x = y - 3$ so that $x = 2 - 3$ or -1. The solution $(-1, 2)$ checks.

SECTION 4.3 SOLVING SYSTEMS OF LINEAR EQUATIONS BY ADDITION

To solve a system of linear equations by the addition method

Step 1. Rewrite each equation in standard form $Ax + By = C$.

Step 2. Multiply one or both equations by a nonzero number so that the coefficients of a variable are opposites.

Step 3. Add the equations.

Step 4. Find the value of one variable by solving the resulting equation.

Step 5. Substitute the value from Step 4 into either original equation to find the value of the other variable.

Step 6. Check the solution in both equations.

If solving a system of linear equations by substitution or addition yields a true statement such as $-2 = -2$, then the graphs of the equations in the system are identical and there is an infinite number of solutions of the system.

Solve by addition.
$$\begin{cases} x - 2y = 8 \\ 3x + y = -4 \end{cases}$$

Multiply both sides of the first equation by -3.
$$\begin{cases} -3x + 6y = -24 \\ 3x + y = -4 \end{cases}$$
$7y = -28$ Add.
$y = -4$ Divide by 7.

To find x, let $y = -4$ in an original equation.
$$x - 2(-4) = 8 \quad \text{First equation}$$
$$x + 8 = 8$$
$$x = 0$$

The solution $(0, -4)$ checks.

Solve: $\begin{cases} 2x - 6y = -2 \\ x = 3y - 1 \end{cases}$

Substitute $3y - 1$ for x in the first equation.
$$2(3y - 1) - 6y = -2$$
$$6y - 2 - 6y = -2$$
$$-2 = -2 \quad \text{True}$$

The system has an infinite number of solutions.

SECTION 4.4 SYSTEMS OF LINEAR EQUATIONS AND PROBLEM SOLVING

Problem-solving steps

1. UNDERSTAND. Read and reread the problem.

Two angles are supplementary if their sum is 180°.

The larger of two supplementary angles is three times the smaller, decreased by twelve. Find the measure of each angle. Let

$$x = \text{measure of smaller angle}$$
$$y = \text{measure of larger angle}$$

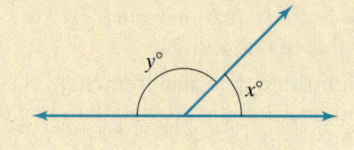

(continued)

DEFINITIONS AND CONCEPTS	EXAMPLES
SECTION 4.4 SYSTEMS OF LINEAR EQUATIONS AND PROBLEM SOLVING (continued)	
2. TRANSLATE.	In words: the sum of supplementary angles is 180° Translate: $x + y = 180$ In words: larger angle is 3 times smaller decreased by 12 Translate: $y = 3x - 12$
3. SOLVE.	Solve the system: $$\begin{cases} x + y = 180 \\ y = 3x - 12 \end{cases}$$ Use the substitution method and replace y with $3x - 12$ in the first equation. $$x + y = 180$$ $$x + (3x - 12) = 180$$ $$4x = 192$$ $$x = 48$$ Since $y = 3x - 12$, then $y = 3 \cdot 48 - 12$ or 132.
4. INTERPRET.	The solution checks. The smaller angle measures 48° and the larger angle measures 132°.
SECTION 4.5 GRAPHING LINEAR INEQUALITIES	
A **linear inequality in two variables** is an inequality that can be written in one of the forms: $$Ax + By < C \quad Ax + By \leq C$$ $$Ax + By > C \quad Ax + By \geq C$$ To graph a linear inequality 1. Graph the boundary line by graphing the related equation. Draw the line solid if the inequality symbol is $\leq$ or $\geq$. Draw the line dashed if the inequality symbol is $<$ or $>$. 2. Choose a test point not on the line. Substitute its coordinates into the original inequality. 3. If the resulting inequality is true, shade the half-plane that contains the test point. If the inequality is not true, shade the half-plane that does not contain the test point.	**Linear Inequalities** $$2x - 5y < 6 \qquad x \geq -5$$ $$y > -8x \qquad y \leq 2$$ Graph $2x - y \leq 4$. 1. Graph $2x - y = 4$. Draw a solid line because the inequality symbol is $\leq$. 2. Check the test point $(0, 0)$ in the inequality $2x - y \leq 4$. $$2 \cdot 0 - 0 \leq 4 \quad \text{Let } x = 0 \text{ and } y = 0.$$ $$0 \leq 4 \quad \text{True}$$ 3. The inequality is true so we shade the half-plane containing $(0, 0)$.

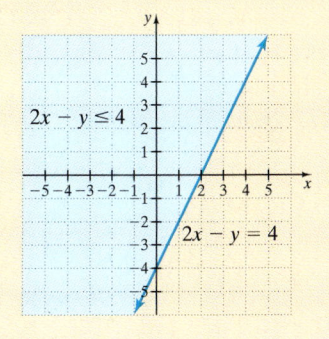

DEFINITIONS AND CONCEPTS

EXAMPLES

SECTION 4.6 SYSTEMS OF LINEAR INEQUALITIES

A system of linear inequalities consists of two or more linear inequalities.

To graph a system of inequalities, graph each inequality in the system. The overlapping region is the solution of the system.

System of Linear Inequalities

$$\begin{cases} x - y \geq 3 \\ y \leq -2x \end{cases}$$

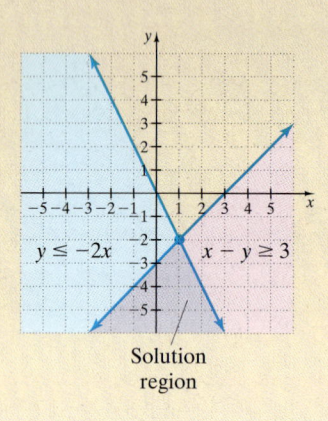

CHAPTER 4 REVIEW

(4.1) Determine whether any of the following ordered pairs satisfy the system of linear equations.

1. $\begin{cases} 2x - 3y = 12 \\ 3x + 4y = 1 \end{cases}$

   a. $(12, 4)$ b. $(3, -2)$ c. $(-3, 6)$

2. $\begin{cases} 4x + y = 0 \\ -8x - 5y = 9 \end{cases}$

   a. $\left(\dfrac{3}{4}, -3\right)$ b. $(-2, 8)$ c. $\left(\dfrac{1}{2}, -2\right)$

3. $\begin{cases} 5x - 6y = 18 \\ 2y - x = -4 \end{cases}$

   a. $(-6, -8)$ b. $\left(3, \dfrac{5}{2}\right)$ c. $\left(3, -\dfrac{1}{2}\right)$

4. $\begin{cases} 2x + 3y = 1 \\ 3y - x = 4 \end{cases}$

   a. $(2, 2)$ b. $(-1, 1)$ c. $(2, -1)$

Solve each system of equations by graphing.

5. $\begin{cases} x + y = 5 \\ x - y = 1 \end{cases}$

6. $\begin{cases} x + y = 3 \\ x - y = -1 \end{cases}$

7. $\begin{cases} x = 5 \\ y = -1 \end{cases}$

8. $\begin{cases} x = -3 \\ y = 2 \end{cases}$

9. $\begin{cases} 2x + y = 5 \\ x = -3y \end{cases}$

10. $\begin{cases} 3x + y = -2 \\ y = -5x \end{cases}$

11. $\begin{cases} y = 3x \\ -6x + 2y = 6 \end{cases}$

12. $\begin{cases} x - 2y = 2 \\ -2x + 4y = -4 \end{cases}$

(4.2) Solve each system of equations by the substitution method.

13. $\begin{cases} y = 2x + 6 \\ 3x - 2y = -11 \end{cases}$

14. $\begin{cases} y = 3x - 7 \\ 2x - 3y = 7 \end{cases}$

15. $\begin{cases} x + 3y = -3 \\ 2x + y = 4 \end{cases}$

16. $\begin{cases} 3x + y = 11 \\ x + 2y = 12 \end{cases}$

17. $\begin{cases} 4y = 2x + 6 \\ x - 2y = -3 \end{cases}$

18. $\begin{cases} 9x = 6y + 3 \\ 6x - 4y = 2 \end{cases}$

19. $\begin{cases} x + y = 6 \\ y = -x - 4 \end{cases}$

20. $\begin{cases} -3x + y = 6 \\ y = 3x + 2 \end{cases}$

(4.3) Solve each system of equations by the addition method.

21. $\begin{cases} 2x + 3y = -6 \\ x - 3y = -12 \end{cases}$

22. $\begin{cases} 4x + y = 15 \\ -4x + 3y = -19 \end{cases}$

23. $\begin{cases} 2x - 3y = -15 \\ x + 4y = 31 \end{cases}$

24. $\begin{cases} x - 5y = -22 \\ 4x + 3y = 4 \end{cases}$

25. $\begin{cases} 2x - 6y = -1 \\ -x + 3y = \dfrac{1}{2} \end{cases}$

26. $\begin{cases} 0.6x - 0.3y = -1.5 \\ 0.04x - 0.02y = -0.1 \end{cases}$

27. $\begin{cases} \dfrac{3}{4}x + \dfrac{2}{3}y = 2 \\ x + \dfrac{y}{3} = 6 \end{cases}$

28. $\begin{cases} 10x + 2y = 0 \\ 3x + 5y = 33 \end{cases}$

(4.4) Solve each problem by writing and solving a system of linear equations.

29. The sum of two numbers is 16. Three times the larger number decreased by the smaller number is 72. Find the two numbers.

30. The Forrest Theater can seat a total of 360 people. They take in $15,150 when every seat is sold. If orchestra section tickets cost $45 and balcony tickets cost $35, find the number of seats in the orchestra section and the number of seats in the balcony.

31. A riverboat can head 340 miles upriver in 19 hours, but the return trip takes only 14 hours. Find the current of the river and find the speed of the riverboat in still water to the nearest tenth of a mile.

	d =	r ·	t
Upriver	340	$x - y$	19
Downriver	340	$x + y$	14

32. Find the amount of a 6% acid solution and the amount of a 14% acid solution Pat Mayfield should combine to prepare 50 cc (cubic centimeters) of a 12% solution.

33. A deli charges $3.80 for a breakfast of three eggs and four strips of bacon. The charge is $2.75 for two eggs and three strips of bacon. Find the cost of each egg and the cost of each strip of bacon.

34. An exercise enthusiast alternates between jogging and walking. He traveled 15 miles during the past 3 hours. He jogs at a rate of 7.5 miles per hour and walks at a rate of 4 miles per hour. Find how much time, to the nearest hundredth of an hour, he actually spent jogging and how much time he spent walking.

(4.5) Graph each inequality.

35. $5x + 4y < 20$

36. $x + 3y > 4$

37. $y \geq -7$

38. $y \leq -4$

39. $-x \leq y$

40. $x \geq -y$

(4.6) Graph the solutions of the following systems of linear inequalities.

41. $\begin{cases} y \geq 2x - 3 \\ y \leq -2x + 1 \end{cases}$

42. $\begin{cases} y \leq -3x - 3 \\ y \leq 2x + 7 \end{cases}$

43. $\begin{cases} -3x + 2y > -1 \\ y < -2 \end{cases}$

44. $\begin{cases} -2x + 3y > -7 \\ x \geq -2 \end{cases}$

MIXED REVIEW

Solve each system of equations by graphing.

45. $\begin{cases} x - 2y = 1 \\ 2x + 3y = -12 \end{cases}$

46. $\begin{cases} 3x - y = -4 \\ 6x - 2y = -8 \end{cases}$

Solve each system of equations.

47. $\begin{cases} x + 4y = 11 \\ 5x - 9y = -3 \end{cases}$

48. $\begin{cases} x + 9y = 16 \\ 3x - 8y = 13 \end{cases}$

49. $\begin{cases} y = -2x \\ 4x + 7y = -15 \end{cases}$

50. $\begin{cases} 3y = 2x + 15 \\ -2x + 3y = 21 \end{cases}$

51. $\begin{cases} 3x - y = 4 \\ 4y = 12x - 16 \end{cases}$

52. $\begin{cases} x + y = 19 \\ x - y = -3 \end{cases}$

53. $\begin{cases} x - 3y = -11 \\ 4x + 5y = -10 \end{cases}$

54. $\begin{cases} -x - 15y = 44 \\ 2x + 3y = 20 \end{cases}$

55. $\begin{cases} 2x + y = 3 \\ 6x + 3y = 9 \end{cases}$

56. $\begin{cases} -3x + y = 5 \\ -3x + y = -2 \end{cases}$

Solve each problem by writing and solving a system of linear equations.

57. The sum of two numbers is 12. Three times the smaller number increased by the larger number is 20. Find the numbers.

58. The difference of two numbers is -18. Twice the smaller decreased by the larger is -23. Find the two numbers.

59. Emma Hodges has a jar containing 65 coins, all of which are either nickels or dimes. The total value of the coins is $5.30. How many of each type does she have?

60. Sarah and Owen Hebert purchased 26 stamps, a mixture of 13¢ and 22¢ stamps. Find the number of each type of stamp if they spent $4.19.

Graph each inequality.

61. $x + 6y < 6$

62. $x + y > -2$

CHAPTER 4 TEST

Answer each question true or false.

1. A system of two linear equations in two variables can have exactly two solutions.

2. Although (1, 4) is not a solution of $x + 2y = 6$, it can still be a solution of the system $\begin{cases} x + 2y = 6 \\ x + y = 5 \end{cases}$

3. If the two equations in a system of linear equations are added and the result is $3 = 0$, the system has no solution.

4. If the two equations in a system of linear equations are added and the result is $3x = 0$, the system has no solution.

Is the ordered pair a solution of the given linear system?

5. $\begin{cases} 2x - 3y = 5 \\ 6x + y = 1 \end{cases}; (1, -1)$

6. $\begin{cases} 4x - 3y = 24 \\ 4x + 5y = -8 \end{cases}; (3, -4)$

Solve each system by graphing.

7. $\begin{cases} x - y = 2 \\ 3x - y = -2 \end{cases}$

8. $\begin{cases} y = -3x \\ 3x + y = 6 \end{cases}$

Solve each system by the substitution method.

9. $\begin{cases} 3x - 2y = -14 \\ y = x + 5 \end{cases}$

10. $\begin{cases} \dfrac{1}{2}x + 2y = -\dfrac{15}{4} \\ 4x = -y \end{cases}$

Solve each system by the addition method.

11. $\begin{cases} x + y = 28 \\ x - y = 12 \end{cases}$

12. $\begin{cases} 4x - 6y = 7 \\ -2x + 3y = 0 \end{cases}$

Solve each system using the substitution method or the addition method.

13. $\begin{cases} 3x + y = 7 \\ 4x + 3y = 1 \end{cases}$

14. $\begin{cases} 3(2x + y) = 4x + 20 \\ x - 2y = 3 \end{cases}$

15. $\begin{cases} \dfrac{x - 3}{2} = \dfrac{2 - y}{4} \\ \dfrac{7 - 2x}{3} = \dfrac{y}{2} \end{cases}$

16. $\begin{cases} 8x - 4y = 12 \\ y = 2x - 3 \end{cases}$

17. $\begin{cases} 0.01x - 0.06y = -0.23 \\ 0.2x + 0.4y = 0.2 \end{cases}$

18. $\begin{cases} x - \dfrac{2}{3}y = 3 \\ -2x + 3y = 10 \end{cases}$

Solve each problem by writing and using a system of linear equations.

19. Two numbers have a sum of 124 and a difference of 32. Find the numbers.

20. Find the amount of a 12% saline solution a lab assistant should add to 80 cc (cubic centimeters) of a 22% saline solution in order to have a 16% solution.

21. Although the number of farms in the United States is still decreasing, small farms are making a comeback. Texas and Missouri are the states with the most number of farms. Texas has 116 thousand more farms than Missouri and the total number of farms for these two states is 336 thousand. Find the number of farms for each state.

22. Two hikers start at opposite ends of the St. Tammany Trails and walk toward each other. The trail is 36 miles long and they meet in 4 hours. If one hiker is twice as fast as the other, find both hiking speeds.

Graph each inequality.

23. $x - y \geq -2$

24. $y > -4x$

25. $2x - 3y > -6$

Graph the solutions of the following systems of linear inequalities.

26. $\begin{cases} y + 2x \leq 4 \\ y \geq 2 \end{cases}$

27. $\begin{cases} 2y - x \geq 1 \\ x + y \geq -4 \end{cases}$

CHAPTER 4 CUMULATIVE REVIEW

1. Insert $<$, $>$, or $=$ in the space between the paired numbers to make each statement true.
   a. $-1 \quad 0$ b. $7 \quad \frac{14}{2}$ c. $-5 \quad -6$

2. Evaluate.
   a. 5^2 b. 2^5

3. Name the property or properties illustrated by each true statement.
   a. $3 \cdot y = y \cdot 3$
   b. $(x + 7) + 9 = x + (7 + 9)$
   c. $(b + 0) + 3 = b + 3$
   d. $0.2 \cdot (z \cdot 5) = 0.2 \cdot (5 \cdot z)$
   e. $-2 \cdot \left(-\frac{1}{2}\right) = 1$
   f. $-2 + 2 = 0$
   g. $-6 \cdot (y \cdot 2) = (-6 \cdot 2) \cdot y$

4. Evaluate $y^2 - 3x$ for $x = 8$ and $y = 5$.

5. Subtract $4x - 2$ from $2x - 3$.

6. Simplify: $7 - 12 + (-5) - 2 + (-2)$

7. Solve $5t - 5 = 6t + 2$ for t.

8. Evaluate $2y^2 - x^2$ for $x = -7$ and $y = -3$.

9. Solve: $\frac{5}{2}x = 15$

10. Simplify: $0.4y - 6.7 + y - 0.3 - 2.6y$

11. Solve: $\frac{x}{2} - 1 = \frac{2}{3}x - 3$

12. Solve: $7(x - 2) - 6(x + 1) = 20$

13. Twice the sum of a number and 4 is the same as four times the number, decreased 12. Find the number.

14. Solve: $5(y - 5) = 5y + 10$

15. Solve $y = mx + b$ for x.

16. Five times the sum of a number and -1 is the same as 6 times the number. Find the number.

17. Solve $-2x \leq -4$. Write the solution set in interval notation.

18. Solve $P = a + b + c$ for b.

19. Graph $x = -2y$ by plotting intercepts.

20. Solve $3x + 7 \geq x - 9$. Write the solution set in interval notation.

21. Find the slope of the line through $(-1, 5)$ and $(2, -3)$.

22. Complete the table of values for $x - 3y = 3$

x	y
	-1
3	
	2

23. Find the slope of the line whose equation is $y = \frac{3}{4}x + 6$.

24. Find the slope of the line parallel to the line passing through $(-1, 3)$ and $(2, -8)$.

25. Find the slope and the y-intercept of the line whose equation is $3x - 4y = 4$.

26. Find the slope and y-intercept of the line whose equation is $y = 7x$.

27. Find an equation of the line passing through $(-1, 5)$ with slope -2. Write the equation in standard form: $Ax + By = C$.

28. Determine whether the lines are parallel, perpendicular or neither.
$$y = 4x - 5$$
$$-4x + y = 7$$

29. Find an equation of the vertical line through $(-1, 5)$.

30. Write an equation of the line with slope -5, through $(-2, 3)$.

31. Find the domain and the range of the relation $\{(0, 2), (3, 3), (-1, 0), (3, -2)\}$.

32. If $f(x) = 5x^2 - 6$, find $f(0)$ and $f(-2)$.

33. Which of the following relations are also functions?
   a. $\{(-1, 1), (2, 3), (7, 3), (8, 6)\}$
   b. $\{(0, -2), (1, 5), (0, 3), (7, 7)\}$

34. Determine which graph(s) are graphs of functions.
   a. b. c.

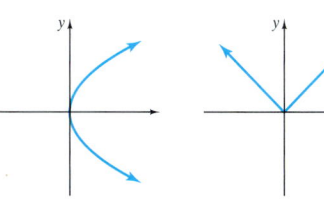

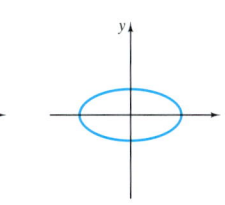

35. Without graphing, determine the number of solutions of the system.
$$\begin{cases} 3x - y = 4 \\ x + 2y = 8 \end{cases}$$

36. Determine whether any ordered pairs satisfy the given system.

$$\begin{cases} 2x - y = 6 \\ 3x + 2y = -5 \end{cases}$$

a. $(1, -4)$ **b.** $(0, 6)$ **c.** $(3, 0)$

Solve each system.

37. $\begin{cases} x + 2y = 7 \\ 2x + 2y = 13 \end{cases}$

38. $\begin{cases} 3x - 4y = 10 \\ y = 2x \end{cases}$

39. $\begin{cases} x + y = 7 \\ x - y = 5 \end{cases}$

40. $\begin{cases} x = 5y - 3 \\ x = 8y + 4 \end{cases}$

41. Find two numbers whose sum is 37 and whose difference is 21.

42. Graph: $x > 1$

43. Graph: $2x - y \geq 3$

44. Graph the solution of the system: $\begin{cases} 2x + 3y < 6 \\ y < 2 \end{cases}$

CHAPTER

5 Exponents and Polynomials

5.1 Exponents

5.2 Adding and Subtracting Polynomials

5.3 Multiplying Polynomials

5.4 Special Products

Integrated Review—Exponents and Operations on Polynomials

5.5 Negative Exponents and Scientific Notation

5.6 Dividing Polynomials

Recall from Chapter 1 that an exponent is a shorthand notation for repeated factors. This chapter explores additional concepts about exponents and exponential expressions. An especially useful type of exponential expression is a polynomial. Polynomials model many real-world phenomena. This chapter will focus on operations on polynomials.

A popular use of the Internet is the World Wide Web. The World Wide Web was invented in 1989–1990 as an environment originally by which scientists could share information. It has grown into a medium containing text, graphics, audio, animation, and video. In Section 5.5, Exercises 91 and 92, you will have the opportunity to estimate the number of visitors to the most popular Web sites.

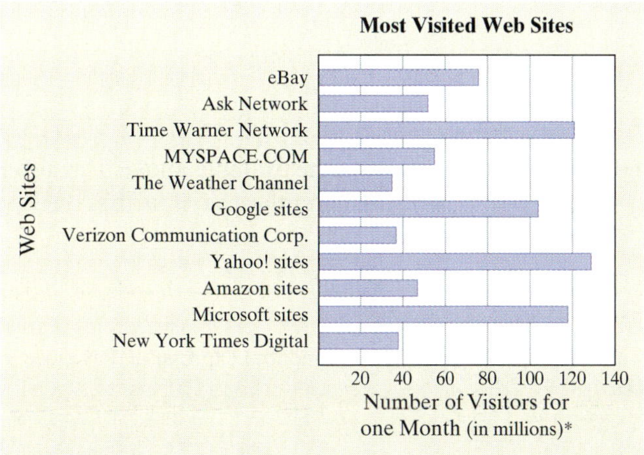

Most Visited Web Sites
Web Sites
eBay
Ask Network
Time Warner Network
MYSPACE.COM
The Weather Channel
Google sites
Verizon Communication Corp.
Yahoo! sites
Amazon sites
Microsoft sites
New York Times Digital
20 40 60 80 100 120 140
Number of Visitors for one Month (in millions)*

* June, 2006 (*Source*: comScore Media Metrix, Inc.)

5.1 EXPONENTS

OBJECTIVES

1. Evaluate exponential expressions.
2. Use the product rule for exponents.
3. Use the power rule for exponents.
4. Use the power rules for products and quotients.
5. Use the quotient rule for exponents, and define a number raised to the 0 power.
6. Decide which rule(s) to use to simplify an expression.

OBJECTIVE 1 ▶ Evaluating exponential expressions. As we reviewed in Section 1.4, an exponent is a shorthand notation for repeated factors. For example, $2 \cdot 2 \cdot 2 \cdot 2 \cdot 2$ can be written as 2^5. The expression 2^5 is called an **exponential expression**. It is also called the fifth **power** of 2, or we say that 2 is **raised** to the fifth power.

$$5^6 = \underbrace{5 \cdot 5 \cdot 5 \cdot 5 \cdot 5 \cdot 5}_{\text{6 factors; each factor is 5}} \quad \text{and} \quad (-3)^4 = \underbrace{(-3) \cdot (-3) \cdot (-3) \cdot (-3)}_{\text{4 factors; each factor is } -3}$$

The **base** of an exponential expression is the repeated factor. The **exponent** is the number of times that the base is used as a factor.

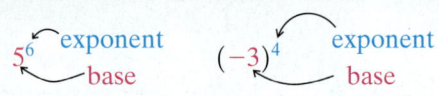

$$5^6 \quad \text{exponent} \atop \text{base} \qquad (-3)^4 \quad \text{exponent} \atop \text{base}$$

EXAMPLE 1 Evaluate each expression.

a. 2^3 **b.** 3^1 **c.** $(-4)^2$ **d.** -4^2 **e.** $\left(\dfrac{1}{2}\right)^4$ **f.** $(0.5)^3$ **g.** $4 \cdot 3^2$

Solution

a. $2^3 = 2 \cdot 2 \cdot 2 = 8$

b. To raise 3 to the first power means to use 3 as a factor only once. Therefore, $3^1 = 3$. Also, when no exponent is shown, the exponent is assumed to be 1.

c. $(-4)^2 = (-4)(-4) = 16$ **d.** $-4^2 = -(4 \cdot 4) = -16$

e. $\left(\dfrac{1}{2}\right)^4 = \dfrac{1}{2} \cdot \dfrac{1}{2} \cdot \dfrac{1}{2} \cdot \dfrac{1}{2} = \dfrac{1}{16}$ **f.** $(0.5)^3 = (0.5)(0.5)(0.5) = 0.125$

g. $4 \cdot 3^2 = 4 \cdot 9 = 36$

PRACTICE 1 Evaluate each expression.

a. 3^3 **b.** 4^1 **c.** $(-8)^2$ **d.** -8^2

e. $\left(\dfrac{3}{4}\right)^3$ **f.** $(0.3)^4$ **g.** $3 \cdot 5^2$

Notice how similar -4^2 is to $(-4)^2$ in the example above. The difference between the two is the parentheses. In $(-4)^2$, the parentheses tell us that the base, or repeated factor, is -4. In -4^2, only 4 is the base.

▶ **Helpful Hint**

Be careful when identifying the base of an exponential expression. Pay close attention to the use of parentheses.

$(-3)^2$ -3^2 $2 \cdot 3^2$

The base is -3.   The base is 3.   The base is 3.

$(-3)^2 = (-3)(-3) = 9$ $-3^2 = -(3 \cdot 3) = -9$ $2 \cdot 3^2 = 2 \cdot 3 \cdot 3 = 18$

An exponent has the same meaning whether the base is a number or a variable. If x is a real number and n is a positive integer, then x^n is the product of n factors, each of which is x.

$$x^n = \underbrace{x \cdot x \cdot x \cdot x \cdot x \cdot \ldots \cdot x}_{n \text{ factors of } x}$$

EXAMPLE 2 Evaluate each expression for the given value of x.

a. $2x^3$; x is 5   **b.** $\dfrac{9}{x^2}$; x is -3

Solution **a.** If x is 5, $2x^3 = 2 \cdot (5)^3$
$= 2 \cdot (5 \cdot 5 \cdot 5)$
$= 2 \cdot 125$
$= 250$

b. If x is -3, $\dfrac{9}{x^2} = \dfrac{9}{(-3)^2}$
$= \dfrac{9}{(-3)(-3)}$
$= \dfrac{9}{9}$
$= 1$

PRACTICE 2 Evaluate each expression for the given value of x.

a. $3x^4$; x is 3   **b.** $\dfrac{6}{x^2}$; x is -4

OBJECTIVE 2 ▶ Using the product rule. Exponential expressions can be multiplied, divided, added, subtracted, and themselves raised to powers. By our definition of an exponent,

$5^4 \cdot 5^3 = \underbrace{(5 \cdot 5 \cdot 5 \cdot 5)}_{\text{4 factors of 5}} \cdot \underbrace{(5 \cdot 5 \cdot 5)}_{\text{3 factors of 5}}$
$= \underbrace{5 \cdot 5 \cdot 5 \cdot 5 \cdot 5 \cdot 5 \cdot 5}_{\text{7 factors of 5}}$
$= 5^7$

Also,

$x^2 \cdot x^3 = (x \cdot x) \cdot (x \cdot x \cdot x)$
$= x \cdot x \cdot x \cdot x \cdot x$
$= x^5$

In both cases, notice that the result is exactly the same if the exponents are added.

$5^4 \cdot 5^3 = 5^{4+3} = 5^7$ and $x^2 \cdot x^3 = x^{2+3} = x^5$

This suggests the following rule.

Product Rule for Exponents

If m and n are positive integers and a is a real number, then

$a^m \cdot a^n = a^{m+n}$ ← Add exponents.
    ↑
    Keep common base.

For example, $3^5 \cdot 3^7 = 3^{5+7} = 3^{12}$ ← Add exponents.
    ↑
    Keep common base.

In other words, to multiply two exponential expressions with a **common base**, keep the base and add the exponents. We call this simplifying the exponential expression.

EXAMPLE 3 Use the product rule to simplify.

a. $4^2 \cdot 4^5$ **b.** $x^4 \cdot x^6$ **c.** $y^3 \cdot y$ **d.** $y^3 \cdot y^2 \cdot y^7$ **e.** $(-5)^7 \cdot (-5)^8$ **f.** $a^2 \cdot b^2$

Solution

a. $4^2 \cdot 4^5 = 4^{2+5} = 4^7$ ← Add exponents.
    ↑
    Keep common base.

b. $x^4 \cdot x^6 = x^{4+6} = x^{10}$

c. $y^3 \cdot y = y^3 \cdot y^1$
$= y^{3+1}$
$= y^4$

d. $y^3 \cdot y^2 \cdot y^7 = y^{3+2+7} = y^{12}$

e. $(-5)^7 \cdot (-5)^8 = (-5)^{7+8} = (-5)^{15}$

f. $a^2 \cdot b^2$ Cannot be simplified because a and b are different bases.

> **Helpful Hint**
> Don't forget that if no exponent is written, it is assumed to be 1.

PRACTICE 3 Use the product rule to simplify.

a. $3^4 \cdot 3^6$ **b.** $y^3 \cdot y^2$
c. $z \cdot z^4$ **d.** $x^3 \cdot x^2 \cdot x^6$
e. $(-2)^5 \cdot (-2)^3$ **f.** $b^3 \cdot t^5$

Concept Check ✓

Where possible, use the product rule to simplify the expression.

a. $z^2 \cdot z^{14}$ **b.** $x^2 \cdot y^{14}$ **c.** $9^8 \cdot 9^3$ **d.** $9^8 \cdot 2^7$

EXAMPLE 4 Use the product rule to simplify $(2x^2)(-3x^5)$.

Solution Recall that $2x^2$ means $2 \cdot x^2$ and $-3x^5$ means $-3 \cdot x^5$.

$(2x^2)(-3x^5) = 2 \cdot x^2 \cdot -3 \cdot x^5$ Remove parentheses.
$= 2 \cdot -3 \cdot x^2 \cdot x^5$ Group factors with common bases.
$= -6x^7$ Simplify.

PRACTICE 4 Use the product rule to simplify $(-5y^3)(-3y^4)$.

EXAMPLE 5 Simplify.

a. $(x^2y)(x^3y^2)$ **b.** $(-a^7b^4)(3ab^9)$

Solution

a. $(x^2y)(x^3y^2) = (x^2 \cdot x^3) \cdot (y^1 \cdot y^2)$ Group like bases and write y as y^1.
$= x^5 \cdot y^3$ or x^5y^3 Multiply.

b. $(-a^7b^4)(3ab^9) = (-1 \cdot 3) \cdot (a^7 \cdot a^1) \cdot (b^4 \cdot b^9)$
$= -3a^8b^{13}$

PRACTICE 5 Simplify.

a. $(y^7z^3)(y^5z)$ **b.** $(-m^4n^4)(7mn^{10})$

> **Helpful Hint**
> These examples will remind you of the difference between adding and multiplying terms.
>
> **Addition**
>
> $5x^3 + 3x^3 = (5 + 3)x^3 = 8x^3$ By the distributive property.
> $7x + 4x^2 = 7x + 4x^2$ Cannot be combined.
>
> **Multiplication**
>
> $(5x^3)(3x^3) = 5 \cdot 3 \cdot x^3 \cdot x^3 = 15x^{3+3} = 15x^6$ By the product rule.
> $(7x)(4x^2) = 7 \cdot 4 \cdot x \cdot x^2 = 28x^{1+2} = 28x^3$ By the product rule.

Answers to Concept Check:
a. z^{16} **b.** cannot be simplified
c. 9^{11} **d.** cannot be simplified

OBJECTIVE 3 ▶ Using the power rule. Exponential expressions can themselves be raised to powers. Let's try to discover a rule that simplifies an expression like $(x^2)^3$. By definition,

$$(x^2)^3 = \underbrace{(x^2)(x^2)(x^2)}_{3 \text{ factors of } x^2}$$

which can be simplified by the product rule for exponents.

$$(x^2)^3 = (x^2)(x^2)(x^2) = x^{2+2+2} = x^6$$

Notice that the result is exactly the same if we multiply the exponents.

$$(x^2)^3 = x^{2 \cdot 3} = x^6$$

The following property states this result.

Power Rule for Exponents

If m and n are positive integers and a is a real number, then

$$(a^m)^n = a^{mn} \quad \leftarrow \text{Multiply exponents.}$$
$$\uparrow$$
$$\text{Keep common base.}$$

For example, $(7^2)^5 = 7^{2 \cdot 5} = 7^{10}$ ← Multiply exponents.
$\quad\quad\quad\quad\quad\quad\quad\quad\uparrow$
$\quad\quad\quad\quad\quad\quad\quad$ Keep common base.

To raise a power to a power, keep the base and multiply the exponents.

EXAMPLE 6 Use the power rule to simplify.

a. $(x^2)^5$ **b.** $(y^8)^2$ **c.** $[(-5)^3]^7$

Solution **a.** $(x^2)^5 = x^{2 \cdot 5} = x^{10}$ **b.** $(y^8)^2 = y^{8 \cdot 2} = y^{16}$ **c.** $[(-5)^3]^7 = (-5)^{21}$

PRACTICE 6 Use the power rule to simplify.

a. $(x^4)^3$ **b.** $(z^3)^7$ **c.** $[(-2)^3]^5$

▶ **Helpful Hint**

Take a moment to make sure that you understand when to apply the product rule and when to apply the power rule.

Product Rule → Add Exponents	Power Rule → Multiply Exponents
$x^5 \cdot x^7 = x^{5+7} = x^{12}$	$(x^5)^7 = x^{5 \cdot 7} = x^{35}$
$y^6 \cdot y^2 = y^{6+2} = y^8$	$(y^6)^2 = y^{6 \cdot 2} = y^{12}$

OBJECTIVE 4 ▶ Using the power rules for products and quotients. When the base of an exponential expression is a product, the definition of x^n still applies. To simplify $(xy)^3$, for example,

$$(xy)^3 = (xy)(xy)(xy) \quad (xy)^3 \text{ means 3 factors of } (xy).$$
$$= x \cdot x \cdot x \cdot y \cdot y \cdot y \quad \text{Group factors with common bases.}$$
$$= x^3 y^3 \quad \text{Simplify.}$$

Notice that to simplify the expression $(xy)^3$, we raise each factor within the parentheses to a power of 3.

$$(xy)^3 = x^3y^3$$

In general, we have the following rule.

> **Power of a Product Rule**
> If n is a positive integer and a and b are real numbers, then
> $$(ab)^n = a^n b^n$$

For example, $(3x)^5 = 3^5 x^5$.

In other words, to raise a product to a power, we raise each factor to the power.

EXAMPLE 7 Simplify each expression.

a. $(st)^4$ **b.** $(2a)^3$ **c.** $\left(\dfrac{1}{3}mn^3\right)^2$ **d.** $(-5x^2y^3z)^2$

Solution

a. $(st)^4 = s^4 \cdot t^4 = s^4 t^4$ Use the power of a product rule.

b. $(2a)^3 = 2^3 \cdot a^3 = 8a^3$ Use the power of a product rule.

c. $\left(\dfrac{1}{3}mn^3\right)^2 = \left(\dfrac{1}{3}\right)^2 \cdot (m)^2 \cdot (n^3)^2 = \dfrac{1}{9}m^2 n^6$ Use the power of a product rule.

d. $(-5x^2y^3z)^2 = (-5)^2 \cdot (x^2)^2 \cdot (y^3)^2 \cdot (z^1)^2$ Use the power rule for exponents.
$= 25x^4 y^6 z^2$

PRACTICE 7 Simplify each expression.

a. $(pr)^5$ **b.** $(6b)^2$ **c.** $\left(\dfrac{1}{4}x^2y\right)^3$ **d.** $(-3a^3b^4c)^4$

Let's see what happens when we raise a quotient to a power. To simplify $\left(\dfrac{x}{y}\right)^3$, for example,

$$\left(\dfrac{x}{y}\right)^3 = \left(\dfrac{x}{y}\right)\left(\dfrac{x}{y}\right)\left(\dfrac{x}{y}\right) \quad \left(\dfrac{x}{y}\right)^3 \text{ means 3 factors of } \left(\dfrac{x}{y}\right)$$

$$= \dfrac{x \cdot x \cdot x}{y \cdot y \cdot y} \quad \text{Multiply fractions.}$$

$$= \dfrac{x^3}{y^3} \quad \text{Simplify.}$$

Notice that to simplify the expression $\left(\dfrac{x}{y}\right)^3$, we raise both the numerator and the denominator to a power of 3.

$$\left(\dfrac{x}{y}\right)^3 = \dfrac{x^3}{y^3}$$

In general, we have the following.

> **Power of a Quotient Rule**
> If n is a positive integer and a and c are real numbers, then
> $$\left(\dfrac{a}{c}\right)^n = \dfrac{a^n}{c^n}, \quad c \neq 0$$

For example, $\left(\dfrac{y}{7}\right)^4 = \dfrac{y^4}{7^4}$.

In other words, to raise a quotient to a power, we raise both the numerator and the denominator to the power.

EXAMPLE 8 Simplify each expression.

a. $\left(\dfrac{m}{n}\right)^7$ b. $\left(\dfrac{x^3}{3y^5}\right)^4$

Solution

a. $\left(\dfrac{m}{n}\right)^7 = \dfrac{m^7}{n^7}, n \neq 0$ Use the power of a quotient rule.

b. $\left(\dfrac{x^3}{3y^5}\right)^4 = \dfrac{(x^3)^4}{3^4 \cdot (y^5)^4}, y \neq 0$ Use the power of a product or quotient rule.

$= \dfrac{x^{12}}{81y^{20}}$ Use the power rule for exponents.

PRACTICE 8 Simplify each expression.

a. $\left(\dfrac{x}{y^2}\right)^5$ b. $\left(\dfrac{2a^4}{b^3}\right)^5$

OBJECTIVE 5 ▶ **Using the quotient rule and defining the zero exponent.** Another pattern for simplifying exponential expressions involves quotients.

To simplify an expression like $\dfrac{x^5}{x^3}$, in which the numerator and the denominator have a common base, we can apply the fundamental principle of fractions and divide the numerator and the denominator by the common base factors. Assume for the remainder of this section that denominators are not 0.

$$\dfrac{x^5}{x^3} = \dfrac{x \cdot x \cdot x \cdot x \cdot x}{x \cdot x \cdot x}$$
$$= \dfrac{x \cdot x \cdot x \cdot x \cdot x}{x \cdot x \cdot x}$$
$$= x \cdot x$$
$$= x^2$$

Notice that the result is exactly the same if we subtract exponents of the common bases.

$$\dfrac{x^5}{x^3} = x^{5-3} = x^2$$

The quotient rule for exponents states this result in a general way.

Quotient Rule for Exponents

If m and n are positive integers and a is a real number, then

$$\dfrac{a^m}{a^n} = a^{m-n}$$

as long as a is not 0.

For example, $\dfrac{x^6}{x^2} = x^{6-2} = x^4$.

In other words, to divide one exponential expression by another with a common base, keep the base and subtract exponents.

EXAMPLE 9 Simplify each quotient.

a. $\dfrac{x^5}{x^2}$ b. $\dfrac{4^7}{4^3}$ c. $\dfrac{(-3)^5}{(-3)^2}$ d. $\dfrac{s^2}{t^3}$ e. $\dfrac{2x^5y^2}{xy}$

Solution

a. $\dfrac{x^5}{x^2} = x^{5-2} = x^3$ Use the quotient rule.

b. $\dfrac{4^7}{4^3} = 4^{7-3} = 4^4 = 256$ Use the quotient rule.

c. $\dfrac{(-3)^5}{(-3)^2} = (-3)^3 = -27$

d. $\dfrac{s^2}{t^3}$ Cannot be simplified because s and t are different bases.

e. Begin by grouping common bases.

$$\dfrac{2x^5y^2}{xy} = 2 \cdot \dfrac{x^5}{x^1} \cdot \dfrac{y^2}{y^1}$$
$$= 2 \cdot (x^{5-1}) \cdot (y^{2-1}) \quad \text{Use the quotient rule.}$$
$$= 2x^4y^1 \quad \text{or} \quad 2x^4y$$

PRACTICE 9 Simplify each quotient.

a. $\dfrac{z^8}{z^4}$ b. $\dfrac{(-5)^5}{(-5)^3}$ c. $\dfrac{8^8}{8^6}$ d. $\dfrac{q^5}{t^2}$ e. $\dfrac{6x^3y^7}{xy^5}$

Concept Check ✓

Suppose you are simplifying each expression. Tell whether you would *add* the exponents, *subtract* the exponents, *multiply* the exponents, *divide* the exponents, or *none of these*.

a. $(x^{63})^{21}$ b. $\dfrac{y^{15}}{y^3}$ c. $z^{16} + z^8$ d. $w^{45} \cdot w^9$

Let's now give meaning to an expression such as x^0. To do so, we will simplify $\dfrac{x^3}{x^3}$ in two ways and compare the results.

$$\dfrac{x^3}{x^3} = x^{3-3} = x^0 \quad \text{Apply the quotient rule.}$$

$$\dfrac{x^3}{x^3} = \dfrac{x \cdot x \cdot x}{x \cdot x \cdot x} = 1 \quad \text{Apply the fundamental principle for fractions.}$$

Since $\dfrac{x^3}{x^3} = x^0$ and $\dfrac{x^3}{x^3} = 1$, we define that $x^0 = 1$ as long as x is not 0.

Zero Exponent

$a^0 = 1$, as long as a is not 0.

In other words, any base raised to the 0 power is 1, as long as the base is not 0.

Answers to Concept Check:
a. multiply   b. subtract
c. none of these   d. add

EXAMPLE 10 Simplify each expression.

a. 3^0 b. $(ab)^0$ c. $(-5)^0$ d. -5^0 e. $\left(\dfrac{3}{100}\right)^0$

Solution

a. $3^0 = 1$

b. Assume that neither a nor b is zero.

$$(ab)^0 = a^0 \cdot b^0 = 1 \cdot 1 = 1$$

c. $(-5)^0 = 1$

d. $-5^0 = -1 \cdot 5^0 = -1 \cdot 1 = -1$

e. $\left(\dfrac{3}{100}\right)^0 = 1$

PRACTICE 10 Simplify the following expressions.

a. -3^0 b. $(-3)^0$ c. 8^0 d. $(0.2)^0$ e. $(xz)^0$

OBJECTIVE 6 ▶ Deciding which rule to use. Let's practice deciding which rule(s) to use to simplify. We will continue this discussion with more examples in Section 5.5.

EXAMPLE 11 Simplify each expression.

a. $\left(\dfrac{st}{2}\right)^4$ b. $(9y^5z^7)^2$ c. $\left(\dfrac{-5x^2}{y^3}\right)^2$

Solution

a. This is a quotient raised to a power, so we use the power of a quotient rule.

$$\left(\dfrac{st}{2}\right)^4 = \dfrac{s^4 t^4}{2^4} = \dfrac{s^4 t^4}{16}$$

b. This is a product raised to a power, so we use the power of a product rule.

$$(9y^5z^7)^2 = 9^2(y^5)^2(z^7)^2 = 81y^{10}z^{14}$$

c. Use the power of a product or quotient rule; then use the power rule for exponents.

$$\left(\dfrac{-5x^2}{y^3}\right)^2 = \dfrac{(-5)^2(x^2)^2}{(y^3)^2} = \dfrac{25x^4}{y^6}$$

PRACTICE 11 Simplify each expression.

a. $\left(\dfrac{5}{xz}\right)^3$ b. $(2z^8x^5)^4$ c. $\left(\dfrac{-3x^3}{y^4}\right)^3$

310 CHAPTER 5 Exponents and Polynomials

VOCABULARY & READINESS CHECK

Use the choices below to fill in each blank. Some choices may be used more than once.

   0   base   add
   1   exponent   multiply

1. Repeated multiplication of the same factor can be written using a(n) _____.
2. In 5^2, the 2 is called the _____ and the 5 is called the _____.
3. To simplify $x^2 \cdot x^7$, keep the base and _____ the exponents.
4. To simplify $(x^3)^6$, keep the base and _____ the exponents.
5. The understood exponent on the term y is _____.
6. If $x^\square = 1$, the exponent is _____.

State the bases and the exponents for each of the following expressions.

7. 3^2
8. $(-3)^6$
9. -4^2
10. $5 \cdot 3^4$
11. $5x^2$
12. $(5x)^2$

5.1 EXERCISE SET

Evaluate each expression. See Example 1.

1. 7^2
2. -3^2
3. $(-5)^1$
4. $(-3)^2$
5. -2^4
6. -4^3
7. $(-2)^4$
8. $(-4)^3$
9. $(0.1)^5$
10. $(0.2)^5$
11. $\left(\dfrac{1}{3}\right)^4$
12. $\left(-\dfrac{1}{9}\right)^2$
13. $7 \cdot 2^5$
14. $9 \cdot 1^7$
15. $-2 \cdot 5^3$
16. $-4 \cdot 3^3$
17. Explain why $(-5)^4 = 625$, while $-5^4 = -625$.
18. Explain why $5 \cdot 4^2 = 80$, while $(5 \cdot 4)^2 = 400$.

Evaluate each expression given the replacement values for x. See Example 2.

19. $x^2; x = -2$
20. $x^3; x = -2$
21. $5x^3; x = 3$
22. $4x^2; x = -1$
23. $2xy^2; x = 3$ and $y = 5$
24. $-4x^2y^3; x = 2$ and $y = -1$
25. $\dfrac{2z^4}{5}; z = -2$
26. $\dfrac{10}{3y^3}; y = 5$

Use the product rule to simplify each expression. Write the results using exponents. See Examples 3 through 5.

27. $x^2 \cdot x^5$
28. $y^2 \cdot y$
29. $(-3)^3 \cdot (-3)^9$
30. $(-5)^7 \cdot (-5)^6$
31. $(5y^4)(3y)$
32. $(-2z^3)(-2z^2)$
33. $(x^9y)(x^{10}y^5)$
34. $(a^2b)(a^{13}b^{17})$
35. $(-8mn^6)(9m^2n^2)$
36. $(-7a^3b^3)(7a^{19}b)$
37. $(4z^{10})(-6z^7)(z^3)$
38. $(12x^5)(-x^6)(x^4)$

△ 39. The rectangle below has width $4x^2$ feet and length $5x^3$ feet. Find its area as an expression in x. ($A = l \cdot w$)

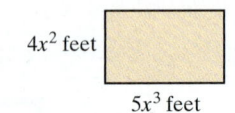

    $4x^2$ feet
    $5x^3$ feet

△ 40. The parallelogram below has base length $9y^7$ meters and height $2y^{10}$ meters. Find its area as an expression in y. ($A = b \cdot h$)

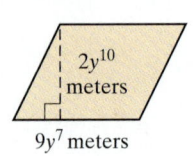

    $2y^{10}$ meters
    $9y^7$ meters

MIXED PRACTICE

Use the power rule and the power of a product or quotient rule to simplify each expression. See Examples 6 through 8.

41. $(x^9)^4$
42. $(y^7)^5$

43. $(pq)^8$
44. $(ab)^6$
45. $(2a^5)^3$
46. $(4x^6)^2$
47. $(x^2y^3)^5$
48. $(a^4b)^7$
49. $(-7a^2b^5c)^2$
50. $(-3x^7yz^2)^3$
51. $\left(\dfrac{r}{s}\right)^9$
52. $\left(\dfrac{q}{t}\right)^{11}$
53. $\left(\dfrac{mp}{n}\right)^5$
54. $\left(\dfrac{xy}{7}\right)^2$
55. $\left(\dfrac{-2xz}{y^5}\right)^2$
56. $\left(\dfrac{xy^4}{-3z^3}\right)^3$

△ 57. The square shown has sides of length $8z^5$ decimeters. Find its area. $(A = s^2)$

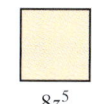

$8z^5$ decimeters

△ 58. Given the circle below with radius $5y$ centimeters, find its area. Do not approximate π. $(A = \pi r^2)$

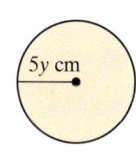

$5y$ cm

△ 59. The vault below is in the shape of a cube. If each side is $3y^4$ feet, find its volume. $(V = s^3)$

$3y^4$ feet
$3y^4$ feet
$3y^4$ feet

△ 60. The silo shown is in the shape of a cylinder. If its radius is $4x$ meters and its height is $5x^3$ meters, find its volume. Do not approximate π. $(V = \pi r^2 h)$

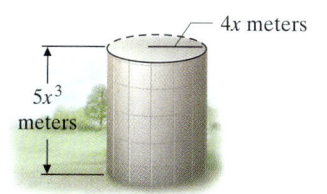

$4x$ meters
$5x^3$ meters

Use the quotient rule and simplify each expression. See Example 9.

61. $\dfrac{x^3}{x}$
62. $\dfrac{y^{10}}{y^9}$
63. $\dfrac{(-4)^6}{(-4)^3}$
64. $\dfrac{(-6)^{13}}{(-6)^{11}}$
65. $\dfrac{p^7q^{20}}{pq^{15}}$
66. $\dfrac{x^8y^6}{xy^5}$
67. $\dfrac{7x^2y^6}{14x^2y^3}$
68. $\dfrac{9a^4b^7}{27ab^2}$

Simplify each expression. See Example 10.

69. 7^0
70. 23^0
71. $(2x)^0$
72. $(4y)^0$
73. $-7x^0$
74. $-2x^0$
75. $5^0 + y^0$
76. $-3^0 + 4^0$

MIXED PRACTICE

Simplify each expression. See Examples 1 through 10.

77. -9^2
78. $(-9)^2$
79. $\left(\dfrac{1}{4}\right)^3$
80. $\left(\dfrac{2}{3}\right)^3$
81. $\left(\dfrac{9}{qr}\right)^2$
82. $\left(\dfrac{pt}{3}\right)^3$
83. a^2a^3a
84. $x^2x^{15}x$
85. $(2x^3)(-8x^4)$
86. $(3y^4)(-5y)$
87. $(a^7b^{12})(a^4b^8)$
88. $(y^2z^2)(y^{15}z^{13})$
89. $(-2mn^6)(-13m^8n)$
90. $(-3s^5t)(-7st^{10})$
91. $(z^4)^{10}$
92. $(t^5)^{11}$
93. $(-6xyz^3)^2$
94. $(-3xy^2a^3)^3$
95. $\dfrac{3x^5}{x^4}$
96. $\dfrac{5x^9}{x^3}$
97. $(9xy)^2$
98. $(2ab)^5$
99. $2^0 + 2^5$
100. $7^2 - 7^0$
101. $\left(\dfrac{3y^5}{6x^4}\right)^3$
102. $\left(\dfrac{2ab}{6yz}\right)^4$
103. $\dfrac{2x^3y^2z}{xyz}$
104. $\dfrac{x^{12}y^{13}}{x^5y^7}$

REVIEW AND PREVIEW

Simplify each expression by combining any like terms. Use the distributive property to remove any parentheses. See Section 2.1.

105. $y - 10 + y$
106. $-6z + 20 - 3z$
107. $7x + 2 - 8x - 6$
108. $10y - 14 - y - 14$
109. $2(x - 5) + 3(5 - x)$
110. $-3(w + 7) + 5(w + 1)$

CONCEPT EXTENSIONS

Solve. See the Concept Checks in this section. For Exercises 111 through 114, match the expression with the operation needed to simplify each. A letter may be used more than once and a letter may not be used at all.

111. $(x^{14})^{23}$
112. $x^{14} \cdot x^{23}$
113. $x^{14} + x^{23}$
114. $\dfrac{x^{35}}{x^{17}}$

a. Add the exponents
b. Subtract the exponents
c. Multiply the exponents
d. Divide the exponents
e. None of these

Fill in the boxes so that each statement is true. (More than one answer is possible for each exercise.)

115. $x^{\square} \cdot x^{\square} = x^{12}$
116. $(x^{\square})^{\square} = x^{20}$
117. $\dfrac{y^{\square}}{y^{\square}} = y^7$
118. $(y^{\square})^{\square} \cdot (y^{\square})^{\square} = y^{30}$

△ 119. The formula $V = x^3$ can be used to find the volume V of a cube with side length x. Find the volume of a cube with side length 7 meters. (Volume is measured in cubic units.)

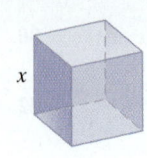

△ 120. The formula $S = 6x^2$ can be used to find the surface area S of a cube with side length x. Find the surface area of a cube with side length 5 meters. (Surface area is measured in square units.)

△ 121. To find the amount of water that a swimming pool in the shape of a cube can hold, do we use the formula for volume of the cube or surface area of the cube? (See Exercises 119 and 120.)

△ 122. To find the amount of material needed to cover an ottoman in the shape of a cube, do we use the formula for volume of the cube or surface area of the cube? (See Exercises 119 and 120.)

123. In your own words, explain why $5^0 = 1$.
124. In your own words, explain when $(-3)^n$ is positive and when it is negative.

Simplify each expression. Assume that variables represent positive integers.

125. $x^{5a} x^{4a}$
126. $b^{9a} b^{4a}$
127. $(a^b)^5$
128. $(2a^{4b})^4$
129. $\dfrac{x^{9a}}{x^{4a}}$
130. $\dfrac{y^{15b}}{y^{6b}}$

131. Suppose you borrow money for 6 months. If the interest rate is compounded monthly, the formula $A = P\left(1 + \dfrac{r}{12}\right)^6$ gives the total amount A to be repaid at the end of 6 months. For a loan of $P = \$1000$ and interest rate of 9% ($r = 0.09$), how much money will you need to pay off the loan?

132. On January 1, 2007, the Federal Reserve discount rate was set at $5\dfrac{1}{4}$%. (*Source:* Federal Reserve Board) The discount rate is the interest rate at which banks can borrow money from the Federal Reserve System. Suppose a bank needs to borrow money from the Federal Reserve System for 3 months. If the interest is compounded monthly, the formula $A = P\left(1 + \dfrac{r}{12}\right)^3$ gives the total amount A to be repaid at the end of 3 months. For a loan of $P = \$500{,}000$ and interest rate of $r = 0.0525$, how much money will the bank repay to the Federal Reserve at the end of 3 months? Round to the nearest dollar.

STUDY SKILLS BUILDER

How Well Do You Know Your Textbook?

The questions below will help determine whether you are familiar with your textbook. For additional information, see Section 1.1 in this text.

1. What does the 💿 icon mean?
2. What does the ＼ icon mean?
3. What does the △ icon mean?
4. Where can you find a review for each chapter? What answers to this review can be found in the back of your text?
5. Each chapter contains an overview of the chapter along with examples. What is this feature called?
6. Each chapter contains a review of vocabulary. What is this feature called?
7. There is a CD in your text. What content is contained on this CD?
8. What is the location of the section that is entirely devoted to study skills?
9. There are practice exercises that are contained in this text. What are they and how can they be used?

5.2 ADDING AND SUBTRACTING POLYNOMIALS

OBJECTIVES

1. Define polynomial, monomial, binomial, trinomial, and degree.
2. Find the value of a polynomial given replacement values for the variables.
3. Simplify a polynomial by combining like terms.
4. Add and subtract polynomials.

OBJECTIVE 1 ▶ Defining polynomial, monomial, binomial, trinomial, and degree. In this section, we introduce a special algebraic expression called a polynomial. Let's first review some definitions presented in Section 2.1.

Recall that a term is a number or the product of a number and variables raised to powers. The terms of the expression $4x^2 + 3x$ are $4x^2$ and $3x$. The terms of the expression $9x^4 - 7x - 1$ are $9x^4$, $-7x$, and -1.

Expression	Terms
$4x^2 + 3x$	$4x^2, 3x$
$9x^4 - 7x - 1$	$9x^4, -7x, -1$
$7y^3$	$7y^3$
5	5

The **numerical coefficient** of a term, or simply the **coefficient,** is the numerical factor of each term. If no numerical factor appears in the term, then the coefficient is understood to be 1. If the term is a number only, it is called a **constant** term or simply a constant.

Term	Coefficient
x^5	1
$3x^2$	3
$-4x$	-4
$-x^2y$	-1
3 (constant)	3

> **Polynomial**
> A **polynomial in x** is a finite sum of terms of the form ax^n, where a is a real number and n is a whole number.

For example,
$$x^5 - 3x^3 + 2x^2 - 5x + 1$$
is a polynomial. Notice that this polynomial is written in **descending powers** of x because the powers of x decrease from left to right. (Recall that the term 1 can be thought of as $1x^0$.)

On the other hand,
$$x^{-5} + 2x - 3$$
is **not** a polynomial because it contains an exponent, -5, that is not a whole number. (We study negative exponents in Section 5.5 of this chapter.)

Some polynomials are given special names.

> **Types of Polynomials**
> A **monomial** is a polynomial with exactly one term.
> A **binomial** is a polynomial with exactly two terms.
> A **trinomial** is a polynomial with exactly three terms.

The following are examples of monomials, binomials, and trinomials. Each of these examples is also a polynomial.

POLYNOMIALS

Monomials	Binomials	Trinomials	None of These
ax^2	$x + y$	$x^2 + 4xy + y^2$	$5x^3 - 6x^2 + 3x - 6$
$-3z$	$3p + 2$	$x^5 + 7x^2 - x$	$-y^5 + y^4 - 3y^3 - y^2 + y$
4	$4x^2 - 7$	$-q^4 + q^3 - 2q$	$x^6 + x^4 - x^3 + 1$

Each term of a polynomial has a **degree**.

> **Degree of a Term**
> The degree of a term is the sum of the exponents on the variables contained in the term.

EXAMPLE 1 Find the degree of each term.

a. $-3x^2$ **b.** $5x^3yz$ **c.** 2

Solution

a. The exponent on x is 2, so the degree of the term is 2.

b. $5x^3yz$ can be written as $5x^3y^1z^1$. The degree of the term is the sum of its exponents, so the degree is $3 + 1 + 1$ or 5.

c. The constant, 2, can be written as $2x^0$ (since $x^0 = 1$). The degree of 2 or $2x^0$ is 0. □

PRACTICE 1 Find the degree of each term.

a. $5y^3$ **b.** $-3a^2b^5c$ **c.** 8

From the preceding, we can say that **the degree of a constant is 0.**
Each polynomial also has a degree.

> **Degree of a Polynomial**
> The degree of a polynomial is the greatest degree of any term of the polynomial.

EXAMPLE 2 Find the degree of each polynomial and tell whether the polynomial is a monomial, binomial, trinomial, or none of these.

a. $-2t^2 + 3t + 6$ **b.** $15x - 10$ **c.** $7x + 3x^3 + 2x^2 - 1$

Solution

a. The degree of the trinomial $-2t^2 + 3t + 6$ is 2, the greatest degree of any of its terms.

b. The degree of the binomial $15x - 10$ or $15x^1 - 10$ is 1.

c. The degree of the polynomial $7x + 3x^3 + 2x^2 - 1$ is 3. □

PRACTICE 2
Find the degree of each polynomial and tell whether the polynomial is a monomial, binomial, trinomial, or none of these.

a. $5b^2 - 3b + 7$ **b.** $7t + 3$

c. $5x^2 + 3x - 6x^3 + 4$

EXAMPLE 3   Complete the table for the polynomial
$$7x^2y - 6xy + x^2 - 3y + 7$$
Use the table to give the degree of the polynomial.

Solution

Term	Numerical Coefficient	Degree of Term
$7x^2y$	7	3
$-6xy$	-6	2
x^2	1	2
$-3y$	-3	1
7	7	0

The degree of the polynomial is 3.

PRACTICE 3
Complete the table for the polynomial $-3x^3y^2 + 4xy^2 - y^2 + 3x - 2$.

Term	Numerical Coefficient	Degree of Term
$-3x^3y^2$		
$4xy^2$		
$-y^2$		
$3x$		
-2		

OBJECTIVE 2 ▶ Evaluating polynomials. Polynomials have different values depending on replacement values for the variables.

EXAMPLE 4   Find the value of the polynomial $3x^2 - 2x + 1$ when $x = -2$.

Solution   Replace x with -2 and simplify.
$$3x^2 - 2x + 1 = 3(-2)^2 - 2(-2) + 1$$
$$= 3(4) + 4 + 1$$
$$= 12 + 4 + 1$$
$$= 17$$

PRACTICE 4
Find the value of the polynomial $2x^2 - 5x + 3$ when $x = -3$.

Many physical phenomena can be modeled by polynomials.

EXAMPLE 5   Finding the Height of a Dropped Object

The Swiss Re Building, in London, is a unique building. Londoners often refer to it as the "pickle building." The building is 592.1 feet tall. An object is dropped from the highest point of this building. Neglecting air resistance, the height in feet of the object above ground at time t seconds is given by the polynomial $-16t^2 + 592.1$. Find the height of the object when $t = 1$ second, and when $t = 6$ seconds.

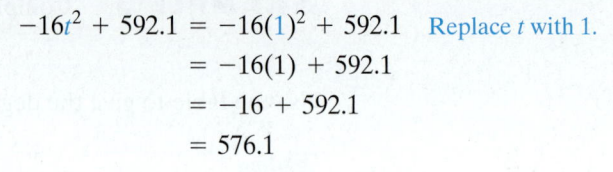

Solution   To find each height, we evaluate the polynomial when $t = 1$ and when $t = 6$.

$$-16t^2 + 592.1 = -16(1)^2 + 592.1 \quad \text{Replace } t \text{ with 1.}$$
$$= -16(1) + 592.1$$
$$= -16 + 592.1$$
$$= 576.1$$

The height of the object at 1 second is 576.1 feet.

$$-16t^2 + 592.1 = -16(6)^2 + 592.1 \quad \text{Replace } t \text{ with 6.}$$
$$= -16(36) + 592.1$$
$$= -576 + 592.1 = 16.1$$

The height of the object at 6 seconds is 16.1 feet.

PRACTICE 5   The cliff divers of Acapulco dive 130 feet into La Quebrada several times a day for the entertainment of the tourists. If a tourist is standing near the diving platform and drops his camera off the cliff, the height of the camera above the water at time t seconds is given by the polynomial $-16t^2 + 130$. Find the height of the camera when $t = 1$ second and when $t = 2$ seconds.

OBJECTIVE 3 ▶ Simplifying polynomials by combining like terms. Polynomials with like terms can be simplified by combining the like terms. Recall that like terms are terms that contain exactly the same variables raised to exactly the same powers.

Like Terms	Unlike Terms
$5x^2, -7x^2$	$3x, 3y$
$y, 2y$	$-2x^2, -5x$
$\frac{1}{2}a^2b, -a^2b$	$6st^2, 4s^2t$

Only like terms can be combined. We combine like terms by applying the distributive property.

EXAMPLE 6   Simplify each polynomial by combining any like terms.

a. $-3x + 7x$

b. $x + 3x^2$

c. $11x^2 + 5 + 2x^2 - 7$

d. $\frac{2}{5}x^4 + \frac{2}{3}x^3 - x^2 + \frac{1}{10}x^4 - \frac{1}{6}x^3$

Solution

a. $-3x + 7x = (-3 + 7)x = 4x$

b. $x + 3x^2$ These terms cannot be combined because x and $3x^2$ are not like terms.

c. $11x^2 + 5 + 2x^2 - 7 = 11x^2 + 2x^2 + 5 - 7$
$$= 13x^2 - 2 \quad \text{Combine like terms.}$$

Section 5.2 Adding and Subtracting Polynomials 317

d. $\dfrac{2}{5}x^4 + \dfrac{2}{3}x^3 - x^2 + \dfrac{1}{10}x^4 - \dfrac{1}{6}x^3$

$= \left(\dfrac{2}{5} + \dfrac{1}{10}\right)x^4 + \left(\dfrac{2}{3} - \dfrac{1}{6}\right)x^3 - x^2$

$= \left(\dfrac{4}{10} + \dfrac{1}{10}\right)x^4 + \left(\dfrac{4}{6} - \dfrac{1}{6}\right)x^3 - x^2$

$= \dfrac{5}{10}x^4 + \dfrac{3}{6}x^3 - x^2$

$= \dfrac{1}{2}x^4 + \dfrac{1}{2}x^3 - x^2$

PRACTICE 6 Simplify each polynomial by combining any like terms.

a. $-4y + 2y$
b. $z + 5z^3$
c. $7a^2 - 5 - 3a^2 - 7$
d. $\dfrac{3}{8}x^3 - x^2 + \dfrac{5}{6}x^4 + \dfrac{1}{12}x^3 - \dfrac{1}{2}x^4$

Concept Check ✓

When combining like terms in the expression, $5x - 8x^2 - 8x$, which of the following is the proper result?

a. $-11x^2$
b. $-8x^2 - 3x$
c. $-11x$
d. $-11x^4$

EXAMPLE 7 Combine like terms to simplify.

$$-9x^2 + 3xy - 5y^2 + 7yx$$

Solution $-9x^2 + 3xy - 5y^2 + 7yx = -9x^2 + (3 + 7)xy - 5y^2$

$= -9x^2 + 10xy - 5y^2$

▶ **Helpful Hint**

This term can be written as $7yx$ or $7xy$.

PRACTICE 7 Combine like terms to simplify: $9xy - 3x^2 - 4yx + 5y^2$.

EXAMPLE 8 Write a polynomial that describes the total area of the squares and rectangles shown below. Then simplify the polynomial.

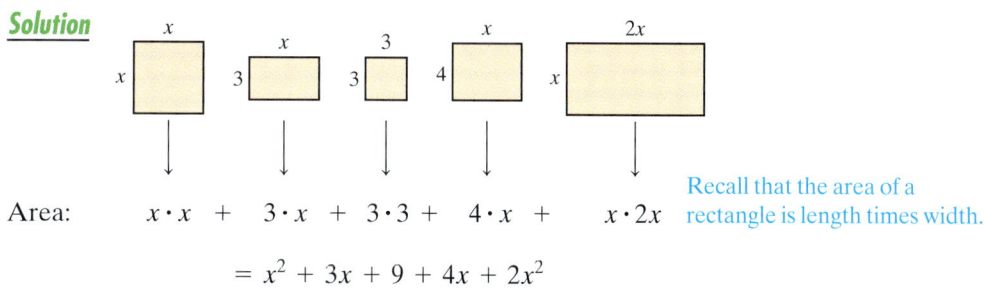

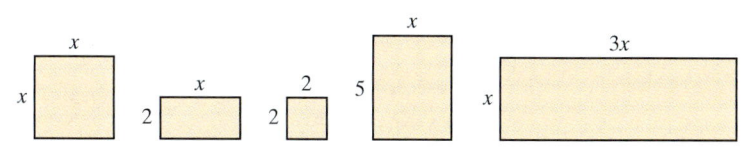

Solution

Area: $\quad x \cdot x \quad + \quad 3 \cdot x \quad + \quad 3 \cdot 3 + \quad 4 \cdot x \quad + \quad x \cdot 2x$

Recall that the area of a rectangle is length times width.

$= x^2 + 3x + 9 + 4x + 2x^2$

$= 3x^2 + 7x + 9 \qquad$ *Combine like terms.*

PRACTICE 8 Write a polynomial that describes the total area of the squares and rectangles shown below. Then simplify the polynomial.

Answer to Concept Check:
b

OBJECTIVE 4 ▶ Adding and subtracting polynomials. We now practice adding and subtracting polynomials.

> **Adding Polynomials**
> To add polynomials, combine all like terms.

EXAMPLE 9 Add $(-2x^2 + 5x - 1)$ and $(-2x^2 + x + 3)$.

Solution

$$(-2x^2 + 5x - 1) + (-2x^2 + x + 3) = -2x^2 + 5x - 1 - 2x^2 + x + 3$$
$$= (-2x^2 - 2x^2) + (5x + 1x) + (-1 + 3)$$
$$= -4x^2 + 6x + 2$$

PRACTICE 9 Add. $(-3x^2 - 4x + 9)$ and $(2x^2 - 2x)$.

EXAMPLE 10 Add: $(4x^3 - 6x^2 + 2x + 7) + (5x^2 - 2x)$.

Solution

$$(4x^3 - 6x^2 + 2x + 7) + (5x^2 - 2x) = 4x^3 - 6x^2 + 2x + 7 + 5x^2 - 2x$$
$$= 4x^3 + (-6x^2 + 5x^2) + (2x - 2x) + 7$$
$$= 4x^3 - x^2 + 7$$

PRACTICE 10 Add. $(-3x^3 + 7x^2 + 3x - 4) + (3x^2 - 9x)$.

Polynomials can be added vertically if we line up like terms underneath one another.

EXAMPLE 11 Add $(7y^3 - 2y^2 + 7)$ and $(6y^2 + 1)$ using the vertical format.

Solution Vertically line up like terms and add.

$$\begin{array}{r} 7y^3 - 2y^2 + 7 \\ 6y^2 + 1 \\ \hline 7y^3 + 4y^2 + 8 \end{array}$$

PRACTICE 11 Add $(5z^3 + 3z^2 + 4z)$ and $(5z^2 + 4z)$ using the vertical format.

To subtract one polynomial from another, recall the definition of subtraction. To subtract a number, we add its opposite: $a - b = a + (-b)$. To subtract a polynomial, we also add its opposite. Just as $-b$ is the opposite of b, $-(x^2 + 5)$ is the opposite of $(x^2 + 5)$.

EXAMPLE 12 Subtract: $(5x - 3) - (2x - 11)$.

Solution From the definition of subtraction, we have

$$(5x - 3) - (2x - 11) = (5x - 3) + [-(2x - 11)] \quad \text{Add the opposite.}$$
$$= (5x - 3) + (-2x + 11) \quad \text{Apply the distributive property.}$$
$$= (5x - 2x) + (-3 + 11)$$
$$= 3x + 8$$

PRACTICE 12 Subtract: $(8x - 7) - (3x - 6)$.

> **Subtracting Polynomials**
> To subtract two polynomials, change the signs of the terms of the polynomial being subtracted and then add.

EXAMPLE 13 Subtract: $(2x^3 + 8x^2 - 6x) - (2x^3 - x^2 + 1)$.

Solution First, change the sign of each term of the second polynomial and then add.

> **Helpful Hint**
> Notice the sign of each term is changed.

$$(2x^3 + 8x^2 - 6x) - (2x^3 - x^2 + 1) = (2x^3 + 8x^2 - 6x) + (-2x^3 + x^2 - 1)$$
$$= 2x^3 - 2x^3 + 8x^2 + x^2 - 6x - 1$$
$$= 9x^2 - 6x - 1 \quad \text{Combine like terms.}$$

PRACTICE 13 Subtract: $(3x^3 - 5x^2 + 4x) - (x^3 - x^2 + 6)$.

EXAMPLE 14 Subtract $(5y^2 + 2y - 6)$ from $(-3y^2 - 2y + 11)$ using the vertical format.

Solution Arrange the polynomials in vertical format, lining up like terms.

$$\begin{array}{r} -3y^2 - 2y + 11 \\ -(5y^2 + 2y - 6) \end{array} \qquad \begin{array}{r} -3y^2 - 2y + 11 \\ -5y^2 - 2y + 6 \\ \hline -8y^2 - 4y + 17 \end{array}$$

PRACTICE 14 Subtract $(6z^2 + 3z - 7)$ from $(-2z^2 - 8z + 5)$ using the vertical format.

EXAMPLE 15 Subtract $(5z - 7)$ from the sum of $(8z + 11)$ and $(9z - 2)$.

Solution Notice that $(5z - 7)$ is to be subtracted **from** a sum. The translation is

$$[(8z + 11) + (9z - 2)] - (5z - 7)$$
$$= 8z + 11 + 9z - 2 - 5z + 7 \quad \text{Remove grouping symbols.}$$
$$= 8z + 9z - 5z + 11 - 2 + 7 \quad \text{Group like terms.}$$
$$= 12z + 16 \quad \text{Combine like terms.}$$

PRACTICE 15 Subtract $(3x + 5)$ from the sum of $(8x - 11)$ and $(2x + 5)$.

EXAMPLE 16 Add or subtract as indicated.
a. $(3x^2 - 6xy + 5y^2) + (-2x^2 + 8xy - y^2)$
b. $(9a^2b^2 + 6ab - 3ab^2) - (5b^2a + 2ab - 3 - 9b^2)$

Solution
a. $(3x^2 - 6xy + 5y^2) + (-2x^2 + 8xy - y^2)$
$$= 3x^2 - 6xy + 5y^2 - 2x^2 + 8xy - y^2$$
$$= x^2 + 2xy + 4y^2 \quad \text{Combine like terms.}$$

Section 5.2 Adding and Subtracting Polynomials **319**

b. $(9a^2b^2 + 6ab - 3ab^2) - (5b^2a + 2ab - 3 - 9b^2)$ Change the sign of each term
$= 9a^2b^2 + 6ab - 3ab^2 - 5b^2a - 2ab + 3 + 9b^2$ of the polynomial being subtracted.
$= 9a^2b^2 + 4ab - 8ab^2 + 3 + 9b^2$ Combine like terms.

PRACTICE 16 Add or subtract as indicated.

a. $(3a^2 - 4ab + 7b^2) + (-8a^2 + 3ab - b^2)$
b. $(5x^2y^2 - 6xy - 4xy^2) - (2x^2y^2 + 4xy - 5 + 6y^2)$

Answers to Concept Check:
a. $3y$ b. $2y^2$ c. $-3y$ d. $2y^2$
e. cannot be simplified

Concept Check ✓

If possible, simplify each expression by performing the indicated operation.

a. $2y + y$ b. $2y \cdot y$ c. $-2y - y$ d. $(-2y)(-y)$ e. $2x + y$

VOCABULARY & READINESS CHECK

Use the choices below to fill in each blank. Not all choices will be used.

least   monomial   trinomial   coefficient
greatest   binomial   constant

1. A(n) _____ is a polynomial with exactly 2 terms.
2. A(n) _____ is a polynomial with exactly one term.
3. A(n) _____ is a polynomial with exactly three terms.
4. The numerical factor of a term is called the _____.
5. A number term is also called a _____.
6. The degree of a polynomial is the _____ degree of any term of the polynomial.

Simplify by combining like terms if possible.

7. $-9y - 5y$
8. $6m^5 + 7m^5$
9. $x + 6x$
10. $7z - z$
11. $5m^2 + 2m$
12. $8p^3 + 3p^2$

5.2 EXERCISE SET   MyMathLab

Find the degree of each of the following polynomials and determine whether it is a monomial, binomial, trinomial, or none of these. See Examples 1 through 3.

1. $x + 2$
2. $-6y + y^2 + 4$
3. $9m^3 - 5m^2 + 4m - 8$
4. $5a^2 + 3a^3 - 4a^4$
5. $12x^4y - x^2y^2 - 12x^2y^4$
6. $7r^2s^2 + 2r - 3s^5$
7. $3zx - 5x^2$
8. $5y + 2$

In the second column, write the degree of the polynomial in the first column. See Examples 1 through 3.

Polynomial	Degree
9. $3xy^2 - 4$	
10. $8x^2y^2$	
11. $5a^2 - 2a + 1$	
12. $4z^6 + 3z^2$	

Find the value of each polynomial when **(a)** $x = 0$ *and* **(b)** $x = -1$. *See Examples 4 and 5.*

13. $x + 6$
14. $2x - 10$

15. $x^2 - 5x - 2$
16. $x^2 - 4$
17. $x^3 - 15$
18. $-2x^3 + 3x^2 - 6$

The CN Tower in Toronto, Ontario, is 1821 feet tall and is the world's tallest self-supporting structure. An object is dropped from the Skypod of the Tower which is at 1150 feet. Neglecting air resistance, the height of the object at time t seconds is given by the polynomial $-16t^2 + 1150$. Find the height of the object at the given times.

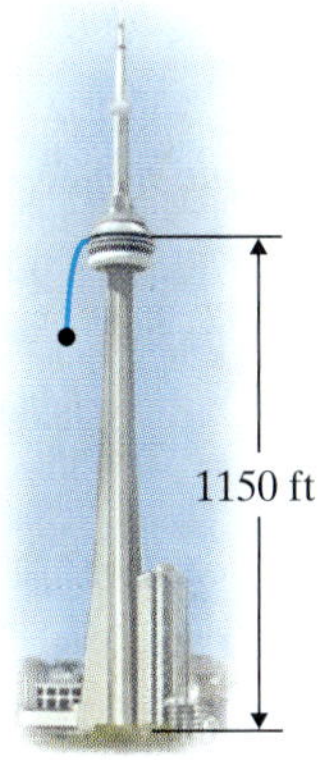

19. $t = 1$ second
20. $t = 7$ seconds
21. $t = 3$ seconds
22. $t = 6$ seconds

Simplify each of the following by combining like terms. See Examples 6 and 7.

23. $14x^2 + 9x^2$
24. $18x^3 - 4x^3$
25. $15x^2 - 3x^2 - y$
26. $12k^3 - 9k^3 + 11$
27. $8s - 5s + 4s$
28. $5y + 7y - 6y$
29. $0.1y^2 - 1.2y^2 + 6.7 - 1.9$
30. $7.6y + 3.2y^2 - 8y - 2.5y^2$
31. $\frac{2}{5}x^2 - \frac{1}{3}x^3 + x^2 - \frac{1}{4}x^3 + 6$
32. $\frac{1}{6}x^4 - \frac{1}{7}x^2 + 5 - \frac{1}{2}x^4 - \frac{3}{7}x^2 + \frac{1}{3}$
33. $6a^2 - 4ab + 7b^2 - a^2 - 5ab + 9b^2$
34. $x^2y + xy - y + 10x^2y - 2y + xy$

Perform the indicated operations. See Examples 9 through 13.

35. $(3x + 7) + (9x + 5)$
36. $(3x^2 + 7) + (3x^2 + 9)$
37. $(-7x + 5) + (-3x^2 + 7x + 5)$
38. $(3x - 8) + (4x^2 - 3x + 3)$
39. $(2x^2 + 5) - (3x^2 - 9)$
40. $(5x^2 + 4) - (-2y^2 + 4)$
41. $3x - (5x - 9)$
42. $4 - (-y - 4)$
43. $(2x^2 + 3x - 9) - (-4x + 7)$
44. $(-7x^2 + 4x + 7) - (-8x + 2)$

Perform the indicated operations. See Examples 11, 14, and 15.

45. $\begin{array}{r} 3t^2 + 4 \\ +5t^2 - 8 \\ \hline \end{array}$

46. $\begin{array}{r} 7x^3 + 3 \\ +2x^3 + 1 \\ \hline \end{array}$

47. $\begin{array}{r} 4z^2 - 8z + 3 \\ -(6z^2 + 8z - 3) \\ \hline \end{array}$

48. $\begin{array}{r} 5u^5 - 4u^2 + 3u - 7 \\ -(3u^5 + 6u^2 - 8u + 2) \\ \hline \end{array}$

49. $\begin{array}{r} 5x^3 - 4x^2 + 6x - 2 \\ -(3x^3 - 2x^2 - x - 4) \\ \hline \end{array}$

50. $\begin{array}{r} 7a^2 - 9a + 6 \\ -(11a^2 - 4a + 2) \\ \hline \end{array}$

51. Subtract $(19x^2 + 5)$ from $(81x^2 + 10)$.
52. Subtract $(2x + xy)$ from $(3x - 9xy)$.
53. Subtract $(2x + 2)$ from the sum of $(8x + 1)$ and $(6x + 3)$.
54. Subtract $(-12x - 3)$ from the sum of $(-5x - 7)$ and $(12x + 3)$.

MIXED PRACTICE

Perform the indicated operations.

55. $(-3y^2 - 4y) + (2y^2 + y - 1)$
56. $(7x^2 + 2x - 9) + (-3x^2 + 5)$
57. $(5x + 8) - (-2x^2 - 6x + 8)$
58. $(-6y^2 + 3y - 4) - (9y^2 - 3y)$
59. $(-8x^4 + 7x) + (-8x^4 + x + 9)$
60. $(6y^5 - 6y^3 + 4) + (-2y^5 - 8y^3 - 7)$
61. $(3x^2 + 5x - 8) + (5x^2 + 9x + 12) - (x^2 - 14)$
62. $(-a^2 + 1) - (a^2 - 3) + (5a^2 - 6a + 7)$
63. Subtract $4x$ from $7x - 3$.
64. Subtract y from $y^2 - 4y + 1$.
65. Subtract $(5x + 7)$ from $(7x^2 + 3x + 9)$.
66. Subtract $(5y^2 + 8y + 2)$ from $(7y^2 + 9y - 8)$.
67. Subtract $(4y^2 - 6y - 3)$ from the sum of $(8y^2 + 7)$ and $(6y + 9)$.
68. Subtract $(5y + 7x^2)$ from the sum of $(8y - x)$ and $(3 + 8x^2)$.
69. Subtract $(-2x^2 + 4x - 12)$ from the sum of $(-x^2 - 2x)$ and $(5x^2 + x + 9)$.
70. Subtract $(4x^2 - 2x + 2)$ from the sum of $(x^2 + 7x + 1)$ and $(7x + 5)$.

Find the area of each figure. Write a polynomial that describes the total area of the rectangles and squares shown in Exercises 71–72. Then simplify the polynomial. See Example 8.

△ 71.

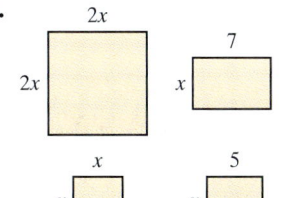

△ 72.

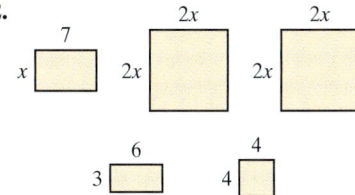

322 CHAPTER 5 Exponents and Polynomials

Recall that the perimeter of a figure such as the ones shown in Exercises 73 through 76 is the sum of the lengths of its sides. Find the perimeter of each polynomial.

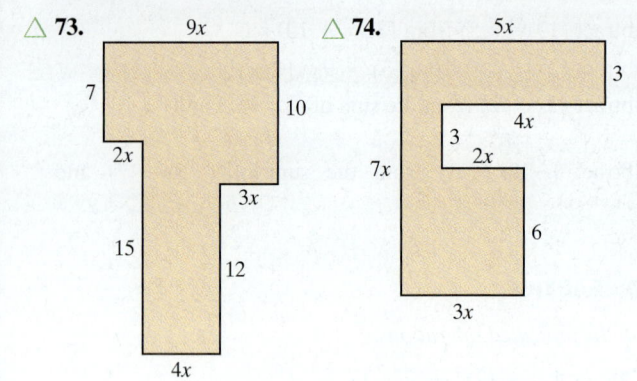

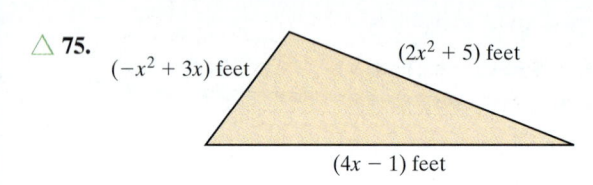

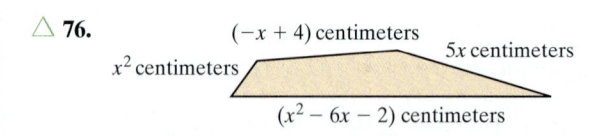

△ **73.** (figure with sides: 9x, 7, 10, 2x, 3x, 15, 12, 4x)

△ **74.** (figure with sides: 5x, 3, 4x, 3, 2x, 7x, 6, 3x)

△ **75.** Triangle with sides $(-x^2 + 3x)$ feet, $(2x^2 + 5)$ feet, $(4x - 1)$ feet

△ **76.** Quadrilateral with sides x^2 centimeters, $(-x + 4)$ centimeters, $5x$ centimeters, $(x^2 - 6x - 2)$ centimeters

△ **77.** A wooden beam is $(4y^2 + 4y + 1)$ meters long. If a piece $(y^2 - 10)$ meters is cut, express the length of the remaining piece of beam as a polynomial in y.

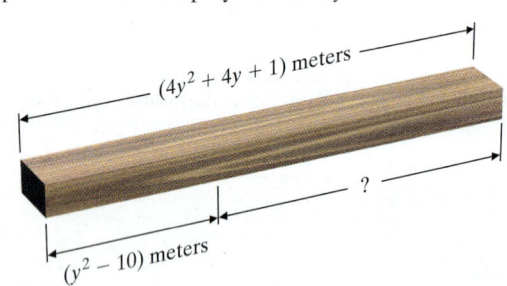

△ **78.** A piece of quarter-round molding is $(13x - 7)$ inches long. If a piece $(2x + 2)$ inches is removed, express the length of the remaining piece of molding as a polynomial in x.

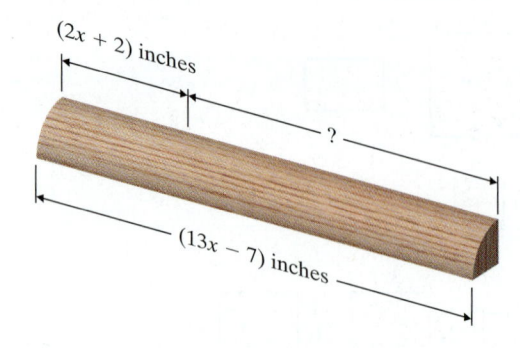

Add or subtract as indicated. See Example 16.

79. $(9a + 6b - 5) + (-11a - 7b + 6)$
80. $(3x - 2 + 6y) + (7x - 2 - y)$
81. $(4x^2 + y^2 + 3) - (x^2 + y^2 - 2)$
82. $(7a^2 - 3b^2 + 10) - (-2a^2 + b^2 - 12)$
83. $(x^2 + 2xy - y^2) + (5x^2 - 4xy + 20y^2)$
84. $(a^2 - ab + 4b^2) + (6a^2 + 8ab - b^2)$
85. $(11r^2s + 16rs - 3 - 2r^2s^2) - (3sr^2 + 5 - 9r^2s^2)$
86. $(3x^2y - 6xy + x^2y^2 - 5) - (11x^2y^2 - 1 + 5yx^2)$

Simplify each polynomial by combining like terms.

87. $7.75x + 9.16x^2 - 1.27 - 14.58x^2 - 18.34$
88. $1.85x^2 - 3.76x + 9.25x^2 + 10.76 - 4.21x$

Perform each indicated operation.

89. $[(7.9y^4 - 6.8y^3 + 3.3y) + (6.1y^3 - 5)] - (4.2y^4 + 1.1y - 1)$
90. $[(1.2x^2 - 3x + 9.1) - (7.8x^2 - 3.1 + 8)] + (1.2x - 6)$

REVIEW AND PREVIEW

Multiply. See Section 5.1.

91. $3x(2x)$
92. $-7x(x)$
93. $(12x^3)(-x^5)$
94. $6r^3(7r^{10})$
95. $10x^2(20xy^2)$
96. $-z^2y(11zy)$

CONCEPT EXTENSIONS

97. Describe how to find the degree of a term.
98. Describe how to find the degree of a polynomial.
99. Explain why xyz is a monomial while $x + y + z$ is a trinomial.
100. Explain why the degree of the term $5y^3$ is 3 and the degree of the polynomial $2y + y + 2y$ is 1.

Match each expression on the left with its simplification on the right. Not all letters on the right must be used and a letter may be used more than once.

101. $10y - 6y^2 - y$ **a.** $3y$
102. $5x + 5x$ **b.** $9y - 6y^2$
103. $(5x - 3) + (5x - 3)$ **c.** $10x$
104. $(15x - 3) - (5x - 3)$ **d.** $25x^2$
    e. $10x - 6$
    f. none of these

Simplify each expression by performing the indicated operation. Explain how you arrived at each answer. See the Concept Check in this section.

105. a. $z + 3z$ **106. a.** $x + x$
   b. $z \cdot 3z$ **b.** $x \cdot x$
   c. $-z - 3z$ **c.** $-x - x$
   d. $(-z)(-3z)$ **d.** $(-x)(-x)$

Fill in the boxes so that the terms in each expression can be combined. Then simplify. Each exercise has more than one solution.

107. $7x^\square + 2x^\square$

108. $(3y^2)^\square + (4y^3)^\square$

Write a polynomial that describes the surface area of each figure. (Recall that the surface area of a solid is the sum of the areas of the faces or sides of the solid.)

△ **109.** (box with dimensions x, x, y)

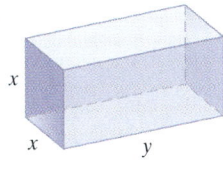

△ **110.** (box with dimensions x, 5, 9)

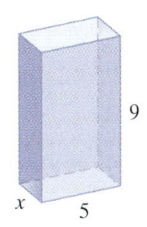

111. The average tuition, fees, and room and board rates charged per year for full-time students in degree-granting two year public colleges is approximated by the polynomial $6.4x^2 + 37.9x + 2856.8$ for the years 1984 through 2006. (*Source:* National Center for Education Statistics & The College Board) Use this model to predict what the costs will be for a student at a public two-year institution in 2010. ($x = 26$). Round to the nearest dollar.

112. The number of wireless telephone subscribers (in millions) x years after 1990 is given by the polynomial $0.74x^2 + 2.6x + 3.2$ for 1990 to 2005. Use this model to predict the number of wireless telephone subscribers in 2010 ($x = 20$). (*Source:* CTIA—The Wireless Association)

113. The polynomial $2.13x^2 + 21.89x + 1190$ represents the sale of electricity (in billion kilowatt-hours) in the U.S. residential sector during 2000–2005. The polynomial $8.71x^2 - 1.46x + 2095$ represents the sale of electricity (in billion kilowatt-hours) in all other U.S. sectors during 2000–2005. In both polynomials, x represents the number of years after 2000. Find a polynomial for the total sales of electricity (in billion kilowatt hours) to all sectors in the United States during this period. (*Source:* Based on data from the Energy Information Administration)

114. The polynomial $-3.5x^2 + 33.3x + 392$ represents the number of prescriptions (in millions) purchased from a supermarket for the years 2000–2005. The polynomial $19x + 141$ represents the number of prescriptions (in millions) purchased through the mail for the same years. In both polynomials, x represents the number of years since 2000. Find a polynomial for the total number of prescriptions purchased from a supermarket or mail order. (*Source*: National Association of Chain Drug Stores)

5.3 MULTIPLYING POLYNOMIALS

OBJECTIVES

1. Use the distributive property to multiply polynomials.
2. Multiply polynomials vertically.

OBJECTIVE 1 ▶ Using the distributive property to multiply polynomials. To multiply polynomials, we apply our knowledge of the rules and definitions of exponents.

Recall from Section 5.1 that to multiply two monomials such as $(-5x^3)$ and $(-2x^4)$, we use the associative and commutative properties and regroup. Remember, also, that to multiply exponential expressions with a common base we use the product rule for exponents and add exponents.

$$(-5x^3)(-2x^4) = (-5)(-2)(x^3)(x^4) = 10x^7$$

EXAMPLES Multiply.

1. $6x \cdot 4x = (6 \cdot 4)(x \cdot x)$ *Use the commutative and associative properties.*
$= 24x^2$ *Multiply.*

2. $-7x^2 \cdot 0.2x^5 = (-7 \cdot 0.2)(x^2 \cdot x^5)$
$= -1.4x^7$

3. $\left(-\dfrac{1}{3}x^5\right)\left(-\dfrac{2}{9}x\right) = \left(-\dfrac{1}{3} \cdot -\dfrac{2}{9}\right) \cdot \left(x^5 \cdot x\right)$
$= \dfrac{2}{27}x^6$

PRACTICES

1–3 Multiply.

1. $5y \cdot 2y$
2. $(5z^3) \cdot (-0.4z^5)$
3. $\left(-\dfrac{1}{9}b^6\right)\left(-\dfrac{7}{8}b^3\right)$

Concept Check ✓

Simplify.

a. $3x \cdot 2x$
b. $3x + 2x$

To multiply polynomials that are not monomials, use the distributive property.

EXAMPLE 4 Use the distributive property to find each product.

a. $5x(2x^3 + 6)$
b. $-3x^2(5x^2 + 6x - 1)$

Solution

a. $5x(2x^3 + 6) = 5x(2x^3) + 5x(6)$ Use the distributive property.
$ = 10x^4 + 30x$ Multiply.

b. $-3x^2(5x^2 + 6x - 1)$
$= (-3x^2)(5x^2) + (-3x^2)(6x) + (-3x^2)(-1)$ Use the distributive property.
$= -15x^4 - 18x^3 + 3x^2$ Multiply.

PRACTICE 4 Use the distributive property to find each product.

a. $3x(5x^5 + 5)$
b. $-5x^3(2x^2 - 9x + 2)$

We also use the distributive property to multiply two binomials. To multiply $(x + 3)$ by $(x + 1)$, distribute the factor $(x + 3)$ first.

$(x + 3)(x + 1) = x(x + 1) + 3(x + 1)$ Distribute $(x + 3)$.
$ = x(x) + x(1) + 3(x) + 3(1)$ Apply distributive property a second time.
$ = x^2 + x + 3x + 3$ Multiply.
$ = x^2 + 4x + 3$ Combine like terms.

This idea can be expanded so that we can multiply any two polynomials.

> **To Multiply Two Polynomials**
> Multiply each term of the first polynomial by each term of the second polynomial, and then combine like terms.

Answers to Concept Check:

a. $6x^2$
b. $5x$

EXAMPLE 5 Multiply $(3x + 2)(2x - 5)$.

Solution Multiply each term of the first binomial by each term of the second.

$$(3x + 2)(2x - 5) = 3x(2x) + 3x(-5) + 2(2x) + 2(-5)$$
$$= 6x^2 - 15x + 4x - 10 \qquad \text{Multiply.}$$
$$= 6x^2 - 11x - 10 \qquad \text{Combine like terms.} \quad \square$$

PRACTICE 5 Multiply $(5x - 2)(2x + 3)$

EXAMPLE 6 Multiply $(2x - y)^2$.

Solution Recall that $a^2 = a \cdot a$, so $(2x - y)^2 = (2x - y)(2x - y)$. Multiply each term of the first polynomial by each term of the second.

$$(2x - y)(2x - y) = 2x(2x) + 2x(-y) + (-y)(2x) + (-y)(-y)$$
$$= 4x^2 - 2xy - 2xy + y^2 \qquad \text{Multiply.}$$
$$= 4x^2 - 4xy + y^2 \qquad \text{Combine like terms.} \quad \square$$

PRACTICE 6 Multiply $(5x - 3y)^2$

Concept Check ✓

Square where indicated. Simplify if possible.

a. $(4a)^2 + (3b)^2$ **b.** $(4a + 3b)^2$

EXAMPLE 7 Multiply $(t + 2)$ by $(3t^2 - 4t + 2)$.

Solution Multiply each term of the first polynomial by each term of the second.

$$(t + 2)(3t^2 - 4t + 2) = t(3t^2) + t(-4t) + t(2) + 2(3t^2) + 2(-4t) + 2(2)$$
$$= 3t^3 - 4t^2 + 2t + 6t^2 - 8t + 4$$
$$= 3t^3 + 2t^2 - 6t + 4 \qquad \text{Combine like terms.} \quad \square$$

PRACTICE 7 Multiply $(y + 4)$ by $(2y^2 - 3y + 5)$

EXAMPLE 8 Multiply $(3a + b)^3$.

Solution Write $(3a + b)^3$ as $(3a + b)(3a + b)(3a + b)$.

$$(3a + b)(3a + b)(3a + b) = (9a^2 + 3ab + 3ab + b^2)(3a + b)$$
$$= (9a^2 + 6ab + b^2)(3a + b)$$
$$= (9a^2 + 6ab + b^2)3a + (9a^2 + 6ab + b^2)b$$
$$= 27a^3 + 18a^2b + 3ab^2 + 9a^2b + 6ab^2 + b^3$$
$$= 27a^3 + 27a^2b + 9ab^2 + b^3 \qquad \square$$

PRACTICE 8 Multiply $(s + 2t)^3$

Answers to Concept Check:
a. $16a^2 + 9b^2$
b. $16a^2 + 24ab + 9b^2$

OBJECTIVE 2 ▶ **Multiplying polynomials vertically.** Another convenient method for multiplying polynomials is to use a vertical format similar to the format used to multiply real numbers. We demonstrate this method by multiplying $(3y^2 - 4y + 1)$ by $(y + 2)$.

EXAMPLE 9 Multiply $(3y^2 - 4y + 1)(y + 2)$. Use a vertical format.

Solution

$$\begin{array}{r} 3y^2 - 4y + 1 \\ \times \quad\quad y + 2 \\ \hline 6y^2 - 8y + 2 \\ 3y^3 - 4y^2 + y \quad\quad \\ \hline 3y^3 + 2y^2 - 7y + 2 \end{array}$$

1st, Multiply $3y^2 - 4y + 1$ by 2.
2nd, Multiply $3y^2 - 4y + 1$ by y. Line up like terms.
3rd, Combine like terms.

▶ **Helpful Hint**
Make sure like terms are lined up.

Thus, $(y + 2)(3y^2 - 4y + 1) = 3y^3 + 2y^2 - 7y + 2$.

PRACTICE 9 Multiply $(5x^2 - 3x + 5)(x - 4)$

When multiplying vertically, be careful if a power is missing, you may want to leave space in the partial products and take care that like terms are lined up.

EXAMPLE 10 Multiply $(2x^3 - 3x + 4)(x^2 + 1)$. Use a vertical format.

Solution

$$\begin{array}{r} 2x^3 \quad\quad - 3x + 4 \\ \times \quad\quad\quad x^2 + 1 \\ \hline 2x^3 \quad\quad - 3x + 4 \\ 2x^5 - 3x^3 + 4x^2 \quad\quad\quad \\ \hline 2x^5 - x^3 + 4x^2 - 3x + 4 \end{array}$$

Leave space for missing powers of x.
← Line up like terms.
Combine like terms.

PRACTICE 10 Multiply $(x^3 - 2x^2 + 1)(x^2 + 2)$.

EXAMPLE 11 Find the product of $(2x^2 - 3x + 4)$ and $(x^2 + 5x - 2)$ using a vertical format.

Solution First, we arrange the polynomials in a vertical format. Then we multiply each term of the second polynomial by each term of the first polynomial.

$$\begin{array}{r} 2x^2 - \quad 3x + 4 \\ x^2 + \quad 5x - 2 \\ \hline -4x^2 + \quad 6x - 8 \\ 10x^3 - 15x^2 + 20x \quad\quad \\ 2x^4 - \quad 3x^3 + \quad 4x^2 \quad\quad\quad\quad \\ \hline 2x^4 + \quad 7x^3 - 15x^2 + 26x - 8 \end{array}$$

Multiply $2x^2 - 3x + 4$ by -2.
Multiply $2x^2 - 3x + 4$ by $5x$.
Multiply $2x^2 - 3x + 4$ by x^2.
Combine like terms.

PRACTICE 11 Find the product of $(5x^2 + 2x - 2)$ and $(x^2 - x + 3)$ using a vertical format.

VOCABULARY & READINESS CHECK

Fill in each blank with the correct choice.

1. The expression $5x(3x + 2)$ equals $5x \cdot 3x + 5x \cdot 2$ by the _____ property.
   a. commutative   b. associative   c. distributive
2. The expression $(x + 4)(7x - 1)$ equals $x(7x - 1) + 4(7x - 1)$ by the _____ property.
   a. commutative   b. associative   c. distributive
3. The expression $(5y - 1)^2$ equals _____.
   a. $2(5y - 1)$ b. $(5y - 1)(5y + 1)$ c. $(5y - 1)(5y - 1)$
4. The expression $9x \cdot 3x$ equals _____.
   a. $27x$ b. $27x^2$ c. $12x$ d. $12x^2$

Perform the indicated operation, if possible.

5. $x^3 \cdot x^5$
6. $x^2 \cdot x^6$
7. $x^3 + x^5$
8. $x^2 + x^6$
9. $x^7 \cdot x^7$
10. $x^{11} \cdot x^{11}$
11. $x^7 + x^7$
12. $x^{11} + x^{11}$

5.3 EXERCISE SET

Multiply. See Examples 1 through 3.

1. $-4n^3 \cdot 7n^7$
2. $9t^6(-3t^5)$
3. $(-3.1x^3)(4x^9)$
4. $(-5.2x^4)(3x^4)$
5. $\left(-\dfrac{1}{3}y^2\right)\left(\dfrac{2}{5}y\right)$
6. $\left(-\dfrac{3}{4}y^7\right)\left(\dfrac{1}{7}y^4\right)$
7. $(2x)(-3x^2)(4x^5)$
8. $(x)(5x^4)(-6x^7)$

Multiply. See Example 4.

9. $3x(2x + 5)$
10. $2x(6x + 3)$
11. $-2a(a + 4)$
12. $-3a(2a + 7)$
13. $3x(2x^2 - 3x + 4)$
14. $4x(5x^2 - 6x - 10)$
15. $-2a^2(3a^2 - 2a + 3)$
16. $-4b^2(3b^3 - 12b^2 - 6)$
17. $-y(4x^3 - 7x^2y + xy^2 + 3y^3)$
18. $-x(6y^3 - 5xy^2 + x^2y - 5x^3)$
19. $\dfrac{1}{2}x^2(8x^2 - 6x + 1)$
20. $\dfrac{1}{3}y^2(9y^2 - 6y + 1)$

Multiply. See Examples 5 and 6.

21. $(x + 4)(x + 3)$
22. $(x + 2)(x + 9)$
23. $(a + 7)(a - 2)$
24. $(y - 10)(y + 11)$
25. $\left(x + \dfrac{2}{3}\right)\left(x - \dfrac{1}{3}\right)$
26. $\left(x + \dfrac{3}{5}\right)\left(x - \dfrac{2}{5}\right)$
27. $(3x^2 + 1)(4x^2 + 7)$
28. $(5x^2 + 2)(6x^2 + 2)$
29. $(2y - 4)^2$
30. $(6x - 7)^2$
31. $(4x - 3)(3x - 5)$
32. $(8x - 3)(2x - 4)$
33. $(3x^2 + 1)^2$
34. $(x^2 + 4)^2$
35. Perform the indicated operations.
    a. $(3x + 5) + (3x + 7)$
    b. $(3x + 5)(3x + 7)$
    c. Explain the difference between the two expressions.
36. Perform the indicated operations.
    a. $9x^2(-10x^2)$
    b. $9x^2 - 10x^2$
    c. Explain the difference between the two expressions.

Multiply. See Example 7.

37. $(x - 2)(x^2 - 3x + 7)$
38. $(x + 3)(x^2 + 5x - 8)$
39. $(x + 5)(x^3 - 3x + 4)$

328 CHAPTER 5 Exponents and Polynomials

40. $(a + 2)(a^3 - 3a^2 + 7)$
41. $(2a - 3)(5a^2 - 6a + 4)$
42. $(3 + b)(2 - 5b - 3b^2)$

Multiply. See Example 8.

43. $(x + 2)^3$
44. $(y - 1)^3$
45. $(2y - 3)^3$
46. $(3x + 4)^3$

Multiply vertically. See Examples 9 through 11.

47. $(2x - 11)(6x + 1)$
48. $(4x - 7)(5x + 1)$
49. $(5x + 1)(2x^2 + 4x - 1)$
50. $(4x - 5)(8x^2 + 2x - 4)$
51. $(x^2 + 5x - 7)(2x^2 - 7x - 9)$
52. $(3x^2 - x + 2)(x^2 + 2x + 1)$

MIXED PRACTICE

Multiply. See Examples 1 through 11.

53. $-1.2y(-7y^6)$
54. $-4.2x(-2x^5)$
55. $-3x(x^2 + 2x - 8)$
56. $-5x(x^2 - 3x + 10)$
57. $(x + 19)(2x + 1)$
58. $(3y + 4)(y + 11)$
59. $\left(x + \frac{1}{7}\right)\left(x - \frac{3}{7}\right)$
60. $\left(m + \frac{2}{9}\right)\left(m - \frac{1}{9}\right)$
61. $(3y + 5)^2$
62. $(7y + 2)^2$
63. $(a + 4)(a^2 - 6a + 6)$
64. $(t + 3)(t^2 - 5t + 5)$
65. $(2x - 5)^3$
66. $(3y - 1)^3$
67. $(4x + 5)(8x^2 + 2x - 4)$
68. $(5x + 4)(x^2 - x + 4)$
69. $(3x^2 + 2x - 4)(2x^2 - 4x + 3)$
70. $(a^2 + 3a - 2)(2a^2 - 5a - 1)$

Express as the product of polynomials. Then multiply.

△ 71. Find the area of the rectangle.

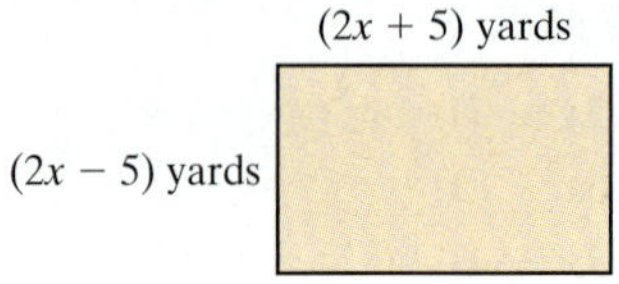

$(2x + 5)$ yards

$(2x - 5)$ yards

△ 72. Find the area of the square field.

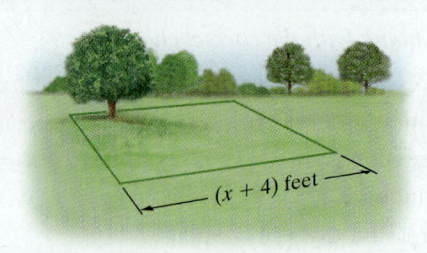

$(x + 4)$ feet

△ 73. Find the area of the triangle.

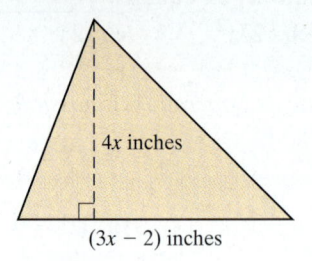

$4x$ inches

$(3x - 2)$ inches

△ 74. Find the volume of the cube.

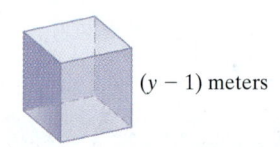

$(y - 1)$ meters

REVIEW AND PREVIEW

Perform the indicated operation. See Section 5.1.

75. $(5x)^2$
76. $(4p)^2$
77. $(-3y^3)^2$
78. $(-7m^2)^2$

CONCEPT EXTENSIONS

△ 79. The area of the larger rectangle below is $x(x + 3)$. Find another expression for this area by finding the sum of the areas of the two smaller rectangles.

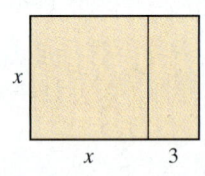

x

x 3

△ 80. Write an expression for the area of the larger rectangle below in two different ways.

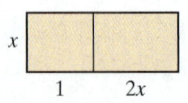

x

1 $2x$

△ 81. The area of the figure below is $(x + 2)(x + 3)$. Find another expression for this area by finding the sum of the areas of the four smaller rectangles.

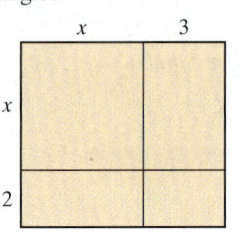

x 3

x

2

△ 82. Write an expression for the area of the figure in two different ways.

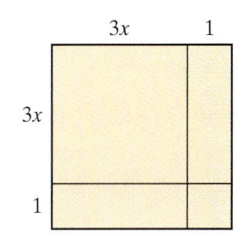

Simplify.

See the Concept Checks in this section.

83. $5a + 6a$ **84.** $5a \cdot 6a$

Square where indicated. Simplify if possible.

85. $(5x)^2 + (2y)^2$ **86.** $(5x + 2y)^2$

MIXED PRACTICE

See Sections 5.2, 5.3. Perform the indicated operations.

87. $(3x - 1) + (10x - 6)$
88. $(2x - 1) + (10x - 7)$
89. $(3x - 1)(10x - 6)$
90. $(2x - 1)(10x - 7)$
91. $(3x - 1) - (10x - 6)$
92. $(2x - 1) - (10x - 7)$
93. Multiply each of the following polynomials.
   a. $(a + b)(a - b)$
   b. $(2x + 3y)(2x - 3y)$
   c. $(4x + 7)(4x - 7)$
   d. Can you make a general statement about all products of the form $(x + y)(x - y)$?

94. Evaluate each of the following.
   a. $(2 + 3)^2$; $2^2 + 3^2$
   b. $(8 + 10)^2$; $8^2 + 10^2$
   Does $(a + b)^2 = a^2 + b^2$ no matter what the values of a and b are? Why or why not?

△ **95.** Write a polynomial that describes the area of the shaded region. (Find the area of the larger square minus the area of the smaller square.)

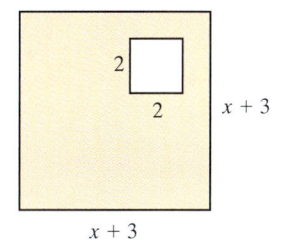

△ **96.** Write a polynomial that describes the area of the shaded region. (See Exercise 95.)

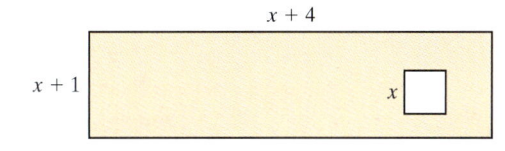

📖 STUDY SKILLS BUILDER

Are You Organized?

Have you ever had trouble finding a completed assignment? When it's time to study for a test, are your notes neat and organized? Have you ever had trouble reading your own mathematics handwriting? (Be honest—I have.)

When any of these things happen, it's time to get organized. Here are a few suggestions:

- Write your notes and complete your homework assignment in a notebook with pockets (spiral or ring binder.)
- Take class notes in this notebook, and then follow the notes with your completed homework assignment.
- When you receive graded papers or handouts, place them in the notebook pocket so that you will not lose them.
- Mark (possibly with an exclamation point) any note(s) that seem extra important to you.
- Mark (possibly with a question mark) any notes or homework that you are having trouble with.
- See your instructor or a math tutor to help you with the concepts or exercises that you are having trouble understanding.
- If you are having trouble reading your own handwriting, *slow down* and write your mathematics work clearly!

Exercises

1. Have you been completing your assignments on time?
2. Have you been correcting any exercises you may be having difficulty with?
3. If you are having trouble with a mathematical concept or correcting any homework exercises, have you visited your instructor, a tutor, or your campus math lab?
4. Are you taking lecture notes in your mathematics course? (By the way, these notes should include worked-out examples solved by your instructor.)
5. Is your mathematics course material (handouts, graded papers, lecture notes) organized?
6. If your answer to Exercise 5 is no, take a moment and review your course material. List at least two ways that you might better organize it. Then read the Study Skills Builder on organizing a notebook in Chapter 2.

330 CHAPTER 5 Exponents and Polynomials

5.4 SPECIAL PRODUCTS

OBJECTIVES

1 Multiply two binomials using the FOIL method.

2 Square a binomial.

3 Multiply the sum and difference of two terms.

OBJECTIVE 1 ▶ **Using the FOIL method.** In this section, we multiply binomials using special products. First, a special order for multiplying binomials called the FOIL order or method is introduced. This method is demonstrated by multiplying $(3x + 1)$ by $(2x + 5)$.

The FOIL Method

F stands for the product of the **First** terms. $(3x + 1)(2x + 5)$
$(3x)(2x) = 6x^2$ **F**

O stands for the product of the **Outer** terms. $(3x + 1)(2x + 5)$
$(3x)(5) = 15x$ **O**

I stands for the product of the **Inner** terms. $(3x + 1)(2x + 5)$
$(1)(2x) = 2x$ **I**

L stands for the product of the **Last** terms. $(3x + 1)(3x + 5)$
$(1)(5) = 5$ **L**

$$\begin{array}{cccc} F & O & I & L \end{array}$$
$$(3x + 1)(2x + 5) = 6x^2 + 15x + 2x + 5$$
$$= 6x^2 + 17x + 5 \quad \text{Combine like terms.}$$

Concept Check ✓

Multiply $(3x + 1)(2x + 5)$ using methods from the last section. Show that the product is still $6x^2 + 17x + 5$.

EXAMPLE 1   Multiply $(x - 3)(x + 4)$ by the FOIL method.

Solution
$$\begin{array}{cccc} & F & O & I & L \end{array}$$
$$(x - 3)(x + 4) = (x)(x) + (x)(4) + (-3)(x) + (-3)(4)$$

$$= x^2 + 4x - 3x - 12$$
$$= x^2 + x - 12 \quad \text{Combine like terms.}$$

PRACTICE
1   Multiply $(x + 2)(x - 5)$ by the FOIL method.

EXAMPLE 2   Multiply $(5x - 7)(x - 2)$ by the FOIL method.

Solution
$$\begin{array}{cccc} & F & O & I & L \end{array}$$
$$(5x - 7)(x - 2) = 5x(x) + 5x(-2) + (-7)(x) + (-7)(-2)$$

$$= 5x^2 - 10x - 7x + 14$$
$$= 5x^2 - 17x + 14 \quad \text{Combine like terms.}$$

Answer to Concept Check:
Multiply and simplify:
$3x(2x + 5) + 1(2x + 5)$

PRACTICE
2   Multiply $(4x - 9)(x - 1)$ by the FOIL method.

EXAMPLE 3 Multiply $2(y + 6)(2y - 1)$.

Solution
$$\begin{align}
2(y + 6)(2y - 1) &= 2(\overset{F}{2y^2} \overset{O}{- 1y} \overset{I}{+ 12y} \overset{L}{- 6}) \\
&= 2(2y^2 + 11y - 6) \quad \text{Simplify inside parentheses.} \\
&= 4y^2 + 22y - 12 \quad \text{Now use the distributive property.} \ \square
\end{align}$$

PRACTICE 3 Multiply $3(x + 5)(3x - 1)$.

OBJECTIVE 2 ▶ Squaring binomials. Now, try squaring a binomial using the FOIL method.

EXAMPLE 4 Multiply $(3y + 1)^2$.

Solution
$$\begin{align}
(3y + 1)^2 &= (3y + 1)(3y + 1) \\
&= \overset{F}{(3y)(3y)} + \overset{O}{(3y)(1)} + \overset{I}{1(3y)} + \overset{L}{1(1)} \\
&= 9y^2 + 3y + 3y + 1 \\
&= 9y^2 + 6y + 1 \quad \square
\end{align}$$

PRACTICE 4 Multiply $(4x - 1)^2$.

Notice the pattern that appears in Example 4.

$(3y + 1)^2 = 9y^2 + 6y + 1$

$9y^2$ is the first term of the binomial squared. $(3y)^2 = 9y^2$.

$6y$ is 2 times the product of both terms of the binomial. $(2)(3y)(1) = 6y$.

1 is the second term of the binomial squared. $(1)^2 = 1$.

This pattern leads to the following, which can be used when squaring a binomial. We call these **special products.**

> **Squaring a Binomial**
> A binomial squared is equal to the square of the first term plus or minus twice the product of both terms plus the square of the second term.
> $$(a + b)^2 = a^2 + 2ab + b^2$$
> $$(a - b)^2 = a^2 - 2ab + b^2$$

This product can be visualized geometrically.

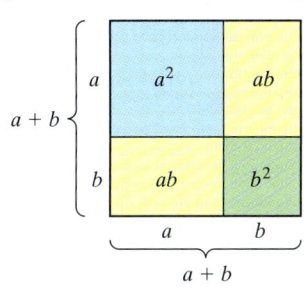

The area of the large square is side · side.

$$\text{Area} = (a + b)(a + b) = (a + b)^2$$

The area of the large square is also the sum of the areas of the smaller rectangles.

$$\text{Area} = a^2 + ab + ab + b^2 = a^2 + 2ab + b^2$$

Thus, $(a + b)^2 = a^2 + 2ab + b^2$.

EXAMPLE 5 Use a special product to square each binomial.

a. $(t + 2)^2$ **b.** $(p - q)^2$ **c.** $(2x + 5)^2$ **d.** $(x^2 - 7y)^2$

Solution

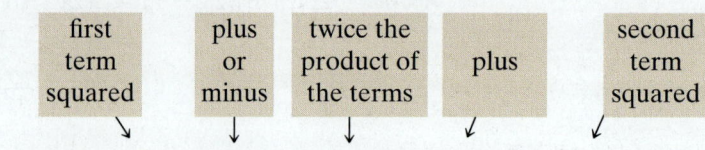

	first term squared	plus or minus	twice the product of the terms	plus	second term squared	

a. $(t + 2)^2 = t^2 + 2(t)(2) + 2^2 = t^2 + 4t + 4$

b. $(p - q)^2 = p^2 - 2(p)(q) + q^2 = p^2 - 2pq + q^2$

c. $(2x + 5)^2 = (2x)^2 + 2(2x)(5) + 5^2 = 4x^2 + 20x + 25$

d. $(x^2 - 7y)^2 = (x^2)^2 - 2(x^2)(7y) + (7y^2) = x^4 - 14x^2y + 49y^2$ □

PRACTICE 5 Use a special product to square each binomial.

a. $(b + 3)^2$ **b.** $(x - y)^2$
c. $(3y + 2)^2$ **d.** $(a^2 - 5b)^2$

▶ **Helpful Hint**

Notice that

$(a + b)^2 \neq a^2 + b^2$ The middle term $2ab$ is missing.
$(a + b)^2 = (a + b)(a + b) = a^2 + 2ab + b^2$

Likewise,

$(a - b)^2 \neq a^2 - b^2$
$(a - b)^2 = (a - b)(a - b) = a^2 - 2ab + b^2$

OBJECTIVE 3 ▶ Multiplying the sum and difference of two terms. Another special product is the product of the sum and difference of the same two terms, such as $(x + y)(x - y)$. Finding this product by the FOIL method, we see a pattern emerge.

$$(x + y)(x - y) = x^2 - xy + xy - y^2$$
$$= x^2 - y^2$$

Notice that the middle two terms subtract out. This is because the **Outer** product is the opposite of the **Inner** product. Only the **difference of squares** remains.

Multiplying the Sum and Difference of Two Terms

The product of the sum and difference of two terms is the square of the first term minus the square of the second term.

$$(a + b)(a - b) = a^2 - b^2$$

Section 5.4 Special Products **333**

EXAMPLE 6 Use a special product to multiply.

a. $4(x + 4)(x - 4)$ **b.** $(6t + 7)(6t - 7)$ **c.** $\left(x - \dfrac{1}{4}\right)\left(x + \dfrac{1}{4}\right)$
d. $(2p - q)(2p + q)$ **e.** $(3x^2 - 5y)(3x^2 + 5y)$

Solution

first term squared minus second term squared

a. $4(x + 4)(x - 4) = 4(x^2 \quad - \quad 4^2) = 4(x^2 - 16) = 4x^2 - 64$
b. $(6t + 7)(6t - 7) = (6t)^2 \quad - \quad 7^2 = 36t^2 - 49$
c. $\left(x - \dfrac{1}{4}\right)\left(x + \dfrac{1}{4}\right) = x^2 - \left(\dfrac{1}{4}\right)^2 = x^2 - \dfrac{1}{16}$
d. $(2p - q)(2p + q) = (2p)^2 - q^2 = 4p^2 - q^2$
e. $(3x^2 - 5y)(3x^2 + 5y) = (3x^2)^2 - (5y)^2 = 9x^4 - 25y^2$

PRACTICE 6 Use a special product to multiply.

a. $3(x + 5)(x - 5)$ **b.** $(4b - 3)(4b + 3)$
c. $\left(x + \dfrac{2}{3}\right)\left(x - \dfrac{2}{3}\right)$ **d.** $(5s + t)(5s - t)$
e. $(2y - 3z^2)(2y + 3z^2)$

Concept Check ✓

Match each expression on the left to the equivalent expression or expressions in the list below.

1. $(a + b)^2$ **2.** $(a + b)(a - b)$

a. $(a + b)(a + b)$ **b.** $a^2 - b^2$ **c.** $a^2 + b^2$ **d.** $a^2 - 2ab + b^2$ **e.** $a^2 + 2ab + b^2$

Let's now practice multiplying polynomials in general. If possible, use a special product.

EXAMPLE 7 Use a special product to multiply, if possible.

a. $(x - 5)(3x + 4)$ **b.** $(7x + 4)^2$ **c.** $(y - 0.6)(y + 0.6)$
d. $(y^4 + 2)(3y^2 - 1)$ **e.** $(a - 3)(a^2 + 2a - 1)$

Solution

a. $(x - 5)(3x + 4) = 3x^2 + 4x - 15x - 20$ FOIL.
$\qquad = 3x^2 - 11x - 20$
b. $(7x + 4)^2 = (7x)^2 + 2(7x)(4) + 4^2$ Squaring a binomial.
$\qquad = 49x^2 + 56x + 16$
c. $(y - 0.6)(y + 0.6) = y^2 - (0.6)^2 = y^2 - 0.36$ Multiplying the sum and difference of 2 terms.
d. $(y^4 + 2)(3y^2 - 1) = 3y^6 - y^4 + 6y^2 - 2$ FOIL.
e. I've inserted this product as a reminder that since it is not a binomial times a binomial, the FOIL order may not be used.

$(a - 3)(a^2 + 2a - 1) = a(a^2 + 2a - 1) - 3(a^2 + 2a - 1)$ Multiplying each term of
$\qquad = a^3 + 2a^2 - a - 3a^2 - 6a + 3$ the binomial by each term
$\qquad = a^3 - a^2 - 7a + 3$ of the trinomial.

Answers to Concept Check:
1. a or e **2.** b

PRACTICE 7 Use a special product to multiply, if possible.

a. $(4x + 3)(x - 6)$
b. $(7b - 2)^2$
c. $(x + 0.4)(x - 0.4)$
d. $(x^2 - 3)(3x^4 + 2)$
e. $(x + 1)(x^2 + 5x - 2)$

> **Helpful Hint**
> - When multiplying two binomials, you may always use the FOIL order or method.
> - When multiplying any two polynomials, you may always use the distributive property to find the product.

VOCABULARY & READINESS CHECK

Answer each exercise true or false.

1. $(x + 4)^2 = x^2 + 16$
2. For $(x + 6)(2x - 1)$ the product of the first terms is $2x^2$.
3. $(x + 4)(x - 4) = x^2 + 16$
4. The product $(x - 1)(x^3 + 3x - 1)$ is a polynomial of degree 5.

5.4 EXERCISE SET

Multiply using the FOIL method. See Examples 1 through 3.

1. $(x + 3)(x + 4)$
2. $(x + 5)(x - 1)$
3. $(x - 5)(x + 10)$
4. $(y - 12)(y + 4)$
5. $(5x - 6)(x + 2)$
6. $(3y - 5)(2y - 7)$
7. $(y - 6)(4y - 1)$
8. $(2x - 9)(x - 11)$
9. $(2x + 5)(3x - 1)$
10. $(6x + 2)(x - 2)$

Multiply. See Examples 4 and 5.

11. $(x - 2)^2$
12. $(x + 7)^2$
13. $(2x - 1)^2$
14. $(7x - 3)^2$
15. $(3a - 5)^2$
16. $(5a + 2)^2$
17. $(5x + 9)^2$
18. $(6s - 2)^2$

19. Using your own words, explain how to square a binomial such as $(a + b)^2$.
20. Explain how to find the product of two binomials using the FOIL method.

Multiply. See Example 6.

21. $(a - 7)(a + 7)$
22. $(b + 3)(b - 3)$
23. $(3x - 1)(3x + 1)$
24. $(4x - 5)(4x + 5)$
25. $\left(3x - \dfrac{1}{2}\right)\left(3x + \dfrac{1}{2}\right)$
26. $\left(10x + \dfrac{2}{7}\right)\left(10x - \dfrac{2}{7}\right)$
27. $(9x + y)(9x - y)$
28. $(2x - y)(2x + y)$
29. $(2x + 0.1)(2x - 0.1)$
30. $(5x - 1.3)(5x + 1.3)$

MIXED PRACTICE

Multiply. See Example 7.

31. $(a + 5)(a + 4)$
32. $(a - 5)(a - 7)$
33. $(a + 7)^2$
34. $(b - 2)^2$
35. $(4a + 1)(3a - 1)$
36. $(6a + 7)(6a + 5)$
37. $(x + 2)(x - 2)$
38. $(x - 10)(x + 10)$
39. $(3a + 1)^2$
40. $(4a - 2)^2$
41. $(x^2 + y)(4x - y^4)$
42. $(x^3 - 2)(5x + y)$
43. $(x + 3)(x^2 - 6x + 1)$
44. $(x - 2)(x^2 - 4x + 2)$

45. $(2a - 3)^2$
46. $(5b - 4x)^2$
47. $(5x - 6z)(5x + 6z)$
48. $(11x - 7y)(11x + 7y)$
49. $(x^5 - 3)(x^5 - 5)$
50. $(a^4 + 5)(a^4 + 6)$
51. $\left(x - \dfrac{1}{3}\right)\left(x + \dfrac{1}{3}\right)$
52. $\left(3x + \dfrac{1}{5}\right)\left(3x - \dfrac{1}{5}\right)$
53. $(a^3 + 11)(a^4 - 3)$
54. $(x^5 + 5)(x^2 - 8)$
55. $3(x - 2)^2$
56. $2(3b + 7)^2$
57. $(3b + 7)(2b - 5)$
58. $(3y - 13)(y - 3)$
59. $(7p - 8)(7p + 8)$
60. $(3s - 4)(3s + 4)$
61. $\left(\dfrac{1}{3}a^2 - 7\right)\left(\dfrac{1}{3}a^2 + 7\right)$
62. $\left(\dfrac{2}{3}a - b^2\right)\left(\dfrac{2}{3}a - b^2\right)$
63. $5x^2(3x^2 - x + 2)$
64. $4x^3(2x^2 + 5x - 1)$
65. $(2r - 3s)(2r + 3s)$
66. $(6r - 2x)(6r + 2x)$
67. $(3x - 7y)^2$
68. $(4s - 2y)^2$
69. $(4x + 5)(4x - 5)$
70. $(3x + 5)(3x - 5)$
71. $(8x + 4)^2$
72. $(3x + 2)^2$
73. $\left(a - \dfrac{1}{2}y\right)\left(a + \dfrac{1}{2}y\right)$
74. $\left(\dfrac{a}{2} + 4y\right)\left(\dfrac{a}{2} - 4y\right)$
75. $\left(\dfrac{1}{5}x - y\right)\left(\dfrac{1}{5}x + y\right)$
76. $\left(\dfrac{y}{6} - 8\right)\left(\dfrac{y}{6} + 8\right)$
77. $(a + 1)(3a^2 - a + 1)$
78. $(b + 3)(2b^2 + b - 3)$

Express each as a product of polynomials in x. Then multiply and simplify.

△ 79. Find the area of the square rug shown if its side is $(2x + 1)$ feet.

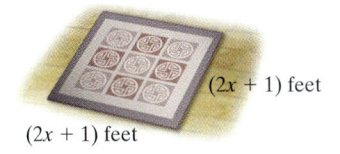

$(2x + 1)$ feet
$(2x + 1)$ feet

△ 80. Find the area of the rectangular canvas if its length is $(3x - 2)$ inches and its width is $(x - 4)$ inches.

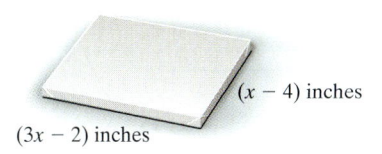

$(x - 4)$ inches
$(3x - 2)$ inches

REVIEW AND PREVIEW

Simplify each expression. See Section 5.1.

81. $\dfrac{50b^{10}}{70b^5}$
82. $\dfrac{x^3 y^6}{xy^2}$
83. $\dfrac{8a^{17}b^{15}}{-4a^7 b^{10}}$
84. $\dfrac{-6a^8 y}{3a^4 y}$
85. $\dfrac{2x^4 y^{12}}{3x^4 y^4}$
86. $\dfrac{-48ab^6}{32ab^3}$

Find the slope of each line. See Section 3.4.

87. 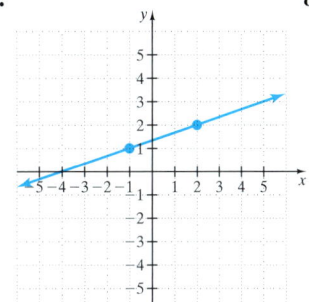
88. 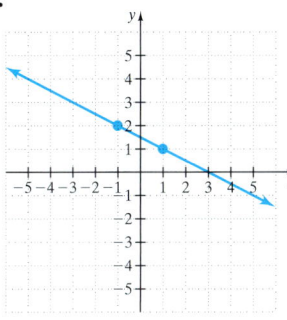
89. 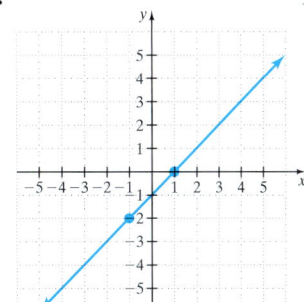
90.

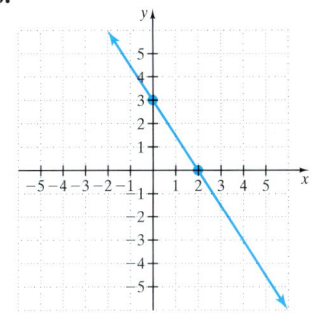

CONCEPT EXTENSIONS

Match each expression on the left to the equivalent expression on the right. See the Concept Check in this section.

91. $(a - b)^2$
92. $(a - b)(a + b)$
93. $(a + b)^2$
94. $(a + b)^2(a - b)^2$

a. $a^2 - b^2$
b. $a^2 + b^2$
c. $a^2 - 2ab + b^2$
d. $a^2 + 2ab + b^2$
e. none of these

Fill in the squares so that a true statement forms.

95. $(x^{\square} + 7)(x^{\square} + 3) = x^4 + 10x^2 + 21$
96. $(5x^{\square} - 2)^2 = 25x^6 - 20x^3 + 4$

336   CHAPTER 5   Exponents and Polynomials

Find the area of each shaded region.

△ **97.**

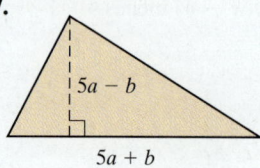

triangle with height $5a - b$ and base $5a + b$

△ **98.**

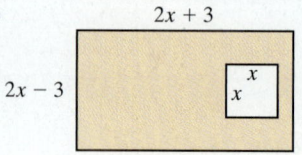

rectangle $2x + 3$ by $2x - 3$ with square of side x removed

△ **99.**

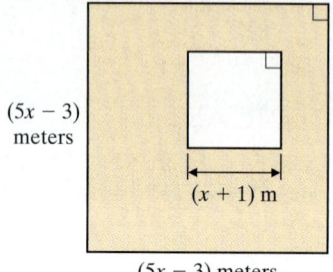

$(5x - 3)$ meters by $(5x - 3)$ meters square with $(x + 1)$ m square removed

△ **100.**

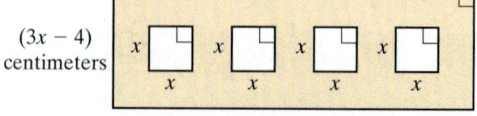

$(3x + 4)$ centimeters by $(3x - 4)$ centimeters rectangle with four squares of side x removed

△ **101.**

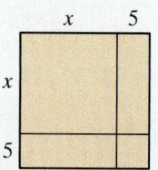

square with sides divided: top x and 5, left x and 5

△ **102.**

square with sides divided: top $2y$ and 11, left $2y$ and 11

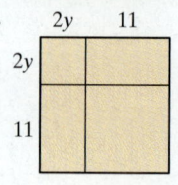

103. In your own words, describe the different methods that can be used to find the product: $(2x - 5)(3x + 1)$.

104. In your own words, describe the different methods that can be used to find the product: $(5x + 1)^2$.

Find each product. For example,

$$[(a + b) - 2][(a + b) + 2] = (a + b)^2 - 2^2$$
$$= a^2 + 2ab + b^2 - 4$$

105. $[(x + y) - 3][(x + y) + 3]$
106. $[(a + c) - 5][(a + c) + 5]$
107. $[(a - 3) + b][(a - 3) - b]$
108. $[(x - 2) + y][(x - 2) - y]$

INTEGRATED REVIEW   EXPONENTS AND OPERATIONS ON POLYNOMIALS

Sections 5.1–5.4

Perform the indicated operations and simplify.

1. $(5x^2)(7x^3)$
2. $(4y^2)(8y^7)$
3. -4^2
4. $(-4)^2$
5. $(x - 5)(2x + 1)$
6. $(3x - 2)(x + 5)$
7. $(x - 5) + (2x + 1)$
8. $(3x - 2) + (x + 5)$
9. $\dfrac{7x^9 y^{12}}{x^3 y^{10}}$
10. $\dfrac{20a^2 b^8}{14a^2 b^2}$
11. $(12m^7 n^6)^2$
12. $(4y^9 z^{10})^3$
13. $3(4y - 3)(4y + 3)$
14. $2(7x - 1)(7x + 1)$
15. $(x^7 y^5)^9$
16. $(3^1 x^9)^3$
17. $(7x^2 - 2x + 3) - (5x^2 + 9)$
18. $(10x^2 + 7x - 9) - (4x^2 - 6x + 2)$
19. $0.7y^2 - 1.2 + 1.8y^2 - 6y + 1$
20. $7.8x^2 - 6.8x + 3.3 + 0.6x^2 - 9$
21. $(x + 4y)^2$
22. $(y - 9z)^2$
23. $(x + 4y) + (x + 4y)$
24. $(y - 9z) + (y - 9z)$
25. $7x^2 - 6xy + 4(y^2 - xy)$
26. $5a^2 - 3ab + 6(b^2 - a^2)$
27. $(x - 3)(x^2 + 5x - 1)$
28. $(x + 1)(x^2 - 3x - 2)$
29. $(2x^3 - 7)(3x^2 + 10)$
30. $(5x^3 - 1)(4x^4 + 5)$
31. $(2x - 7)(x^2 - 6x + 1)$
32. $(5x - 1)(x^2 + 2x - 3)$

Perform exercises and simplify, if possible.

33. $5x^3 + 5y^3$
34. $(5x^3)(5y^3)$
35. $(5x^3)^3$
36. $\dfrac{5x^3}{5y^3}$
37. $x + x$
38. $x \cdot x$

5.5 NEGATIVE EXPONENTS AND SCIENTIFIC NOTATION

OBJECTIVES

1. Simplify expressions containing negative exponents.
2. Use all the rules and definitions for exponents to simplify exponential expressions.
3. Write numbers in scientific notation.
4. Convert numbers from scientific notation to standard form.

OBJECTIVE 1 ▶ Simplifying expressions containing negative exponents. Our work with exponential expressions so far has been limited to exponents that are positive integers or 0. Here we expand to give meaning to an expression like x^{-3}.

Suppose that we wish to simplify the expression $\dfrac{x^2}{x^5}$. If we use the quotient rule for exponents, we subtract exponents:

$$\dfrac{x^2}{x^5} = x^{2-5} = x^{-3}, \quad x \neq 0$$

But what does x^{-3} mean? Let's simplify $\dfrac{x^2}{x^5}$ using the definition of x^n.

$$\dfrac{x^2}{x^5} = \dfrac{x \cdot x}{x \cdot x \cdot x \cdot x \cdot x}$$

$$= \dfrac{x \cdot x}{x \cdot x \cdot x \cdot x \cdot x} \quad \text{Divide numerator and denominator by common factors by applying the fundamental principle for fractions.}$$

$$= \dfrac{1}{x^3}$$

If the quotient rule is to hold true for negative exponents, then x^{-3} must equal $\dfrac{1}{x^3}$. From this example, we state the definition for negative exponents.

Negative Exponents

If a is a real number other than 0 and n is an integer, then

$$a^{-n} = \dfrac{1}{a^n}$$

For example, $x^{-3} = \dfrac{1}{x^3}$.

In other words, another way to write a^{-n} is to take its reciprocal and change the sign of its exponent.

EXAMPLE 1 Simplify by writing each expression with positive exponents only.

a. 3^{-2} **b.** $2x^{-3}$ **c.** $2^{-1} + 4^{-1}$ **d.** $(-2)^{-4}$ **e.** $\dfrac{1}{y^{-4}}$ **f.** $\dfrac{1}{7^{-2}}$

Solution

a. $3^{-2} = \dfrac{1}{3^2} = \dfrac{1}{9}$ Use the definition of negative exponents.

b. $2x^{-3} = 2 \cdot \dfrac{1}{x^3} = \dfrac{2}{x^3}$ Use the definition of negative exponents.

c. $2^{-1} + 4^{-1} = \dfrac{1}{2} + \dfrac{1}{4} = \dfrac{2}{4} + \dfrac{1}{4} = \dfrac{3}{4}$

d. $(-2)^{-4} = \dfrac{1}{(-2)^4} = \dfrac{1}{(-2)(-2)(-2)(-2)} = \dfrac{1}{16}$

e. $\dfrac{1}{y^{-4}} = \dfrac{1}{\dfrac{1}{y^4}} = y^4$ **f.** $\dfrac{1}{7^{-2}} = \dfrac{1}{\dfrac{1}{7^2}} = \dfrac{7^2}{1}$ or 49

▶ **Helpful Hint**
Don't forget that since there are no parentheses, only x is the base for the exponent -3.

CHAPTER 5 Exponents and Polynomials

> **PRACTICE**
> **1** Simplify by writing each expression with positive exponents only.
>
> a. 5^{-3} b. $3y^{-4}$ c. $3^{-1} + 2^{-1}$
>
> d. $(-5)^{-2}$ e. $\dfrac{1}{x^{-5}}$ f. $\dfrac{1}{4^{-3}}$

> ▶ **Helpful Hint**
>
> A negative exponent *does not affect* the sign of its base.
> Remember: Another way to write a^{-n} is to take its reciprocal and change the sign of its exponent: $a^{-n} = \dfrac{1}{a^n}$. For example,
>
> $$x^{-2} = \dfrac{1}{x^2}, \qquad 2^{-3} = \dfrac{1}{2^3} \text{ or } \dfrac{1}{8}$$
>
> $$\dfrac{1}{y^{-4}} = \dfrac{1}{\frac{1}{y^4}} = y^4, \qquad \dfrac{1}{5^{-2}} = 5^2 \text{ or } 25$$

From the preceding Helpful Hint, we know that $x^{-2} = \dfrac{1}{x^2}$ and $\dfrac{1}{y^{-4}} = y^4$. We can use this to include another statement in our definition of negative exponents.

> **Negative Exponents**
>
> If a is a real number other than 0 and n is an integer, then
>
> $$a^{-n} = \dfrac{1}{a^n} \quad \text{and} \quad \dfrac{1}{a^{-n}} = a^n$$

EXAMPLE 2 Simplify each expression. Write results using positive exponents only.

a. $\dfrac{1}{x^{-3}}$ b. $\dfrac{1}{3^{-4}}$ c. $\dfrac{p^{-4}}{q^{-9}}$ d. $\dfrac{5^{-3}}{2^{-5}}$

Solution

a. $\dfrac{1}{x^{-3}} = \dfrac{x^3}{1} = x^3$ b. $\dfrac{1}{3^{-4}} = \dfrac{3^4}{1} = 81$ c. $\dfrac{p^{-4}}{q^{-9}} = \dfrac{q^9}{p^4}$ d. $\dfrac{5^{-3}}{2^{-5}} = \dfrac{2^5}{5^3} = \dfrac{32}{125}$

> **PRACTICE**
> **2** Simplify each expression. Write results using positive exponents only.
>
> a. $\dfrac{1}{s^{-5}}$ b. $\dfrac{1}{2^{-3}}$ c. $\dfrac{x^{-7}}{y^{-5}}$ d. $\dfrac{4^{-3}}{3^{-2}}$

EXAMPLE 3 Simplify each expression. Write answers with positive exponents.

a. $\dfrac{y}{y^{-2}}$ b. $\dfrac{3}{x^{-4}}$ c. $\dfrac{x^{-5}}{x^7}$

Solution

a. $\dfrac{y}{y^{-2}} = \dfrac{y^1}{y^{-2}} = y^{1-(-2)} = y^3$ Remember that $\dfrac{a^m}{a^n} = a^{m-n}$.

b. $\dfrac{3}{x^{-4}} = 3 \cdot \dfrac{1}{x^{-4}} = 3 \cdot x^4$ or $3x^4$

c. $\dfrac{x^{-5}}{x^7} = x^{-5-7} = x^{-12} = \dfrac{1}{x^{12}}$

PRACTICE 3 Simplify each expression. Write answers with positive exponents.

a. $\dfrac{x^{-3}}{x^2}$ **b.** $\dfrac{5}{y^{-7}}$ **c.** $\dfrac{z}{z^{-4}}$

OBJECTIVE 2 ▶ Simplifying exponential expressions. All the previously stated rules for exponents apply for negative exponents also. Here is a summary of the rules and definitions for exponents.

> **Summary of Exponent Rules**
> If m and n are integers and a, b, and c are real numbers, then:
> Product rule for exponents: $a^m \cdot a^n = a^{m+n}$
> Power rule for exponents: $(a^m)^n = a^{m \cdot n}$
> Power of a product: $(ab)^n = a^n b^n$
> Power of a quotient: $\left(\dfrac{a}{c}\right)^n = \dfrac{a^n}{c^n},\ c \neq 0$
> Quotient rule for exponents: $\dfrac{a^m}{a^n} = a^{m-n},\ a \neq 0$
> Zero exponent: $a^0 = 1,\ a \neq 0$
> Negative exponent: $a^{-n} = \dfrac{1}{a^n},\ a \neq 0$

EXAMPLE 4 Simplify the following expressions. Write each result using positive exponents only.

a. $\left(\dfrac{2}{3}\right)^{-3}$ **b.** $\dfrac{(x^3)^4 x}{x^7}$ **c.** $\left(\dfrac{3a^2}{b}\right)^{-3}$ **d.** $\dfrac{4^{-1} x^{-3} y}{4^{-3} x^2 y^{-6}}$ **e.** $(y^{-3} z^6)^{-6}$ **f.** $\left(\dfrac{-2x^3 y}{xy^{-1}}\right)^3$

Solution

a. $\left(\dfrac{2}{3}\right)^{-3} = \dfrac{2^{-3}}{3^{-3}} = \dfrac{3^3}{2^3} = \dfrac{27}{8}$

b. $\dfrac{(x^3)^4 x}{x^7} = \dfrac{x^{12} \cdot x}{x^7} = \dfrac{x^{12+1}}{x^7} = \dfrac{x^{13}}{x^7} = x^{13-7} = x^6$ Use the power rule.

c. $\left(\dfrac{3a^2}{b}\right)^{-3} = \dfrac{3^{-3}(a^2)^{-3}}{b^{-3}}$ Raise each factor in the numerator and the denominator to the -3 power.

$= \dfrac{3^{-3} a^{-6}}{b^{-3}}$ Use the power rule.

$= \dfrac{b^3}{3^3 a^6}$ Use the negative exponent rule.

$= \dfrac{b^3}{27 a^6}$ Write 3^3 as 27.

d. $\dfrac{4^{-1}x^{-3}y}{4^{-3}x^2y^{-6}} = 4^{-1-(-3)}x^{-3-2}y^{1-(-6)} = 4^2x^{-5}y^7 = \dfrac{4^2y^7}{x^5} = \dfrac{16y^7}{x^5}$

e. $(y^{-3}z^6)^{-6} = y^{18} \cdot z^{-36} = \dfrac{y^{18}}{z^{36}}$

f. $\left(\dfrac{-2x^3y}{xy^{-1}}\right)^3 = \dfrac{(-2)^3x^9y^3}{x^3y^{-3}} = \dfrac{-8x^9y^3}{x^3y^{-3}} = -8x^{9-3}y^{3-(-3)} = -8x^6y^6$

PRACTICE
4 Simplify the following expression. Write each result using positive exponents only.

a. $\left(\dfrac{3}{4}\right)^{-2}$
b. $\dfrac{x^2(x^5)^3}{x^7}$
c. $\left(\dfrac{5p^8}{q}\right)^{-2}$
d. $\dfrac{6^{-2}x^{-4}y^{-7}}{6^{-3}x^3y^{-9}}$
e. $(a^4b^{-3})^{-5}$
f. $\left(\dfrac{-3x^4y}{x^2y^{-2}}\right)^3$

OBJECTIVE 3 ▶ Writing numbers in scientific notation. Both very large and very small numbers frequently occur in many fields of science. For example, the distance between the sun and the dwarf planet Pluto is approximately 5,906,000,000 kilometers, and the mass of a proton is approximately 0.00000000000000000000000165 gram. It can be tedious to write these numbers in this standard decimal notation, so **scientific notation** is used as a convenient shorthand for expressing very large and very small numbers.

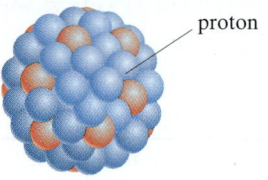

Mass of proton is approximately
0.000 000 000 000 000 000 000 001 65 gram

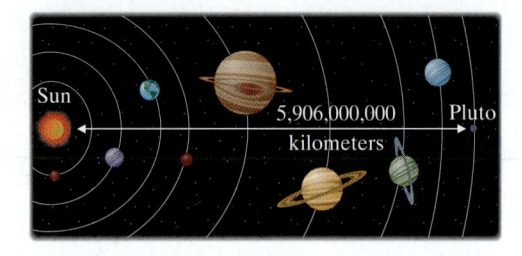

Scientific Notation
A positive number is written in scientific notation if it is written as the product of a number a, where $1 \leq a < 10$, and an integer power r of 10:

$$a \times 10^r$$

The numbers below are written in scientific notation. The × sign for multiplication is used as part of the notation.

2.03×10^2 $\quad 7.362 \times 10^7 \quad$ 5.906×10^9 (Distance between the sun and Pluto)
1×10^{-3} $\quad 8.1 \times 10^{-5} \quad$ 1.65×10^{-24} (Mass of a proton)

The following steps are useful when writing numbers in scientific notation.

To Write a Number in Scientific Notation
STEP 1. Move the decimal point in the original number to the left or right so that the new number has a value between 1 and 10.

STEP 2. Count the number of decimal places the decimal point is moved in Step 1. If the original number is 10 or greater, the count is positive. If the original number is less than 1, the count is negative.

STEP 3. Multiply the new number in Step 1 by 10 raised to an exponent equal to the count found in Step 2.

EXAMPLE 5 Write each number in scientific notation.

a. 367,000,000 **b.** 0.000003 **c.** 20,520,000,000 **d.** 0.00085

Solution

a. STEP 1. Move the decimal point until the number is between 1 and 10.
367,000,000.
8 places

STEP 2. The decimal point is moved 8 places, and the original number is 10 or greater, so the count is positive 8.

STEP 3. $367,000,000 = 3.67 \times 10^8$.

b. STEP 1. Move the decimal point until the number is between 1 and 10.
0.000003
6 places

STEP 2. The decimal point is moved 6 places, and the original number is less than 1, so the count is -6.

STEP 3. $0.000003 = 3.0 \times 10^{-6}$

c. $20,520,000,000 = 2.052 \times 10^{10}$

d. $0.00085 = 8.5 \times 10^{-4}$

PRACTICE 5 Write each number in scientific notation.

a. 0.000007 **b.** 20,700,000 **c.** 0.0043 **d.** 812,000,000

OBJECTIVE 4 ▶ Converting numbers to standard form. A number written in scientific notation can be rewritten in standard form. For example, to write 8.63×10^3 in standard form, recall that $10^3 = 1000$.

$$8.63 \times 10^3 = 8.63(1000) = 8630$$

Notice that the exponent on the 10 is positive 3, and we moved the decimal point 3 places to the right.

To write 7.29×10^{-3} in standard form, recall that $10^{-3} = \dfrac{1}{10^3} = \dfrac{1}{1000}$.

$$7.29 \times 10^{-3} = 7.29\left(\dfrac{1}{1000}\right) = \dfrac{7.29}{1000} = 0.00729$$

The exponent on the 10 is negative 3, and we moved the decimal to the left 3 places.

In general, **to write a scientific notation number in standard form,** move the decimal point the same number of places as the exponent on 10. If the exponent is positive, move the decimal point to the right; if the exponent is negative, move the decimal point to the left.

EXAMPLE 6 Write each number in standard notation, without exponents.

a. 1.02×10^5 **b.** 7.358×10^{-3} **c.** 8.4×10^7 **d.** 3.007×10^{-5}

Solution

a. Move the decimal point 5 places to the right.

$$1.02 \times 10^5 = 102,000.$$

b. Move the decimal point 3 places to the left.

$$7.358 \times 10^{-3} = 0.007358$$

c. $8.4 \times 10^7 = 84{,}000{,}000.$ 7 places to the right

d. $3.007 \times 10^{-5} = 0.00003007$ 5 places to the left

PRACTICE

6 Write each number in standard notation, without exponents.

a. 3.67×10^{-4} b. 8.954×10^6 c. 2.009×10^{-5} d. 4.054×10^3

Concept Check ✓

Which number in each pair is larger?

a. 7.8×10^3 or 2.1×10^5 b. 9.2×10^{-2} or 2.7×10^4 c. 5.6×10^{-4} or 6.3×10^{-5}

Performing operations on numbers written in scientific notation makes use of the rules and definitions for exponents.

EXAMPLE 7   Perform each indicated operation. Write each result in standard decimal notation.

a. $(8 \times 10^{-6})(7 \times 10^3)$ b. $\dfrac{12 \times 10^2}{6 \times 10^{-3}}$

Solution

a. $(8 \times 10^{-6})(7 \times 10^3) = (8 \cdot 7) \times (10^{-6} \cdot 10^3)$
$= 56 \times 10^{-3}$
$= 0.056$

b. $\dfrac{12 \times 10^2}{6 \times 10^{-3}} = \dfrac{12}{6} \times 10^{2-(-3)} = 2 \times 10^5 = 200{,}000$

PRACTICE

7 Perform each indicated operation. Write each result in standard decimal notation.

a. $(5 \times 10^{-4})(8 \times 10^6)$ b. $\dfrac{64 \times 10^3}{32 \times 10^{-7}}$

Answers to Concept Check:
a. 2.1×10^5 b. 2.7×10^4
c. 5.6×10^{-4}

Calculator Explorations

Scientific Notation

To enter a number written in scientific notation on a scientific calculator, locate the scientific notation key, which may be marked $\boxed{\text{EE}}$ or $\boxed{\text{EXP}}$. To enter 3.1×10^7, press $\boxed{3.1}$ $\boxed{\text{EE}}$ $\boxed{7}$. The display should read $\boxed{3.1 \quad 07}$.

Enter each number written in scientific notation on your calculator.

1. 5.31×10^3
2. -4.8×10^{14}
3. 6.6×10^{-9}
4. -9.9811×10^{-2}

Multiply each of the following on your calculator. Notice the form of the result.

5. $3{,}000{,}000 \times 5{,}000{,}000$
6. $230{,}000 \times 1000$

Multiply each of the following on your calculator. Write the product in scientific notation.

7. $(3.26 \times 10^6)(2.5 \times 10^{13})$
8. $(8.76 \times 10^{-4})(1.237 \times 10^9)$

VOCABULARY & READINESS CHECK

Fill in each blank with the correct choice.

1. The expression x^{-3} equals _____.
   a. $-x^3$ b. $\dfrac{1}{x^3}$ c. $\dfrac{-1}{x^3}$ d. $\dfrac{1}{x^{-3}}$

2. The expression 5^{-4} equals _____.
   a. -20 b. -625 c. $\dfrac{1}{20}$ d. $\dfrac{1}{625}$

3. The number 3.021×10^{-3} is written in _____.
   a. standard form   b. expanded form
   c. scientific notation

4. The number 0.0261 is written in _____.
   a. standard form   b. expanded form
   c. scientific notation

Write each expression using positive exponents only.

5. $5x^{-2}$ 6. $3x^{-3}$ 7. $\dfrac{1}{y^{-6}}$ 8. $\dfrac{1}{x^{-3}}$ 9. $\dfrac{4}{y^{-3}}$ 10. $\dfrac{16}{y^{-7}}$

5.5 EXERCISE SET

Simplify each expression. Write each result using positive exponents only. See Examples 1 through 3.

1. 4^{-3}
2. 6^{-2}
3. $(-2)^{-4}$
4. $(-3)^{-5}$
5. $7x^{-3}$
6. $(7x)^{-3}$
7. $\left(\dfrac{1}{2}\right)^{-5}$
8. $\left(\dfrac{1}{8}\right)^{-2}$
9. $\left(-\dfrac{1}{4}\right)^{-3}$
10. $\left(-\dfrac{1}{8}\right)^{-2}$
11. $3^{-1}+2^{-1}$
12. $4^{-1}+4^{-2}$
13. $\dfrac{1}{p^{-3}}$
14. $\dfrac{1}{q^{-5}}$
15. $\dfrac{p^{-5}}{q^{-4}}$
16. $\dfrac{r^{-5}}{s^{-2}}$
17. $\dfrac{x^{-2}}{x}$
18. $\dfrac{y}{y^{-3}}$
19. $\dfrac{z^{-4}}{z^{-7}}$
20. $\dfrac{x^{-4}}{x^{-1}}$
21. $3^{-2}+3^{-1}$
22. $4^{-2}-4^{-3}$
23. $\dfrac{-1}{p^{-4}}$
24. $\dfrac{-1}{y^{-6}}$
25. -2^0-3^0
26. $5^0+(-5)^0$

MIXED PRACTICE

Simplify each expression. Write each result using positive exponents only. See Examples 1 through 4.

27. $\dfrac{x^2 x^5}{x^3}$
28. $\dfrac{y^4 y^5}{y^6}$
29. $\dfrac{p^2 p}{p^{-1}}$
30. $\dfrac{y^3 y}{y^{-2}}$
31. $\dfrac{(m^5)^4 m}{m^{10}}$
32. $\dfrac{(x^2)^8 x}{x^9}$
33. $\dfrac{r}{r^{-3}r^{-2}}$
34. $\dfrac{p}{p^{-3}q^{-5}}$
35. $(x^5 y^3)^{-3}$
36. $(z^5 x^5)^{-3}$
37. $\dfrac{(x^2)^3}{x^{10}}$
38. $\dfrac{(y^4)^2}{y^{12}}$
39. $\dfrac{(a^5)^2}{(a^3)^4}$
40. $\dfrac{(x^2)^5}{(x^4)^3}$
41. $\dfrac{8k^4}{2k}$
42. $\dfrac{27r^4}{3r^6}$
43. $\dfrac{-6m^4}{-2m^3}$
44. $\dfrac{15a^4}{-15a^5}$
45. $\dfrac{-24a^6 b}{6ab^2}$
46. $\dfrac{-5x^4 y^5}{15x^4 y^2}$
47. $(-2x^3 y^{-4})(3x^{-1}y)$
48. $(-5a^4 b^{-7})(-a^{-4}b^3)$
49. $(a^{-5}b^2)^{-6}$
50. $(4^{-1}x^5)^{-2}$
51. $\left(\dfrac{x^{-2}y^4}{x^3 y^7}\right)^2$
52. $\left(\dfrac{a^5 b}{a^7 b^{-2}}\right)^{-3}$
53. $\dfrac{4^2 z^{-3}}{4^3 z^{-5}}$
54. $\dfrac{3^{-1}x^4}{3^3 x^{-7}}$
55. $\dfrac{2^{-3}x^{-4}}{2^2 x}$
56. $\dfrac{5^{-1}z^7}{5^{-2}z^9}$
57. $\dfrac{7ab^{-4}}{7^{-1}a^{-3}b^2}$
58. $\dfrac{6^{-5}x^{-1}y^2}{6^{-2}x^{-4}y^4}$
59. $\left(\dfrac{a^{-5}b}{ab^3}\right)^{-4}$
60. $\left(\dfrac{r^{-2}s^{-3}}{r^{-4}s^{-3}}\right)^{-3}$
61. $\dfrac{(xy^3)^5}{(xy)^{-4}}$
62. $\dfrac{(rs)^{-3}}{(r^2 s^3)^2}$
63. $\dfrac{(-2xy^{-3})^{-3}}{(xy^{-1})^{-1}}$
64. $\dfrac{(-3x^2 y^2)^{-2}}{(xyz)^{-2}}$
65. $\dfrac{6x^2 y^3}{-7xy^5}$
66. $\dfrac{-8xa^2 b}{-5xa^5 b}$
67. $\dfrac{(a^4 b^{-7})^{-5}}{(5a^2 b^{-1})^{-2}}$
68. $\dfrac{(a^6 b^{-2})^4}{(4a^{-3}b^{-3})^3}$

Write each number in scientific notation. See Example 5.

69. 78,000
70. 9,300,000,000
71. 0.00000167
72. 0.00000017
73. 0.00635
74. 0.00194
75. 1,160,000
76. 700,000
77. More than 2,000,000,000 pencils are manufactured in the United States annually. Write this number in scientific notation (*Source*: AbsoluteTrivia.com)
78. The temperature at the interior of the Earth is 20,000,000 degrees Celsius. Write 20,000,000 in scientific notation.

344 CHAPTER 5 Exponents and Polynomials

79. The Cassini-Huygens Space Mission to Saturn was launched October 15, 1997, with a goal of reaching and orbiting Saturn and its moons. When the Cassini spacecraft disconnected the Huygens probe, which landed on the surface of the moon Titan on January 14, 2005, it was approximately 1,212,000,000 km from earth. Write 1,212,000,000 in scientific notation. (*Source:* Jet Propulsion Laboratory, California Institute of Technology)

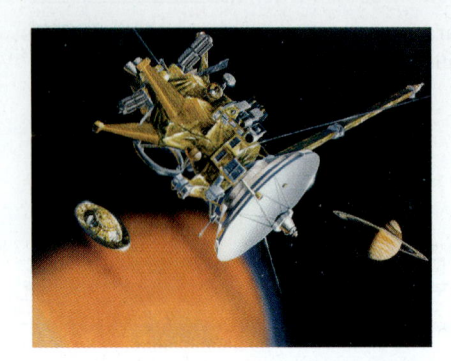

80. At this writing, the world's largest optical telescopes are the twin Keck Telescopes located near the summit of Mauna Kea in Hawaii. The elevation of the Keck Telescopes is about 13,600 feet above sea level. Write 13,600 in scientific notation. (*Source:* W.M. Keck Observatory)

Write each number in standard notation. See Example 6.

81. 8.673×10^{-10}

82. 9.056×10^{-4}

83. 3.3×10^{-2}

84. 4.8×10^{-6}

85. 2.032×10^{4}

86. 9.07×10^{10}

87. Each second, the Sun converts 7.0×10^8 tons of hydrogen into helium and energy in the form of gamma rays. Write this number in standard notation. (*Source:* Students for the Exploration and Development of Space)

88. In chemistry, Avogadro's number is the number of atoms in one mole of an element. Avogadro's number is $6.02214199 \times 10^{23}$. Write this number in standard notation. (*Source:* National Institute of Standards and Technology)

89. The distance light travels in 1 year is 9.460×10^{12} kilometers. Write this number in standard notation.

90. The population of the world is 6.067×10^9. Write this number in standard notation. (*Source:* U.S. Bureau of the Census)

MIXED PRACTICE

See Examples 5 and 6. Below are some interesting facts about the Internet. If a number is written in standard form, write it in scientific notation. If a number is written in scientific notation, write it in standard form.

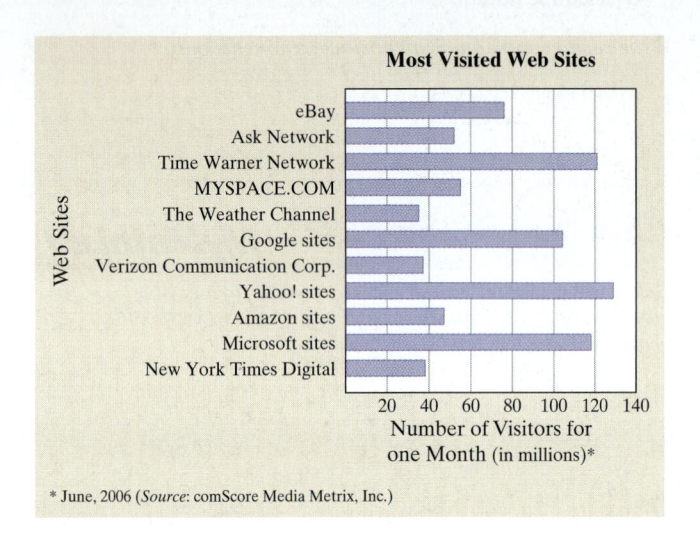

The bar graph above shows the most visited Web sites on the computer.

91. Estimate the length of the longest bar. Then write the number in scientific notation.

92. Estimate the length of the shortest bar. Then write the number in scientific notation.

93. The total number of Internet users exceeds 1,000,000,000. (*Source:* Computer Industry, Almanac)

94. In a recent year, the retail sales generated by the Internet was 1.08×10^{11} dollars. (*Source:* U.S. Census Bureau)

95. An estimated 5.7×10^7 American adults read online weblogs (blogs). (*Source:* PEW Internet & American Life Project)

96. Junk e-mail (SPAM) costs consumers and businesses an estimated $23,000,000,000.

Evaluate each expression using exponential rules. Write each result in standard notation. See Example 7.

97. $(1.2 \times 10^{-3})(3 \times 10^{-2})$

98. $(2.5 \times 10^6)(2 \times 10^{-6})$

99. $(4 \times 10^{-10})(7 \times 10^{-9})$

100. $(5 \times 10^6)(4 \times 10^{-8})$

101. $\dfrac{8 \times 10^{-1}}{16 \times 10^5}$

102. $\dfrac{25 \times 10^{-4}}{5 \times 10^{-9}}$

103. $\dfrac{1.4 \times 10^{-2}}{7 \times 10^{-8}}$

104. $\dfrac{0.4 \times 10^5}{0.2 \times 10^{11}}$

REVIEW AND PREVIEW

Simplify the following. See Section 5.1.

105. $\dfrac{5x^7}{3x^4}$
106. $\dfrac{27y^{14}}{3y^7}$
107. $\dfrac{15z^4y^3}{21zy}$
108. $\dfrac{18a^7b^{17}}{30a^7b}$

Use the distributive property and multiply. See Sections 5.3 and 5.5.

109. $\dfrac{1}{y}(5y^2 - 6y + 5)$
110. $\dfrac{2}{x}(3x^5 + x^4 - 2)$

CONCEPT EXTENSIONS

△ 111. Find the volume of the cube.

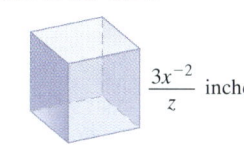

$\dfrac{3x^{-2}}{z}$ inches

△ 112. Find the area of the triangle.

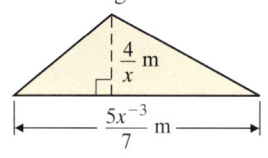

$\dfrac{4}{x}$ m

$\dfrac{5x^{-3}}{7}$ m

Simplify.

113. $(2a^3)^3 a^4 + a^5 a^8$
114. $(2a^3)^3 a^{-3} + a^{11} a^{-5}$

Fill in the boxes so that each statement is true. (More than one answer is possible for these exercises.)

115. $x^\square = \dfrac{1}{x^5}$
116. $7^\square = \dfrac{1}{49}$
117. $z^\square \cdot z^\square = z^{-10}$
118. $(x^\square)^\square = x^{-15}$

119. Which is larger? See the Concept Check in this section.
   a. 9.7×10^{-2} or 1.3×10^1
   b. 8.6×10^5 or 4.4×10^7
   c. 6.1×10^{-2} or 5.6×10^{-4}

120. It was stated earlier that for an integer n,
$$x^{-n} = \dfrac{1}{x^n}, \quad x \neq 0$$
   Explain why x may not equal 0.

121. Determine whether each statement is true or false.
   a. $5^{-1} < 5^{-2}$
   b. $\left(\dfrac{1}{5}\right)^{-1} < \left(\dfrac{1}{5}\right)^{-2}$
   c. $a^{-1} < a^{-2}$ for all nonzero numbers.

Simplify each expression. Assume that variables represent positive integers.

122. $a^{-4m} \cdot a^{5m}$
123. $(x^{-3s})^3$
124. $(3y^{2z})^3$
125. $a^{4m+1} \cdot a^4$

Simplify each expression. Write each result in standard notation.

126. $(2.63 \times 10^{12})(-1.5 \times 10^{-10})$
127. $(6.785 \times 10^{-4})(4.68 \times 10^{10})$

Light travels at a rate of 1.86×10^5 miles per second. Use this information and the distance formula $d = r \cdot t$ to answer Exercises 128 and 129.

128. If the distance from the moon to the Earth is 238,857 miles, find how long it takes the reflected light of the moon to reach the Earth. (Round to the nearest tenth of a second.)

129. If the distance from the sun to the Earth is 93,000,000 miles, find how long it takes the light of the sun to reach the Earth. (Round to the nearest tenth of a second.)

5.6 DIVIDING POLYNOMIALS

OBJECTIVES

1. Divide a polynomial by a monomial.
2. Use long division to divide a polynomial by another polynomial.

OBJECTIVE 1 ▶ Dividing by a monomial. Now that we know how to add, subtract, and multiply polynomials, we practice dividing polynomials.

To divide a polynomial by a monomial, recall addition of fractions. Fractions that have a common denominator are added by adding the numerators:

$$\dfrac{a}{c} + \dfrac{b}{c} = \dfrac{a+b}{c}$$

If we read this equation from right to left and let a, b, and c be monomials, $c \neq 0$, we have the following:

> **Dividing a Polynomial By a Monomial**
> Divide each term of the polynomial by the monomial.
> $$\dfrac{a+b}{c} = \dfrac{a}{c} + \dfrac{b}{c}, \quad c \neq 0$$

Throughout this section, we assume that denominators are not 0.

EXAMPLE 1 Divide $6m^2 + 2m$ by $2m$.

Solution We begin by writing the quotient in fraction form. Then we divide each term of the polynomial $6m^2 + 2m$ by the monomial $2m$.

$$\frac{6m^2 + 2m}{2m} = \frac{6m^2}{2m} + \frac{2m}{2m}$$

$$= 3m + 1 \quad \text{Simplify.}$$

Check: We know that if $\dfrac{6m^2 + 2m}{2m} = 3m + 1$, then $2m \cdot (3m + 1)$ must equal $6m^2 + 2m$. Thus, to check, we multiply.

$$2m(3m + 1) = 2m(3m) + 2m(1) = 6m^2 + 2m$$

The quotient $3m + 1$ checks. □

PRACTICE 1 Divide $8t^3 + 4t^2$ by $4t^2$.

EXAMPLE 2 Divide $\dfrac{9x^5 - 12x^2 + 3x}{3x^2}$.

Solution
$$\frac{9x^5 - 12x^2 + 3x}{3x^2} = \frac{9x^5}{3x^2} - \frac{12x^2}{3x^2} + \frac{3x}{3x^2} \quad \text{Divide each term by } 3x^2.$$

$$= 3x^3 - 4 + \frac{1}{x} \quad \text{Simplify.}$$

Notice that the quotient is not a polynomial because of the term $\dfrac{1}{x}$. This expression is called a rational expression—we will study rational expressions further in Chapter 7. Although the quotient of two polynomials is not always a polynomial, we may still check by multiplying.

Check:
$$3x^2\left(3x^3 - 4 + \frac{1}{x}\right) = 3x^2(3x^3) - 3x^2(4) + 3x^2\left(\frac{1}{x}\right)$$

$$= 9x^5 - 12x^2 + 3x \quad \square$$

PRACTICE 2 Divide $\dfrac{16x^6 + 20x^3 - 12x}{4x^2}$.

EXAMPLE 3 Divide $\dfrac{8x^2y^2 - 16xy + 2x}{4xy}$.

Solution
$$\frac{8x^2y^2 - 16xy + 2x}{4xy} = \frac{8x^2y^2}{4xy} - \frac{16xy}{4xy} + \frac{2x}{4xy} \quad \text{Divide each term by } 4xy.$$

$$= 2xy - 4 + \frac{1}{2y} \quad \text{Simplify.}$$

Check: $4xy\left(2xy - 4 + \dfrac{1}{2y}\right) = 4xy(2xy) - 4xy(4) + 4xy\left(\dfrac{1}{2y}\right)$

$= 8x^2y^2 - 16xy + 2x$ ☐

PRACTICE 3 Divide $\dfrac{15x^4y^4 - 10xy + y}{5xy}$.

Concept Check ✓

In which of the following is $\dfrac{x+5}{5}$ simplified correctly?

a. $\dfrac{x}{5} + 1$ **b.** x **c.** $x + 1$

OBJECTIVE 2 ▶ Using long division to divide by a polynomial. To divide a polynomial by a polynomial other than a monomial, we use a process known as long division. Polynomial long division is similar to number long division, so we review long division by dividing 13 into 3660.

$$\begin{array}{r} 281 \\ 13\overline{)3660} \\ \underline{26\downarrow} \\ 106 \\ \underline{104\downarrow} \\ 20 \\ \underline{13} \\ 7 \end{array}$$

▶ **Helpful Hint**
Recall that 3660 is called the dividend.

$2 \cdot 13 = 26$
Subtract and bring down the next digit in the dividend.
$8 \cdot 13 = 104$
Subtract and bring down the next digit in the dividend.
$1 \cdot 13 = 13$
Subtract. There are no more digits to bring down, so the remainder is 7.

The quotient is 281 R 7, which can be written as $281\dfrac{7}{13}$ ← remainder
← divisor

Recall that division can be checked by multiplication. To check a division problem such as this one, we see that

$$13 \cdot 281 + 7 = 3660$$

Now we demonstrate long division of polynomials.

EXAMPLE 4 Divide $x^2 + 7x + 12$ by $x + 3$ using long division.

Solution

To subtract, change the signs of these terms and add.

$$\begin{array}{r} x \\ x + 3\overline{)x^2 + 7x + 12} \\ \underline{\overline{x^2 \mp 3x}}\downarrow \\ 4x + 12 \end{array}$$

How many times does x divide x^2? $\dfrac{x^2}{x} = x$.
Multiply: $x(x+3)$.
Subtract and bring down the next term.

Now we repeat this process.

$$\begin{array}{r} x + 4 \\ x + 3\overline{)x^2 + 7x + 12} \\ \underline{\overline{x^2 \mp 3x}} \\ 4x + 12 \\ \underline{\overline{4x \mp 12}} \\ 0 \end{array}$$

How many times does x divide $4x$? $\dfrac{4x}{x} = 4$.

To subtract, change the signs of these terms and add.

Multiply: $4(x+3)$.
Subtract. The remainder is 0.

The quotient is $x + 4$.

Answer to Concept Check: a

348 CHAPTER 5 Exponents and Polynomials

Check: We check by multiplying.

$$\underset{\downarrow}{\text{divisor}} \cdot \underset{\downarrow}{\text{quotient}} + \underset{\downarrow}{\text{remainder}} = \underset{\downarrow}{\text{dividend}}$$

or

$$(x + 3) \cdot (x + 4) + 0 = x^2 + 7x + 12$$

The quotient checks. □

PRACTICE 4 Divide $x^2 + 5x + 6$ by $x + 2$ using long division.

EXAMPLE 5 Divide $6x^2 + 10x - 5$ by $3x - 1$ using long division.

Solution

$$\begin{array}{r} 2x + 4 \\ 3x - 1 \overline{) 6x^2 + 10x - 5} \\ \underline{-6x^2 \mp 2x} \downarrow \\ 12x - 5 \\ \underline{-12x \mp 4} \\ -1 \end{array}$$

$\dfrac{6x^2}{3x} = 2x$, so $2x$ is a term of the quotient.
Multiply $2x(3x - 1)$.
Subtract and bring down the next term.
$\dfrac{12x}{3x} = 4$, multiply $4(3x - 1)$
Subtract. The remainder is -1.

Thus $(6x^2 + 10x - 5)$ divided by $(3x - 1)$ is $(2x + 4)$ with a remainder of -1. This can be written as

$$\dfrac{6x^2 + 10x - 5}{3x - 1} = 2x + 4 + \dfrac{-1}{3x - 1} \quad \begin{array}{l} \leftarrow \text{remainder} \\ \leftarrow \text{divisor} \end{array}$$

Check: To check, we multiply $(3x - 1)(2x + 4)$. Then we add the remainder, -1, to this product.

$$(3x - 1)(2x + 4) + (-1) = (6x^2 + 12x - 2x - 4) - 1$$
$$= 6x^2 + 10x - 5$$

The quotient checks. □

PRACTICE 5 Divide $4x^2 + 8x - 7$ by $2x + 1$ using long division.

In Example 5, the degree of the divisor, $3x - 1$, is 1 and the degree of the remainder, -1, is 0. The division process is continued until the degree of the remainder polynomial is less than the degree of the divisor polynomial.

EXAMPLE 6 Divide $\dfrac{4x^2 + 7 + 8x^3}{2x + 3}$.

Solution Before we begin the division process, we rewrite

$$4x^2 + 7 + 8x^3 \quad \text{as} \quad 8x^3 + 4x^2 + 0x + 7$$

Notice that we have written the polynomial in descending order and have represented the missing x term by $0x$.

$$
\begin{array}{r}
4x^2 - 4x + 6 \\
2x + 3 \overline{\smash{)}\, 8x^3 + 4x^2 + 0x + 7} \\
\underline{8x^3 + 12x^2} \\
-8x^2 + 0x \\
\underline{-8x^2 - 12x} \\
12x + 7 \\
\underline{12x + 18} \\
-11 \quad \text{Remainder}
\end{array}
$$

Thus, $\dfrac{4x^2 + 7 + 8x^3}{2x + 3} = 4x^2 - 4x + 6 + \dfrac{-11}{2x + 3}$.

PRACTICE 6 Divide $\dfrac{11x - 3 + 9x^3}{3x + 2}$.

EXAMPLE 7 Divide $\dfrac{2x^4 - x^3 + 3x^2 + x - 1}{x^2 + 1}$.

Solution Before dividing, rewrite the divisor polynomial

$$x^2 + 1 \quad \text{as} \quad x^2 + 0x + 1$$

The $0x$ term represents the missing x^1 term in the divisor.

$$
\begin{array}{r}
2x^2 - x + 1 \\
x^2 + 0x + 1 \overline{\smash{)}\, 2x^4 - x^3 + 3x^2 + x - 1} \\
\underline{2x^4 + 0x^3 + 2x^2} \\
-x^3 + x^2 + x \\
\underline{-x^3 - 0x^2 - x} \\
x^2 + 2x - 1 \\
\underline{x^2 + 0x + 1} \\
2x - 1 \quad \text{Remainder}
\end{array}
$$

Thus, $\dfrac{2x^4 - x^3 + 3x^2 + x - 1}{x^2 + 1} = 2x^2 - x + 1 + \dfrac{2x - 2}{x^2 + 1}$.

PRACTICE 7 Divide $\dfrac{3x^4 - 2x^3 - 3x^2 + x + 4}{x^2 + 2}$.

VOCABULARY & READINESS CHECK

Use the choices below to fill in each blank. Choices may be used more than once.

dividend   divisor   quotient

1. In $6\overline{\smash{)}\, 18}^{\,3}$, the 18 is the _____, the 3 is the _____ and the 6 is the _____.

2. In $x + 1 \overline{\smash{)}\, x^2 + 3x + 2}^{\,x + 2}$, the $x + 1$ is the _____, the $x^2 + 3x + 2$ is the _____ and the $x + 2$ is the _____.

Simplify each expression mentally.

3. $\dfrac{a^6}{a^4}$ **4.** $\dfrac{p^8}{p^3}$ **5.** $\dfrac{y^2}{y}$ **6.** $\dfrac{a^3}{a}$

5.6 EXERCISE SET

Perform each division. See Examples 1 through 3.

1. $\dfrac{12x^4 + 3x^2}{x}$
2. $\dfrac{15x^2 - 9x^5}{x}$
3. $\dfrac{20x^3 - 30x^2 + 5x + 5}{5}$
4. $\dfrac{8x^3 - 4x^2 + 6x + 2}{2}$
5. $\dfrac{15p^3 + 18p^2}{3p}$
6. $\dfrac{14m^2 - 27m^3}{7m}$
7. $\dfrac{-9x^4 + 18x^5}{6x^5}$
8. $\dfrac{6x^5 + 3x^4}{3x^4}$
9. $\dfrac{-9x^5 + 3x^4 - 12}{3x^3}$
10. $\dfrac{6a^2 - 4a + 12}{-2a^2}$
11. $\dfrac{4x^4 - 6x^3 + 7}{-4x^4}$
12. $\dfrac{-12a^3 + 36a - 15}{3a}$

Find each quotient using long division. See Examples 4 and 5.

13. $\dfrac{x^2 + 4x + 3}{x + 3}$
14. $\dfrac{x^2 + 7x + 10}{x + 5}$
15. $\dfrac{2x^2 + 13x + 15}{x + 5}$
16. $\dfrac{3x^2 + 8x + 4}{x + 2}$
17. $\dfrac{2x^2 - 7x + 3}{x - 4}$
18. $\dfrac{3x^2 - x - 4}{x - 1}$
19. $\dfrac{9a^3 - 3a^2 - 3a + 4}{3a + 2}$
20. $\dfrac{4x^3 + 12x^2 + x - 14}{2x + 3}$
21. $\dfrac{8x^2 + 10x + 1}{2x + 1}$
22. $\dfrac{3x^2 + 17x + 7}{3x + 2}$
23. $\dfrac{2x^3 + 2x^2 - 17x + 8}{x - 2}$
24. $\dfrac{4x^3 + 11x^2 - 8x - 10}{x + 3}$

Find each quotient using long division. Don't forget to write the polynomials in descending order and fill in any missing terms. See Examples 6 and 7.

25. $\dfrac{x^2 - 36}{x - 6}$
26. $\dfrac{a^2 - 49}{a - 7}$
27. $\dfrac{x^3 - 27}{x - 3}$
28. $\dfrac{x^3 + 64}{x + 4}$
29. $\dfrac{1 - 3x^2}{x + 2}$
30. $\dfrac{7 - 5x^2}{x + 3}$
31. $\dfrac{-4b + 4b^2 - 5}{2b - 1}$
32. $\dfrac{-3y + 2y^2 - 15}{2y + 5}$

MIXED PRACTICE

Divide. If the divisor contains 2 or more terms, use long division. See Examples 1 through 7.

33. $\dfrac{a^2b^2 - ab^3}{ab}$
34. $\dfrac{m^3n^2 - mn^4}{mn}$
35. $\dfrac{8x^2 + 6x - 27}{2x - 3}$
36. $\dfrac{18w^2 + 18w - 8}{3w + 4}$
37. $\dfrac{2x^2y + 8x^2y^2 - xy^2}{2xy}$
38. $\dfrac{11x^3y^3 - 33xy + x^2y^2}{11xy}$
39. $\dfrac{2b^3 + 9b^2 + 6b - 4}{b + 4}$
40. $\dfrac{2x^3 + 3x^2 - 3x + 4}{x + 2}$
41. $\dfrac{5x^2 + 28x - 10}{x + 6}$
42. $\dfrac{2x^2 + x - 15}{x + 3}$
43. $\dfrac{10x^3 - 24x^2 - 10x}{10x}$
44. $\dfrac{2x^3 + 12x^2 + 16}{4x^2}$
45. $\dfrac{6x^2 + 17x - 4}{x + 3}$
46. $\dfrac{2x^2 - 9x + 15}{x - 6}$
47. $\dfrac{30x^2 - 17x + 2}{5x - 2}$
48. $\dfrac{4x^2 - 13x - 12}{4x + 3}$
49. $\dfrac{3x^4 - 9x^3 + 12}{-3x}$
50. $\dfrac{8y^6 - 3y^2 - 4y}{4y}$

51. $\dfrac{x^3 + 6x^2 + 18x + 27}{x + 3}$

52. $\dfrac{x^3 - 8x^2 + 32x - 64}{x - 4}$

53. $\dfrac{y^3 + 3y^2 + 4}{y - 2}$

54. $\dfrac{3x^3 + 11x + 12}{x + 4}$

55. $\dfrac{5 - 6x^2}{x - 2}$

56. $\dfrac{3 - 7x^2}{x - 3}$

Divide.

57. $\dfrac{x^5 + x^2}{x^2 + x}$

58. $\dfrac{x^6 - x^4}{x^3 + 1}$

REVIEW AND PREVIEW

Multiply each expression. See Section 5.3.

59. $2a(a^2 + 1)$
60. $-4a(3a^2 - 4)$
61. $2x(x^2 + 7x - 5)$
62. $4y(y^2 - 8y - 4)$
63. $-3xy(xy^2 + 7x^2y + 8)$
64. $-9xy(4xyz + 7xy^2z + 2)$
65. $9ab(ab^2c + 4bc - 8)$
66. $-7sr(6s^2r + 9sr^2 + 9rs + 8)$

Use the bar graph below to answer Exercises 67 through 70. See Section 1.9.

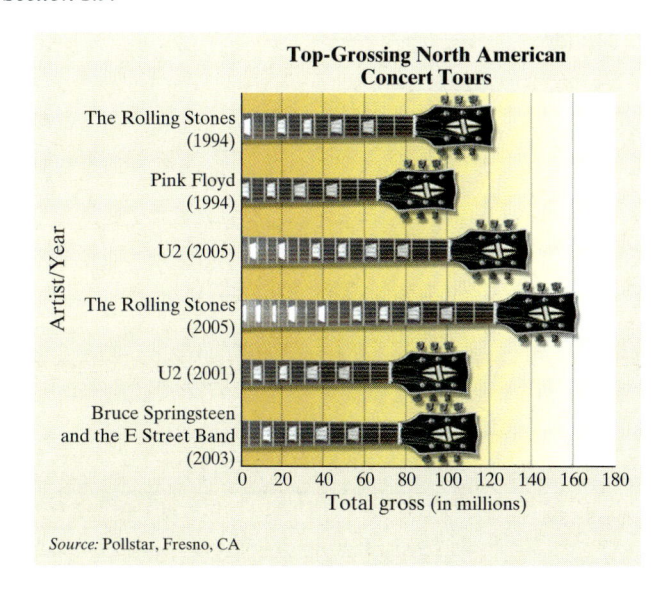

Top-Grossing North American Concert Tours

Source: Pollstar, Fresno, CA

67. Which artist has grossed the most money on one tour?
68. Estimate the amount of money made by the 2005 concert tour of The Rolling Stones.
69. Estimate the amount of money made by the 2005 concert tour of U2.
70. Which artist shown has grossed the least amount of money on a tour?

CONCEPT EXTENSIONS

△ 71. The perimeter of a square is $(12x^3 + 4x - 16)$ feet. Find the length of its side.

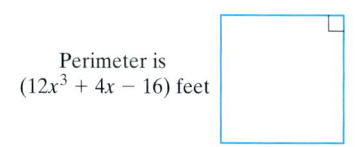

Perimeter is $(12x^3 + 4x - 16)$ feet

△ 72. The volume of the swimming pool shown is $(36x^5 - 12x^3 + 6x^2)$ cubic feet. If its height is $2x$ feet and its width is $3x$ feet, find its length.

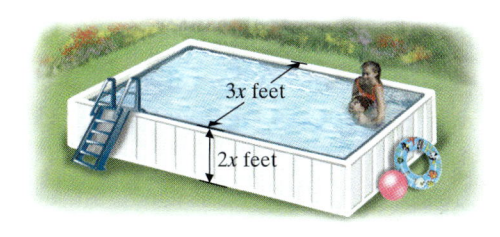

3x feet

2x feet

73. In which of the following is $\dfrac{a + 7}{7}$ simplified correctly? See the Concept Check in this section.

   a. $a + 1$ b. a c. $\dfrac{a}{7} + 1$

74. Explain how to check a polynomial long division result when the remainder is 0.

75. Explain how to check a polynomial long division result when the remainder is not 0.

△ 76. The area of the following parallelogram is $(10x^2 + 31x + 15)$ square meters. If its base is $(5x + 3)$ meters, find its height.

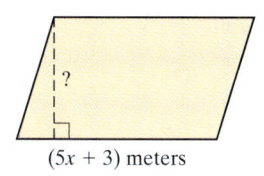

$(5x + 3)$ meters

△ 77. The area of the top of the Ping-Pong table is $(49x^2 + 70x - 200)$ square inches. If its length is $(7x + 20)$ inches, find its width.

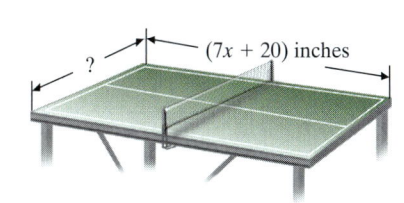

$(7x + 20)$ inches

78. $(18x^{10a} - 12x^{8a} + 14x^{5a} - 2x^{3a}) \div 2x^{3a}$
79. $(25y^{11b} + 5y^{6b} - 20y^{3b} + 100y^b) \div 5y^b$

THE BIGGER PICTURE SIMPLIFYING EXPRESSIONS AND SOLVING EQUATIONS

Now we continue our outline from Sections 1.7 and 2.9. Although suggestions are given, this outline should be in your own words. Once you complete this new portion, try the exercises below.

I. Simplifying Expressions
   A. Real Numbers
      1. Add (Section 1.5)
      2. Subtract (Section 1.6)
      3. Multiply or Divide (Section 1.7)
   B. Exponents— $x^7 \cdot x^5 = x^{12}; (x^7)^5 = x^{35}; \dfrac{x^7}{x^5} = x^2;$
      $x^0 = 1; 8^{-2} = \dfrac{1}{8^2} = \dfrac{1}{64}$
   C. Polynomials
      1. Add: Combine like terms.
         $(3y^2 + 6y + 7) + (9y^2 - 11y - 15)$
         $= 3y^2 + 6y + 7 + 9y^2 - 11y - 15$
         $= 12y^2 - 5y - 8$
      2. Subtract: Change the sign of the terms of the polynomial being subtracted, then add.
         $(3y^2 + 6y + 7) - (9y^2 - 11y - 15)$
         $= 3y^2 + 6y + 7 - 9y^2 + 11y + 15$
         $= -6y^2 + 17y + 22$
      3. Multiply: Multiply each term of one polynomial by each term of the other polynomial.
         $(x + 5)(2x^2 - 3x + 4)$
         $= x(2x^2 - 3x + 4) + 5(2x^2 - 3x + 4)$
         $= 2x^3 - 3x^2 + 4x + 10x^2 - 15x + 20$
         $= 2x^3 + 7x^2 - 11x + 20$
      4. Divide:
         a. To divide by a monomial, divide each term of the polynomial by the monomial.
            $\dfrac{8x^2 + 2x - 6}{2x} = \dfrac{8x^2}{2x} + \dfrac{2x}{2x} - \dfrac{6}{2x}$
            $= 4x + 1 - \dfrac{3}{x}$
         b. To divide by a polynomial other than a monomial, use long division.

$$\begin{array}{r} x - 6 + \dfrac{40}{2x+5} \\ 2x + 5 \overline{\smash{\big)}\, 2x^2 - 7x + 10} \\ \underline{2x^2 + 5x} \\ -12x + 10 \\ \underline{-12x - 30} \\ 40 \end{array}$$

II. Solving Equations and Inequalities
   A. Linear Equations (Section 2.4)
   B. Linear Inequalities (Section 2.9)

Simplify the expressions.

1. $-5.7 + (-0.23)$
2. $\dfrac{1}{2} - \dfrac{9}{10}$
3. $(-5x^2y^3)(-x^7y)$
4. $2^{-3}a^{-7}a^3$
5. $(7y^3 - 6y + 2) - (y^3 + 2y^2 + 2)$
6. Subtract $(y^2 + 7)$ from $(9y^2 - 3y)$
7. Multiply: $(x - 3)(4x^2 - x + 7)$
8. Multiply: $(6m - 5)^2$
9. Divide: $\dfrac{20n^2 - 5n + 10}{5n}$
10. Divide: $\dfrac{6x^2 - 20x + 20}{3x - 1}$

Solve the equations or inequalities.

11. $-6x = 3.6$
12. $-6x < 3.6$
13. $6x + 6 \geq 8x + 2$
14. $7y + 3(y - 1) = 4(y + 1) - 3$

CHAPTER 5 GROUP ACTIVITY

Modeling with Polynomials

The polynomial $-2.02x^2 + 51.60x + 674.60$ dollars represents consumer spending per person per year on all U.S. media from 2000 to 2006. This includes spending on subscription TV services, recorded music, newspapers, magazines, books, home video, theater movies, video games, and educational software. The polynomial model $-1.39x^2 + 22.61x + 206.53$ dollars represents consumer spending per person per year on subscription TV services alone during this same period. In both models, x is the number of years after 2000. (*Source:* Based on data from *Statistical Abstract of the United States, 2007*).

In this project, you will have the opportunity to investigate these polynomial models numerically, algebraically, and graphically. This project may be completed by working in groups or individually.

1. Use the polynomials to complete the following table showing the annual consumer spending per person over the period 2000–2006 by evaluating each polynomial at the given values of x. Then subtract each value in the fourth column from the corresponding value in the third column. Record the result in the last column, "Difference." What do you think these values represent? What trends do you notice in the data?

Year	x	Consumer Spending per Person per Year on All U.S. Media	Consumer Spending per Person per Year on Subscription TV	Difference
2000	0			
2002	2			
2004	4			
2006	6			

2. Use the polynomial models to find a new polynomial model representing the amount of consumer spending per person on U.S. media other than subscription TV services (such as recorded music, newspapers, magazines, books, home video, theater movies, video games, and educational software). Then use this new polynomial to complete the following table.

Year	x	Consumer Spending per Person per Year on Media Other Than Subscription TV
2000	0	
2002	2	
2004	4	
2006	6	

3. Compare the values in the last column of the table in Question 1 to the values in the last column of the table in Question 2. What do you notice? What can you conclude?

4. Use the polynomial models to estimate consumer spending on
   a. all U.S. media,
   b. subscription TV,
   c. media other than subscription TV for the year 2008.

5. Use the polynomial models to estimate consumer spending on
   a. all U.S. media,
   b. subscription TV,
   c. media other than subscription TV for the year 2010.

6. Create a bar graph that represents the data for consumer spending on all U.S. media in the years 2000, 2002, 2004, and 2006 along with your estimates for 2008 and 2010. Study your bar graph. Discuss what the graph implies about the future.

CHAPTER 5 VOCABULARY CHECK

Fill in each blank with one of the words or phrases listed below.

term   coefficient   monomial   binomial   trinomial
polynomials   degree of a term   degree of a polynomial   FOIL

1. A _____ is a number or the product of numbers and variables raised to powers.
2. The _____ method may be used when multiplying two binomials.
3. A polynomial with exactly 3 terms is called a _____.
4. The _____ is the greatest degree of any term of the polynomial.
5. A polynomial with exactly 2 terms is called a _____.
6. The _____ of a term is its numerical factor.
7. The _____ is the sum of the exponents on the variables in the term.
8. A polynomial with exactly 1 term is called a _____.
9. Monomials, binomials, and trinomials are all examples of _____.

STUDY SKILLS BUILDER

Are You Preparing for a Test on Chapter 5?

Below is a list of some *common trouble areas* for topics covered in Chapter 5. After studying for your test—but before taking your test—read these.

- Do you know that a negative exponent does not make the base a negative number? For example,

$$3^{-2} = \frac{1}{3^2} = \frac{1}{9}$$

354 CHAPTER 5 Exponents and Polynomials

- Make sure you remember that x has an understood coefficient of 1 and an understood exponent of 1. For example,

$$2x + x = 2x + 1x = 3x; \quad x^5 \cdot x = x^5 \cdot x^1 = x^6$$

- Do you know the difference between $5x^2$ and $(5x)^2$?

$$5x^2 \text{ is } 5 \cdot x^2; \quad (5x)^2 = 5^2 \cdot x^2 \text{ or } 25 \cdot x^2$$

- Can you evaluate $x^2 - x$ when $x = -2$?

$$x^2 - x = (-2)^2 - (-2) = 4 - (-2) = 4 + 2 = 6$$

- Can you subtract $5x^2 + 1$ from $3x^2 - 6$?

$$(3x^2 - 6) - (5x^2 + 1) = 3x^2 - 6 - 5x^2 - 1 = -2x^2 - 7$$

- Make sure you are familiar with squaring a binomial and other special products.

$$(3x - 4)^2 = (3x)^2 - 2(3x)(4) + 4^2 = 9x^2 - 24x + 16$$

or

$$(3x - 4)^2 = (3x - 4)(3x - 4) = 9x^2 - 24x + 16$$
$$(2x^2 + 1)(2x^2 - 1) = (2x^2)^2 - 1^2 = 4x^4 - 1$$

Remember: This is simply a checklist of common trouble areas. For a review of Chapter 5, see the Highlights and Chapter Review.

> **Helpful Hint**
> Are you preparing for your test? Don't forget to take the Chapter 5 Test on page 359. Then check your answers at the back of the text and use the Chapter Test Prep Video CD to see the fully worked-out solutions to any of the exercises you want to review.

CHAPTER 5 HIGHLIGHTS

DEFINITIONS AND CONCEPTS	EXAMPLES
SECTION 5.1 EXPONENTS	
a^n means the product of n factors, each of which is a.	$3^2 = 3 \cdot 3 = 9$ $(-5)^3 = (-5)(-5)(-5) = -125$ $\left(\dfrac{1}{2}\right)^4 = \dfrac{1}{2} \cdot \dfrac{1}{2} \cdot \dfrac{1}{2} \cdot \dfrac{1}{2} = \dfrac{1}{16}$
If m and n are integers and no denominators are 0,	
Product Rule: $a^m \cdot a^n = a^{m+n}$	$x^2 \cdot x^7 = x^{2+7} = x^9$
Power Rule: $(a^m)^n = a^{mn}$	$(5^3)^8 = 5^{3 \cdot 8} = 5^{24}$
Power of a Product Rule: $(ab)^n = a^n b^n$	$(7y)^4 = 7^4 y^4$
Power of a Quotient Rule: $\left(\dfrac{a}{b}\right)^n = \dfrac{a^n}{b^n}$	$\left(\dfrac{x}{8}\right)^3 = \dfrac{x^3}{8^3}$
Quotient Rule: $\dfrac{a^m}{a^n} = a^{m-n}$	$\dfrac{x^9}{x^4} = x^{9-4} = x^5$
Zero Exponent: $a^0 = 1, a \neq 0$.	$5^0 = 1, x^0 = 1, x \neq 0$
SECTION 5.2 ADDING AND SUBTRACTING POLYNOMIALS	
A **term** is a number or the product of numbers and variables raised to powers.	**Terms** $-5x, 7a^2b, \dfrac{1}{4}y^4, 0.2$
The **numerical coefficient** or **coefficient** of a term is its numerical factor.	**Term** **Coefficient** $7x^2$ 7 y 1 $-a^2b$ -1

DEFINITIONS AND CONCEPTS	EXAMPLES

SECTION 5.2 ADDING AND SUBTRACTING POLYNOMIALS (continued)

A **polynomial** is a term or a finite sum of terms in which variables may appear in the numerator raised to whole number powers only.	**Polynomials** $3x^2 - 2x + 1$ (Trinomial) $-0.2a^2b - 5b^2$ (Binomial)
A **monomial** is a polynomial with exactly 1 term. A **binomial** is a polynomial with exactly 2 terms. A **trinomial** is a polynomial with exactly 3 terms.	$\frac{5}{6}y^3$ (Monomial)
The **degree of a term** is the sum of the exponents on the variables in the term.	Term Degree $-5x^3$ 3 3 (or $3x^0$) 0 $2a^2b^2c$ 5
The **degree of a polynomial** is the greatest degree of any term of the polynomial.	Polynomial Degree $5x^2 - 3x + 2$ 2 $7y + 8y^2z^3 - 12$ $2 + 3 = 5$
To add polynomials, add or combine like terms.	Add: $(7x^2 - 3x + 2) + (-5x - 6) = 7x^2 - 3x + 2 - 5x - 6$ $= 7x^2 - 8x - 4$
To subtract two polynomials, change the signs of the terms of the second polynomial, then add.	Subtract: $(17y^2 - 2y + 1) - (-3y^3 + 5y - 6)$ $= (17y^2 - 2y + 1) + (3y^3 - 5y + 6)$ $= 17y^2 - 2y + 1 + 3y^3 - 5y + 6$ $= 3y^3 + 17y^2 - 7y + 7$

SECTION 5.3 MULTIPLYING POLYNOMIALS

To multiply two polynomials, multiply each term of one polynomial by each term of the other polynomial, and then combine like terms.	Multiply: $(2x + 1)(5x^2 - 6x + 2)$ $= 2x(5x^2 - 6x + 2) + 1(5x^2 - 6x + 2)$ $= 10x^3 - 12x^2 + 4x + 5x^2 - 6x + 2$ $= 10x^3 - 7x^2 - 2x + 2$

SECTION 5.4 SPECIAL PRODUCTS

The **FOIL method** may be used when multiplying two binomials.	Multiply: $(5x - 3)(2x + 3)$ F O I L $(5x - 3)(2x + 3) = (5x)(2x) + (5x)(3) + (-3)(2x) + (-3)(3)$ $= 10x^2 + 15x - 6x - 9$ $= 10x^2 + 9x - 9$
Squaring a Binomial $(a + b)^2 = a^2 + 2ab + b^2$ $(a - b)^2 = a^2 - 2ab + b^2$	Square each binomial. $(x + 5)^2 = x^2 + 2(x)(5) + 5^2$ $= x^2 + 10x + 25$ $(3x - 2y)^2 = (3x)^2 - 2(3x)(2y) + (2y)^2$ $= 9x^2 - 12xy + 4y^2$
Multiplying the Sum and Difference of Two Terms $(a + b)(a - b) = a^2 - b^2$	Multiply: $(6y + 5)(6y - 5) = (6y)^2 - 5^2$ $= 36y^2 - 25$

DEFINITIONS AND CONCEPTS	EXAMPLES

SECTION 5.5 NEGATIVE EXPONENTS AND SCIENTIFIC NOTATION

If $a \neq 0$ and n is an integer,

$$a^{-n} = \frac{1}{a^n}$$

Rules for exponents are true for positive and negative integers.

$3^{-2} = \frac{1}{3^2} = \frac{1}{9}$; $5x^{-2} = \frac{5}{x^2}$

Simplify: $\left(\dfrac{x^{-2}y}{x^5}\right)^{-2} = \dfrac{x^4 y^{-2}}{x^{-10}}$

$= x^{4-(-10)} y^{-2}$

$= \dfrac{x^{14}}{y^2}$

A positive number is written in scientific notation if it is as the product of a number a, $1 \leq a < 10$, and an integer power r of 10.

$$a \times 10^r$$

Write each number in scientific notation.

$12{,}000 = 1.2 \times 10^4$

$0.00000568 = 5.68 \times 10^{-6}$

SECTION 5.6 DIVIDING POLYNOMIALS

To divide a polynomial by a monomial:

$$\frac{a+b}{c} = \frac{a}{c} + \frac{b}{c}$$

To divide a polynomial by a polynomial other than a monomial, use long division.

Divide:

$\dfrac{15x^5 - 10x^3 + 5x^2 - 2x}{5x^2} = \dfrac{15x^5}{5x^2} - \dfrac{10x^3}{5x^2} + \dfrac{5x^2}{5x^2} - \dfrac{2x}{5x^2}$

$= 3x^3 - 2x + 1 - \dfrac{2}{5x}$

$5x - 1 + \dfrac{-4}{2x+3}$

$2x+3 \overline{) 10x^2 + 13x - 7}$

CHAPTER 5 REVIEW

(5.1) State the base and the exponent for each expression.

1. 7^9
2. $(-5)^4$
3. -5^4
4. x^6

Evaluate each expression.

5. 8^3
6. $(-6)^2$
7. -6^2
8. $-4^3 - 4^0$
9. $(3b)^0$
10. $\dfrac{8b}{8b}$

Simplify each expression.

11. $y^2 \cdot y^7$
12. $x^9 \cdot x^5$
13. $(2x^5)(-3x^6)$
14. $(-5y^3)(4y^4)$
15. $(x^4)^2$
16. $(y^3)^5$
17. $(3y^6)^4$
18. $(2x^3)^3$
19. $\dfrac{x^9}{x^4}$
20. $\dfrac{z^{12}}{z^5}$
21. $\dfrac{a^5 b^4}{ab}$
22. $\dfrac{x^4 y^6}{xy}$

23. $\dfrac{12xy^6}{3x^4 y^{10}}$
24. $\dfrac{2x^7 y^8}{8xy^2}$
25. $5a^7(2a^4)^3$
26. $(2x)^2(9x)$
27. $(-5a)^0 + 7^0 + 8^0$
28. $8x^0 + 9^0$

Simplify the given expression and choose the correct result.

29. $\left(\dfrac{3x^4}{4y}\right)^3$

 a. $\dfrac{27x^{64}}{64y^3}$ b. $\dfrac{27x^{12}}{64y^3}$ c. $\dfrac{9x^{12}}{12y^3}$ d. $\dfrac{3x^{12}}{4y^3}$

30. $\left(\dfrac{5a^6}{b^3}\right)^2$

 a. $\dfrac{10a^{12}}{b^6}$ b. $\dfrac{25a^{36}}{b^9}$ c. $\dfrac{25a^{12}}{b^6}$ d. $25a^{12}b^6$

(5.2) Find the degree of each term.

31. $-5x^4 y^3$
32. $10x^3 y^2 z$
33. $35a^5 bc^2$
34. $95xyz$

Find the degree of each polynomial.

35. $y^5 + 7x - 8x^4$
36. $9y^2 + 30y + 25$
37. $-14x^2y - 28x^2y^3 - 42x^2y^2$
38. $6x^2y^2z^2 + 5x^2y^3 - 12xyz$
39. **a.** Complete the table for the polynomial $x^2y^2 + 5x^2 - 7y^2 + 11xy - 1$.

Term	Numerical Coefficient	Degree of Term
x^2y^2		
$5x^2$		
$-7y^2$		
$11xy$		
-1		

b. What is the degree of the polynomial?

40. The surface area of a box with a square base and a height of 5 units is given by the polynomial $2x^2 + 20x$. Fill in the table below by evaluating $2x^2 + 20x$ for the given values of x.

x	1	3	5.1	10
$2x^2 + 20x$				

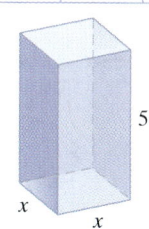

Combine like terms in each expression.

41. $7a^2 - 4a^2 - a^2$
42. $9y + y - 14y$
43. $6a^2 + 4a + 9a^2$
44. $21x^2 + 3x + x^2 + 6$
45. $4a^2b - 3b^2 - 8q^2 - 10a^2b + 7q^2$
46. $2s^{14} + 3s^{13} + 12s^{12} - s^{10}$

Add or subtract as indicated.

47. $(3x^2 + 2x + 6) + (5x^2 + x)$
48. $(2x^5 + 3x^4 + 4x^3 + 5x^2) + (4x^2 + 7x + 6)$
49. $(-5y^2 + 3) - (2y^2 + 4)$
50. $(3x^2 - 7xy + 7y^2) - (4x^2 - xy + 9y^2)$
51. Add $(-9x^2 + 6x + 2)$ and $(4x^2 - x - 1)$.
52. Subtract $(3x - y)$ from $(7x - 14y)$.
53. Subtract $(4x^2 + 8x - 7)$ from the sum of $(x^2 + 7x + 9)$ and $(x^2 + 4)$.
54. With the ownership of computers growing rapidly, the market for new software is also increasing. The revenue for software publishers (in millions of dollars) in the United States from 2001 to 2006 can be represented by the polynomial $754x^2 - 228x + 80,134$ where x is the number of years since 2001. Use this model to predict the revenues from software sales in 2009. (*Source:* Software & Information Industry Association)

(5.3) Multiply each expression.

55. $4(2a + 7)$
56. $9(6a - 3)$
57. $-7x(x^2 + 5)$
58. $-8y(4y^2 - 6)$
59. $(3a^3 - 4a + 1)(-2a)$
60. $(6b^3 - 4b + 2)(7b)$
61. $(2x + 2)(x - 7)$
62. $(2x - 5)(3x + 2)$
63. $(x - 9)^2$
64. $(x - 12)^2$
65. $(4a - 1)(a + 7)$
66. $(6a - 1)(7a + 3)$
67. $(5x + 2)^2$
68. $(3x + 5)^2$
69. $(x + 7)(x^3 + 4x - 5)$
70. $(x + 2)(x^5 + x + 1)$
71. $(x^2 + 2x + 4)(x^2 + 2x - 4)$
72. $(x^3 + 4x + 4)(x^3 + 4x - 4)$
73. $(x + 7)^3$
74. $(2x - 5)^3$

(5.4) Use special products to multiply each of the following.

75. $(x + 7)^2$
76. $(x - 5)^2$
77. $(3x - 7)^2$
78. $(4x + 2)^2$
79. $(5x - 9)^2$
80. $(5x + 1)(5x - 1)$
81. $(7x + 4)(7x - 4)$
82. $(a + 2b)(a - 2b)$
83. $(2x - 6)(2x + 6)$
84. $(4a^2 - 2b)(4a^2 + 2b)$

Express each as a product of polynomials in x. Then multiply and simplify.

85. Find the area of the square if its side is $(3x - 1)$ meters.

86. Find the area of the rectangle. $(x - 1)$ miles by $(5x + 2)$ miles

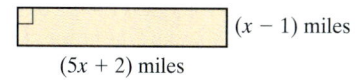

(5.5) Simplify each expression.

87. 7^{-2}
88. -7^{-2}
89. $2x^{-4}$
90. $(2x)^{-4}$
91. $\left(\dfrac{1}{5}\right)^{-3}$
92. $\left(\dfrac{-2}{3}\right)^{-2}$
93. $2^0 + 2^{-4}$
94. $6^{-1} - 7^{-1}$

Simplify each expression. Write each answer using positive exponents only.

95. $\dfrac{x^5}{x^{-3}}$

96. $\dfrac{z^4}{z^{-4}}$

97. $\dfrac{r^{-3}}{r^{-4}}$

98. $\dfrac{y^{-2}}{y^{-5}}$

99. $\left(\dfrac{bc^{-2}}{bc^{-3}}\right)^4$

100. $\left(\dfrac{x^{-3}y^{-4}}{x^{-2}y^{-5}}\right)^{-3}$

101. $\dfrac{x^{-4}y^{-6}}{x^2 y^7}$

102. $\dfrac{a^5 b^{-5}}{a^{-5} b^5}$

103. $a^{6m} a^{5m}$

104. $\dfrac{(x^{5+h})^3}{x^5}$

105. $(3xy^{2z})^3$

106. $a^{m+2} a^{m+3}$

Write each number in scientific notation.

107. 0.00027

108. 0.8868

109. 80,800,000

110. 868,000

111. Google.com is an Internet search engine that handles 91,000,000 searches every day. Write 91,000,000 in scientific notation. (*Source:* Google, Inc.)

112. The approximate diameter of the Milky Way galaxy is 150,000 light years. Write this number in scientific notation. (*Source:* NASA IMAGE/POETRY Education and Public Outreach Program)

Write each number in standard form.

113. 8.67×10^5

114. 3.86×10^{-3}

115. 8.6×10^{-4}

116. 8.936×10^5

117. The volume of the planet Jupiter is 1.43128×10^{15} cubic kilometers. Write this number in standard notation. (*Source:* National Space Science Data Center)

118. An angstrom is a unit of measure, equal to 1×10^{-10} meter, used for measuring wavelengths or the diameters of atoms. Write this number in standard notation. (*Source:* National Institute of Standards and Technology)

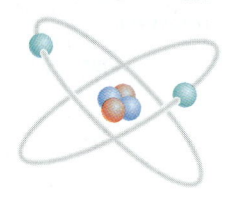

Simplify. Express each result in standard form.

119. $(8 \times 10^4)(2 \times 10^{-7})$

120. $\dfrac{8 \times 10^4}{2 \times 10^{-7}}$

(5.6) *Divide.*

121. $\dfrac{x^2 + 21x + 49}{7x^2}$

122. $\dfrac{5a^3 b - 15ab^2 + 20ab}{-5ab}$

123. $(a^2 - a + 4) \div (a - 2)$

124. $(4x^2 + 20x + 7) \div (x + 5)$

125. $\dfrac{a^3 + a^2 + 2a + 6}{a - 2}$

126. $\dfrac{9b^3 - 18b^2 + 8b - 1}{3b - 2}$

127. $\dfrac{4x^4 - 4x^3 + x^2 + 4x - 3}{2x - 1}$

128. $\dfrac{-10x^2 - x^3 - 21x + 18}{x - 6}$

△ 129. The area of the rectangle below is $(15x^3 - 3x^2 + 60)$ square feet. If its length is $3x^2$ feet, find its width.

Area is $(15x^3 - 3x^2 + 60)$ sq feet

△ 130. The perimeter of the equilateral triangle below is $(21a^3 b^6 + 3a - 3)$ units. Find the length of a side.

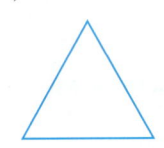

Perimeter is $(21a^3 b^6 + 3a - 3)$ units

MIXED REVIEW

Evaluate.

131. $\left(-\dfrac{1}{2}\right)^3$

Simplify each expression. Write each answer using positive exponents only.

132. $(4xy^2)(x^3 y^5)$

133. $\dfrac{18x^9}{27x^3}$

134. $\left(\dfrac{3a^4}{b^2}\right)^3$

135. $(2x^{-4} y^3)^{-4}$

136. $\dfrac{a^{-3} b^6}{9^{-1} a^{-5} b^{-2}}$

Perform the indicated operations and simplify.

137. $(6x + 2) + (5x - 7)$

138. $(-y^2 - 4) + (3y^2 - 6)$

139. $(8y^2 - 3y + 1) - (3y^2 + 2)$

140. $(5x^2 + 2x - 6) - (-x - 4)$

141. $4x(7x^2 + 3)$

142. $(2x + 5)(3x - 2)$

143. $(x - 3)(x^2 + 4x - 6)$

144. $(7x - 2)(4x - 9)$

Use special products to multiply.

145. $(5x + 4)^2$

146. $(6x + 3)(6x - 3)$

Divide.

147. $\dfrac{8a^4 - 2a^3 + 4a - 5}{2a^3}$

148. $\dfrac{x^2 + 2x + 10}{x + 5}$

149. $\dfrac{4x^3 + 8x^2 - 11x + 4}{2x - 3}$

CHAPTER 5 TEST

Remember to use your Chapter Test Prep Video CD to see the fully worked-out solutions to any of the exercises you want to review.

Evaluate each expression.

1. 2^5 **2.** $(-3)^4$ **3.** -3^4 **4.** 4^{-3}

Simplify each exponential expression. Write the result using only positive exponents.

5. $(3x^2)(-5x^9)$

6. $\dfrac{y^7}{y^2}$

7. $\dfrac{r^{-8}}{r^{-3}}$

8. $\left(\dfrac{x^2 y^3}{x^3 y^{-4}}\right)^2$

9. $\dfrac{6^2 x^{-4} y^{-1}}{6^3 x^{-3} y^7}$

Express each number in scientific notation.

10. 563,000

11. 0.0000863

Write each number in standard form.

12. 1.5×10^{-3}

13. 6.23×10^4

14. Simplify. Write the answer in standard form.

$(1.2 \times 10^5)(3 \times 10^{-7})$

15. a. Complete the table for the polynomial $4xy^2 + 7xyz + x^3 y - 2$.

Term	Numerical Coefficient	Degree of Term
$4xy^2$		
$7xyz$		
$x^3 y$		
-2		

b. What is the degree of the polynomial?

16. Simplify by combining like terms.

$5x^2 + 4xy - 7x^2 + 11 + 8xy$

Perform each indicated operation.

17. $(8x^3 + 7x^2 + 4x - 7) + (8x^3 - 7x - 6)$

18. $5x^3 + x^2 + 5x - 2 - (8x^3 - 4x^2 + x - 7)$

19. Subtract $(4x + 2)$ from the sum of $(8x^2 + 7x + 5)$ and $(x^3 - 8)$.

Multiply.

20. $(3x + 7)(x^2 + 5x + 2)$

21. $3x^2(2x^2 - 3x + 7)$

22. $(x + 7)(3x - 5)$

23. $\left(3x - \dfrac{1}{5}\right)\left(3x + \dfrac{1}{5}\right)$

24. $(4x - 2)^2$

25. $(8x + 3)^2$

26. $(x^2 - 9b)(x^2 + 9b)$

27. The height of the Bank of China in Hong Kong is 1001 feet. Neglecting air resistance, the height of an object dropped from this building at time t seconds is given by the polynomial $-16t^2 + 1001$. Find the height of the object at the given times below.

t	0 seconds	1 second	3 seconds	5 seconds
$-16t^2 + 1001$				

△ **28.** Find the area of the top of the table. Express the area as a product, then multiply and simplify.

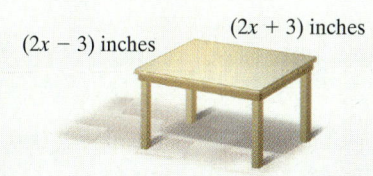

$(2x - 3)$ inches $(2x + 3)$ inches

Divide.

29. $\dfrac{4x^2 + 24xy - 7x}{8xy}$

30. $(x^2 + 7x + 10) \div (x + 5)$

31. $\dfrac{27x^3 - 8}{3x + 2}$

CHAPTER 5 CUMULATIVE REVIEW

1. Tell whether each statement is true or false.
   a. $8 \geq 8$
   b. $8 \leq 8$
   c. $23 \leq 0$
   d. $23 \geq 0$

2. Find the absolute value of each number.
   a. $|-7.2|$
   b. $|0|$
   c. $\left|-\frac{1}{2}\right|$

3. Divide. Write all quotients in lowest terms.
   a. $\frac{4}{5} \div \frac{5}{16}$
   b. $\frac{7}{10} \div 14$
   c. $\frac{3}{8} \div \frac{3}{10}$

4. Multiply. Write products in lowest terms.
   a. $\frac{3}{4} \cdot \frac{7}{21}$
   b. $\frac{1}{2} \cdot 4\frac{5}{6}$

5. Evaluate the following:
   a. 3^2
   b. 5^3
   c. 2^4
   d. 7^1
   e. $\left(\frac{3}{7}\right)^2$

6. Evaluate $\frac{2x - 7y}{x^2}$ for $x = 5$ and $y = 1$.

7. Add.
   a. $-3 + (-7)$
   b. $-1 + (-20)$
   c. $-2 + (-10)$

8. Simplify: $8 + 3(2 \cdot 6 - 1)$

9. Subtract 8 from -4.

10. Is $x = 1$ a solution of $5x^2 + 2 = x - 8$.

11. Find the reciprocal of each number.
    a. 22
    b. $\frac{3}{16}$
    c. -10
    d. $-\frac{9}{13}$

12. Subtract:
    a. $7 - 40$
    b. $-5 - (-10)$

13. Use an associative property to complete each statement.
    a. $5 + (4 + 6) =$ _____
    b. $(-1 \cdot 2) \cdot 5 =$ _____

14. Simplify: $\frac{4(-3) + (-8)}{5 + (-5)}$

15. Simplify each expression.
    a. $10 + (x + 12)$
    b. $-3(7x)$

16. Use the distributive property to write $-2(x + 3y - z)$ without parentheses.

17. Find each product by using the distributive property to remove parentheses.
    a. $5(x + 2)$
    b. $-2(y + 0.3z - 1)$
    c. $-(x + y - 2z + 6)$

18. Simplify: $2(6x - 1) - (x - 7)$

19. Solve $x - 7 = 10$ for x.

20. Write the phrase as an algebraic expression: double a number, subtracted from the sum of a number and seven

21. Solve: $\frac{5}{2}x = 15$

22. Solve: $2x + \frac{1}{8} = x - \frac{3}{8}$

23. Twice a number, added to seven, is the same as three subtracted from the number. Find the number.

24. Solve: $10 = 5j - 2$

25. Twice the sum of a number and 4 is the same as four times the number, decreased by 12. Find the number.

26. Solve: $\frac{7x + 5}{3} = x + 3$

27. The length of a rectangular road sign is 2 feet less than three times its width. Find the dimensions if the perimeter is 28 feet.

28. Graph $x < 5$ and write in interval notation.

29. Solve $F = \frac{9}{5}C + 32$ for C.

30. Find the slope of each line.
    a. $x = -1$
    b. $y = 7$

31. Graph $2 < x \leq 4$.

32. Recall that the grade of a road is its slope written as a percent. Find the grade of the road shown.

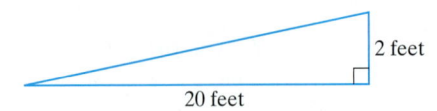

33. Complete the following ordered-pair solutions for the equation $3x + y = 12$.
    a. $(0, \ )$
    b. $(\ , 6)$
    c. $(-1, \ )$

34. Solve the system: $\begin{cases} 3x + 2y = -8 \\ 2x - 6y = -9 \end{cases}$

35. Graph the linear equation $2x + y = 5$.

36. Solve the system: $\begin{cases} x = -3y + 3 \\ 2x + 9y = 5 \end{cases}$

37. Graph $x = 2$.

38. Evaluate.
    a. $(-5)^2$
    b. -5^2
    c. $2 \cdot 5^2$

39. Find the slope of the line $x = 5$.

40. Simplify: $\dfrac{(z^2)^3 \cdot z^7}{z^9}$

41. Graph $x + y < 7$.

42. Subtract $(5y^2 - 6) - (y^2 + 2)$.

43. Use the product rule to simplify $(2x^2)(-3x^5)$.

44. Find the value of $-x^2$ when
    a. $x = 2$ b. $x = -2$

45. Add $(-2x^2 + 5x - 1)$ and $(-2x^2 + x + 3)$.

46. Multiply $(10x^2 - 3)(10x^2 + 3)$.

47. Multiply $(2x - y)^2$.

48. Multiply $(10x^2 + 3)^2$.

49. Divide $6m^2 + 2m$ by $2m$.

50. Evaluate.
    a. 5^{-1}
    b. 7^{-2}

CHAPTER

6 Factoring Polynomials

6.1 The Greatest Common Factor and Factoring by Grouping

6.2 Factoring Trinomials of the Form $x^2 + bx + c$

6.3 Factoring Trinomials of the Form $ax^2 + bx + c$ and Perfect Square Trinomials

6.4 Factoring Trinomials of the Form $ax^2 + bx + c$ by Grouping

6.5 Factoring Binomials

Integrated Review—Choosing a Factoring Strategy

6.6 Solving Quadratic Equations by Factoring

6.7 Quadratic Equations and Problem Solving

In Chapter 5, you learned how to multiply polynomials. This chapter deals with an operation that is the reverse process of multiplying, called *factoring*. Factoring is an important algebraic skill because this process allows us to write a sum as a product.

At the end of this chapter, we use factoring to help us solve equations other than linear equations, and in Chapter 7 we use factoring to simplify and perform arithmetic operations on rational expressions.

The majority of college students hold credit cards. According to the Nellie May Corporation, 56% of final year undergraduate students carry four or more cards with an average balance of $2864. The circle graph below shows when students with credit cards obtained a card.

American Consumer Credit Counseling (ACCC) released the following guidelines to help protect college students from the lasting effects of credit card debt.

- Pay at least the minimum payment on all bills before the due date.
- Keep a budget. Record all income and record all outgoings for at least two months so you know where your money is going.
- Avoid borrowing money to pay off other creditors.
- Only carry as much cash as your weekly budget allows.

In Section 6.5, Exercises 77 through 80, you will have the opportunity to calculate percents of students with credit cards during selected years.

76% of Undergraduate College Students Have Credit Cards—When Do These Students Obtain a Card?

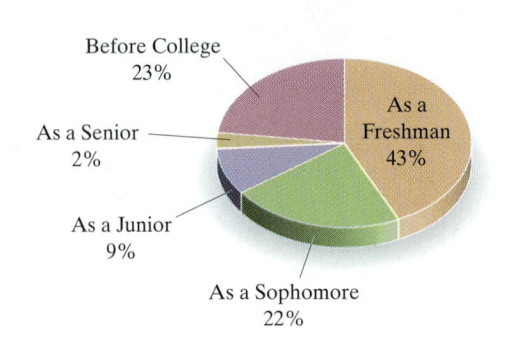

- Before College 23%
- As a Senior 2%
- As a Junior 9%
- As a Sophomore 22%
- As a Freshman 43%

6.1 THE GREATEST COMMON FACTOR AND FACTORING BY GROUPING

OBJECTIVES

1. Find the greatest common factor of a list of integers.
2. Find the greatest common factor of a list of terms.
3. Factor out the greatest common factor from a polynomial.
4. Factor a polynomial by grouping.

In the product $2 \cdot 3 = 6$, the numbers 2 and 3 are called **factors** of 6 and $2 \cdot 3$ is a **factored form** of 6. This is true of polynomials also. Since $(x + 2)(x + 3) = x^2 + 5x + 6$, then $(x + 2)$ and $(x + 3)$ are factors of $x^2 + 5x + 6$, and $(x + 2)(x + 3)$ is a factored form of the polynomial.

a factored form of 6

$$2 \cdot 3 = 6$$
factor factor product

a factored form of x^5

$$x^2 \cdot x^3 = x^5$$
factor factor product

a factored form of $x^2 + 5x + 6$

$$(x + 2)(x + 3) = x^2 + 5x + 6$$
factor factor product

Do you see that factoring is the reverse process of multiplying?

$$x^2 + 5x + 6 \xrightleftharpoons[\text{multiplying}]{\text{factoring}} (x + 2)(x + 3)$$

Concept Check ✓

Multiply: $2(x - 4)$

What do you think the result of factoring $2x - 8$ would be? Why?

The first step in factoring a polynomial is to see whether the terms of the polynomial have a common factor. If there is one, we can write the polynomial as a product by **factoring out** the common factor. We will usually factor out the **greatest common factor (GCF)**.

OBJECTIVE 1 ▶ Finding the greatest common factor of a list of integers. The GCF of a list of integers is the largest integer that is a factor of all the integers in the list. For example, the GCF of 12 and 20 is 4 because 4 is the largest integer that is a factor of both 12 and 20. With large integers, the GCF may not be easily found by inspection. When this happens, use the following steps.

> **Finding the GCF of a List of Integers**
>
> **STEP 1.** Write each number as a product of prime numbers.
>
> **STEP 2.** Identify the common prime factors.
>
> **STEP 3.** The product of all common prime factors found in Step 2 is the greatest common factor. If there are no common prime factors, the greatest common factor is 1.

Recall from Section 1.3 that a prime number is a whole number other than 1, whose only factors are 1 and itself.

Answers to Concept Check:

$2x - 8$; The result would be $2(x - 4)$ because factoring is the reverse process of multiplying.

EXAMPLE 1 Find the GCF of each list of numbers.

a. 28 and 40   **b.** 55 and 21   **c.** 15, 18, and 66

Solution

a. Write each number as a product of primes.

$$28 = 2 \cdot 2 \cdot 7 = 2^2 \cdot 7$$
$$40 = 2 \cdot 2 \cdot 2 \cdot 5 = 2^3 \cdot 5$$

There are two common factors, each of which is 2, so the GCF is

$$\text{GCF} = 2 \cdot 2 = 4$$

b. $55 = 5 \cdot 11$
$21 = 3 \cdot 7$

There are no common prime factors; thus, the GCF is 1.

c. $15 = 3 \cdot 5$
$18 = 2 \cdot 3 \cdot 3 = 2 \cdot 3^2$
$66 = 2 \cdot 3 \cdot 11$

The only prime factor common to all three numbers is 3, so the GCF is

$$\text{GCF} = 3$$

PRACTICE 1 Find the GCF of each list of numbers.

a. 36 and 42   **b.** 35 and 44   **c.** 12, 16, and 40

OBJECTIVE 2 ▶ Finding the greatest common factor of a list of terms. The greatest common factor of a list of variables raised to powers is found in a similar way. For example, the GCF of x^2, x^3, and x^5 is x^2 because each term contains a factor of x^2 and no higher power of x is a factor of each term.

$$x^2 = x \cdot x$$
$$x^3 = x \cdot x \cdot x$$
$$x^5 = x \cdot x \cdot x \cdot x \cdot x$$

There are two common factors, each of which is x, so the GCF = $x \cdot x$ or x^2.

From this example, we see that **the GCF of a list of common variables raised to powers is the variable raised to the smallest exponent in the list.**

EXAMPLE 2 Find the GCF of each list of terms.

a. x^3, x^7, and x^5 **b.** y, y^4, and y^7

Solution

a. The GCF is x^3, since 3 is the smallest exponent to which x is raised.
b. The GCF is y^1 or y, since 1 is the smallest exponent on y.

PRACTICE 2 Find the GCF of each list of terms.

a. y^7, y^4, and y^6 **b.** x, x^4, and x^2

In general, the **greatest common factor (GCF) of a list of terms** is the product of the GCF of the numerical coefficients and the GCF of the variable factors.

Section 6.1 The Greatest Common Factor and Factoring by Grouping 365

EXAMPLE 3 Find the GCF of each list of terms.

a. $6x^2$, $10x^3$, and $-8x$ **b.** $-18y^2$, $-63y^3$, and $27y^4$ **c.** a^3b^2, a^5b, and a^6b^2

Solution

a.
$6x^2 = 2 \cdot 3 \cdot x^2$
$10x^3 = 2 \cdot 5 \cdot x^3$
$-8x = -1 \cdot 2 \cdot 2 \cdot 2 \cdot x^1$ $\Big\}$ → The GCF of x^2, x^3, and x^1 is x^1 or x.
GCF $= 2 \cdot x^1$ or $2x$

b.
$-18y^2 = -1 \cdot 2 \cdot 3 \cdot 3 \cdot y^2$
$-63y^3 = -1 \cdot 3 \cdot 3 \cdot 7 \cdot y^3$
$27y^4 = 3 \cdot 3 \cdot 3 \cdot y^4$ $\Big\}$ → The GCF of y^2, y^3, and y^4 is y^2.
GCF $= 3 \cdot 3 \cdot y^2$ or $9y^2$

c. The GCF of a^3, a^5, and a^6 is a^3.
The GCF of b^2, b, and b^2 is b. Thus,
the GCF of a^3b^2, a^5b, and a^6b^2 is a^3b.

> **Helpful Hint**
> Remember that the GCF of a list of terms contains the smallest exponent on each common variable.
>
> Smallest exponent on x.
>
> The GCF of x^5y^6, x^2y^7 and x^3y^4 is x^2y^4.
>
> Smallest exponent on y.

PRACTICE 3 Find the GCF of each list of terms.

a. $5y^4$, $15y^2$, and $-20y^3$ **b.** $4x^2$, x^3, and $3x^8$ **c.** a^4b^2, a^3b^5, and a^2b^3

OBJECTIVE 3 ▶ Factoring out the greatest common factor. The first step in factoring a polynomial is to find the GCF of its terms. Once we do so, we can write the polynomial as a product by **factoring out** the GCF.

The polynomial $8x + 14$, for example, contains two terms: $8x$ and 14. The GCF of these terms is 2. We factor out 2 from each term by writing each term as a product of 2 and the term's remaining factors.

$$8x + 14 = 2 \cdot 4x + 2 \cdot 7$$

Using the distributive property, we can write

$$8x + 14 = 2 \cdot 4x + 2 \cdot 7$$
$$= 2(4x + 7)$$

Thus, a factored form of $8x + 14$ is $2(4x + 7)$. We can check by multiplying:

$$2(4x + 7) = 2 \cdot 4x + 2 \cdot 7 = 8x + 14.$$

> **Helpful Hint**
> A factored form of $8x + 14$ is *not*
>
> $$2 \cdot 4x + 2 \cdot 7$$
>
> Although the *terms* have been factored (written as a product), the *polynomial* $8x + 14$ has not been factored (written as a product). A factored form of $8x + 14$ is the *product* $2(4x + 7)$.

Concept Check ✓

Which of the following is/are factored form(s) of $7t + 21$?

a. 7 **b.** $7 \cdot t + 7 \cdot 3$ **c.** $7(t + 3)$ **d.** $7(t + 21)$

Answer to Concept Check: c

366 CHAPTER 6 Factoring Polynomials

EXAMPLE 4 Factor each polynomial by factoring out the GCF.

a. $6t + 18$ **b.** $y^5 - y^7$

Solution

a. The GCF of terms $6t$ and 18 is 6.

$$6t + 18 = 6 \cdot t + 6 \cdot 3$$
$$= 6(t + 3) \quad \text{Apply the distributive property.}$$

Our work can be checked by multiplying 6 and $(t + 3)$.

$$6(t + 3) = 6 \cdot t + 6 \cdot 3 = 6t + 18, \text{ the original polynomial.}$$

b. The GCF of y^5 and y^7 is y^5. Thus,

$$y^5 - y^7 = y^5(1) - y^5(y^2)$$
$$= y^5(1 - y^2)$$

▶ **Helpful Hint**
Don't forget the 1.

PRACTICE 4 Factor each polynomial by factoring out the GCF.

a. $4t + 12$ **b.** $y^8 + y^4$

EXAMPLE 5 Factor: $-9a^5 + 18a^2 - 3a$

Solution

$$-9a^5 + 18a^2 - 3a = (3a)(-3a^4) + (3a)(6a) + (3a)(-1)$$
$$= 3a(-3a^4 + 6a - 1)$$

▶ **Helpful Hint**
Don't forget the -1.

PRACTICE 5 Factor $-8b^6 + 16b^4 - 8b^2$.

In Example 5 we could have chosen to factor out a $-3a$ instead of $3a$. If we factor out a $-3a$, we have

$$-9a^5 + 18a^2 - 3a = (-3a)(3a^4) + (-3a)(-6a) + (-3a)(1)$$
$$= -3a(3a^4 - 6a + 1)$$

▶ **Helpful Hint**
Notice the changes in signs when factoring out $-3a$.

EXAMPLES Factor.

6. $6a^4 - 12a = 6a(a^3 - 2)$

7. $\dfrac{3}{7}x^4 + \dfrac{1}{7}x^3 - \dfrac{5}{7}x^2 = \dfrac{1}{7}x^2(3x^2 + x - 5)$

8. $15p^2q^4 + 20p^3q^5 + 5p^3q^3 = 5p^2q^3(3q + 4pq^2 + p)$

PRACTICES 6–8 Factor.

6. $5x^4 - 20x$ **7.** $\dfrac{5}{9}z^5 + \dfrac{1}{9}z^4 - \dfrac{2}{9}z^3$ **8.** $8a^2b^4 - 20a^3b^3 + 12ab^3$

Section 6.1 The Greatest Common Factor and Factoring by Grouping 367

EXAMPLE 9 Factor: $5(x + 3) + y(x + 3)$

Solution The binomial $(x + 3)$ is the greatest common factor. Use the distributive property to factor out $(x + 3)$.

$$5(x + 3) + y(x + 3) = (x + 3)(5 + y)$$

PRACTICE 9 Factor $8(y - 2) + x(y - 2)$.

EXAMPLE 10 Factor: $3m^2n(a + b) - (a + b)$

Solution The greatest common factor is $(a + b)$.

$$3m^2n(a + b) - 1(a + b) = (a + b)(3m^2n - 1)$$

PRACTICE 10 Factor $7xy^3(p + q) - (p + q)$

OBJECTIVE 4 ▶ Factoring by grouping. Once the GCF is factored out, we can often continue to factor the polynomial, using a variety of techniques. We discuss here a technique for factoring polynomials called **grouping**.

EXAMPLE 11 Factor $xy + 2x + 3y + 6$ by grouping. Check by multiplying.

Solution The GCF of the first two terms is x, and the GCF of the last two terms is 3.

$$xy + 2x + 3y + 6 = (xy + 2x) + (3y + 6) \quad \text{Group terms.}$$
$$= x(y + 2) + 3(y + 2) \quad \text{Factor out GCF from each grouping.}$$

▶ **Helpful Hint**
Notice that this form, $x(y + 2) + 3(y + 2)$, is *not* a factored form of the original polynomial. It is a sum, not a product.

Next we factor out the common binomial factor, $(y + 2)$.

$$x(y + 2) + 3(y + 2) = (y + 2)(x + 3)$$

Now the result is a factored form because it is a product. We were able to write the polynomial as a product because of the common binomial factor, $(y + 2)$, that appeared. If this does not happen, try rearranging the terms of the original polynomial.

Check: Multiply $(y + 2)$ by $(x + 3)$.

$$(y + 2)(x + 3) = xy + 2x + 3y + 6,$$

the original polynomial.
Thus, the factored form of $xy + 2x + 3y + 6$ is the product $(y + 2)(x + 3)$.

PRACTICE 11 Factor $xy + 3y + 4x + 12$ by grouping. Check by multiplying.

You may want to try these steps when factoring by grouping.

To Factor a Four-Term Polynomial by Grouping
STEP 1. Group the terms in two groups of two terms so that each group has a common factor.
STEP 2. Factor out the GCF from each group.
STEP 3. If there is now a common binomial factor in the groups, factor it out.
STEP 4. If not, rearrange the terms and try these steps again.

368 CHAPTER 6 Factoring Polynomials

EXAMPLES Factor by grouping.

12. $3x^2 + 4xy - 3x - 4y$
$= (3x^2 + 4xy) + (-3x - 4y)$
$= x(3x + 4y) - 1(3x + 4y)$ Factor each group. A -1 is factored from the second pair of terms so that there is a common factor, $(3x + 4y)$.
$= (3x + 4y)(x - 1)$ Factor out the common factor, $(3x + 4y)$.

13. $2a^2 + 5ab + 2a + 5b$
$= (2a^2 + 5ab) + (2a + 5b)$ Factor each group. An understood 1 is written before
$= a(2a + 5b) + 1(2a + 5b)$ $(2a + 5b)$ to help remember that $(2a + 5b)$ is $1(2a + 5b)$.
$= (2a + 5b)(a + 1)$ Factor out the common factor, $(2a + 5b)$.

> **Helpful Hint**
> Notice the factor of 1 is written when $(2a + 5b)$ is factored out.

PRACTICES 12–13

12. Factor $2xy + 3y^2 - 2x - 3y$ by grouping.

13. Factor $7a^3 + 5a^2 + 7a + 5$ by grouping.

EXAMPLES Factor by grouping.

14. $3xy + 2 - 3x - 2y$

Notice that the first two terms have no common factor other than 1. However, if we rearrange these terms, a grouping emerges that does lead to a common factor.

$3xy + 2 - 3x - 2y$
$= (3xy - 3x) + (-2y + 2)$
$= 3x(y - 1) - 2(y - 1)$ Factor -2 from the second group so that there is a common factor $(y - 1)$.
$= (y - 1)(3x - 2)$ Factor out the common factor, $(y - 1)$.

15. $5x - 10 + x^3 - x^2 = 5(x - 2) + x^2(x - 1)$

There is no common binomial factor that can now be factored out. No matter how we rearrange the terms, no grouping will lead to a common factor. Thus, this polynomial is not factorable by grouping.

PRACTICES 14–15

14. Factor $4xy + 15 - 12x - 5y$ by grouping.

15. Factor $9y - 18 + y^3 - 4y^2$.

> **Helpful Hint**
> One more reminder: When **factoring** a polynomial, make sure the polynomial is written as a **product**. For example, it is true that
> $$3x^2 + 4xy - 3x - 4y = \underbrace{x(3x + 4y) - 1(3x + 4y)}_{\text{but is not a factored form}},$$
> since it is a **sum (difference)**, not a **product**. A factored form of $3x^2 + 4xy - 3x - 4y$ is the product $(3x + 4y)(x - 1)$.

Factoring out a greatest common factor first makes factoring by any method easier, as we see in the next example.

EXAMPLE 16 Factor: $4ax - 4ab - 2bx + 2b^2$

Solution First, factor out the common factor 2 from all four terms.

$4ax - 4ab - 2bx + 2b^2$
$= 2(2ax - 2ab - bx + b^2)$ Factor out 2 from all four terms.
$= 2[2a(x - b) - b(x - b)]$ Factor each pair of terms. A "$-b$" is factored from the second pair so that there is a common factor, $x - b$.
$= 2(x - b)(2a - b)$ Factor out the common binomial.

PRACTICE 16 Factor $3xy - 3ay - 6ax + 6a^2$

> **Helpful Hint**
> Throughout this chapter, we will be factoring polynomials. Even when the instructions do not so state, it is always a good idea to check your answers by multiplying.

VOCABULARY & READINESS CHECK

Use the choices below to fill in each blank. Some choices may be used more than once and some may not be used at all.

greatest common factor   factors   factoring   true   false   least   greatest

1. Since $5 \cdot 4 = 20$, the numbers 5 and 4 are called _____ of 20.
2. The _____ of a list of integers is the largest integer that is a factor of all the integers in the list.
3. The greatest common factor of a list of common variables raised to powers is the variable raised to the _____ exponent in the list.
4. The process of writing a polynomial as a product is called _____.
5. True or false: A factored form of $7x + 21 + xy + 3y$ is $7(x + 3) + y(x + 3)$. _____
6. True or false: A factored form of $3x^3 + 6x + x^2 + 2$ is $3x(x^2 + 2)$. _____

Write the prime factorization of the following integers.

7. 14
8. 15

Write the GCF of the following pairs of integers.

9. 18, 3
10. 7, 35
11. 20, 15
12. 6, 15

6.1 EXERCISE SET

Find the GCF for each list. See Examples 1 through 3.

1. 32, 36
2. 36, 90
3. 18, 42, 84
4. 30, 75, 135
5. 24, 14, 21
6. 15, 25, 27
7. y^2, y^4, y^7
8. x^3, x^2, x^5
9. z^7, z^9, z^{11}
10. y^8, y^{10}, y^{12}
11. $x^{10}y^2, xy^2, x^3y^3$
12. p^7q, p^8q^2, p^9q^3
13. $14x, 21$
14. $20y, 15$
15. $12y^4, 20y^3$
16. $32x^5, 18x^2$
17. $-10x^2, 15x^3$
18. $-21x^3, 14x$
19. $12x^3, -6x^4, 3x^5$
20. $15y^2, 5y^7, -20y^3$
21. $-18x^2y, 9x^3y^3, 36x^3y$
22. $7x^3y^3, -21x^2y^2, 14xy^4$
23. $20a^6b^2c^8, 50a^7b$
24. $40x^7y^2z, 64x^9y$

Factor out the GCF from each polynomial. See Examples 4 through 10.

25. $3a + 6$
26. $18a + 12$
27. $30x - 15$
28. $42x - 7$
29. $x^3 + 5x^2$
30. $y^5 + 6y^4$
31. $6y^4 + 2y^3$
32. $5x^2 + 10x^6$
33. $4x - 8y + 4$
34. $7x + 21y - 7$
35. $6x^3 - 9x^2 + 12x$
36. $12x^3 + 16x^2 - 8x$
37. $a^7b^6 - a^3b^2 + a^2b^5 - a^2b^2$
38. $x^9y^6 + x^3y^5 - x^4y^3 + x^3y^3$
39. $8x^5 + 16x^4 - 20x^3 + 12$
40. $9y^6 - 27y^4 + 18y^2 + 6$

41. $\dfrac{1}{3}x^4 + \dfrac{2}{3}x^3 - \dfrac{4}{3}x^5 + \dfrac{1}{3}x$

42. $\dfrac{2}{5}y^7 - \dfrac{4}{5}y^5 + \dfrac{3}{5}y^2 - \dfrac{2}{5}y$

43. $y(x^2 + 2) + 3(x^2 + 2)$
44. $x(y^2 + 1) - 3(y^2 + 1)$
45. $z(y + 4) - 3(y + 4)$
46. $8(x + 2) - y(x + 2)$
47. $r(z^2 - 6) + (z^2 - 6)$
48. $q(b^3 - 5) + (b^3 - 5)$

Factor a negative number or a GCF with a negative coefficient from each polynomial. See Example 5.

49. $-2x - 14$
50. $-7y - 21$
51. $-2x^5 + x^7$
52. $-5y^3 + y^6$
53. $-6a^4 + 9a^3 - 3a^2$
54. $-5m^6 + 10m^5 - 5m^3$

Factor each four-term polynomial by grouping. See Examples 11 through 16.

55. $x^3 + 2x^2 + 5x + 10$
56. $x^3 + 4x^2 + 3x + 12$
57. $5x + 15 + xy + 3y$
58. $xy + y + 2x + 2$
59. $6x^3 - 4x^2 + 15x - 10$
60. $16x^3 - 28x^2 + 12x - 21$
61. $5m^3 + 6mn + 5m^2 + 6n$
62. $8w^2 + 7wv + 8w + 7v$
63. $2y - 8 + xy - 4x$
64. $6x - 42 + xy - 7y$
65. $2x^3 - x^2 + 8x - 4$
66. $2x^3 - x^2 - 10x + 5$
67. $4x^2 - 8xy - 3x + 6y$
68. $5xy - 15x - 6y + 18$
69. $5q^2 - 4pq - 5q + 4p$
70. $6m^2 - 5mn - 6m + 5n$
71. $2x^4 + 5x^3 + 2x^2 + 5x$
72. $4y^4 + y^2 + 20y^3 + 5y$
73. $12x^2y - 42x^2 - 4y + 14$
74. $90 + 15y^2 - 18x - 3xy^2$

MIXED PRACTICE

Factor. See Examples 4 through 16.

75. $32xy - 18x^2$
76. $10xy - 15x^2$
77. $y(x + 2) - 3(x + 2)$
78. $z(y - 4) + 3(y - 4)$
79. $14x^3y + 7x^2y - 7xy$
80. $5x^3y - 15x^2y + 10xy$
81. $28x^3 - 7x^2 + 12x - 3$
82. $15x^3 + 5x^2 - 6x - 2$
83. $-40x^8y^6 - 16x^9y^5$
84. $-21x^3y - 49x^2y^2$
85. $6a^2 + 9ab^2 + 6ab + 9b^3$
86. $16x^2 + 4xy^2 + 8xy + 2y^3$

REVIEW AND PREVIEW

Multiply. See Section 5.3.

87. $(x + 2)(x + 5)$
88. $(y + 3)(y + 6)$
89. $(b + 1)(b - 4)$
90. $(x - 5)(x + 10)$

Fill in the chart by finding two numbers that have the given product and sum. The first column is filled in for you.

		91.	92.	93.	94.	95.	96.
Two Numbers	4, 7						
Their Product	28	12	20	8	16	−10	−24
Their Sum	11	8	9	−9	−10	3	−5

CONCEPT EXTENSIONS

See the Concept Checks in this section.

97. Which of the following is/are factored form(s) of $8a - 24$?
   a. $8 \cdot a - 24$
   b. $8(a - 3)$
   c. $4(2a - 12)$
   d. $8 \cdot a - 2 \cdot 12$

Which of the following expressions are factored?

98. $(a + 6)(a + 2)$
99. $(x + 5)(x + y)$
100. $5(2y + z) - b(2y + z)$
101. $3x(a + 2b) + 2(a + 2b)$

102. Construct a binomial whose greatest common factor is $5a^3$. (*Hint:* Multiply $5a^3$ by a binomial whose terms contain no common factor other than 1. $5a^3(\square + \square)$.)

103. Construct a trinomial whose greatest common factor is $2x^2$. See the hint for Exercise 102.

104. Explain how you can tell whether a polynomial is written in factored form.

105. Construct a four-term polynomial that can be factored by grouping.

106. The number (in millions) of single digital downloads annually in the United States each year during 2003–2005 can be modeled by the polynomial $45x^2 + 95x$, where x is the number of years since 2003. (*Source:* Recording Industry Association of America)
   a. Find the number of single digital downloads in 2005. To do so, let $x = 2$ and evaluate $45x^2 + 95x$.
   b. Use this expression to predict the number of single digital downloads in 2009.
   c. Factor the polynomial $45x^2 + 95x$.

107. The number (in thousands) of students who graduated from U.S. high schools each year during 2003–2005 can be modeled by $-8x^2 + 50x + 3020$, where x is the number of years since 2003. (*Source:* National Center for Education Statistics)

   a. Find the number of students who graduated from U.S. high schools in 2005. To do so, let $x = 2$ and evaluate $-8x^2 + 50x + 3020$.

   b. Use this expression to predict the number of students who graduated from U.S. high schools in 2007.

   c. Factor the polynomial $-8x^2 + 50x + 3020$.

Write an expression for the area of each shaded region. Then write the expression as a factored polynomial.

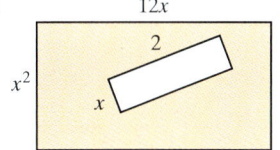

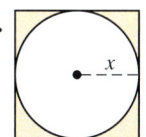

△ **108.** (rectangle x^2 by $12x$ with inner rectangle 2 by x)

△ **109.** (square with inscribed circle of radius x)

Write an expression for the length of each rectangle. (**Hint:** *Factor the area binomial and recall that Area = width · length.*)

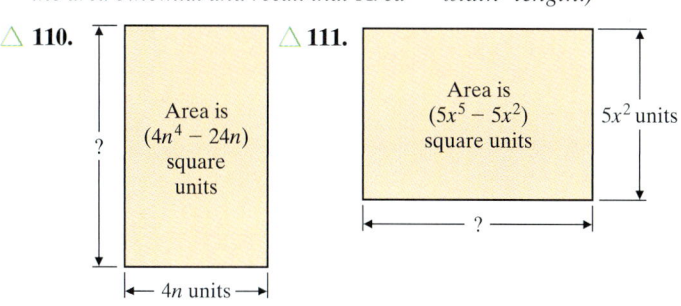

△ **110.** Area is $(4n^4 - 24n)$ square units; width is $4n$ units; length ?

△ **111.** Area is $(5x^5 - 5x^2)$ square units; height is $5x^2$ units; length ?

Factor each polynomial by grouping.

112. $x^{2n} + 2x^n + 3x^n + 6$
   (**Hint:** Don't forget that $x^{2n} = x^n \cdot x^n$.)

113. $x^{2n} + 6x^n + 10x^n + 60$

114. $3x^{2n} + 21x^n - 5x^n - 35$

115. $12x^{2n} - 10x^n - 30x^n + 25$

6.2 FACTORING TRINOMIALS OF THE FORM $x^2 + bx + c$

OBJECTIVES

1. Factor trinomials of the form $x^2 + bx + c$.
2. Factor out the greatest common factor and then factor a trinomial of the form $x^2 + bx + c$.

OBJECTIVE 1 ▶ Factoring trinomials of the form $x^2 + bx + c$. In this section, we factor trinomials of the form $x^2 + bx + c$, such as

$$x^2 + 4x + 3, \quad x^2 - 8x + 15, \quad x^2 + 4x - 12, \quad r^2 - r - 42$$

Notice that for these trinomials, the coefficient of the squared variable is 1.

Recall that factoring means to write as a product and that factoring and multiplying are reverse processes. Using the FOIL method of multiplying binomials, we have that

$$\overset{F\quad O\quad I\quad L}{(x+3)(x+1) = x^2 + 1x + 3x + 3}$$
$$= x^2 + 4x + 3$$

Thus, a factored form of $x^2 + 4x + 3$ is $(x + 3)(x + 1)$.

Notice that the product of the first terms of the binomials is $x \cdot x = x^2$, the first term of the trinomial. Also, the product of the last two terms of the binomials is $3 \cdot 1 = 3$, the third term of the trinomial. The sum of these same terms is $3 + 1 = 4$, the coefficient of the middle term, x, of the trinomial.

The product of these numbers is 3.

$$x^2 + 4x + 3 = (x + 3)(x + 1)$$

The sum of these numbers is 4.

Many trinomials, such as the one above, factor into two binomials. To factor $x^2 + 7x + 10$, let's assume that it factors into two binomials and begin by writing two pairs of parentheses. The first term of the trinomial is x^2, so we use x and x as the first terms of the binomial factors.

$$x^2 + 7x + 10 = (x + \square)(x + \square)$$

To determine the last term of each binomial factor, we look for two integers whose product is 10 and whose sum is 7. Since our numbers must have a positive product and a positive sum, we list pairs of positive integer factors of 10 only.

Positive Factors of 10	Sum of Factors
1, 10	$1 + 10 = 11$
2, 5	$2 + 5 = 7$

The correct pair of numbers is 2 and 5 because their product is 10 and their sum is 7. Now we can fill in the last terms of the binomial factors.

$$x^2 + 7x + 10 = (x + 2)(x + 5)$$

Check: To see if we have factored correctly, multiply.

$$(x + 2)(x + 5) = x^2 + 5x + 2x + 10$$
$$= x^2 + 7x + 10 \qquad \text{Combine like terms.}$$

> **Helpful Hint**
> Since multiplication is commutative, the factored form of $x^2 + 7x + 10$ can be written as either $(x + 2)(x + 5)$ or $(x + 5)(x + 2)$.

Factoring a Trinomial of the Form $x^2 + bx + c$

The factored form of $x^2 + bx + c$ is

The product of these numbers is c.

$$x^2 + bx + c = (x + \square)(x + \square)$$

The sum of these numbers is b.

EXAMPLE 1 Factor: $x^2 + 7x + 12$

Solution We begin by writing the first terms of the binomial factors.

$$(x + \square)(x + \square)$$

Next we look for two numbers whose product is 12 and whose sum is 7. Since our numbers must have a positive product and a positive sum, we look at pairs of positive factors of 12 only.

Positive Factors of 12	Sum of Factors	
1, 12	13	
2, 6	8	
3, 4	7	Correct sum, so the numbers are 3 and 4.

Thus, $x^2 + 7x + 12 = (x + 3)(x + 4)$

Check: $(x + 3)(x + 4) = x^2 + 4x + 3x + 12 = x^2 + 7x + 12.$ □

PRACTICE 1 Factor $x^2 + 5x + 6$.

EXAMPLE 2 Factor: $x^2 - 12x + 35$

Solution Again, we begin by writing the first terms of the binomials.

$$(x + \Box)(x + \Box)$$

Now we look for two numbers whose product is 35 and whose sum is -12. Since our numbers must have a positive product and a negative sum, we look at pairs of negative factors of 35 only.

Negative Factors of 35	Sum of Factors
$-1, -35$	-36
$-5, -7$	-12

Correct sum, so the numbers are -5 and -7.

Thus, $x^2 - 12x + 35 = (x - 5)(x - 7)$

Check: To check, multiply $(x - 5)(x - 7)$.

PRACTICE 2 Factor $x^2 - 17x + 70$.

EXAMPLE 3 Factor: $x^2 + 4x - 12$

Solution $x^2 + 4x - 12 = (x + \Box)(x + \Box)$

We look for two numbers whose product is -12 and whose sum is 4. Since our numbers must have a negative product, we look at pairs of factors with opposite signs.

Factors of -12	Sum of Factors
$-1, 12$	11
$1, -12$	-11
$-2, 6$	4
$2, -6$	-4
$-3, 4$	1
$3, -4$	-1

Correct sum, so the numbers are -2 and 6.

Thus, $x^2 + 4x - 12 = (x - 2)(x + 6)$

PRACTICE 3 Factor $x^2 + 5x - 14$.

EXAMPLE 4 Factor: $r^2 - r - 42$

Solution Because the variable in this trinomial is r, the first term of each binomial factor is r.

$$r^2 - r - 42 = (r + \Box)(r + \Box)$$

Now we look for two numbers whose product is -42 and whose sum is -1, the numerical coefficient of r. The numbers are 6 and -7. Therefore,

$$r^2 - r - 42 = (r + 6)(r - 7)$$

PRACTICE 4 Factor $p^2 - 2p - 63$.

EXAMPLE 5 Factor: $a^2 + 2a + 10$

Solution Look for two numbers whose product is 10 and whose sum is 2. Neither 1 and 10 nor 2 and 5 give the required sum, 2. We conclude that $a^2 + 2a + 10$ is not factorable with integers. A polynomial such as $a^2 + 2a + 10$ is called a **prime polynomial.**

PRACTICE 5 Factor $b^2 + 5b + 1$.

EXAMPLE 6 Factor: $x^2 + 7xy + 6y^2$

Solution $$x^2 + 7xy + 6y^2 = (x + \square)(x + \square)$$

Recall that the middle term $7xy$ is the same as $7yx$. Thus, we can see that $7y$ is the "coefficient" of x. We then look for two terms whose product is $6y^2$ and whose sum is $7y$. The terms are $6y$ and $1y$ or $6y$ and y because $6y \cdot y = 6y^2$ and $6y + y = 7y$. Therefore,

$$x^2 + 7xy + 6y^2 = (x + 6y)(x + y)$$

PRACTICE 6 Factor $x^2 + 7xy + 12y^2$.

EXAMPLE 7 Factor: $x^4 + 5x^2 + 6$

Solution As usual, we begin by writing the first terms of the binomials. Since the greatest power of x in this polynomial is x^4, we write

$$(x^2 + \square)(x^2 + \square) \quad \text{since } x^2 \cdot x^2 = x^4$$

Now we look for two factors of 6 whose sum is 5. The numbers are 2 and 3. Thus,

$$x^4 + 5x^2 + 6 = (x^2 + 2)(x^2 + 3)$$

PRACTICE 7 Factor $x^4 + 13x^2 + 12$.

If the terms of a polynomial are not written in descending powers of the variable, you may want to do so before factoring.

EXAMPLE 8 Factor: $40 - 13t + t^2$

Solution First, we rearrange terms so that the trinomial is written in descending powers of t.

$$40 - 13t + t^2 = t^2 - 13t + 40$$

Next, try to factor.

$$t^2 - 13t + 40 = (t + \square)(t + \square)$$

Now we look for two factors of 40 whose sum is -13. The numbers are -8 and -5. Thus,

$$t^2 - 13t + 40 = (t - 8)(t - 5)$$

PRACTICE 8 Factor $48 - 14x + x^2$.

The following sign patterns may be useful when factoring trinomials.

> **Helpful Hint**
> A positive constant in a trinomial tells us to look for two numbers with the same sign. The sign of the coefficient of the middle term tells us whether the signs are both positive or both negative.
>
> both positive — same sign
> $x^2 + 10x + 16 = (x + 2)(x + 8)$
>
> both negative — same sign
> $x^2 - 10x + 16 = (x - 2)(x - 8)$
>
> A negative constant in a trinomial tells us to look for two numbers with opposite signs.
>
> opposite signs
> $x^2 + 6x - 16 = (x + 8)(x - 2)$
>
> opposite signs
> $x^2 - 6x - 16 = (x - 8)(x + 2)$

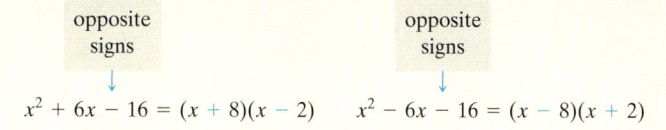

OBJECTIVE 2 ▶ Factoring out the greatest common factor. Remember that the first step in factoring any polynomial is to factor out the greatest common factor (if there is one other than 1 or −1).

EXAMPLE 9 Factor: $3m^2 - 24m - 60$

Solution First we factor out the greatest common factor, 3, from each term.

$$3m^2 - 24m - 60 = 3(m^2 - 8m - 20)$$

Now we factor $m^2 - 8m - 20$ by looking for two factors of −20 whose sum is −8. The factors are −10 and 2. Therefore, the complete factored form is

$$3m^2 - 24m - 60 = 3(m + 2)(m - 10)$$

> **Helpful Hint**
> Remember to write the common factor 3 as part of the factored form.

PRACTICE 9 Factor $4x^2 - 24x + 36$.

EXAMPLE 10 Factor: $2x^4 - 26x^3 + 84x^2$

Solution

$$2x^4 - 26x^3 + 84x^2 = 2x^2(x^2 - 13x + 42) \quad \text{Factor out common factor, } 2x^2.$$
$$= 2x^2(x - 6)(x - 7) \quad \text{Factor } x^2 - 13x + 42.$$

PRACTICE 10 Factor $3y^4 - 18y^3 - 21y^2$.

VOCABULARY & READINESS CHECK

Fill in each blank with "true or false."

1. To factor $x^2 + 7x + 6$, we look for two numbers whose product is 6 and whose sum is 7. _____
2. We can write the factorization $(y + 2)(y + 4)$ also as $(y + 4)(y + 2)$. _____
3. The factorization $(4x - 12)(x - 5)$ is completely factored. _____
4. The factorization $(x + 2y)(x + y)$ may also be written as $(x + 2y)^2$. _____

Complete each factored form.

5. $x^2 + 9x + 20 = (x + 4)(x \quad)$
6. $x^2 + 12x + 35 = (x + 5)(x \quad)$
7. $x^2 - 7x + 12 = (x - 4)(x \quad)$
8. $x^2 - 13x + 22 = (x - 2)(x \quad)$
9. $x^2 + 4x + 4 = (x + 2)(x \quad)$
10. $x^2 + 10x + 24 = (x + 6)(x \quad)$

6.2 EXERCISE SET

Factor each trinomial completely. If a polynomial can't be factored, write "prime." See Examples 1 through 8.

1. $x^2 + 7x + 6$
2. $x^2 + 6x + 8$
3. $y^2 - 10y + 9$
4. $y^2 - 12y + 11$
5. $x^2 - 6x + 9$
6. $x^2 - 10x + 25$
7. $x^2 - 3x - 18$
8. $x^2 - x - 30$
9. $x^2 + 3x - 70$
10. $x^2 + 4x - 32$
11. $x^2 + 5x + 2$
12. $x^2 - 7x + 5$
13. $x^2 + 8xy + 15y^2$
14. $x^2 + 6xy + 8y^2$
15. $a^4 - 2a^2 - 15$
16. $y^4 - 3y^2 - 70$
17. $13 + 14m + m^2$
18. $17 + 18n + n^2$
19. $10t - 24 + t^2$
20. $6q - 27 + q^2$
21. $a^2 - 10ab + 16b^2$
22. $a^2 - 9ab + 18b^2$

MIXED PRACTICE

Factor each trinomial completely. Some of these trinomials contain a greatest common factor (other than 1). Don't forget to factor out the GCF first. See Examples 1 through 10.

23. $2z^2 + 20z + 32$
24. $3x^2 + 30x + 63$
25. $2x^3 - 18x^2 + 40x$
26. $3x^3 - 12x^2 - 36x$
27. $x^2 - 3xy - 4y^2$
28. $x^2 - 4xy - 77y^2$
29. $x^2 + 15x + 36$
30. $x^2 + 19x + 60$
31. $x^2 - x - 2$
32. $x^2 - 5x - 14$
33. $r^2 - 16r + 48$
34. $r^2 - 10r + 21$
35. $x^2 + xy - 2y^2$
36. $x^2 - xy - 6y^2$
37. $3x^2 + 9x - 30$
38. $4x^2 - 4x - 48$
39. $3x^2 - 60x + 108$
40. $2x^2 - 24x + 70$
41. $x^2 - 18x - 144$
42. $x^2 + x - 42$
43. $r^2 - 3r + 6$
44. $x^2 + 4x - 10$
45. $x^2 - 8x + 15$
46. $x^2 - 9x + 14$
47. $6x^3 + 54x^2 + 120x$
48. $3x^3 + 3x^2 - 126x$
49. $4x^2y + 4xy - 12y$
50. $3x^2y - 9xy + 45y$
51. $x^2 - 4x - 21$
52. $x^2 - 4x - 32$
53. $x^2 + 7xy + 10y^2$
54. $x^2 - 3xy - 4y^2$
55. $64 + 24t + 2t^2$
56. $50 + 20t + 2t^2$
57. $x^3 - 2x^2 - 24x$
58. $x^3 - 3x^2 - 28x$
59. $2t^5 - 14t^4 + 24t^3$
60. $3x^6 + 30x^5 + 72x^4$
61. $5x^3y - 25x^2y^2 - 120xy^3$
62. $7a^3b - 35a^2b^2 + 42ab^3$
63. $162 - 45m + 3m^2$
64. $48 - 20n + 2n^2$
65. $-x^2 + 12x - 11$ (Factor out -1 first.)
66. $-x^2 + 8x - 7$ (Factor out -1 first.)
67. $\frac{1}{2}y^2 - \frac{9}{2}y - 11$ (Factor out $\frac{1}{2}$ first.)
68. $\frac{1}{3}y^2 - \frac{5}{3}y - 8$ (Factor out $\frac{1}{3}$ first.)
69. $x^3y^2 + x^2y - 20x$
70. $a^2b^3 + ab^2 - 30b$

REVIEW AND PREVIEW

Multiply. See Section 5.4.

71. $(2x + 1)(x + 5)$
72. $(3x + 2)(x + 4)$
73. $(5y - 4)(3y - 1)$
74. $(4z - 7)(7z - 1)$
75. $(a + 3b)(9a - 4b)$
76. $(y - 5x)(6y + 5x)$

CONCEPT EXTENSIONS

77. Write a polynomial that factors as $(x - 3)(x + 8)$.
78. To factor $x^2 + 13x + 42$, think of two numbers whose _____ is 42 and whose _____ is 13.

Complete each sentence in your own words.

79. If $x^2 + bx + c$ is factorable and c is negative, then the signs of the last-term factors of the binomials are opposite because . . . .
80. If $x^2 + bx + c$ is factorable and c is positive, then the signs of the last-term factors of the binomials are the same because . . . .

Remember that perimeter means distance around. Write the perimeter of each rectangle as a simplified polynomial. Then factor the polynomial.

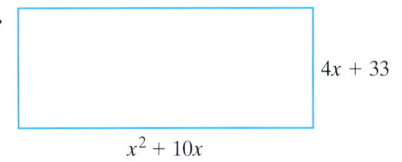

△ 81.

$4x + 33$

$x^2 + 10x$

△ 82.

$12x^2$

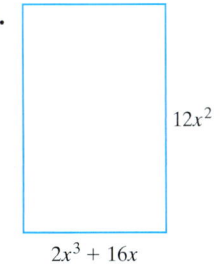

$2x^3 + 16x$

83. An object is thrown upward from the top of an 80-foot building with an initial velocity of 64 feet per second. The height of the object after t seconds is given by $-16t^2 + 64t + 80$. Factor this polynomial.

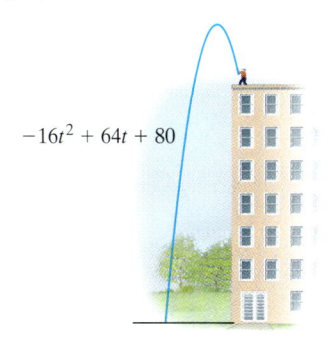

$-16t^2 + 64t + 80$

Factor each trinomial completely.

84. $x^2 + x + \dfrac{1}{4}$
85. $x^2 + \dfrac{1}{2}x + \dfrac{1}{16}$
86. $y^2(x + 1) - 2y(x + 1) - 15(x + 1)$
87. $z^2(x + 1) - 3z(x + 1) - 70(x + 1)$

Factor each trinomial. (**Hint:** *Notice that* $x^{2n} + 4x^n + 3$ *factors as* $(x^n + 1)(x^n + 3)$. **Remember:** $x^n \cdot x^n = x^{n+n}$ *or* x^{2n}.)

88. $x^{2n} + 5x^n + 6$
89. $x^{2n} + 8x^n - 20$

Find a positive value of c so that each trinomial is factorable.

90. $x^2 + 6x + c$
91. $t^2 + 8t + c$
92. $y^2 - 4y + c$
93. $n^2 - 16n + c$

Find a positive value of b so that each trinomial is factorable.

94. $x^2 + bx + 15$
95. $y^2 + by + 20$
96. $m^2 + bm - 27$
97. $x^2 + bx - 14$

6.3 FACTORING TRINOMIALS OF THE FORM $ax^2 + bx + c$ AND PERFECT SQUARE TRINOMIALS

OBJECTIVES

1. Factor trinomials of the form $ax^2 + bx + c$, where $a \neq 1$.
2. Factor out a GCF before factoring a trinomial of the form $ax^2 + bx + c$.
3. Factor perfect square trinomials.

OBJECTIVE 1 ▶ **Factoring trinomials of the form $ax^2 + bx + c$.** In this section, we factor trinomials of the form $ax^2 + bx + c$, such as

$$3x^2 + 11x + 6, \quad 8x^2 - 22x + 5, \quad \text{and} \quad 2x^2 + 13x - 7$$

Notice that the coefficient of the squared variable in these trinomials is a number other than 1. We will factor these trinomials using a trial-and-check method based on our work in the last section.

To begin, let's review the relationship between the numerical coefficients of the trinomial and the numerical coefficients of its factored form. For example, since

$$(2x + 1)(x + 6) = 2x^2 + 13x + 6,$$

a factored form of $2x^2 + 13x + 6$ is $(2x + 1)(x + 6)$

Notice that $2x$ and x are factors of $2x^2$, the first term of the trinomial. Also, 6 and 1 are factors of 6, the last term of the trinomial, as shown:

$$2x^2 + 13x + 6 = (2x + 1)(x + 6)$$

Also notice that $13x$, the middle term, is the sum of the following products:

$$2x^2 + 13x + 6 = (2x + 1)(x + 6)$$

$1x$
$+12x$
$13x$ Middle term

Let's use this pattern to factor $5x^2 + 7x + 2$. First, we find factors of $5x^2$. Since all numerical coefficients in this trinomial are positive, we will use factors with positive numerical coefficients only. Thus, the factors of $5x^2$ are $5x$ and x. Let's try these factors as first terms of the binomials. Thus far, we have

$$5x^2 + 7x + 2 = (5x + \square)(x + \square)$$

Next, we need to find positive factors of 2. Positive factors of 2 are 1 and 2. Now we try possible combinations of these factors as second terms of the binomials until we obtain a middle term of $7x$.

$$(5x + 1)(x + 2) = 5x^2 + 11x + 2$$

$1x$
$+10x$
$11x$ ⟶ Incorrect middle term

Let's try switching factors 2 and 1.

$$(5x + 2)(x + 1) = 5x^2 + 7x + 2$$

$2x$
$+5x$
$7x$ ⟶ Correct middle term

Thus the factored form of $5x^2 + 7x + 2$ is $(5x + 2)(x + 1)$. To check, we multiply $(5x + 2)$ and $(x + 1)$. The product is $5x^2 + 7x + 2$.

EXAMPLE 1   Factor: $3x^2 + 11x + 6$

Solution   Since all numerical coefficients are positive, we use factors with positive numerical coefficients. We first find factors of $3x^2$.

Factors of $3x^2$: $3x^2 = 3x \cdot x$

If factorable, the trinomial will be of the form

$$3x^2 + 11x + 6 = (3x + \square)(x + \square)$$

Next we factor 6.

Factors of 6: $6 = 1 \cdot 6$, $6 = 2 \cdot 3$

Now we try combinations of factors of 6 until a middle term of $11x$ is obtained. Let's try 1 and 6 first.

$$(3x + 1)(x + 6) = 3x^2 + 19x + 6$$

$$\frac{\begin{array}{c}1x\\+18x\end{array}}{19x} \longrightarrow \text{Incorrect middle term}$$

Now let's next try 6 and 1.

$$(3x + 6)(x + 1)$$

Before multiplying, notice that the terms of the factor $3x + 6$ have a common factor of 3. The terms of the original trinomial $3x^2 + 11x + 6$ have no common factor other than 1, so the terms of its factors will also contain no common factor other than 1. This means that $(3x + 6)(x + 1)$ is not a factored form.

Next let's try 2 and 3 as last terms.

$$(3x + 2)(x + 3) = 3x^2 + 11x + 6$$

$$\frac{\begin{array}{c}2x\\+9x\end{array}}{11x} \longrightarrow \text{Correct middle term}$$

Thus a factored form of $3x^2 + 11x + 6$ is $(3x + 2)(x + 3)$. □

PRACTICE
1 Factor: $2x^2 + 11x + 15$.

> ▶ **Helpful Hint**
> If the terms of a trinomial have no common factor (other than 1), then the terms of neither of its binomial factors will contain a common factor (other than 1).

Concept Check ✓

Do the terms of $3x^2 + 29x + 18$ have a common factor? Without multiplying, decide which of the following factored forms could not be a factored form of $3x^2 + 29x + 18$.

a. $(3x + 18)(x + 1)$ **b.** $(3x + 2)(x + 9)$
c. $(3x + 6)(x + 3)$ **d.** $(3x + 9)(x + 2)$

EXAMPLE 2   Factor: $8x^2 - 22x + 5$

Solution   Factors of $8x^2$: $8x^2 = 8x \cdot x$, $8x^2 = 4x \cdot 2x$

We'll try $8x$ and x.

$$8x^2 - 22x + 5 = (8x + \square)(x + \square)$$

Since the middle term, $-22x$, has a negative numerical coefficient, we factor 5 into negative factors.

Factors of 5: $5 = -1 \cdot -5$

Let's try -1 and -5.

$$(8x - 1)(x - 5) = 8x^2 - 41x + 5$$

$$\frac{\begin{array}{c}-1x\\+(-40x)\end{array}}{-41x} \longrightarrow \text{Incorrect middle term}$$

Answers to Concept Check:
no; a, c, d

Now let's try -5 and -1.

$$(8x - 5)(x - 1) = 8x^2 - 13x + 5$$

$$\begin{array}{r} -5x \\ +(-8x) \\ \hline -13x \end{array} \longrightarrow \text{Incorrect middle term}$$

Don't give up yet! We can still try other factors of $8x^2$. Let's try $4x$ and $2x$ with -1 and -5.

$$(4x - 1)(2x - 5) = 8x^2 - 22x + 5$$

$$\begin{array}{r} -2x \\ +(-20x) \\ \hline -22x \end{array} \longrightarrow \text{Correct middle term}$$

A factored form of $8x^2 - 22x + 5$ is $(4x - 1)(2x - 5)$. ☐

PRACTICE 2 Factor: $15x^2 - 22x + 8$.

EXAMPLE 3 Factor: $2x^2 + 13x - 7$

Solution Factors of $2x^2$: $\quad 2x^2 = 2x \cdot x$

Factors of -7: $\quad -7 = -1 \cdot 7, \quad -7 = 1 \cdot -7$

We try possible combinations of these factors:

$(2x + 1)(x - 7) = 2x^2 - 13x - 7 \quad$ Incorrect middle term
$(2x - 1)(x + 7) = 2x^2 + 13x - 7 \quad$ Correct middle term

A factored form of $2x^2 + 13x - 7$ is $(2x - 1)(x + 7)$. ☐

PRACTICE 3 Factor: $4x^2 + 11x - 3$.

EXAMPLE 4 Factor: $10x^2 - 13xy - 3y^2$

Solution Factors of $10x^2$: $\quad 10x^2 = 10x \cdot x, \quad 10x^2 = 2x \cdot 5x$

Factors of $-3y^2$: $\quad -3y^2 = -3y \cdot y, \quad -3y^2 = 3y \cdot -y$

We try some combinations of these factors:

$\qquad\qquad\qquad\qquad\quad$ Correct $\quad\;\,$ Correct
$\qquad\qquad\qquad\qquad\quad\;\; \downarrow \qquad\qquad \downarrow$
$(10x - 3y)(x + y) = 10x^2 + 7xy - 3y^2$
$(x + 3y)(10x - y) = 10x^2 + 29xy - 3y^2$
$(5x + 3y)(2x - y) = 10x^2 + xy - 3y^2$
$(2x - 3y)(5x + y) = 10x^2 - 13xy - 3y^2 \quad$ Correct middle term

A factored form of $10x^2 - 13xy - 3y^2$ is $(2x - 3y)(5x + y)$. ☐

PRACTICE 4 Factor: $21x^2 + 11xy - 2y^2$.

Section 6.3 Factoring Trinomials of the Form $ax^2 + bx + c$ and Perfect Square Trinomials 381

EXAMPLE 5 Factor: $3x^4 - 5x^2 - 8$

Solution Factors of $3x^4$: $3x^4 = 3x^2 \cdot x^2$
Factors of -8: $-8 = -2 \cdot 4, 2 \cdot -4, -1 \cdot 8, 1 \cdot -8$

Try combinations of these factors:

$\qquad\qquad\qquad$ Correct $\qquad$ Correct
$\qquad\qquad\qquad\quad \downarrow \qquad\qquad\;\; \downarrow$

$(3x^2 - 2)(x^2 + 4) = 3x^4 + 10x^2 - 8$
$(3x^2 + 4)(x^2 - 2) = 3x^4 - 2x^2 - 8$
$(3x^2 + 8)(x^2 - 1) = 3x^4 + 5x^2 - 8$ $\quad$ Incorrect sign on middle term, so switch signs in binomial factors.
$(3x^2 - 8)(x^2 + 1) = 3x^4 - 5x^2 - 8$ $\quad$ Correct middle term.

A factored form of $3x^4 - 5x^2 - 8$ is $(3x^2 - 8)(x^2 + 1)$.

PRACTICE 5 Factor: $2x^4 - 5x^2 - 7$.

> **Helpful Hint**
> Study the last two lines of Example 5. If a factoring attempt gives you a middle term whose numerical coefficient is the opposite of the desired numerical coefficient, try switching the signs of the last terms in the binomials.
>
> Switched signs $\begin{cases} (3x^2 + 8)(x^2 - 1) = 3x^4 + 5x^2 - 8 \quad \text{Middle term: } +5x \\ (3x^2 - 8)(x^2 + 1) = 3x^4 - 5x^2 - 8 \quad \text{Middle term: } -5x \end{cases}$

OBJECTIVE 2 ▶ **Factoring out the greatest common factor.** Don't forget that the first step in factoring any polynomial is to look for a common factor to factor out.

EXAMPLE 6 Factor: $24x^4 + 40x^3 + 6x^2$

Solution Notice that all three terms have a common factor of $2x^2$. Thus we factor out $2x^2$ first.

$$24x^4 + 40x^3 + 6x^2 = 2x^2(12x^2 + 20x + 3)$$

Next we factor $12x^2 + 20x + 3$.

Factors of $12x^2$: $\;\; 12x^2 = 4x \cdot 3x, \quad 12x^2 = 12x \cdot x, \quad 12x^2 = 6x \cdot 2x$

Since all terms in the trinomial have positive numerical coefficients, we factor 3 using positive factors only.

Factors of 3: $\;\; 3 = 1 \cdot 3$

We try some combinations of the factors.

$2x^2(4x + 3)(3x + 1) = 2x^2(12x^2 + 13x + 3)$
$2x^2(12x + 1)(x + 3) = 2x^2(12x^2 + 37x + 3)$
$2x^2(2x + 3)(6x + 1) = 2x^2(12x^2 + 20x + 3)$ $\quad$ Correct middle term

A factored form of $24x^4 + 40x^3 + 6x^2$ is $2x^2(2x + 3)(6x + 1)$.

> **Helpful Hint**
> Don't forget to include the common factor in the factored form.

PRACTICE 6 Factor: $3x^3 + 17x^2 + 10x$

When the term containing the squared variable has a negative coefficient, you may want to first factor out a common factor of -1.

EXAMPLE 7 Factor: $-6x^2 - 13x + 5$

Solution We begin by factoring out a common factor of -1.

$-6x^2 - 13x + 5 = -1(6x^2 + 13x - 5)$ Factor out -1.
$= -1(3x - 1)(2x + 5)$ Factor $6x^2 + 13x - 5$.

PRACTICE 7 Factor: $-8x^2 + 2x + 3$

OBJECTIVE 3 ▶ Factoring perfect square trinomials. A trinomial that is the square of a binomial is called a **perfect square trinomial.** For example,

$$(x + 3)^2 = (x + 3)(x + 3)$$
$$= x^2 + 6x + 9$$

Thus $x^2 + 6x + 9$ is a perfect square trinomial.

In Chapter 5, we discovered special product formulas for squaring binomials.

$$(a + b)^2 = a^2 + 2ab + b^2 \quad \text{and} \quad (a - b)^2 = a^2 - 2ab + b^2$$

Because multiplication and factoring are reverse processes, we can now use these special products to help us factor perfect square trinomials. If we reverse these equations, we have the following.

> **Factoring Perfect Square Trinomials**
>
> $$a^2 + 2ab + b^2 = (a + b)^2$$
> $$a^2 - 2ab + b^2 = (a - b)^2$$

▶ **Helpful Hint**

Notice that for both given forms of a perfect square trinomial, the last term is positive. This is because the last term is a square.

To use these equations to help us factor, we must first be able to recognize a perfect square trinomial. A trinomial is a perfect square when

1. two terms, a^2 and b^2, are squares and
2. another term is $2 \cdot a \cdot b$ or $-2 \cdot a \cdot b$. That is, this term is twice the product of a and b, or its opposite.

When a trinomial fits this description, its factored form is $(a + b)^2$.

EXAMPLE 8 Factor: $x^2 + 12x + 36$

Solution First, is this a perfect square trinomial?

$$x^2 + 12x + 36$$

1. $x^2 = (x)^2$ and $36 = 6^2$.
2. Is $2 \cdot x \cdot 6$ the middle term? Yes, $2 \cdot x \cdot 6 = 12x$.

Thus, $x^2 + 12x + 36$ factors as $(x + 6)^2$.

PRACTICE 8 Factor: $x^2 + 14x + 49$

85. $-27t + 7t^2 - 4$
86. $-3t + 4t^2 - 7$
87. $49p^2 - 7p - 2$
88. $3r^2 + 10r - 8$
89. $m^3 + 18m^2 + 81m$
90. $y^3 + 12y^2 + 36y$
91. $5x^2y^2 + 20xy + 1$
92. $3a^2b^2 + 12ab + 1$
93. $6a^5 + 37a^3b^2 + 6ab^4$
94. $5m^5 + 26m^3h^2 + 5mh^4$

REVIEW AND PREVIEW

Multiply the following. See Section 5.4.

95. $(x - 2)(x + 2)$
96. $(y^2 + 3)(y^2 - 3)$
97. $(a + 3)(a^2 - 3a + 9)$
98. $(z - 2)(z^2 + 2z + 4)$

As of 2006, approximately 80% of U.S. households have access to the Internet. The following graph shows the percent of households having Internet access grouped according to household income. See Section 3.1.

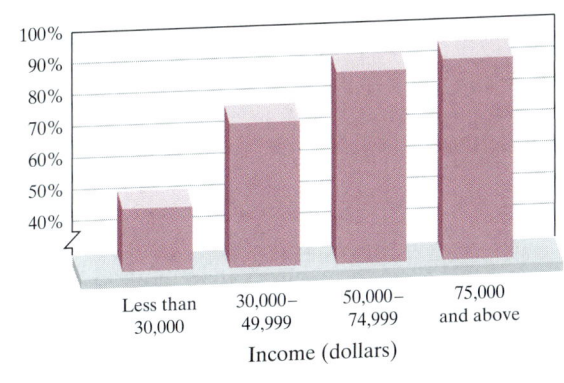

99. Which range of household income corresponds to the highest percent of households having access to the Internet?
100. Which range of household income corresponds to the greatest increase in percent of households having access to the Internet?
101. Describe any trend you see.
102. Why don't the percents shown in the graph add to 100%?

CONCEPT EXTENSIONS

See the Concept Check in this section.

103. Do the terms of $4x^2 + 19x + 12$ have a common factor (other than 1)?
104. Without multiplying, decide which of the following factored forms is not a factored form of $4x^2 + 19x + 12$.
    a. $(2x + 4)(2x + 3)$ b. $(4x + 4)(x + 3)$
    c. $(4x + 3)(x + 4)$ d. $(2x + 2)(2x + 6)$

105. Describe a perfect square trinomial.
106. Write the perfect square trinomial that factors as $(x + 3y)^2$.

Write the perimeter of each figure as a simplified polynomial. Then factor the polynomial.

107. Triangle with sides $3x^2 + 1$, $6x + 4$, $x^2 + 15x$.
108. Square with side $3y^2$ and $-22y + 7$.

Factor each trinomial completely.

109. $4x^2 + 2x + \dfrac{1}{4}$
110. $27x^2 + 2x - \dfrac{1}{9}$
111. $4x^2(y - 1)^2 + 10x(y - 1)^2 + 25(y - 1)^2$
112. $3x^2(a + 3)^3 - 10x(a + 3)^3 + 25(a + 3)^3$
113. Fill in the blank so that $x^2 + $ _____ $x + 16$ is a perfect square trinomial.
114. Fill in the blank so that $9x^2 + $ _____ $x + 25$ is a perfect square trinomial.

The area of the largest square in the figure is $(a + b)^2$. Use this figure to answer Exercises 115 and 116.

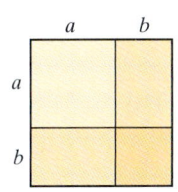

115. Write the area of the largest square as the sum of the areas of the smaller squares and rectangles.
116. What factoring formula from this section is visually represented by this square?

Find a positive value of b so that each trinomial is factorable.

117. $3x^2 + bx - 5$ 118. $2y^2 + by + 3$

Find a positive value of c so that each trinomial is factorable.

119. $5x^2 + 7x + c$ 120. $11y^2 - 40y + c$

Factor completely. Don't forget to first factor out the greatest common factor.

121. $-12x^3y^2 + 3x^2y^2 + 15xy^2$
122. $-12r^3x^2 + 38r^2x^2 + 14rx^2$
123. $4x^2(y - 1)^2 + 20x(y - 1)^2 + 25(y - 1)^2$
124. $3x^2(a + 3)^3 - 28x(a + 3)^3 + 25(a + 3)^3$

Factor.

125. $3x^{2n} + 17x^n + 10$
126. $2x^{2n} + 5x^n - 12$
127. In your own words, describe the steps you will use to factor a trinomial.

STUDY SKILLS BUILDER

Are You Satisfied with Your Performance on a Particular Quiz or Exam?

If not, don't forget to analyze your quiz or exam and look for common errors. Were most of your errors a result of:

- *Carelessness?* Did you turn in your quiz or exam before the allotted time expired? If so, resolve to use any extra time to check your work.
- *Running out of time?* Try completing any questions that you are unsure of last and delay checking your work until all questions have been answered.
- *Not understanding a concept?* If so, review that concept and correct your work so that you make sure it doesn't happen before the next quiz or the final exam.
- *Test conditions?* When studying for a quiz or exam, make sure you place yourself in conditions similar to test conditions. For example, before your next quiz or exam, take a sample test without the aid of your notes or text.

(For a sample test, see your instructor or use the Chapter Test at the end of each chapter.)

Exercises

1. Have you corrected all your previous quizzes and exams?
2. List any errors you have found common to two or more of your graded papers.
3. Is one of your common errors not understanding a concept? If so, are you making sure you understand all the concepts for the next quiz or exam?
4. Is one of your common errors making careless mistakes? If so, are you now taking all the time allotted to check over your work so that you can minimize the number of careless mistakes?
5. Are you satisfied with your grades thus far on quizzes and tests?
6. If your answer to Exercise 5 is no, are there any more suggestions you can make to your instructor or yourself to help? If so, list them here and share these with your instructor.

6.4 FACTORING TRINOMIALS OF THE FORM $ax^2 + bx + c$ BY GROUPING

OBJECTIVE

1. Use the grouping method to factor trinomials of the form $ax^2 + bx + c$.

OBJECTIVE 1 ▶ Using the grouping method. There is an alternative method that can be used to factor trinomials of the form $ax^2 + bx + c$, $a \neq 1$. This method is called the **grouping method** because it uses factoring by grouping as we learned in Section 6.1.

To see how this method works, recall from Section 6.1 that to factor a trinomial such as $x^2 + 11x + 30$, we find two numbers such that

Product is 30
↓
$x^2 + 11x + 30$
↓
Sum is 11.

To factor a trinomial such as $2x^2 + 11x + 12$ by grouping, we use an extension of the method in Section 6.1. Here we look for two numbers such that

Product is $2 \cdot 12 = 24$
↓
$2x^2 + 11x + 12$
↓
Sum is 11.

This time, we use the two numbers to write

$2x^2 + 11x + 12$ as

$= 2x^2 + \Box x + \Box x + 12$

Section 6.4 Factoring Trinomials of the Form $ax^2 + bx + c$ by Grouping   387

Then we factor by grouping. Since we want a positive product, 24, and a positive sum, 11, we consider pairs of positive factors of 24 only.

Factors of 24	Sum of Factors	
1, 24	25	
2, 12	14	
3, 8	11	Correct sum

The factors are 3 and 8. Now we use these factors to write the middle term $11x$ as $3x + 8x$ (or $8x + 3x$). We replace $11x$ with $3x + 8x$ in the original trinomial and then we can factor by grouping.

$$\begin{aligned}
2x^2 + 11x + 12 &= 2x^2 + 3x + 8x + 12 \\
&= (2x^2 + 3x) + (8x + 12) \quad \text{Group the terms.} \\
&= x(2x + 3) + 4(2x + 3) \quad \text{Factor each group.} \\
&= (2x + 3)(x + 4) \quad \text{Factor out } (2x + 3).
\end{aligned}$$

In general, we have the following procedure.

To Factor Trinomials by Grouping

STEP 1. Factor out a greatest common factor, if there is one other than 1.

STEP 2. For the resulting trinomial $ax^2 + bx + c$, find two numbers whose product is $a \cdot c$ and whose sum is b.

STEP 3. Write the middle term, bx, using the factors found in Step 2.

STEP 4. Factor by grouping.

EXAMPLE 1 Factor $3x^2 + 31x + 10$ by grouping.

Solution

STEP 1. The terms of this trinomial contain no greatest common factor other than 1 (or -1).

STEP 2. In $3x^2 + 31x + 10$, $a = 3$, $b = 31$, and $c = 10$.

Let's find two numbers whose product is $a \cdot c$ or $3(10) = 30$ and whose sum is b or 31. The numbers are 1 and 30.

Factors of 30	Sum of factors	
5, 6	11	
3, 10	13	
2, 15	17	
1, 30	31	Correct sum

STEP 3. Write $31x$ as $1x + 30x$ so that $3x^2 + 31x + 10 = 3x^2 + 1x + 30x + 10$.

STEP 4. Factor by grouping.

$$\begin{aligned}
3x^2 + 1x + 30x + 10 &= x(3x + 1) + 10(3x + 1) \\
&= (3x + 1)(x + 10)
\end{aligned}$$

PRACTICE 1 Factor $5x^2 + 61x + 12$ by grouping.

EXAMPLE 2 Factor $8x^2 - 14x + 5$ by grouping.

Solution

STEP 1. The terms of this trinomial contain no greatest common factor other than 1.

STEP 2. This trinomial is of the form $ax^2 + bx + c$ with $a = 8, b = -14,$ and $c = 5$. Find two numbers whose product is $a \cdot c$ or $8 \cdot 5 = 40$, and whose sum is b or -14.

The numbers are -4 and -10.

Factors of 40	Sum of Factors
$-40, -1$	-41
$-20, -2$	-22
$-10, -4$	-14

Correct sum

STEP 3. Write $-14x$ as $-4x - 10x$ so that
$$8x^2 - 14x + 5 = 8x^2 - 4x - 10x + 5$$

STEP 4. Factor by grouping.
$$8x^2 - 4x - 10x + 5 = 4x(2x - 1) - 5(2x - 1)$$
$$= (2x - 1)(4x - 5)$$

PRACTICE 2 Factor $12x^2 - 19x + 5$ by grouping.

EXAMPLE 3 Factor $6x^2 - 2x - 20$ by grouping.

Solution

STEP 1. First factor out the greatest common factor, 2.
$$6x^2 - 2x - 20 = 2(3x^2 - x - 10)$$

STEP 2. Next notice that $a = 3, b = -1,$ and $c = -10$ in the resulting trinomial. Find two numbers whose product is $a \cdot c$ or $3(-10) = -30$ and whose sum is $b, -1$. The numbers are -6 and 5.

STEP 3. $3x^2 - x - 10 = 3x^2 - 6x + 5x - 10$

STEP 4. $3x^2 - 6x + 5x - 10 = 3x(x - 2) + 5(x - 2)$
$$= (x - 2)(3x + 5)$$

The factored form of $6x^2 - 2x - 20 = 2(x - 2)(3x + 5)$.

↑ Don't forget to include the common factor of 2.

PRACTICE 3 Factor $30x^2 - 14x - 4$ by grouping.

EXAMPLE 4 Factor $18y^4 + 21y^3 - 60y^2$ by grouping.

Solution

STEP 1. First factor out the greatest common factor, $3y^2$.
$$18y^4 + 21y^3 - 60y^2 = 3y^2(6y^2 + 7y - 20)$$

STEP 2. Notice that $a = 6, b = 7,$ and $c = -20$ in the resulting trinomial. Find two numbers whose product is $a \cdot c$ or $6(-20) = -120$ and whose sum is 7. It may help to factor -120 as a product of primes and -1.
$$-120 = 2 \cdot 2 \cdot 2 \cdot 3 \cdot 5 \cdot (-1)$$

Then choose pairings of factors until you have two pairings whose sum is 7.

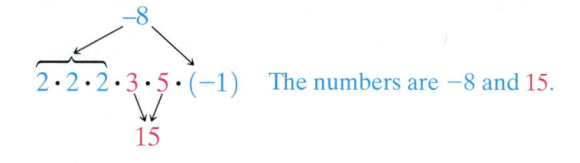

$2 \cdot 2 \cdot 2 \cdot 3 \cdot 5 \cdot (-1)$ The numbers are -8 and 15.

STEP 3. $6y^2 + 7y - 20 = 6y^2 - 8y + 15y - 20$

STEP 4. $6y^2 - 8y + 15y - 20 = 2y(3y - 4) + 5(3y - 4)$
$= (3y - 4)(2y + 5)$

The factored form of $18y^4 + 21y^3 - 60y^2$ is $3y^2(3y - 4)(2y + 5)$.

Don't forget to include the common factor of $3y^2$.

PRACTICE 4 Factor $40m^4 + 5m^3 - 35m^2$ by grouping.

EXAMPLE 5 Factor $4x^2 + 20x + 25$ by grouping.

Solution

STEP 1. The terms of this trinomial contain no greatest common factor other than 1 (or -1).

STEP 2. In $4x^2 + 20x + 25$, $a = 4$, $b = 20$, and $c = 25$. Find two numbers whose product is $a \cdot c$ or $4 \cdot 25 = 100$ and whose sum is 20. The numbers are 10 and 10.

STEP 3. Write $20x$ as $10x + 10x$ so that
$$4x^2 + 20x + 25 = 4x^2 + 10x + 10x + 25$$

STEP 4. Factor by grouping.
$$4x^2 + 10x + 10x + 25 = 2x(2x + 5) + 5(2x + 5)$$
$$= (2x + 5)(2x + 5)$$

The factored form of $4x^2 + 20x + 25$ is $(2x + 5)(2x + 5)$ or $(2x + 5)^2$.

PRACTICE 5 Factor $16x^2 + 24x + 9$ by grouping.

A trinomial that is the square of a binomial, such as the trinomial in Example 5, is called a **perfect square trinomial.** From Chapter 5, there are special product formulas we can use to help us recognize and factor these trinomials. To study these formulas further, see Section 6.3, Objective 3. **Remember:** A perfect square trinomial, such as the one in Example 5, may be factored by special product formulas or by other methods of factoring trinomials, such as by grouping.

6.4 EXERCISE SET

Factor each polynomial by grouping. Notice that Step 3 has already been done in these exercises. See Examples 1 through 5.

1. $x^2 + 3x + 2x + 6$
2. $x^2 + 5x + 3x + 15$
3. $y^2 + 8y - 2y - 16$
4. $z^2 + 10z - 7z - 70$
5. $8x^2 - 5x - 24x + 15$
6. $4x^2 - 9x - 32x + 72$
7. $5x^4 - 3x^2 + 25x^2 - 15$
8. $2y^4 - 10y^2 + 7y^2 - 35$

MIXED PRACTICE

Factor each trinomial by grouping. Exercises 9–12 are broken into parts to help you get started. See Examples 1 through 5.

9. $6x^2 + 11x + 3$
   a. Find two numbers whose product is $6 \cdot 3 = 18$ and whose sum is 11.
   b. Write $11x$ using the factors from part (a).
   c. Factor by grouping.

10. $8x^2 + 14x + 3$
    a. Find two numbers whose product is $8 \cdot 3 = 24$ and whose sum is 14.
    b. Write $14x$ using the factors from part (a).
    c. Factor by grouping.

11. $15x^2 - 23x + 4$
   a. Find two numbers whose product is $15 \cdot 4 = 60$ and whose sum is -23.
   b. Write $-23x$ using the factors from part (a).
   c. Factor by grouping.

12. $6x^2 - 13x + 5$
   a. Find two numbers whose product is $6 \cdot 5 = 30$ and whose sum is -13.
   b. Write $-13x$ using the factors from part (a).
   c. Factor by grouping.

13. $21y^2 + 17y + 2$

14. $15x^2 + 11x + 2$

15. $7x^2 - 4x - 11$

16. $8x^2 - x - 9$

17. $10x^2 - 9x + 2$

18. $30x^2 - 23x + 3$

19. $2x^2 - 7x + 5$

20. $2x^2 - 7x + 3$

21. $12x + 4x^2 + 9$

22. $20x + 25x^2 + 4$

23. $4x^2 - 8x - 21$

24. $6x^2 - 11x - 10$

25. $10x^2 - 23x + 12$

26. $21x^2 - 13x + 2$

27. $2x^3 + 13x^2 + 15x$

28. $3x^3 + 8x^2 + 4x$

29. $16y^2 - 34y + 18$

30. $4y^2 - 2y - 12$

31. $-13x + 6 + 6x^2$

32. $-25x + 12 + 12x^2$

33. $54a^2 - 9a - 30$

34. $30a^2 + 38a - 20$

35. $20a^3 + 37a^2 + 8a$

36. $10a^3 + 17a^2 + 3a$

37. $12x^3 - 27x^2 - 27x$

38. $30x^3 - 155x^2 + 25x$

39. $3x^2y + 4xy^2 + y^3$

40. $6r^2t + 7rt^2 + t^3$

41. $20z^2 + 7z + 1$

42. $36z^2 + 6z + 1$

43. $5x^2 + 50xy + 125y^2$

44. $3x^2 + 42xy + 147y^2$

45. $24a^2 - 6ab - 30b^2$

46. $30a^2 + 5ab - 25b^2$

47. $15p^4 + 31p^3q + 2p^2q^2$

48. $20s^4 + 61s^3t + 3s^2t^2$

49. $162a^4 - 72a^2 + 8$

50. $32n^4 - 112n^2 + 98$

51. $35 + 12x + x^2$

52. $33 + 14x + x^2$

53. $6 - 11x + 5x^2$

54. $5 - 12x + 7x^2$

REVIEW AND PREVIEW

Multiply. See Sections 5.3 and 5.4.

55. $(x - 2)(x + 2)$

56. $(y - 5)(y + 5)$

57. $(y + 4)(y + 4)$

58. $(x + 7)(x + 7)$

59. $(9z + 5)(9z - 5)$

60. $(8y + 9)(8y - 9)$

61. $(x - 3)(x^2 + 3x + 9)$

62. $(2z - 1)(4z^2 + 2z + 1)$

CONCEPT EXTENSIONS

Write the perimeter of each figure as a simplified polynomial. Then factor the polynomial.

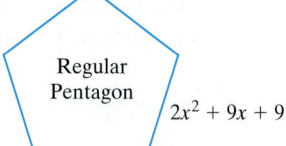

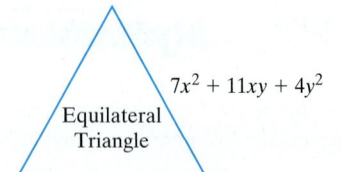

63. Regular Pentagon, side $2x^2 + 9x + 9$

64. Equilateral Triangle, side $7x^2 + 11xy + 4y^2$

Factor each polynomial by grouping.

65. $x^{2n} + 2x^n + 3x^n + 6$
   (***Hint:*** Don't forget that $x^{2n} = x^n \cdot x^n$.)

66. $x^{2n} + 6x^n + 10x^n + 60$

67. $3x^{2n} + 16x^n - 35$

68. $12x^{2n} - 40x^n + 25$

69. In your own words, explain how to factor a trinomial by grouping.

6.5 FACTORING BINOMIALS

OBJECTIVES

1. Factor the difference of two squares.
2. Factor the sum or difference of two cubes.

OBJECTIVE 1 ▶ **Factoring the difference of two squares.** When learning to multiply binomials in Chapter 5, we studied a special product, the product of the sum and difference of two terms, a and b:

$$(a + b)(a - b) = a^2 - b^2$$

For example, the product of $x + 3$ and $x - 3$ is

$$(x + 3)(x - 3) = x^2 - 9$$

The binomial $x^2 - 9$ is called a **difference of squares**. In this section, we use the pattern for the product of a sum and difference to factor the binomial difference of squares.

Factoring the Difference of Two Squares

$$a^2 - b^2 = (a + b)(a - b)$$

▶ **Helpful Hint**

Since multiplication is commutative, remember that the order of factors does not matter. In other words,

$$a^2 - b^2 = (a + b)(a - b) \text{ or } (a - b)(a + b)$$

EXAMPLE 1 Factor: $x^2 - 25$

Solution $x^2 - 25$ is the difference of two squares since $x^2 - 25 = x^2 - 5^2$. Therefore,

$$x^2 - 25 = x^2 - 5^2 = (x + 5)(x - 5)$$

Multiply to check.

PRACTICE 1 Factor $x^2 - 81$.

EXAMPLE 2 Factor each difference of squares.

a. $4x^2 - 1$ **b.** $25a^2 - 9b^2$ **c.** $y^2 - \dfrac{4}{9}$

Solution

a. $4x^2 - 1 = (2x)^2 - 1^2 = (2x + 1)(2x - 1)$
b. $25a^2 - 9b^2 = (5a)^2 - (3b)^2 = (5a + 3b)(5a - 3b)$
c. $y^2 - \dfrac{4}{9} = y^2 - \left(\dfrac{2}{3}\right)^2 = \left(y + \dfrac{2}{3}\right)\left(y - \dfrac{2}{3}\right)$

PRACTICE 2 Factor each difference of squares.

a. $9x^2 - 1$ **b.** $36a^2 - 49b^2$ **c.** $p^2 - \dfrac{25}{36}$

EXAMPLE 3 Factor: $x^4 - y^6$

Solution This is a difference of squares since $x^4 = (x^2)^2$ and $y^6 = (y^3)^2$. Thus,
$$x^4 - y^6 = (x^2)^2 - (y^3)^2 = (x^2 + y^3)(x^2 - y^3)$$

PRACTICE 3 Factor $p^4 - q^{10}$.

EXAMPLE 4 Factor each binomial.

a. $y^4 - 16$ **b.** $x^2 + 4$

Solution

a. $y^4 - 16 = (y^2)^2 - 4^2$
$= (y^2 + 4)(y^2 - 4)$.   Factor the difference of two squares.
                This binomial can be factored further since it is the difference of two squares.
$= (y^2 + 4)(y + 2)(y - 2)$ Factor the difference of two squares.

b. $x^2 + 4$

Note that the binomial $x^2 + 4$ is the *sum* of two squares since we can write $x^2 + 4$ as $x^2 + 2^2$. We might try to factor using $(x + 2)(x + 2)$ or $(x - 2)(x - 2)$. But when we multiply to check, we find that neither factoring is correct.

$$(x + 2)(x + 2) = x^2 + 4x + 4$$
$$(x - 2)(x - 2) = x^2 - 4x + 4$$

In both cases, the product is a trinomial, not the required binomial. In fact, $x^2 + 4$ is a prime polynomial.

PRACTICE 4 Factor each binomial.

a. $z^4 - 81$ **b.** $m^2 + 49$

> **Helpful Hint**
>
> When factoring, don't forget:
>
> - See whether the terms have a greatest common factor (GCF) (other than 1) that can be factored out.
> - Other than a GCF, the **sum** of two squares cannot be factored using real numbers.
> - Factor completely. Always check to see whether any factors can be factored further.

EXAMPLES Factor each difference of two squares.

5. $4x^3 - 49x = x(4x^2 - 49)$ Factor out the common factor, x.
              $= x[(2x)^2 - 7^2]$
              $= x(2x + 7)(2x - 7)$ Factor the difference of two squares.

6. $162x^4 - 2 = 2(81x^4 - 1)$ Factor out the common factor, 2.
              $= 2(9x^2 + 1)(9x^2 - 1)$ Factor the difference of two squares.
              $= 2(9x^2 + 1)(3x + 1)(3x - 1)$ Factor the difference of two squares.

PRACTICES 5–6 Factor each difference of two squares.

5. $36y^3 - 25y$ **6.** $80y^4 - 5$

Section 6.5 Factoring Binomials 393

EXAMPLE 7   Factor: $-49x^2 + 16$

Solution   Factor as is, or, if you like, rearrange terms.

Factor as is: $-49x^2 + 16 = -1(49x^2 - 16)$ Factor out -1.
$= -1(7x + 4)(7x - 4)$ Factor the difference of two squares.

Rewrite binomial: $-49x^2 + 16 = 16 - 49x^2 = 4^2 - (7x)^2$
$= (4 + 7x)(4 - 7x)$

Both factorizations are correct and are equal. To see this, factor -1 from $(4 - 7x)$ in the second factorization.

PRACTICE 7   Factor: $-9x^2 + 100$

OBJECTIVE 2 ▶ Factoring the sum or difference of two cubes. Although the sum of two squares usually does not factor, the sum or difference of two cubes can be factored and reveals factoring patterns. The pattern for the sum of cubes is illustrated by multiplying the binomial $x + y$ and the trinomial $x^2 - xy + y^2$.

$$\begin{array}{r} x^2 - xy + y^2 \\ \underline{x + y} \\ x^2y - xy^2 + y^3 \\ \underline{x^3 - x^2y + xy^2 } \\ x^3 + y^3 \end{array}$$

Thus, $(x + y)(x^2 - xy + y^2) = x^3 + y^3$ Sum of cubes

The pattern for the difference of two cubes is illustrated by multiplying the binomial $x - y$ by the trinomial $x^2 + xy + y^2$. The result is

$(x - y)(x^2 + xy + y^2) = x^3 - y^3$ Difference of cubes

Factoring the Sum or Difference of Two Cubes

$a^3 + b^3 = (a + b)(a^2 - ab + b^2)$
$a^3 - b^3 = (a - b)(a^2 + ab + b^2)$

Recall that "factor" means "to write as a product." Above are patterns for writing sums and differences as products.

EXAMPLE 8   Factor: $x^3 + 8$

Solution   First, write the binomial in the form $a^3 + b^3$.

$x^3 + 8 = x^3 + 2^3$ Write in the form $a^3 + b^3$.

If we replace a with x and b with 2 in the formula above, we have

$x^3 + 2^3 = (x + 2)[x^2 - (x)(2) + 2^2]$
$= (x + 2)(x^2 - 2x + 4)$

PRACTICE 8   Factor $x^3 + 64$.

Helpful Hint

When factoring sums or differences of cubes, notice the sign patterns.

$$x^3 + y^3 = (x + y)(x^2 - xy + y^2)$$

same sign; opposite signs; always positive

$$x^3 - y^3 = (x - y)(x^2 + xy + y^2)$$

same sign; opposite signs; always positive

EXAMPLE 9 Factor: $y^3 - 27$

Solution
$$y^3 - 27 = y^3 - 3^3 \quad \text{Write in the form } a^3 - b^3.$$
$$= (y - 3)[y^2 + (y)(3) + 3^2]$$
$$= (y - 3)(y^2 + 3y + 9)$$

PRACTICE 9 Factor $x^3 - 125$.

EXAMPLE 10 Factor: $64x^3 + 1$

Solution
$$64x^3 + 1 = (4x)^3 + 1^3$$
$$= (4x + 1)[(4x)^2 - (4x)(1) + 1^2]$$
$$= (4x + 1)(16x^2 - 4x + 1)$$

PRACTICE 10 Factor $27y^3 + 1$.

EXAMPLE 11 Factor: $54a^3 - 16b^3$

Solution Remember to factor out common factors first before using other factoring methods.

$$54a^3 - 16b^3 = 2(27a^3 - 8b^3) \quad \text{Factor out the GCF 2.}$$
$$= 2[(3a)^3 - (2b)^3] \quad \text{Difference of two cubes}$$
$$= 2(3a - 2b)[(3a)^2 + (3a)(2b) + (2b)^2]$$
$$= 2(3a - 2b)(9a^2 + 6ab + 4b^2)$$

PRACTICE 11 Factor $32x^3 - 500y^3$.

Calculator Explorations

Graphing

A graphing calculator is a convenient tool for evaluating an expression at a given replacement value. For example, let's evaluate $x^2 - 6x$ when $x = 2$. To do so, store the value 2 in the variable x and then enter and evaluate the algebraic expression.

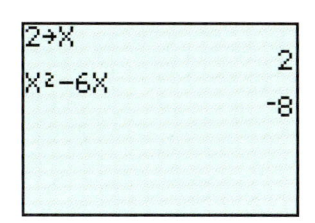

```
2→X
         2
X²-6X
        -8
```

The value of $x^2 - 6x$ when $x = 2$ is -8. You may want to use this method for evaluating expressions as you explore the following.

We can use a graphing calculator to explore factoring patterns numerically. Use your calculator to evaluate $x^2 - 2x + 1$, $x^2 - 2x - 1$, and $(x - 1)^2$ for each value of x given in the table. What do you observe?

	$x^2 - 2x + 1$	$x^2 - 2x - 1$	$(x - 1)^2$
$x = 5$			
$x = -3$			
$x = 2.7$			
$x = -12.1$			
$x = 0$			

Notice in each case that $x^2 - 2x - 1 \neq (x - 1)^2$. Because for each x in the table the value of $x^2 - 2x + 1$ and the value of $(x - 1)^2$ are the same, we might guess that $x^2 - 2x + 1 = (x - 1)^2$. We can verify our guess algebraically with multiplication:

$$(x - 1)(x - 1) = x^2 - x - x + 1 = x^2 - 2x + 1$$

VOCABULARY & READINESS CHECK

Use the choices below to fill in each blank. Some choices may be used more than once and some choices may not be used at all.

| true | difference of two squares | sum of two cubes |
| false | difference of two cubes | |

1. The expression $x^3 - 27$ is called a(n) _____.
2. The expression $x^2 - 49$ is called a(n) _____.
3. The expression $z^3 + 1$ is called a(n) _____.
4. True or false: The binomial $y^2 + 9$ factors as $(y + 3)^2$. _____

Write each number or term as a square.

5. 64   **6.** 100   **7.** $49x^2$ **8.** $25y^4$

Write each number or term as a cube.

9. 64   **10.** 1   **11.** $8y^3$ **12.** x^6

6.5 EXERCISE SET

Factor each binomial completely. See Examples 1 through 7.

1. $x^2 - 4$
2. $x^2 - 36$
3. $81p^2 - 1$
4. $49m^2 - 1$
5. $25y^2 - 9$
6. $49a^2 - 16$
7. $121m^2 - 100n^2$
8. $169a^2 - 49b^2$
9. $x^2y^2 - 1$
10. $a^2b^2 - 16$
11. $x^2 - \dfrac{1}{4}$
12. $y^2 - \dfrac{1}{16}$
13. $-4r^2 + 1$
14. $-9t^2 + 1$
15. $16r^2 + 1$
16. $49y^2 + 1$
17. $-36 + x^2$
18. $-1 + y^2$
19. $m^4 - 1$
20. $n^4 - 16$
21. $m^4 - n^{18}$
22. $n^4 - r^6$

Factor the sum or difference of two cubes. See Examples 8 through 11.

23. $x^3 + 125$
24. $p^3 + 1$
25. $8a^3 - 1$
26. $27y^3 - 1$
27. $m^3 + 27n^3$
28. $y^3 + 64z^3$
29. $5k^3 + 40$
30. $6r^3 + 162$
31. $x^3y^3 - 64$
32. $a^3b^3 - 8$
33. $250r^3 - 128t^3$
34. $24x^3 - 81y^3$

MIXED PRACTICE

Factor each binomial completely. See Examples 1 through 11.

35. $r^2 - 64$
36. $q^2 - 121$
37. $x^2 - 169y^2$
38. $x^2 - 225y^2$
39. $27 - t^3$
40. $125 - r^3$
41. $18r^2 - 8$
42. $32t^2 - 50$
43. $9xy^2 - 4x$
44. $36x^2y - 25y$
45. $8m^3 + 64$
46. $2x^3 + 54$
47. $xy^3 - 9xyz^2$
48. $x^3y - 4xy^3$
49. $36x^2 - 64y^2$
50. $225a^2 - 81b^2$
51. $144 - 81x^2$
52. $12x^2 - 27$
53. $x^3y^3 - z^6$
54. $a^3b^3 - c^9$
55. $49 - \dfrac{9}{25}m^2$
56. $100 - \dfrac{4}{81}n^2$
57. $t^3 + 343$
58. $s^3 + 216$
59. $n^3 + 49n$
60. $y^3 + 64y$
61. $x^6 - 81x^2$
62. $n^9 - n^5$
63. $64p^3q - 81pq^3$
64. $100x^3y - 49xy^3$
65. $27x^2y^3 + xy^2$
66. $8x^3y^3 + x^3y$
67. $125a^4 - 64ab^3$
68. $64m^4 - 27mn^3$
69. $16x^4 - 64x^2$
70. $25y^4 - 100y^2$

REVIEW AND PREVIEW

Solve each equation. See Section 2.3.

71. $x - 6 = 0$
72. $y + 5 = 0$
73. $2m + 4 = 0$

74. $3x - 9 = 0$

75. $5z - 1 = 0$

76. $4a + 2 = 0$

Solve. See Section 6.1. The percent of undergraduate college students who have credit cards each year from 2000 through 2006 can be approximately modeled by the polynomial $-1.2x^2 + 4x + 80$, where x is the number of years since 2000.

77. Find the percent of college students who had credit cards in 2003.

78. Find the percent of college students who had credit cards in 2006.

79. Write a factored form of $-1.2x^2 + 4x + 80$ by factoring -4 from the terms of this polynomial.

80. Use your answers to Exercises 77 and 78 to write down any trends.

CONCEPT EXTENSIONS

Factor each expression completely.

81. $(x + 2)^2 - y^2$
82. $(y - 6)^2 - z^2$
83. $a^2(b - 4) - 16(b - 4)$
84. $m^2(n + 8) - 9(n + 8)$
85. $(x^2 + 6x + 9) - 4y^2$ (***Hint:*** Factor the trinomial in parentheses first.)
86. $(x^2 + 2x + 1) - 36y^2$
87. $x^{2n} - 100$
88. $x^{2n} - 81$
89. What binomial multiplied by $(x - 6)$ gives the difference of two squares?
90. What binomial multiplied by $(5 + y)$ gives the difference of two squares?
91. In your own words, explain how to tell whether a binomial is a difference of squares. Then explain how to factor a difference of squares.
92. In your own words, explain how to tell whether a binomial is a sum of cubes. Then explain how to factor a sum of cubes.

93. An object is dropped from the top of Pittsburgh's USX Tower, which is 841 feet tall. (*Source: World Almanac* research) The height of the object after *t* seconds is given by the expression $841 - 16t^2$.
   a. Find the height of the object after 2 seconds.
   b. Find the height of the object after 5 seconds.
   c. To the nearest whole second, estimate when the object hits the ground.
   d. Factor $841 - 16t^2$.

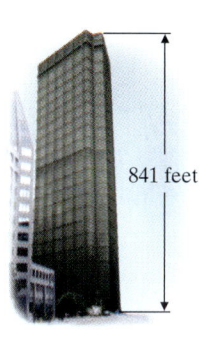

94. A worker on the top of the Aetna Life Building in San Francisco accidentally drops a bolt. The Aetna Life Building is 529 feet tall. (*Source: World Almanac* research) The height of the bolt after *t* seconds is given by the expression $529 - 16t^2$.
   a. Find the height of the bolt after 1 second.
   b. Find the height of the bolt after 4 seconds.
   c. To the nearest whole second, estimate when the bolt hits the ground.
   d. Factor $529 - 16t^2$.

95. At this writing, the world's tallest building is the Taipei 101 in Taipei, Taiwan, at a height of 1671 feet. (*Source:* Council on Tall Buildings and Urban Habitat) Suppose a worker is suspended 71 feet below the top of the pinnacle atop the building, at a height of 1600 feet above the ground. If the worker accidentally drops a bolt, the height of the bolt after *t* seconds is given by the expression $1600 - 16t^2$.
   a. Find the height of the bolt after 3 seconds.
   b. Find the height of the bolt after 7 seconds.
   c. To the nearest whole second, estimate when the bolt hits the ground.
   d. Factor $1600 - 16t^2$.

96. A performer with the Moscow Circus is planning a stunt involving a free fall from the top of the Moscow State University building, which is 784 feet tall. (*Source:* Council on Tall

Buildings and Urban Habitat) Neglecting air resistance, the performer's height above gigantic cushions positioned at ground level after t seconds is given by the expression $784 - 16t^2$.

a. Find the performer's height after 2 seconds.

b. Find the performer's height after 5 seconds.

c. To the nearest whole second, estimate when the performer reaches the cushions positioned at ground level.

d. Factor $784 - 16t^2$.

STUDY SKILLS BUILDER

Are You Getting All the Mathematics Help That You Need?

Remember that, in addition to your instructor, there are many places to get help with your mathematics course. For example:

- This text has an accompanying video lesson for every section and the CD in this text contains worked out solutions to every Chapter Test exercise.
- The back of the book contains answers to odd-numbered exercises and selected solutions.
- A student *Solutions Manual* is available that contains worked-out solutions to odd-numbered exercises as well as solutions to every exercise in the Integrated Reviews, Chapter Reviews, Chapter Tests, and Cumulative Reviews.
- Don't forget to check with your instructor for other local resources available to you, such as a tutor center.

Exercises

1. List items you find helpful in the text and all student supplements to this text.
2. List all the campus help that is available to you for this course.
3. List any help (besides the textbook) from Exercises 1 and 2 above that you are using.
4. List any help (besides the textbook) that you feel you should try.
5. Write a goal for yourself that includes trying anything you listed in Exercise 4 during the next week.

INTEGRATED REVIEW   CHOOSING A FACTORING STRATEGY

Sections 6.1–6.5

The following steps may be helpful when factoring polynomials.

Factoring a Polynomial

STEP 1. Are there any common factors? If so, factor out the GCF.

STEP 2. How many terms are in the polynomial?

    a. If there are **two** terms, decide if one of the following can be applied.

        i. Difference of two squares: $a^2 - b^2 = (a + b)(a - b)$.

        ii. Difference of two cubes: $a^3 - b^3 = (a - b)(a^2 + ab + b^2)$.

        iii. Sum of two cubes: $a^3 + b^3 = (a + b)(a^2 - ab + b^2)$.

    b. If there are **three** terms, try one of the following.

        i. Perfect square trinomial: $a^2 + 2ab + b^2 = (a + b)^2$
$$a^2 - 2ab + b^2 = (a - b)^2.$$

        ii. If not a perfect square trinomial, factor using the methods presented in Sections 6.2 through 6.4.

    c. If there are **four** or more terms, try factoring by grouping.

STEP 3. See if any factors in the factored polynomial can be factored further.

STEP 4. Check by multiplying.

Study the next five examples to help you use the steps on the previous page.

EXAMPLE 1 Factor $10t^2 - 17t + 3$.

Solution

STEP 1. The terms of this polynomial have no common factor (other than 1).

STEP 2. There are three terms, so this polynomial is a trinomial. This trinomial is not a perfect square trinomial, so factor using methods from earlier sections.

$$\text{Factors of } 10t^2: \quad 10t^2 = 2t \cdot 5t, \quad 10t^2 = t \cdot 10t$$

Since the middle term, $-17t$, has a negative numerical coefficient, find negative factors of 3.

$$\text{Factors of 3:} \quad 3 = -1 \cdot -3$$

Try different combinations of these factors. The correct combination is

$$(2t - 3)(5t - 1) = 10t^2 - 17t + 3$$

$$\underline{} \quad -15t$$
$$ \quad -2t$$
$$ \quad -17t \quad \text{Correct middle term}$$

STEP 3. No factor can be factored further, so we have factored completely.

STEP 4. To check, multiply $2t - 3$ and $5t - 1$.

$$(2t - 3)(5t - 1) = 10t^2 - 2t - 15t + 3 = 10t^2 - 17t + 3$$

The factored form of $10t^2 - 17t + 3$ is $(2t - 3)(5t - 1)$.

PRACTICE 1 Factor $6x^2 - 11x + 3$.

EXAMPLE 2 Factor $2x^3 + 3x^2 - 2x - 3$.

Solution

STEP 1. There are no factors common to all terms.

STEP 2. Try factoring by grouping since this polynomial has four terms.

$$2x^3 + 3x^2 - 2x - 3 = x^2(2x + 3) - 1(2x + 3) \quad \text{Factor out the greatest common factor for each pair of terms.}$$
$$= (2x + 3)(x^2 - 1) \quad \text{Factor out } 2x + 3.$$

STEP 3. The binomial $x^2 - 1$ can be factored further. It is the difference of two squares.

$$= (2x + 3)(x + 1)(x - 1) \quad \text{Factor } x^2 - 1 \text{ as a difference of squares.}$$

STEP 4. Check by finding the product of the three binomials. The polynomial factored completely is $(2x + 3)(x + 1)(x - 1)$.

PRACTICE 2 Factor $3x^3 + x^2 - 12x - 4$.

EXAMPLE 3 Factor $12m^2 - 3n^2$.

Solution

STEP 1. The terms of this binomial contain a greatest common factor of 3.

$$12m^2 - 3n^2 = 3(4m^2 - n^2) \quad \text{Factor out the greatest common factor.}$$

STEP 2. The binomial $4m^2 - n^2$ is a difference of squares.

$$= 3(2m + n)(2m - n) \quad \text{Factor the difference of squares.}$$

STEP 3. No factor can be factored further.

STEP 4. We check by multiplying.

$$3(2m + n)(2m - n) = 3(4m^2 - n^2) = 12m^2 - 3n^2$$

The factored form of $12m^2 - 3n^2$ is $3(2m + n)(2m - n)$. □

PRACTICE 3 Factor $27x^2 - 3y^2$.

EXAMPLE 4 Factor $x^3 + 27y^3$.

Solution

STEP 1. The terms of this binomial contain no common factor (other than 1).

STEP 2. This binomial is the sum of two cubes.

$$x^3 + 27y^3 = (x)^3 + (3y)^3$$
$$= (x + 3y)[x^2 - x(3y) + (3y)^2]$$
$$= (x + 3y)(x^2 - 3xy + 9y^2)$$

STEP 3. No factor can be factored further.

STEP 4. We check by multiplying.

$$(x + 3y)(x^2 - 3xy + 9y^2) = x(x^2 - 3xy + 9y^2) + 3y(x^2 - 3xy + 9y^2)$$
$$= x^3 - 3x^2y + 9xy^2 + 3x^2y - 9xy^2 + 27y^3$$
$$= x^3 + 27y^3$$

Thus, $x^3 + 27y^3$ factored completely is $(x + 3y)(x^2 - 3xy + 9y^2)$. □

PRACTICE 4 Factor $8a^3 + b^3$.

EXAMPLE 5 Factor $30a^2b^3 + 55a^2b^2 - 35a^2b$.

Solution

STEP 1. $30a^2b^3 + 55a^2b^2 - 35a^2b = 5a^2b(6b^2 + 11b - 7)$ Factor out the GCF.

STEP 2. $\qquad\qquad\qquad\qquad\qquad\qquad = 5a^2b(2b - 1)(3b + 7)$ Factor the resulting trinomial.

STEP 3. No factor can be factored further.

STEP 4. Check by multiplying.

The trinomial factored completely is $5a^2b(2b - 1)(3b + 7)$. □

PRACTICE 5 Factor $60x^3y^2 - 66x^2y^2 - 36xy^2$.

Factor the following completely.

1. $x^2 + 2xy + y^2$
2. $x^2 - 2xy + y^2$
3. $a^2 + 11a - 12$
4. $a^2 - 11a + 10$
5. $a^2 - a - 6$
6. $a^2 - 2a + 1$
7. $x^2 + 2x + 1$
8. $x^2 + x - 2$
9. $x^2 + 4x + 3$
10. $x^2 + x - 6$
11. $x^2 + 7x + 12$
12. $x^2 + x - 12$
13. $x^2 + 3x - 4$
14. $x^2 - 7x + 10$
15. $x^2 + 2x - 15$
16. $x^2 + 11x + 30$
17. $x^2 - x - 30$
18. $x^2 + 11x + 24$
19. $2x^2 - 98$
20. $3x^2 - 75$
21. $x^2 + 3x + xy + 3y$

22. $3y - 21 + xy - 7x$
23. $x^2 + 6x - 16$
24. $x^2 - 3x - 28$
25. $4x^3 + 20x^2 - 56x$
26. $6x^3 - 6x^2 - 120x$
27. $12x^2 + 34x + 24$
28. $8a^2 + 6ab - 5b^2$
29. $4a^2 - b^2$
30. $28 - 13x - 6x^2$
31. $20 - 3x - 2x^2$
32. $x^2 - 2x + 4$
33. $a^2 + a - 3$
34. $6y^2 + y - 15$
35. $4x^2 - x - 5$
36. $x^2y - y^3$
37. $4t^2 + 36$
38. $x^2 + x + xy + y$
39. $ax + 2x + a + 2$
40. $18x^3 - 63x^2 + 9x$
41. $12a^3 - 24a^2 + 4a$
42. $x^2 + 14x - 32$
43. $x^2 - 14x - 48$
44. $16a^2 - 56ab + 49b^2$
45. $25p^2 - 70pq + 49q^2$
46. $7x^2 + 24xy + 9y^2$
47. $125 - 8y^3$
48. $64x^3 + 27$
49. $-x^2 - x + 30$
50. $-x^2 + 6x - 8$
51. $14 + 5x - x^2$
52. $3 - 2x - x^2$
53. $3x^4y + 6x^3y - 72x^2y$
54. $2x^3y + 8x^2y^2 - 10xy^3$
55. $5x^3y^2 - 40x^2y^3 + 35xy^4$
56. $4x^4y - 8x^3y - 60x^2y$
57. $12x^3y + 243xy$
58. $6x^3y^2 + 8xy^2$
59. $4 - x^2$
60. $9 - y^2$
61. $3rs - s + 12r - 4$
62. $x^3 - 2x^2 + 3x - 6$
63. $4x^2 - 8xy - 3x + 6y$
64. $4x^2 - 2xy - 7yz + 14xz$
65. $6x^2 + 18xy + 12y^2$
66. $12x^2 + 46xy - 8y^2$
67. $xy^2 - 4x + 3y^2 - 12$
68. $x^2y^2 - 9x^2 + 3y^2 - 27$
69. $5(x + y) + x(x + y)$
70. $7(x - y) + y(x - y)$
71. $14t^2 - 9t + 1$
72. $3t^2 - 5t + 1$
73. $3x^2 + 2x - 5$
74. $7x^2 + 19x - 6$
75. $x^2 + 9xy - 36y^2$
76. $3x^2 + 10xy - 8y^2$
77. $1 - 8ab - 20a^2b^2$
78. $1 - 7ab - 60a^2b^2$
79. $9 - 10x^2 + x^4$
80. $36 - 13x^2 + x^4$
81. $x^4 - 14x^2 - 32$
82. $x^4 - 22x^2 - 75$
83. $x^2 - 23x + 120$
84. $y^2 + 22y + 96$
85. $6x^3 - 28x^2 + 16x$
86. $6y^3 - 8y^2 - 30y$
87. $27x^3 - 125y^3$
88. $216y^3 - z^3$
89. $x^3y^3 + 8z^3$
90. $27a^3b^3 + 8$
91. $2xy - 72x^3y$
92. $2x^3 - 18x$
93. $x^3 + 6x^2 - 4x - 24$
94. $x^3 - 2x^2 - 36x + 72$
95. $6a^3 + 10a^2$
96. $4n^2 - 6n$
97. $a^2(a + 2) + 2(a + 2)$
98. $a - b + x(a - b)$
99. $x^3 - 28 + 7x^2 - 4x$
100. $a^3 - 45 - 9a + 5a^2$

CONCEPT EXTENSIONS

Factor.

101. $(x - y)^2 - z^2$
102. $(x + 2y)^2 - 9$
103. $81 - (5x + 1)^2$
104. $b^2 - (4a + c)^2$
105. Explain why it makes good sense to factor out the GCF first, before using other methods of factoring.
106. The sum of two squares usually does not factor. Is the sum of two squares $9x^2 + 81y^2$ factorable?
107. Which of the following are equivalent to $(x + 10)(x - 7)$?
    a. $(x - 7)(x + 10)$
    b. $-1(x + 10)(x - 7)$
    c. $-1(x + 10)(7 - x)$
    d. $-1(-x - 10)(7 - x)$

6.6 SOLVING QUADRATIC EQUATIONS BY FACTORING

OBJECTIVES

1. Solve quadratic equations by factoring.
2. Solve equations with degree greater than 2 by factoring.
3. Find the x-intercepts of the graph of a quadratic equation in two variables.

In this section, we introduce a new type of equation—the **quadratic equation.**

> **Quadratic Equation**
> A quadratic equation is one that can be written in the form
> $$ax^2 + bx + c = 0$$
> where $a, b,$ and c are real numbers and $a \neq 0$.

Some examples of quadratic equations are shown below.

$$x^2 - 9x - 22 = 0 \qquad 4x^2 - 28 = -49 \qquad x(2x - 7) = 4$$

The form $ax^2 + bx + c = 0$ is called the **standard form** of a quadratic equation. The quadratic equation $x^2 - 9x - 22 = 0$ is the only equation above that is in standard form.

Quadratic equations model many real-life situations. For example, let's suppose we want to know how long before a person diving from a 144-foot cliff reaches the ocean. The answer to this question is found by solving the quadratic equation $-16t^2 + 144 = 0$. (See Example 1 in Section 6.7.)

144 feet

OBJECTIVE 1 ▶ Solving quadratic equations by factoring. Some quadratic equations can be solved by making use of factoring and the **zero factor property.**

> **Zero Factor Theorem**
> If a and b are real numbers and if $ab = 0$, then $a = 0$ or $b = 0$.

This theorem states that if the product of two numbers is 0 then at least one of the numbers must be 0.

EXAMPLE 1 Solve: $(x - 3)(x + 1) = 0$.

Solution If this equation is to be a true statement, then either the factor $x - 3$ must be 0 or the factor $x + 1$ must be 0. In other words, either

$$x - 3 = 0 \quad \text{or} \quad x + 1 = 0$$

If we solve these two linear equations, we have

$$x = 3 \quad \text{or} \quad x = -1$$

Thus, 3 and -1 are both solutions of the equation $(x - 3)(x + 1) = 0$. To check, we replace x with 3 in the original equation. Then we replace x with -1 in the original equation.

Check: Let $x = 3$.

$(x - 3)(x + 1) = 0$
$(3 - 3)(3 + 1) \stackrel{?}{=} 0$ Replace x with 3.
$0(4) = 0$ True

Let $x = -1$.

$(x - 3)(x + 1) = 0$
$(-1 - 3)(-1 + 1) \stackrel{?}{=} 0$ Replace x with -1.
$(-4)(0) = 0$ True

The solutions are 3 and -1, or we say that the solution set is $\{-1, 3\}$.

PRACTICE 1 Solve: $(x + 4)(x - 5) = 0$.

Helpful Hint

The zero factor property says that *if a product is 0, then a factor is 0*.

If $a \cdot b = 0$, then $a = 0$ or $b = 0$.
If $x(x + 5) = 0$, then $x = 0$ or $x + 5 = 0$.
If $(x + 7)(2x - 3) = 0$, then $x + 7 = 0$ or $2x - 3 = 0$.

Use this property only when the product is 0. For example, if $a \cdot b = 8$, we do not know the value of a or b. The values may be $a = 2, b = 4$ or $a = 8, b = 1$, or any other two numbers whose product is 8.

EXAMPLE 2 Solve: $x(5x - 2) = 0$

Solution
$$x(5x - 2) = 0$$
$$x = 0 \quad \text{or} \quad 5x - 2 = 0 \quad \text{Use the zero factor property.}$$
$$5x = 2$$
$$x = \frac{2}{5}$$

Check: Let $x = 0$.

$x(5x - 2) = 0$
$0(5 \cdot 0 - 2) \stackrel{?}{=} 0$ Replace x with 0.
$0(-2) \stackrel{?}{=} 0$
$0 = 0$ True

Let $x = \frac{2}{5}$.

$x(5x - 2) = 0$
$\frac{2}{5}\left(5 \cdot \frac{2}{5} - 2\right) \stackrel{?}{=} 0$ Replace x with $\frac{2}{5}$.
$\frac{2}{5}(2 - 2) \stackrel{?}{=} 0$
$\frac{2}{5}(0) \stackrel{?}{=} 0$
$0 = 0$ True

The solutions are 0 and $\frac{2}{5}$.

PRACTICE 2 Solve: $x(7x - 6) = 0$.

EXAMPLE 3 Solve: $x^2 - 9x - 22 = 0$

Solution One side of the equation is 0. However, to use the zero factor property, one side of the equation must be 0 *and* the other side must be written as a product (must be factored). Thus, we must first factor this polynomial.

$$x^2 - 9x - 22 = 0$$
$$(x - 11)(x + 2) = 0 \quad \text{Factor.}$$

Now we can apply the zero factor property.

$$x - 11 = 0 \quad \text{or} \quad x + 2 = 0$$
$$x = 11 \qquad\qquad x = -2$$

Check: Let $x = 11$.

$x^2 - 9x - 22 = 0$
$11^2 - 9 \cdot 11 - 22 \stackrel{?}{=} 0$
$121 - 99 - 22 \stackrel{?}{=} 0$
$22 - 22 \stackrel{?}{=} 0$
$0 = 0$ True

Let $x = -2$.

$x^2 - 9x - 22 = 0$
$(-2)^2 - 9(-2) - 22 \stackrel{?}{=} 0$
$4 + 18 - 22 \stackrel{?}{=} 0$
$22 - 22 \stackrel{?}{=} 0$
$0 = 0$ True

The solutions are 11 and -2.

PRACTICE 3 Solve: $x^2 - 8x - 48 = 0$.

EXAMPLE 4 Solve: $4x^2 - 28x = -49$

Solution First we rewrite the equation in standard form so that one side is 0. Then we factor the polynomial.

$$4x^2 - 28x = -49$$
$$4x^2 - 28x + 49 = 0 \quad \text{Write in standard form by adding 49 to both sides.}$$
$$(2x - 7)(2x - 7) = 0 \quad \text{Factor.}$$

Next we use the zero factor property and set each factor equal to 0. Since the factors are the same, the related equations will give the same solution.

$$2x - 7 = 0 \quad \text{or} \quad 2x - 7 = 0 \quad \text{Set each factor equal to 0.}$$
$$2x = 7 \qquad\qquad 2x = 7 \quad \text{Solve.}$$
$$x = \frac{7}{2} \qquad\qquad x = \frac{7}{2}$$

Check: Although $\frac{7}{2}$ occurs twice, there is a single solution. Check this solution in the original equation. The solution is $\frac{7}{2}$.

PRACTICE 4 Solve: $9x^2 - 24x = -16$.

The following steps may be used to solve a quadratic equation by factoring.

> **To Solve Quadratic Equations by Factoring**
> **STEP 1.** Write the equation in standard form so that one side of the equation is 0.
> **STEP 2.** Factor the quadratic expression completely.
> **STEP 3.** Set each factor containing a variable equal to 0.
> **STEP 4.** Solve the resulting equations.
> **STEP 5.** Check each solution in the original equation.

Since it is not always possible to factor a quadratic polynomial, not all quadratic equations can be solved by factoring. Other methods of solving quadratic equations are presented in Chapter 9.

EXAMPLE 5 Solve: $x(2x - 7) = 4$

Solution First we write the equation in standard form; then we factor.

$$x(2x - 7) = 4$$
$$2x^2 - 7x = 4 \qquad \text{Multiply.}$$
$$2x^2 - 7x - 4 = 0 \qquad \text{Write in standard form.}$$
$$(2x + 1)(x - 4) = 0 \qquad \text{Factor.}$$
$$2x + 1 = 0 \quad \text{or} \quad x - 4 = 0 \quad \text{Set each factor equal to zero.}$$
$$2x = -1 \qquad\qquad x = 4 \quad \text{Solve.}$$
$$x = -\frac{1}{2}$$

Check the solutions in the original equation. The solutions are $-\frac{1}{2}$ and 4.

PRACTICE 5 Solve: $x(3x + 7) = 6$.

Helpful Hint

To solve the equation $x(2x - 7) = 4$, do **not** set each factor equal to 4. Remember that to apply the zero factor property, one side of the equation must be 0 and the other side of the equation must be in factored form.

Concept Check ✓

Explain the error and solve the equation correctly.

$$(x - 3)(x + 1) = 5$$
$$x - 3 = 5 \quad \text{or} \quad x + 1 = 5$$
$$x = 8 \quad \text{or} \quad x = 4$$

EXAMPLE 6 Solve: $-2x^2 - 4x + 30 = 0$.

Solution The equation is in standard form so we begin by factoring out a common factor of -2.

$$-2x^2 - 4x + 30 = 0$$
$$-2(x^2 + 2x - 15) = 0 \quad \text{Factor out } -2.$$
$$-2(x + 5)(x - 3) = 0 \quad \text{Factor the quadratic.}$$

Next, set each factor **containing a variable** equal to 0.

$$x + 5 = 0 \quad \text{or} \quad x - 3 = 0 \quad \text{Set each factor containing a variable equal to 0.}$$
$$x = -5 \quad \text{or} \quad x = 3 \quad \text{Solve.}$$

Note: The factor -2 is a constant term containing no variables and can never equal 0. The solutions are -5 and 3. □

PRACTICE 6 Solve: $-3x^2 - 6x + 72 = 0$.

OBJECTIVE 2 ▶ **Solving equations with degree greater than two by factoring.** Some equations involving polynomials of degree higher than 2 may also be solved by factoring and then applying the zero factor theorem.

EXAMPLE 7 Solve: $3x^3 - 12x = 0$.

Solution Factor the left side of the equation. Begin by factoring out the common factor of $3x$.

$$3x^3 - 12x = 0$$
$$3x(x^2 - 4) = 0 \quad \text{Factor out the GCF } 3x.$$
$$3x(x + 2)(x - 2) = 0 \quad \text{Factor } x^2 - 4, \text{ a difference of squares.}$$

$$3x = 0 \quad \text{or} \quad x + 2 = 0 \quad \text{or} \quad x - 2 = 0 \quad \text{Set each factor equal to 0.}$$
$$x = 0 \quad \text{or} \quad x = -2 \quad \text{or} \quad x = 2 \quad \text{Solve.}$$

Thus, the equation $3x^3 - 12x = 0$ has three solutions: 0, -2, and 2. To check, replace x with each solution in the original equation.

Let $x = 0$.	Let $x = -2$.	Let $x = 2$.
$3(0)^3 - 12(0) \stackrel{?}{=} 0$	$3(-2)^3 - 12(-2) \stackrel{?}{=} 0$	$3(2)^3 - 12(2) \stackrel{?}{=} 0$
$0 = 0$	$3(-8) + 24 \stackrel{?}{=} 0$	$3(8) - 24 \stackrel{?}{=} 0$
	$0 = 0$	$0 = 0$

Substituting 0, -2, or 2 into the original equation results each time in a true equation. The solutions are 0, -2, and 2. □

PRACTICE 7 Solve: $7x^3 - 63x = 0$.

Answer to Concept Check:
To use the zero factor property, one side of the equation must be 0, not 5. Correctly, $(x - 3)(x + 1) = 5$, $x^2 - 2x - 3 = 5$, $x^2 - 2x - 8 = 0$, $(x - 4)(x + 2) = 0$, $x - 4 = 0$ or $x + 2 = 0$, $x = 4$ or $x = -2$.

EXAMPLE 8 Solve: $(5x - 1)(2x^2 + 15x + 18) = 0$.

Solution

$(5x - 1)(2x^2 + 15x + 18) = 0$

$(5x - 1)(2x + 3)(x + 6) = 0$ Factor the trinomial.

$5x - 1 = 0$ or $2x + 3 = 0$ or $x + 6 = 0$ Set each factor equal to 0.

$5x = 1$ or $2x = -3$ or $x = -6$ Solve.

$x = \dfrac{1}{5}$ or $x = -\dfrac{3}{2}$

The solutions are $\dfrac{1}{5}, -\dfrac{3}{2}$, and -6. Check by replacing x with each solution in the original equation. The solutions are $-6, -\dfrac{3}{2}$, and $\dfrac{1}{5}$. □

PRACTICE 8 Solve: $(3x - 2)(2x^2 - 13x + 15) = 0$.

EXAMPLE 9 Solve: $2x^3 - 4x^2 - 30x = 0$.

Solution Begin by factoring out the GCF $2x$.

$2x^3 - 4x^2 - 30x = 0$

$2x(x^2 - 2x - 15) = 0$ Factor out the GCF $2x$.

$2x(x - 5)(x + 3) = 0$ Factor the quadratic.

$2x = 0$ or $x - 5 = 0$ or $x + 3 = 0$ Set each factor containing a variable equal to 0.

$x = 0$ or $x = 5$ or $x = -3$ Solve.

Check by replacing x with each solution in the cubic equation. The solutions are $-3, 0$, and 5. □

PRACTICE 9 Solve: $5x^3 + 5x^2 - 30x = 0$.

OBJECTIVE 3 ▶ Finding x-intercepts of the graph of a quadratic equation. In Chapter 3, we graphed linear equations in two variables, such as $y = 5x - 6$. Recall that to find the x-intercept of the graph of a linear equation, let $y = 0$ and solve for x. This is also how to find the x-intercepts of the graph of a **quadratic equation in two variables,** such as $y = x^2 - 5x + 4$.

EXAMPLE 10 Find the x-intercepts of the graph of $y = x^2 - 5x + 4$.

Solution Let $y = 0$ and solve for x.

$y = x^2 - 5x + 4$

$0 = x^2 - 5x + 4$ Let $y = 0$.

$0 = (x - 1)(x - 4)$ Factor.

$x - 1 = 0$ or $x - 4 = 0$ Set each factor equal to 0.

$x = 1$ or $x = 4$ Solve.

The x-intercepts of the graph of $y = x^2 - 5x + 4$ are $(1, 0)$ and $(4, 0)$.

The graph of $y = x^2 - 5x + 4$ is shown in the margin. □

PRACTICE 10 Find the x-intercepts of the graph of $y = x^2 - 6x + 8$.

In general, a quadratic equation in two variables is one that can be written in the form $y = ax^2 + bx + c$ where $a \neq 0$. The graph of such an equation is called a **parabola** and will open up or down depending on the sign of a.

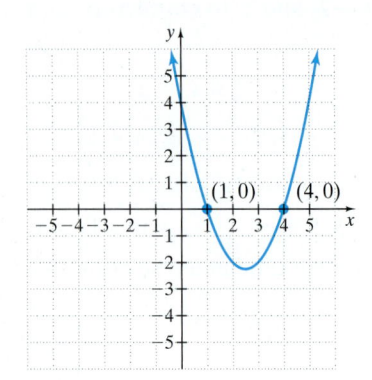

Notice that the x-intercepts of the graph of $y = ax^2 + bx + c$ are the real number solutions of $0 = ax^2 + bx + c$. Also, the real number solutions of $0 = ax^2 + bx + c$ are the x-intercepts of the graph of $y = ax^2 + bx + c$. We study more about graphs of quadratic equations in two variables in Chapter 9.

Graph of $y = ax^2 + bx + c$
x-intercepts are solutions of $0 = ax^2 + bx + c$

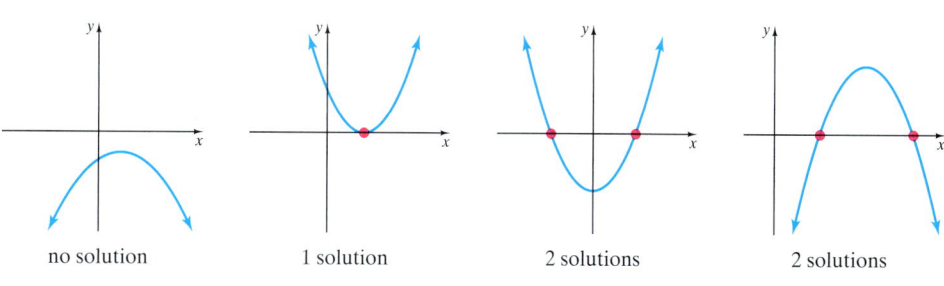

no solution   1 solution   2 solutions   2 solutions

Graphing Calculator Explorations

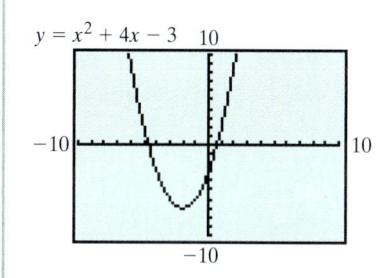

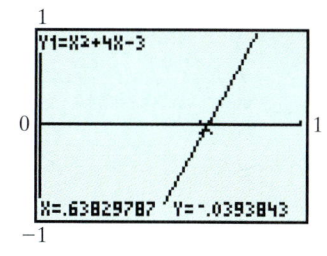

A grapher may be used to find solutions of a quadratic equation whether the related quadratic polynomial is factorable or not. For example, let's use a grapher to approximate the solutions of $0 = x^2 + 4x - 3$. To do so, graph $y_1 = x^2 + 4x - 3$. Recall that the x-intercepts of this graph are the solutions of $0 = x^2 + 4x - 3$.

Notice that the graph appears to have an x-intercept between -5 and -4 and one between 0 and 1. Many graphers contain a TRACE feature. This feature activates a graph cursor that can be used to *trace* along a graph while the corresponding x- and y-coordinates are shown on the screen. Use the TRACE feature to confirm that x-intercepts lie between -5 and -4 and also 0 and 1. To approximate the x-intercepts to the nearest tenth, use a ROOT or a ZOOM feature on your grapher or redefine the viewing window. (A ROOT feature calculates the x-intercept. A ZOOM feature magnifies the viewing window around a specific location such as the graph cursor.) If we redefine the window to $[0, 1]$ on the x-axis and $[-1, 1]$ on the y-axis, the following graph is generated.

By using the TRACE feature, we can conclude that one x-intercept is approximately 0.6 to the nearest tenth. By repeating these steps for the other x-intercept, we find that it is approximately -4.6.

Use a grapher to approximate the real number solutions to the nearest tenth. If an equation has no real number solution, state so.

1. $3x^2 - 4x - 6 = 0$
2. $x^2 - x - 9 = 0$
3. $2x^2 + x + 2 = 0$
4. $-4x^2 - 5x - 4 = 0$
5. $-x^2 + x + 5 = 0$
6. $10x^2 + 6x - 3 = 0$

VOCABULARY & READINESS CHECK

Use the choices below to fill in each blank. Not all choices will be used.

    $-3, 5$ $a = 0$ or $b = 0$ 0   linear
    $3, -5$ quadratic   1

1. An equation that can be written in the form $ax^2 + bx + c = 0$, (with $a \neq 0$), is called a(n) _____ equation.
2. If the product of two numbers is 0, then at least one of the numbers must be _____.
3. The solutions to $(x - 3)(x + 5) = 0$ are _____.
4. If $a \cdot b = 0$, then _____.

Solve each equation by inspection.

5. $(a - 3)(a - 7) = 0$
6. $(a - 5)(a - 2) = 0$
7. $(x + 8)(x + 6) = 0$
8. $(x + 2)(x + 3) = 0$
9. $(x + 1)(x - 3) = 0$
10. $(x - 1)(x + 2) = 0$

6.6 EXERCISE SET

Solve each equation. See Examples 1 and 2.

1. $(x - 2)(x + 1) = 0$
2. $(x + 4)(x - 10) = 0$
3. $(x + 9)(x + 17) = 0$
4. $(x + 11)(x + 1) = 0$
5. $x(x + 6) = 0$
6. $x(x - 7) = 0$
7. $3x(x - 8) = 0$
8. $2x(x + 12) = 0$
9. $(2x + 3)(4x - 5) = 0$
10. $(3x - 2)(5x + 1) = 0$
11. $(2x - 7)(7x + 2) = 0$
12. $(9x + 1)(4x - 3) = 0$
13. $\left(x - \frac{1}{2}\right)\left(x + \frac{1}{3}\right) = 0$
14. $\left(x + \frac{2}{9}\right)\left(x - \frac{1}{4}\right) = 0$
15. $(x + 0.2)(x + 1.5) = 0$
16. $(x + 1.7)(x + 2.3) = 0$
17. Write a quadratic equation that has two solutions, 6 and −1. Leave the polynomial in the equation in factored form.
18. Write a quadratic equation that has two solutions, 0 and −2. Leave the polynomial in the equation in factored form.

Solve. See Examples 3 through 6.

19. $x^2 - 13x + 36 = 0$
20. $x^2 + 2x - 63 = 0$
21. $x^2 + 2x - 8 = 0$
22. $x^2 - 5x + 6 = 0$
23. $x^2 - 7x = 0$
24. $x^2 - 3x = 0$
25. $x^2 - 4x = 32$
26. $x^2 - 5x = 24$
27. $x^2 = 16$
28. $x^2 = 9$
29. $(x + 4)(x - 9) = 4x$
30. $(x + 3)(x + 8) = x$
31. $x(3x - 1) = 14$
32. $x(4x - 11) = 3$
33. $-3x^2 + 75 = 0$
34. $-2y^2 + 72 = 0$
35. $24x^2 + 44x = 8$
36. $6x^2 + 57x = 30$

Solve each equation. See Examples 7 through 9.

37. $x^3 - 12x^2 + 32x = 0$
38. $x^3 - 14x^2 + 49x = 0$
39. $(4x - 3)(16x^2 - 24x + 9) = 0$
40. $(2x + 5)(4x^2 + 20x + 25) = 0$
41. $4x^3 - x = 0$
42. $4y^3 - 36y = 0$
43. $32x^3 - 4x^2 - 6x = 0$
44. $15x^3 + 24x^2 - 63x = 0$

MIXED PRACTICE

Solve each equation. See Examples 1 through 9. (A few exercises are linear equations.)

45. $(x + 3)(x - 2) = 0$
46. $(x - 6)(x + 7) = 0$
47. $x^2 + 20x = 0$
48. $x^2 + 15x = 0$
49. $4(x - 7) = 6$
50. $5(3 - 4x) = 9$
51. $4y^2 - 1 = 0$
52. $4y^2 - 81 = 0$
53. $(2x + 3)(2x^2 - 5x - 3) = 0$
54. $(2x - 9)(x^2 + 5x - 36) = 0$
55. $x^2 - 15 = -2x$
56. $x^2 - 26 = -11x$
57. $30x^2 - 11x - 30 = 0$
58. $12x^2 + 7x - 12 = 0$
59. $5x^2 - 6x - 8 = 0$
60. $9x^2 + 7x = 2$
61. $6y^2 - 22y - 40 = 0$
62. $3x^2 - 6x - 9 = 0$
63. $(y - 2)(y + 3) = 6$
64. $(y - 5)(y - 2) = 28$
65. $3x^3 + 19x^2 - 72x = 0$
66. $36x^3 + x^2 - 21x = 0$
67. $x^2 + 14x + 49 = 0$
68. $x^2 + 22x + 121 = 0$
69. $12y = 8y^2$
70. $9y = 6y^2$
71. $7x^3 - 7x = 0$
72. $3x^3 - 27x = 0$
73. $3x^2 + 8x - 11 = 13 - 6x$
74. $2x^2 + 12x - 1 = 4 + 3x$
75. $3x^2 - 20x = -4x^2 - 7x - 6$
76. $4x^2 - 20x = -5x^2 - 6x - 5$

Find the x-intercepts of the graph of each equation. See Example 10.

77. $y = (3x + 4)(x - 1)$
78. $y = (5x - 3)(x - 4)$
79. $y = x^2 - 3x - 10$
80. $y = x^2 + 7x + 6$
81. $y = 2x^2 + 11x - 6$
82. $y = 4x^2 + 11x + 6$

For Exercises 83 through 88, match each equation with its graph. See Example 10.

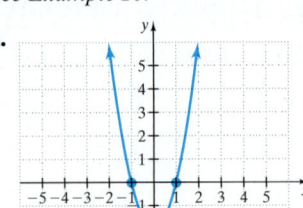

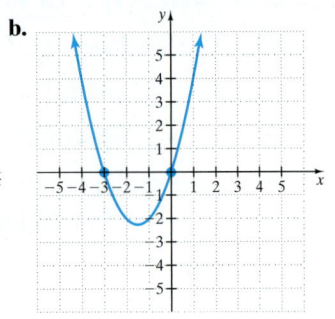

a.
b.

c.

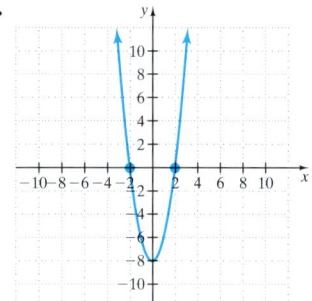

d.

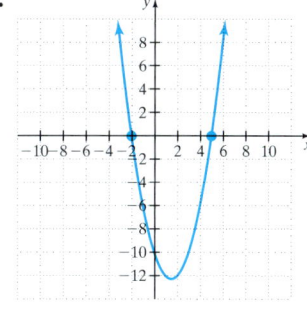

e.

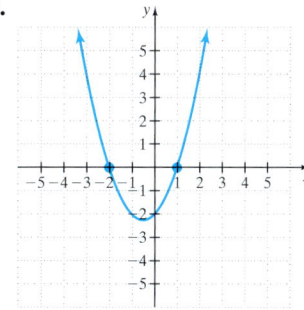

f.

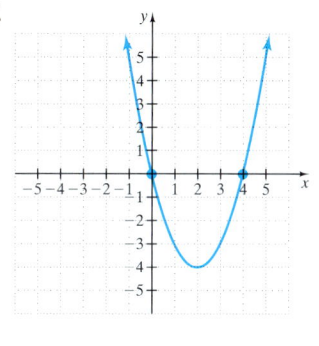

83. $y = (x + 2)(x - 1)$
84. $y = (x - 5)(x + 2)$
85. $y = x(x + 3)$
86. $y = x(x - 4)$
87. $y = 2x^2 - 8$
88. $y = 2x^2 - 2$

REVIEW AND PREVIEW

Perform the following operations. Write all results in lowest terms. See Section 1.3.

89. $\dfrac{3}{5} + \dfrac{4}{9}$

90. $\dfrac{2}{3} + \dfrac{3}{7}$

91. $\dfrac{7}{10} - \dfrac{5}{12}$

92. $\dfrac{5}{9} - \dfrac{5}{12}$

93. $\dfrac{7}{8} \div \dfrac{7}{15}$

94. $\dfrac{5}{12} - \dfrac{3}{10}$

95. $\dfrac{4}{5} \cdot \dfrac{7}{8}$

96. $\dfrac{3}{7} \cdot \dfrac{12}{17}$

CONCEPT EXTENSIONS

For Exercises 97 and 98, see the Concept Check in this section.

97. Explain the error and solve correctly:

$$x(x - 2) = 8$$
$$x = 8 \quad \text{or} \quad x - 2 = 8$$
$$x = 10$$

98. Explain the error and solve correctly:

$$(x - 4)(x + 2) = 0$$
$$x = -4 \quad \text{or} \quad x = 2$$

99. Write a quadratic equation in standard form that has two solutions, 5 and 7.

100. Write an equation that has three solutions, 0, 1, and 2.

101. A compass is accidentally thrown upward and out of an air balloon at a height of 300 feet. The height, y, of the compass at time x in seconds is given by the equation

$$y = -16x^2 + 20x + 300$$

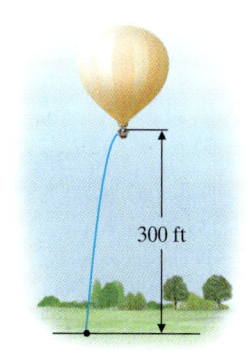

a. Find the height of the compass at the given times by filling in the table below.

time, x	0	1	2	3	4	5	6
height, y							

b. Use the table to determine when the compass strikes the ground.

c. Use the table to approximate the maximum height of the compass.

d. Plot the points (x, y) on a rectangular coordinate system and connect them with a smooth curve. Explain your results.

102. A rocket is fired upward from the ground with an initial velocity of 100 feet per second. The height, y, of the rocket at any time x is given by the equation

$$y = -16x^2 + 100x$$

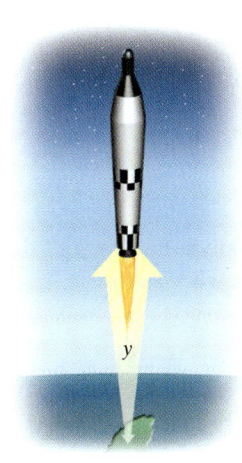

a. Find the height of the rocket at the given times by filling in the table below.

time, x	0	1	2	3	4	5	6	7
height, y								

b. Use the table to approximate when the rocket strikes the ground to the nearest second.

c. Use the table to approximate the maximum height of the rocket.

d. Plot the points (x, y) on a rectangular coordinate system and connect them with a smooth curve. Explain your results.

103. $(x - 3)(3x + 4) = (x + 2)(x - 6)$

104. $(2x - 3)(x + 6) = (x - 9)(x + 2)$

105. $(2x - 3)(x + 8) = (x - 6)(x + 4)$

106. $(x + 6)(x - 6) = (2x - 9)(x + 4)$

Solve each equation. First, multiply the binomial.

To solve $(x - 6)(2x - 3) = (x + 2)(x + 9)$, see below.

$$(x - 6)(2x - 3) = (x + 2)(x + 9)$$
$$2x^2 - 15x + 18 = x^2 + 11x + 18$$
$$x^2 - 26x = 0$$
$$x(x - 26) = 0$$
$$x = 0 \quad \text{or} \quad x - 26 = 0$$
$$x = 26$$

THE BIGGER PICTURE

Simplifying Expressions and Solving Equations

Now we continue our outline from Sections 1.7, 2.9, and 5.6. Although suggestions are given, this outline should be in your own words. Once you complete this new portion, try the exercises below.

I. Simplifying Expressions
  A. Real Numbers
   1. Add (Section 1.5)
   2. Subtract (Section 1.6)
   3. Multiply or Divide (Section 1.7)
  B. Exponents (Section 5.1)
  C. Polynomials
   1. Add (Section 5.2)
   2. Subtract (Section 5.2)
   3. Multiply (Section 5.3)
   4. Divide (Section 5.6)
  D. Factoring Polynomials—see the Chapter 6 Integrated Review for steps.
   $3x^4 - 78x^2 + 75$
   $= 3(x^4 - 26x^2 + 25)$ Factor out GCF—always first step.
   $= 3(x^2 - 25)(x^2 - 1)$ Factor trinomial.
   $= 3(x + 5)(x - 5)(x + 1)(x - 1)$ Factor further—each difference of squares.

II. Solving Equations and Inequalities
  A. Linear Equations (Section 2.4)
  B. Linear Inequalities (Section 2.9)
  C. Quadratic & Higher Degree Equations (Solving by Factoring)—highest power on variable is at least 2 when equation is written in standard form (set equal to 0).

   $x^2 + x = 6$
   $x^2 + x - 6 = 0$ Write the equation in standard form (set it equal to 0).
   $(x - 2)(x + 3) = 0$ Factor.
   $x = 2 \quad \text{or} \quad x = -3$ Set each factor equal to 0 and solve.

Simplify each expression.

1. $-7 + (-27)$ **2.** $\dfrac{(x^3)^4}{(x^{-2})^5}$

3. $(x^3 - 6x^2 + 2) - (5x^3 - 6)$

4. $\dfrac{3y^3 - 3y^2 + 9}{3y^2}$

Factor each expression.

5. $10x^3 - 250x$

6. $x^2 - 36x + 35$

7. $6xy + 15x - 6y - 15$

8. $5xy^2 - 2xy - 7x$

Solve each equation. Remember to use your outline to determine whether the equation is linear or quadratic and how to proceed with solving.

9. $(x - 5)(2x + 1) = 0$

10. $5x - 5 = 0$

11. $x(x - 12) = 28$

12. $7(x - 3) + 2(5x + 1) = 14$

6.7 QUADRATIC EQUATIONS AND PROBLEM SOLVING

OBJECTIVE

1. Solve problems that can be modeled by quadratic equations.

OBJECTIVE 1 ▶ Solving problems modeled by quadratic equations. Some problems may be modeled by quadratic equations. To solve these problems, we use the same problem-solving steps that were introduced in Section 2.5. When solving these problems, keep in mind that a solution of an equation that models a problem may not be a solution to the problem. For example, a person's age or the length of a rectangle is always a positive number. Discard solutions that do not make sense as solutions of the problem.

EXAMPLE 1 Finding Free-Fall Time

Since the 1940s, one of the top tourist attractions in Acapulco, Mexico is watching the cliff divers off the La Quebrada. The divers' platform is about 144 feet above the sea. These divers must time their descent just right, since they land in the crashing Pacific, in an inlet that is at most $9\frac{1}{2}$ feet deep. Neglecting air resistance, the height h in feet of a cliff diver above the ocean after t seconds is given by the quadratic equation $h = -16t^2 + 144$.

Find out how long it takes the diver to reach the ocean.

Solution

1. UNDERSTAND. Read and reread the problem. Then draw a picture of the problem.
   The equation $h = -16t^2 + 144$ models the height of the falling diver at time t. Familiarize yourself with this equation by find the height of the diver at time $t = 1$ second and $t = 2$ seconds.

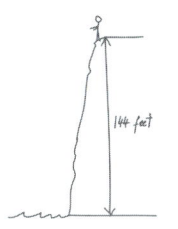

   When $t = 1$ second, the height of the diver is $h = -16(1)^2 + 144 = 128$ feet.
   When $t = 2$ seconds, the height of the diver is $h = -16(2)^2 + 144 = 80$ feet.

2. TRANSLATE. To find out how long it takes the diver to reach the ocean, we want to know the value of t for which $h = 0$.

   $0 = -16t^2 + 144$
   $0 = -16(t^2 - 9)$ Factor out -16.
   $0 = -16(t - 3)(t + 3)$ Factor completely.
   $t - 3 = 0$ or $t + 3 = 0$ Set each factor containing a variable equal to 0.
   $t = 3$ or $t = -3$ Solve.

3. INTERPRET. Since the time t cannot be negative, the proposed solution is 3 seconds.

Check: Verify that the height of the diver when t is 3 seconds is 0.

   When $t = 3$ seconds, $h = -16(3)^2 + 144 = -144 + 144 = 0$.

State: It takes the diver 3 seconds to reach the ocean.

PRACTICE

1 Cliff divers also frequent the falls at Waimea Falls Park in Oahu, Hawaii. One of the popular diving spots is 64 feet high. Neglecting air resistance, the height of a diver above the pool after t seconds is $h = -16t^2 + 64$. Find how long it takes a diver to reach the pool.

EXAMPLE 2   Finding an Unknown Number

The square of a number plus three times the number is 70. Find the number.

Solution

1. **UNDERSTAND.** Read and reread the problem. Suppose that the number is 5. The square of 5 is 5^2 or 25. Three times 5 is 15. Then $25 + 15 = 40$, not 70, so the number must be greater than 5. Remember, the purpose of proposing a number, such as 5, is to better understand the problem. Now that we do, we will let $x =$ the number.

2. **TRANSLATE.**

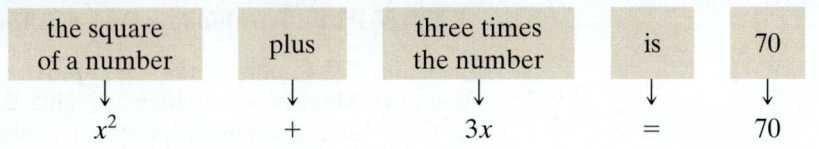

the square of a number	plus	three times the number	is	70
↓	↓	↓	↓	↓
x^2	$+$	$3x$	$=$	70

3. **SOLVE.**

$$x^2 + 3x = 70$$
$$x^2 + 3x - 70 = 0 \quad \text{Subtract 70 from both sides.}$$
$$(x + 10)(x - 7) = 0 \quad \text{Factor.}$$
$$x + 10 = 0 \quad \text{or} \quad x - 7 = 0 \quad \text{Set each factor equal to 0.}$$
$$x = -10 \qquad\qquad x = 7 \quad \text{Solve.}$$

4. **INTERPRET.**

Check: The square of -10 is $(-10)^2$, or 100. Three times -10 is $3(-10)$ or -30. Then $100 + (-30) = 70$, the correct sum, so -10 checks.

The square of 7 is 7^2 or 49. Three times 7 is $3(7)$, or 21. Then $49 + 21 = 70$, the correct sum, so 7 checks.

State: There are two numbers. They are -10 and 7.

> **PRACTICE 2**   The square of a number minus eight times the number is equal to forty-eight. Find the number.

EXAMPLE 3   Finding the Dimensions of a Sail

The height of a triangular sail is 2 meters less than twice the length of the base. If the sail has an area of 30 square meters, find the length of its base and the height.

Solution

1. **UNDERSTAND.** Read and reread the problem. Since we are finding the length of the base and the height, we let

$$x = \text{the length of the base}$$

and since the height is 2 meters less than twice the base,

$$2x - 2 = \text{the height}$$

An illustration is shown to the right.

2. **TRANSLATE.** We are given that the area of the triangle is 30 square meters, so we use the formula for area of a triangle.

$$\text{area of triangle} = \frac{1}{2} \cdot \text{base} \cdot \text{height}$$

$$30 = \frac{1}{2} \cdot x \cdot (2x - 2)$$

3. **SOLVE.** Now we solve the quadratic equation.

$$30 = \frac{1}{2}x(2x - 2)$$

$$30 = x^2 - x \qquad \text{Multiply.}$$
$$x^2 - x - 30 = 0 \qquad \text{Write in standard form.}$$
$$(x - 6)(x + 5) = 0 \qquad \text{Factor.}$$
$$x - 6 = 0 \quad \text{or} \quad x + 5 = 0 \qquad \text{Set each factor equal to 0.}$$
$$x = 6 \qquad\qquad x = -5$$

4. **INTERPRET.** Since x represents the length of the base, we discard the solution -5. The base of a triangle cannot be negative. The base is then 6 meters and the height is $2(6) - 2 = 10$ meters.

Check: To check this problem, we recall that $\frac{1}{2}$ base $\cdot$ height $=$ area, or

$$\frac{1}{2}(6)(10) = 30 \quad \text{The required area}$$

State: The base of the triangular sail is 6 meters and the height is 10 meters.

PRACTICE 3 An engineering team from Georgia Tech earned second place in a recent flight competition, with their triangular shaped paper hang glider. The base of their prize-winning entry was 1 foot less than three times the height. If the area of the triangular glider wing was 210 square feet, find the dimensions of the wing. (*Source: The Technique* [Georgia Tech's newspaper], April 18, 2003)

The next examples make use of the **Pythagorean theorem** and consecutive integers. Before we review this theorem, recall that a **right triangle** is a triangle that contains a 90° or right angle. The **hypotenuse** of a right triangle is the side opposite the right angle and is the longest side of the triangle. The **legs** of a right triangle are the other sides of the triangle.

Pythagorean Theorem
In a right triangle, the sum of the squares of the lengths of the two legs is equal to the square of the length of the hypotenuse.

$$(\text{leg})^2 + (\text{leg})^2 = (\text{hypotenuse})^2 \quad \text{or} \quad a^2 + b^2 = c^2$$

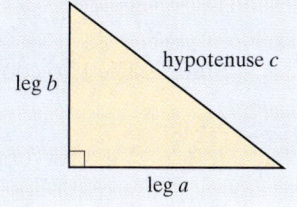

▶ **Helpful Hint**
If you use this formula, don't forget that c represents the length of the hypotenuse.

414 CHAPTER 6 Factoring Polynomials

Study the following diagrams for a review of consecutive integers.

Examples

If x is the first integer, then consecutive integers are
$x, x + 1, x + 2, \ldots$

If x is the first even integer, then consecutive even integers are
$x, x + 2, x + 4, \ldots$

If x is the first odd integer, then consecutive odd integers are
$x, x + 2, x + 4, \ldots$

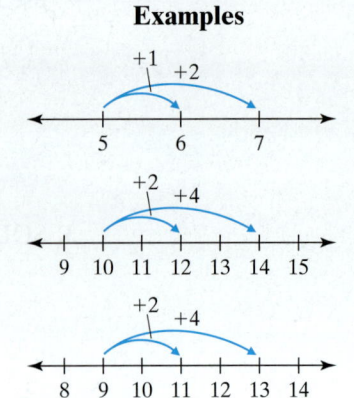

EXAMPLE 4 Finding Consecutive Even Integers

Find two consecutive even integers whose product is 34 more than their sum.

Solution

1. UNDERSTAND. Read and reread the problem. Let's just choose two consecutive even integers to help us better understand the problem. Let's choose 10 and 12. Their product is $10(12) = 120$ and their sum is $10 + 12 = 22$. The product is $120 - 22$, or 98 greater than the sum. Thus our guess is incorrect, but we have a better understanding of this example.

    Let's let x and $x + 2$ be the consecutive even integers.

2. TRANSLATE.

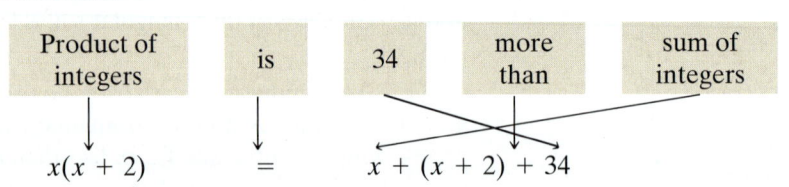

Product of integers	is	34	more than	sum of integers
$x(x + 2)$	=		$x + (x + 2) + 34$	

3. SOLVE. Now we solve the equation.

    $\begin{aligned} x(x + 2) &= x + (x + 2) + 34 \\ x^2 + 2x &= x + x + 2 + 34 &&\text{Multiply.} \\ x^2 + 2x &= 2x + 36 &&\text{Combine like terms.} \\ x^2 - 36 &= 0 &&\text{Write in standard form.} \\ (x + 6)(x - 6) &= 0 &&\text{Factor.} \\ x + 6 = 0 \quad &\text{or} \quad x - 6 = 0 &&\text{Set each factor equal to 0.} \\ x = -6 \quad &\phantom{\text{or}} \quad x = 6 &&\text{Solve.} \end{aligned}$

4. INTERPRET. If $x = -6$, then $x + 2 = -6 + 2$, or -4.
   If $x = 6$, then $x + 2 = 6 + 2$, or 8.

Check: $-6, -4$ $\qquad\qquad\qquad$ 6, 8

$\begin{aligned} -6(-4) &\stackrel{?}{=} -6 + (-4) + 34 & 6(8) &\stackrel{?}{=} 6 + 8 + 34 \\ 24 &\stackrel{?}{=} -10 + 34 & 48 &\stackrel{?}{=} 14 + 34 \\ 24 &= 24 \qquad\text{True} & 48 &= 48 \qquad\text{True} \end{aligned}$

State: The two consecutive even integers are -6 and -4 or 6 and 8.

PRACTICE 4 Find two consecutive integers whose product is 41 more than their sum.

EXAMPLE 5   Finding the Dimensions of a Triangle

Find the lengths of the sides of a right triangle if the lengths can be expressed as three consecutive even integers.

Solution

1. **UNDERSTAND.** Read and reread the problem. Let's suppose that the length of one leg of the right triangle is 4 units. Then the other leg is the next even integer, or 6 units, and the hypotenuse of the triangle is the next even integer, or 8 units. Remember that the hypotenuse is the longest side. Let's see if a triangle with sides of these lengths forms a right triangle. To do this, we check to see whether the Pythagorean theorem holds true.

$$4^2 + 6^2 \stackrel{?}{=} 8^2$$
$$16 + 36 \stackrel{?}{=} 64$$
$$52 = 64 \quad \text{False}$$

Our proposed numbers do not check, but we now have a better understanding of the problem.

We let x, $x + 2$, and $x + 4$ be three consecutive even integers. Since these integers represent lengths of the sides of a right triangle, we have the following.

x = one leg
$x + 2$ = other leg
$x + 4$ = hypotenuse (longest side)

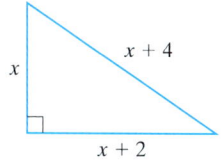

2. **TRANSLATE.** By the Pythagorean theorem, we have that

$$(\text{leg})^2 + (\text{leg})^2 = (\text{hypotenuse})^2$$
$$(x)^2 + (x + 2)^2 = (x + 4)^2$$

3. **SOLVE.** Now we solve the equation.

$$x^2 + (x + 2)^2 = (x + 4)^2$$
$$x^2 + x^2 + 4x + 4 = x^2 + 8x + 16 \quad \text{Multiply.}$$
$$2x^2 + 4x + 4 = x^2 + 8x + 16 \quad \text{Combine like terms.}$$
$$x^2 - 4x - 12 = 0 \quad \text{Write in standard form.}$$
$$(x - 6)(x + 2) = 0 \quad \text{Factor.}$$
$$x - 6 = 0 \quad \text{or} \quad x + 2 = 0 \quad \text{Set each factor equal to 0.}$$
$$x = 6 \quad\quad\quad\quad x = -2$$

4. **INTERPRET.** We discard $x = -2$ since length cannot be negative. If $x = 6$, then $x + 2 = 8$ and $x + 4 = 10$.

Check: Verify that

$$(\text{leg})^2 + (\text{leg})^2 = (\text{hypotenuse})^2$$
$$6^2 + 8^2 \stackrel{?}{=} 10^2$$
$$36 + 64 \stackrel{?}{=} 100$$
$$100 = 100 \quad \text{True}$$

State: The sides of the right triangle have lengths 6 units, 8 units, and 10 units.

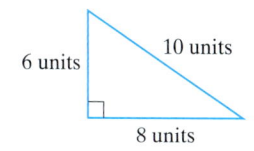

PRACTICE

5 Find the dimensions of a right triangle where the second leg is 1 unit less than double the first leg, and the hypotenuse is 1 unit more than double the length of the first leg.

6.7 EXERCISE SET

MIXED PRACTICE

See Examples 1 through 5 for all exercises. For Exercises 1 through 6, represent each given condition using a single variable, x.

1. The length and width of a rectangle whose length is 4 centimeters more than its width

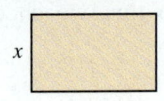

2. The length and width of a rectangle whose length is twice its width

3. Two consecutive odd integers

4. Two consecutive even integers

5. The base and height of a triangle whose height is one more than four times its base

6. The base and height of a trapezoid whose base is three less than five times its height

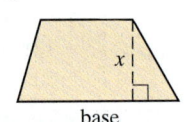

Use the information given to find the dimensions of each figure.

7. The *area* of the square is 121 square units. Find the length of its sides.

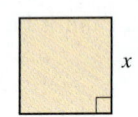

8. The *area* of the rectangle is 84 square inches. Find its length and width.

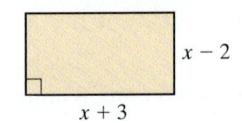

9. The *perimeter* of the quadrilateral is 120 centimeters. Find the lengths of the sides.

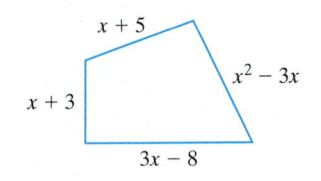

10. The *perimeter* of the triangle is 85 feet. Find the lengths of its sides.

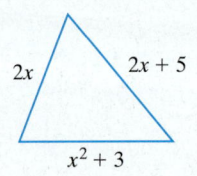

11. The *area* of the parallelogram is 96 square miles. Find its base and height.

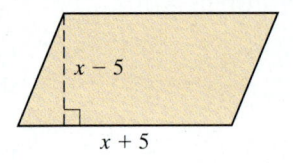

12. The *area* of the circle is 25π square kilometers. Find its radius.

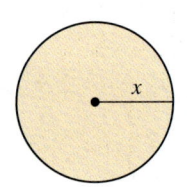

Solve.

13. An object is thrown upward from the top of an 80-foot building with an initial velocity of 64 feet per second. The height h of the object after t seconds is given by the quadratic equation $h = -16t^2 + 64t + 80$. When will the object hit the ground?

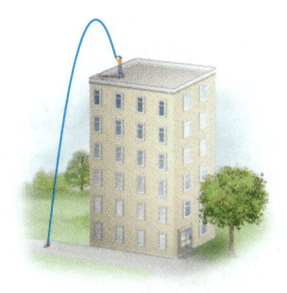

14. A hang glider pilot accidentally drops her compass from the top of a 400-foot cliff. The height h of the compass after t seconds is given by the quadratic equation $h = -16t^2 + 400$. When will the compass hit the ground?

15. The length of a rectangle is 7 centimeters less than twice its width. Its area is 30 square centimeters. Find the dimensions of the rectangle.

16. The length of a rectangle is 9 inches more than its width. Its area is 112 square inches. Find the dimensions of the rectangle.

The equation $D = \frac{1}{2}n(n-3)$ gives the number of diagonals D for a polygon with n sides. For example, a polygon with 6 sides has $D = \frac{1}{2} \cdot 6(6-3)$ or $D = 9$ diagonals. (See if you can count all 9 diagonals. Some are shown in the figure.) Use this equation, $D = \frac{1}{2}n(n-3)$, for Exercises 17 through 20.

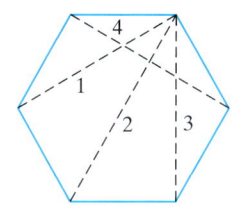

17. Find the number of diagonals for a polygon that has 12 sides.

18. Find the number of diagonals for a polygon that has 15 sides.

19. Find the number of sides n for a polygon that has 35 diagonals.

20. Find the number of sides n for a polygon that has 14 diagonals.

Solve.

21. The sum of a number and its square is 132. Find the number(s).

22. The sum of a number and its square is 182. Find the number(s).

23. The product of two consecutive room numbers is 210. Find the room numbers.

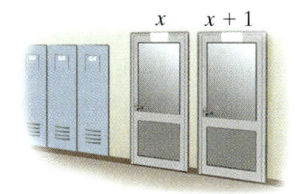

24. The product of two consecutive page numbers is 420. Find the page numbers.

25. A ladder is leaning against a building so that the distance from the ground to the top of the ladder is one foot less than the length of the ladder. Find the length of the ladder if the distance from the bottom of the ladder to the building is 5 feet.

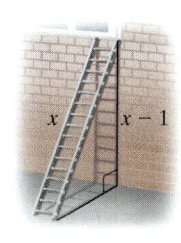

26. Use the given figure to find the length of the guy wire.

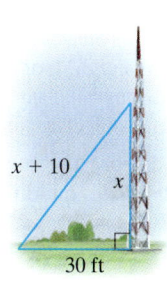

27. If the sides of a square are increased by 3 inches, the area becomes 64 square inches. Find the length of the sides of the original square.

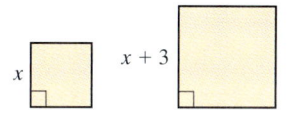

28. If the sides of a square are increased by 5 meters, the area becomes 100 square meters. Find the length of the sides of the original square.

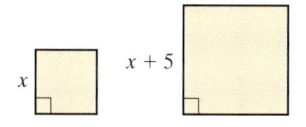

29. One leg of a right triangle is 4 millimeters longer than the smaller leg and the hypotenuse is 8 millimeters longer than the smaller leg. Find the lengths of the sides of the triangle.

30. One leg of a right triangle is 9 centimeters longer than the other leg and the hypotenuse is 45 centimeters. Find the lengths of the legs of the triangle.

31. The length of the base of a triangle is twice its height. If the area of the triangle is 100 square kilometers, find the height.

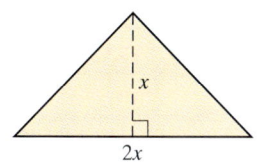

32. The height of a triangle is 2 millimeters less than the base. If the area is 60 square millimeters, find the base.

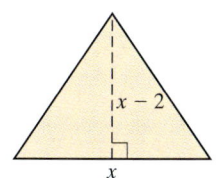

33. Find the length of the shorter leg of a right triangle if the longer leg is 12 feet more than the shorter leg and the hypotenuse is 12 feet less than twice the shorter leg.

34. Find the length of the shorter leg of a right triangle if the longer leg is 10 miles more than the shorter leg and the hypotenuse is 10 miles less than twice the shorter leg.

35. An object is dropped from 39 feet below the tip of the pinnacle atop one of the 1483-foot-tall Petronas Twin Towers in Kuala Lumpur, Malaysia. (*Source:* Council on Tall Buildings and Urban Habitat) The height h of the object after t seconds is given by the equation $h = -16t^2 + 1444$. Find how many seconds pass before the object reaches the ground.

36. An object is dropped from the top of 311 South Wacker Drive, a 961-foot-tall office building in Chicago. (*Source:* Council on Tall Buildings and Urban Habitat) The height h of the object after t seconds is given by the equation $h = -16t^2 + 961$. Find how many seconds pass before the object reaches the ground.

37. At the end of 2 years, P dollars invested at an interest rate r compounded annually increases to an amount, A dollars, given by
$$A = P(1 + r)^2$$
Find the interest rate if $100 increased to $144 in 2 years. Write your answer as a percent.

38. At the end of 2 years, P dollars invested at an interest rate r compounded annually increases to an amount, A dollars, given by
$$A = P(1 + r)^2$$
Find the interest rate if $2000 increased to $2420 in 2 years. Write your answer as a percent.

△ 39. Find the dimensions of a rectangle whose width is 7 miles less than its length and whose area is 120 square miles.

△ 40. Find the dimensions of a rectangle whose width is 2 inches less than half its length and whose area is 160 square inches.

41. If the cost, C, for manufacturing x units of a certain product is given by $C = x^2 - 15x + 50$, find the number of units manufactured at a cost of $9500.

42. If a switchboard handles n telephones, the number C of telephone connections it can make simultaneously is given by the equation $C = \dfrac{n(n-1)}{2}$. Find how many telephones are handled by a switchboard making 120 telephone connections simultaneously.

REVIEW AND PREVIEW

The following double line graph shows a comparison of the number of farms in the United States and the size of the average farm. Use this graph to answer Exercises 43–49. See Section 3.1.

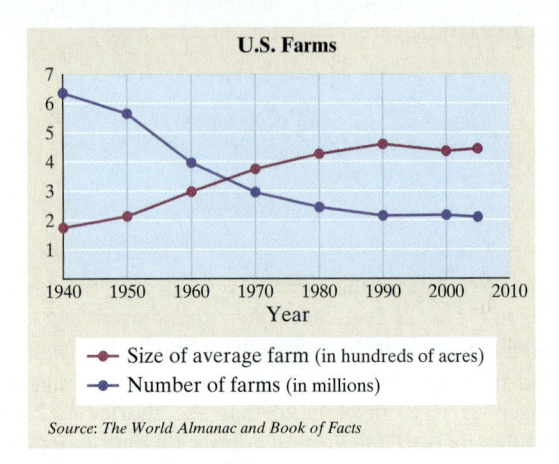

U.S. Farms

— Size of average farm (in hundreds of acres)
— Number of farms (in millions)

Source: The World Almanac and Book of Facts

△ 43. Approximate the size of the average farm in 1940.

△ 44. Approximate the size of the average farm in 2005.

45. Approximate the number of farms in 1940.

46. Approximate the number of farms in 2005.

47. Approximate the year that the colored lines in this graph intersect.

✎ 48. In your own words, explain the meaning of the point of intersection in the graph.

✎ 49. Describe the trends shown in this graph and speculate as to why these trends have occurred.

Write each fraction in simplest form. See Section 1.3.

50. $\dfrac{20}{35}$

51. $\dfrac{24}{32}$

52. $\dfrac{27}{18}$

53. $\dfrac{15}{27}$

54. $\dfrac{14}{42}$

55. $\dfrac{45}{50}$

CONCEPT EXTENSIONS

△ 56. Two boats travel at right angles to each other after leaving the same dock at the same time. One hour later the boats are 17 miles apart. If one boat travels 7 miles per hour faster than the other boat, find the rate of each boat.

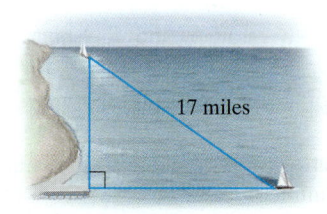

△ 57. The side of a square equals the width of a rectangle. The length of the rectangle is 6 meters longer than its width. The sum of the areas of the square and the rectangle is 176 square meters. Find the side of the square.

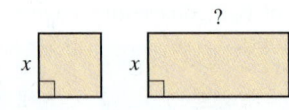

58. The sum of two numbers is 20, and the sum of their squares is 218. Find the numbers.

59. The sum of two numbers is 25, and the sum of their squares is 325. Find the numbers.

△ 60. According to the International America's Cup Class (IACC) rule, a sailboat competing in the America's Cup match must have a 110-foot-tall mast and a combined mainsail and jib sail area of 3000 square feet. (*Source:* America's Cup Organizing Committee) A design for an IACC-class sailboat calls for the mainsail to be 60% of the combined sail area. If the height of the triangular mainsail is 28 feet more than twice the length of the boom, find the length of the boom and the height of the mainsail.

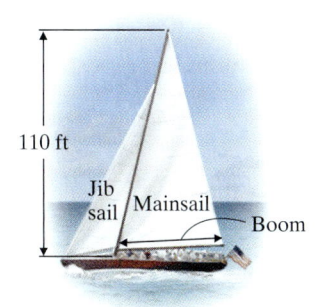

△ **61.** A rectangular pool is surrounded by a walk 4 meters wide. The pool is 6 meters longer than its width. If the total area of the pool and walk is 576 square meters more than the area of the pool, find the dimensions of the pool.

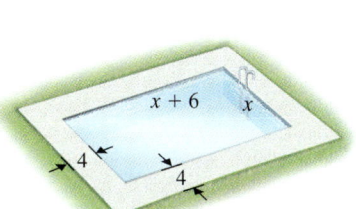

△ **62.** A rectangular garden is surrounded by a walk of uniform width. The area of the garden is 180 square yards. If the dimensions of the garden plus the walk are 16 yards by 24 yards, find the width of the walk.

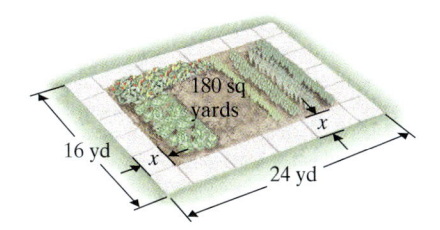

63. Write down two numbers whose sum is 10. Square each number and find the sum of the squares. Use this work to write a word problem like Exercise 59. Then give the word problem to a classmate to solve.

CHAPTER 6 GROUP ACTIVITY

Choosing Among Building Options

Whether putting in a new floor, hanging new wallpaper, or retiling a bathroom, it may be necessary to choose among several different materials with different pricing schemes. If a fixed amount of money is available for projects like these, it can be helpful to compare the choices by calculating how much area can be covered by a fixed dollar-value of material.

In this project, you will have the opportunity to choose among three different choices of materials for building a patio around a swimming pool. This project may be completed by working in groups or individually.

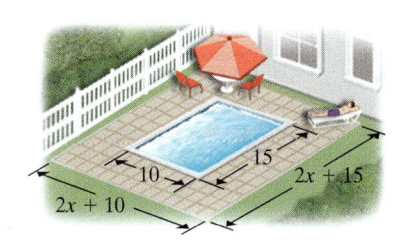

Situation: Suppose you have just had a 10-foot-by-15-foot in-ground swimming pool installed in your backyard. You have $3000 left from the building project that you would like to spend on surrounding the pool with a patio, equally wide on all sides (see figure). You have talked to several local suppliers about options for building this patio and must choose among the following.

Option	Material	Price
A	Poured cement	$5 per square foot
B	Brick	$7.50 per square foot plus a $30 flat fee for delivering the bricks
C	Outdoor carpeting	$4.50 per square foot plus $10.86 per foot of the pool's perimeter to install edging

△ **1.** Find the area of the swimming pool.

△ **2.** Write an algebraic expression for the total area of the region containing both the pool and the patio.

△ **3.** Use subtraction to find an algebraic expression for the area of just the patio (not including the area of the pool).

△ **4.** Find the perimeter of the swimming pool alone.

5. For each patio material option, write an algebraic expression for the total cost of installing the patio based on its area and the given price information.

6. If you plan to spend the entire $3000 on the patio, how wide would the patio in option A be?

7. If you plan to spend the entire $3000 on the patio, how wide would the patio in option B be?

8. If you plan to spend the entire $3000 on the patio, how wide would the patio in option C be?

9. Which option would you choose? Why? Discuss the pros and cons of each option.

CHAPTER 6 VOCABULARY CHECK

Fill in each blank with one of the words or phrases listed below. Not all choices will be used.

factoring	quadratic equation	perfect square trinomial
greatest common factor	0	sum of two cubes
difference of two cubes	difference of two squares	1

1. An equation that can be written in the form $ax^2 + bx + c = 0$ (with a not 0) is called a _____.
2. _____ is the process of writing an expression as a product.
3. The _____ of a list of terms is the product of all common factors.
4. A trinomial that is the square of some binomial is called a _____.
5. The expression $a^2 - b^2$ is called a(n) _____.
6. The expression $a^3 - b^3$ is called a(n) _____.
7. The expression $a^3 + b^3$ is called a(n) _____.
8. By the zero factor property, if the product of two numbers is 0, then at least one of the numbers must be ____.

> **Helpful Hint**
> Are you preparing for your test? Don't forget to take the Chapter 6 Test on page 426. Then check your answers at the back of the text and use the Chapter Test Prep Video CD to see the fully worked-out solutions to any of the exercises you want to review.

CHAPTER 6 HIGHLIGHTS

DEFINITIONS AND CONCEPTS	EXAMPLES
SECTION 6.1 THE GREATEST COMMON FACTOR AND FACTORING BY GROUPING	
Factoring is the process of writing an expression as a product.	Factor: $6 = 2 \cdot 3$ $$x^2 + 5x + 6 = (x + 2)(x + 3)$$
To Find the GCF of a List of Integers **Step 1.** Write each number as a product of primes. **Step 2.** Identify the common prime factors. **Step 3.** The product of all common factors is the greatest common factor. If there are no common prime factors, the GCF is 1.	Find the GCF of 12, 36, and 48. $12 = 2 \cdot 2 \cdot 3$ $36 = 2 \cdot 2 \cdot 3 \cdot 3$ $48 = 2 \cdot 2 \cdot 2 \cdot 2 \cdot 3$ $\text{GCF} = 2 \cdot 2 \cdot 3 = 12$
The GCF of a list of common variables raised to powers is the variable raised to the smallest exponent in the list.	The GCF of z^5, z^3, and z^{10} is z^3.
The GCF of a list of terms is the product of all common factors.	Find the GCF of $8x^2y$, $10x^3y^2$, and $26x^2y^3$. The GCF of 8, 10, and 26 is 2. The GCF of x^2, x^3, and x^2 is x^2. The GCF of y, y^2, and y^3 is y. The GCF of the terms is $2x^2y$.

DEFINITIONS AND CONCEPTS	EXAMPLES
SECTION 6.1 THE GREATEST COMMON FACTOR AND FACTORING BY GROUPING (continued)	

To Factor by Grouping	Factor $10ax + 15a - 6xy - 9y$.
Step 1. Arrange the terms so that the first two terms have a common factor and the last two have a common factor.	**Step 1.** $10ax + 15a - 6xy - 9y$
Step 2. For each pair of terms, factor out the pair's GCF.	**Step 2.** $5a(2x + 3) - 3y(2x + 3)$
Step 3. If there is now a common binomial factor, factor it out.	**Step 3.** $(2x + 3)(5a - 3y)$
Step 4. If there is no common binomial factor, begin again, rearranging the terms differently. If no rearrangement leads to a common binomial factor, the polynomial cannot be factored.	

SECTION 6.2 FACTORING TRINOMIALS OF THE FORM $x^2 + bx + c$	
To factor a trinomial of the form $x^2 + bx + c$, look for two numbers whose product is c and whose sum is b. The factored form is $(x + \text{one number})(x + \text{other number})$	Factor: $x^2 + 7x + 12$ $3 + 4 = 7 \qquad 3 \cdot 4 = 12$ $(x + 3)(x + 4)$

SECTION 6.3 FACTORING TRINOMIALS OF THE FORM $ax^2 + bx + c$	
To factor $ax^2 + bx + c$, try various combinations of factors of ax^2 and c until a middle term of bx is obtained when checking.	Factor: $3x^2 + 14x - 5$ Factors of $3x^2$: $3x, x$ Factors of -5: $-1, 5$ and $1, -5$. $(3x - 1)(x + 5)$ $-1x$ $15x$ $14x$ Correct middle term
A **perfect square trinomial** is a trinomial that is the square of some binomial.	Perfect square trinomial = square of binomial $x^2 + 4x + 4 = (x + 2)^2$ $25x^2 - 10x + 1 = (5x - 1)^2$
Factoring Perfect Square Trinomials: $a^2 + 2ab + b^2 = (a + b)^2$ $a^2 - 2ab + b^2 = (a - b)^2$	Factor: $x^2 + 6x + 9 = x^2 + 2 \cdot x \cdot 3 + 3^2 = (x + 3)^2$ $4x^2 - 12x + 9 = (2x)^2 - 2 \cdot 2x \cdot 3 + 3^2 = (2x - 3)^2$

SECTION 6.4 FACTORING TRINOMIALS OF THE FORM $ax^2 + bx + c$ BY GROUPING	
To Factor $ax^2 + bx + c$ by Grouping	Factor: $3x^2 + 14x - 5$
Step 1. Find two numbers whose product is $a \cdot c$ and whose sum is b.	**Step 1.** Find two numbers whose product is $3 \cdot (-5)$ or -15 and whose sum is 14. They are 15 and -1.
Step 2. Rewrite bx, using the factors found in Step 1.	**Step 2.** $3x^2 + 14x - 5$ $= 3x^2 + 15x - 1x - 5$
Step 3. Factor by grouping.	**Step 3.** $= 3x(x + 5) - 1(x + 5)$ $= (x + 5)(3x - 1)$

DEFINITIONS AND CONCEPTS	EXAMPLES
SECTION 6.5 FACTORING BINOMIALS	
Difference of Squares $$a^2 - b^2 = (a + b)(a - b)$$ **Sum or Difference of Cubes** $$a^3 + b^3 = (a + b)(a^2 - ab + b^2)$$ $$a^3 - b^3 = (a - b)(a^2 + ab + b^2)$$	Factor: $$x^2 - 9 = x^2 - 3^2 = (x + 3)(x - 3)$$ $$y^3 + 8 = y^3 + 2^3 = (y + 2)(y^2 - 2y + 4)$$ $$125z^3 - 1 = (5z)^3 - 1^3 = (5z - 1)(25z^2 + 5z + 1)$$
INTEGRATED REVIEW—CHOOSING A FACTORING STRATEGY	
To Factor a Polynomial, **Step 1.** Factor out the GCF. **Step 2.** **a.** If two terms, **i.** $a^2 - b^2 = (a + b)(a - b)$ **ii.** $a^3 - b^3 = (a - b)(a^2 + ab + b^2)$ **iii.** $a^3 + b^3 = (a + b)(a^2 - ab + b^2)$ **b.** If three terms, **i.** $a^2 + 2ab + b^2 = (a + b)^2$ **ii.** Methods in Sections 6.2 and 6.3 **c.** If four or more terms, try factoring by grouping. **Step 3.** See if any factors can be factored further. **Step 4.** Check by multiplying.	Factor: $2x^4 - 6x^2 - 8$ **Step 1.** $2x^4 - 6x^2 - 8 = 2(x^4 - 3x^2 - 4)$ **Step 2. b. ii.** $= 2(x^2 + 1)(x^2 - 4)$ **Step 3.** $= 2(x^2 + 1)(x + 2)(x - 2)$ **Step 4.** Check by multiplying. $$2(x^2 + 1)(x + 2)(x - 2) = 2(x^2 + 1)(x^2 - 4)$$ $$= 2(x^4 - 3x^2 - 4)$$ $$= 2x^4 - 6x^2 - 8$$
SECTION 6.6 SOLVING QUADRATIC EQUATIONS BY FACTORING	
A **quadratic equation** is an equation that can be written in the form $ax^2 + bx + c = 0$ with a not 0. The form $ax^2 + bx + c = 0$ is called the **standard form** of a quadratic equation.	**Quadratic Equation** **Standard Form** $x^2 = 16$ $x^2 - 16 = 0$ $y = -2y^2 + 5$ $2y^2 + y - 5 = 0$
Zero Factor Theorem If a and b are real numbers and if $ab = 0$, then $a = 0$ or $b = 0$.	If $(x + 3)(x - 1) = 0$, then $x + 3 = 0$ or $x - 1 = 0$
To solve quadratic equations by factoring, **Step 1.** Write the equation in standard form: $ax^2 + bx + c = 0$. **Step 2.** Factor the quadratic. **Step 3.** Set each factor containing a variable equal to 0. **Step 4.** Solve the equations. **Step 5.** Check in the original equation.	Solve: $3x^2 = 13x - 4$ **Step 1.** $3x^2 - 13x + 4 = 0$ **Step 2.** $(3x - 1)(x - 4) = 0$ **Step 3.** $3x - 1 = 0$ or $x - 4 = 0$ **Step 4.** $3x = 1$ or $x = 4$ $x = \dfrac{1}{3}$ **Step 5.** Check both $\dfrac{1}{3}$ and 4 in the original equation.

DEFINITIONS AND CONCEPTS	EXAMPLES
SECTION 6.7 QUADRATIC EQUATIONS AND PROBLEM SOLVING	

Problem-Solving Steps

A garden is in the shape of a rectangle whose length is two feet more than its width. If the area of the garden is 35 square feet, find its dimensions.

1. UNDERSTAND the problem.

1. Read and reread the problem. Guess a solution and check your guess.
   Let x be the width of the rectangular garden. Then $x + 2$ is the length.

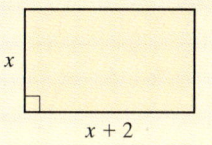

2. TRANSLATE.

2. In words: length · width = area
   Translate: $(x + 2) \cdot x = 35$

3. SOLVE.

3. $(x + 2)x = 35$
   $x^2 + 2x - 35 = 0$
   $(x - 5)(x + 7) = 0$
   $x - 5 = 0$ or $x + 7 = 0$
   $x = 5$ or $x = -7$

4. INTERPRET.

4. Discard the solution of -7 since x represents width.
   Check: If x is 5 feet then $x + 2 = 5 + 2 = 7$ feet. The area of a rectangle whose width is 5 feet and whose length is 7 feet is (5 feet)(7 feet) or 35 square feet.
   State: The garden is 5 feet by 7 feet.

STUDY SKILLS BUILDER

Are You Prepared for a Test on Chapter 6?

Below is a list of some *common trouble areas* for students in Chapter 6. After studying for your test—but before taking your test—read these.

- The difference of two squares such as $x^2 - 25$ factors as $x^2 - 25 = (x + 5)(x - 5)$.
- The sum of two squares, for example, $x^2 + 25$, cannot be factored using real numbers.
- Don't forget that the first step to factor any polynomial is to first factor out any common factors.
  $$9x^2 - 36 = 9(x^2 - 4) = 9(x + 2)(x - 2)$$
- Can you completely factor $x^4 - 24x^2 - 25$?
  $$x^4 - 24x^2 - 25 = (x^2 - 25)(x^2 + 1)$$
  $$= (x + 5)(x - 5)(x^2 + 1)$$

- Remember that to use the zero factor property to solve a quadratic equation, one side of the equation must be 0 and the other side must be a factored polynomial.

$x(x - 2) = 3$ Cannot use zero factor property.
$x^2 - 2x - 3 = 0$
$(x - 3)(x + 1) = 0$ Now we can use zero factor property.
$x - 3 = 0$ or $x + 1 = 0$
$x = 3$ or $x = -1$

Remember: This is simply a sampling of selected topics given to check your understanding. For a review of Chapter 6 in your text, see the material at the end of this chapter.

CHAPTER 6 REVIEW

(6.1) *Complete the factoring.*

1. $6x^2 - 15x = 3x(\quad)$
2. $2x^3y - 6x^2y^2 - 8xy^3 = 2xy(\quad)$

Factor the GCF from each polynomial.

3. $20x^2 + 12x$
4. $6x^2y^2 - 3xy^3$
5. $-8x^3y + 6x^2y^2$
6. $3x(2x + 3) - 5(2x + 3)$
7. $5x(x + 1) - (x + 1)$

Factor.

8. $3x^2 - 3x + 2x - 2$
9. $6x^2 + 10x - 3x - 5$
10. $3a^2 + 9ab + 3b^2 + ab$

(6.2) *Factor each trinomial.*

11. $x^2 + 6x + 8$
12. $x^2 - 11x + 24$
13. $x^2 + x + 2$
14. $x^2 - 5x - 6$
15. $x^2 + 2x - 8$
16. $x^2 + 4xy - 12y^2$
17. $x^2 + 8xy + 15y^2$
18. $3x^2y + 6xy^2 + 3y^3$
19. $72 - 18x - 2x^2$
20. $32 + 12x - 4x^2$

(6.3) or **(6.4)** *Factor each trinomial.*

21. $2x^2 + 11x - 6$
22. $4x^2 - 7x + 4$
23. $4x^2 + 4x - 3$
24. $6x^2 + 5xy - 4y^2$
25. $6x^2 - 25xy + 4y^2$
26. $18x^2 - 60x + 50$
27. $2x^2 - 23xy - 39y^2$
28. $4x^2 - 28xy + 49y^2$
29. $18x^2 - 9xy - 20y^2$
30. $36x^3y + 24x^2y^2 - 45xy^3$

(6.5) *Factor each binomial.*

31. $4x^2 - 9$
32. $9t^2 - 25s^2$
33. $16x^2 + y^2$
34. $x^3 - 8y^3$
35. $8x^3 + 27$
36. $2x^3 + 8x$
37. $54 - 2x^3y^3$
38. $9x^2 - 4y^2$
39. $16x^4 - 1$
40. $x^4 + 16$

(6.6) *Solve the following equations.*

41. $(x + 6)(x - 2) = 0$
42. $3x(x + 1)(7x - 2) = 0$
43. $4(5x + 1)(x + 3) = 0$
44. $x^2 + 8x + 7 = 0$
45. $x^2 - 2x - 24 = 0$
46. $x^2 + 10x = -25$
47. $x(x - 10) = -16$
48. $(3x - 1)(9x^2 + 3x + 1) = 0$
49. $56x^2 - 5x - 6 = 0$
50. $20x^2 - 7x - 6 = 0$
51. $5(3x + 2) = 4$
52. $6x^2 - 3x + 8 = 0$
53. $12 - 5t = -3$
54. $5x^3 + 20x^2 + 20x = 0$
55. $4t^3 - 5t^2 - 21t = 0$

56. Write a quadratic equation that has the two solutions 4 and 5.

(6.7) *Use the given information to choose the correct dimensions.*

△ 57. The perimeter of a rectangle is 24 inches. The length is twice the width. Find the dimensions of the rectangle.
 a. 5 inches by 7 inches
 b. 5 inches by 10 inches
 c. 4 inches by 8 inches
 d. 2 inches by 10 inches

△ 58. The area of a rectangle is 80 meters. The length is one more than three times the width. Find the dimensions of the rectangle.
 a. 8 meters by 10 meters
 b. 4 meters by 13 meters
 c. 4 meters by 20 meters
 d. 5 meters by 16 meters

Use the given information to find the dimensions of each figure.

△ 59. The *area* of the square is 81 square units. Find the length of a side.

△ 60. The *perimeter* of the quadrilateral is 47 units. Find the lengths of the sides.

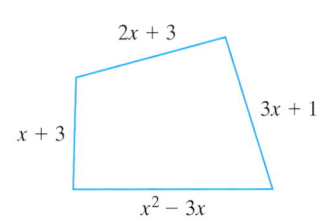

△ 61. A flag for a local organization is in the shape of a rectangle whose length is 15 inches less than twice its width. If the area of the flag is 500 square inches, find its dimensions.

△ 62. The base of a triangular sail is four times its height. If the area of the triangle is 162 square yards, find the base.

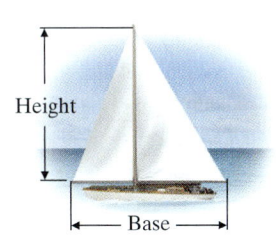

63. Find two consecutive positive integers whose product is 380.

64. A rocket is fired from the ground with an initial velocity of 440 feet per second. Its height h after t seconds is given by the equation
$$h = -16t^2 + 440t$$

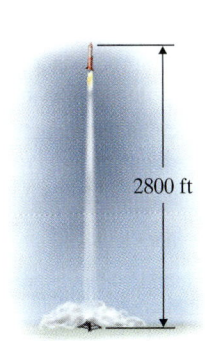

a. Find how many seconds pass before the rocket reaches a height of 2800 feet. Explain why two answers are obtained.

b. Find how many seconds pass before the rocket reaches the ground again.

65. An object is dropped from the top of the 625-foot-tall Waldorf-Astoria Hotel on Park Avenue in New York City. (*Source: World Almanac* research) The height h of the object after t seconds is given by the equation $h = -16t^2 + 625$. Find how many seconds pass before the object reaches the ground.

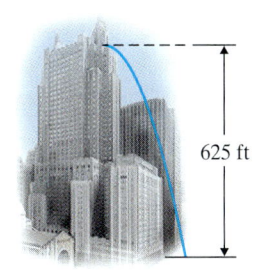

△ 66. An architect's squaring instrument is in the shape of a right triangle. Find the length of the long leg of the right triangle if the hypotenuse is 8 centimeters longer than the long leg and the short leg is 8 centimeters shorter than the long leg.

MIXED REVIEW

Factor completely.

67. $7x - 63$

68. $11x(4x - 3) - 6(4x - 3)$

69. $m^2 - \dfrac{4}{25}$

70. $3x^3 - 4x^2 + 6x - 8$

71. $xy + 2x - y - 2$

72. $2x^2 + 2x - 24$

73. $3x^3 - 30x^2 + 27x$

74. $4x^2 - 81$

75. $2x^2 - 18$

76. $16x^2 - 24x + 9$

77. $5x^2 + 20x + 20$

78. $2x^2 + 5x - 12$

79. $4x^2y - 6xy^2$

80. $8x^2 - 15x - x^3$

81. $125x^3 + 27$

82. $24x^2 - 3x - 18$

83. $(x + 7)^2 - y^2$

84. $x^2(x + 3) - 4(x + 3)$

85. $54a^3b - 2b$

86. To factor $x^2 + 2x - 48$, think of two numbers whose product is _____ and whose sum is _____.

87. What is the first step to factoring $3x^2 + 15x + 30$?

Write the perimeter of each figure as a simplified polynomial. Then factor each polynomial.

△ 88.

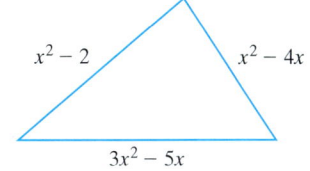

△ 89. [rectangle with width $2x^2 + 3$ and length $6x^2 - 14x$]

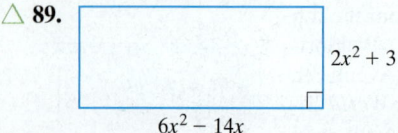

Solve.

90. $2x^2 - x - 28 = 0$
91. $x^2 - 2x = 15$
92. $2x(x + 7)(x + 4) = 0$
93. $x(x - 5) = -6$
94. $x^2 = 16x$

Solve.

95. The perimeter of the following triangle is 48 inches. Find the lengths of its sides. [triangle with sides $x^2 + 3$, $4x + 5$, $2x$]

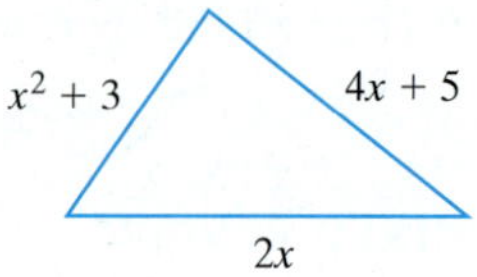

96. The width of a rectangle is 4 inches less than its length. Its area is 12 square inches. Find the dimensions of the rectangle.

97. A 6-foot-tall person drops an object from the top of the Westin Peachtree Plaza in Atlanta, Georgia. The Westin building is 723 feet tall. (*Source: World Almanac* research) The height h of the object after t seconds is given by the equation $h = -16t^2 + 729$. Find how many seconds pass before the object reaches the ground.

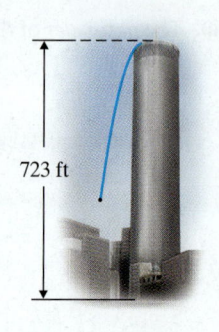

Write an expression for the area of the shaded region. Then write the expression as a factored polynomial.

△ 98. [figure showing shaded region with dimensions $6x$, $4x$, x, and a circle of radius x]

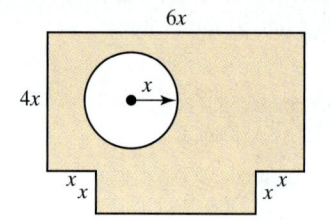

CHAPTER 6 TEST

Remember to use the Chapter Test Prep Video CD to see the fully worked-out solutions to any of the exercises you want to review.

Factor each polynomial completely. If a polynomial cannot be factored, write "prime."

1. $x^2 + 11x + 28$
2. $49 - m^2$
3. $y^2 + 22y + 121$
4. $4(a + 3) - y(a + 3)$
5. $x^2 + 4$
6. $y^2 - 8y - 48$
7. $x^2 + x - 10$
8. $9x^3 + 39x^2 + 12x$
9. $3a^2 + 3ab - 7a - 7b$
10. $3x^2 - 5x + 2$
11. $x^2 + 14xy + 24y^2$
12. $180 - 5x^2$
13. $6t^2 - t - 5$
14. $xy^2 - 7y^2 - 4x + 28$
15. $x - x^5$
16. $-xy^3 - x^3y$
17. $64x^3 - 1$
18. $8y^3 - 64$

Solve each equation.

19. $(x - 3)(x + 9) = 0$
20. $x^2 + 5x = 14$
21. $x(x + 6) = 7$
22. $3x(2x - 3)(3x + 4) = 0$
23. $5t^3 - 45t = 0$
24. $t^2 - 2t - 15 = 0$
25. $6x^2 = 15x$

Solve each problem.

△ 26. A deck for a home is in the shape of a triangle. The length of the base of the triangle is 9 feet longer than its altitude. If the area of the triangle is 68 square feet, find the length of the base.

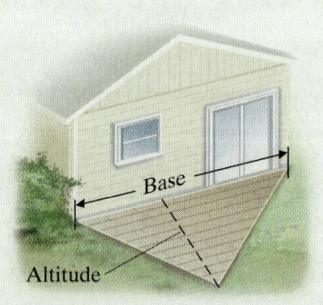

27. The sum of two numbers is 17 and the sum of their squares is 145. Find the numbers.

28. An object is dropped from the top of the Woolworth Building on Broadway in New York City. The height h of the object after t seconds is given by the equation

$$h = -16t^2 + 784$$

Find how many seconds pass before the object reaches the ground.

△ 29. Find the lengths of the sides of a right triangle if the hypotenuse is 10 centimeters longer than the shorter leg and 5 centimeters longer than the longer leg.

CHAPTER 6 CUMULATIVE REVIEW

1. Translate each sentence into a mathematical statement.
   a. Nine is less than or equal to eleven.
   b. Eight is greater than one.
   c. Three is not equal to four.
2. Insert < or > in the space to make each statement true.
   a. $|-5|$ ___ $|-3|$
   b. $|0|$ ___ $|-2|$
3. Write each fraction in lowest terms.
   a. $\frac{42}{49}$ b. $\frac{11}{27}$ c. $\frac{88}{20}$
4. Evaluate $\frac{x}{y} + 5x$ if $x = 20$ and $y = 10$.
5. Simplify: $\frac{8 + 2 \cdot 3}{2^2 - 1}$
6. Evaluate $\frac{x}{y} + 5x$ if $x = -20$ and $y = 10$.
7. Add.
   a. $3 + (-7) + (-8)$
   b. $[7 + (-10)] + [-2 + |-4|]$
8. Evaluate $\frac{x}{y} + 5x$ if $x = -20$ and $y = -10$.
9. Multiply.
   a. $(-6)(4)$ b. $2(-1)$
   c. $(-5)(-10)$
10. Simplify: $5 - 2(3x - 7)$
11. Simplify each expression by combining like terms.
    a. $7x - 3x$ b. $10y^2 + y^2$
    c. $8x^2 + 2x - 3x$
12. Solve: $0.8y + 0.2(y - 1) = 1.8$

Solve.

13. $3 - x = 7$
14. $\frac{x}{-7} = -4$
15. $-3x = 33$
16. $-\frac{2}{3}x = -22$
17. $8(2 - t) = -5t$
18. $-z = \frac{7z + 3}{5}$
19. Balsa wood sticks are commonly used to build models (for example, bridge models). A 48-inch Balsa wood stick is to be cut into two pieces so that the longer piece is 3 times the shorter. Find the length of each piece.
20. Solve $3x + 9 \leq 5(x - 1)$. Write the solution set using interval notation.
21. Graph the linear equation $y = -\frac{1}{3}x + 2$.
22. Is the ordered pair $(-1, 2)$ a solution of $-7x - 8y = -9$?
23. Find the slope and y-intercept of the line whose equation is $3x - 4y = 4$.
24. Find the slope of the line through $(5, -6)$ and $(5, 2)$.
25. Evaluate each expression for the given value of x.
    a. $2x^3$; x is 5   b. $\frac{9}{x^2}$; x is -3
26. Find the slope and y-intercept of the line whose equation is $7x - 3y = 2$.
27. Find the degree of each term.
    a. $-3x^2$ b. $5x^3yz$ c. 2
28. Find an equation of the vertical line through $(0, 7)$.
29. Subtract: $(2x^3 + 8x^2 - 6x) - (2x^3 - x^2 + 1)$
30. Find an equation of the line with slope 4 and y-intercept $\left(0, \frac{1}{2}\right)$. Write the equation in standard form.
31. Multiply $(3x + 2)(2x - 5)$.
32. Write an equation of the line through $(-4, 0)$ and $(6, -1)$. Write the equation in standard form.
33. Multiply $(3y + 1)^2$.
34. Solve the system: $\begin{cases} -x + 3y = 18 \\ -3x + 2y = 19 \end{cases}$
35. Simplify by writing each expression with positive exponents only.
    a. 3^{-2} b. $2x^{-3}$
    c. $2^{-1} + 4^{-1}$ d. $(-2)^{-4}$
    e. $\frac{1}{y^{-4}}$ f. $\frac{1}{7^{-2}}$
36. Simplify: $\frac{(5a^7)^2}{a^5}$
37. Write each number in scientific notation.
    a. 367,000,000   b. 0.000003
    c. 20,520,000,000
    d. 0.00085
38. Multiply: $(3x - 7y)^2$
39. Divide $x^2 + 7x + 12$ by $x + 3$ using long division.
40. Simplify: $\frac{(xy)^{-3}}{(x^5y^6)^3}$
41. Find the GCF of each list of terms.
    a. $x^3, x^7,$ and x^5 b. $y, y^4,$ and y^7

Factor.

42. $z^3 + 7z + z^2 + 7$
43. $x^2 + 7x + 12$
44. $2x^3 + 2x^2 - 84x$
45. $8x^2 - 22x + 5$
46. $-4x^2 - 23x + 6$
47. $25a^2 - 9b^2$
48. $9xy^2 - 16x$
49. Solve $(x - 3)(x + 1) = 0$.
50. Solve $x^2 - 13x = -36$.

CHAPTER

7 Rational Expressions

- 7.1 Simplifying Rational Expressions
- 7.2 Multiplying and Dividing Rational Expressions
- 7.3 Adding and Subtracting Rational Expressions with Common Denominators and Least Common Denominator
- 7.4 Adding and Subtracting Rational Expressions with Unlike Denominators
- 7.5 Solving Equations Containing Rational Expressions

  Integrated Review—Summary on Rational Expressions

- 7.6 Proportion and Problem Solving with Rational Equations
- 7.7 Variation and Problem Solving
- 7.8 Simplifying Complex Fractions

In this chapter, we expand our knowledge of algebraic expressions to include another category called rational expressions, such as $\frac{x+1}{x}$. We explore the operations of addition, subtraction, multiplication, and division for these algebraic fractions, using principles similar to the principles for number fractions.

Cephalic index is the ratio of the maximum width of the head to its maximum length, sometimes multiplied by 100. It is used by anthropologists and forensic scientists on human skulls, but it is used especially on animal skulls to categorize animals such as dogs and cats. In Section 7.1, Exercise 95, you will have the opportunity to calculate this index for a human skull.

Cephalic Index Formula: $C = \dfrac{100W}{L}$

where W is width of the skull and L is length of the skull.

Cephalic Index for Dogs		
Value	Scientific Term	Meaning
<80 or <75	*dolichocephalic*	"long-headed"
	mesocephalic	"medium-headed"
>80	*brachycephalic*	"short-headed"

A **brachycephalic** skull is relatively broad and short, as in the Pug.
A **mesocephalic** skull is of intermediate length and width, as in the Cocker Spaniel.
A **dolichocephalic** skull is relatively long, as in the Afghan Hound.

Section 7.1 Simplifying Rational Expressions 429

7.1 SIMPLIFYING RATIONAL EXPRESSIONS

OBJECTIVES

1. Find the value of a rational expression given a replacement number.
2. Identify values for which a rational expression is undefined.
3. Simplify or write rational expressions in lowest terms.
4. Write equivalent rational expressions of the form $-\dfrac{a}{b} = \dfrac{-a}{b} = \dfrac{a}{-b}$.

OBJECTIVE 1 ▶ **Evaluating rational expressions.** As we reviewed in Chapter 1, a rational number is a number that can be written as a quotient of integers. A **rational expression** is also a quotient; it is a quotient of polynomials.

Rational Expression
A rational expression is an expression that can be written in the form $\dfrac{P}{Q}$, where P and Q are polynomials and $Q \neq 0$.

Rational Expressions

$$\dfrac{2}{3} \qquad \dfrac{3y^3}{8} \qquad \dfrac{-4p}{p^3 + 2p + 1} \qquad \dfrac{5x^2 - 3x + 2}{3x + 7}$$

Rational expressions have different values depending on what value replaces the variable. Next, we review the standard order of operations by finding values of rational expressions for given replacement values of the variable.

EXAMPLE 1 Find the value of $\dfrac{x + 4}{2x - 3}$ for the given replacement values.

a. $x = 5$ **b.** $x = -2$

Solution

a. Replace each x in the expression with 5 and then simplify.

$$\dfrac{x + 4}{2x - 3} = \dfrac{5 + 4}{2(5) - 3} = \dfrac{9}{10 - 3} = \dfrac{9}{7}$$

b. Replace each x in the expression with -2 and then simplify.

$$\dfrac{x + 4}{2x - 3} = \dfrac{-2 + 4}{2(-2) - 3} = \dfrac{2}{-7} \text{ or } -\dfrac{2}{7}$$

PRACTICE 1 Find the value of $\dfrac{x + 6}{3x - 2}$ for the given replacement values.

a. $x = 3$ **b.** $x = -3$

In the example above, we wrote $\dfrac{2}{-7}$ as $-\dfrac{2}{7}$. For a negative fraction such as $\dfrac{2}{-7}$, recall from Section 1.7 that

$$\dfrac{2}{-7} = \dfrac{-2}{7} = -\dfrac{2}{7}$$

In general, for any fraction,

$$\dfrac{-a}{b} = \dfrac{a}{-b} = -\dfrac{a}{b}, \qquad b \neq 0$$

This is also true for rational expressions. For example,

$$\dfrac{\underbrace{-(x + 2)}_{\uparrow}}{x} = \dfrac{x + 2}{-x} = -\dfrac{x + 2}{x}$$

Notice the parentheses.

▶ **Helpful Hint**

Do you recall why division by 0 is not defined? Remember, for example, that $\frac{8}{4} = 2$ because $2 \cdot 4 = 8$. Thus, if $\frac{8}{0} = $ *a number*, then *the number* $\cdot 0 = 8$. There is no number that when multiplied by 0 equals 8; thus $\frac{8}{0}$ is undefined. This is true in general for fractions and rational expressions.

OBJECTIVE 2 ▶ **Identifying when a rational expression is undefined.** In the definition of rational expression (first "box" in this section), notice that we wrote $Q \neq 0$ for the denominator Q. This is because the denominator of a rational expression must not equal 0 since division by 0 is not defined. (See the margin Helpful Hint.) This means we must be careful when replacing the variable in a rational expression by a number. For example, suppose we replace x with 5 in the rational expression $\frac{3+x}{x-5}$. The expression becomes

$$\frac{3+x}{x-5} = \frac{3+5}{5-5} = \frac{8}{0}$$

But division by 0 is undefined. Therefore, in this rational expression we can allow x to be any real number *except* 5. **A rational expression is undefined for values that make the denominator 0.** Thus, to find values for which a rational expression is undefined, find values for which the denominator is 0.

EXAMPLE 2 Are there any values for x for which each rational expression is undefined?

a. $\dfrac{x}{x-3}$ b. $\dfrac{x^2+2}{3x^2-5x+2}$ c. $\dfrac{x^3-6x^2-10x}{3}$ d. $\dfrac{2}{x^2+1}$

Solution To find values for which a rational expression is undefined, find values that make the *denominator* 0.

a. The denominator of $\dfrac{x}{x-3}$ is 0 when $x - 3 = 0$ or when $x = 3$. Thus, when $x = 3$, the expression $\dfrac{x}{x-3}$ is undefined.

b. Set the denominator equal to zero.

$$3x^2 - 5x + 2 = 0$$
$$(3x - 2)(x - 1) = 0 \quad \text{Factor.}$$
$$3x - 2 = 0 \quad \text{or} \quad x - 1 = 0 \quad \text{Set each factor equal to zero.}$$
$$3x = 2 \quad \text{or} \quad x = 1 \quad \text{Solve.}$$
$$x = \frac{2}{3}$$

Thus, when $x = \dfrac{2}{3}$ or $x = 1$, the denominator $3x^2 - 5x + 2$ is 0. So the rational expression $\dfrac{x^2+2}{3x^2-5x+2}$ is undefined when $x = \dfrac{2}{3}$ or when $x = 1$.

c. The denominator of $\dfrac{x^3-6x^2-10x}{3}$ is never 0, so there are no values of x for which this expression is undefined.

d. No matter which real number x is replaced by, the denominator $x^2 + 1$ does not equal 0, so there are no real numbers for which this expression is undefined. ☐

PRACTICE 2 Are there any values of x for which each rational expression is undefined?

a. $\dfrac{x}{x+6}$ b. $\dfrac{x^4-3x^2+7x}{7}$ c. $\dfrac{x^2-5}{x^2+6x+8}$ d. $\dfrac{3}{x^4+5}$

Note: Unless otherwise stated, we will now assume that variables in rational expressions are only replaced by values for which the expressions are defined.

OBJECTIVE 3 ▶ **Simplifying rational expressions.** A fraction is said to be written in lowest terms or simplest form when the numerator and denominator have no common factors other than 1 (or -1). For example, the fraction $\dfrac{7}{10}$ is in lowest terms since the numerator and denominator have no common factors other than 1 (or -1).

The process of writing a rational expression in lowest terms or simplest form is called **simplifying** a rational expression.

Section 7.1 Simplifying Rational Expressions 431

Simplifying a rational expression is similar to simplifying a fraction. Recall that to simplify a fraction, we essentially "remove factors of 1." Our ability to do this comes from these facts:

- Any nonzero number over itself simplifies to 1 $\left(\dfrac{5}{5} = 1, \dfrac{-7.26}{-7.26} = 1, \text{ or } \dfrac{c}{c} = 1 \text{ as long as } c \text{ is not } 0\right)$, and

- The product of any number and 1 is that number $\left(19 \cdot 1 = 19, -8.9 \cdot 1 = -8.9, \dfrac{a}{b} \cdot 1 = \dfrac{a}{b}\right)$.

In other words, we have the following:

Simplify: $\dfrac{15}{20}$

$\dfrac{15}{20} = \dfrac{3 \cdot 5}{2 \cdot 2 \cdot 5}$ Factor the numerator and the denominator.

$= \dfrac{3 \cdot 5}{2 \cdot 2 \cdot 5}$ Look for common factors.

$= \dfrac{3}{2 \cdot 2} \cdot \dfrac{5}{5}$ Common factors in the numerator and denominator form factors of 1.

$= \dfrac{3}{2 \cdot 2} \cdot 1$ Write $\dfrac{5}{5}$ as 1.

$= \dfrac{3}{2 \cdot 2} = \dfrac{3}{4}$ Multiply to remove a factor of 1.

$\dfrac{a \cdot c}{b \cdot c} = \dfrac{a}{b} \cdot \dfrac{c}{c} = \dfrac{a}{b}$

Since $\dfrac{a}{b} \cdot 1 = \dfrac{a}{b}$

Before we use the same technique to simplify a rational expression, remember that as long as the denominator is not 0, $\dfrac{a^3 b}{a^3 b} = 1, \dfrac{x+3}{x+3} = 1,$ and $\dfrac{7x^2 + 5x - 100}{7x^2 + 5x - 100} = 1.$

Simplify: $\dfrac{x^2 - 9}{x^2 + x - 6}$

$\dfrac{x^2 - 9}{x^2 + x - 6} = \dfrac{(x-3)(x+3)}{(x-2)(x+3)}$ Factor the numerator and the denominator.

$= \dfrac{(x-3)\,(x+3)}{(x-2)\,(x+3)}$ Look for common factors.

$= \dfrac{x-3}{x-2} \cdot \dfrac{x+3}{x+3}$

$= \dfrac{x-3}{x-2} \cdot 1$ Write $\dfrac{x+3}{x+3}$ as 1.

$= \dfrac{x-3}{x-2}$ Multiply to remove a factor of 1.

Just as for numerical fractions, we can use a shortcut notation. Remember that as long as exact factors in both the numerator and denominator are divided out, we are "removing a factor of 1." We will use the following notation to show this:

$\dfrac{x^2 - 9}{x^2 + x - 6} = \dfrac{(x-3)\,(x+3)}{(x-2)\,(x+3)}$ A factor of 1 is identified by the shading.

$= \dfrac{x-3}{x-2}$ Remove a factor of 1.

Thus, the rational expression $\dfrac{x^2 - 9}{x^2 + x - 6}$ has the same value as the rational expression $\dfrac{x-3}{x-2}$ for all values of x except 2 and -3. (Remember that when x is 2, the denominator of both rational expressions is 0 and when x is -3, the original rational expression has a denominator of 0.)

As we simplify rational expressions, we will assume that the simplified rational expression is equal to the original rational expression for all real numbers except those

432 CHAPTER 7 Rational Expressions

for which either denominator is 0. The following steps may be used to simplify rational expressions.

> **To Simplify a Rational Expression**
> **STEP 1.** Completely factor the numerator and denominator.
> **STEP 2.** Divide out factors common to the numerator and denominator. (This is the same as "removing a factor of 1.")

EXAMPLE 3 Simplify: $\dfrac{5x - 5}{x^3 - x^2}$

Solution To begin, we factor the numerator and denominator if possible. Then we look for common factors.

$$\dfrac{5x - 5}{x^3 - x^2} = \dfrac{5\,(x - 1)}{x^2\,(x - 1)} = \dfrac{5}{x^2}$$

PRACTICE 3 Simplify: $\dfrac{x^6 - x^5}{6x - 6}$

EXAMPLE 4 Simplify: $\dfrac{x^2 + 8x + 7}{x^2 - 4x - 5}$

Solution We factor the numerator and denominator and then look for common factors.

$$\dfrac{x^2 + 8x + 7}{x^2 - 4x - 5} = \dfrac{(x + 7)\,(x + 1)}{(x - 5)\,(x + 1)} = \dfrac{x + 7}{x - 5}$$

PRACTICE 4 Simplify: $\dfrac{x^2 + 5x + 4}{x^2 + 2x - 8}$

EXAMPLE 5 Simplify: $\dfrac{x^2 + 4x + 4}{x^2 + 2x}$

Solution We factor the numerator and denominator and then look for common factors.

$$\dfrac{x^2 + 4x + 4}{x^2 + 2x} = \dfrac{(x + 2)\,(x + 2)}{x\,(x + 2)} = \dfrac{x + 2}{x}$$

PRACTICE 5 Simplify: $\dfrac{x^3 + 9x^2}{x^2 + 18x + 81}$

> **▶ Helpful Hint**
> When simplifying a rational expression, we look for **common *factors*, not common *terms*.**
>
> $\dfrac{x \cdot (x + 2)}{x \cdot x} = \dfrac{x + 2}{x}$ $\quad\Big|\quad$ $\dfrac{x + 2}{x}$
>
> Common factors. These can be divided out. $\quad\Big|\quad$ Common terms. There is no factor of 1 that can be generated.

Concept Check ✓

Recall that we can only remove *factors* of 1. Which of the following are *not* true? Explain why.

a. $\dfrac{3 - 1}{3 + 5}$ simplifies to $-\dfrac{1}{5}$?

b. $\dfrac{2x + 10}{2}$ simplifies to $x + 5$?

c. $\dfrac{37}{72}$ simplifies to $\dfrac{3}{2}$?

d. $\dfrac{2x + 3}{2}$ simplifies to $x + 3$?

Answers to Concept Check:
a, c, d

EXAMPLE 6 Simplify: $\dfrac{x+9}{x^2-81}$

Solution We factor and then divide out common factors.

$$\dfrac{x+9}{x^2-81} = \dfrac{x+9}{(x+9)(x-9)} = \dfrac{1}{x-9}$$

PRACTICE 6 Simplify: $\dfrac{x-7}{x^2-49}$

EXAMPLE 7 Simplify each rational expression.

a. $\dfrac{x+y}{y+x}$ **b.** $\dfrac{x-y}{y-x}$

Solution

a. The expression $\dfrac{x+y}{y+x}$ can be simplified by using the commutative property of addition to rewrite the denominator $y+x$ as $x+y$.

$$\dfrac{x+y}{y+x} = \dfrac{x+y}{x+y} = 1$$

b. The expression $\dfrac{x-y}{y-x}$ can be simplified by recognizing that $y-x$ and $x-y$ are opposites. In other words, $y-x = -1(x-y)$. We proceed as follows:

$$\dfrac{x-y}{y-x} = \dfrac{1 \cdot (x-y)}{-1 \cdot (x-y)} = \dfrac{1}{-1} = -1$$

PRACTICE 7 Simplify each rational expression.

a. $\dfrac{s-t}{t-s}$ **b.** $\dfrac{2c+d}{d+2c}$

EXAMPLE 8 Simplify: $\dfrac{4-x^2}{3x^2-5x-2}$

Solution

$$\dfrac{4-x^2}{3x^2-5x-2} = \dfrac{(2-x)(2+x)}{(x-2)(3x+1)} \quad \text{Factor.}$$

$$= \dfrac{(-1)(x-2)(2+x)}{(x-2)(3x+1)} \quad \text{Write } 2-x \text{ as } -1(x-2).$$

$$= \dfrac{(-1)(2+x)}{3x+1} \text{ or } \dfrac{-2-x}{3x+1} \quad \text{Simplify.}$$

PRACTICE 8 Simplify: $\dfrac{2x^2-5x-12}{16-x^2}$

OBJECTIVE 4 ▶ **Writing equivalent forms of rational expressions.** From Example 7a, we have $y+x = x+y$. $\quad y+x$ and $x+y$ are equivalent.

From Example 7b, we have $y-x = -1(x-y)$. $y-x$ and $x-y$ are opposites.

Thus, $\dfrac{x+y}{y+x} = \dfrac{x+y}{x+y} = 1$ and $\dfrac{x-y}{y-x} = \dfrac{x-y}{-1(x-y)} = \dfrac{1}{-1} = -1$.

When performing operations on rational expressions, equivalent forms of answers often result. For this reason, it is very important to be able to recognize equivalent answers.

EXAMPLE 9 List some equivalent forms of $-\dfrac{5x-1}{x+9}$.

Solution To do so, recall that $-\dfrac{a}{b} = \dfrac{-a}{b} = \dfrac{a}{-b}$. Thus

$$-\dfrac{5x-1}{x+9} = \dfrac{-(5x-1)}{x+9} = \dfrac{-5x+1}{x+9} \text{ or } \dfrac{1-5x}{x+9}$$

Also,

$$-\dfrac{5x-1}{x+9} = \dfrac{5x-1}{-(x+9)} = \dfrac{5x-1}{-x-9} \text{ or } \dfrac{5x-1}{-9-x}$$

Thus $-\dfrac{5x-1}{x+9} = \dfrac{-(5x-1)}{x+9} = \dfrac{-5x+1}{x+9} = \dfrac{5x-1}{-(x+9)} = \dfrac{5x-1}{-x-9}$

> **Helpful Hint**
> Remember, a negative sign in front of a fraction or rational expression may be moved to the numerator or the denominator, but *not* both.

PRACTICE 9 List some equivalent forms of $-\dfrac{x+3}{6x-11}$.

Keep in mind that many rational expressions may look different, but in fact be equivalent.

VOCABULARY & READINESS CHECK

Use the choices below to fill in each blank. Not all choices will be used.

$-1 \quad\quad 0 \quad\quad$ simplifying $\quad\quad \dfrac{-a}{-b} \quad\quad \dfrac{-a}{b} \quad\quad \dfrac{a}{-b}$

$1 \quad\quad 2 \quad\quad$ rational expression

1. A _____ is an expression that can be written in the form $\dfrac{P}{Q}$ where P and Q are polynomials and $Q \neq 0$.
2. The expression $\dfrac{x+3}{3+x}$ simplifies to _____.
3. The expression $\dfrac{x-3}{3-x}$ simplifies to _____.
4. A rational expression is undefined for values that make the denominator _____.
5. The expression $\dfrac{7x}{x-2}$ is undefined for $x =$ _____.
6. The process of writing a rational expression in lowest terms is called _____.
7. For a rational expression, $-\dfrac{a}{b} =$ _____ = _____.

Decide which rational expression can be simplified. (Do not actually simplify.)

8. $\dfrac{x}{x+7}$
9. $\dfrac{3+x}{x+3}$
10. $\dfrac{5-x}{x-5}$
11. $\dfrac{x+2}{x+8}$

7.1 EXERCISE SET

Find the value of the following expressions when $x = 2$, $y = -2$, and $z = -5$. See Example 1.

1. $\dfrac{x+5}{x+2}$
2. $\dfrac{x+8}{x+1}$
3. $\dfrac{4z-1}{z-2}$
4. $\dfrac{7y-1}{y-1}$
5. $\dfrac{y^3}{y^2-1}$
6. $\dfrac{z}{z^2-5}$

7. $\dfrac{x^2+8x+2}{x^2-x-6}$
8. $\dfrac{x+5}{x^2+4x-8}$

Find any numbers for which each rational expression is undefined. See Example 2.

9. $\dfrac{7}{2x}$
10. $\dfrac{3}{5x}$
11. $\dfrac{x+3}{x+2}$
12. $\dfrac{5x+1}{x-9}$

13. $\dfrac{x-4}{2x-5}$

14. $\dfrac{x+1}{5x-2}$

15. $\dfrac{x^2-5x-2}{4}$

16. $\dfrac{9y^5+y^3}{9}$

17. $\dfrac{3x^2+9}{x^2-5x-6}$

18. $\dfrac{11x^2+1}{x^2-5x-14}$

19. $\dfrac{9x^3+4}{x^2+36}$

20. $\dfrac{19x^3+2}{x^2+4}$

21. $\dfrac{x}{3x^2+13x+14}$

22. $\dfrac{x}{2x^2+15x+27}$

Study Example 9. Then list four equivalent forms for each rational expression.

23. $-\dfrac{x-10}{x+8}$

24. $-\dfrac{x+11}{x-4}$

25. $-\dfrac{5y-3}{y-12}$

26. $-\dfrac{8y-1}{y-15}$

Simplify each expression. See Examples 3 through 8.

27. $\dfrac{x+7}{7+x}$

28. $\dfrac{y+9}{9+y}$

29. $\dfrac{x-7}{7-x}$

30. $\dfrac{y-9}{9-y}$

31. $\dfrac{2}{8x+16}$

32. $\dfrac{3}{9x+6}$

33. $\dfrac{x-2}{x^2-4}$

34. $\dfrac{x+5}{x^2-25}$

35. $\dfrac{2x-10}{3x-30}$

36. $\dfrac{3x-9}{4x-16}$

37. $\dfrac{-5a-5b}{a+b}$

38. $\dfrac{-4x-4y}{x+y}$

39. $\dfrac{7x+35}{x^2+5x}$

40. $\dfrac{9x+99}{x^2+11x}$

41. $\dfrac{x+5}{x^2-4x-45}$

42. $\dfrac{x-3}{x^2-6x+9}$

43. $\dfrac{5x^2+11x+2}{x+2}$

44. $\dfrac{12x^2+4x-1}{2x+1}$

45. $\dfrac{x^3+7x^2}{x^2+5x-14}$

46. $\dfrac{x^4-10x^3}{x^2-17x+70}$

47. $\dfrac{14x^2-21x}{2x-3}$

48. $\dfrac{4x^2+24x}{x+6}$

49. $\dfrac{x^2+7x+10}{x^2-3x-10}$

50. $\dfrac{2x^2+7x-4}{x^2+3x-4}$

51. $\dfrac{3x^2+7x+2}{3x^2+13x+4}$

52. $\dfrac{4x^2-4x+1}{2x^2+9x-5}$

53. $\dfrac{2x^2-8}{4x-8}$

54. $\dfrac{5x^2-500}{35x+350}$

55. $\dfrac{4-x^2}{x-2}$

56. $\dfrac{49-y^2}{y-7}$

57. $\dfrac{x^2-1}{x^2-2x+1}$

58. $\dfrac{x^2-16}{x^2-8x+16}$

59. $\dfrac{m^2-6m+9}{m^2-m-6}$

60. $\dfrac{m^2-4m+4}{m^2+m-6}$

61. $\dfrac{11x^2-22x^3}{6x-12x^2}$

62. $\dfrac{24y^2-8y^3}{15y-5y^2}$

Simplify. These expressions contain 4-term polynomials and sums and differences of cubes.

63. $\dfrac{x^2+xy+2x+2y}{x+2}$

64. $\dfrac{ab+ac+b^2+bc}{b+c}$

65. $\dfrac{5x+15-xy-3y}{2x+6}$

66. $\dfrac{xy-6x+2y-12}{y^2-6y}$

67. $\dfrac{x^3+8}{x+2}$

68. $\dfrac{x^3+64}{x+4}$

69. $\dfrac{x^3-1}{1-x}$

70. $\dfrac{3-x}{x^3-27}$

71. $\dfrac{2xy+5x-2y-5}{3xy+4x-3y-4}$

72. $\dfrac{2xy+2x-3y-3}{2xy+4x-3y-6}$

MIXED PRACTICE

Simplify each expression. Then determine whether the given answer is correct. See Examples 3 through 9.

73. $\dfrac{9-x^2}{x-3}$; Answer: $-3-x$

74. $\dfrac{100-x^2}{x-10}$; Answer: $-10-x$

75. $\dfrac{7-34x-5x^2}{25x^2-1}$; Answer: $\dfrac{x+7}{-5x-1}$

76. $\dfrac{2-15x-8x^2}{64x^2-1}$; Answer: $\dfrac{x+2}{-8x-1}$

REVIEW AND PREVIEW

Perform each indicated operation. See Section 1.3.

77. $\dfrac{1}{3} \cdot \dfrac{9}{11}$

78. $\dfrac{5}{27} \cdot \dfrac{2}{5}$

79. $\dfrac{1}{3} \div \dfrac{1}{4}$

80. $\dfrac{7}{8} \div \dfrac{1}{2}$

81. $\dfrac{13}{20} \div \dfrac{2}{9}$

82. $\dfrac{8}{15} \div \dfrac{5}{8}$

CONCEPT EXTENSIONS

Which of the following are incorrect and why? See the Concept Check in this section.

83. $\dfrac{5a-15}{5}$ simplifies to $a-3$?

84. $\dfrac{7m-9}{7}$ simplifies to $m-9$?

85. $\dfrac{1+2}{1+3}$ simplifies to $\dfrac{2}{3}$?

86. $\dfrac{46}{54}$ simplifies to $\dfrac{6}{5}$?

87. Explain how to write a fraction in lowest terms.
88. Explain how to write a rational expression in lowest terms.
89. Explain why the denominator of a fraction or a rational expression must not equal 0.
90. Does $\dfrac{(x-3)(x+3)}{x-3}$ have the same value as $x+3$ for all real numbers? Explain why or why not.
91. The total revenue R from the sale of a popular music compact disc is approximately given by the equation
$$R = \dfrac{150x^2}{x^2 + 3}$$
where x is the number of years since the CD has been released and revenue R is in millions of dollars.
    a. Find the total revenue generated by the end of the first year.
    b. Find the total revenue generated by the end of the second year.
    c. Find the total revenue generated in the second year only.
92. For a certain model fax machine, the manufacturing cost C per machine is given by the equation
$$C = \dfrac{250x + 10{,}000}{x}$$
where x is the number of fax machines manufactured and cost C is in dollars per machine.
    a. Find the cost per fax machine when manufacturing 100 fax machines.
    b. Find the cost per fax machine when manufacturing 1000 fax machines.
    c. Does the cost per machine decrease or increase when more machines are manufactured? Explain why this is so.

Solve.

93. The dose of medicine prescribed for a child depends on the child's age A in years and the adult dose D for the medication. Young's Rule is a formula used by pediatricians that gives a child's dose C as
$$C = \dfrac{DA}{A + 12}$$ 
Suppose that an 8-year-old child needs medication, and the normal adult dose is 1000 mg. What size dose should the child receive?

94. Calculating body-mass index is a way to gauge whether a person should lose weight. Doctors recommend that body-mass index values fall between 18.5 and 25. The formula for body-mass index B is
$$B = \dfrac{703w}{h^2}$$
where w is weight in pounds and h is height in inches. Should a 148-pound person who is 5 feet 6 inches tall lose weight?

95. Anthropologists and forensic scientists use a measure called the cephalic index to help classify skulls. The cephalic index of a skull with width W and length L from front to back is given by the formula 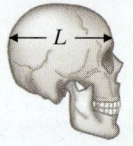
$$C = \dfrac{100W}{L}$$
A long skull has an index value less than 75, a medium skull has an index value between 75 and 85, and a broad skull has an index value over 85. Find the cephalic index of a skull that is 5 inches wide and 6.4 inches long. Classify the skull.

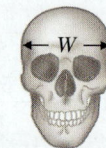

96. During a storm, water treatment engineers monitor how quickly rain is falling. If too much rain comes too fast, there is a danger of sewers backing up. A formula that gives the rainfall intensity i in millimeters per hour for a certain strength storm in eastern Virginia is
$$i = \dfrac{5840}{t + 29}$$
where t is the duration of the storm in minutes. What rainfall intensity should engineers expect for a storm of this strength in eastern Virginia that lasts for 80 minutes? Round your answer to one decimal place.

97. To calculate a quarterback's rating in football, you may use the formula $\left[\dfrac{20C + 0.5A + Y + 80T - 100I}{A}\right]\left(\dfrac{25}{6}\right)$, where C = the number of completed passes, A = the number of attempted passes, Y = total yards thrown for passes, T = the number of touchdown passes, and I = the number of interceptions. For the 2006 season, Peyton Manning, of the Indianapolis Colts, had final season totals of 557 attempts, 362 completions, 4397 yards, 31 touchdown passes, and 9 interceptions. Calculate Manning's quarterback rating for the 2006 season. Round the answer to the nearest tenth. (*Source:* The NFL)

98. A baseball player's slugging percent S can be calculated by the following formula: $S = \dfrac{h + d + 2t + 3r}{b}$, where h = number of hits, d = number of doubles, t = number of triples, r = number of home runs, and b = number at bats. During the 2006 season, David Ortiz of the Boston Red Sox had 558 at bats, 160 hits, 29 doubles, 2 triples, and 54 home runs. Calculate Ortiz's 2006 slugging percent. Round to the nearest tenth of a percent. (*Source:* Major League Baseball)

99. A company's gross profit margin P can be computed with the formula $P = \dfrac{R - C}{R}$, where R = the company's revenue and C = cost of goods sold. For fiscal year 2006, consumer electronics retailer Best Buy had revenues of $30.8 billion and cost of goods sold of $23.1 billion. What was Best Buy's gross profit margin in 2006? Express the answer as a percent, rounded to the nearest tenth of a percent. (*Source:* Best Buy Company, Inc.)

How does the graph of $y = \frac{x^2 - 9}{x - 3}$ compare to the graph of $y = x + 3$? Recall that $\frac{x^2 - 9}{x - 3} = \frac{(x + 3)(x - 3)}{x - 3} = x + 3$ as long as x is not 3. This means that the graph of $y = \frac{x^2 - 9}{x - 3}$ is the same as the graph of $y = x + 3$ with $x \neq 3$. To graph $y = \frac{x^2 - 9}{x - 3}$, then, graph the linear equation $y = x + 3$ and place an open dot on the graph at 3. This open dot or interruption of the line at 3 means $x \neq 3$.

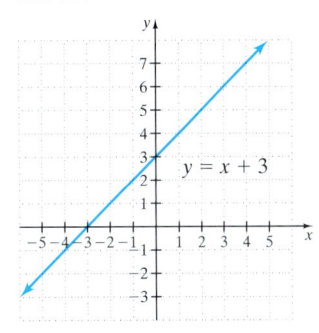

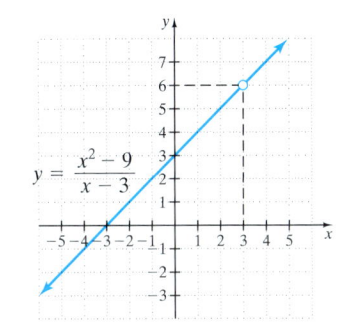

100. Graph $y = \frac{x^2 - 25}{x + 5}$.

101. Graph $y = \frac{x^2 - 16}{x - 4}$.

102. Graph $y = \frac{x^2 + x - 12}{x + 4}$.

103. Graph $y = \frac{x^2 - 6x + 8}{x - 2}$.

STUDY SKILLS BUILDER

Is Your Notebook Still Organized?

It's never too late to organize your material in a course. Let's see how you are doing.

1. Are all your graded papers in one place in your math notebook or binder?
2. Flip through the pages of your notebook. Are your notes neat and readable?
3. Are your notes complete with no sections missing?
4. Are important notes marked in some way (like an exclamation point) so that you will know to review them before a quiz or test?
5. Are your assignments complete?
6. Do exercises that have given you trouble have a mark (like a question mark) so that you will remember to talk to your instructor or a tutor about them?
7. Describe your attitude toward this course.
8. List ways your attitude can improve and make a commitment to work on at least one of those during the next week.

7.2 MULTIPLYING AND DIVIDING RATIONAL EXPRESSIONS

OBJECTIVES

1. Multiply rational expressions.
2. Divide rational expressions.
3. Multiply or divide rational expressions.

OBJECTIVE 1 ▶ Multiplying rational expressions. Just as simplifying rational expressions is similar to simplifying number fractions, multiplying and dividing rational expressions is similar to multiplying and dividing number fractions.

Fractions

Multiply: $\frac{3}{5} \cdot \frac{10}{11}$

Rational Expressions

Multiply: $\frac{x - 3}{x + 5} \cdot \frac{2x + 10}{x^2 - 9}$

Multiply numerators and then multiply denominators.

$\frac{3}{5} \cdot \frac{10}{11} = \frac{3 \cdot 10}{5 \cdot 11}$ $\quad$ $\frac{x - 3}{x + 5} \cdot \frac{2x + 10}{x^2 - 9} = \frac{(x - 3) \cdot (2x + 10)}{(x + 5) \cdot (x^2 - 9)}$

Simplify by factoring numerators and denominators.

$= \frac{3 \cdot 2 \cdot 5}{5 \cdot 11}$ $\quad$ $= \frac{(x - 3) \cdot 2(x + 5)}{(x + 5)(x + 3)(x - 3)}$

Apply the fundamental principle.

$= \frac{3 \cdot 2}{11}$ or $\frac{6}{11}$ $\quad$ $= \frac{2}{x + 3}$

CHAPTER 7 Rational Expressions

> **Multiplying Rational Expressions**
>
> If $\dfrac{P}{Q}$ and $\dfrac{R}{S}$ are rational expressions, then
>
> $$\dfrac{P}{Q} \cdot \dfrac{R}{S} = \dfrac{PR}{QS}$$
>
> To multiply rational expressions, multiply the numerators and then multiply the denominators.

Note: Recall that for Sections 7.1 through 7.4, we assume variables in rational expressions have only those replacement values for which the expressions are defined.

EXAMPLE 1 Multiply.

a. $\dfrac{25x}{2} \cdot \dfrac{1}{y^3}$ b. $\dfrac{-7x^2}{5y} \cdot \dfrac{3y^5}{14x^2}$

Solution To multiply rational expressions, multiply the numerators and then multiply the denominators of both expressions. Then simplify if possible.

a. $\dfrac{25x}{2} \cdot \dfrac{1}{y^3} = \dfrac{25x \cdot 1}{2 \cdot y^3} = \dfrac{25x}{2y^3}$

The expression $\dfrac{25x}{2y^3}$ is in simplest form.

b. $\dfrac{-7x^2}{5y} \cdot \dfrac{3y^5}{14x^2} = \dfrac{-7x^2 \cdot 3y^5}{5y \cdot 14x^2}$ Multiply.

The expression $\dfrac{-7x^2 \cdot 3y^5}{5y \cdot 14x^2}$ is not in simplest form, so we factor the numerator and the denominator and divide out common factors.

$= \dfrac{-1 \cdot \boxed{7} \cdot 3 \cdot \boxed{x^2} \cdot \boxed{y} \cdot y^4}{5 \cdot 2 \cdot \boxed{7} \cdot \boxed{x^2} \cdot \boxed{y}}$

$= -\dfrac{3y^4}{10}$

PRACTICE 1 Multiply.

a. $\dfrac{4a}{5} \cdot \dfrac{3}{b^2}$ b. $\dfrac{-3p^4}{q^2} \cdot \dfrac{2q^3}{9p^4}$

When multiplying rational expressions, it is usually best to factor each numerator and denominator. This will help us when we divide out common factors to write the product in lowest terms.

EXAMPLE 2 Multiply: $\dfrac{x^2 + x}{3x} \cdot \dfrac{6}{5x + 5}$

Solution $\dfrac{x^2 + x}{3x} \cdot \dfrac{6}{5x + 5} = \dfrac{x(x+1)}{3x} \cdot \dfrac{2 \cdot 3}{5(x+1)}$ Factor numerators and denominators.

$= \dfrac{x(x+1) \cdot 2 \cdot 3}{3x \cdot 5 \, \boxed{(x+1)}}$ Multiply.

$= \dfrac{2}{5}$ Simplify by dividing out common factors.

PRACTICE 2 Multiply: $\dfrac{x^2 - x}{5x} \cdot \dfrac{15}{x^2 - 1}$.

The following steps may be used to multiply rational expressions.

> **Multiplying Rational Expressions**
>
> **STEP 1.** Completely factor numerators and denominators.
>
> **STEP 2.** Multiply numerators and multiply denominators.
>
> **STEP 3.** Simplify or write the product in lowest terms by dividing out common factors.

Concept Check ✓

Which of the following is a true statement?

a. $\dfrac{1}{3} \cdot \dfrac{1}{2} = \dfrac{1}{5}$ b. $\dfrac{2}{x} \cdot \dfrac{5}{x} = \dfrac{10}{x}$ c. $\dfrac{3}{x} \cdot \dfrac{1}{2} = \dfrac{3}{2x}$ d. $\dfrac{x}{7} \cdot \dfrac{x+5}{4} = \dfrac{2x+5}{28}$

EXAMPLE 3 Multiply: $\dfrac{3x+3}{5x-5x^2} \cdot \dfrac{2x^2+x-3}{4x^2-9}$

Solution

$$\dfrac{3x+3}{5x-5x^2} \cdot \dfrac{2x^2+x-3}{4x^2-9} = \dfrac{3(x+1)}{5x(1-x)} \cdot \dfrac{(2x+3)(x-1)}{(2x-3)(2x+3)} \quad \text{Factor.}$$

$$= \dfrac{3(x+1)(2x+3)(x-1)}{5x(1-x)(2x-3)(2x+3)} \quad \text{Multiply.}$$

$$= \dfrac{3(x+1)(x-1)}{5x(1-x)(2x-3)} \quad \text{Divide out common factors.}$$

Next, recall that $x-1$ and $1-x$ are opposites so that $x-1 = -1(1-x)$.

$$= \dfrac{3(x+1)(-1)(1-x)}{5x(1-x)(2x-3)} \quad \text{Write } x-1 \text{ as } -1(1-x).$$

$$= \dfrac{-3(x+1)}{5x(2x-3)} \quad \text{or} \quad -\dfrac{3(x+1)}{5x(2x-3)} \quad \text{Divide out common factors.} \quad \square$$

PRACTICE 3 Multiply: $\dfrac{6-3x}{6x+6x^2} \cdot \dfrac{3x^2-2x-5}{x^2-4}$.

OBJECTIVE 2 ▶ Dividing rational expressions. We can divide by a rational expression in the same way we divide by a fraction. To divide by a fraction, multiply by its reciprocal.

> **▶ Helpful Hint**
>
> Don't forget how to find reciprocals. The reciprocal of $\dfrac{a}{b}$ is $\dfrac{b}{a}$, $a \neq 0, b \neq 0$.

For example, to divide $\dfrac{3}{2}$ by $\dfrac{7}{8}$, multiply $\dfrac{3}{2}$ by $\dfrac{8}{7}$.

$$\dfrac{3}{2} \div \dfrac{7}{8} = \dfrac{3}{2} \cdot \dfrac{8}{7} = \dfrac{3 \cdot 4 \cdot 2}{2 \cdot 7} = \dfrac{12}{7}$$

Answer to Concept Check: c

440 CHAPTER 7 Rational Expressions

> **Dividing Rational Expressions**
> If $\dfrac{P}{Q}$ and $\dfrac{R}{S}$ are rational expressions and $\dfrac{R}{S}$ is not 0, then
> $$\dfrac{P}{Q} \div \dfrac{R}{S} = \dfrac{P}{Q} \cdot \dfrac{S}{R} = \dfrac{PS}{QR}$$
> To divide two rational expressions, multiply the first rational expression by the reciprocal of the second rational expression.

EXAMPLE 4 Divide: $\dfrac{3x^3y^7}{40} \div \dfrac{4x^3}{y^2}$

Solution
$$\dfrac{3x^3y^7}{40} \div \dfrac{4x^3}{y^2} = \dfrac{3x^3y^7}{40} \cdot \dfrac{y^2}{4x^3} \quad \text{Multiply by the reciprocal of } \dfrac{4x^3}{y^2}.$$
$$= \dfrac{3x^3y^9}{160x^3}$$
$$= \dfrac{3y^9}{160} \quad \text{Simplify.}$$

PRACTICE 4 Divide: $\dfrac{5a^3b^2}{24} \div \dfrac{10a^5}{6}$.

EXAMPLE 5 Divide: $\dfrac{(x-1)(x+2)}{10}$ by $\dfrac{2x+4}{5}$.

Solution
$$\dfrac{(x-1)(x+2)}{10} \div \dfrac{2x+4}{5} = \dfrac{(x-1)(x+2)}{10} \cdot \dfrac{5}{2x+4} \quad \text{Multiply by the reciprocal of } \dfrac{2x+4}{5}.$$
$$= \dfrac{(x-1)(x+2)\cdot 5}{5\cdot 2\cdot 2\cdot (x+2)} \quad \text{Factor and multiply.}$$
$$= \dfrac{x-1}{4} \quad \text{Simplify.}$$

PRACTICE 5 Divide $\dfrac{(3x+1)(x-5)}{3}$ by $\dfrac{4x-20}{9}$.

The following may be used to divide by a rational expression.

> **Dividing by a Rational Expression**
> Multiply by its reciprocal.

EXAMPLE 6 Divide: $\dfrac{6x+2}{x^2-1} \div \dfrac{3x^2+x}{x-1}$

Solution

$$\frac{6x+2}{x^2-1} \div \frac{3x^2+x}{x-1} = \frac{6x+2}{x^2-1} \cdot \frac{x-1}{3x^2+x}$$ Multiply by the reciprocal.

$$= \frac{2(3x+1)(x-1)}{(x+1)(x-1) \cdot x(3x+1)}$$ Factor and multiply.

$$= \frac{2}{x(x+1)}$$ Simplify.

PRACTICE 6 Divide $\dfrac{10x-2}{x^2-9} \div \dfrac{5x^2-x}{x+3}$.

EXAMPLE 7 Divide: $\dfrac{2x^2-11x+5}{5x-25} \div \dfrac{4x-2}{10}$

Solution

$$\frac{2x^2-11x+5}{5x-25} \div \frac{4x-2}{10} = \frac{2x^2-11x+5}{5x-25} \cdot \frac{10}{4x-2}$$ Multiply by the reciprocal.

$$= \frac{(2x-1)(x-5) \cdot 2 \cdot 5}{5(x-5) \cdot 2(2x-1)}$$ Factor and multiply.

$$= \frac{1}{1} \text{ or } 1$$ Simplify.

PRACTICE 7 Divide $\dfrac{3x^2-11x-4}{2x-8} \div \dfrac{9x+3}{6}$.

OBJECTIVE 3 ▶ Multiplying or dividing rational expressions. Let's make sure that we understand the difference between multiplying and dividing rational expressions.

Rational Expressions	
Multiplication	Multiply the numerators and multiply the denominators.
Division	Multiply by the reciprocal of the divisor.

EXAMPLE 8 Multiply or divide as indicated.

a. $\dfrac{x-4}{5} \cdot \dfrac{x}{x-4}$ b. $\dfrac{x-4}{5} \div \dfrac{x}{x-4}$ c. $\dfrac{x^2-4}{2x+6} \cdot \dfrac{x^2+4x+3}{2-x}$

Solution

a. $\dfrac{x-4}{5} \cdot \dfrac{x}{x-4} = \dfrac{(x-4) \cdot x}{5 \cdot (x-4)} = \dfrac{x}{5}$

b. $\dfrac{x-4}{5} \div \dfrac{x}{x-4} = \dfrac{x-4}{5} \cdot \dfrac{x-4}{x} = \dfrac{(x-4)^2}{5x}$

c. $\dfrac{x^2-4}{2x+6} \cdot \dfrac{x^2+4x+3}{2-x} = \dfrac{(x-2)(x+2) \cdot (x+1)(x+3)}{2(x+3) \cdot (2-x)}$ Factor and multiply.

$= \dfrac{(x-2)(x+2) \cdot (x+1)(x+3)}{2(x+3) \cdot (2-x)}$

$= \dfrac{-1(x+2)(x+1)}{2}$ Divide out common factors. Recall that $\dfrac{x-2}{2-x} = -1$.

$= -\dfrac{(x+2)(x+1)}{2}$

PRACTICE 8 Multiply or divide as indicated.

a. $\dfrac{y+9}{8x} \cdot \dfrac{y+9}{2x}$ b. $\dfrac{y+9}{8x} \div \dfrac{y+9}{2}$ c. $\dfrac{35x-7x^2}{x^2-25} \cdot \dfrac{x^2+3x-10}{x^2+4x}$

VOCABULARY & READINESS CHECK

Use one of the choices below to fill in the blank.

    opposites   reciprocals

1. The expressions $\dfrac{x}{2y}$ and $\dfrac{2y}{x}$ are called _____.

Multiply or divide as indicated.

2. $\dfrac{a}{b} \cdot \dfrac{c}{d} =$ _____

3. $\dfrac{a}{b} \div \dfrac{c}{d} =$ _____

4. $\dfrac{x}{7} \cdot \dfrac{x}{6} =$ _____

5. $\dfrac{x}{7} \div \dfrac{x}{6} =$ _____

7.2 EXERCISE SET

Find each product and simplify if possible. See Examples 1 through 3.

1. $\dfrac{3x}{y^2} \cdot \dfrac{7y}{4x}$

2. $\dfrac{9x^2}{y} \cdot \dfrac{4y}{3x^3}$

3. $\dfrac{8x}{2} \cdot \dfrac{x^5}{4x^2}$

4. $\dfrac{6x^2}{10x^3} \cdot \dfrac{5x}{12}$

5. $-\dfrac{5a^2b}{30a^2b^2} \cdot b^3$

6. $-\dfrac{9x^3y^2}{18xy^5} \cdot y^3$

7. $\dfrac{x}{2x-14} \cdot \dfrac{x^2-7x}{5}$

8. $\dfrac{4x-24}{20x} \cdot \dfrac{5}{x-6}$

9. $\dfrac{6x+6}{5} \cdot \dfrac{10}{36x+36}$

10. $\dfrac{x^2+x}{8} \cdot \dfrac{16}{x+1}$

11. $\dfrac{(m+n)^2}{m-n} \cdot \dfrac{m}{m^2+mn}$

12. $\dfrac{(m-n)^2}{m+n} \cdot \dfrac{m}{m^2-mn}$

13. $\dfrac{x^2-25}{x^2-3x-10} \cdot \dfrac{x+2}{x}$

14. $\dfrac{a^2-4a+4}{a^2-4} \cdot \dfrac{a+3}{a-2}$

15. $\dfrac{x^2+6x+8}{x^2+x-20} \cdot \dfrac{x^2+2x-15}{x^2+8x+16}$

16. $\dfrac{x^2+9x+20}{x^2-15x+44} \cdot \dfrac{x^2-11x+28}{x^2+12x+35}$

Find each quotient and simplify. See Examples 4 through 7.

17. $\dfrac{5x^7}{2x^5} \div \dfrac{15x}{4x^3}$

18. $\dfrac{9y^4}{6y} \div \dfrac{y^2}{3}$

19. $\dfrac{8x^2}{y^3} \div \dfrac{4x^2y^3}{6}$

20. $\dfrac{7a^2b}{3ab^2} \div \dfrac{21a^2b^2}{14ab}$

21. $\dfrac{(x-6)(x+4)}{4x} \div \dfrac{2x-12}{8x^2}$

22. $\dfrac{(x+3)^2}{5} \div \dfrac{5x+15}{25}$

23. $\dfrac{3x^2}{x^2-1} \div \dfrac{x^5}{(x+1)^2}$

24. $\dfrac{9x^5}{a^2 - b^2} \div \dfrac{27x^2}{3b - 3a}$

25. $\dfrac{m^2 - n^2}{m + n} \div \dfrac{m}{m^2 + nm}$

26. $\dfrac{(m - n)^2}{m + n} \div \dfrac{m^2 - mn}{m}$

27. $\dfrac{x + 2}{7 - x} \div \dfrac{x^2 - 5x + 6}{x^2 - 9x + 14}$

28. $\dfrac{x - 3}{2 - x} \div \dfrac{x^2 + 3x - 18}{x^2 + 2x - 8}$

29. $\dfrac{x^2 + 7x + 10}{x - 1} \div \dfrac{x^2 + 2x - 15}{x - 1}$

30. $\dfrac{x + 1}{(x + 1)(2x + 3)} \div \dfrac{20x + 100}{2x + 3}$

MIXED PRACTICE

Multiply or divide as indicated. See Examples 1 through 8.

31. $\dfrac{5x - 10}{12} \div \dfrac{4x - 8}{8}$

32. $\dfrac{6x + 6}{5} \div \dfrac{9x + 9}{10}$

33. $\dfrac{x^2 + 5x}{8} \cdot \dfrac{9}{3x + 15}$

34. $\dfrac{3x^2 + 12x}{6} \cdot \dfrac{9}{2x + 8}$

35. $\dfrac{7}{6p^2 + q} \div \dfrac{14}{18p^2 + 3q}$

36. $\dfrac{3x + 6}{20} \div \dfrac{4x + 8}{8}$

37. $\dfrac{3x + 4y}{x^2 + 4xy + 4y^2} \cdot \dfrac{x + 2y}{2}$

38. $\dfrac{x^2 - y^2}{3x^2 + 3xy} \cdot \dfrac{3x^2 + 6x}{3x^2 - 2xy - y^2}$

39. $\dfrac{(x + 2)^2}{x - 2} \div \dfrac{x^2 - 4}{2x - 4}$

40. $\dfrac{x + 3}{x^2 - 9} \div \dfrac{5x + 15}{(x - 3)^2}$

41. $\dfrac{x^2 - 4}{24x} \div \dfrac{2 - x}{6xy}$

42. $\dfrac{3y}{3 - x} \div \dfrac{12xy}{x^2 - 9}$

43. $\dfrac{a^2 + 7a + 12}{a^2 + 5a + 6} \cdot \dfrac{a^2 + 8a + 15}{a^2 + 5a + 4}$

44. $\dfrac{b^2 + 2b - 3}{b^2 + b - 2} \cdot \dfrac{b^2 - 4}{b^2 + 6b + 8}$

45. $\dfrac{5x - 20}{3x^2 + x} \cdot \dfrac{3x^2 + 13x + 4}{x^2 - 16}$

46. $\dfrac{9x + 18}{4x^2 - 3x} \cdot \dfrac{4x^2 - 11x + 6}{x^2 - 4}$

47. $\dfrac{8n^2 - 18}{2n^2 - 5n + 3} \div \dfrac{6n^2 + 7n - 3}{n^2 - 9n + 8}$

48. $\dfrac{36n^2 - 64}{3n^2 + 10n + 8} \div \dfrac{3n^2 - 13n + 12}{n^2 - 5n - 14}$

49. Find the quotient of $\dfrac{x^2 - 9}{2x}$ and $\dfrac{x + 3}{8x^4}$.

50. Find the quotient of $\dfrac{4x^2 + 4x + 1}{4x + 2}$ and $\dfrac{4x + 2}{16}$.

Multiply or divide as indicated. Some of these expressions contain 4-term polynomials and sums and differences of cubes. See Examples 1 through 8.

51. $\dfrac{a^2 + ac + ba + bc}{a - b} \div \dfrac{a + c}{a + b}$

52. $\dfrac{x^2 + 2x - xy - 2y}{x^2 - y^2} \div \dfrac{2x + 4}{x + y}$

53. $\dfrac{3x^2 + 8x + 5}{x^2 + 8x + 7} \cdot \dfrac{x + 7}{x^2 + 4}$

54. $\dfrac{16x^2 + 2x}{16x^2 + 10x + 1} \cdot \dfrac{1}{4x^2 + 2x}$

55. $\dfrac{x^3 + 8}{x^2 - 2x + 4} \cdot \dfrac{4}{x^2 - 4}$

56. $\dfrac{9y}{3y - 3} \cdot \dfrac{y^3 - 1}{y^3 + y^2 + y}$

57. $\dfrac{a^2 - ab}{6a^2 + 6ab} \div \dfrac{a^3 - b^3}{a^2 - b^2}$

58. $\dfrac{x^3 + 27y^3}{6x} \div \dfrac{x^2 - 9y^2}{x^2 - 3xy}$

REVIEW AND PREVIEW

Perform each indicated operation. See Section 1.3.

59. $\dfrac{1}{5} + \dfrac{4}{5}$

60. $\dfrac{3}{15} + \dfrac{6}{15}$

61. $\dfrac{9}{9} - \dfrac{19}{9}$

62. $\dfrac{4}{3} - \dfrac{8}{3}$

63. $\dfrac{6}{5} + \left(\dfrac{1}{5} - \dfrac{8}{5}\right)$

64. $-\dfrac{3}{2} + \left(\dfrac{1}{2} - \dfrac{3}{2}\right)$

Graph each linear equation. See Section 3.2.

65. $x - 2y = 6$

66. $5x - y = 10$

CONCEPT EXTENSIONS

Identify each statement as true or false. If false, correct the multiplication. See the Concept Check in this section.

67. $\dfrac{4}{a} \cdot \dfrac{1}{b} = \dfrac{4}{ab}$

68. $\dfrac{2}{3} \cdot \dfrac{2}{4} = \dfrac{2}{7}$

69. $\dfrac{x}{5} \cdot \dfrac{x + 3}{4} = \dfrac{2x + 3}{20}$

70. $\dfrac{7}{a} \cdot \dfrac{3}{a} = \dfrac{21}{a}$

444 CHAPTER 7 Rational Expressions

△ **71.** Find the area of the rectangle.

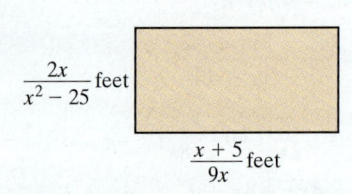

$\frac{2x}{x^2-25}$ feet

$\frac{x+5}{9x}$ feet

△ **72.** Find the area of the square.

$\frac{2x}{5x+3}$ meters

Multiply or divide as indicated.

73. $\left(\dfrac{x^2-y^2}{x^2+y^2} \div \dfrac{x^2-y^2}{3x}\right) \cdot \dfrac{x^2+y^2}{6}$

74. $\left(\dfrac{x^2-9}{x^2-1} \cdot \dfrac{x^2+2x+1}{2x^2+9x+9}\right) \div \dfrac{2x+3}{1-x}$

75. $\left(\dfrac{2a+b}{b^2} \cdot \dfrac{3a^2-2ab}{ab+2b^2}\right) \div \dfrac{a^2-3ab+2b^2}{5ab-10b^2}$

76. $\left(\dfrac{x^2y^2-xy}{4x-4y} \div \dfrac{3y-3x}{8x-8y}\right) \cdot \dfrac{y-x}{8}$

✎ **77.** In your own words, explain how you multiply rational expressions.

✎ **78.** Explain how dividing rational expressions is similar to dividing rational numbers.

7.3 ADDING AND SUBTRACTING RATIONAL EXPRESSIONS WITH COMMON DENOMINATORS AND LEAST COMMON DENOMINATOR

OBJECTIVES

1. Add and subtract rational expressions with the same denominator.
2. Find the least common denominator of a list of rational expressions.
3. Write a rational expression as an equivalent expression whose denominator is given.

OBJECTIVE 1 ▶ Adding and subtracting rational expressions with the same denominator. Like multiplication and division, addition and subtraction of rational expressions is similar to addition and subtraction of rational numbers. In this section, we add and subtract rational expressions with a common (or the same) denominator.

Add: $\dfrac{6}{5} + \dfrac{2}{5}$ | Add: $\dfrac{9}{x+2} + \dfrac{3}{x+2}$

Add the numerators and place the sum over the common denominator.

$\dfrac{6}{5} + \dfrac{2}{5} = \dfrac{6+2}{5}$ | $\dfrac{9}{x+2} + \dfrac{3}{x+2} = \dfrac{9+3}{x+2}$

$= \dfrac{8}{5}$ Simplify. | $= \dfrac{12}{x+2}$ Simplify.

Adding and Subtracting Rational Expressions with Common Denominators

If $\dfrac{P}{R}$ and $\dfrac{Q}{R}$ are rational expressions, then

$$\dfrac{P}{R} + \dfrac{Q}{R} = \dfrac{P+Q}{R} \quad \text{and} \quad \dfrac{P}{R} - \dfrac{Q}{R} = \dfrac{P-Q}{R}$$

To add or subtract rational expressions, add or subtract numerators and place the sum or difference over the common denominator.

Section 7.3 Adding and Subtracting Rational Expressions with Common Denominators and Least Common Denominator 445

EXAMPLE 1 Add: $\dfrac{5m}{2n} + \dfrac{m}{2n}$

Solution
$\dfrac{5m}{2n} + \dfrac{m}{2n} = \dfrac{5m + m}{2n}$ Add the numerators.

$= \dfrac{6m}{2n}$ Simplify the numerator by combining like terms.

$= \dfrac{3m}{n}$ Simplify by applying the fundamental principle.

PRACTICE 1 Add: $\dfrac{7a}{4b} + \dfrac{a}{4b}$.

EXAMPLE 2 Subtract: $\dfrac{2y}{2y - 7} - \dfrac{7}{2y - 7}$

Solution
$\dfrac{2y}{2y - 7} - \dfrac{7}{2y - 7} = \dfrac{2y - 7}{2y - 7}$ Subtract the numerators.

$= \dfrac{1}{1}$ or 1 Simplify.

PRACTICE 2 Subtract: $\dfrac{3x}{3x - 2} - \dfrac{2}{3x - 2}$.

EXAMPLE 3 Subtract: $\dfrac{3x^2 + 2x}{x - 1} - \dfrac{10x - 5}{x - 1}$.

Solution
$\dfrac{3x^2 + 2x}{x - 1} - \dfrac{10x - 5}{x - 1} = \dfrac{(3x^2 + 2x) - (10x - 5)}{x - 1}$ Subtract the numerators. Notice the parentheses.

$= \dfrac{3x^2 + 2x - 10x + 5}{x - 1}$ Use the distributive property.

$= \dfrac{3x^2 - 8x + 5}{x - 1}$ Combine like terms.

$= \dfrac{(x - 1)(3x - 5)}{x - 1}$ Factor.

$= 3x - 5$ Simplify.

> **Helpful Hint**
> Parentheses are inserted so that the entire numerator, $10x - 5$, is subtracted.

PRACTICE 3 Subtract: $\dfrac{4x^2 + 15x}{x + 3} - \dfrac{8x + 15}{x + 3}$.

> **Helpful Hint**
> Notice how the numerator $10x - 5$ has been subtracted in Example 3.
> This $-$ sign applies to the entire numerator of $10x - 5$.
> So parentheses are inserted here to indicate this.
>
> $\dfrac{3x^2 + 2x}{x - 1} - \dfrac{10x - 5}{x - 1} = \dfrac{3x^2 + 2x - (10x - 5)}{x - 1}$

OBJECTIVE 2 ▶ Finding the least common denominator. To add and subtract fractions with **unlike** denominators, first find a least common denominator (LCD), and then write all fractions as equivalent fractions with the LCD.

For example, suppose we add $\frac{8}{3}$ and $\frac{2}{5}$. The LCD of denominators 3 and 5 is 15, since 15 is the least common multiple (LCM) of 3 and 5. That is, 15 is the smallest number that both 3 and 5 divide into evenly.

Next, rewrite each fraction so that its denominator is 15.

$$\frac{8}{3} + \frac{2}{5} = \frac{8(5)}{3(5)} + \frac{2(3)}{5(3)} = \frac{40}{15} + \frac{6}{15} = \frac{40+6}{15} = \frac{46}{15}$$

We are multiplying by 1.

To add or subtract rational expressions with unlike denominators, we also first find an LCD and then write all rational expressions as equivalent expressions with the LCD. The **least common denominator (LCD) of a list of rational expressions** is a polynomial of least degree whose factors include all the factors of the denominators in the list.

Finding the Least Common Denominator (LCD)

STEP 1. Factor each denominator completely.

STEP 2. The least common denominator (LCD) is the product of all unique factors found in Step 1, each raised to a power equal to the greatest number of times that the factor appears in any one factored denominator.

EXAMPLE 4 Find the LCD for each pair.

a. $\frac{1}{8}, \frac{3}{22}$ **b.** $\frac{7}{5x}, \frac{6}{15x^2}$

Solution

a. Start by finding the prime factorization of each denominator.

$$8 = 2 \cdot 2 \cdot 2 = 2^3 \quad \text{and}$$
$$22 = 2 \cdot 11$$

Next, write the product of all the unique factors, each raised to a power equal to the greatest number of times that the factor appears in any denominator.

The greatest number of times that the factor 2 appears is 3.

The greatest number of times that the factor 11 appears is 1.

$$\text{LCD} = 2^3 \cdot 11^1 = 8 \cdot 11 = 88$$

b. Factor each denominator.

$$5x = 5 \cdot x \quad \text{and}$$
$$15x^2 = 3 \cdot 5 \cdot x^2$$

The greatest number of times that the factor 5 appears is 1.
The greatest number of times that the factor 3 appears is 1.
The greatest number of times that the factor x appears is 2.

$$\text{LCD} = 3^1 \cdot 5^1 \cdot x^2 = 15x^2$$

PRACTICE 4 Find the LCD for each pair.

a. $\frac{3}{14}, \frac{5}{21}$ **b.** $\frac{4}{9y}, \frac{11}{15y^3}$

Section 7.3 Adding and Subtracting Rational Expressions with Common Denominators and Least Common Denominator 447

EXAMPLE 5 Find the LCD of

a. $\dfrac{7x}{x+2}$ and $\dfrac{5x^2}{x-2}$

b. $\dfrac{3}{x}$ and $\dfrac{6}{x+4}$

Solution

a. The denominators $x + 2$ and $x - 2$ are completely factored already. The factor $x + 2$ appears once and the factor $x - 2$ appears once.

$$\text{LCD} = (x+2)(x-2)$$

b. The denominators x and $x + 4$ cannot be factored further. The factor x appears once and the factor $x + 4$ appears once.

$$\text{LCD} = x(x+4)$$

PRACTICE 5 Find the LCD of

a. $\dfrac{16}{y-5}$ and $\dfrac{3y^3}{y-4}$

b. $\dfrac{8}{a}$ and $\dfrac{5}{a+2}$

EXAMPLE 6 Find the LCD of $\dfrac{6m^2}{3m+15}$ and $\dfrac{2}{(m+5)^2}$.

Solution We factor each denominator.

$$3m + 15 = 3(m+5)$$

$$(m+5)^2 = (m+5)^2 \quad \text{This denominator is already factored.}$$

The greatest number of times that the factor 3 appears is 1.

The greatest number of times that the factor $m + 5$ appears *in any one denominator* is 2.

$$\text{LCD} = 3(m+5)^2$$

PRACTICE 6 Find the LCD of $\dfrac{2x^3}{(2x-1)^2}$ and $\dfrac{5x}{6x-3}$.

Concept Check ✓

Choose the correct LCD of $\dfrac{x}{(x+1)^2}$ and $\dfrac{5}{x+1}$.

a. $x + 1$ b. $(x+1)^2$ c. $(x+1)^3$ d. $5x(x+1)^2$

EXAMPLE 7 Find the LCD of $\dfrac{t-10}{t^2-t-6}$ and $\dfrac{t+5}{t^2+3t+2}$.

Solution Start by factoring each denominator.

$$t^2 - t - 6 = (t-3)(t+2)$$

$$t^2 + 3t + 2 = (t+1)(t+2)$$

$$\text{LCD} = (t-3)(t+2)(t+1)$$

PRACTICE 7 Find the LCD of $\dfrac{x-5}{x^2+5x+4}$ and $\dfrac{x+8}{x^2-16}$.

Answer to Concept Check: b

448 CHAPTER 7 Rational Expressions

EXAMPLE 8 Find the LCD of $\dfrac{2}{x-2}$ and $\dfrac{10}{2-x}$.

Solution The denominators $x-2$ and $2-x$ are opposites. That is, $2-x = -1(x-2)$. Use $x-2$ or $2-x$ as the LCD.

$$\text{LCD} = x-2 \quad \text{or} \quad \text{LCD} = 2-x$$

PRACTICE 8 Find the LCD of $\dfrac{5}{3-x}$ and $\dfrac{4}{x-3}$.

OBJECTIVE 3 ▶ Writing equivalent rational expressions. Next we practice writing a rational expression as an equivalent rational expression with a given denominator. To do this, we multiply by a form of 1. Recall that multiplying an expression by 1 produces an equivalent expression. In other words,

$$\frac{P}{Q} = \frac{P}{Q} \cdot 1 = \frac{P}{Q} \cdot \frac{R}{R} = \frac{PR}{QR}.$$

EXAMPLE 9 Write each rational expression as an equivalent rational expression with the given denominator.

a. $\dfrac{4b}{9a} = \dfrac{}{27a^2b}$ **b.** $\dfrac{7x}{2x+5} = \dfrac{}{6x+15}$

Solution

a. We can ask ourselves: "What do we multiply $9a$ by to get $27a^2b$?" The answer is $3ab$, since $9a(3ab) = 27a^2b$. So we multiply by 1 in the form of $\dfrac{3ab}{3ab}$.

$$\frac{4b}{9a} = \frac{4b}{9a} \cdot 1 = \frac{4b}{9a} \cdot \frac{3ab}{3ab}$$

$$= \frac{4b(3ab)}{9a(3ab)} = \frac{12ab^2}{27a^2b}$$

b. First, factor the denominator on the right.

$$\frac{7x}{2x+5} = \frac{}{3(2x+5)}$$

To obtain the denominator on the right from the denominator on the left, we multiply by 1 in the form of $\dfrac{3}{3}$.

$$\frac{7x}{2x+5} = \frac{7x}{2x+5} \cdot \frac{3}{3} = \frac{7x \cdot 3}{(2x+5) \cdot 3} = \frac{21x}{3(2x+5)} \text{ or } \frac{21x}{6x+15}$$

PRACTICE 9 Write each rational expression as an equivalent fraction with the given denominator.

a. $\dfrac{3x}{5y} = \dfrac{}{35xy^2}$ **b.** $\dfrac{9x}{4x+7} = \dfrac{}{8x+14}$

EXAMPLE 10 Write the rational expression as an equivalent rational expression with the given denominator.

$$\frac{5}{x^2-4} = \frac{}{(x-2)(x+2)(x-4)}$$

Section 7.3 Adding and Subtracting Rational Expressions with Common Denominators and Least Common Denominator 449

Solution First, factor the denominator $x^2 - 4$ as $(x - 2)(x + 2)$.

If we multiply the original denominator $(x - 2)(x + 2)$ by $x - 4$, the result is the new denominator $(x + 2)(x - 2)(x - 4)$. Thus, we multiply by 1 in the form of $\dfrac{x - 4}{x - 4}$.

$$\frac{5}{\underbrace{x^2 - 4}_{\text{Factored denominator}}} = \frac{5}{(x - 2)(x + 2)} = \frac{5}{(x - 2)(x + 2)} \cdot \frac{x - 4}{x - 4}$$

$$= \frac{5(x - 4)}{(x - 2)(x + 2)(x - 4)}$$

$$= \frac{5x - 20}{(x - 2)(x + 2)(x - 4)}$$

PRACTICE 10 Write the rational expression as an equivalent rational expression with the given denominator.

$$\frac{3}{x^2 - 2x - 15} = \frac{}{(x - 2)(x + 3)(x - 5)}$$

VOCABULARY & READINESS CHECK

Use the choices below to fill in each blank. Not all choices will be used.

$\dfrac{9}{22}$ $\dfrac{5}{22}$ $\dfrac{9}{11}$ $\dfrac{5}{11}$ $\dfrac{ac}{b}$ $\dfrac{a - c}{b}$ $\dfrac{a + c}{b}$ $\dfrac{5 - 6 + x}{x}$ $\dfrac{5 - (6 + x)}{x}$

1. $\dfrac{7}{11} + \dfrac{2}{11} = $ _____
2. $\dfrac{7}{11} - \dfrac{2}{11} = $ _____
3. $\dfrac{a}{b} + \dfrac{c}{b} = $ _____
4. $\dfrac{a}{b} - \dfrac{c}{b} = $ _____
5. $\dfrac{5}{x} - \dfrac{6 + x}{x} = $ _____

7.3 EXERCISE SET

Add or subtract as indicated. Simplify the result if possible. See Examples 1 through 3.

1. $\dfrac{a + 1}{13} + \dfrac{8}{13}$
2. $\dfrac{x + 1}{7} + \dfrac{6}{7}$
3. $\dfrac{4m}{3n} + \dfrac{5m}{3n}$
4. $\dfrac{3p}{2q} + \dfrac{11p}{2q}$
5. $\dfrac{4m}{m - 6} - \dfrac{24}{m - 6}$
6. $\dfrac{8y}{y - 2} - \dfrac{16}{y - 2}$
7. $\dfrac{9}{3 + y} + \dfrac{y + 1}{3 + y}$
8. $\dfrac{9}{y + 9} + \dfrac{y - 5}{y + 9}$
9. $\dfrac{5x^2 + 4x}{x - 1} - \dfrac{6x + 3}{x - 1}$
10. $\dfrac{x^2 + 9x}{x + 7} - \dfrac{4x + 14}{x + 7}$
11. $\dfrac{4a}{a^2 + 2a - 15} - \dfrac{12}{a^2 + 2a - 15}$
12. $\dfrac{3y}{y^2 + 3y - 10} - \dfrac{6}{y^2 + 3y - 10}$
13. $\dfrac{2x + 3}{x^2 - x - 30} - \dfrac{x - 2}{x^2 - x - 30}$
14. $\dfrac{3x - 1}{x^2 + 5x - 6} - \dfrac{2x - 7}{x^2 + 5x - 6}$
15. $\dfrac{2x + 1}{x - 3} + \dfrac{3x + 6}{x - 3}$
16. $\dfrac{4p - 3}{2p + 7} + \dfrac{3p + 8}{2p + 7}$
17. $\dfrac{2x^2}{x - 5} - \dfrac{25 + x^2}{x - 5}$
18. $\dfrac{6x^2}{2x - 5} - \dfrac{25 + 2x^2}{2x - 5}$
19. $\dfrac{5x + 4}{x - 1} - \dfrac{2x + 7}{x - 1}$
20. $\dfrac{7x + 1}{x - 4} - \dfrac{2x + 21}{x - 4}$

Find the LCD for each list of rational expressions. See Examples 4 through 8.

21. $\dfrac{19}{2x}, \dfrac{5}{4x^3}$

22. $\dfrac{17x}{4y^5}, \dfrac{2}{8y}$

23. $\dfrac{9}{8x}, \dfrac{3}{2x+4}$

24. $\dfrac{1}{6y}, \dfrac{3x}{4y+12}$

25. $\dfrac{2}{x+3}, \dfrac{5}{x-2}$

26. $\dfrac{-6}{x-1}, \dfrac{4}{x+5}$

27. $\dfrac{x}{x+6}, \dfrac{10}{3x+18}$

28. $\dfrac{12}{x+5}, \dfrac{x}{4x+20}$

29. $\dfrac{8x^2}{(x-6)^2}, \dfrac{13x}{5x-30}$

30. $\dfrac{9x^2}{7x-14}, \dfrac{6x}{(x-2)^2}$

31. $\dfrac{1}{3x+3}, \dfrac{8}{2x^2+4x+2}$

32. $\dfrac{19x+5}{4x-12}, \dfrac{3}{2x^2-12x+18}$

33. $\dfrac{5}{x-8}, \dfrac{3}{8-x}$

34. $\dfrac{2x+5}{3x-7}, \dfrac{5}{7-3x}$

35. $\dfrac{5x+1}{x^2+3x-4}, \dfrac{3x}{x^2+2x-3}$

36. $\dfrac{4}{x^2+4x+3}, \dfrac{4x-2}{x^2+10x+21}$

37. $\dfrac{2x}{3x^2+4x+1}, \dfrac{7}{2x^2-x-1}$

38. $\dfrac{3x}{4x^2+5x+1}, \dfrac{5}{3x^2-2x-1}$

39. $\dfrac{1}{x^2-16}, \dfrac{x+6}{2x^3-8x^2}$

40. $\dfrac{5}{x^2-25}, \dfrac{x+9}{3x^3-15x^2}$

Rewrite each rational expression as an equivalent rational expression with the given denominator. See Examples 9 and 10.

41. $\dfrac{3}{2x} = \dfrac{}{4x^2}$

42. $\dfrac{3}{9y^5} = \dfrac{}{72y^9}$

43. $\dfrac{6}{3a} = \dfrac{}{12ab^2}$

44. $\dfrac{5}{4y^2x} = \dfrac{}{32y^3x^2}$

45. $\dfrac{9}{2x+6} = \dfrac{}{2y(x+3)}$

46. $\dfrac{4x+1}{3x+6} = \dfrac{}{3y(x+2)}$

47. $\dfrac{9a+2}{5a+10} = \dfrac{}{5b(a+2)}$

48. $\dfrac{5+y}{2x^2+10} = \dfrac{}{4(x^2+5)}$

49. $\dfrac{x}{x^3+6x^2+8x} = \dfrac{}{x(x+4)(x+2)(x+1)}$

50. $\dfrac{5x}{x^3+2x^2-3x} = \dfrac{}{x(x-1)(x-5)(x+3)}$

51. $\dfrac{9y-1}{15x^2-30} = \dfrac{}{30x^2-60}$

52. $\dfrac{6m-5}{3x^2-9} = \dfrac{}{12x^2-36}$

MIXED PRACTICE

Perform the indicated operations.

53. $\dfrac{5x}{7}+\dfrac{9x}{7}$

54. $\dfrac{5x}{7}\cdot\dfrac{9x}{7}$

55. $\dfrac{x+3}{4}\div\dfrac{2x-1}{4}$

56. $\dfrac{x+3}{4}-\dfrac{2x-1}{4}$

57. $\dfrac{x^2}{x-6}-\dfrac{5x+6}{x-6}$

58. $\dfrac{x^2+5x}{x^2-25}\cdot\dfrac{3x-15}{x^2}$

59. $\dfrac{-2x}{x^3-8x}+\dfrac{3x}{x^3-8x}$

60. $\dfrac{-2x}{x^3-8x}\div\dfrac{3x}{x^3-8x}$

61. $\dfrac{12x-6}{x^2+3x}\cdot\dfrac{4x^2+13x+3}{4x^2-1}$

62. $\dfrac{x^3+7x^2}{3x^3-x^2}\div\dfrac{5x^2+36x+7}{9x^2-1}$

REVIEW AND PREVIEW

Perform each indicated operation. See Section 1.3.

63. $\dfrac{2}{3}+\dfrac{5}{7}$

64. $\dfrac{9}{10}-\dfrac{3}{5}$

65. $\dfrac{2}{6}-\dfrac{3}{4}$

66. $\dfrac{11}{15}+\dfrac{5}{9}$

67. $\dfrac{1}{12}+\dfrac{3}{20}$

68. $\dfrac{7}{30}+\dfrac{3}{18}$

CONCEPT EXTENSIONS

69. Choose the correct LCD of $\dfrac{11a^3}{4a-20}$ and $\dfrac{15a^3}{(a-5)^2}$. See the Concept Check in this section.

   a. $4a(a-5)(a+5)$
   b. $a-5$
   c. $(a-5)^2$
   d. $4(a-5)^2$
   e. $(4a-20)(a-5)^2$

70. An algebra student approaches you with a problem. He's tried to subtract two rational expressions, but his result does not match the book's. Check to see if the student has made an error. If so, correct his work shown below.

$$\dfrac{2x-6}{x-5}-\dfrac{x+4}{x-5}$$
$$=\dfrac{2x-6-x+4}{x-5}$$
$$=\dfrac{x-2}{x-5}$$

Section 7.3 Adding and Subtracting Rational Expressions with Common Denominators and Least Common Denominator 451

Multiple choice. Select the correct result.

71. $\dfrac{3}{x} + \dfrac{y}{x} =$

  a. $\dfrac{3+y}{x^2}$ b. $\dfrac{3+y}{2x}$ c. $\dfrac{3+y}{x}$

72. $\dfrac{3}{x} - \dfrac{y}{x} =$

  a. $\dfrac{3-y}{x^2}$ b. $\dfrac{3-y}{2x}$ c. $\dfrac{3-y}{x}$

73. $\dfrac{3}{x} \cdot \dfrac{y}{x} =$

  a. $\dfrac{3y}{x}$ b. $\dfrac{3y}{x^2}$ c. $3y$

74. $\dfrac{3}{x} \div \dfrac{y}{x} =$

  a. $\dfrac{3}{y}$ b. $\dfrac{y}{3}$ c. $\dfrac{3}{x^2 y}$

Write each rational expression as an equivalent expression with a denominator of $x - 2$.

75. $\dfrac{5}{2-x}$

76. $\dfrac{8y}{2-x}$

77. $-\dfrac{7+x}{2-x}$

78. $\dfrac{x-3}{-(x-2)}$

△ 79. A square has a side of length $\dfrac{5}{x-2}$ meters. Express its perimeter as a rational expression.

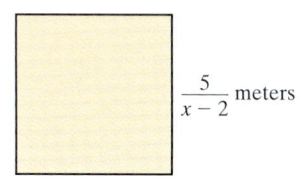

$\dfrac{5}{x-2}$ meters

△ 80. A trapezoid has sides of the indicated lengths. Find its perimeter.

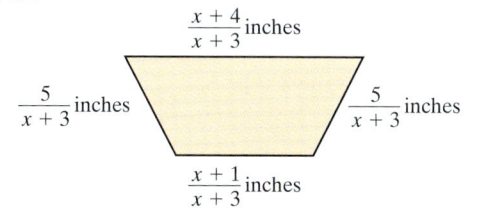

$\dfrac{x+4}{x+3}$ inches

$\dfrac{5}{x+3}$ inches $\dfrac{5}{x+3}$ inches

$\dfrac{x+1}{x+3}$ inches

81. Write two rational expressions with the same denominator whose sum is $\dfrac{5}{3x-1}$.

82. Write two rational expressions with the same denominator whose difference is $\dfrac{x-7}{x^2+1}$.

83. The planet Mercury revolves around the sun in 88 Earth days. It takes Jupiter 4332 Earth days to make one revolution around the sun. (*Source:* National Space Science Data Center) If the two planets are aligned as shown in the figure, how long will it take for them to align again?

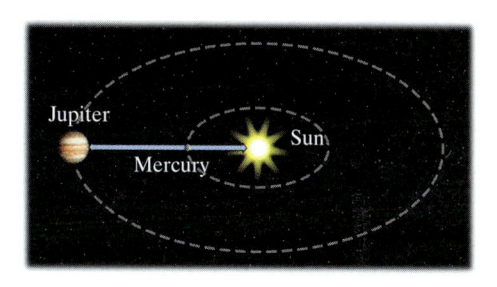

84. You are throwing a barbecue and you want to make sure that you purchase the same number of hot dogs as hot dog buns. Hot dogs come 8 to a package and hot dog buns come 12 to a package. What is the least number of each type of package you should buy?

85. Write some instructions to help a friend who is having difficulty finding the LCD of two rational expressions.

86. Explain why the LCD of the rational expressions $\dfrac{7}{x+1}$ and $\dfrac{9x}{(x+1)^2}$ is $(x+1)^2$ and not $(x+1)^3$.

87. In your own words, describe how to add or subtract two rational expressions with the same denominators.

88. Explain the similarities between subtracting $\dfrac{3}{8}$ from $\dfrac{7}{8}$ and subtracting $\dfrac{6}{x+3}$ from $\dfrac{9}{x+3}$.

STUDY SKILLS BUILDER

How Are You Doing?

If you haven't done so yet, take a few moments and think about how you are doing in this course. Are you working toward your goal of successfully completing this course? Is your performance on homework, quizzes, and tests satisfactory? If not, you might want to see your instructor to see if he/she has any suggestions on how you can improve your performance. Reread Section 1.1 for ideas on places to get help with your mathematics course.

Answer the following.

1. List any textbook supplements you are using to help you through this course.

2. List any campus resources you are using to help you through this course.

3. Write a short paragraph describing how you are doing in your mathematics course.

4. If improvement is needed, list ways that you can work toward improving your situation as described in Exercise 3.

7.4 ADDING AND SUBTRACTING RATIONAL EXPRESSIONS WITH UNLIKE DENOMINATORS

OBJECTIVE

1. Add and subtract rational expressions with unlike denominators.

OBJECTIVE 1 ▶ Adding and subtracting rational expressions with unlike denominators. In the previous section, we practiced all the skills we need to add and subtract rational expressions with unlike or different denominators. We add or subtract rational expressions the same way as we add or subtract fractions. You may want to use the steps below.

Adding or Subtracting Rational Expressions with Unlike Denominators

STEP 1. Find the LCD of the rational expressions.

STEP 2. Rewrite each rational expression as an equivalent expression whose denominator is the LCD found in Step 1.

STEP 3. Add or subtract numerators and write the sum or difference over the common denominator.

STEP 4. Simplify or write the rational expression in simplest form.

EXAMPLE 1 Perform each indicated operation.

a. $\dfrac{a}{4} - \dfrac{2a}{8}$ **b.** $\dfrac{3}{10x^2} + \dfrac{7}{25x}$

Solution

a. First, we must find the LCD. Since $4 = 2^2$ and $8 = 2^3$, the LCD $= 2^3 = 8$. Next we write each fraction as an equivalent fraction with the denominator 8, then we subtract.

$$\dfrac{a}{4} - \dfrac{2a}{8} = \dfrac{a(2)}{4(2)} - \dfrac{2a}{8} = \dfrac{2a}{8} - \dfrac{2a}{8} = \dfrac{2a - 2a}{8} = \dfrac{0}{8} = 0$$

Multiplying the numerator and denominator by 2 is the same as multiplying by $\dfrac{2}{2}$ or 1.

b. Since $10x^2 = 2 \cdot 5 \cdot x \cdot x$ and $25x = 5 \cdot 5 \cdot x$, the LCD $= 2 \cdot 5^2 \cdot x^2 = 50x^2$. We write each fraction as an equivalent fraction with a denominator of $50x^2$.

$$\dfrac{3}{10x^2} + \dfrac{7}{25x} = \dfrac{3(5)}{10x^2(5)} + \dfrac{7(2x)}{25x(2x)}$$

$$= \dfrac{15}{50x^2} + \dfrac{14x}{50x^2}$$

$$= \dfrac{15 + 14x}{50x^2} \qquad \text{Add numerators. Write the sum over the common denominator.} \quad \square$$

PRACTICE 1 Perform each indicated operation.

a. $\dfrac{2x}{5} - \dfrac{6x}{15}$ **b.** $\dfrac{7}{8a} + \dfrac{5}{12a^2}$

Section 7.4 Adding and Subtracting Rational Expressions with Unlike Denominators   453

EXAMPLE 2   Subtract: $\dfrac{6x}{x^2 - 4} - \dfrac{3}{x + 2}$

Solution   Since $x^2 - 4 = (x + 2)(x - 2)$, the LCD $= (x - 2)(x + 2)$. We write equivalent expressions with the LCD as denominators.

$$\dfrac{6x}{x^2 - 4} - \dfrac{3}{x + 2} = \dfrac{6x}{(x - 2)(x + 2)} - \dfrac{3(x - 2)}{(x + 2)(x - 2)}$$

$$= \dfrac{6x - 3(x - 2)}{(x + 2)(x - 2)} \quad \text{Subtract numerators. Write the difference over the common denominator.}$$

$$= \dfrac{6x - 3x + 6}{(x + 2)(x - 2)} \quad \text{Apply the distributive property in the numerator.}$$

$$= \dfrac{3x + 6}{(x + 2)(x - 2)} \quad \text{Combine like terms in the numerator.}$$

Next we factor the numerator to see if this rational expression can be simplified.

$$= \dfrac{3(x + 2)}{(x + 2)(x - 2)} \quad \text{Factor.}$$

$$= \dfrac{3}{x - 2} \quad \text{Divide out common factors to simplify.} \quad \square$$

PRACTICE 2   Subtract: $\dfrac{12x}{x^2 - 25} - \dfrac{6}{x + 5}$

EXAMPLE 3   Add: $\dfrac{2}{3t} + \dfrac{5}{t + 1}$

Solution   The LCD is $3t(t + 1)$. We write each rational expression as an equivalent rational expression with a denominator of $3t(t + 1)$.

$$\dfrac{2}{3t} + \dfrac{5}{t + 1} = \dfrac{2(t + 1)}{3t(t + 1)} + \dfrac{5(3t)}{(t + 1)(3t)}$$

$$= \dfrac{2(t + 1) + 5(3t)}{3t(t + 1)} \quad \text{Add numerators. Write the sum over the common denominator.}$$

$$= \dfrac{2t + 2 + 15t}{3t(t + 1)} \quad \text{Apply the distributive property in the numerator.}$$

$$= \dfrac{17t + 2}{3t(t + 1)} \quad \text{Combine like terms in the numerator.} \quad \square$$

PRACTICE 3   Add: $\dfrac{3}{5y} + \dfrac{2}{y + 1}$

EXAMPLE 4   Subtract: $\dfrac{7}{x - 3} - \dfrac{9}{3 - x}$

Solution   To find a common denominator, we notice that $x - 3$ and $3 - x$ are opposites. That is, $3 - x = -(x - 3)$. We write the denominator $3 - x$ as $-(x - 3)$ and simplify.

$$\frac{7}{x-3} - \frac{9}{3-x} = \frac{7}{x-3} - \frac{9}{-(x-3)}$$

$$= \frac{7}{x-3} - \frac{-9}{x-3} \qquad \text{Apply } \frac{a}{-b} = \frac{-a}{b}.$$

$$= \frac{7-(-9)}{x-3} \qquad \text{Subtract numerators. Write the difference over the common denominator.}$$

$$= \frac{16}{x-3}$$

PRACTICE 4 Subtract: $\dfrac{6}{x-5} - \dfrac{7}{5-x}$

EXAMPLE 5 Add: $1 + \dfrac{m}{m+1}$

Solution Recall that 1 is the same as $\dfrac{1}{1}$. The LCD of $\dfrac{1}{1}$ and $\dfrac{m}{m+1}$ is $m+1$.

$$1 + \frac{m}{m+1} = \frac{1}{1} + \frac{m}{m+1} \qquad \text{Write 1 as } \frac{1}{1}.$$

$$= \frac{1(m+1)}{1(m+1)} + \frac{m}{m+1} \qquad \text{Multiply both the numerator and the denominator of } \frac{1}{1} \text{ by } m+1.$$

$$= \frac{m+1+m}{m+1} \qquad \text{Add numerators. Write the sum over the common denominator.}$$

$$= \frac{2m+1}{m+1} \qquad \text{Combine like terms in the numerator.}$$

PRACTICE 5 Add: $2 + \dfrac{b}{b+3}$

EXAMPLE 6 Subtract: $\dfrac{3}{2x^2+x} - \dfrac{2x}{6x+3}$

Solution First, we factor the denominators.

$$\frac{3}{2x^2+x} - \frac{2x}{6x+3} = \frac{3}{x(2x+1)} - \frac{2x}{3(2x+1)}$$

The LCD is $3x(2x+1)$. We write equivalent expressions with denominators of $3x(2x+1)$.

$$= \frac{3(3)}{x(2x+1)(3)} - \frac{2x(x)}{3(2x+1)(x)}$$

$$= \frac{9-2x^2}{3x(2x+1)} \qquad \text{Subtract numerators. Write the difference over the common denominator.}$$

PRACTICE 6 Subtract: $\dfrac{5}{2x^2+3x} - \dfrac{3x}{4x+6}$

Section 7.4 Adding and Subtracting Rational Expressions with Unlike Denominators 455

EXAMPLE 7 Add: $\dfrac{2x}{x^2 + 2x + 1} + \dfrac{x}{x^2 - 1}$

Solution First we factor the denominators.

$$\dfrac{2x}{x^2 + 2x + 1} + \dfrac{x}{x^2 - 1} = \dfrac{2x}{(x + 1)(x + 1)} + \dfrac{x}{(x + 1)(x - 1)}$$

Now we write the rational expressions as equivalent expressions with denominators of $(x + 1)(x + 1)(x - 1)$, the LCD.

$$= \dfrac{2x(x - 1)}{(x + 1)(x + 1)(x - 1)} + \dfrac{x(x + 1)}{(x + 1)(x - 1)(x + 1)}$$

$$= \dfrac{2x(x - 1) + x(x + 1)}{(x + 1)^2(x - 1)} \quad \text{Add numerators. Write the sum over the common denominator.}$$

$$= \dfrac{2x^2 - 2x + x^2 + x}{(x + 1)^2(x - 1)} \quad \text{Apply the distributive property in the numerator.}$$

$$= \dfrac{3x^2 - x}{(x + 1)^2(x - 1)} \quad \text{or} \quad \dfrac{x(3x - 1)}{(x + 1)^2(x - 1)}$$

PRACTICE 7 Add: $\dfrac{2x}{x^2 + 7x + 12} + \dfrac{3x}{x^2 - 9}$

The numerator was factored as a last step to see if the rational expression could be simplified further. Since there are no factors common to the numerator and the denominator, we can't simplify further.

VOCABULARY & READINESS CHECK

Match each exercise with the first step needed to perform the operation. Do not actually perform the operation.

1. $\dfrac{3}{4} - \dfrac{y}{4}$ 2. $\dfrac{2}{a} \cdot \dfrac{3}{(a + 6)}$ 3. $\dfrac{x + 1}{x} \div \dfrac{x - 1}{x}$ 4. $\dfrac{9}{x - 2} - \dfrac{x}{x + 2}$

a. Multiply the first rational expression by the reciprocal of the second rational expression.
b. Find the LCD. Write each expression as an equivalent expression with the LCD as denominator.
c. Multiply numerators, then multiply denominators.
d. Subtract numerators. Place the difference over a common denominator.

7.4 EXERCISE SET

MIXED PRACTICE

Perform each indicated operation. Simplify if possible. See Examples 1 through 7.

1. $\dfrac{4}{2x} + \dfrac{9}{3x}$

2. $\dfrac{15}{7a} + \dfrac{8}{6a}$

3. $\dfrac{15a}{b} - \dfrac{6b}{5}$

4. $\dfrac{4c}{d} - \dfrac{8d}{5}$

5. $\dfrac{3}{x} + \dfrac{5}{2x^2}$

6. $\dfrac{14}{3x^2} + \dfrac{6}{x}$

7. $\dfrac{6}{x + 1} + \dfrac{10}{2x + 2}$

8. $\dfrac{8}{x + 4} - \dfrac{3}{3x + 12}$

9. $\dfrac{3}{x + 2} - \dfrac{2x}{x^2 - 4}$

10. $\dfrac{5}{x - 4} + \dfrac{4x}{x^2 - 16}$

11. $\dfrac{3}{4x} + \dfrac{8}{x - 2}$

12. $\dfrac{5}{y^2} - \dfrac{y}{2y + 1}$

13. $\dfrac{6}{x - 3} + \dfrac{8}{3 - x}$

14. $\dfrac{15}{y - 4} + \dfrac{20}{4 - y}$

15. $\dfrac{9}{x-3} + \dfrac{9}{3-x}$

16. $\dfrac{5}{a-7} + \dfrac{5}{7-a}$

17. $\dfrac{-8}{x^2-1} - \dfrac{7}{1-x^2}$

18. $\dfrac{-9}{25x^2-1} + \dfrac{7}{1-25x^2}$

19. $\dfrac{5}{x} + 2$

20. $\dfrac{7}{x^2} - 5x$

21. $\dfrac{5}{x-2} + 6$

22. $\dfrac{6y}{y+5} + 1$

23. $\dfrac{y+2}{y+3} - 2$

24. $\dfrac{7}{2x-3} - 3$

25. $\dfrac{-x+2}{x} - \dfrac{x-6}{4x}$

26. $\dfrac{-y+1}{y} - \dfrac{2y-5}{3y}$

27. $\dfrac{5x}{x+2} - \dfrac{3x-4}{x+2}$

28. $\dfrac{7x}{x-3} - \dfrac{4x+9}{x-3}$

29. $\dfrac{3x^4}{7} - \dfrac{4x^2}{21}$

30. $\dfrac{5x}{6} + \dfrac{11x^2}{2}$

31. $\dfrac{1}{x+3} - \dfrac{1}{(x+3)^2}$

32. $\dfrac{5x}{(x-2)^2} - \dfrac{3}{x-2}$

33. $\dfrac{4}{5b} + \dfrac{1}{b-1}$

34. $\dfrac{1}{y+5} + \dfrac{2}{3y}$

35. $\dfrac{2}{m} + 1$

36. $\dfrac{6}{x} - 1$

37. $\dfrac{2x}{x-7} - \dfrac{x}{x-2}$

38. $\dfrac{9x}{x-10} - \dfrac{x}{x-3}$

39. $\dfrac{6}{1-2x} - \dfrac{4}{2x-1}$

40. $\dfrac{10}{3n-4} - \dfrac{5}{4-3n}$

41. $\dfrac{7}{(x+1)(x-1)} + \dfrac{8}{(x+1)^2}$

42. $\dfrac{5}{(x+1)(x+5)} - \dfrac{2}{(x+5)^2}$

43. $\dfrac{x}{x^2-1} - \dfrac{2}{x^2-2x+1}$

44. $\dfrac{x}{x^2-4} - \dfrac{5}{x^2-4x+4}$

45. $\dfrac{3a}{2a+6} - \dfrac{a-1}{a+3}$

46. $\dfrac{1}{x+y} - \dfrac{y}{x^2-y^2}$

47. $\dfrac{y-1}{2y+3} + \dfrac{3}{(2y+3)^2}$

48. $\dfrac{x-6}{5x+1} + \dfrac{6}{(5x+1)^2}$

49. $\dfrac{5}{2-x} + \dfrac{x}{2x-4}$

50. $\dfrac{-1}{a-2} + \dfrac{4}{4-2a}$

51. $\dfrac{15}{x^2+6x+9} + \dfrac{2}{x+3}$

52. $\dfrac{2}{x^2+4x+4} + \dfrac{1}{x+2}$

53. $\dfrac{13}{x^2-5x+6} - \dfrac{5}{x-3}$

54. $\dfrac{-7}{y^2-3y+2} - \dfrac{2}{y-1}$

55. $\dfrac{70}{m^2-100} + \dfrac{7}{2(m+10)}$

56. $\dfrac{27}{y^2-81} + \dfrac{3}{2(y+9)}$

57. $\dfrac{x+8}{x^2-5x-6} + \dfrac{x+1}{x^2-4x-5}$

58. $\dfrac{x+4}{x^2+12x+20} + \dfrac{x+1}{x^2+8x-20}$

59. $\dfrac{5}{4n^2-12n+8} - \dfrac{3}{3n^2-6n}$

60. $\dfrac{6}{5y^2-25y+30} - \dfrac{2}{4y^2-8y}$

MIXED PRACTICE

Perform the indicated operations. Addition, subtraction, multiplication, and division of rational expressions are included here.

61. $\dfrac{15x}{x+8} \cdot \dfrac{2x+16}{3x}$

62. $\dfrac{9z+5}{15} \cdot \dfrac{5z}{81z^2-25}$

63. $\dfrac{8x+7}{3x+5} - \dfrac{2x-3}{3x+5}$

Section 7.4 Adding and Subtracting Rational Expressions with Unlike Denominators 457

64. $\dfrac{2z^2}{4z-1} - \dfrac{z-2z^2}{4z-1}$

65. $\dfrac{5a+10}{18} \div \dfrac{a^2-4}{10a}$

66. $\dfrac{9}{x^2-1} \div \dfrac{12}{3x+3}$

67. $\dfrac{5}{x^2-3x+2} + \dfrac{1}{x-2}$

68. $\dfrac{4}{2x^2+5x-3} + \dfrac{2}{x+3}$

REVIEW AND PREVIEW

Solve the following linear and quadratic equations. See Sections 2.4 and 6.5.

69. $3x + 5 = 7$

70. $5x - 1 = 8$

71. $2x^2 - x - 1 = 0$

72. $4x^2 - 9 = 0$

73. $4(x+6) + 3 = -3$

74. $2(3x+1) + 15 = -7$

CONCEPT EXTENSIONS

Perform each indicated operation.

75. $\dfrac{3}{x} - \dfrac{2x}{x^2-1} + \dfrac{5}{x+1}$

76. $\dfrac{5}{x-2} + \dfrac{7x}{x^2-4} - \dfrac{11}{x}$

77. $\dfrac{5}{x^2-4} + \dfrac{2}{x^2-4x+4} - \dfrac{3}{x^2-x-6}$

78. $\dfrac{8}{x^2+6x+5} - \dfrac{3x}{x^2+4x-5} + \dfrac{2}{x^2-1}$

79. $\dfrac{9}{x^2+9x+14} - \dfrac{3x}{x^2+10x+21} + \dfrac{x+4}{x^2+5x+6}$

80. $\dfrac{x+10}{x^2-3x-4} - \dfrac{8}{x^2+6x+5} - \dfrac{9}{x^2+x-20}$

81. A board of length $\dfrac{3}{x+4}$ inches was cut into two pieces. If one piece is $\dfrac{1}{x-4}$ inches, express the length of the other piece as a rational expression.

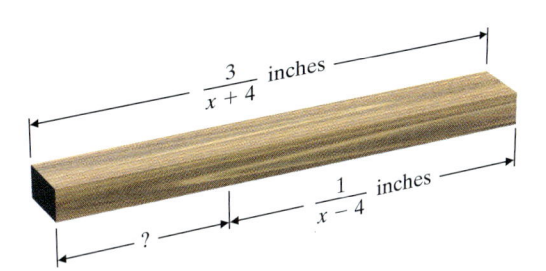

82. The length of a rectangle is $\dfrac{3}{y-5}$ feet, while its width is $\dfrac{2}{y}$ feet. Find its perimeter and then find its area.

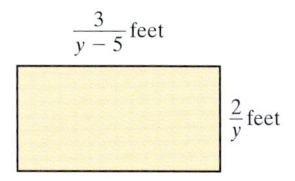

83. In ice hockey, penalty killing percentage is a statistic calculated as $1 - \dfrac{G}{P}$, where G = opponent's power play goals and P = opponent's power play opportunities. Simplify this expression.

84. The dose of medicine prescribed for a child depends on the child's age A in years and the adult dose D for the medication. Two expressions that give a child's dose are Young's Rule, $\dfrac{DA}{A+12}$, and Cowling's Rule, $\dfrac{D(A+1)}{24}$. Find an expression for the difference in the doses given by these expressions.

85. Explain when the LCD of the rational expressions in a sum is the product of the denominators.

86. Explain when the LCD is the same as one of the denominators of a rational expression to be added or subtracted.

87. Two angles are said to be complementary if the sum of their measures is 90°. If one angle measures $\dfrac{40}{x}$ degrees, find the measure of its complement.

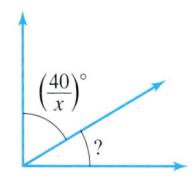

88. Two angles are said to be supplementary if the sum of their measures is 180°. If one angle measures $\dfrac{x+2}{x}$ degrees, find the measure of its supplement.

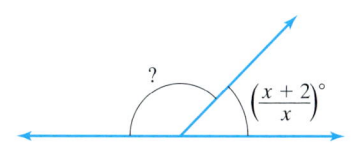

89. In your own words, explain how to add two rational expressions with different denominators.

90. In your own words, explain how to subtract two rational expressions with different denominators.

THE BIGGER PICTURE SIMPLIFYING EXPRESSIONS AND SOLVING EQUATIONS

Now we continue our outline from Sections 1.7, 2.9, 5.6, and 6.6. Although suggestions are given, this outline should be in your own words. Once you complete this new portion, try the exercises below.

I. Simplifying Expressions

   A. Real Numbers
      1. Add (Section 1.5)
      2. Subtract (Section 1.6)
      3. Multiply or Divide (Section 1.7)
   B. Exponents (Section 5.1)
   C. Polynomials
      1. Add (Section 5.2)
      2. Subtract (Section 5.2)
      3. Multiply (Section 5.3)
      4. Divide (Section 5.6)
   D. Factoring Polynomials (Chapter 6 Integrated Review)
   E. Rational Expressions
      1. Simplify: Factor the numerator and denominator. Then divide out factors of 1 by dividing out common factors in the numerator and denominator.

      $$\frac{x^2 - 9}{7x^2 - 21x} = \frac{(x+3)(x-3)}{7x(x-3)} = \frac{x+3}{7x}$$

      2. Multiply: Multiply numerators, then multiply denominators.

      $$\frac{5z}{2z^2 - 9z - 18} \cdot \frac{22z + 33}{10z}$$
      $$= \frac{5 \cdot z}{(2z+3)(z-6)} \cdot \frac{11(2z+3)}{2 \cdot 5 \cdot z} = \frac{11}{2(z-6)}$$

      3. Divide: First fraction times the reciprocal of the second fraction.

      $$\frac{14}{x+5} \div \frac{x+1}{2} = \frac{14}{x+5} \cdot \frac{2}{x+1}$$
      $$= \frac{28}{(x+5)(x+1)}$$

      4. Add or Subtract: Must have same denominator. If not find the LCD and write each fraction as an equivalent fraction with the LCD as denominator.

      $$\frac{9}{10} - \frac{x+1}{x+5} = \frac{9(x+5)}{10(x+5)} - \frac{10(x+1)}{10(x+5)}$$
      $$= \frac{9x + 45 - 10x - 10}{10(x+5)}$$
      $$= \frac{-x + 35}{10(x+5)}$$

II. Solving Equations and Inequalities

   A. Linear Equations (Section 2.4)
   B. Linear Inequalities (Section 2.9)
   C. Quadratic & Higher Degree Equations (Section 6.6)

Perform indicated operations and simplify.

1. $-8.6 + (-9.1)$
2. $(-8.6)(-9.1)$
3. $14 - (-14)$
4. $3x^4 - 7 + x^4 - x^2 - 10$
5. $\dfrac{5x^2 - 5}{25x + 25}$
6. $\dfrac{7x}{x^2 + 4x + 3} \div \dfrac{x}{2x + 6}$
7. $\dfrac{2}{9} - \dfrac{5}{6}$
8. $\dfrac{x}{9} - \dfrac{x+3}{5}$

Factor.

9. $9x^3 - 2x^2 - 11x$
10. $12xy - 21x + 4y - 7$

Solve.

11. $7x - 14 = 5x + 10$
12. $\dfrac{-x + 2}{5} < \dfrac{3}{10}$
13. $1 + 4(x + 4) = 3^2 + x$
14. $x(x - 2) = 24$

7.5 SOLVING EQUATIONS CONTAINING RATIONAL EXPRESSIONS

OBJECTIVES

1. Solve equations containing rational expressions.
2. Solve equations containing rational expressions for a specified variable.

OBJECTIVE 1 ▶ Solving equations containing rational expressions. In Chapter 2, we solved equations containing fractions. In this section, we continue the work we began in Chapter 2 by solving equations containing rational expressions.

Examples of Equations Containing Rational Expressions

$$\frac{x}{2} + \frac{8}{3} = \frac{1}{6} \quad \text{and} \quad \frac{4x}{x^2 + x - 30} + \frac{2}{x - 5} = \frac{1}{x + 6}$$

To solve equations such as these, use the multiplication property of equality to clear the equation of fractions by multiplying both sides of the equation by the LCD.

EXAMPLE 1 Solve: $\frac{x}{2} + \frac{8}{3} = \frac{1}{6}$

Solution The LCD of denominators 2, 3, and 6 is 6, so we multiply both sides of the equation by 6.

$$6\left(\frac{x}{2} + \frac{8}{3}\right) = 6\left(\frac{1}{6}\right)$$

$$6\left(\frac{x}{2}\right) + 6\left(\frac{8}{3}\right) = 6\left(\frac{1}{6}\right) \quad \text{Use the distributive property.}$$

$$3 \cdot x + 16 = 1 \quad \text{Multiply and simplify.}$$

$$3x = -15 \quad \text{Subtract 16 from both sides.}$$

$$x = -5 \quad \text{Divide both sides by 3.}$$

▶ **Helpful Hint**
Make sure that *each* term is multiplied by the LCD, 6.

Check: To check, we replace x with -5 in the original equation.

$$\frac{x}{2} + \frac{8}{3} = \frac{1}{6}$$

$$\frac{-5}{2} + \frac{8}{3} \stackrel{?}{=} \frac{1}{6} \quad \text{Replace } x \text{ with } -5.$$

$$\frac{1}{6} = \frac{1}{6} \quad \text{True}$$

This number checks, so the solution is -5.

PRACTICE 1 Solve: $\frac{x}{3} + \frac{4}{5} = \frac{2}{15}$

EXAMPLE 2 Solve: $\frac{t - 4}{2} - \frac{t - 3}{9} = \frac{5}{18}$

Solution The LCD of denominators 2, 9, and 18 is 18, so we multiply both sides of the equation by 18.

$$18\left(\frac{t - 4}{2} - \frac{t - 3}{9}\right) = 18\left(\frac{5}{18}\right)$$

$$18\left(\frac{t - 4}{2}\right) - 18\left(\frac{t - 3}{9}\right) = 18\left(\frac{5}{18}\right) \quad \text{Use the distributive property.}$$

$$9(t - 4) - 2(t - 3) = 5 \quad \text{Simplify.}$$

$$9t - 36 - 2t + 6 = 5 \quad \text{Use the distributive property.}$$

$$7t - 30 = 5 \quad \text{Combine like terms.}$$

$$7t = 35$$

$$t = 5 \quad \text{Solve for } t.$$

▶ **Helpful Hint**
Multiply *each* term by 18.

460 CHAPTER 7 Rational Expressions

Check:
$$\frac{t-4}{2} - \frac{t-3}{9} = \frac{5}{18}$$
$$\frac{5-4}{2} - \frac{5-3}{9} \stackrel{?}{=} \frac{5}{18} \quad \text{Replace } t \text{ with 5.}$$
$$\frac{1}{2} - \frac{2}{9} \stackrel{?}{=} \frac{5}{18} \quad \text{Simplify.}$$
$$\frac{5}{18} = \frac{5}{18} \quad \text{True}$$

The solution is 5.

PRACTICE 2 Solve: $\dfrac{x+4}{4} - \dfrac{x-3}{3} = \dfrac{11}{12}$

Recall from Section 7.1 that a rational expression is defined for all real numbers except those that make the denominator of the expression 0. This means that if an equation contains *rational expressions with variables in the denominator,* we must be certain that the proposed solution does not make the denominator 0. If replacing the variable with the proposed solution makes the denominator 0, the rational expression is undefined and this proposed solution must be rejected.

EXAMPLE 3 Solve: $3 - \dfrac{6}{x} = x + 8$

Solution In this equation, 0 cannot be a solution because if x is 0, the rational expression $\dfrac{6}{x}$ is undefined. The LCD is x, so we multiply both sides of the equation by x.

$$x\left(3 - \frac{6}{x}\right) = x(x + 8)$$

▶ **Helpful Hint**
Multiply *each* term by x.

$$x(3) - x\left(\frac{6}{x}\right) = x \cdot x + x \cdot 8 \quad \text{Use the distributive property.}$$
$$3x - 6 = x^2 + 8x \quad \text{Simplify.}$$

Now we write the quadratic equation in standard form and solve for x.

$$0 = x^2 + 5x + 6$$
$$0 = (x + 3)(x + 2) \quad \text{Factor.}$$
$$x + 3 = 0 \quad \text{or} \quad x + 2 = 0 \quad \text{Set each factor equal to 0 and solve.}$$
$$x = -3 \qquad\qquad x = -2$$

Notice that neither -3 nor -2 makes the denominator in the original equation equal to 0.

Check: To check these solutions, we replace x in the original equation by -3, and then by -2.

If $x = -3$:
$$3 - \frac{6}{x} = x + 8$$
$$3 - \frac{6}{-3} \stackrel{?}{=} -3 + 8$$
$$3 - (-2) \stackrel{?}{=} 5$$
$$5 = 5 \quad \text{True}$$

If $x = -2$:
$$3 - \frac{6}{x} = x + 8$$
$$3 - \frac{6}{-2} \stackrel{?}{=} -2 + 8$$
$$3 - (-3) \stackrel{?}{=} 6$$
$$6 = 6 \quad \text{True}$$

Both -3 and -2 are solutions.

PRACTICE 3 Solve: $8 + \dfrac{7}{x} = x + 2$

Section 7.5 Solving Equations Containing Rational Expressions 461

The following steps may be used to solve an equation containing rational expressions.

> **Solving an Equation Containing Rational Expressions**
> **STEP 1.** Multiply both sides of the equation by the LCD of all rational expressions in the equation.
> **STEP 2.** Remove any grouping symbols and solve the resulting equation.
> **STEP 3.** Check the solution in the original equation.

EXAMPLE 4 Solve: $\dfrac{4x}{x^2 + x - 30} + \dfrac{2}{x - 5} = \dfrac{1}{x + 6}$

Solution

The denominator $x^2 + x - 30$ factors as $(x + 6)(x - 5)$. The LCD is then $(x + 6)(x - 5)$, so we multiply both sides of the equation by this LCD.

$(x + 6)(x - 5)\left(\dfrac{4x}{x^2 + x - 30} + \dfrac{2}{x - 5}\right) = (x + 6)(x - 5)\left(\dfrac{1}{x + 6}\right)$ Multiply by the LCD.

$(x + 6)(x - 5) \cdot \dfrac{4x}{x^2 + x - 30} + (x + 6)(x - 5) \cdot \dfrac{2}{x - 5}$ Apply the distributive property.

$= (x + 6)(x - 5) \cdot \dfrac{1}{x + 6}$

$4x + 2(x + 6) = x - 5$ Simplify.
$4x + 2x + 12 = x - 5$ Apply the distributive property.
$6x + 12 = x - 5$ Combine like terms.
$5x = -17$
$x = -\dfrac{17}{5}$ Divide both sides by 5.

Check: Check by replacing x with $-\dfrac{17}{5}$ in the original equation. The solution is $-\dfrac{17}{5}$.

PRACTICE 4 Solve: $\dfrac{6x}{x^2 - 5x - 14} - \dfrac{3}{x + 2} = \dfrac{1}{x - 7}$

EXAMPLE 5 Solve: $\dfrac{2x}{x - 4} = \dfrac{8}{x - 4} + 1$

Solution Multiply both sides by the LCD, $x - 4$.

$(x - 4)\left(\dfrac{2x}{x - 4}\right) = (x - 4)\left(\dfrac{8}{x - 4} + 1\right)$ Multiply by the LCD. Notice that 4 cannot be a solution.

$(x - 4) \cdot \dfrac{2x}{x - 4} = (x - 4) \cdot \dfrac{8}{x - 4} + (x - 4) \cdot 1$ Use the distributive property.

$2x = 8 + (x - 4)$ Simplify.
$2x = 4 + x$
$x = 4$

Notice that 4 makes the denominator 0 in the original equation. Therefore, 4 is *not* a solution.

This equation has *no solution*.

PRACTICE 5 Solve: $\dfrac{7}{x - 2} = \dfrac{3}{x - 2} + 4$

> **Helpful Hint**
> As we can see from Example 5, it is important to check the proposed solution(s) in the *original* equation.

Concept Check ✓

When can we clear fractions by multiplying through by the LCD?

a. When adding or subtracting rational expressions
b. When solving an equation containing rational expressions
c. Both of these
d. Neither of these

EXAMPLE 6 Solve: $x + \dfrac{14}{x-2} = \dfrac{7x}{x-2} + 1$

Solution Notice the denominators in this equation. We can see that 2 can't be a solution. The LCD is $x - 2$, so we multiply both sides of the equation by $x - 2$.

$$(x-2)\left(x + \frac{14}{x-2}\right) = (x-2)\left(\frac{7x}{x-2} + 1\right)$$

$$(x-2)(x) + (x-2)\left(\frac{14}{x-2}\right) = (x-2)\left(\frac{7x}{x-2}\right) + (x-2)(1)$$

$x^2 - 2x + 14 = 7x + x - 2$	Simplify.
$x^2 - 2x + 14 = 8x - 2$	Combine like terms.
$x^2 - 10x + 16 = 0$	Write the quadratic equation in standard form.
$(x-8)(x-2) = 0$	Factor.
$x - 8 = 0 \quad \text{or} \quad x - 2 = 0$	Set each factor equal to 0.
$x = 8 \qquad\qquad x = 2$	Solve.

As we have already noted, 2 can't be a solution of the original equation. So we need only replace x with 8 in the original equation. We find that 8 is a solution; the only solution is 8. ☐

PRACTICE 6 Solve: $x + \dfrac{x}{x-5} = \dfrac{5}{x-5} - 7$

OBJECTIVE 2 ▶ Solving equations for a specified variable. The last example in this section is an equation containing several variables, and we are directed to solve for one of the variables. The steps used in the preceding examples can be applied to solve equations for a specified variable as well.

EXAMPLE 7 Solve: $\dfrac{1}{a} + \dfrac{1}{b} = \dfrac{1}{x}$ for x.

Solution (This type of equation often models a work problem, as we shall see in Section 7.6.) The LCD is abx, so we multiply both sides by abx.

$$abx\left(\frac{1}{a} + \frac{1}{b}\right) = abx\left(\frac{1}{x}\right)$$

$$abx\left(\frac{1}{a}\right) + abx\left(\frac{1}{b}\right) = abx \cdot \frac{1}{x}$$

$bx + ax = ab$	Simplify.
$x(b + a) = ab$	Factor out x from each term on the left side.

Answer to Concept Check: b

$$\frac{x(b+a)}{b+a} = \frac{ab}{b+a} \quad \text{Divide both sides by } b+a.$$

$$x = \frac{ab}{b+a} \quad \text{Simplify.}$$

This equation is now solved for x.

PRACTICE 7 Solve: $\dfrac{1}{a} + \dfrac{1}{b} = \dfrac{1}{x}$ for b

Graphing Calculator Explorations

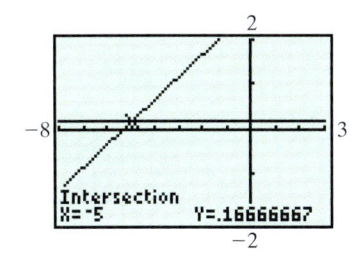

A graphing calculator may be used to check solutions of equations containing rational expressions. For example, to check the solution of Example 1, $\dfrac{x}{2} + \dfrac{8}{3} = \dfrac{1}{6}$, graph $y_1 = \dfrac{x}{2} + \dfrac{8}{3}$ and $y_2 = \dfrac{1}{6}$.

Use TRACE and ZOOM, or use INTERSECT, to find the point of intersection. The point of intersection has an x-value of -5, so the solution of the equation is -5.

Use a graphing calculator to check the examples of this section.

1. Example 2
2. Example 3
3. Example 5
4. Example 6

7.5 EXERCISE SET

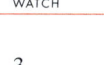

Solve each equation and check each solution. See Examples 1 through 3.

1. $\dfrac{x}{5} + 3 = 9$
2. $\dfrac{x}{5} - 2 = 9$
3. $\dfrac{x}{2} + \dfrac{5x}{4} = \dfrac{x}{12}$
4. $\dfrac{x}{6} + \dfrac{4x}{3} = \dfrac{x}{18}$
5. $2 - \dfrac{8}{x} = 6$
6. $5 + \dfrac{4}{x} = 1$
7. $2 + \dfrac{10}{x} = x + 5$
8. $6 + \dfrac{5}{y} = y - \dfrac{2}{y}$
9. $\dfrac{a}{5} = \dfrac{a-3}{2}$
10. $\dfrac{b}{5} = \dfrac{b+2}{6}$
11. $\dfrac{x-3}{5} + \dfrac{x-2}{2} = \dfrac{1}{2}$
12. $\dfrac{a+5}{4} + \dfrac{a+5}{2} = \dfrac{a}{8}$

Solve each equation and check each proposed solution. See Examples 4 through 6.

13. $\dfrac{3}{2a-5} = -1$
14. $\dfrac{6}{4-3x} = -3$
15. $\dfrac{4y}{y-4} + 5 = \dfrac{5y}{y-4}$
16. $\dfrac{2a}{a+2} - 5 = \dfrac{7a}{a+2}$

17. $2 + \dfrac{3}{a-3} = \dfrac{a}{a-3}$
18. $\dfrac{2y}{y-2} - \dfrac{4}{y-2} = 4$
19. $\dfrac{1}{x+3} + \dfrac{6}{x^2-9} = 1$
20. $\dfrac{1}{x+2} + \dfrac{4}{x^2-4} = 1$
21. $\dfrac{2y}{y+4} + \dfrac{4}{y+4} = 3$
22. $\dfrac{5y}{y+1} - \dfrac{3}{y+1} = 4$
23. $\dfrac{2x}{x+2} - 2 = \dfrac{x-8}{x-2}$
24. $\dfrac{4y}{y-3} - 3 = \dfrac{3y-1}{y+3}$

MIXED PRACTICE

Solve each equation. See Examples 1 through 6.

25. $\dfrac{2}{y} + \dfrac{1}{2} = \dfrac{5}{2y}$
26. $\dfrac{6}{3y} + \dfrac{3}{y} = 1$
27. $\dfrac{a}{a-6} = \dfrac{-2}{a-1}$
28. $\dfrac{5}{x-6} = \dfrac{x}{x-2}$
29. $\dfrac{11}{2x} + \dfrac{2}{3} = \dfrac{7}{2x}$
30. $\dfrac{5}{3} - \dfrac{3}{2x} = \dfrac{3}{2}$
31. $\dfrac{2}{x-2} + 1 = \dfrac{x}{x+2}$
32. $1 + \dfrac{3}{x+1} = \dfrac{x}{x-1}$
33. $\dfrac{x+1}{3} - \dfrac{x-1}{6} = \dfrac{1}{6}$
34. $\dfrac{3x}{5} - \dfrac{x-6}{3} = -\dfrac{2}{5}$

464 CHAPTER 7 Rational Expressions

35. $\dfrac{t}{t-4} = \dfrac{t+4}{6}$

36. $\dfrac{15}{x+4} = \dfrac{x-4}{x}$

37. $\dfrac{y}{2y+2} + \dfrac{2y-16}{4y+4} = \dfrac{2y-3}{y+1}$

38. $\dfrac{1}{x+2} = \dfrac{4}{x^2-4} - \dfrac{1}{x-2}$

39. $\dfrac{4r-4}{r^2+5r-14} + \dfrac{2}{r+7} = \dfrac{1}{r-2}$

40. $\dfrac{3}{x+3} = \dfrac{12x+19}{x^2+7x+12} - \dfrac{5}{x+4}$

41. $\dfrac{x+1}{x+3} = \dfrac{x^2-11x}{x^2+x-6} - \dfrac{x-3}{x-2}$

42. $\dfrac{2t+3}{t-1} - \dfrac{2}{t+3} = \dfrac{5-6t}{t^2+2t-3}$

Solve each equation for the indicated variable. See Example 7.

43. $R = \dfrac{E}{I}$ for I (Electronics: resistance of a circuit)

44. $T = \dfrac{V}{Q}$ for Q (Water purification: settling time)

45. $T = \dfrac{2U}{B+E}$ for B (Merchandising: stock turnover rate)

46. $i = \dfrac{A}{t+B}$ for t (Hydrology: rainfall intensity)

47. $B = \dfrac{705w}{h^2}$ for w (Health: body-mass index)

48. $\dfrac{A}{W} = L$ for W (Geometry: area of a rectangle)

49. $N = R + \dfrac{V}{G}$ for G (Urban forestry: tree plantings per year)

50. $C = \dfrac{D(A+1)}{24}$ for A (Medicine: Cowling's Rule for child's dose)

51. $\dfrac{C}{\pi r} = 2$ for r (Geometry: circumference of a circle)

52. $W = \dfrac{CE^2}{2}$ for C (Electronics: energy stored in a capacitor)

53. $\dfrac{1}{y} + \dfrac{1}{3} = \dfrac{1}{x}$ for x

54. $\dfrac{1}{5} + \dfrac{2}{y} = \dfrac{1}{x}$ for x

REVIEW AND PREVIEW

Write each phrase as an expression.

55. The reciprocal of x

56. The reciprocal of $x + 1$

57. The reciprocal of x, added to the reciprocal of 2

58. The reciprocal of x, subtracted from the reciprocal of 5

Answer each question.

59. If a tank is filled in 3 hours, what fractional part of the tank is filled in 1 hour?

60. If a strip of beach is cleaned in 4 hours, what fractional part of the beach is cleaned in 1 hour?

Identify the x- and y-intercepts. See Section 3.3.

61.

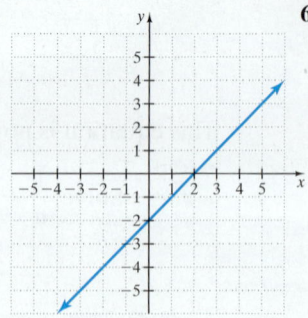

62.

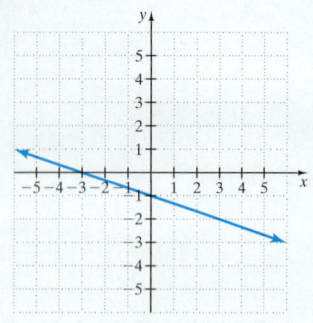

63.

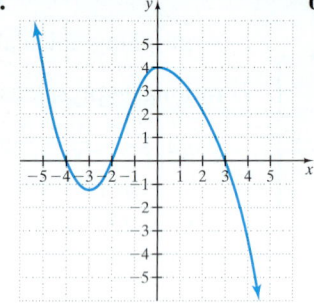

64.

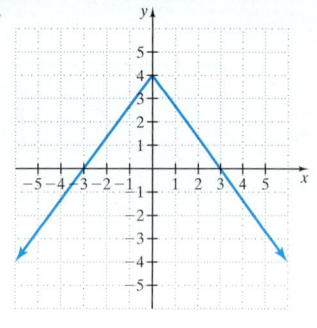

CONCEPT EXTENSIONS

65. Explain the difference between solving an equation such as $\dfrac{x}{2} + \dfrac{3}{4} = \dfrac{x}{4}$ for x and performing an operation such as adding $\dfrac{x}{2} + \dfrac{3}{4}$.

66. When solving an equation such as $\dfrac{y}{4} = \dfrac{y}{2} - \dfrac{1}{4}$, we may multiply all terms by 4. When subtracting two rational expressions such as $\dfrac{y}{2} - \dfrac{1}{4}$, we may not. Explain why.

Determine whether each of the following is an equation or an expression. If it is an equation, then solve it for its variable. If it is an expression, perform the indicated operation.

67. $\dfrac{1}{x} + \dfrac{5}{9}$

68. $\dfrac{1}{x} + \dfrac{5}{9} = \dfrac{2}{3}$

69. $\dfrac{5}{x-1} - \dfrac{2}{x} = \dfrac{5}{x(x-1)}$

70. $\dfrac{5}{x-1} - \dfrac{2}{x}$

Recall that two angles are supplementary if the sum of their measures is 180°. Find the measures of the following supplementary angles.

71. $\left(\dfrac{20x}{3}\right)°$, $\left(\dfrac{32x}{6}\right)°$

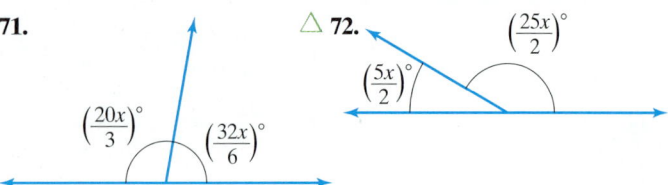

72. $\left(\dfrac{25x}{2}\right)°$, $\left(\dfrac{5x}{2}\right)°$

Recall that two angles are complementary if the sum of their measures is 90°. Find the measures of the following complementary angles.

△ 73.

$\left(\frac{450}{x}\right)°$
$\left(\frac{150}{x}\right)°$

△ 74.

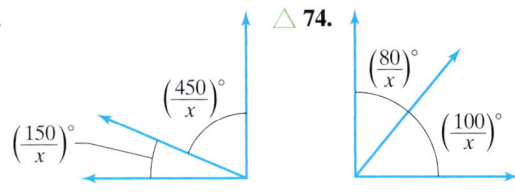

$\left(\frac{80}{x}\right)°$
$\left(\frac{100}{x}\right)°$

Solve each equation.

75. $\dfrac{5}{a^2 + 4a + 3} + \dfrac{2}{a^2 + a - 6} - \dfrac{3}{a^2 - a - 2} = 0$

76. $\dfrac{-2}{a^2 + 2a - 8} + \dfrac{1}{a^2 + 9a + 20} = \dfrac{-4}{a^2 + 3a - 10}$

INTEGRATED REVIEW SUMMARY ON RATIONAL EXPRESSIONS
Sections 7.1–7.5

It is important to know the difference between performing operations with rational expressions and solving an equation containing rational expressions. Study the examples below.

PERFORMING OPERATIONS WITH RATIONAL EXPRESSIONS

Adding: $\dfrac{1}{x} + \dfrac{1}{x + 5} = \dfrac{1 \cdot (x + 5)}{x(x + 5)} + \dfrac{1 \cdot x}{x(x + 5)} = \dfrac{x + 5 + x}{x(x + 5)} = \dfrac{2x + 5}{x(x + 5)}$

Subtracting: $\dfrac{3}{x} - \dfrac{5}{x^2 y} = \dfrac{3 \cdot xy}{x \cdot xy} - \dfrac{5}{x^2 y} = \dfrac{3xy - 5}{x^2 y}$

Multiplying: $\dfrac{2}{x} \cdot \dfrac{5}{x - 1} = \dfrac{2 \cdot 5}{x(x - 1)} = \dfrac{10}{x(x - 1)}$

Dividing: $\dfrac{4}{2x + 1} \div \dfrac{x - 3}{x} = \dfrac{4}{2x + 1} \cdot \dfrac{x}{x - 3} = \dfrac{4x}{(2x + 1)(x - 3)}$

SOLVING AN EQUATION CONTAINING RATIONAL EXPRESSIONS
To solve an equation containing rational expressions, we clear the equation of fractions by multiplying both sides by the LCD.

$\dfrac{3}{x} - \dfrac{5}{x - 1} = \dfrac{1}{x(x - 1)}$ Note that x can't be 0 or 1.

$x(x - 1)\left(\dfrac{3}{x}\right) - x(x - 1)\left(\dfrac{5}{x - 1}\right) = x(x - 1) \cdot \dfrac{1}{x(x - 1)}$ Multiply both sides by the LCD.

$3(x - 1) - 5x = 1$ Simplify.
$3x - 3 - 5x = 1$ Use the distributive property.
$-2x - 3 = 1$ Combine like terms.
$-2x = 4$ Add 3 to both sides.
$x = -2$ Divide both sides by -2.

Determine whether each of the following is an equation or an expression. If it is an equation, solve it for its variable. If it is an expression, perform the indicated operation.

1. $\dfrac{1}{x} + \dfrac{2}{3}$

2. $\dfrac{3}{a} + \dfrac{5}{6}$

3. $\dfrac{1}{x} + \dfrac{2}{3} = \dfrac{3}{x}$

4. $\dfrac{3}{a} + \dfrac{5}{6} = 1$

5. $\dfrac{2}{x - 1} - \dfrac{1}{x}$

6. $\dfrac{4}{x - 3} - \dfrac{1}{x}$

7. $\dfrac{2}{x + 1} - \dfrac{1}{x} = 1$

8. $\dfrac{4}{x - 3} - \dfrac{1}{x} = \dfrac{6}{x(x - 3)}$

9. $\dfrac{15x}{x+8} \cdot \dfrac{2x+16}{3x}$

10. $\dfrac{9z+5}{15} \cdot \dfrac{5z}{81z^2-25}$

11. $\dfrac{2x+1}{x-3} + \dfrac{3x+6}{x-3}$

12. $\dfrac{4p-3}{2p+7} + \dfrac{3p+8}{2p+7}$

13. $\dfrac{x+5}{7} = \dfrac{8}{2}$

14. $\dfrac{1}{2} = \dfrac{x-1}{8}$

15. $\dfrac{5a+10}{18} \div \dfrac{a^2-4}{10a}$

16. $\dfrac{9}{x^2-1} + \dfrac{12}{3x+3}$

17. $\dfrac{x+2}{3x-1} + \dfrac{5}{(3x-1)^2}$

18. $\dfrac{4}{(2x-5)^2} + \dfrac{x+1}{2x-5}$

19. $\dfrac{x-7}{x} - \dfrac{x+2}{5x}$

20. $\dfrac{9}{x^2-4} + \dfrac{2}{x+2} = \dfrac{-1}{x-2}$

21. $\dfrac{3}{x+3} = \dfrac{5}{x^2-9} - \dfrac{2}{x-3}$

22. $\dfrac{10x-9}{x} - \dfrac{x-4}{3x}$

7.6 PROPORTION AND PROBLEM SOLVING WITH RATIONAL EQUATIONS

OBJECTIVES

1. Solve proportions.
2. Use proportions to solve problems.
3. Solve problems about numbers.
4. Solve problems about work.
5. Solve problems about distance.

OBJECTIVE 1 ▶ Solving proportions. A **ratio** is the quotient of two numbers or two quantities. For example, the ratio of 2 to 5 can be written as $\dfrac{2}{5}$, the quotient of 2 and 5.

If two ratios are equal, we say the ratios are **in proportion** to each other. A **proportion** is a mathematical statement that two ratios are equal.

For example, the equation $\dfrac{1}{2} = \dfrac{4}{8}$ is a proportion, as is $\dfrac{x}{5} = \dfrac{8}{10}$, because both sides of the equations are ratios. When we want to emphasize the equation as a proportion, we read the proportion $\dfrac{1}{2} = \dfrac{4}{8}$ as "one is to two as four is to eight"

In a proportion, cross products are equal. To understand cross products, let's start with the proportion

$$\dfrac{a}{b} = \dfrac{c}{d}$$

and multiply both sides by the LCD, bd.

$bd\left(\dfrac{a}{b}\right) = bd\left(\dfrac{c}{d}\right)$ Multiply both sides by the LCD, bd.

$\underbrace{ad}_{\text{Cross product}} = \underbrace{bc}_{\text{Cross product}}$ Simplify.

Notice why ad and bc are called cross products.

Cross Products

If $\dfrac{a}{b} = \dfrac{c}{d}$, then $ad = bc$.

For example, if

$$\frac{1}{2} = \frac{4}{8}, \text{ then } \begin{aligned} 1 \cdot 8 &= 2 \cdot 4 \text{ or} \\ 8 &= 8 \end{aligned}$$

Notice that a proportion contains four numbers (or expressions). If any three numbers are known, we can solve and find the fourth number.

EXAMPLE 1 Solve for x: $\dfrac{45}{x} = \dfrac{5}{7}$

Solution This is an equation with rational expressions, and also a proportion. Below are two ways to solve.

Since this is a rational equation, we can use the methods of the previous section.

$$\frac{45}{x} = \frac{5}{7}$$

$7x \cdot \dfrac{45}{x} = 7x \cdot \dfrac{5}{7}$ Multiply both sides by LCD $7x$.

$7 \cdot 45 = x \cdot 5$ Divide out common factors.

$315 = 5x$ Multiply.

$\dfrac{315}{5} = \dfrac{5x}{5}$ Divide both sides by 5.

$63 = x$ Simplify.

Since this is also a proportion, we may set cross products equal.

$$\frac{45}{x} = \frac{5}{7}$$

$45 \cdot 7 = x \cdot 5$ Set cross products equal.

$315 = 5x$ Multiply.

$\dfrac{315}{5} = \dfrac{5x}{5}$ Divide both sides by 5.

$63 = x$ Simplify.

Check: Both methods give us a solution of 63. To check, substitute 63 for x in the original proportion. The solution is 63.

PRACTICE 1 Solve for x: $\dfrac{36}{x} = \dfrac{4}{11}$

In this section, if the rational equation is a proportion, we will use cross products to solve.

EXAMPLE 2 Solve for x: $\dfrac{x-5}{3} = \dfrac{x+2}{5}$

Solution

$$\frac{x-5}{3} = \frac{x+2}{5}$$

$5(x - 5) = 3(x + 2)$ Set cross products equal.

$5x - 25 = 3x + 6$ Multiply.

$5x = 3x + 31$ Add 25 to both sides.

$2x = 31$ Subtract $3x$ from both sides.

$\dfrac{2x}{2} = \dfrac{31}{2}$ Divide both sides by 2.

$x = \dfrac{31}{2}$

Check: Verify that $\dfrac{31}{2}$ is the solution.

PRACTICE 2 Solve for x: $\dfrac{3x+2}{9} = \dfrac{x-1}{2}$

OBJECTIVE 2 ▶ Using proportions to solve problems. Proportions can be used to model and solve many real-life problems. When using proportions in this way, it is important to judge whether the solution is reasonable. Doing so helps us to decide if the proportion has been formed correctly. We use the same problem-solving steps that were introduced in Section 2.4.

EXAMPLE 3   Calculating the Cost of Recordable Compact Discs

Three boxes of CD-Rs (recordable compact discs) cost $37.47. How much should 5 boxes cost?

Solution

1. **UNDERSTAND.** Read and reread the problem. We know that the cost of 5 boxes is more than the cost of 3 boxes, or $37.47, and less than the cost of 6 boxes, which is double the cost of 3 boxes, or 2($37.47) = $74.94. Let's suppose that 5 boxes cost $60.00. To check, we see if 3 boxes is to 5 boxes as the *price* of 3 boxes is to the *price* of 5 boxes. In other words, we see if

$$\frac{3 \text{ boxes}}{5 \text{ boxes}} = \frac{\text{price of 3 boxes}}{\text{price of 5 boxes}}$$

or

$$\frac{3}{5} = \frac{37.47}{60.00}$$

$3(60.00) = 5(37.47)$ Set cross products equal.

or

$180.00 = 187.35$ Not a true statement.

Thus, $60 is not correct, but we now have a better understanding of the problem.

Let x = price of 5 boxes of CD-Rs.

2. **TRANSLATE.**

$$\frac{3 \text{ boxes}}{5 \text{ boxes}} = \frac{\text{price of 3 boxes}}{\text{price of 5 boxes}}$$

$$\frac{3}{5} = \frac{37.47}{x}$$

3. **SOLVE.**

$$\frac{3}{5} = \frac{37.47}{x}$$

$3x = 5(37.47)$ Set cross products equal.

$3x = 187.35$

$x = 62.45$ Divide both sides by **3**.

4. **INTERPRET.**

Check: Verify that 3 boxes is to 5 boxes as $37.47 is to $62.45. Also, notice that our solution is a reasonable one as discussed in Step 1.

State: Five boxes of CD-Rs cost $62.45.

PRACTICE 3   Four 2-liter bottles of Diet Pepsi cost $5.16. How much will seven 2-liter bottles cost?

> **Helpful Hint**
> The proportion $\dfrac{5 \text{ boxes}}{3 \text{ boxes}} = \dfrac{\text{price of 5 boxes}}{\text{price of 3 boxes}}$ could also have been used to solve Example 3. Notice that the cross products are the same.

Similar triangles have the same shape but not necessarily the same size. In similar triangles, the measures of corresponding angles are equal, and corresponding sides are in proportion.

If triangle ABC and triangle XYZ shown are similar, then we know that the measure of angle A = the measure of angle X, the measure of angle B = the measure of angle Y, and the measure of angle C = the measure of angle Z. We also know that corresponding sides are in proportion: $\dfrac{a}{x} = \dfrac{b}{y} = \dfrac{c}{z}$.

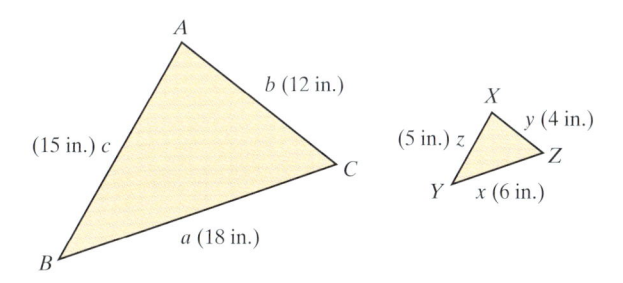

In this section, we will position similar triangles so that they have the same orientation.

To show that corresponding sides are in proportion for the triangles above, we write the ratios of the corresponding sides.

$$\dfrac{a}{x} = \dfrac{18}{6} = 3 \qquad \dfrac{b}{y} = \dfrac{12}{4} = 3 \qquad \dfrac{c}{z} = \dfrac{15}{5} = 3$$

EXAMPLE 4 **Finding the Length of a Side of a Triangle**

If the following two triangles are similar, find the missing length x.

Solution

1. UNDERSTAND. Read the problem and study the figure.
2. TRANSLATE. Since the triangles are similar, their corresponding sides are in proportion and we have

$$\dfrac{2}{3} = \dfrac{10}{x}$$

3. SOLVE. To solve, we multiply both sides by the LCD, $3x$, or cross multiply.

$$2x = 30$$
$$x = 15 \quad \text{Divide both sides by 2.}$$

4. INTERPRET.

Check: To check, replace x with 15 in the original proportion and see that a true statement results.

State: The missing length is 15 yards.

PRACTICE 4 If the following two triangles are similar, find x.

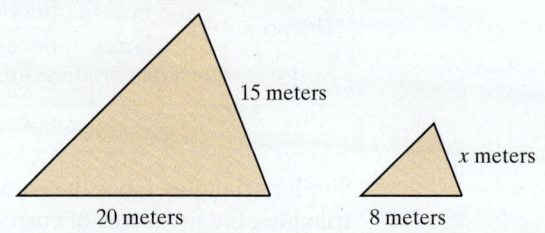

15 meters
20 meters
x meters
8 meters

OBJECTIVE 3 ▶ Solving problems about numbers. Let's continue to solve problems. The remaining problems are all modeled by rational equations.

EXAMPLE 5 Finding an Unknown Number

The quotient of a number and 6, minus $\frac{5}{3}$, is the quotient of the number and 2. Find the number.

Solution

1. UNDERSTAND. Read and reread the problem. Suppose that the unknown number is 2, then we see if the quotient of 2 and 6, or $\frac{2}{6}$, minus $\frac{5}{3}$ is equal to the quotient of 2 and 2, or $\frac{2}{2}$.

$$\frac{2}{6} - \frac{5}{3} = \frac{1}{3} - \frac{5}{3} = -\frac{4}{3}, \text{ not } \frac{2}{2}$$

Don't forget that the purpose of a proposed solution is to better understand the problem.

Let x = the unknown number.

2. TRANSLATE.

In words:	the quotient of x and 6	minus	$\frac{5}{3}$	is	the quotient of x and 2
Translate:	$\frac{x}{6}$	$-$	$\frac{5}{3}$	$=$	$\frac{x}{2}$

3. SOLVE. Here, we solve the equation $\frac{x}{6} - \frac{5}{3} = \frac{x}{2}$. We begin by multiplying both sides of the equation by the LCD, 6.

$$6\left(\frac{x}{6} - \frac{5}{3}\right) = 6\left(\frac{x}{2}\right)$$

$6\left(\frac{x}{6}\right) - 6\left(\frac{5}{3}\right) = 6\left(\frac{x}{2}\right)$ Apply the distributive property.

$x - 10 = 3x$ Simplify.

$-10 = 2x$ Subtract x from both sides.

$\frac{-10}{2} = \frac{2x}{2}$ Divide both sides by 2.

$-5 = x$ Simplify.

4. INTERPRET.

Check: To check, we verify that "the quotient of -5 and 6 minus $\frac{5}{3}$ is the quotient of -5 and 2," or $-\frac{5}{6} - \frac{5}{3} = -\frac{5}{2}$.

State: The unknown number is -5.

PRACTICE 5 The quotient of a number and 5, minus $\frac{3}{2}$, is the quotient of the number and 10.

OBJECTIVE 4 ▶ Solving problems about work. The next example is often called a work problem. Work problems usually involve people or machines doing a certain task.

EXAMPLE 6   Finding Work Rates

Sam Waterton and Frank Schaffer work in a plant that manufactures automobiles. Sam can complete a quality control tour of the plant in 3 hours while his assistant, Frank, needs 7 hours to complete the same job. The regional manager is coming to inspect the plant facilities, so both Sam and Frank are directed to complete a quality control tour together. How long will this take?

Solution

1. UNDERSTAND. Read and reread the problem. The key idea here is the relationship between the **time** (hours) it takes to complete the job and the **part of the job** completed in 1 unit of time (hour). For example, if the **time** it takes Sam to complete the job is 3 hours, the **part of the job** he can complete in 1 hour is $\frac{1}{3}$. Similarly, Frank can complete $\frac{1}{7}$ of the job in 1 hour.

   Let x = the **time** in hours it takes Sam and Frank to complete the job together. Then $\frac{1}{x}$ = the **part of the job** they complete in 1 hour.

	Hours to Complete Total Job	Part of Job Completed in 1 Hour
Sam	3	$\frac{1}{3}$
Frank	7	$\frac{1}{7}$
Together	x	$\frac{1}{x}$

2. TRANSLATE.

In words:	part of job Sam completed in 1 hour	added to	part of job Frank completed in 1 hour	is equal to	part of job they completed together in 1 hour
Translate:	$\frac{1}{3}$	$+$	$\frac{1}{7}$	$=$	$\frac{1}{x}$

3. SOLVE. Here, we solve the equation $\frac{1}{3} + \frac{1}{7} = \frac{1}{x}$. We begin by multiplying both sides of the equation by the LCD, $21x$.

$$21x\left(\frac{1}{3}\right) + 21x\left(\frac{1}{7}\right) = 21x\left(\frac{1}{x}\right)$$
$$7x + 3x = 21 \qquad \text{Simplify.}$$
$$10x = 21$$
$$x = \frac{21}{10} \quad \text{or} \quad 2\frac{1}{10} \text{ hours}$$

4. INTERPRET.

Check: Our proposed solution is $2\frac{1}{10}$ hours. This proposed solution is reasonable since $2\frac{1}{10}$ hours is more than half of Sam's time and less than half of Frank's time. Check this solution in the originally *stated* problem.

State: Sam and Frank can complete the quality control tour in $2\frac{1}{10}$ hours. □

PRACTICE

6 Cindy Liu and Mary Beckwith own a landscaping company. Cindy can complete a certain garden planting in 3 hours, while Mary takes 4 hours to complete the same Job. If both of them work together, how long will it take to plant the garden?

Concept Check ✓

Solve $E = mc^2$

a. for m. **b.** for c^2.

OBJECTIVE 5 ▶ Solving problems about distance. Next we look at a problem solved by the distance formula,

$$d = r \cdot t$$

EXAMPLE 7 Finding Speeds of Vehicles

A car travels 180 miles in the same time that a truck travels 120 miles. If the car's speed is 20 miles per hour faster than the truck's, find the car's speed and the truck's speed.

Solution

1. UNDERSTAND. Read and reread the problem. Suppose that the truck's speed is 45 miles per hour. Then the car's speed is 20 miles per hour more, or 65 miles per hour.

We are given that the car travels 180 miles in the same time that the truck travels 120 miles. To find the time it takes the car to travel 180 miles, remember that since $d = rt$, we know that $\frac{d}{r} = t$.

$$\text{Car's Time} \qquad\qquad \text{Truck's Time}$$

$$t = \frac{d}{r} = \frac{180}{65} = 2\frac{50}{65} = 2\frac{10}{13} \text{ hours} \qquad t = \frac{d}{r} = \frac{120}{45} = 2\frac{30}{45} = 2\frac{2}{3} \text{ hours}$$

Since the times are not the same, our proposed solution is not correct. But we have a better understanding of the problem.

Let x = the speed of the truck.

Since the car's speed is 20 miles per hour faster than the truck's, then

$$x + 20 = \text{the speed of the car}$$

Use the formula $d = r \cdot t$ or **d**istance = **r**ate · **t**ime. Prepare a chart to organize the information in the problem.

▶ **Helpful Hint**

If $d = r \cdot t$,

then $t = \frac{d}{r}$

or $time = \frac{distance}{rate}$.

	Distance	=	Rate	·	Time
Truck	120		x		$\frac{120}{x}$ ← distance / ← rate
Car	180		$x + 20$		$\frac{180}{x+20}$ ← distance / ← rate

Answers to Concept Check:

a. $m = \dfrac{E}{c^2}$ **b.** $c^2 = \dfrac{E}{m}$

2. **TRANSLATE.** Since the car and the truck traveled the same amount of time, we have that

In words: car's time = truck's time

Translate: $\dfrac{180}{x+20} = \dfrac{120}{x}$

3. **SOLVE.** We begin by multiplying both sides of the equation by the LCD, $x(x+20)$, or cross multiplying.

$$\dfrac{180}{x+20} = \dfrac{120}{x}$$

$180x = 120(x+20)$

$180x = 120x + 2400$ Use the distributive property.

$60x = 2400$ Subtract $120x$ from both sides.

$x = 40$ Divide both sides by 60.

4. **INTERPRET.** The speed of the truck is 40 miles per hour. The speed of the car must then be $x + 20$ or 60 miles per hour.

Check: Find the time it takes the car to travel 180 miles and the time it takes the truck to travel 120 miles.

Car's Time

$t = \dfrac{d}{r} = \dfrac{180}{60} = 3$ hours

Truck's Time

$t = \dfrac{d}{r} = \dfrac{120}{40} = 3$ hours

Since both travel the same amount of time, the proposed solution is correct.

State: The car's speed is 60 miles per hour and the truck's speed is 40 miles per hour.

PRACTICE 7 A bus travels 180 miles in the same time that a car travels 240 miles. If the car's speed is 15 mph faster than the speed of the bus, find the speed of the car and the speed of the bus.

VOCABULARY & READINESS CHECK

Without solving algebraically, select the best choice for each exercise.

1. One person can complete a job in 7 hours. A second person can complete the same job in 5 hours. How long will it take them to complete the job if they work together?
   a. more than 7 hours
   b. between 5 and 7 hours
   c. less than 5 hours

2. One inlet pipe can fill a pond in 30 hours. A second inlet pipe can fill the same pond in 25 hours. How long before the pond is filled if both inlet pipes are on?
   a. less than 25 hours
   b. between 25 and 30 hours
   c. more than 30 hours

7.6 EXERCISE SET

Solve each proportion. See Examples 1 and 2. For additional exercises on proportion and proportion applications, see Appendix C.

1. $\dfrac{2}{3} = \dfrac{x}{6}$

2. $\dfrac{x}{2} = \dfrac{16}{6}$

3. $\dfrac{x}{10} = \dfrac{5}{9}$

4. $\dfrac{9}{4x} = \dfrac{6}{2}$

5. $\dfrac{x+1}{2x+3} = \dfrac{2}{3}$

6. $\dfrac{x+1}{x+2} = \dfrac{5}{3}$

7. $\dfrac{9}{5} = \dfrac{12}{3x+2}$

8. $\dfrac{6}{11} = \dfrac{27}{3x-2}$

474   CHAPTER 7   Rational Expressions

Solve. See Example 3.

9. The ratio of the weight of an object on Earth to the weight of the same object on Pluto is 100 to 3. If an elephant weighs 4100 pounds on Earth, find the elephant's weight on Pluto.

10. If a 170-pound person weighs approximately 65 pounds on Mars, about how much does a 9000-pound satellite weigh? Round your answer to the nearest pound.

11. There are 110 calories per 28.8 grams of Frosted Flakes cereal. Find how many calories are in 43.2 grams of this cereal.

12. On an architect's blueprint, 1 inch corresponds to 4 feet. Find the length of a wall represented by a line that is $3\frac{7}{8}$ inches long on the blueprint.

Find the unknown length x or y in the following pairs of similar triangles. See Example 4.

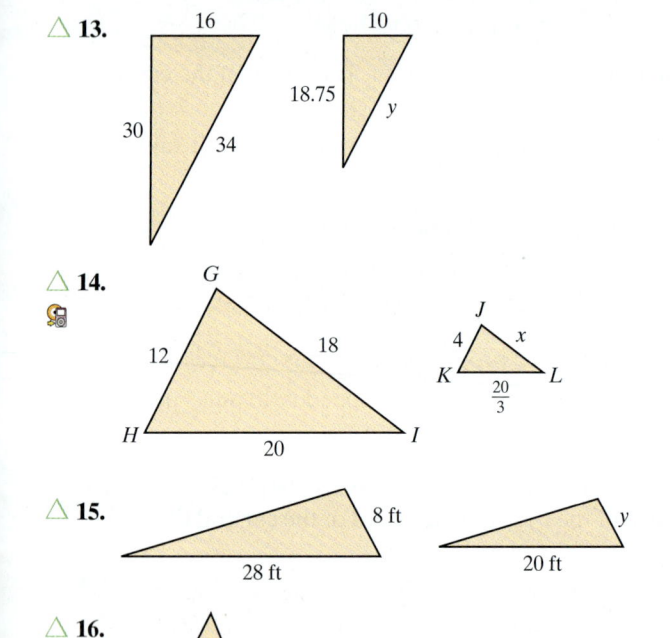

△ 13.

△ 14.

△ 15.

△ 16.

Solve the following. See Example 5.

17. Three times the reciprocal of a number equals 9 times the reciprocal of 6. Find the number.

18. Twelve divided by the sum of x and 2 equals the quotient of 4 and the difference of x and 2. Find x.

19. If twice a number added to 3 is divided by the number plus 1, the result is three halves. Find the number.

20. A number added to the product of 6 and the reciprocal of the number equals -5. Find the number.

See Example 6.

21. Smith Engineering found that an experienced surveyor surveys a roadbed in 4 hours. An apprentice surveyor needs 5 hours to survey the same stretch of road. If the two work together, find how long it takes them to complete the job.

22. An experienced bricklayer constructs a small wall in 3 hours. The apprentice completes the job in 6 hours. Find how long it takes if they work together.

23. In 2 minutes, a conveyor belt moves 300 pounds of recyclable aluminum from the delivery truck to a storage area. A smaller belt moves the same quantity of cans the same distance in 6 minutes. If both belts are used, find how long it takes to move the cans to the storage area.

24. Find how long it takes the conveyor belts described in Exercise 23 to move 1200 pounds of cans. (*Hint:* Think of 1200 pounds as four 300-pound jobs.)

See Example 7.

25. A jogger begins her workout by jogging to the park, a distance of 12 miles. She then jogs home at the same speed but along a different route. This return trip is 18 miles and her time is one hour longer. Find her jogging speed. Complete the accompanying chart and use it to find her jogging speed.

	Distance	=	Rate	·	Time
Trip to Park	12				
Return Trip	18				

26. A boat can travel 9 miles upstream in the same amount of time it takes to travel 11 miles downstream. If the current of the river is 3 miles per hour, complete the chart below and use it to find the speed of the boat in still water.

	Distance	=	Rate	·	Time
Upstream	9		$r - 3$		
Downstream	11		$r + 3$		

27. A cyclist rode the first 20-mile portion of his workout at a constant speed. For the 16-mile cooldown portion of his workout, he reduced his speed by 2 miles per hour. Each portion of the workout took the same time. Find the cyclist's speed during the first portion and find his speed during the cooldown portion.

28. A semi-truck travels 300 miles through the flatland in the same amount of time that it travels 180 miles through mountains. The rate of the truck is 20 miles per hour slower in the mountains than in the flatland. Find both the flatland rate and mountain rate.

MIXED PRACTICE

Solve the following. See Examples 1 through 7. (Note: Some exercises can be modeled by equations without rational expressions.)

29. A human factors expert recommends that there be at least 9 square feet of floor space in a college classroom for every student in the class. Find the minimum floor space that 40 students need.

30. Due to space problems at a local university, a 20-foot by 12-foot conference room is converted into a classroom. Find the maximum number of students the room can accommodate. (See Exercise 29.)

31. One-fourth equals the quotient of a number and 8. Find the number.

32. Four times a number added to 5 is divided by 6. The result is $\frac{7}{2}$. Find the number.

33. Marcus and Tony work for Lombardo's Pipe and Concrete. Mr. Lombardo is preparing an estimate for a customer. He knows that Marcus lays a slab of concrete in 6 hours. Tony lays the same size slab in 4 hours. If both work on the job and the cost of labor is $45.00 per hour, decide what the labor estimate should be.

34. Mr. Dodson can paint his house by himself in 4 days. His son needs an additional day to complete the job if he works by himself. If they work together, find how long it takes to paint the house.

35. A pilot can travel 400 miles with the wind in the same amount of time as 336 miles against the wind. Find the speed of the wind if the pilot's speed in still air is 230 miles per hour.

36. A fisherman on Pearl River rows 9 miles downstream in the same amount of time he rows 3 miles upstream. If the current is 6 miles per hour, find how long it takes him to cover the 12 miles.

37. Find the unknown length y.

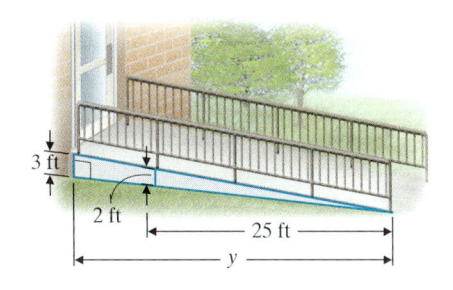

△ 38. Find the unknown length y.

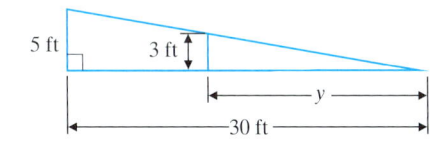

39. Ken Hall, a tailback, holds the high school sports record for total yards rushed in a season. In 1953, he rushed for 4045 total yards in 12 games. Find his average rushing yards per game. Round your answer to the nearest whole yard.

40. To estimate the number of people in Jackson, population 50,000, who have no health insurance, 250 people were polled. Of those polled, 39 had no insurance. How many people in the city might we expect to be uninsured?

41. Two divided by the difference of a number and 3 minus 4 divided by a number plus 3, equals 8 times the reciprocal of the difference of the number squared and 9. What is the number?

42. If 15 times the reciprocal of a number is added to the ratio of 9 times a number minus 7 and the number plus 2, the result is 9. What is the number?

43. A pilot flies 630 miles with a tail wind of 35 miles per hour. Against the wind, he flies only 455 miles in the same amount of time. Find the rate of the plane in still air.

44. A marketing manager travels 1080 miles in a corporate jet and then an additional 240 miles by car. If the car ride takes one hour longer than the jet ride takes, and if the rate of the jet is 6 times the rate of the car, find the time the manager travels by jet and find the time the manager travels by car.

45. To mix weed killer with water correctly, it is necessary to mix 8 teaspoons of weed killer with 2 gallons of water. Find how many gallons of water are needed to mix with the entire box if it contains 36 teaspoons of weed killer.

46. The directions for a certain bug spray concentrate is to mix 3 ounces of concentrate with 2 gallons of water. How many ounces of concentrate are needed to mix with 5 gallons of water?

47. A boater travels 16 miles per hour on the water on a still day. During one particular windy day, he finds that he travels 48 miles with the wind behind him in the same amount of time that he travels 16 miles into the wind. Find the rate of the wind.

Let x be the rate of the wind.

	r	×	t	=	d
with wind	16 + x				48
into wind	16 − x				16

48. The current on a portion of the Mississippi River is 3 miles per hour. A barge can go 6 miles upstream in the same amount of time it takes to go 10 miles downstream. Find the speed of the boat in still water.

Let x be the speed of the boat in still water.

	r	×	t	=	d
upstream	x − 3				6
downstream	x + 3				10

49. The best-selling two-seater sports car is the Mazda Miata. A driver of this car took a day-trip around the California coastline driving at two different speeds. He drove 70 miles at a slower speed and 300 miles at a speed 40 miles per hour faster. If the time spent during the faster speed was twice that spent at a slower speed, find the two speeds during the trip. (*Source: Guinness World Records*)

50. Currently, the Toyota Corolla is the most produced car in the world. Suppose that during a drive test of two Corollas, one car travels 224 miles in the same time that the second car travels 175 miles. If the speed of one car is 14 miles per hour faster than the speed of the second car, find the speed of both cars. (*Source: Guinness World Records*)

476 CHAPTER 7 Rational Expressions

51. One custodian cleans a suite of offices in 3 hours. When a second worker is asked to join the regular custodian, the job takes only $1\frac{1}{2}$ hours. How long does it take the second worker to do the same job alone?

52. One person proofreads a copy for a small newspaper in 4 hours. If a second proofreader is also employed, the job can be done in $2\frac{1}{2}$ hours. How long does it take for the second proofreader to do the same job alone?

△ 53. An architect is completing the plans for a triangular deck. Use the diagram below to find the missing dimension.

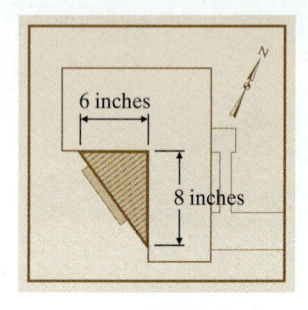

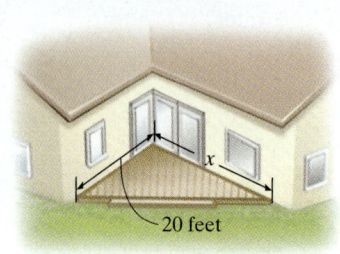

△ 54. A student wishes to make a small model of a triangular mainsail in order to study the effects of wind on the sail. The smaller model will be the same shape as a regular-size sailboat's mainsail. Use the following diagram to find the missing dimensions.

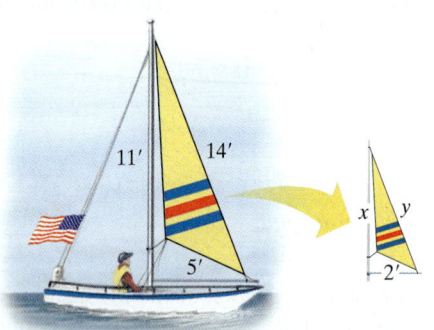

55. The manufacturers of cans of salted mixed nuts state that the ratio of peanuts to other nuts is 3 to 2. If 324 peanuts are in a can, find how many other nuts should also be in the can.

56. There are 1280 calories in a 14-ounce portion of Eagle Brand Milk. Find how many calories are in 2 ounces of Eagle Brand Milk.

57. A pilot can fly an MD-11 2160 miles with the wind in the same time as she can fly 1920 miles against the wind. If the speed of the wind is 30 mph, find the speed of the plane in still air. (*Source*: Air Transport Association of America)

58. A pilot can fly a DC-10 1365 miles against the wind in the same time as he can fly 1575 miles with the wind. If the speed of the plane in still air is 490 miles per hour, find the speed of the wind. (*Source*: Air Transport Association of America)

59. One pipe fills a storage pool in 20 hours. A second pipe fills the same pool in 15 hours. When a third pipe is added and all three are used to fill the pool, it takes only 6 hours. Find how long it takes the third pipe to do the job.

60. One pump fills a tank 2 times as fast as another pump. If the pumps work together, they fill the tank in 18 minutes. How long does it take for each pump to fill the tank?

61. A car travels 280 miles in the same time that a motorcycle travels 240 miles. If the car's speed is 10 miles per hour more than the motorcycle's, find the speed of the car and the speed of the motorcycle.

62. A walker travels 3.6 miles in the same time that a jogger travels 6 miles. If the walker's speed is 2 miles per hour less than the jogger's, find the speed of the walker and the speed of the jogger.

63. In 6 hours, an experienced cook prepares enough pies to supply a local restaurant's daily order. Another cook prepares the same number of pies in 7 hours. Together with a third cook, they prepare the pies in 2 hours. Find how long it takes the third cook to prepare the pies alone.

64. It takes 9 hours for pump A to fill a tank alone. Pump B takes 15 hours to fill the same tank alone. If pumps A, B, and C are used, the tank fills in 5 hours. How long does it take pump C to fill the tank alone?

65. One pump fills a tank 3 times as fast as another pump. If the pumps work together, they fill the tank in 21 minutes. How long does it take for each pump to fill the tank?

66. Mrs. Smith balances the company books in 8 hours. It takes her assistant 12 hours to do the same job. If they work together, find how long it takes them to balance the books.

Given that the following pairs of triangles are similar, find each missing length.

△ 67.

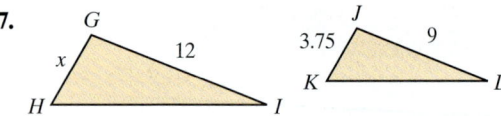

△ 68.

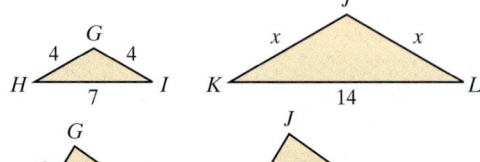

△ 69.

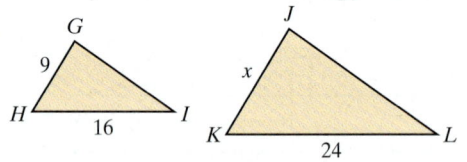

△ 70.

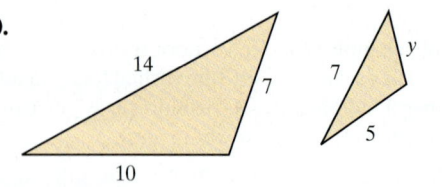

REVIEW AND PREVIEW

Find the slope of the line through each pair of points. Use the slope to determine whether the line is vertical, horizontal, or moves upward or downward from left to right. See Section 3.4.

71. $(-2, 5), (4, -3)$
72. $(0, 4), (2, 10)$
73. $(-3, -6), (1, 5)$
74. $(-2, 7), (3, -2)$
75. $(3, 7), (3, -2)$
76. $(0, -4), (2, -4)$

CONCEPT EXTENSIONS

The following bar graph shows the capacity of the United States to generate electricity from the wind in the years shown. Use this graph for Exercises 77 and 78.

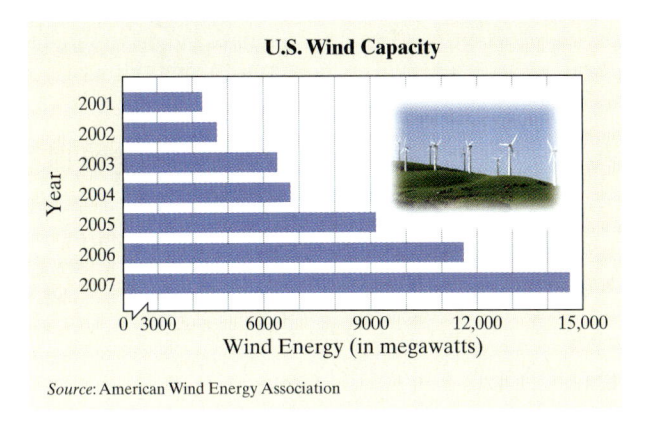

U.S. Wind Capacity

Source: American Wind Energy Association

77. Find the approximate increase in megawatt capacity during the 2-year period from 2001 to 2003.

78. Find the approximate increase in megawatt capacity during the 2-year period from 2004 to 2006.

In general, 1000 megawatts will serve the average electricity needs of 560,000 people. Use this fact and the preceding graph to answer Exercises 79 and 80.

79. In 2007, the number of megawatts that were generated from wind would serve the electricity needs of how many people? (Round to the nearest ten-thousand.)

80. How many megawatts of electricity are needed to serve the city or town in which you live?

81. Person A can complete a job in 5 hours, and person B can complete the same job in 3 hours. Without solving algebraically, discuss reasonable and unreasonable answers for how long it would take them to complete the job together.

82. For which of the following equations can we immediately use cross products to solve for x?

    a. $\dfrac{2-x}{5} = \dfrac{1+x}{3}$ b. $\dfrac{2}{5} - x = \dfrac{1+x}{3}$

83. For what value of x is $\dfrac{x}{x-1}$ in proportion to $\dfrac{x+1}{x}$? Explain your result.

84. If x is 10, is $\dfrac{2}{x}$ in proportion to $\dfrac{x}{50}$? Explain why or why not.

One of the great algebraists of ancient times was a man named Diophantus. Little is known of his life other than that he lived and worked in Alexandria. Some historians believe he lived during the first century of the Christian era, about the time of Nero. The only clue to his personal life is the following epigram found in a collection called the Palatine Anthology.

God granted him youth for a sixth of his life and added a twelfth part to this. He clothed his cheeks in down. He lit him the light of wedlock after a seventh part and five years after his marriage, He granted him a son. Alas, lateborn wretched child. After attaining the measure of half his father's life, cruel fate overtook him, thus leaving Diophantus during the last four years of his life only such consolation as the science of numbers. How old was Diophantus at his death?*

We are looking for Diophantus' age when he died, so let x represent that age. If we sum the parts of his life, we should get the total age.

Parts of his life
$\begin{cases} \dfrac{1}{6}x + \dfrac{1}{12}x \text{ is the time of his youth.} \\ \dfrac{1}{7}x \text{ is the time between his youth and when he married.} \\ 5 \text{ years is the time between his marriage and the birth of his son.} \\ \dfrac{1}{2}x \text{ is the time Diophantus had with his son.} \\ 4 \text{ years is the time between his son's death and his own.} \end{cases}$

The sum of these parts should equal Diophantus' age when he died.

$$\dfrac{1}{6} \cdot x + \dfrac{1}{12} \cdot x + \dfrac{1}{7} \cdot x + 5 + \dfrac{1}{2} \cdot x + 4 = x$$

85. Solve the epigram.

86. How old was Diophantus when his son was born? How old was the son when he died?

87. Solve the following epigram:

    I was four when my mother packed my lunch and sent me off to school. Half my life was spent in school and another sixth was spent on a farm. Alas, hard times befell me. My crops and cattle fared poorly and my land was sold. I returned to school for 3 years and have spent one tenth of my life teaching. How old am I?

88. Write an epigram describing your life. Be sure that none of the time periods in your epigram overlap.

89. A hyena spots a giraffe 0.5 mile away and begins running toward it. The giraffe starts running away from the hyena just as the hyena begins running toward it. A hyena can run at a speed of 40 mph and a giraffe can run at 32 mph. How long will it take for the hyena to overtake the giraffe? (*Source: World Almanac* and *Book of Facts*)

*From *The Nature and Growth of Modern Mathematics*, Edna Kramer, 1970, Fawcett Premier Books, Vol. 1, pages 107–108.

478 CHAPTER 7 Rational Expressions

> **THE BIGGER PICTURE   SIMPLIFYING EXPRESSIONS AND SOLVING EQUATIONS**

Now we continue our outline from Sections 1.7, 2.9, 5.6, 6.6, and 7.4. Although suggestions are given, this outline should be in your own words. Once you complete this new portion, try the exercises below.

I. Simplifying Expressions
   A. Real Numbers
      1. Add (Section 1.5)
      2. Subtract (Section 1.6)
      3. Multiply or Divide (Section 1.7)
   B. Exponents (Section 5.1)
   C. Polynomials
      1. Add (Section 5.2)
      2. Subtract (Section 5.2)
      3. Multiply (Section 5.3)
      4. Divide (Section 5.6)
   D. Factoring Polynomials (Chapter 6 Integrated Review)
   E. Rational Expressions
      1. Simplify (Section 7.1)
      2. Multiply (Section 7.2)
      3. Divide (Section 7.2)
      4. Add or Subtract (Section 7.4)

II. Solving Equations and Inequalities
   A. Linear Equations (Section 2.4)
   B. Linear Inequalities (Section 2.9)
   C. Quadratic and Higher Degree Equations (Section 6.6)
   D. Equations with Rational Expressions—solving equations with rational expressions

$$\frac{3}{x} - \frac{1}{x-1} = \frac{4}{x-1} \quad \text{Equation with rational expressions.}$$

$$x(x-1) \cdot \frac{3}{x} - x(x-1)\frac{1}{x-1} \quad \text{Multiply through by } x(x-1).$$

$$= x(x-1)\frac{4}{x-1}$$

$$3(x-1) - x \cdot 1 = x \cdot 4 \quad \text{Simplify.}$$
$$3x - 3 - x = 4x \quad \text{Use the distributive property.}$$
$$-3 = 2x \quad \text{Simplify and move variable terms to right side.}$$
$$-\frac{3}{2} = x \quad \text{Divide both sides by 2.}$$

   E. Proportions—an equation with two ratios equal. Set cross products equal, then solve.

$$\frac{5}{x} = \frac{9}{2x-3}, \text{ or } 5(2x-3) = 9 \cdot x$$

or $10x - 15 = 9x$ or $x = 15$

Multiply.
1. $(3x - 2)(4x^2 - x - 5)$
2. $(2x - y)^2$

Factor.
3. $8y^3 - 20y^5$
4. $9m^2 - 11mn + 2n^2$

Simplify or solve.

If an expression, perform indicated operations and simplify. If an equation or inequality, solve it.

5. $\dfrac{7}{x} = \dfrac{9}{x-10}$

6. $\dfrac{7}{x} + \dfrac{9}{x-10}$

7. $(-3x^5)\left(\dfrac{1}{2}x^7\right)(8x)$

8. $5x - 1 = |-4| + |-5|$

9. $\dfrac{8-12}{12 \div 3 \cdot 2}$

10. $-2(3y - 4) \leq 5y - 7 - 7y - 1$

11. $\dfrac{7}{x} + \dfrac{5}{2x+3} = \dfrac{-2}{x}$

12. $\dfrac{(a^{-3}b^2)^{-5}}{ab^4}$

7.7 VARIATION AND PROBLEM SOLVING

OBJECTIVES

1. Solve problems involving direct variation.
2. Solve problems involving inverse variation.
3. Other types of direct and inverse variation.
4. Variation and problem solving.

In Chapter 3, we studied linear equations in two variables. Recall that such an equation can be written in the form $Ax + By = C$, where A and B are not both 0.

Also recall that the graph of a linear equation in two variables is a line. In this section, we begin by looking at a particular family of linear equations—those that can be written in the form

$$y = kx,$$

where k is a constant. This family of equations is called *direct variation*.

Section 7.7 Variation and Problem Solving 479

OBJECTIVE 1 ▶ **Solving direct variation problems.** Let's suppose that you are earning $7.25 per hour at a part-time job. The amount of money you earn depends on the number of hours you work. This is illustrated by the following table:

Hours Worked	0	1	2	3	4
Money Earned (before deductions)	0	7.25	14.50	21.75	29.00

and so on

In general, to calculate your earnings (before deductions) multiply the constant $7.25 by the number of hours you work. If we let y represent the amount of money earned and x represent the number of hours worked, we get the direct variation equation

$$y = 7.25 \cdot x$$

earnings = $7.25 · hours worked

Notice that in this direct variation equation, as the number of hours increases, the pay increases as well.

Direct Variation

y varies directly as x, or **y is directly proportional to x,** if there is a nonzero constant k such that

$$y = kx$$

The number k is called the **constant of variation** or the **constant of proportionality.**

In our direct variation example: $y = 7.25x$, the constant of variation is 7.25.

Let's use the previous table to graph $y = 7.25x$. We begin our graph at the ordered-pair solution (0, 0). Why? We assume that the least amount of hours worked is 0. If 0 hours are worked, then the pay is $0.

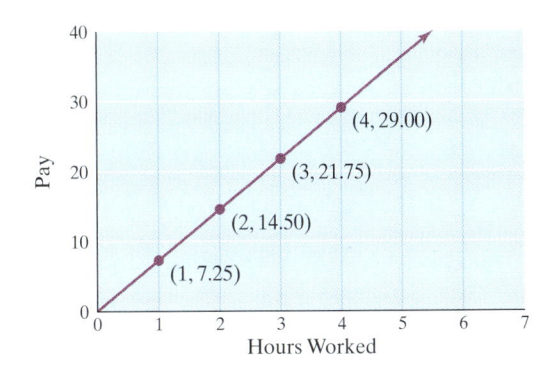

As illustrated in this graph, a direct variation equation $y = kx$ is linear. Also notice that $y = 7.25x$ is a function since its graph passes the vertical line test.

EXAMPLE 1 Write a direct variation equation of the form $y = kx$ that satisfies the ordered pairs in the table below.

x	2	9	1.5	−1
y	6	27	4.5	−3

Solution We are given that there is a direct variation relationship between x and y. This means that

$$y = kx$$

By studying the given values, you may be able to mentally calculate k. If not, to find k, we simply substitute one given ordered pair into this equation and solve for k. We'll use the given pair $(2, 6)$.

$$y = kx$$
$$6 = k \cdot 2$$
$$\frac{6}{2} = \frac{k \cdot 2}{2}$$
$$3 = k \qquad \text{Solve for } k.$$

Since $k = 3$, we have the equation $y = 3x$.

To check, see that each given y is 3 times the given x. □

PRACTICE

1 Write a direct variation of the form $y = kx$ that satisfies the ordered pairs in the table below.

x	2	8	-4	1.3
y	10	40	-20	6.5

Let's try another type of direct variation example.

EXAMPLE 2 Suppose that y varies directly as x. If y is 17 when x is 34, find the constant of variation and the direct variation equation. Then find y when x is 12.

Solution Let's use the same method as in Example 1 to find x. Since we are told that y varies directly as x, we know the relationship is of the form

$$y = kx$$

Let $y = 17$ and $x = 34$ and solve for k.

$$17 = k \cdot 34$$
$$\frac{17}{34} = \frac{k \cdot 34}{34}$$
$$\frac{1}{2} = k \qquad \text{Solve for } k.$$

Thus, the constant of variation is $\frac{1}{2}$ and the equation is $y = \frac{1}{2}x$.

To find y when $x = 12$, use $y = \frac{1}{2}x$ and replace x with 12.

$$y = \frac{1}{2}x$$
$$y = \frac{1}{2} \cdot 12 \qquad \text{Replace } x \text{ with 12.}$$
$$y = 6$$

Thus, when x is 12, y is 6. □

PRACTICE

2 If y varies directly as x and y is 12 when x is 48, find the constant of variation and the direct variation equation. Then find y when x is 20.

Let's review a few facts about linear equations of the form $y = kx$.

> **Direct Variation: y = kx**
> - There is a direct variation relationship between x and y.
> - The graph is a line.
> - The line will always go through the origin $(0, 0)$. Why?
>   Let $x = 0$. Then $y = k \cdot 0$ or $y = 0$.
> - The slope of the graph of $y = kx$ is k, the constant of variation. Why? Remember that the slope of an equation of the form $y = mx + b$ is m, the coefficient of x.
> - The equation $y = kx$ describes a function. Each x has a unique y and its graph passes the vertical line test.

EXAMPLE 3 The line is the graph of a direct variation equation. Find the constant of variation and the direct variation equation.

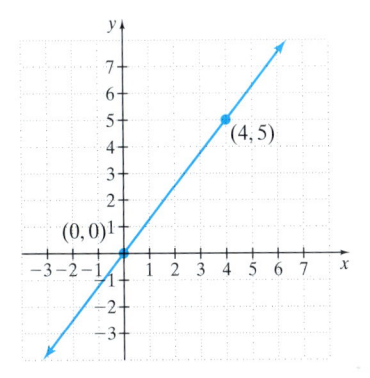

Solution Recall that k, the constant of variation is the same as the slope of the line. Thus, to find k, we use the slope formula and find slope.

Using the given points $(0, 0)$, and $(4, 5)$, we have

$$\text{slope} = \frac{5 - 0}{4 - 0} = \frac{5}{4}.$$

Thus, $k = \frac{5}{4}$ and the variation equation is $y = \frac{5}{4}x$.

PRACTICE 3 Find the constant of variation and the direct variation equation for the line below.

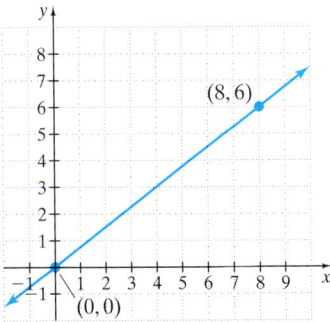

OBJECTIVE 2 ▶ Solving inverse variation problems. In this section, we will introduce another type of variation, called inverse variation.

Let's suppose you need to drive a distance of 40 miles. You know that the faster you drive the distance, the sooner you arrive at your destination. Recall that there is a mathematical relationship between distance, rate, and time. It is $d = r \cdot t$. In our example, distance is a constant 40 miles, so we have $40 = r \cdot t$ or $t = \dfrac{40}{r}$.

For example, if you drive 10 mph, the time to drive the 40 miles is

$$t = \dfrac{40}{r} = \dfrac{40}{10} = 4 \text{ hours}$$

If you drive 20 mph, the time is

$$t = \dfrac{40}{r} = \dfrac{40}{20} = 2 \text{ hours}$$

Again, notice that as speed increases, time decreases. Below are some ordered-pair solutions of $t = \dfrac{40}{r}$ and its graph.

Rate (mph)	r	5	10	20	40	60	80
Time (hr)	t	8	4	2	1	$\dfrac{2}{3}$	$\dfrac{1}{2}$

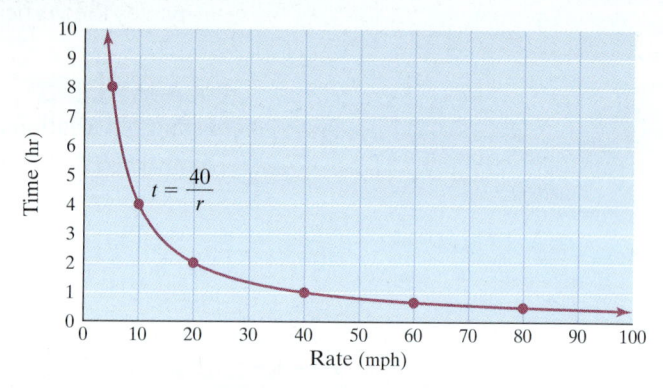

Notice that the graph of this variation is not a line, but it passes the vertical line test so $t = \dfrac{40}{r}$ does describe a function. This is an example of inverse variation.

Inverse Variation

y varies inversely as x, or **y is inversely proportional to x,** if there is a nonzero constant k such that

$$y = \dfrac{k}{x}$$

The number k is called the **constant of variation** or the **constant of proportionality.**

In our inverse variation example, $t = \dfrac{40}{r}$ or $y = \dfrac{40}{x}$, the constant of variation is 40.

We can immediately see differences and similarities in direct variation and inverse variation.

Direct variation	$y = kx$	linear equation	both
Inverse variation	$y = \dfrac{k}{x}$	rational equation	functions

Remember that $y = \dfrac{k}{x}$ is a rational equation and not a linear equation. Also notice that because x is in the denominator, x can be any value except 0.

We can still derive an inverse variation equation from a table of values.

EXAMPLE 4 Write an inverse variation equation of the form $y = \dfrac{k}{x}$ that satisfies the ordered pairs in the table below.

x	2	4	$\dfrac{1}{2}$
y	6	3	24

Solution Since there is an inverse variation relationship between x and y, we know that $y = \dfrac{k}{x}$. To find k, choose one given ordered pair and substitute the values into the equation. We'll use $(2, 6)$.

$$y = \frac{k}{x}$$

$$6 = \frac{k}{2}$$

$$2 \cdot 6 = 2 \cdot \frac{k}{2} \quad \text{Multiply both sides by 2.}$$

$$12 = k \quad \text{Solve for } k.$$

Since $k = 12$, we have the equation $y = \dfrac{12}{x}$. □

PRACTICE 4 Write an inverse variation equation of the form $y = \dfrac{k}{x}$ that satisfies the ordered pairs in the table below.

x	2	-1	$\dfrac{1}{3}$
y	4	-8	24

▶ **Helpful Hint**

Multiply both sides of the inverse variation relationship equation $y = \dfrac{k}{x}$ by x (as long as x is not 0), and we have $xy = k$. This means that if y varies inversely as x, their product is always the constant of variation k. For an example of this, check the table from Example 4.

x	2	4	$\dfrac{1}{2}$
y	6	3	24

$$2 \cdot 6 = 12 \qquad 4 \cdot 3 = 12 \qquad \frac{1}{2} \cdot 24 = 12$$

EXAMPLE 5 Suppose that y varies inversely as x. If $y = 0.02$ when $x = 75$, find the constant of variation and the inverse variation equation. Then find y when x is 30.

Solution Since y varies inversely as x, the constant of variation may be found by simply finding the product of the given x and y.

$$k = xy = 75(0.02) = 1.5$$

To check, we will use the inverse variation equation

$$y = \frac{k}{x}.$$

Let $y = 0.02$ and $x = 75$ and solve for k.

$$0.02 = \frac{k}{75}$$

$$75(0.02) = 75 \cdot \frac{k}{75} \quad \text{Multiply both sides by 75.}$$

$$1.5 = k \quad \text{Solve for } k.$$

Thus, the constant of variation is 1.5 and the equation is $y = \dfrac{1.5}{x}$.

To find y when $x = 30$ use $y = \dfrac{1.5}{x}$ and replace x with 30.

$$y = \frac{1.5}{x}$$

$$y = \frac{1.5}{30} \quad \text{Replace } x \text{ with 30.}$$

$$y = 0.05$$

Thus, when x is 30, y is 0.05.

PRACTICE 5 If y varies inversely as x and y is 0.05 when x is 42, find the constant of variation and the inverse variation equation. Then find y when x is 70.

OBJECTIVE 3 ▶ **Solving other types of direct and inverse variation problems.** It is possible for y to vary directly or inversely as powers of x.

> **Direct and Inverse Variation as nth Powers of x**
>
> **y varies directly as a power of x** if there is a nonzero constant k and a natural number n such that
>
> $$y = kx^n$$
>
> **y varies inversely as a power of x** if there is a nonzero constant k and a natural number n such that
>
> $$y = \frac{k}{x^n}$$

EXAMPLE 6 The surface area of a cube A varies directly as the square of a length of its side s. If A is 54 when s is 3, find A when $s = 4.2$.

Solution Since the surface area A varies directly as the square of side s, we have

$$A = ks^2.$$

To find k, let $A = 54$ and $s = 3$.

$$A = k \cdot s^2$$
$$54 = k \cdot 3^2 \quad \text{Let } A = 54 \text{ and } s = 3.$$
$$54 = 9k \quad 3^2 = 9.$$
$$6 = k \quad \text{Divide by 9.}$$

The formula for surface area of a cube is then

$$A = 6s^2 \text{ where } s \text{ is the length of a side.}$$

To find the surface area when $s = 4.2$, substitute.

$$A = 6s^2$$
$$A = 6 \cdot (4.2)^2$$
$$A = 105.84$$

The surface area of a cube whose side measures 4.2 units is 105.84 square units.

PRACTICE 6 The area of an isosceles right triangle A varies directly as the square of one of its legs x. If A is 32 when x is 8, find A when $x = 3.6$.

OBJECTIVE 4 ▶ **Solving applications of variation.** There are many real-life applications of direct and inverse variation.

EXAMPLE 7 The weight of a body w varies inversely with the square of its distance from the center of Earth d. If a person weighs 160 pounds on the surface of Earth, what is the person's weight 200 miles above the surface? (Assume that the radius of Earth is 4000 miles.)

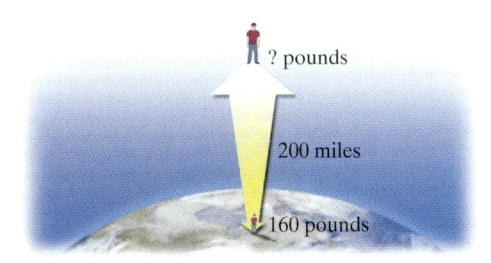

Solution

1. UNDERSTAND. Make sure you read and reread the problem.
2. TRANSLATE. Since we are told that weight w varies inversely with the square of its distance from the center of Earth, d, we have

$$w = \frac{k}{d^2}.$$

3. SOLVE. To solve the problem, we first find k. To do so, use the fact that the person weighs 160 pounds on Earth's surface, which is a distance of 4000 miles from Earth's center.

$$w = \frac{k}{d^2}$$

$$160 = \frac{k}{(4000)^2}$$

$$2{,}560{,}000{,}000 = k$$

Thus, we have $w = \dfrac{2{,}560{,}000{,}000}{d^2}$

Since we want to know the person's weight 200 miles above the Earth's surface, we let $d = 4200$ and find w.

$$w = \frac{2{,}560{,}000{,}000}{d^2}$$

$$w = \frac{2{,}560{,}000{,}000}{(4200)^2} \quad \text{A person 200 miles above the Earth's surface is 4200 miles from the Earth's center.}$$

$$w \approx 145 \quad \text{Simplify.}$$

4. INTERPRET.

Check: Your answer is reasonable since the farther a person is from Earth, the less the person weighs.

State: Thus, 200 miles above the surface of the Earth, a 160-pound person weighs approximately 145 pounds.

PRACTICE

7 Robert Boyle investigated the relationship between volume of a gas and its pressure. He developed Boyle's law, which states that the volume of a gas varies inversely with pressure if the temperature is held constant. If 50 ml of oxygen is at a pressure of 20 atmospheres, what will the volume of the oxygen be at a pressure of 40 atmospheres?

486 CHAPTER 7 Rational Expressions

VOCABULARY & READINESS CHECK

State whether each equation represents direct or inverse variation.

1. $y = \dfrac{k}{x}$, where k is a constant. _____
2. $y = kx$, where k is a constant. _____
3. $y = 5x$ _____
4. $y = \dfrac{5}{x}$ _____
5. $y = \dfrac{7}{x^2}$ _____
6. $y = 6.5x^4$ _____
7. $y = \dfrac{11}{x}$ _____
8. $y = 18x$ _____
9. $y = 12x^2$ _____
10. $y = \dfrac{20}{x^3}$ _____

7.7 EXERCISE SET

Write a direct variation equation, $y = kx$, that satisfies the ordered pairs in each table. See Example 1.

1.
x	0	6	10
y	0	3	5

2.
x	0	2	−1	3
y	0	14	−7	21

3.
x	−2	2	4	5
y	−12	12	24	30

4.
x	3	9	−2	12
y	1	3	$-\dfrac{2}{3}$	4

Write a direct variation equation, $y = kx$, that describes each graph. See Example 3.

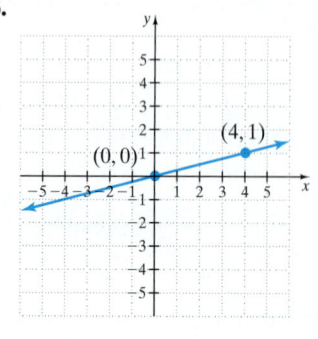

5. [graph with points (0,0) and (1,3)]
6. [graph with points (0,0) and (4,1)]
7. [graph with points (0,0) and (3,2)]
8. [graph with points (0,0) and (2,5)]

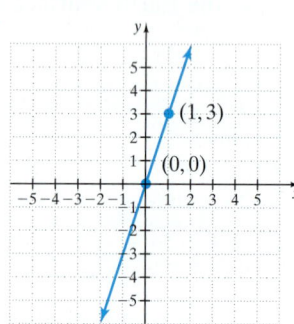

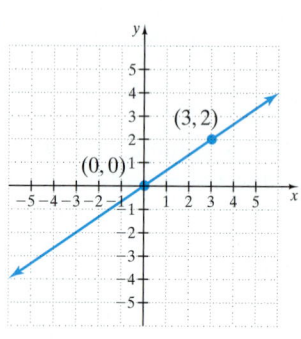

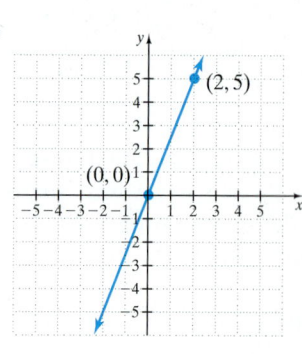

Write an inverse variation equation, $y = \dfrac{k}{x}$, that satisfies the ordered pairs in each table. See Example 4.

9.
x	1	−7	3.5	−2
y	7	−1	2	−3.5

10.
x	2	−11	4	−4
y	11	−2	5.5	−5.5

11.
x	10	$\dfrac{1}{2}$	$-\dfrac{3}{2}$
y	0.05	1	$-\dfrac{1}{3}$

12.
x	4	$\dfrac{1}{5}$	−8
y	0.1	2	−0.05

MIXED PRACTICE

Write an equation to describe each variation. Use k for the constant of proportionality. See Examples 1 through 6.

13. y varies directly as x
14. a varies directly as b
15. h varies inversely as t
16. s varies inversely as t
17. z varies directly as x^2
18. p varies inversely as x^2
19. y varies inversely as z^3
20. x varies directly as y^4
21. x varies inversely as $\sqrt{y}$
22. y varies directly as d^2

Solve. See Examples 2, 5, and 6.

23. y varies directly as x. If $y = 20$ when $x = 5$, find y when x is 10.
24. y varies directly as x. If $y = 27$ when $x = 3$, find y when x is 2.
25. y varies inversely as x. If $y = 5$ when $x = 60$, find y when x is 100.

26. y varies inversely as x. If $y = 200$ when $x = 5$, find y when x is 4.
27. z varies directly as x^2. If $z = 96$ when $x = 4$, find z when $x = 3$.
28. s varies directly as t^3. If $s = 270$ when $t = 3$, find s when $x = 1$.
29. a varies inversely as b^3. If $a = \frac{3}{2}$ when $b = 2$, find a when b is 3.
30. p varies inversely as q^2. If $p = \frac{5}{16}$ when $q = 8$, find p when $q = \frac{1}{2}$.

Solve. See Examples 1 through 7.

31. Your paycheck (before deductions) varies directly as the number of hours you work. If your paycheck is $112.50 for 18 hours, find your pay for 10 hours.
32. If your paycheck (before deductions) is $244.50 for 30 hours, find your pay for 34 hours. See Exercise 31.
33. The cost of manufacturing a certain type of headphone varies inversely as the number of headphones increases. If 5000 headphones can be manufactured for $9.00 each, find the cost to manufacture 7500 headphones.

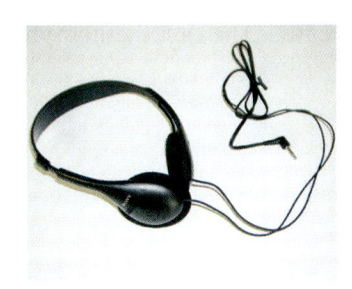

34. The cost of manufacturing a certain composition notebook varies inversely as the number of notebooks increases. If 10,000 notebooks can be manufactured for $0.50 each, find the cost to manufacture 18,000 notebooks.
35. The distance a spring stretches varies directly with the weight attached to the spring. If a 60-pound weight stretches the spring 4 inches, find the distance that an 80-pound weight stretches the spring.

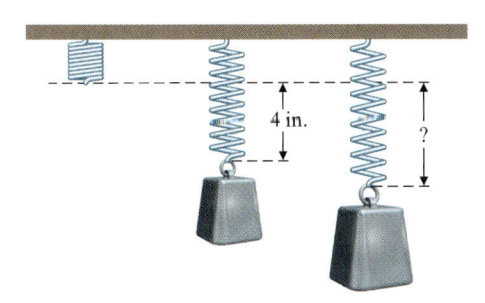

36. If a 30-pound weight stretches a spring 10 inches, find the distance a 20-pound weight stretches the spring. (See Exercise 35.)
37. The weight of an object varies inversely as the square of its distance from the *center* of the Earth. If a person weighs 180 pounds on Earth's surface, what is his weight 10 miles above the surface of the Earth? (Assume that the Earth's radius is 4000 miles.)
38. For a constant distance, the rate of travel varies inversely as the time traveled. If a family travels 55 mph and arrives at a destination in 4 hours, how long will the return trip take traveling at 60 mph?
39. The distance d that an object falls is directly proportional to the square of the time of the fall, t. A person who is parachuting for the first time is told to wait 10 seconds before opening the parachute. If the person falls 64 feet in 2 seconds, find how far he falls in 10 seconds.

40. The distance needed for a car to stop, d is directly proportional to the square of its rate of travel, r. Under certain driving conditions, a car traveling 60 mph needs 300 feet to stop. With these same driving conditions, how long does it take a car to stop if the car is traveling 30 mph when the brakes are applied?

REVIEW AND PREVIEW

Simplify. Follow the circled steps in the order shown.

41. $\dfrac{\frac{3}{4} + \frac{1}{4}}{\frac{3}{8} + \frac{13}{8}}$ ① Add. ③ Divide. ② Add.

42. $\dfrac{\frac{9}{5} + \frac{6}{5}}{\frac{17}{6} + \frac{7}{6}}$ ① Add. ③ Divide. ② Add.

43. $\dfrac{\frac{2}{5} + \frac{1}{5}}{\frac{7}{10} + \frac{7}{10}}$ ① Add. ③ Divide. ② Add.

44. $\dfrac{\frac{1}{4} + \frac{5}{4}}{\frac{3}{8} + \frac{7}{8}}$ ① Add. ③ Divide. ② Add.

CONCEPT EXTENSIONS

45. Suppose that y varies directly as x. If x is tripled, what is the effect on y?
46. Suppose that y varies directly as x^2. If x is tripled, what is the effect on y?

488 CHAPTER 7 Rational Expressions

47. The period, P, of a pendulum (the time of one complete back and forth swing) varies directly with the square root of its length, l. If the length of the pendulum is quadrupled, what is the effect on the period, P?

48. For a constant distance, the rate of travel r varies inversely with the time traveled, t. If a car traveling 100 mph completes a test track in 6 minutes, find the rate needed to complete the same test track in 4 minutes. (*Hint:* Convert minutes to hours.)

7.8 SIMPLIFYING COMPLEX FRACTIONS

OBJECTIVES
1. Simplify complex fractions using method 1.
2. Simplify complex fractions using method 2.

A rational expression whose numerator or denominator or both numerator and denominator contain fractions is called a **complex rational expression** or a **complex fraction**. Some examples are

$$\frac{4}{2-\frac{1}{2}}, \quad \frac{\frac{3}{2}}{\frac{4}{7}-x}, \quad \frac{\frac{1}{x+2}}{x+2-\frac{1}{x}} \begin{matrix} \leftarrow \text{Numerator of complex fraction} \\ \leftarrow \text{Main fraction bar} \\ \leftarrow \text{Denominator of complex fraction} \end{matrix}$$

Our goal in this section is to write complex fractions in simplest form. A complex fraction is in simplest form when it is in the form $\frac{P}{Q}$, where P and Q are polynomials that have no common factors.

OBJECTIVE 1 ▶ Simplifying complex fractions—method 1. In this section, two methods of simplifying complex fractions are presented. The first method presented uses the fact that the main fraction bar indicates division.

Method 1: Simplifying a Complex Fraction

STEP 1. Add or subtract fractions in the numerator or denominator so that the numerator is a single fraction and the denominator is a single fraction.

STEP 2. Perform the indicated division by multiplying the numerator of the complex fraction by the reciprocal of the denominator of the complex fraction.

STEP 3. Write the rational expression in simplest form.

EXAMPLE 1 Simplify the complex fraction $\dfrac{\frac{5}{8}}{\frac{2}{3}}$.

Solution Since the numerator and denominator of the complex fraction are already single fractions, we proceed to step 2: perform the indicated division by multiplying the numerator $\frac{5}{8}$ by the reciprocal of the denominator $\frac{2}{3}$.

$$\frac{\frac{5}{8}}{\frac{2}{3}} = \frac{5}{8} \div \frac{2}{3} = \frac{5}{8} \cdot \frac{3}{2} = \frac{15}{16}$$

The reciprocal of $\frac{2}{3}$ is $\frac{3}{2}$.

PRACTICE 1 Simplify the complex fraction $\dfrac{\frac{3}{4}}{\frac{6}{11}}$.

Section 7.8 Simplifying Complex Fractions 489

EXAMPLE 2 Simplify: $\dfrac{\dfrac{2}{3}+\dfrac{1}{5}}{\dfrac{2}{3}-\dfrac{2}{9}}$.

Solution Simplify above and below the main fraction bar separately. First, add $\dfrac{2}{3}$ and $\dfrac{1}{5}$ to obtain a single fraction in the numerator; then subtract $\dfrac{2}{9}$ from $\dfrac{2}{3}$ to obtain a single fraction in the denominator.

$$\dfrac{\dfrac{2}{3}+\dfrac{1}{5}}{\dfrac{2}{3}-\dfrac{2}{9}} = \dfrac{\dfrac{2(5)}{3(5)}+\dfrac{1(3)}{5(3)}}{\dfrac{2(3)}{3(3)}-\dfrac{2}{9}} \quad \text{The LCD of the numerator's fractions is 15.}$$

The LCD of the denominator's fractions is 9.

$$= \dfrac{\dfrac{10}{15}+\dfrac{3}{15}}{\dfrac{6}{9}-\dfrac{2}{9}} \quad \text{Simplify.}$$

$$= \dfrac{\dfrac{13}{15}}{\dfrac{4}{9}} \quad \begin{array}{l}\text{Add the numerator's fractions.}\\ \text{Subtract the denominator's fractions.}\end{array}$$

Next, perform the indicated division by multiplying the numerator of the complex fraction by the reciprocal of the denominator of the complex fraction.

$$\dfrac{\dfrac{13}{15}}{\dfrac{4}{9}} = \dfrac{13}{15} \cdot \dfrac{9}{4} \quad \text{The reciprocal of } \dfrac{4}{9} \text{ is } \dfrac{9}{4}.$$

$$= \dfrac{13 \cdot 3 \cdot 3}{3 \cdot 5 \cdot 4} = \dfrac{39}{20} \quad \text{Simplify.} \quad \square$$

PRACTICE 2 Simplify: $\dfrac{\dfrac{3}{4}+\dfrac{2}{3}}{\dfrac{3}{4}-\dfrac{1}{5}}$

EXAMPLE 3 Simplify: $\dfrac{\dfrac{1}{z}-\dfrac{1}{2}}{\dfrac{1}{3}-\dfrac{z}{6}}$

Solution Subtract to get a single fraction in the numerator and a single fraction in the denominator of the complex fraction.

$$\dfrac{\dfrac{1}{z}-\dfrac{1}{2}}{\dfrac{1}{3}-\dfrac{z}{6}} = \dfrac{\dfrac{2}{2z}-\dfrac{z}{2z}}{\dfrac{2}{6}-\dfrac{z}{6}} \quad \begin{array}{l}\text{The LCD of the numerator's fractions is } 2z.\\ \\ \text{The LCD of the denominator's fractions is 6.}\end{array}$$

$$= \dfrac{\dfrac{2-z}{2z}}{\dfrac{2-z}{6}}$$

$$= \dfrac{2-z}{2z} \cdot \dfrac{6}{2-z} \quad \text{Multiply by the reciprocal of } \dfrac{2-z}{6}.$$

490 CHAPTER 7   Rational Expressions

$$= \frac{2 \cdot 3 \cdot (2-z)}{2 \cdot z \cdot (2-z)} \qquad \text{Factor.}$$

$$= \frac{3}{z} \qquad \text{Write in simplest form.} \qquad \square$$

PRACTICE 3 Simplify: $\dfrac{\dfrac{4}{x} - \dfrac{1}{2}}{\dfrac{1}{5} - \dfrac{x}{10}}$

OBJECTIVE 2 ▶ Simplifying complex fractions—method 2. Next we study a second method for simplifying complex fractions. In this method, we multiply the numerator and the denominator of the complex fraction by the LCD of all fractions in the complex fraction.

Method 2: Simplifying a Complex Fraction
STEP 1. Find the LCD of all the fractions in the complex fraction.
STEP 2. Multiply both the numerator and the denominator of the complex fraction by the LCD from Step 1.
STEP 3. Perform the indicated operations and write the result in simplest form.

We use method 2 to rework Example 2.

EXAMPLE 4 Simplify: $\dfrac{\dfrac{2}{3} + \dfrac{1}{5}}{\dfrac{2}{3} - \dfrac{2}{9}}$

Solution The LCD of $\dfrac{2}{3}, \dfrac{1}{5}, \dfrac{2}{3}$ and $\dfrac{2}{9}$ is 45, so we multiply the numerator and the denominator of the complex fraction by 45. Then we perform the indicated operations, and write in simplest form.

$$\frac{\dfrac{2}{3} + \dfrac{1}{5}}{\dfrac{2}{3} - \dfrac{2}{9}} = \frac{45\left(\dfrac{2}{3} + \dfrac{1}{5}\right)}{45\left(\dfrac{2}{3} - \dfrac{2}{9}\right)}$$

$$= \frac{45\left(\dfrac{2}{3}\right) + 45\left(\dfrac{1}{5}\right)}{45\left(\dfrac{2}{3}\right) - 45\left(\dfrac{2}{9}\right)} \qquad \text{Apply the distributive property.}$$

$$= \frac{30 + 9}{30 - 10} = \frac{39}{20} \qquad \text{Simplify.} \qquad \square$$

PRACTICE 4 Simplify: $\dfrac{\dfrac{3}{4} + \dfrac{2}{3}}{\dfrac{3}{4} - \dfrac{1}{5}}$

▶ **Helpful Hint**
The same complex fraction was simplified using two different methods in Examples 2 and 4. Notice that the simplified results are the same.

EXAMPLE 5   Simplify: $\dfrac{\dfrac{x+1}{y}}{\dfrac{x}{y}+2}$

Solution   The LCD of $\dfrac{x+1}{y}, \dfrac{x}{y},$ and $\dfrac{2}{1}$ is y, so we multiply the numerator and the denominator of the complex fraction by y.

$$\dfrac{\dfrac{x+1}{y}}{\dfrac{x}{y}+2} = \dfrac{y\left(\dfrac{x+1}{y}\right)}{y\left(\dfrac{x}{y}+2\right)}$$

$$= \dfrac{y\left(\dfrac{x+1}{y}\right)}{y\left(\dfrac{x}{y}\right) + y \cdot 2} \quad \text{Apply the distributive property in the denominator.}$$

$$= \dfrac{x+1}{x+2y} \quad \text{Simplify.} \qquad \square$$

PRACTICE 5   Simplify: $\dfrac{\dfrac{a-b}{b}}{\dfrac{a}{b}+4}$

EXAMPLE 6   Simplify: $\dfrac{\dfrac{x}{y}+\dfrac{3}{2x}}{\dfrac{x}{2}+y}$

Solution   The LCD of $\dfrac{x}{y}, \dfrac{3}{2x}, \dfrac{x}{2},$ and $\dfrac{y}{1}$ is $2xy$, so we multiply both the numerator and the denominator of the complex fraction by $2xy$.

$$\dfrac{\dfrac{x}{y}+\dfrac{3}{2x}}{\dfrac{x}{2}+y} = \dfrac{2xy\left(\dfrac{x}{y}+\dfrac{3}{2x}\right)}{2xy\left(\dfrac{x}{2}+y\right)}$$

$$= \dfrac{2xy\left(\dfrac{x}{y}\right) + 2xy\left(\dfrac{3}{2x}\right)}{2xy\left(\dfrac{x}{2}\right) + 2xy(y)} \quad \text{Apply the distributive property.}$$

$$= \dfrac{2x^2 + 3y}{x^2y + 2xy^2}$$

$$\text{or } \dfrac{2x^2 + 3y}{xy(x+2y)} \qquad \square$$

PRACTICE 6   Simplify: $\dfrac{\dfrac{4}{3b}+\dfrac{b}{a}}{\dfrac{a}{3}-b}$

VOCABULARY & READINESS CHECK

Complete the steps by writing the simplified complex fraction.

1. $\dfrac{\frac{y}{2}}{\frac{5x}{2}} = \dfrac{2\left(\frac{y}{2}\right)}{2\left(\frac{5x}{2}\right)} = \dfrac{?}{?}$

2. $\dfrac{\frac{10}{x}}{\frac{z}{x}} = \dfrac{x\left(\frac{10}{x}\right)}{x\left(\frac{z}{x}\right)} = \dfrac{?}{?}$

3. $\dfrac{\frac{3}{x}}{\frac{5}{x^2}} = \dfrac{x^2\left(\frac{3}{x}\right)}{x^2\left(\frac{5}{x^2}\right)} = \dfrac{?}{?}$

4. $\dfrac{\frac{a}{10}}{\frac{b}{20}} = \dfrac{20\left(\frac{a}{10}\right)}{20\left(\frac{b}{20}\right)} = \dfrac{?}{?}$

7.8 EXERCISE SET

MIXED PRACTICE

Simplify each complex fraction. See Examples 1 through 6.

1. $\dfrac{\frac{1}{2}}{\frac{3}{4}}$

2. $\dfrac{\frac{1}{8}}{-\frac{5}{12}}$

3. $\dfrac{\frac{4x}{9}}{-\frac{2x}{3}}$

4. $\dfrac{-\frac{6y}{11}}{\frac{4y}{9}}$

5. $\dfrac{\frac{1+x}{6}}{\frac{1+x}{3}}$

6. $\dfrac{\frac{6x-3}{5x^2}}{\frac{2x-1}{10x}}$

7. $\dfrac{\frac{1}{2}+\frac{2}{3}}{\frac{5}{9}-\frac{5}{6}}$

8. $\dfrac{\frac{3}{4}-\frac{1}{2}}{\frac{3}{8}+\frac{1}{6}}$

9. $\dfrac{2+\frac{7}{10}}{1+\frac{3}{5}}$

10. $\dfrac{4-\frac{11}{12}}{5+\frac{1}{4}}$

11. $\dfrac{\frac{1}{3}}{\frac{1}{2}-\frac{1}{4}}$

12. $\dfrac{\frac{7}{10}-\frac{3}{5}}{\frac{1}{2}}$

13. $\dfrac{-\frac{2}{9}}{-\frac{14}{3}}$

14. $\dfrac{\frac{3}{8}}{\frac{4}{15}}$

15. $\dfrac{-\frac{5}{12x^2}}{\frac{25}{16x^3}}$

16. $\dfrac{-\frac{7}{8y}}{\frac{21}{4y}}$

17. $\dfrac{\frac{m}{n}-1}{\frac{m}{n}+1}$

18. $\dfrac{\frac{x}{2}+2}{\frac{x}{2}-2}$

19. $\dfrac{\frac{1}{5}-\frac{1}{x}}{\frac{7}{10}+\frac{1}{x^2}}$

20. $\dfrac{\frac{1}{y^2}+\frac{2}{3}}{\frac{1}{y}-\frac{5}{6}}$

21. $\dfrac{1+\frac{1}{y-2}}{y+\frac{1}{y-2}}$

22. $\dfrac{x-\frac{1}{2x+1}}{1-\frac{x}{2x+1}}$

23. $\dfrac{\frac{4y-8}{16}}{\frac{6y-12}{4}}$

24. $\dfrac{\frac{7y+21}{3}}{\frac{3y+9}{8}}$

25. $\dfrac{\frac{x}{y}+1}{\frac{x}{y}-1}$

26. $\dfrac{\frac{3}{5y}+8}{\frac{3}{5y}-8}$

27. $\dfrac{1}{2+\frac{1}{3}}$

28. $\dfrac{3}{1-\frac{4}{3}}$

29. $\dfrac{\frac{ax+ab}{x^2-b^2}}{\frac{x+b}{x-b}}$

30. $\dfrac{\frac{m+2}{m-2}}{\frac{2m+4}{m^2-4}}$

31. $\dfrac{\frac{-3+y}{4}}{\frac{8+y}{28}}$

32. $\dfrac{\frac{-x+2}{18}}{\frac{8}{9}}$

33. $\dfrac{3+\frac{12}{x}}{1-\frac{16}{x^2}}$

34. $\dfrac{2+\frac{6}{x}}{1-\frac{9}{x^2}}$

35. $\dfrac{\frac{8}{x+4}+2}{\frac{12}{x+4}-2}$

36. $\dfrac{\frac{25}{x+5}+5}{\frac{3}{x+5}-5}$

37. $\dfrac{\frac{s}{r}+\frac{r}{s}}{\frac{s}{r}-\frac{r}{s}}$

38. $\dfrac{\frac{2}{x}+\frac{x}{2}}{\frac{2}{x}-\frac{x}{2}}$

39. $\dfrac{\frac{6}{x-5}+\frac{x}{x-2}}{\frac{3}{x-6}-\frac{2}{x-5}}$

40. $\dfrac{\frac{4}{x}+\frac{x}{x+1}}{\frac{1}{2x}+\frac{1}{x+6}}$

REVIEW AND PREVIEW

Simplify.

41. $\sqrt{81}$

42. $\sqrt{16}$

43. $\sqrt{1}$

44. $\sqrt{0}$

45. $\sqrt{\dfrac{1}{25}}$

46. $\sqrt{\dfrac{1}{49}}$

47. $\sqrt{\dfrac{4}{9}}$

48. $\sqrt{\dfrac{121}{100}}$

CONCEPT EXTENSIONS

49. Explain how to simplify a complex fraction using method 1.
50. Explain how to simplify a complex fraction using method 2.

To find the average of two numbers, we find their sum and divide by 2. For example, the average of 65 and 81 is found by simplifying $\dfrac{65+81}{2}$. *This simplifies to* $\dfrac{146}{2} = 73$.

51. Find the average of $\dfrac{1}{3}$ and $\dfrac{3}{4}$.

52. Write the average of $\dfrac{3}{n}$ and $\dfrac{5}{n^2}$ as a simplified rational expression.

53. In electronics, when two resistors R_1 (read R sub 1) and R_2 (read R sub 2) are connected in parallel, the total resistance is given by the complex fraction

$$\frac{1}{\dfrac{1}{R_1}+\dfrac{1}{R_2}}.$$

Simplify this expression.

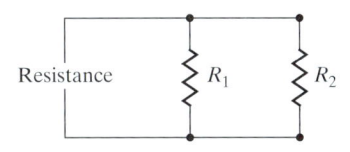

54. Astronomers occasionally need to know the day of the week a particular date fell on. The complex fraction

$$\frac{J+\dfrac{3}{2}}{7}$$

where J is the *Julian day number,* is used to make this calculation. Simplify this expression.

Simplify each of the following. First, write each expression with positive exponents. Then simplify the complex fraction. The first step has been completed for Exercise 55.

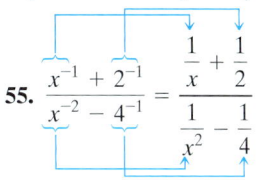

55. $\dfrac{x^{-1}+2^{-1}}{x^{-2}-4^{-1}}=\dfrac{\dfrac{1}{x}+\dfrac{1}{2}}{\dfrac{1}{x^2}-\dfrac{1}{4}}$

56. $\dfrac{3^{-1}-x^{-1}}{9^{-1}-x^{-2}}$

57. $\dfrac{y^{-2}}{1-y^{-2}}$

58. $\dfrac{4+x^{-1}}{3+x^{-1}}$

59. If the distance formula $d = r \cdot t$ is solved for t, then $t = \dfrac{d}{r}$. Use this formula to find t if distance d is $\dfrac{20x}{3}$ miles and rate r is $\dfrac{5x}{9}$ miles per hour. Write t in simplified form.

△ 60. If the formula for area of a rectangle, $A = l \cdot w$, is solved for w, then $w = \dfrac{A}{l}$. Use this formula to find w if area A is $\dfrac{4x-2}{3}$ square meters and length l is $\dfrac{6x-3}{5}$ meters. Write w in simplified form.

CHAPTER 7 GROUP ACTIVITY

Comparing Dosage Formulas

In this project, you will have the opportunity to investigate two well-known formulas for predicting the correct doses of medication for children. This project may be completed by working in groups or individually.

Young's Rule and Cowling's Rule are dose formulas for prescribing medicines to children. Unlike formulas for, say area or distance, these dose formulas describe only an approximate relationship. The formulas relate a child's age A in years and an adult dose D of medication to the proper child's dose C. The formulas are most accurate when applied to children between the ages of 2 and 13.

$$\text{Young's Rule:}\quad C = \frac{DA}{A+12}$$

$$\text{Cowling's Rule:}\quad C = \frac{D(A+1)}{24}$$

1. Let the adult dose $D = 1000$ mg. Complete the Young's Rule and Cowling's Rule columns of the following table comparing the doses predicted by both formulas for ages 2 through 13.

2. Use the data from the table in Question 1 to form sets of ordered pairs of the form (age, child's dose) for each formula. Graph the ordered pairs for each formula on the same graph. Describe the shapes of the graphed data.

3. Use your table, graph, or both, to decide whether either formula will consistently predict a larger dose than the other. If so, which one? If not, is there an age at which the doses predicted by one becomes greater than the doses predicted by the other? If so, estimate that age.

4. Use your graph to estimate for what age the difference in the two predicted doses is greatest.

5. Return to the table in Question 1 and complete the last column, titled "Difference," by finding the absolute value of the difference between the Young's dose and the Cowling's dose for each age. Use this column in the table to verify your graphical estimate found in Question 4.

6. Does Cowling's Rule ever predict exactly the adult dose? If so, at what age? Explain. Does Young's Rule ever predict exactly the adult dose? If so, at what age? Explain.

7. Many doctors prefer to use formulas that relate doses to factors other than a child's age. Why is age not necessarily the most important factor when predicting a child's dose? What other factors might be used?

Age A	Young's Rule	Cowling's Rule	Difference	Age A	Young's Rule	Cowling's Rule	Difference
2				8			
3				9			
4				10			
5				11			
6				12			
7				13			

CHAPTER 7 VOCABULARY CHECK

Fill in each blank with one of the words or phrases listed below.

rational expression   complex fraction   ratio   proportion
cross products   direct variation   inverse variation

1. A _____ is the quotient of two numbers.
2. $\frac{x}{2} = \frac{7}{16}$ is an example of a _____.
3. If $\frac{a}{b} = \frac{c}{d}$, then ad and bc are called _____.
4. A _____ is an expression that can be written in the form $\frac{P}{Q}$, where P and Q are polynomials and Q is not 0.
5. In a _____, the numerator or denominator or both may contain fractions.
6. The equation $y = \frac{k}{x}$ is an example of _____.
7. The equation $y = kx$ is an example of _____.

> **Helpful Hint**
> Are you preparing for your test? Don't forget to take the Chapter 7 Test on page 501. Then check your answers at the back of the text and use the Chapter Test Prep Video CD to see the fully worked-out solutions to any of the exercises you want to review.

CHAPTER 7 HIGHLIGHTS

DEFINITIONS AND CONCEPTS	EXAMPLES
SECTION 7.1 SIMPLIFYING RATIONAL EXPRESSIONS	
A **rational expression** is an expression that can be written in the form $\frac{P}{Q}$, where P and Q are polynomials and Q does not equal 0. To find values for which a rational expression is undefined, find values for which the denominator is 0. **To Simplify a Rational Expression** **Step 1.** Factor the numerator and denominator. **Step 2.** Divide out factors common to the numerator and denominator. (This is the same as removing a factor of 1.)	$\frac{7y^3}{4}, \frac{x^2 + 6x + 1}{x - 3}, \frac{-5}{s^3 + 8}$ Find any values for which the expression $\frac{5y}{y^2 - 4y + 3}$ is undefined. $\quad y^2 - 4y + 3 = 0$ Set the denominator equal to 0. $\quad (y - 3)(y - 1) = 0$ Factor. $\quad y - 3 = 0$ or $y - 1 = 0$ Set each factor equal to 0. $\quad y = 3 \qquad\qquad y = 1$ Solve. The expression is undefined when y is 3 and when y is 1. Simplify: $\frac{4x + 20}{x^2 - 25}$ $\frac{4x + 20}{x^2 - 25} = \frac{4(x + 5)}{(x + 5)(x - 5)} = \frac{4}{x - 5}$

DEFINITIONS AND CONCEPTS	EXAMPLES

SECTION 7.2 MULTIPLYING AND DIVIDING RATIONAL EXPRESSIONS

To multiply rational expressions,

Step 1. Factor numerators and denominators.

Step 2. Multiply numerators and multiply denominators.

Step 3. Write the product in simplest form.

$$\frac{P}{Q} \cdot \frac{R}{S} = \frac{PR}{QS}$$

Multiply: $\dfrac{4x+4}{2x-3} \cdot \dfrac{2x^2+x-6}{x^2-1}$

$$\frac{4x+4}{2x-3} \cdot \frac{2x^2+x-6}{x^2-1} = \frac{4(x+1)}{2x-3} \cdot \frac{(2x-3)(x+2)}{(x+1)(x-1)}$$
$$= \frac{4(x+1)(2x-3)(x+2)}{(2x-3)(x+1)(x-1)}$$
$$= \frac{4(x+2)}{x-1}$$

To divide by a rational expression, multiply by the reciprocal.

$$\frac{P}{Q} \div \frac{R}{S} = \frac{P}{Q} \cdot \frac{S}{R} = \frac{PS}{QR}$$

Divide: $\dfrac{15x+5}{3x^2-14x-5} \div \dfrac{15}{3x-12}$

$$\frac{15x+5}{3x^2-14x-5} \div \frac{15}{3x-12} = \frac{5(3x+1)}{(3x+1)(x-5)} \cdot \frac{3(x-4)}{3 \cdot 5}$$
$$= \frac{x-4}{x-5}$$

SECTION 7.3 ADDING AND SUBTRACTING RATIONAL EXPRESSIONS WITH COMMON DENOMINATORS AND LEAST COMMON DENOMINATOR

To add or subtract rational expressions with the same denominator, add or subtract numerators, and place the sum or difference over a common denominator.

$$\frac{P}{R} + \frac{Q}{R} = \frac{P+Q}{R}$$
$$\frac{P}{R} - \frac{Q}{R} = \frac{P-Q}{R}$$

Perform indicated operations.

$$\frac{5}{x+1} + \frac{x}{x+1} = \frac{5+x}{x+1}$$

$$\frac{2y+7}{y^2-9} - \frac{y+4}{y^2-9} = \frac{(2y+7)-(y+4)}{y^2-9}$$
$$= \frac{2y+7-y-4}{y^2-9}$$
$$= \frac{y+3}{(y+3)(y-3)}$$
$$= \frac{1}{y-3}$$

To find the least common denominator (LCD),

Step 1. Factor the denominators.

Step 2. The LCD is the product of all unique factors, each raised to a power equal to the greatest number of times that it appears in any one factored denominator.

Find the LCD for

$$\frac{7x}{x^2+10x+25} \text{ and } \frac{11}{3x^2+15x}$$

$$x^2+10x+25 = (x+5)(x+5)$$
$$3x^2+15x = 3x(x+5)$$

LCD is $3x(x+5)(x+5)$ or $3x(x+5)^2$

DEFINITIONS AND CONCEPTS	EXAMPLES

SECTION 7.4 ADDING AND SUBTRACTING RATIONAL EXPRESSIONS WITH UNLIKE DENOMINATORS

To add or subtract rational expressions with unlike denominators,

Step 1. Find the LCD.

Step 2. Rewrite each rational expression as an equivalent expression whose denominator is the LCD.

Step 3. Add or subtract numerators and place the sum or difference over the common denominator.

Step 4. Write the result in simplest form.

Perform the indicated operation.

$$\frac{9x+3}{x^2-9} - \frac{5}{x-3}$$

$$= \frac{9x+3}{(x+3)(x-3)} - \frac{5}{x-3}$$

LCD is $(x+3)(x-3)$.

$$= \frac{9x+3}{(x+3)(x-3)} - \frac{5(x+3)}{(x-3)(x+3)}$$

$$= \frac{9x+3-5(x+3)}{(x+3)(x-3)}$$

$$= \frac{9x+3-5x-15}{(x+3)(x-3)}$$

$$= \frac{4x-12}{(x+3)(x-3)}$$

$$= \frac{4(x-3)}{(x+3)(x-3)} = \frac{4}{x+3}$$

SECTION 7.5 SOLVING EQUATIONS CONTAINING RATIONAL EXPRESSIONS

To solve an equation containing rational expressions,

Step 1. Multiply both sides of the equation by the LCD of all rational expressions in the equation.

Step 2. Remove any grouping symbols and solve the resulting equation.

Step 3. Check the solution in the original equation.

Solve: $\dfrac{5x}{x+2} + 3 = \dfrac{4x-6}{x+2}$

$$(x+2)\left(\frac{5x}{x+2} + 3\right) = (x+2)\left(\frac{4x-6}{x+2}\right)$$

$$(x+2)\left(\frac{5x}{x+2}\right) + (x+2)(3) = (x+2)\left(\frac{4x-6}{x+2}\right)$$

$$5x + 3x + 6 = 4x - 6$$
$$4x = -12$$
$$x = -3$$

The solution checks and the solution is -3.

SECTION 7.6 PROPORTIONS AND PROBLEM SOLVING WITH RATIONAL EQUATIONS

A **ratio** is the quotient of two numbers or two quantities. A **proportion** is a mathematical statement that two ratios are equal.

Cross products:

If $\dfrac{a}{b} = \dfrac{c}{d}$, then $ad = bc$.

Proportions

$$\frac{2}{3} = \frac{8}{12} \qquad \frac{x}{7} = \frac{15}{35}$$

Cross Products

$2 \cdot 12$ or $24 \qquad\qquad 3 \cdot 8$ or 24

$$\frac{2}{3} = \frac{8}{12}$$

DEFINITIONS AND CONCEPTS	EXAMPLES
SECTION 7.6 PROPORTIONS AND PROBLEM SOLVING WITH RATIONAL EQUATIONS (continued)	
	Solve: $\dfrac{3}{4} = \dfrac{x}{x-1}$
	$\dfrac{3}{4} \bowtie \dfrac{x}{x-1}$
	$3(x-1) = 4x$ Set cross products equal.
	$3x - 3 = 4x$
	$-3 = x$
Problem-Solving Steps 1. UNDERSTAND. Read and reread the problem.	A small plane and a car leave Kansas City, Missouri, and head for Minneapolis, Minnesota, a distance of 450 miles. The speed of the plane is 3 times the speed of the car, and the plane arrives 6 hours ahead of the car. Find the speed of the car. Let x = the speed of the car. Then $3x$ = the speed of the plane.

	Distance	=	*Rate* · *Time*
Car	450	x	$\dfrac{450}{x} \left(\dfrac{\text{distance}}{\text{rate}} \right)$
Plane	450	$3x$	$\dfrac{450}{3x} \left(\dfrac{\text{distance}}{\text{rate}} \right)$

2. TRANSLATE. 3. SOLVE. 4. INTERPRET.	In words: plane's time + 6 hours = car's time Translate: $\dfrac{450}{3x} + 6 = \dfrac{450}{x}$ $\dfrac{450}{3x} + 6 = \dfrac{450}{x}$ $3x\left(\dfrac{450}{3x}\right) + 3x(6) = 3x\left(\dfrac{450}{x}\right)$ $450 + 18x = 1350$ $18x = 900$ $x = 50$ **Check** this solution in the originally stated problem. **State** the conclusion: The speed of the car is 50 miles per hour.

SECTION 7.7 VARIATION AND PROBLEM SOLVING	
y **varies directly as** x, or y is **directly proportional to** x, if there is a nonzero constant k such that $y = kx$	The circumference of a circle C varies directly as its radius r. $C = \underbrace{2\pi}_{k} r$
y **varies inversely as** x, or y is **inversely proportional to** x, if there is a nonzero constant k such that $y = \dfrac{k}{x}$	Pressure P varies inversely with volume V. $P = \dfrac{k}{V}$

498 CHAPTER 7 Rational Expressions

DEFINITIONS AND CONCEPTS	EXAMPLES

SECTION 7.8 SIMPLIFYING COMPLEX FRACTIONS

Method 1: To Simplify a Complex Fraction

Step 1. Add or subtract fractions in the numerator and the denominator of the complex fraction.

Step 2. Perform the indicated division.

Step 3. Write the result in lowest terms.

Simplify:

$$\dfrac{\dfrac{1}{x} + 2}{\dfrac{1}{x} - \dfrac{1}{y}} = \dfrac{\dfrac{1}{x} + \dfrac{2x}{x}}{\dfrac{y}{xy} - \dfrac{x}{xy}}$$

$$= \dfrac{\dfrac{1 + 2x}{x}}{\dfrac{y - x}{xy}}$$

$$= \dfrac{1 + 2x}{x} \cdot \dfrac{x\,y}{y - x}$$

$$= \dfrac{y(1 + 2x)}{y - x}$$

Method 2: To Simplify a Complex Fraction

Step 1. Find the LCD of all fractions in the complex fraction.

Step 2. Multiply the numerator and the denominator of the complex fraction by the LCD.

Step 3. Perform the indicated operations and write the result in lowest terms.

$$\dfrac{\dfrac{1}{x} + 2}{\dfrac{1}{x} - \dfrac{1}{y}} = \dfrac{xy\left(\dfrac{1}{x} + 2\right)}{xy\left(\dfrac{1}{x} - \dfrac{1}{y}\right)}$$

$$= \dfrac{xy\left(\dfrac{1}{x}\right) + xy(2)}{xy\left(\dfrac{1}{x}\right) - xy\left(\dfrac{1}{y}\right)}$$

$$= \dfrac{y + 2xy}{y - x} \text{ or } \dfrac{y(1 + 2x)}{y - x}$$

📖 STUDY SKILLS BUILDER

Are You Prepared for a Test on Chapter 7?

Below I have listed *a common trouble* area for students in Chapter 7. After studying for your test, but before taking your test, read this.

Do you know the differences between how to perform operations such as $\dfrac{4}{x} + \dfrac{2}{3}$ or $\dfrac{4}{x} \div \dfrac{2}{x}$ and how to solve an equation such as $\dfrac{4}{x} + \dfrac{2}{3} = 1$?

$$\dfrac{4}{x} + \dfrac{2}{3} = \dfrac{4 \cdot 3}{x \cdot 3} + \dfrac{2 \cdot x}{3 \cdot x}$$

Addition—write each expression as an equivalent expression with the same LCD denominator.

$$= \dfrac{12}{3x} + \dfrac{2x}{3x} = \dfrac{12 + 2x}{3x} \text{ or } \dfrac{2(6 + x)}{3x}, \text{ the sum.}$$

$$\dfrac{4}{x} \div \dfrac{2}{x} = \dfrac{4}{x} \cdot \dfrac{x}{2} = \dfrac{4 \cdot x}{x \cdot 2} = \dfrac{4}{2} = 2, \text{ the quotient.}$$

Division—multiply the first rational expression by the reciprocal of the second.

$$\dfrac{4}{x} + \dfrac{2}{3} = 1 \qquad \text{Equation to be solved.}$$

$$3x\left(\dfrac{4}{x} + \dfrac{2}{3}\right) = 3x \cdot 1 \qquad \text{Multiply both sides of the equation by the LCD, } 3x.$$

$$3x\left(\dfrac{4}{x}\right) + 3x\left(\dfrac{2}{3}\right) = 3x \cdot 1 \qquad \text{Use the distributive property.}$$

$$12 + 2x = 3x \qquad \text{Multiply and simplify.}$$

$$12 = x \qquad \text{Subtract } 2x \text{ from both sides.}$$

The solution is 12.

For more examples and exercises, see the Chapter 7 Integrated Review.

CHAPTER 7 REVIEW

(7.1) *Find any real number for which each rational expression is undefined.*

1. $\dfrac{x+5}{x^2-4}$
2. $\dfrac{5x+9}{4x^2-4x-15}$

Find the value of each rational expression when $x=5$, $y=7$, and $z=-2$.

3. $\dfrac{2-z}{z+5}$
4. $\dfrac{x^2+xy-y^2}{x+y}$

Simplify each rational expression.

5. $\dfrac{2x+6}{x^2+3x}$
6. $\dfrac{3x-12}{x^2-4x}$
7. $\dfrac{x+2}{x^2-3x-10}$
8. $\dfrac{x+4}{x^2+5x+4}$
9. $\dfrac{x^3-4x}{x^2+3x+2}$
10. $\dfrac{5x^2-125}{x^2+2x-15}$
11. $\dfrac{x^2-x-6}{x^2-3x-10}$
12. $\dfrac{x^2-2x}{x^2+2x-8}$

Simplify each expression. This section contains four-term polynomials and sums and differences of two cubes.

13. $\dfrac{x^2+xa+xb+ab}{x^2-xc+bx-bc}$
14. $\dfrac{x^2+5x-2x-10}{x^2-3x-2x+6}$
15. $\dfrac{4-x}{x^3-64}$
16. $\dfrac{x^2-4}{x^3+8}$

(7.2) *Perform each indicated operation and simplify.*

17. $\dfrac{15x^3y^2}{z} \cdot \dfrac{z}{5xy^3}$
18. $\dfrac{-y^3}{8} \cdot \dfrac{9x^2}{y^3}$
19. $\dfrac{x^2-9}{x^2-4} \cdot \dfrac{x-2}{x+3}$
20. $\dfrac{2x+5}{x-6} \cdot \dfrac{2x}{-x+6}$
21. $\dfrac{x^2-5x-24}{x^2-x-12} \div \dfrac{x^2-10x+16}{x^2+x-6}$
22. $\dfrac{4x+4y}{xy^2} \div \dfrac{3x+3y}{x^2y}$
23. $\dfrac{x^2+x-42}{x-3} \cdot \dfrac{(x-3)^2}{x+7}$
24. $\dfrac{2a+2b}{3} \cdot \dfrac{a-b}{a^2-b^2}$
25. $\dfrac{2x^2-9x+9}{8x-12} \div \dfrac{x^2-3x}{2x}$
26. $\dfrac{x^2-y^2}{x^2+xy} \div \dfrac{3x^2-2xy-y^2}{3x^2+6x}$
27. $\dfrac{x-y}{4} \div \dfrac{y^2-2y-xy+2x}{16x+24}$
28. $\dfrac{5+x}{7} \div \dfrac{xy+5y-3x-15}{7y-35}$

(7.3) *Perform each indicated operation and simplify.*

29. $\dfrac{x}{x^2+9x+14} + \dfrac{7}{x^2+9x+14}$
30. $\dfrac{x}{x^2+2x-15} + \dfrac{5}{x^2+2x-15}$

31. $\dfrac{4x-5}{3x^2} - \dfrac{2x+5}{3x^2}$
32. $\dfrac{9x+7}{6x^2} - \dfrac{3x+4}{6x^2}$

Find the LCD of each pair of rational expressions.

33. $\dfrac{x+4}{2x}, \dfrac{3}{7x}$
34. $\dfrac{x-2}{x^2-5x-24}, \dfrac{3}{x^2+11x+24}$

Rewrite each rational expression as an equivalent expression whose denominator is the given polynomial.

35. $\dfrac{5}{7x} = \dfrac{}{14x^3y}$
36. $\dfrac{9}{4y} = \dfrac{}{16y^3x}$
37. $\dfrac{x+2}{x^2+11x+18} = \dfrac{}{(x+2)(x-5)(x+9)}$
38. $\dfrac{3x-5}{x^2+4x+4} = \dfrac{}{(x+2)^2(x+3)}$

(7.4) *Perform each indicated operation and simplify.*

39. $\dfrac{4}{5x^2} - \dfrac{6}{y}$
40. $\dfrac{2}{x-3} - \dfrac{4}{x-1}$
41. $\dfrac{4}{x+3} - 2$
42. $\dfrac{3}{x^2+2x-8} + \dfrac{2}{x^2-3x+2}$
43. $\dfrac{2x-5}{6x+9} - \dfrac{4}{2x^2+3x}$
44. $\dfrac{x-1}{x^2-2x+1} - \dfrac{x+1}{x-1}$

Find the perimeter and the area of each figure.

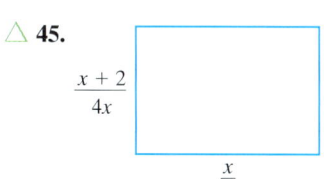

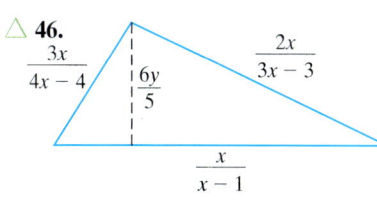

△ 45. rectangle with sides $x+2$ and $4x$

△ 46. triangle with sides $3x$, $4x-4$, $\dfrac{2x}{3x-3}$, height $\dfrac{6y}{5}$, base $\dfrac{x}{x-1}$, and $\dfrac{x}{8}$

(7.5) *Solve each equation.*

47. $\dfrac{n}{10} = 9 - \dfrac{n}{5}$
48. $\dfrac{2}{x+1} - \dfrac{1}{x-2} = -\dfrac{1}{2}$
49. $\dfrac{y}{2y+2} + \dfrac{2y-16}{4y+4} = \dfrac{y-3}{y+1}$
50. $\dfrac{2}{x-3} - \dfrac{4}{x+3} = \dfrac{8}{x^2-9}$
51. $\dfrac{x-3}{x+1} - \dfrac{x-6}{x+5} = 0$
52. $x+5 = \dfrac{6}{x}$

Solve the equation for the indicated variable.

53. $\dfrac{4A}{5b} = x^2$, for b
54. $\dfrac{x}{7} + \dfrac{y}{8} = 10$, for y

(7.6) *Solve each proportion.*

55. $\dfrac{x}{2} = \dfrac{12}{4}$

56. $\dfrac{20}{1} = \dfrac{x}{25}$

57. $\dfrac{2}{x-1} = \dfrac{3}{x+3}$

58. $\dfrac{4}{y-3} = \dfrac{2}{y-3}$

Solve.

59. A machine can process 300 parts in 20 minutes. Find how many parts can be processed in 45 minutes.

60. As his consulting fee, Mr. Visconti charges $90.00 per day. Find how much he charges for 3 hours of consulting. Assume an 8-hour work day.

61. Five times the reciprocal of a number equals the sum of $\dfrac{3}{2}$ the reciprocal of the number and $\dfrac{7}{6}$. What is the number?

62. The reciprocal of a number equals the reciprocal of the difference of 4 and the number. Find the number.

63. A car travels 90 miles in the same time that a car traveling 10 miles per hour slower travels 60 miles. Find the speed of each car.

64. The current in a bayou near Lafayette, Louisiana, is 4 miles per hour. A paddle boat travels 48 miles upstream in the same amount of time it takes to travel 72 miles downstream. Find the speed of the boat in still water.

65. When Mark and Maria manicure Mr. Stergeon's lawn, it takes them 5 hours. If Mark works alone, it takes 7 hours. Find how long it takes Maria alone.

66. It takes pipe A 20 days to fill a fish pond. Pipe B takes 15 days. Find how long it takes both pipes together to fill the pond.

Given that the pairs of triangles are similar, find each missing length x.

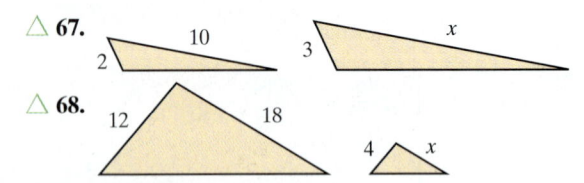

67.

68.

(7.7) *Solve.*

69. y varies directly as x. If $y = 40$ when $x = 4$, find y when x is 11.

70. y varies inversely as x. If $y = 4$ when $x = 6$, find y when x is 48.

71. y varies inversely as x^3. If $y = 12.5$ when $x = 2$, find y when x is 3.

72. y varies directly as x^2. If $y = 175$ when $x = 5$, find y when $x = 10$.

73. The cost of manufacturing a certain medicine varies inversely as the amount of medicine manufactured increases. If 3000 milliliters can be manufactured for $6600, find the cost to manufacture 5000 milliliters.

74. The distance a spring stretches varies directly with the weight attached to the spring. If a 150-pound weight stretches the spring 8 inches, find the distance that a 90-pound weight stretches the spring.

(7.8) *Simplify each complex fraction.*

75. $\dfrac{\dfrac{5x}{27}}{\dfrac{10xy}{21}}$

76. $\dfrac{\dfrac{3}{5} + \dfrac{2}{7}}{\dfrac{1}{5} + \dfrac{5}{6}}$

77. $\dfrac{3 - \dfrac{1}{y}}{2 - \dfrac{1}{y}}$

78. $\dfrac{\dfrac{6}{x+2} + 4}{\dfrac{8}{x+2} - 4}$

MIXED REVIEW

Simplify each rational expression.

79. $\dfrac{4x + 12}{8x^2 + 24x}$

80. $\dfrac{x^3 - 6x^2 + 9x}{x^2 + 4x - 21}$

Perform the indicated operations and simplify.

81. $\dfrac{x^2 + 9x + 20}{x^2 - 25} \cdot \dfrac{x^2 - 9x + 20}{x^2 + 8x + 16}$

82. $\dfrac{x^2 - x - 72}{x^2 - x - 30} \div \dfrac{x^2 + 6x - 27}{x^2 - 9x + 18}$

83. $\dfrac{x}{x^2 - 36} + \dfrac{6}{x^2 - 36}$

84. $\dfrac{5x - 1}{4x} - \dfrac{3x - 2}{4x}$

85. $\dfrac{4}{3x^2 + 8x - 3} + \dfrac{2}{3x^2 - 7x + 2}$

86. $\dfrac{3x}{x^2 + 9x + 14} - \dfrac{6x}{x^2 + 4x - 21}$

Solve.

87. $\dfrac{4}{a-1} + 2 = \dfrac{3}{a-1}$

88. $\dfrac{x}{x+3} + 4 = \dfrac{x}{x+3}$

Solve.

89. The quotient of twice a number and three, minus one-sixth is the quotient of the number and two. Find the number.

90. Mr. Crocker can paint his house by himself in three days. His son will need an additional day to complete the job if he works alone. If they work together, find how long it takes to paint the house.

Given that the following pairs of triangles are similar, find each missing length.

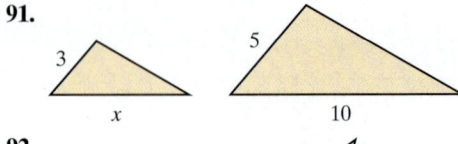

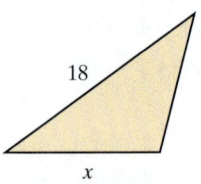

91.

92.

Simplify each complex fraction.

93. $\dfrac{\dfrac{1}{4}}{\dfrac{1}{3} + \dfrac{1}{2}}$

94. $\dfrac{4 + \dfrac{2}{x}}{6 + \dfrac{3}{x}}$

CHAPTER 7 TEST

1. Find any real numbers for which the following expression is undefined.
$$\frac{x+5}{x^2+4x+3}$$

2. For a certain computer desk, the average cost C (in dollars) per desk manufactured is
$$C = \frac{100x + 3000}{x}$$
where x is the number of desks manufactured.

   a. Find the average cost per desk when manufacturing 200 computer desks.
   b. Find the average cost per desk when manufacturing 1000 computer desks.

Simplify each rational expression.

3. $\dfrac{3x-6}{5x-10}$

4. $\dfrac{x+6}{x^2+12x+36}$

5. $\dfrac{x+3}{x^3+27}$

6. $\dfrac{2m^3-2m^2-12m}{m^2-5m+6}$

7. $\dfrac{ay+3a+2y+6}{ay+3a+5y+15}$

8. $\dfrac{y-x}{x^2-y^2}$

Perform the indicated operation and simplify if possible.

9. $\dfrac{3}{x-1} \cdot (5x-5)$

10. $\dfrac{y^2-5y+6}{2y+4} \cdot \dfrac{y+2}{2y-6}$

11. $\dfrac{15x}{2x+5} - \dfrac{6-4x}{2x+5}$

12. $\dfrac{5a}{a^2-a-6} - \dfrac{2}{a-3}$

13. $\dfrac{6}{x^2-1} + \dfrac{3}{x+1}$

14. $\dfrac{x^2-9}{x^2-3x} \div \dfrac{xy+5x+3y+15}{2x+10}$

15. $\dfrac{x+2}{x^2+11x+18} + \dfrac{5}{x^2-3x-10}$

Solve each equation.

16. $\dfrac{4}{y} - \dfrac{5}{3} = \dfrac{-1}{5}$

17. $\dfrac{5}{y+1} = \dfrac{4}{y+2}$

18. $\dfrac{a}{a-3} = \dfrac{3}{a-3} - \dfrac{3}{2}$

19. $x - \dfrac{14}{x-1} = 4 - \dfrac{2x}{x-1}$

20. $\dfrac{10}{x^2-25} = \dfrac{3}{x+5} + \dfrac{1}{x-5}$

Simplify each complex fraction.

21. $\dfrac{\dfrac{5x^2}{yz^2}}{\dfrac{10x}{z^3}}$

22. $\dfrac{5 - \dfrac{1}{y^2}}{\dfrac{1}{y} + \dfrac{2}{y^2}}$

23. y varies directly as x. If $y = 10$ when $x = 15$, find y when x is 42.

24. y varies inversely as x^2. If $y = 8$ when $x = 5$, find y when x is 15.

25. In a sample of 85 fluorescent bulbs, 3 were found to be defective. At this rate, how many defective bulbs should be found in 510 bulbs?

26. One number plus five times its reciprocal is equal to six. Find the number.

27. A pleasure boat traveling down the Red River takes the same time to go 14 miles upstream as it takes to go 16 miles downstream. If the current of the river is 2 miles per hour, find the speed of the boat in still water.

28. An inlet pipe can fill a tank in 12 hours. A second pipe can fill the tank in 15 hours. If both pipes are used, find how long it takes to fill the tank.

△ 29. Given that the two triangles are similar, find x.

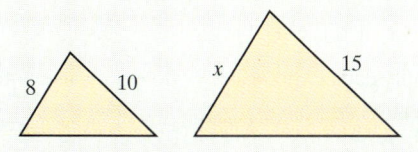

CHAPTER 7 CUMULATIVE REVIEW

1. Write each sentence as an equation. Let x represent the unknown number.
   a. The quotient of 15 and a number is 4.
   b. Three subtracted from 12 is a number.
   c. Four times a number, added to 17, is not equal to 21.
   d. Triple a number is less than 48.

2. Write each sentence as an equation. Let x represent the unknown number.
   a. The difference of 12 and a number is -45.
   b. The product of 12 and a number is -45.
   c. A number less 10 is twice the number.

3. Rajiv Puri invested part of his $20,000 inheritance in a mutual funds account that pays 7% simple interest yearly and the rest in a certificate of deposit that pays 9% simple interest yearly. At the end of one year, Rajiv's investments earned $1550. Find the amount he invested at each rate.

4. The number of non-business bankruptcies has increased over the years. In 2002, the number of non-business bankruptcies was 80,000 less than twice the number in 1994. If the total of non-business bankruptcies for these two years is 2,290,000 find the number of non-business bankruptcies for each year. (*Source:* American Bankruptcy Institute)

5. Graph $x - 3y = 6$ by finding and plotting intercepts.

6. Find the slope of the line whose equation is $7x + 2y = 9$.

7. Use the product rule to simplify each expression.
   a. $4^2 \cdot 4^5$
   b. $x^4 \cdot x^6$
   c. $y^3 \cdot y$
   d. $y^3 \cdot y^2 \cdot y^7$
   e. $(-5)^7 \cdot (-5)^8$
   f. $a^2 \cdot b^2$

8. Simplify.
   a. $\dfrac{x^9}{x^7}$
   b. $\dfrac{x^{19}y^5}{xy}$
   c. $(x^5y^2)^3$
   d. $(-3a^2b)(5a^3b)$

9. Subtract $(5z - 7)$ from the sum of $(8z + 11)$ and $(9z - 2)$.

10. Subtract $(9x^2 - 6x + 2)$ from $(x + 1)$.

11. Multiply: $(3a + b)^3$

12. Multiply: $(2x + 1)(5x^2 - x + 2)$

13. Use a special product to square each binomial.
    a. $(t + 2)^2$
    b. $(p - q)^2$
    c. $(2x + 5)^2$
    d. $(x^2 - 7y)^2$

14. Multiply.
    a. $(x + 9)^2$
    b. $(2x + 1)(2x - 1)$
    c. $8x(x^2 + 1)(x^2 - 1)$

15. Simplify each expression. Write results using positive exponents only.
    a. $\dfrac{1}{x^{-3}}$
    b. $\dfrac{1}{3^{-4}}$
    c. $\dfrac{p^{-4}}{q^{-9}}$
    d. $\dfrac{5^{-3}}{2^{-5}}$

16. Simplify. Write results with positive exponents.
    a. 5^{-3}
    b. $\dfrac{9}{x^{-7}}$
    c. $\dfrac{11^{-1}}{7^{-2}}$

17. Divide: $\dfrac{4x^2 + 7 + 8x^3}{2x + 3}$

18. Divide $(4x^3 - 9x + 2)$ by $(x - 4)$.

19. Find the GCF of each list of numbers.
    a. 28 and 40
    b. 55 and 21
    c. 15, 18, and 66

20. Find the GCF of $9x^2$, $6x^3$, and $21x^5$.

Factor.

21. $-9a^5 + 18a^2 - 3a$
22. $7x^6 - 7x^5 + 7x^4$
23. $3m^2 - 24m - 60$
24. $-2a^2 + 10a + 12$
25. $3x^2 + 11x + 6$
26. $10m^2 - 7m + 1$
27. $x^2 + 12x + 36$
28. $4x^2 + 12x + 9$
29. $x^2 + 4$
30. $x^2 - 4$
31. $x^3 + 8$
32. $27y^3 - 1$
33. $2x^3 + 3x^2 - 2x - 3$
34. $3x^3 + 5x^2 - 12x - 20$
35. $12m^2 - 3n^2$
36. $x^5 - x$
37. Solve: $x(2x - 7) = 4$
38. Solve: $3x^2 + 5x = 2$
39. Find the x-intercepts of the graph of $y = x^2 - 5x + 4$.
40. Find the x-intercepts of the graph of $y = x^2 - x - 6$.

41. The height of a triangular sail is 2 meters less than twice the length of the base. If the sail has an area of 30 square meters, find the length of its base and the height.

42. The height of a parallelogram is 5 feet more than three times its base. If the area of the parallelogram is 182 square feet, find the length of its base and height.

43. Simplify: $\dfrac{5x - 5}{x^3 - x^2}$

44. Simplify: $\dfrac{2x^2 - 50}{4x^4 - 20x^3}$

45. Divide: $\dfrac{6x + 2}{x^2 - 1} \div \dfrac{3x^2 + x}{x - 1}$

46. Multiply: $\dfrac{6x^2 - 18x}{3x^2 - 2x} \cdot \dfrac{15x - 10}{x^2 - 9}$

47. Simplify: $\dfrac{\dfrac{x+1}{y}}{\dfrac{x}{y} + 2}$

48. Simplify: $\dfrac{\dfrac{m}{3} + \dfrac{n}{6}}{\dfrac{m+n}{12}}$

CHAPTER

8 Roots and Radicals

8.1 Introduction to Radicals
8.2 Simplifying Radicals
8.3 Adding and Subtracting Radicals
8.4 Multiplying and Dividing Radicals

Integrated Review—Simplifying Radicals

8.5 Solving Equations Containing Radicals
8.6 Radical Equations and Problem Solving
8.7 Rational Exponents

When we think of pendulums, we often think of grandfather clocks. In fact, pendulums can be used to provide accurate timekeeping. But, did you know that pendulums can also be used to demonstrate that the earth rotates on its axis?

In 1851, French physicist Léon Foucault developed a special pendulum in an experiment to demonstrate that the Earth rotated on its axis. He connected his tall pendulum, capable of running for many hours, to the roof of the Paris Observatory. The pendulum's bob was able to swing back and forth in one plane, but not to twist in other directions. So, when the pendulum bob appeared to move in a circle over time, he demonstrated that it was not the pendulum but the building that moved. And since the building was firmly attached to the earth, it must be the earth rotating which created the apparent circular motion of the bob. In Section 8.1, Exercise 85 on page 510, roots are used to explore the time it takes Foucault's pendulum to complete one swing of its bob.

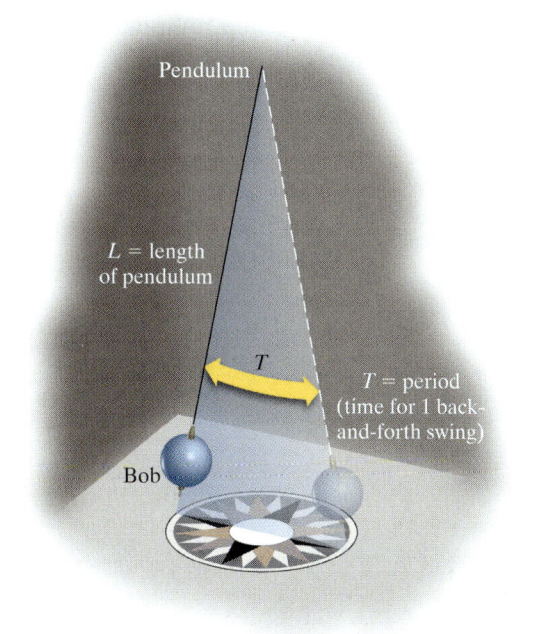

Having spent the last chapter studying equations, we return now to algebraic expressions. We expand on our skills of operating on expressions—adding, subtracting, multiplying, dividing, and raising to powers—to include finding roots. Just as subtraction is defined by addition and division by multiplication, finding roots is defined by raising to powers. As we master finding roots, we will work with equations that contain roots and solve problems that can be modeled by such equations.

8.1 INTRODUCTION TO RADICALS

OBJECTIVES

1. Find square roots.
2. Find cube roots.
3. Find nth roots.
4. Approximate square roots.
5. Simplify radicals containing variables.

OBJECTIVE 1 ▶ Finding square roots. In this section, we define finding the **root** of a number by its reverse operation, raising a number to a power. We begin with squares and square roots.

$$\text{The } square \text{ of } 5 \text{ is } 5^2 = 25.$$
$$\text{The } square \text{ of } -5 \text{ is } (-5)^2 = 25.$$
$$\text{The } square \text{ of } \frac{1}{2} \text{ is } \left(\frac{1}{2}\right)^2 = \frac{1}{4}.$$

The reverse operation of squaring a number is finding the **square root** of a number. For example,

A *square root* of 25 is 5, because $5^2 = 25$.

A *square root* of 25 is also -5, because $(-5)^2 = 25$.

A *square root* of $\frac{1}{4}$ is $\frac{1}{2}$, because $\left(\frac{1}{2}\right)^2 = \frac{1}{4}$.

> **In general, a number b is a square root of a number a if $b^2 = a$.**

Notice that both 5 and -5 are square roots of 25. The symbol $\sqrt{}$ is used to denote the **positive** or **principal square root** of a number. For example,

$$\sqrt{25} = 5 \text{ since } 5^2 = 25 \text{ and } 5 \text{ is positive}.$$

The symbol $-\sqrt{}$ is used to denote the **negative square root**. For example,

$$-\sqrt{25} = -5$$

The symbol $\sqrt{}$ is called a **radical** or **radical sign.** The expression within or under a radical sign is called the **radicand**. An expression containing a radical is called a **radical expression.**

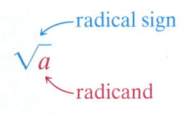

$$\sqrt{a}$$

radical sign

radicand

Square Root

If a is a positive number, then

$\sqrt{a}$ is the **positive square root** of a and

$-\sqrt{a}$ is the **negative square root** of a.

$\sqrt{a} = b$ only if $b^2 = a$ and $b > 0$

Also, $\sqrt{0} = 0$.

EXAMPLE 1 Find each square root.

a. $\sqrt{36}$ b. $-\sqrt{16}$ c. $\sqrt{\dfrac{9}{100}}$ d. $\sqrt{0}$ e. $\sqrt{0.64}$

Solution

a. $\sqrt{36} = 6$, because $6^2 = 36$ and 6 is positive.

b. $-\sqrt{16} = -4$. The negative sign in front of the radical indicates the negative square root of 16.

c. $\sqrt{\dfrac{9}{100}} = \dfrac{3}{10}$ because $\left(\dfrac{3}{10}\right)^2 = \dfrac{9}{100}$ and $\dfrac{3}{10}$ is positive.

d. $\sqrt{0} = 0$ because $0^2 = 0$.

e. $\sqrt{0.64} = 0.8$, because $(0.8)^2 = 0.64$ and 0.8 is positive.

PRACTICE 1 Find each square root.

a. $\sqrt{\dfrac{4}{81}}$ b. $-\sqrt{25}$ c. $\sqrt{144}$ d. $\sqrt{0.49}$ e. $-\sqrt{1}$

Is the square root of a negative number a real number? For example, is $\sqrt{-4}$ a real number? To answer this question, we ask ourselves, is there a real number whose square is -4? Since there is no real number whose square is -4, we say that $\sqrt{-4}$ is not a real number. In general,

> A square root of a negative number is not a real number.

We will discuss numbers such as $\sqrt{-4}$ in Chapter 9.

OBJECTIVE 2 ▶ Finding cube roots. We can find roots other than square roots. For example, since $2^3 = 8$, we call 2 the **cube root** of 8. In symbols, we write

$$\sqrt[3]{8} = 2 \quad \text{The number 3 is called the \textbf{index}.}$$

Also,

$$\sqrt[3]{27} = 3 \quad \text{Since } 3^3 = 27$$
$$\sqrt[3]{-64} = -4 \quad \text{Since } (-4)^3 = -64$$

Notice that unlike the square root of a negative number, the cube root of a negative number is a real number. This is so because while we cannot find a real number whose **square** is negative, we **can** find a real number whose **cube** is negative. In fact, the cube of a negative number is a negative number. Therefore, the cube root of a negative number is a negative number.

EXAMPLE 2 Find each cube root.

a. $\sqrt[3]{1}$ b. $\sqrt[3]{-27}$ c. $\sqrt[3]{\dfrac{1}{125}}$

Solution

a. $\sqrt[3]{1} = 1$ because $1^3 = 1$.

b. $\sqrt[3]{-27} = -3$ because $(-3)^3 = -27$.

c. $\sqrt[3]{\dfrac{1}{125}} = \dfrac{1}{5}$ because $\left(\dfrac{1}{5}\right)^3 = \dfrac{1}{125}$.

PRACTICE 2 Find each cube root.

a. $\sqrt[3]{0}$ b. $\sqrt[3]{-64}$ c. $\sqrt[3]{\dfrac{1}{8}}$

OBJECTIVE 3 ▶ Finding nth roots. Just as we can raise a real number to powers other than 2 or 3, we can find roots other than square roots and cube roots. In fact, we can take the nth root of a number where n is any natural number. An **nth root** of a number a is a number whose nth power is a. The natural number n is called the **index**.

In symbols, the *n*th root of *a* is written as $\sqrt[n]{a}$. The index 2 is usually omitted for square roots.

> **Helpful Hint**
> If the index is even, such as $\sqrt{}, \sqrt[4]{}, \sqrt[6]{}$, and so on, the radicand must be nonnegative for the root to be a real number. For example,
>
> $\sqrt[4]{16} = 2$ but $\sqrt[4]{-16}$ is not a real number
> $\sqrt[6]{64} = 2$ but $\sqrt[6]{-64}$ is not a real number

Concept Check ✓
Which of the following is a real number?

a. $\sqrt{-64}$ b. $\sqrt[4]{-64}$ c. $\sqrt[5]{-64}$ d. $\sqrt[6]{-64}$

EXAMPLE 3 Find each root.

a. $\sqrt[4]{16}$ b. $\sqrt[5]{-32}$ c. $-\sqrt[3]{8}$ d. $\sqrt[4]{-81}$

Solution

a. $\sqrt[4]{16} = 2$ because $2^4 = 16$ and 2 is positive.
b. $\sqrt[5]{-32} = -2$ because $(-2)^5 = -32$.
c. $-\sqrt[3]{8} = -2$ since $\sqrt[3]{8} = 2$.
d. $\sqrt[4]{-81}$ is not a real number since the index 4 is even and the radicand -81 is negative.

PRACTICE 3 Find each root.

a. $\sqrt[4]{81}$ b. $\sqrt[5]{100,000}$ c. $\sqrt[6]{-64}$ d. $\sqrt[3]{-125}$

OBJECTIVE 4 ▶ Approximating square roots. Recall that numbers such as $1, 4, 9, 25,$ and $\frac{4}{25}$ are called **perfect squares**, since $1^2 = 1, 2^2 = 4, 3^2 = 9, 5^2 = 25,$ and $\left(\frac{2}{5}\right)^2 = \frac{4}{25}$.
Square roots of perfect square radicands simplify to rational numbers. What happens when we try to simplify a root such as $\sqrt{3}$? Since 3 is not a perfect square, $\sqrt{3}$ is not a rational number. It cannot be written as a quotient of integers. It is called an **irrational number** and we can find a decimal **approximation** of it. To find decimal approximations, use a calculator or an appendix. (For calculator help, see the box at the end of this section.)

EXAMPLE 4 Use a calculator or an appendix to approximate $\sqrt{3}$ to three decimal places.

Solution We may use an appendix or a calculator to approximate $\sqrt{3}$. To use a calculator, find the square root key $\boxed{\sqrt{}}$.

$$\sqrt{3} \approx 1.732050808$$

To three decimal places, $\sqrt{3} \approx 1.732$.

PRACTICE 4 Use a calculator or an appendix to approximate $\sqrt{17}$ to three decimal places.

Answer to Concept Check: c

OBJECTIVE 5 ▶ Simplifying radicals containing variables. Radicals can also contain variables. To simplify radicals containing variables, special care must be taken. To see how we simplify $\sqrt{x^2}$, let's look at a few examples in this form.

If $x = 3$, we have $\sqrt{3^2} = \sqrt{9} = 3$, or x.
If x is 5, we have $\sqrt{5^2} = \sqrt{25} = 5$, or x.

From these two examples, you may think that $\sqrt{x^2}$ simplifies to x. Let's now look at an example where x is a negative number. If $x = -3$, we have $\sqrt{(-3)^2} = \sqrt{9} = 3$, not -3, our original x. To make sure that $\sqrt{x^2}$ simplifies to a nonnegative number, we have the following.

> For any real number a,
> $$\sqrt{a^2} = |a|.$$

Thus,
$$\sqrt{x^2} = |x|,$$
$$\sqrt{(-8)^2} = |-8| = 8$$
$$\sqrt{(7y)^2} = |7y|, \text{ and so on.}$$

To avoid this, for the rest of the chapter we assume that **if a variable appears in the radicand of a radical expression, it represents positive numbers only.** Then

$\sqrt{x^2} = |x| = x$ since x is a positive number.
$\sqrt{y^2} = y$ Because $(y)^2 = y^2$
$\sqrt{x^8} = x^4$ Because $(x^4)^2 = x^8$
$\sqrt{9x^2} = 3x$ Because $(3x)^2 = 9x^2$
$\sqrt[3]{8z^{12}} = 2z^4$ Because $(2z^4)^3 = 8z^{12}$

EXAMPLE 5 Simplify each expression. Assume that all variables represent positive numbers.

a. $\sqrt{z^2}$ **b.** $\sqrt{x^6}$ **c.** $\sqrt[3]{27y^6}$ **d.** $\sqrt{16x^{16}}$ **e.** $\sqrt{\dfrac{x^4}{25}}$ **f.** $\sqrt[3]{-125a^{12}b^{15}}$

Solution

a. $\sqrt{z^2} = z$ because $(z)^2 = z^2$.
b. $\sqrt{x^6} = x^3$ because $(x^3)^2 = x^6$.
c. $\sqrt[3]{27y^6} = 3y^2$ because $(3y^2)^3 = 27y^6$.
d. $\sqrt{16x^{16}} = 4x^8$ because $(4x^8)^2 = 16x^{16}$.
e. $\sqrt{\dfrac{x^4}{25}} = \dfrac{x^2}{5}$ because $\left(\dfrac{x^2}{5}\right)^2 = \dfrac{x^4}{25}$.
f. $\sqrt[3]{-125a^{12}b^{15}} = -5a^4b^5$ because $(-5a^4b^5)^3 = -125a^{12}b^{15}$.

PRACTICE 5 Simplify each expression. Assume that all variables represent positive numbers.

a. $\sqrt{x^{10}}$ **b.** $\sqrt{y^{14}}$ **c.** $\sqrt[3]{125z^9}$ **d.** $\sqrt{49x^2}$ **e.** $\sqrt{\dfrac{z^4}{36}}$ **f.** $\sqrt[3]{-8a^6b^{12}}$

508 CHAPTER 8 Roots and Radicals

Calculator Explorations

To simplify or approximate square roots using a calculator, locate the key marked $\sqrt{}$. To simplify $\sqrt{25}$ using a scientific calculator, press $\boxed{25}$ $\boxed{\sqrt{}}$. The display should read $\boxed{5}$. To simplify $\sqrt{25}$ using a graphing calculator, press $\boxed{\sqrt{}}$ $\boxed{25}$ $\boxed{\text{ENTER}}$.

To approximate $\sqrt{30}$, press $\boxed{30}$ $\boxed{\sqrt{}}$ (or $\boxed{\sqrt{}}$ $\boxed{30}$ $\boxed{\text{ENTER}}$). The display should read $\boxed{5.4772256}$. This is an approximation for $\sqrt{30}$. A three-decimal-place approximation is

$$\sqrt{30} \approx 5.477$$

Is this answer reasonable? Since 30 is between perfect squares 25 and 36, $\sqrt{30}$ is between $\sqrt{25} = 5$ and $\sqrt{36} = 6$. The calculator result is then reasonable since 5.4772256 is between 5 and 6.

Use a calculator to approximate each expression to three decimal places. Decide whether each result is reasonable.

1. $\sqrt{7}$
2. $\sqrt{14}$
3. $\sqrt{11}$
4. $\sqrt{200}$
5. $\sqrt{82}$
6. $\sqrt{46}$

Many scientific calculators have a key, such as $\boxed{\sqrt[x]{y}}$, that can be used to approximate roots other than square roots. To approximate these roots using a graphing calculator, look under the $\boxed{\text{MATH}}$ menu or consult your manual.

Use a calculator to approximate each expression to three decimal places. Decide whether each result is reasonable.

7. $\sqrt[3]{40}$
8. $\sqrt[3]{71}$
9. $\sqrt[4]{20}$
10. $\sqrt[4]{15}$
11. $\sqrt[5]{18}$
12. $\sqrt[6]{2}$

VOCABULARY & READINESS CHECK

Use the choices below to fill in each blank.

principal   radical sign   index   radicand

1. In the expression $\sqrt[4]{16}$, the number 4 is called the _____, the number 16 is called the _____ and $\sqrt{}$ called the _____.

2. The symbol $\sqrt{}$ is used to denote the positive, or _____, square root.

Answer each exercise true or false.

3. $\sqrt{-16}$ simplifies to a real number.

4. $\sqrt{64} = 8$ while $\sqrt[3]{64} = 4$.

5. The number 9 has two square roots.

6. $\sqrt{0} = 0$ and $\sqrt{1} = 1$.

7. If x is a positive number, $\sqrt{x^{10}} = x^5$.

8. If x is a positive number, $\sqrt{x^{16}} = x^4$.

8.1 EXERCISE SET

Find each square root. See Example 1.

1. $\sqrt{16}$
2. $\sqrt{64}$
3. $\sqrt{\dfrac{1}{25}}$
4. $\sqrt{\dfrac{1}{64}}$
5. $-\sqrt{100}$
6. $-\sqrt{36}$
7. $\sqrt{-4}$
8. $\sqrt{-25}$
9. $-\sqrt{121}$
10. $-\sqrt{49}$
11. $\sqrt{\dfrac{9}{25}}$
12. $\sqrt{\dfrac{4}{81}}$
13. $\sqrt{900}$
14. $\sqrt{400}$
15. $\sqrt{144}$
16. $\sqrt{169}$
17. $\sqrt{\dfrac{1}{100}}$
18. $\sqrt{\dfrac{1}{121}}$
19. $\sqrt{0.25}$
20. $\sqrt{0.49}$

Find each cube root. See Example 2.

21. $\sqrt[3]{125}$
22. $\sqrt[3]{64}$
23. $\sqrt[3]{-64}$
24. $\sqrt[3]{-27}$
25. $-\sqrt[3]{8}$
26. $-\sqrt[3]{27}$
27. $\sqrt[3]{\dfrac{1}{8}}$
28. $\sqrt[3]{\dfrac{1}{64}}$
29. $\sqrt[3]{-125}$
30. $\sqrt[3]{-1}$

MIXED PRACTICE

Find each root. See Examples 1 through 3.

31. $\sqrt[5]{32}$
32. $\sqrt[4]{81}$
33. $\sqrt{81}$
34. $\sqrt{49}$
35. $\sqrt[4]{-16}$
36. $\sqrt{-9}$
37. $\sqrt[3]{\dfrac{27}{64}}$
38. $\sqrt[3]{-\dfrac{8}{27}}$
39. $-\sqrt[4]{625}$
40. $-\sqrt[5]{32}$
41. $\sqrt[6]{1}$
42. $\sqrt[5]{1}$

Approximate each square root to three decimal places. See Example 4.

43. $\sqrt{7}$
44. $\sqrt{10}$
45. $\sqrt{37}$
46. $\sqrt{27}$
47. $\sqrt{136}$
48. $\sqrt{8}$

49. A standard baseball diamond is a square with 90-foot sides connecting the bases. The distance from home plate to second base is $90 \cdot \sqrt{2}$ feet. Approximate $\sqrt{2}$ to two decimal places and use your result to approximate the distance $90 \cdot \sqrt{2}$ feet.

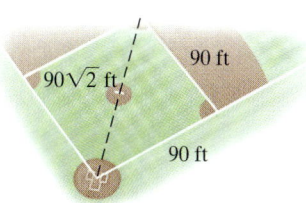

50. The roof of the warehouse shown needs to be shingled. The total area of the roof is exactly $240 \cdot \sqrt{41}$ square feet. Approximate $\sqrt{41}$ to two decimal places and use your result to approximate the area $240 \cdot \sqrt{41}$ square feet. Approximate this area to the nearest whole number.

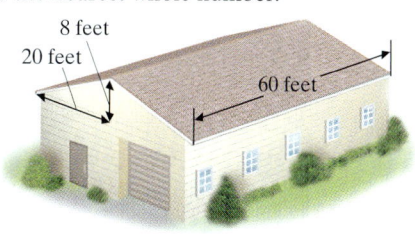

Find each root. Assume that all variables represent positive numbers. See Example 5.

51. $\sqrt{x^4}$
52. $\sqrt{y^{10}}$
53. $\sqrt{9x^8}$
54. $\sqrt{36x^{12}}$
55. $\sqrt{81x^2}$
56. $\sqrt{100z^4}$
57. $\sqrt{\dfrac{x^6}{36}}$
58. $\sqrt{\dfrac{y^8}{49}}$
59. $\sqrt{\dfrac{25y^2}{9}}$
60. $\sqrt{\dfrac{4x^2}{81}}$
61. $\sqrt{16a^6b^4}$
62. $\sqrt{4m^{14}n^2}$
63. $\sqrt[3]{a^6b^{18}}$
64. $\sqrt[3]{x^{12}y^{18}}$
65. $\sqrt[3]{-8x^3y^{27}}$
66. $\sqrt[3]{-27a^6b^{30}}$

REVIEW AND PREVIEW

Write each integer as a product of two integers such that one of the factors is a perfect square. For example, we can write $18 = 9 \cdot 2$, where 9 is a perfect square.

67. 50
68. 8
69. 32
70. 75
71. 28
72. 44
73. 27
74. 90

CONCEPT EXTENSIONS

Solve. See the Concept Check in this section.

75. Which of the following is a real number?
    a. $\sqrt[7]{-1}$
    b. $\sqrt[3]{-125}$
    c. $\sqrt[6]{-128}$
    d. $\sqrt[8]{-1}$

76. a. $\sqrt{-1}$
    b. $\sqrt[3]{-1}$
    c. $\sqrt[4]{-1}$
    d. $\sqrt[5]{-1}$

The length of a side of a square in given by the expression $\sqrt{A}$, where A is the square's area. Use this expression for Exercises 77 through 80. Be sure to attach the appropriate units.

77. The area of a square is 49 square miles. Find the length of a side of the square.

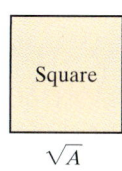

510 CHAPTER 8 Roots and Radicals

△ **78.** The area of a square is $\frac{1}{81}$ square meters. Find the length of a side of the square.

△ **79.** Sony makes the current smallest mini disc player. It is in the shape of a square with area of 9.0601 square inches. Find the length of a side. (*Source:* SONY)

△ **80.** A parking lot is in the shape of a square with area 2500 square yards. Find the length of a side.

81. Simplify $\sqrt{\sqrt{81}}$.

82. Simplify $\sqrt[3]{\sqrt[3]{1}}$.

83. Simplify $\sqrt{\sqrt{10,000}}$.

84. Simplify $\sqrt{\sqrt{1,600,000,000}}$.

85. The formula for calculating the period (one back and forth swing) of a pendulum is $T = 2\pi\sqrt{\frac{L}{g}}$, where T is time of the period of the swing, L is the length of the pendulum, and g is the acceleration of gravity. At the California Academy of Sciences, one can see a Foucault's pendulum with a length = 30 ft, and $g = 32$ ft/sec^2. Using $\pi = 3.14$, find the period of this pendulum. (Round to the nearest tenth of a second.)

86. If the amount of gold discovered by humankind could be assembled in one place, it would make a cube with a volume of 195,112 cubic feet. Each side of the cube would be $\sqrt[3]{195,112}$ feet long. How long would one side of the cube be? (*Source: Reader's Digest*)

87. Explain why the square root of a negative number is not a real number.

88. Explain why the cube root of a negative number is a real number.

89. Graph $y = \sqrt{x}$. (Complete the table below, plot the ordered pair solutions, and draw a smooth curve through the points. Remember that since the radicand cannot be negative, this particular graph begins at the point with coordinates (0, 0).)

x	y
0	0
1	
3	(approximate)
4	
9	

90. Graph $y = \sqrt[3]{x}$. (Complete the table below, plot the ordered pair solutions, and draw a smooth curve through the points.)

x	y
-8	
-2	(approximate)
-1	
0	
1	
2	(approximate)
8	

Recall from this section that $\sqrt{a^2} = |a|$ for any real number a. Simplify the following given that x represents any real number.

91. $\sqrt{x^2}$

92. $\sqrt{4x^2}$

93. $\sqrt{(x+2)^2}$

94. $\sqrt{x^2 + 6x + 9}$ (**Hint:** First factor $x^2 + 6x + 9$.)

Using a graphing calculator, graph each function. Observe the graph from left to right and give the ordered pair that corresponds to the "beginning" of the graph. Then tell why the graph starts at that point.

95. $y = \sqrt{x - 2}$

96. $y = \sqrt{x + 3}$

97. $y = \sqrt{x + 4}$

98. $y = \sqrt{x - 5}$

> **STUDY SKILLS BUILDER**

How Well Do You Know Your Textbook?

Let's check to see whether you are familiar with your textbook yet. Remember, for help, see Section 1.1 in this text.

1. What does the 🖳 icon mean?
2. What does the ✎ icon mean?
3. What does the △ icon mean?
4. Where can you find a review for each chapter? What answers to this review can be found in the back of your text?
5. Each chapter contains an overview of the chapter along with examples. What is this feature called?
6. Each chapter contains a review of vocabulary. What is this feature called?
7. There is a CD in your text. What content is contained on this CD?
8. What is the location of the section that is entirely devoted to study skills?
9. There are Practice exercises that are contained in the text. What are they and how can they be used?

8.2 SIMPLIFYING RADICALS

OBJECTIVES

1. Use the product rule to simplify square roots.
2. Use the quotient rule to simplify square roots.
3. Simplify radicals containing variables.
4. Simplify higher roots.

OBJECTIVE 1 ▶ Simplifying radicals using the product rule. A square root is simplified when the radicand contains no perfect square factors (other than 1). For example, $\sqrt{20}$ is not simplified because $\sqrt{20} = \sqrt{4 \cdot 5}$ and 4 is a perfect square.

To begin simplifying square roots, we notice the following pattern.

$$\sqrt{9 \cdot 16} = \sqrt{144} = 12$$
$$\sqrt{9} \cdot \sqrt{16} = 3 \cdot 4 = 12$$

Since both expressions simplify to 12, we can write

$$\sqrt{9 \cdot 16} = \sqrt{9} \cdot \sqrt{16}$$

This suggests the following product rule for square roots.

> **Product Rule for Square Roots**
> If $\sqrt{a}$ and $\sqrt{b}$ are real numbers, then
> $$\sqrt{a \cdot b} = \sqrt{a} \cdot \sqrt{b}$$

In other words, the square root of a product is equal to the product of the square roots.

To simplify $\sqrt{20}$, for example, we factor 20 so that one of its factors is a perfect square factor.

$$\sqrt{20} = \sqrt{4 \cdot 5} \quad \text{Factor 20.}$$
$$= \sqrt{4} \cdot \sqrt{5} \quad \text{Use the product rule.}$$
$$= 2\sqrt{5} \quad \text{Write } \sqrt{4} \text{ as 2.}$$

The notation $2\sqrt{5}$ means $2 \cdot \sqrt{5}$. Since the radicand 5 has no perfect square factor other than 1 then $2\sqrt{5}$ is in simplest form.

> ▶ **Helpful Hint**
> A radical expression in simplest form does *not* mean a decimal approximation. The simplest form of a radical expression is an exact form and may still contain a radical.
>
> $$\sqrt{20} = 2\sqrt{5} \qquad \sqrt{20} \approx 4.47$$
> exact $\qquad\qquad$ decimal approximation

512 CHAPTER 8 Roots and Radicals

EXAMPLE 1 Simplify.
a. $\sqrt{54}$ b. $\sqrt{12}$ c. $\sqrt{200}$ d. $\sqrt{35}$

Solution

a. Try to factor 54 so that at least one of the factors is a perfect square. Since 9 is a perfect square and $54 = 9 \cdot 6$,

$$\begin{aligned}\sqrt{54} &= \sqrt{9 \cdot 6} &&\text{Factor 54 so that one factor is a perfect square.}\\ &= \sqrt{9} \cdot \sqrt{6} &&\text{Apply the product rule.}\\ &= 3\sqrt{6} &&\text{Write } \sqrt{9} \text{ as 3.}\end{aligned}$$

b. $$\begin{aligned}\sqrt{12} &= \sqrt{4 \cdot 3} &&\text{Factor 12 so that one factor is a perfect square.}\\ &= \sqrt{4} \cdot \sqrt{3} &&\text{Apply the product rule.}\\ &= 2\sqrt{3} &&\text{Write } \sqrt{4} \text{ as 2.}\end{aligned}$$

c. $$\begin{aligned}\sqrt{200} &= \sqrt{100 \cdot 2} &&\text{Factor 200 so that one factor is a perfect square.}\\ &= \sqrt{100} \cdot \sqrt{2} &&\text{Apply the product rule.}\\ &= 10\sqrt{2} &&\text{Write } \sqrt{100} \text{ as 10.}\end{aligned}$$

d. The radicand 35 contains no perfect square factors other than 1. Thus $\sqrt{35}$ is in simplest form.

PRACTICE 1 Simplify.
a. $\sqrt{24}$ b. $\sqrt{60}$ c. $\sqrt{42}$ d. $\sqrt{300}$

In Example 1, part (c), 100 is the largest perfect square factor of 200. What happens if we don't use the largest perfect square factor? Although using the largest perfect square factor saves time, the result is the same no matter what perfect square factor is used. For example, it is also true that $200 = 4 \cdot 50$. Then

$$\begin{aligned}\sqrt{200} &= \sqrt{4} \cdot \sqrt{50}\\ &= 2 \cdot \sqrt{50}\end{aligned}$$

Since $\sqrt{50}$ is not in simplest form, we continue.

$$\begin{aligned}\sqrt{200} &= 2 \cdot \sqrt{50}\\ &= 2 \cdot \sqrt{25 \cdot 2}\\ &= 2 \cdot \sqrt{25} \cdot \sqrt{2}\\ &= 2 \cdot 5 \cdot \sqrt{2}\\ &= 10\sqrt{2}\end{aligned}$$

EXAMPLE 2 Simplify $3\sqrt{8}$.

Solution Remember that $3\sqrt{8}$ means $3 \cdot \sqrt{8}$.

$$\begin{aligned}3 \cdot \sqrt{8} &= 3 \cdot \sqrt{4 \cdot 2} &&\text{Factor 8 so that one factor is a perfect square.}\\ &= 3 \cdot \sqrt{4} \cdot \sqrt{2} &&\text{Use the product rule.}\\ &= 3 \cdot 2 \cdot \sqrt{2} &&\text{Write } \sqrt{4} \text{ as 2.}\\ &= 6 \cdot \sqrt{2} \text{ or } 6\sqrt{2} &&\text{Write } 3 \cdot 2 \text{ as 6.}\end{aligned}$$

PRACTICE 2 Simplify $5\sqrt{40}$.

OBJECTIVE 2 ▶ Simplifying radicals using the quotient rule. Next, let's examine the square root of a quotient.

$$\sqrt{\frac{16}{4}} = \sqrt{4} = 2$$

Also,

$$\frac{\sqrt{16}}{\sqrt{4}} = \frac{4}{2} = 2$$

Since both expressions equal 2, we can write

$$\sqrt{\frac{16}{4}} = \frac{\sqrt{16}}{\sqrt{4}}$$

This suggests the following quotient rule.

Quotient Rule for Square Roots
If $\sqrt{a}$ and $\sqrt{b}$ are real numbers and $b \neq 0$, then

$$\sqrt{\frac{a}{b}} = \frac{\sqrt{a}}{\sqrt{b}}$$

In other words, the square root of a quotient is equal to the quotient of the square roots.

EXAMPLE 3 Simplify.

a. $\sqrt{\frac{25}{36}}$ b. $\sqrt{\frac{3}{64}}$ c. $\sqrt{\frac{40}{81}}$

Use the quotient rule.

Solution

a. $\sqrt{\frac{25}{36}} = \frac{\sqrt{25}}{\sqrt{36}} = \frac{5}{6}$

b. $\sqrt{\frac{3}{64}} = \frac{\sqrt{3}}{\sqrt{64}} = \frac{\sqrt{3}}{8}$

c. $\sqrt{\frac{40}{81}} = \frac{\sqrt{40}}{\sqrt{81}}$ Use the quotient rule.

$= \frac{\sqrt{4} \cdot \sqrt{10}}{9}$ Apply the product rule and write $\sqrt{81}$ as 9.

$= \frac{2\sqrt{10}}{9}$ Write $\sqrt{4}$ as 2.

PRACTICE 3 Simplify.

a. $\sqrt{\frac{5}{49}}$ b. $\sqrt{\frac{9}{100}}$ c. $\sqrt{\frac{18}{25}}$

OBJECTIVE 3 ▶ Simplifying radicals containing variables. Recall that $\sqrt{x^6} = x^3$ because $(x^3)^2 = x^6$. If an odd exponent occurs, we write the exponential expression so that one factor is the greatest even power contained in the expression. Then we use the product rule to simplify.

EXAMPLE 4 Simplify. Assume that all variables represent positive numbers.

a. $\sqrt{x^5}$ b. $\sqrt{8y^2}$ c. $\sqrt{\dfrac{45}{x^6}}$ d. $\sqrt{\dfrac{5p^3}{9}}$

Solution

a. $\sqrt{x^5} = \sqrt{x^4 \cdot x} = \sqrt{x^4} \cdot \sqrt{x} = x^2\sqrt{x}$

b. $\sqrt{8y^2} = \sqrt{4 \cdot 2 \cdot y^2} = \sqrt{4y^2 \cdot 2} = \sqrt{4y^2} \cdot \sqrt{2} = 2y\sqrt{2}$

c. $\sqrt{\dfrac{45}{x^6}} = \dfrac{\sqrt{45}}{\sqrt{x^6}} = \dfrac{\sqrt{9 \cdot 5}}{x^3} = \dfrac{\sqrt{9} \cdot \sqrt{5}}{x^3} = \dfrac{3\sqrt{5}}{x^3}$

d. $\sqrt{\dfrac{5p^3}{9}} = \dfrac{\sqrt{5p^3}}{\sqrt{9}} = \dfrac{\sqrt{p^2 \cdot 5p}}{3} = \dfrac{\sqrt{p^2} \cdot \sqrt{5p}}{3} = \dfrac{p\sqrt{5p}}{3}$

PRACTICE 4 Simplify. Assume that all variables represent positive numbers.

a. $\sqrt{x^7}$ b. $\sqrt{12a^4}$ c. $\sqrt{\dfrac{98}{z^8}}$ d. $\sqrt{\dfrac{11y^9}{49}}$

OBJECTIVE 4 ▶ Simplifying higher roots. The product and quotient rules also apply to roots other than square roots. In general, we have the following product and quotient rules for radicals.

Product Rule for Radicals
If $\sqrt[n]{a}$ and $\sqrt[n]{b}$ are real numbers, then

$$\sqrt[n]{a \cdot b} = \sqrt[n]{a} \cdot \sqrt[n]{b}$$

Quotient Rule for Radicals
If $\sqrt[n]{a}$ and $\sqrt[n]{b}$ are real numbers and $b \neq 0$, then

$$\sqrt[n]{\dfrac{a}{b}} = \dfrac{\sqrt[n]{a}}{\sqrt[n]{b}}$$

To simplify cube roots, look for perfect cube factors of the radicand. For example, 8 is a perfect cube, since $2^3 = 8$.

To simplify $\sqrt[3]{48}$, factor 48 as $8 \cdot 6$.

$\sqrt[3]{48} = \sqrt[3]{8 \cdot 6}$ Factor 48.
$\phantom{\sqrt[3]{48}} = \sqrt[3]{8} \cdot \sqrt[3]{6}$ Apply the product rule.
$\phantom{\sqrt[3]{48}} = 2\sqrt[3]{6}$ Write $\sqrt[3]{8}$ as 2.

$2\sqrt[3]{6}$ is in simplest form since the radicand 6 contains no perfect cube factors other than 1.

EXAMPLE 5 Simplify.

a. $\sqrt[3]{54}$ b. $\sqrt[3]{18}$ c. $\sqrt[3]{\dfrac{7}{8}}$ d. $\sqrt[3]{\dfrac{40}{27}}$

Solution

a. $\sqrt[3]{54} = \sqrt[3]{27 \cdot 2} = \sqrt[3]{27} \cdot \sqrt[3]{2} = 3\sqrt[3]{2}$

b. The number 18 contains no perfect cube factors, so $\sqrt[3]{18}$ cannot be simplified further.

c. $\sqrt[3]{\dfrac{7}{8}} = \dfrac{\sqrt[3]{7}}{\sqrt[3]{8}} = \dfrac{\sqrt[3]{7}}{2}$

d. $\sqrt[3]{\dfrac{40}{27}} = \dfrac{\sqrt[3]{40}}{\sqrt[3]{27}} = \dfrac{\sqrt[3]{8 \cdot 5}}{3} = \dfrac{\sqrt[3]{8} \cdot \sqrt[3]{5}}{3} = \dfrac{2\sqrt[3]{5}}{3}$

PRACTICE 5 Simplify.

a. $\sqrt[3]{24}$ b. $\sqrt[3]{38}$ c. $\sqrt[3]{\dfrac{5}{27}}$ d. $\sqrt[3]{\dfrac{15}{64}}$

To simplify fourth roots, look for perfect fourth powers of the radicand. For example, 16 is a perfect fourth power since $2^4 = 16$.

To simplify $\sqrt[4]{32}$, factor 32 as $16 \cdot 2$.

$\sqrt[4]{32} = \sqrt[4]{16 \cdot 2}$ Factor 32.

$\phantom{\sqrt[4]{32}} = \sqrt[4]{16} \cdot \sqrt[4]{2}$ Apply the product rule.

$\phantom{\sqrt[4]{32}} = 2\sqrt[4]{2}$ Write $\sqrt[4]{16}$ as 2.

EXAMPLE 6 Simplify.

a. $\sqrt[4]{243}$ b. $\sqrt[4]{\dfrac{3}{16}}$ c. $\sqrt[5]{64}$

Solution

a. $\sqrt[4]{243} = \sqrt[4]{81 \cdot 3} = \sqrt[4]{81} \cdot \sqrt[4]{3} = 3\sqrt[4]{3}$ b. $\sqrt[4]{\dfrac{3}{16}} = \dfrac{\sqrt[4]{3}}{\sqrt[4]{16}} = \dfrac{\sqrt[4]{3}}{2}$

c. $\sqrt[5]{64} = \sqrt[5]{32 \cdot 2} = \sqrt[5]{32} \cdot \sqrt[5]{2} = 2\sqrt[5]{2}$

PRACTICE 6 Simplify.

a. $\sqrt[4]{32}$ b. $\sqrt[4]{\dfrac{5}{81}}$ c. $\sqrt[5]{96}$

VOCABULARY & READINESS CHECK

Use the choices below to fill in the blanks. Not all choices will be used.

$a \cdot b$ $\dfrac{a}{b}$ $\dfrac{\sqrt{a}}{\sqrt{b}}$ $\sqrt{a} \cdot \sqrt{b}$

1. If $\sqrt{a}$ and $\sqrt{b}$ are real numbers, then $\sqrt{a \cdot b} =$ _____ .

2. If $\sqrt{a}$ and $\sqrt{b}$ are real numbers, then $\sqrt{\dfrac{a}{b}} =$ _____ .

For Exercises 3 and 4, fill in the blanks using the example: $\sqrt{4 \cdot 9} = \sqrt{\underline{4}} \cdot \sqrt{\underline{9}} = \underline{2} \cdot \underline{3} = \underline{6}$.

3. $\sqrt{16 \cdot 25} = \sqrt{\underline{}} \cdot \sqrt{\underline{}} = \underline{} \cdot \underline{} = \underline{}$ **4.** $\sqrt{36 \cdot 3} = \sqrt{\underline{}} \cdot \sqrt{\underline{}} = \underline{} \cdot \sqrt{\underline{}} = \underline{}$

8.2 EXERCISE SET

Use the product rule to simplify each radical. See Examples 1 and 2.

1. $\sqrt{20}$
2. $\sqrt{44}$
3. $\sqrt{50}$
4. $\sqrt{28}$
5. $\sqrt{33}$
6. $\sqrt{21}$
7. $\sqrt{98}$
8. $\sqrt{125}$
9. $\sqrt{60}$
10. $\sqrt{90}$
11. $\sqrt{180}$
12. $\sqrt{150}$
13. $\sqrt{52}$
14. $\sqrt{75}$
15. $3\sqrt{25}$
16. $9\sqrt{36}$
17. $7\sqrt{63}$
18. $11\sqrt{99}$
19. $-5\sqrt{27}$
20. $-6\sqrt{75}$

Use the quotient rule and the product rule to simplify each radical. See Example 3.

21. $\sqrt{\dfrac{8}{25}}$
22. $\sqrt{\dfrac{63}{16}}$
23. $\sqrt{\dfrac{27}{121}}$
24. $\sqrt{\dfrac{24}{169}}$
25. $\sqrt{\dfrac{9}{4}}$
26. $\sqrt{\dfrac{100}{49}}$
27. $\sqrt{\dfrac{125}{9}}$
28. $\sqrt{\dfrac{27}{100}}$
29. $\sqrt{\dfrac{11}{36}}$
30. $\sqrt{\dfrac{30}{49}}$
31. $-\sqrt{\dfrac{27}{144}}$
32. $-\sqrt{\dfrac{84}{121}}$

Simplify each radical. Assume that all variables represent positive numbers. See Example 4.

33. $\sqrt{x^7}$
34. $\sqrt{y^3}$
35. $\sqrt{x^{13}}$
36. $\sqrt{y^{17}}$
37. $\sqrt{36a^3}$
38. $\sqrt{81b^5}$
39. $\sqrt{96x^4}$
40. $\sqrt{40y^{10}}$
41. $\sqrt{\dfrac{12}{m^2}}$
42. $\sqrt{\dfrac{63}{p^2}}$
43. $\sqrt{\dfrac{9x}{y^{10}}}$
44. $\sqrt{\dfrac{6y^2}{z^{16}}}$
45. $\sqrt{\dfrac{88}{x^{12}}}$
46. $\sqrt{\dfrac{500}{y^{22}}}$

MIXED PRACTICE

Simplify each radical. See Examples 1 through 4.

47. $8\sqrt{4}$
48. $6\sqrt{49}$
49. $\sqrt{\dfrac{36}{121}}$
50. $\sqrt{\dfrac{25}{144}}$
51. $\sqrt{175}$
52. $\sqrt{700}$
53. $\sqrt{\dfrac{20}{9}}$
54. $\sqrt{\dfrac{45}{64}}$
55. $\sqrt{24m^7}$
56. $\sqrt{50n^{13}}$
57. $\sqrt{\dfrac{23y^3}{4x^6}}$
58. $\sqrt{\dfrac{41x^5}{9y^8}}$

Simplify each radical. See Example 5.

59. $\sqrt[3]{24}$
60. $\sqrt[3]{81}$
61. $\sqrt[3]{250}$
62. $\sqrt[3]{56}$
63. $\sqrt[3]{\dfrac{5}{64}}$
64. $\sqrt[3]{\dfrac{32}{125}}$
65. $\sqrt[3]{\dfrac{23}{8}}$
66. $\sqrt[3]{\dfrac{37}{27}}$
67. $\sqrt[3]{\dfrac{15}{64}}$
68. $\sqrt[3]{\dfrac{4}{27}}$
69. $\sqrt[3]{80}$
70. $\sqrt[3]{108}$

Simplify. See Example 6.

71. $\sqrt[4]{48}$
72. $\sqrt[4]{405}$
73. $\sqrt[4]{\dfrac{8}{81}}$
74. $\sqrt[4]{\dfrac{25}{256}}$
75. $\sqrt[5]{96}$
76. $\sqrt[5]{128}$
77. $\sqrt[5]{\dfrac{5}{32}}$
78. $\sqrt[5]{\dfrac{16}{243}}$

Simplify.

△ 79. If a cube is to have a volume of 80 cubic inches, then each side must be $\sqrt[3]{80}$ inches long. Simplify the radical representing the side length.

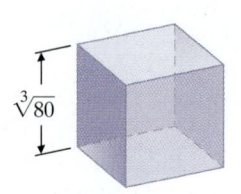

△ 80. Jeannie Boswell is swimming across a 40-foot-wide river, trying to head straight across to the opposite shore. However, the current is strong enough to move her downstream 100 feet by the time she reaches land. (See the figure.) Because of the current, the actual distance she swam is $\sqrt{11{,}600}$ feet. Simplify this radical.

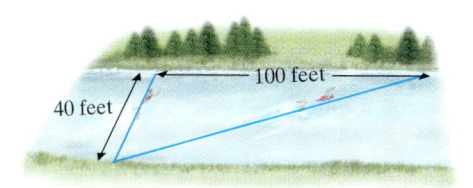

REVIEW AND PREVIEW

Perform the following operations. See Sections 5.2 and 5.3.

81. $6x + 8x$
82. $(6x)(8x)$
83. $(2x + 3)(x - 5)$
84. $(2x + 3) + (x - 5)$
85. $9y^2 - 9y^2$
86. $(9y^2)(-8y^2)$

CONCEPT EXTENSIONS

Simplify each radical. Assume that all variables represent positive numbers.

87. $\sqrt{x^6 y^3}$
88. $\sqrt{a^{13} b^{14}}$
89. $\sqrt{98 x^5 y^4}$
90. $\sqrt{27 x^8 y^{11}}$
91. $\sqrt[3]{-8x^6}$
92. $\sqrt[3]{27 x^{12}}$

93. By using replacement values for a and b, show that $\sqrt{a^2 + b^2}$ does not equal $a + b$.

94. By using replacement values for a and b, show that $\sqrt{a + b}$ does not equal $\sqrt{a} + \sqrt{b}$.

The length of a side of a cube is given by the expression $\sqrt{\dfrac{A}{6}}$ units where A square units is the cube's surface area. Use this expression for Exercises 95 through 98. Be sure to attach the appropriate units.

△ **95.** The surface area of a cube is 120 square inches. Find the exact length of a side of the cube.

△ **96.** The surface area of a cube is 594 square feet. Find the exact length of a side of the cube.

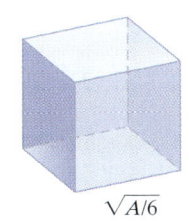

97. A Guinness World record was set in December 2004, when an electrical engineering student from Johannesburg, South Africa, solved 42 Rubik's cubes in one hour, the most ever in that time. Rubik's cube, named after its inventor, Erno Rubik, was first imagined by him in 1974, and by 1980 was a worldwide phenomenon. These cubes have remained unchanged in size, and a standard Rubik's cube has a surface area of 30.375 square inches. Find the length of one side of a Rubik's cube. (*Source: Guinness Book of World Records*)

△ **98.** The Borg spaceship on *Star Trek: The Next Generation* is in the shape of a cube. Suppose a model of this ship has a surface area of 121 square inches. Find the length of a side of the ship.

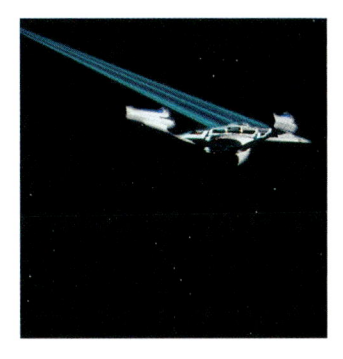

The cost C in dollars per day to operate a small delivery service is given by $C = 100\sqrt[3]{n} + 700$, where n is the number of deliveries per day.

99. Find the cost if the number of deliveries is 1000.

100. Approximate the cost if the number of deliveries is 500.

The Mosteller formula for calculating body surface area is $B = \sqrt{\dfrac{hw}{3600}}$, where B is an individual's body surface area in square meters, h is the individual's height in centimeters, and w is the individual's weight in kilograms. Use this formula in Exercises 101 and 102. Round answers to the nearest tenth.

101. Find the body surface area of a person who is 169 cm tall and weighs 64 kilograms.

102. Approximate the body surface area of a person who is 183 cm tall and weighs 85 kilograms.

8.3 ADDING AND SUBTRACTING RADICALS

OBJECTIVES

1. Add or subtract like radicals.
2. Simplify radical expressions, and then add or subtract any like radicals.

OBJECTIVE 1 ▶ **Adding and subtracting like radicals.** To combine like terms, we use the distributive property.

$$5x + 3x = (5 + 3)x = 8x$$

The distributive property can also be applied to expressions containing radicals. For example,

$$5\sqrt{2} + 3\sqrt{2} = (5 + 3)\sqrt{2} = 8\sqrt{2}$$

Also,

$$9\sqrt{5} - 6\sqrt{5} = (9 - 6)\sqrt{5} = 3\sqrt{5}$$

Radical terms $5\sqrt{2}$ and $3\sqrt{2}$ are **like radicals**, as are $9\sqrt{5}$ and $6\sqrt{5}$.

> **Like Radicals**
> Like radicals are radical expressions that have the same index and the same radicand.

From the examples above, we can see that **only like radicals can be combined** in this way. For example, the expression $2\sqrt{3} + 3\sqrt{2}$ cannot be simplified further since the radicals are not like radicals. Also, the expression $4\sqrt{7} + 4\sqrt[3]{7}$ cannot be simplified further because the radicals are not like radicals since the indices are different.

EXAMPLE 1 Simplify by combining like radical terms.

a. $4\sqrt{5} + 3\sqrt{5}$ b. $\sqrt{10} - 6\sqrt{10}$ c. $\sqrt[3]{7} + \sqrt[3]{7} - 4\sqrt[3]{5}$ d. $2\sqrt{6} + 2\sqrt[3]{6}$

Solution

a. $4\sqrt{5} + 3\sqrt{5} = (4 + 3)\sqrt{5} = 7\sqrt{5}$

b. $\sqrt{10} - 6\sqrt{10} = 1\sqrt{10} - 6\sqrt{10} = (1 - 6)\sqrt{10} = -5\sqrt{10}$

c. $\sqrt[3]{7} + \sqrt[3]{7} - 4\sqrt[3]{5} = 1\sqrt[3]{7} + 1\sqrt[3]{7} - 4\sqrt[3]{5} = (1 + 1)\sqrt[3]{7} - 4\sqrt[3]{5} = 2\sqrt[3]{7} - 4\sqrt[3]{5}$

This expression cannot be simplified further since the radicands are not the same.

d. $2\sqrt{6} + 2\sqrt[3]{6}$ cannot be simplified further since the indices are not the same.

PRACTICE 1 Simplify by combining like radical terms.

a. $3\sqrt{2} + 5\sqrt{2}$ b. $\sqrt{6} - 8\sqrt{6}$
c. $6\sqrt[4]{5} - 2\sqrt[4]{5} + 11\sqrt[4]{7}$ d. $4\sqrt{13} - 5\sqrt[3]{13}$

Concept Check ✓

Which is true?

a. $2 + 3\sqrt{5} = 5\sqrt{5}$ b. $2\sqrt{3} + 2\sqrt{7} = 2\sqrt{10}$ c. $\sqrt{3} + \sqrt{5} = \sqrt{8}$
d. $\sqrt{3} + \sqrt{3} = 3$ e. None of the above is true.

Answer to Concept Check:
e (a, b, and c are not true since each left side cannot be simplified further. For d, $\sqrt{3} + \sqrt{3} = 2\sqrt{3}$.)

OBJECTIVE 2 ▶ **Simplifying radicals, then adding or subtracting.** At first glance, it appears that the expression $\sqrt{50} + \sqrt{8}$ cannot be simplified further because the radicands are different. However, the product rule can be used to simplify each radical, and then further simplification might be possible.

Section 8.3 Adding and Subtracting Radicals 519

EXAMPLE 2 Add or subtract by first simplifying each radical.

a. $\sqrt{50} + \sqrt{8}$ **b.** $7\sqrt{12} - \sqrt{75}$ **c.** $\sqrt{25} - \sqrt{27} - 2\sqrt{18} - \sqrt{16}$

Solution

a. First simplify each radical.

$$\begin{aligned}
\sqrt{50} + \sqrt{8} &= \sqrt{25 \cdot 2} + \sqrt{4 \cdot 2} && \text{Factor radicands.} \\
&= \sqrt{25} \cdot \sqrt{2} + \sqrt{4} \cdot \sqrt{2} && \text{Apply the product rule.} \\
&= 5\sqrt{2} + 2\sqrt{2} && \text{Simplify } \sqrt{25} \text{ and } \sqrt{4}. \\
&= 7\sqrt{2} && \text{Add like radicals.}
\end{aligned}$$

b. $\begin{aligned}
7\sqrt{12} - \sqrt{75} &= 7\sqrt{4 \cdot 3} - \sqrt{25 \cdot 3} && \text{Factor radicands.} \\
&= 7\sqrt{4} \cdot \sqrt{3} - \sqrt{25} \cdot \sqrt{3} && \text{Apply the product rule.} \\
&= 7 \cdot 2\sqrt{3} - 5\sqrt{3} && \text{Simplify } \sqrt{4} \text{ and } \sqrt{25}. \\
&= 14\sqrt{3} - 5\sqrt{3} && \text{Multiply.} \\
&= 9\sqrt{3} && \text{Subtract like radicals.}
\end{aligned}$

c. $\begin{aligned}
\sqrt{25} &- \sqrt{27} - 2\sqrt{18} - \sqrt{16} \\
&= 5 - \sqrt{9 \cdot 3} - 2\sqrt{9 \cdot 2} - 4 && \text{Factor radicands.} \\
&= 5 - \sqrt{9} \cdot \sqrt{3} - 2\sqrt{9} \cdot \sqrt{2} - 4 && \text{Apply the product rule.} \\
&= 5 - 3\sqrt{3} - 2 \cdot 3\sqrt{2} - 4 && \text{Simplify.} \\
&= 1 - 3\sqrt{3} - 6\sqrt{2} && \text{Write } 5 - 4 \text{ as 1 and } 2 \cdot 3 \text{ as 6.}
\end{aligned}$

PRACTICE 2 Add or subtract by first simplifying each radical.

a. $\sqrt{45} + \sqrt{20}$ **b.** $\sqrt{36} + 3\sqrt{24} - \sqrt{40} - \sqrt{150}$
c. $\sqrt{98} - 5\sqrt{8}$

If radical expressions contain variables, we proceed in a similar way. Simplify radicals using the product and quotient rules. Then add or subtract any like radicals.

EXAMPLE 3 Simplify $2\sqrt{x^2} - \sqrt{25x^5} + \sqrt{x^5}$. Assume variables represent positive numbers.

Solution $\begin{aligned}
2\sqrt{x^2} &- \sqrt{25x^5} + \sqrt{x^5} \\
&= 2x - \sqrt{25x^4 \cdot x} + \sqrt{x^4 \cdot x} && \text{Factor radicands so that one factor is a perfect square. Simplify } \sqrt{x^2}. \\
&= 2x - \sqrt{25x^4} \cdot \sqrt{x} + \sqrt{x^4} \cdot \sqrt{x} && \text{Use the product rule.} \\
&= 2x - 5x^2\sqrt{x} + x^2\sqrt{x} && \text{Write } \sqrt{25x^4} \text{ as } 5x^2 \text{ and } \sqrt{x^4} \text{ as } x^2. \\
&= 2x - 4x^2\sqrt{x} && \text{Add like radicals.}
\end{aligned}$

PRACTICE 3 Simplify $\sqrt{x^3} - 8x\sqrt{x} + 3\sqrt{x^2}$. Assume variables represent positive numbers.

EXAMPLE 4 Simplify the radical expression: $5\sqrt[3]{16x^3} - \sqrt[3]{54x^3}$

Solution $\begin{aligned}
5\sqrt[3]{16x^3} &- \sqrt[3]{54x^3} \\
&= 5\sqrt[3]{8x^3 \cdot 2} - \sqrt[3]{27x^3 \cdot 2} && \text{Factor radicands so that one factor is a perfect cube.} \\
&= 5 \cdot \sqrt[3]{8x^3} \cdot \sqrt[3]{2} - \sqrt[3]{27x^3} \cdot \sqrt[3]{2} && \text{Use the product rule.}
\end{aligned}$

$$= 5 \cdot 2x \cdot \sqrt[3]{2} - 3x \cdot \sqrt[3]{2}$$ Write $\sqrt[3]{8x^3}$ as $2x$ and $\sqrt[3]{27x^3}$ as $3x$.
$$= 10x\sqrt[3]{2} - 3x\sqrt[3]{2}$$ Write $5 \cdot 2x$ as $10x$.
$$= 7x\sqrt[3]{2}$$ Subtract like radicands.

PRACTICE 4 Simplify the radical expression: $4\sqrt[3]{81x^6} - \sqrt[3]{24x^6}$

VOCABULARY & READINESS CHECK

Fill in each blank.

1. Radicals that have the same index and same radicand are called _____.
2. The expressions $7\sqrt[3]{2x}$ and $-\sqrt[3]{2x}$ are called _____.
3. $11\sqrt{2} + 6\sqrt{2} =$ _____.
   a. $66\sqrt{2}$ b. $17\sqrt{2}$ c. $17\sqrt{4}$
4. $\sqrt{5}$ is the same as _____.
   a. $0\sqrt{5}$ b. $1\sqrt{5}$ c. $5\sqrt{5}$
5. $\sqrt{5} + \sqrt{5} =$ _____
   a. $\sqrt{10}$ b. 5 c. $2\sqrt{5}$
6. $9\sqrt{7} - \sqrt{7} =$ _____
   a. $8\sqrt{7}$ b. 9 c. 0

8.3 EXERCISE SET

Simplify each expression by combining like radicals where possible. See Example 1.

1. $4\sqrt{3} - 8\sqrt{3}$
2. $2\sqrt{5} - 9\sqrt{5}$
3. $3\sqrt{6} + 8\sqrt{6} - 2\sqrt{6} - 5$
4. $12\sqrt{2} - 3\sqrt{2} + 8\sqrt{2} + 10$
5. $\sqrt{11} + \sqrt{11} + 11$
6. $\sqrt{13} + 13 + \sqrt{13}$
7. $6\sqrt{5} - 5\sqrt{5} + \sqrt{2}$
8. $4\sqrt{3} + \sqrt{5} - 3\sqrt{3}$
9. $\sqrt[3]{16} + \sqrt[3]{16} - 4\sqrt[3]{16}$
10. $\sqrt[3]{49} - 8\sqrt[3]{49} + \sqrt[3]{49}$
11. $2\sqrt[3]{3} + 5\sqrt[3]{3} - \sqrt[3]{3}$
12. $8\sqrt[3]{5} + 2\sqrt[3]{5} + \sqrt{5}$
13. $2\sqrt[3]{2} - 7\sqrt[3]{2} - 6$
14. $5\sqrt[3]{9} + 2 - 11\sqrt[3]{9}$

MIXED PRACTICE

Add or subtract by first simplifying each radical and then combining any like radical terms. Assume that all variables represent positive real numbers. See Examples 2 and 3.

15. $\sqrt{12} + \sqrt{27}$
16. $\sqrt{50} + \sqrt{18}$
17. $\sqrt{45} + 3\sqrt{20}$
18. $\sqrt{28} + \sqrt{63}$
19. $2\sqrt{54} - \sqrt{20} + \sqrt{45} - \sqrt{24}$
20. $2\sqrt{8} - \sqrt{128} + \sqrt{48} + \sqrt{18}$
21. $4x - 3\sqrt{x^2} + \sqrt{x}$
22. $x - 6\sqrt{x^2} + 2\sqrt{x}$
23. $\sqrt{25x} + \sqrt{36x} - 11\sqrt{x}$
24. $3\sqrt{x^3} - x\sqrt{4x}$
25. $\sqrt{16x} - \sqrt{x^3}$
26. $\sqrt{8x^3} - \sqrt{x^2}$
27. $12\sqrt{5} - \sqrt{5} - 4\sqrt{5}$
28. $7\sqrt{3} + 2\sqrt{3} - 13\sqrt{3}$
29. $\sqrt{5} + \sqrt[3]{5}$
30. $\sqrt{5} + \sqrt{5}$
31. $4 + 8\sqrt{2} - 9$
32. $6 - 2\sqrt{3} - \sqrt{3}$
33. $8 - \sqrt{2} - 5\sqrt{2}$
34. $\sqrt{75} + \sqrt{48}$
35. $5\sqrt{32} - \sqrt{72}$
36. $2\sqrt{80} - \sqrt{45}$
37. $\sqrt{8} + \sqrt{9} + \sqrt{18} + \sqrt{81}$
38. $\sqrt{6} + \sqrt{16} + \sqrt{24} + \sqrt{25}$
39. $\sqrt{\dfrac{5}{9}} + \sqrt{\dfrac{5}{81}}$
40. $\sqrt{\dfrac{3}{64}} + \sqrt{\dfrac{3}{16}}$
41. $\sqrt{\dfrac{3}{4}} - \sqrt{\dfrac{3}{64}}$
42. $\sqrt{\dfrac{7}{25}} - \sqrt{\dfrac{7}{100}}$
43. $2\sqrt{45} - 2\sqrt{20}$
44. $5\sqrt{18} + 2\sqrt{32}$
45. $\sqrt{35} - \sqrt{140}$
46. $\sqrt{6} - \sqrt{600}$
47. $5\sqrt{2x} + \sqrt{98x}$
48. $3\sqrt{9x} + 2\sqrt{x}$

Section 8.3 Adding and Subtracting Radicals 521

49. $5\sqrt{x} + 4\sqrt{4x} - 13\sqrt{x}$
50. $\sqrt{9x} + \sqrt{81x} - 11\sqrt{x}$
51. $\sqrt{3x^3} + 3x\sqrt{x}$
52. $x\sqrt{4x} + \sqrt{9x^3}$

Add or subtract by first simplifying each radical and then combining any like radical terms. Assume that all variables represent positive real numbers. See Example 4.

53. $\sqrt[3]{81} + \sqrt[3]{24}$
54. $\sqrt[3]{32} - \sqrt[3]{4}$
55. $4\sqrt[3]{9} - \sqrt[3]{243}$
56. $7\sqrt[3]{6} - \sqrt[3]{48}$
57. $2\sqrt[3]{8} + 2\sqrt[3]{16}$
58. $3\sqrt[3]{27} + 3\sqrt[3]{81}$
59. $\sqrt[3]{8} + \sqrt[3]{54} - 5$
60. $\sqrt[3]{64} + \sqrt[3]{14} - 9$
61. $\sqrt{32x^2} + \sqrt[3]{32} + \sqrt{4x^2}$
62. $\sqrt{18x^2} + \sqrt[3]{24} + \sqrt{2x^2}$
63. $\sqrt{40x} + \sqrt[3]{40} - 2\sqrt{10x} - \sqrt[3]{5}$
64. $\sqrt{72x^2} + \sqrt[3]{54} - x\sqrt{50} - 3\sqrt[3]{2}$

REVIEW AND PREVIEW

Square each binomial. See Section 5.4.

65. $(x + 6)^2$
66. $(3x + 2)^2$
67. $(2x - 1)^2$
68. $(x - 5)^2$

Solve each system of linear equations. See Section 4.2.

69. $\begin{cases} x = 2y \\ x + 5y = 14 \end{cases}$
70. $\begin{cases} y = -5x \\ x + y = 16 \end{cases}$

CONCEPT EXTENSIONS

71. In your own words, describe like radicals.
72. In the expression $\sqrt{5} + 2 - 3\sqrt{5}$, explain why 2 and -3 cannot be combined.
73. Find the perimeter of the rectangular picture frame.

$\sqrt{5}$ inches

$3\sqrt{5}$ inches

74. Find the perimeter of the plot of land.

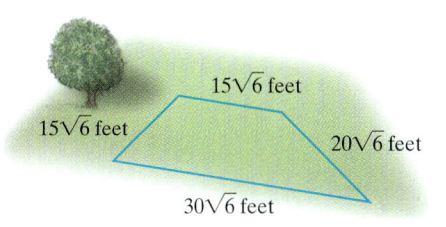

$15\sqrt{6}$ feet
$15\sqrt{6}$ feet
$20\sqrt{6}$ feet
$30\sqrt{6}$ feet

75. An 8-foot-long water trough is to be made of wood. Each of the two triangular end pieces has an area of $\dfrac{3\sqrt{27}}{4}$ square feet. The two side panels are both rectangular. In simplest radical form, find the total area of the wood needed.

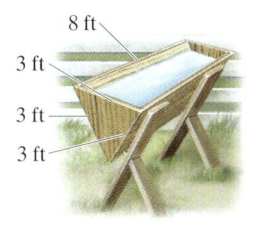

8 ft
3 ft
3 ft
3 ft

76. Eight wooden braces are to be attached along the diagonals of the vertical sides of a storage bin. Each of four of these diagonals has a length of $\sqrt{52}$ feet, while each of the other four has a length of $\sqrt{80}$ feet. In simplest radical form, find the total length of the wood needed for these braces.

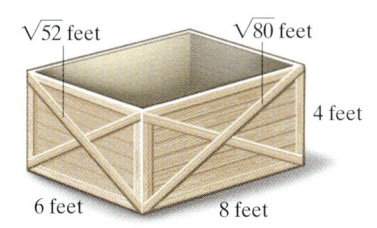

$\sqrt{52}$ feet
$\sqrt{80}$ feet
4 feet
6 feet
8 feet

Simplify.

77. $\sqrt{\dfrac{x^3}{16}} - x\sqrt{\dfrac{9x}{25}} + \dfrac{\sqrt{81x^3}}{2}$

78. $7\sqrt{x^{11}y^7} - x^2y\sqrt{25x^7y^5} + \sqrt{8x^8y^2}$

STUDY SKILLS BUILDER

Learning New Terms?

By now, you have encountered many new terms. It's never too late to make a list of new terms and review them frequently. Remember that placing these new terms (including page references) on 3 × 5 index cards might help you later when you're preparing for a quiz.

Answer the following.

1. How do new terms stand out in this text so that they can be found?
2. Name one way placing a word and its definition on a 3 × 5 card might be helpful.

522 CHAPTER 8 Roots and Radicals

8.4 MULTIPLYING AND DIVIDING RADICALS

OBJECTIVES
1. Multiply radicals.
2. Divide radicals.
3. Rationalize denominators.
4. Rationalize using conjugates.

OBJECTIVE 1 ▶ Multiplying radicals. In Section 8.2 we used the product and quotient rules for radicals to help us simplify radicals. In this section, we use these rules to simplify products and quotients of radicals.

Product Rule for Radicals

If $\sqrt[n]{a}$ and $\sqrt[n]{b}$ are real numbers, then

$$\sqrt[n]{a} \cdot \sqrt[n]{b} = \sqrt[n]{a \cdot b}$$

This property says that the product of the nth roots of two numbers is the nth root of the product of the two numbers. For example,

$$\sqrt{3} \cdot \sqrt{2} = \sqrt{3 \cdot 2} = \sqrt{6} \quad \text{Also,} \quad \sqrt[3]{5} \cdot \sqrt[3]{7} = \sqrt[3]{5 \cdot 7} = \sqrt[3]{35}$$

EXAMPLE 1 Multiply. Then simplify if possible.

a. $\sqrt{7} \cdot \sqrt{3}$ b. $\sqrt{3} \cdot \sqrt{3}$ c. $\sqrt{3} \cdot \sqrt{15}$ d. $2\sqrt{3} \cdot 5\sqrt{2}$ e. $\sqrt{2x^3} \cdot \sqrt{6x}$

Solution

a. $\sqrt{7} \cdot \sqrt{3} = \sqrt{7 \cdot 3} = \sqrt{21}$
b. $\sqrt{3} \cdot \sqrt{3} = \sqrt{3 \cdot 3} = \sqrt{9} = 3$
c. $\sqrt{3} \cdot \sqrt{15} = \sqrt{45}$. Next, simplify $\sqrt{45}$.
   $\sqrt{45} = \sqrt{9 \cdot 5} = \sqrt{9} \cdot \sqrt{5} = 3\sqrt{5}$
d. $2\sqrt{3} \cdot 5\sqrt{2} = 2 \cdot 5\sqrt{3 \cdot 2} = 10\sqrt{6}$
e. $\sqrt{2x^3} \cdot \sqrt{6x} = \sqrt{2x^3 \cdot 6x}$ Use the product rule.
   $= \sqrt{12x^4}$ Multiply.
   $= \sqrt{4x^4 \cdot 3}$ Write $12x^4$ so that one factor is a perfect square.
   $= \sqrt{4x^4} \cdot \sqrt{3}$ Use the product rule.
   $= 2x^2\sqrt{3}$ Simplify.

PRACTICE 1 Multiply. Then simplify if possible.

a. $\sqrt{11} \cdot \sqrt{7}$ b. $9\sqrt{10} \cdot 8\sqrt{3}$ c. $\sqrt{5} \cdot \sqrt{10}$
d. $\sqrt{17} \cdot \sqrt{17}$ e. $\sqrt{15y} \cdot \sqrt{5y^3}$

From Example 1b, we found that

$$\sqrt{3} \cdot \sqrt{3} = 3 \quad \text{or} \quad (\sqrt{3})^2 = 3$$

This is true in general.

If a is a positive number,

$$\sqrt{a} \cdot \sqrt{a} = a \quad \text{or} \quad (\sqrt{a})^2 = a$$

Concept Check ✓
Identify the true statement(s).

a. $\sqrt{7} \cdot \sqrt{7} = 7$ b. $\sqrt{2} \cdot \sqrt{3} = 6$
c. $(\sqrt{131})^2 = 131$ d. $\sqrt{5x} \cdot \sqrt{5x} = 5x$ (Here x is a positive number.)

Answer to Concept Check:
a, c, d

Section 8.4 Multiplying and Dividing Radicals

EXAMPLE 2 Find $(3\sqrt{2})^2$

Solution $(3\sqrt{2})^2 = 3^2 \cdot (\sqrt{2})^2 = 9 \cdot 2 = 18$

PRACTICE 2 Find $(2\sqrt{7})^2$.

EXAMPLE 3 Multiply $\sqrt[3]{4} \cdot \sqrt[3]{18}$. Then simplify if possible.

Solution $\sqrt[3]{4} \cdot \sqrt[3]{18} = \sqrt[3]{4 \cdot 18} = \sqrt[3]{4 \cdot 2 \cdot 9} = \sqrt[3]{8 \cdot 9} = \sqrt[3]{8} \cdot \sqrt[3]{9} = 2\sqrt[3]{9}$

PRACTICE 3 Multiply $\sqrt[3]{10} \cdot \sqrt[3]{50}$. Then simplify if possible.

When multiplying radical expressions containing more than one term, use the same techniques we use to multiply other algebraic expressions with more than one term.

EXAMPLE 4 Multiply. Then simplify if possible.

a. $\sqrt{5}(\sqrt{5} - \sqrt{2})$ **b.** $\sqrt{3x}(\sqrt{x} - 5\sqrt{3})$
c. $(\sqrt{x} + \sqrt{2})(\sqrt{3} - \sqrt{2})$

Solution

a. Using the distributive property, we have

$\sqrt{5}(\sqrt{5} - \sqrt{2}) = \sqrt{5} \cdot \sqrt{5} - \sqrt{5} \cdot \sqrt{2}$

$= 5 - \sqrt{10}$ Since $\sqrt{5} \cdot \sqrt{5} = 5$ and $\sqrt{5} \cdot \sqrt{2} = \sqrt{10}$

b. $\sqrt{3x}(\sqrt{x} - 5\sqrt{3}) = \sqrt{3x} \cdot \sqrt{x} - \sqrt{3x} \cdot 5\sqrt{3}$ Use the distributive property.
$= \sqrt{3x \cdot x} - 5\sqrt{3x \cdot 3}$ Use the product rule.
$= \sqrt{3 \cdot x^2} - 5\sqrt{9 \cdot x}$ Factor each radicand so that one factor is a perfect square.
$= \sqrt{3} \cdot \sqrt{x^2} - 5 \cdot \sqrt{9} \cdot \sqrt{x}$ Use the product rule.
$= x\sqrt{3} - 5 \cdot 3 \cdot \sqrt{x}$ Simplify.
$= x\sqrt{3} - 15\sqrt{x}$ Simplify.

c. Use the FOIL method of multiplication.

$$\overset{F}{} \quad \overset{O}{} \quad \overset{I}{} \quad \overset{L}{}$$
$(\sqrt{x} + \sqrt{2})(\sqrt{3} - \sqrt{2}) = \sqrt{x} \cdot \sqrt{3} - \sqrt{x} \cdot \sqrt{2} + \sqrt{2} \cdot \sqrt{3} - \sqrt{2} \cdot \sqrt{2}$
$= \sqrt{3x} - \sqrt{2x} + \sqrt{6} - \sqrt{4}$ Apply the product rule.
$= \sqrt{3x} - \sqrt{2x} + \sqrt{6} - 2$ Simplify.

PRACTICE 4 Multiply. Then simplify if possible.

a. $\sqrt{3}(\sqrt{3} - \sqrt{5})$ **b.** $\sqrt{2z}(\sqrt{z} + 7\sqrt{2})$
c. $(\sqrt{x} - \sqrt{7})(\sqrt{x} + \sqrt{2})$

The special product formulas can be used to multiply expressions containing radicals.

EXAMPLE 5 Multiply. Then simplify if possible.

a. $(\sqrt{5} - 7)(\sqrt{5} + 7)$ **b.** $(\sqrt{7x} + 2)^2$

Solution

a. Recall from Chapter 5 that $(a - b)(a + b) = a^2 - b^2$. Then

$$(\sqrt{5} - 7)(\sqrt{5} + 7) = (\sqrt{5})^2 - 7^2$$
$$= 5 - 49$$
$$= -44$$

b. Recall that $(a + b)^2 = a^2 + 2ab + b^2$. Then

$$(\sqrt{7x} + 2)^2 = (\sqrt{7x})^2 + 2(\sqrt{7x})(2) + (2)^2$$
$$= 7x + 4\sqrt{7x} + 4$$

PRACTICE 5 Multiply. Then simplify if possible.

a. $(\sqrt{7} + 4)(\sqrt{7} - 4)$ **b.** $(\sqrt{3x} - 5)^2$

OBJECTIVE 2 ▶ Dividing radicals. To simplify quotients of radical expressions, we use the quotient rule.

Quotient Rule for Radicals
If $\sqrt[n]{a}$ and $\sqrt[n]{b}$ are real numbers and $b \neq 0$, then

$$\frac{\sqrt[n]{a}}{\sqrt[n]{b}} = \sqrt[n]{\frac{a}{b}}, \text{ providing } b \neq 0$$

EXAMPLE 6 Divide. Then simplify if possible.

a. $\dfrac{\sqrt{14}}{\sqrt{2}}$ **b.** $\dfrac{\sqrt{100}}{\sqrt{5}}$ **c.** $\dfrac{\sqrt{12x^3}}{\sqrt{3x}}$

Solution Use the quotient rule and then simplify the resulting radicand.

a. $\dfrac{\sqrt{14}}{\sqrt{2}} = \sqrt{\dfrac{14}{2}} = \sqrt{7}$

b. $\dfrac{\sqrt{100}}{\sqrt{5}} = \sqrt{\dfrac{100}{5}} = \sqrt{20} = \sqrt{4 \cdot 5} = \sqrt{4} \cdot \sqrt{5} = 2\sqrt{5}$

c. $\dfrac{\sqrt{12x^3}}{\sqrt{3x}} = \sqrt{\dfrac{12x^3}{3x}} = \sqrt{4x^2} = 2x$

PRACTICE 6 Divide. Then simplify if possible.

a. $\dfrac{\sqrt{21}}{\sqrt{7}}$ **b.** $\dfrac{\sqrt{48}}{\sqrt{6}}$ **c.** $\dfrac{\sqrt{45y^5}}{\sqrt{5y}}$

EXAMPLE 7 Divide $\dfrac{\sqrt[3]{32}}{\sqrt[3]{4}}$. Then simplify if possible.

Solution $\dfrac{\sqrt[3]{32}}{\sqrt[3]{4}} = \sqrt[3]{\dfrac{32}{4}} = \sqrt[3]{8} = 2$

PRACTICE 7 Divide $\dfrac{\sqrt[3]{625}}{\sqrt[3]{5}}$. Then simplify if possible.

OBJECTIVE 3 ▶ Rationalizing denominators. It is sometimes easier to work with radical expressions if the denominator does not contain a radical. To rewrite the expression so that the denominator does not contain a radical expression, we use the fact that we can multiply the numerator and the denominator of a fraction by the same nonzero number without changing the value of the expression. This is the same as multiplying the fraction by 1. For example, to get rid of the radical in the denominator of $\dfrac{\sqrt{5}}{\sqrt{2}}$, we multiply by 1 in the form of $\dfrac{\sqrt{2}}{\sqrt{2}}$. Then

$$\dfrac{\sqrt{5}}{\sqrt{2}} = \dfrac{\sqrt{5}}{\sqrt{2}} \cdot 1 = \dfrac{\sqrt{5}}{\sqrt{2}} \cdot \dfrac{\sqrt{2}}{\sqrt{2}} = \dfrac{\sqrt{5} \cdot \sqrt{2}}{\sqrt{2} \cdot \sqrt{2}} = \dfrac{\sqrt{10}}{2}$$

This process is called **rationalizing** the denominator.

EXAMPLE 8 Rationalize each denominator.

a. $\dfrac{2}{\sqrt{7}}$ **b.** $\dfrac{\sqrt{5}}{\sqrt{12}}$ **c.** $\sqrt{\dfrac{1}{18x}}$

Solution

a. To rewrite $\dfrac{2}{\sqrt{7}}$ so that there is no radical in the denominator, we multiply by 1 in the form of $\dfrac{\sqrt{7}}{\sqrt{7}}$.

$$\dfrac{2}{\sqrt{7}} = \dfrac{2}{\sqrt{7}} \cdot \dfrac{\sqrt{7}}{\sqrt{7}} = \dfrac{2 \cdot \sqrt{7}}{\sqrt{7} \cdot \sqrt{7}} = \dfrac{2\sqrt{7}}{7}$$

b. We can multiply by $\dfrac{\sqrt{12}}{\sqrt{12}}$, but see what happens if we simplify first.

$$\dfrac{\sqrt{5}}{\sqrt{12}} = \dfrac{\sqrt{5}}{\sqrt{4 \cdot 3}} = \dfrac{\sqrt{5}}{2\sqrt{3}}$$

To rationalize the denominator now, we multiply by $\dfrac{\sqrt{3}}{\sqrt{3}}$.

$$\dfrac{\sqrt{5}}{2\sqrt{3}} = \dfrac{\sqrt{5}}{2\sqrt{3}} \cdot \dfrac{\sqrt{3}}{\sqrt{3}} = \dfrac{\sqrt{5} \cdot \sqrt{3}}{2\sqrt{3} \cdot \sqrt{3}} = \dfrac{\sqrt{15}}{2 \cdot 3} = \dfrac{\sqrt{15}}{6}$$

c. First we simplify.

$$\sqrt{\dfrac{1}{18x}} = \dfrac{\sqrt{1}}{\sqrt{18x}} = \dfrac{1}{\sqrt{9} \cdot \sqrt{2x}} = \dfrac{1}{3\sqrt{2x}}$$

Now to rationalize the denominator, we multiply by $\dfrac{\sqrt{2x}}{\sqrt{2x}}$.

$$\dfrac{1}{3\sqrt{2x}} = \dfrac{1}{3\sqrt{2x}} \cdot \dfrac{\sqrt{2x}}{\sqrt{2x}} = \dfrac{1 \cdot \sqrt{2x}}{3\sqrt{2x} \cdot \sqrt{2x}} = \dfrac{\sqrt{2x}}{3 \cdot 2x} = \dfrac{\sqrt{2x}}{6x}$$

PRACTICE 8 Rationalize each denominator.

a. $\dfrac{4}{\sqrt{5}}$ b. $\dfrac{\sqrt{3}}{\sqrt{18}}$ c. $\sqrt{\dfrac{3}{14x}}$

As a general rule, simplify a radical expression first and then rationalize the denominator.

EXAMPLE 9 Rationalize each denominator.

a. $\dfrac{5}{\sqrt[3]{4}}$ b. $\dfrac{\sqrt[3]{7}}{\sqrt[3]{3}}$

Solution

a. Since the denominator contains a cube root, we multiply the numerator and the denominator by a factor that gives the **cube root of a perfect cube** in the denominator. Recall that $\sqrt[3]{8} = 2$ and that the denominator $\sqrt[3]{4}$ multiplied by $\sqrt[3]{2}$ is $\sqrt[3]{4 \cdot 2}$ or $\sqrt[3]{8}$.

$$\dfrac{5}{\sqrt[3]{4}} = \dfrac{5 \cdot \sqrt[3]{2}}{\sqrt[3]{4} \cdot \sqrt[3]{2}} = \dfrac{5\sqrt[3]{2}}{\sqrt[3]{8}} = \dfrac{5\sqrt[3]{2}}{2}$$

b. Recall that $\sqrt[3]{27} = 3$. Multiply the denominator $\sqrt[3]{3}$ by $\sqrt[3]{9}$ and the result is $\sqrt[3]{3 \cdot 9}$ or $\sqrt[3]{27}$.

$$\dfrac{\sqrt[3]{7}}{\sqrt[3]{3}} = \dfrac{\sqrt[3]{7} \cdot \sqrt[3]{9}}{\sqrt[3]{3} \cdot \sqrt[3]{9}} = \dfrac{\sqrt[3]{63}}{\sqrt[3]{27}} = \dfrac{\sqrt[3]{63}}{3}$$

PRACTICE 9 Rationalize each denominator.

a. $\dfrac{3}{\sqrt[3]{25}}$ b. $\dfrac{\sqrt[3]{6}}{\sqrt[3]{5}}$

OBJECTIVE 4 ▶ Rationalizing denominators using conjugates. To rationalize a denominator that is a sum, such as the denominator in

$$\dfrac{2}{4 + \sqrt{3}}$$

we multiply the numerator and the denominator by $4 - \sqrt{3}$. The expressions $4 + \sqrt{3}$ and $4 - \sqrt{3}$ are called **conjugates** of each other. When a radical expression such as $4 + \sqrt{3}$ is multiplied by its conjugate $4 - \sqrt{3}$, the product simplifies to an expression that contains no radicals.

$$(a + b)(a - b) = a^2 - b^2$$
$$(4 + \sqrt{3})(4 - \sqrt{3}) = 4^2 - (\sqrt{3})^2 = 16 - 3 = 13$$

Section 8.4 Multiplying and Dividing Radicals **527**

Then

$$\frac{2}{4 + \sqrt{3}} = \frac{2(4 - \sqrt{3})}{(4 + \sqrt{3})(4 - \sqrt{3})} = \frac{2(4 - \sqrt{3})}{13}$$

EXAMPLE 10   Rationalize each denominator and simplify.

a. $\dfrac{2}{1 + \sqrt{3}}$ b. $\dfrac{\sqrt{5} + 4}{\sqrt{5} - 1}$ c. $\dfrac{3}{1 + \sqrt{x}}$

Solution

a. Multiply the numerator and the denominator of this fraction by the conjugate of $1 + \sqrt{3}$, that is, by $1 - \sqrt{3}$.

$$\frac{2}{1 + \sqrt{3}} = \frac{2(1 - \sqrt{3})}{(1 + \sqrt{3})(1 - \sqrt{3})}$$

$$= \frac{2(1 - \sqrt{3})}{1^2 - (\sqrt{3})^2}$$

> **Helpful Hint**
> Don't forget that $(\sqrt{3})^2 = 3$.

$$= \frac{2(1 - \sqrt{3})}{1 - 3}$$

$$= \frac{2(1 - \sqrt{3})}{-2}$$

$$= -\frac{2(1 - \sqrt{3})}{2} \qquad \frac{a}{-b} = -\frac{a}{b}$$

$$= -1(1 - \sqrt{3}) \qquad \text{Simplify.}$$

$$= -1 + \sqrt{3} \qquad \text{Multiply.}$$

b. $\dfrac{\sqrt{5} + 4}{\sqrt{5} - 1} = \dfrac{(\sqrt{5} + 4)(\sqrt{5} + 1)}{(\sqrt{5} - 1)(\sqrt{5} + 1)}$ Multiply the numerator and denominator by $\sqrt{5} + 1$, the conjugate of $\sqrt{5} - 1$.

$$= \frac{5 + \sqrt{5} + 4\sqrt{5} + 4}{5 - 1} \qquad \text{Multiply.}$$

$$= \frac{9 + 5\sqrt{5}}{4} \qquad \text{Simplify.}$$

c. $\dfrac{3}{1 + \sqrt{x}} = \dfrac{3(1 - \sqrt{x})}{(1 + \sqrt{x})(1 - \sqrt{x})}$ Multiply the numerator and denominator by $1 - \sqrt{x}$, the conjugate of $1 + \sqrt{x}$.

$$= \frac{3(1 - \sqrt{x})}{1 - x}$$

PRACTICE
10   Rationalize each denominator and simplify.

a. $\dfrac{4}{1 + \sqrt{5}}$ b. $\dfrac{\sqrt{3} + 2}{\sqrt{3} - 1}$ c. $\dfrac{8}{5 - \sqrt{x}}$

EXAMPLE 11 Simplify $\dfrac{12 - \sqrt{18}}{9}$.

Solution First simplify $\sqrt{18}$.

$$\dfrac{12 - \sqrt{18}}{9} = \dfrac{12 - \sqrt{9 \cdot 2}}{9} = \dfrac{12 - 3\sqrt{2}}{9}$$

Next, factor out a common factor of 3 from the terms in the numerator and the denominator and simplify.

$$\dfrac{12 - 3\sqrt{2}}{9} = \dfrac{3(4 - \sqrt{2})}{3 \cdot 3} = \dfrac{4 - \sqrt{2}}{3}$$

PRACTICE 11 Simplify $\dfrac{14 - \sqrt{28}}{6}$.

VOCABULARY & READINESS CHECK

Fill in each blank.

1. $\sqrt{7} \cdot \sqrt{3} = $ _____
2. $\sqrt{10} \cdot \sqrt{10} = $ _____
3. $\dfrac{\sqrt{15}}{\sqrt{3}} = $ _____

4. The process of eliminating the radical in the denominator of a radical expression is called _____.

5. The conjugate of $2 + \sqrt{3}$ is _____.

8.4 EXERCISE SET

Multiply and simplify. Assume that all variables represent positive real numbers. See Examples 1, 2, 4, and 5.

1. $\sqrt{8} \cdot \sqrt{2}$
2. $\sqrt{3} \cdot \sqrt{12}$
3. $\sqrt{10} \cdot \sqrt{5}$
4. $\sqrt{2} \cdot \sqrt{14}$
5. $(\sqrt{6})^2$
6. $(\sqrt{10})^2$
7. $\sqrt{2x} \cdot \sqrt{2x}$
8. $\sqrt{5y} \cdot \sqrt{5y}$
9. $(2\sqrt{5})^2$
10. $(3\sqrt{10})^2$
11. $(6\sqrt{x})^2$
12. $(8\sqrt{y})^2$
13. $\sqrt{3x^5} \cdot \sqrt{6x}$
14. $\sqrt{21y^7} \cdot \sqrt{3y}$
15. $\sqrt{2xy^2} \cdot \sqrt{8xy}$
16. $\sqrt{18x^2y^2} \cdot \sqrt{2x^2y}$
17. $\sqrt{6}(\sqrt{5} + \sqrt{7})$
18. $\sqrt{10}(\sqrt{3} - \sqrt{7})$
19. $\sqrt{10}(\sqrt{2} + \sqrt{5})$
20. $\sqrt{6}(\sqrt{3} + \sqrt{2})$
21. $\sqrt{7y}(\sqrt{y} - 2\sqrt{7})$
22. $\sqrt{5b}(2\sqrt{b} + \sqrt{5})$
23. $(\sqrt{3} + 6)(\sqrt{3} - 6)$
24. $(\sqrt{5} + 2)(\sqrt{5} - 2)$
25. $(\sqrt{3} + \sqrt{5})(\sqrt{2} - \sqrt{5})$
26. $(\sqrt{7} + \sqrt{5})(\sqrt{2} - \sqrt{5})$
27. $(2\sqrt{11} + 1)(\sqrt{11} - 6)$
28. $(5\sqrt{3} + 2)(\sqrt{3} - 1)$
29. $(\sqrt{x} + 6)(\sqrt{x} - 6)$
30. $(\sqrt{y} + 5)(\sqrt{y} - 5)$
31. $(\sqrt{x} - 7)^2$
32. $(\sqrt{x} + 4)^2$
33. $(\sqrt{6y} + 1)^2$
34. $(\sqrt{3y} - 2)^2$

Divide and simplify. Assume that all variables represent positive real numbers. See Example 6.

35. $\dfrac{\sqrt{32}}{\sqrt{2}}$
36. $\dfrac{\sqrt{40}}{\sqrt{10}}$
37. $\dfrac{\sqrt{21}}{\sqrt{3}}$
38. $\dfrac{\sqrt{55}}{\sqrt{5}}$
39. $\dfrac{\sqrt{90}}{\sqrt{5}}$
40. $\dfrac{\sqrt{96}}{\sqrt{8}}$
41. $\dfrac{\sqrt{75y^5}}{\sqrt{3y}}$
42. $\dfrac{\sqrt{24x^7}}{\sqrt{6x}}$
43. $\dfrac{\sqrt{150}}{\sqrt{2}}$

Section 8.4 Multiplying and Dividing Radicals   529

44. $\dfrac{\sqrt{120}}{\sqrt{3}}$ 45. $\dfrac{\sqrt{72y^5}}{\sqrt{3y^3}}$

46. $\dfrac{\sqrt{54x^3}}{\sqrt{2x}}$ 47. $\dfrac{\sqrt{24x^3y^4}}{\sqrt{2xy}}$

48. $\dfrac{\sqrt{96x^5y^3}}{\sqrt{3x^2y}}$

Rationalize each denominator and simplify. Assume that all variables represent positive real numbers. See Example 8.

49. $\dfrac{\sqrt{3}}{\sqrt{5}}$ 50. $\dfrac{\sqrt{2}}{\sqrt{3}}$ 51. $\dfrac{7}{\sqrt{2}}$

52. $\dfrac{8}{\sqrt{11}}$ 53. $\dfrac{1}{\sqrt{6y}}$ 54. $\dfrac{1}{\sqrt{10z}}$

55. $\sqrt{\dfrac{3}{x}}$ 56. $\sqrt{\dfrac{5}{x}}$ 57. $\sqrt{\dfrac{1}{8}}$

58. $\sqrt{\dfrac{1}{27}}$ 59. $\sqrt{\dfrac{2}{15}}$ 60. $\sqrt{\dfrac{11}{14}}$

61. $\dfrac{8y}{\sqrt{5}}$ 62. $\dfrac{7x}{\sqrt{2}}$ 63. $\sqrt{\dfrac{y}{12x}}$

64. $\sqrt{\dfrac{x}{20y}}$

Multiply or divide as indicated. See Examples 3 and 7.

85. $\sqrt[3]{12} \cdot \sqrt[3]{4}$ 86. $\sqrt[3]{9} \cdot \sqrt[3]{6}$

87. $2\sqrt[3]{5} \cdot 6\sqrt[3]{2}$ 88. $8\sqrt[3]{4} \cdot 7\sqrt[3]{7}$

89. $\sqrt[3]{15} \cdot \sqrt[3]{25}$ 90. $\sqrt[3]{4} \cdot \sqrt[3]{4}$

91. $\dfrac{\sqrt[3]{54}}{\sqrt[3]{2}}$ 92. $\dfrac{\sqrt[3]{80}}{\sqrt[3]{10}}$

93. $\dfrac{\sqrt[3]{120}}{\sqrt[3]{5}}$ 94. $\dfrac{\sqrt[3]{270}}{\sqrt[3]{5}}$

Rationalize each denominator. See Example 9.

95. $\sqrt[3]{\dfrac{5}{4}}$ 96. $\sqrt[3]{\dfrac{7}{9}}$

97. $\dfrac{6}{\sqrt[3]{2}}$ 98. $\dfrac{3}{\sqrt[3]{5}}$

99. $\sqrt[3]{\dfrac{1}{9}}$ 100. $\sqrt[3]{\dfrac{8}{11}}$

101. $\sqrt[3]{\dfrac{2}{9}}$ 102. $\sqrt[3]{\dfrac{3}{4}}$

Rationalize each denominator and simplify. Assume that all variables represent positive real numbers. See Example 10.

65. $\dfrac{3}{\sqrt{2}+1}$ 66. $\dfrac{6}{\sqrt{5}+2}$

67. $\dfrac{\sqrt{5}+1}{\sqrt{6}-\sqrt{5}}$ 68. $\dfrac{\sqrt{3}+1}{\sqrt{3}-\sqrt{2}}$

69. $\dfrac{3}{\sqrt{x}-4}$ 70. $\dfrac{4}{\sqrt{x}-1}$

MIXED PRACTICE

Rationalize each denominator and simplify.

71. $\sqrt{\dfrac{3}{20}}$ 72. $\sqrt{\dfrac{3}{50}}$

73. $\dfrac{4}{2-\sqrt{5}}$ 74. $\dfrac{2}{1-\sqrt{2}}$

75. $\dfrac{3x}{\sqrt{2x}}$ 76. $\dfrac{5y}{\sqrt{3y}}$

77. $\dfrac{5}{2+\sqrt{x}}$ 78. $\dfrac{9}{3+\sqrt{x}}$

Simplify the following. See Example 11.

79. $\dfrac{6+2\sqrt{3}}{2}$ 80. $\dfrac{9+6\sqrt{2}}{3}$

81. $\dfrac{18-12\sqrt{5}}{6}$ 82. $\dfrac{8-20\sqrt{3}}{4}$

83. $\dfrac{15\sqrt{3}+5}{5}$ 84. $\dfrac{8+16\sqrt{2}}{8}$

REVIEW AND PREVIEW

Solve each equation. See Sections 2.4 and 5.3.

103. $x + 5 = 7^2$
104. $2y - 1 = 3^2$
105. $4z^2 + 6z - 12 = (2z)^2$
106. $16x^2 + x + 9 = (4x)^2$
107. $9x^2 + 5x + 4 = (3x+1)^2$
108. $x^2 + 3x + 4 = (x+2)^2$

CONCEPT EXTENSIONS

△ 109. Find the area of a rectangle whose length is $13\sqrt{2}$ meters and width is $5\sqrt{6}$ meters.

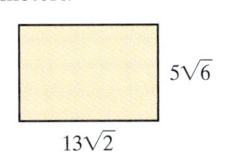

△ 110. Find the volume of a cube whose length is $\sqrt{3}$ feet, width is $\sqrt{2}$ feet, and height is $\sqrt{2}$ feet.

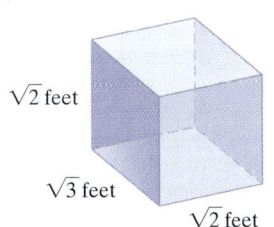

△ 111. If a circle has area A, then the formula for the radius r of the circle is

$$r = \sqrt{\dfrac{A}{\pi}}$$

Simplify this expression by rationalizing the denominator.

△ 112. If a round ball has volume V, then the formula for the radius r of the ball is

$$r = \sqrt[3]{\frac{3V}{4\pi}}$$

Simplify this expression by rationalizing the denominator.

Identify each statement as true or false. See the Concept Check in this section.

113. $\sqrt{5} \cdot \sqrt{5} = 5$
114. $\sqrt{5} \cdot \sqrt{3} = 15$
115. $\sqrt{3x} \cdot \sqrt{3x} = 2\sqrt{3x}$
116. $\sqrt{3x} + \sqrt{3x} = 2\sqrt{3x}$
117. $\sqrt{11} + \sqrt{2} = \sqrt{13}$
118. $\sqrt{11} \cdot \sqrt{2} = \sqrt{22}$

119. When rationalizing the denominator of $\dfrac{\sqrt{2}}{\sqrt{3}}$, explain why both the numerator and the denominator must be multiplied by $\sqrt{3}$.

120. In your own words, explain why $\sqrt{6} + \sqrt{2}$ cannot be simplified further, but $\sqrt{6} \cdot \sqrt{2}$ can be.

121. When rationalizing the denominator of $\dfrac{\sqrt[3]{2}}{\sqrt[3]{3}}$, explain why both the numerator and the denominator must be multiplied by $\sqrt[3]{9}$.

122. When rationalizing the denominator of $\dfrac{5}{1 + \sqrt{2}}$, explain why multiplying by $\dfrac{\sqrt{2}}{\sqrt{2}}$ will not accomplish this, but multiplying by $\dfrac{1 - \sqrt{2}}{1 - \sqrt{2}}$ will.

It is often more convenient to work with a radical expression whose numerator is rationalized. Rationalize the numerator of each expression by multiplying the numerator and denominator by the conjugate of the numerator.

123. $\dfrac{\sqrt{3} + 1}{\sqrt{2} - 1}$

124. $\dfrac{\sqrt{2} - 2}{2 - \sqrt{3}}$

INTEGRATED REVIEW — SIMPLIFYING RADICALS

Sections 8.1–8.4

Simplify. Assume that all variables represent positive numbers.

1. $\sqrt{36}$
2. $\sqrt{48}$
3. $\sqrt{x^4}$
4. $\sqrt{y^7}$
5. $\sqrt{16x^2}$
6. $\sqrt{18x^{11}}$
7. $\sqrt[3]{8}$
8. $\sqrt[4]{81}$
9. $\sqrt[3]{-27}$
10. $\sqrt{-4}$
11. $\sqrt{\dfrac{11}{9}}$
12. $\sqrt[3]{\dfrac{7}{64}}$
13. $-\sqrt{16}$
14. $-\sqrt{25}$
15. $\sqrt{\dfrac{9}{49}}$
16. $\sqrt{\dfrac{1}{64}}$
17. $\sqrt{a^8 b^2}$
18. $\sqrt{x^{10} y^{20}}$
19. $\sqrt{25m^6}$
20. $\sqrt{9n^{16}}$

Add or subtract as indicated.

21. $5\sqrt{7} + \sqrt{7}$
22. $\sqrt{50} - \sqrt{8}$
23. $5\sqrt{2} - 5\sqrt{3}$
24. $2\sqrt{x} + \sqrt{25x} - \sqrt{36x} + 3x$

Multiply and simplify if possible.

25. $\sqrt{2} \cdot \sqrt{15}$
26. $\sqrt{3} \cdot \sqrt{3}$
27. $(2\sqrt{7})^2$
28. $(3\sqrt{5})^2$
29. $\sqrt{3}(\sqrt{11} + 1)$
30. $\sqrt{6}(\sqrt{3} - 2)$
31. $\sqrt{8y} \cdot \sqrt{2y}$
32. $\sqrt{15x^2} \cdot \sqrt{3x^2}$
33. $(\sqrt{x} - 5)(\sqrt{x} + 2)$
34. $(3 + \sqrt{2})^2$

Divide and simplify if possible.

35. $\dfrac{\sqrt{8}}{\sqrt{2}}$
36. $\dfrac{\sqrt{45}}{\sqrt{15}}$
37. $\dfrac{\sqrt{24x^5}}{\sqrt{2x}}$
38. $\dfrac{\sqrt{75a^4b^5}}{\sqrt{5ab}}$

Rationalize each denominator.

39. $\sqrt{\dfrac{1}{6}}$
40. $\dfrac{x}{\sqrt{20}}$
41. $\dfrac{4}{\sqrt{6}+1}$
42. $\dfrac{\sqrt{2}+1}{\sqrt{x}-5}$

8.5 SOLVING EQUATIONS CONTAINING RADICALS

OBJECTIVES

1. Solve radical equations by using the squaring property of equality once.
2. Solve radical equations by using the squaring property of equality twice.

OBJECTIVE 1 ▸ Using the squaring property once. In this section, we solve **radical equations** such as

$$\sqrt{x+3} = 5 \quad \text{and} \quad \sqrt{2x+1} = \sqrt{3x}$$

Radical equations contain variables in the radicand. To solve these equations, we rely on the following squaring property.

> **The Squaring Property of Equality**
> If $a = b$, then $a^2 = b^2$.

Unfortunately, this squaring property does not guarantee that all solutions of the new equation are solutions of the original equation. For example, if we square both sides of the equation

$$x = 2$$

we have

$$x^2 = 4$$

This new equation has two solutions, 2 and -2, while the original equation $x = 2$ has only one solution. Thus, squaring both sides of the original equation resulted in an equation that has an **extraneous solution** that isn't a solution of the original equation. For this reason, we must **always check proposed solutions of radical equations in the original equation.** If a proposed solution does not work, we call that value an **extraneous solution.**

EXAMPLE 1 Solve: $\sqrt{x+3} = 5$

Solution To solve this radical equation, we use the squaring property of equality and square both sides of the equation.

$$\sqrt{x+3} = 5$$
$$(\sqrt{x+3})^2 = 5^2 \quad \text{Square both sides.}$$
$$x + 3 = 25 \quad \text{Simplify.}$$
$$x = 22 \quad \text{Subtract 3 from both sides.}$$

Check: We replace x with 22 in the original equation.

▸ **Helpful Hint**
Don't forget to check the proposed solutions of radical equations in the original equation.

$$\sqrt{x+3} = 5 \quad \text{Original equation}$$
$$\sqrt{22+3} \stackrel{?}{=} 5 \quad \text{Let } x = 22.$$
$$\sqrt{25} \stackrel{?}{=} 5$$
$$5 = 5 \quad \text{True}$$

Since a true statement results, 22 is the solution.

PRACTICE 1 Solve: $\sqrt{x-5} = 2$

EXAMPLE 2 Solve: $\sqrt{x} + 6 = 4$

Solution First we set the radical by itself on one side of the equation. Then we square both sides.

$$\sqrt{x} + 6 = 4$$
$$\sqrt{x} = -2 \quad \text{Subtract 6 from both sides to get the radical by itself.}$$

Recall that $\sqrt{x}$ is the principal or nonnegative square root of x so that $\sqrt{x}$ cannot equal -2 and thus this equation has no solution. We arrive at the same conclusion if we continue by applying the squaring property.

$$\sqrt{x} = -2$$
$$(\sqrt{x})^2 = (-2)^2 \quad \text{Square both sides.}$$
$$x = 4 \quad \text{Simplify.}$$

Check: We replace x with 4 in the original equation.

$$\sqrt{x} + 6 = 4 \quad \text{Original equation}$$
$$\sqrt{4} + 6 \stackrel{?}{=} 4 \quad \text{Let } x = 4.$$
$$2 + 6 = 4 \quad \text{False}$$

Since 4 *does not* satisfy the original equation, this equation has no solution.

PRACTICE 2 Solve: $\sqrt{x} + 5 = 3$

Example 2 makes it very clear that we *must* check proposed solutions in the original equation to determine if they are truly solutions. Remember, if a proposed solution is not an actual solution, we say that the value is an **extraneous solution.**

The following steps can be used to solve radical equations containing square roots.

Solving a Radical Equation Containing Square Roots

STEP 1. Arrange terms so that one radical is by itself on one side of the equation. That is, isolate a radical.

STEP 2. Square both sides of the equation.

STEP 3. Simplify both sides of the equation.

STEP 4. If the equation still contains a radical term, repeat steps 1 through 3.

STEP 5. Solve the equation.

STEP 6. Check all solutions in the original equation for extraneous solutions.

EXAMPLE 3 Solve: $\sqrt{x} = \sqrt{5x - 2}$.

Solution Each of the radicals is already isolated, since each is by itself on one side of the equation. So we begin solving by squaring both sides.

$$\sqrt{x} = \sqrt{5x - 2} \quad \text{Original equation}$$
$$(\sqrt{x})^2 = (\sqrt{5x - 2})^2 \quad \text{Square both sides.}$$
$$x = 5x - 2 \quad \text{Simplify.}$$
$$-4x = -2 \quad \text{Subtract } 5x \text{ from both sides.}$$
$$x = \frac{-2}{-4} = \frac{1}{2} \quad \text{Divide both sides by } -4 \text{ and simplify.}$$

Check: We replace x with $\dfrac{1}{2}$ in the original equation.

$$\sqrt{x} = \sqrt{5x-2} \qquad \text{Original equation}$$
$$\sqrt{\dfrac{1}{2}} \stackrel{?}{=} \sqrt{5 \cdot \dfrac{1}{2} - 2} \qquad \text{Let } x = \dfrac{1}{2}.$$
$$\sqrt{\dfrac{1}{2}} \stackrel{?}{=} \sqrt{\dfrac{5}{2} - 2} \qquad \text{Multiply.}$$
$$\sqrt{\dfrac{1}{2}} \stackrel{?}{=} \sqrt{\dfrac{5}{2} - \dfrac{4}{2}} \qquad \text{Write 2 as } \dfrac{4}{2}.$$
$$\sqrt{\dfrac{1}{2}} = \sqrt{\dfrac{1}{2}} \qquad \text{True}$$

This statement is true, so the solution is $\dfrac{1}{2}$.

PRACTICE 3 Solve: $\sqrt{7x-4} = \sqrt{x}$

EXAMPLE 4 Solve: $\sqrt{4y^2 + 5y - 15} = 2y$

Solution The radical is already isolated, so we start by squaring both sides.

$$\sqrt{4y^2 + 5y - 15} = 2y$$
$$\left(\sqrt{4y^2 + 5y - 15}\right)^2 = (2y)^2 \qquad \text{Square both sides.}$$
$$4y^2 + 5y - 15 = 4y^2 \qquad \text{Simplify.}$$
$$5y - 15 = 0 \qquad \text{Subtract } 4y^2 \text{ from both sides.}$$
$$5y = 15 \qquad \text{Add 15 to both sides.}$$
$$y = 3 \qquad \text{Divide both sides by 5.}$$

Check: We replace y with 3 in the original equation.

$$\sqrt{4y^2 + 5y - 15} = 2y \qquad \text{Original equation}$$
$$\sqrt{4 \cdot 3^2 + 5 \cdot 3 - 15} \stackrel{?}{=} 2 \cdot 3 \qquad \text{Let } y = 3.$$
$$\sqrt{4 \cdot 9 + 15 - 15} \stackrel{?}{=} 6 \qquad \text{Simplify.}$$
$$\sqrt{36} \stackrel{?}{=} 6$$
$$6 = 6 \qquad \text{True}$$

This statement is true, so the solution is 3.

PRACTICE 4 Solve: $\sqrt{16y^2 + 4y - 28} = 4y$

EXAMPLE 5 Solve: $\sqrt{x+3} - x = -3$

Solution First we isolate the radical by adding x to both sides. Then we square both sides.

$$\sqrt{x+3} - x = -3$$
$$\sqrt{x+3} = x - 3 \qquad \text{Add } x \text{ to both sides.}$$
$$\left(\sqrt{x+3}\right)^2 = (x-3)^2 \qquad \text{Square both sides.}$$
$$x + 3 = \underbrace{x^2 - 6x + 9}$$

> ▶ **Helpful Hint**
> Don't forget that $(x-3)^2 = (x-3)(x-3) = x^2 - 6x + 9$

To solve the resulting quadratic equation, we write the equation in standard form by subtracting x and 3 from both sides.

$$3 = x^2 - 7x + 9 \quad \text{Subtract } x \text{ from both sides.}$$
$$0 = x^2 - 7x + 6 \quad \text{Subtract 3 from both sides.}$$
$$0 = (x - 6)(x - 1) \quad \text{Factor.}$$
$$0 = x - 6 \quad \text{or} \quad 0 = x - 1 \quad \text{Set each factor equal to zero.}$$
$$6 = x \qquad\qquad 1 = x \qquad \text{Solve for } x.$$

Check: We replace x with 6 and then x with 1 in the original equation.

Let $x = 6$.
$$\sqrt{x + 3} - x = -3$$
$$\sqrt{6 + 3} - 6 \stackrel{?}{=} -3$$
$$\sqrt{9} - 6 \stackrel{?}{=} -3$$
$$3 - 6 \stackrel{?}{=} -3$$
$$-3 = -3 \quad \text{True}$$

Let $x = 1$.
$$\sqrt{x + 3} - x = -3$$
$$\sqrt{1 + 3} - 1 \stackrel{?}{=} -3$$
$$\sqrt{4} - 1 \stackrel{?}{=} -3$$
$$2 - 1 \stackrel{?}{=} -3$$
$$1 = -3 \quad \text{False}$$

Since replacing x with 1 resulted in a false statement, 1 is an extraneous solution. The only solution is 6.

PRACTICE 5 Solve: $\sqrt{x + 15} - x = -5$

OBJECTIVE 2 ▶ Using the squaring property twice. If a radical equation contains two radicals, we may need to use the squaring property twice.

EXAMPLE 6 Solve: $\sqrt{x - 4} = \sqrt{x} - 2$

Solution
$$\sqrt{x - 4} = \sqrt{x} - 2$$
$$(\sqrt{x - 4})^2 = (\sqrt{x} - 2)^2 \quad \text{Square both sides.}$$
$$x - 4 = x - 4\sqrt{x} + 4$$
$$-8 = -4\sqrt{x}$$
$$2 = \sqrt{x} \quad \text{Divide both sides by } -4.$$
$$4 = x \quad \text{Square both sides again.}$$

> **Helpful Hint**
> $(\sqrt{x} - 2)^2 = (\sqrt{x} - 2)(\sqrt{x} - 2)$
> $= \sqrt{x} \cdot \sqrt{x} - 2\sqrt{x} - 2\sqrt{x} + 4$
> $= x - 4\sqrt{x} + 4$

Check the proposed solution in the original equation. The solution is 4.

PRACTICE 6 Solve: $\sqrt{x - 4} = \sqrt{x} - 16$

8.5 EXERCISE SET

Solve each equation. See Examples 1 through 3.

1. $\sqrt{x} = 9$
2. $\sqrt{x} = 4$
3. $\sqrt{x + 5} = 2$
4. $\sqrt{x + 12} = 3$
5. $\sqrt{x - 2} = 5$
6. $4\sqrt{x - 7} = 5$
7. $3\sqrt{x + 5} = 2$
8. $3\sqrt{x + 8} = 5$
9. $\sqrt{x} = \sqrt{3x - 8}$
10. $\sqrt{x} = \sqrt{4x - 3}$
11. $\sqrt{4x - 3} = \sqrt{x + 3}$
12. $\sqrt{5x - 4} = \sqrt{x + 8}$

Solve each equation. See Examples 4 and 5.

13. $\sqrt{9x^2 + 2x - 4} = 3x$
14. $\sqrt{4x^2 + 3x - 9} = 2x$
15. $\sqrt{x} = x - 6$
16. $\sqrt{x} = x - 2$

17. $\sqrt{x+7} = x+5$
18. $\sqrt{x+5} = x-1$
19. $\sqrt{3x+7} - x = 3$
20. $x = \sqrt{4x-7} + 1$
21. $\sqrt{16x^2 + 2x + 2} = 4x$
22. $\sqrt{4x^2 + 3x + 2} = 2x$
23. $\sqrt{2x^2 + 6x + 9} = 3$
24. $\sqrt{3x^2 + 6x + 4} = 2$

Solve each equation. See Example 6.

25. $\sqrt{x-7} = \sqrt{x} - 1$
26. $\sqrt{x-8} = \sqrt{x} - 2$
27. $\sqrt{x} + 2 = \sqrt{x+24}$
28. $\sqrt{x} + 5 = \sqrt{x+55}$
29. $\sqrt{x+8} = \sqrt{x} + 2$
30. $\sqrt{x+1} = \sqrt{x+15}$

MIXED PRACTICE

Solve each equation. See Examples 1 through 6.

31. $\sqrt{2x+6} = 4$
32. $\sqrt{3x+7} = 5$
33. $\sqrt{x+6} + 1 = 3$
34. $\sqrt{x+5} + 2 = 5$
35. $\sqrt{x+6} + 5 = 3$
36. $\sqrt{2x-1} + 7 = 1$
37. $\sqrt{16x^2 - 3x + 6} = 4x$
38. $\sqrt{9x^2 - 2x + 8} = 3x$
39. $-\sqrt{x} = -6$
40. $-\sqrt{y} = -8$
41. $\sqrt{x+9} = \sqrt{x} - 3$
42. $\sqrt{x-6} = \sqrt{x+36}$
43. $\sqrt{2x+1} + 3 = 5$
44. $\sqrt{3x-1} + 1 = 4$
45. $\sqrt{x} + 3 = 7$
46. $\sqrt{x} + 5 = 10$
47. $\sqrt{4x} = \sqrt{2x+6}$
48. $\sqrt{5x+6} = \sqrt{8x}$
49. $\sqrt{2x+1} = x-7$
50. $\sqrt{2x+5} = x-5$
51. $x = \sqrt{2x-2} + 1$
52. $\sqrt{2x-4} + 2 = x$
53. $\sqrt{1-8x} - x = 4$
54. $\sqrt{2x+5} - 1 = x$

REVIEW AND PREVIEW

Translate each sentence into an equation and then solve. See Section 2.5.

55. If 8 is subtracted from the product of 3 and x, the result is 19. Find x.
56. If 3 more than x is subtracted from twice x, the result is 11. Find x.
57. The length of a rectangle is twice the width. The perimeter is 24 inches. Find the length.
58. The length of a rectangle is 2 inches longer than the width. The perimeter is 24 inches. Find the length.

CONCEPT EXTENSIONS

Solve each equation.

59. $\sqrt{x-3} + 3 = \sqrt{3x+4}$
60. $\sqrt{2x+3} = \sqrt{x-2} + 2$

61. Explain why proposed solutions of radical equations must be checked in the original equation.

62. Is 8 a solution of the equation $\sqrt{x-4} - 5 = \sqrt{x+1}$? Explain why or why not.

△ 63. The formula $b = \sqrt{\dfrac{V}{2}}$ can be used to determine the length b of a side of the base of a square-based pyramid with height 6 units and volume V cubic units.

a. Find the length of the side of the base that produces a pyramid with each volume. (Round to the nearest tenth of a unit.)

V	20	200	2000
b			

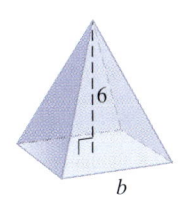

b. Notice in the table that volume V has been increased by a factor of 10 each time. Does the corresponding length b of a side increase by a factor of 10 each time also?

△ 64. The formula $r = \sqrt{\dfrac{V}{2\pi}}$ can be used to determine the radius r of a cylinder with height 2 units and volume V cubic units.

a. Find the radius needed to manufacture a cylinder with each volume. (Round to the nearest tenth of a unit.)

V	10	100	1000
r			

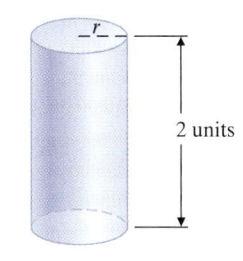

b. Notice in the table that volume V has been increased by a factor of 10 each time. Does the corresponding radius increase by a factor of 10 each time also?

Graphing calculators can be used to solve equations. To solve $\sqrt{x-2} = x-5$, *for example, graph* $y_1 = \sqrt{x-2}$ *and* $y_2 = x-5$ *on the same set of axes. Use the Trace and Zoom features or an Intersect feature to find the point of intersection of the graphs. The x-value of the point is the solution of the equation. Use a graphing calculator to solve the equations below. Approximate solutions to the nearest hundredth.*

65. $\sqrt{x-2} = x-5$
66. $\sqrt{x+1} = 2x-3$
67. $-\sqrt{x+4} = 5x-6$
68. $-\sqrt{x+5} = -7x+1$

THE BIGGER PICTURE — SIMPLIFYING EXPRESSIONS AND SOLVING EQUATIONS

Now we continue our outline from Sections 1.7, 2.9, 5.6, 6.6, 7.4, and 7.6. Although suggestions are given, this outline should be in your own words. Once you complete this new portion, try the exercises below.

I. **Simplifying Expressions**
   A. **Real Numbers**
      1. Add (Section 1.5)
      2. Subtract (Section 1.6)
      3. Multiply or Divide (Section 1.7)
   B. **Exponents** (Section 5.1)
   C. **Polynomials**
      1. Add (Section 5.2)
      2. Subtract (Section 5.2)
      3. Multiply (Section 5.3)
      4. Divide (Section 5.6)
   D. **Factoring Polynomials** (Chapter 6 Integrated Review)
   E. **Rational Expressions**
      1. Simplify (Section 7.1)
      2. Multiply (Section 7.2)
      3. Divide (Section 7.2)
      4. Add or Subtract (Section 7.4)
   F. **Radicals**
      1. Simplify square roots: If possible, factor the radicand so that one factor is a perfect square. Then use the product rule, and simplify.
      $$\sqrt{75} = \sqrt{25 \cdot 3} = \sqrt{25} \cdot \sqrt{3} = 5\sqrt{3}$$
      2. Add or subtract: Only like radicals (same index and radicand) can be added or subtracted.
      $$8\sqrt{10} - \sqrt{40} + \sqrt{5}$$
      $$= 8\sqrt{10} - 2\sqrt{10} + \sqrt{5}$$
      $$= 6\sqrt{10} + \sqrt{5}$$
      3. Multiply or divide:
      $$\sqrt{a} \cdot \sqrt{b} = \sqrt{ab}; \quad \frac{\sqrt{a}}{\sqrt{b}} = \sqrt{\frac{a}{b}}.$$
      $$\sqrt{11} \cdot \sqrt{3} = \sqrt{33};$$
      $$\frac{\sqrt{140}}{\sqrt{7}} = \sqrt{\frac{140}{7}} = \sqrt{20} = \sqrt{4 \cdot 5} = 2\sqrt{5}$$
      4. Rationalizing the denominator:
         a. If the denominator is one term,
         $$\frac{5}{\sqrt{11}} = \frac{5 \cdot \sqrt{11}}{\sqrt{11} \cdot \sqrt{11}} = \frac{5\sqrt{11}}{11}$$
         b. If the denominator has two terms, multiply by 1 in the form of $\dfrac{\text{conjugate of denominator}}{\text{conjugate of denominator}}$.

$$\frac{13}{3 + \sqrt{2}} = \frac{13}{3 + \sqrt{2}} \cdot \frac{3 - \sqrt{2}}{3 - \sqrt{2}}$$
$$= \frac{13(3 - \sqrt{2})}{9 - 2} = \frac{13(3 - \sqrt{2})}{7}$$

II. **Solving Equations and Inequalities**
   A. **Linear Equations** (Section 2.4)
   B. **Linear Inequalities** (Section 2.9)
   C. **Quadratic and Higher Degree Equations** (Section 6.6)
   D. **Equations with Rational Expressions** (Section 7.5)
   E. **Proportions** (Section 7.6)
   F. **Equations with Radicals** To solve, isolate a radical, then square both sides. You may have to repeat this. Check possible solution in the original equation.

$\sqrt{x + 49} + 7 = x$	
$\sqrt{x + 49} = x - 7$	Subtract 7 from both sides.
$x + 49 = x^2 - 14x + 49$	Square both sides.
$0 = x^2 - 15x$	Set terms equal to 0.
$0 = x(x - 15)$	Factor.
$x\not= 0$ or $x = 15$	Set each factor equal to 0 and solve.

Perform indicated operations and simplify. If necessary, rationalize the denominator.

1. $\sqrt{56}$
2. $\sqrt{\dfrac{20x^5}{49}}$
3. $(-5x^{12}y^{-3})(3x^{-7}y^{14})$
4. $\sqrt{\dfrac{10}{11}}$
5. $\dfrac{8}{\sqrt{5} - 1}$
6. $\dfrac{1}{2}(6x^2 - 4) + \dfrac{1}{3}(6x^2 - 9) - 14$

Solve each equation or inequality.

7. $9x - 7 = 7x - 9$
8. $\dfrac{x}{5} = \dfrac{x - 3}{11}$
9. $-5(2y + 1) \le 3y - 2 - 2y + 1$
10. $x(x + 1) = 42$
11. $\dfrac{-6}{x - 7} + \dfrac{8}{x} = \dfrac{-4}{x - 7}$
12. $1 + 4(x - 2) = x(x - 6) - x^2 + 13$

8.6 RADICAL EQUATIONS AND PROBLEM SOLVING

OBJECTIVES

1. Use the Pythagorean formula to solve problems.
2. Use the distance formula.
3. Solve problems using formulas containing radicals.

OBJECTIVE 1 ▶ Using the Pythagorean formula. Applications of radicals can be found in geometry, finance, science, and other areas of technology. Our first application involves the Pythagorean theorem, giving a formula that relates the lengths of the three sides of a right triangle. We first studied the Pythagorean theorem in Chapter 6 and we review it here.

> **The Pythagorean Theorem**
> If a and b are lengths of the legs of a right triangle and c is the length of the hypotenuse, then $a^2 + b^2 = c^2$.

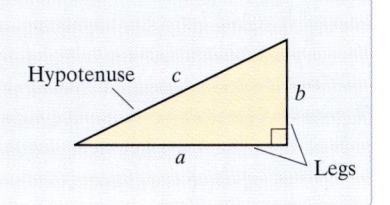

That is, the square of the length of the hypotenuse is equal to the sum of the squares of the lengths of the legs.

EXAMPLE 1 Find the length of the hypotenuse of a right triangle whose legs are 6 inches and 8 inches long.

Solution Because this is a right triangle, we use the Pythagorean theorem. We let $a = 6$ inches and $b = 8$ inches. Length c must be the length of the hypotenuse.

$$a^2 + b^2 = c^2 \quad \text{Use the Pythagorean theorem.}$$
$$6^2 + 8^2 = c^2 \quad \text{Substitute the lengths of the legs.}$$
$$36 + 64 = c^2 \quad \text{Simplify.}$$
$$100 = c^2$$

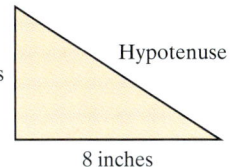

Since c represents a length, we know that c is positive and is the principal square root of 100.

$$100 = c^2$$
$$\sqrt{100} = c \quad \text{Use the definition of principal square root.}$$
$$10 = c \quad \text{Simplify.}$$

The hypotenuse has a length of 10 inches.

PRACTICE
1 Find the length of the hypotenuse of a right triangle whose legs are 5 inches and 12 inches long.

EXAMPLE 2 Find the length of the leg of the right triangle shown. Give the exact length and a two-decimal-place approximation.

Solution We let $a = 2$ meters and b be the unknown length of the other leg. The hypotenuse is $c = 5$ meters.

$$a^2 + b^2 = c^2 \quad \text{Use the Pythagorean theorem.}$$
$$2^2 + b^2 = 5^2 \quad \text{Let } a = 2 \text{ and } c = 5.$$
$$4 + b^2 = 25$$
$$b^2 = 21$$
$$b = \sqrt{21} \approx 4.58 \text{ meters}$$

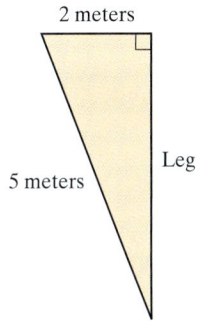

The length of the leg is exactly $\sqrt{21}$ meters or approximately 4.58 meters.

538 CHAPTER 8 Roots and Radicals

PRACTICE 2 Find the length of the leg of the right triangle shown. Give the exact length and a two-decimal-place approximation.

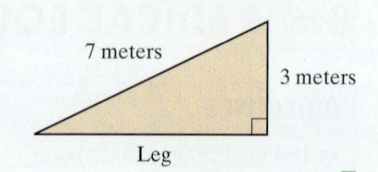

7 meters
3 meters
Leg

EXAMPLE 3 Finding a Distance

A surveyor must determine the distance across a lake at points P and Q as shown in the figure. To do this, she finds a third point R perpendicular to line PQ. If the length of $\overline{PR}$ is 320 feet and the length of $\overline{QR}$ is 240 feet, what is the distance across the lake? Approximate this distance to the nearest whole foot.

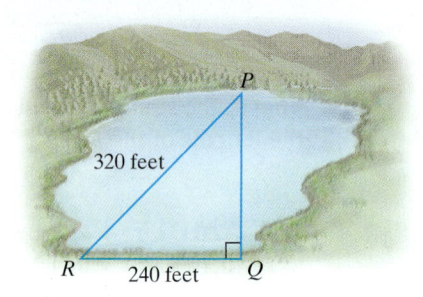

Solution

1. UNDERSTAND. Read and reread the problem. We will set up the problem using the Pythagorean theorem. By creating a line perpendicular to line PQ, the surveyor deliberately constructed a right triangle. The hypotenuse, $\overline{PR}$, has a length of 320 feet, so we let $c = 320$ in the Pythagorean theorem. The side $\overline{QR}$ is one of the legs, so we let $a = 240$ and $b =$ the unknown length.

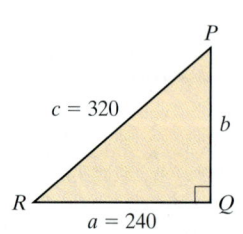

2. TRANSLATE.

$$a^2 + b^2 = c^2 \quad \text{Use the Pythagorean theorem.}$$
$$240^2 + b^2 = 320^2 \quad \text{Let } a = 240 \text{ and } c = 320.$$

3. SOLVE.

$$57{,}600 + b^2 = 102{,}400$$
$$b^2 = 44{,}800 \quad \text{Subtract 57,600 from both sides.}$$
$$b = \sqrt{44{,}800} \quad \text{Use the definition of principal square root.}$$

4. INTERPRET.

Check: See that $240^2 + \left(\sqrt{44{,}800}\right)^2 = 320^2$.

State: The distance across the lake is **exactly** $\sqrt{44{,}800}$ feet. The surveyor can now use a calculator to find that $\sqrt{44{,}800}$ feet is **approximately** 211.6601 feet, so the distance across the lake is roughly 212 feet.

PRACTICE 3 Find the length of a bridge, to the nearest whole foot, to be constructed from point A to point B across Little Marsh. Use the distances shown in the diagram.

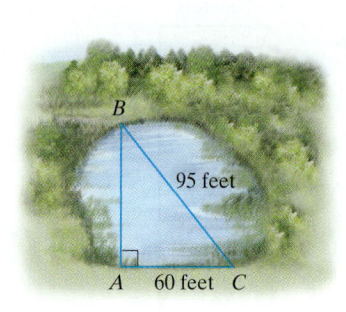

B
95 feet
A 60 feet C

Section 8.6 Radical Equations and Problem Solving

OBJECTIVE 2 ▶ Using the distance formula. A second important application of radicals is in finding the distance between two points in the plane. By using the Pythagorean theorem, the following formula can be derived.

> **Distance Formula**
>
> The distance d between two points with coordinates (x_1, y_1) and (x_2, y_2) is given by
>
> $$d = \sqrt{(x_2 - x_1)^2 + (y_2 - y_1)^2}$$

EXAMPLE 4 Find the distance between $(-1, 9)$ and $(-3, -5)$.

Solution Use the distance formula with $(x_1, y_1) = (-1, 9)$ and $(x_2, y_2) = (-3, -5)$.

$$\begin{aligned} d &= \sqrt{(x_2 - x_1)^2 + (y_2 - y_1)^2} & \text{The distance formula.} \\ &= \sqrt{[-3 - (-1)]^2 + (-5 - 9)^2} & \text{Substitute known values.} \\ &= \sqrt{(-2)^2 + (-14)^2} & \text{Simplify.} \\ &= \sqrt{4 + 196} \\ &= \sqrt{200} = 10\sqrt{2} & \text{Simplify the radical.} \end{aligned}$$

The distance is **exactly** $10\sqrt{2}$ units or **approximately** 14.1 units.

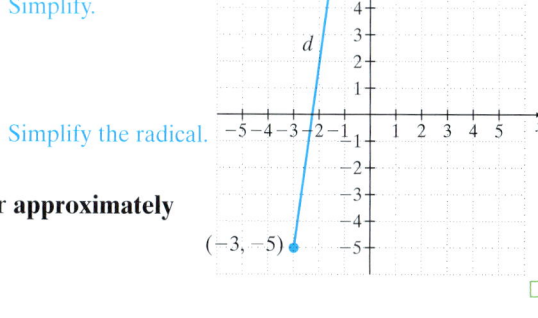

PRACTICE 4 Find the distance between $(-2, 5)$ and $(-4, -7)$.

OBJECTIVE 3 ▶ Using formulas containing radicals. The Pythagorean theorem is an extremely important result in mathematics and should be memorized. But there are other applications involving formulas containing radicals that are not quite as well known, such as the velocity formula used in the next example.

EXAMPLE 5 Determining Velocity

A formula used to determine the velocity v, in feet per second, of an object (neglecting air resistance) after it has fallen a certain height is $v = \sqrt{2gh}$, where g is the acceleration due to gravity, and h is the height the object has fallen. On Earth, the acceleration g due to gravity is approximately 32 feet per second per second. Find the velocity of a person after falling 5 feet.

Solution We are told that $g = 32$ feet per second per second. To find the velocity v when $h = 5$ feet, we use the velocity formula.

$$\begin{aligned} v &= \sqrt{2gh} & \text{Use the velocity formula.} \\ &= \sqrt{2 \cdot 32 \cdot 5} & \text{Substitute known values.} \\ &= \sqrt{320} \\ &= 8\sqrt{5} & \text{Simplify the radicand.} \end{aligned}$$

The velocity of the person after falling 5 feet is **exactly** $8\sqrt{5}$ feet per second, or **approximately** 17.9 feet per second.

PRACTICE 5 Use the formula in Example 5 to find the velocity of an object after it has fallen 12 feet.

8.6 EXERCISE SET

Use the Pythagorean theorem to find the unknown side of each right triangle. See Examples 1 and 2.

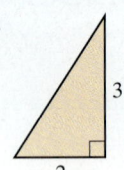

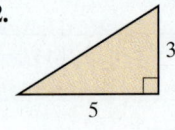

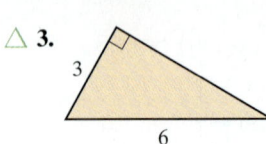

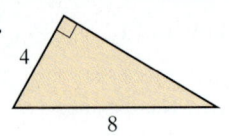

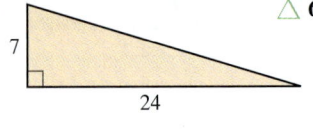

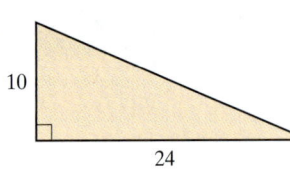

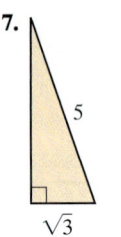

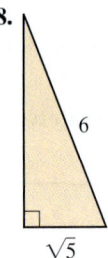

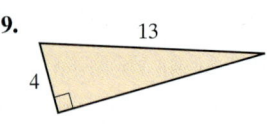

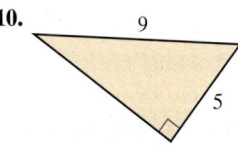

1. (legs 3 and 2)

2. (leg 3, hypotenuse 5)

3. (leg 3, hypotenuse 6)

4. (leg 4, hypotenuse 8)

5. (leg 7, leg 24)

6. (leg 10, leg 24)

7. (leg 5, leg $\sqrt{3}$)

8. (leg 6, leg $\sqrt{5}$)

9. (leg 4, hypotenuse 13)

10. (leg 9, hypotenuse 5) — with legs 9 and 5

Find the length of the unknown side of each right triangle with sides a, b, and c, where c is the hypotenuse. See Examples 1 and 2.

11. $a = 4, b = 5$

12. $a = 2, b = 7$

13. $b = 2, c = 6$

14. $b = 1, c = 5$

15. $a = \sqrt{10}, c = 10$

16. $a = \sqrt{7}, c = \sqrt{35}$

Solve. See Examples 1 through 5.

17. Evan and Noah Saacks want to determine the distance at certain points across a pond on their property. They are able to measure the distances shown on the following diagram. Find how wide the pond is to the nearest tenth of a foot.

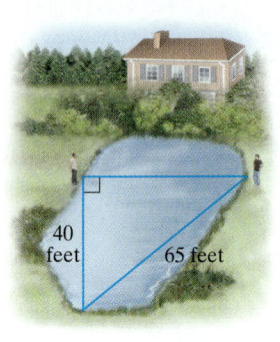

(40 feet, 65 feet)

18. Use the formula from Example 5 and find the velocity of an object after it has fallen 20 feet.

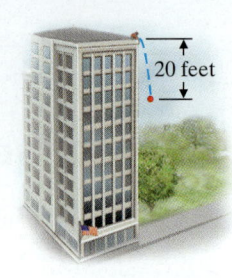

19. A wire is used to anchor a 20-foot-high pole. One end of the wire is attached to the top of the pole. The other end is fastened to a stake five feet away from the bottom of the pole. Find the length of the wire, to the nearest tenth of a foot.

20. Jim Spivey needs to connect two underground pipelines, which are offset by 3 feet, as pictured in the diagram. Neglecting the joints needed to join the pipes, find the length of the shortest possible connecting pipe rounded to the nearest hundredth of a foot.

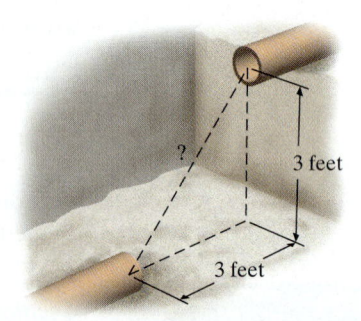

21. Robert Weisman needs to attach a diagonal brace to a rectangular frame in order to make it structurally sound. If the framework is 6 feet by 10 feet, find how long the brace needs to be to the nearest tenth of a foot.

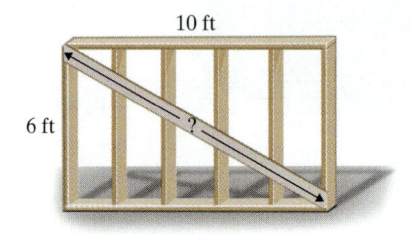

△ **22.** Elizabeth Kaster is flying a kite. She let out 80 feet of string and attached the string to a stake in the ground. The kite is now directly above her brother Mike, who is 32 feet away from Elizabeth. Find the height of the kite to the nearest foot.

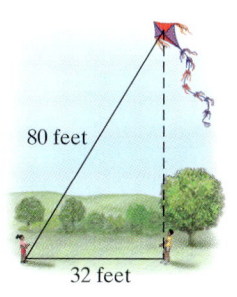

Use the distance formula to find the distance between the points given. See Example 4.

23. $(3, 6), (5, 11)$

24. $(2, 3), (9, 7)$

25. $(-3, 1), (5, -2)$

26. $(-2, 6), (3, -2)$

27. $(3, -2), (1, -8)$

28. $(-5, 8), (-2, 2)$

29. $\left(\frac{1}{2}, 2\right), (2, -1)$

30. $\left(\frac{1}{3}, 1\right), (1, -1)$

31. $(3, -2), (5, 7)$

32. $(-2, -3), (-1, 4)$

Solve each problem. See Example 5.

△ **33.** For a square-based pyramid, the formula $b = \sqrt{\dfrac{3V}{h}}$ describes the relationship between the length b of one side of the base, the volume V, and the height h. Find the volume if each side of the base is 6 feet long, and the pyramid is 2 feet high.

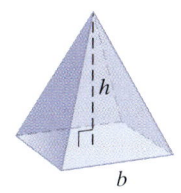

34. The formula $t = \dfrac{\sqrt{d}}{4}$ relates the distance d, in feet, that an object falls in t seconds, assuming that air resistance does not slow down the object. Find how long, to the nearest hundredth of a second, it takes an object to reach the ground from the top of the Sears Tower in Chicago, a distance of 1450 feet. (*Source: World Almanac and Book of Facts*)

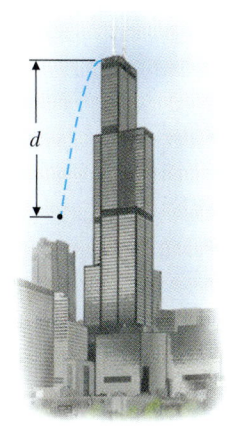

35. Police use the formula $s = \sqrt{30fd}$ to estimate the speed s of a car in miles per hour. In this formula, d represents the distance the car skidded in feet and f represents the coefficient of friction. The value of f depends on the type of road surface, and for wet concrete f is 0.35. Find how fast a car was moving if it skidded 280 feet on wet concrete, to the nearest mile per hour.

36. The coefficient of friction of a certain dry road is 0.95. Use the formula in Exercise 35 to find how far a car will skid on this dry road if it is traveling at a rate of 60 mph. Round the length to the nearest foot.

37. The formula $v = \sqrt{2.5r}$ can be used to estimate the maximum safe velocity, v, in miles per hour, at which a car can travel if it is driven along a curved road with a **radius of curvature**, r, in feet. To the nearest whole number, find the maximum safe speed if a cloverleaf exit on an expressway has a radius of curvature of 300 feet.

38. Use the formula from Exercise 37 to find the radius of curvature if the safe velocity is 30 mph.

39. The maximum distance d in kilometers that you can see from a height of h meters is given by $d = 3.5\sqrt{h}$. Find how far you can see from the top of the Texas Commerce Tower in Houston, a height of 305.4 meters. Round to the nearest tenth of a kilometer. (*Source: World Almanac and Book of Facts*)

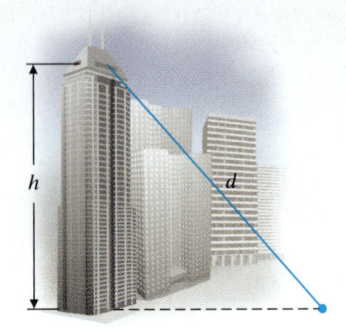

40. Use the formula from Exercise 39 to determine how high above the ground you need to be to see 40 kilometers. Round to the nearest tenth of a meter.

REVIEW AND PREVIEW

Simplify using rules for exponents. See Section 5.1.

41. 2^5

42. $(-3)^3$

43. $\left(-\dfrac{1}{5}\right)^2$

44. $\left(\dfrac{2}{7}\right)^3$

45. $x^2 \cdot x^3$

46. $x^4 \cdot x^2$

47. $y^3 \cdot y$

48. $x \cdot x^7$

CONCEPT EXTENSIONS

For each triangle, find the length of y, then x.

49.

50.

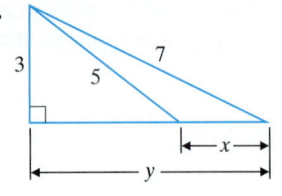

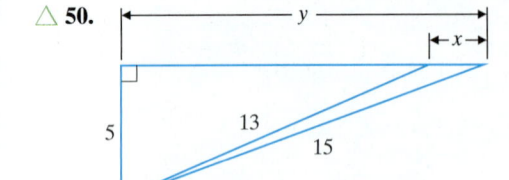

Solve.

51. Mike and Sandra Hallahan leave the seashore at the same time. Mike drives northward at a rate of 30 miles per hour, while Sandra drives west at 60 mph. Find how far apart they are after 3 hours to the nearest mile.

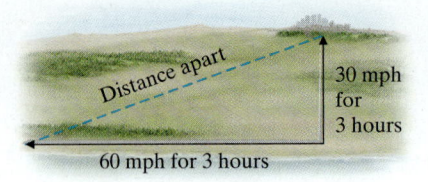

52. Railroad tracks are invariably made up of relatively short sections of rail connected by expansion joints. To see why this construction is necessary, consider a single rail 100 feet long (or 1200 inches). On an extremely hot day, suppose it expands 1 inch in the hot sun to a new length of 1201 inches. Theoretically, the track would bow upward as pictured.

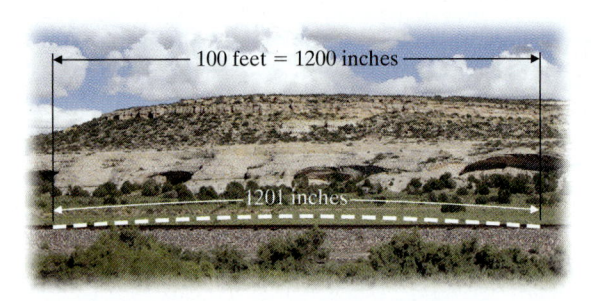

Let us approximate the bulge in the railroad this way. Calculate the height h of the bulge.

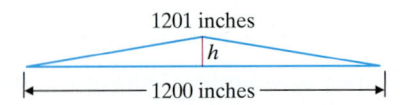

53. Based on the results of Exercise 52, explain why railroads use short sections of rail connected by expansion joints.

8.7 RATIONAL EXPONENTS

OBJECTIVES

1. Evaluate exponential expressions of the form $a^{1/n}$.
2. Evaluate exponential expressions of the form $a^{m/n}$.
3. Evaluate exponential expressions of the form $a^{-m/n}$.
4. Use rules for exponents to simplify expressions containing fractional exponents.

OBJECTIVE 1 ▶ Evaluating $a^{1/n}$. Radical notation is widely used, as we've seen. In this section, we study an alternate notation, one that proves to be more efficient and compact. This alternate notation makes use of expressions containing an exponent that is a rational number but not necessarily an integer, for example,

$$3^{1/2}, 2^{-3/4}, \text{ and } y^{5/6}$$

In giving meaning to rational exponents, keep in mind that we want the rules for operating with them to be the same as the rules for operating with integer exponents. For this to be true,

$$(3^{1/2})^2 = 3^{1/2 \cdot 2} = 3^1 = 3$$

Also, we know that

$$(\sqrt{3})^2 = 3$$

Since the square of both $3^{1/2}$ and $\sqrt{3}$ is 3, it would be reasonable to say that

$$3^{1/2} \text{ means } \sqrt{3}$$

In general, we have the following.

Definition of $a^{1/n}$

If n is a positive integer and $\sqrt[n]{a}$ is a real number, then

$$a^{1/n} = \sqrt[n]{a}$$

Notice that the denominator of the rational exponent is the same as the index of the corresponding radical.

EXAMPLE 1 Write in radical notation. Then simplify.

a. $25^{1/2}$ **b.** $8^{1/3}$ **c.** $-16^{1/4}$ **d.** $(-27)^{1/3}$ **e.** $\left(\dfrac{1}{9}\right)^{1/2}$

Solution

a. $25^{1/2} = \sqrt{25} = 5$
b. $8^{1/3} = \sqrt[3]{8} = 2$
c. In $-16^{1/4}$, the base of the exponent is 16. Thus **the negative sign is not affected by the exponent;** so $-16^{1/4} = -\sqrt[4]{16} = -2$.
d. The parentheses show that -27 is the base. $(-27)^{1/3} = \sqrt[3]{-27} = -3$.
e. $\left(\dfrac{1}{9}\right)^{1/2} = \sqrt{\dfrac{1}{9}} = \dfrac{1}{3}$

PRACTICE 1 Write in radical notation. Then simplify.

a. $36^{1/2}$ **b.** $125^{1/3}$ **c.** $-\left(\dfrac{1}{81}\right)^{1/4}$ **d.** $(-1000)^{1/3}$ **e.** $32^{1/5}$

OBJECTIVE 2 ▶ **Evaluating $a^{m/n}$.** In Example 1, each rational exponent has a numerator of 1. What happens if the numerator is some other positive integer? Consider $8^{2/3}$. Since $\frac{2}{3}$ is the same $\frac{1}{3} \cdot 2$, we reason that

$$8^{2/3} = 8^{(1/3)2} = (8^{1/3})^2 = (\sqrt[3]{8})^2 = 2^2 = 4$$

The denominator 3 of the rational exponent is the same as the index of the radical. The numerator 2 of the fractional exponent indicates that the radical base is to be squared.

Definition of $a^{m/n}$
If m and n are integers with $n > 0$ and if a is a positive number, then
$$a^{m/n} = (a^{1/n})^m = (\sqrt[n]{a})^m$$
Also,
$$a^{m/n} = (a^m)^{1/n} = \sqrt[n]{a^m}$$

EXAMPLE 2 Simplify each expression.

a. $4^{3/2}$ **b.** $27^{2/3}$ **c.** $-16^{3/4}$

Solution

a. $4^{3/2} = (4^{1/2})^3 = (\sqrt{4})^3 = 2^3 = 8$

b. $27^{2/3} = (27^{1/3})^2 = (\sqrt[3]{27})^2 = 3^2 = 9$

c. The negative sign is **not** affected by the exponent since the base of the exponent is 16. Thus, $-16^{3/4} = -(16^{1/4})^3 = -(\sqrt[4]{16})^3 = -2^3 = -8$.

PRACTICE 2 Simplify each expression.

a. $9^{3/2}$ **b.** $8^{5/3}$ **c.** $-625^{1/4}$

▶ **Helpful Hint**
Recall that
$$-3^2 = -(3 \cdot 3) = -9$$
and
$$(-3)^2 = (-3)(-3) = 9$$
In other words, without parentheses the exponent 2 applies to the base 3, **not** -3. The same is true of rational exponents. For example,
$$-16^{1/2} = -\sqrt{16} = -4$$
and
$$(-27)^{1/3} = \sqrt[3]{-27} = -3$$

OBJECTIVE 3 ▶ **Evaluating $a^{-m/n}$.** If the exponent is a negative rational number, use the following definition.

Definition of $a^{-m/n}$
If $a^{-m/n}$ is a nonzero real number, then
$$a^{-m/n} = \frac{1}{a^{m/n}}$$

Section 8.7 Rational Exponents **545**

EXAMPLE 3 Write each expression with a positive exponent and then simplify.

a. $36^{-1/2}$ **b.** $16^{-3/4}$ **c.** $-9^{1/2}$ **d.** $32^{-4/5}$

Solution

a. $36^{-1/2} = \dfrac{1}{36^{1/2}} = \dfrac{1}{\sqrt{36}} = \dfrac{1}{6}$

b. $16^{-3/4} = \dfrac{1}{16^{3/4}} = \dfrac{1}{(\sqrt[4]{16})^3} = \dfrac{1}{2^3} = \dfrac{1}{8}$

c. $-9^{1/2} = -\sqrt{9} = -3$

d. $32^{-4/5} = \dfrac{1}{32^{4/5}} = \dfrac{1}{(\sqrt[5]{32})^4} = \dfrac{1}{2^4} = \dfrac{1}{16}$

PRACTICE 3 Write each expression with a positive exponent and then simplify.

a. $25^{-1/2}$ **b.** $1000^{-2/3}$ **c.** $-49^{1/2}$ **d.** $1024^{-2/5}$

OBJECTIVE 4 ▶ Using exponent rules with fractional exponents. It can be shown that the properties of integer exponents hold for rational exponents. By using these properties and definitions, we can now simplify products and quotients of expressions containing rational exponents.

EXAMPLE 4 Simplify each expression. Write results with positive exponents only. Assume that all variables represent positive numbers.

a. $3^{1/2} \cdot 3^{3/2}$ **b.** $\dfrac{5^{1/3}}{5^{2/3}}$ **c.** $(x^{1/4})^{12}$ **d.** $\dfrac{x^{1/5}}{x^{-4/5}}$ **e.** $\left(\dfrac{y^{3/5}}{z^{1/4}}\right)^2$

Solution

a. $3^{1/2} \cdot 3^{3/2} = 3^{(1/2)+(3/2)} = 3^{4/2} = 3^2 = 9$

b. $\dfrac{5^{1/3}}{5^{2/3}} = 5^{(1/3)-(2/3)} = 5^{-1/3} = \dfrac{1}{5^{1/3}}$

c. $(x^{1/4})^{12} = x^{(1/4)12} = x^3$

d. $\dfrac{x^{1/5}}{x^{-4/5}} = x^{(1/5)-(-4/5)} = x^{5/5} = x^1$ or x

e. $\left(\dfrac{y^{3/5}}{z^{1/4}}\right)^2 = \dfrac{y^{(3/5)2}}{z^{(1/4)2}} = \dfrac{y^{6/5}}{z^{1/2}}$

PRACTICE 4 Simplify each expression. Write results with positive exponents only. Assume that all variables represent positive numbers.

a. $6^{3/5} \cdot 6^{7/5}$ **b.** $\dfrac{7^{1/6}}{7^{3/6}}$ **c.** $(z^{3/8})^{16}$ **d.** $\dfrac{a^{3/7}}{a^{-4/7}}$ **e.** $\left(\dfrac{x^{5/8}}{y^{2/3}}\right)^{12}$

8.7 EXERCISE SET

Simplify each expression. See Examples 1 and 2.

1. $8^{1/3}$
2. $16^{1/4}$
3. $9^{1/2}$
4. $16^{1/2}$
5. $16^{3/4}$
6. $27^{4/3}$
7. $32^{2/5}$
8. $-64^{5/6}$

Simplify each expression. See Example 3.

9. $-16^{-1/4}$
10. $-8^{-1/3}$
11. $16^{-3/2}$
12. $27^{-4/3}$
13. $81^{-3/2}$
14. $32^{-2/5}$
15. $\left(\dfrac{4}{25}\right)^{-1/2}$
16. $\left(\dfrac{8}{27}\right)^{-1/3}$

17. Explain the meaning of the numbers 2, 3, and 4 in the exponential expression $4^{3/2}$.
18. Explain why $-4^{1/2}$ is a real number but $(-4)^{1/2}$ is not.

Simplify each expression. Write each answer with positive exponents. Assume that all variables represent positive numbers. See Example 4.

19. $2^{1/3} \cdot 2^{2/3}$
20. $4^{2/5} \cdot 4^{3/5}$
21. $\dfrac{4^{3/4}}{4^{1/4}}$
22. $\dfrac{9^{7/2}}{9^{3/2}}$
23. $\dfrac{x^{1/6}}{x^{5/6}}$
24. $\dfrac{x^{1/4}}{x^{3/4}}$
25. $(x^{1/2})^6$
26. $(x^{1/3})^6$

27. Explain how simplifying $x^{1/2} \cdot x^{1/3}$ is similar to simplifying $x^2 \cdot x^3$.
28. Explain how simplifying $(x^{1/2})^{1/3}$ is similar to simplifying $(x^2)^3$.

MIXED PRACTICE

Simplify each expression.

29. $81^{1/2}$
30. $(-32)^{6/5}$
31. $(-8)^{1/3}$
32. $36^{1/2}$
33. $-81^{1/4}$
34. $-64^{1/3}$
35. $\left(\dfrac{1}{81}\right)^{1/2}$
36. $\left(\dfrac{9}{16}\right)^{1/2}$
37. $\left(\dfrac{27}{64}\right)^{1/3}$
38. $\left(\dfrac{16}{81}\right)^{1/4}$
39. $9^{3/2}$
40. $16^{3/2}$
41. $64^{3/2}$
42. $64^{2/3}$
43. $-8^{2/3}$
44. $8^{2/3}$
45. $4^{5/2}$
46. $9^{4/2}$
47. $\left(\dfrac{4}{9}\right)^{3/2}$
48. $\left(\dfrac{8}{27}\right)^{2/3}$
49. $\left(\dfrac{1}{81}\right)^{3/4}$
50. $\left(\dfrac{1}{32}\right)^{3/5}$

51. $4^{-1/2}$
52. $9^{-1/2}$
53. $125^{-1/3}$
54. $216^{-1/3}$
55. $625^{-3/4}$
56. $256^{-5/8}$

Simplify each expression. Write each answer with positive exponents. Assume that all variables represent positive numbers.

57. $3^{4/3} \cdot 3^{2/3}$
58. $2^{5/4} \cdot 2^{3/4}$
59. $\dfrac{6^{2/3}}{6^{1/3}}$
60. $\dfrac{3^{3/5}}{3^{1/5}}$
61. $(x^{2/3})^9$
62. $(x^6)^{3/4}$
63. $\dfrac{6^{1/3}}{6^{-5/3}}$
64. $\dfrac{2^{-3/4}}{2^{5/4}}$
65. $\dfrac{3^{-3/5}}{3^{2/5}}$
66. $\dfrac{5^{1/4}}{5^{-3/4}}$
67. $\left(\dfrac{x^{1/3}}{y^{3/4}}\right)^2$
68. $\left(\dfrac{x^{1/2}}{y^{2/3}}\right)^6$
69. $\left(\dfrac{x^{2/5}}{y^{3/4}}\right)^8$
70. $\left(\dfrac{x^{3/4}}{y^{1/6}}\right)^3$

REVIEW AND PREVIEW

Solve each system of linear inequalities by graphing on a single coordinate system. See Section 4.5.

71. $\begin{cases} x + y < 6 \\ y \geq 2x \end{cases}$
72. $\begin{cases} 2x - y \geq 3 \\ x < 5 \end{cases}$

Solve each quadratic equation. See Section 6.5.

73. $x^2 - 4 = 3x$
74. $x^2 + 2x = 8$
75. $2x^2 - 5x - 3 = 0$
76. $3x^2 + x - 2 = 0$

CONCEPT EXTENSIONS

77. If a population grows at a rate of 8% annually, the formula $P = P_O(1.08)^N$ can be used to estimate the total population P after N years have passed, assuming the original population is P_O. Find the population after $1\frac{1}{2}$ years if the original population of 10,000 people is growing at a rate of 8% annually.

78. Money grows in a certain savings account at a rate of 4% compounded annually. The amount of money A in the account at time t is given by the formula
$$A = P(1.04)^t$$
where P is the original amount deposited in the account. Find the amount of money in the account after $3\frac{3}{4}$ years if $200 was initially deposited.

Use a calculator and approximate each to three decimal places.

79. $5^{3/4}$
80. $20^{1/8}$
81. $18^{3/5}$
82. $42^{3/10}$

CHAPTER 8 GROUP ACTIVITY

Investigating the Dimensions of Cylinders

The volume V (in cubic units) of a cylinder is given by the formula $V = \pi r^2 h$, where r is the radius of the cylinder and h is its height. In this project, you will investigate the radii of several cylinders by completing the table below.

For this project, you will need several empty cans of different sizes, a 2-cup (16-fluid-ounce) transparent measuring cup with metric markings (in milliliters), a metric ruler, and water. This project may be completed by working in groups or individually.

1. For each can, measure its volume by filling it with water and pouring the water into the measuring cup. Find the volume of the water in milliliters (ml). Record the volumes of the cans in the table. (Remember that 1 ml = 1 cm³.)

2. Use a ruler to measure the height of each can in centimeters (cm). Record the heights in the table.

3. Solve the formula $V = \pi r^2 h$ for the radius r.

4. Use your formula from Question 3 to calculate an estimate of each can's radius based on the volume and height measurements recorded in the table. Record these calculated radii in the table.

5. Try to measure the radius of each can and record these measured radii in the table. (Remember that radius = $\frac{1}{2}$ diameter.)

6. How close are the values of the calculated radius and the measured radius of each can? What factors could account for the differences?

Can	Volume (ml)	Height (cm)	Calculated Radius (cm)	Measured Radius (cm)
A				
B				
C				
D				

CHAPTER 8 VOCABULARY CHECK

Fill in each blank with one of the words or phrases listed below.

| index | rationalizing the denominator | principal square root | radicand |
| conjugate | radical | like radicals | |

1. The expressions $5\sqrt{x}$ and $7\sqrt{x}$ are examples of _____.
2. In the expression $\sqrt[3]{45}$ the number 3 is the _____, the number 45 is the _____, and $\sqrt{}$ is called the _____ sign.
3. The _____ of $(a + b)$ is $(a - b)$.
4. The _____ of 25 is 5.
5. The process of eliminating the radical in the denominator of a radical expression is called _____.

> **Helpful Hint**
> Are you preparing for your test? Don't forget to take the Chapter 8 Test on page 553. Then check your answers at the back of the text and use the Chapter Test Prep Video CD to see the fully worked-out solutions to any of the exercises you want to review.

CHAPTER 8 HIGHLIGHTS

DEFINITIONS AND CONCEPTS	EXAMPLES

SECTION 8.1 INTRODUCTION TO RADICALS

The **positive or principal square root** of a positive number a is written as $\sqrt{a}$. The **negative square root** of a is written as $-\sqrt{a}$. $\sqrt{a} = b$ only if $b^2 = a$ and $b > 0$.	$\sqrt{25} = 5 \qquad \sqrt{100} = 10$ $-\sqrt{9} = -3 \qquad \sqrt{\dfrac{4}{49}} = \dfrac{2}{7}$

(continued)

DEFINITIONS AND CONCEPTS	EXAMPLES

SECTION 8.1 INTRODUCTION TO RADICALS (continued)

A square root of a negative number is not a real number.	$\sqrt{-4}$ is not a real number.
The **cube root** of a real number a is written as $\sqrt[3]{a}$ and $\sqrt[3]{a} = b$ only if $b^3 = a$.	$\sqrt[3]{64} = 4 \qquad \sqrt[3]{-8} = -2$
The **nth root** of a number a is written as $\sqrt[n]{a}$ and $\sqrt[n]{a} = b$ only if $b^n = a$.	$\sqrt[4]{81} = 3$
In $\sqrt[n]{a}$, the natural number n is called the **index**, the symbol $\sqrt{}$ is called a **radical**, and the expression within the radical is called the **radicand**. (*Note*: If the index is even, the radicand must be non-negative for the root to be a real number.)	$\sqrt[5]{-32} = -2$ index $\downarrow$ $\sqrt[n]{a}$ $\uparrow$ radicand

SECTION 8.2 SIMPLIFYING RADICALS

Product rule for radicals If $\sqrt[n]{a}$ and $\sqrt[n]{b}$ are real numbers, then $\sqrt[n]{a} \cdot \sqrt[n]{b} = \sqrt[n]{a \cdot b}$. A square root is in **simplified form** if the radicand contains no perfect square factors other than 1. To simplify a square root, factor the radicand so that one of its factors is a perfect square factor. To simplify cube roots, factor the radicand so that one of its factors is a perfect cube.	$\sqrt{2} \cdot \sqrt{3} = \sqrt{6}$ $\sqrt[3]{7} \cdot \sqrt[3]{2} = \sqrt[3]{14}$ $\sqrt{45} = \sqrt{9 \cdot 5}$ $\phantom{\sqrt{45}} = \sqrt{9} \cdot \sqrt{5}$ $\phantom{\sqrt{45}} = 3\sqrt{5} \qquad$ in simplest form. $\sqrt[3]{48} = \sqrt[3]{8 \cdot 6}$ $\phantom{\sqrt[3]{48}} = \sqrt[3]{8} \cdot \sqrt[3]{6}$ $\phantom{\sqrt[3]{48}} = 2\sqrt[3]{6}$
Quotient rule for radicals If $\sqrt[n]{a}$ and $\sqrt[n]{b}$ are real numbers and $b \neq 0$, then $$\sqrt[n]{\frac{a}{b}} = \frac{\sqrt[n]{a}}{\sqrt[n]{b}}$$	$\sqrt{\dfrac{18}{x^6}} = \dfrac{\sqrt{9 \cdot 2}}{\sqrt{x^6}} = \dfrac{\sqrt{9} \cdot \sqrt{2}}{x^3} = \dfrac{3\sqrt{2}}{x^3}$ $\sqrt[3]{\dfrac{18}{x^6}} = \dfrac{\sqrt[3]{18}}{\sqrt[3]{x^6}} = \dfrac{\sqrt[3]{18}}{x^2}$

SECTION 8.3 ADDING AND SUBTRACTING RADICALS

Like radicals are radical expressions that have the same index and the same radicand. To combine like radicals, use the distributive property.	***Like Radicals*** $5\sqrt{2}, -7\sqrt{2}, \sqrt{2} \quad$ Also, $-\sqrt[3]{11}, 3\sqrt[3]{11}$ $2\sqrt{7} - 13\sqrt{7} = (2 - 13)\sqrt{7} = -11\sqrt{7}$ $\sqrt[3]{24} + \sqrt[3]{8} + \sqrt[3]{81}$ $= \sqrt[3]{8 \cdot 3} + 2 + \sqrt[3]{27 \cdot 3}$ $= \sqrt[3]{8} \cdot \sqrt[3]{3} + 2 + \sqrt[3]{27} \cdot \sqrt[3]{3}$ $= 2\sqrt[3]{3} + 2 + 3\sqrt[3]{3}$ $= (2 + 3)\sqrt[3]{3} + 2$ $= 5\sqrt[3]{3} + 2$

DEFINITIONS AND CONCEPTS	EXAMPLES
SECTION 8.4 MULTIPLYING AND DIVIDING RADICALS	
The product and quotient rules for radicals may be used to simplify products and quotients of radicals.	Perform indicated operations and simplify. $$\sqrt{3} \cdot \sqrt{11} = \sqrt{33}$$ $$(2\sqrt{5})^2 = 2^2 \cdot (\sqrt{5})^2 = 4 \cdot 5 = 20$$ Multiply. $$(\sqrt{3x} + 1)(\sqrt{5} - \sqrt{3})$$ $$= \sqrt{15x} - \sqrt{9x} + \sqrt{5} - \sqrt{3}$$ $$= \sqrt{15x} - 3\sqrt{x} + \sqrt{5} - \sqrt{3}$$ $$\frac{\sqrt[3]{56x^4}}{\sqrt[3]{7x}} = \sqrt[3]{\frac{56x^4}{7x}} = \sqrt[3]{8x^3} = 2x$$
The process of eliminating the radical in the denominator of a radical expression is called **rationalizing the denominator**.	Rationalize the denominator. $$\frac{5}{\sqrt{11}} = \frac{5 \cdot \sqrt{11}}{\sqrt{11} \cdot \sqrt{11}} = \frac{5\sqrt{11}}{11}$$
The **conjugate** of $a + b$ is $a - b$.	The conjugate of $2 + \sqrt{3}$ is $2 - \sqrt{3}$.
To rationalize a denominator that is a sum or difference of radicals, multiply the numerator and the denominator by the conjugate of the denominator.	Rationalize the denominator. $$\frac{5}{6 - \sqrt{5}} = \frac{5(6 + \sqrt{5})}{(6 - \sqrt{5})(6 + \sqrt{5})}$$ $$= \frac{5(6 + \sqrt{5})}{36 + 6\sqrt{5} - 6\sqrt{5} - 5}$$ $$= \frac{5(6 + \sqrt{5})}{31}$$
SECTION 8.5 SOLVING EQUATIONS CONTAINING RADICALS	

To solve a radical equation containing square roots

Step 1. Get one radical by itself on one side of the equation.
Step 2. Square both sides of the equation.
Step 3. Simplify both sides of the equation.
Step 4. If the equation still contains a radical term, repeat steps 1 through 3.
Step 5. Solve the equation.
Step 6. Check solutions in the original equation.

Solve $\sqrt{2x - 1} - x = -2$.

$$\sqrt{2x - 1} = x - 2$$
$$(\sqrt{2x - 1})^2 = (x - 2)^2 \quad \text{Square both sides.}$$
$$2x - 1 = x^2 - 4x + 4$$
$$0 = x^2 - 6x + 5$$
$$0 = (x - 1)(x - 5) \quad \text{Factor.}$$
$$x - 1 = 0 \quad \text{or} \quad x - 5 = 0$$
$$x = 1 \quad \text{or} \quad x = 5 \quad \text{Solve.}$$

Check both proposed solutions in the original equation. 5 checks but 1 does not. The only solution is 5.

DEFINITIONS AND CONCEPTS	EXAMPLES

SECTION 8.6 RADICAL EQUATIONS AND PROBLEM SOLVING

Problem-solving steps

1. UNDERSTAND. Read and reread the problem.

A gutter is mounted on the eaves of a house 15 feet above the ground. A garden is adjacent to the house so that the closest a ladder can be placed to the house is 6 feet. How long a ladder is needed for installing the gutter? Let x = the length of the ladder.

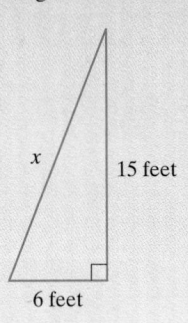

2. TRANSLATE.

Here, we use the Pythagorean theorem. The unknown length x is the hypotenuse.

In words:
$$(\text{leg})^2 + (\text{leg})^2 = (\text{hypotenuse})^2$$

3. SOLVE.

Translate:
$$6^2 + 15^2 = x^2$$
$$36 + 225 = x^2$$
$$261 = x^2$$
$$\sqrt{261} = x \quad \text{or} \quad x = 3\sqrt{29}$$

4. INTERPRET.

Check and state. The ladder needs to be $3\sqrt{29}$ feet or approximately 16.2 feet long.

SECTION 8.7 RATIONAL EXPONENTS

If n is a positive integer and $\sqrt[n]{a}$ is a real number, then $a^{1/n} = \sqrt[n]{a}$.

If m and n are integers with $n > 0$ and a is positive, then
$$a^{m/n} = (a^{1/n})^m = (\sqrt[n]{a})^m$$

Also
$$a^{m/n} = (a^m)^{1/n} = \sqrt[n]{a^m}$$

If $a^{m/n}$ is a nonzero real number, then
$$a^{-m/n} = \frac{1}{a^{m/n}}$$

Properties for integer exponents hold for rational exponents also.

$$a^m \cdot a^n = a^{m+n}$$
$$(a^m)^n = a^{mn}$$
$$\frac{a^m}{a^n} = a^{m-n}$$

$$9^{1/2} = \sqrt{9} = 3$$
$$(-8)^{1/3} = \sqrt[3]{-8} = -2$$

$$25^{3/2} = (25^{1/2})^3 = (\sqrt{25})^3 = 5^3 \text{ or } 125$$

$$81^{-3/4} = \frac{1}{81^{3/4}} = \frac{1}{(\sqrt[4]{81})^3} = \frac{1}{3^3} = \frac{1}{27}$$

$$x^{1/2} \cdot x^{1/4} = x^{(1/2)+(1/4)} = x^{(2/4)+(1/4)} = x^{3/4}$$
$$(x^{2/3})^{1/5} = x^{(2/3) \cdot (1/5)} = x^{2/15}$$
$$\frac{x^{5/6}}{x^{1/6}} = x^{(5/6)-(1/6)} = x^{4/6} = x^{2/3}$$

CHAPTER 8 REVIEW

(8.1) Find the root.

1. $\sqrt{81}$
2. $-\sqrt{49}$
3. $\sqrt[3]{27}$
4. $\sqrt[4]{16}$
5. $-\sqrt{\dfrac{9}{64}}$
6. $\sqrt{\dfrac{36}{81}}$
7. $\sqrt[4]{\dfrac{16}{81}}$
8. $\sqrt[3]{-\dfrac{27}{64}}$

9. Which radical(s) is not a real number?
   a. $\sqrt{4}$ b. $-\sqrt{4}$ c. $\sqrt{-4}$ d. $\sqrt[3]{-4}$

10. Which radical(s) is not a real number?
    a. $\sqrt{-5}$ b. $\sqrt[3]{-5}$ c. $\sqrt[4]{-5}$ d. $\sqrt[5]{-5}$

Find the following roots. Assume that variables represent positive numbers only.

11. $\sqrt{x^{12}}$
12. $\sqrt{x^8}$
13. $\sqrt{9x^6}$
14. $\sqrt{25x^4}$
15. $\sqrt{\dfrac{16}{y^{10}}}$
16. $\sqrt{\dfrac{y^{12}}{49}}$

(8.2) Simplify each expression using the product rule. Assume that variables represent nonnegative real numbers.

17. $\sqrt{54}$
18. $\sqrt{88}$
19. $\sqrt{150x^3}$
20. $\sqrt{92y^5}$
21. $\sqrt[3]{54}$
22. $\sqrt[3]{88}$
23. $\sqrt[4]{48}$
24. $\sqrt[4]{162}$

Simplify each expression using the quotient rule. Assume that variables represent positive real numbers.

25. $\sqrt{\dfrac{18}{25}}$
26. $\sqrt{\dfrac{75}{64}}$
27. $\sqrt{\dfrac{45y^2}{4x^4}}$
28. $\sqrt{\dfrac{20x^5}{9x^2}}$
29. $\sqrt[4]{\dfrac{9}{16}}$
30. $\sqrt[3]{\dfrac{40}{27}}$
31. $\sqrt[3]{\dfrac{3}{8}}$
32. $\sqrt[4]{\dfrac{5}{81}}$

(8.3) Add or subtract by combining like radicals.

33. $3\sqrt[3]{2} + 2\sqrt[3]{3} - 4\sqrt[3]{2}$
34. $5\sqrt{2} + 2\sqrt[3]{2} - 8\sqrt{2}$
35. $\sqrt{6} + 2\sqrt[3]{6} - 4\sqrt[3]{6} + 5\sqrt{6}$
36. $3\sqrt{5} - \sqrt[3]{5} - 2\sqrt{5} + 3\sqrt[3]{5}$

Add or subtract by simplifying each radical and then combining like terms. Assume that variables represent nonnegative real numbers.

37. $\sqrt{28x} + \sqrt{63x} + \sqrt[3]{56}$
38. $\sqrt{75y} + \sqrt{48y} - \sqrt[4]{16}$
39. $\sqrt{\dfrac{5}{9}} - \sqrt{\dfrac{5}{36}}$
40. $\sqrt{\dfrac{11}{25}} + \sqrt{\dfrac{11}{16}}$
41. $2\sqrt[3]{125} - 5\sqrt[3]{8}$
42. $3\sqrt[3]{16} - 2\sqrt[3]{2}$

(8.4) Find the product and simplify if possible.

43. $3\sqrt{10} \cdot 2\sqrt{5}$
44. $2\sqrt[3]{4} \cdot 5\sqrt[3]{6}$
45. $\sqrt{3}(2\sqrt{6} - 3\sqrt{12})$
46. $4\sqrt{5}(2\sqrt{10} - 5\sqrt{5})$
47. $(\sqrt{3} + 2)(\sqrt{6} - 5)$
48. $(2\sqrt{5} + 1)(4\sqrt{5} - 3)$
49. $(\sqrt{x} - 2)^2$
50. $(\sqrt{y} + 4)^2$

Divide and simplify if possible. Assume that all variables represent positive numbers.

51. $\dfrac{\sqrt{27}}{\sqrt{3}}$
52. $\dfrac{\sqrt{20}}{\sqrt{5}}$
53. $\dfrac{\sqrt{160}}{\sqrt{8}}$
54. $\dfrac{\sqrt{96}}{\sqrt{3}}$
55. $\dfrac{\sqrt{30x^6}}{\sqrt{2x^3}}$
56. $\dfrac{\sqrt{54x^5y^2}}{\sqrt{3xy^2}}$

Rationalize each denominator and simplify.

57. $\dfrac{\sqrt{2}}{\sqrt{11}}$
58. $\dfrac{\sqrt{3}}{\sqrt{13}}$
59. $\sqrt{\dfrac{5}{6}}$
60. $\sqrt{\dfrac{7}{10}}$
61. $\dfrac{1}{\sqrt{5x}}$
62. $\dfrac{5}{\sqrt{3y}}$
63. $\sqrt{\dfrac{3}{x}}$
64. $\sqrt{\dfrac{6}{y}}$
65. $\dfrac{3}{\sqrt{5} - 2}$
66. $\dfrac{8}{\sqrt{10} - 3}$
67. $\dfrac{\sqrt{2} + 1}{\sqrt{3} - 1}$
68. $\dfrac{\sqrt{3} - 2}{\sqrt{5} + 2}$
69. $\dfrac{10}{\sqrt{x} + 5}$
70. $\dfrac{8}{\sqrt{x} - 1}$
71. $\sqrt[3]{\dfrac{7}{9}}$
72. $\sqrt[3]{\dfrac{3}{4}}$
73. $\sqrt[3]{\dfrac{3}{2}}$
74. $\sqrt[3]{\dfrac{5}{4}}$

(8.5) Solve each radical equations.

75. $\sqrt{2x} = 6$
76. $\sqrt{x + 3} = 4$
77. $\sqrt{x + 3} = 8$
78. $\sqrt{x + 8} = 3$
79. $\sqrt{2x + 1} = x - 7$
80. $\sqrt{3x + 1} = x - 1$
81. $\sqrt{x + 3} = \sqrt{x + 15}$
82. $\sqrt{x - 5} = \sqrt{x - 1}$

552 CHAPTER 8 Roots and Radicals

(8.6) *Use the Pythagorean theorem to find the length of each unknown side. Give an exact answer and a two-decimal-place approximation.*

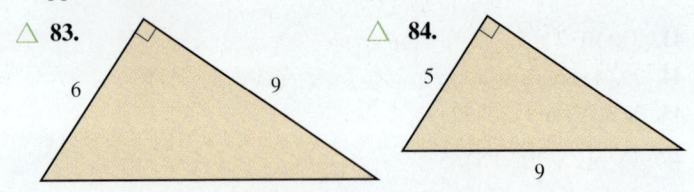

△ **83.** (triangle with legs 6 and 9)

△ **84.** (triangle with leg 5 and hypotenuse 9)

Solve. Give an exact answer and a two-decimal-place approximation.

△ **85.** Romeo is standing 20 feet away from the wall below Juliet's balcony during a school play. Juliet is on the balcony, 12 feet above the ground. Find how far apart Romeo and Juliet are.

△ **86.** The diagonal of a rectangle is 10 inches long. If the width of the rectangle is 5 inches, find the length of the rectangle.

Use the distance formula to find the distance between the points.

87. $(6, -2)$ and $(-3, 5)$

88. $(2, 8)$ and $(-6, 10)$

Use the formula $r = \sqrt{\dfrac{S}{4\pi}}$, where $r =$ the radius of a sphere and $S =$ the surface area of the sphere, for Exercises 89 and 90.

89. Find the radius of a sphere to the nearest tenth of an inch if the area is 72 square inches.

△ **90.** Find the exact surface area of a sphere if its radius is 6 inches. (Do not approximate π.)

(8.7) *Write each of the following with fractional exponents and simplify if possible. Assume that variables represent nonnegative real numbers.*

91. $\sqrt{a^5}$ **92.** $\sqrt[5]{a^3}$

93. $\sqrt[6]{x^{15}}$ **94.** $\sqrt[4]{x^{12}}$

Evaluate each of the following expressions.

95. $16^{1/2}$ **96.** $36^{1/2}$

97. $(-8)^{1/3}$ **98.** $(-32)^{1/5}$

99. $-64^{3/2}$ **100.** $-8^{2/3}$

101. $\left(\dfrac{16}{81}\right)^{3/4}$ **102.** $\left(\dfrac{9}{25}\right)^{3/2}$

Simplify each expression using positive exponents only. Assume that variables represent positive real numbers.

103. $8^{1/3} \cdot 8^{4/3}$ **104.** $4^{3/2} \cdot 4^{1/2}$

105. $\dfrac{3^{1/6}}{3^{5/6}}$ **106.** $\dfrac{2^{1/4}}{2^{-3/5}}$

107. $(x^{-1/3})^6$ **108.** $\left(\dfrac{x^{1/2}}{y^{1/3}}\right)^2$

MIXED REVIEW

Find each root. Assume all variables represent positive numbers.

109. $\sqrt{144}$ **110.** $-\sqrt[3]{64}$

111. $\sqrt{16x^{16}}$ **112.** $\sqrt{4x^{24}}$

Simplify each expression. Assume all variables represent positive numbers.

113. $\sqrt{18x^7}$ **114.** $\sqrt{48y^6}$

115. $25^{-\frac{1}{2}}$ **116.** $64^{-2/3}$

117. $\sqrt{\dfrac{y^4}{81}}$ **118.** $\sqrt{\dfrac{x^9}{9}}$

Add or subtract by simplifying and then combining like terms. Assume all variables represent positive numbers.

119. $\sqrt{12} + \sqrt{75}$ **120.** $\sqrt{63} + \sqrt{28} - \sqrt[3]{27}$

121. $\sqrt{\dfrac{3}{16}} - \sqrt{\dfrac{3}{4}}$ **122.** $\sqrt{45x^3} + x\sqrt{20x} - \sqrt{5x^3}$

Multiply and simplify if possible. Assume all variables represent positive numbers.

123. $\sqrt{7} \cdot \sqrt{14}$ **124.** $\sqrt{3}(\sqrt{9} - \sqrt{2})$

125. $(\sqrt{2} + 4)(\sqrt{5} - 1)$ **126.** $(\sqrt{x} + 3)^2$

Divide and simplify if possible. Assume all variables represent positive numbers.

127. $\dfrac{\sqrt{120}}{\sqrt{5}}$ **128.** $\dfrac{\sqrt{60x^9}}{\sqrt{15x^7}}$

Rationalize each denominator and simplify.

129. $\sqrt{\dfrac{2}{7}}$ **130.** $\dfrac{3}{\sqrt{2x}}$

131. $\dfrac{3}{\sqrt{x} - 6}$ **132.** $\dfrac{\sqrt{7} - 5}{\sqrt{5} + 3}$

Solve each radical equation.

133. $\sqrt{4x} = 2$ **134.** $\sqrt{x - 4} = 3$

135. $\sqrt{4x + 8} + 6 = x$ **136.** $\sqrt{x - 8} = \sqrt{x} - 2$

137. Use the Pythagorean theorem to find the length of the unknown side. Give an exact answer and a two-decimal-place approximation.

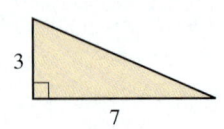

(right triangle with legs 3 and 7)

138. The diagonal of a rectangle is 6 inches long. If the width of the rectangle is 2 inches, find the length of the rectangle.

139. The Apple (computer) store on Fifth Avenue in New York City, which opened in 2006, features a distinctive glass cube. If the glass surface area of this cube (which includes only the four walls and the roof) is 5120 square feet, find the length of a side of the cube.

STUDY SKILLS BUILDER

Are You Prepared for a Test on Chapter 8?

Below I have listed some *common trouble areas* for students in Chapter 8. After studying for your test—but before taking your test—read these.

- Do you understand the difference between $\sqrt{3} \cdot \sqrt{2}$ and $\sqrt{3} + \sqrt{2}$?

$$\sqrt{3} \cdot \sqrt{2} = \sqrt{3 \cdot 2} = \sqrt{6}$$

$\sqrt{3} + \sqrt{2}$ cannot be simplified further. The terms are unlike terms.

- Do you understand the difference between rationalizing the denominator of $\dfrac{\sqrt{3}}{\sqrt{7}}$ and rationalizing the denominator of $\dfrac{\sqrt{3}}{\sqrt{7}+1}$?

$$\frac{\sqrt{3}}{\sqrt{7}} = \frac{\sqrt{3}\cdot\sqrt{7}}{\sqrt{7}\cdot\sqrt{7}} = \frac{\sqrt{21}}{7}$$

$$\frac{\sqrt{3}}{\sqrt{7}+1} = \frac{\sqrt{3}(\sqrt{7}-1)}{(\sqrt{7}+1)(\sqrt{7}-1)}$$
$$= \frac{\sqrt{3}(\sqrt{7}-1)}{7-1} = \frac{\sqrt{3}(\sqrt{7}-1)}{6}$$

- To solve an equation containing a radical, don't forget to first isolate the radical.

$$\sqrt{x} - 10 = -4$$
$$\sqrt{x} = 6 \quad \text{Isolate the radical.}$$
$$(\sqrt{x})^2 = 6^2 \quad \text{Square both sides.}$$
$$x = 36 \quad \text{Simplify.}$$

Make sure you check the proposed solution in the original equation.

Remember: This is simply a listing of a few common trouble areas. For a review of Chapter 8, see the Highlights and Chapter Review at the end of the chapter.

CHAPTER 8 TEST

Remember to use the Chapter Test Prep Video CD to see the fully worked-out solutions to any of the exercises you want to review.

Simplify the following. Indicate if the expression is not a real number.

1. $\sqrt{16}$
2. $\sqrt[3]{-125}$
3. $16^{3/4}$
4. $\left(\dfrac{9}{16}\right)^{1/2}$
5. $\sqrt[4]{-81}$
6. $27^{-2/3}$

Simplify each radical expression. Assume that variables represent positive numbers only.

7. $\sqrt{54}$
8. $\sqrt{92}$
9. $\sqrt{3x^6}$
10. $\sqrt{8x^4y^7}$
11. $\sqrt{9x^9}$
12. $\sqrt[3]{8}$
13. $\sqrt[3]{40}$
14. $\sqrt{x^{10}}$
15. $\sqrt{y^7}$
16. $\sqrt{\dfrac{5}{16}}$
17. $\sqrt{\dfrac{y^3}{25}}$
18. $\sqrt[3]{\dfrac{2}{27}}$
19. $3\sqrt{8x}$

Perform each indicated operation. Assume that all variables represent positive numbers.

20. $\sqrt{13} + \sqrt{13} - 4\sqrt{13}$
21. $\sqrt{12} - 2\sqrt{75}$
22. $\sqrt{2x^2} + \sqrt[3]{54} - x\sqrt{18}$
23. $\sqrt{\dfrac{3}{4}} + \sqrt{\dfrac{3}{25}}$
24. $\sqrt{7} \cdot \sqrt{14}$
25. $\sqrt{2}(\sqrt{6} - \sqrt{5})$
26. $(\sqrt{x}+2)(\sqrt{x}-3)$
27. $\dfrac{\sqrt{50}}{\sqrt{10}}$
28. $\dfrac{\sqrt{40x^4}}{\sqrt{2x}}$

Rationalize the denominator.

29. $\sqrt{\dfrac{2}{3}}$
30. $\sqrt[3]{\dfrac{5}{9}}$
31. $\sqrt{\dfrac{5}{12x^2}}$
32. $\dfrac{2\sqrt{3}}{\sqrt{3}-3}$

Solve each of the following radical equations.

33. $\sqrt{x+8} = 11$
34. $\sqrt{3x-6} = \sqrt{x+4}$
35. $\sqrt{2x-2} = x-5$
△ 36. Find the length of the unknown leg of a right triangle if the other leg is 8 inches long and the hypotenuse is 12 inches long.
37. Find the distance between $(-3, 6)$ and $(-2, 8)$.

Simplify each expression using positive exponents only.

38. $16^{-3/4} \cdot 16^{-1/4}$
39. $\left(\dfrac{x^{2/3}}{y^{2/5}}\right)^5$

CHAPTER 8 CUMULATIVE REVIEW

1. Simplify each expression.
   a. $\dfrac{(-12)(-3) + 3}{-7 - (-2)}$
   b. $\dfrac{2(-3)^2 - 20}{-5 + 4}$

2. Simplify each expression.
   a. $\dfrac{4(-3) - (-6)}{-8 + 4}$
   b. $\dfrac{3 + (-3)(-2)^3}{-1 - (-4)}$

3. Solve. $2x + 3x - 5 + 7 = 10x + 3 - 6x - 4$
4. Solve. $6y - 11 + 4 + 2y = 8 + 15y - 8y$
5. Complete the table for the equation $y = 3x$.

x	y
-1	
	0
	-9

6. Complete the table for the equation $2x + y = 6$.

x	y
0	
	-2
3	

7. Find an equation of the line with y-intercept $(0, -3)$ and slope of $\dfrac{1}{4}$.

8. Find an equation of a line with y-intercept $(0, 4)$ and slope of -2.

9. Find an equation of the line parallel to the line $y = 5$ and passing through $(-2, -3)$.

10. Find an equation of the line perpendicular to $y = 2x + 4$ and passing through $(1, 5)$.

11. Which of the following linear equations are functions?
    a. $y = x$
    b. $y = 2x + 1$
    c. $y = 5$
    d. $x = -1$

12. Which of the following linear equations are functions?
    a. $2x + 3 = y$
    b. $x + 4 = 0$
    c. $\dfrac{1}{2}y = 2x$
    d. $y = 0$

13. Determine whether $(12, 6)$ is a solution of the given system
    $\begin{cases} 2x - 3y = 6 \\ x = 2y \end{cases}$

14. Which of the following ordered pairs is a solution of the given system?
    $\begin{cases} 2x + y = 4 \\ x + y = 2 \end{cases}$
    a. $(1, 1)$
    b. $(2, 0)$

15. Solve the system:
    $\begin{cases} 2x + y = 10 \\ x = y + 2 \end{cases}$

16. Solve the system:
    $\begin{cases} 3y = x + 10 \\ 2x + 5y = 24 \end{cases}$

17. Solve the system:
    $\begin{cases} -x - \dfrac{y}{2} = \dfrac{5}{2} \\ -\dfrac{x}{2} + \dfrac{y}{4} = 0 \end{cases}$

18. Solve the system:
    $\begin{cases} \dfrac{x}{2} + y = \dfrac{5}{6} \\ 2x - y = \dfrac{5}{6} \end{cases}$

19. Eric Daly, a chemistry teaching assistant, needs 10 liters of a 20% saline solution (salt water) for his 2 P.M. laboratory class. Unfortunately, the only mixtures on hand are a 5% saline solution and a 25% saline solution. How much of each solution should he mix to produce the 20% solution?

20. Two streetcars are 11 miles apart and traveling toward each other on parallel tracks. They meet in 12 minutes. Find the speed of each streetcar if one travels 15 miles per hour faster than the other.

21. Graph the solution of the system:
    $\begin{cases} 3x \geq y \\ x + 2y \leq 8 \end{cases}$

22. Graph the solution of the system:
    $\begin{cases} x + y \leq 1 \\ 2x - y \geq 2 \end{cases}$

23. Combine like terms to simplify. $-9x^2 + 3xy - 5y^2 + 7yx$

24. Combine like terms to simplify.
    $4a^2 + 3a - 2a^2 + 7a - 5$

25. Factor: $x^2 + 7xy + 6y^2$
26. Factor: $3x^2 + 15x + 18$
27. Simplify: $\dfrac{4 - x^2}{3x^2 - 5x - 2}$
28. Simplify: $\dfrac{2x^2 + 7x + 3}{x^2 - 9}$
29. Divide: $\dfrac{3x^3 y^7}{40} \div \dfrac{4x^3}{y^2}$
30. Divide: $\dfrac{12x^2 y^3}{5} \div \dfrac{3y^3}{x}$
31. Subtract: $\dfrac{2y}{2y - 7} - \dfrac{7}{2y - 7}$
32. Subtract: $\dfrac{-4x^2}{x + 1} - \dfrac{4x}{x + 1}$
33. Add: $\dfrac{2x}{x^2 + 2x + 1} + \dfrac{x}{x^2 - 1}$
34. Add: $\dfrac{3x}{x^2 + 5x + 6} + \dfrac{1}{x^2 + 2x - 3}$
35. Solve: $\dfrac{x}{2} + \dfrac{8}{3} = \dfrac{1}{6}$
36. Solve: $\dfrac{1}{21} + \dfrac{x}{7} = \dfrac{5}{3}$

37. If the following two triangles are similar, find the missing length x.

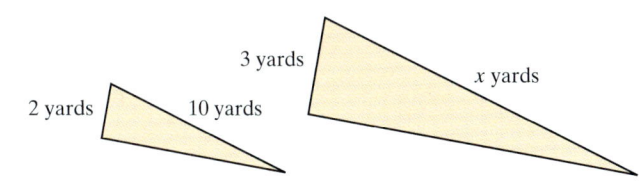

38. If the following two triangles are similar, find the missing length.

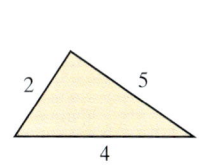

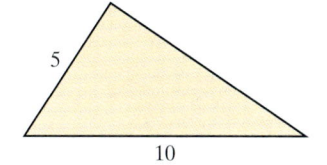

39. Simplify: $\dfrac{\dfrac{1}{z} - \dfrac{1}{2}}{\dfrac{1}{3} - \dfrac{z}{6}}$

40. Simplify: $\dfrac{x + 3}{\dfrac{1}{x} + \dfrac{1}{3}}$

41. Simplify.
  a. $\sqrt{54}$
  b. $\sqrt{12}$
  c. $\sqrt{200}$
  d. $\sqrt{35}$

42. Simplify.
  a. $\sqrt{40}$
  b. $\sqrt{500}$
  c. $\sqrt{63}$
  d. $\sqrt{169}$

43. Multiply. Then simplify, if possible.
  a. $(\sqrt{5} - 7)(\sqrt{5} + 7)$
  b. $(\sqrt{7x} + 2)^2$

44. Multiply. Then simplify, if possible.
  a. $(\sqrt{6} + 2)^2$
  b. $(\sqrt{x} + 5)(\sqrt{x} - 5)$

45. Solve. $\sqrt{x} + 6 = 4$

46. Solve. $\sqrt{x + 4} = \sqrt{3x - 1}$

47. Find the length of the hypotenuse of a right triangle whose legs are 6 inches and 8 inches long.

48. Find the length of the unknown leg of a right triangle whose other leg is 9 and whose hypotenuse is 13.

49. Simplify each expression.
  a. $4^{3/2}$
  b. $27^{2/3}$
  c. $-16^{3/4}$

50. Simplify each expression.
  a. $9^{5/2}$
  b. $-81^{1/4}$
  c. $(-64)^{2/3}$

CHAPTER

9 Quadratic Equations

9.1 Solving Quadratic Equations by the Square Root Property

9.2 Solving Quadratic Equations by Completing the Square

9.3 Solving Quadratic Equations by the Quadratic Formula

Integrated Review—Summary on Solving Quadratic Equations

9.4 Complex Solutions of Quadratic Equations

9.5 Graphing Quadratic Equations

An important part of the study of algebra is learning to use methods for solving equations. In Chapter 2, we presented techniques for solving linear equations in one variable. In Chapter 6, we solved quadratic equations in one variable by factoring the quadratic expressions. We now present other methods for solving quadratic equations in one variable.

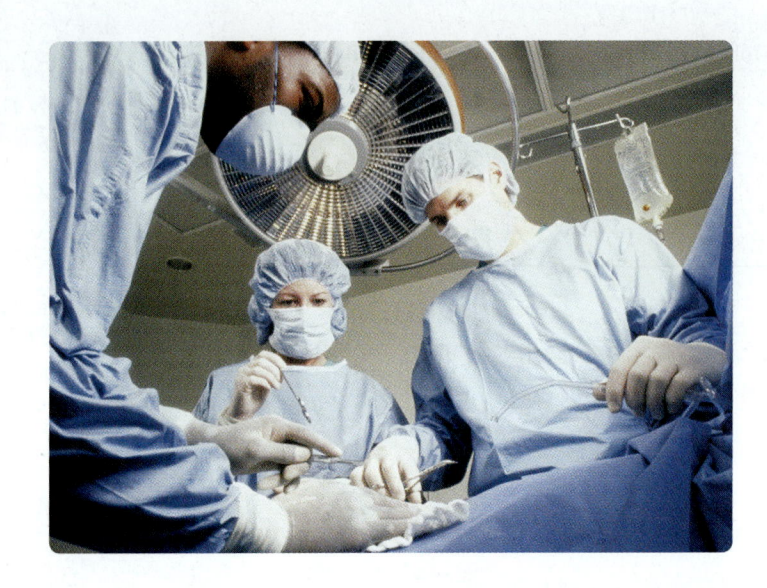

There are currently over 90,000 people living in the United States with a functioning kidney transplant, and that number is increasing every year. The bars on the graph below show the yearly number of kidney transplants performed, and the broken line shows the number of patients waiting for a kidney transplant. Notice that the gap between the bars and the broken line is increasing, meaning that the gap between the number of patients waiting for a transplant and the number receiving a transplant is increasing. Physicians use these data to help plan for the future. In Section 9.3, Exercises 97 and 98, we will use a formula called the quadratic formula to predict the number of kidney transplants and the number of patients on the waiting list at a future date.

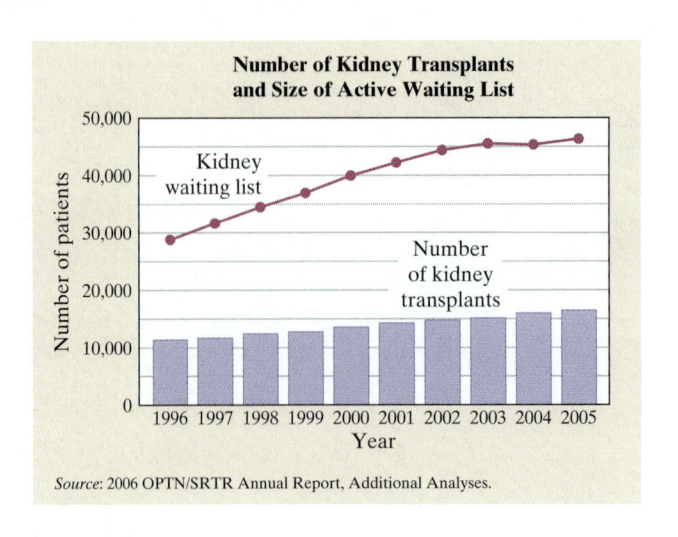

Number of Kidney Transplants and Size of Active Waiting List

Source: 2006 OPTN/SRTR Annual Report, Additional Analyses.

9.1 SOLVING QUADRATIC EQUATIONS BY THE SQUARE ROOT PROPERTY

OBJECTIVES

1. Use the square root property to solve quadratic equations.
2. Solve problems modeled by quadratic equations.

OBJECTIVE 1 ▶ Using the square root property. Recall that a quadratic equation is an equation that can be written in the form

$$ax^2 + bx + c = 0$$

where a, b, and c are real numbers and $a \neq 0$.

To solve quadratic equations by factoring, use the **zero factor theorem:** If the product of two numbers is zero, then at least one of the two numbers is zero. For example, to solve $x^2 - 4 = 0$, we first factor the left side of the equation and then set each factor equal to 0.

$$x^2 - 4 = 0$$
$$(x + 2)(x - 2) = 0 \quad \text{Factor.}$$
$$x + 2 = 0 \quad \text{or} \quad x - 2 = 0 \quad \text{Apply the zero factor theorem.}$$
$$x = -2 \quad \text{or} \quad x = 2 \quad \text{Solve each equation.}$$

The solutions are -2 and 2.

Now let's solve $x^2 - 4 = 0$ another way. First, add 4 to both sides of the equation.

$$x^2 - 4 = 0$$
$$x^2 = 4 \quad \text{Add 4 both sides.}$$

Now we see that the value for x must be a number whose square is 4. Therefore $x = \sqrt{4} = 2$ or $x = -\sqrt{4} = -2$. This reasoning is an example of the square root property.

Square Root Property

If $x^2 = a$ for $a \geq 0$, then

$$x = \sqrt{a} \quad \text{or} \quad x = -\sqrt{a}$$

EXAMPLE 1 Use the square root property to solve $x^2 - 9 = 0$.

Solution First we solve for x^2 by adding 9 to both sides.

$$x^2 - 9 = 0$$
$$x^2 = 9 \quad \text{Add 9 to both sides.}$$

Next we use the square root property.

$$x = \sqrt{9} \quad \text{or} \quad x = -\sqrt{9}$$
$$x = 3 \qquad\qquad x = -3$$

Check:

$$x^2 - 9 = 0 \quad \text{Original equation} \qquad x^2 - 9 = 0 \quad \text{Original equation}$$
$$3^2 - 9 \stackrel{?}{=} 0 \quad \text{Let } x = 3. \qquad (-3)^2 - 9 \stackrel{?}{=} 0 \quad \text{Let } x = -3.$$
$$0 = 0 \quad \text{True} \qquad\qquad 0 = 0 \quad \text{True}$$

The solutions are 3 and -3.

PRACTICE 1 Use the square root property to solve $x^2 - 16 = 0$.

558 CHAPTER 9 Quadratic Equations

EXAMPLE 2 Use the square root property to solve $2x^2 = 7$.

Solution First we solve for x^2 by dividing both sides by 2. Then we use the square root property.

$$2x^2 = 7$$

$$x^2 = \frac{7}{2} \qquad \text{Divide both sides by 2.}$$

$$x = \sqrt{\frac{7}{2}} \quad \text{or} \quad x = -\sqrt{\frac{7}{2}} \qquad \text{Use the square root property.}$$

If the denominators are rationalized, we have

$$x = \frac{\sqrt{7} \cdot \sqrt{2}}{\sqrt{2} \cdot \sqrt{2}} \quad \text{or} \quad x = -\frac{\sqrt{7} \cdot \sqrt{2}}{\sqrt{2} \cdot \sqrt{2}} \qquad \text{Rationalize the denominator.}$$

$$x = \frac{\sqrt{14}}{2} \qquad\qquad x = -\frac{\sqrt{14}}{2} \qquad \text{Simplify.}$$

Remember to check both solutions in the original equation. The solutions are $\frac{\sqrt{14}}{2}$ and $-\frac{\sqrt{14}}{2}$.

PRACTICE 2 Use the square root property to solve $5x^2 = 13$.

EXAMPLE 3 Use the square root property to solve $(x - 3)^2 = 16$.

Solution Instead of x^2, here we have $(x - 3)^2$. But the square root property can still be used.

$$(x - 3)^2 = 16$$

$$x - 3 = \sqrt{16} \quad \text{or} \quad x - 3 = -\sqrt{16} \qquad \text{Use the square root property.}$$

$$x - 3 = 4 \qquad\qquad x - 3 = -4 \qquad \text{Write } \sqrt{16} \text{ as 4 and } -\sqrt{16} \text{ as } -4.$$

$$x = 7 \qquad\qquad\quad x = -1 \qquad\quad \text{Solve.}$$

Check:

$(x - 3)^2 = 16$ Original equation $\qquad (x - 3)^2 = 16$ Original equation
$(7 - 3)^2 \stackrel{?}{=} 16$ Let $x = 7$. $\qquad\qquad (-1 - 3)^2 \stackrel{?}{=} 16$ Let $x = -1$.
$4^2 \stackrel{?}{=} 16$ Simplify. $\qquad\qquad\qquad (-4)^2 \stackrel{?}{=} 16$ Simplify.
$16 = 16$ True $\qquad\qquad\qquad\qquad 16 = 16$ True

Both 7 and -1 are solutions.

PRACTICE 3 Use the square root property to solve $(x - 5)^2 = 36$.

EXAMPLE 4 Use the square root property to solve $(x + 1)^2 = 8$.

Solution $(x + 1)^2 = 8$

$$x + 1 = \sqrt{8} \quad \text{or} \quad x + 1 = -\sqrt{8} \qquad \text{Use the square root property.}$$

$$x + 1 = 2\sqrt{2} \qquad\qquad x + 1 = -2\sqrt{2} \qquad \text{Simplify the radical.}$$

$$x = -1 + 2\sqrt{2} \qquad\quad x = -1 - 2\sqrt{2} \qquad \text{Solve for } x.$$

Check both solutions in the original equation. The solutions are $-1 + 2\sqrt{2}$ and $-1 - 2\sqrt{2}$. This can be written compactly as $-1 \pm 2\sqrt{2}$. The notation $\pm$ is read as "plus or minus."

▶ **Helpful Hint**

read "plus or minus"
↓
The notation $-1 \pm \sqrt{5}$, for example, is just a shorthand notation for both $-1 + \sqrt{5}$ and $-1 - \sqrt{5}$.

PRACTICE 4 Use the square root property to solve $(x + 2)^2 = 12$.

EXAMPLE 5 Use the square root property to solve $(x - 1)^2 = -2$.

Solution This equation has no real solution because the square root of -2 is not a real number. □

PRACTICE 5 Use the square root property to solve $(x - 8)^2 = -5$.

EXAMPLE 6 Use the square root property to solve $(5x - 2)^2 = 10$.

Solution $(5x - 2)^2 = 10$

$5x - 2 = \sqrt{10}$ or $5x - 2 = -\sqrt{10}$ Use the square root property.
$5x = 2 + \sqrt{10}$ $5x = 2 - \sqrt{10}$ Add 2 to both sides.
$x = \dfrac{2 + \sqrt{10}}{5}$ $x = \dfrac{2 - \sqrt{10}}{5}$ Divide both sides by 5.

Check both solutions in the original equation. The solutions are $\dfrac{2 + \sqrt{10}}{5}$ and $\dfrac{2 - \sqrt{10}}{5}$, which can be written as $\dfrac{2 \pm \sqrt{10}}{5}$. □

PRACTICE 6 Use the square root property to solve $(3x - 5)^2 = 17$.

> **Helpful Hint**
> For some applications and graphing purposes, decimal approximations of exact solutions to quadratic equations may be desired.
>
> | Exact Solutions from Example 6 | Decimal Approximations |
> |---|---|
> | $\dfrac{2 + \sqrt{10}}{5}$ | $\approx$ 1.032 |
> | $\dfrac{2 - \sqrt{10}}{5}$ | $\approx$ -0.232 |

OBJECTIVE 2 ▶ Solving problems modeled by quadratic equations. Many real-world applications are modeled by quadratic equations.

EXAMPLE 7 Finding the Length of Time of a Dive

The record for the highest dive into a lake was made by Harry Froboess of Switzerland. In 1936 he dove 394 feet from the airship Hindenburg into Lake Constance. To the nearest tenth of a second, how long did his dive take? (*Source: The Guiness Book of Records*)

Solution

1. UNDERSTAND. To approximate the time of the dive, we use the formula $h = 16t^2$ *
   where t is time in seconds and h is the distance in feet, traveled by a free-falling body or object. For example, to find the distance traveled in 1 second, or 3 seconds, we let $t = 1$ and then $t = 3$.

   If $t = 1, h = 16(1)^2 = 16 \cdot 1 = 16$ feet
   If $t = 3, h = 16(3)^2 = 16 \cdot 9 = 144$ feet

   Since a body travels 144 feet in 3 seconds, we now know the dive of 394 feet lasted longer than 3 seconds.

*The formula $h = 16t^2$ does not take into account air resistance.

2. **TRANSLATE.** Use the formula $h = 16t^2$, let the distance $h = 394$, and we have the equation $394 = 16t^2$.

3. **SOLVE.** To solve $394 = 16t^2$ for t, we will use the square root property.

$$394 = 16t^2$$

$$\frac{394}{16} = t^2 \qquad \text{Divide both sides by 16.}$$

$$24.625 = t^2 \qquad \text{Simplify.}$$

$$\sqrt{24.625} = t \quad \text{or} \quad -\sqrt{24.625} = t \qquad \text{Use the square root property.}$$

$$5.0 \approx t \quad \text{or} \quad -5.0 \approx t \qquad \text{Approximate.}$$

4. **INTERPRET.**

   Check: We reject the solution -5.0 since the length of the dive is not a negative number.

   State: The dive lasted approximately 5 seconds.

PRACTICE 7 On August 16, 1960, as part of an Air Force research program, Air Force Captain Joseph W. Kittinger, Jr. stepped off a platform lifted to 102,800 feet above the New Mexico desert by a hot air balloon. As part of the research, Captain Kittinger free fell for 84,700 feet before opening his parachute. To the nearest tenth of a second, how long did Captain Kittinger free fall before he opened his chute? (*Source: PBS, Nova*)

9.1 EXERCISE SET

Use the square root property to solve each quadratic equation. See Examples 1, 2, and 5.

1. $x^2 = 64$
2. $x^2 = 121$
3. $x^2 = 21$
4. $x^2 = 22$
5. $x^2 = \frac{1}{25}$
6. $x^2 = \frac{1}{16}$
7. $x^2 = -4$
8. $x^2 = -25$
9. $3x^2 = 13$
10. $5x^2 = 2$
11. $7x^2 = 4$
12. $2x^2 = 9$
13. $x^2 - 2 = 0$
14. $x^2 - 15 = 0$
15. $2x^2 - 10 = 0$
16. $7x^2 - 21 = 0$
17. Explain why the equation $x^2 = -9$ has no real solution.
18. Explain why the equation $x^2 = 9$ has two solutions.

Use the square root property to solve each quadratic equation. See Examples 3 through 6.

19. $(x - 5)^2 = 49$
20. $(x + 2)^2 = 25$
21. $(x + 2)^2 = 7$
22. $(x - 7)^2 = 2$
23. $\left(m - \frac{1}{2}\right)^2 = \frac{1}{4}$
24. $\left(m + \frac{1}{3}\right)^2 = \frac{1}{9}$
25. $(p + 2)^2 = 10$
26. $(p - 7)^2 = 13$
27. $(3y + 2)^2 = 100$
28. $(4y - 3)^2 = 81$
29. $(z - 4)^2 = -9$
30. $(z + 7)^2 = -20$
31. $(2x - 11)^2 = 50$
32. $(3x - 17)^2 = 28$
33. $(3x - 7)^2 = 32$
34. $(5x - 11)^2 = 54$
35. $(2p - 5)^2 = 121$
36. $(3p - 1)^2 = 4$

MIXED PRACTICE

Use the square root property to solve. See Examples 3 through 6.

37. $x^2 - 2 = 0$
38. $x^2 - 15 = 0$
39. $(x + 6)^2 = 24$
40. $(x + 5)^2 = 20$
41. $\frac{1}{2}n^2 = 5$
42. $\frac{1}{5}y^2 = 2$
43. $(4x - 1)^2 = 5$
44. $(7x - 2)^2 = 11$
45. $3z^2 = 36$
46. $3z^2 = 24$
47. $(8 - 3x)^2 - 45 = 0$
48. $(10 - 9x)^2 - 75 = 0$

The formula for area of a square is $A = s^2$ where s is the length of a side. Use this formula for Exercises 49 through 52. For each exercise, give an exact answer and a two-decimal-place approximation.

△ **49.** If the area of a square is 20 square inches, find the length of a side.

△ **50.** If the area of a square is 32 square meters, find the length of a side.

△ **51.** The "Water Cube" National Swimming Center is being constructed in Beijing for the 2008 Summer Olympics. Its square base has an area of 31,329 sq meters. Find the length of a side of this building. (*Source:* ARUP East Asia)

△ **52.** The Washington Monument has a square base whose area is approximately 3039 square feet. Find the length of a side. (*Source: The World Almanac*)

Solve. See Example 7. For Exercises 53 through 58, use the formula from $h = 16t^2$ and round answers to the nearest tenth of a second. (Recall that this formula does not take into account any air resistance.)

53. If a sandblaster drops his goggles from a bridge 400 feet from the water below, find how long it takes for the goggles to hit the water.

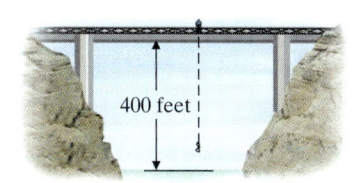

54. In 1988, Eddie Turner saved Frank Fanan, who became unconscious after an injury while jumping out of an airplane. Fanan fell 11,136 feet before Turner pulled his ripcord. Determine the time of Fanan's unconscious free-fall.

55. The highest regularly performed dives are made by professional divers from La Quebrada, in Acapulco, Mexico. The performer dives head first from a height of 115 feet (into 12 feet of water). Determine the time of a dive. (*Source: The Guinness Book of Records*)

56. In 1962, Vesna Vulovic, a flight attendant from Yugoslavia, survived a fall from 33,300 feet when the DC-9 airplane in which she was traveling blew up. She fell still strapped into her flight attendant's seat in the tail section of the plane. To the nearest tenth of a second, determine the time it took her to reach the ground. (*Source: Aviation Security*)

57. In March 2007, the Hualapai Indian Tribe allowed the Grand Canyon Skywalk to be built over the rim of the Grand Canyon on its tribal land. The skywalk extends 70 feet beyond the canyon's edge and is 4000 feet above the canyon floor. Determine the time, to the nearest tenth of a second, it would take an object, dropped off the skywalk, to land at the bottom of the Grand Canyon. (*Source: Boston Globe;* 03/21/07)

△ **58.** In 1997, stuntman Stig Gunther of Denmark jumped from a height of 343 feet off a crane onto an airbag. Determine the time of Gunther's stunt fall. (*Source: Guinness Book of World Records*)

△ **59.** The area of a circle is found by the equation $A = \pi r^2$. If the area A of a certain circle is 36π square inches, find its radius r.

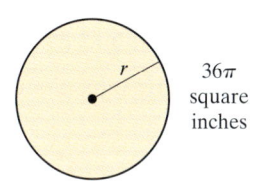

60. If the area of the circle below is 10π square units, find its exact radius. (See Exercise 59.)

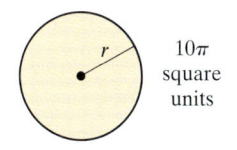

REVIEW AND PREVIEW

Factor each perfect square trinomial. See Section 6.3.

61. $x^2 + 6x + 9$ **62.** $y^2 + 10y + 25$
63. $x^2 - 4x + 4$ **64.** $x^2 - 20x + 100$

CONCEPT EXTENSIONS

Solve each quadratic equation by first factoring the perfect square trinomial on the left side. Then apply the square root property.

65. $x^2 + 4x + 4 = 16$ **66.** $z^2 - 6z + 9 = 25$
67. $x^2 + 14x + 49 = 31$ **68.** $y^2 - 10y + 25 = 11$

For Exercises 69 through 72, solve each quadratic equation by using the square root property. If necessary, use a calculator and round each solution to the nearest hundredth.

69. $x^2 = 1.78$ **70.** $(x - 1.37)^2 = 5.71$

71. The number y of CVS stores open for business from 2003 through 2006 is given by the equation $y = -120(x - 4)^2 + 6200$, where $x = 0$ represents the year 2003. Assume that this trend continues and find the year after 2006 in which there will be 6080 stores open. (*Hint*: Replace y with 6080 and solve for x.) (*Source:* Based on Data from CVS Corporation)

72. U.S. soybean production y (in billion bushels) from 2004 through 2006 is given by the equation $y = 0.09(x - 1)^2 + 3.06$, where $x = 0$ represents the year 2004. Assume that this trend continues and find the year in which there will be 4.5 billion bushels. (*Hint:* Replace y with 4.5 and solve for x.) (*Source:* Based on Data from U.S. Department of Agriculture)

73. The number of cattle y (in thousands) on farms in Kansas from 2003 through 2006 is approximated by the equation $y = 75x^2 + 6400$, where $x = 0$ represents the year 2003. Assume that this trend continues and find the year in which there are 7600 cattle. (*Hint:* Replace y with 7600 and solve for x.) (*Source:* Based on Data from U.S. Department of Agriculture)

STUDY SKILLS BUILDER

Are You Preparing for Your Final Exam?

To prepare for your final exam, try the following study techniques:

- Review the material that you will be responsible for on your exam. This includes material from your textbook, your notebook, and any handouts from your instructor.
- Review any formulas that you may need to memorize.
- Check to see if your instructor or mathematics department will be conducting a final exam review.
- Check with your instructor to see whether final exams from previous semesters/quarters are available to students for review.
- Use your previously taken exams as a practice final exam. To do so, rewrite the test questions in mixed order on blank sheets of paper. This will help you prepare for exam conditions.
- If you are unsure of a few concepts, see your instructor or visit a learning lab for assistance. Also, view the video segment of any troublesome sections.
- If you need further exercises to work, try the Cumulative Reviews at the end of the chapters.

Once again, good luck! I hope you are enjoying this textbook and your mathematics course.

9.2 SOLVING QUADRATIC EQUATIONS BY COMPLETING THE SQUARE

OBJECTIVES

1. Write perfect square trinomials.
2. Solve quadratic equations of the form $x^2 + bx + c = 0$ by completing the square.
3. Solve quadratic equations of the form $ax^2 + bx + c = 0$ by completing the square.

OBJECTIVE 1 ▶ Writing perfect square trinomials. In the last section, we used the square root property to solve equations such as

$$(x + 1)^2 = 8 \quad \text{and} \quad (5x - 2)^2 = 3$$

Notice that one side of each equation is a quantity squared and that the other side is a constant. To solve

$$x^2 + 2x = 4$$

Section 9.2 Solving Quadratic Equations by Completing the Square 563

notice that if we add 1 to both sides of the equation, the left side is a perfect square trinomial that can be factored.

$$x^2 + 2x + 1 = 4 + 1 \quad \text{Add 1 to both sides.}$$
$$(x + 1)^2 = 5 \quad \text{Factor.}$$

Now we can solve this equation as we did in the previous section by using the square root property.

$$x + 1 = \sqrt{5} \quad \text{or} \quad x + 1 = -\sqrt{5} \quad \text{Use the square root property.}$$
$$x = -1 + \sqrt{5} \quad x = -1 - \sqrt{5} \quad \text{Solve.}$$

The solutions are $-1 \pm \sqrt{5}$.

Adding a number to $x^2 + 2x$ to form a perfect square trinomial is called **completing the square** on $x^2 + 2x$.

In general, we have the following.

Completing the Square

To complete the square on $x^2 + bx$, add $\left(\dfrac{b}{2}\right)^2$. To find $\left(\dfrac{b}{2}\right)^2$, **find half the coefficient of x, then square the result.**

EXAMPLE 1 Complete the square for each expression and then factor the resulting perfect square trinomial.

a. $x^2 + 10x$ **b.** $m^2 - 6m$ **c.** $x^2 + x$

Solution

a. The coefficient of the x-term is 10. Half of 10 is 5, and $5^2 = 25$. Add 25.

$$x^2 + 10x + 25 = (x + 5)^2$$

b. Half the coefficient of m is -3, and $(-3)^2$ is 9. Add 9.

$$m^2 - 6m + 9 = (m - 3)^2$$

c. Half the coefficient of x is $\dfrac{1}{2}$ and $\left(\dfrac{1}{2}\right)^2 = \dfrac{1}{4}$. Add $\dfrac{1}{4}$.

$$x^2 + x + \dfrac{1}{4} = \left(x + \dfrac{1}{2}\right)^2$$

PRACTICE 1 Complete the square for each expression and then factor the resulting perfect square trinomial.

a. $z^2 + 8z$ **b.** $x^2 - 12x$ **c.** $b^2 + 5b$

OBJECTIVE 2 ▶ Completing the square to solve $x^2 + bx + c = 0$. By completing the square, a quadratic equation can be solved using the square root property.

EXAMPLE 2 Solve $x^2 + 6x + 3 = 0$ by completing the square.

Solution First we get the variable terms alone by subtracting 3 from both sides of the equation.

$$x^2 + 6x + 3 = 0$$
$$x^2 + 6x = -3 \quad \text{Subtract 3 from both sides.}$$

Next we find half the coefficient of the x-term, then square it. Add this result to **both sides** of the equation. This will make the left side a perfect square trinomial. The coefficient of x is 6, and half of 6 is 3. So we add 3^2 or 9 to both sides.

$$x^2 + 6x + 9 = -3 + 9 \quad \text{Complete the square.}$$
$$(x + 3)^2 = 6 \quad \text{Factor the trinomial } x^2 + 6x + 9.$$
$$x + 3 = \sqrt{6} \quad \text{or} \quad x + 3 = -\sqrt{6} \quad \text{Use the square root property.}$$
$$x = -3 + \sqrt{6} \quad\quad x = -3 - \sqrt{6} \quad \text{Subtract 3 from both sides.}$$

Check by substituting $-3 + \sqrt{6}$ and $-3 - \sqrt{6}$ in the original equation. The solutions are $-3 \pm \sqrt{6}$.

PRACTICE 2 Solve $x^2 + 2x - 5 = 0$ by completing the square.

> **Helpful Hint**
> Remember, when solving an equation by completing the square, add the number that completes the square to **both sides of the equation.**

EXAMPLE 3 Solve $x^2 - 10x = -14$ by completing the square.

Solution The variable terms are already alone on one side of the equation. The coefficient of x is -10. Half of -10 is -5, and $(-5)^2 = 25$. So we add 25 to both sides.

> **Helpful Hint**
> Add 25 to *both* sides of the equation.

$$x^2 - 10x = -14$$
$$x^2 - 10x + 25 = -14 + 25$$
$$(x - 5)^2 = 11 \quad \text{Factor the trinomial and simplify } -14 + 25.$$
$$x - 5 = \sqrt{11} \quad \text{or} \quad x - 5 = -\sqrt{11} \quad \text{Use the square root property.}$$
$$x = 5 + \sqrt{11} \quad\quad x = 5 - \sqrt{11} \quad \text{Add 5 to both sides.}$$

The solutions are $5 \pm \sqrt{11}$.

PRACTICE 3 Solve $x^2 - 8x = -8$ by completing the square.

OBJECTIVE 3 ▶ Completing the square to solve $ax^2 + bx + c = 0$. The method of completing the square can be used to solve *any* quadratic equation whether the coefficient of the squared variable is 1 or not. When the coefficient of the squared variable is not 1, we first divide both sides of the equation by the coefficient of the squared variable so that the new coefficient is 1. Then we complete the square.

EXAMPLE 4 Solve $4x^2 - 8x - 5 = 0$ by completing the square.

Solution
$$4x^2 - 8x - 5 = 0$$
$$x^2 - 2x - \frac{5}{4} = 0 \quad \text{Divide both sides by 4.}$$
$$x^2 - 2x = \frac{5}{4} \quad \text{Get the variable terms alone on one side of the equation.}$$

The coefficient of x is -2. Half of -2 is -1, and $(-1)^2 = 1$. So we add 1 to both sides.

$$x^2 - 2x + 1 = \frac{5}{4} + 1$$
$$(x - 1)^2 = \frac{9}{4} \quad \text{Factor } x^2 - 2x + 1 \text{ and simplify } \frac{5}{4} + 1.$$

$$x - 1 = \sqrt{\frac{9}{4}} \quad \text{or} \quad x - 1 = -\sqrt{\frac{9}{4}} \quad \text{Use the square root property.}$$

$$x = 1 + \frac{3}{2} \qquad\qquad x = 1 - \frac{3}{2} \quad \text{Add 1 to both sides and simplify the radical.}$$

$$x = \frac{5}{2} \qquad\qquad x = -\frac{1}{2} \quad \text{Simplify.}$$

Both $\frac{5}{2}$ and $-\frac{1}{2}$ are solutions.

PRACTICE 4 Solve $9x^2 - 36x - 13 = 0$ by completing the square.

The following steps may be used to solve a quadratic equation in x by completing the square.

> **Solving a Quadratic Equation in x by Completing the Square**
> **STEP 1.** If the coefficient of x^2 is 1, go to Step 2. If not, divide both sides of the equation by the coefficient of x^2.
> **STEP 2.** Get all terms with variables on one side of the equation and constants on the other side.
> **STEP 3.** Find half the coefficient of x and then square the result. Add this number to both sides of the equation.
> **STEP 4.** Factor the resulting perfect square trinomial.
> **STEP 5.** Use the square root property to solve the equation.

EXAMPLE 5 Solve $2x^2 + 6x = -7$ by completing the square.

Solution The coefficient of x^2 is not 1. We divide both sides by 2, the coefficient of x^2.

$$2x^2 + 6x = -7$$

$$x^2 + 3x = -\frac{7}{2} \qquad \text{Divide both sides by 2.}$$

$$x^2 + 3x + \frac{9}{4} = -\frac{7}{2} + \frac{9}{4} \qquad \text{Add } \left(\frac{3}{2}\right)^2 \text{ or } \frac{9}{4} \text{ to both sides.}$$

$$\left(x + \frac{3}{2}\right)^2 = -\frac{5}{4} \qquad \text{Factor the left side and simplify the right.}$$

There is no real solution to this equation since the square root of a negative number is not a real number.

PRACTICE 5 Solve $2x^2 + 12x = -20$ by completing the square.

EXAMPLE 6 Solve $2x^2 = 10x + 1$ by completing the square.

Solution First we divide both sides of the equation by 2, the coefficient of x^2.

$$2x^2 = 10x + 1$$

$$x^2 = 5x + \frac{1}{2} \qquad \text{Divide both sides by 2.}$$

Next we get the variable terms alone by subtracting $5x$ from both sides.

$$x^2 - 5x = \frac{1}{2}$$

$$x^2 - 5x + \frac{25}{4} = \frac{1}{2} + \frac{25}{4} \qquad \text{Add } \left(-\frac{5}{2}\right)^2 \text{ or } \frac{25}{4} \text{ to both sides.}$$

$$\left(x - \frac{5}{2}\right)^2 = \frac{27}{4} \qquad \text{Factor the left side and simplify the right side.}$$

$$x - \frac{5}{2} = \sqrt{\frac{27}{4}} \quad \text{or} \quad x - \frac{5}{2} = -\sqrt{\frac{27}{4}} \qquad \text{Use the square root property.}$$

$$x - \frac{5}{2} = \frac{3\sqrt{3}}{2} \qquad x - \frac{5}{2} = -\frac{3\sqrt{3}}{2} \qquad \text{Simplify.}$$

$$x = \frac{5}{2} + \frac{3\sqrt{3}}{2} \qquad x = \frac{5}{2} - \frac{3\sqrt{3}}{2}$$

The solutions are $\dfrac{5 \pm 3\sqrt{3}}{2}$.

PRACTICE 6 Solve $2x^2 = 6x - 3$ by completing the square.

VOCABULARY & READINESS CHECK

Use the choices below to fill in each blank. Not all choices will be used, and these exercises come from Sections 9.1 and 9.2.

| $\sqrt{a}$ | linear equation | zero | $\left(\frac{b}{2}\right)^2$ | $\frac{b}{2}$ | 6 |
| $\pm\sqrt{a}$ | quadratic equation | one | completing the square | 9 | 3 |

1. By the zero factor property, if the product of two numbers is zero, then at least one of these two numbers must be _____.
2. If a is a positive number, and if $x^2 = a$, then $x = $ _____.
3. An equation that can be written in the form $ax^2 + bx + c = 0$ where $a, b,$ and c are real numbers and a is not zero is called a(n) _____.
4. The process of solving a quadratic equation by writing it in the form $(x + a)^2 = c$ is called _____.
5. To complete the square on $x^2 + 6x$, add _____.
6. To complete the square on $x^2 + bx$, add _____.

Fill in the blank with the number needed to make each expression a perfect square trinomial. See Example 1.

7. $p^2 + 8p + $ _____
8. $p^2 + 6p + $ _____
9. $x^2 + 20x + $ _____
10. $x^2 + 18x + $ _____
11. $y^2 + 14y + $ _____
12. $y^2 + 2y + $ _____

9.2 EXERCISE SET

Complete the square for each expression and then factor the resulting perfect square trinomial. See Example 1.

1. $x^2 + 4x$
2. $x^2 + 6x$
3. $k^2 - 12k$
4. $k^2 - 16k$
5. $x^2 - 3x$
6. $x^2 - 5x$
7. $m^2 - m$
8. $y^2 + y$

Solve each quadratic equation by completing the square. See Examples 2 and 3.

9. $x^2 + 8x = -12$
10. $x^2 - 10x = -24$
11. $x^2 + 2x - 7 = 0$
12. $z^2 + 6z - 9 = 0$
13. $x^2 - 6x = 0$
14. $y^2 + 4y = 0$

15. $z^2 + 5z = 7$

16. $x^2 - 7x = 5$

17. $x^2 - 2x - 1 = 0$

18. $x^2 - 4x + 2 = 0$

19. $y^2 + 5y + 4 = 0$

20. $y^2 - 5y + 6 = 0$

Solve each quadratic equation by completing the square. See Examples 4 through 6.

21. $3x^2 - 6x = 24$

22. $2x^2 + 18x = -40$

23. $5x^2 + 10x + 6 = 0$

24. $3x^2 - 12x + 14 = 0$

25. $2x^2 = 6x + 5$

26. $4x^2 = -20x + 3$

27. $2y^2 + 8y + 5 = 0$

28. $4z^2 - 8z + 1 = 0$

MIXED PRACTICE

Solve each quadratic equation by completing the square. See Examples 1 through 6.

29. $x^2 + 6x - 25 = 0$

30. $x^2 - 6x + 7 = 0$

31. $x^2 - 3x - 3 = 0$

32. $x^2 - 9x + 3 = 0$

33. $2y^2 - 3y + 1 = 0$

34. $2y^2 - y - 1 = 0$

35. $x(x + 3) = 18$

36. $x(x - 3) = 18$

37. $3z^2 + 6z + 4 = 0$

38. $2y^2 + 8y + 9 = 0$

39. $4x^2 + 16x = 48$

40. $6x^2 - 30x = -36$

REVIEW AND PREVIEW

Simplify each expression. See Section 8.3.

41. $\dfrac{3}{4} - \sqrt{\dfrac{25}{16}}$

42. $\dfrac{3}{5} + \sqrt{\dfrac{16}{25}}$

43. $\dfrac{1}{2} - \sqrt{\dfrac{9}{4}}$

44. $\dfrac{9}{10} - \sqrt{\dfrac{49}{100}}$

Simplify each expression. See Section 8.4.

45. $\dfrac{6 + 4\sqrt{5}}{2}$

46. $\dfrac{10 - 20\sqrt{3}}{2}$

47. $\dfrac{3 - 9\sqrt{2}}{6}$

48. $\dfrac{12 - 8\sqrt{7}}{16}$

CONCEPT EXTENSIONS

49. In your own words, describe a perfect square trinomial.

50. Describe how to find the number to add to $x^2 - 7x$ to make a perfect square trinomial.

51. Write your own quadratic equation to be solved by completing the square. Write it in the form

perfect square trinomial = a number that is not a perfect square

$$x^2 + 6x + 9 = 11$$

For example,

a. Solve $x^2 + 6x + 9 = 11$.

b. Write and solve your quadratic equation by completing the square.

52. Follow the directions of Exercise 51, except write your equation in the form

perfect square trinomial = negative number

Solve your quadratic equation by completing the square.

53. Find a value of k that will make $x^2 + kx + 16$ a perfect square trinomial.

54. Find a value of k that will make $x^2 + kx + 25$ a perfect square trinomial.

55. Retail sales y (in millions of dollars) for bookstores in the United States from 2003 through 2005 can be represented by the equation $y = 250x^2 - 750x + 7800$. In this equation x is the number of years after 2002. Assume that this trend continues and predict the years after 2002 in which the retail sales for U.S. bookstores will be $8800 million. (*Source: Based on data from the Statistical Abstract of the United States*)

56. The average price of gold y (in dollars per ounce) from 2000 through 2006 is given by the equation $y = 10x^2 - 6x + 280$. Assume that this trend continues and find the year after 2000 in which the price of gold will be $1036 per ounce. (*Source: Based on data from U.S. Geological survey, Minerals Information*)

Recall that a graphing calculator may be used to solve an equation. For example, to solve $x^2 + 8x = -12$ (Exercise 1), graph

$y_1 = x^2 + 8x$ *(left side of equation) and*

$y_2 = -12$ *(right side of equation)*

The x-coordinate of the point of intersection of the graphs is the solution. Use a graphing calculator and solve each equation. Round solutions to the nearest hundredth.

57. Exercise 9

58. Exercise 10

59. Exercise 25

60. Exercise 26

9.3 SOLVING QUADRATIC EQUATIONS BY THE QUADRATIC FORMULA

OBJECTIVES

1. Use the quadratic formula to solve quadratic equations.
2. Approximate solutions to quadratic equations.
3. Determine the number of solutions of a quadratic equation by using the discriminant.

OBJECTIVE 1 ▶ Using the quadratic formula. We can use the technique of completing the square to develop a formula to find solutions of any quadratic equation. We develop and use the **quadratic formula** in this section.

Recall that a quadratic equation in **standard form** is

$$ax^2 + bx + c = 0, \quad a \neq 0$$

To develop the quadratic formula, let's complete the square for this quadratic equation in standard form.

First we divide both sides of the equation by the coefficient of x^2 and then get the variable terms alone on one side of the equation.

$$x^2 + \frac{b}{a}x + \frac{c}{a} = 0 \qquad \text{Divide by } a; \text{ recall that } a \text{ cannot be 0.}$$

$$x^2 + \frac{b}{a}x = -\frac{c}{a} \qquad \text{Get the variable terms alone on one side of the equation.}$$

The coefficient of x is $\frac{b}{a}$. Half of $\frac{b}{a}$ is $\frac{b}{2a}$ and $\left(\frac{b}{2a}\right)^2 = \frac{b^2}{4a^2}$. So we add $\frac{b^2}{4a^2}$ to both sides of the equation.

$$x^2 + \frac{b}{a}x + \frac{b^2}{4a^2} = -\frac{c}{a} + \frac{b^2}{4a^2} \qquad \text{Add } \frac{b^2}{4a^2} \text{ to both sides.}$$

$$\left(x + \frac{b}{2a}\right)^2 = -\frac{c}{a} + \frac{b^2}{4a^2} \qquad \text{Factor the left side.}$$

$$\left(x + \frac{b}{2a}\right)^2 = -\frac{4ac}{4a^2} + \frac{b^2}{4a^2} \qquad \text{Multiply } -\frac{c}{a} \text{ by } \frac{4a}{4a} \text{ so that both terms on the right side have a common denominator.}$$

$$\left(x + \frac{b}{2a}\right)^2 = \frac{b^2 - 4ac}{4a^2} \qquad \text{Simplify the right side.}$$

Now we use the square root property.

$$x + \frac{b}{2a} = \sqrt{\frac{b^2 - 4ac}{4a^2}} \quad \text{or} \quad x + \frac{b}{2a} = -\sqrt{\frac{b^2 - 4ac}{4a^2}} \qquad \text{Use the square root property.}$$

$$x + \frac{b}{2a} = \frac{\sqrt{b^2 - 4ac}}{2a} \qquad x + \frac{b}{2a} = -\frac{\sqrt{b^2 - 4ac}}{2a} \qquad \text{Simplify the radical.}$$

$$x = -\frac{b}{2a} + \frac{\sqrt{b^2 - 4ac}}{2a} \qquad x = -\frac{b}{2a} - \frac{\sqrt{b^2 - 4ac}}{2a} \qquad \text{Subtract } \frac{b}{2a} \text{ from both sides.}$$

$$x = \frac{-b + \sqrt{b^2 - 4ac}}{2a} \qquad x = \frac{-b - \sqrt{b^2 - 4ac}}{2a} \qquad \text{Simplify.}$$

The solutions are $\dfrac{-b \pm \sqrt{b^2 - 4ac}}{2a}$. This final equation is called the **quadratic formula** and gives the solutions of any quadratic equation.

Quadratic Formula

If a, b, and c are real numbers and $a \neq 0$, a quadratic equation written in the form $ax^2 + bx + c = 0$ has solutions

$$x = \frac{-b \pm \sqrt{b^2 - 4ac}}{2a}$$

Section 9.3 Solving Quadratic Equations by the Quadratic Formula

> **Helpful Hint**
> Don't forget that to correctly identify a, b, and c in the quadratic formula, you should write the equation in standard form.
>
> **Quadratic Equations in Standard Form**
> $5x^2 - 6x + 2 = 0 \quad a = 5, b = -6, c = 2$
> $4y^2 - 9 = 0 \quad a = 4, b = 0, c = -9$
> $x^2 + x = 0 \quad a = 1, b = 1, c = 0$
> $\sqrt{2}x^2 + \sqrt{5}x + \sqrt{3} = 0 \quad a = \sqrt{2}, b = \sqrt{5}, c = \sqrt{3}$

EXAMPLE 1 Solve $3x^2 + x - 3 = 0$ using the quadratic formula.

Solution This equation is in standard form with $a = 3$, $b = 1$, and $c = -3$. By the quadratic formula, we have

$$x = \frac{-b \pm \sqrt{b^2 - 4ac}}{2a}$$

$$x = \frac{-1 \pm \sqrt{1^2 - 4 \cdot 3 \cdot (-3)}}{2 \cdot 3} \quad \text{Let } a = 3, b = 1, \text{ and } c = -3.$$

$$= \frac{-1 \pm \sqrt{1 + 36}}{6} \quad \text{Simplify.}$$

$$= \frac{-1 \pm \sqrt{37}}{6}$$

Check both solutions in the original equation. The solutions are $\dfrac{-1 + \sqrt{37}}{6}$ and $\dfrac{-1 - \sqrt{37}}{6}$.

PRACTICE 1 Solve $5x^2 + x - 2 = 0$ using the quadratic formula.

EXAMPLE 2 Solve $2x^2 - 9x = 5$ using the quadratic formula.

Solution First we write the equation in standard form by subtracting 5 from both sides.
$$2x^2 - 9x = 5$$
$$2x^2 - 9x - 5 = 0$$

Next we note that $a = 2$, $b = -9$, and $c = -5$. We substitute these values into the quadratic formula.

> **Helpful Hint**
> Notice that the fraction bar is under the entire numerator of $-b \pm \sqrt{b^2 - 4ac}$.

$$x = \frac{-b \pm \sqrt{b^2 - 4ac}}{2a}$$

$$x = \frac{-(-9) \pm \sqrt{(-9)^2 - 4 \cdot 2 \cdot (-5)}}{2 \cdot 2} \quad \text{Substitute in the formula.}$$

$$= \frac{9 \pm \sqrt{81 + 40}}{4} \quad \text{Simplify.}$$

$$= \frac{9 \pm \sqrt{121}}{4} = \frac{9 \pm 11}{4}$$

Then,
$$x = \frac{9 - 11}{4} = -\frac{1}{2} \quad \text{or} \quad x = \frac{9 + 11}{4} = 5$$

Check $-\dfrac{1}{2}$ and 5 in the original equation. Both $-\dfrac{1}{2}$ and 5 are solutions.

PRACTICE 2 Solve $3x^2 + 2x = 8$ using the quadratic formula.

The following steps may be useful when solving a quadratic equation by the quadratic formula.

> **Solving a Quadratic Equation by the Quadratic Formula**
> **STEP 1.** Write the quadratic equation in standard form: $ax^2 + bx + c = 0$.
> **STEP 2.** If necessary, clear the equation of fractions to simplify calculations.
> **STEP 3.** Identify a, b, and c.
> **STEP 4.** Replace a, b, and c in the quadratic formula with the identified values, and simplify.

Concept Check ✓

For the quadratic equation $2x^2 - 5 = 7x$, if $a = 2$ and $c = -5$ in the quadratic formula, the value of b is which of the following?

a. $\dfrac{7}{2}$ b. 7   c. -5 d. -7

EXAMPLE 3   Solve $7x^2 = 1$ using the quadratic formula.

Solution   First we write the equation in standard form by subtracting 1 from both sides.

$$7x^2 = 1$$
$$7x^2 - 1 = 0$$

> **Helpful Hint**
> $7x^2 - 1 = 0$ can be written as $7x^2 + 0x - 1 = 0$. This form helps you see that $b = 0$.

Next we replace a, b, and c with the identified values: $a = 7$, $b = 0$ and $c = -1$.

$$x = \frac{0 \pm \sqrt{0^2 - 4 \cdot 7 \cdot (-1)}}{2 \cdot 7} \quad \text{Substitute in the formula.}$$

$$= \frac{\pm\sqrt{28}}{14} \quad \text{Simplify.}$$

$$= \frac{\pm 2\sqrt{7}}{14}$$

$$= \pm\frac{\sqrt{7}}{7}$$

The solutions are $\dfrac{\sqrt{7}}{7}$ and $-\dfrac{\sqrt{7}}{7}$.

PRACTICE 3   Solve $3x^2 = 5$ using the quadratic formula.

Notice that the equation in Example 3, $7x^2 = 1$, could have been easily solved by dividing both sides by 7 and then using the square root property. We solved the equation by the quadratic formula to show that this formula can be used to solve any quadratic equation.

Answer to Concept Check: d

EXAMPLE 4 Solve $x^2 = -x - 1$ using the quadratic formula.

Solution First we write the equation in standard form.

$$x^2 + x + 1 = 0$$

Next we replace a, b, and c in the quadratic formula with $a = 1$, $b = 1$, and $c = 1$.

$$x = \frac{-1 \pm \sqrt{1^2 - 4 \cdot 1 \cdot 1}}{2 \cdot 1} \quad \text{Substitute in the formula.}$$

$$= \frac{-1 \pm \sqrt{-3}}{2} \quad \text{Simplify.}$$

There is no real number solution because $\sqrt{-3}$ is not a real number.

PRACTICE 4 Solve $x^2 = 3x - 4$ using the quadratic formula.

EXAMPLE 5 Solve $\frac{1}{2}x^2 - x = 2$ by using the quadratic formula.

Solution We write the equation in standard form and then clear the equation of fractions by multiplying both sides by the LCD, 2.

$$\frac{1}{2}x^2 - x = 2$$

$$\frac{1}{2}x^2 - x - 2 = 0 \quad \text{Write in standard form.}$$

$$x^2 - 2x - 4 = 0 \quad \text{Multiply both sides by 2.}$$

Here, $a = 1$, $b = -2$, and $c = -4$, so we substitute these values into the quadratic formula.

$$x = \frac{-(-2) \pm \sqrt{(-2)^2 - 4 \cdot 1 \cdot (-4)}}{2 \cdot 1}$$

$$= \frac{2 \pm \sqrt{20}}{2} = \frac{2 \pm 2\sqrt{5}}{2} \quad \text{Simplify.}$$

$$= \frac{2(1 \pm \sqrt{5})}{2} = 1 \pm \sqrt{5} \quad \text{Factor and simplify.}$$

The solutions are $1 - \sqrt{5}$ and $1 + \sqrt{5}$.

PRACTICE 5 Solve $\frac{1}{5}x^2 - x = 1$ using the quadratic formula.

Notice that in Example 5, although we cleared the equation of fractions, the coefficients $a = \frac{1}{2}$, $b = -1$, and $c = -2$ will give the same results.

> **Helpful Hint**
> When simplifying an expression such as
> $$\frac{3 \pm 6\sqrt{2}}{6}$$
> first factor out a common factor from the terms of the numerator and then simplify.
> $$\frac{3 \pm 6\sqrt{2}}{6} = \frac{3(1 \pm 2\sqrt{2})}{2 \cdot 3} = \frac{1 \pm 2\sqrt{2}}{2}$$

OBJECTIVE 2 ▶ **Approximate solutions to quadratic equations.** Sometimes approximate solutions for quadratic equations are appropriate.

EXAMPLE 6 Approximate the exact solutions of the quadratic equation in Example 1. Round the approximations to the nearest tenth.

Solution From Example 1, we have exact solutions $\frac{-1 \pm \sqrt{37}}{6}$. Thus,

$$\frac{-1 + \sqrt{37}}{6} \approx 0.847127088 \approx 0.8 \text{ to the nearest tenth.}$$

$$\frac{-1 - \sqrt{37}}{6} \approx -1.180460422 \approx -1.2 \text{ to the nearest tenth.}$$

Thus approximate solutions to the quadratic equation in Example 1 are 0.8 and −1.2. □

PRACTICE 6 Approximate the exact solutions of the quadratic equation in Practice 1. Round the approximations to the nearest tenth.

OBJECTIVE 3 ▶ **Using the discriminant.** In the quadratic formula, $x = \frac{-b \pm \sqrt{b^2 - 4ac}}{2a}$, the radicand $b^2 - 4ac$ is called the **discriminant** because, by knowing its value, we can **discriminate** among the possible number and type of solutions of a quadratic equation. Possible values of the discriminant and their meanings are summarized next.

Discriminant

The following table corresponds the discriminant $b^2 - 4ac$ of a quadratic equation of the form $ax^2 + bx + c = 0$ with the number of solutions of the equation.

$b^2 - 4ac$	Number of Solutions
Positive	Two distinct real solutions
Zero	One real solution
Negative	No real solution*

*In this case, the quadratic equation will have two complex (but not real) solutions. See Section 9.4 for a discussion of complex numbers.

EXAMPLE 7 Use the discriminant to determine the number of solutions of $3x^2 + x - 3 = 0$.

Solution In $3x^2 + x - 3 = 0$, $a = 3$, $b = 1$, and $c = -3$. Then

$$b^2 - 4ac = (1)^2 - 4(3)(-3) = 1 + 36 = 37$$

Since the discriminant is 37, a positive number, this equation has two distinct real solutions.

We solved this equation in Example 1 of this section, and the solutions are $\dfrac{-1 + \sqrt{37}}{6}$ and $\dfrac{-1 - \sqrt{37}}{6}$, two distinct real solutions.

PRACTICE 7 Use the discriminant to determine the number of solutions of $5x^2 + x - 2 = 0$.

EXAMPLE 8 Use the discriminant to determine the number of solutions of each quadratic equation.

a. $x^2 - 6x + 9 = 0$ **b.** $5x^2 + 4 = 0$

Solution

a. In $x^2 - 6x + 9 = 0$, $a = 1$, $b = -6$, and $c = 9$.

$$b^2 - 4ac = (-6)^2 - 4(1)(9) = 36 - 36 = 0$$

Since the discriminant is 0, this equation has one real solution.

b. In $5x^2 + 4 = 0$, $a = 5$, $b = 0$, and $c = 4$.

$$b^2 - 4ac = 0^2 - 4(5)(4) = 0 - 80 = -80$$

Since the discriminant is -80, a negative number, this equation has no real solution.

PRACTICE 8 Use the discriminant to determine the number of solutions of each quadratic equation.

a. $x^2 - 10x + 35 = 0$ **b.** $5x^2 + 3x = 0$

VOCABULARY & READINESS CHECK

Fill in each blank.

1. The quadratic formula is _____ .

Identify the values of a, b, and c in each quadratic equation.

2. $5x^2 - 7x + 1 = 0$; $a = \_\_$, $b = \_\_\_$, $c = \_\_$
3. $x^2 + 3x - 7 = 0$; $a = \_\_$, $b = \_\_$, $c = \_\_\_$
4. $x^2 - 6 = 0$; $a = \_\_$, $b = \_\_$, $c = \_\_\_$
5. $x^2 + x - 1 = 0$; $a = \_\_\_$, $b = \_\_$, $c = \_\_\_$
6. $9x^2 - 4 = 0$; $a = \_\_\_$, $b = \_\_$, $c = \_\_\_$

9.3 EXERCISE SET

Simplify the following.

1. $\dfrac{-1 \pm \sqrt{1^2 - 4(1)(-2)}}{2(1)}$

2. $\dfrac{-(-5) \pm \sqrt{(-5)^2 - 4(2)(3)}}{2(2)}$

3. $\dfrac{-5 \pm \sqrt{5^2 - 4(1)(2)}}{2(1)}$

4. $\dfrac{-7 \pm \sqrt{7^2 - 4(2)(1)}}{2(2)}$

5. $\dfrac{-(-4) \pm \sqrt{(-4)^2 - 4(2)(1)}}{2(2)}$

6. $\dfrac{-6 \pm \sqrt{6^2 - 4(3)(1)}}{2(3)}$

Use the quadratic formula to solve each quadratic equation. See Examples 1 through 4.

7. $x^2 - 3x + 2 = 0$
8. $x^2 - 5x - 6 = 0$
9. $3k^2 + 7k + 1 = 0$
10. $7k^2 + 3k - 1 = 0$
11. $49x^2 - 4 = 0$
12. $25x^2 - 15 = 0$
13. $5z^2 - 4z + 3 = 0$
14. $3z^2 + 2z + 1 = 0$
15. $y^2 = 7y + 30$
16. $y^2 = 5y + 36$
17. $2x^2 = 10$
18. $5x^2 = 15$
19. $m^2 - 12 = m$
20. $m^2 - 14 = 5m$
21. $3 - x^2 = 4x$
22. $10 - x^2 = 2x$
23. $2a^2 - 7a + 3 = 0$
24. $3a^2 - 7a + 2 = 0$
25. $x^2 - 5x - 2 = 0$
26. $x^2 - 2x - 5 = 0$
27. $3x^2 - x - 14 = 0$
28. $5x^2 - 13x - 6 = 0$
29. $6x^2 + 9x = 2$
30. $3x^2 - 9x = 8$
31. $7p^2 + 2 = 8p$
32. $11p^2 + 2 = 10p$
33. $a^2 - 6a + 2 = 0$
34. $a^2 - 10a + 19 = 0$
35. $2x^2 - 6x + 3 = 0$
36. $5x^2 - 8x + 2 = 0$
37. $3x^2 = 1 - 2x$
38. $5y^2 = 4 - y$
39. $20y^2 = 3 - 11y$
40. $2z^2 = z + 3$
41. $x^2 + x + 1 = 0$
42. $k^2 + 2k + 5 = 0$
43. $4y^2 = 6y + 1$
44. $6z^2 + 3z + 2 = 0$

Use the quadratic formula to solve each quadratic equation. See Example 5.

45. $3p^2 - \dfrac{2}{3}p + 1 = 0$
46. $\dfrac{5}{2}p^2 - p + \dfrac{1}{2} = 0$

47. $\dfrac{m^2}{2} = m + \dfrac{1}{2}$
48. $\dfrac{m^2}{2} = 3m - 1$
49. $4p^2 + \dfrac{3}{2} = -5p$
50. $4p^2 + \dfrac{3}{2} = 5p$
51. $5x^2 = \dfrac{7}{2}x + 1$
52. $2x^2 = \dfrac{5}{2}x + \dfrac{7}{2}$
53. $28x^2 + 5x + \dfrac{11}{4} = 0$
54. $\dfrac{2}{3}x^2 - 2x - \dfrac{2}{3} = 0$
55. $5z^2 - 2z = \dfrac{1}{5}$
56. $9z^2 + 12z = -1$
57. $x^2 + 3\sqrt{2}x - 5 = 0$
58. $y^2 - 2\sqrt{5}y - 1 = 0$

MIXED PRACTICE

Use the quadratic formula to solve each quadratic equation. Find the exact solutions; then approximate these solutions to the nearest tenth. See Examples 1 through 6.

59. $3x^2 = 21$
60. $2x^2 = 26$
61. $x^2 + 6x + 1 = 0$
62. $x^2 + 4x + 2 = 0$
63. $x^2 = 9x + 4$
64. $x^2 = 7x + 5$
65. $3x^2 - 2x - 2 = 0$
66. $5x^2 - 3x - 1 = 0$

Use the discriminant to determine the number of solutions of each quadratic equation. See Examples 7 and 8.

67. $x^2 + 3x - 1 = 0$
68. $x^2 - 5x - 3 = 0$
69. $3x^2 + x + 5 = 0$
70. $2x^2 + x + 4 = 0$
71. $4x^2 + 4x = -1$
72. $7x^2 - x = 0$
73. $9x^2 + 2x = 0$
74. $x^2 + 10x = -25$
75. $5x^2 + 1 = 0$
76. $4x^2 + 9 = 12x$
77. $x^2 + 36 = -12x$
78. $10x^2 + 2 = 0$

REVIEW AND PREVIEW

Simplify each radical. See Section 8.2.

79. $\sqrt{48}$
80. $\sqrt{104}$
81. $\sqrt{50}$
82. $\sqrt{80}$

Solve the following. See Section 2.6.

△ **83.** The height of a triangle is 4 times the length of the base. The area of the triangle is 18 square feet. Find the height and base of the triangle.

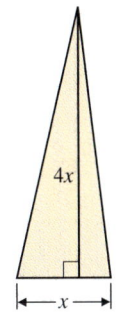

△ **84.** The length of a rectangle is 6 inches more than its width. The area of the rectangle is 391 square inches. Find the dimensions of the rectangle.

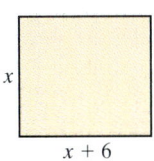

CONCEPT EXTENSIONS

Solve. See the Concept Check in this section. For the quadratic equation $5x^2 + 2 = x$, if $a = 5$,

85. What is the value of b?
 a. $\frac{1}{5}$ b. 0   c. -1 d. 1

86. What is the value of c?
 a. 5   b. x c. -2 d. 2

For the quadratic equation $7y^2 = 3y$, if $b = 3$,

87. What is the value of a?
 a. 7   b. -7 c. 0   d. 1

88. What is the value of c?
 a. 7   b. 3   c. 0   d. 1

△ **89.** The largest chocolate bar was a 5026-lb scaled-up model of a Novi chocolate bar, made by the Elah-Dufour United Food Company in 2000. The bar had a base area of 50.8 square feet and its length was 0.5 feet longer than twice its width. Find the length and the width of the bar, rounded to one decimal place. (*Source: Guinness Book of World Records*, 2005)

△ **90.** The area of a rectangular conference room table is 95 square feet. If its length is six feet longer than its width, find the dimensions of the table. Round each dimension to the nearest tenth.

91. $1.2x^2 - 5.2x - 3.9 = 0$

92. $7.3z^2 + 5.4z - 1.1 = 0$

A rocket is launched from the top of an 80-foot cliff with an initial velocity of 120 feet per second. The height, h, of the rocket after t seconds is given by the equation $h = -16t^2 + 120t + 80$. Use this for Exercises 93 and 94.

93. How long after the rocket is launched will it be 30 feet from the ground? Round to the nearest tenth of a second.

94. How long after the rocket is launched will it strike the ground? Round to the nearest tenth of a second. (*Hint:* The rocket will strike the ground when its height $h = 0$.)

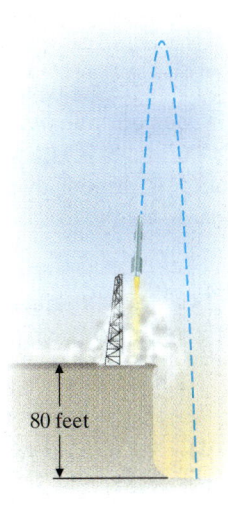

95. Explain how the quadratic formula is developed and why it is useful.

96. The gross profit y (in millions of dollars) of eBay from 2004 through 2006 is given by the equation $y = -50x^2 + 1128x + 2656$, where $x = 0$ represents 2004. Assume that this trend continues and predict the first year in which eBay's gross profit will be $6368 million. (*Source:* Based on data from eBay)

97. The number of yearly kidney transplants from 1996 to 2005 can be modeled by the equation $y = 3.6x^2 + 578x + 13{,}538$, where $x = 0$ represents the year 2000. Assume that this trend continues and predict the year in which 23,018 kidney transplants will be performed.

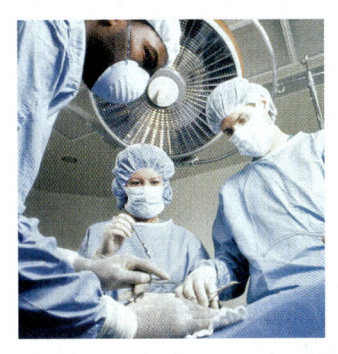

98. The number of patients on the kidney waiting list from 1996 to 2005 can be modeled by the equation $y = -181.5x^2 + 2202.2x + 40{,}000$, where $x = 0$ represents the year 2000. Assume that this trend continues and predict the year in which 43,872 patients will be on the waiting list.

99. The average annual salary y for Microsoft information technology (IT) professionals for the years 2003 through 2006 is given by the equation $y = -1100x^2 + 11{,}800x + 46{,}769$, where $x = 0$ represents 2003. Assume that this trend continues and predict the year in which the average Microsoft IT professional salary will be $78,269. (*Source:* Microsoft Corporation)

100. The number of Target stores y operating in the United States from 2003 through 2006 is given by the equation $y = 6x^2 + 75x + 1225$, where x is the number of years after 2003. Assume that this trend continues and predict the year after 2003 in which the number of Target stores will be 1891. (*Source:* Target Corporation)

THE BIGGER PICTURE   SIMPLIFYING EXPRESSIONS AND SOLVING EQUATIONS

Now we continue our outline from Sections 1.7, 2.9, 5.6, 6.6, 7.4, 7.6, and 8.5. Although suggestions are given, this outline should be in your own words. Once you complete this new portion, try the exercises below.

I. Simplifying Expressions
   A. **Real Numbers**
      1. Add (Section 1.5)
      2. Subtract (Section 1.6)
      3. Multiply or Divide (Section 1.7)
   B. **Exponents** (Section 5.1)
   C. **Polynomials**
      1. Add (Section 5.2)
      2. Subtract (Section 5.2)
      3. Multiply (Section 5.3)
      4. Divide (Section 5.6)
   D. **Factoring Polynomials** (Chapter 6 Integrated Review)
   E. **Rational Expressions**
      1. Simplify (Section 7.1)
      2. Multiply (Section 7.2)
      3. Divide (Section 7.2)
      4. Add or Subtract (Section 7.4)
   F. **Radicals**
      1. Simplify (Section 8.2)
      2. Add or Subtract (Section 8.3)
      3. Multiply or Divide (Section 8.4)
      4. Rationalize the denominator (Section 8.4)

II. Solving Equations and Inequalities
   A. **Linear Equations** (Section 2.4)
   B. **Linear Inequalities** (Section 2.9)
   C. **Quadratic and Higher Degree Equations** (Sections 6.6 and 9.3)
      1. If in the form $x^2 = a$, solve by the Square Root Property. If not, write the equation in standard form (one side is 0).
      2. If the polynomial on one side factors, solve by factoring.
      3. If the polynomial does not factor, solve by the quadratic formula.
   D. **Equations with Rational Expressions** (Section 7.5)
   E. **Proportions** (Section 7.6)
   F. **Equations with Radicals** (Section 8.5)

Perform indicated operations and simplify.

1. $7.9 - 9.7$
2. $5 + (-3) + (-7)$
3. $(-4)^2 - 5^2$
4. $7x - 2 + \dfrac{1}{3}(9x - 3) + 5$
5. $\left(\dfrac{1}{2}x + 5\right)\left(\dfrac{1}{2}x - 5\right)$
6. $\dfrac{9x^2y + 3xy - 12y}{3xy}$
7. $\dfrac{x^2}{(x-5)(x-4)} - \dfrac{3x + 10}{(x-5)(x-4)}$
8. $\dfrac{x}{x - 10} + \dfrac{5}{x + 3}$
9. $\sqrt{50}$
10. $\dfrac{\sqrt{30a^2b^3}}{\sqrt{3ab}}$
11. $\sqrt{\dfrac{2}{3}}$
12. $\dfrac{7x - 14}{x^2 - 4} \cdot \dfrac{x^2 + 5x + 6}{49}$

Solve.

13. $x^2 + 3x - 5 = 0$
14. $x^2 + x = x^2 + 6$
15. $-2x \le 5.6$
16. $2x^2 + 15x = 8$
17. $\sqrt{x + 2} + 4 = x$
18. $\dfrac{5}{x} - \dfrac{3}{x - 4} = \dfrac{7 + x}{x(x - 4)}$

By Factoring	By Square Root Property	By Quadratic Formula
$x^2 + x = 6$	$9x^2 = 2$	$x^2 + x = 5$
$x^2 + x - 6 = 0$	$x^2 = \dfrac{2}{9}$	$x^2 + x - 5 = 0$
$(x - 2)(x + 3) = 0$		$a = 1, b = 1, c = -5$
$x - 2 = 0$ or $x + 3 = 0$	$x = \pm\sqrt{\dfrac{2}{9}} = \dfrac{\pm 2}{3}$	$x = \dfrac{-1 \pm \sqrt{1^2 - 4(1)(-5)}}{2 \cdot 1}$
$x = 2$ or $x = -3$		$x = \dfrac{-1 \pm \sqrt{21}}{2}$

INTEGRATED REVIEW SUMMARY ON SOLVING QUADRATIC EQUATIONS

Sections 9.1–9.3

An important skill in mathematics is learning when to use one technique in favor of another. We now practice this by deciding which method to use when solving quadratic equations. Although both the quadratic formula and completing the square can be used to solve any quadratic equation, the quadratic formula is usually less tedious and thus preferred. The following steps may be used to solve a quadratic equation.

> **To Solve a Quadratic Equation**
>
> **STEP 1.** If the equation is in the form $ax^2 = c$ or $(ax + b)^2 = c$, use the square root property and solve. If not, go to Step 2.
>
> **STEP 2.** Write the equation in standard form: $ax^2 + bx + c = 0$.
>
> **STEP 3.** Try to solve the equation by the factoring method. If not possible, go to Step 4.
>
> **STEP 4.** Solve the equation by the quadratic formula.

Study the examples below to help you review these steps.

EXAMPLE 1 Solve $m^2 - 2m - 7 = 0$.

Solution The equation is in standard form, but the quadratic expression $m^2 - 2m - 7$ is not factorable, so use the quadratic formula with $a = 1$, $b = -2$, and $c = -7$.

$$m^2 - 2m - 7 = 0$$

$$m = \frac{-(-2) \pm \sqrt{(-2)^2 - 4 \cdot 1 \cdot (-7)}}{2 \cdot 1} = \frac{2 \pm \sqrt{32}}{2}$$

$$m = \frac{2 \pm 4\sqrt{2}}{2} = \frac{2(1 \pm 2\sqrt{2})}{2} = 1 \pm 2\sqrt{2}$$

The solutions are $1 - 2\sqrt{2}$ and $1 + 2\sqrt{2}$.

PRACTICE 1 Solve $y^2 - 3y - 4 = 0$.

EXAMPLE 2 Solve $(3x + 1)^2 = 20$.

Solution This equation is in a form that makes the square root property easy to apply.

$$(3x + 1)^2 = 20$$

$$3x + 1 = \pm\sqrt{20} \quad \text{Apply the square root property.}$$

$$3x + 1 = \pm 2\sqrt{5} \quad \text{Simplify } \sqrt{20}.$$

$$3x = -1 \pm 2\sqrt{5}$$

$$x = \frac{-1 \pm 2\sqrt{5}}{3}$$

The solutions are $\dfrac{-1 - 2\sqrt{5}}{3}$ and $\dfrac{-1 + 2\sqrt{5}}{3}$.

PRACTICE 2 Solve $(2x + 5)^2 = 45$.

EXAMPLE 3 Solve $x^2 - \dfrac{11}{2}x = -\dfrac{5}{2}$.

Solution The fractions make factoring more difficult and also complicate the calculations for using the quadratic formula. Clear the equation of fractions by multiplying both sides of the equation by the LCD 2.

$$x^2 - \frac{11}{2}x = -\frac{5}{2}$$

$$x^2 - \frac{11}{2}x + \frac{5}{2} = 0 \qquad \text{Write in standard form.}$$

$$2x^2 - 11x + 5 = 0 \qquad \text{Multiply both sides by 2.}$$

$$(2x - 1)(x - 5) = 0 \qquad \text{Factor.}$$

$$2x - 1 = 0 \quad \text{or} \quad x - 5 = 0 \qquad \text{Apply the zero factor theorem.}$$

$$2x = 1 \qquad\qquad x = 5$$

$$x = \frac{1}{2} \qquad\qquad x = 5$$

The solutions are $\dfrac{1}{2}$ and 5.

PRACTICE 3 Solve $x^2 - \dfrac{5}{2}x = -\dfrac{3}{2}$.

Choose and use a method to solve each equation.

1. $5x^2 - 11x + 2 = 0$
2. $5x^2 + 13x - 6 = 0$
3. $x^2 - 1 = 2x$
4. $x^2 + 7 = 6x$
5. $a^2 = 20$
6. $a^2 = 72$
7. $x^2 - x + 4 = 0$
8. $x^2 - 2x + 7 = 0$
9. $3x^2 - 12x + 12 = 0$
10. $5x^2 - 30x + 45 = 0$
11. $9 - 6p + p^2 = 0$
12. $49 - 28p + 4p^2 = 0$
13. $4y^2 - 16 = 0$
14. $3y^2 - 27 = 0$
15. $x^4 - 3x^3 + 2x^2 = 0$
16. $x^3 + 7x^2 + 12x = 0$
17. $(2z + 5)^2 = 25$
18. $(3z - 4)^2 = 16$
19. $30x = 25x^2 + 2$
20. $12x = 4x^2 + 4$
21. $\dfrac{2}{3}m^2 - \dfrac{1}{3}m - 1 = 0$
22. $\dfrac{5}{8}m^2 + m - \dfrac{1}{2} = 0$
23. $x^2 - \dfrac{1}{2}x - \dfrac{1}{5} = 0$
24. $x^2 + \dfrac{1}{2}x - \dfrac{1}{8} = 0$
25. $4x^2 - 27x + 35 = 0$
26. $9x^2 - 16x + 7 = 0$
27. $(7 - 5x)^2 = 18$
28. $(5 - 4x)^2 = 75$
29. $3z^2 - 7z = 12$
30. $6z^2 + 7z = 6$
31. $x = x^2 - 110$
32. $x = 56 - x^2$
33. $\dfrac{3}{4}x^2 - \dfrac{5}{2}x - 2 = 0$
34. $x^2 - \dfrac{6}{5}x - \dfrac{8}{5} = 0$
35. $x^2 - 0.6x + 0.05 = 0$
36. $x^2 - 0.1x - 0.06 = 0$
37. $10x^2 - 11x + 2 = 0$
38. $20x^2 - 11x + 1 = 0$
39. $\dfrac{1}{2}z^2 - 2z + \dfrac{3}{4} = 0$
40. $\dfrac{1}{5}z^2 - \dfrac{1}{2}z - 2 = 0$

41. Explain how you will decide what method to use when solving quadratic equations.

9.4 COMPLEX SOLUTIONS OF QUADRATIC EQUATIONS

OBJECTIVES

1. Write complex numbers using i notation.
2. Add and subtract complex numbers.
3. Multiply complex numbers.
4. Divide complex numbers.
5. Solve quadratic equations that have complex solutions.

In Chapter 8, we learned that $\sqrt{-4}$, for example, is not a real number because there is no real number whose square is -4. However, our real number system can be extended to include numbers like $\sqrt{-4}$. This extended number system is called the **complex number** system. The complex number system includes the **imaginary unit i,** which is defined next.

Imaginary Unit i

The imaginary unit, written i, is the number whose square is -1. That is,
$$i^2 = -1 \quad \text{and} \quad i = \sqrt{-1}$$

OBJECTIVE 1 ▶ **Writing complex numbers using i notation.** We use i to write numbers like $\sqrt{-6}$ as the product of a real number and i. Since $i = \sqrt{-1}$, we have
$$\sqrt{-6} = \sqrt{-1 \cdot 6} = \sqrt{-1} \cdot \sqrt{6} = i\sqrt{6}$$

EXAMPLE 1 Write each radical as the product of a real number and i.

a. $\sqrt{-4}$ b. $\sqrt{-11}$ c. $\sqrt{-20}$

Solution Write each negative radicand as a product of a positive number and -1. Then write $\sqrt{-1}$ as i.

a. $\sqrt{-4} = \sqrt{-1 \cdot 4} = \sqrt{-1} \cdot \sqrt{4} = i \cdot 2 = 2i$
b. $\sqrt{-11} = \sqrt{-1 \cdot 11} = \sqrt{-1} \cdot \sqrt{11} = i\sqrt{11}$
c. $\sqrt{-20} = \sqrt{-1 \cdot 20} = \sqrt{-1} \cdot \sqrt{20} = i \cdot 2\sqrt{5} = 2i\sqrt{5}$

PRACTICE 1 Write each radical as the product of a real number and i.

a. $\sqrt{-36}$ b. $\sqrt{-15}$ c. $\sqrt{-48}$

The numbers $2i$, $i\sqrt{11}$, and $2i\sqrt{5}$ are called **pure imaginary numbers.** Both real numbers and pure imaginary numbers are complex numbers.

Complex Numbers and Pure Imaginary Numbers

A complex number is a number that can be written in the form
$$a + bi$$
where a and b are real numbers. A complex number that can be written in the form
$$0 + bi$$
$b \neq 0$, is also called a pure imaginary number.

A complex number written in the form $a + bi$ is in **standard form.** We call a the real part and bi the imaginary part of the complex number $a + bi$.

EXAMPLE 2 Identify each number as a complex number by writing it in standard form $a + bi$.

a. 7 b. 0 c. $\sqrt{20}$ d. $\sqrt{-27}$ e. $2 + \sqrt{-4}$

Solution

a. 7 is a complex number since $7 = 7 + 0i$.
b. 0 is a complex number since $0 = 0 + 0i$.
c. $\sqrt{20}$ is a complex number since $\sqrt{20} = 2\sqrt{5} = 2\sqrt{5} + 0i$.
d. $\sqrt{-27}$ is a complex number since $\sqrt{-27} = i \cdot 3\sqrt{3} = 0 + 3i\sqrt{3}$.
e. $2 + \sqrt{-4}$ is a complex number since $2 + \sqrt{-4} = 2 + 2i$. □

PRACTICE 2 Identify each number as a complex number by writing it in standard form $a + bi$.

a. 6 b. 0 c. $\sqrt{24}$
d. $\sqrt{-1}$ e. $5 + \sqrt{-9}$

OBJECTIVE 2 ▶ Adding and subtracting complex numbers. We now present arithmetic operations—addition, subtraction, multiplication, and division—for the complex number system. Complex numbers are added and subtracted in the same way as we add and subtract polynomials.

EXAMPLE 3 Simplify the sum or difference. Write the result in standard form.

a. $(2 + 3i) + (-6 - i)$ b. $-i + (3 + 7i)$ c. $(5 - i) - 4$

Solution Add the real parts and then add the imaginary parts.

a. $(2 + 3i) + (-6 - i) = [2 + (-6)] + (3i - i) = -4 + 2i$
b. $-i + (3 + 7i) = 3 + (-i + 7i) = 3 + 6i$
c. $(5 - i) - 4 = (5 - 4) - i = 1 - i$ □

PRACTICE 3 Simplify the sum or difference. Write the result in standard form.

a. $(4 + 3i) + (-8 - 2i)$ b. $(5i) + (6 - 9i)$
c. $(3 - 2i) - 4$

EXAMPLE 4 Subtract $(11 - i)$ from $(1 + i)$.

Solution $(1 + i) - (11 - i) = 1 + i - 11 + i = (1 - 11) + (i + i) = -10 + 2i$ □

PRACTICE 4 Subtract $(13 - 5i)$ from $(3i)$.

OBJECTIVE 3 ▶ Multiplying complex numbers. Use the distributive property and the FOIL method to multiply complex numbers.

EXAMPLE 5 Find the following products and write in standard form.

a. $5i(2 - i)$ b. $(7 - 3i)(4 + 2i)$ c. $(2 + 3i)(2 - 3i)$

Solution

a. By the distributive property, we have

$5i(2 - i) = 5i \cdot 2 - 5i \cdot i$ Apply the distributive property.
$= 10i - 5i^2$
$= 10i - 5(-1)$ Write i^2 as -1.
$= 10i + 5$
$= 5 + 10i$ Write in standard form.

$$\begin{array}{cccc} & \text{F} & \text{O} & \text{I} & \text{L} \end{array}$$

b. $(7 - 3i)(4 + 2i) = 28 + 14i - 12i - 6i^2$
$ = 28 + 2i - 6(-1)$ Write i^2 as -1.
$ = 28 + 2i + 6$
$ = 34 + 2i$

c. $(2 + 3i)(2 - 3i) = 4 - 6i + 6i - 9i^2$
$ = 4 - 9(-1)$ Write i^2 as -1.
$ = 13$

PRACTICE 5   Find the following products and write in standard form.

a. $2i(3 - 4i)$ **b.** $(3 + i)(2 - 3i)$ **c.** $(5 - 2i)(5 + 2i)$

The product in part (c) is the real number 13. Notice that one factor is the sum of 2 and $3i$, and the other factor is the difference of 2 and $3i$. When complex number factors are related as these two are, their product is a real number. In general,

$$(a + bi)(a - bi) = a^2 + b^2$$
     sum   difference   real number

The complex numbers $a + bi$ and $a - bi$ are called **complex conjugates** of each other. For example, $2 - 3i$ is the conjugate of $2 + 3i$, and $2 + 3i$ is the conjugate of $2 - 3i$. Also,

The conjugate of $3 - 10i$ is $3 + 10i$.
The conjugate of 5 is 5. (Note that $5 = 5 + 0i$ and its conjugate is $5 - 0i = 5$.)
The conjugate of $4i$ is $-4i$. ($0 - 4i$ is the conjugate of $0 + 4i$.)

OBJECTIVE 4 ▶ Dividing complex numbers. The fact that the product of a complex number and its conjugate is a real number provides a method for dividing by a complex number and for simplifying fractions whose denominators are complex numbers.

EXAMPLE 6   Write $\dfrac{4 + i}{3 - 4i}$ in standard form.

Solution   To write this quotient as a complex number in the standard form $a + bi$, we need to find an equivalent fraction whose denominator is a real number. By multiplying both numerator and denominator by the denominator's conjugate, we obtain a new fraction that is an equivalent fraction with a real number denominator.

$\dfrac{4 + i}{3 - 4i} = \dfrac{(4 + i)}{(3 - 4i)} \cdot \dfrac{(3 + 4i)}{(3 + 4i)}$ Multiply numerator and denominator by $3 + 4i$.

$\phantom{\dfrac{4 + i}{3 - 4i}} = \dfrac{12 + 16i + 3i + 4i^2}{9 - 16i^2}$

$\phantom{\dfrac{4 + i}{3 - 4i}} = \dfrac{12 + 19i + 4(-1)}{9 - 16(-1)}$ Recall that $i^2 = -1$.

$\phantom{\dfrac{4 + i}{3 - 4i}} = \dfrac{12 + 19i - 4}{9 + 16} = \dfrac{8 + 19i}{25}$

$\phantom{\dfrac{4 + i}{3 - 4i}} = \dfrac{8}{25} + \dfrac{19}{25}i$ Write in standard form.

Note that our last step was to write $\dfrac{4 + i}{3 - 4i}$ in standard form $a + bi$, where a and b are real numbers.

PRACTICE 6   Write $\dfrac{3 - i}{2 + 5i}$ in standard form.

OBJECTIVE 5 ▶ Solving quadratic equations with complex solutions. Some quadratic equations have complex solutions.

EXAMPLE 7 Solve $(x + 2)^2 = -25$.

Solution Begin by applying the square root property.

$$(x + 2)^2 = -25$$
$$x + 2 = \pm\sqrt{-25} \quad \text{Apply the square root property.}$$
$$x + 2 = \pm 5i \quad \text{Write } \sqrt{-25} \text{ as } 5i.$$
$$x = -2 \pm 5i$$

The solutions are $-2 + 5i$ and $-2 - 5i$.

PRACTICE 7 Solve $(x - 3)^2 = -16$.

EXAMPLE 8 Solve $m^2 = 4m - 5$.

Solution Write the equation in standard form and use the quadratic formula to solve.

$$m^2 = 4m - 5$$
$$m^2 - 4m + 5 = 0 \quad \text{Write the equation in standard form.}$$

Apply the quadratic formula with $a = 1, b = -4,$ and $c = 5$.

$$m = \frac{4 \pm \sqrt{16 - 4 \cdot 1 \cdot 5}}{2 \cdot 1}$$
$$= \frac{4 \pm \sqrt{-4}}{2}$$
$$= \frac{4 \pm 2i}{2} \quad \text{Write } \sqrt{-4} \text{ as } 2i.$$
$$= \frac{2(2 \pm i)}{2} = 2 \pm i$$

The solutions are $2 - i$ and $2 + i$.

PRACTICE 8 Solve $y^2 = 3y - 5$.

EXAMPLE 9 Solve $x^2 + x = -1$.

Solution
$$x^2 + x = -1$$
$$x^2 + x + 1 = 0 \quad \text{Write in standard form.}$$
$$x = \frac{-1 \pm \sqrt{1 - 4 \cdot 1 \cdot 1}}{2 \cdot 1} \quad \text{Apply the quadratic formula with } a = 1, b = 1, \text{ and } c = 1.$$
$$= \frac{-1 \pm \sqrt{-3}}{2}$$
$$= \frac{-1 \pm i\sqrt{3}}{2}$$

The solutions are $\dfrac{-1 - i\sqrt{3}}{2}$ and $\dfrac{-1 + i\sqrt{3}}{2}$.

PRACTICE 9 Solve $x^2 + x = -4$.

VOCABULARY & READINESS CHECK

Use the choices below to fill in each blank.

real   complex   standard
imaginary   conjugate

1. A number that can be written in the form $a + bi$ is called a(n) _____ number.
2. A complex number that can be written in the form $0 + bi$ is also called a(n) _____ number.
3. A complex number that can be written in the form $a + 0i$ is also called a(n) _____ number.
4. The _____ of $a + bi$ is $a - bi$.
5. The form $a + bi$ is called _____ form.

9.4 EXERCISE SET

Write each expression in i notation. See Example 1.

1. $\sqrt{-9}$
2. $\sqrt{-64}$
3. $\sqrt{-100}$
4. $\sqrt{-16}$
5. $\sqrt{-50}$
6. $\sqrt{-98}$
7. $\sqrt{-63}$
8. $\sqrt{-44}$

Add or subtract as indicated. See Examples 2 through 4.

9. $(2 - i) + (-5 + 10i)$
10. $(-7 + 2i) + (5 - 3i)$
11. $(-11 + 3i) - (1 - 3i)$
12. $(1 + i) - (-6 - 8i)$
13. $(3 - 4i) - (2 - i)$
14. $(-6 + i) - (3 + i)$
15. $(16 + 2i) + (-7 - 6i)$
16. $(-3 - i) + (-4 - i)$

Multiply. See Example 5.

17. $4i(3 - 2i)$
18. $-2i(5 + 4i)$
19. $(6 - 2i)(4 + i)$
20. $(6 + 2i)(4 - i)$
21. $(3 + 8i)(3 - 8i)$
22. $(-9 + 2i)(-9 - 2i)$

23. Earlier in this text, we learned that $\sqrt{-4}$ is not a real number. Explain what that means and explain what type of number $\sqrt{-4}$ is.

24. Describe how to find the conjugate of a complex number.

Divide. Write each answer in standard form. See Example 6.

25. $\dfrac{8 - 12i}{4}$
26. $\dfrac{14 + 28i}{-7}$
27. $\dfrac{7 - i}{4 - 3i}$
28. $\dfrac{4 - 3i}{7 - i}$

Solve the following quadratic equations for complex solutions. See Example 7.

29. $(x + 1)^2 = -9$
30. $(y - 2)^2 = -25$
31. $(2z - 3)^2 = -12$
32. $(3p + 5)^2 = -18$

Solve the following quadratic equations for complex solutions. See Examples 8 and 9.

33. $y^2 + 6y + 13 = 0$
34. $y^2 - 2y + 5 = 0$
35. $4x^2 + 7x + 4 = 0$
36. $8x^2 - 7x + 2 = 0$
37. $2m^2 - 4m + 5 = 0$
38. $5m^2 - 6m + 7 = 0$

MIXED PRACTICE

Perform the indicated operations. Write results in standard form.

39. $3 + (12 - 7i)$
40. $(-14 + 5i) + 3i$
41. $-9i(5i - 7)$
42. $10i(4i - 1)$
43. $(2 - i) - (3 - 4i)$
44. $(3 + i) - (-6 + i)$
45. $\dfrac{15 + 10i}{5i}$
46. $\dfrac{-18 + 12i}{-6i}$

47. Subtract $(2 + 3i)$ from $(-5 + i)$.
48. Subtract $(-8 - i)$ from $(7 - 4i)$.

49. $(4 - 3i)(4 + 3i)$
50. $(12 - 5i)(12 + 5i)$
51. $\dfrac{4 - i}{1 + 2i}$
52. $\dfrac{9 - 2i}{-3 + i}$
53. $(5 + 2i)^2$
54. $(9 - 7i)^2$

Solve the following quadratic equations for complex solutions.

55. $(y - 4)^2 = -64$
56. $(x + 7)^2 = -1$
57. $4x^2 = -100$
58. $7x^2 = -28$
59. $z^2 + 6z + 10 = 0$
60. $z^2 + 4z + 13 = 0$
61. $2a^2 - 5a + 9 = 0$
62. $4a^2 + 3a + 2 = 0$
63. $(2x + 8)^2 = -20$
64. $(6z - 4)^2 = -24$
65. $3m^2 + 108 = 0$
66. $5m^2 + 80 = 0$
67. $x^2 + 14x + 50 = 0$
68. $x^2 + 8x + 25 = 0$

584 CHAPTER 9 Quadratic Equations

REVIEW AND PREVIEW

Graph the following linear equations in two variables. See Section 3.2.

69. $y = -3$
70. $x = 4$
71. $y = 3x - 2$
72. $y = 2x + 3$

The line graph shows the percent of U.S. households with computers. Use this graph for Exercises 73 through 75. See Sections 3.5 and 7.2.

73. Estimate the percent of households with computers in 2000.

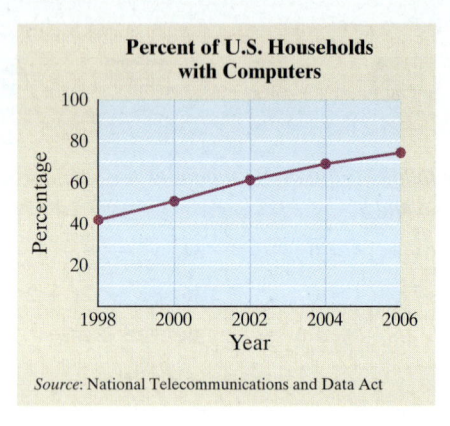

Percent of U.S. Households with Computers

Source: National Telecommunications and Data Act

74. The percent growth of computers in households was almost linear from 1998 to 2006. Approximate this growth with a linear equation. To do so, find an equation of the line through the ordered pairs (0, 42) and (8, 74.4), where x is the number of years since 1998 and y is the percent of households that have computers. Write the equation in slope-intercept form.

75. Use the equation found in Exercise 74 to predict the percent of U.S. households with computers in 2008.

CONCEPT EXTENSIONS

Answer the following true or false.

76. Every real number is a complex number.
77. Every complex number is a real number.
78. If a complex number such as $2 + 3i$ is a solution of a quadratic equation, then its conjugate $2 - 3i$ is also a solution.
79. Some pure imaginary numbers are real numbers.

9.5 GRAPHING QUADRATIC EQUATIONS

OBJECTIVES

1. Graph quadratic equations of the form $y = ax^2$.
2. Graph quadratic equations of the form $y = ax^2 + bx + c$.
3. Use the vertex formula to determine the vertex of a parabola.

OBJECTIVE 1 ▶ Graphing $y = ax^2$. Recall from Section 3.2 that the graph of a linear equation in two variables $Ax + By = C$ is a straight line. Also recall from Section 6.5 that the graph of a quadratic equation in two variables $y = ax^2 + bx + c$ is a parabola. In this section, we further investigate the graph of a quadratic equation.

To graph the quadratic equation $y = x^2$, select a few values for x and find the corresponding y-values. Make a table of values to keep track. Then plot the points corresponding to these solutions.

If $x = -3$, then $y = (-3)^2$, or 9.
If $x = -2$, then $y = (-2)^2$, or 4.
If $x = -1$, then $y = (-1)^2$, or 1.
If $x = 0$, then $y = 0^2$, or 0.
If $x = 1$, then $y = 1^2$, or 1.
If $x = 2$, then $y = 2^2$, or 4.
If $x = 3$, then $y = 3^2$, or 9.

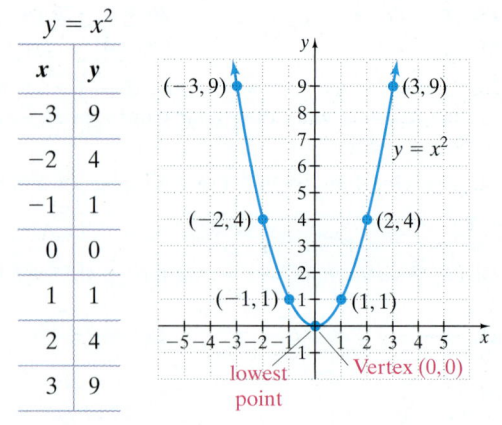

$y = x^2$

x	y
−3	9
−2	4
−1	1
0	0
1	1
2	4
3	9

Clearly, these points are not on one straight line. As we saw in Chapter 6, the graph of $y = x^2$ is a smooth curve through the plotted points. This curve is called a **parabola**. The lowest point on a parabola opening upward is called the **vertex**. The vertex is $(0, 0)$ for the parabola $y = x^2$. If we fold the graph paper along the y-axis, the two pieces of the parabola match perfectly. For this reason, we say the graph is **symmetric about the y-axis,** and we call the y-axis the **axis of symmetry.**

Notice that the parabola that corresponds to the equation $y = x^2$ opens upward. This happens when the coefficient of x^2 is positive. In the equation $y = x^2$, the coefficient of x^2 is 1. Example 1 shows the graph of a quadratic equation whose coefficient of x^2 is negative.

EXAMPLE 1 Graph $y = -2x^2$.

Solution Select x-values and calculate the corresponding y-values. Plot the ordered pairs found. Then draw a smooth curve through those points. When the coefficient of x^2 is negative, the corresponding parabola opens downward. When a parabola opens downward, the vertex is the highest point of the parabola. The vertex of this parabola is (0, 0) and the axis of symmetry is again the y-axis.

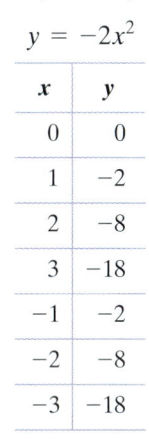

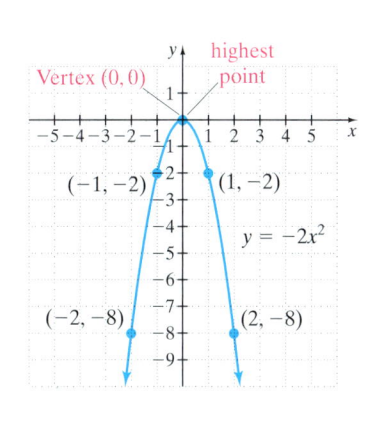

$y = -2x^2$

x	y
0	0
1	-2
2	-8
3	-18
-1	-2
-2	-8
-3	-18

PRACTICE
1 Graph $y = -\dfrac{1}{2}x^2$.

OBJECTIVE 2 ▶ **Graphing $y = ax^2 + bx + c$.** Just as for linear equations, we can use x- and y-intercepts to help graph quadratic equations. Recall from Chapter 3 that an x-intercept is the point where the graph intersects the x-axis. A y-intercept is the point where the graph intersects the y-axis. We find intercepts just as we did in Chapter 3.

▶ **Helpful Hint**
Recall that:
To find x-intercepts, let $y = 0$ and solve for x.
To find y-intercepts, let $x = 0$ and solve for y.

EXAMPLE 2 Graph $y = x^2 - 4$.

Solution First, find intercepts. To find the y-intercept, let $x = 0$. Then

$$y = 0^2 - 4 = -4$$

To find x-intercepts, we let $y = 0$.

$$0 = x^2 - 4$$
$$0 = (x - 2)(x + 2)$$
$$x - 2 = 0 \quad \text{or} \quad x + 2 = 0$$
$$x = 2 \qquad\qquad x = -2$$

Thus far, we have the *y*-intercept $(0, -4)$ and the *x*-intercepts $(2, 0)$ and $(-2, 0)$. Now we can select additional *x*-values, find the corresponding *y*-values, plot the points, and draw a smooth curve through the points.

$y = x^2 - 4$

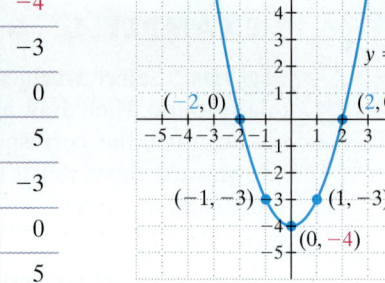

x	y
0	−4
1	−3
2	0
3	5
−1	−3
−2	0
−3	5

Notice that the vertex of this parabola is $(0, -4)$.

PRACTICE 2 Graph $y = x^2 + 1$.

Concept Check ✓
Tell whether the graph of each equation opens upward or downward.
a. $y = 2x^2$ **b.** $y = 3x^2 + 4x - 5$ **c.** $y = -5x^2 + 2$

> ▶ **Helpful Hint**
> For the graph of $y = ax^2 + bx + c$,
> If *a* is positive, the parabola opens upward.
> If *a* is negative, the parabola opens downward.

Concept Check ✓
For which of the following graphs of $y = ax^2 + bx + c$ would the value of *a* be negative?

a. **b.**

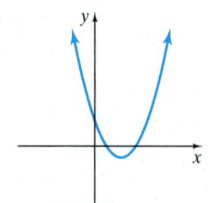

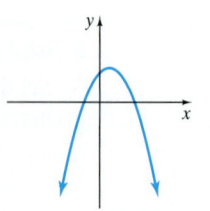

OBJECTIVE 3 ▶ Using the vertex formula. Thus far, we have accidentally stumbled upon the vertex of each parabola that we have graphed. However, our choice of values for *x* may not yield an ordered pair for the vertex of the parabola. It would be helpful if we could first find the vertex of a parabola, next determine whether the parabola opens upward or downward, and finally calculate additional points such as *x*- and *y*-intercepts as needed. In fact, there is a formula that may be used to find the vertex of a parabola.

One way to develop this formula is to notice that the *x*-value of the vertex of the parabolas that we are considering lies halfway between its *x*-intercepts. We can use this fact to find a formula for the vertex.

Answers to Concept Checks: **a.** upward **b.** upward **c.** downward; b

Recall that the x-intercepts of a parabola may be found by solving $0 = ax^2 + bx + c$. These solutions, by the quadratic formula, are

$$x = \frac{-b - \sqrt{b^2 - 4ac}}{2a}, \quad x = \frac{-b + \sqrt{b^2 - 4ac}}{2a}$$

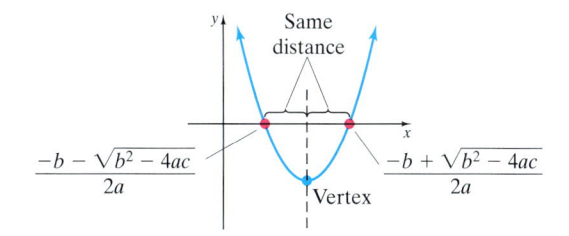

The x-coordinate of the vertex of a parabola is halfway between its x-intercepts, so the x-value of the vertex may be found by computing the average, or $\frac{1}{2}$ of the sum of the intercepts.

$$\begin{aligned} x &= \frac{1}{2}\left(\frac{-b - \sqrt{b^2 - 4ac}}{2a} + \frac{-b + \sqrt{b^2 - 4ac}}{2a}\right) \\ &= \frac{1}{2}\left(\frac{-b - \sqrt{b^2 - 4ac} - b + \sqrt{b^2 - 4ac}}{2a}\right) \\ &= \frac{1}{2}\left(\frac{-2b}{2a}\right) \\ &= \frac{-b}{2a} \end{aligned}$$

Vertex Formula

The vertex of the parabola $y = ax^2 + bx + c$ has x-coordinate

$$\frac{-b}{2a}$$

The corresponding y-coordinate of the vertex is found by substituting the x-coordinate into the equation and evaluating y.

EXAMPLE 3 Graph $y = x^2 - 6x + 8$.

Solution In the equation $y = x^2 - 6x + 8$, $a = 1$ and $b = -6$. The x-coordinate of the vertex is

$$x = \frac{-b}{2a} = \frac{-(-6)}{2 \cdot 1} = 3 \quad \text{Use the vertex formula, } x = \frac{-b}{2a}.$$

To find the corresponding y-coordinate, we let $x = 3$ in the original equation.

$$y = x^2 - 6x + 8 = 3^2 - 6 \cdot 3 + 8 = -1$$

The vertex is $(3, -1)$ and the parabola opens upward since a is positive. We now find and plot the intercepts.

To find the x-intercepts, we let $y = 0$.

$$0 = x^2 - 6x + 8$$

We factor the expression $x^2 - 6x + 8$ to find $(x - 4)(x - 2) = 0$. The x-intercepts are $(4, 0)$ and $(2, 0)$.

If we let $x = 0$ in the original equation, then $y = 8$ and the y-intercept is $(0, 8)$. Now we plot the vertex $(3, -1)$ and the intercepts $(4, 0)$, $(2, 0)$, and $(0, 8)$. Then we can sketch the parabola. These and two additional points are shown in the table.

$y = x^2 - 6x + 8$

x	y
3	−1
4	0
2	0
0	8
1	3
5	3
6	8

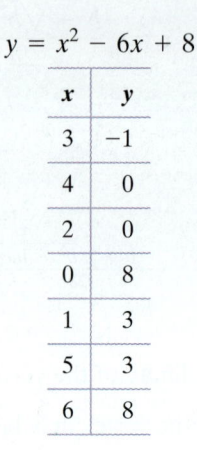

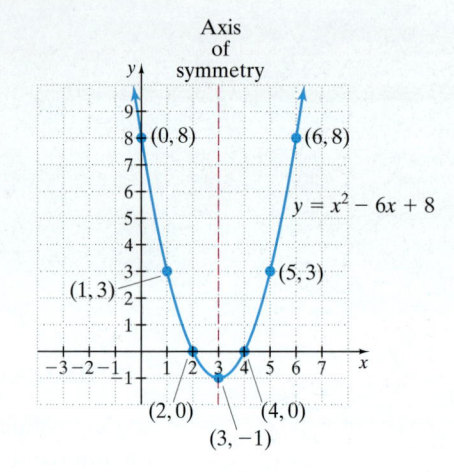

PRACTICE 3 Graph $y = x^2 - 3x - 4$.

Study Example 3 and let's use it to write down a general procedure for graphing quadratic equations.

Graphing Parabolas Defined by $y = ax^2 + bx + c$

1. **Find the vertex by using the formula $x = -\dfrac{b}{2a}$.** Don't forget to find the y-value of the vertex.
2. **Find the intercepts.**
   - Let $x = 0$ and solve for y to find the y-intercept. There will be only one.
   - Let $y = 0$ and solve for x to find any x-intercepts. There may be 0, 1, or 2.
3. **Plot the vertex and the intercepts.**
4. **Find and plot additional points on the graph.** Then draw a smooth curve through the plotted points. Keep in mind if $a > 0$, the parabola opens up and if $a < 0$, the parabola opens down.

EXAMPLE 4 Graph $y = x^2 + 2x - 5$.

Solution In the equation $y = x^2 + 2x - 5$, $a = 1$ and $b = 2$. Using the vertex formula, we find that the x-coordinate of the vertex is

$$x = \frac{-b}{2a} = \frac{-2}{2 \cdot 1} = -1$$

The y-coordinate of the vertex is

$$y = (-1)^2 + 2(-1) - 5 = -6$$

Thus the vertex is $(-1, -6)$.

To find the x-intercepts, we let $y = 0$.

$$0 = x^2 + 2x - 5$$

This cannot be solved by factoring, so we use the quadratic formula.

$$x = \frac{-2 \pm \sqrt{2^2 - 4(1)(-5)}}{2 \cdot 1} \qquad \text{Let } a = 1, b = 2, \text{ and } c = -5.$$

$$x = \frac{-2 \pm \sqrt{24}}{2}$$

$$x = \frac{-2 \pm 2\sqrt{6}}{2}$$ Simplify the radical.

$$x = \frac{2(-1 \pm \sqrt{6})}{2} = -1 \pm \sqrt{6}$$

The x-intercepts are $(-1 + \sqrt{6}, 0)$ and $(-1 - \sqrt{6}, 0)$. We use a calculator to approximate these so that we can easily graph these intercepts.

$$-1 + \sqrt{6} \approx 1.4 \quad \text{and} \quad -1 - \sqrt{6} \approx -3.4$$

To find the y-intercept, we let $x = 0$ in the original equation and find that $y = -5$. Thus the y-intercept is $(0, -5)$.

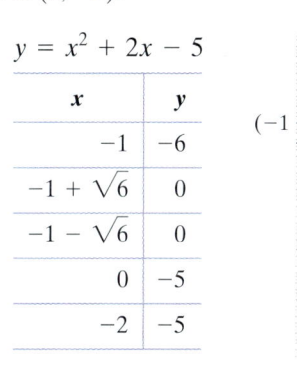

$y = x^2 + 2x - 5$

x	y
-1	-6
$-1 + \sqrt{6}$	0
$-1 - \sqrt{6}$	0
0	-5
-2	-5

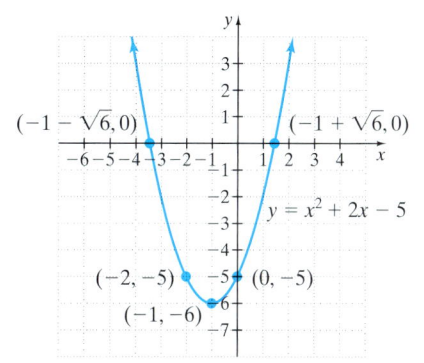

PRACTICE 4 Graph $y = x^2 + 4x - 7$.

> **Helpful Hint**
>
> Notice that the number of x-intercepts of the graph of the parabola $y = ax^2 + bx + c$ is the same as the number of real solutions of $0 = ax^2 + bx + c$. 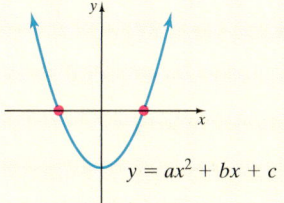
>
> Two x-intercepts
> Two real solutions of
> $0 = ax^2 + bx + c$ 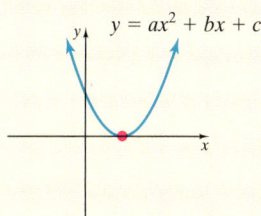
>
> One x-intercept
> One real solution of
> $0 = ax^2 + bx + c$ 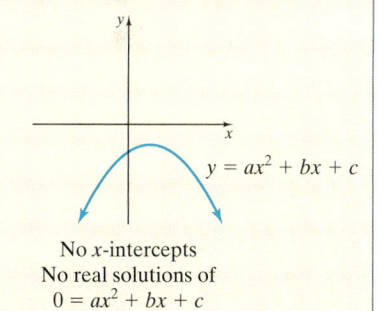
>
> No x-intercepts
> No real solutions of
> $0 = ax^2 + bx + c$

Graphing Calculator Explorations

Recall that a graphing calculator may be used to solve quadratic equations. The x-intercepts of the graph of $y = ax^2 + bx + c$ are solutions of $0 = ax^2 + bx + c$. To solve $x^2 - 7x - 3 = 0$, for example, graph $y_1 = x^2 - 7x - 3$. The x-intercepts of the graph are the solutions of the equation.

Use a graphing calculator to solve each quadratic equation. Round solutions to two decimal places.

1. $x^2 - 7x - 3 = 0$
2. $2x^2 - 11x - 1 = 0$
3. $-1.7x^2 + 5.6x - 3.7 = 0$
4. $-5.8x^2 + 2.3x - 3.9 = 0$
5. $5.8x^2 - 2.6x - 1.9 = 0$
6. $7.5x^2 - 3.7x - 1.1 = 0$

9.5 EXERCISE SET

Graph each quadratic equation by finding and plotting ordered pair solutions. See Example 1.

1. $y = 2x^2$
2. $y = 3x^2$
3. $y = -x^2$
4. $y = -4x^2$

Sketch the graph of each equation. Label the vertex and the intercepts. See Examples 2 through 4.

5. $y = x^2 - 1$
6. $y = x^2 - 16$
7. $y = x^2 + 4$
8. $y = x^2 + 9$
9. $y = -x^2 + 4x - 4$
10. $y = -x^2 - 2x - 1$
11. $y = x^2 + 5x + 4$
12. $y = x^2 + 7x + 10$
13. $y = x^2 - 4x + 5$
14. $y = x^2 - 6x + 10$
15. $y = 2 - x^2$
16. $y = 3 - x^2$

MIXED PRACTICE

Sketch the graph of each equation. Label the vertex and the intercepts. See Examples 1 through 4.

17. $y = \frac{1}{3}x^2$
18. $y = \frac{1}{2}x^2$
19. $y = x^2 + 6x$
20. $y = x^2 - 4x$
21. $y = x^2 + 2x - 8$
22. $y = x^2 - 2x - 3$
23. $y = -\frac{1}{2}x^2$
24. $y = -\frac{1}{3}x^2$
25. $y = 2x^2 - 11x + 5$
26. $y = 2x^2 + x - 3$
27. $y = -x^2 + 4x - 3$
28. $y = -x^2 + 6x - 8$
29. $y = x^2 + 2x - 2$
30. $y = x^2 - 4x - 3$
31. $y = x^2 - 3x + 1$
32. $y = x^2 - 2x - 5$

REVIEW AND PREVIEW

Simplify the following complex fractions. See Section 7.8.

33. $\dfrac{\frac{1}{7}\frac{2}{5}}{\frac{2}{5}}$

34. $\dfrac{\frac{3}{8}}{\frac{1}{7}}$

35. $\dfrac{\frac{1}{x}\frac{2}{x^2}}{\frac{2}{x^2}}$

36. $\dfrac{\frac{x}{5}}{\frac{2}{x}}$

37. $\dfrac{2x}{1 - \frac{1}{x}}$

38. $\dfrac{x}{x - \frac{1}{x}}$

39. $\dfrac{\frac{a-b}{2b}}{\frac{b-a}{8b^2}}$

40. $\dfrac{\frac{2a^2}{a-3}}{\frac{a}{3-a}}$

CONCEPT EXTENSIONS

The graph of a quadratic equation that takes the form $y = ax^2 + bx + c$ is the graph of a function. Write the domain and the range of each of the functions graphed.

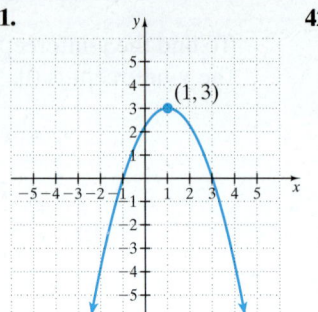

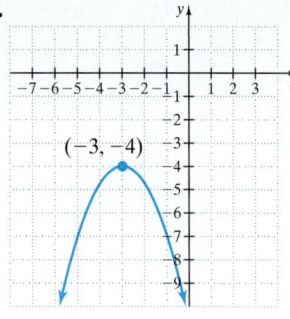

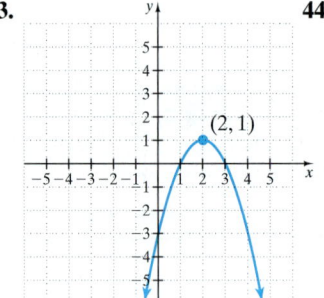

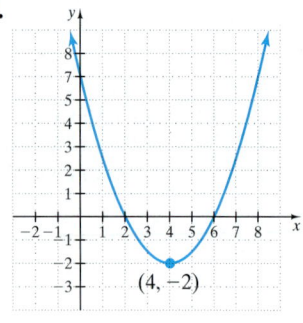

41. [graph with vertex (1, 3)]

42. [graph with vertex (-3, -4)]

43. [graph with vertex (2, 1)]

44. [graph with vertex (4, -2)]

45. The height h of a fireball launched from a Roman candle with an initial velocity of 128 feet per second is given by the equation

$$h = -16t^2 + 128t$$

where t is time in seconds after launch. Use the graph of this function to answer the questions.

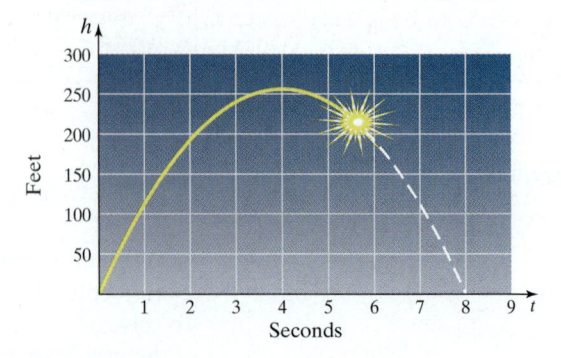

a. Estimate the maximum height of the fireball.

b. Estimate the time when the fireball is at its maximum height.

c. Estimate the time when the fireball returns to the ground.

46. Determine the maximum number and the minimum number of x-intercepts for a parabola. Explain your answer.

Match the values given with the correct graph of each quadratic equation of the form $y = a(x - h)^2 + k$.

47. $a > 0, h > 0, k > 0$ **48.** $a < 0, h > 0, k > 0$ **49.** $a > 0, h > 0, k < 0$ **50.** $a < 0, h > 0, k < 0$

A B 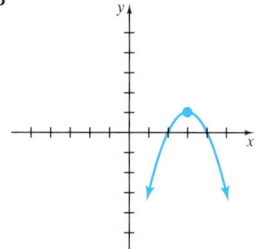 C 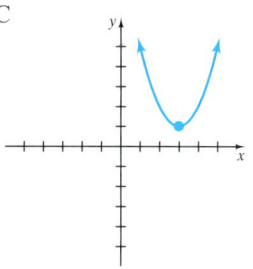 D

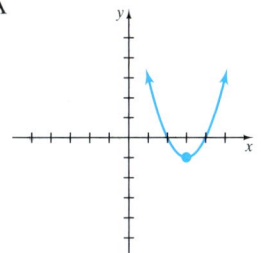

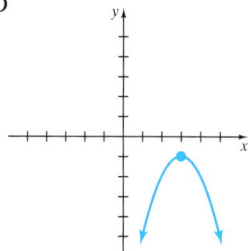

CHAPTER 9 GROUP ACTIVITY

Modeling a Physical Situation

When water comes out of a water fountain, it initially heads upward, but then gravity causes the water to fall. The curve formed by the stream of water can be modeled using a quadratic equation (parabola).

In this project, you will have the opportunity to model the parabolic path of water as it leaves a drinking fountain. This project may be completed by working in groups or individually.

1. Using the figure above, collect data for the x-intercepts of the parabolic path. Let points A and B in the figure be on the x-axis and let the coordinates of point A be $(0, 0)$. Use a ruler to measure the distance between points A and B **on the figure** to the nearest even one-tenth centimeter, and use this information to determine the coordinates of point B. Record this data in the data table. (*Hint:* If the distance from A to B measures 8 one-tenth centimeters, then the coordinates of point B are $(8, 0)$.)

2. Next, collect data for the vertex V of the parabolic path. What is the relationship between the x-coordinate of the vertex and the x-intercepts found in Question 1? What is the line of symmetry? To locate point V in the figure, find the midpoint of the line segment joining points A and B and mark point V on the path of water directly above the midpoint. To approximate the y-coordinate of the vertex, use a ruler to measure its distance from the x-axis to the nearest one-tenth centimeter. Record this data in the data table.

3. Plot the points from the data table on a rectangular coordinate system. Sketch the parabola through your points A, B, and V.

4. Which of the following models best fits the data you collected? Explain your reasoning.

   a. $y = 16x + 18$
   b. $y = -13x^2 + 20x$
   c. $y = 0.13x^2 - 2.6x$
   d. $y = -0.13x^2 + 2.6x$

5. (Optional) Enter your data into a graphing calculator and use the quadratic curve-fitting feature to find a model for your data. How does the model compare with your selection from Question 4?

Data Table

	x	y
Point A		
Point B		
Point V		

CHAPTER 9 VOCABULARY CHECK

Fill in each blank with one of the words listed below.

square root   complex   imaginary i
completing the square   quadratic   conjugate   vertex

1. If $x^2 = a$, then $x = \sqrt{a}$ or $x = -\sqrt{a}$. This property is called the _____ property.
2. A number that can be written in the form $a + bi$ is called a(n) _____ number.
3. The formula $\frac{-b}{2a}$ where $y = ax^2 + bx + c$ is called the _____ formula.
4. A complex number that can be written in the form $0 + bi$ is also called a(n) _____ number.
5. The _____ of $2 + 3i$ is $2 - 3i$.
6. $\sqrt{-1} = $ ___.
7. The process of solving a quadratic equation by writing it in the form $(x + a)^2 = c$ is called _____.
8. The formula $x = \frac{-b \pm \sqrt{b^2 - 4ac}}{2a}$ is called the _____ formula.

> **Helpful Hint**
>
> Are you preparing for your test? Don't forget to take the Chapter 9 Test on page 598. Then check your answers at the back of the text and use the Chapter Test Prep Video CD to see the fully worked-out solutions to any of the exercises you want to review.

CHAPTER 9 HIGHLIGHTS

DEFINITIONS AND CONCEPTS	EXAMPLES
SECTION 9.1 SOLVING QUADRATIC EQUATIONS BY THE SQUARE ROOT PROPERTY	
Square Root Property If $x^2 = a$ for $a \geq 0$, then $x = \pm\sqrt{a}$	*Solve the equation.* $(x - 1)^2 = 15$ $x - 1 = \pm\sqrt{15}$ $x = 1 \pm \sqrt{15}$
SECTION 9.2 SOLVING QUADRATIC EQUATIONS BY COMPLETING THE SQUARE	
To Solve a Quadratic Equation by Completing the Square **Step 1.** If the coefficient of x^2 is not 1, divide both sides of the equation by the coefficient. **Step 2.** Isolate all terms with variables on one side. **Step 3.** Complete the square by adding the square of half of the coefficient of x to both sides. **Step 4.** Factor the perfect square trinomial. **Step 5.** Apply the square root property to solve.	Solve $2x^2 + 12x - 10 = 0$ by completing the square. $\frac{2x^2}{2} + \frac{12x}{2} - \frac{10}{2} = \frac{0}{2}$ Divide by 2. $x^2 + 6x - 5 = 0$ Simplify. $x^2 + 6x = 5$ Add 5. The coefficient of x is 6. Half of 6 is 3 and $3^2 = 9$. Add 9 to both sides. $x^2 + 6x + 9 = 5 + 9$ $(x + 3)^2 = 14$ Factor. $x + 3 = \pm\sqrt{14}$ $x = -3 \pm \sqrt{14}$

DEFINITIONS AND CONCEPTS	EXAMPLES

SECTION 9.3 SOLVING QUADRATIC EQUATIONS BY THE QUADRATIC FORMULA

Quadratic Formula

If a, b, and c are real numbers and $a \neq 0$, the quadratic equation $ax^2 + bx + c = 0$ has solutions

$$x = \frac{-b \pm \sqrt{b^2 - 4ac}}{2a}$$

To Solve a Quadratic Equation by the Quadratic Formula

Step 1. Write the equation in standard form: $ax^2 + bx + c = 0$.
Step 2. If necessary, clear the equation of fractions.
Step 3. Identify a, b, and c.
Step 4. Replace a, b, and c in the quadratic formula by known values, and simplify.

Identify a, b, and c in the quadratic equation

$$4x^2 - 6x = 5$$

First, subtract 5 from both sides.

$$4x^2 - 6x - 5 = 0$$

$$a = 4, b = -6, \text{ and } c = -5$$

Solve $3x^2 - 2x - 2 = 0$.

In this equation, $a = 3$, $b = -2$, and $c = -2$.

$$x = \frac{-(-2) \pm \sqrt{(-2)^2 - 4(3)(-2)}}{2 \cdot 3}$$

$$= \frac{2 \pm \sqrt{4 - (-24)}}{6}$$

$$= \frac{2 \pm \sqrt{28}}{6} = \frac{2 \pm \sqrt{4 \cdot 7}}{6} = \frac{2 \pm 2\sqrt{7}}{6}$$

$$= \frac{2(1 \pm \sqrt{7})}{2 \cdot 3} = \frac{1 \pm \sqrt{7}}{3}$$

To Solve a Quadratic Equation

Step 1. If the equation is in the form $(ax + b)^2 = c$, use the square root property and solve. If not, go to step 2.
Step 2. Write the equation in standard form: $ax^2 + bx + c = 0$.
Step 3. Try to solve by factoring. If not, go to step 4.
Step 4. Solve by the quadratic formula.

Solve $(3x - 1)^2 = 10$.

$3x - 1 = \pm\sqrt{10}$ Square root property.
$3x = 1 \pm \sqrt{10}$ Add 1.
$x = \dfrac{1 \pm \sqrt{10}}{3}$ Divide by 3.

Solve $x(2x + 9) = 5$.
$2x^2 + 9x - 5 = 0$
$(2x - 1)(x + 5) = 0$
$2x - 1 = 0$ or $x + 5 = 0$
$2x = 1$ $x = -5$
$x = \dfrac{1}{2}$

SECTION 9.4 COMPLEX SOLUTIONS OF QUADRATIC EQUATIONS

The **imaginary unit**, written i, is the number whose square is -1. That is,

$$i^2 = -1 \quad \text{and} \quad i = \sqrt{-1}$$

A **complex number** is a number that can be written in the form

$$a + bi$$

Write $\sqrt{-10}$ as the product of a real number and i.
$$\sqrt{-10} = \sqrt{-1 \cdot 10} = \sqrt{-1} \cdot \sqrt{10} = i\sqrt{10}$$

Identify each number as a complex number by writing it in **standard form** $a + bi$.

$7 = 7 + 0i$
$\sqrt{-5} = 0 + i\sqrt{5}$ Also an imaginary number
$1 - \sqrt{-9} = 1 - 3i$

(continued)

DEFINITIONS AND CONCEPTS	EXAMPLES

SECTION 9.4 COMPLEX SOLUTIONS OF QUADRATIC EQUATIONS (continued)

where a and b are real numbers. A complex number that can be written in the form $0 + bi, b \neq 0$, is also called an **imaginary number**.

Complex numbers are added and subtracted in the same way as polynomials are added and subtracted.

Simplify the sum or difference.

$$(2 + 3i) - (1 - 6i) = 2 + 3i - 1 + 6i$$
$$= 1 + 9i$$

$$2i + (5 - 3i) = 2i + 5 - 3i$$
$$= 5 - i$$

Use the distributive property to multiply complex numbers.

Multiply: $(4 - i)(-2 + 3i)$
$$= -8 + 12i + 2i - 3i^2$$
$$= -8 + 14i + 3$$
$$= -5 + 14i$$

The complex numbers $a + bi$ and $a - bi$ are called **complex conjugates**.

The conjugate of $5 - 6i$ is $5 + 6i$.

To write a quotient of complex numbers in standard form $a + bi$, multiply numerator and denominator by the denominator's conjugate.

Write $\dfrac{3 - 2i}{2 + i}$ in standard form.

$$\frac{3 - 2i}{2 + i} \cdot \frac{(2 - i)}{(2 - i)} = \frac{6 - 7i + 2i^2}{4 - i^2}$$
$$= \frac{6 - 7i + 2(-1)}{4 - (-1)}$$
$$= \frac{4 - 7i}{5} \text{ or } \frac{4}{5} - \frac{7}{5}i$$

Some quadratic equations have complex solutions.

Solve $x^2 - 3x = -3$
$$x^2 - 3x + 3 = 0$$
$$a = 1, b = -3, c = 3$$
$$x = \frac{-(-3) \pm \sqrt{(-3)^2 - 4(1)(3)}}{2(1)}$$
$$= \frac{3 \pm \sqrt{-3}}{2} = \frac{3 \pm i\sqrt{3}}{2}$$

SECTION 9.5 GRAPHING QUADRATIC EQUATIONS

The graph of a quadratic equation $y = ax^2 + bx + c, a \neq 0$, is called a **parabola**. The lowest point on a parabola opening upward or the highest point on a parabola opening downward is called the **vertex**. The vertical line through the vertex is the **axis of symmetry**.

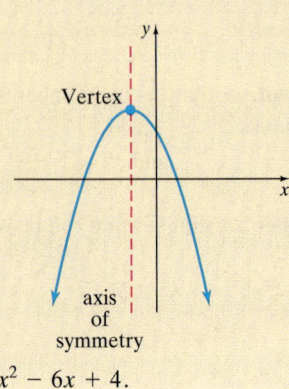

Graph $y = 2x^2 - 6x + 4$.

DEFINITIONS AND CONCEPTS	EXAMPLES

SECTION 9.5 GRAPHING QUADRATIC EQUATIONS (continued)

The vertex of the parabola $y = ax^2 + bx + c$ has x-value $\dfrac{-b}{2a}$.	The x-value of the vertex is $$x = \frac{-b}{2a} = \frac{-(-6)}{2(2)} = \frac{6}{4} = \frac{3}{2}$$ The y-value is $$y = 2\left(\frac{3}{2}\right)^2 - 6\left(\frac{3}{2}\right) + 4 = -\frac{1}{2}$$ The vertex is $\left(\dfrac{3}{2}, -\dfrac{1}{2}\right)$. The y-intercept is $$y = 2 \cdot 0^2 - 6 \cdot 0 + 4 = 4$$ The x-intercepts are found by solving $$0 = 2x^2 - 6x + 4$$ $$0 = 2(x^2 - 3x + 2)$$ $$0 = 2(x-2)(x-1)$$ $x - 2 = 0$ or $x - 1 = 0$ $x = 2$ or $x = 1$
Find more ordered pair solutions as needed.	

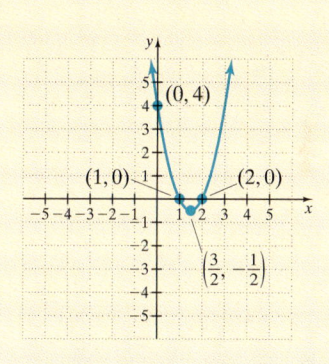

CHAPTER 9 REVIEW

(9.1) Use the square root property to solve each quadratic equation.

1. $x^2 = 36$
2. $x^2 = 81$
3. $k^2 = 50$
4. $k^2 = 45$
5. $(x - 11)^2 = 49$
6. $(x + 3)^2 = 100$
7. $(4p + 5)^2 = 41$
8. $(3p + 7)^2 = 37$

Solve. For Exercises 9 and 10, use the formula $h = 16t^2$, where h is the height in feet at time t seconds.

9. If Kara Washington dives from a height of 100 feet, how long before she hits the water?

10. How long does a 5-mile free-fall take? Round your result to the nearest tenth of a second. (*Hint:* 1 mi = 5280 ft)

(9.2) Complete the square for the following expressions and then factor the resulting perfect square trinomial.

11. $a^2 + 4a$
12. $a^2 - 12a$
13. $m^2 - 3m$
14. $m^2 + 5m$

Solve each quadratic equation by completing the square.

15. $x^2 - 9x = -8$
16. $x^2 + 8x = 20$
17. $x^2 + 4x = 1$
18. $x^2 - 8x = 3$
19. $x^2 - 6x + 7 = 0$
20. $x^2 + 6x + 7 = 0$
21. $2y^2 + y - 1 = 0$
22. $y^2 + 3y - 1 = 0$

(9.3) Use the quadratic formula to solve each quadratic equation.

23. $9x^2 + 30x + 25 = 0$
24. $16x^2 - 72x + 81 = 0$
25. $7x^2 = 35$
26. $11x^2 = 33$
27. $x^2 - 10x + 7 = 0$
28. $x^2 + 4x - 7 = 0$
29. $3x^2 + x - 1 = 0$
30. $x^2 + 3x - 1 = 0$
31. $2x^2 + x + 5 = 0$
32. $7x^2 - 3x + 1 = 0$

For the Exercise numbers given, approximate the exact solutions to the nearest tenth.

33. Exercise 29
34. Exercise 30

Use the discriminant to determine the number of solutions of each quadratic equation.

35. $x^2 - 7x - 1 = 0$
36. $x^2 + x + 5 = 0$
37. $9x^2 + 1 = 6x$
38. $x^2 + 6x = 5$
39. $5x^2 + 4 = 0$
40. $x^2 + 25 = 10x$

41. The average price of gold (in dollars per troy ounce) from 2001 to 2005 is given by the equation $y = -x^2 + 51x + 218$. In this equation, x is the number of years since 2000. Assume that this trend continues and find the first year after 2000 in which the price of gold will be $658 per troy ounce. (A troy ounce is a little over 15 grams.) (*Source:* National Mining Association)

42. The number of combined kidney-liver transplants is increasing. From 2000 to 2005, the number of these transplants is modeled by the equation $y = 2.1x^2 + 31x + 125$, where x is the number of years since 2000. Assume that this trend continues and predict the year that the number of kidney-liver transplants will be 645.

(9.4) Perform the indicated operations. Write the resulting complex number in standard form.

43. $\sqrt{-144}$
44. $\sqrt{-36}$
45. $\sqrt{-108}$
46. $\sqrt{-500}$
47. $2i(3 - 5i)$
48. $i(-7 - i)$
49. $(7 - i) + (14 - 9i)$
50. $(10 - 4i) + (9 - 21i)$
51. $3 - (11 + 2i)$
52. $(-4 - 3i) + 5i$
53. $(2 - 3i)(3 - 2i)$
54. $(2 + 5i)(5 - i)$
55. $(3 - 4i)(3 + 4i)$
56. $(7 - 2i)(7 - 2i)$
57. $\dfrac{2 - 6i}{4i}$
58. $\dfrac{5 - i}{2i}$
59. $\dfrac{4 - i}{1 + 2i}$
60. $\dfrac{1 + 3i}{2 - 7i}$

Solve each quadratic equation.

61. $3x^2 = -48$
62. $5x^2 = -125$
63. $x^2 - 4x + 13 = 0$
64. $x^2 + 4x + 11 = 0$

(9.5) Graph each quadratic equation and find and plot any intercept points.

65. $y = 5x^2$
66. $y = -\dfrac{1}{2}x^2$

Graph each quadratic equation. Label the vertex and the intercept points with their coordinates.

67. $y = x^2 - 25$
68. $y = x^2 - 36$
69. $y = x^2 + 3$
70. $y = x^2 + 8$
71. $y = -4x^2 + 8$
72. $y = -3x^2 + 9$
73. $y = x^2 + 3x - 10$
74. $y = x^2 + 3x - 4$
75. $y = -x^2 - 5x - 6$
76. $y = 3x^2 - x - 2$
77. $y = 2x^2 - 11x - 6$
78. $y = -x^2 + 4x + 8$

Match each quadratic equation with its graph.

79. $y = 2x^2$
80. $y = -x^2$

A

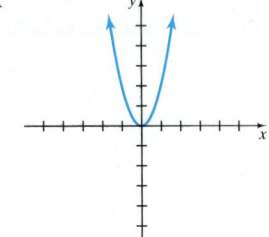

B

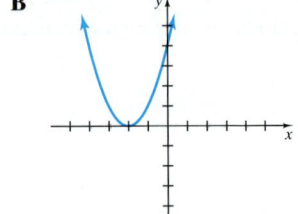

81. $y = x^2 + 4x + 4$ **82.** $y = x^2 + 5x + 4$ **85.**   **86.**

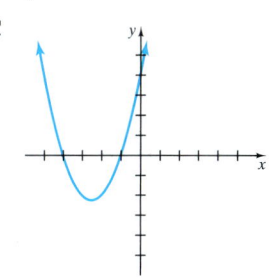

C   D

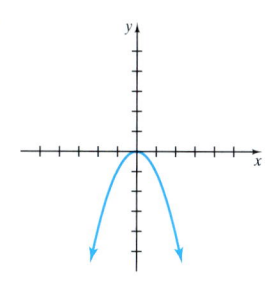

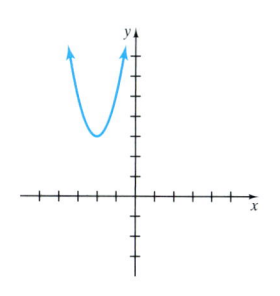

Quadratic equations in the form $y = ax^2 + bx + c$ are graphed below. Determine the number of real solutions for the related equation $0 = ax^2 + bx + c$ from each graph.

83.   **84.**

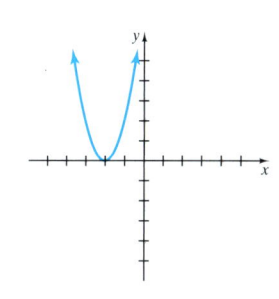

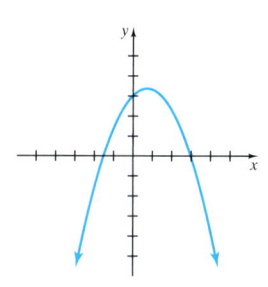

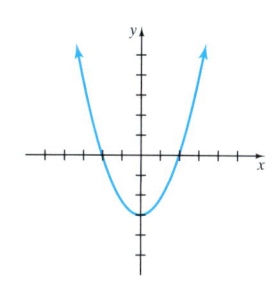

MIXED REVIEW

Use the square root property to solve each quadratic equation.

87. $x^2 = 49$ **88.** $y^2 = 75$ **89.** $(x - 7)^2 = 64$

Solve each quadratic equation by completing the square.

90. $x^2 + 4x = 6$ **91.** $3x^2 + x = 2$ **92.** $4x^2 - x - 2 = 0$

Use the quadratic formula to solve each quadratic equation.

93. $4x^2 - 3x - 2 = 0$ **94.** $5x^2 + x - 2 = 0$
95. $4x^2 + 12x + 9 = 0$ **96.** $2x^2 + x + 4 = 0$

Graph each quadratic equation. Label the vertex and the intercept points with their coordinates.

97. $y = 4 - x^2$ **98.** $y = x^2 + 4$
99. $y = x^2 + 6x + 8$ **100.** $y = x^2 - 2x - 4$

STUDY SKILLS BUILDER

Are You Prepared for a Test on Chapter 9?

Below I have listed some common trouble areas for students in Chapter 9. After studying for your test—but before taking your test—read these.

- Don't forget that to use the square root property, one side of your equation should be a squared variable or variable expression.

   Solve: $3x^2 = 15$
   $x^2 = 5$ Divide both sides by 3 to isolate x^2.
   $x = \sqrt{5}$ or $x = -\sqrt{5}$ Use the square root property.

- Remember that to identify a, b, and c for the quadratic formula, write the quadratic equation in standard form: $ax^2 + bx + c = 0$

   Solve: $x^2 = -x + 1$
   $x^2 + x - 1 = 0$ Write in standard form.

   Here, $a = 1$, $b = 1$, and $c = -1$.

   $$x = \frac{-1 \pm \sqrt{1^2 - 4(1)(-1)}}{2(1)} = \frac{-1 \pm \sqrt{5}}{2}$$

Remember: This is simply a listing of a few common trouble areas. For a review of Chapter 9, see the Highlights and Chapter Review at the end of this chapter.

CHAPTER 9 TEST

Solve using the square root property.

1. $5k^2 = 80$
2. $(3m - 5)^2 = 8$

Solve by completing the square.

3. $x^2 - 26x + 160 = 0$
4. $3x^2 + 12x - 4 = 0$

Solve using the quadratic formula.

5. $x^2 - 3x - 10 = 0$
6. $p^2 - \frac{5}{3}p - \frac{1}{3} = 0$

Solve by the most appropriate method.

7. $(3x - 5)(x + 2) = -6$
8. $(3x - 1)^2 = 16$
9. $3x^2 - 7x - 2 = 0$
10. $x^2 - 4x + 5 = 0$
11. $3x^2 - 7x + 2 = 0$
12. $2x^2 - 6x + 1 = 0$
13. $9x^3 = x$

Perform the indicated operations. Write the resulting complex number in standard form.

14. $\sqrt{-25}$
15. $\sqrt{-200}$
16. $(3 + 2i) + (5 - i)$
17. $(3 + 2i) - (3 - 2i)$
18. $(3 + 2i)(3 - 2i)$
19. $\frac{3 - i}{1 + 2i}$

Graph each quadratic equation. Label the vertex and the intercept points with their coordinates.

20. $y = -5x^2$
21. $y = x^2 - 4$
22. $y = x^2 - 7x + 10$
23. $y = 2x^2 + 4x - 1$

Solve.

24. The height of a triangle is 4 times the length of the base. The area of the triangle is 18 square feet. Find the height and base of the triangle.

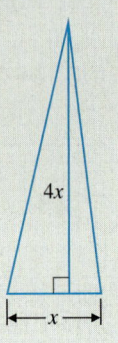

25. The number of diagonals d that a polygon with n sides has is given by the formula
$$d = \frac{n^2 - 3n}{2}$$
Find the number of sides of a polygon if it has 9 diagonals.

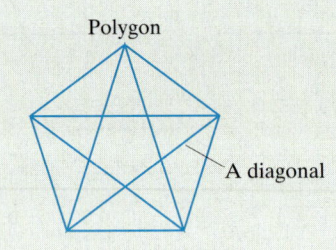

Solve.

26. The highest dive from a diving board by a woman was made by Lucy Wardle of the United States. She dove from a height of 120.75 feet at Ocean Park, Hong Kong, in 1985. To the nearest tenth of a second, how long did the dive take? Use the formula $h = 16t^2$.

CHAPTER 9 CUMULATIVE REVIEW

1. Find the value of each expression when $x = 2$ and $y = -5$.
   a. $\frac{x - y}{12 + x}$
   b. $x^2 - 3y$
   c. $2 + 8.1a + a - 6$
   d. $2x^2 - 2x$

2. Find the value of each expression when $x = -4$ and $y = 7$.
   a. $\frac{x - y}{7 - x}$
   b. $x^2 + 2y$

3. Simplify each expression by combining like terms.
   a. $2x + 3x + 5 + 2$
   b. $-5a - 3 + a + 2$
   c. $4y - 3y^2$
   d. $2.3x + 5x - 6$
   e. $-\frac{1}{2}b + b$

4. Simplify each expression by combining like terms.
   a. $4x - 3 + 7 - 5x$
   b. $-6y + 3y - 8 + 8y$

5. Identify the x- and y-intercepts.
   a. 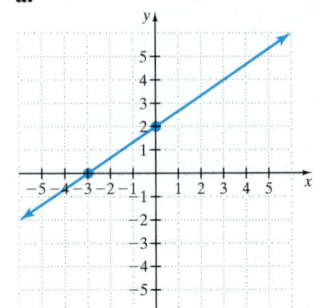
   b.

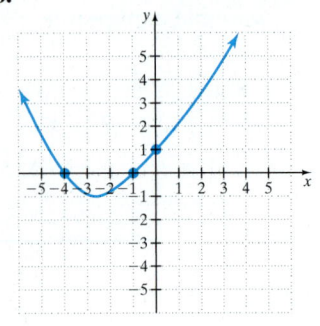

c.

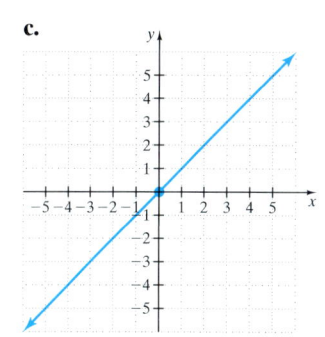

d.

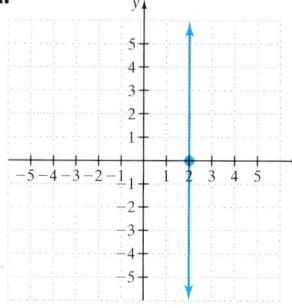

e.

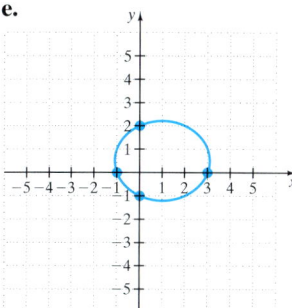

6. Identify the x- and y-intercepts.

   a.

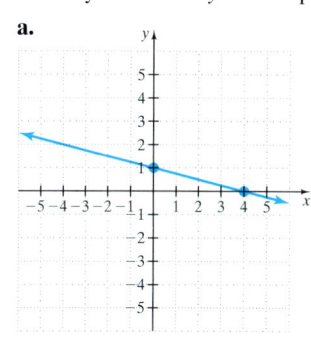

   b.

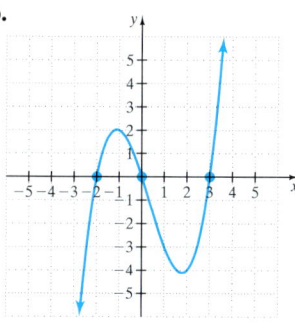

   c.

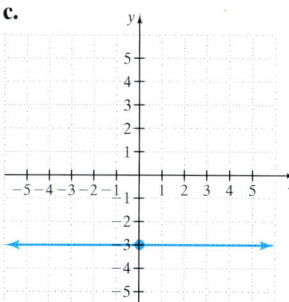

   d.

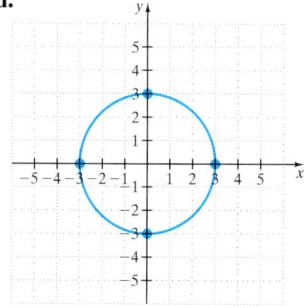

7. Determine whether the graphs of $y = -\frac{1}{5}x + 1$ and $2x + 10y = 30$ are parallel lines, perpendicular lines, or neither.

8. Determine whether the graphs of $y = 3x + 7$ and $x + 3y = -15$ are parallel lines, perpendicular lines, or neither.

9. Solve the following system of equations by graphing.
$$\begin{cases} 2x + y = 7 \\ 2y = -4x \end{cases}$$

10. Solve the following system by graphing.
$$\begin{cases} y = x + 2 \\ 2x + y = 5 \end{cases}$$

11. Solve the system.
$$\begin{cases} 7x - 3y = -14 \\ -3x + y = 6 \end{cases}$$

12. Solve the system.
$$\begin{cases} 5x + y = 3 \\ y = -5x \end{cases}$$

13. Solve the system.
$$\begin{cases} 3x - 2y = 2 \\ -9x + 6y = -6 \end{cases}$$

14. Solve the system.
$$\begin{cases} -2x + y = 7 \\ 6x - 3y = -21 \end{cases}$$

15. As part of an exercise program, Albert and Louis started walking each morning. They live 15 miles away from each other and decided to meet one day by walking toward one another. After 2 hours they meet. If Louis walks one mile per hour faster than Albert, find both walking speeds.

16. A coin purse contains dimes and quarters only. There are 15 coins totalling $2.85. How many dimes and how many quarters are in the purse?

17. Graph the solution of the system
$$\begin{cases} -3x + 4y < 12 \\ x \geq 2 \end{cases}$$

18. Graph the solution of the system.
$$\begin{cases} 2x - y \leq 6 \\ y \geq 2 \end{cases}$$

19. Simplify each expression.

   a. $\left(\dfrac{st}{2}\right)^4$ **b.** $(9y^5z^7)^2$ **c.** $\left(\dfrac{-5x^2}{y^3}\right)^2$

20. Simplify.

   a. $\left(\dfrac{-6x}{y^3}\right)^3$ **b.** $\dfrac{a^2b^7}{(2b^2)^5}$

   c. $\dfrac{(3y)^2}{y^2}$ **d.** $\dfrac{(x^2y^4)^2}{xy^3}$

21. Solve $(5x - 1)(2x^2 + 15x + 18) = 0$.

22. Solve $(x + 1)(2x^2 - 3x - 5) = 0$.

23. Solve $\dfrac{45}{x} = \dfrac{5}{7}$.

24. Solve $\dfrac{2x + 7}{3} = \dfrac{x - 6}{2}$.

25. Find each root.

   a. $\sqrt[4]{16}$ **b.** $\sqrt[5]{-32}$ **c.** $-\sqrt[3]{8}$

   d. $\sqrt[4]{-81}$

26. Find each root.

   a. $\sqrt[3]{27}$ **b.** $\sqrt[4]{256}$ **c.** $\sqrt[3]{-125}$ **d.** $\sqrt[5]{1}$

27. Simplify the following expressions.

   a. $\sqrt{\dfrac{25}{36}}$ b. $\sqrt{\dfrac{3}{64}}$ c. $\sqrt{\dfrac{40}{81}}$

28. Simplify the following expressions.

   a. $\sqrt{\dfrac{4}{25}}$ b. $\sqrt{\dfrac{16}{121}}$ c. $\sqrt{\dfrac{2}{49}}$

29. Add or subtract by first simplifying each radical.

   a. $\sqrt{50} + \sqrt{8}$
   b. $7\sqrt{12} - \sqrt{75}$
   c. $\sqrt{25} - \sqrt{27} - 2\sqrt{18} - \sqrt{16}$

30. Add or subtract by first simplifying each radical.

   a. $\sqrt{80} + \sqrt{20}$
   b. $2\sqrt{98} - 2\sqrt{18}$
   c. $\sqrt{32} + \sqrt{121} - \sqrt{12}$

31. Multiply. Then simplify if possible.

   a. $\sqrt{7} \cdot \sqrt{3}$ b. $\sqrt{3} \cdot \sqrt{3}$ c. $\sqrt{3} \cdot \sqrt{15}$
   d. $2\sqrt{3} \cdot 5\sqrt{2}$ e. $\sqrt{2x^3} \cdot \sqrt{6x}$

32. Multiply. Then simplify if possible.

   a. $\sqrt{2} \cdot \sqrt{5}$ b. $\sqrt{56} \cdot \sqrt{7}$
   c. $(4\sqrt{3})^2$ d. $3\sqrt{8} \cdot 7\sqrt{2}$

33. Solve $\sqrt{x} = \sqrt{5x - 2}$.

34. Solve $\sqrt{x - 4} + 7 = 2$.

35. A surveyor must determine the distance across a lake at points P and Q. To do this, she finds a third point R perpendicular to line PQ. If the length of $\overline{PR}$ is 320 feet and the length of $\overline{QR}$ is 240 feet, what is the distance across the lake? Approximate this distance to the nearest whole foot.

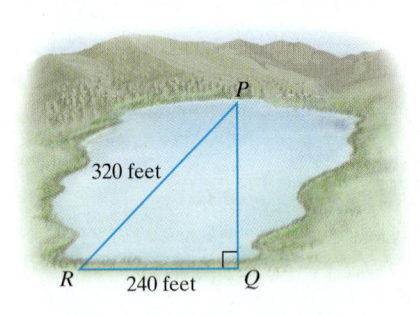

36. Find the distance between $(-7, 4)$ and $(2, 5)$.

37. Write in radical notation. Then simplify.

   a. $25^{1/2}$ b. $8^{1/3}$
   c. $-16^{1/4}$ d. $(-27)^{1/3}$
   e. $\left(\dfrac{1}{9}\right)^{1/2}$

38. Write in radical notation, then simplify.

   a. $-49^{1/2}$ b. $256^{1/4}$
   c. $(-64)^{1/3}$ d. $\left(\dfrac{25}{36}\right)^{1/2}$
   e. $(32)^{1/5}$

39. Use the square root property to solve $2x^2 = 7$.

40. Use the square root property to solve $3(x - 4)^2 = 9$.

41. Solve $x^2 - 10x = -14$ by completing the square.

42. Solve $x^2 + 4x = 8$ by completing the square.

43. Solve $2x^2 - 9x = 5$ using the quadratic formula.

44. Solve $2x^2 + 5x = 7$ using the quadratic formula.

45. Write each radical as the product of a real number and i.

   a. $\sqrt{-4}$
   b. $\sqrt{-11}$
   c. $\sqrt{-20}$

46. Write each radical as the product of a real number and i.

   a. $\sqrt{-7}$
   b. $\sqrt{-16}$
   c. $\sqrt{-27}$

47. Graph $y = x^2 - 4$.

48. Graph $y = x^2 + 2x + 3$.

Appendix A

The Bigger Picture/Practice Final Exam

A.1 THE BIGGER PICTURE: SIMPLIFYING EXPRESSIONS AND SOLVING EQUATIONS

I. Simplifying Expressions

A. Real Numbers

1. **Add:** (Sec. 1.5)

   $-1.7 + (-0.21) = -1.91$ Adding like signs.
   Add absolute values. Attach common sign.

   $-7 + 3 = -4$ Adding different signs.
   Subtract absolute values. Attach the sign of the number with the larger absolute value.

2. **Subtract:** Add the first number to the opposite of the second number. (Sec. 1.6)

   $$17 - 25 = 17 + (-25) = -8$$

3. **Multiply or divide:** Multiply or divide the two numbers as usual. If the signs are the same, the answer is positive. If the signs are different, the answer is negative. (Sec. 1.7)

   $$-10 \cdot 3 = -30, \quad -81 \div (-3) = 27$$

B. Exponents (Sec. 5.1)

$$x^7 \cdot x^5 = x^{12}; \quad (x^7)^5 = x^{35}; \quad \frac{x^7}{x^5} = x^2; \quad x^0 = 1; \quad 8^{-2} = \frac{1}{8^2} = \frac{1}{64}$$

C. Polynomials

1. **Add:** Combine like terms. (Sec. 5.2)

   $$(3y^2 + 6y + 7) + (9y^2 - 11y - 15) = 3y^2 + 6y + 7 + 9y^2 - 11y - 15$$
   $$= 12y^2 - 5y - 8$$

2. **Subtract:** Change the sign of the terms of the polynomial being subtracted, then add. (Sec. 5.2)

   $$(3y^2 + 6y + 7) - (9y^2 - 11y - 15) = 3y^2 + 6y + 7 - 9y^2 + 11y + 15$$
   $$= -6y^2 + 17y + 22$$

3. **Multiply:** Multiply each term of one polynomial by each term of the other polynomial. (Sec. 5.3)

   $$(x + 5)(2x^2 - 3x + 4) = x(2x^2 - 3x + 4) + 5(2x^2 - 3x + 4)$$
   $$= 2x^3 - 3x^2 + 4x + 10x^2 - 15x + 20$$
   $$= 2x^3 + 7x^2 - 11x + 20$$

4. **Divide:** (Sec. 5.6)
   a. To divide by a monomial, divide each term of the polynomial by the monomial.
   $$\frac{8x^2 + 2x - 6}{2x} = \frac{8x^2}{2x} + \frac{2x}{2x} - \frac{6}{2x} = 4x + 1 - \frac{3}{x}$$
   b. To divide by a polynomial other than a monomial, use long division.

   $$\begin{array}{r} x - 6 + \frac{40}{2x+5} \\ 2x+5\overline{)2x^2 - 7x + 10} \\ \underline{2x^2 + 5x} \\ -12x + 10 \\ \underline{-12x - 30} \\ 40 \end{array}$$

D. **Factoring Polynomials**
   See the Chapter 6 Integrated Review for steps.

   $3x^4 - 78x^2 + 75 = 3(x^4 - 26x^2 + 25)$ Factor out GCF—always first step.
   $\qquad\qquad\qquad\quad = 3(x^2 - 25)(x^2 - 1)$ Factor trinomial.
   $\qquad\qquad\qquad\quad = 3(x + 5)(x - 5)(x + 1)(x - 1)$ Factor further—each difference of squares.

E. **Rational Expressions**
   1. **Simplify:** Factor the numerator and denominator. Then divide out factors of 1 by dividing out common factors in the numerator and denominator. (Sec. 7.1)
   $$\frac{x^2 - 9}{7x^2 - 21x} = \frac{(x + 3)(x - 3)}{7x(x - 3)} = \frac{x + 3}{7x}$$

   2. **Multiply:** Multiply numerators, then multiply denominators. (Sec. 7.2)
   $$\frac{5z}{2z^2 - 9z - 18} \cdot \frac{22z + 33}{10z} = \frac{5 \cdot z}{(2z + 3)(z - 6)} \cdot \frac{11(2z + 3)}{2 \cdot 5 \cdot z} = \frac{11}{2(z - 6)}$$

   3. **Divide:** First fraction times the reciprocal of the second fraction. (Sec. 7.2)
   $$\frac{14}{x + 5} \div \frac{x + 1}{2} = \frac{14}{x + 5} \cdot \frac{2}{x + 1} = \frac{28}{(x + 5)(x + 1)}$$

   4. **Add or subtract:** Must have same denominator. If not, find the LCD and write each fraction as an equivalent fraction with the LCD as denominator. (Sec. 7.4)
   $$\frac{9}{10} - \frac{x + 1}{x + 5} = \frac{9(x + 5)}{10(x + 5)} - \frac{10(x + 1)}{10(x + 5)}$$
   $$= \frac{9x + 45 - 10x - 10}{10(x + 5)} = \frac{-x + 35}{10(x + 5)}$$

F. **Radicals**
   1. **Simplify square roots:** If possible, factor the radicand so that one factor is a perfect square. Then use the product rule and simplify. (Sec. 8.2)
   $$\sqrt{75} = \sqrt{25 \cdot 3} = \sqrt{25} \cdot \sqrt{3} = 5\sqrt{3}$$

   2. **Add or subtract:** Only like radicals (same index and radicand) can be added or subtracted. (Sec. 8.3)
   $$8\sqrt{10} - \sqrt{40} + \sqrt{5} = 8\sqrt{10} - 2\sqrt{10} + \sqrt{5} = 6\sqrt{10} + \sqrt{5}$$

3. **Multiply or divide:** $\sqrt{a} \cdot \sqrt{b} = \sqrt{ab}$; $\dfrac{\sqrt{a}}{\sqrt{b}} = \sqrt{\dfrac{a}{b}}$. (Sec. 8.4)

$$\sqrt{11} \cdot \sqrt{3} = \sqrt{33}; \quad \dfrac{\sqrt{140}}{\sqrt{7}} = \sqrt{\dfrac{140}{7}} = \sqrt{20} = \sqrt{4 \cdot 5} = 2\sqrt{5}$$

4. **Rationalizing the denominator:** (Sec. 8.4)
   a. If denominator is one term,

   $$\dfrac{5}{\sqrt{11}} = \dfrac{5 \cdot \sqrt{11}}{\sqrt{11} \cdot \sqrt{11}} = \dfrac{5\sqrt{11}}{11}$$

   b. If denominator is two terms, multiply by 1 in the form of $\dfrac{\text{conjugate of denominator}}{\text{conjugate of denominator}}$.

   $$\dfrac{13}{3 + \sqrt{2}} = \dfrac{13}{3 + \sqrt{2}} \cdot \dfrac{3 - \sqrt{2}}{3 - \sqrt{2}} = \dfrac{13(3 - \sqrt{2})}{9 - 2} = \dfrac{13(3 - \sqrt{2})}{7}$$

II. Solving Equations and Inequalities

A. Linear Equations: Power on variable is 1 and there are no variables in denominator. (Sec. 2.4)

$7(x - 3) = 4x + 6$	Linear equation (If fractions, multiply by LCD.)
$7x - 21 = 4x + 6$	Use the distributive property.
$7x = 4x + 27$	Add 21 to both sides.
$3x = 27$	Subtract $4x$ from both sides.
$x = 9$	Divide both sides by 3.

B. Linear Inequalities: Same as linear equation except if you multiply or divide by a negative number, then reverse direction of inequality. (Sec. 2.9)

$-4x + 11 \leq -1$	Linear inequality
$-4x \leq -12$	Subtract 11 from both sides.
$\dfrac{-4x}{-4} \geq \dfrac{-12}{-4}$	Divide both sides by -4 and reverse the direction of the inequality symbol.
$x \geq 3$	Simplify.

C. Quadratic and Higher Degree Equations: Solve: first write the equation in standard form (one side is 0.)

1. If the polynomial on one side factors, solve by factoring. (Sec. 6.6)
2. If the polynomial does not factor, solve by the quadratic formula. (Sec. 9.3)

By factoring:	**By quadratic formula:**
$x^2 + x = 6$	$x^2 + x = 5$
$x^2 + x - 6 = 0$	$x^2 + x - 5 = 0$
$(x - 2)(x + 3) = 0$	
$x - 2 = 0$ or $x + 3 = 0$	$a = 1, b = 1, c = -5$
$x = 2$ or $x = -3$	$x = \dfrac{-1 \pm \sqrt{1^2 - 4(1)(-5)}}{2 \cdot 1}$
	$x = \dfrac{-1 \pm \sqrt{21}}{2}$

D. Equations with Rational Expressions: Make sure the proposed solution does not make the denominator 0. (Sec. 7.5)

$$\frac{3}{x} - \frac{1}{x-1} = \frac{4}{x-1} \quad \text{Equation with rational expressions.}$$

$$x(x-1) \cdot \frac{3}{x} - x(x-1) \cdot \frac{1}{x-1} = x(x-1) \cdot \frac{4}{x-1} \quad \text{Multiply through by } x(x-1).$$

$$3(x-1) - x \cdot 1 = x \cdot 4 \quad \text{Simplify.}$$
$$3x - 3 - x = 4x \quad \text{Use the distributive property.}$$
$$-3 = 2x \quad \text{Simplify and move variable terms to right side.}$$
$$-\frac{3}{2} = x \quad \text{Divide both sides by 2.}$$

E. Proportions: An equation with two ratios equal. Set cross products equal, then solve. Make sure the proposed solution does not make the denominator 0. (Sec. 7.6)

$$\frac{5}{x} = \frac{9}{2x-3}$$
$$5(2x-3) = 9 \cdot x \quad \text{Set cross products equal.}$$
$$10x - 15 = 9x \quad \text{Multiply.}$$
$$x = 15 \quad \text{Write equation with variable terms on one side and constants on the other.}$$

F. Equations with Radicals: To solve, isolate a radical, then square both sides. You may have to repeat this. Check possible solution in the original equation. (Sec. 8.5)

$$\sqrt{x+49} + 7 = x$$
$$\sqrt{x+49} = x - 7 \quad \text{Subtract 7 from both sides.}$$
$$x + 49 = x^2 - 14x + 49 \quad \text{Square both sides.}$$
$$0 = x^2 - 15x \quad \text{Set terms equal to 0.}$$
$$0 = x(x-15) \quad \text{Factor.}$$
$$\cancel{x=0} \text{ or } x = 15 \quad \text{Set each factor equal to 0 and solve.}$$

A.2 PRACTICE FINAL EXAM TEST PREP VIDEO

Preparing for your Final Exam? Take this Practice Final and watch the full video solutions to any of the exercises you want to review. You will find the Practice Final video in the Video Lecture Series. The video also provides you with an overview to help you approach different problem types just as you will need to do on a Final Exam. To build your own study guide, use the Bigger Picture feature in the text. See Appendix A.1 for an example.

Evaluate.

1. -3^4
2. 4^{-3}
3. $6[5 + 2(3-8) - 3]$

Perform the indicated operations and simplify if possible.

4. $(5x^3 + x^2 + 5x - 2) - (8x^3 - 4x^2 + x - 7)$
5. $(4x - 2)^2$
6. $(3x + 7)(x^2 + 5x + 2)$

Factor.

7. $y^2 - 8y - 48$
8. $9x^3 + 39x^2 + 12x$
9. $180 - 5x^2$
10. $3a^2 + 3ab - 7a - 7b$
11. $8y^3 - 64$

Simplify. Write answers with positive exponents only.

12. $\left(\dfrac{x^2 y^3}{x^3 y^{-4}}\right)^2$

13. $\dfrac{5 - \dfrac{1}{y^2}}{\dfrac{1}{y} + \dfrac{2}{y^2}}$

Perform the indicated operations and simplify if possible.

14. $\dfrac{x^2 - 9}{x^2 - 3x} \div \dfrac{xy + 5x + 3y + 15}{2x + 10}$

15. $\dfrac{5a}{a^2 - a - 6} - \dfrac{2}{a - 3}$

Solve each equation or inequality.

16. $4(n - 5) = -(4 - 2n)$
17. $x(x + 6) = 7$
18. $3x - 5 \geq 7x + 3$
19. $2x^2 - 6x + 1 = 0$
20. $\dfrac{4}{y} - \dfrac{5}{3} = -\dfrac{1}{5}$
21. $\dfrac{5}{y + 1} = \dfrac{4}{y + 2}$
22. $\dfrac{a}{a - 3} = \dfrac{3}{a - 3} - \dfrac{3}{2}$
23. $\sqrt{2x - 2} = x - 5$

Graph the following.

24. $5x - 7y = 10$
25. $x - 3 = 0$
26. $y > -4x$

Find the slope of each line.

27. through $(6, -5)$ and $(-1, 2)$
28. $-3x + y = 5$

Write equations of the following lines. Write each equation in standard form.

29. through $(2, -5)$ and $(1, 3)$
30. through $(-5, -1)$ and parallel to $x = 7$

Solve each system of equations.

31. $\begin{cases} 3x - 2y = -14 \\ y = x + 5 \end{cases}$

32. $\begin{cases} 4x - 6y = 7 \\ -2x + 3y = 0 \end{cases}$

Answer the questions about functions.

33. If $h(x) = x^3 - x$, find
    a. $h(-1)$ b. $h(0)$ c. $h(4)$

34. Find the domain and range of the function graphed.

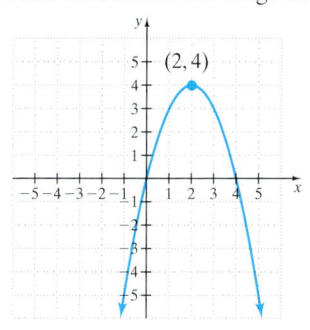

Evaluate.

35. $\sqrt{16}$
36. $27^{-2/3}$
37. $\left(\dfrac{9}{16}\right)^{1/2}$

Simplify.

38. $\sqrt{54}$
39. $\sqrt{9x^9}$

Perform the indicated operations and simplify if possible.

40. $\sqrt{12} - 2\sqrt{75}$
41. $\dfrac{\sqrt{40x^4}}{\sqrt{2x}}$
42. $\sqrt{2}\left(\sqrt{6} - \sqrt{5}\right)$

Rationalize each denominator.

43. $\sqrt{\dfrac{5}{12x^2}}$
44. $\dfrac{2\sqrt{3}}{\sqrt{3} - 3}$

Solve each application.

45. One number plus five times its reciprocal is equal to six. Find the number.

46. Some states have a single area code for the entire state. Two such states have area codes where one is double the other. If the sum of these integers is 1203, find the two area codes.

47. Two trains leave Los Angeles simultaneously traveling on the same track in opposite directions at speeds of 50 and 64 mph. How long will it take before they are 285 miles apart?

48. Find the amount of a 12% saline solution a lab assistant should add to 80 cc (cubic centimeters) of a 22% saline solution in order to have a 16% solution.

Appendix B

Geometry

B.1 GEOMETRIC FORMULAS

Rectangle 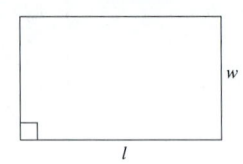
Perimeter: $P = 2l + 2w$
Area: $A = lw$

Square 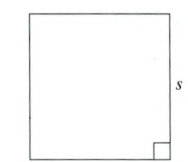
Perimeter: $P = 4s$
Area: $A = s^2$

Triangle 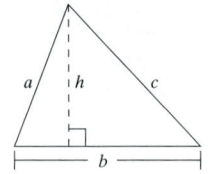
Perimeter: $P = a + b + c$
Area: $A = \dfrac{1}{2}bh$

Sum of Angles of Triangle 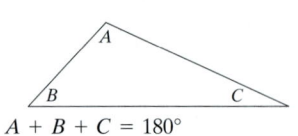
$A + B + C = 180°$
The sum of the measures of the three angles is 180°.

Right Triangles 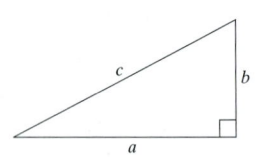
Perimeter: $P = a + b + c$
Area: $A = \dfrac{1}{2}ab$
One 90° (right) angle

Pythagorean Theorem (for right triangles) 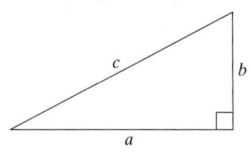
$a^2 + b^2 = c^2$

Isosceles Triangle 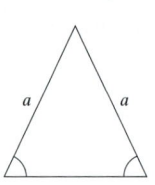
Triangle has:
two equal sides and two equal angles.

Equilateral Triangle 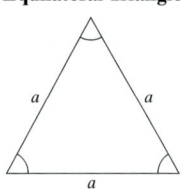
Triangle has:
three equal sides and three equal angles.
Measure of each angle is 60°.

Trapezoid 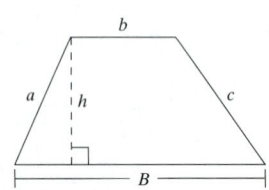
Perimeter: $P = a + b + c + B$
Area: $A = \dfrac{1}{2}h(B + b)$

Parallelogram 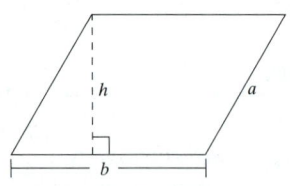
Perimeter: $P = 2a + 2b$
Area: $A = bh$

Circle 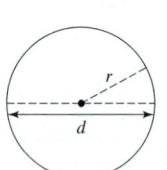
Circumference: $C = \pi d$
$C = 2\pi r$
Area: $A = \pi r^2$

Rectangular Solid 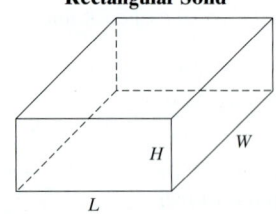
Volume: $V = LWH$
Surface Area:
$S = 2LW + 2HL + 2HW$

Cube 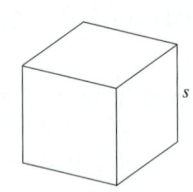
Volume: $V = s^3$
Surface Area: $S = 6s^2$

Cone 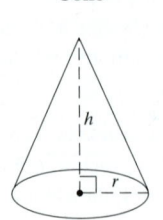
Volume: $V = \dfrac{1}{3}\pi r^2 h$

Right Circular Cylinder 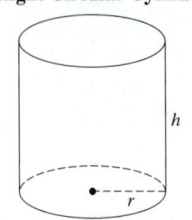
Volume: $V = \pi r^2 h$
Surface Area: $S = 2\pi r^2 + 2\pi rh$

Sphere 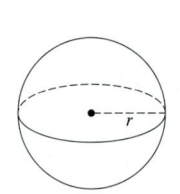
Volume: $V = \dfrac{4}{3}\pi r^3$
Surface Area: $S = 4\pi r^2$

Other Formulas
Distance: $d = rt$ (r = rate, t = time)
Percent: $p = br$ (p = percentage, b = base, r = rate)

Temperature: $F = \dfrac{9}{5}C + 32$ $C = \dfrac{5}{9}(F - 32)$

Simple Interest: $I = Prt$
(P = principal, r = annual interest rate, t = time in years)

B.2 REVIEW OF GEOMETRIC FIGURES

Plane figures have length and width but no thickness or depth

Name	Description
Polygon	Union of three or more coplanar line segments that intersect with each other only at each end point, with each end point shared by two segments.
Triangle	Polygon with three sides (sum of measures of three angles is 180°).
Scalene Triangle	Triangle with no sides of equal length.
Isosceles Triangle	Triangle with two sides of equal length.
Equilateral Triangle	Triangle with all sides of equal length.
Right Triangle	Triangle that contains a right angle.
Quadrilateral	Polygon with four sides (sum of measures of four angles is 360°).
Trapezoid	Quadrilateral with exactly one pair of opposite sides parallel.
Isosceles Trapezoid	Trapezoid with legs of equal length.
Parallelogram	Quadrilateral with both pairs of opposite sides parallel and equal in length.

Plane figures have length and width but no thickness or depth

Name	Description	Figure
Rhombus	Parallelogram with all sides of equal length.	
Rectangle	Parallelogram with four right angles.	
Square	Rectangle with all sides of equal length.	
Circle	All points in a plane the same distance from a fixed point called the **center.**	
Rectangular Solid	A solid with six sides, all of which are rectangles.	
Cube	A rectangular solid whose six sides are squares.	
Sphere	All points the same distance from a fixed point, called the center.	
Right Circular Cylinder	A cylinder consisting of two circular bases that are perpendicular to its altitude.	
Right Circular Cone	A cone with a circular base that is perpendicular to its altitude.	

B.3 REVIEW OF ANGLES, LINES, AND SPECIAL TRIANGLES

The word **geometry** is formed from the Greek words, **geo,** meaning earth, and **metron,** meaning measure. Geometry literally means to measure the earth.

This section contains a review of some basic geometric ideas. It will be assumed that fundamental ideas of geometry such as point, line, ray, and angle are known. In this appendix, the notation $\angle 1$ is read "angle 1" and the notation $m\angle 1$ is read "the measure of angle 1."

We first review types of angles.

Angles

A **right angle** is an angle whose measure is 90°. A right angle can be indicated by a square drawn at the vertex of the angle, as shown below.

An angle whose measure is more than 0° but less than 90° is called an **acute angle.**

An angle whose measure is greater than 90° but less than 180° is called an **obtuse angle.**

An angle whose measure is 180° is called a **straight angle.**

Two angles are said to be **complementary** if the sum of their measures is 90°. Each angle is called the **complement** of the other.

Two angles are said to be **supplementary** if the sum of their measures is 180°. Each angle is called the **supplement** of the other.

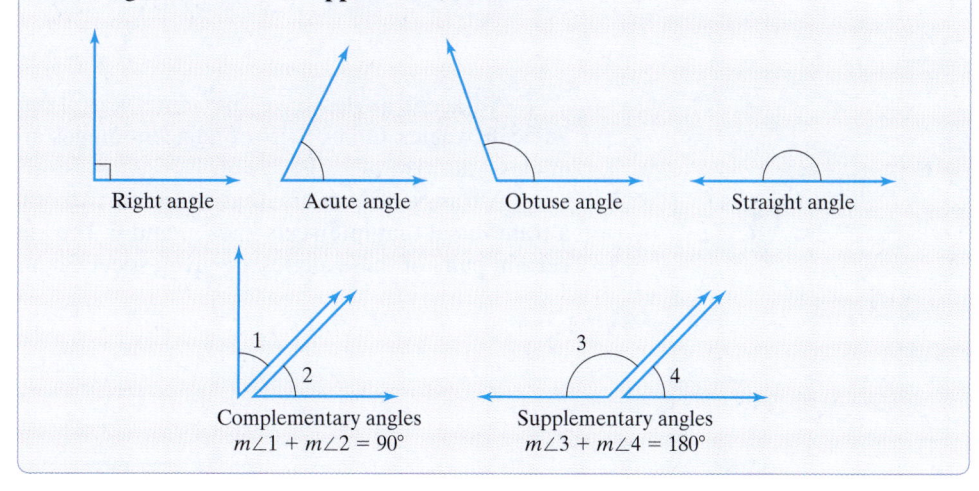

Right angle   Acute angle   Obtuse angle   Straight angle

Complementary angles
$m\angle 1 + m\angle 2 = 90°$

Supplementary angles
$m\angle 3 + m\angle 4 = 180°$

EXAMPLE 1 If an angle measures 28°, find its complement.

Solution Two angles are complementary if the sum of their measures is 90°. The complement of a 28° angle is an angle whose measure is $90° - 28° = 62°$. To check, notice that $28° + 62° = 90°$. □

Plane is an undefined term that we will describe. A plane can be thought of as a flat surface with infinite length and width, but no thickness. A plane is two dimensional. The arrows in the following diagram indicate that a plane extends indefinitely and has no boundaries.

Figures that lie on a plane are called **plane figures.** (See the description of common plane figures in Appendix B.2.) Lines that lie in the same plane are called **coplanar.**

> **Lines**
>
> Two lines are **parallel** if they lie in the same plane but never meet.
>
> **Intersecting lines** meet or cross in one point.
>
> Two lines that form right angles when they intersect are said to be **perpendicular.**
>
> Parallel lines   Intersecting lines   Intersecting lines that are perpendicular

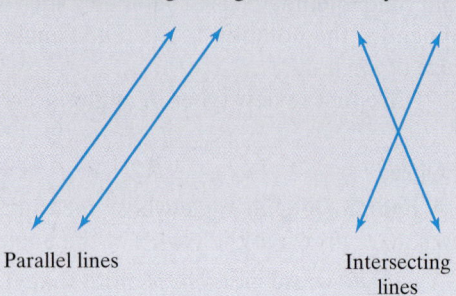

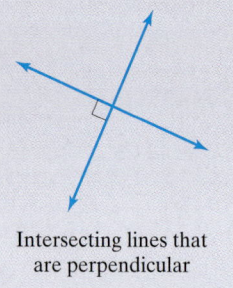

Two intersecting lines form **vertical angles.** Angles 1 and 3 are vertical angles. Also angles 2 and 4 are vertical angles. It can be shown that **vertical angles have equal measures.**

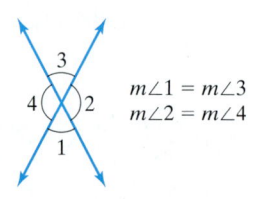

$$m\angle 1 = m\angle 3$$
$$m\angle 2 = m\angle 4$$

Adjacent angles have the same vertex and share a side. Angles 1 and 2 are adjacent angles. Other pairs of adjacent angles are angles 2 and 3, angles 3 and 4, and angles 4 and 1.

A **transversal** is a line that intersects two or more lines in the same plane. Line l is a transversal that intersects lines m and n. The eight angles formed are numbered and certain pairs of these angles are given special names.

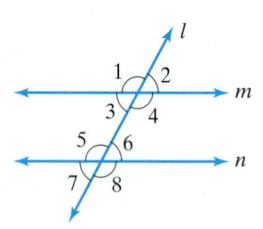

Corresponding angles: $\angle 1$ and $\angle 5, \angle 3$ and $\angle 7, \angle 2$ and $\angle 6$, and $\angle 4$ and $\angle 8$.
Exterior angles: $\angle 1, \angle 2, \angle 7$, and $\angle 8$.
Interior angles: $\angle 3, \angle 4, \angle 5$, and $\angle 6$.
Alternate interior angles: $\angle 3$ and $\angle 6, \angle 4$ and $\angle 5$.

These angles and parallel lines are related in the following manner.

> **Parallel Lines Cut by a Transversal**
> 1. If two parallel lines are cut by a transversal, then
>   a. corresponding angles are equal and
>   b. alternate interior angles are equal.
> 2. If corresponding angles formed by two lines and a transversal are equal, then the lines are parallel.
> 3. If alternate interior angles formed by two lines and a transversal are equal, then the lines are parallel.

EXAMPLE 2 Given that lines m and n are parallel and that the measure of angle 1 is 100°, find the measures of angles 2, 3, and 4.

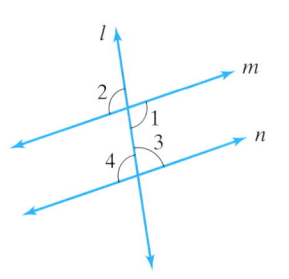

Solution $m\angle 2 = 100°$, since angles 1 and 2 are vertical angles.

$m\angle 4 = 100°$, since angles 1 and 4 are alternate interior angles.

$m\angle 3 = 180° - 100° = 80°$, since angles 4 and 3 are supplementary angles.

A **polygon** is the union of three or more coplanar line segments that intersect each other only at each end point, with each end point shared by exactly two segments.

A **triangle** is a polygon with three sides. The sum of the measures of the three angles of a triangle is 180°. In the following figure, $m\angle 1 + m\angle 2 + m\angle 3 = 180°$.

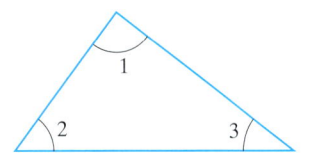

EXAMPLE 3 Find the measure of the third angle of the triangle shown.

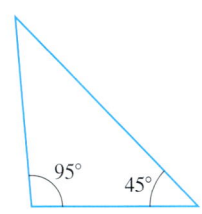

Solution The sum of the measures of the angles of a triangle is 180°. Since one angle measures 45° and the other angle measures 95°, the third angle measures $180° - 45° - 95° = 40°$.

Two triangles are **congruent** if they have the same size and the same shape. In congruent triangles, the measures of corresponding angles are equal and the lengths of corresponding sides are equal. The following triangles are congruent.

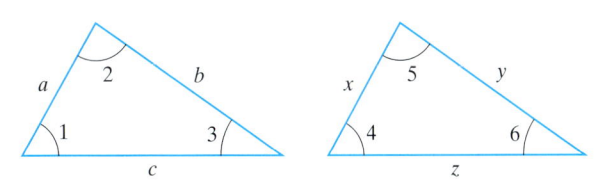

Corresponding angles are equal: $m\angle 1 = m\angle 4$, $m\angle 2 = m\angle 5$, and $m\angle 3 = m\angle 6$. Also, lengths of corresponding sides are equal: $a = x$, $b = y$, and $c = z$.

Any one of the following may be used to determine whether two triangles are congruent.

Congruent Triangles

1. If the measures of two angles of a triangle equal the measures of two angles of another triangle and the lengths of the sides between each pair of angles are equal, the triangles are congruent.

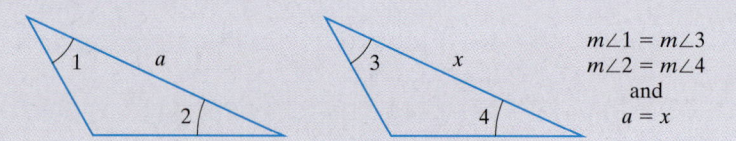

$m\angle 1 = m\angle 3$
$m\angle 2 = m\angle 4$
and
$a = x$

2. If the lengths of the three sides of a triangle equal the lengths of corresponding sides of another triangle, the triangles are congruent.

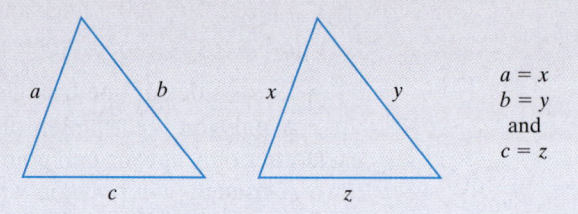

$a = x$
$b = y$
and
$c = z$

3. If the lengths of two sides of a triangle equal the lengths of corresponding sides of another triangle, and the measures of the angles between each pair of sides are equal, the triangles are congruent.

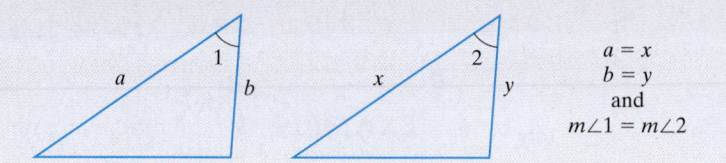

$a = x$
$b = y$
and
$m\angle 1 = m\angle 2$

Two triangles are **similar** if they have the same shape. In similar triangles, the measures of corresponding angles are equal and corresponding sides are in proportion. The following triangles are similar. (All similar triangles drawn in this appendix will be oriented the same.)

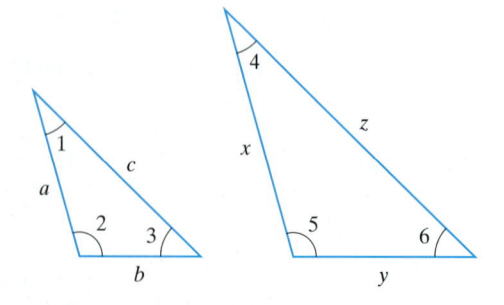

Corresponding angles are equal: $m\angle 1 = m\angle 4$, $m\angle 2 = m\angle 5$, and $m\angle 3 = m\angle 6$. Also, corresponding sides are proportional: $\dfrac{a}{x} = \dfrac{b}{y} = \dfrac{c}{z}$.

Any one of the following may be used to determine whether two triangles are similar.

> **Similar Triangles**
> 1. If the measures of two angles of a triangle equal the measures of two angles of another triangle, the triangles are similar. 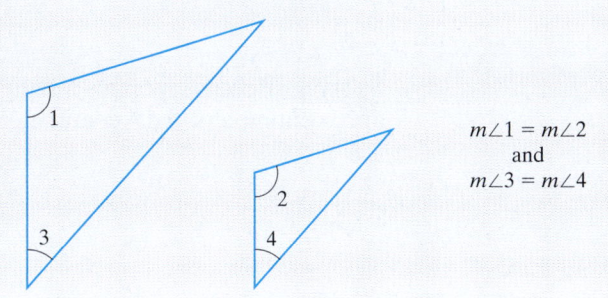
>
> $m\angle 1 = m\angle 2$
> and
> $m\angle 3 = m\angle 4$
>
> 2. If three sides of one triangle are proportional to three sides of another triangle, the triangles are similar. 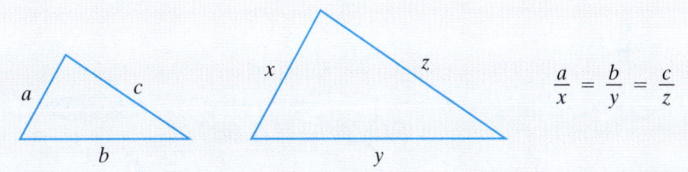
>
> $\dfrac{a}{x} = \dfrac{b}{y} = \dfrac{c}{z}$
>
> 3. If two sides of a triangle are proportional to two sides of another triangle and the measures of the included angles are equal, the triangles are similar. 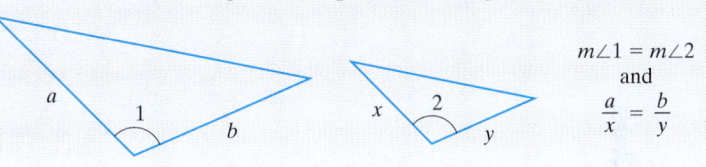
>
> $m\angle 1 = m\angle 2$
> and
> $\dfrac{a}{x} = \dfrac{b}{y}$

EXAMPLE 4 Given that the following triangles are similar, find the missing length x.

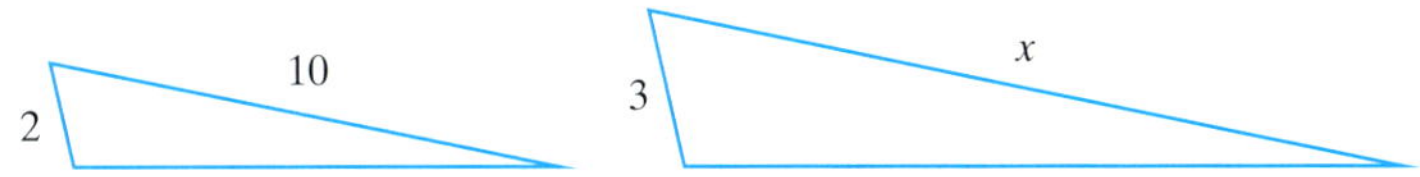

Solution Since the triangles are similar, corresponding sides are in proportion. Thus, $\dfrac{2}{3} = \dfrac{10}{x}$. To solve this equation for x, we multiply both sides by the LCD, $3x$.

$$3x\left(\dfrac{2}{3}\right) = 3x\left(\dfrac{10}{x}\right)$$
$$2x = 30$$
$$x = 15$$

The missing length is 15 units.   □

A **right triangle** contains a right angle. The side opposite the right angle is called the **hypotenuse**, and the other two sides are called the **legs**. The **Pythagorean theorem** gives a formula that relates the lengths of the three sides of a right triangle.

> **The Pythagorean Theorem**
> If a and b are the lengths of the legs of a right triangle, and c is the length of the hypotenuse, then $a^2 + b^2 = c^2$.

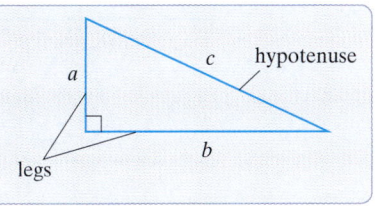

EXAMPLE 5 Find the length of the hypotenuse of a right triangle whose legs have lengths of 3 centimeters and 4 centimeters.

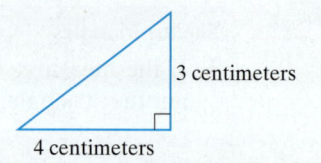

Solution Because we have a right triangle, we use the Pythagorean theorem. The legs are 3 centimeters and 4 centimeters, so let $a = 3$ and $b = 4$ in the formula.

$$a^2 + b^2 = c^2$$
$$3^2 + 4^2 = c^2$$
$$9 + 16 = c^2$$
$$25 = c^2$$

Since c represents a length, we assume that c is positive. Thus, if c^2 is 25, c must be 5. The hypotenuse has a length of 5 centimeters.

B.3 EXERCISE SET

Find the complement of each angle. See Example 1.

1. $19°$
2. $65°$
3. $70.8°$
4. $45\frac{2}{3}°$
5. $11\frac{1}{4}°$
6. $19.6°$

Find the supplement of each angle.

7. $150°$
8. $90°$
9. $30.2°$
10. $81.9°$
11. $79\frac{1}{2}°$
12. $165\frac{8}{9}°$

13. If lines m and n are parallel, find the measures of angles 1 through 7. See Example 2.

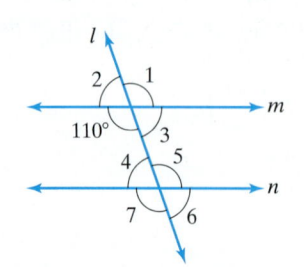

14. If lines m and n are parallel, find the measures of angles 1 through 5. See Example 2.

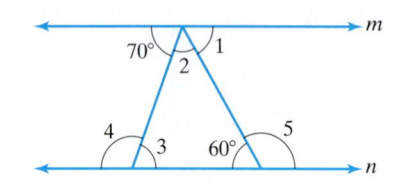

In each of the following, the measures of two angles of a triangle are given. Find the measure of the third angle. See Example 3.

15. $11°, 79°$
16. $8°, 102°$
17. $25°, 65°$
18. $44°, 19°$
19. $30°, 60°$
20. $67°, 23°$

In each of the following, the measure of one angle of a right triangle is given. Find the measures of the other two angles.

21. $45°$
22. $60°$
23. $17°$
24. $30°$
25. $39\frac{3}{4}°$
26. $72.6°$

Given that each of the following pairs of triangles is similar, find the missing lengths. See Example 4.

27.

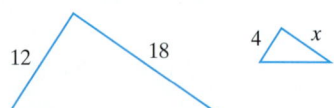

28.

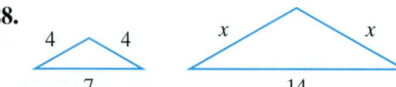

29.

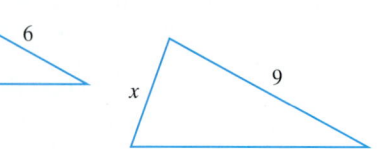

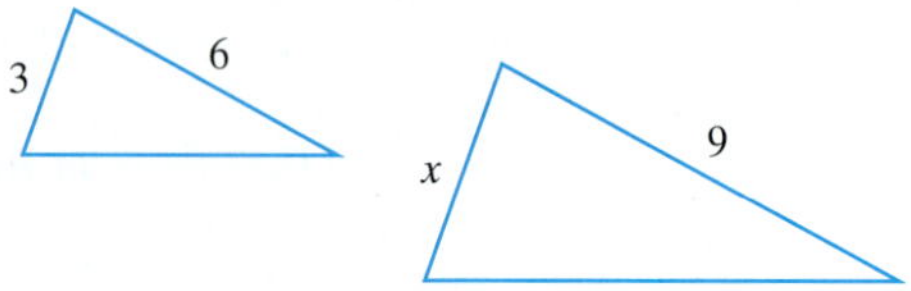

30.

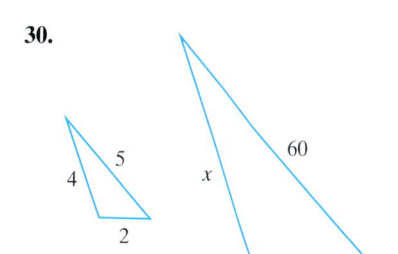

Use the Pythagorean theorem to find the missing lengths in the right triangles. See Example 5.

31.

32.

33.

34.

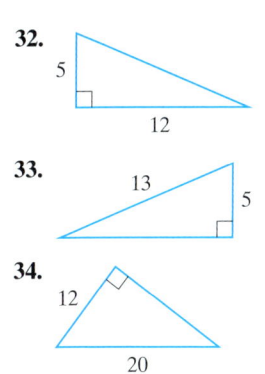

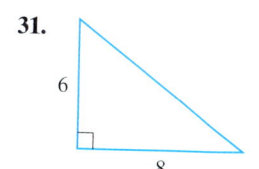

B.4 REVIEW OF VOLUME AND SURFACE AREA

A **convex solid** is a set of points, S, not all in one plane, such that for any two points A and B in S, all points between A and B are also in S. In this appendix, we will find the volume and surface area of special types of solids called polyhedrons. A solid formed by the intersection of a finite number of planes is called a **polyhedron.** The box below is an example of a polyhedron.

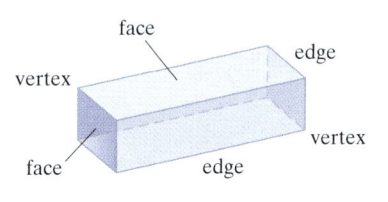

Polyhedron

Each of the plane regions of the polyhedron is called a **face** of the polyhedron. If the intersection of two faces is a line segment, this line segment is an **edge** of the polyhedron. The intersections of the edges are the **vertices** of the polyhedron.

Volume is a measure of the space of a solid. The volume of a box or can, for example, is the amount of space inside. Volume can be used to describe the amount of juice in a pitcher or the amount of concrete needed to pour a foundation for a house.

The volume of a solid is the number of **cubic units** in the solid. A cubic centimeter and a cubic inch are illustrated.

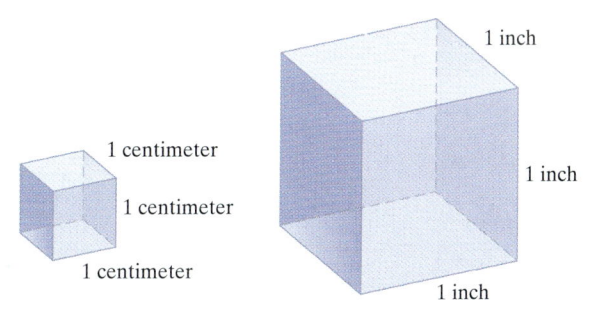

1 cubic centimeter   **1 cubic inch**

The **surface area** of a polyhedron is the sum of the areas of the faces of the polyhedron. For example, each face of the cube to the left above has an area of 1 square centimeter. Since there are 6 faces of the cube, the sum of the areas of the faces is 6 square centimeters. Surface area can be used to describe the amount of material needed to cover or form a solid. Surface area is measured in square units.

Formulas for finding the volumes, V, and surface areas, SA, of some common solids are given next.

Volume and Surface Area Formulas of Common Solids

Solid	Formulas
RECTANGULAR SOLID 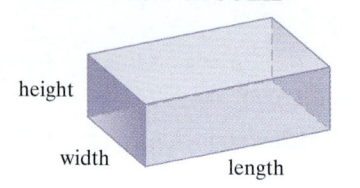	$V = lwh$ $SA = 2lh + 2wh + 2lw$ where h = height, w = width, l = length
CUBE 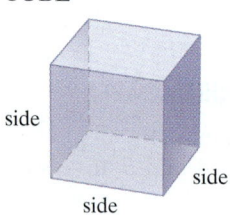	$V = s^3$ $SA = 6s^2$ where s = side
SPHERE 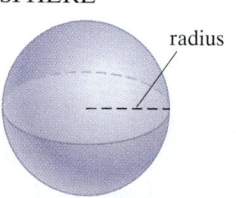	$V = \frac{4}{3}\pi r^3$ $SA = 4\pi r^2$ where r = radius
CIRCULAR CYLINDER 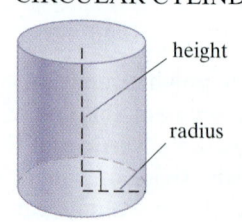	$V = \pi r^2 h$ $SA = 2\pi rh + 2\pi r^2$ where h = height, r = radius
CONE 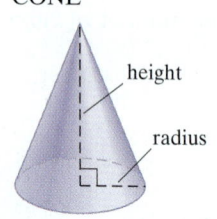	$V = \frac{1}{3}\pi r^2 h$ $SA = \pi r\sqrt{r^2 + h^2} + \pi r^2$ where h = height, r = radius
SQUARE-BASED PYRAMID 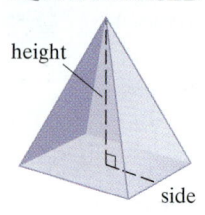	$V = \frac{1}{3}s^2 h$ $SA = B + \frac{1}{2}pl$ where B = area of base; p = perimeter of base, h = height, s = side, l = slant height

> **Helpful Hint**
> Volume is measured in cubic units. Surface area is measured in square units.

EXAMPLE 1 Find the volume and surface area of a rectangular box that is 12 inches long, 6 inches wide, and 3 inches high.

Solution Let $h = 3$ in., $l = 12$ in., and $w = 6$ in.

$$V = lwh$$
$$V = 12 \text{ inches} \cdot 6 \text{ inches} \cdot 3 \text{ inches} = 216 \text{ cubic inches}$$

The volume of the rectangular box is 216 cubic inches.

$$SA = 2lh + 2wh + 2lw$$
$$= 2(12 \text{ in.})(3 \text{ in.}) + 2(6 \text{ in.})(3 \text{ in.}) + 2(12 \text{ in.})(6 \text{ in.})$$
$$= 72 \text{ sq in.} + 36 \text{ sq in.} + 144 \text{ sq in.}$$
$$= 252 \text{ sq in.}$$

The surface area of rectangular box is 252 square inches. □

EXAMPLE 2 Find the volume and surface area of a ball of radius 2 inches. Give the exact volume and surface area and then use the approximation $\frac{22}{7}$ for π.

Solution

$$V = \frac{4}{3}\pi r^3 \qquad \text{Formula for volume of a sphere}$$

$$V = \frac{4}{3} \cdot \pi (2 \text{ in.})^3 \qquad \text{Let } r = 2 \text{ inches.}$$

$$= \frac{32}{3}\pi \text{ cu in.} \qquad \text{Simplify.}$$

$$\approx \frac{32}{3} \cdot \frac{22}{7} \text{ cu in.} \qquad \text{Approximate } \pi \text{ with } \frac{22}{7}.$$

$$= \frac{704}{21} \text{ or } 33\frac{11}{21} \text{ cu in.}$$

The volume of the sphere is exactly $\frac{32}{3}\pi$ cubic inches or approximately $33\frac{11}{21}$ cubic inches.

$$SA = 4\pi r^2 \qquad \text{Formula for surface area}$$
$$SA = 4 \cdot \pi (2 \text{ in.})^2 \qquad \text{Let } r = 2 \text{ inches.}$$
$$= 16\pi \text{ sq in.} \qquad \text{Simplify.}$$
$$\approx 16 \cdot \frac{22}{7} \text{ sq in.} \qquad \text{Approximate } \pi \text{ with } \frac{22}{7}.$$
$$= \frac{352}{7} \text{ or } 50\frac{2}{7} \text{ sq in.}$$

The surface area of the sphere is exactly 16π square inches or approximately $50\frac{2}{7}$ square inches.

□

B.4 EXERCISE SET

Find the volume and surface area of each solid. See Examples 1 and 2. For formulas that contain π, give an exact answer and then approximate using $\frac{22}{7}$ for π.

1. (Rectangular box: 4 in. × 3 in. × 6 in.)

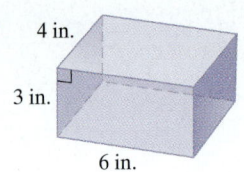

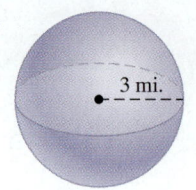

2. (Sphere with radius 3 mi.)

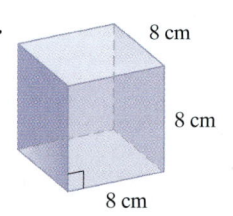

3. (Cube: 8 cm × 8 cm × 8 cm)

4. (Rectangular box: 4 cm × 4 cm × 8 cm)

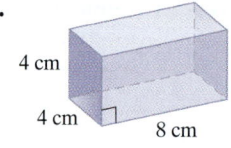

5. (For surface area, use 3.14 for π and round to two decimal places.) (Cone: height 3 yd, radius 2 yd)

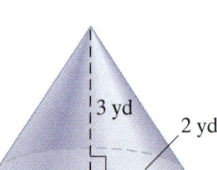

6. (Cylinder: height 10 ft, radius 6 ft)

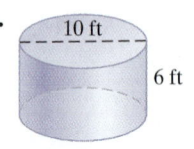

7. (Sphere with diameter 10 in.)

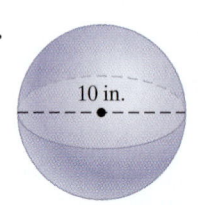

8. Find the volume only. (Cone: radius $1\frac{3}{4}$ in., height 9 in.)

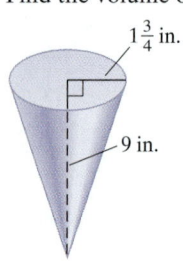

9. (Square pyramid: slant 5 cm, height 4 cm, base 6 cm)

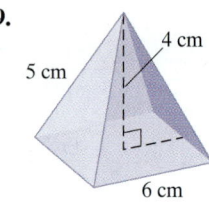

10. (Cube: 1 ft on each side)

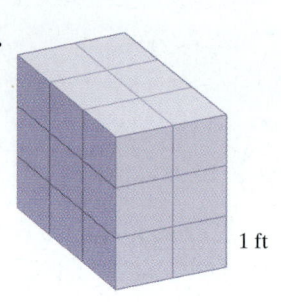

Solve.

11. Find the volume of a cube with edges of $1\frac{1}{3}$ inches.

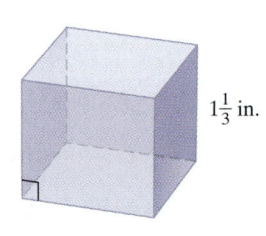

12. A water storage tank is in the shape of a cone with the pointed end down. If the radius is 14 ft and the depth of the tank is 15 ft, approximate the volume of the tank in cubic feet. Use $\frac{22}{7}$ for π.

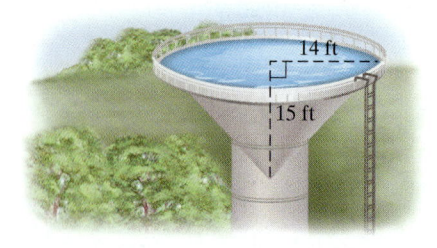

13. Find the surface area of a rectangular box 2 ft by 1.4 ft by 3 ft.

14. Find the surface area of a box in the shape of a cube that is 5 ft on each side.

15. Find the volume of a pyramid with a square base 5 in. on a side and a height of $1\frac{3}{10}$ in.

16. Approximate to the nearest hundredth the volume of a sphere with a radius of 2 cm. Use 3.14 for π.

17. A paperweight is in the shape of a square-based pyramid 20 cm tall. If an edge of the base is 12 cm, find the volume of the paperweight.

18. A bird bath is made in the shape of a hemisphere (half-sphere). If its radius is 10 in., approximate the volume. Use $\frac{22}{7}$ for π.

19. Find the exact surface area of a sphere with a radius of 7 in.

20. A tank is in the shape of a cylinder 8 ft tall and 3 ft in radius. Find the exact surface area of the tank.

21. Find the volume of a rectangular block of ice 2 ft by $2\frac{1}{2}$ ft by $1\frac{1}{2}$ ft.

22. Find the capacity (volume in cubic feet) of a rectangular ice chest with inside measurements of 3 ft by $1\frac{1}{2}$ ft by $1\frac{3}{4}$ ft.

23. An ice cream cone with a 4-cm diameter and 3-cm depth is filled exactly level with the top of the cone. Approximate how much ice cream (in cubic centimeters) is in the cone. Use $\frac{22}{7}$ for π.

24. A child's toy is in the shape of a square-based pyramid 10 in. tall. If an edge of the base is 7 in., find the volume of the toy.

Appendix C

Additional Exercises on Proportion and Proportion Applications

1. $\dfrac{30}{10} = \dfrac{15}{y}$
2. $\dfrac{16}{20} = \dfrac{z}{35}$
3. $\dfrac{8}{15} = \dfrac{z}{6}$
4. $\dfrac{12}{10} = \dfrac{z}{16}$
5. $\dfrac{-3.5}{12.5} = \dfrac{-7}{n}$
6. $\dfrac{-0.2}{0.7} = \dfrac{-8}{n}$
7. $\dfrac{n}{0.6} = \dfrac{0.05}{12}$
8. $\dfrac{7.8}{13} = \dfrac{n}{2.6}$
9. $\dfrac{8}{\frac{2}{3}} = \dfrac{24}{n}$
10. $\dfrac{12}{\frac{3}{4}} = \dfrac{48}{n}$
11. $\dfrac{7}{9} = \dfrac{35}{3x}$
12. $\dfrac{5}{6} = \dfrac{40}{3x}$
13. $\dfrac{7x}{18} = \dfrac{5}{3}$
14. $\dfrac{6x}{5} = \dfrac{19}{10}$
15. $\dfrac{11}{7} = \dfrac{4}{x+1}$
16. $\dfrac{12}{5} = \dfrac{7}{x-1}$
17. $\dfrac{x-3}{2x+1} = \dfrac{4}{9}$
18. $\dfrac{x+1}{3x-4} = \dfrac{3}{8}$
19. $\dfrac{2x+1}{4} = \dfrac{6x-1}{5}$
20. $\dfrac{11x+2}{3} = \dfrac{10x-5}{2}$

Solve. A self-tanning lotion advertises that a 3-oz bottle will provide four applications.

21. Jen Haddad found a great deal on a 14-oz bottle of the self-tanning lotion she had been using. Based on the advertising claims, how many applications of the self-tanner should Jen expect? Round down to the whole number.

22. The Community College thespians need fake tans for a play they are doing. If the play has a cast of 35, how many ounces of self-tanning lotion should the cast purchase? Round up to the next whole number onces.

The school's computer lab goes through 5 reams of printer paper every 3 weeks.

23. Find out how long a case of printer paper is likely to last (a case of paper holds 8 reams of paper). Round to the nearest week.

24. How many cases of printer paper should be purchased to last the entire semester of 15 weeks? Round to the next case.

A recipe for pancakes calls for 2 cups flour and $1\dfrac{1}{2}$ cup milk to make a serving for four people.

25. Ming has plenty of flour, but only 4 cups milk. How many servings can he make?

26. The swim team has a weekly breakfast after early practice. How much flour will it take to make pancakes for 18 swimmers?

Solve.

27. A 16-oz grande Tazo Black Iced Tea at Starbucks has 80 calories. How many calories are there in a 24-oz venti Tazo Black Iced Tea? (*Source:* Starbucks Coffee Company)

28. A 16-oz nonfat Caramel Macchiato at Starbucks has 220 calories. How many calories are there in a 12-oz nonfat Caramel Macchiato? (*Source:* Starbucks Coffee Company)

29. Mosquitos are annoying insects. To eliminate mosquito larvae, a certain granular substance can be applied to standing water in a ratio of 1 tsp per 25 sq ft of standing water.
    a. At this rate, find how many teaspoons of granules must be used for 450 square feet.
    b. If 3 tsp = 1 tbsp, how many tablespoons of granule must be used?

30. Another type of mosquito control is liquid, where 3 oz of pesticide is mixed with 100 oz of water. This mixture is sprayed on roadsides to control mosquito breeding grounds hidden by tall grass.
    a. If one mixture of water with this pesticide can treat 150 feet of roadway, how many ounces of pesticide are needed to treat one mile? (*Hint:* 1 mile = 5280 feet)
    b. If 8 liquid ounces equals one cup, write your answer to part a in cups. Round to the nearest cup.

31. The daily supply of oxygen for one person is provided by 625 square feet of lawn. A total of 3750 square feet of lawn would provide the daily supply of oxygen for how many people? (*Source:* Professional Lawn Care Association of America)

32. In the United States, approximately 71 million of the 200 million cars and light trucks in service have driver-side air bags. In a parking lot containing 800 cars and light trucks, how many would be expected to have driver-side air bags? (*Source:* Insurance Institute for Highway Safety)

33. A student would like to estimate the height of the Statue of Liberty in New York City's harbor. The length of the Statue of Liberty's right arm is 42 feet. The student's right arm is 2 feet long and her height is $5\frac{1}{3}$ feet. Use this information to estimate the height of the Statue of Liberty. How close is your estimate to the statue's actual height of 111 feet, 1 inch from heel to top of head? (*Source:* National Park Service)

34. The length of the Statue of Liberty's index finger is 8 feet while the height to the top of the head is about 111 feet. Suppose your measurements are proportionaly the same as this statue and your height is 5 feet.

  a. Use this information to find the proposed length of your index finger. Give an exact measurement and then a decimal rounded to the nearest hundredth.

  b. Measure your index finger and write it as decimal in feet rounded to the nearest hundredth. How close is the length of your index finger to the answer to **a**? Explain why.

35. There are 72 milligrams of cholesterol in a 3.5 ounce serving of lobster. How much cholesterol is in 5 ounces of lobster? Round to the nearest tenth of a milligram. (*Source:* The National Institute of Health)

36. There are 76 milligrams of cholesterol in a 3-ounce serving of skinless chicken. How much cholesterol is in 8 ounces of chicken? (*Source:* USDA)

37. Medication is prescribed in 7 out of every 10 hospital emergency room visits that involve an injury. If a large urban hospital had 620 emergency room visits involving an injury in the past month, how many of these visits would you expect included a prescription for medication? (*Source:* National Center for Health Statistics)

38. Currently in the American population of people aged 65 years old and older, there are 145 women for every 100 men. In a nursing home with 280 male residents over the age of 65, how many female residents over the age of 65 would be expected? (*Source:* U.S. Bureau of the Census)

39. One out of three American adults has worked in the restaurant industry at some point during his or her life. In an office of 84 workers, how many of these people would you expect to have worked in the restaurant industry at some point? (*Source:* National Restaurant Association)

40. One pound of firmly packed brown sugar yields $2\frac{1}{4}$ cups. How many pounds of brown sugar will be required in a recipe that calls for 6 cups of firmly packed brown sugar? (*Source:* Based on data from *Family Circle* magazine)

Appendix D

Operations on Decimals

To **add** or **subtract** decimals, write the numbers vertically with decimal points lined up. Add or subtract as with whole numbers and place the decimal point in the answer directly below the decimal points in the problem.

EXAMPLE 1   Add $5.87 + 23.279 + 0.003$.

Solution
$$\begin{array}{r} 5.87 \\ 23.279 \\ +0.003 \\ \hline 29.152 \end{array}$$

EXAMPLE 2   Subtract $32.15 - 11.237$.

Solution
$$\begin{array}{r} 3\overset{1}{2}.\overset{11}{\cancel{1}}\overset{4}{\cancel{5}}\overset{10}{\cancel{0}} \\ -11.237 \\ \hline 20.913 \end{array}$$

To **multiply** decimals, multiply the numbers as if they were whole numbers. The decimal point in the product is placed so that the number of decimal places in the product is the same as the sum of the number of decimal places in the factors.

EXAMPLE 3   Multiply 0.072×3.5.

Solution
$$\begin{array}{r} 0.072 \quad \text{3 decimal places} \\ \times 3.5 \quad \text{1 decimal place} \\ \hline 360 \\ 216 \\ \hline 0.2520 \quad \text{4 decimal places} \end{array}$$

To **divide** decimals, move the decimal point in the divisor to the right of the last digit. Move the decimal point in the dividend the same number of places that the decimal point in the divisor was moved. The decimal point in the quotient lies directly above the decimal point in the dividend.

EXAMPLE 4   Divide $9.46 \div 0.04$.

Solution
$$\begin{array}{r} 236.5 \\ 04.\overline{)946.0} \\ \underline{-8} \\ 14 \\ \underline{-12} \\ 26 \\ \underline{-24} \\ 20 \\ \underline{-20} \end{array}$$

APPENDIX D | EXERCISE SET

Perform the indicated operations.

1. $9.076 + 8.004$
2. 6.3
   $\times 0.05$
3. 27.004
   -14.2
4. 0.0036
   7.12
   32.502
   $+0.05$
5. 107.92
   $+3.04$
6. $7.2 \div 4$
7. $10 - 7.6$
8. $40 \div 0.25$
9. $126.32 - 97.89$
10. 3.62
    7.11
    12.36
    4.15
    $+2.29$
11. 3.25
    $\times 70$
12. 26.014
    -7.8
13. $8.1 \div 3$
14. 1.2366
    0.005
    15.17
    $+0.97$
15. $55.405 - 6.1711$
16. $8.09 + 0.22$
17. $60 \div 0.75$
18. $20 - 12.29$
19. $7.612 \div 100$
20. 8.72
    1.12
    14.86
    3.98
    $+1.99$
21. $12.312 \div 2.7$
22. $0.443 \div 100$
23. 569.2
    71.25
    $+8.01$
24. $3.706 - 2.91$
25. $768 - 0.17$
26. $63 \div 0.28$
27. $12 + 0.062$
28. $0.42 + 18$
29. $76 - 14.52$
30. $1.1092 \div 0.47$
31. $3.311 \div 0.43$
32. $7.61 + 0.0004$
33. 762.12
    89.7
    $+11.55$
34. $444 \div 0.6$
35. $23.4 - 0.821$
36. $3.7 + 5.6$
37. $476.12 - 112.97$
38. $19.872 \div 0.54$
39. $0.007 + 7$
40. 51.77
    $+3.6$

Appendix E

Mean, Median, and Mode

It is sometimes desirable to be able to describe a set of data, or a set of numbers, by a single "middle" number. Three such **measures of central tendency** are the mean, the median, and the mode.

The most common measure of central tendency is the mean (sometimes called the arithmetic mean or the average). The **mean** of a set of data items, denoted by $\bar{x}$, is the sum of the items divided by the number of items.

EXAMPLE 1 Seven students in a psychology class conducted an experiment on mazes. Each student was given a pencil and asked to successfully complete the same maze. The timed results are below.

Student	Ann	Thanh	Carlos	Jesse	Melinda	Ramzi	Dayni
Time (Seconds)	13.2	11.8	10.7	16.2	15.9	13.8	18.5

a. Who completed the maze in the shortest time? Who completed the maze in the longest time?
b. Find the mean.
c. How many students took longer than the mean time? How many students took shorter than the mean time?

Solution

a. Carlos completed the maze in 10.7 seconds, the shortest time. Dayni completed the maze in 18.5 seconds, the longest time.
b. To find the mean, $\bar{x}$, find the sum of the data items and divide by 7, the number of items.

$$\bar{x} = \frac{13.2 + 11.8 + 10.7 + 16.2 + 15.9 + 13.8 + 18.5}{7} = \frac{100.1}{7} = 14.3$$

c. Three students, Jesse, Melinda, and Dayni, had times longer than the mean time. Four students, Ann, Thanh, Carlos, and Ramzi, had times shorter than the mean time. □

Two other measures of central tendency are the median and the mode.

The **median** of an ordered set of numbers is the middle number. If the number of items is even, the median is the mean of the two middle numbers. The **mode** of a set of numbers is the number that occurs most often. It is possible for a data set to have no mode or more than one mode.

EXAMPLE 2 Find the median and the mode of the following list of numbers. These numbers were high temperatures for fourteen consecutive days in a city in Montana.

76, 80, 85, 86, 89, 87, 82, 77, 76, 79, 82, 89, 89, 92

Solution

First, write the numbers in order.

76, 76, 77, 79, 80, 82, 82, 85, 86, 87, 89, 89, 89, 92

Since there are an even number of items, the median is the mean of the two middle numbers.

$$\text{median} = \frac{82 + 85}{2} = 83.5$$

The mode is 89, since 89 occurs most often.

APPENDIX E | EXERCISE SET

For each of the following data sets, find the mean, the median, and the mode. If necessary, round the mean to one decimal place.

1. 21, 28, 16, 42, 38
2. 42, 35, 36, 40, 50
3. 7.6, 8.2, 8.2, 9.6, 5.7, 9.1
4. 4.9, 7.1, 6.8, 6.8, 5.3, 4.9
5. 0.2, 0.3, 0.5, 0.6, 0.6, 0.9, 0.2, 0.7, 1.1
6. 0.6, 0.6, 0.8, 0.4, 0.5, 0.3, 0.7, 0.8, 0.1
7. 231, 543, 601, 293, 588, 109, 334, 268
8. 451, 356, 478, 776, 892, 500, 467, 780

The eight tallest buildings in the United States are listed below. Use this table for Exercises 9 through 12.

Building	Height (feet)
Sears Tower, Chicago, IL	1454
Empire State, New York, NY	1250
Amoco, Chicago, IL	1136
John Hancock Center, Chicago, IL	1127
First Interstate World Center, Los Angeles, CA	1107
Chrysler, New York, NY	1046
NationsBank Tower, Atlanta, GA	1023
Texas Commerce Tower, Houston, TX	1002

9. Find the mean height for the five tallest buildings.
10. Find the median height for the five tallest buildings.
11. Find the median height for the eight tallest buildings.
12. Find the mean height for the eight tallest buildings.

During an experiment, the following times (in seconds) were recorded: 7.8, 6.9, 7.5, 4.7, 6.9, 7.0.

13. Find the mean. Round to the nearest tenth.
14. Find the median.
15. Find the mode.

In a mathematics class, the following test scores were recorded for a student: 86, 95, 91, 74, 77, 85.

16. Find the mean. Round to the nearest hundredth.
17. Find the median.
18. Find the mode.

The following pulse rates were recorded for a group of fifteen students: 78, 80, 66, 68, 71, 64, 82, 71, 70, 65, 70, 75, 77, 86, 72.

19. Find the mean.
20. Find the median.
21. Find the mode.
22. How many rates were higher than the mean?
23. How many rates were lower than the mean?
24. Have each student in your algebra class take his/her pulse rate. Record the data and find the mean, the median, and the mode.

Find the missing numbers in each list of numbers. (These numbers are not necessarily in numerical order.)

25. __, __, 16, 18, __
    The mode is 21.   The mean is 20.
26. __, __, __, __, 40
    The mode is 35.   The median is 37.   The mean is 38.

Appendix F

Tables

F.1 TABLE OF SQUARES AND SQUARE ROOTS

n	n^2	$\sqrt{n}$	n	n^2	$\sqrt{n}$
1	1	1.000	51	2601	7.141
2	4	1.414	52	2704	7.211
3	9	1.732	53	2809	7.280
4	16	2.000	54	2916	7.348
5	25	2.236	55	3025	7.416
6	36	2.449	56	3136	7.483
7	49	2.646	57	3249	7.550
8	64	2.828	58	3364	7.616
9	81	3.000	59	3481	7.681
10	100	3.162	60	3600	7.746
11	121	3.317	61	3721	7.810
12	144	3.464	62	3844	7.874
13	169	3.606	63	3969	7.937
14	196	3.742	64	4096	8.000
15	225	3.873	65	4225	8.062
16	256	4.000	66	4356	8.124
17	289	4.123	67	4489	8.185
18	324	4.243	68	4624	8.246
19	361	4.359	69	4761	8.307
20	400	4.472	70	4900	8.367
21	441	4.583	71	5041	8.426
22	484	4.690	72	5184	8.485
23	529	4.796	73	5329	8.544
24	576	4.899	74	5476	8.602
25	625	5.000	75	5625	8.660
26	676	5.099	76	5776	8.718
27	729	5.196	77	5929	8.775
28	784	5.292	78	6084	8.832
29	841	5.385	79	6241	8.888
30	900	5.477	80	6400	8.944
31	961	5.568	81	6561	9.000
32	1024	5.657	82	6724	9.055
33	1089	5.745	83	6889	9.110
34	1156	5.831	84	7056	9.165
35	1225	5.916	85	7225	9.220
36	1296	6.000	86	7396	9.274
37	1369	6.083	87	7569	9.327
38	1444	6.164	88	7744	9.381
39	1521	6.245	89	7921	9.434
40	1600	6.325	90	8100	9.487
41	1681	6.403	91	8281	9.539
42	1764	6.481	92	8464	9.592
43	1849	6.557	93	8649	9.644
44	1936	6.633	94	8836	9.695
45	2025	6.708	95	9025	9.747
46	2116	6.782	96	9216	9.798
47	2209	6.856	97	9409	9.849
48	2304	6.928	98	9604	9.899
49	2401	7.000	99	9801	9.950
50	2500	7.071	100	10,000	10.000

F.2 TABLE OF PERCENT, DECIMAL, AND FRACTION EQUIVALENTS

Percent, Decimal, and Fraction Equivalents		
Percent	**Decimal**	**Fraction**
1%	0.01	$\frac{1}{100}$
5%	0.05	$\frac{1}{20}$
10%	0.1	$\frac{1}{10}$
12.5% or $12\frac{1}{2}$%	0.125	$\frac{1}{8}$
$16.\overline{6}$% or $16\frac{2}{3}$%	$0.1\overline{6}$	$\frac{1}{6}$
20%	0.2	$\frac{1}{5}$
25%	0.25	$\frac{1}{4}$
30%	0.3	$\frac{3}{10}$
$33.\overline{3}$% or $33\frac{1}{3}$%	$0.\overline{3}$	$\frac{1}{3}$
37.5% or $37\frac{1}{2}$%	0.375	$\frac{3}{8}$
40%	0.4	$\frac{2}{5}$
50%	0.5	$\frac{1}{2}$
60%	0.6	$\frac{3}{5}$
62.5% or $62\frac{1}{2}$%	0.625	$\frac{5}{8}$
$66.\overline{6}$% or $66\frac{2}{3}$%	$0.\overline{6}$	$\frac{2}{3}$
70%	0.7	$\frac{7}{10}$
75%	0.75	$\frac{3}{4}$
80%	0.8	$\frac{4}{5}$
$83.\overline{3}$% or $83\frac{1}{3}$%	$0.8\overline{3}$	$\frac{5}{6}$
87.5% or $87\frac{1}{2}$%	0.875	$\frac{7}{8}$
90%	0.9	$\frac{9}{10}$
100%	1.0	1
110%	1.1	$1\frac{1}{10}$
125%	1.25	$1\frac{1}{4}$
$133.\overline{3}$% or $133\frac{1}{3}$%	$1.\overline{3}$	$1\frac{1}{3}$
150%	1.5	$1\frac{1}{2}$
$166.\overline{6}$% or $166\frac{2}{3}$%	$1.\overline{6}$	$1\frac{2}{3}$
175%	1.75	$1\frac{3}{4}$
200%	2.0	2

Answers to Selected Exercises

CHAPTER 1 REVIEW OF REAL NUMBERS

Section 1.2
Practice Exercises
1. a. $<$ b. $>$ c. $<$ 2. a. True   b. False   c. True   d. True   3. a. $3 < 8$ b. $15 \geq 9$ c. $6 \neq 7$ 4. -52 5. a. 25   b. 25   c. $25, -15, -99$ d. $25, \frac{7}{3}, -15, -\frac{3}{4}, -3.7, 8.8, -99$ e. $\sqrt{5}$ f. $25, \frac{7}{3}, -15, -\frac{3}{4}, \sqrt{5}, -3.7, 8.8, -99$ 6. a. $<$ b. $>$ c. $=$ 7. a. 8   b. 9   c. 2.5   d. $\frac{5}{11}$ e. $\sqrt{3}$ 8. a. $=$ b. $>$ c. $<$ d. $>$ e. $<$

Vocabulary and Readiness Check 1.2
1. whole   3. inequality   5. real   7. irrational

Exercise Set 1.2
1. $>$ 3. $=$ 5. $<$ 7. $<$ 9. $32 < 212$ 11. $2631 > 2456$ 13. true   15. false   17. false   19. true   21. $30 \leq 45$ 23. $8 < 12$ 25. $5 \geq 4$ 27. $15 \neq -2$ 29. $535; -8$ 31. $-21{,}350$ 33. $350; -126$ 35. 1998, 1999   37. 1998, 1999, 2000   39. 279 million $>$ 273 million   41. whole, integers, rational, real   43. integers, rational, real   45. natural, whole, integers, rational, real   47. rational, real   49. irrational, real   51. false   53. true   55. true   57. true   59. false   61. $>$ 63. $>$ 65. $<$ 67. $<$ 69. $>$ 71. $=$ 73. $<$ 75. $<$ 77. $-0.04 > -26.7$ 79. sun   81. sun   83. $20 \leq 25$ 85. $6 > 0$ 87. $-12 < -10$ 89. answers may vary

Section 1.3
Practice Exercises
1. a. $2 \cdot 2 \cdot 3 \cdot 3$ b. $3 \cdot 5 \cdot 5$ 2. a. $\frac{7}{8}$ b. $\frac{16}{3}$ c. $\frac{7}{25}$ 3. $\frac{7}{24}$ 4. a. $\frac{27}{16}$ b. $\frac{1}{36}$ c. $\frac{5}{2}$ 5. a. 1   b. $\frac{6}{5}$ c. $\frac{4}{5}$ d. $\frac{1}{2}$ 6. $\frac{14}{21}$ 7. a. $\frac{46}{77}$ b. $3\frac{15}{26}$ c. $\frac{1}{2}$

Vocabulary and Readiness Check 1.3
1. fraction   3. product   5. factors, product   7. equivalent   9. $\frac{1}{4}$ 11. $\frac{2}{5}$

Exercise Set 1.3
1. $3 \cdot 11$ 3. $2 \cdot 7 \cdot 7$ 5. $2 \cdot 2 \cdot 5$ 7. $3 \cdot 5 \cdot 5$ 9. $3 \cdot 3 \cdot 5$ 11. $\frac{1}{2}$ 13. $\frac{2}{3}$ 15. $\frac{3}{7}$ 17. $\frac{3}{5}$ 19. $\frac{3}{8}$ 21. $\frac{1}{2}$ 23. $\frac{6}{7}$ 25. 15   27. $\frac{1}{6}$ 29. $\frac{25}{27}$ 31. $\frac{11}{20}$ sq mi   33. $\frac{3}{5}$ 35. 1   37. $\frac{1}{3}$ 39. $\frac{9}{35}$ 41. $\frac{21}{30}$ 43. $\frac{4}{18}$ 45. $\frac{16}{20}$ 47. $\frac{23}{21}$ 49. $1\frac{2}{3}$ 51. $\frac{5}{66}$ 53. $\frac{7}{5}$ 55. $\frac{1}{5}$ 57. $\frac{3}{8}$ 59. $\frac{1}{9}$ 61. $\frac{5}{7}$ 63. $\frac{65}{21}$ 65. $\frac{2}{5}$ 67. $\frac{9}{7}$ 69. $\frac{3}{4}$ 71. $\frac{17}{3}$ 73. $\frac{7}{26}$ 75. 1   77. $\frac{1}{5}$ 79. $5\frac{1}{6}$ 81. $\frac{17}{18}$ 83. $55\frac{1}{4}$ ft   85. $6\frac{7}{50}$ m   87. answers may vary   89. $3\frac{3}{8}$ mi   91. $\frac{7}{50}$ 93. $\frac{1}{4}$ 95. $\frac{160}{509}$ 97. $\frac{7}{36}$ sq ft

Section 1.4
Practice Exercises
1. a. 1   b. 25   c. $\frac{1}{100}$ d. 9   e. $\frac{8}{125}$ 2. a. 33   b. 11   c. $\frac{32}{9}$ or $3\frac{5}{9}$ d. 36   e. $\frac{3}{16}$ 3. $\frac{31}{11}$ 4. 4   5. $\frac{9}{22}$ 6. a. 9   b. $\frac{8}{15}$ c. $\frac{19}{10}$ d. 33   7. No   8. a. $6x$ b. $x - 8$ c. $x \cdot 9$ or $9x$ d. $2x + 3$ e. $7 + x$ 9. a. $x + 7 = 13$ b. $x - 2 = 11$ c. $2x + 9 \neq 25$ d. $5(11) \geq x$

Calculator Explorations 1.4
1. 625   3. 59,049   5. 30   7. 9857   9. 2376

Vocabulary and Readiness Check 1.4
1. base, exponent   3. variable   5. equation   7. solving   9. add   11. divide

Exercise Set 1.4
1. 243   3. 27   5. 1   7. 5   9. $\frac{1}{125}$ 11. $\frac{16}{81}$ 13. 49   15. 16   17. 1.44   19. 17   21. 20   23. 10   25. 21   27. 45   29. 0   31. $\frac{2}{7}$ 33. 30   35. 2   37. $\frac{7}{18}$ 39. $\frac{27}{10}$ 41. $\frac{7}{5}$ 43. no   45. a. 64   b. 43   c. 19   d. 22   47. 9   49. 1   51. 1   53. 11

A1

A2 Answers to Selected Exercises

55. 45 **57.** 27 **59.** 132 **61.** $\frac{37}{18}$ **63.** 16, 64, 144, 256 **65.** yes **67.** no **69.** no **71.** yes **73.** no **75.** $x + 15$ **77.** $x - 5$
79. $3x + 22$ **81.** $1 + 2 = 9 \div 3$ **83.** $3 \neq 4 \div 2$ **85.** $5 + x = 20$ **87.** $13 - 3x = 13$ **89.** $\frac{12}{x} = \frac{1}{2}$ **91.** answers may vary
93. $(20 - 4) \cdot 4 \div 2$ **95.** 28 m **97.** 12,000 sq ft **99.** 6.5% **101.** $13.08

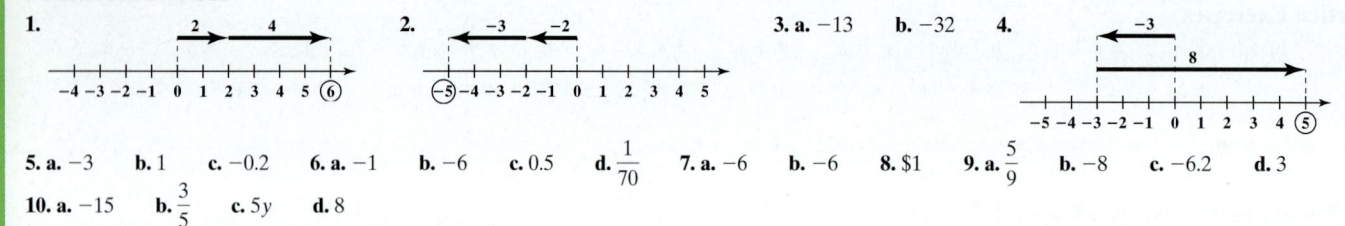

Section 1.5
Practice Exercises
1. **2.** **3. a.** -13 **b.** -32 **4.**
5. a. -3 **b.** 1 **c.** -0.2 **6. a.** -1 **b.** -6 **c.** 0.5 **d.** $\frac{1}{70}$ **7. a.** -6 **b.** -6 **8.** $1 **9. a.** $\frac{5}{9}$ **b.** -8 **c.** -6.2 **d.** 3
10. a. -15 **b.** $\frac{3}{5}$ **c.** $5y$ **d.** 8

Vocabulary and Readiness Check 1.5
1. opposites **3.** n **5.** positive number **7.** negative number **9.** 0

Exercise Set 1.5
1. 9 **3.** -14 **5.** 1 **7.** -12 **9.** -5 **11.** -12 **13.** -4 **15.** 7 **17.** -2 **19.** 0 **21.** -19 **23.** 31 **25.** -47 **27.** -2.1
29. -8 **31.** 38 **33.** -13.1 **35.** $\frac{2}{8} = \frac{1}{4}$ **37.** $-\frac{3}{16}$ **39.** $-\frac{13}{10}$ **41.** -8 **43.** -59 **45.** -9 **47.** 5 **49.** 11 **51.** -18 **53.** 19
55. -0.7 **57.** $-6°$ **59.** $-16,427$ ft **61.** $-$9250 million **63.** -9 **65.** -6 **67.** 2 **69.** 0 **71.** -6 **73.** answers may vary **75.** -2
77. 0 **79.** $-\frac{2}{3}$ **81.** answers may vary **83.** yes **85.** no **87.** July **89.** October **91.** 4.7°F **93.** negative **95.** positive

Section 1.6
Practice Exercises
1. a. -13 **b.** -7 **c.** 12 **d.** -2 **2. a.** 10.9 **b.** $-\frac{1}{2}$ **c.** $-\frac{19}{20}$ **3.** -7 **4. a.** -6 **b.** 6.1 **5. a.** -20 **b.** 13 **6. a.** 2 **b.** 13
7. $357 **8. a.** 28° **b.** 137°

Vocabulary and Readiness Check 1.6
1. $7 - x$ **3.** $x - 7$ **5.** $7 - x$

Exercise Set 1.6
1. -10 **3.** -5 **5.** 19 **7.** $\frac{1}{6}$ **9.** 2 **11.** -11 **13.** 11 **15.** 5 **17.** 37 **19.** -6.4 **21.** -71 **23.** 0 **25.** 4.1 **27.** $\frac{2}{11}$
29. $-\frac{11}{12}$ **31.** 8.92 **33.** 13 **35.** -5 **37.** -1 **39.** -23 **41.** answers may vary **43.** -26 **45.** -24 **47.** 3 **49.** -45 **51.** -4
53. 13 **55.** 6 **57.** 9 **59.** -9 **61.** -7 **63.** $\frac{7}{5}$ **65.** 21 **67.** $\frac{1}{4}$ **69.** 100° **71.** -23 yd or 23 yd loss **73.** -569 or 569 B.C.
75. -308 ft **77.** 19,852 ft **79.** 130° **81.** 30° **83.** no **85.** no **87.** yes **89.** $-4.4°$; $2.6°$; $12°$; $23.5°$; $15.3°$; $3.9°$; $-0.3°$; $-6.3°$; $-18.2°$; $-15.7°$; $-10.3°$ **91.** October **93.** true **95.** true **97.** negative, -2.6466

Integrated Review—Operations on Real Numbers
1. negative **2.** negative **3.** positive **4.** 0 **5.** positive **6.** 0 **7.** positive **8.** positive **9.** $-\frac{1}{7}; \frac{1}{7}$ **10.** $\frac{12}{5}; \frac{12}{5}$ **11.** 3; 3
12. $-\frac{9}{11}; \frac{9}{11}$ **13.** -42 **14.** 10 **15.** 2 **16.** -18 **17.** -7 **18.** -39 **19.** -2 **20.** -9 **21.** -3.4 **22.** -9.8 **23.** $-\frac{25}{28}$ **24.** $-\frac{5}{24}$
25. -4 **26.** -24 **27.** 6 **28.** 20 **29.** 6 **30.** 61 **31.** -6 **32.** -16 **33.** -19 **34.** -13 **35.** -4 **36.** -1 **37.** $\frac{13}{20}$ **38.** $-\frac{29}{40}$
39. 4 **40.** 9 **41.** -1 **42.** -3 **43.** 8 **44.** 10 **45.** 47 **46.** $\frac{2}{3}$

Section 1.7
Practice Exercises
1. a. -40 **b.** 12 **c.** -54 **2. a.** -30 **b.** 24 **c.** 0 **d.** 26 **3. a.** -0.046 **b.** $-\frac{4}{15}$ **c.** 14 **4. a.** 36 **b.** -36 **c.** -64 **d.** -64
5. a. $\frac{3}{8}$ **b.** $\frac{1}{15}$ **c.** $-\frac{7}{2}$ **d.** $-\frac{1}{5}$ **6. a.** -8 **b.** -4 **c.** 5 **7. a.** 3 **b.** -16 **c.** $-\frac{6}{5}$ **d.** $-\frac{1}{18}$ **8. a.** 0 **b.** undefined
c. undefined **9. a.** $\frac{-84}{5}$ **b.** 11 **10. a.** -9 **b.** 33 **c.** $\frac{5}{3}$

Calculator Explorations 1.7
1. 38   **3.** −441   **5.** 163.$\overline{3}$ **7.** 54,499   **9.** 15,625

Vocabulary and Readiness Check 1.7
1. 0, 0   **3.** positive   **5.** negative   **7.** positive

Exercise Set 1.7
1. −24   **3.** −2   **5.** 50   **7.** −12   **9.** 0   **11.** −18   **13.** $\frac{3}{10}$ **15.** $\frac{2}{3}$ **17.** −7   **19.** 0.14   **21.** −800   **23.** −28   **25.** 25   **27.** −$\frac{8}{27}$
29. −121   **31.** −$\frac{1}{4}$ **33.** −30   **35.** 23   **37.** −7   **39.** true   **41.** false   **43.** 16   **45.** −1   **47.** 25   **49.** −49   **51.** $\frac{1}{9}$ **53.** $\frac{3}{2}$
55. −$\frac{1}{14}$ **57.** −$\frac{11}{3}$ **59.** $\frac{1}{0.2}$ **61.** −6.3   **63.** −9   **65.** 4   **67.** −4   **69.** 0   **71.** −5   **73.** undefined   **75.** 3   **77.** −15   **79.** −$\frac{18}{7}$
81. $\frac{20}{27}$ **83.** −1   **85.** −$\frac{9}{2}$ **87.** −4   **89.** 16   **91.** −3   **93.** −$\frac{16}{7}$ **95.** 2   **97.** $\frac{6}{5}$ **99.** −5   **101.** $\frac{3}{2}$ **103.** −21   **105.** 41
107. −134   **109.** 3   **111.** 0   **113.** −$24,812 million   **115.** yes   **117.** no   **119.** yes   **121.** answers may vary   **123.** 1, −1
125. positive   **127.** not possible   **129.** negative   **131.** −2 + $\frac{-15}{3}$; −7   **133.** 2[−5 + (−3)]; −16

The Bigger Picture
1. −5   **2.** −14   **3.** −$\frac{26}{35}$ **4.** 8   **5.** 49   **6.** −49   **7.** −21   **8.** undefined   **9.** 0   **10.** −12   **11.** −16.6   **12.** $\frac{1}{6}$ **13.** −79
14. $\frac{10}{13}$ **15.** 50   **16.** −12

Section 1.8
Practice Exercises
1. a. 8 · x   **b.** 17 + x   **2. a.** 2 + (9 + 7)   **b.** (−4 · 2) · 7   **3. a.** x + 14   **b.** −30x   **4. a.** 5x − 5y   **b.** −24 − 12t   **c.** 6x − 8y − 2z
d. −3 + y   **e.** −x + 7 − 2s   **f.** 14x + 14   **5. a.** 5(w + 3)   **b.** 9(w + z)   **6. a.** commutative property of multiplication   **b.** associative
property of addition   **c.** identity element for addition   **d.** multiplicative inverse property   **e.** commutative property of addition   **f.** additive
inverse property   **g.** commutative and associative properties of multiplication

Vocabulary and Readiness Check 1.8
1. commutative property of addition   **3.** distributive property   **5.** associative property of addition   **7.** opposites or additive inverses

Exercise Set 1.8
1. 16 + x   **3.** y · (−4)   **5.** yx   **7.** 13 + 2x   **9.** x · (yz)   **11.** (2 + a) + b   **13.** (4a) · b   **15.** a + (b + c)   **17.** 17 + b   **19.** 24y
21. y   **23.** 26 + a   **25.** −72x   **27.** s   **29.** answers may vary   **31.** 4x + 4y   **33.** 9x − 54   **35.** 6x + 10   **37.** 28x − 21   **39.** 18 + 3x
41. −2y + 2z   **43.** −21y − 35   **45.** 5x + 20m + 10   **47.** −4 + 8m − 4n   **49.** −5x − 2   **51.** −r + 3 + 7p   **53.** 3x + 4   **55.** −x + 3y
57. 6r + 8   **59.** −36x − 70   **61.** −16x − 25   **63.** 4(1 + y)   **65.** 11(x + y)   **67.** −1(5 + x)   **69.** 30(a + b)   **71.** commutative property
of multiplication   **73.** associative property of addition   **75.** distributive property   **77.** associative property of multiplication   **79.** identity
element of addition   **81.** distributive property   **83.** commutative and associative properties of multiplication   **85.** −8; $\frac{1}{8}$ **87.** −x; $\frac{1}{x}$
89. 2x; −2x   **91.** no   **93.** yes   **95.** answers may vary

Chapter 1 Vocabulary Check
1. inequality symbols   **2.** equation   **3.** absolute value   **4.** variable   **5.** opposites   **6.** numerator   **7.** solution   **8.** reciprocals
9. exponent   **10.** denominator   **11.** grouping symbols   **12.** set

Chapter 1 Review
1. <   **3.** >   **5.** <   **7.** =   **9.** >   **11.** 4 ≥ −3   **13.** 0.03 < 0.3   **15. a.** 1, 3   **b.** 0, 1, 3   **c.** −6, 0, 1, 3   **d.** −6, 0, 1, 1$\frac{1}{2}$, 3, 9.62
e. π   **f.** −6, 0, 1, 1$\frac{1}{2}$, 3, π, 9.62   **17.** Friday   **19.** 2 · 2 · 3 · 3   **21.** $\frac{12}{25}$ **23.** $\frac{13}{10}$ **25.** 9$\frac{3}{8}$ **27.** 15   **29.** $\frac{7}{12}$ **31.** $A = \frac{34}{121}$ sq in.; $P = 2\frac{4}{11}$ in.
33. 2$\frac{15}{16}$ lb   **35.** 11$\frac{5}{16}$ lb   **37.** Odera   **39.** 3$\frac{7}{8}$ lb   **41.** 16   **43.** $\frac{4}{49}$ **45.** 70   **47.** 37   **49.** $\frac{18}{7}$ **51.** 20 − 12 = 2 · 4   **53.** 18   **55.** 5
57. 63°   **59.** no   **61.** −$\frac{2}{3}$ **63.** 7   **65.** −17   **67.** −5   **69.** 3.9   **71.** −14   **73.** 5   **75.** −19   **77.** 15   **79.** −$\frac{1}{6}$ **81.** −48   **83.** 3
85. undefined   **87.** undefined   **89.** −12   **91.** 9   **93.** −5   **95.** commutative property of addition   **97.** distributive property
99. associative property of addition   **101.** distributive property   **103.** multiplicative inverse   **105.** 5y − 10   **107.** −7 + x − 4z
109. −12z − 27   **111.** <   **113.** −15.3   **115.** −80   **117.** −$\frac{1}{4}$ **119.** 16   **121.** −5   **123.** −$\frac{5}{6}$

Chapter 1 Test

1. $|-7| > 5$ **2.** $9 + 5 \geq 4$ **3.** -5 **4.** -11 **5.** -3 **6.** -39 **7.** 12 **8.** -2 **9.** undefined **10.** -8 **11.** $-\dfrac{1}{3}$ **12.** $4\dfrac{5}{8}$
13. $-\dfrac{5}{2}$ or $-2\dfrac{1}{2}$ **14.** -32 **15.** -48 **16.** 3 **17.** 0 **18.** $>$ **19.** $>$ **20.** $<$ **21.** $=$ **22.** $2221 < 10{,}993$ or $10{,}993 > 2221$
23. a. $1, 7$ **b.** $0, 1, 7$ **c.** $-5, -1, 0, 1, 7$ **d.** $-5, -1, 0, \dfrac{1}{4}, 1, 7, 11.6$ **e.** $\sqrt{7}, 3\pi$ **f.** $-5, -1, 0, \dfrac{1}{4}, 1, 7, 11.6, \sqrt{7}, 3\pi$ **24.** 40 **25.** 12
26. 22 **27.** -1 **28.** associative property of addition **29.** commutative property of multiplication **30.** distributive property
31. multiplicative inverse property **32.** 9 **33.** -3 **34.** second down **35.** yes **36.** $17°$ **37.** \$650 million **38.** \$420

CHAPTER 2 EQUATIONS, INEQUALITIES, AND PROBLEM SOLVING

Section 2.1

Practice Exercises

1. a. 1 **b.** -7 **c.** $-\dfrac{1}{5}$ **d.** 43 **e.** -1 **2. a.** like terms **b.** unlike terms **c.** like terms **d.** like terms **3. a.** $7x^2$ **b.** $-2y$
c. $5x + 5x^2$ **4. a.** $11y - 5$ **b.** $5x - 6$ **c.** $-\dfrac{1}{4}t$ **d.** $12.2y + 13$ **e.** $5z - 3z^4$ **5. a.** $6x - 21$ **b.** $-15x + 20z + 25$
c. $-2x + y - z + 2$ **6. a.** $36x + 10$ **b.** $-11x + 1$ **c.** $-30x - 17$ **7.** $-5x + 4$ **8. a.** $3 + 2x$ **b.** $x - 1$ **c.** $2x + 10$ **d.** $\dfrac{13}{2}x$

Vocabulary and Readiness Check 2.1

1. expression; term **3.** numerical coefficient **5.** numerical coefficient **7.** -7 **9.** 1 **11.** $-\dfrac{5}{3}$ **13.** like **15.** unlike

Exercise Set 2.1

1. $15y$ **3.** $13w$ **5.** $-7b - 9$ **7.** $-m - 6$ **9.** -8 **11.** $7.2x - 5.2$ **13.** $4x - 3$ **15.** $5x^2$ **17.** $1.3x + 3.5$ **19.** answers may vary
21. $5y - 20$ **23.** $-2x - 4$ **25.** $7d - 11$ **27.** $-10x + 15y - 30$ **29.** $-3x + 2y - 1$ **31.** $2x + 14$ **33.** $10x - 3$ **35.** $-4x - 9$
37. $-4m - 3$ **39.** $k - 6$ **41.** $-15x + 18$ **43.** 16 **45.** $x + 5$ **47.** $x + 2$ **49.** $2k + 10$ **51.** $-3x + 5$ **53.** -11 **55.** $3y + \dfrac{5}{6}$
57. $-22 + 24x$ **59.** $0.9m + 1$ **61.** $10 - 6x - 9y$ **63.** $-x - 38$ **65.** $5x - 7$ **67.** $2x - 4$ **69.** $2x + 7$ **71.** $\dfrac{3}{4}x + 12$
73. $-2 + 12x$ **75.** $8(x + 6)$ or $8x + 48$ **77.** $x - 10$ **79.** $\dfrac{7x}{6}$ **81.** $7x - 7$ **83.** 2 **85.** -23 **87.** -25 **89.** $(18x - 2)$ ft
91. balanced **93.** balanced **95.** answers may vary **97.** $(15x + 23)$ in. **99.** $5b^2c^3 + b^3c^2$ **101.** $5x^2 + 9x$ **103.** $-7x^2y$

Section 2.2

Practice Exercises

1. -8 **2.** -1.8 **3.** $\dfrac{1}{10}$ **4.** 10 **5.** -10 **6.** 18 **7.** -8 **8. a.** 7 **b.** $9 - x$ **c.** $(9 - x)$ ft **9.** $(s - 3.8)$ mph

Vocabulary and Readiness Check 2.2

1. equation; expression **3.** solution **5.** addition

Exercise Set 2.2

1. 3 **3.** -2 **5.** 3 **7.** 0.5 **9.** $\dfrac{5}{12}$ **11.** -0.7 **13.** 3 **15.** answers may vary **17.** -3 **19.** -9 **21.** -10 **23.** 2 **25.** -7
27. -1 **29.** -9 **31.** -12 **33.** $-\dfrac{1}{2}$ **35.** 11 **37.** 21 **39.** 25 **41.** -14 **43.** $\dfrac{1}{4}$ **45.** 11 **47.** 13 **49.** -30 **51.** -0.4
53. -7 **55.** $-\dfrac{1}{3}$ **57.** -17.9 **59.** $-\dfrac{3}{4}$ **61.** 0 **63.** 1.83 **65.** $20 - p$ **67.** $(10 - x)$ ft **69.** $(180 - x)°$ **71.** $\left(m + 1\dfrac{1}{2}\right)$ ft
73. $7x$ sq mi **75.** $\dfrac{8}{5}$ **77.** $\dfrac{1}{2}$ **79.** -9 **81.** x **83.** y **85.** x **87.** $(173 - 3x)°$ **89.** answers may vary **91.** 4
93. answers may vary **95.** 250 ml **97.** answers may vary **99.** solution **101.** not a solution

Section 2.3

Practice Exercises

1. 20 **2.** -12 **3.** 65 **4.** 1.5 **5.** $-\dfrac{12}{35}$ **6.** 11 **7.** -3 **8.** 6 **9.** $\dfrac{11}{12}$ **10.** $3x + 6$

Vocabulary and Readiness Check 2.3

1. multiplication **3.** false **5.** 9 **7.** 2

Exercise Set 2.3

1. 4 **3.** 0 **5.** 12 **7.** −12 **9.** 3 **11.** 2 **13.** 0 **15.** 6.3 **17.** 10 **19.** −20 **21.** 0 **23.** −9 **25.** −3 **27.** −30 **29.** 3 **31.** $\frac{10}{9}$ **33.** −1 **35.** −4 **37.** $-\frac{1}{2}$ **39.** 0 **41.** 4 **43.** $-\frac{1}{14}$ **45.** 0.21 **47.** 5 **49.** 6 **51.** −5.5 **53.** −5 **55.** 0 **57.** −3 **59.** $-\frac{9}{28}$ **61.** $\frac{14}{3}$ **63.** −9 **65.** −2 **67.** $\frac{11}{2}$ **69.** $-\frac{1}{4}$ **71.** $\frac{9}{10}$ **73.** $-\frac{17}{20}$ **75.** −16 **77.** $2x + 2$ **79.** $2x + 2$ **81.** $5x + 20$ **83.** $7x - 12$ **85.** 1 **87.** > **89.** = **91.** < **93.** −48 **95.** answers may vary **97.** answers may vary **99.** $\frac{700}{3}$ mg **101.** −2.95 **103.** 0.02

Section 2.4
Practice Exercises

1. 3 **2.** $\frac{21}{13}$ **3.** −15 **4.** 3 **5.** 0 **6.** no solution **7.** all real numbers

Calculator Explorations 2.4

1. solution **3.** not a solution **5.** solution

Vocabulary and Readiness Check 2.4

1. equation **3.** expression **5.** expression **7.** equation

Exercise Set 2.4

1. −6 **3.** 3 **5.** 1 **7.** $\frac{3}{2}$ **9.** 0 **11.** −1 **13.** 4 **15.** −4 **17.** −3 **19.** 2 **21.** 50 **23.** 1 **25.** $\frac{7}{3}$ **27.** 0.2 **29.** all real numbers **31.** no solution **33.** no solution **35.** all real numbers **37.** 18 **39.** $\frac{19}{9}$ **41.** $\frac{14}{3}$ **43.** 13 **45.** 4 **47.** all real numbers **49.** $-\frac{3}{5}$ **51.** −5 **53.** 10 **55.** no solution **57.** 3 **59.** −17 **61.** −4 **63.** 3 **65.** all real numbers **67.** $-8 - x$ **69.** $-3 + 2x$ **71.** $9(x + 20)$ **73.** $(6x - 8)$ m **75. a.** all real numbers **b.** answers may vary **c.** answers may vary **77.** a **79.** b **81.** c **83.** answers may vary **85. a.** $x + x + x + 2x + 2x = 28$ **b.** $x = 4$ **c.** $x = 4$ cm; $2x = 8$ cm **87.** answers may vary **89.** 15.3 **91.** −0.2 **93.** $-\frac{7}{8}$ **95.** no solution

Integrated Review

1. 6 **2.** −17 **3.** 12 **4.** −26 **5.** −3 **6.** −1 **7.** $\frac{27}{2}$ **8.** $\frac{25}{2}$ **9.** 8 **10.** −64 **11.** 2 **12.** −3 **13.** no solution **14.** no solution **15.** −2 **16.** −2 **17.** $-\frac{5}{6}$ **18.** $\frac{1}{6}$ **19.** 1 **20.** 6 **21.** 4 **22.** 1 **23.** $\frac{9}{5}$ **24.** $-\frac{6}{5}$ **25.** all real numbers **26.** all real numbers **27.** 0 **28.** −1.6 **29.** $\frac{4}{19}$ **30.** $-\frac{5}{19}$ **31.** $\frac{7}{2}$ **32.** $-\frac{1}{4}$ **33.** no solution **34.** no solution **35.** $\frac{7}{6}$ **36.** $\frac{1}{15}$

Section 2.5
Practice Exercises

1. 9 **2.** 2 **3.** 9 in. and 36 in. **4.** 22 Republican and 28 Democratic Governors **5.** 25°, 75°, 80° **6.** 46, 48, 50

Vocabulary and Readiness Check 2.5

1. $2x; 2x - 31$ **3.** $x + 5; 2(x + 5)$ **5.** $20 - y; \frac{20 - y}{3}$ or $(20 - y) \div 3$

Exercise Set 2.5

1. $2x + 7 = x + 6; -1$ **3.** $3x - 6 = 2x + 8; 14$ **5.** $2(x - 8) = 3(x + 3); -25$ **7.** $4(-2 + x) = 5x + \frac{1}{2}; -\frac{17}{2}$ **9.** 5 ft, 12 ft **11.** Armanty: 22 tons; Hoba West: 66 tons **13.** China: 42,400; U.S.: 36,594 **15.** 1st angle: 37.5°; 2nd angle: 37.5°; 3rd angle: 105° **17.** $3x + 3$ **19.** $x + 2, x + 4, 2x + 4$ **21.** $x + 1; x + 2; x + 3; 4x + 6$ **23.** $x + 2; x + 4; 2x + 6$ **25.** 234, 235 **27.** Belgium: 32; France: 33; Spain: 34 **29.** 3 in.; 6 in.; 16 in. **31.** $\frac{5}{4}$ **33.** Botswana: 32,000,000 carats; Angola: 8,000,000 carats **35.** 58°, 60°, 62° **37.** Russia: 22; Austria: 23; Canada: 24; United States: 25 **39.** −16 **41.** Weller: 108, 375; Pavich: 88, 179 **43.** 43°, 137° **45.** 1 **47.** 1st angle: 65°; 2nd angle: 115° **49.** Maglev: 361 mph; TGV: 357.2 mph **51.** $\frac{5}{2}$ **53.** California: 58; Montana: 56 **55.** Colts: 29; Bears: 17 **57.** 34.5°, 34.5°, 111° **59.** 1st piece: 5 in.; 2nd piece: 10 in.; 3rd piece: 25 in. **61.** Hawaii **63.** Texas: $31.1 million; Florida: $29.4 million **65.** answers may vary **67.** 34 **69.** 225π **71.** answers may vary

Section 2.6
Practice Exercises
1. 116 sec or 1 min 56 sec
2. width: 9 ft
3. 46.4°F
4. length: 28 in.; width: 5 in.
5. $r = \dfrac{I}{Pt}$
6. $s = \dfrac{H - 10a}{5a}$
7. $d = \dfrac{N - F}{n - 1}$
8. $B = \dfrac{2A - ab}{a}$

Exercise Set 2.6
1. $h = 3$
3. $h = 3$
5. $h = 20$
7. $c = 12$
9. $r \approx 2.5$
11. $T = 3$
13. $h \approx 15$
15. $h = \dfrac{f}{5g}$
17. $w = \dfrac{V}{lh}$
19. $y = 7 - 3x$
21. $R = \dfrac{A - P}{PT}$
23. $A = \dfrac{3V}{h}$
25. $a = P - b - c$
27. $h = \dfrac{S - 2\pi r^2}{2\pi r}$
29. a. area: 103.5 sq ft; perimeter: 41 ft
b. baseboard: perimeter; carpet: area
31. a. area: 480 sq in.; perimeter: 120 in.
b. frame: perimeter; glass: area
33. 70 ft
35. −10°C
37. 6.25 hr
39. length: 78 ft; width: 52 ft
41. 18 ft, 36 ft, 48 ft
43. 55.2 mph
45. 96 piranhas
47. 2 bags
49. one 16-in. pizza
51. $x = 6$ m, $2.5x = 15$ m
53. 0.2 hr or 12 min
55. 13 in.
57. 2.25 hr
59. 12,090 ft
61. 50°C
63. 515,509.5 cu in.
65. 449 cu in.
67. 332.6°F
69. $\dfrac{9}{x + 5}$
71. $3(x + 4)$
73. $3(x - 12)$
75. ● = ■ - ▲
77. −109.3°F
79. 500 sec or $8\dfrac{1}{3}$ min
81. 608.33 ft
83. 565.5 cu in.
85. It multiplies the area by 4.

Section 2.7
Practice Exercises
1. 62.5%
2. 360
3. a. 4%
b. 87%
c. 13 people
4. discount: $408; new price: $72
5. 50.7%
6. 520 new films
7. 2 liters of 5% eyewash; 4 liters of 2% eyewash

Vocabulary and Readiness Check 2.7
1. no
3. yes

Exercise Set 2.7
1. 11.2
3. 55%
5. 180
7. 69%
9. 1896.3 million bushels or 1,896,300,000 bushels
11. discount: $1480; new price: $17,020
13. $46.58
15. 35%
17. 30%
19. $104
21. $42,500
23. 2 gal
25. 7 lb
27. 4.6
29. 50
31. 30%
33. 71%
35. 178,778
37. 46%; 28%; 9%; 6%; total: 100%
39. decrease: $64; sale price: $192
41. 115% increase
43. 230 million
45. 400 oz
47. markup: $18.90; adult ticket price: $45.90
49. 300%
51. 120 employees
53. 5 lb
55. 4.6%
57. 335 decisions
59. 854 thousand Scoville units
61. 361 college students
63. >
65. =
67. >
69. no; answers may vary
71. no; answers may vary
73. 9.6%
75. 26.9%; yes
77. 17.1%

Section 2.8
Practice Exercises
1. 2.2 hr
2. eastbound: 62 mph; westbound: 52 mph
3. 106 $5 bills; 59 $20 bills
4. $18,000 at 11.5%; $12,000 at 6%

Exercise Set 2.8
1. $666\dfrac{2}{3}$ mi
3. 55 mph
5. 0.10 y
7. 0.05(x + 7)
9. 20(4y) or 80y
11. 50(35 − x)
13. 12 $10 bills; 32 $5 bills
15. $11,500 at 8%; $13,500 at 9%
17. $7000 at 11% profit; $3000 at 4% loss
19. 187 adult tickets; 313 child tickets
21. $30,000 at 8%; $24,000 at 10%
23. 2 hr $37\dfrac{1}{2}$ min
25. 483 dimes; 161 nickels
27. $4500
29. 2.2 mph; 3.3 mph
31. 27.5 mi
33. −4
35. $\dfrac{9}{16}$
37. −4
39. 25 $100 bills; 71 $50 bills; 175 $20 bills
41. 25 skateboards
43. 800 books
45. answers may vary

Section 2.9
Practice Exercises
1. (−∞, 5)
2. [−5, ∞)
3. (−∞, 3]
4. (−3, ∞)
5. [7, ∞)
6. (−∞, −7]
7. $\left(-\infty, \dfrac{9}{4}\right)$
8. [0, ∞)
9. [−3, 1)

10. (−2, 2] **11.** $\left(-\frac{16}{3}, \frac{4}{3}\right)$ **12.** $x \leq 3.2$; Kasonga can afford at most 3 classes.

Vocabulary and Readiness Check 2.9
1. expression **3.** inequality **5.** equation **7.** −5 **9.** 4.1

Exercise Set 2.9
1. $x \geq 2$ **3.** $x < -5$ **5.** $(-\infty, -1]$
7. $\left(-\infty, \frac{1}{2}\right)$ **9.** $[5, \infty)$
11. $x < -3$, $(-\infty, -3)$ **13.** $x \geq -5$, $[-5, \infty)$
15. $x \geq -2$, $[2, \infty)$ **17.** $x > -3$, $(-3, \infty)$
19. $x \leq 1$, $(-\infty, 1]$ **21.** $x > -5$, $(-5, \infty)$
23. $x \leq -2$, $(-\infty, -2]$ **25.** $x \leq -8$, $(-\infty, -8]$
27. $x > 4$, $(4, \infty)$ **29.** $x \geq 20$, $[20, \infty)$
31. $x > 16$, $(16, \infty)$ **33.** $x > -3$, $(-3, \infty)$
35. $x \leq -\frac{2}{3}$, $\left(-\infty, -\frac{2}{3}\right]$ **37.** $x > \frac{8}{3}$, $\left(\frac{8}{3}, \infty\right)$
39. $x > -13$, $(-13, \infty)$ **41.** $x < 0$, $(-\infty, 0)$
43. $x \leq 0$, $(-\infty, 0]$ **45.** $x > 3$, $(3, \infty)$
47. $x \leq 0$, $(-\infty, 0]$ **49.** answers may vary **51.** $(-1, 3)$
53. $[0, 2)$ **55.** $(-1, 2)$
57. $[4, 5]$ **59.** $(1, 5]$
61. $(1, 4)$ **63.** $\left(0, \frac{14}{3}\right]$
65. answers may vary **67.** $x > -10$ **69.** 86 people **71.** $x \leq 35$ **73.** at least 10% **75.** at least 193
77. $x < 200$ recommended; $200 \leq x \leq 240$ borderline; $x > 240$ high **79.** $-3 < x < 3$ **81.** $-38.2° \leq F \leq 113°$ **83.** 8
85. 1 **87.** $\frac{16}{49}$ **89.** $52.70 **91.** 2005 **93.** $0.924 \leq d \leq 0.987$ **95.** $(1, \infty)$
97. $\left(-\infty, \frac{5}{8}\right)$

The Bigger Picture
1. −3 **2.** $(-\infty, -3)$ **3.** $\frac{2}{9}$ **4.** $-\frac{1}{4}$ **5.** $[-15, \infty)$ **6.** no solution **7.** 7 **8.** $(-\infty, 37)$ **9.** all real numbers **10.** $\frac{41}{29}$

Chapter 2 Vocabulary Check
1. like terms **2.** linear equation in one variable **3.** equivalent equations **4.** compound inequalities **5.** formula
6. linear inequality in one variable **7.** numerical coefficient

Answers to Selected Exercises

Chapter 2 Review
1. $6x$ 3. $4x - 2$ 5. $3n - 18$ 7. $-6x + 7$ 9. $3x - 7$ 11. 4 13. 6 15. 0 17. -23 19. 5; 5 21. b 23. b 25. -12
27. 0 29. 0.75 31. -6 33. -1 35. $-\frac{1}{5}$ 37. $3x + 3$ 39. -4 41. 2 43. no solution 45. $\frac{3}{4}$ 47. 20 49. $\frac{23}{7}$ 51. 102
53. 6665.5 in. 55. Kellogg: 35 plants; Keebler: 18 plants 57. 3 59. $w = 9$ 61. $m = \frac{y - b}{x}$ 63. $x = \frac{2y - 7}{5}$ 65. $\pi = \frac{C}{D}$ 67. 15 m
69. 1 hr 20 min 71. 20% 73. 110 75. mark-up: $209; new price: $2109 77. 40% solution: 10 gal; 10% solution: 20 gal 79. 18%
81. 966 customers 83. 50 km 85. 80 nickels 87. $(0, \infty)$
89. $[0.5, 1.5]$ 91. $(-\infty, -4)$
93. $(-\infty, 4]$ 95. $\left(-\frac{1}{2}, \frac{3}{4}\right)$
97. $\left(-\infty, \frac{19}{3}\right]$ 99. $2500 101. $x = 4$ 103. $a = -\frac{3}{2}$ 105. all real numbers
107. -13 109. $h = \frac{3V}{A}$ 111. 160 113. $(9, \infty)$ 115. $(-\infty, 0]$

Chapter 2 Test
1. $y - 10$ 2. $5.9x + 1.2$ 3. $-2x + 10$ 4. $10y + 1$ 5. -5 6. 8 7. $\frac{7}{10}$ 8. 0 9. 27 10. 3 11. 0.25 12. $\frac{25}{7}$
13. no solution 14. 21 15. 7 gal 16. 401, 802 17. $8500 @ 10%; $17,000 @ 12% 18. $2\frac{1}{2}$ hr 19. $x = 6$ 20. $h = \frac{V}{\pi r^2}$
21. $y = \frac{3x - 10}{4}$ 22. $(-\infty, -2]$ 23. $(-\infty, 4)$
24. $\left(-1, \frac{7}{3}\right)$ 25. $\left(\frac{2}{5}, \infty\right)$

Chapter 2 Cumulative Review
1. a. 11, 112 b. 0, 11, 112 c. $-3, -2, 0, 11, 112$ d. $-3, -2, -1.5, 0, \frac{1}{4}, 11, 112$ e. $\sqrt{2}$ f. all numbers in the given set; Sec. 1.2, Ex. 5
3. a. 4 b. 5 c. 0 d. $\frac{1}{2}$ e. 5.6; Sec. 1.2, Ex. 7 5. a. $2 \cdot 2 \cdot 2 \cdot 5$ b. $3 \cdot 3 \cdot 7$; Sec. 1.3, Ex. 1 7. $\frac{8}{20}$; Sec. 1.3, Ex. 6 9. 66; Sec. 1.4, Ex. 4
11. 2 is a solution; Sec. 1.4, Ex. 7 13. -3; Sec. 1.5, Ex. 2 15. 2; Sec. 1.5, Ex. 4 17. a. 10 b. $\frac{1}{2}$ c. $2x$ d. -6; Sec. 1.5, Ex. 10
19. a. 9.9 b. $-\frac{4}{5}$ c. $\frac{2}{15}$; Sec. 1.6, Ex. 2 21. a. 52° b. 118°; Sec. 1.6, Ex. 8 23. a. -0.06 b. $-\frac{7}{15}$ c. 16; Sec. 1.7, Ex. 3 25. a. 6
b. -12 c. $-\frac{8}{15}$ d. $-\frac{1}{6}$; Sec. 1.7, Ex. 7 27. a. $5 + x$ b. $x \cdot 3$; Sec. 1.8, Ex. 1 29. a. $8(2 + x)$ b. $7(s + t)$; Sec. 1.8, Ex. 5
31. $-2x - 1$; Sec. 2.1, Ex. 7 33. $\frac{5}{4}$; Sec. 2.2, Ex. 3 35. 19; Sec. 2.2, Ex. 6 37. 140; Sec. 2.3, Ex. 3 39. 2; Sec. 2.4, Ex. 1 41. 10; Sec. 2.5, Ex. 2
43. $\frac{V}{wh} = l$; Sec. 2.6, Ex. 5 45. $(-\infty, -10]$; Sec. 2.9, Ex. 2

CHAPTER 3 GRAPHING
Section 3.1
Practice Exercises
1. a. Germany, 45 million Internet users b. approximately 5 million Internet users 2. a. 70 beats per minute b. 60 beats per minute
c. 5 minutes after lighting 3. 4. a. (2000, 92), (2001, 84), (2002, 73), (2003, 64), (2004, 65), (2005, 67), (2006, 96)

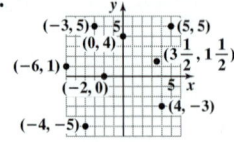

5. a. true b. true c. false 6. a. $(0, -8)$ b. $(6, 4)$ c. $(-3, -14)$

b. Wildfires

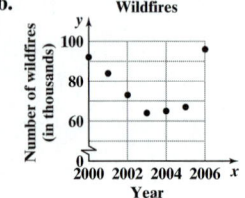

Answers to Selected Exercises   A9

7.	x	y
a.	−2	8
b.	3	−12
c.	0	0

8.	x	y
a.	−10	−4
b.	0	−2
c.	10	0

9. x	0	1	2	3	4
y	12,000	10,200	8400	6600	4800

Vocabulary and Readiness Check 3.1
1. x-axis   **3.** origin   **5.** x-coordinate; y-coordinate   **7.** solution

Exercise Set 3.1
1. France   **3.** France, U.S., Spain, and China   **5.** 30 million   **7.** 72,600   **9.** 2007; 104,000   **11.** 15.9   **13.** from 1996 to 1998   **15.** 2002
17. (1, 5) and (3.7, 2.2) are in quadrant I, $\left(-1, 4\frac{1}{2}\right)$ is in quadrant II, (−5, −2) is in quadrant III, (2, −4) and $\left(\frac{1}{2}, -3\right)$ are in quadrant IV, (−3, 0) lies on the x-axis, (0, −1) lies on the y-axis   **19.** (0, 0)   **21.** (3, 2)   **23.** (−2, −2)
25. (2, −1)   **27.** (0, −3)   **29.** (1, 3)   **31.** (−3, −1)   **33. a.** (2002, 12), (2003, 14), (2004, 14), (2005, 11), (2006, 12)
35. a. (2001, 1770), (2003, 2800), (2005, 3904), (2007, 7500), (2009, 10,800)
b. Regular Season Games Won by Super Bowl Winner
b. Ethanol Fuel Production in the U.S. 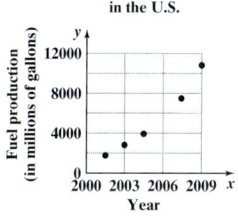
c. The ethanol production is increasing as the years increase.   **37. a.** (2313, 2), (2085, 1), (2711, 21), (2869, 39), (2920, 42), (4038, 99), (1783, 0), (2493, 9)
b. Average Annual Snowfall for Selected U.S. Cities   **c.** The farther from the equator, the more snowfall.   **39.** yes; no; yes   **41.** yes; yes   **43.** no; yes; yes
45. (−4, −2), (4, 0)   **47.** (−8, −5), (16, 1)   **49.** 0; 7; $-\frac{2}{7}$
51. 2; 2; 5   **53.** 0; −3; 2   **55.** 2; 6; 3   **57.** −12; 5; −6   **59.** $\frac{5}{7}; \frac{5}{2}; -1$ **61.** 0; −5; −2   **63.** 2; 1; −6
65. a. 13,000; 21,000; 29,000   **b.** 45 desks   **67. a.** 53.57; 48.87; 44.17   **b.** year 4: 2005   **69.** In 2004, there were 1308 Target stores.
71. year 1: 75 stores; year 2: 100 stores; year 3: 75 stores   **73.** $a = b$ **75.** $y = 5 - x$ **77.** $y = -\frac{1}{2}x + \frac{5}{4}$ **79.** $y = -2x$ **81.** $y = \frac{1}{3}x - 2$
83. false   **85.** true   **87.** negative; negative   **89.** positive; negative   **91.** 0; 0   **93.** y   **95.** no; answers may vary   **97.** answers may vary
99. (4, −7)   **101. a.** (−2, 6)   **b.** 28 units   **c.** 45 sq units

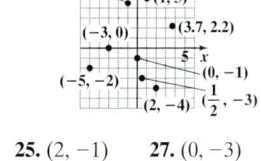

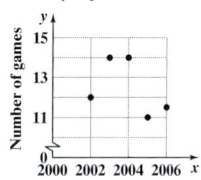

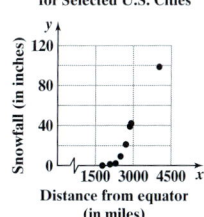

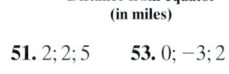

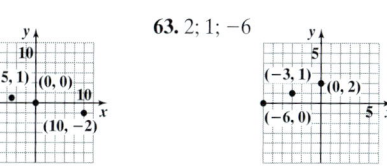

Section 3.2
Practice Exercises
1. a. yes   **b.** no   **c.** yes   **d.** yes   **2.**   **3.** 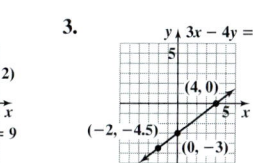   **4.** 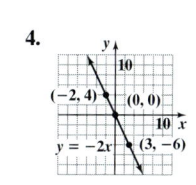   **5.**

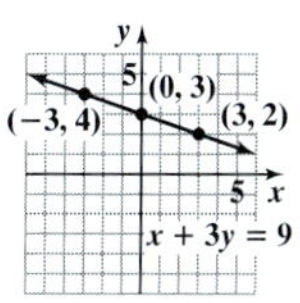

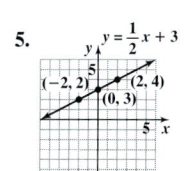

A10 Answers to Selected Exercises

6. The graph of $y = -2x + 3$ is the same as the graph of $y = -2x$ except that the graph of $y = -2x + 3$ is moved 3 units upward.

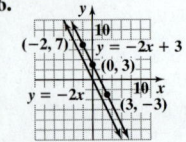

7. a. [graph showing points (0, 371) and (10, 593), Number of computer software engineers (in thousands) vs. Year after 2000] **b.** We predict 700 thousand computer software application engineers in the year 2015.

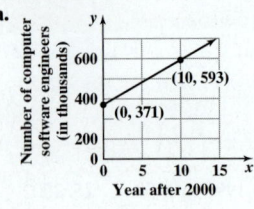

Calculator Explorations 3.2

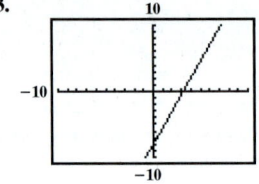

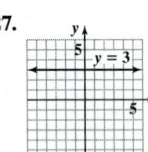

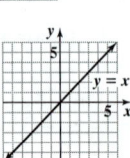

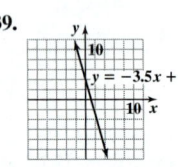

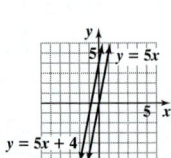

1. [graph] **3.** [graph] **5.** [graph]

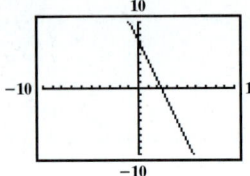

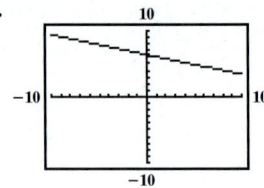

Exercise Set 3.2

1. yes **3.** yes **5.** no **7.** yes

9.
x	y
6	0
4	-2
5	-1

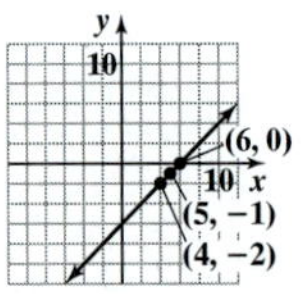

11.
x	y
1	-4
0	0
-1	4

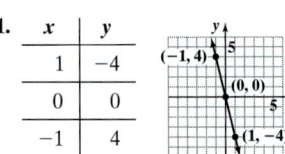

13.
x	y
0	0
6	2
-3	-1

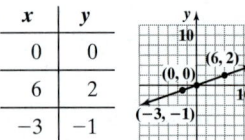

15.
x	y
0	3
1	-1
2	-5

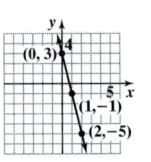

17. [graph: $x + y = 1$] **19.** [graph: $x - y = -2$] **21.** [graph: $x - 2y = 6$] **23.** [graph: $y = 6x + 3$]

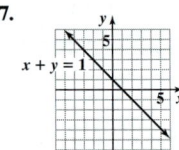

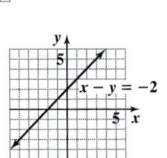

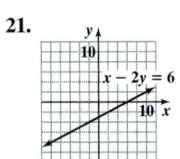

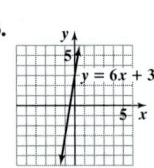

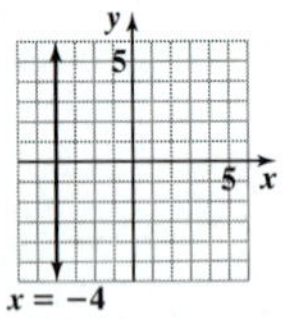

25. [graph: $x = -4$] **27.** [graph: $y = 3$] **29.** [graph: $y = x$] **31.** [graph: $x = -3y$] **33.** [graph: $x + 3y = 9$] **35.** [graph: $y = \frac{1}{2}x + 2$]

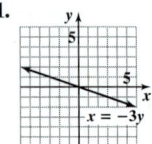

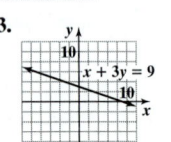

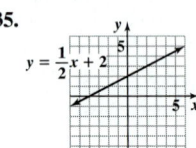

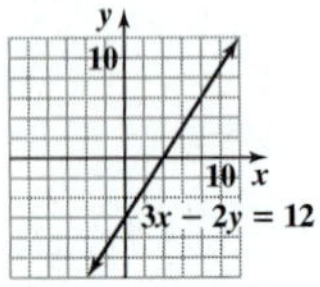

37. [graph: $3x - 2y = 12$] **39.** [graph: $y = -3.5x + 4$] **41.** [graph: $y = 5x + 4$, $y = 5x$] **43.** [graph: $y = -2x$, $y = -2x - 3$] **45.** [graph: $y = \frac{1}{2}x + 2$, $y = \frac{1}{2}x$] **47.** c

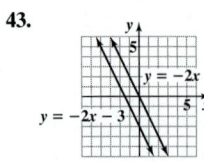

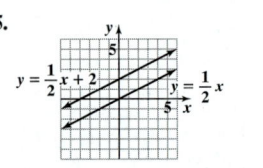

49. d **51. a.** (8, 7) **b.** In 2005, there were 7 million snowboarders. **c.** 10.5 million snowboarders **53.** The expected minimum salary after 5 years experience is $545 thousand. **55.** (4, -1) **57.** -5 **59.** $-\frac{1}{10}$ **61.** $y = x + 5$ [graph] **63.** $2x + 3y = 6$ [graph]

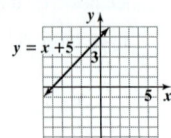

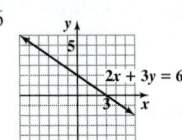

65. $x + y = 12$; $y = 9$ cm **67.** answers may vary **69.** 0, 1, 1, 4, 4 [graph]

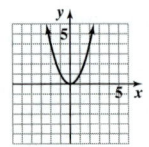

Section 3.3
Practice Exercises
1. x-intercept: $(-4, 0)$; y-intercept: $(0, -6)$
2. x-intercepts: $(-1, 0)$, $(-0.5, 0)$; y-intercept: $(0, 1)$
3. x-intercept: $(0, 0)$; y-intercept: $(0, 0)$
4. x-intercept: none; y-intercept: $(0, 3)$
5. x-intercepts: $(-1, 0)$, $(5, 0)$; y-intercepts: $(0, 2)$, $(0, -2)$

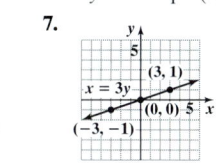

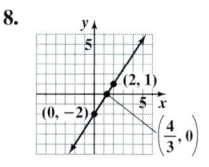

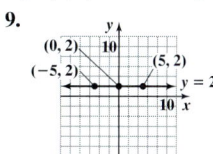

6. – 10. [graphs]

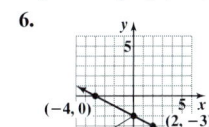

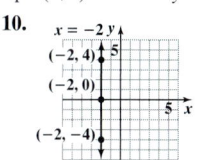

Calculator Explorations 3.3
1. [graph: $x = 3.78y$]
3. [graph: $3x + 7y = 21$]
5. [graph: $-2.2x + 6.8y = 15.5$]

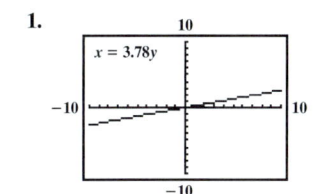

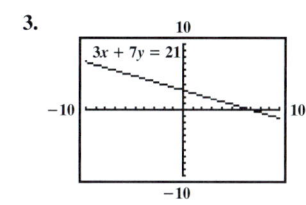

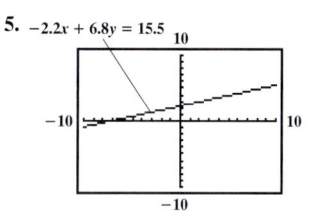

Vocabulary and Readiness Check 3.3
1. linear 3. horizontal 5. y-intercept 7. y; x 9. false 11. true

Exercise Set 3.3
1. $(-1, 0)$; $(0, 1)$ 3. $(-2, 0)$; $(2, 0)$; $(0, -2)$ 5. $(-2, 0)$; $(1, 0)$; $(3, 0)$; $(0, 3)$ 7. $(-1, 0)$; $(1, 0)$; $(0, 1)$; $(0, -2)$ 9. infinite 11. 0

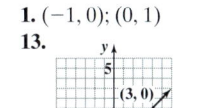

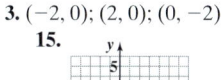

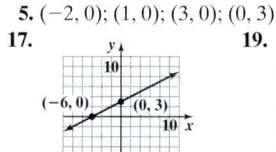

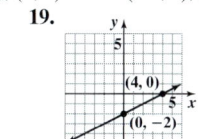

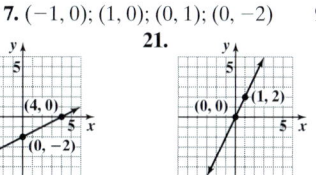

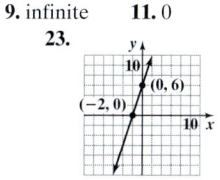

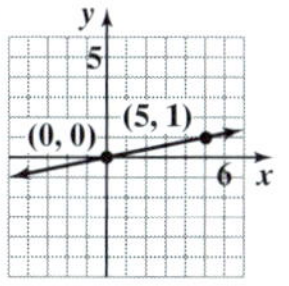

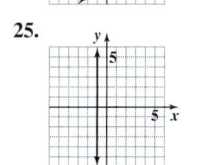

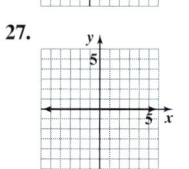

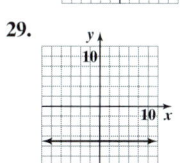

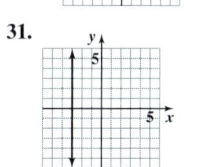

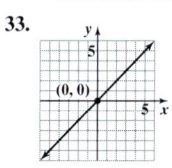

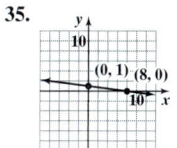

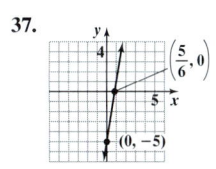

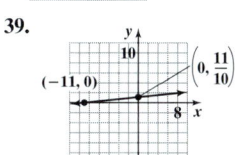

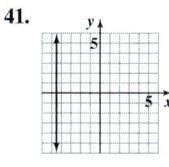

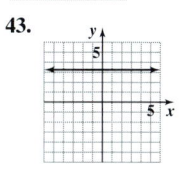

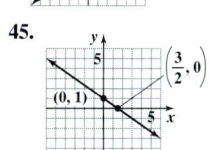

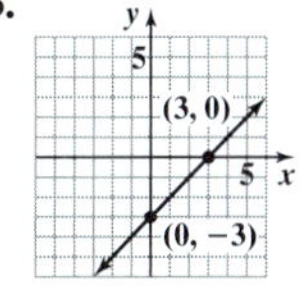

13. – 47. [graphs]

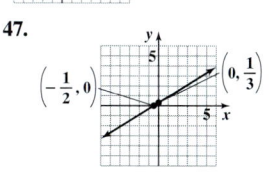

49. C 51. E 53. B 55. $\dfrac{3}{2}$ 57. 6 59. $-\dfrac{6}{5}$ 61. a. $(0, 6505)$ b. In 2003, the revenue for Disney Parks and Resorts was about $6505 million. 63. a. $(22, 0)$ b. 22 years after 2002, 0 people will attend the movies at the theater. c. answer may vary 65. a. $(0, 200)$; no chairs and 200 computer desks are manufactured. b. $(400, 0)$; 400 chairs and no computer desks are manufactured. c. [graph: $3x + 6y = 1200$] d. 300 chairs 67. $y = -4$ 69. answers may vary 71. answers may vary

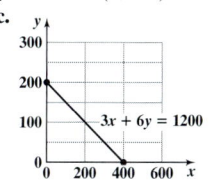

Section 3.4
Practice Exercises
1. -1 2. $\dfrac{1}{3}$ 3. $m = \dfrac{2}{3}$; y-intercept: $(0, -2)$ 4. $m = 6$; y-intercept: $(0, -5)$ 5. $m = -\dfrac{5}{2}$; y-intercept: $(0, 4)$ 6. $m = 0$ 7. slope is undefined 8. a. perpendicular b. neither c. parallel 9. 25% 10. $m = \dfrac{0.75 \text{ dollar}}{1 \text{ pound}}$; The Wash-n-Fold charges $0.75 per pound of laundry.

A12 Answers to Selected Exercises

Calculator Explorations 3.4
1. [graph: $y = -3.8x + 9$, $y = -3.8x$, $y = -3.8x - 3$]
3. [graph: $y = \frac{1}{4}x + 5$, $y = \frac{1}{4}x$, $y = \frac{1}{4}x - 8$]

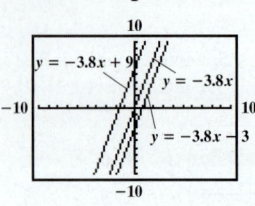

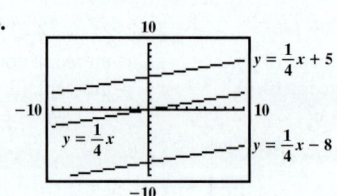

Vocabulary and Readiness Check 3.4
1. slope **3.** 0 **5.** positive **7.** $y; x$ **9.** positive **11.** 0 **13.** downward **15.** vertical

Exercise Set 3.4
1. -1 **3.** undefined **5.** $-\frac{2}{3}$ **7.** 0 **9.** $m = -\frac{4}{3}$ **11.** undefined slope **13.** $m = \frac{5}{2}$ **15.** line 1 **17.** line 2 **19.** D **21.** B **23.** E
25. undefined slope **27.** $m = 0$ **29.** undefined slope **31.** $m = 0$ **33.** $m = 5$ **35.** $m = -0.3$ **37.** $m = -2$ **39.** $m = \frac{2}{3}$
41. undefined slope **43.** $m = \frac{1}{2}$ **45.** $m = 0$ **47.** $m = -\frac{3}{4}$ **49.** $m = 4$ **51.** neither **53.** neither **55.** parallel **57.** perpendicular
59. $\frac{3}{5}$ **61.** 12.5% **63.** 40% **65.** 79% **67.** $m = 3$; Every 1 year, there are/should be 3 million more U.S. households with personal computers.
69. $m = 0.42$; It costs $0.42 per 1 mile to own and operate a compact car. **71.** $y = 2x - 14$ **73.** $y = -6x - 11$ **75. a.** 1 **b.** -1 **77. a.** $\frac{9}{11}$
b. $-\frac{11}{9}$ **79.** $m = \frac{1}{2}$ **81.** answers may vary **83.** 28.5 mi per gal **85.** 2000; 28.1 mi per gal **87.** from 2000 to 2001 **89.** $x = 6$
91. a. (2001, 1132), (2006, 1657) **b.** 105 **c.** For the years 2001 through 2006, the price per acre of U.S. farmland rose approximately $105 per year.
93. The slope through $(-3, 0)$ and $(1, 1)$ is $\frac{1}{4}$. The slope through $(-3, 0)$ and $(-4, 4)$ is -4. The product of the slopes is -1, so the sides are perpendicular.
95. -0.25 **97.** 0.875 **99.** The line becomes steeper.

Integrated Review
1. $m = 2$ **2.** $m = 0$ **3.** $m = -\frac{2}{3}$ **4.** undefined slope **5.** [graph $y = -2x$] **6.** [graph $x + y = 3$, (0, 3)] **7.** [graph $x = -1$]
8. [graph $y = 4$] **9.** [graph $x - 2y = 6$, (6, 0), (0, -3), (4, -1)] **10.** [graph $y = 3x + 2$] **11.** [graph $5x + 3y = 15$] **12.** [graph $2x - 4y = 8$, (4, 0), (0, -2)]
13. parallel **14.** neither **15. a.** (0, 1650) **b.** In 2002, there were 1650 million admissions to movie theaters in the U.S. **c.** -75
d. For the years 2002 through 2005, the number of movie theater admissions decreased at a rate of 75 million per year. **16. a.** (9, 26.6)
b. In 2009, the predicted revenue for online advertising is $26.6 billion.

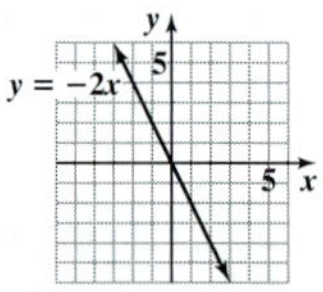

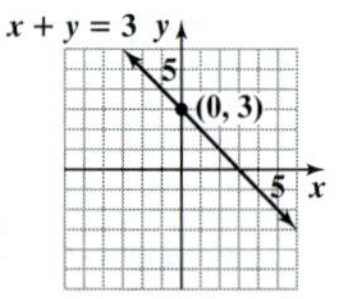

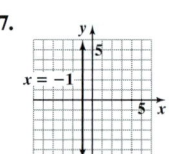

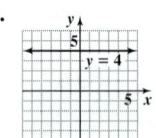

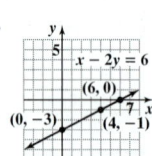

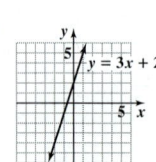

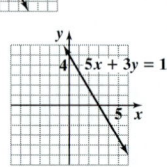

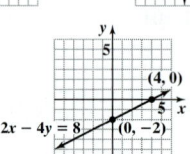

Section 3.5
Practice Exercises
1. $y = \frac{1}{2}x + 7$ **2.** [graph (0, -5), (3, -3)] **3.** [graph (1, 1), (0, -2)] **4.** $4x - y = 5$ **5.** $5x + 4y = 19$ **6.** $x = 3$ **7.** $y = 3$

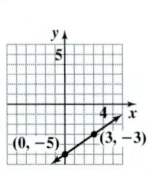

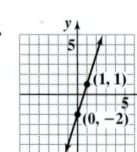

8. a. $y = -1500x + 195{,}000$ **b.** $105{,}000

Calculator Explorations 3.5
1. [graph $y = -6x$, $y = 6x$, $y = x$]
3. [graph $y = x + 2$, $y = \frac{3}{4}x + 2$, $y = \frac{1}{2}x + 2$]
5. [graph $y = 7x + 5$, $y = -7x + 5$]

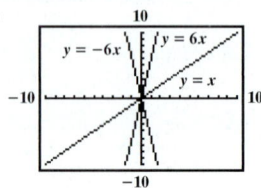

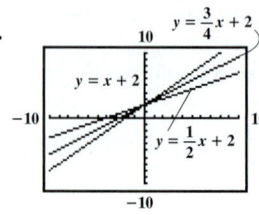

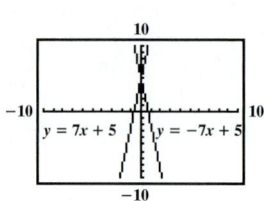

Vocabulary and Readiness Check 3.5
1. slope-intercept; $m; b$ **3.** point-slope   **5.** horizontal   **7.** slope-intercept

Exercise Set 3.5
1. $y = 5x + 3$ **3.** $y = -4x - \frac{1}{6}$ **5.** $y = \frac{2}{3}x$ **7.** $y = -8$ **9.** $y = -\frac{1}{5}x + \frac{1}{9}$ **11.** [graph with (0,1)]   **13.** [graph with (0,5)]
15. [graph with (0,0)]   **17.** [graph with (0,6)]   **19.** [graph with (0,2)]   **21.** [graph with (0,0)]   **23.** $-6x + y = -10$ **25.** $8x + y = -13$
27. $3x - 2y = 27$ **29.** $x + 2y = -3$ **31.** $2x - y = 4$ **33.** $8x - y = -11$ **35.** $4x - 3y = -1$ **37.** $8x + 13y = 0$ **39.** $x = 0$
41. $y = 3$ **43.** $x = -\frac{7}{3}$ **45.** $y = 2$ **47.** $y = 5$ **49.** $x = 6$ **51.** $y = -\frac{1}{2}x + \frac{5}{3}$ **53.** $y = -x + 17$ **55.** $x = -\frac{3}{4}$ **57.** $y = x + 16$
59. $y = -5x + 7$ **61.** $y = 7$ **63.** $y = \frac{3}{2}x$ **65.** $y = -3$ **67.** $y = -\frac{4}{7}x - \frac{18}{7}$ **69. a.** $s = 32t$ **b.** 128 ft/sec   **71. a.** $y = 14,000x + 29,000$
b. 113,000 hybrids   **73. a.** $y = 0.9x + 79.6$ **b.** 88.6 persons per sq mi   **75. a.** $(0, 14.7), (10, 14.14)$ **b.** $y = -0.056x + 14.7$ **c.** 13.58 births per thousand population   **77. a.** $(0, 5), (3, 20)$ **b.** $y = 5x + 5$ **c.** 50 thousand, or 50,000   **79.** -1 **81.** 5   **83.** no   **85.** yes
87. answers may vary   **89. a.** $3x - y = -5$ **b.** $x + 3y = 5$ **91. a.** $3x + 2y = -1$ **b.** $2x - 3y = 21$

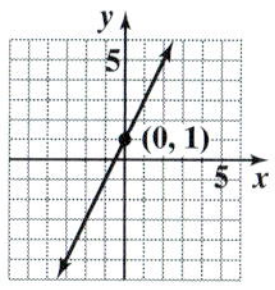

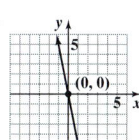

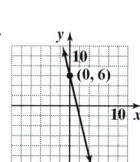

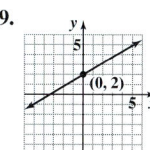

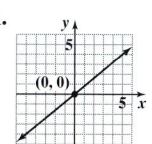

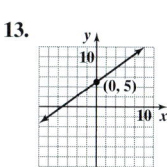

Section 3.6
Practice Exercises
1. Domain: $\{0, 1, 5\}$; Range: $\{-2, 0, 3, 4\}$ **2. a.** function   **b.** not a function   **3. a.** not a function   **b.** function   **4. a.** function
b. function   **c.** function   **d.** not a function   **5. a.** function   **b.** function   **c.** function   **d.** not a function   **6. a.** 69°F   **b.** November
c. yes   **7. a.** $h(2) = 9; (2, 9)$ **b.** $h(-5) = 30; (-5, 30)$ **c.** $h(0) = 5; (0, 5)$ **8. a.** domain: $(-\infty, \infty)$ **b.** domain: $(-\infty, 0) \cup (0, \infty)$
9. a. domain: $[-4, 6]$; range: $[-2, 3]$ **b.** domain: $(-\infty, \infty)$; range: $(-\infty, 3]$

Vocabulary and Readiness Check 3.6
1. relation   **3.** range   **5.** vertical   **7.** $(3, 7)$

Exercise Set 3.6
1. $\{-7, 0, 2, 10\}$; $\{-7, 0, 4, 10\}$ **3.** $\{0, 1, 5\}$; $\{-2\}$ **5.** yes   **7.** no   **9.** no   **11.** yes   **13.** yes   **15.** no   **17.** yes   **19.** yes   **21.** yes
23. no   **25.** no   **27.** 9:30 p.m.   **29.** January 1 and December 1   **31.** yes; it passes the vertical line test   **33.** $4.25 per hour   **35.** 2009
37. yes; answers may vary   **39.** $-9, -5, 1$ **41.** $6, 2, 11$ **43.** $-6, 0, 9$ **45.** $2, 0, 3$ **47.** $5, 0, -20$ **49.** $5, 3, 35$ **51.** $(3, 6)$
53. $\left(0, -\frac{1}{2}\right)$ **55.** $(-2, 9)$ **57.** $(-\infty, \infty)$ **59.** all real number except -5 or $(-\infty, -5) \cup (-5, \infty)$ **61.** $(-\infty, \infty)$ **63.** domain: $(-\infty, \infty)$;
range: $[-4, \infty)$ **65.** domain: $(-\infty, \infty)$; range: $(-\infty, \infty)$ **67.** domain: $(-\infty, \infty)$; range: $\{2\}$ **69.** $(-2, 1)$ **71.** $(-3, -1)$ **73.** $f(-5) = 12$
75. $(3, -4)$ **77.** $f(5) = 0$ **79. a.** 166.38 cm   **b.** 148.25 cm   **81.** answers may vary   **83.** $f(x) = x + 7$ **85. a.** $-3s + 12$ **b.** $-3r + 12$
87. a. 132   **b.** $a^2 - 12$

Chapter 3 Vocabulary Check
1. solution   **2.** y-axis   **3.** linear   **4.** x-intercept   **5.** standard   **6.** y-intercept   **7.** slope-intercept   **8.** point-slope   **9.** y **10.** x-axis
11. x **12.** slope   **13.** function   **14.** domain   **15.** range   **16.** relation

Chapter 3 Review
1. [graph with $(-7,0)$]   **3.** [graph with $(-2,-5)$]   **5.** [graph with $(0.7, 0.7)$]   **7. a.** $(8.00, 1); (7.50, 10); (6.50, 25); (5.00, 50); (2.00, 100)$ **b.** [scatter plot: Number of boards purchased vs Price per board (in dollars)]

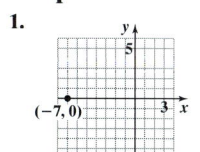

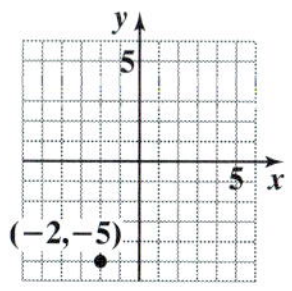

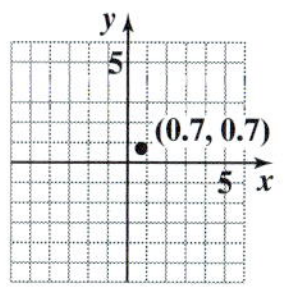

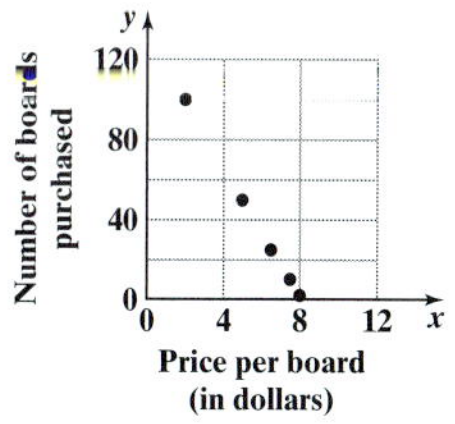

9. no; yes   **11.** yes; yes   **13.** $(7, 44)$ **15.** $(-3, 0); (1, 3); (9, 9)$ **17.** $(0, 0); (10, 5); (-10, -5)$

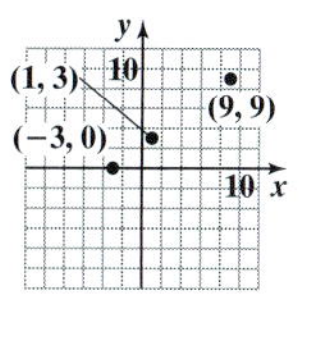

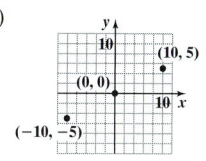

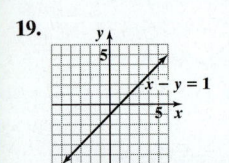

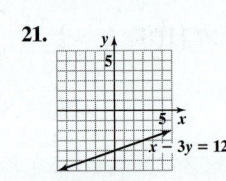

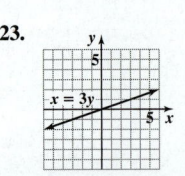

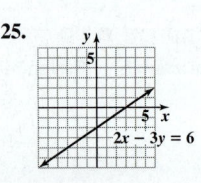

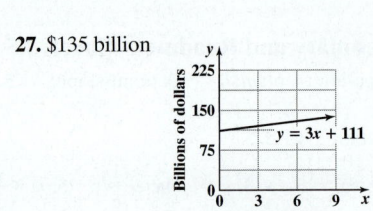

19. [graph: $x - y = 1$] **21.** [graph: $x - 3y = 12$] **23.** [graph: $x = 3y$] **25.** [graph: $2x - 3y = 6$] **27.** $135 billion [graph: $y = 3x + 111$, Billions of dollars vs Years after 1999]

29. $(0, -3)$ **31.** $(-1, 0); (2, 0); (3, 0); (0, -2)$ **33.** [graph: $-4x + y = 8$, $(0, 8)$, $(-2, 0)$] **35.** [graph: $x = 5$, $(5, 0)$] **37.** [graph: $x = 5y$, $(0, 0)$] **39.** [graph: $y + 6 = 0$, $(0, -6)$]

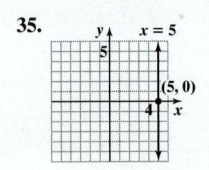

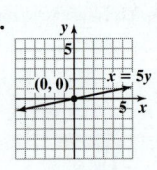

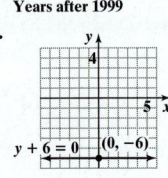

41. $m = \dfrac{1}{5}$ **43.** b **45.** a **47.** $\dfrac{3}{4}$ **49.** 4 **51.** 3 **53.** 0 **55.** perpendicular **57.** neither **59.** Every 1 year, monthly day care costs increase by $17.90 **61.** $m = -3; (0, 7)$ **63.** $m = 0; (0, 2)$ **65.** $y = -5x + \dfrac{1}{2}$ **67.** [graph: $y = 3x - 1$] **69.** [graph: $5x - 3y = 15$]

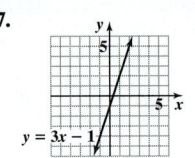

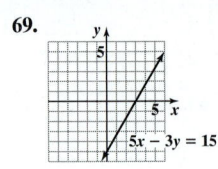

71. c **73.** b **75.** $3x + y = -5$ **77.** $y = -3$ **79.** $6x + y = 11$ **81.** $x + y = 6$ **83.** $x = 5$ **85.** $x = 6$ **87.** no **89.** yes **91.** no **93.** no **95. a.** 6 **b.** 10 **c.** 5 **97. a.** 45 **b.** -35 **c.** 0 **99.** $(-\infty, \infty)$ **101.** domain: $[-3, 5]$ range: $[-4, 2]$ **103.** domain: $\{3\}$ range: $(-\infty, \infty)$ **105.** $7; -1; -3$ **107.** $(3, 0); (0, -2)$ **109.** [graph: $(10, 0), (0, -2)$] **111.** [graph: $(0, 0), (1, -4)$] **113.** [graph]

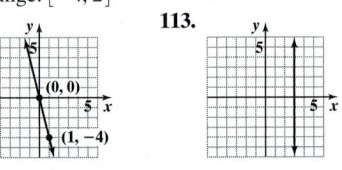

115. $m = -1$ **117.** $m = 2$ **119.** $m = \dfrac{2}{3}; (0, -5)$ **121.** $5x + y = 8$ **123.** $4x + y = -3$ **125.** 2002; 27.1 billion lb **127.** 2002, 2003, 2006

Chapter 3 Test

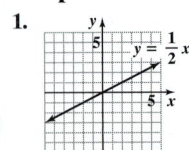

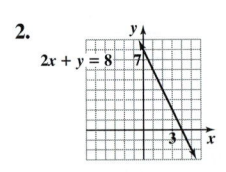

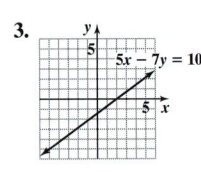

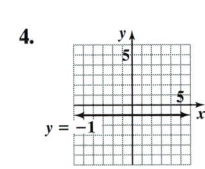

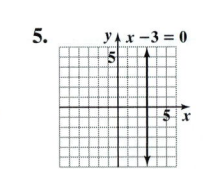

1. [graph: $y = \dfrac{1}{2}x$] **2.** [graph: $2x + y = 8$] **3.** [graph: $5x - 7y = 10$] **4.** [graph: $y = -1$] **5.** [graph: $x - 3 = 0$]

6. $\dfrac{2}{5}$ **7.** 0 **8.** -1 **9.** 3 **10.** undefined **11.** $m = \dfrac{7}{3}; \left(0, -\dfrac{2}{3}\right)$ **12.** neither **13.** $x + 4y = 10$ **14.** $7x + 6y = 0$ **15.** $8x + y = 11$ **16.** $x = -5$ **17.** $x - 8y = -96$ **18.** yes **19.** no **20. a.** 0 **b.** 0 **c.** 60 **21.** all real numbers except -1 or $(-\infty, -1) \cup (-1, \infty)$ **22.** domain: $(-\infty, \infty)$; range: $(-\infty, 4]$ **23.** domain: $(-\infty, \infty)$; range: $(-\infty, \infty)$ **24.** $(7, 20)$ **25.** 210 liters **26.** 490 liters **27.** July **28.** 63°F **29.** January, February, March, November, December

Chapter 3 Cumulative Review

1. a. $<$ **b.** $>$ **c.** $>$; Sec. 1.2, Ex. 1 **3.** $\dfrac{2}{39}$; Sec. 1.3, Ex. 3 **5.** $\dfrac{8}{3}$; Sec. 1.4, Ex. 3 **7. a.** -19 **b.** 30 **c.** -0.5 **d.** $-\dfrac{4}{5}$ **e.** 6.7 **f.** $\dfrac{1}{40}$; Sec. 1.5, Ex. 6 **9. a.** -6; **b.** 6.3; Sec. 1.6, Ex. 4 **11. a.** -6 **b.** 0 **c.** $\dfrac{3}{4}$; Sec. 1.7, Ex. 10 **13. a.** $22 + x$ **b.** $-21x$; Sec. 1.8, Ex. 3 **15. a.** -3 **b.** 22 **c.** 1 **d.** -1 **e.** $\dfrac{1}{7}$; Sec. 2.1, Ex. 1 **17.** -1.6; Sec. 2.2, Ex. 2 **19.** $\dfrac{15}{4}$; Sec. 2.3, Ex. 5 **21.** $3x + 3$; Sec. 2.3, Ex. 10 **23.** 0; Sec. 2.4, Ex. 4 **25.** 202 Republicans, 235 Democrats; Sec. 2.5, Ex. 4 **27.** 40 ft; Sec. 2.6, Ex. 2 **29.** $\dfrac{y - b}{m} = x$; Sec. 2.6, Ex. 6 **31.** 40% solution: 8 liters; 70% solution: 4 liters; Sec. 2.7, Ex. 7 **33.** [number line, -1]; Sec. 2.9, Ex. 1 **35.** [number line, $1, 4$] $[1, 4)$; Sec. 2.9, Ex. 10 **37. a.** solution **b.** not a solution **c.** solution; Sec. 3.1, Ex. 5 **39. a.** yes **b.** yes **c.** no **d.** yes; Sec. 3.2, Ex. 1 **41.** 0; Sec. 3.4, Ex. 6 **43.** $y = \dfrac{1}{4}x - 3$; Sec. 3.5, Ex. 1

CHAPTER 4 SOLVING SYSTEMS OF LINEAR EQUATIONS AND INEQUALITIES

Section 4.1
Practice Exercises
1. no   **2.** yes   **3.** (8, 5)   **4.** (−3, −5) 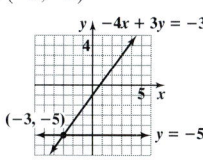   **5.** no solution; inconsistent, independent

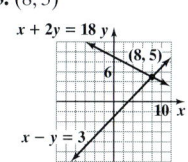

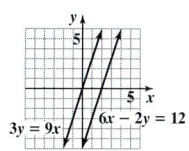

6. infinite number of solutions; consistent, dependent 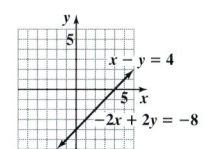   **7.** one solution   **8.** no solution

Calculator Explorations 4.1
1. (0.37, 0.23)   **3.** (0.03, −1.89)

Vocabulary and Readiness Check 4.1
1. dependent   **3.** consistent   **5.** inconsistent   **7.** one solution, (−1, 3)   **9.** infinite number of solutions

Exercise Set 4.1
1. a. no **b.** yes   **3. a.** yes **b.** no   **5. a.** yes **b.** yes   **7. a.** no **b.** no   **9.**   **11.**   **13.**

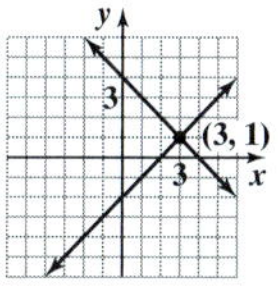

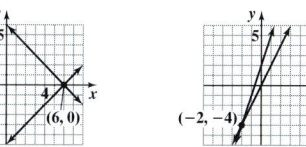

15. 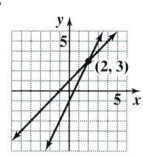   **17.** 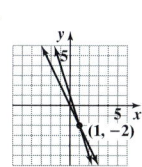   **19.** 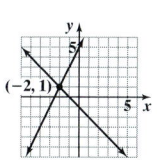   **21.** no solution 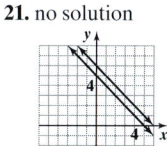   **23.** 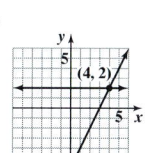   **25.**

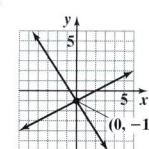

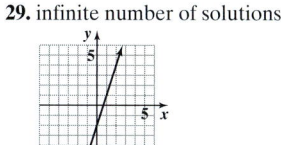

27. no solution 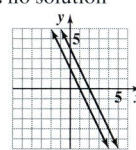   **29.** infinite number of solutions   **31.** 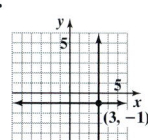   **33.** 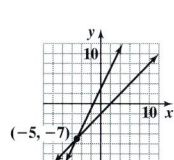   **35.**

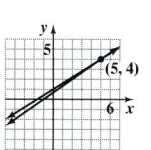

37. infinite number of solutions 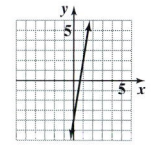   **39.** intersecting, one solution   **41.** parallel, no solution   **43.** identical lines, infinite number of solutions

45. intersecting, one solution   **47.** intersecting, one solution   **49.** identical lines, infinite number of solutions   **51.** parallel, no solution
53. 2   **55.** $-\dfrac{2}{5}$ **57.** 2   **59.** answers may vary; possible answer   **61.** answers may vary; possible answer 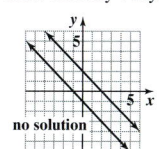   **63.** answers may vary

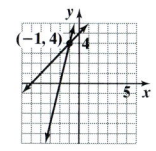

A16 Answers to Selected Exercises

65. 2000, 2001, 2002   **67.** 2001, 2002, 2003   **69.** answers may vary   **71.** answers may vary   **73. a.** (4, 9)   **b.**   **c.** yes
75. answers may vary

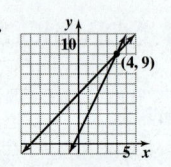

Section 4.2
Practice Exercises
1. (8, 7)   **2.** (−3, −6)   **3.** $\left(4, \dfrac{2}{3}\right)$ **4.** (−3, 2)   **5.** infinite number of solutions   **6.** no solution

Vocabulary and Readiness Check 4.2
1. (1, 4)   **3.** infinite number of solutions   **5.** (0, 0)

Exercise Set 4.2
1. (2, 1)   **3.** (−3, 9)   **5.** (2, 7)   **7.** $\left(-\dfrac{1}{5}, \dfrac{43}{5}\right)$ **9.** (2, −1)   **11.** (−2, 4)   **13.** (4, 2)   **15.** (−2, −1)   **17.** no solution   **19.** (3, −1)
21. (3, 5)   **23.** $\left(\dfrac{2}{3}, -\dfrac{1}{3}\right)$ **25.** (−1, −4)   **27.** (−6, 2)   **29.** (2, 1)   **31.** no solution   **33.** infinite number of solutions   **35.** $\left(\dfrac{1}{2}, 2\right)$
37. (1, −3)   **39.** $-6x - 4y = -12$ **41.** $-12x + 3y = 9$ **43.** $5n$ **45.** $-15b$ **47.** answers may vary   **49.** no; answers may vary
51. c; answers may vary   **53. a.** (13, 492)   **b.** In 1970 + 13 = 1983, the number of men and women receiving bachelor's degrees was the same.
c. answers may vary   **55.** (−2.6, 1.3)   **57.** (3.28, 2.1)

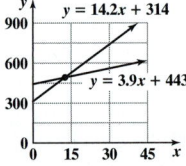

Section 4.3
Practice Exercises
1. (5, 3)   **2.** (3, −4)   **3.** no solution   **4.** infinite number of solutions   **5.** (2, 2)   **6.** $\left(-\dfrac{8}{5}, \dfrac{6}{5}\right)$

Exercise Set 4.3
1. (1, 2)   **3.** (2, −3)   **5.** (−2, −5)   **7.** (5, −2)   **9.** (−7, 5)   **11.** (6, 0)   **13.** no solution   **15.** infinite number of solutions   **17.** $\left(2, -\dfrac{1}{2}\right)$
19. (−2, 0)   **21.** (1, −1)   **23.** infinite number of solutions   **25.** $\left(\dfrac{12}{11}, -\dfrac{4}{11}\right)$ **27.** $\left(\dfrac{3}{2}, 3\right)$ **29.** infinite number of solutions   **31.** (1, 6)
33. $\left(-\dfrac{1}{2}, -2\right)$ **35.** infinite number of solutions   **37.** $\left(-\dfrac{2}{3}, \dfrac{2}{5}\right)$ **39.** (2, 4)   **41.** (−0.5, 2.5)   **43.** (2, 5)   **45.** (−3, 2)   **47.** (0, 3)   **49.** (5, 7)
51. $\left(\dfrac{1}{3}, 1\right)$ **53.** infinite number of solutions   **55.** (−8.9, 10.6)   **57.** $2x + 6 = x - 3$ **59.** $20 - 3x = 2$ **61.** $4(n + 6) = 2n$
63. 2; $6x - 2y = -24$ **65.** b; answers may vary   **67.** answers may vary   **69. a.** $b = 15$ **b.** any real number except 15   **71.** (−4.2, 9.6)
73. a. (5, 294) or (5, 295) or (5, 296)   **b.** In 2009 (2004 + 5), the number of pharmacy technician jobs equals the number of network and data analyst jobs.
c. 294–296 thousand

Integrated Review
1. (2, 5)   **2.** (4, 2)   **3.** (5, −2)   **4.** (6, −14)   **5.** (−3, 2)   **6.** (−4, 3)   **7.** (0, 3)   **8.** (−2, 4)   **9.** (5, 7)   **10.** (−3, −23)   **11.** $\left(\dfrac{1}{3}, 1\right)$
12. $\left(-\dfrac{1}{4}, 2\right)$ **13.** no solution   **14.** infinite number of solutions   **15.** (0.5, 3.5)   **16.** (−0.75, 1.25)   **17.** infinite number of solutions
18. no solution   **19.** (7, −3)   **20.** (−1, −3)   **21.** answers may vary   **22.** answers may vary

Section 4.4
Practice Exercises
1. 18, 12   **2. a.** Adult: $19   **b.** Child: $6   **c.** No, the regular rates are less than the group rate.   **3.** 1.75 mph and 3.75 mph
4. 15 pounds of Kona and 5 pounds of Blue Mountain

Exercise Set 4.4
1. c   **3.** b   **5.** a   **7.** $\begin{cases} x + y = 15 \\ x - y = 7 \end{cases}$ **9.** $\begin{cases} x + y = 6500 \\ x = y + 800 \end{cases}$ **11.** 33 and 50   **13.** 14 and −3   **15.** Taurasi: 860 points; Augustus: 744 points
17. child's ticket: $18; adult's ticket: $29   **19.** quarters: 53; nickels: 27   **21.** Apple: $87.97; Microsoft: $27.29   **23.** daily fee: $32; mileage

Answers to Selected Exercises   **A17**

charge: $0.25 per mi   **25.** distance downstream = distance upstream = 18 mi; time downstream: 2 hr; time upstream: $4\frac{1}{2}$ hr; still water: 6.5 mph; current: 2.5 mph   **27.** still air: 455 mph; wind: 65 mph   **29.** $4\frac{1}{2}$ hr   **31.** 12% solution: $7\frac{1}{2}$ oz; 4% solution: $4\frac{1}{2}$ oz   **33.** $4.95 beans: 113 lb; $2.65 beans: 87 lb   **35.** 60°, 30°   **37.** 20°, 70°   **39.** number sold at $9.50: 23; number sold at $7.50: 67   **41.** $2\frac{1}{4}$ mph and $2\frac{3}{4}$ mph   **43.** 30%: 50 gal; 60%: 100 gal   **45.** length: 42 in.; width: 30 in.   **47.** (3, ∞)   **49.** $\left[\frac{1}{2}, \infty\right)$ **51.** a   **53.** width: 9 ft; length: 15 ft   **55. a.** (27.1, 39.5)   **b.** For viewers 27.1 years over 18 (or 45.1 years old) the percent who watch cable news and network news is the same, or 39.5%.   **c.** answers may vary

Section 4.5
Practice Exercises

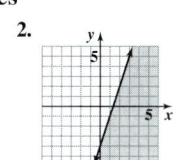

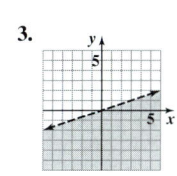

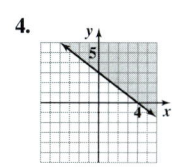

1.–**5.** [graphs]

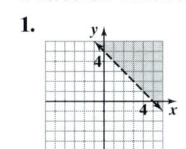

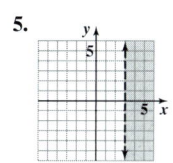

Vocabulary and Readiness Check 4.5
1. linear inequality in two variables   **3.** false   **5.** true   **7.** yes   **9.** yes   **11.** yes   **13.** no

Exercise Set 4.5
1. no; yes   **3.** no; no   **5.** no; yes   **7.**–**41.** [graphs]   **43.** e   **45.** c   **47.** f   **49.** 8   **51.** −32   **53.** 48   **55.** 25   **57.** −2   **59.** yes   **61.** yes   **63.** $x + y \geq 13$ [graph]   **65.** answers may vary   **67.** [graph: $2.5x + 0.25y \leq 20$]   **69.** answers may vary   **71. a.** $30x + 0.15y \leq 500$ **b.** [graph]   **c.** answers may vary

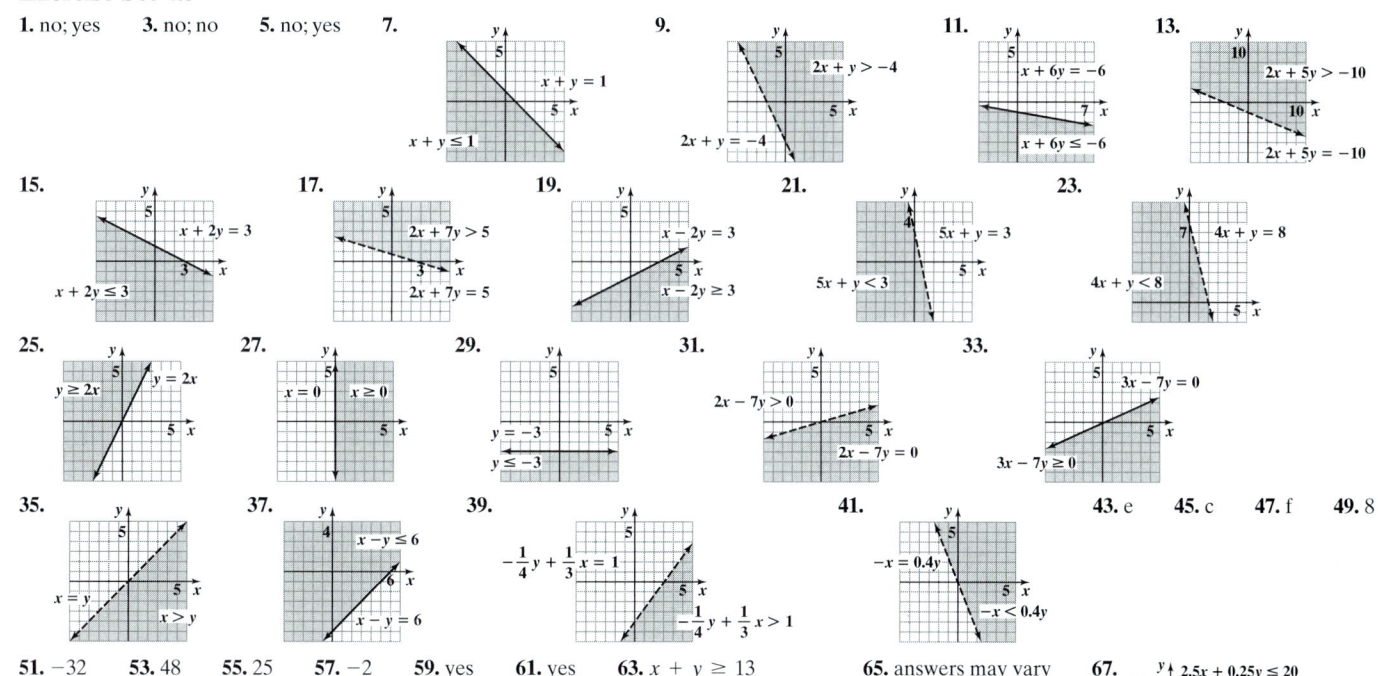

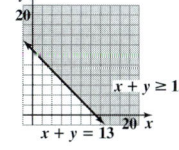

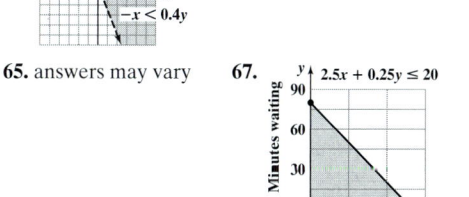

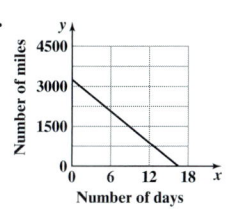

Section 4.6
Practice Exercises
1. 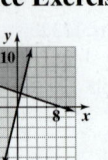 [graph] 2. 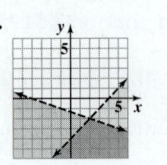 [graph] 3. 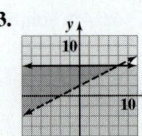 [graph]

Exercise Set 4.6
1. 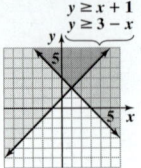 $y \geq x + 1$, $y \geq 3 - x$ [graph]
3. 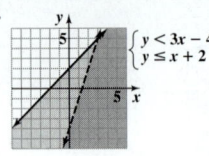 $\begin{cases} y < 3x - 4 \\ y \leq x + 2 \end{cases}$ [graph]
5. 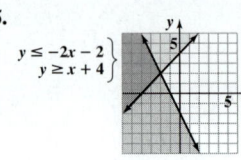 $\begin{cases} y \leq -2x - 2 \\ y \geq x + 4 \end{cases}$ [graph]
7. 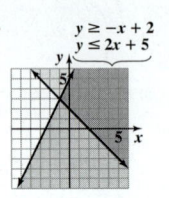 $y \geq -x + 2$, $y \leq 2x + 5$ [graph]
9. 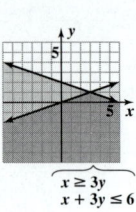 [graph] $x \geq 3y$, $x + 3y \leq 6$
11. 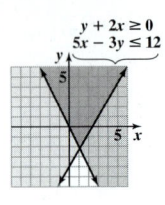 $y + 2x \geq 0$, $5x - 3y \leq 12$ [graph]
13. 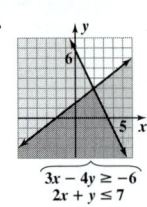 [graph] $3x - 4y \geq -6$, $2x + y \leq 7$
15. 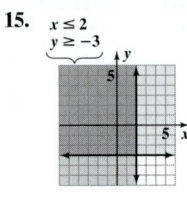 $x \leq 2$, $y \geq -3$ [graph]
17. 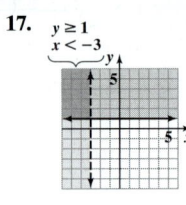 $y \geq 1$, $x < -3$ [graph]
19. 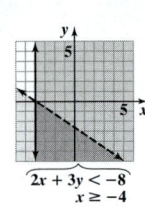 [graph] $2x + 3y < -8$, $x \geq -4$
21. 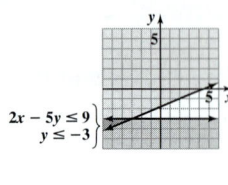 [graph] $2x - 5y \leq 9$, $y \leq -3$
23. 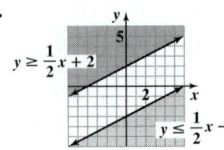 $y \geq \frac{1}{2}x + 2$, $y \leq \frac{1}{2}x - 3$ [graph]
25. 16   27. $36x^2$ 29. $100y^6$ 31. C   33. D   35. answers may vary   37. 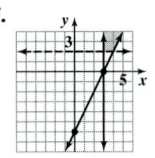 [graph]

Chapter 4 Vocabulary Check
1. dependent   2. system of linear equations   3. consistent   4. solution   5. addition; substitution   6. inconsistent   7. independent   8. system of linear inequalities

Chapter 4 Review
1. a. no b. yes c. no   3. a. no b. no c. yes   5. 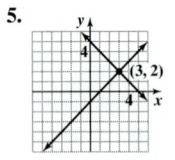 [graph (3, 2)]   7. 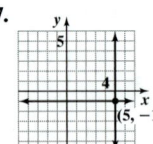 [graph (5, −1)]   9. 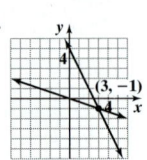 [graph (3, −1)]
11. no solution [graph]   13. $(-1, 4)$ 15. $(3, -2)$ 17. infinite number of solutions   19. no solution   21. $(-6, 2)$ 23. $(3, 7)$
25. infinite number of solutions   27. $(8, -6)$ 29. -6 and 22 31. current of river: 3.2 mph; speed in still water: 21.1 mph
33. egg: $0.40; strip of bacon: $0.65   35. 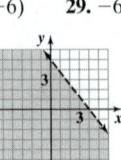 [graph]   37. 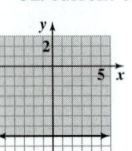 [graph]   39. 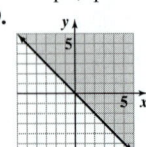 [graph]   41. 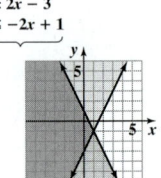 $y \geq 2x - 3$, $y \leq -2x + 1$ [graph]

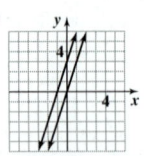

Answers to Selected Exercises   A19

43. 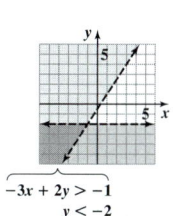 [graph: $-3x+2y>-1$, $y<-2$]   **45.** [graph]   **47.** $(3,2)$ **49.** $\left(1\frac{1}{2},-3\right)$ **51.** infinite number of solutions   **53.** $(-5,2)$

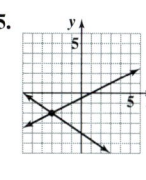

55. infinite number of solutions   **57.** 4 and 8   **59.** 24 nickels and 41 dimes   **61.** 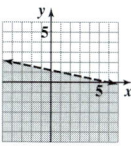 [graph]

Chapter 4 Test

1. false   **2.** false   **3.** true   **4.** false   **5.** no   **6.** yes   **7.** 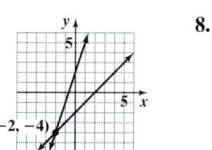 [graph through $(-2,-4)$]   **8.** 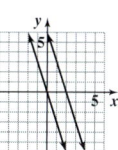 [graph]   **9.** $(-4,1)$ **10.** $\left(\frac{1}{2},-2\right)$

11. $(20,8)$ **12.** no solution   **13.** $(4,-5)$ **14.** $(7,2)$ **15.** $(5,-2)$ **16.** infinite number of solutions   **17.** $(-5,3)$ **18.** $\left(\frac{47}{5},\frac{48}{5}\right)$

19. 78, 46   **20.** 120 cc   **21.** Texas: 226 thousand; Missouri: 110 thousand   **22.** 3 mph; 6 mph   **23.** 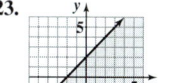 [graph]   **24.** 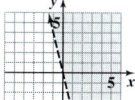 [graph]

25. 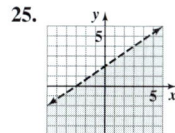 [graph]   **26.** 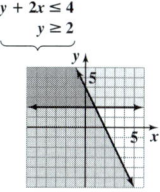 [graph: $y+2x\leq 4$, $y\geq 2$]   **27.** 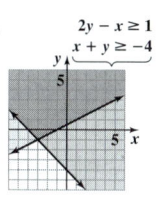 [graph: $2y-x\geq 1$, $x+y\geq -4$]

Chapter 4 Cumulative Review

1. a. $<$ **b.** $=$ **c.** $>$; Sec. 1.2, Ex. 6   **3. a.** commutative property of multiplication   **b.** associative property of addition   **c.** identity element for addition   **d.** commutative property of multiplication   **e.** multiplicative inverse property   **f.** additive inverse property   **g.** commutative and associative properties of multiplication; Sec. 1.8, Ex. 6   **5.** $-2x-1$; Sec. 2.1, Ex. 7   **7.** -7; Sec. 2.2, Ex. 4   **9.** 6; Sec. 2.3, Ex. 1
11. 12; Sec. 2.4, Ex. 3   **13.** 10; Sec. 2.5, Ex. 2   **15.** $x=\dfrac{y-b}{m}$; Sec. 2.6, Ex. 6   **17.** $[2,\infty)$; Sec. 2.9, Ex. 3   **19.** [graph through $(0,0)$, $x=-2y$]; Sec. 3.3, Ex. 7

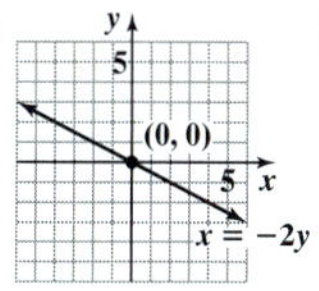

21. $-\dfrac{8}{3}$; Sec. 3.4, Ex. 1   **23.** $\dfrac{3}{4}$; Sec. 3.4, Ex. 3   **25.** slope: $\dfrac{3}{4}$; y-intercept: $(0,-1)$; Sec. 3.4, Ex. 5   **27.** $2x+y=3$; Sec. 3.5, Ex. 4
29. $x=-1$; Sec. 3.5, Ex. 6   **31.** domain: $\{-1,0,3\}$; range: $\{-2,0,2,3\}$; Sec. 3.6, Ex. 1   **33. a.** function   **b.** not a function; Sec. 3.6, Ex. 2
35. one solution; Sec. 4.1, Ex. 8   **37.** $\left(6,\dfrac{1}{2}\right)$; Sec. 4.2, Ex. 3   **39.** $(6,1)$; Sec. 4.3, Ex. 1   **41.** 29 and 8; Sec. 4.4, Ex. 1
43. 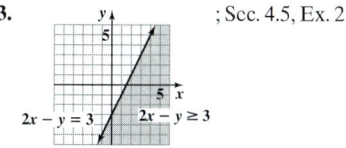 [graph: $2x-y=3$, $2x-y\geq 3$]; Sec. 4.5, Ex. 2

A20 Answers to Selected Exercises

CHAPTER 5 EXPONENTS AND POLYNOMIALS

Section 5.1
Practice Exercises
1. a. 27 b. 4 c. 64 d. −64 e. $\frac{27}{64}$ f. 0.0081 g. 75 2. a. 243 b. $\frac{3}{8}$ 3. a. 3^{10} b. y^5 c. z^5 d. x^{11} e. $(-2)^8$ f. $b^3 \cdot t^5$
4. $15y^7$ 5. a. $y^{12}z^4$ b. $-7m^5n^{14}$ 6. a. x^{12} b. z^{21} c. $(-2)^{15}$ 7. a. p^5r^5 b. $36b^2$ c. $\frac{1}{64}x^6y^3$ d. $81a^{12}b^{16}c^4$ 8. a. $\frac{x^5}{y^{10}}$ b. $\frac{32a^{20}}{b^{15}}$
9. a. z^4 b. 25 c. 64 d. $\frac{q^5}{t^2}$ e. $6x^2y^2$ 10. a. −1 b. 1 c. 1 d. 1 e. 1 11. a. $\frac{125}{x^3z^3}$ b. $16z^{32}x^{20}$ c. $\frac{-27x^9}{y^{12}}$

Vocabulary and Readiness Check 5.1
1. exponent 3. add 5. 1 7. base: 3; exponent: 2 9. base: 4; exponent: 2 11. base: 5; exponent: 1; base: x; exponent: 2

Exercise Set 5.1
1. 49 3. −5 5. −16 7. 16 9. 0.00001 11. $\frac{1}{81}$ 13. 224 15. −250 17. answers may vary 19. 4 21. 135 23. 150
25. $\frac{32}{5}$ 27. x^7 29. $(-3)^{12}$ 31. $15y^5$ 33. $x^{19}y^6$ 35. $-72m^3n^8$ 37. $-24z^{20}$ 39. $20x^5$ sq ft 41. x^{36} 43. p^8q^8 45. $8a^{15}$
47. $x^{10}y^{15}$ 49. $49a^4b^{10}c^2$ 51. $\frac{r^9}{s^9}$ 53. $\frac{m^5p^5}{n^5}$ 55. $\frac{4x^2z^2}{y^{10}}$ 57. $64z^{10}$ sq dm 59. $27y^{12}$ cu ft 61. x^2 63. −64 65. p^6q^5 67. $\frac{y^3}{2}$
69. 1 71. 1 73. −7 75. 2 77. −81 79. $\frac{1}{64}$ 81. $\frac{81}{q^2r^2}$ 83. a^6 85. $-16x^7$ 87. $a^{11}b^{20}$ 89. $26m^9n^7$ 91. z^{40} 93. $36x^2y^2z^6$
95. $3x$ 97. $81x^2y^2$ 99. 33 101. $\frac{y^{15}}{8x^{12}}$ 103. $2x^2y$ 105. $2y - 10$ 107. $-x - 4$ 109. $-x + 5$ 111. c 113. e
115. answers may vary 117. answers may vary 119. 343 cu m 121. volume 123. answers may vary 125. x^{9a} 127. a^{5b}
129. x^{5a} 131. $1045.85

Section 5.2
Practice Exercises
1. a. degree 3 b. degree 8 c. degree 0 2. a. trinomial, degree 2 b. binomial, degree 1 c. none of these, degree 3

3.
Term	Numerical Coefficient	Degree of Term
$-3x^3y^2$	−3	5
$4xy^2$	4	3
$-y^2$	−1	2
$3x$	3	1
−2	−2	0

4. 36 5. 114 ft; 66 ft 6. a. $-2y$ b. $z + 5z^3$ c. $4a^2 - 12$ d. $\frac{1}{3}x^4 + \frac{11}{24}x^3 - x^2$ 7. $-3x^2 + 5xy + 5y^2$ 8. $4x^2 + 7x + 4$
9. $-x^2 - 6x + 9$ 10. $-3x^3 + 10x^2 - 6x - 4$ 11. $5z^3 + 8z^2 + 8z$
12. $5x - 1$ 13. $2x^3 - 4x^2 + 4x - 6$ 14. $-8z^2 - 11z + 12$ 15. $7x - 11$
16. a. $-5a^2 - ab + 6b^2$ b. $3x^2y^2 - 10xy - 4xy^2 - 6y^2 + 5$

Vocabulary and Readiness Check 5.2
1. binomial 3. trinomial 5. constant 7. $-14y$ 9. $7x$ 11. $5m^2 + 2m$

Exercise Set 5.2
1. 1; binomial 3. 3; none of these 5. 6; trinomial 7. 2; binomial 9. 3 11. 2 13. a. 6 b. 5 15. a. −2 b. 4
17. a. −15 b. −16 19. 1134 ft 21. 1006 ft 23. $23x^2$ 25. $12x^2 - y$ 27. $7s$ 29. $-1.1y^2 + 4.8$ 31. $-\frac{7}{12}x^3 + \frac{7}{5}x^2 + 6$
33. $5a^2 - 9ab + 16b^2$ 35. $12x + 12$ 37. $-3x^2 + 10$ 39. $-x^2 + 14$ 41. $-2x + 9$ 43. $2x^2 + 7x - 16$ 45. $8t^2 - 4$
47. $-2z^2 - 16z + 6$ 49. $2x^3 - 2x^2 + 7x + 2$ 51. $62x^2 + 5$ 53. $12x + 2$ 55. $-y^2 - 3y - 1$ 57. $2x^2 + 11x$ 59. $-16x^4 + 8x + 9$
61. $7x^2 + 14x + 18$ 63. $3x - 3$ 65. $7x^2 - 2x + 2$ 67. $4y^2 + 12y + 19$ 69. $6x^2 - 5x + 21$ 71. $4x^2 + 7x + x^2 + 5x; 5x^2 + 12x$
73. $18x + 44$ 75. $(x^2 + 7x + 4)$ ft 77. $(3y^2 + 4y + 11)$ m 79. $-2a - b + 1$ 81. $3x^2 + 5$ 83. $6x^2 - 2xy + 19y^2$
85. $8r^2s + 16rs - 8 + 7r^2s^2$ 87. $-5.42x^2 + 7.75x - 19.61$ 89. $3.7y^4 - 0.7y^3 + 2.2y - 4$ 91. $6x^2$ 93. $-12x^8$ 95. $200x^3y^2$
97., 99. answers may vary 101. b 103. e 105. a. $4z$ b. $3z^2$ c. $-4z$ d. $3z^2$; answers may vary 107. answers may vary
109. $2x^2 + 4xy$ 111. $8169 113. $10.84x^2 + 20.43x + 3285$

Section 5.3
Practice Exercises
1. $10y^2$ 2. $-2z^8$ 3. $\frac{7}{72}b^9$ 4. a. $15x^6 + 15x$ b. $-10x^5 + 45x^4 - 10x^3$ 5. $10x^2 + 11x - 6$ 6. $25x^2 - 30xy + 9y^2$
7. $2y^3 + 5y^2 - 7y + 20$ 8. $s^3 + 6s^2t + 12st^2 + 8t^3$ 9. $5x^3 - 23x^2 + 17x - 20$ 10. $x^5 - 2x^4 + 2x^3 - 3x^2 + 2$
11. $5x^4 - 3x^3 + 11x^2 + 8x - 6$

Vocabulary and Readiness Check 5.3
1. distributive **3.** $(5y - 1)(5y - 1)$ **5.** x^8 **7.** cannot simplify **9.** x^{14} **11.** $2x^7$

Exercise Set 5.3
1. $-28n^{10}$ **3.** $-12.4x^{12}$ **5.** $-\frac{2}{15}y^3$ **7.** $-24x^8$ **9.** $6x^2 + 15x$ **11.** $-2a^2 - 8a$ **13.** $6x^3 - 9x^2 + 12x$ **15.** $-6a^4 + 4a^3 - 6a^2$
17. $-4x^3y + 7x^2y^2 - xy^3 - 3y^4$ **19.** $4x^4 - 3x^3 + \frac{1}{2}x^2$ **21.** $x^2 + 7x + 12$ **23.** $a^2 + 5a - 14$ **25.** $x^2 + \frac{1}{3}x - \frac{2}{9}$ **27.** $12x^4 + 25x^2 + 7$
29. $4y^2 - 16y + 16$ **31.** $12x^2 - 29x + 15$ **33.** $9x^4 + 6x^2 + 1$ **35. a.** $6x + 12$ **b.** $9x^2 + 36x + 35$ **c.** answers may vary
37. $x^3 - 5x^2 + 13x - 14$ **39.** $x^4 + 5x^3 - 3x^2 - 11x + 20$ **41.** $10a^3 - 27a^2 + 26a - 12$ **43.** $x^3 + 6x^2 + 12x + 8$
45. $8y^3 - 36y^2 + 54y - 27$ **47.** $12x^2 - 64x - 11$ **49.** $10x^3 + 22x^2 - x - 1$ **51.** $2x^4 + 3x^3 - 58x^2 + 4x + 63$ **53.** $8.4y^7$
55. $-3x^3 - 6x^2 + 24x$ **57.** $2x^2 + 39x + 19$ **59.** $x^2 - \frac{2}{7}x - \frac{3}{49}$ **61.** $9y^2 + 30y + 25$ **63.** $a^3 - 2a^2 - 18a + 24$
65. $8x^3 - 60x^2 + 150x - 125$ **67.** $32x^3 + 48x^2 - 6x - 20$ **69.** $6x^4 - 8x^3 - 7x^2 + 22x - 12$ **71.** $(4x^2 - 25)$ sq yd
73. $(6x^2 - 4x)$ sq in. **75.** $25x^2$ **77.** $9y^6$ **79.** $x^2 + 3x$ **81.** $x^2 + 5x + 6$ **83.** $11a$ **85.** $25x^2 + 4y^2$ **87.** $13x - 7$
89. $30x^2 - 28x + 6$ **91.** $-7x + 5$ **93. a.** $a^2 - b^2$ **b.** $4x^2 - 9y^2$ **c.** $16x^2 - 49$ **d.** answers may vary **95.** $(x^2 + 6x + 5)$ sq units

Section 5.4
Practice Exercises
1. $x^2 - 3x - 10$ **2.** $4x^2 - 13x + 9$ **3.** $9x^2 + 42x - 15$ **4.** $16x^2 - 8x + 1$ **5. a.** $b^2 + 6b + 9$ **b.** $x^2 - 2xy + y^2$ **c.** $9y^2 + 12y + 4$
d. $a^4 - 10a^2b + 25b^2$ **6. a.** $3x^2 - 75$ **b.** $16b^2 - 9$ **c.** $x^2 - \frac{4}{9}$ **d.** $25s^2 - t^2$ **e.** $4y^2 - 9z^4$ **7. a.** $4x^2 - 21x - 18$ **b.** $49b^2 - 28b + 4$
c. $x^2 - 0.16$ **d.** $3x^6 - 9x^4 + 2x^2 - 6$ **e.** $x^3 + 6x^2 + 3x - 2$

Vocabulary and Readiness Check 5.4
1. false **3.** false

Exercise Set 5.4
1. $x^2 + 7x + 12$ **3.** $x^2 + 5x - 50$ **5.** $5x^2 + 4x - 12$ **7.** $4y^2 - 25y + 6$ **9.** $6x^2 + 13x - 5$ **11.** $x^2 - 4x + 4$ **13.** $4x^2 - 4x + 1$
15. $9a^2 - 30a + 25$ **17.** $25x^2 + 90x + 81$ **19.** answers may vary **21.** $a^2 - 49$ **23.** $9x^2 - 1$ **25.** $9x^2 - \frac{1}{4}$ **27.** $81x^2 - y^2$
29. $4x^2 - 0.01$ **31.** $a^2 + 9a + 20$ **33.** $a^2 + 14a + 49$ **35.** $12a^2 - a - 1$ **37.** $x^2 - 4$ **39.** $9a^2 + 6a + 1$ **41.** $4x^3 - x^2y^4 + 4xy - y^5$
43. $x^3 - 3x^2 - 17x + 3$ **45.** $4a^2 - 12a + 9$ **47.** $25x^2 - 36z^2$ **49.** $x^{10} - 8x^5 + 15$ **51.** $x^2 - \frac{1}{9}$ **53.** $a^7 - 3a^3 + 11a^4 - 33$
55. $3x^2 - 12x + 12$ **57.** $6b^2 - b - 35$ **59.** $49p^2 - 64$ **61.** $\frac{1}{9}a^4 - 49$ **63.** $15x^4 - 5x^3 + 10x^2$ **65.** $4r^2 - 9s^2$ **67.** $9x^2 - 42xy + 49y^2$
69. $16x^2 - 25$ **71.** $64x^2 + 64x + 16$ **73.** $a^2 - \frac{1}{4}y^2$ **75.** $\frac{1}{25}x^2 - y^2$ **77.** $3a^3 + 2a^2 + 1$ **79.** $(2x + 1)(2x + 1)$ sq ft or $(4x^2 + 4x + 1)$ sq ft
81. $\frac{5b^5}{7}$ **83.** $-2a^{10}b^5$ **85.** $\frac{2y^8}{3}$ **87.** $\frac{1}{3}$ **89.** 1 **91.** c **93.** d **95.** 2 **97.** $\left(\frac{25}{2}a^2 - \frac{1}{2}b^2\right)$ sq units **99.** $(24x^2 - 32x + 8)$ sq m
101. $(x^2 + 10x + 25)$ sq units **103.** answers may vary **105.** $x^2 + 2xy + y^2 - 9$ **107.** $a^2 - 6a + 9 - b^2$

Integrated Review
1. $35x^5$ **2.** $32y^9$ **3.** -16 **4.** 16 **5.** $2x^2 - 9x - 5$ **6.** $3x^2 + 13x - 10$ **7.** $3x - 4$ **8.** $4x + 3$ **9.** $7x^6y^2$ **10.** $\frac{10b^6}{7}$
11. $144m^{14}n^{12}$ **12.** $64y^{27}z^{30}$ **13.** $48y^2 - 27$ **14.** $98x^2 - 2$ **15.** $x^{63}y^{45}$ **16.** $27x^{27}$ **17.** $2x^2 - 2x - 6$ **18.** $6x^2 + 13x - 11$
19. $2.5y^2 - 6y - 0.2$ **20.** $8.4x^2 - 6.8x - 5.7$ **21.** $x^2 + 8xy + 16y^2$ **22.** $y^2 - 18yz + 81z^2$ **23.** $2x + 8y$ **24.** $2y - 18z$
25. $7x^2 - 10xy + 4y^2$ **26.** $-a^2 - 3ab + 6b^2$ **27.** $x^3 + 2x^2 - 16x + 3$ **28.** $x^3 - 2x^2 - 5x - 2$ **29.** $6x^5 + 20x^3 - 21x^2 - 70$
30. $20x^7 + 25x^3 - 4x^4 - 5$ **31.** $2x^3 - 19x^2 + 44x - 7$ **32.** $5x^3 + 9x^2 - 17x + 3$ **33.** cannot simplify **34.** $25x^3y^3$ **35.** $125x^9$
36. $\frac{r^3}{y^3}$ **37.** $2x$ **38.** x^2

Section 5.5
Practice Exercises
1. a. $\frac{1}{125}$ **b.** $\frac{3}{y^4}$ **c.** $\frac{5}{6}$ **d.** $\frac{1}{25}$ **e.** x^5 **f.** 64 **2. a.** s^5 **b.** 8 **c.** $\frac{y^5}{x^7}$ **d.** $\frac{9}{64}$ **3. a.** $\frac{1}{x^5}$ **b.** $5y^7$ **c.** z^5 **4. a.** $\frac{16}{9}$ **b.** x^{10}
c. $\frac{q^2}{25p^{16}}$ **d.** $\frac{6y^2}{x^7}$ **e.** $\frac{b^{15}}{a^{20}}$ **f.** $-27x^6y^9$ **5. a.** 7×10^{-6} **b.** 2.07×10^7 **c.** 4.3×10^{-3} **d.** 8.12×10^8 **6. a.** 0.000367 **b.** 8,954,000
c. 0.00002009 **d.** 4054 **7. a.** 4000 **b.** 20,000,000,000

Calculator Explorations 5.5
1. 5.31 EE 3 **3.** 6.6 EE −9 **5.** 1.5×10^{13} **7.** 8.15×10^{19}

Vocabulary and Readiness Check 5.5

1. $\dfrac{1}{x^3}$ 3. scientific notation 5. $\dfrac{5}{x^2}$ 7. y^6 9. $4y^3$

Exercise Set 5.5

1. $\dfrac{1}{64}$ 3. $\dfrac{1}{16}$ 5. $\dfrac{7}{x^3}$ 7. 32 9. -64 11. $\dfrac{5}{6}$ 13. p^3 15. $\dfrac{q^4}{p^5}$ 17. $\dfrac{1}{x^3}$ 19. z^3 21. $\dfrac{4}{9}$ 23. $-p^4$ 25. -2 27. x^4 29. p^4
31. m^{11} 33. r^6 35. $\dfrac{1}{x^{15}y^9}$ 37. $\dfrac{1}{x^4}$ 39. $\dfrac{1}{a^2}$ 41. $4k^3$ 43. $3m$ 45. $-\dfrac{4a^5}{b}$ 47. $-\dfrac{6x^2}{y^3}$ 49. $\dfrac{a^{30}}{b^{12}}$ 51. $\dfrac{1}{x^{10}y^6}$ 53. $\dfrac{z^2}{4}$ 55. $\dfrac{1}{32x^5}$
57. $\dfrac{49a^4}{b^6}$ 59. $a^{24}b^8$ 61. x^9y^{19} 63. $-\dfrac{y^8}{8x^2}$ 65. $-\dfrac{6x}{7y^2}$ 67. $\dfrac{25b^{33}}{a^{16}}$ 69. 7.8×10^4 71. 1.67×10^{-6} 73. 6.35×10^{-3} 75. 1.16×10^6
77. 2×10^9 79. 1.212×10^9 81. 0.0000000008673 83. 0.033 85. 20,320 87. 700,000,000 89. 9,460,000,000,000
91. Yahoo! sites, 130,000,000, 1.3×10^8 93. 1×10^9 95. 57,000,000 97. 0.000036 99. 0.00000000000000028 101. 0.0000005
103. 200,000 105. $\dfrac{5x^3}{3}$ 107. $\dfrac{5z^3y^2}{7}$ 109. $5y - 6 + \dfrac{5}{y}$ 111. $\dfrac{27}{x^6z^3}$ cu in. 113. $9a^{13}$ 115. -5 117. answers may vary
119. a. 1.3×10^1 b. 4.4×10^7 c. 6.1×10^{-2} 121. a. false b. true c. false 123. $\dfrac{1}{x^{9s}}$ 125. a^{4m+5} 127. 31,753,800 129. 500 sec

Section 5.6
Practice Exercises

1. $2t + 1$ 2. $4x^4 + 5x - \dfrac{3}{x}$ 3. $3x^3y^3 - 2 + \dfrac{1}{5x}$ 4. $x + 3$ 5. $2x + 3 + \dfrac{-10}{2x+1}$ 6. $3x^2 - 2x + 5 + \dfrac{-13}{3x+2}$ 7. $3x^2 - 2x - 9 + \dfrac{5x+22}{x^2+2}$

Vocabulary and Readiness Check 5.6

1. dividend, quotient, divisor 3. a^2 5. y

Exercise Set 5.6

1. $12x^3 + 3x$ 3. $4x^3 - 6x^2 + x + 1$ 5. $5p^2 + 6p$ 7. $-\dfrac{3}{2x} + 3$ 9. $-3x^2 + x - \dfrac{4}{x^3}$ 11. $-1 + \dfrac{3}{2x} - \dfrac{7}{4x^4}$ 13. $x + 1$ 15. $2x + 3$
17. $2x + 1 + \dfrac{7}{x-4}$ 19. $3a^2 - 3a + 1 + \dfrac{2}{3a+2}$ 21. $4x + 3 - \dfrac{2}{2x+1}$ 23. $2x^2 + 6x - 5 - \dfrac{2}{x-2}$ 25. $x + 6$ 27. $x^2 + 3x + 9$
29. $-3x + 6 - \dfrac{11}{x+2}$ 31. $2b - 1 - \dfrac{6}{2b-1}$ 33. $ab - b^2$ 35. $4x + 9$ 37. $x + 4xy - \dfrac{y}{2}$ 39. $2b^2 + b + 2 - \dfrac{12}{b+4}$
41. $5x - 2 + \dfrac{2}{x+6}$ 43. $x^2 - \dfrac{12x}{5} - 1$ 45. $6x - 1 - \dfrac{1}{x+3}$ 47. $6x - 1$ 49. $-x^3 + 3x^2 - \dfrac{4}{x}$ 51. $x^2 + 3x + 9$
53. $y^2 + 5y + 10 + \dfrac{24}{y-2}$ 55. $-6x - 12 - \dfrac{19}{x-2}$ 57. $x^3 - x^2 + x$ 59. $2a^3 + 2a$ 61. $2x^3 + 14x^2 - 10x$ 63. $-3x^2y^3 - 21x^3y^2 - 24xy$
65. $9a^2b^3c + 36ab^2c - 72ab$ 67. The Rolling Stones (2005) 69. $139 million 71. $(3x^3 + x - 4)$ ft 73. c 75. answers may vary
77. $(7x - 10)$ in. 79. $5y^{10b} + y^{5b} - 4y^{2b} + 20$

The Bigger Picture

1. -5.93 2. $-\dfrac{2}{5}$ 3. $5x^9y^4$ 4. $\dfrac{1}{8a^4}$ 5. $6y^3 - 2y^2 - 6y$ 6. $8y^2 - 3y - 7$ 7. $4x^3 - 13x^2 + 10x - 21$ 8. $36m^2 - 60m + 25$
9. $4n - 1 + \dfrac{2}{n}$ 10. $2x - 6 + \dfrac{14}{3x-1}$ 11. -0.6 12. $(-0.6, \infty)$ 13. $(-\infty, 2]$ 14. $\dfrac{2}{3}$

Chapter 5 Vocabulary Check

1. term 2. FOIL 3. trinomial 4. degree of a polynomial 5. binomial 6. coefficient 7. degree of a term 8. monomial
9. polynomials

Chapter 5 Review

1. base: 7; exponent: 9 3. base: 5; exponent: 4 5. 512 7. -36 9. 1 11. y^9 13. $-6x^{11}$ 15. x^8 17. $81y^{24}$ 19. x^5 21. a^4b^3
23. $\dfrac{4}{x^3y^4}$ 25. $40a^{19}$ 27. 3 29. b 31. 7 33. 8 35. 5 37. 5 39. a. $1, 4; 5, 2; -7, 2; 11, 2; -1, 0$ b. 4 41. $2a^2$
43. $15a^2 + 4a$ 45. $-6a^2b - 3b^2 - q^2$ 47. $8x^2 + 3x + 6$ 49. $-7y^2 - 1$ 51. $-5x^2 + 5x + 1$ 53. $-2x^2 - x + 20$ 55. $8a + 28$
57. $-7x^3 - 35x$ 59. $-6a^4 + 8a^2 - 2a$ 61. $2x^2 - 12x - 14$ 63. $x^2 - 18x + 81$ 65. $4a^2 + 27a - 7$ 67. $25x^2 + 20x + 4$
69. $x^4 + 7x^3 + 4x^2 + 23x - 35$ 71. $x^4 + 4x^3 + 4x^2 - 16$ 73. $x^3 + 21x^2 + 147x + 343$ 75. $x^2 + 14x + 49$ 77. $9x^2 - 42x + 49$
79. $25x^2 - 90x + 81$ 81. $49x^2 - 16$ 83. $4x^2 - 36$ 85. $(9x^2 - 6x + 1)$ sq m 87. $\dfrac{1}{49}$ 89. $\dfrac{2}{x^4}$ 91. 125 93. $\dfrac{17}{16}$ 95. x^8 97. r

99. c^4 **101.** $\dfrac{1}{x^6 y^{13}}$ **103.** a^{11m} **105.** $27x^3 y^{6z}$ **107.** 2.7×10^{-4} **109.** 8.08×10^7 **111.** 9.1×10^7 **113.** $867{,}000$ **115.** 0.00086
117. $1{,}431{,}280{,}000{,}000{,}000$ **119.** 0.016 **121.** $\dfrac{1}{7} + \dfrac{3}{x} + \dfrac{7}{x^2}$ **123.** $a + 1 + \dfrac{6}{a-2}$ **125.** $a^2 + 3a + 8 + \dfrac{22}{a-2}$
127. $2x^3 - x^2 + 2 - \dfrac{1}{2x-1}$ **129.** $\left(5x - 1 + \dfrac{20}{x^2}\right)$ ft **131.** $-\dfrac{1}{8}$ **133.** $\dfrac{2x^6}{3}$ **135.** $\dfrac{x^{16}}{16 y^{12}}$ **137.** $11x - 5$ **139.** $5y^2 - 3y - 1$
141. $28x^3 + 12x$ **143.** $x^3 + x^2 - 18x + 18$ **145.** $25x^2 + 40x + 16$ **147.** $4a - 1 + \dfrac{2}{a^2} - \dfrac{5}{2a^3}$ **149.** $2x^2 + 7x + 5 + \dfrac{19}{2x-3}$

Chapter 5 Test
1. 32 **2.** 81 **3.** -81 **4.** $\dfrac{1}{64}$ **5.** $-15x^{11}$ **6.** y^5 **7.** $\dfrac{1}{r^5}$ **8.** $\dfrac{y^{14}}{x^2}$ **9.** $\dfrac{1}{6xy^8}$ **10.** 5.63×10^5 **11.** 8.63×10^{-5} **12.** 0.0015
13. $62{,}300$ **14.** 0.036 **15. a.** $4, 3; 7, 3; 1, 4; -2, 0$ **b.** 4 **16.** $-2x^2 + 12xy + 11$ **17.** $16x^3 + 7x^2 - 3x - 13$ **18.** $-3x^3 + 5x^2 + 4x + 5$
19. $x^3 + 8x^2 + 3x - 5$ **20.** $3x^3 + 22x^2 + 41x + 14$ **21.** $6x^4 - 9x^3 + 21x^2$ **22.** $3x^2 + 16x - 35$ **23.** $9x^2 - \dfrac{1}{25}$
24. $16x^2 - 16x + 4$ **25.** $64x^2 + 48x + 9$ **26.** $x^4 - 81b^2$ **27.** 1001 ft; 985 ft; 857 ft; 601 ft **28.** $(4x^2 - 9)$ sq in. **29.** $\dfrac{x}{2y} + 3 - \dfrac{7}{8y}$
30. $x + 2$ **31.** $9x^2 - 6x + 4 - \dfrac{16}{3x+2}$

Chapter 5 Cumulative Review
1. a. true **b.** true **c.** false **d.** true; Sec. 1.2, Ex. 2 **3. a.** $\dfrac{64}{25}$ **b.** $\dfrac{1}{20}$ **c.** $\dfrac{5}{4}$; Sec. 1.3, Ex. 4 **5. a.** 9 **b.** 125 **c.** 16 **d.** 7
e. $\dfrac{9}{49}$; Sec. 1.4, Ex. 1 **7. a.** -10 **b.** -21 **c.** -12; Sec. 1.5, Ex. 3 **9.** -12; Sec. 1.6, Ex. 3 **11. a.** $\dfrac{1}{22}$ **b.** $\dfrac{16}{3}$ **c.** $-\dfrac{1}{10}$
d. $-\dfrac{13}{9}$; Sec. 1.7, Ex. 5 **13. a.** $(5 + 4) + 6$ **b.** $-1 \cdot (2 \cdot 5)$; Sec. 1.8, Ex. 2 **15. a.** $22 + x$ **b.** $-21x$; Sec. 1.8, Ex. 3 **17. a.** $5x + 10$
b. $-2y - 0.6z + 2$ **c.** $-x - y + 2z - 6$; Sec. 2.1, Ex. 5 **19.** 17; Sec. 2.2, Ex. 1 **21.** 6; Sec. 2.3, Ex. 1 **23.** -10; Sec. 2.5, Ex. 1
25. 10; Sec. 2.5, Ex. 2 **27.** width: 4 ft; length: 10 ft; Sec. 2.6, Ex. 4 **29.** $\dfrac{5F - 160}{9} = C$; Sec. 2.6, Ex. 8
31. ⟵——|———|⟶ ; Sec. 2.9, Ex. 9 **33. a.** $(0, 12)$ **b.** $(2, 6)$ **c.** $(-1, 15)$; Sec. 3.1, Ex. 6
35. ; Sec. 3.2, Ex. 2 **37.** ; Sec. 3.3, Ex. 9 **39.** undefined slope; Sec. 3.4, Ex. 7
41. ; Sec. 4.5, Ex. 1 **43.** $-6x^7$; Sec. 5.1, Ex. 4 **45.** $-4x^2 + 6x + 2$; Sec. 5.2, Ex. 9 **47.** $4x^2 - 4xy + y^2$; Sec. 5.3, Ex. 6
49. $3m + 1$; Sec. 5.6, Ex. 1

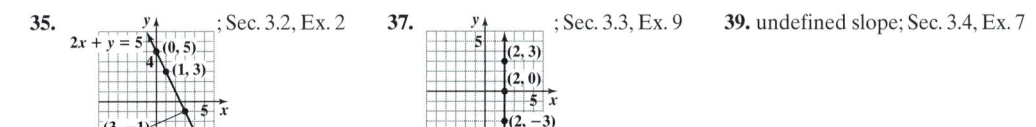

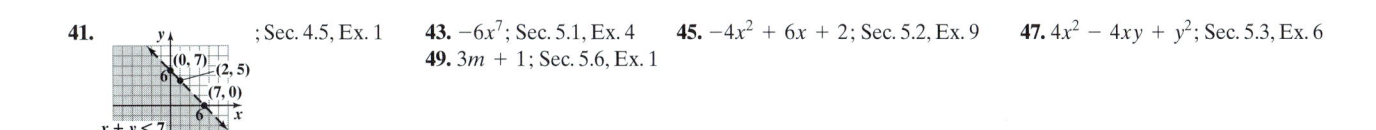

CHAPTER 6 FACTORING POLYNOMIALS

Section 6.1
Practice Exercises
1. a. 6 **b.** 1 **c.** 4 **2. a.** y^4 **b.** x **3. a.** $5y^2$ **b.** x^2 **c.** $a^2 b^2$ **4. a.** $4(t + 3)$ **b.** $y^4(y^4 + 1)$
5. $8b^2(-b^4 + 2b^2 - 1)$ or $-8b^2(b^4 - 2b^2 + 1)$ **6.** $5x(x^3 - 4)$ **7.** $\dfrac{1}{9}z^3(5z^2 + z - 2)$ **8.** $4ab^3(2ab - 5a^2 + 3)$ **9.** $(y - 2)(8 + x)$
10. $(p + q)(7xy^3 - 1)$ **11.** $(x + 3)(y + 4)$ **12.** $(2x + 3y)(y - 1)$ **13.** $(7a + 5)(a^2 + 1)$ **14.** $(y - 3)(4x - 5)$
15. cannot be factored by grouping **16.** $3(x - a)(y - 2a)$

Vocabulary and Readiness Check 6.1
1. factors **3.** least **5.** false **7.** $2 \cdot 7$ **9.** 3 **11.** 5

Exercise Set 6.1
1. 4 **3.** 6 **5.** 1 **7.** y^2 **9.** z^7 **11.** xy^2 **13.** 7 **15.** $4y^3$ **17.** $5x^2$ **19.** $3x^3$ **21.** $9x^2 y$ **23.** $10a^6 b$ **25.** $3(a + 2)$ **27.** $15(2x - 1)$
29. $x^2(x + 5)$ **31.** $2y^3(3y + 1)$ **33.** $4(x - 2y + 1)$ **35.** $3x(2x^2 - 3x + 4)$ **37.** $a^2 b^2(a^5 b^4 - a + b^3 - 1)$ **39.** $4(2x^5 + 4x^4 - 5x^3 + 3)$
41. $\dfrac{1}{3}x(x^3 + 2x^2 - 4x^4 + 1)$ **43.** $(x^2 + 2)(y + 3)$ **45.** $(y + 4)(z - 3)$ **47.** $(z^2 - 6)(r + 1)$ **49.** $-2(x + 7)$ **51.** $-x^5(2 - x^2)$

53. $-3a^2(2a^2 - 3a + 1)$ **55.** $(x + 2)(x^2 + 5)$ **57.** $(x + 3)(5 + y)$ **59.** $(3x - 2)(2x^2 + 5)$ **61.** $(5m^2 + 6n)(m + 1)$ **63.** $(y - 4)(2 + x)$
65. $(2x - 1)(x^2 + 4)$ **67.** $(x - 2y)(4x - 3)$ **69.** $(5q - 4p)(q - 1)$ **71.** $x(x^2 + 1)(2x + 5)$ **73.** $2(2y - 7)(3x^2 - 1)$ **75.** $2x(16y - 9x)$
77. $(x + 2)(y - 3)$ **79.** $7xy(2x^2 + x - 1)$ **81.** $(4x - 1)(7x^2 + 3)$ **83.** $-8x^8y^5(5y + 2x)$ **85.** $3(2a + 3b^2)(a + b)$ **87.** $x^2 + 7x + 10$
89. $b^2 - 3b - 4$ **91.** 2, 6 **93.** $-1, -8$ **95.** $-2, 5$ **97.** b **99.** factored **101.** not factored **103.** answers may vary
105. answers may vary **107. a.** 3088 thousand or 3,088,000 **b.** 3092 thousand or 3,092,000 **c.** $-2(4x^2 - 25x - 1510)$ **109.** $4x^2 - \pi x^2; x^2(4 - \pi)$
111. $(x^3 - 1)$ units **113.** $(x^n + 6)(x^n + 10)$ **115.** $(2x^n - 5)(6x^n - 5)$

Section 6.2

Practice Exercises
1. $(x + 2)(x + 3)$ **2.** $(x - 10)(x - 7)$ **3.** $(x + 7)(x - 2)$ **4.** $(p - 9)(p + 7)$ **5.** prime polynomial **6.** $(x + 3y)(x + 4y)$
7. $(x^2 + 12)(x^2 + 1)$ **8.** $(x - 6)(x - 8)$ **9.** $4(x - 3)(x - 3)$ **10.** $3y^2(y - 7)(y + 1)$

Vocabulary and Readiness Check 6.2
1. true **3.** false **5.** $+5$ **7.** -3 **9.** $+2$

Exercise Set 6.2
1. $(x + 6)(x + 1)$ **3.** $(y - 9)(y - 1)$ **5.** $(x - 3)(x - 3)$ or $(x - 3)^2$ **7.** $(x - 6)(x + 3)$ **9.** $(x + 10)(x - 7)$ **11.** prime
13. $(x + 5y)(x + 3y)$ **15.** $(a^2 - 5)(a^2 + 3)$ **17.** $(m + 13)(m + 1)$ **19.** $(t - 2)(t + 12)$ **21.** $(a - 2b)(a - 8b)$ **23.** $2(z + 8)(z + 2)$
25. $2x(x - 5)(x - 4)$ **27.** $(x - 4y)(x + y)$ **29.** $(x + 12)(x + 3)$ **31.** $(x - 2)(x + 1)$ **33.** $(r - 12)(r - 4)$ **35.** $(x + 2y)(x - y)$
37. $3(x + 5)(x - 2)$ **39.** $3(x - 18)(x - 2)$ **41.** $(x - 24)(x + 6)$ **43.** prime **45.** $(x - 5)(x - 3)$ **47.** $6x(x + 4)(x + 5)$
49. $4y(x^2 + x - 3)$ **51.** $(x - 7)(x + 3)$ **53.** $(x + 5y)(x + 2y)$ **55.** $2(t + 8)(t + 4)$ **57.** $x(x - 6)(x + 4)$ **59.** $2t^3(t - 4)(t - 3)$
61. $5xy(x - 8y)(x + 3y)$ **63.** $3(m - 9)(m - 6)$ **65.** $-1(x - 11)(x - 1)$ **67.** $\frac{1}{2}(y - 11)(y + 2)$ **69.** $x(xy - 4)(xy + 5)$
71. $2x^2 + 11x + 5$ **73.** $15y^2 - 17y + 4$ **75.** $9a^2 + 23ab - 12b^2$ **77.** $x^2 + 5x - 24$ **79.** answers may vary
81. $2x^2 + 28x + 66; 2(x + 3)(x + 11)$ **83.** $-16(t - 5)(t + 1)$ **85.** $\left(x + \frac{1}{4}\right)\left(x + \frac{1}{4}\right)$ or $\left(x + \frac{1}{4}\right)^2$ **87.** $(x + 1)(z - 10)(z + 7)$
89. $(x^n + 10)(x^n - 2)$ **91.** 7; 12; 15; 16 **93.** 15; 28; 39; 48; 55; 60; 63; 64 **95.** 9; 12; 21 **97.** 5; 13

Section 6.3

Practice Exercises
1. $(2x + 5)(x + 3)$ **2.** $(5x - 4)(3x - 2)$ **3.** $(4x - 1)(x + 3)$ **4.** $(7x - y)(3x + 2y)$ **5.** $(2x^2 - 7)(x^2 + 1)$ **6.** $x(3x + 2)(x + 5)$
7. $-1(4x - 3)(2x + 1)$ **8.** $(x + 7)^2$ **9.** $(2x + 9y)(2x + y)$ **10.** $(6n^2 - 1)^2$ **11.** $3x(2x - 7)^2$

Vocabulary and Readiness Check 6.3
1. perfect square trinomial **3.** perfect square trinomial **5.** no **7.** 8^2 **9.** $(11a)^2$ **11.** $(6p^2)^2$

Exercise Set 6.3
1. $x + 4$ **3.** $10x - 1$ **5.** $5x - 2$ **7.** $(2x + 3)(x + 5)$ **9.** $(y - 1)(8y - 9)$ **11.** $(2x + 1)(x - 5)$ **13.** $(4r - 1)(5r + 8)$
15. $(10x + 1)(x + 3)$ **17.** prime **19.** $(3x - 5y)(2x - y)$ **21.** $(3m - 5)(5m + 3)$ **23.** $x(3x + 2)(4x + 1)$ **25.** $3(7b + 5)(b - 3)$
27. $(3z + 4)(4z - 3)$ **29.** $2y^2(3x - 10)(x + 3)$ **31.** $(2x - 7)(2x + 3)$ **33.** $-1(x - 6)(x + 4)$ **35.** $x(4x + 3)(x - 3)$ **37.** $(4x - 9)(6x - 1)$
39. $(x + 11)^2$ **41.** $(x - 8)^2$ **43.** $(4a - 3)^2$ **45.** $(x^2 + 2)^2$ **47.** $2(n - 7)^2$ **49.** $(4y + 5)^2$ **51.** $(2x + 11)(x - 9)$ **53.** $(8x + 3)(3x + 4)$
55. $(3a + b)(a + 3b)$ **57.** $(x - 4)(x - 5)$ **59.** $(p + 6q)^2$ **61.** $(xy - 5)^2$ **63.** $b(8a - 3)(5a + 3)$ **65.** $2x(3x + 2)(5x + 3)$
67. $2y(3y + 5)(y - 3)$ **69.** $5x^2(2x - y)(x + 3y)$ **71.** $-1(2x - 5)(7x - 2)$ **73.** $p^2(4p - 5)(4p - 5)$ or $p^2(4p - 5)^2$ **75.** $(3x - 2)(x + 1)$
77. $(4x + 9y)(2x - 3y)$ **79.** prime **81.** $(3x - 4y)^2$ **83.** $(6x - 7)(3x + 2)$ **85.** $(7t + 1)(t - 4)$ **87.** $(7p + 1)(7p - 2)$ **89.** $m(m + 9)^2$
91. prime **93.** $a(6a^2 + b^2)(a^2 + 6b^2)$ **95.** $x^2 - 4$ **97.** $a^3 + 27$ **99.** \$75,000 and above **101.** answers may vary **103.** no
105. answers may vary **107.** $4x^2 + 21x + 5; (4x + 1)(x + 5)$ **109.** $\left(2x + \frac{1}{2}\right)\left(2x + \frac{1}{2}\right)$ or $\left(2x + \frac{1}{2}\right)^2$ **111.** $(y - 1)^2(4x^2 + 10x + 25)$ **113.** 8
115. $a^2 + 2ab + b^2$ **117.** 2; 14 **119.** 2 **121.** $-3xy^2(4x - 5)(x + 1)$ **123.** $(y - 1)^2(2x + 5)^2$ **125.** $(3x^n + 2)(x^n + 5)$ **127.** answers may vary

Section 6.4

Practice Exercises
1. $(5x + 1)(x + 12)$ **2.** $(4x - 5)(3x - 1)$ **3.** $2(5x + 1)(3x - 2)$ **4.** $5m^2(8m - 7)(m + 1)$ **5.** $(4x + 3)^2$

Exercise Set 6.4
1. $(x + 3)(x + 2)$ **3.** $(y + 8)(y - 2)$ **5.** $(8x - 5)(x - 3)$ **7.** $(5x^2 - 3)(x^2 + 5)$ **9. a.** 9, 2 **b.** $9x + 2x$ **c.** $(2x + 3)(3x + 1)$
11. a. $-20, -3$ **b.** $-20x - 3x$ **c.** $(3x - 4)(5x - 1)$ **13.** $(3y + 2)(7y + 1)$ **15.** $(7x - 11)(x + 1)$ **17.** $(5x - 2)(2x - 1)$
19. $(2x - 5)(x - 1)$ **21.** $(2x + 3)(2x + 3)$ or $(2x + 3)^2$ **23.** $(2x + 3)(2x - 7)$ **25.** $(5x - 4)(2x - 3)$ **27.** $x(2x + 3)(x + 5)$
29. $2(8y - 9)(y - 1)$ **31.** $(2x - 3)(3x - 2)$ **33.** $3(3a + 2)(6a - 5)$ **35.** $a(4a + 1)(5a + 8)$ **37.** $3x(4x + 3)(x - 3)$ **39.** $y(3x + y)(x + y)$
41. prime **43.** $5(x + 5y)^2$ **45.** $6(a + b)(4a - 5b)$ **47.** $p^2(15p + q)(p + 2q)$ **49.** $2(9a^2 - 2)^2$ **51.** $(7 + x)(5 + x)$ or $(x + 7)(x + 5)$
53. $(6 - 5x)(1 - x)$ or $(5x - 6)(x - 1)$ **55.** $x^2 - 4$ **57.** $y^2 + 8y + 16$ **59.** $81z^2 - 25$ **61.** $x^3 - 27$
63. $10x^2 + 45x + 45; 5(2x + 3)(x + 3)$ **65.** $(x^n + 2)(x^n + 3)$ **67.** $(3x^n - 5)(x^n + 7)$ **69.** answers may vary

Section 6.5
Practice Exercises
1. $(x + 9)(x - 9)$ **2. a.** $(3x - 1)(3x + 1)$ **b.** $(6a - 7b)(6a + 7b)$ **c.** $\left(p + \frac{5}{6}\right)\left(p - \frac{5}{6}\right)$ **3.** $(p^2 - q^5)(p^2 + q^5)$
4. a. $(z^2 + 9)(z + 3)(z - 3)$ **b.** prime polynomial **5.** $y(6y + 5)(6y - 5)$ **6.** $5(4y^2 + 1)(2y + 1)(2y - 1)$
7. $-1(3x + 10)(3x - 10)$ or $(10 + 3x)(10 - 3x)$ **8.** $(x + 4)(x^2 - 4x + 16)$ **9.** $(x - 5)(x^2 + 5x + 25)$ **10.** $(3y + 1)(9y^2 - 3y + 1)$
11. $4(2x - 5y)(4x^2 + 10xy + 25y^2)$

Calculator Explorations 6.5

	$x^2 - 2x + 1$	$x^2 - 2x - 1$	$(x - 1)^2$
$x = 5$	16	14	16
$x = -3$	16	14	16
$x = 2.7$	2.89	0.89	2.89
$x = -12.1$	171.61	169.61	171.61
$x = 0$	1	-1	1

Vocabulary and Readiness Check 6.5
1. difference of two cubes   **3.** sum of two cubes   **5.** 8^2 **7.** $(7x)^2$ **9.** 4^3 **11.** $(2y)^3$

Exercise Set 6.5
1. $(x + 2)(x - 2)$ **3.** $(9p + 1)(9p - 1)$ **5.** $(5y - 3)(5y + 3)$ **7.** $(11m + 10n)(11m - 10n)$ **9.** $(xy - 1)(xy + 1)$ **11.** $\left(x - \frac{1}{2}\right)\left(x + \frac{1}{2}\right)$
13. $-1(2r + 1)(2r - 1)$ **15.** prime **17.** $(-6 + x)(6 + x)$ or $-1(6 + x)(6 - x)$ **19.** $(m^2 + 1)(m + 1)(m - 1)$ **21.** $(m^2 + n^9)(m^2 - n^9)$
23. $(x + 5)(x^2 - 5x + 25)$ **25.** $(2a - 1)(4a^2 + 2a + 1)$ **27.** $(m + 3n)(m^2 - 3mn + 9n^2)$ **29.** $5(k + 2)(k^2 - 2k + 4)$
31. $(xy - 4)(x^2y^2 + 4xy + 16)$ **33.** $2(5r - 4t)(25r^2 + 20rt + 16t^2)$ **35.** $(r + 8)(r - 8)$ **37.** $(x + 13y)(x - 13y)$ **39.** $(3 - t)(9 + 3t + t^2)$
41. $2(3r + 2)(3r - 2)$ **43.** $x(3y + 2)(3y - 2)$ **45.** $8(m + 2)(m^2 - 2m + 4)$ **47.** $xy(y - 3z)(y + 3z)$ **49.** $4(3x - 4y)(3x + 4y)$
51. $9(4 - 3x)(4 + 3x)$ **53.** $(xy - z^2)(x^2y^2 + xyz^2 + z^4)$ **55.** $\left(7 - \frac{3}{5}m\right)\left(7 + \frac{3}{5}m\right)$ **57.** $(t + 7)(t^2 - 7t + 49)$ **59.** $n(n^2 + 49)$
61. $x^2(x^2 + 9)(x + 3)(x - 3)$ **63.** $pq(8p + 9q)(8p - 9q)$ **65.** $xy^2(27xy + 1)$ **67.** $a(5a - 4b)(25a^2 + 20ab + 16b^2)$ **69.** $16x^2(x + 2)(x - 2)$
71. 6 **73.** -2 **75.** $\frac{1}{5}$ **77.** 81.2% **79.** $-4(0.3x^2 - x - 20)$ **81.** $(x + 2 + y)(x + 2 - y)$ **83.** $(a + 4)(a - 4)(b - 4)$
85. $(x + 3 + 2y)(x + 3 - 2y)$ **87.** $(x^n + 10)(x^n - 10)$ **89.** $(x + 6)$ **91.** answers may vary **93. a.** 777 ft **b.** 441 ft **c.** 7 sec
d. $(29 + 4t)(29 - 4t)$ **95. a.** 1456 ft **b.** 816 ft **c.** 10 sec **d.** $16(10 + t)(10 - t)$

Integrated Review
Practice Exercises 6.1–6.5
1. $(3x - 1)(2x - 3)$ **2.** $(3x + 1)(x - 2)(x + 2)$ **3.** $3(3x - y)(3x + y)$ **4.** $(2a + b)(4a^2 - 2ab + b^2)$ **5.** $6xy^2(5x + 2)(2x - 3)$

Exercise Set 6.1–6.5
1. $(x + y)^2$ **2.** $(x - y)^2$ **3.** $(a + 12)(a - 1)$ **4.** $(a - 5)(a - 2)$ **5.** $(a + 2)(a - 3)$ **6.** $(a + 1)^2$ **7.** $(x + 1)^2$ **8.** $(x + 2)(x - 1)$
9. $(x + 1)(x + 3)$ **10.** $(x + 3)(x - 2)$ **11.** $(x + 3)(x + 4)$ **12.** $(x + 4)(x - 3)$ **13.** $(x + 4)(x - 1)$ **14.** $(x - 5)(x - 2)$
15. $(x + 5)(x - 3)$ **16.** $(x + 6)(x + 5)$ **17.** $(x - 6)(x + 5)$ **18.** $(x + 8)(x + 3)$ **19.** $2(x + 7)(x - 7)$ **20.** $3(x + 5)(x - 5)$
21. $(x + 3)(x + y)$ **22.** $(y - 7)(3 + x)$ **23.** $(x + 8)(x - 2)$ **24.** $(x - 7)(x + 4)$ **25.** $4x(x + 7)(x - 2)$ **26.** $6x(x - 5)(x + 4)$
27. $2(3x + 4)(2x + 3)$ **28.** $(2a - b)(4a + 5b)$ **29.** $(2a + b)(2a - b)$ **30.** $(4 - 3x)(7 + 2x)$ **31.** $(5 - 2x)(4 + x)$ **32.** prime **33.** prime
34. $(3y + 5)(2y - 3)$ **35.** $(4x - 5)(x + 1)$ **36.** $y(x + y)(x - y)$ **37.** $4(t^2 + 9)$ **38.** $(x + 1)(x + y)$ **39.** $(x + 1)(a + 2)$
40. $9x(2x^2 - 7x + 1)$ **41.** $4a(3a^2 - 6a + 1)$ **42.** $(x + 16)(x - 2)$ **43.** prime **44.** $(4a - 7b)^2$ **45.** $(5p - 7q)^2$ **46.** $(7x + 3y)(x + 3y)$
47. $(5 - 2y)(25 + 10y + 4y^2)$ **48.** $(4x + 3)(16x^2 - 12x + 9)$ **49.** $-(x - 5)(x + 6)$ **50.** $-(x - 2)(x - 4)$ **51.** $(7 - x)(2 + x)$
52. $(3 + x)(1 - x)$ **53.** $3x^2y(x + 6)(x - 4)$ **54.** $2xy(x + 5y)(x - y)$ **55.** $5xy^2(x - 7y)(x - y)$ **56.** $4x^2y(x - 5)(x + 3)$
57. $3xy(4x^2 + 81)$ **58.** $2xy^2(3x^2 + 4)$ **59.** $(2 + x)(2 - x)$ **60.** $(3 + y)(3 - y)$ **61.** $(s + 4)(3r - 1)$ **62.** $(x - 2)(x^2 + 3)$
63. $(4x - 3)(x - 2y)$ **64.** $(2x - y)(2x + 7z)$ **65.** $6(x + 2y)(x + y)$ **66.** $2(x + 4y)(6x - y)$ **67.** $(x + 3)(y + 2)(y - 2)$
68. $(y + 3)(y - 3)(x^2 + 3)$ **69.** $(5 + x)(x + y)$ **70.** $(x - y)(7 + y)$ **71.** $(7t - 1)(2t - 1)$ **72.** prime **73.** $(3x + 5)(x - 1)$
74. $(7x - 2)(x + 3)$ **75.** $(x + 12y)(x - 3y)$ **76.** $(3x - 2y)(x + 4y)$ **77.** $(1 - 10ab)(1 + 2ab)$ **78.** $(1 + 5ab)(1 - 12ab)$
79. $(3 + x)(3 - x)(1 + x)(1 - x)$ **80.** $(3 + x)(3 - x)(2 + x)(2 - x)$ **81.** $(x + 4)(x - 4)(x^2 + 2)$ **82.** $(x + 5)(x - 5)(x^2 + 3)$
83. $(x - 15)(x - 8)$ **84.** $(y + 16)(y + 6)$ **85.** $2x(3x - 2)(x - 4)$ **86.** $2y(3y + 5)(y - 3)$ **87.** $(3x - 5y)(9x^2 + 15xy + 25y^2)$
88. $(6y - z)(36y^2 + 6yz + z^2)$ **89.** $(xy + 2z)(x^2y^2 - 2xyz + 4z^2)$ **90.** $(3ab + 2)(9a^2b^2 - 6ab + 4)$ **91.** $2xy(1 + 6x)(1 - 6x)$
92. $2x(x + 3)(x - 3)$ **93.** $(x + 2)(x - 2)(x + 6)$ **94.** $(x - 2)(x + 6)(x - 6)$ **95.** $2a^2(3a + 5)$ **96.** $2n(2n - 3)$ **97.** $(a^2 + 2)(a + 2)$
98. $(a - b)(1 + x)$ **99.** $(x + 2)(x - 2)(x + 7)$ **100.** $(a + 3)(a - 3)(a + 5)$ **101.** $(x - y + z)(x - y - z)$
102. $(x + 2y + 3)(x + 2y - 3)$ **103.** $(9 + 5x + 1)(9 - 5x - 1)$ **104.** $(b + 4a + c)(b - 4a - c)$ **105.** answers may vary
106. yes; $9(x^2 + 9y^2)$ **107.** a, c

Section 6.6
Practice Exercises
1. $-4, 5$ **2.** $0, \frac{6}{7}$ **3.** $-4, 12$ **4.** $\frac{4}{3}$ **5.** $-3, \frac{2}{3}$ **6.** $-6, 4$ **7.** $-3, 0, 3$ **8.** $\frac{2}{3}, \frac{3}{2}, 5$ **9.** $-3, 0, 2$ **10.** The x-intercepts are $(2, 0)$ and $(4, 0)$.

Calculator Explorations 6.6
1. $-0.9, 2.2$ **3.** no real solution **5.** $-1.8, 2.8$

Vocabulary and Readiness Check 6.6
1. quadratic **3.** $3, -5$ **5.** $3, 7$ **7.** $-8, -6$ **9.** $-1, 3$

Exercise Set 6.6
1. $2, -1$ **3.** $-9, -17$ **5.** $0, -6$ **7.** $0, 8$ **9.** $-\frac{3}{2}, \frac{5}{4}$ **11.** $\frac{7}{2}, -\frac{2}{7}$ **13.** $\frac{1}{2}, -\frac{1}{3}$ **15.** $-0.2, -1.5$
17. answers may vary; for example, $(x-6)(x+1)=0$ **19.** $9, 4$ **21.** $-4, 2$ **23.** $0, 7$ **25.** $8, -4$ **27.** $4, -4$ **29.** $-3, 12$ **31.** $\frac{7}{3}, -2$
33. $-5, 5$ **35.** $-2, \frac{1}{6}$ **37.** $0, 4, 8$ **39.** $\frac{3}{4}$ **41.** $-\frac{1}{2}, 0, \frac{1}{2}$ **43.** $-\frac{3}{8}, 0, \frac{1}{2}$ **45.** $-3, 2$ **47.** $-20, 0$ **49.** $\frac{17}{2}$ **51.** $-\frac{1}{2}, \frac{1}{2}$
53. $-\frac{3}{2}, -\frac{1}{2}, 3$ **55.** $-5, 3$ **57.** $-\frac{5}{6}, \frac{6}{5}$ **59.** $2, -\frac{4}{5}$ **61.** $-\frac{4}{3}, 5$ **63.** $-4, 3$ **65.** $\frac{8}{3}, -9, 0$ **67.** -7 **69.** $0, \frac{3}{2}$ **71.** $0, 1, -1$
73. $-6, \frac{4}{3}$ **75.** $\frac{6}{7}, 1$ **77.** $\left(-\frac{4}{3}, 0\right), (1, 0)$ **79.** $(-2, 0), (5, 0)$ **81.** $(-6, 0), \left(\frac{1}{2}, 0\right)$ **83.** e **85.** b **87.** c **89.** $\frac{47}{45}$ **91.** $\frac{17}{60}$ **93.** $\frac{15}{8}$
95. $\frac{7}{10}$ **97.** didn't write equation in standard form; should be $x = 4$ or $x = -2$ **99.** answers may vary; for example, $x^2 - 12x + 35 = 0$
101. a. 300; 304; 276; 216; 124; 0; -156 **b.** 5 sec **c.** 304 ft **d.** [graph of $y = -16x^2 + 20x + 300$] **103.** $0, \frac{1}{2}$
105. $0, -15$

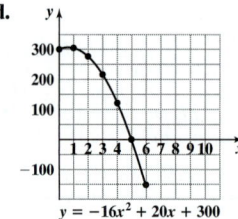

The Bigger Picture
1. -34 **2.** x^{22} **3.** $-4x^3 - 6x^2 + 8$ **4.** $y - 1 + \frac{3}{y^2}$ **5.** $10x(x+5)(x-5)$ **6.** $(x-1)(x-35)$ **7.** $3(2y+5)(x-1)$
8. $x(5y-7)(y+1)$ **9.** $5, -\frac{1}{2}$ **10.** 1 **11.** $-2, 14$ **12.** $\frac{33}{17}$

Section 6.7
Practice Exercises
1. 2 sec **2.** There are 2 numbers. They are -4 and 12. **3.** base: 35 ft; height: 12 ft **4.** 7 and 8 or -6 and -5
5. leg: 8 units; leg: 15 units; hypotenuse: 17 units

Exercise Set 6.7
1. width $= x$; length $= x + 4$ **3.** x and $x + 2$ if x is an odd integer **5.** base $= x$; height $= 4x + 1$ **7.** 11 units
9. 15 cm, 13 cm, 70 cm, 22 cm **11.** base $= 16$ mi; height $= 6$ mi **13.** 5 sec **15.** length $= 5$ cm; width $= 6$ cm **17.** 54 diagonals
19. 10 sides **21.** -12 or 11 **23.** 14, 15 **25.** 13 feet **27.** 5 in. **29.** 12 mm, 16 mm, 20 mm **31.** 10 km **33.** 36 ft **35.** 9.5 sec
37. 20% **39.** length: 15 mi; width: 8 mi **41.** 105 units **43.** 175 acres **45.** 6.25 million **47.** 1966 **49.** answers may vary **51.** $\frac{3}{4}$
53. $\frac{5}{9}$ **55.** $\frac{9}{10}$ **57.** 8 m **59.** 10 and 15 **61.** width: 29 m; length: 35 m **63.** answers may vary

Chapter 6 Vocabulary Check
1. quadratic equation **2.** factoring **3.** greatest common factor **4.** perfect square trinomial **5.** difference of two squares
6. difference of two cubes **7.** sum of two cubes **8.** 0

Chapter 6 Review
1. $2x - 5$ **3.** $4x(5x + 3)$ **5.** $-2x^2y(4x - 3y)$ **7.** $(x+1)(5x-1)$ **9.** $(2x-1)(3x+5)$ **11.** $(x+4)(x+2)$ **13.** prime
15. $(x+4)(x-2)$ **17.** $(x+5y)(x+3y)$ **19.** $2(3-x)(12+x)$ **21.** $(2x-1)(x+6)$ **23.** $(2x+3)(2x-1)$ **25.** $(6x-y)(x-4y)$
27. $(2x+3y)(x-13y)$ **29.** $(6x+5y)(3x-4y)$ **31.** $(2x+3)(2x-3)$ **33.** prime **35.** $(2x+3)(4x^2-6x+9)$
37. $2(3-xy)(9+3xy+x^2y^2)$ **39.** $(4x^2+1)(2x+1)(2x-1)$ **41.** $-6, 2$ **43.** $-\frac{1}{5}, -3$ **45.** $-4, 6$ **47.** $2, 8$ **49.** $-\frac{2}{7}, \frac{3}{8}$ **51.** $-\frac{2}{5}$

53. 3   **55.** $0, -\dfrac{7}{4}, 3$ **57.** c   **59.** 9 units   **61.** width: 20 in.; length: 25 in.   **63.** 19 and 20   **65.** 6.25 sec   **67.** $7(x - 9)$
69. $\left(m + \dfrac{2}{5}\right)\left(m - \dfrac{2}{5}\right)$ **71.** $(y + 2)(x - 1)$ **73.** $3x(x - 9)(x - 1)$ **75.** $2(x + 3)(x - 3)$ **77.** $5(x + 2)^2$ **79.** $2xy(2x - 3y)$
81. $(5x + 3)(25x^2 - 15x + 9)$ **83.** $(x + 7 + y)(x + 7 - y)$ **85.** $2b(3a - 1)(9a^2 + 3a + 1)$ **87.** factor out the GCF, 3
89. $16x^2 - 28x + 6$; $2(4x - 1)(2x - 3)$ **91.** $-3, 5$ **93.** $3, 2$ **95.** 19 in., 8 in., 21 in.   **97.** 6.75 sec

Chapter 6 Test
1. $(x + 7)(x + 4)$ **2.** $(7 - m)(7 + m)$ **3.** $(y + 11)^2$ **4.** $(a + 3)(4 - y)$ **5.** prime   **6.** $(y - 12)(y + 4)$ **7.** prime
8. $3x(3x + 1)(x + 4)$ **9.** $(3a - 7)(a + b)$ **10.** $(3x - 2)(x - 1)$ **11.** $(x + 12y)(x + 2y)$ **12.** $5(6 + x)(6 - x)$ **13.** $(6t + 5)(t - 1)$
14. $(y + 2)(y - 2)(x - 7)$ **15.** $x(1 + x^2)(1 + x)(1 - x)$ **16.** $-xy(y^2 + x^2)$ **17.** $(4x - 1)(16x^2 + 4x + 1)$ **18.** $8(y - 2)(y^2 + 2y + 4)$
19. $-9, 3$ **20.** $-7, 2$ **21.** $-7, 1$ **22.** $0, \dfrac{3}{2}, -\dfrac{4}{3}$ **23.** $0, 3, -3$ **24.** $-3, 5$ **25.** $0, \dfrac{5}{2}$ **26.** 17 ft   **27.** 8 and 9   **28.** 7 sec
29. hypotenuse: 25 cm; legs: 15 cm, 20 cm

Chapter 6 Cumulative Review
1. a. $9 \leq 11$ **b.** $8 > 1$ **c.** $3 \neq 4$; Sec. 1.2, Ex. 3   **3. a.** $\dfrac{6}{7}$ **b.** $\dfrac{11}{27}$ **c.** $\dfrac{22}{5}$; Sec. 1.3, Ex. 2   **5.** $\dfrac{14}{3}$; Sec. 1.4, Ex. 5
7. a. -12 **b.** -1; Sec. 1.5, Ex. 7   **9. a.** -24 **b.** -2 **c.** 50; Sec. 1.7, Ex. 1   **11. a.** $4x$ **b.** $11y^2$ **c.** $8x^2 - x$; Sec. 2.1, Ex. 3
13. -4; Sec. 2.2, Ex. 7   **15.** -11; Sec. 2.3, Ex. 2   **17.** $\dfrac{16}{3}$; Sec. 2.4, Ex. 2   **19.** shorter: 12 in.; longer: 36 in.; Sec. 2.5, Ex. 3
21. Sec. 3.2, Ex. 5   **23.** $m = \dfrac{3}{4}$; y-intercept: $(0, -1)$; Sec. 3.4, Ex. 5   **25. a.** 250   **b.** 1; Sec. 5.1, Ex. 2   **27. a.** 2   **b.** 5
c. 0; Sec. 5.2, Ex. 1   **29.** $9x^2 - 6x - 1$; Sec. 5.2, Ex. 13   **31.** $6x^2 - 11x - 10$; Sec. 5.3, Ex. 5
33. $9y^2 + 6y + 1$; Sec. 5.4, Ex. 4   **35. a.** $\dfrac{1}{9}$ **b.** $\dfrac{2}{x^3}$ **c.** $\dfrac{3}{4}$ **d.** $\dfrac{1}{16}$ **e.** y^4 **f.** 49; Sec. 5.5, Ex. 1
37. a. 3.67×10^6 **b.** 3.0×10^{-6} **c.** 2.052×10^{10} **d.** 8.5×10^{-4}; Sec. 5.5, Ex. 5   **39.** $x + 4$; Sec. 5.6, Ex. 4
41. a. x^3 **b.** y; Sec. 6.1, Ex. 2   **43.** $(x + 3)(x + 4)$; Sec. 6.2, Ex. 1   **45.** $(4x - 1)(2x - 5)$; Sec. 6.3, Ex. 2   **47.** $(5a + 3b)(5a - 3b)$; Sec. 6.5, Ex. 2b
49. $3, -1$; Sec. 6.6, Ex. 1

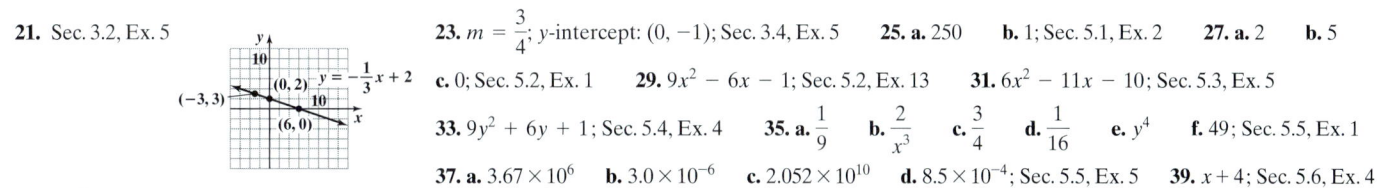

CHAPTER 7 RATIONAL EXPRESSIONS

Section 7.1
Practice Exercises
1. a. $\dfrac{9}{7}$ **b.** $-\dfrac{3}{11}$ **2. a.** $x = -6$ **b.** none   **c.** $x = -2$ or $x = -4$ **d.** none   **3.** $\dfrac{x^5}{6}$ **4.** $\dfrac{x + 1}{x - 2}$ **5.** $\dfrac{x^2}{x + 9}$ **6.** $\dfrac{1}{x + 7}$ **7. a.** -1 **b.** 1
8. $-\dfrac{2x + 3}{x + 4}$ or $\dfrac{-2x - 3}{x + 4}$ **9.** $\dfrac{-(x + 3)}{6x - 11}$; $\dfrac{-x - 3}{6x - 11}$; $\dfrac{x + 3}{-(6x - 11)}$; $\dfrac{x + 3}{-6x + 11}$; $\dfrac{x + 3}{11 - 6x}$

Vocabulary and Readiness Check 7.1
1. rational expression   **3.** -1 **5.** 2   **7.** $\dfrac{-a}{b}, \dfrac{a}{-b}$ **9.** yes   **11.** no

Exercise Set 7.1
1. $\dfrac{7}{4}$ **3.** 3   **5.** $-\dfrac{8}{3}$ **7.** $-\dfrac{11}{2}$ **9.** $x = 0$ **11.** $x = -2$ **13.** $x = \dfrac{5}{2}$ **15.** none   **17.** $x = 6, x = -1$ **19.** none   **21.** $x = -2, x = -\dfrac{7}{3}$
23. $\dfrac{-(x - 10)}{x + 8}$; $\dfrac{-x + 10}{x + 8}$; $\dfrac{x - 10}{-(x + 8)}$; $\dfrac{x - 10}{-x - 8}$ **25.** $\dfrac{-(5y - 3)}{y - 12}$; $\dfrac{-5y + 3}{y - 12}$; $\dfrac{5y - 3}{-(y - 12)}$; $\dfrac{5y - 3}{-y + 12}$ **27.** 1   **29.** -1 **31.** $\dfrac{1}{4(x + 2)}$ **33.** $\dfrac{1}{x + 2}$
35. can't simplify   **37.** -5 **39.** $\dfrac{7}{x}$ **41.** $\dfrac{1}{x - 9}$ **43.** $5x + 1$ **45.** $\dfrac{x^2}{x - 2}$ **47.** $7x$ **49.** $\dfrac{x + 5}{x - 5}$ **51.** $\dfrac{x + 2}{x + 4}$ **53.** $\dfrac{x + 2}{2}$
55. $-(x + 2)$ or $-x - 2$ **57.** $\dfrac{x + 1}{x - 1}$ **59.** $\dfrac{m - 3}{m + 2}$ **61.** $\dfrac{11x}{6}$ **63.** $x + y$ **65.** $\dfrac{5 - y}{2}$ **67.** $x^2 - 2x + 4$ **69.** $-x^2 - x - 1$ **71.** $\dfrac{2y + 5}{3y + 4}$
73. correct   **75.** correct   **77.** $\dfrac{3}{11}$ **79.** $\dfrac{4}{3}$ **81.** $\dfrac{117}{40}$ **83.** correct   **85.** incorrect; $\dfrac{1 + 2}{1 + 3} = \dfrac{3}{4}$ **87.** answers may vary   **89.** answers may vary
91. a. $37.5 million   **b.** $\approx$85.7 million   **c.** $\approx$48.2 million   **93.** 400 mg   **95.** $C = 78.125$; medium   **97.** 101.0   **99.** 25.0%
101. $y = \dfrac{x^2 - 16}{x - 4}$ **103.** $y = \dfrac{x^2 - 6x + 8}{x - 2}$

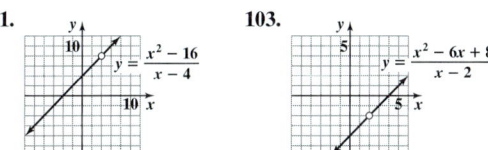

Section 7.2
Practice Exercises
1. a. $\dfrac{12a}{5b^2}$ b. $-\dfrac{2q}{3}$ 2. $\dfrac{3}{x+1}$ 3. $-\dfrac{3x-5}{2x(x+2)}$ 4. $\dfrac{b^2}{8a^2}$ 5. $\dfrac{3(3x+1)}{4}$ 6. $\dfrac{2}{x(x-3)}$ 7. 1 8. a. $\dfrac{(y+9)^2}{16x^2}$ b. $\dfrac{1}{4x}$ c. $-\dfrac{7(x-2)}{x+4}$

Vocabulary and Readiness Check 7.2
1. reciprocals 3. $\dfrac{a \cdot d}{b \cdot c}$ or $\dfrac{ad}{bc}$ 5. $\dfrac{6}{7}$

Exercise Set 7.2
1. $\dfrac{21}{4y}$ 3. x^4 5. $-\dfrac{b^2}{6}$ 7. $\dfrac{x^2}{10}$ 9. $\dfrac{1}{3}$ 11. $\dfrac{m+n}{m-n}$ 13. $\dfrac{x+5}{x}$ 15. $\dfrac{(x+2)(x-3)}{(x-4)(x+4)}$ 17. $\dfrac{2x^4}{3}$ 19. $\dfrac{12}{y^6}$ 21. $x(x+4)$ 23. $\dfrac{3(x+1)}{x^3(x-1)}$
25. $m^2 - n^2$ 27. $-\dfrac{x+2}{x-3}$ 29. $\dfrac{x+2}{x-3}$ 31. $\dfrac{5}{6}$ 33. $\dfrac{3x}{8}$ 35. $\dfrac{3}{2}$ 37. $\dfrac{3x+4y}{2(x+2y)}$ 39. $\dfrac{2(x+2)}{x-2}$ 41. $-\dfrac{y(x+2)}{4}$ 43. $\dfrac{(a+5)(a+3)}{(a+2)(a+1)}$
45. $\dfrac{5}{x}$ 47. $\dfrac{2(n-8)}{3n-1}$ 49. $4x^3(x-3)$ 51. $\dfrac{(a+b)^2}{a-b}$ 53. $\dfrac{3x+5}{x^2+4}$ 55. $\dfrac{4}{x-2}$ 57. $\dfrac{a-b}{6(a^2+ab+b^2)}$ 59. 1 61. $-\dfrac{10}{9}$ 63. $-\dfrac{1}{5}$
65. [graph of $x - 2y = 6$] 67. true 69. false; $\dfrac{x^2 + 3x}{20}$ 71. $\dfrac{2}{9(x-5)}$ sq ft 73. $\dfrac{x}{2}$ 75. $\dfrac{5a(2a+b)(3a-2b)}{b^2(a-b)(a+2b)}$ 77. answers may vary

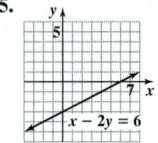

Section 7.3
Practice Exercises
1. $\dfrac{2a}{b}$ 2. 1 3. $4x - 5$ 4. a. 42 b. $45y^3$ 5. a. $(y-5)(y-4)$ b. $a(a+2)$ 6. $3(2x-1)^2$ 7. $(x+4)(x-4)(x+1)$
8. $(3-x)$ or $(x-3)$ 9. a. $\dfrac{21x^2y}{35xy^2}$ b. $\dfrac{18x}{8x+14}$ 10. $\dfrac{3x-6}{(x-2)(x+3)(x-5)}$

Vocabulary and Readiness Check 7.3
1. $\dfrac{9}{11}$ 3. $\dfrac{a+c}{b}$ 5. $\dfrac{5-(6+x)}{x}$

Exercise Set 7.3
1. $\dfrac{a+9}{13}$ 3. $\dfrac{3m}{n}$ 5. 4 7. $\dfrac{y+10}{3+y}$ 9. $5x+3$ 11. $\dfrac{4}{a+5}$ 13. $\dfrac{1}{x-6}$ 15. $\dfrac{5x+7}{x-3}$ 17. $x+5$ 19. 3 21. $4x^3$
23. $8x(x+2)$ 25. $(x+3)(x-2)$ 27. $3(x+6)$ 29. $5(x-6)^2$ 31. $6(x+1)^2$ 33. $x - 8$ or $8 - x$ 35. $(x-1)(x+4)(x+3)$
37. $(3x+1)(x+1)(x-1)(2x+1)$ 39. $2x^2(x+4)(x-4)$ 41. $\dfrac{6x}{4x^2}$ 43. $\dfrac{24b^2}{12ab^2}$ 45. $\dfrac{9y}{2y(x+3)}$ 47. $\dfrac{9ab+2b}{5b(a+2)}$
49. $\dfrac{x^2+x}{x(x+4)(x+2)(x+1)}$ 51. $\dfrac{18y-2}{30x^2-60}$ 53. $2x$ 55. $\dfrac{x+3}{2x-1}$ 57. $x+1$ 59. $\dfrac{1}{x^2-8}$ 61. $\dfrac{6(4x+1)}{x(2x+1)}$ 63. $\dfrac{29}{21}$ 65. $-\dfrac{5}{12}$
67. $\dfrac{7}{30}$ 69. d 71. c 73. b 75. $-\dfrac{5}{x-2}$ 77. $\dfrac{7+x}{x-2}$ 79. $\dfrac{20}{x-2}$ m 81. answers may vary 83. 95,304 Earth days
85. answers may vary 87. answers may vary

Section 7.4
Practice Exercises
1. a. 0 b. $\dfrac{21a+10}{24a^2}$ 2. $\dfrac{6}{x-5}$ 3. $\dfrac{13y+3}{5y(y+1)}$ 4. $\dfrac{13}{x-5}$ 5. $\dfrac{3b+6}{b+3}$ or $\dfrac{3(b+2)}{b+3}$ 6. $\dfrac{10-3x^2}{2x(2x+3)}$ 7. $\dfrac{x(5x+6)}{(x+4)(x+3)(x-3)}$

Vocabulary and Readiness Check 7.4
1. d 3. a

Exercise Set 7.4

1. $\dfrac{5}{x}$ **3.** $\dfrac{75a - 6b^2}{5b}$ **5.** $\dfrac{6x + 5}{2x^2}$ **7.** $\dfrac{11}{x + 1}$ **9.** $\dfrac{x - 6}{(x - 2)(x + 2)}$ **11.** $\dfrac{35x - 6}{4x(x - 2)}$ **13.** $-\dfrac{2}{x - 3}$ **15.** 0 **17.** $-\dfrac{1}{x^2 - 1}$ **19.** $\dfrac{5 + 2x}{x}$

21. $\dfrac{6x - 7}{x - 2}$ **23.** $-\dfrac{y + 4}{y + 3}$ **25.** $\dfrac{-5x + 14}{4x}$ or $-\dfrac{5x - 14}{4x}$ **27.** 2 **29.** $\dfrac{9x^4 - 4x^2}{21}$ **31.** $\dfrac{x + 2}{(x + 3)^2}$ **33.** $\dfrac{9b - 4}{5b(b - 1)}$ **35.** $\dfrac{2 + m}{m}$

37. $\dfrac{x^2 + 3x}{(x - 7)(x - 2)}$ or $\dfrac{x(x + 3)}{(x - 7)(x - 2)}$ **39.** $\dfrac{10}{1 - 2x}$ **41.** $\dfrac{15x - 1}{(x + 1)^2(x - 1)}$ **43.** $\dfrac{x^2 - 3x - 2}{(x - 1)^2(x + 1)}$ **45.** $\dfrac{a + 2}{2(a + 3)}$ **47.** $\dfrac{y(2y + 1)}{(2y + 3)^2}$

49. $\dfrac{x - 10}{2(x - 2)}$ **51.** $\dfrac{2x + 21}{(x + 3)^2}$ **53.** $\dfrac{-5x + 23}{(x - 2)(x - 3)}$ **55.** $\dfrac{7}{2(m - 10)}$ **57.** $\dfrac{2x^2 - 2x - 46}{(x + 1)(x - 6)(x - 5)}$ or $\dfrac{2(x^2 - x - 23)}{(x + 1)(x - 6)(x - 5)}$

59. $\dfrac{n + 4}{4n(n - 1)(n - 2)}$ **61.** 10 **63.** 2 **65.** $\dfrac{25a}{9(a - 2)}$ **67.** $\dfrac{x + 4}{(x - 2)(x - 1)}$ **69.** $x = \dfrac{2}{3}$ **71.** $x = -\dfrac{1}{2}, x = 1$ **73.** $x = -\dfrac{15}{2}$

75. $\dfrac{6x^2 - 5x - 3}{x(x + 1)(x - 1)}$ **77.** $\dfrac{4x^2 - 15x + 6}{(x - 2)^2(x + 2)(x - 3)}$ **79.** $\dfrac{-2x^2 + 14x + 55}{(x + 2)(x + 7)(x + 3)}$ **81.** $\dfrac{2x - 16}{(x + 4)(x - 4)}$ in. **83.** $\dfrac{P - G}{P}$ **85.** answers may vary

87. $\left(\dfrac{90x - 40}{x}\right)°$ **89.** answers may vary

The Bigger Picture

1. -17.7 **2.** 78.26 **3.** 28 **4.** $4x^4 - x^2 - 17$ **5.** $\dfrac{x - 1}{5}$ **6.** $\dfrac{14}{x + 1}$ **7.** $-\dfrac{11}{18}$ **8.** $\dfrac{-4x - 27}{45}$ or $-\dfrac{4x + 27}{45}$ **9.** $x(9x - 11)(x + 1)$

10. $(4y - 7)(3x + 1)$ **11.** 12 **12.** $\left(\tfrac{1}{2}, \infty\right)$ **13.** $-\dfrac{8}{3}$ **14.** $-4, 6$

Section 7.5
Practice Exercises

1. -2 **2.** 13 **3.** $-1, 7$ **4.** $-\dfrac{19}{2}$ **5.** 3 **6.** -8 **7.** $b = \dfrac{ax}{a - x}$

Graphing Calculator Explorations 7.5

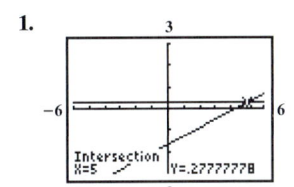

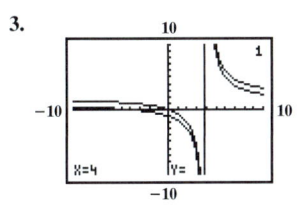

1.

3.

Exercise Set 7.5

1. 30 **3.** 0 **5.** -2 **7.** $-5, 2$ **9.** 5 **11.** 3 **13.** 1 **15.** 5 **17.** no solution **19.** 4 **21.** -8 **23.** $6, -4$ **25.** 1 **27.** $3, -4$
29. -3 **31.** 0 **33.** -2 **35.** $8, -2$ **37.** no solution **39.** 3 **41.** $-11, 1$ **43.** $I = \dfrac{E}{R}$ **45.** $B = \dfrac{2U - TE}{T}$ **47.** $W = \dfrac{Bh^2}{705}$
49. $G = \dfrac{V}{N - R}$ **51.** $r = \dfrac{C}{2\pi}$ **53.** $x = \dfrac{3y}{3 + y}$ **55.** $\dfrac{1}{x}$ **57.** $\dfrac{1}{x} + \dfrac{1}{2}$ **59.** $\dfrac{1}{3}$ **61.** $(2, 0), (0, -2)$ **63.** $(-4, 0), (-2, 0), (3, 0), (0, 4)$
65. answers may vary **67.** $\dfrac{5x + 9}{9x}$ **69.** no solution **71.** $100°, 80°$ **73.** $22.5°, 67.5°$ **75.** $\dfrac{17}{4}$

Integrated Review

1. expression; $\dfrac{3 + 2x}{3x}$ **2.** expression; $\dfrac{18 + 5a}{6a}$ **3.** equation; 3 **4.** equation; 18 **5.** expression; $\dfrac{x + 1}{x(x - 1)}$ **6.** expression; $\dfrac{3(x + 1)}{x(x - 3)}$

7. equation; no solution **8.** equation; 1 **9.** expression; 10 **10.** expression; $\dfrac{z}{3(9z - 5)}$ **11.** expression; $\dfrac{5x + 7}{x - 3}$ **12.** expression; $\dfrac{7p + 5}{2p + 7}$

13. equation; 23 **14.** equation; 5 **15.** expression; $\dfrac{25a}{9(a - 2)}$ **16.** expression; $\dfrac{4x + 5}{(x + 1)(x - 1)}$ **17.** expression; $\dfrac{3x^2 + 5x + 3}{(3x - 1)^2}$

18. expression; $\dfrac{2x^2 - 3x - 1}{(2x - 5)^2}$ **19.** expression; $\dfrac{4x - 37}{5x}$ **20.** equation; $-\dfrac{7}{3}$ **21.** equation; $\dfrac{8}{5}$ **22.** expression; $\dfrac{29x - 23}{3x}$

Section 7.6
Practice Exercises
1. 99 **2.** $\dfrac{13}{3}$ **3.** $9.03 **4.** 6 **5.** 15 **6.** $1\dfrac{5}{7}$ hr **7.** bus: 45 mph; car: 60 mph

Vocabulary and Readiness Check 7.6
1. c

Exercise Set 7.6
1. 4 **3.** $\dfrac{50}{9}$ **5.** −3 **7.** $\dfrac{14}{9}$ **9.** 123 lb **11.** 165 cal **13.** $y = 21.25$ **15.** $y = 5\dfrac{5}{7}$ ft **17.** 2 **19.** −3 **21.** $2\dfrac{2}{9}$ hr **23.** $1\dfrac{1}{2}$ min
25. trip to park rate: r; to park time: $\dfrac{12}{r}$; return trip rate: r; return time: $\dfrac{18}{r} = \dfrac{12}{r} + 1$; $r = 6$ mph **27.** 1st portion: 10 mph; cooldown: 8 mph
29. 360 sq ft **31.** 2 **33.** $108.00 **35.** 20 mph **37.** $y = 37\dfrac{1}{2}$ ft **39.** 337 yd/game **41.** 5 **43.** 217 mph **45.** 9 gal **47.** 8 mph
49. 35 mph; 75 mph **51.** 3 hr **53.** $26\dfrac{2}{3}$ ft **55.** 216 nuts **57.** 510 mph **59.** 20 hr **61.** car: 70 mph; motorcycle: 60 mph **63.** $5\dfrac{1}{4}$ hr
65. first pump: 28 min; second pump: 84 min **67.** $x = 5$ **69.** $x = 13.5$ **71.** $-\dfrac{4}{3}$; downward **73.** $\dfrac{11}{4}$; upward **75.** undefined slope; vertical
77. 2000 megawatts **79.** 8,190,000 people **81.** answers may vary **83.** none; answers may vary **85.** 84 yr **87.** 30 yr **89.** 3.75 min

The Bigger Picture
1. $12x^3 - 11x^2 - 13x + 10$ **2.** $4x^2 - 4xy + y^2$ **3.** $4y^3(2 - 5y^2)$ **4.** $(9m - 2n)(m - n)$ **5.** −35 **6.** $\dfrac{16x - 70}{x(x - 10)}$ or $\dfrac{2(8x - 35)}{x(x - 10)}$
7. $-12x^{13}$ **8.** 2 **9.** $-\dfrac{1}{2}$ **10.** $[4, \infty)$ **11.** $-\dfrac{27}{23}$ **12.** $\dfrac{a^{14}}{b^{14}}$

Section 7.7
Practice Exercises
1. $y = 5x$ **2.** $y = \dfrac{1}{4}x$; when x is 20, $y = 5$. **3.** $k = \dfrac{3}{4}$; $y = \dfrac{3}{4}x$ **4.** $y = \dfrac{8}{x}$ **5.** $k = 2.1$; $y = \dfrac{2.1}{x}$; when x is 70, $y = 0.03$. **6.** 6.48 sq units
7. 25 ml

Vocabulary and Readiness Check 7.7
1. inverse **3.** direct **5.** inverse **7.** inverse **9.** direct

Exercise Set 7.7
1. $y = \dfrac{1}{2}x$ **3.** $y = 6x$ **5.** $y = 3x$ **7.** $y = \dfrac{2}{3}x$ **9.** $y = \dfrac{7}{x}$ **11.** $y = \dfrac{0.5}{x}$ **13.** $y = kx$ **15.** $h = \dfrac{k}{t}$ **17.** $z = kx^2$ **19.** $y = \dfrac{k}{z^3}$
21. $x = \dfrac{k}{\sqrt{y}}$ **23.** $y = 40$ **25.** $y = 3$ **27.** $z = 54$ **29.** $a = \dfrac{4}{9}$ **31.** $62.50 **33.** $6 **35.** $5\dfrac{1}{3}$ in. **37.** 179.1 lb **39.** 1600 ft **41.** $\dfrac{1}{2}$
43. $\dfrac{3}{7}$ **45.** multiplied by 3 **47.** it is doubled

Section 7.8
Practice Exercises
1. $\dfrac{11}{8}$ **2.** $\dfrac{85}{33}$ **3.** $\dfrac{5(8 - x)}{x(2 - x)}$ **4.** $\dfrac{85}{33}$ **5.** $\dfrac{a - b}{a + 4b}$ **6.** $\dfrac{4a + 3b^2}{ab(a - 3b)}$

Vocabulary and Readiness Check 7.8
1. $\dfrac{y}{5x}$ **3.** $\dfrac{3x}{5}$

Exercise Set 7.8
1. $\dfrac{2}{3}$ **3.** $\dfrac{2}{3}$ **5.** $\dfrac{1}{2}$ **7.** $-\dfrac{21}{5}$ **9.** $\dfrac{27}{16}$ **11.** $\dfrac{4}{3}$ **13.** $\dfrac{1}{21}$ **15.** $-\dfrac{4x}{15}$ **17.** $\dfrac{m - n}{m + n}$ **19.** $\dfrac{2x(x - 5)}{7x^2 + 10}$ **21.** $\dfrac{1}{y - 1}$ **23.** $\dfrac{1}{6}$ **25.** $\dfrac{x + y}{x - y}$
27. $\dfrac{3}{7}$ **29.** $\dfrac{a}{x + b}$ **31.** $\dfrac{7(y - 3)}{8 + y}$ **33.** $\dfrac{3x}{x - 4}$ **35.** $-\dfrac{x + 8}{x - 2}$ **37.** $\dfrac{s^2 + r^2}{s^2 - r^2}$ **39.** $\dfrac{(x - 6)(x + 4)}{x - 2}$ **41.** 9 **43.** 1 **45.** $\dfrac{1}{5}$ **47.** $\dfrac{2}{3}$
49. answers may vary **51.** $\dfrac{13}{24}$ **53.** $\dfrac{R_1 R_2}{R_2 + R_1}$ **55.** $\dfrac{2x}{2 - x}$ **57.** $\dfrac{1}{y^2 - 1}$ **59.** 12 hr

Chapter 7 Vocabulary Check
1. ratio   **2.** proportion   **3.** cross products   **4.** rational expression   **5.** complex fraction   **6.** inverse variation   **7.** direct variation

Chapter 7 Review
1. $x = 2, x = -2$ **3.** $\dfrac{4}{3}$ **5.** $\dfrac{2}{x}$ **7.** $\dfrac{1}{x-5}$ **9.** $\dfrac{x(x-2)}{x+1}$ **11.** $\dfrac{x-3}{x-5}$ **13.** $\dfrac{x+a}{x-c}$ **15.** $-\dfrac{1}{x^2+4x+16}$ **17.** $\dfrac{3x^2}{y}$ **19.** $\dfrac{x-3}{x+2}$
21. $\dfrac{x+3}{x-4}$ **23.** $(x-6)(x-3)$ **25.** $\dfrac{1}{2}$ **27.** $-\dfrac{2(2x+3)}{y-2}$ **29.** $\dfrac{1}{x+2}$ **31.** $\dfrac{2x-10}{3x^2}$ **33.** $14x$ **35.** $\dfrac{10x^2y}{14x^3y}$ **37.** $\dfrac{x^2-3x-10}{(x+2)(x-5)(x+9)}$
39. $\dfrac{4y-30x^2}{5x^2y}$ **41.** $\dfrac{-2x-2}{x+3}$ **43.** $\dfrac{x-4}{3x}$ **45.** $\dfrac{x^2+2x+4}{4x}; \dfrac{x+2}{32}$ **47.** 30   **49.** no solution   **51.** $\dfrac{9}{7}$ **53.** $b = \dfrac{4A}{5x^2}$ **55.** $x = 6$
57. $x = 9$ **59.** 675 parts   **61.** 3   **63.** fast car speed: 30 mph; slow car speed: 20 mph   **65.** $17\dfrac{1}{2}$ hr   **67.** $x = 15$ **69.** $y = 110$ **71.** $y = \dfrac{100}{27}$
73. $3960   **75.** $-\dfrac{7}{18y}$ **77.** $\dfrac{3y-1}{2y-1}$ **79.** $\dfrac{1}{2x}$ **81.** $\dfrac{x-4}{x+4}$ **83.** $\dfrac{1}{x-6}$ **85.** $\dfrac{2}{(x+3)(x-2)}$ **87.** $\dfrac{1}{2}$ **89.** 1   **91.** $x = 6$ **93.** $\dfrac{3}{10}$

Chapter 7 Test
1. $x = -1, x = -3$ **2. a.** $115   **b.** $103   **3.** $\dfrac{3}{5}$ **4.** $\dfrac{1}{x+6}$ **5.** $\dfrac{1}{x^2-3x+9}$ **6.** $\dfrac{2m(m+2)}{m-2}$ **7.** $\dfrac{a+2}{a+5}$ **8.** $-\dfrac{1}{x+y}$ **9.** 15
10. $\dfrac{y-2}{4}$ **11.** $\dfrac{19x-6}{2x+5}$ **12.** $\dfrac{3a-4}{(a-3)(a+2)}$ **13.** $\dfrac{3}{x-1}$ **14.** $\dfrac{2(x+5)}{x(y+5)}$ **15.** $\dfrac{x^2+2x+35}{(x+9)(x+2)(x-5)}$ **16.** $\dfrac{30}{11}$ **17.** -6
18. no solution   **19.** $-2, 5$ **20.** no solution   **21.** $\dfrac{xz}{2y}$ **22.** $\dfrac{5y^2-1}{y+2}$ **23.** 28   **24.** $\dfrac{8}{9}$ **25.** 18 bulbs   **26.** 5 or 1   **27.** 30 mph
28. $6\dfrac{2}{3}$ hr   **29.** $x = 12$

Chapter 7 Cumulative Review
1. a. $\dfrac{15}{x} = 4$ **b.** $12 - 3 = x$ **c.** $4x + 17 \neq 21$ **d.** $3x < 48$; Sec. 1.4, Ex. 9   **3.** amount at 7%: $12,500; amount at 9%: $7500; Sec. 2.8, Ex. 4
5. ; Sec. 3.3, Ex. 6   **7. a.** 4^7 **b.** x^{10} **c.** y^4 **d.** y^{12} **e.** $(-5)^{15}$;   **f.** a^2b^2; Sec. 5.1, Ex. 3   **9.** $12z + 16$; Sec. 5.2, Ex. 15
11. $27a^3 + 27a^2b + 9ab^2 + b^3$; Sec. 5.3, Ex. 8   **13. a.** $t^2 + 4t + 4$ **b.** $p^2 - 2pq + q^2$ **c.** $4x^2 + 20x + 25$ **d.** $x^4 - 14x^2y + 49y^2$; Sec. 5.4, Ex. 5
15. a. x^3 **b.** 81   **c.** $\dfrac{q^9}{p^4}$ **d.** $\dfrac{32}{125}$; Sec. 5.5, Ex. 2   **17.** $4x^2 - 4x + 6 + \dfrac{-11}{2x+3}$; Sec. 5.6, Ex. 6   **19. a.** 4   **b.** 1   **c.** 3; Sec. 6.1, Ex. 1
21. $-3a(3a^4 - 6a + 1)$; Sec. 6.1, Ex. 5   **23.** $3(m+2)(m-10)$; Sec. 6.2, Ex. 9   **25.** $(3x+2)(x+3)$; Sec. 6.3, Ex. 1   **27.** $(x+6)^2$; Sec. 6.3, Ex. 8
29. prime polynomial; Sec. 6.5, Ex. 4b   **31.** $(x+2)(x^2 - 2x + 4)$; Sec. 6.5, Ex. 8   **33.** $(2x+3)(x+1)(x-1)$; Sec. 6.6, Ex. 2
35. $3(2m+n)(2m-n)$; Int. Rev., Ex. 3   **37.** $-\dfrac{1}{2}, 4$; Sec. 6.6, Ex. 5   **39.** $(1, 0), (4, 0)$; Sec. 6.6, Ex. 10   **41.** base: 6 m; height: 10 m; Sec. 6.7, Ex. 3
43. $\dfrac{5}{x^2}$; Sec. 7.1, Ex. 3   **45.** $\dfrac{2}{x(x+1)}$; Sec. 7.2, Ex. 6   **47.** $\dfrac{x+1}{x+2y}$; Sec. 7.8, Ex. 5

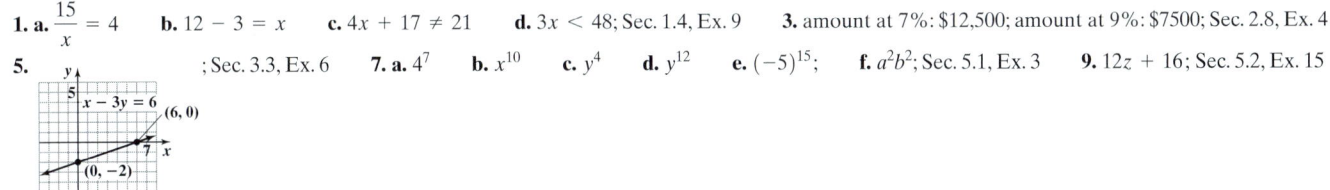

CHAPTER 8 ROOTS AND RADICALS

Section 8.1
Practice Exercises
1. a. $\dfrac{2}{9}$ **b.** -5 **c.** 12   **d.** 0.7   **e.** -1 **2. a.** 0   **b.** -4 **c.** $\dfrac{1}{2}$ **3. a.** 3   **b.** 10   **c.** not a real number   **d.** -5 **4.** 4.123
5. a. x^5 **b.** y^7 **c.** $5z^3$ **d.** $7x$ **e.** $\dfrac{z^2}{6}$ **f.** $-2a^2b^4$

Calculator Explorations 8.1
1. 2.646; yes   **3.** 3.317; yes   **5.** 9.055; yes   **7.** 3.420; yes   **9.** 2.115; yes   **11.** 1.783; yes

Vocabulary and Readiness Check 8.1
1. index, radicand, radical sign   **3.** false   **5.** true   **7.** true

Exercise Set 8.1

1. 4 **3.** $\frac{1}{5}$ **5.** -10 **7.** not a real number **9.** -11 **11.** $\frac{3}{5}$ **13.** 30 **15.** 12 **17.** $\frac{1}{10}$ **19.** 0.5 **21.** 5 **23.** -4 **25.** -2 **27.** $\frac{1}{2}$ **29.** -5 **31.** 2 **33.** 9 **35.** not a real number **37.** $-\frac{3}{4}$ **39.** -5 **41.** 1 **43.** 2.646 **45.** 6.083 **47.** 11.662 **49.** $\sqrt{2} \approx 1.41$; 126.90 ft **51.** x^2 **53.** $3x^4$ **55.** $9x$ **57.** $\frac{x^3}{6}$ **59.** $\frac{5y}{3}$ **61.** $4a^3b^2$ **63.** a^2b^6 **65.** $-2xy^9$ **67.** $25 \cdot 2$ **69.** $16 \cdot 2$ or $4 \cdot 8$ **71.** $4 \cdot 7$ **73.** $9 \cdot 3$ **75.** a, b **77.** 7 mi **79.** 3.01 in. **81.** 3 **83.** 10 **85.** $T = 6.1$ seconds **87.** answers may vary **89.** 1; 1.7; 2; 3 **91.** $|x|$ **93.** $|x + 2|$ **95.** $(2, 0)$ **97.** $(-4, 0)$

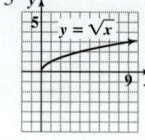

Section 8.2
Practice Exercises

1. a. $2\sqrt{6}$ **b.** $2\sqrt{15}$ **c.** $\sqrt{42}$ **d.** $10\sqrt{3}$ **2.** $10\sqrt{10}$ **3. a.** $\frac{\sqrt{5}}{7}$ **b.** $\frac{3}{10}$ **c.** $\frac{3\sqrt{2}}{5}$ **4. a.** $x^3\sqrt{x}$ **b.** $2a^2\sqrt{3}$ **c.** $\frac{7\sqrt{2}}{z^4}$ **d.** $\frac{y^4\sqrt{11y}}{7}$ **5. a.** $2\sqrt[3]{3}$ **b.** $\sqrt[3]{38}$ **c.** $\frac{\sqrt[3]{5}}{3}$ **d.** $\frac{\sqrt[3]{15}}{4}$ **6. a.** $2\sqrt[4]{2}$ **b.** $\frac{\sqrt[4]{5}}{3}$ **c.** $2\sqrt[5]{3}$

Vocabulary and Readiness Check 8.2

1. $\sqrt{a} \cdot \sqrt{b}$ **3.** 16; 25; 4; 5; 20

Exercise Set 8.2

1. $2\sqrt{5}$ **3.** $5\sqrt{2}$ **5.** $\sqrt{33}$ **7.** $7\sqrt{2}$ **9.** $2\sqrt{15}$ **11.** $6\sqrt{5}$ **13.** $2\sqrt{13}$ **15.** 15 **17.** $21\sqrt{7}$ **19.** $-15\sqrt{3}$ **21.** $\frac{2\sqrt{2}}{5}$ **23.** $\frac{3\sqrt{3}}{11}$ **25.** $\frac{3}{2}$ **27.** $\frac{5\sqrt{5}}{3}$ **29.** $\frac{\sqrt{11}}{6}$ **31.** $-\frac{\sqrt{3}}{4}$ **33.** $x^3\sqrt{x}$ **35.** $x^6\sqrt{x}$ **37.** $6a\sqrt{a}$ **39.** $4x^2\sqrt{6}$ **41.** $\frac{2\sqrt{3}}{m}$ **43.** $\frac{3\sqrt{x}}{y^5}$ **45.** $\frac{2\sqrt{22}}{x^6}$ **47.** 16 **49.** $\frac{6}{11}$ **51.** $5\sqrt{7}$ **53.** $\frac{2\sqrt{5}}{3}$ **55.** $2m^3\sqrt{6m}$ **57.** $\frac{y\sqrt{23y}}{2x^3}$ **59.** $2\sqrt[3]{3}$ **61.** $5\sqrt[3]{2}$ **63.** $\frac{\sqrt[3]{5}}{4}$ **65.** $\frac{\sqrt[3]{23}}{2}$ **67.** $\frac{\sqrt[3]{15}}{4}$ **69.** $2\sqrt[3]{10}$ **71.** $2\sqrt[4]{3}$ **73.** $\frac{\sqrt[4]{8}}{3}$ **75.** $2\sqrt[5]{3}$ **77.** $\frac{\sqrt[5]{5}}{2}$ **79.** $2\sqrt[3]{10}$ **81.** $14x$ **83.** $2x^2 - 7x - 15$ **85.** 0 **87.** $x^3y\sqrt{y}$ **89.** $7x^2y^2\sqrt{2x}$ **91.** $-2x^2$ **93.** answers may vary **95.** $2\sqrt{5}$ in. **97.** 2.25 in. **99.** $1700 **101.** 1.7 sq m

Section 8.3
Practice Exercises

1. a. $8\sqrt{2}$ **b.** $-7\sqrt{6}$ **c.** $4\sqrt[4]{5} + 11\sqrt[4]{7}$ **d.** $4\sqrt{13} - 5\sqrt[3]{13}$ **2. a.** $5\sqrt{5}$ **b.** $6 - 2\sqrt{10} + \sqrt{6}$ **c.** $-3\sqrt{2}$ **3.** $-7x\sqrt{x} + 3x$ **4.** $10x^2\sqrt[3]{3}$

Vocabulary and Readiness Check 8.3

1. like radicals **3.** $17\sqrt{2}$ **5.** $2\sqrt{5}$

Exercise Set 8.3

1. $-4\sqrt{3}$ **3.** $9\sqrt{6} - 5$ **5.** $2\sqrt{11} + 11$ **7.** $\sqrt{5} + \sqrt{2}$ **9.** $-2\sqrt[3]{16}$ **11.** $7\sqrt[3]{3} - \sqrt{3}$ **13.** $-5\sqrt[3]{2} - 6$ **15.** $5\sqrt{3}$ **17.** $9\sqrt{5}$ **19.** $4\sqrt{6} + \sqrt{5}$ **21.** $x + \sqrt{x}$ **23.** 0 **25.** $4\sqrt{x} - x\sqrt{x}$ **27.** $7\sqrt{5}$ **29.** $\sqrt{5} + \sqrt[3]{5}$ **31.** $-5 + 8\sqrt{2}$ **33.** $8 - 6\sqrt{2}$ **35.** $14\sqrt{2}$ **37.** $5\sqrt{2} + 12$ **39.** $\frac{4\sqrt{5}}{9}$ **41.** $\frac{3\sqrt{3}}{8}$ **43.** $2\sqrt{5}$ **45.** $-\sqrt{35}$ **47.** $12\sqrt{2x}$ **49.** 0 **51.** $x\sqrt{3x} + 3x\sqrt{x}$ **53.** $5\sqrt[3]{3}$ **55.** $\sqrt[3]{9}$ **57.** $4 + 4\sqrt[3]{2}$ **59.** $-3 + 3\sqrt[3]{2}$ **61.** $4x\sqrt{2} + 2\sqrt[4]{4} + 2x$ **63.** $\sqrt[5]{5}$ **65.** $x^2 + 12x + 36$ **67.** $4x^2 - 4x + 1$ **69.** $(4, 2)$ **71.** answers may vary **73.** $8\sqrt{5}$ in. **75.** $\left(48 + \frac{9\sqrt{3}}{2}\right)$ sq ft **77.** $\frac{83x\sqrt{x}}{20}$

Section 8.4
Practice Exercises
1. a. $\sqrt{77}$ b. $72\sqrt{30}$ c. $5\sqrt{2}$ d. 17 e. $5y^2\sqrt{3}$ 2. 28   3. $5\sqrt[3]{4}$ 4. a. $3 - \sqrt{15}$ b. $z\sqrt{2} + 14\sqrt{z}$ c. $x + \sqrt{2x} - \sqrt{7x} - \sqrt{14}$
5. a. -9 b. $3x - 10\sqrt{3x} + 25$ 6. a. $\sqrt{3}$ b. $2\sqrt{2}$ c. $3y^2$ 7. 5   8. a. $\dfrac{4\sqrt{5}}{5}$ b. $\dfrac{\sqrt{6}}{6}$ c. $\dfrac{\sqrt{42x}}{14x}$ 9. a. $\dfrac{3\sqrt[3]{5}}{5}$ b. $\dfrac{\sqrt[3]{150}}{5}$
10. a. $-1 + \sqrt{5}$ b. $\dfrac{5 + 3\sqrt{3}}{2}$ c. $\dfrac{8(5 + \sqrt{x})}{25 - x}$ 11. $\dfrac{7 - \sqrt{7}}{3}$

Vocabulary and Readiness Check 8.4
1. $\sqrt{21}$ 3. $\sqrt{\dfrac{15}{3}}$ or $\sqrt{5}$ 5. $2 - \sqrt{3}$

Exercise Set 8.4
1. 4   3. $5\sqrt{2}$ 5. 6   7. $2x$ 9. 20   11. $36x$ 13. $3x^3\sqrt{2}$ 15. $4xy\sqrt{y}$ 17. $\sqrt{30} + \sqrt{42}$ 19. $2\sqrt{5} + 5\sqrt{2}$ 21. $y\sqrt{7} - 14\sqrt{y}$
23. -33 25. $\sqrt{6} - \sqrt{15} + \sqrt{10} - 5$ 27. $16 - 11\sqrt{11}$ 29. $x - 36$ 31. $x - 14\sqrt{x} + 49$ 33. $6y + 2\sqrt{6y} + 1$ 35. 4   37. $\sqrt{7}$
39. $3\sqrt{2}$ 41. $5y^2$ 43. $5\sqrt{3}$ 45. $2y\sqrt{6}$ 47. $2xy\sqrt{3y}$ 49. $\dfrac{\sqrt{15}}{5}$ 51. $\dfrac{7\sqrt{2}}{2}$ 53. $\dfrac{\sqrt{6y}}{6y}$ 55. $\dfrac{\sqrt{3x}}{x}$ 57. $\dfrac{\sqrt{2}}{4}$ 59. $\dfrac{\sqrt{30}}{15}$
61. $\dfrac{8y\sqrt{5}}{5}$ 63. $\dfrac{\sqrt{3xy}}{6x}$ 65. $3\sqrt{2} - 3$ 67. $5 + \sqrt{30} + \sqrt{6} + \sqrt{5}$ 69. $\dfrac{3\sqrt{x} + 12}{x - 16}$ 71. $\dfrac{\sqrt{15}}{10}$ 73. $-8 - 4\sqrt{5}$ 75. $\dfrac{3\sqrt{2x}}{2}$
77. $\dfrac{10 - 5\sqrt{x}}{4 - x}$ 79. $3 + \sqrt{3}$ 81. $3 - 2\sqrt{5}$ 83. $3\sqrt{3} + 1$ 85. $2\sqrt[3]{6}$ 87. $12\sqrt[3]{10}$ 89. $5\sqrt[3]{3}$ 91. 3   93. $2\sqrt[3]{3}$ 95. $\dfrac{\sqrt[3]{10}}{2}$
97. $3\sqrt[3]{4}$ 99. $\dfrac{\sqrt[3]{3}}{3}$ 101. $\dfrac{\sqrt[3]{6}}{3}$ 103. 44   105. 2   107. 3   109. $130\sqrt{3}$ sq m   111. $\dfrac{\sqrt{A\pi}}{\pi}$ 113. true   115. false   117. false
119. answer may vary   121. answer may vary   123. $\dfrac{2}{\sqrt{6} - \sqrt{2} - \sqrt{3} + 1}$

Integrated Review
1. 6   2. $4\sqrt{3}$ 3. x^2 4. $y^3\sqrt{y}$ 5. $4x$ 6. $3x^5\sqrt{2x}$ 7. 2   8. 3   9. -3 10. not a real number   11. $\dfrac{\sqrt{11}}{3}$ 12. $\dfrac{\sqrt[3]{7}}{4}$ 13. -4
14. -5 15. $\dfrac{3}{7}$ 16. $\dfrac{1}{8}$ 17. a^4b 18. x^5y^{10} 19. $5m^3$ 20. $3n^8$ 21. $6\sqrt{7}$ 22. $3\sqrt{2}$ 23. cannot be simplified   24. $\sqrt{x} + 3x$
25. $\sqrt{30}$ 26. 3   27. 28   28. 45   29. $\sqrt{33} + \sqrt{3}$ 30. $3\sqrt{2} - 2\sqrt{6}$ 31. $4y$ 32. $3x^2\sqrt{5}$ 33. $x - 3\sqrt{x} - 10$ 34. $11 + 6\sqrt{2}$
35. 2   36. $\sqrt{3}$ 37. $2x^2\sqrt{3}$ 38. $ab^2\sqrt{15a}$ 39. $\dfrac{\sqrt{6}}{6}$ 40. $\dfrac{x\sqrt{5}}{10}$ 41. $\dfrac{4\sqrt{6} - 4}{5}$ 42. $\dfrac{\sqrt{2x} + 5\sqrt{2} + \sqrt{x} + 5}{x - 25}$

Section 8.5
Practice Exercises
1. 9   2. no solution   3. $\dfrac{2}{3}$ 4. 7   5. 10   6. 16

Exercise Set 8.5
1. 81   3. -1 5. 49   7. no solution   9. 4   11. 2   13. 2   15. 9   17. -3 19. $-1, -2$ 21. no solution   23. $0, -3$ 25. 16
27. 25   29. 1   31. 5   33. -2 35. no solution   37. 2   39. 36   41. no solution   43. $\dfrac{3}{2}$ 45. 16   47. 3   49. 12   51. 3, 1   53. -1
55. $3x - 8 = 19; x = 9$ 57. $2(2x + x) = 24$; length $= 8$ in.   59. 4, 7   61. answers may vary   63. a. $3.2, 10, 31.6$ b. no   65. 7.30 67. 0.76

The Bigger Picture
1. $2\sqrt{14}$ 2. $\dfrac{2x^2\sqrt{5x}}{7}$ 3. $-15x^5y^{11}$ 4. $\dfrac{\sqrt{110}}{11}$ 5. $2(\sqrt{5} + 1)$ or $2\sqrt{5} + 2$ 6. $5x^2 - 19$ 7. -1
8. $-\dfrac{5}{2}$ 9. $\left[-\dfrac{4}{11}, \infty\right)$ 10. $6, -7$ 11. $\dfrac{28}{3}$ 12. 2

Section 8.6
Practice Exercises
1. 13 in.   2. $2\sqrt{10}$ m ≈ 6.32 m   3. $\sqrt{5425} \approx 74$ ft   4. $2\sqrt{37}$ units   5. $16\sqrt{3}$ ft/sec ≈ 27.71 ft/sec

A34 Answers to Selected Exercises

Exercise Set 8.6
1. $\sqrt{13}$ 3. $3\sqrt{3}$ 5. 25   7. $\sqrt{22}$ 9. $3\sqrt{17}$ 11. $\sqrt{41}$ 13. $4\sqrt{2}$ 15. $3\sqrt{10}$ 17. 51.2 ft   19. 20.6 ft   21. 11.7 ft
23. $\sqrt{29}$ 25. $\sqrt{73}$ 27. $2\sqrt{10}$ 29. $\dfrac{3\sqrt{5}}{2}$ 31. $\sqrt{85}$ 33. 24 cu ft   35. 54 mph   37. 27 mph   39. 61.2 km   41. 32   43. $\dfrac{1}{25}$
45. x^5 47. y^4 49. $y = 2\sqrt{10}, x = 2\sqrt{10} - 4$ 51. 201 mi   53. answers may vary

Section 8.7
Practice Exercises
1. a. $\sqrt{36} = 6$ b. $\sqrt[3]{125} = 5$ c. $-\sqrt[4]{\dfrac{1}{81}} = -\dfrac{1}{3}$ d. $\sqrt[3]{-1000} = -10$ e. $\sqrt[5]{32} = 2$ 2. a. 27   b. 32   c. -5 3. a. $\dfrac{1}{25^{1/2}} = \dfrac{1}{5}$
b. $\dfrac{1}{1000^{2/3}} = \dfrac{1}{100}$ c. $-\sqrt{49} = -7$ d. $\dfrac{1}{1024^{2/5}} = \dfrac{1}{16}$ 4. a. 36   b. $\dfrac{1}{7^{1/3}}$ c. z^6 d. a e. $\dfrac{x^{15/2}}{y^8}$

Exercise Set 8.7
1. 2   3. 3   5. 8   7. 4   9. $-\dfrac{1}{2}$ 11. $\dfrac{1}{64}$ 13. $\dfrac{1}{729}$ 15. $\dfrac{5}{2}$ 17. answers may vary   19. 2   21. 2   23. $\dfrac{1}{x^{2/3}}$ 25. x^3
27. answers may vary   29. 9   31. -2 33. -3 35. $\dfrac{1}{9}$ 37. $\dfrac{3}{4}$ 39. 27   41. 512   43. -4 45. 32   47. $\dfrac{8}{27}$ 49. $\dfrac{1}{27}$ 51. $\dfrac{1}{2}$
53. $\dfrac{1}{5}$ 55. $\dfrac{1}{125}$ 57. 9   59. $6^{1/3}$ 61. x^6 63. 36   65. $\dfrac{1}{3}$ 67. $\dfrac{x^{2/3}}{y^{3/2}}$ 69. $\dfrac{x^{16/5}}{y^6}$ 71. [graph: $x+y<6$, $y \geq 2x$]   73. $-1, 4$ 75. $-\dfrac{1}{2}, 3$
77. 11,224 people   79. 3.344   81. 5.665

Chapter 8 Vocabulary Check
1. like radicals   2. index, radicand, radical   3. conjugate   4. principal square root   5. rationalizing the denominator

Chapter 8 Review
1. 9   3. 3   5. $-\dfrac{3}{8}$ 7. $\dfrac{2}{3}$ 9. c   11. x^6 13. $3x^3$ 15. $\dfrac{4}{y^5}$ 17. $3\sqrt{6}$ 19. $5x\sqrt{6x}$ 21. $3\sqrt[3]{2}$ 23. $2\sqrt[4]{3}$ 25. $\dfrac{3\sqrt{2}}{5}$
27. $\dfrac{3y\sqrt{5}}{2x^2}$ 29. $\dfrac{\sqrt[4]{9}}{2}$ 31. $\dfrac{\sqrt[3]{3}}{2}$ 33. $2\sqrt[3]{3} - \sqrt[3]{2}$ 35. $6\sqrt{6} - 2\sqrt[3]{6}$ 37. $5\sqrt{7x} + 2\sqrt[3]{7}$ 39. $\dfrac{\sqrt{5}}{6}$ 41. 0   43. $30\sqrt{2}$ 45. $6\sqrt{2} - 18$
47. $3\sqrt{2} - 5\sqrt{3} + 2\sqrt{6} - 10$ 49. $x - 4\sqrt{x} + 4$ 51. 3   53. $2\sqrt{5}$ 55. $x\sqrt{15x}$ 57. $\dfrac{\sqrt{22}}{11}$ 59. $\dfrac{\sqrt{30}}{6}$ 61. $\dfrac{\sqrt{5x}}{5x}$ 63. $\dfrac{\sqrt{3x}}{x}$
65. $3\sqrt{5} + 6$ 67. $\dfrac{\sqrt{6} + \sqrt{2} + \sqrt{3} + 1}{2}$ 69. $\dfrac{10\sqrt{x} - 50}{x - 25}$ 71. $\dfrac{\sqrt[3]{21}}{3}$ 73. $\dfrac{\sqrt[3]{12}}{2}$ 75. 18   77. 25   79. 12   81. 1   83. $3\sqrt{13}; 10.82$
85. $4\sqrt{34}$ ft; 23.32 ft   87. $\sqrt{130}$ 89. 2.4 in.   91. $a^{5/2}$ 93. $x^{5/2}$ 95. 4   97. -2 99. -512 101. $\dfrac{8}{27}$ 103. 32   105. $\dfrac{1}{3^{2/3}}$
107. $\dfrac{1}{x^2}$ 109. 12   111. $4x^8$ 113. $3x^3\sqrt{2x}$ 115. $\dfrac{1}{5}$ 117. $\dfrac{y^2}{9}$ 119. $7\sqrt{3}$ 121. $-\dfrac{\sqrt{3}}{4}$ 123. $7\sqrt{2}$ 125. $\sqrt{10} - \sqrt{2} + 4\sqrt{5} - 4$
127. $2\sqrt{6}$ 129. $\dfrac{\sqrt{14}}{7}$ 131. $\dfrac{3\sqrt{x} + 18}{x - 36}$ 133. 1   135. 14   137. $\sqrt{58}; 7.62$ 139. 32 ft

Chapter 8 Test
1. 4   2. -5 3. 8   4. $\dfrac{3}{4}$ 5. not a real number   6. $\dfrac{1}{9}$ 7. $3\sqrt{6}$ 8. $2\sqrt{23}$ 9. $x^3\sqrt{3}$ 10. $2x^2y^3\sqrt{2y}$ 11. $3x^4\sqrt{x}$ 12. 2
13. $2\sqrt[3]{5}$ 14. x^5 15. $y^3\sqrt{y}$ 16. $\dfrac{\sqrt{5}}{4}$ 17. $\dfrac{y\sqrt{y}}{5}$ 18. $\dfrac{\sqrt[3]{2}}{3}$ 19. $6\sqrt{2x}$ 20. $-2\sqrt{13}$ 21. $-8\sqrt{3}$ 22. $3\sqrt[3]{2} - 2x\sqrt{2}$ 23. $\dfrac{7\sqrt{3}}{10}$
24. $7\sqrt{2}$ 25. $2\sqrt{3} - \sqrt{10}$ 26. $x - \sqrt{x} - 6$ 27. $\sqrt{5}$ 28. $2x\sqrt{5x}$ 29. $\dfrac{\sqrt{6}}{3}$ 30. $\dfrac{\sqrt[3]{15}}{3}$ 31. $\dfrac{\sqrt{15}}{6x}$ 32. $-1 - \sqrt{3}$ 33. 9
34. 5   35. 9   36. $4\sqrt{5}$ in.   37. $\sqrt{5}$ 38. $\dfrac{1}{16}$ 39. $\dfrac{x^{10/3}}{y^2}$

Chapter 8 Cumulative Review

1. a. $-\dfrac{39}{5}$ **b.** 2; Sec. 1.7, Ex. 9   **3.** -3; Sec. 2.2, Ex. 5

5.

x	y
-1	-3
0	0
-3	-9; Sec. 3.1, Ex. 7

7. $y = \dfrac{1}{4}x - 3$; (Sec. 3.5, Ex. 1)   **9.** $y = -3$; Sec. 3.5, Ex. 7   **11.** a, b, c; Sec. 3.6, Ex. 5   **13.** yes; Sec. 4.1, Ex. 1   **15.** $(4, 2)$; Sec. 4.2, Ex. 1   **17.** $\left(-\dfrac{5}{4}, -\dfrac{5}{2}\right)$; Sec. 4.3, Ex. 6   **19.** 5% saline solution: 2.5 L; 25% saline solution: 7.5 L; Sec. 4.4, Ex. 4

21. ;Sec. 4.6, Ex. 1   **23.** $-9x^2 + 10xy - 5y^2$; Sec. 5.2, Ex. 7   **25.** $(x + 6y)(x + y)$; Sec. 6.2, Ex. 6   **27.** $\dfrac{-2 - x}{3x + 1}$; Sec. 7.1, Ex. 8   **29.** $\dfrac{3y^9}{160}$; Sec. 7.2, Ex. 4   **31.** 1; Sec. 7.3, Ex. 2   **33.** $\dfrac{x(3x - 1)}{(x + 1)^2(x - 1)}$; Sec. 7.4, Ex. 7   **35.** -5; Sec. 7.5, Ex. 1   **37.** 15 yd; Sec. 7.6, Ex. 4   **39.** $\dfrac{3}{z}$; Sec. 7.8, Ex. 3   **41. a.** $3\sqrt{6}$ **b.** $2\sqrt{3}$ **c.** $10\sqrt{2}$ **d.** $\sqrt{35}$; Sec. 8.2, Ex. 1   **43. a.** -44 **b.** $7x + 4\sqrt{7x} + 4$; Sec. 8.4, Ex. 5   **45.** no solution; Sec. 8.5, Ex. 2   **47.** 10 in.; Sec. 8.6, Ex. 1   **49. a.** 8   **b.** 9   **c.** -8; Sec. 8.7, Ex. 2

CHAPTER 9 SOLVING QUADRATIC EQUATIONS

Section 9.1
Practice Exercises

1. $-4, 4$ **2.** $-\dfrac{\sqrt{65}}{5}, \dfrac{\sqrt{65}}{5}$ **3.** $-1, 11$ **4.** $-2 \pm 2\sqrt{3}$ **5.** no real solution   **6.** $\dfrac{5 \pm \sqrt{17}}{3}$ **7.** 72.8 sec

Exercise Set 9.1

1. ± 8 **3.** $\pm\sqrt{21}$ **5.** $\pm\dfrac{1}{5}$ **7.** no real solution   **9.** $\pm\dfrac{\sqrt{39}}{3}$ **11.** $\pm\dfrac{2\sqrt{7}}{7}$ **13.** $\pm\sqrt{2}$ **15.** $\pm\sqrt{5}$ **17.** answers may vary   **19.** $12, -2$ **21.** $-2 \pm \sqrt{7}$ **23.** $1, 0$ **25.** $-2 \pm \sqrt{10}$ **27.** $\dfrac{8}{3}, -4$ **29.** no real solution   **31.** $\dfrac{11 \pm 5\sqrt{2}}{2}$ **33.** $\dfrac{7 \pm 4\sqrt{2}}{3}$ **35.** $8, -3$ **37.** $\pm\sqrt{2}$ **39.** $-6 \pm 2\sqrt{6}$ **41.** $\pm\sqrt{10}$ **43.** $\dfrac{1 \pm \sqrt{5}}{4}$ **45.** $\pm 2\sqrt{3}$ **47.** $\dfrac{-8 \pm 3\sqrt{5}}{-3}$ or $\dfrac{8 \pm 3\sqrt{5}}{3}$ **49.** $2\sqrt{5}$ in. ≈ 4.47 in.   **51.** 177 m   **53.** 5 sec   **55.** 2.7 sec   **57.** 15.8 sec   **59.** 6 in.   **61.** $(x + 3)^2$ **63.** $(x - 2)^2$ **65.** $2, -6$ **67.** $-7 \pm \sqrt{31}$ **69.** ± 1.33 **71.** 2008   **73.** 2007

Section 9.2
Practice Exercises

1. a. $z^2 + 8z + 16 = (z + 4)^2$ **b.** $x^2 - 12x + 36 = (x - 6)^2$ **c.** $b^2 + 5b + \dfrac{25}{4} = \left(b + \dfrac{5}{2}\right)^2$ **2.** $-1 \pm \sqrt{6}$ **3.** $4 \pm 2\sqrt{2}$ **4.** $\dfrac{13}{3}, \dfrac{-1}{3}$ **5.** no real solution   **6.** $\dfrac{3 \pm \sqrt{3}}{2}$

Vocabulary and Readiness Check 9.2

1. zero   **3.** quadratic equation   **5.** 9   **7.** 16   **9.** 100   **11.** 49

Exercise Set 9.2

1. $x^2 + 4x + 4 = (x + 2)^2$ **3.** $k^2 - 12k + 36 = (k - 6)^2$ **5.** $x^2 - 3x + \dfrac{9}{4} = \left(x - \dfrac{3}{2}\right)^2$ **7.** $m^2 - m + \dfrac{1}{4} = \left(m - \dfrac{1}{2}\right)^2$ **9.** $-6, -2$ **11.** $-1 \pm 2\sqrt{2}$ **13.** $0, 6$ **15.** $\dfrac{-5 \pm \sqrt{53}}{2}$ **17.** $1 \pm \sqrt{2}$ **19.** $-1, -4$ **21.** $-2, 4$ **23.** no real solution   **25.** $\dfrac{3 \pm \sqrt{19}}{2}$

A36 Answers to Selected Exercises

27. $-2 \pm \dfrac{\sqrt{6}}{2}$ **29.** $-3 \pm \sqrt{34}$ **31.** $\dfrac{3 \pm \sqrt{21}}{2}$ **33.** $\dfrac{1}{2}, 1$ **35.** $-6, 3$ **37.** no real solution **39.** $2, -6$ **41.** $-\dfrac{1}{2}$ **43.** -1
45. $3 + 2\sqrt{5}$ **47.** $\dfrac{1 - 3\sqrt{2}}{2}$ **49.** answers may vary **51. a.** $-3 \pm \sqrt{11}$ **b.** answers may vary **53.** $k = 8$ or $k = -8$ **55.** 4 years, or 2007
57. $-6, -2$ **59.** $\approx -0.68, 3.68$

Section 9.3
Practice Exercises
1. $\dfrac{-1 \pm \sqrt{41}}{10}$ **2.** $\dfrac{4}{3}, -2$ **3.** $\dfrac{\pm\sqrt{15}}{3}$ **4.** no real solution **5.** $\dfrac{5 \pm 3\sqrt{5}}{2}$ **6.** $\dfrac{-1 + \sqrt{41}}{10} \approx 0.5; \dfrac{-1 - \sqrt{41}}{10} \approx -0.7$
7. two distinct real solutions **8. a.** no real solutions **b.** two distinct real solutions

Vocabulary and Readiness Check 9.3
1. $x = \dfrac{-b \pm \sqrt{b^2 - 4ac}}{2a}$ **3.** $1, 3, -7$ **5.** $1, 1, -1$

Exercise Set 9.3
1. $-2, 1$ **3.** $\dfrac{-5 \pm \sqrt{17}}{2}$ **5.** $\dfrac{2 \pm \sqrt{2}}{2}$ **7.** $1, 2$ **9.** $\dfrac{-7 \pm \sqrt{37}}{6}$ **11.** $\pm\dfrac{2}{7}$ **13.** no real solution **15.** $-3, 10$ **17.** $\pm\sqrt{5}$
19. $-3, 4$ **21.** $-2 \pm \sqrt{7}$ **23.** $\dfrac{1}{2}, 3$ **25.** $\dfrac{5 \pm \sqrt{33}}{2}$ **27.** $-2, \dfrac{7}{3}$ **29.** $\dfrac{-9 \pm \sqrt{129}}{12}$ **31.** $\dfrac{4 \pm \sqrt{2}}{7}$ **33.** $3 \pm \sqrt{7}$ **35.** $\dfrac{3 \pm \sqrt{3}}{2}$
37. $-1, \dfrac{1}{3}$ **39.** $-\dfrac{3}{4}, \dfrac{1}{5}$ **41.** no real solution **43.** $\dfrac{3 \pm \sqrt{13}}{4}$ **45.** no real solution **47.** $1 \pm \sqrt{2}$ **49.** $-\dfrac{3}{4}, -\dfrac{1}{2}$ **51.** $\dfrac{7 \pm \sqrt{129}}{20}$
53. no real solution **55.** $\dfrac{1 \pm \sqrt{2}}{5}$ **57.** $\dfrac{-3\sqrt{2} \pm \sqrt{38}}{2}$ **59.** $\pm\sqrt{7}; -2.6, 2.6$ **61.** $-3 \pm 2\sqrt{2}; -5.8, -0.2$ **63.** $\dfrac{9 \pm \sqrt{97}}{2}; 9.4, -0.4$
65. $\dfrac{1 \pm \sqrt{7}}{3}; 1.2, -0.5$ **67.** 2 real solutions **69.** no real solution **71.** 1 real solution **73.** 2 real solutions **75.** no real solution
77. 1 real solution **79.** $4\sqrt{3}$ **81.** $5\sqrt{2}$ **83.** base: 3 ft; height: 12 ft **85.** c **87.** b **89.** 10.3 ft by 4.9 ft **91.** $-0.7, 5.0$
93. 7.9 sec **95.** answers may vary **97.** 2015 **99.** 2008

The Bigger Picture
1. -1.8 **2.** -5 **3.** -9 **4.** $10x + 2$ **5.** $\dfrac{1}{4}x^2 - 25$ **6.** $3x + 1 - \dfrac{4}{x}$ **7.** $\dfrac{x + 2}{x - 4}$ **8.** $\dfrac{x^2 + 8x - 50}{(x - 10)(x + 3)}$ **9.** $5\sqrt{2}$ **10.** $b\sqrt{10a}$
11. $\dfrac{\sqrt{6}}{3}$ **12.** $\dfrac{x + 3}{7}$ **13.** $\dfrac{-3 \pm \sqrt{29}}{2}$ **14.** 6 **15.** $\{x | x \geq -2.8\}$ **16.** $\dfrac{1}{2}, -8$ **17.** 7 **18.** 27

Integrated Review 9.1–9.3
Practice Exercises
1. $-1, 4$ **2.** $\dfrac{-5 \pm 3\sqrt{5}}{2}$ **3.** $\dfrac{3}{2}, 1$

Integrated Review Exercises
1. $2, \dfrac{1}{5}$ **2.** $\dfrac{2}{5}, -3$ **3.** $1 \pm \sqrt{2}$ **4.** $3 \pm \sqrt{2}$ **5.** $\pm 2\sqrt{5}$ **6.** $\pm 6\sqrt{2}$ **7.** no real solution **8.** no real solution **9.** 2 **10.** 3 **11.** 3
12. $\dfrac{7}{2}$ **13.** ± 2 **14.** ± 3 **15.** $0, 1, 2$ **16.** $0, -3, -4$ **17.** $0, -5$ **18.** $\dfrac{8}{3}, 0$ **19.** $\dfrac{3 \pm \sqrt{7}}{5}$ **20.** $\dfrac{3 \pm \sqrt{5}}{2}$ **21.** $\dfrac{3}{2}, -1$ **22.** $\dfrac{2}{5}, -2$
23. $\dfrac{5 \pm \sqrt{105}}{20}$ **24.** $\dfrac{-1 \pm \sqrt{3}}{4}$ **25.** $5, \dfrac{7}{4}$ **26.** $1, \dfrac{7}{9}$ **27.** $\dfrac{7 \pm 3\sqrt{2}}{5}$ **28.** $\dfrac{5 \pm 5\sqrt{3}}{4}$ **29.** $\dfrac{7 \pm \sqrt{193}}{6}$ **30.** $\dfrac{-7 \pm \sqrt{193}}{12}$ **31.** $11, -10$
32. $7, -8$ **33.** $4, -\dfrac{2}{3}$ **34.** $2, -\dfrac{4}{5}$ **35.** $0.5, 0.1$ **36.** $0.3, -0.2$ **37.** $\dfrac{11 \pm \sqrt{41}}{20}$ **38.** $\dfrac{11 \pm \sqrt{41}}{40}$ **39.** $\dfrac{4 \pm \sqrt{10}}{2}$ **40.** $\dfrac{5 \pm \sqrt{185}}{4}$
41. answers may vary

Section 9.4
Practice Exercises
1. a. $6i$ b. $i\sqrt{15}$ c. $4i\sqrt{3}$ 2. a. $6 + 0i$ b. $0 + 0i$ c. $2\sqrt{6} + 0i$ d. $0 + i$ e. $5 + 3i$ 3. a. $-4 + i$ b. $6 - 4i$ c. $-1 - 2i$
4. $-13 + 8i$ 5. a. $8 + 6i$ b. $9 - 7i$ c. $29 + 0i$ 6. $\dfrac{1}{29} - \dfrac{17}{29}i$ 7. $3 \pm 4i$ 8. $\dfrac{3 \pm i\sqrt{11}}{2}$ 9. $\dfrac{-1 \pm i\sqrt{15}}{2}$

Vocabulary and Readiness Check 9.4
1. complex   3. real   5. standard

Exercise Set 9.4
1. $3i$ 3. $10i$ 5. $5i\sqrt{2}$ 7. $3i\sqrt{7}$ 9. $-3 + 9i$ 11. $-12 + 6i$ 13. $1 - 3i$ 15. $9 - 4i$ 17. $8 + 12i$ 19. $26 - 2i$ 21. 73
23. answers may vary   25. $2 - 3i$ 27. $\dfrac{31}{25} + \dfrac{17}{25}i$ 29. $-1 \pm 3i$ 31. $\dfrac{3 \pm 2i\sqrt{3}}{2}$ 33. $-3 \pm 2i$ 35. $\dfrac{-7 \pm i\sqrt{15}}{8}$ 37. $\dfrac{2 \pm i\sqrt{6}}{2}$
39. $15 - 7i$ 41. $45 + 63i$ 43. $-1 + 3i$ 45. $2 - 3i$ 47. $-7 - 2i$ 49. 25 51. $\dfrac{2}{5} - \dfrac{9}{5}i$ 53. $21 + 20i$ 55. $4 \pm 8i$ 57. $\pm 5i$
59. $-3 \pm i$ 61. $\dfrac{5 \pm i\sqrt{47}}{4}$ 63. $-4 \pm i\sqrt{5}$ 65. $\pm 6i$ 67. $-7 \pm i$ 69. [graph: $y = -3$]   71. [graph: $y = 3x - 2$]   73. 51%
75. 82.5% 77. false   79. false

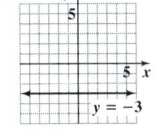

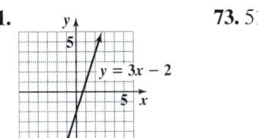

Section 9.5
Practice Exercises
1. [graph with points $(0,0)$, $(-2,-2)$, $(2,-2)$, $(-4,-8)$, $(4,-8)$]
2. [graph with points $(-2,5)$, $(2,5)$, $(-1,2)$, $(1,2)$, $(0,1)$]
3. [graph with points $(-2,6)$, $(4,0)$, $(-1,0)$, $(0,-4)$, $(3,-4)$, $(1.5,-6.25)$]
4. [graph with points $(-2-\sqrt{11},0)$, $(-2+\sqrt{11},0)$, $(-4,-7)$, $(0,-7)$, $(-2,-11)$]

Calculator Explorations 9.5
1. $x = -0.41, 7.41$ 3. $x = 0.91, 2.38$ 5. $x = -0.39, 0.84$

Exercise Set 9.5
1. [graph: $(-2,8)$, $(2,8)$, $(-1,2)$, $(1,2)$, $(0,0)$]
3. [graph: $(0,0)$, $(1,-1)$, $(-1,-1)$, $(-2,-4)$, $(2,-4)$, $(-3,-9)$, $(3,-9)$]
5. [graph: $(-1,0)$, $(1,0)$, $(0,-1)$]
7. [graph: $(0,4)$]
9. [graph: $(2,0)$, $(0,-4)$]
11. [graph: $(0,4)$, $(-1,0)$, $(-4,0)$, $\left(-\dfrac{5}{2}, -\dfrac{9}{4}\right)$]
13. [graph: $(0,5)$, $(2,1)$]
15. [graph: $(-\sqrt{2},0)$, $(\sqrt{2},0)$, $(0,2)$]
17. [graph]
19. [graph: $(-6,0)$, $(0,0)$, $(-3,-9)$]
21. [graph: $(-4,0)$, $(2,0)$, $(0,-8)$, $(-1,-9)$]
23. [graph: $(0,0)$]
25. $\left(\dfrac{1}{2}, 0\right)$, [graph: $(0,5)$, $(5,0)$, $\left(\dfrac{11}{4}, -\dfrac{81}{8}\right)$]
27. [graph: $(1,0)$, $(3,0)$, $(2,1)$, $(0,-3)$]
29. [graph: $(-1-\sqrt{3},0)$, $(-1+\sqrt{3},0)$, $(0,-2)$, $(-1,-3)$]
31. [graph: $\left(\dfrac{3-\sqrt{5}}{2}, 0\right)$, $\left(\dfrac{3+\sqrt{5}}{2}, 0\right)$, $(0,1)$, $\left(1\dfrac{1}{2}, -1\dfrac{1}{4}\right)$]
33. $\dfrac{5}{14}$

A38 Answers to Selected Exercises

35. $\dfrac{x}{2}$ **37.** $\dfrac{2x^2}{x-1}$ **39.** $-4b$ **41.** domain: $(-\infty, \infty)$; range: $(-\infty, 3]$ **43.** domain: $(-\infty, \infty)$; range: $(-\infty, 1]$ **45. a.** 256 ft **b.** $t = 4$ sec **c.** $t = 8$ sec **47.** C **49.** A

Chapter 9 Vocabulary Check

1. square root **2.** complex **3.** vertex **4.** imaginary **5.** conjugate **6.** i **7.** completing the square **8.** quadratic

Chapter 9 Review

1. ± 6 **3.** $\pm 5\sqrt{2}$ **5.** 4, 18 **7.** $\dfrac{-5 \pm \sqrt{41}}{4}$ **9.** 2.5 sec **11.** $a^2 + 4a + 4 = (a+2)^2$ **13.** $m^2 - 3m + \dfrac{9}{4} = \left(m - \dfrac{3}{2}\right)^2$ **15.** 1, 8
17. $-2 \pm \sqrt{5}$ **19.** $3 \pm \sqrt{2}$ **21.** $\dfrac{1}{2}, -1$ **23.** $-\dfrac{5}{3}$ **25.** $\pm\sqrt{5}$ **27.** $5 \pm 3\sqrt{2}$ **29.** $\dfrac{-1 \pm \sqrt{13}}{6}$ **31.** no real solution
33. 0.4, -0.8 **35.** 2 real solutions **37.** 1 real solution **39.** no real solution **41.** 2011 **43.** $12i$ **45.** $6i\sqrt{3}$ **47.** $10 + 6i$
49. $21 - 10i$ **51.** $-8 - 2i$ **53.** $-13i$ **55.** 25 **57.** $-\dfrac{3}{2} - \dfrac{1}{2}i$ **59.** $\dfrac{2}{5} - \dfrac{9}{5}i$ **61.** $\pm 4i$ **63.** $2 \pm 3i$

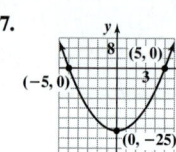

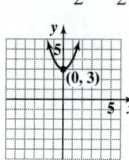

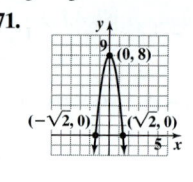

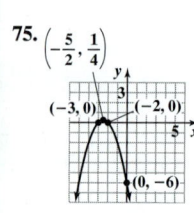

65. [graph: vertex $(0,0)$] **67.** [graph: $(-5,0)$, $(5,0)$, $(0,-25)$] **69.** [graph: $(0,3)$] **71.** [graph: $(0,8)$, $(-\sqrt{2},0)$, $(\sqrt{2},0)$] **73.** [graph: $(-5,0)$, $(2,0)$, $(0,-10)$, vertex $\left(-\dfrac{3}{2}, -\dfrac{49}{4}\right)$]

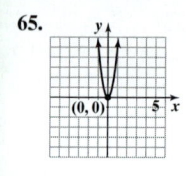

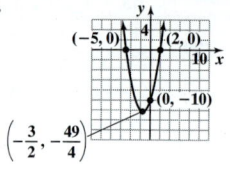

75. $\left(-\dfrac{5}{2}, \dfrac{1}{4}\right)$ [graph: $(-3,0)$, $(-2,0)$] **77.** $\left(-\dfrac{1}{2}, 0\right)$ [graph: $(6,0)$, $(0,-6)$, vertex $\left(\dfrac{11}{4}, -\dfrac{169}{8}\right)$] **79.** A **81.** B **83.** one real solution **85.** no real solution **87.** ± 7
89. 15, -1 **91.** $\dfrac{2}{3}, -1$ **93.** $\dfrac{3 \pm \sqrt{41}}{8}$ **95.** $-\dfrac{3}{2}$

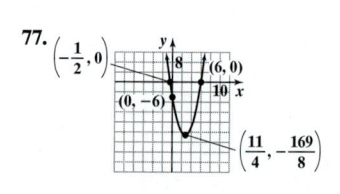

97. [graph: $(0,4)$, $(-2,0)$, $(2,0)$] **99.** [graph: $(0,8)$, $(-2,0)$, $(-4,0)$, vertex $(-3,-1)$]

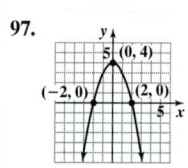

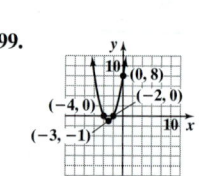

Chapter 9 Test

1. ± 4 **2.** $\dfrac{5 \pm 2\sqrt{2}}{3}$ **3.** 10, 16 **4.** $\dfrac{-6 \pm 4\sqrt{3}}{3}$ **5.** $-2, 5$ **6.** $\dfrac{5 \pm \sqrt{37}}{6}$ **7.** $-\dfrac{4}{3}, 1$ **8.** $-1, \dfrac{5}{3}$ **9.** $\dfrac{7 \pm \sqrt{73}}{6}$ **10.** $2 \pm i$
11. $\dfrac{1}{3}, 2$ **12.** $\dfrac{3 \pm \sqrt{7}}{2}$ **13.** $0, \pm\dfrac{1}{3}$ **14.** $5i$ **15.** $10i\sqrt{2}$ **16.** $8 + i$ **17.** $4i$ **18.** 13 **19.** $\dfrac{1}{5} - \dfrac{7}{5}i$

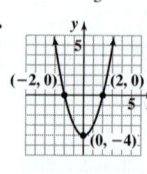

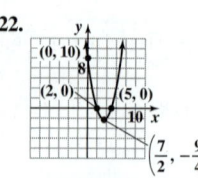

20. [graph: $(0,0)$] **21.** [graph: $(-2,0)$, $(2,0)$] **22.** [graph: $(0,10)$, $(2,0)$, $(5,0)$, vertex $\left(\dfrac{7}{2}, -\dfrac{9}{4}\right)$] **23.** [graph: $\left(\dfrac{-2-\sqrt{6}}{2}, 0\right)$, $\left(\dfrac{-2+\sqrt{6}}{2}, 0\right)$, $(-1,-3)$]

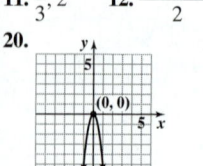

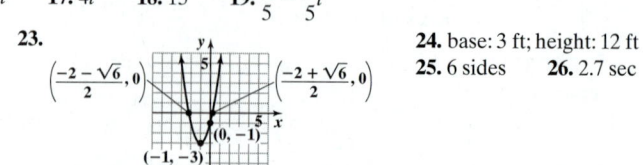

24. base: 3 ft; height: 12 ft **25.** 6 sides **26.** 2.7 sec

Chapter 9 Cumulative Review

1. a. $\dfrac{1}{2}$ **b.** 19; Sec. 1.6, Ex. 6 **3. a.** $5x + 7$ **b.** $-4a - 1$ **c.** $4y - 3y^2$ **d.** $7.3x - 6$; **e.** $\dfrac{1}{2}b$; Sec. 2.1, Ex. 4 **5. a.** x-int: $(-3, 0)$; y-int: $(0, 2)$
b. x-int: $(-4, 0)$, $(-1, 0)$; y-int: $(0, 1)$ **c.** x-int and y-int: $(0, 0)$ **d.** x-int: $(2, 0)$; y-int: none **e.** x-int: $(-1, 0)$, $(3, 0)$; y-int: $(0, -1)$, $(0, 2)$; Sec. 3.3, Ex. 1–5
7. parallel; Sec. 3.4 Ex. 8a **9.** no solution; Sec. 4.1, Ex. 5 **11.** $(-2, 0)$; Sec. 4.2, Ex. 4 **13.** infinite number of solutions; Sec. 4.3, Ex. 4
15. Albert: 3.25 mph; Louis 4.25 mph; Sec. 4.4, Ex. 3 **17.** [graph of $\begin{cases} -3x + 4y < 12 \\ x \geq 2 \end{cases}$]; Sec. 4.6, Ex. 3 **19. a.** $\dfrac{s^4 t^4}{16}$ **b.** $81y^{10}z^{14}$ **c.** $\dfrac{25x^4}{y^6}$; Sec. 5.1, Ex. 11
21. $-6, -\dfrac{3}{2}, \dfrac{1}{5}$; Sec. 6.6, Ex. 8 **23.** 63; Sec. 7.6, Ex. 1 **25. a.** 2 **b.** -2 **c.** -2 **d.** not a real number; Sec. 8.1, Ex. 5

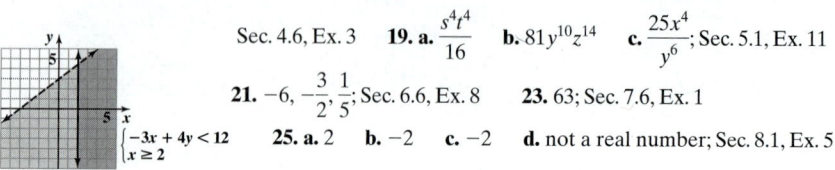

27. a. $\dfrac{5}{6}$ **b.** $\dfrac{\sqrt{3}}{8}$ **c.** $\dfrac{2\sqrt{10}}{9}$; Sec. 8.2, Ex. 3 **29. a.** $7\sqrt{2}$ **b.** $9\sqrt{3}$ **c.** $1 - 3\sqrt{3} - 6\sqrt{2}$; Sec. 8.3, Ex. 2 **31. a.** $\sqrt{21}$ **b.** 3 **c.** $3\sqrt{5}$ **d.** $10\sqrt{6}$ **e.** $2x^2\sqrt{3}$; Sec. 8.4, Ex. 1 **33.** $\dfrac{1}{2}$; Sec. 8.5, Ex. 3 **35.** $\sqrt{44{,}800} \approx 212$ ft; Sec. 8.6, Ex. 3 **37. a.** $\sqrt{25} = 5$ **b.** $\sqrt[3]{8} = 2$ **c.** $-\sqrt[4]{16} = -2$ **d.** $\sqrt[3]{-27} = -3$ **e.** $\sqrt{\dfrac{1}{9}} = \dfrac{1}{3}$; Sec. 8.7, Ex. 1 **39.** $\dfrac{\sqrt{14}}{2}, -\dfrac{\sqrt{14}}{2}$; Sec. 9.1, Ex. 2 **41.** $5 \pm \sqrt{11}$; Sec. 9.2, Ex. 3 **43.** $-\dfrac{1}{2}, 5$; Sec. 9.3, Ex. 2 **45. a.** $2i$ **b.** $i\sqrt{11}$ **c.** $2i\sqrt{5}$; Sec. 9.4, Ex. 1 **47.** ; Sec. 9.5, Ex. 2

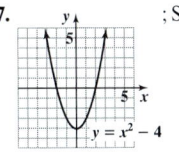

APPENDIX A.2 PRACTICE FINAL EXAM

1. -81 **2.** $\dfrac{1}{64}$ **3.** -48 **4.** $-3x^3 + 5x^2 + 4x + 5$ **5.** $16x^2 - 16x + 4$ **6.** $3x^3 + 22x^2 + 41x + 14$ **7.** $(y - 12)(y + 4)$ **8.** $3x(3x + 1)(x + 4)$ **9.** $5(6 + x)(6 - x)$ **10.** $(3a - 7)(a + b)$ **11.** $8(y - 2)(y^2 + 2y + 4)$ **12.** $\dfrac{y^{14}}{x^2}$ **13.** $\dfrac{5y^2 - 1}{y + 2}$ **14.** $\dfrac{2(x + 5)}{x(y + 5)}$ **15.** $\dfrac{3a - 4}{(a - 3)(a + 2)}$ **16.** 8 **17.** $-7, 1$ **18.** $(-\infty, -2]$ **19.** $\dfrac{3 \pm \sqrt{7}}{2}$ **20.** $\dfrac{30}{11}$ **21.** -6 **22.** no solution **23.** 9 **24.** **25.** 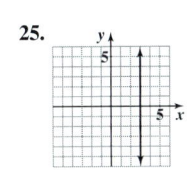 **26.** 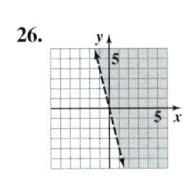 **27.** $m = -1$ **28.** $m = 3$ **29.** $8x + y = 11$ **30.** $x = -5$ **31.** $(-4, 1)$ **32.** no solution **33. a.** 0 **b.** 0 **c.** 60 **34.** Domain: $(-\infty, \infty)$; Range: $(-\infty, 4]$ **35.** 4 **36.** $\dfrac{1}{9}$ **37.** $\dfrac{3}{4}$ **38.** $3\sqrt{6}$ **39.** $3x^4\sqrt{x}$ **40.** $-8\sqrt{3}$ **41.** $2x\sqrt{5x}$ **42.** $2\sqrt{3} - \sqrt{10}$ **43.** $\dfrac{\sqrt{15}}{6x}$ **44.** $-1 - \sqrt{3}$ **45.** 5 or 1 **46.** Rhode Island: 401; Vermont: 802 **47.** $2\dfrac{1}{2}$ hr **48.** 120 cc

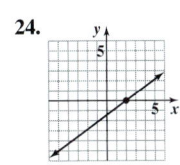

APPENDIX B.3 REVIEW OF ANGLES, LINES, AND SPECIAL TRIANGLES

1. $71°$ **3.** $19.2°$ **5.** $78\dfrac{3}{4}°$ **7.** $30°$ **9.** $149.8°$ **11.** $100\dfrac{1}{2}°$ **13.** $m\angle 1 = m\angle 5 = m\angle 7 = 110°$; $m\angle 2 = m\angle 3 = m\angle 4 = m\angle 6 = 70°$ **15.** $90°$ **17.** $90°$ **19.** $90°$ **21.** $45°, 90°$ **23.** $73°, 90°$ **25.** $50\dfrac{1}{4}°, 90°$ **27.** $x = 6$ **29.** $x = 4.5$ **31.** 10 **33.** 12

APPENDIX B.4 REVIEW OF VOLUME AND SURFACE AREA

1. $V = 72$ cu in.; $SA = 108$ sq in. **3.** $V = 512$ cu cm; $SA = 384$ sq cm **5.** $V = 4\pi$ cu yd $\approx 12\dfrac{4}{7}$ cu yd.; $SA = (2\sqrt{13}\pi + 4\pi)$ sq yd ≈ 35.20 sq yd **7.** $V = \dfrac{500}{3}\pi$ cu in. $\approx 523\dfrac{17}{21}$ cu in.; $SA = 100\pi$ sq in. $\approx 314\dfrac{2}{7}$ sq in. **9.** $V = 48$ cu cm; $SA = 96$ sq cm **11.** $2\dfrac{10}{27}$ cu in. **13.** 26 sq ft **15.** $10\dfrac{5}{6}$ cu in. **17.** 960 cu cm **19.** 196π sq in. **21.** $7\dfrac{1}{2}$ cu ft **23.** $12\dfrac{4}{7}$ cu cm

APPENDIX C

1. 5 **3.** 3.2 or $\dfrac{16}{5}$ **5.** 25 **7.** 0.0025 **9.** 2 **11.** 15 **13.** $\dfrac{90}{21}$ **15.** $\dfrac{17}{11}$ **17.** 31 **19.** $\dfrac{9}{14}$ **21.** 18 applications **23.** 5 weeks **25.** $10\dfrac{2}{3}$ servings **27.** 120 calories **29. a.** 18 tsp **b.** 6 tbsp **31.** 6 people **33.** 112 ft; 11-in. difference **35.** 102.9 mg **37.** 434 emergency room visits **39.** 28 workers

APPENDIX D OPERATIONS ON DECIMALS

1. 17.08 **3.** 12.804 **5.** 110.96 **7.** 2.4 **9.** 28.43 **11.** 227.5 **13.** 2.7 **15.** 49.2339 **17.** 80 **19.** 0.07612 **21.** 4.56 **23.** 648.46
25. 767.83 **27.** 12.062 **29.** 61.48 **31.** 7.7 **33.** 863.37 **35.** 22.579 **37.** 363.15 **39.** 7.007

APPENDIX E MEAN, MEDIAN, AND MODE

1. mean: 29, median: 28, no mode **3.** mean: 8.1, median: 8.2, mode: 8.2 **5.** mean: 0.6, median: 0.6, mode: 0.2 and 0.6
7. mean: 370.9, median: 313.5, no mode **9.** 1214.8 ft **11.** 1117 ft **13.** 6.8 **15.** 6.9 **17.** 85.5 **19.** 73 **21.** 70 and 71
23. 9 **25.** 21, 21, 24

Index

A
Absolute value, 13
Acute angle, 609
Addition method (of solving systems of linear equations), 265–269
Addition property of equality, 82–86, 92–94, 147
Addition
  associative property of, 59–60
  commutative property of, 58–59
  of complex numbers, 580
  of decimals, 622
  of fractions, 19–21
  of polynomials, 318
  of radicals, 518–520
  of rational expressions, 444–448, 452–455
  of real numbers, 34–39
Additive inverses, 38–39, 62–63
Adjacent angles, 610
Algebraic expressions. *See also* Polynomials
  definition, 27
  evaluating, 27–28, 44
    using real numbers, 54–55
  simplifying, 74–78
  writing word phrases as, 29–31, 77–78, 86–87, 94–95
Alternate interior angles, 610
Angles, 45, 609–611
Associative property
  of addition, 59–60
  of multiplication, 59–60
Axis of symmetry, 585

B
Bar graph(s), 73, 170
Base
  of exponent, 24
  of exponential expression, 302

Binomials, 313, 314
  factoring the difference of two squares, 391–393
  factoring the sum or difference of two cubes, 393–394
  multiplying sum and difference of two terms, 332–334
  multiplying, 330–334
  squaring, 331–332
Boundary (line), 282
Braces, 24–26
Brackets, 24–26
Broken line graphs, 170

C
Calculators. *See also under* Graphing calculators; Scientific calculators
  approximating square roots, 508
  checking equations, 102
  evaluating exponents, 31
  negative numbers, 55
  scientific notation and, 342
Cephalic index, 428
Circle graphs, 128–129
Coefficient
  definition, 74, 313
  numerical, 74
Commutative property
  of addition, 58–59
  of multiplication, 58–59
Complementary angles, 45, 609
Complex conjugates, 581
Complex fractions
  definition, 488
  simplifying, 490–491
Complex numbers
  adding and subtracting, 580
  definition, 579

Polynomials (*continued*)
  trinomials. *See* Trinomials
  types of, 313
Positive integers, 9–10
Positive numbers, 11
Positive square root, 504
Power of a product rule, 306–307
Power of a quotient rule, 306–307
Power rule for exponents, 305
Prime numbers, 16–17
Principle square root, 504
Problem-solving
  applications with linear inequalities, 152–153
  by adding real numbers, 37
  calculating break-even point, 291–292
  consecutive integer problems, 110, 414
  dimensions of a sail, 412–413
  dimensions of triangles, 415
  direct and inverse variation problems, 478–485
  direct translation problems, 106–107
  direct variation problems, 478–481
  distance problems, 138–140, 472–473, 538–539
  finding height of a dropped object, 315–316
  formulas and, 115–121
  free fall times, 411
  interest problems, 142–143
  length of time of a dive, 559–560
  using linear equations in two variables, 221–222
  mixture problems, 132–133
  money problems, 140–142
  number problems, 470–471
  percent equations, 127–132
  using polynomials, 315–316
  price problems, 272–275
  problems modeled by quadratic equations, 559–560
  using proportions, 466–473
  quadratic equations and, 411
  radical equations and, 537–539
  rate problems, 275–276
  road sign dimensions, 119–120
  similar triangle problems, 469
  solution/mixture problems, 276–277
  steps in, 105, 127
  by subtracting real numbers, 44–45
  using systems of linear equations, 272–277
  time given rate and distance, 116–117
  unknown number problems, 106–107, 272–273, 412, 470–471
  unknown quantities problems, 107–109
  velocity problems, 538–539
  work rate problems, 471–472
  writing equations that model a problem, 105–110
Product rule for exponents, 303–304
Product rule for square roots, 511–513
Proportions, 620–621
  definition, 466
  problem-solving using, 468–473
  solving problems with rational equations, 466–467
Pure imaginary numbers, 579
Pythagorean theorem, 413–415
  radical equations and, 537–538

Q

Quadrants, 172
Quadratic equations
  approximating solutions to, 572–573
  complex solutions of, 579–582
  definition, 402
  finding x-intercepts of, 406–407
  on graphing calculator, 407
  graphs of, 584–589
  solving
    $ax^2 + bx + c = 0$, 564–566
    by completing the square, 563–566
    by factoring, 403–406
    by the quadratic formula, 568–572
    problems modeled by, 411–415, 559–560
    by square root method, 557–559
    $x^2 + bx + c = 0$, 563–564
  standard form, 402, 568–569
  in two variables, 406–407
Quadratic formula, 568–572
Quotient rule for exponents, 307–309
Quotient rule for radicals, 524
Quotient rule for square roots, 513
Quotient, of two real numbers, 51–55

R

Radical(s), 504. *See also* Cubed roots; *n*th roots; Radical expressions; Square roots
  adding and subtracting, 518–520
  multiplying and dividing, 522–525
  product rule for, 511–515

quotient rule for, 513–515
rationalizing denominators of, 525–528
simplifying higher roots, 514–515
with variables, 507
Radical equations
containing square roots, 532
definition, 531
problem-solving using, 537–539
solving, 531–534
Radical expressions. *See* Radical(s); Radical equations
Radical sign, 504
Radicand, 504
Radius of curvature formula, 541
Radius of sphere, 552
Range (of relation), 227
Ratio, 466
Rational equations, 459–463
solving proportions with, 466–467
Rational exponents
evaluating $a^{1/n}$, 543
evaluating $a^{-m/n}$, 543
evaluating $a^{m/n}$, 544
Rational expressions
adding and subtracting, 444–448, 452–455
complex, 488–491
definition, 429
dividing, 439–442
equations containing, 459–463
evaluating, 429
multiplying, 437–439, 441–442
simplifying, 430–433
undefined, 430
writing equivalent forms of, 433–434
Rational numbers, 10–11
Real numbers, 11–14
absolute value of, 13
adding, 34–39
dividing, 51–55
evaluating algebraic expressions using, 54–55
identity properties of, 61–62
multiplying, 49–52
operations on calculator, 55
order property for, 12–13
properties of, 58–63
subtracting, 41–45
Reciprocals, 18–19, 51–52, 62–63
Rectangular coordinate system, 172–174

Relation(s), 227
domain of, 227
as functions, 227. *See* Functions
range of, 227
Right angle, 609
Right triangles, 413, 613–614
Rise (of line), 203
Run (of line), 203

S

Scatter diagram, 174
Scientific calculator, 31, 55
Scientific notation, 340–342
converting numbers to standard form, 341–342
writing a number in, 340–341
Set, 7
Signed numbers, 11
Similar triangles, 612–613
Simple inequalities, 151–152
Simplified fractions, 16
Slope. *See also* Slope-intercept form
definition, 203
equations for, 208
finding
given its equation, 205–206
given two points of the line, 203–205
of horizontal lines, 207–208
of parallel lines, 208–211
of perpendicular lines, 208–211
rise, 203
run, 203
undefined, 207
of vertical lines, 207–208
zero, 207
Slope-intercept form, 206, 218
graphing equations using, 218–219
Solution, 28, 82
ordered pair as, 175–176
Special products, 331–334
definition, 331
FOIL method, 330–334
Square root(s). *See also* Radical expressions; Radicals approximating, 506
using a calculator, 508
definition, 504
finding, 504–505
perfect squares, 506
product rule for, 511–515
quotient rule for, 513

Square root(s) (*continued*)
   radical equations containing, 532
   table of, 626
Square root property, 557–559
Squares table, 626
Squaring property of equality, 531–534
Straight angle, 609
Study skills
   familiarity with textbook and supplements, 202, 312, 511
   getting mathematics help, 398
   goals for success in mathematics course, 81
   homework assignments, 97, 264
   learning new terms, 521
   notebook organization, 126
   organizational skills, 329
   preparing for final exam, 562
   self-evaluation, 451
      of academic performance, 184
      of quiz/exam performance, 386
   studying for an exam, 288
   taking an exam, 145
Substitution method, 258–262
Subtraction
   of complex numbers, 580
   of decimals, 622
   of fractions, 19–21
   of polynomials, 319–320
   of radicals, 518–520
   of rational expressions, 444–448, 452–455
   of real numbers, 41–45
Sum of two cubes, 393–394
Supplementary angles, 45, 609
Surface area,
   formulas for, 616
   of polyhedron, 615
   of sphere, 552
Symbols
   absolute value, 13
   braces, 7, 24–26
   brackets, 24–26
   equal, 7
   fraction bar, 24–26
   for function, 232
   greater than or equal to, 8
   grouping, 24–26
   imaginary unit, 579
   inequality, 7–8
   less than or equal to, 8
   negative infinity, 146
   parentheses, 24–26
   radical sign, 504
Symmetric about the y-axis, 585
Systems of linear equations. *See* Linear equations

T

Table of values, 177–179
Term(s)
   coefficient of, 74
   constant, 313
   definition, 74
   degree of, 314
   like, 74–78
   unlike, 74–78
Tips for success in mathematics. *See also* Study skills
   general tips, 2–3
   getting help, 4
   preparation, 2
   studying for and taking exams, 4–5
   time management, 5
   using this text, 3–4
Transversal, 610
Triangles, 611–614
Trinomials
   of the form $ax^2 + bx + c$, 377–381, 386–389
   of the form $x^2 + bx + c$, 371–375
   grouping method to factor, 386
   perfect square, 382–383, 389, 562–563

U

Undefined quotient, 53
Unknown number problems, 471–472

V

Variables, 27
Velocity problems, 538
Vertex/vertices
   formula, 586–589
   of parabolas, 584–589
   of polyhedrons, 615
Vertical angles, 610
Vertical line(s), 197–199
   finding equations of, 220–221
   slope of, 207–208
Vertical line test, 228–231

Volume, 615
   formulas for, 616

W

Whole numbers, 7
Work rate problems, 471–472

X

x-axis, 172
x-coordinate, 172
x-intercept, 194–197
   of quadratic equation graphs, 406–407

Y

y-axis, 172
y-coordinate, 172
y-intercept, 194–197

Z

Zero
   as divisor or dividend, 53
   as a factor, 49–50
Zero exponent, 308–309
Zero factor property, 402
Zero factor theorem, 402–405

Photo Credits

Chapter 1

Page 1 Julian Baum/Photo Researchers, Inc.

Page 2 Sepp Seitz/Woodfin Camp & Associates, Inc.

Page 3 © Rachel Epstein/PhotoEdit

Page 10 Tim Davis/Corbis RF

Page 16 Photo Researchers, Inc.

Page 27 Joseph McBride/Stone/Getty Images

Page 40 © Alison Wright/The Image Works

Page 40 © ZEFA/Masterfile

Page 47 © Sharna Balfour/Gallo Images/Corbis

Page 47 Bishop Airport

Page 64 Guildhall Library, Corporation of London, UK/The Bridgeman Art Library

Chapter 2

Page 73 © G. Bowater/CORBIS All Rights Reserved

Page 87 Laurence R. Lowry/Stock Boston

Page 87 Ed Pritchard/Stone/Getty Images

Page 89 Hugh Sitton/Stone/Getty Images

Page 89 Klaus G. Hinkelmann

Page 97 © Spencer Grant/PhotoEdit

Page 107 © Editorial Image, LLC/Alamy

Page 108 U.S. Congress, Office of Technology Assessment

Page 109 AP Wide World Photos

Page 110 © Rolf Bruderer/Masterfile

Page 111 © Forestier Yves/CORBIS SYGMA

Page 114 © Liu Jin/Agence France Presse/Getty Images

Page 114 © G. Bowater/CORBIS All Rights Reserved

Page 116 Sean Reid/Alaska Stock

Page 123 © Buddy Mays/CORBIS All Rights Reserved

Page 125 Colin Braley/Corbis/Reuters America LLC

Page 125 © 2005 Norbert Wu/www.norbertwu.com

Page 126 John Elk III/Stock Boston

Page 130 Blend Images/Alamy

Page 131 Image State/International Stock Photography Ltd.

Page 136 © vario images GmbH & Co. KG/Alamy

Page 136 Neal Ulevich/Bloomberg News/Landov LLC

Page 153 © Jeff Greenberg/The Image Works

Page 164 Catherine Karnow/Woodfin Camp & Associates, Inc.

Chapter 3

Page 169 © Editorial Image, LLC/Alamy

Page 170 Getty Images/Stockbyte

Page 178 David Young-Wolff/Stone/Getty Images

Page 190 © Michael Newman/PhotoEdit

Page 192 Tom Stillo/Omni-Photo Communications, Inc.

Page 193 Amy C. Etra/PhotoEdit Inc.

Page 201 © Tony Freeman/PhotoEdit

Page 202 © Bill Aron/PhotoEdit

Page 211 © John Neubaur/PhotoEdit

Page 217 © Editorial Image, LLC/Alamy

Page 222 © Ian Shaw/Alamy

Page 225 © Jim West/The Image Works

Page 225 Geri Eingberg Photography

Page 226 © Editorial Image, LLC/Alamy

Chapter 4

Page 249 © Image100/CORBIS All Rights Reserved

Page 264 © Peter Hvizdak/The Image Works

Page 271 © Image100/CORBIS All Rights Reserved

Page 271 © Bo Bridges/CORBIS All Rights Reserved

Page 273 © Alberto E. Rodriguez/Getty Images for Davidson & Choy Publicity

Page 278 Barry Gossage/NBAE via Getty Images, Inc.

Page 278 Photo by Mitchell Layton/Getty Images, Inc.

Page 279 AP Wide World Photos

Page 298 AP Wide World Photos

Photo Credits **P3**

Chapter 5

Page 301 David Young-Wolff/PhotoEdit

Page 323 © Becky Olstad/The Christian Science Monitor via Getty Images

Page 323 © George B. Diebold/CORBIS All Rights Reserved

Page 344 AP Wide World Photos

Page 344 G. Brad Lewis/Omjalla Images

Chapter 6

Page 362 © Ariel Skelley/CORBIS

Page 397 © Ariel Skelley/CORBIS

Page 397 AP Wide World Photos

Page 398 John Freeman/Stone/Getty Images

Page 411 AP Wide World Photos

Chapter 7

Page 428 © Design Pics Inc/Alamy

Page 428 © Herbert Spichtinger/ZEFA/Corbis

Page 428 Jerry Young/Getty Images

Page 436 Chuck Keeler/Frozen Images

Page 436 © Ron Vesely/MLB Photos via Getty Images, Inc.

Page 471 © Charles O'Rear/CORBIS

Page 479 © Spencer Grant/PhotoEdit

Page 487 © Susan Van Etten/PhotoEdit

Page 487 Brian Erler/Taxi/Getty Images

Page 501 Photodisc/Getty Images

Chapter 8

Page 503 Javier Larrea/AGE Fotostock America, Inc.

Page 510 Alamy Images

Page 510 Dong Lin/California Academy of Sciences—Geology

Page 517 Jessica Wecker/Photo Researchers, Inc.

Page 517 Photofest

Page 517 Peter Poulides/Getty Images Inc.—Stone Allstock

Page 539 Andy Belcher/ImageState/International Stock Photography Ltd.

Chapter 9

Page 556 Jim Pickerell/The Stock Connection

Page 559 Hulton Archive/Getty Images

Page 561 © CSPA/NewSport/CORBIS All Rights Reserved

Page 561 © David Kadlubowski/CORBIS All Rights Reserved

Page 562 © David R. Frazier Photolibrary, Inc./Alamy

Page 575 Jim Pickerell/The Stock Connection

Page 591 © Karen Preuss/The Image Works

Page 596 Horizon International Images Limited/Alamy

Martin-Gay's VIDEO RESOURCES
Help Students Succeed

MARTIN-GAY'S CHAPTER TEST PREP VIDEO (AVAILABLE WITH THIS TEXT)

- Provides students with help during their most "teachable moment"—while they are studying for a test.
- Text author Elayn Martin-Gay presents step-by-step solutions to the exact exercises found in each Chapter Test in the book.
- Easy video navigation allows students to instantly access the worked-out solutions to the exercises they want to review.
- Close captioned in English and Spanish.

NEW MARTIN-GAY'S INTERACTIVE DVD/CD LECTURE SERIES

Martin-Gay's video series has been comprehensively updated to address the way today's students study and learn. The new videos offer students active learning at their pace, with the following resources and more:

- **A complete lecture** for each section of the text, presented by Elayn Martin-Gay. Students can easily review a section or a specific topic before a homework assignment, quiz, or test. Exercises in the text marked with the are worked on the video.
- A **new interface** with menu and navigation features helps students quickly find and focus on the examples and exercises they need to review.
- Martin-Gay's "pop-ups" reinforce key terms and definitions and are a great support for multiple learning styles.
- A new **Practice Final Exam Video** helps students prepare for the final exam. This Practice Final Exam is included in the text in Appendix A.2. At the click of a button, students can watch the full solutions to each exercise on the exam when they need help. Overview clips provide a brief overview on how to approach different problem types—just as they will need to do on a Final Exam.
- **Interactive Concept Checks** allow students to check their understanding of essential concepts. Like the concept checks in the text, these multiple choice exercises focus on common misunderstandings. After making their answer selection, students are told whether they're correct or not, and why! Elayn also presents the full solution.
- **Study Skills Builders** help students develop effective study habits and reinforce the advice provided in Section 1.1, Tips for Success in Mathematics, found in the text and video.
- **Close-captioned in Spanish and English**
- Ask your bookstore for information about Martin-Gay's *Beginning Algebra,* Fifth Edition Interactive DVD/CD Lecture Series or visit www.mypearsonstore.com.

You will find Interactive Concept Checks and Study Skills Builders in the following sections on the Interactive DVD/CD Lecture Series:

Interactive Concept Checks Section		Study Skills Builders Section
1.3	5.6	1.2 Time Management
1.4	6.1	2.5 Are You Familiar with the Resources Available with Your Textbook?
1.5	6.2	
1.6	6.3	2.6 Have You Decided to Complete This Course Successfully?
1.7	6.5	
1.8	6.6	2.7 How Well Do You Know Your Textbook?
2.1	Ch 6 Integrated Review	2.8 Tips for Studying for an Exam
2.2	7.1	3.1 How Are Your Homework Assignments Going?
2.3	7.2	
2.4	7.3	3.2 Doing Your Homework Online
2.8	7.5	
3.3	7.6	4.6 What to Do the Day of an Exam?
3.4	7.7	5.1 Are You Satisfied with Your Performance on a Particular Quiz?
3.5	7.8	
3.6	8.1	5.2 Are You Organized?
4.1	8.3	
4.2	8.4	5.5 Are You Familiar with the Resources Available with Your Textbook?
4.3	8.7	
4.4	9.2	6.4 Have You Decided to Complete This Course Successfully?
4.5	9.3	
5.3	9.4	6.7 How Well Do You Know Your Textbook?
5.4	9.5	7.4 Tips for Studying for an Exam
		8.2 Are You Satisfied with Your Performance on a Particular Quiz or Exam?
		8.5 What to Do the Day of an Exam?
		8.6 How Are Your Homework Assignment Going?
		9.1 Preparing for Your Final Exam

$$-1 \cdot \frac{(x-3)(x+1)}{(3-x)(3+x)} \quad 3-x$$

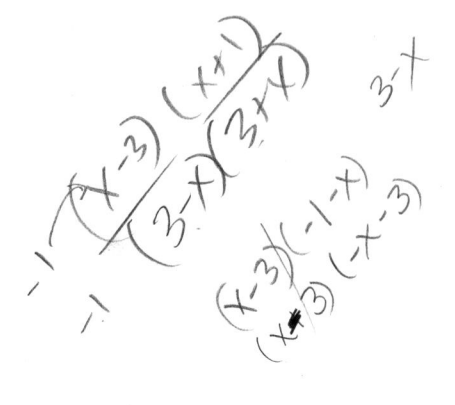

$$-1 \cdot \frac{(x-3)(-1-x)}{(x+3)(-x-3)}$$